Phylogeny of Primates

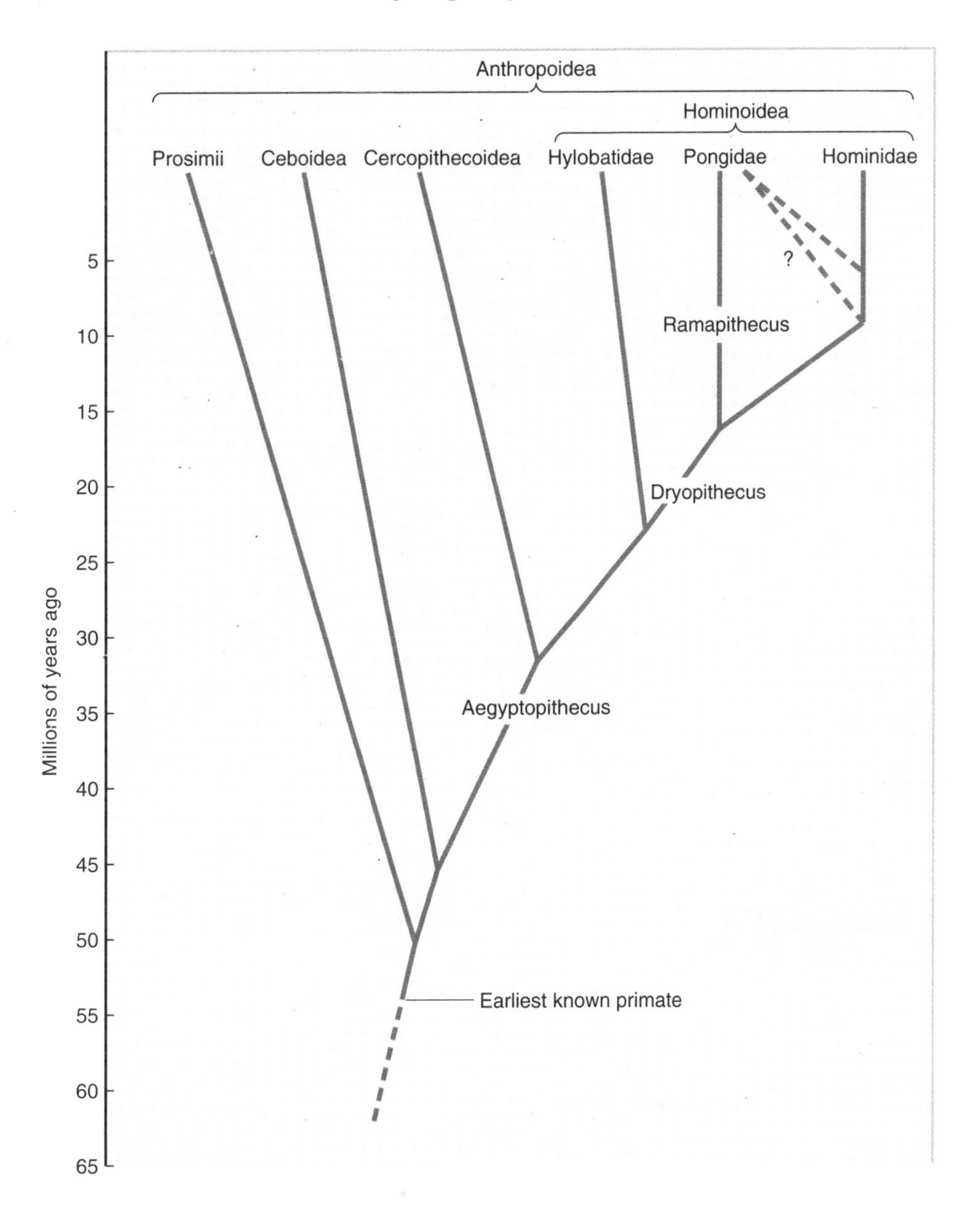

This "family tree" of the primates shows the estimated dates of splits in the lineage and the approximate position of several of the more important fossils. The orangutan lineage (Pongidae), with *Ramapithecus* in the middle, split from the rest of the great apes (Hominoidea) about 18 million years ago. There is controversy over the proper positioning and time of origin of the gorilla and chimpanzee lineages (dashed lines). Genetically, chimpanzees and gorillas are more like humans (Hominidae) than orangutans (Pongidae), but we are inclined to insulate ourselves from the rest of the animal kingdom and therefore we feel safer with the chimpanzees and gorillas placed, as they traditionally are, with the orangutans.

Human Biology

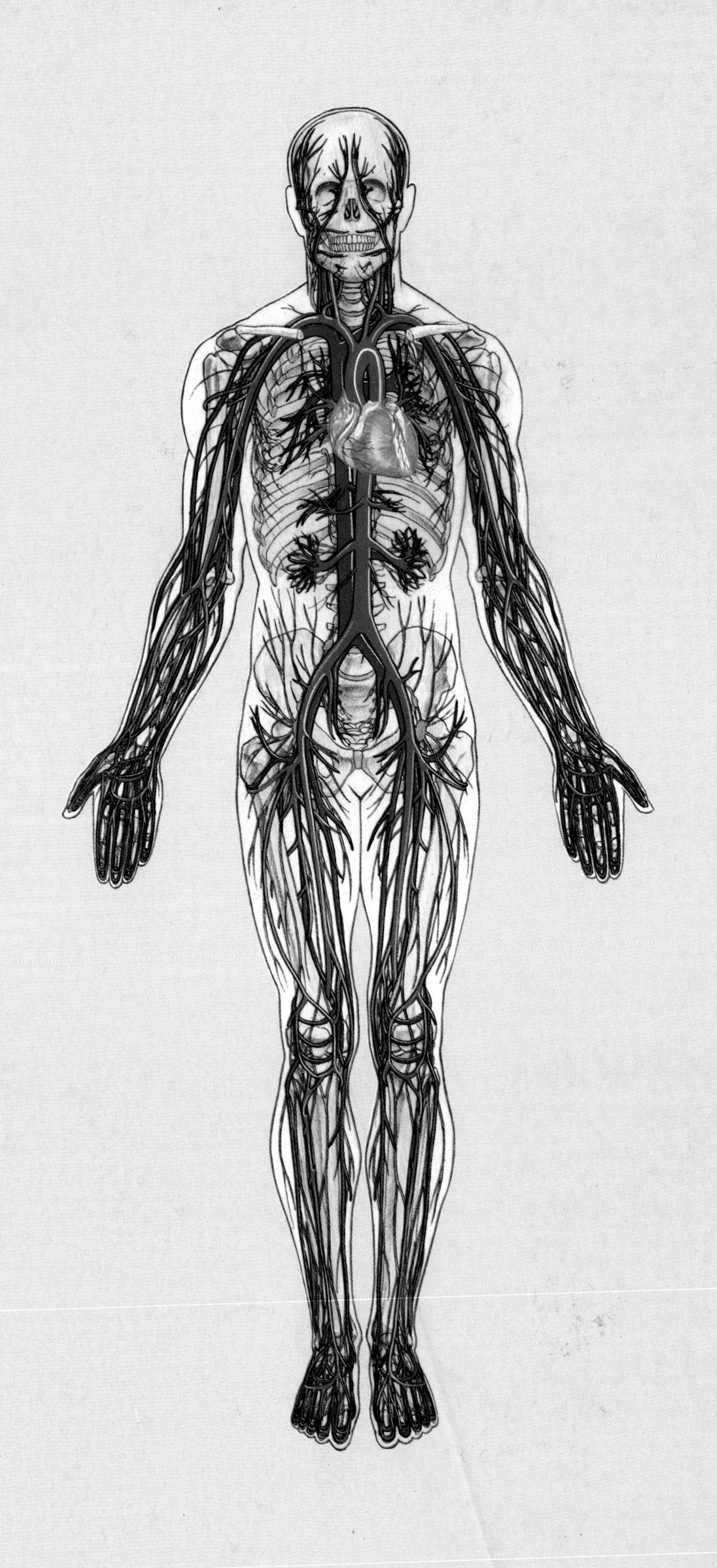

Human Biology

Donald J. Farish

Sonoma State University
Rohnert Park, California

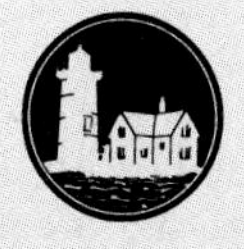

Jones and Bartlett Publishers

Boston London

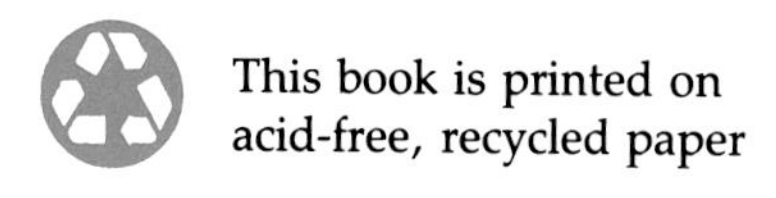

This book is printed on
acid-free, recycled paper

Editorial, Sales, and Customer Service Offices
Jones and Bartlett Publishers
One Exeter Plaza
Boston, MA 02116

Jones and Bartlett Publishers International
PO Box 1498
London W6 7RS
England

Portions of this book first appeared in *Introduction to Biology: A Human Perspective,* by Donald J. Farish, © 1986 by Jones and Bartlett Publishers (Original © 1984 by PWS Publishers).

Library of Congress Cataloging-in-Publication

Farish, Donald J., 1942–
Human biology / Donald J. Farish.
Includes index.
ISBN 0-86720-114-2
1. Human biology. I. Title.
QP36.F37 1993 92-30478
612—dc20 CIP

Interior Designer: Debbie Schneck
Cover Designer: Design Ad Cetera
Cover Photographer: Dave Nagel/The Gamma Liaison Network
Cover Artist: Vincent Perez
Chapter/Part Opening Photographer: Mark Philbrick
Typesetter: Modern Graphics, Inc.
Separator: Pre-Press Co.
Cover Printer: Henry N. Sawyer

Printed in the United States of America
97 96 95 94 93 10 9 8 7 6 5 4 3 2 1

CHAPTER 1: Fig. 1.1 p. 6: Top, © Hank Morgan, Science Source/Photo Researchers; middle, bottom: Supplied by Carolina Biological Supply Company. Fig. 1.2 p. 7: The Bettmann Archive. Fig. 1.4 p. 9: The Bettmann Archive. Fig. 1.5 p. 10: From Singer, *A History of Biology to about the Year 1900, Third Edition.* © 1959 by Abelard-Schuman. Fig. 1.6 p. 11: From Singer, *A History of Biology to about the Year 1900, Third Edition.* © 1959 by Abelard-Schuman. Fig. 1.7 p. 12: Reproduced by courtesy of the Trustees, The National Gallery, London. Fig. 1.8 p. 12: The Bettmann Archive. Fig. 1.9 p. 12: The Bettmann Archive. Fig. 1.10 p. 13: The Bettmann Archive. Fig. 1.11 p. 13: The Bettmann Archive. Fig. 1.12 p. 13: The Bettmann Archive. Fig. 1.14 p. 16: From Singer, *A History of Biology to about the Year 1900, Third Edition.* © 1959 by Abelard-Schuman. Fig. 1.15 p. 16: From Singer, *A History of Biology to about the Year 1900, Third Edition.* © 1959 by Abelard-Schuman. Fig. 1.16 p. 17 (a): The Bettmann Archive; (b): From Singer, *A History of Biology to about the Year 1900, Third Edition.* © 1959 by Abelard-Schuman. Fig. 1.17 p. 17 (a): North Wind Picture Archives; (b): From Singer, *A History of Biology to about the Year 1900, Third Edition.* © 1959 by Abelard-Schuman. Fig. 1.18 p. 18: From Singer, *A History of Biology to about the year 1900, Third Edition.* © 1959 by Abelard-Schuman. Fig. 1.19 p. 19: From Singer, *A History of Biology to about the year 1900, Third Edition.* © 1959 by Abelard-Schuman. Fig. 1.20 p. 20: The Bettmann Archive.

CHAPTER 2: Fig. 2.1 p. 27: The Bettmann Archive. Fig. 2.2 p. 27: The Bettmann Archive. Fig. 2.3 p. 28: The Bettmann Archive. Fig. 2.5 p. 29: The Bettmann Archive. Fig. 2.7 p. 30: The Bettmann Archive. Fig. 2.10 p. 32: The Bettmann Archive.

CHAPTER 3: Fig. 3.10 p. 55: Supplied by Carolina Biological Supplied Company.

CHAPTER 4: Fig. 4.1 p. 69: From Strickberger, *Evolution* © 1990 by Jones and Bartlett Publishers. Fig., Box 4.2 p. 72: (a), © EM Unit, CVL Weybridge/Science Photo Library, Photo Researchers; (b): © Omikron/Science Source, Photo Researchers; (c): © Lee D. Simon/Science Source, Photo Researchers. Fig. 4.4 p. 75: © Dr. Donald Lundgren and Dr. Ralph Slepecky, Syracuse University. Fig. 4.5 p. 78: © Bruce Iverson, BSc. Fig. 4.6 p. 79: © Don Fawcett, MD, Science Source/Photo Researchers. Fig. 4.8, top: © Don Fawcett, MD, Science Source/Photo Researchers. Fig. 4.11 p. 82: Photo © Hugh E. Huxley, Medical Research Council, Cambridge. Drawing from "The Mechanism of Muscular Contraction,: by H.E Huxley. Copyright © 1965 by Scientific American, Inc. All rights reserved. Fig. 4.13 p. 82 (a): Photo © Peter Satir, Albert Einstein College of Medicine. Fig. 4.14 p. 83 (a): © Photo Researchers. Fig. 4.15 p. 84 (b): © Alexander Rich, Photo Researchers. Fig. 4.16 p. 84 (a), (b): © Daniel

continued on page 630

To My Biological Mentors:

R.C. Axtell, North Carolina State University
H.E. Evans, late of Harvard University
G.G.E. Scudder, University of British Columbia
E.O. Wilson, Harvard University

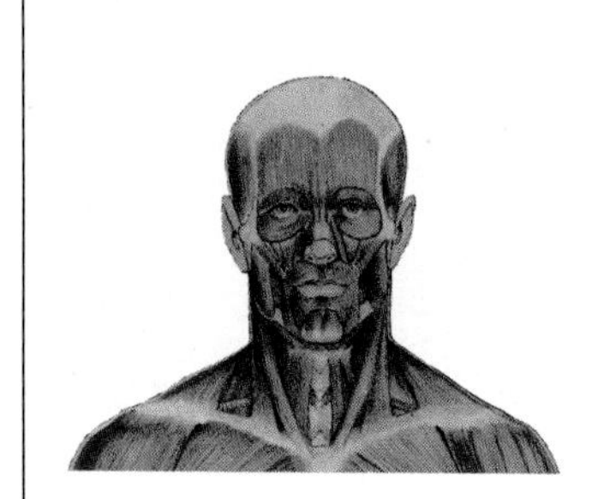
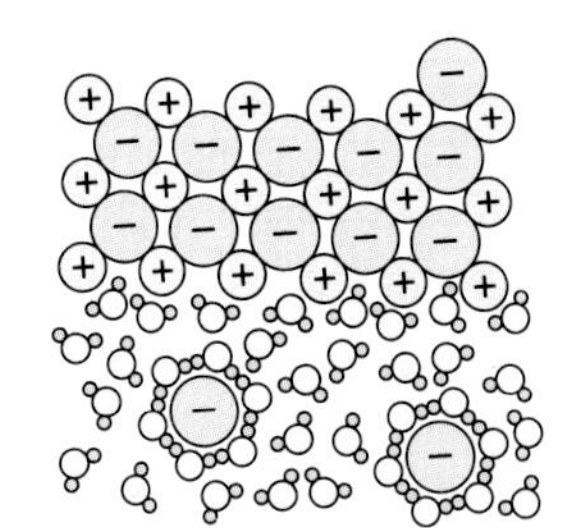
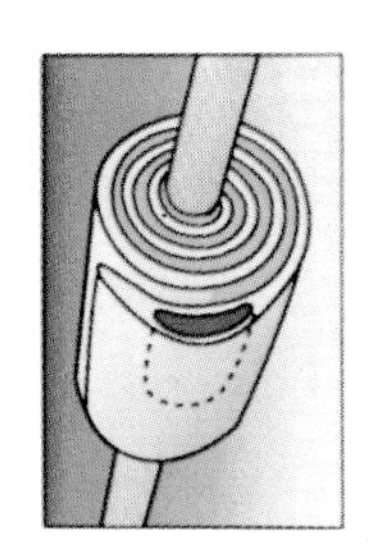
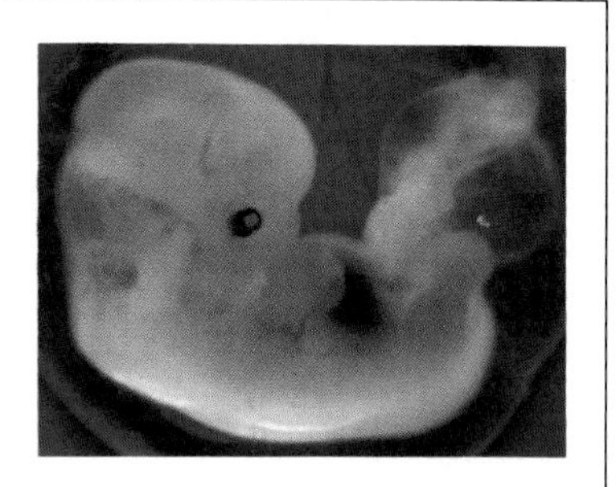

Brief Contents

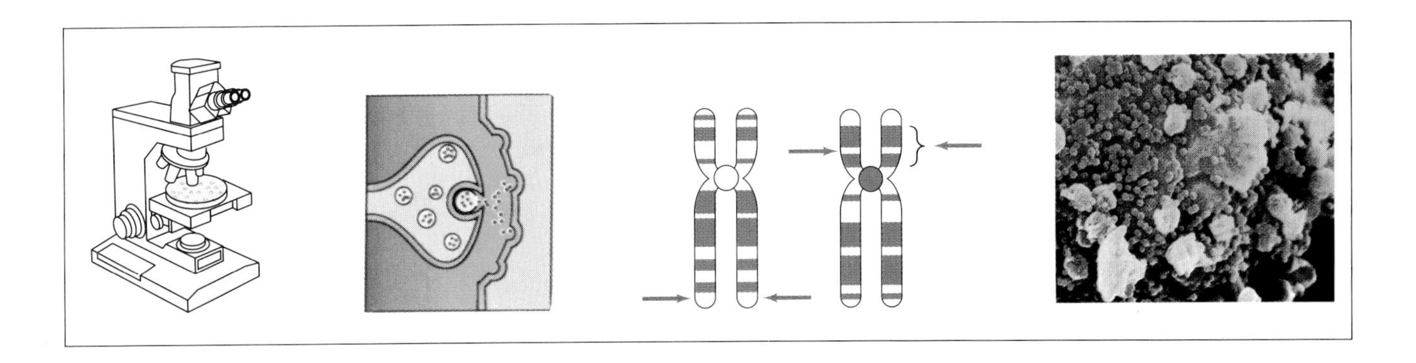

Contents

Preface to the Instructor

The present text is an outgrowth of my 20 years of teaching introductory biology to more than 5,000 students at three different universities. It represents the culmination of more than a decade of text writing and refining, all of which I have continually tested in my own classes.

Obviously, each of us as teachers must evolve our own individual approach to the presentation of a discipline which we love enough to have made it the center of our professional lives, and this text will not work for every instructor in every course. Let me mention, however, what I perceive to be the virtues of this book:

1. *The Audience.* I have endeavored to write to the audience I teach, which is to say, to non-science majors for whom biology is a general education course (and one which many are none-too-anxious to take). Most of the texts I see are, in my view, written more with an eye to impressing the instructor with their completeness and currency than to meeting the needs of the students, especially GE students. I have attempted to serve the students adequately, but without sacrificing completeness and currency, by including explanation along with the traditional description.

2. *Orientation.* Most texts jump rather promptly into a small-to-large discussion of biology, beginning with a mini-course in chemistry. A few offer a brief discussion of unifying principles in what amounts to an extended Preface. I rarely see an effort to position Biology in the academic firmament. However, I believe that GE students appreciate knowing something of the underpinnings of a discipline, its values, and its history. Therefore, I have tried to address those needs with a two-chapter overview of Biology that covers major historical developments from the Greeks to the beginning of the 20th century. My own students have responded very well to text drafts of these materials, even before illustrations were available.

3. *Focus.* My focus is the human organism. For obvious reasons, GE students are more interested in how their own bodies work than in how, for example, an earthworm's body works. Since the principles associated with how humans and earthworms make a living are the same, it simply makes sense to me to illustrate those principles in a manner that maximizes attention and interest—and, as a consequence, learning.

4. *Style.* I realized that text writers had gone too far the day I saw a student's text in which every word in one chapter was highlighted. The outline style of writing is great for packing a lot of material into a small space; it is deadly to read, however, and is generally lacking in explanation. I have adopted more of the narrative style favored by GE students who are bright and literate students despite their often expressed aversion to biology (and to science generally). Ideally, they should enjoy reading their biology textbook. That might be too much to expect, but I certainly do expect them to understand and appreciate biological concepts and principles, as opposed to merely memorizing a directory of terms.

5. *Principles.* The danger, of course, is that one can go too far with the narrative approach, and end up with a purely descriptive anatomy and physiology text that resembles a set of Reader's Digest offprints ("I am Joe's Tongue", etc). I have tried to avoid that danger. Of the 25 chapters, only 10 directly discuss human organ systems, and in each of those the discussion begins with a section on "The Nature of the Problem," which is a presentation of the governing principles that shape the particular organ system. Homeostasis and evolution are the dominant themes, and they are introduced at the outset and reinforced throughout the text.

6. *Pedagogical features.* There are various ways in which I have endeavored to capture attention or to enhance the learning process. Each Part begins with a Part Lead, which presents an overview of that Part. Each Chapter begins with a "teaser"—a set of examples ending in questions to be addressed in that chapter—and ends with a narrative, not an outline-style, summary. In recognition that understanding how things work sometimes is most easily accomplished by examining what happens when things go wrong, I provide many examples of diseases and similar problems in each chapter, but I do so in the form of information boxes, so as to avoid breaking up the flow of the main text. There is a list of Key Terms at the end of each chapter, and a set of chapter questions.

7. *Art Program.* My editors have helped me to assemble a fine collection of photographs and illustrations. The special section containing the award-winning anatomical artwork of Vincent Perez is especially noteworthy. Drawn exclusively for Jones and Bartlett Publishers, it serves as an excellent reference for students.

8. *Supplements.* To facilitate the teaching process and to provide opportunities for course enrichment, we have developed the following ancillary materials to accompany *Human Biology:*

- Instructor's Edition. This special edition combines an instructor's section with the regular student edition. It consists of chapter overviews, learning objectives, topics for discussion, critical thinking/essay questions, and sources for lecture materials.
- Test Bank. Available to adopters, this resource is also available in computerized form. It contains approximately 1200 multiple-choice questions organized by chapter with answers.
- Student Study Guide. My goal in preparing this guide has been to help students master the topics and information presented in the text, and to hone their critical thinking skills. The contents include chapter overviews and outlines, learning objectives, reproductions of selected figures, answers to text questions, and self-tests.
- Transparencies. A set of full color transparency acetates of key illustrations from the text is available from Jones and Bartlett for class use by adopters.
- Videos. An award-winning video library on the human body is available for use by adopting institutions. Please contact the marketing department at Jones and Bartlett for details.
- Instructor Updates. These updates on some of the more significant developments in biology will be referenced to the text and mailed to adopters at the beginning of every semester.
- AIDS I SmartBook. This Macintosh-based, interactive software program is a learning and self-assessment tool for AIDS education. A site license is available to adopters. Contact Jones and Bartlett for details.
- *The Biology of Aids, Second Edition,* by Fan, Conner, and Villarreal. This brief paperbound textbook provides the nonspecialist with a firm scientific overview of AIDS. It is useful to instructors desiring expanded coverage of this topic. Contact the Jones and Bartlett marketing department about a special combined package with *Human Biology.*

9. *Length.* I have written the book to be somewhat longer than some instructors can easily handle in one semester, and have included more chapters than most instructors will want to handle in one course in recognition of the fact that not everyone wants to treat the same topics every semester. However, I have not intended that each topic be addressed in some ultimate, complete way. Instead, my interest was to limit the overall size of the book to something under the size of a Sears Roebuck Catalog (a feat less and less commonly achieved among introductory books these days), so as not to seem quite so daunting to students.

10. *The Author.* Finally, and for what it's worth, unlike a surprisingly large number of recent texts, this book was actually written by a classroom instructor with a doctorate in biology (a fact which may also account for its rough edges), rather than by a professional writer. As such, its concepts and foci have been tested in the classroom while the book was in development. I am personally convinced that makes for a stronger book, but I leave it to you, my fellow instructors, to affirm or rebut that conclusion.

Preface to the Student

As a subject for study, biology can be both easy and difficult. Its ease stems largely from its familiar subject matter—all of you have "studied" biology ever since you first watched worms wriggling on a sidewalk after a rainstorm, or collected fall leaves, or wondered at the colors of flowers. Biology's difficulty comes partly from its enormous and interlocked scope, and partly from the phase through which it is now passing.

Biology is an interlocked discipline, unlike fields in the humanities. It is possible to specialize in American literature, for example, without once looking at French literature. It is possible to become an expert in the history of the United States without paying more than passing attention to Oriental or ancient history. It is no longer possible to study one of the subdisciplines of biology and remain ignorant of the rest. As recently as 40 years ago, that might have been possible, for much of biology was still descriptive at that point, and one could study "natural history" without reference to laboratory studies. The union of genetics and evolution into what is called "the modern synthesis," and the advent of molecular biology has changed our outlook, for now life in all its diversity is seen as having both a common origin and a remarkably large number of shared cellular processes. Knowledge of these processes is indispensable in understanding biology as a discipline today, but because they are primarily chemical in nature, some knowledge of chemistry is also required—and so it builds.

Biology is rapidly moving from a descriptive to a theoretical science, just as physics and chemistry did before it (and as perhaps some of the social sciences will ultimately do). At the moment, however, its theoretical base is still embryonic, and the descriptive part is still vitally important. Thus, biology cannot be reduced to a handful of laws and principles as can physics—yet neither is it any longer simply an assemblage of facts from which one can randomly sample for awhile, as from a smorgasbord table, and come away both replete and wise in the science.

The task of a professional biologist writing a biology textbook in the last decade of the 20th century, then, is to provide a balance of theory and description that tells the reader where biology has been, what it now is, where it is going, and why—and all in a space concise enough to be covered in the confines of a one- or two-semester course! That is a daunting challenge, and one which no one has yet entirely met.

The approach followed in this text is to focus on the processes of homeostasis and evolution by natural selection, terms with which you will become intimately familiar as you read through the book. These processes are exemplified primarily by a detailed examination of the human as a representative species. The human species has the virtue not only of being relatively well studied, but it also tends to be of greater interest than are other species to those students for whom this course will represent both a beginning and an end to their formal study of biology.

I have also written a Student Study Guide to accompany this text. My goal in preparing the study guide has been to help you master the topics and information presented in this textbook. The contents include chapter overviews and outlines, learning objectives, reproductions of selected figures, answers to text questions, and self-tests.

Some knowledge of biology is central in any educated person's competencies. I hope you find this text worthy of your attention.

Donald J. Farish

Santa Rosa, California
January, 1993

Acknowledgments

First and foremost, I wish to acknowledge my wife, Ellen, without whose assistance this book could not have been written. She rendered most of the conceptual art and some of the representational art, tirelessly proofread manuscript, delivered countless parcels to Federal Express, and provided encouragement whenever it was needed. I could not have asked for a more supportive partner.

I also wish to acknowledge Vincent Perez and his associates, who are responsible for what I believe to be the finest anatomical artwork ever to grace the pages of an introductory biology text.

My thanks as well to the staff of York Graphic Services, especially Mary Jo Gregory and Marilyn James, and Judy Songdahl and Art Bartlett of Jones and Bartlett Publishers, all of whom pushed and prodded at just the right times.

Finally, I wish to thank those individuals, friends and strangers alike, who critiqued manuscript and art, provided references, and in general helped to make this a much better book than I could have ever accomplished on my own. They include Paul Benko, James Christmann, Galen Clothier, Wesley W. Ebert, and David Hanes, all from the Department of Biology at Sonoma State University; and James Gale, Matthew James, Donald D. Marshall, Susan McKillop, and Duncan Poland, also of Sonoma State. As well as:

Ray Barnett
California State University, Chico

Edmund E. Bedecarrax
City College of San Francisco

Michael Bell
Richland College
Dallas, Texas

Robert P. Breitenbach
University of Missouri

Jean DeSaix
University of North Carolina

Sheldon R. Gordon
Oakland University
Rochester, Michigan

Gene Kalland
California State University, Dominguez Hills

Richard Klotz
State University of New York, Cortland

Mary Ann Klouda
California State University, Sacramento

William H. Leonard
Louisiana State University

William D. O'Dell
University of Nebraska, Omaha

Gerald Summers
University of Missouri

Jeff Thompson
California State University, San Bernardino

PART

I

The Origin of Modern Biological Thought

What is biology? How did it come into being as a science? What are its relationships with the other natural sciences? What are its governing principles, and how did they arise?

The following two chapters are devoted to a brief review of the history of biology and its unifying theme—evolution. Spending time reviewing the history of biology does much more than simply provide a set of names, dates, and facts for use on examinations. It also provides an appropriate sense of perspective. We would be missing the point if we simply looked at what the ancients believed, and then shook our heads at their naivete and our superior wisdom. Learning *why* they believed as they did allows us to understand why it is that we have come to believe as we do.

As we begin, keep two things in mind. First, biology is not merely a staggering array of facts from which a number of underlying principles have been derived. Biology, like all the natural sciences, is also a process—a set of rules and procedures—by which these facts and principles have been determined. These rules and procedures have changed profoundly over the centuries. Any temptation to sneer at the hodgepodge of facts, fears, and myths that constituted ancient biology should be mitigated by an understanding that the ancients were following different rules.

Second, the history of biology represents a wonderful example of cultural evolution—the passing on of findings and beliefs from one generation to another—which itself is peculiarly human.

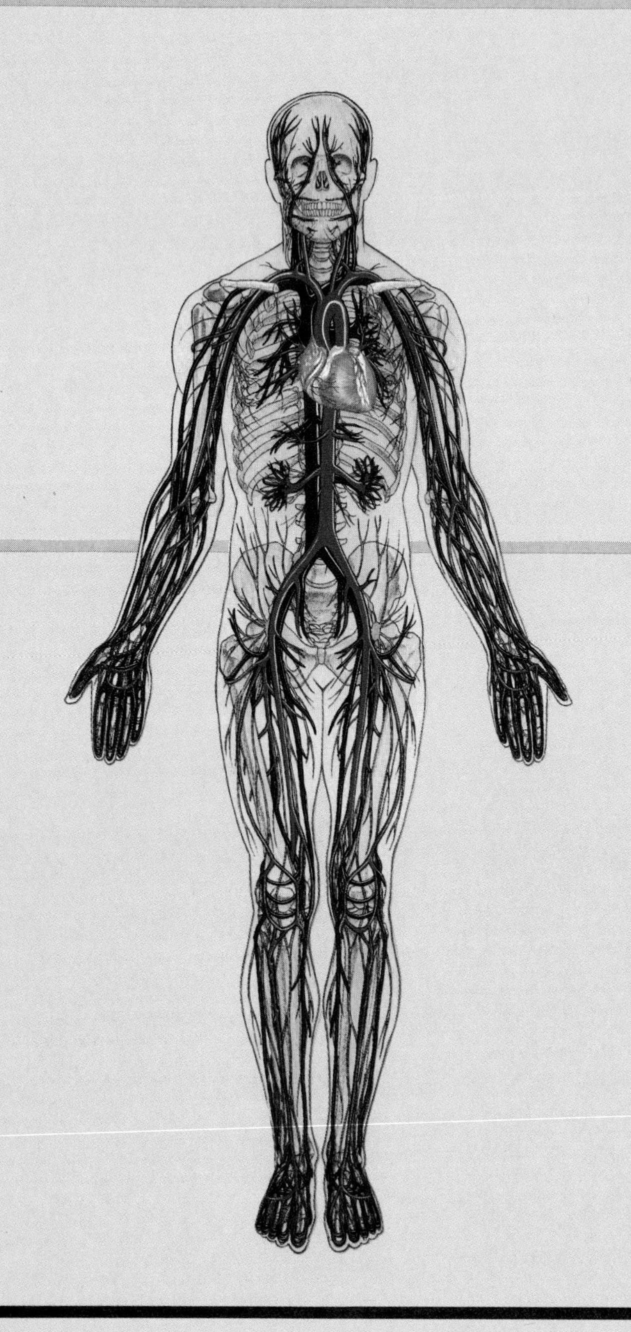

Some of the major organs of the human body. Correctly identifying their function has been a long and protracted process.

We are simply the most recent recipients of the accumulated wisdom of the human race, a wisdom to which each generation has made its contributions. Consider the way in which you have learned basic biology. You all know some of the functions of most of the large organs of the body, for example—but *when* did you first learn that the heart pumps blood, or that the brain is the seat of intelligence, or that the lungs provide oxygen to the body? More to the point, how many of these facts have you *verified* by direct observation? If you are like most people, you cannot remember when and how you learned these facts, and you have not verified any of them experimentally but believe they are true nonetheless. You will have assumed, quite rightly, that if these facts were not true, someone would long ago have proved otherwise.

Of course, the notion that facts need to be discovered and proved before they can be regarded as facts is precisely the point. Not only did each of the facts mentioned above have to be discovered, but frequently preexisting beliefs had to be repudiated and replaced. Scientists had to question prevailing wisdom and to test their doubts and beliefs. As a consequence, our heritage of biological information is now so huge, of such long duration, and so thoroughly inculcated into our society, that only with difficulty can we look backward to a time when it did not exist.

Cultural evolution is very much an ongoing process. For example, it was not until the 1970s that the proportion of young women going to college equaled that of young men, a legacy of the belief of many people that higher education was wasted on women since they just got married and had children anyway—a belief that most of you presumably do not share. However, it is all but certain that some of *your* more cherished beliefs will be rejected, if not ridiculed, by the generation that follows you. Therefore, we should resist the temptation to be smug about *what* the Greeks or Romans believed a hundred or more generations ago, and instead concentration on *why* they believed as they did, and *how* it was that the beliefs we now hold came into being.

A review of the history of biology also has pragmatic value because, for the beginning student, it provides the opportunity for a quick overview of the entire discipline. We could, of course, bypass these topics and plunge directly into the heart of biology, but that would be like trying to assemble a jigsaw puzzle without peeking at the cover of the box to see how the completed puzzle is supposed to look. Even if we study each piece in great detail, we would likely be no further ahead in deducing the appearance of the entire puzzle. Far better that we first stand well back from the pieces of the puzzle and get a rough sense of how they are supposed to look when assembled.

CHAPTER

1

The Beginnings of Biology

Most educated people recognize the names of Aristotle and Hippocrates and can correctly identify them as Greek naturalists who lived several hundred years before the birth of Jesus. If biology had such an early beginning, and given that progress in the sciences is generally additive, why has our understanding of biology been so slow to develop? Why, for example, did people believe (until the seventeenth century) that blood was converted to sweat, as opposed to being circulated through the heart? Why was it widely believed from the seventeenth to the nineteenth centuries that sperm cells contained tiny people?

We shall consider these questions, and others, as we explore the development of biology as a science.

WHAT IS BIOLOGY?

The first thing that biology teaches us is the shortcomings of a dictionary. A dictionary tells us that "biology" is from the Greek word *bios,* which means "life." What the dictionary does not tell us, however, is how enormously varied are the forms of this study. For some biologists, biology is the study of living organisms—how they are constructed, how they function. For others, it is the way in which organisms interact with each other or with the nonliving environment. For still others, biology is the study of life itself—what it is that distinguishes life from nonlife, how life first arose, how it replicates itself (see Figure 1.1).

Biology was not always this complex.

BIOLOGY, GREEKS, AND ROMANS

In the most literal sense of the word, humans have always been biologists. From the earliest glimmerings of our existence, we have been dependent on plants and animals for our food. Blessed with a large brain yet cursed by the absence of great physical prowess, knowledge of the habits and habitats of the organisms that we consume has been indispensible for our survival as a species. More than once, so-called "primitive" societies have pointed out to professional biologists species of birds or fish that the biologists had overlooked. That type of knowledge must have been present from our very beginnings as a species.

Early Greek Naturalists

From the Western perspective, the existing written record traces the study of biology to the ancient Greeks. (There are smatterings of writings from the Babylonians and the Egyptians, but they do not permit a definitive statement to be made about the level of their biological studies. Records from the Chinese and Indian cultures are more complete, but neither became integrated into the subsequent development of biology in Western Europe.)

The quality of early Greek biology was frequently impressive. For example, during the sixth century B.C., Alcmaeon, a Greek living in southern Italy, studied the embryonic development of the chicken, and through dissections described the nerves of the eye and discovered what we now know as the eustachian tube, which links the middle ear with the throat. He also distinguished between arteries and veins, and identified the brain as the seat of intelligence. Xenophanes, a contemporary of Alcmaeon, correctly interpreted fossils he found in the mountains as the

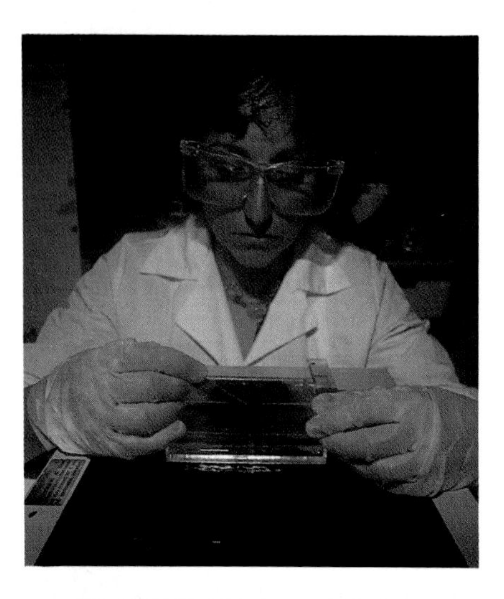

FIGURE 1.1

A Variety of Biologists

Biologists undertake the study of their discipline in a wide variety of ways. Each is a legitimate approach to this very broad area of science.

remains of marine organisms, and concluded that the mountains must at one time have been under water.

Hippocrates (460–377 B.C.) has been called the father of both biology and medicine (biology having first been studied because of its medical importance). To most of the ancient Greeks, illnesses were commonly ascribed to the actions of gods or demons. Hippocrates differed. In writing about epilepsy, which had been termed the "sacred disease," Hippocrates said (in what sounds like a very modern point of view):

> It seems to me that the disease called sacred is no more divine than any other. It has a natural cause, just as other diseases have. Men think it divine merely because they do not understand it. But if they called everything divine which they did not understand, why there would be no end of divine things!

The greatest of the Greek naturalists was Aristotle (384–322 B.C.; see Figure 1.2). After studying with Plato for about 20 years, Aristotle became the tutor of Alexander the Great. He later moved to the island of Lesbos, where he conducted most of his biological investigations, and then to Athens, where he established a school.

FIGURE 1.2

Aristotle

Aristotle, the student of Plato and the teacher of Alexander, was the greatest of the Greek naturalists. Although his scientific observations were excellent, he validated none of his assumptions experimentally and consequently came to many erroneous conclusions.

We know from the surviving body of his works that Aristotle not only made many observations but also derived deductions, or what we would now call hypotheses, from these observations. Aristotle's importance to the development of biology is so great that it is difficult to overstate. From dissections, he described in detail the urinary and reproductive systems of mammals, and provided accompanying drawings. He made extensive and impressively accurate observations of a wide variety of marine organisms. However, perhaps his most important work was to group organisms of similar type, apparently the earliest serious attempt at classification (see Figure 1.3).

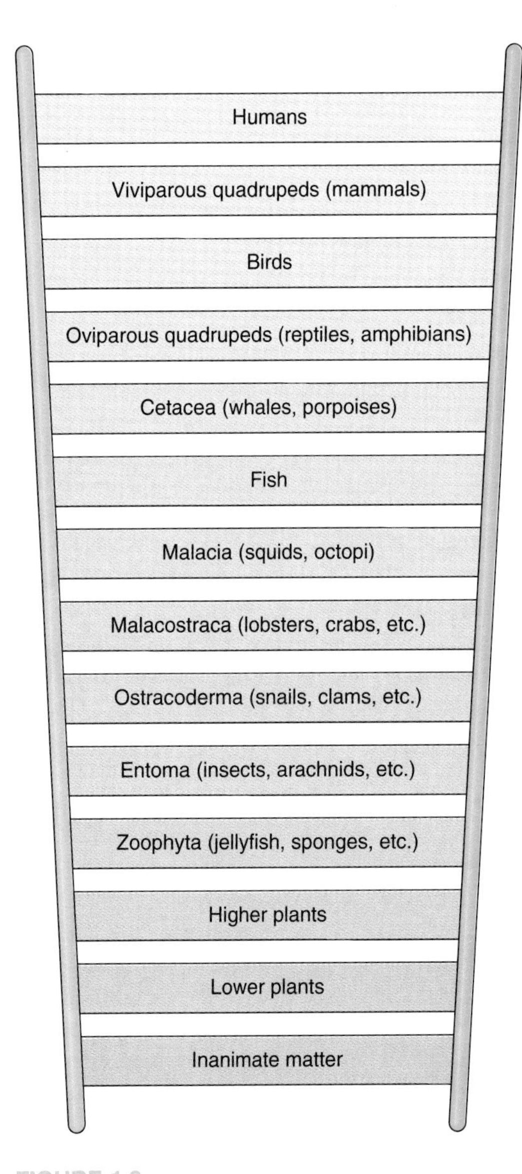

FIGURE 1.3

Aristotle's *scala naturae*

Aristotle saw living organisms as occupying distinct steps on the path to perfection. Are you surprised to see what organism he saw as being closest to perfect?

Although the categories to which Aristotle assigned animals have in most instances been redesignated, he so accurately characterized mammals that he recognized porpoises and whales to be mammals rather than fish. He correctly pointed out that whales had lungs and breathed air, that they were warm-blooded, and that they gave birth to live young and nursed their young—traits also possessed by land mammals. He also correctly distinguished between bony and cartilaginous fish.

Aristotle's arrangement of organisms in a single graded *scala naturae* ("ladder of life") combined acute observation with metaphysics. He used the degree of "perfection" possessed by organisms in assigning them to particular rungs of the ladder. In illustrating what he meant by "perfection," Aristotle included such things as the degree of development at birth and "powers of soul" (whether vegetative, as in plants, sentient, as in animals, or rational, as in humans). Aristotle's views prompted speculation about evolution—but not until 2000 years after his death!

Socrates, Plato, and Alexandria

By the time of Aristotle, in the fourth century B.C., Greek biology was beginning to blossom—indeed, as the above examples show, it had already reached an impressively advanced level. Because of the pervasive influence of Greek culture on Western civilization, we might have expected that biological knowledge would have continued to grow, perhaps even rapidly, long after Aristotle's death. Yet such was not the case. Indeed, not only was there not continued growth but much of what was known by Aristotle was soon forgotten and essentially lost for more than 1500 years. Why was this emerging flower of knowledge nipped in the bud?

Although Aristotle was the student of Plato, who in turn was the student of Socrates, in many respects he followed the philosophy of their predecessors. Socrates and Plato, in contrast, emphasized ethics and abstract thought rather than natural history. With the breakup of Alexander's empire, and the gradual decline of Greek democracy into highly politicized city states governed by tyrants, ethics and politics became more important than natural history. Moreover, in the slave society of ancient Greece, there was a strong predilection against hands-on work by free men and women, and even the scientists therefore tended to be thinkers rather than doers.

Thus, although Aristotle had followers, most notably the botanist Theophrastus (380–287 B.C.), Greek biology essentially peaked during the fourth century B.C. To be sure, the subsequent decline was long and slow, and there were occasional bursts of creativity. Indeed, Ptolemy,[1] one of Alexander's generals, was himself a student of Aristotle, and when he took over the Egyptian portion of Alexander's empire, he established a magnificent museum in Alexandria, which acted as a magnet for the most prominent of the Greek scientists. The museum was more like a modern research institute, with working facilities, grants for scholars, and a library that, by the time of Julius Caesar, comprised over 700,000 scrolls and contained the collected wisdom of the Western world at that time.

Ptolemy's motives were not entirely noble. Science became the servant of religion, for religion was a way of maintaining power over the masses. However, human dissection was permitted during the reigns of the early Ptolemies, and knowledge of human anatomy was substantially enhanced. In the third century B.C., Herophilus correctly determined the role of the brain (Aristotle had disagreed with Alcmaeon and erroneously attributed intelligence to the heart), and distinguished motor from sensory nerves. He also added new information on the digestive system and liver, and correctly distinguished between tendons and nerves (which Aristotle had confused). Erasistratus, a contemporary and rival of Herophilus, observed and identified the valves of the heart, but erroneously believed that the arteries contained air (they are frequently devoid of blood in dead animals).

Roman Conquest

The Romans conquered Greece in the second century B.C. and occupied Egypt a century later. Their interest in biology was limited to the use of plants in medicine. The Romans disseminated Greek biological knowledge but added little to it themselves—and most of what was added was done by Greeks living under Roman domination.

An important exception was Pliny the Elder (A.D. 23–79; see Figure 1.4), a Roman civil servant, who took it upon himself to catalogue virtually all that was known of natural history in a 37-volume treatise that was, for hundreds of years to follow, the bible of science. Unfortunately, Pliny showed very little inclination to separate fact from fable, and included a great deal of nonsense intermingled with accurate observations. For instance, he reported that grain diseases could be prevented if a toad were carried around a field at night and then buried in a pot. In addition, he catalogued strange races of humans who had only one eye, or a single giant foot, or feet that faced backwards. Regrettably, Pliny's encylopedia became the

[1]Ptolemy's descendants adopted his name and ruled Egypt for 300 years. Cleopatra and her brothers were the last of the Ptolemies.

FIGURE 1.4

Pliny the Elder
The most famous of the Roman naturalists, Pliny compiled all that was known of natural history at that time into his renowned multivolume work. Although Pliny combined fact as well as fantasy, his work became the standard scientific reference for hundreds of years.

font of biological wisdom during the Middle Ages, and his statements were accepted as absolute and unchallengeable.

Galen (A.D. 131–200), a Greek living under Roman domination, was the last of the great biologists of antiquity. He was particularly interested in human anatomy and physiology, but the culture of his day prevented him from dissecting humans. Thus he was forced to extrapolate from dissections of other animals, often with devastating results.

Galen was an early proponent of experimentation (or more correctly, demonstration). By tying the ureters, he showed that the kidneys, not the bladder, produce urine. By tying an artery at two points and then cutting it open, he showed that arteries carry blood, not air. He also demonstrated the role of the spinal cord by severing it in living animals and assessing the extent of paralysis that resulted.

In other areas, most notably blood circulation, he was very wrong indeed. Galen was strongly influenced by his belief in a divine and purposeful creator (although he was not himself a Christian), and these beliefs biased his scientific work. Thus he felt obliged to integrate his beliefs of how air and spirits were infused into the body with his views on blood circulation and, in a very arbitrary way, he assigned functions to various organs so as to unite his anatomical findings with his philosophical beliefs. In so doing, he forged a very strange, but as it turned out, enormously persistent, physiology.

BIOLOGY IN THE MIDDLE AGES

Although Rome did not fall until the fifth century A.D., biological investigation virtually ceased with the death of Galen. The decline in biological investigation that had commenced with the death of Aristotle, interrupted only occasionally by flashes of brilliance and scholarship, finally reached its nadir. Indeed, in most respects, Western civilization was "on hold" from about 200 to 1200 A.D. What brought about this long period of scientific dormancy? What caused the reawakening?

Biology and the Church

The causes of the decline and fall of biology were many and complex. The absence of printing and poor communications hindered the dissemination of knowledge. Epidemic diseases, particularly malaria and bubonic plague, and the warring nature of Roman society did nothing to foster a climate in which biology could thrive. The dominant philosophies of the Stoics and Epicureans emphasized subjectivity over objectivity, and speculation over observation. Astrology and superstition, the enemies of science, were in the forefront.

The rise of Christianity has often been linked to the decline of science, but that decline began long before Christianity emerged as the dominant religion of Europe. The attitudes of the early Church, however, hardly promoted interest and growth in science. Christians were told to read Genesis for the creation of the earth, the Prophets for philosophy, and Kings for history. Tertullian, an early Christian writer, said there was no need for curiosity or inquiry. Science was to be sacrificed to the authority of scripture. Learning was predominantly theological and moral, because the focus was on the hereafter, and saving the soul became much more important than saving the body. Indeed, in 1130, monks were prohibited from practicing medicine (they were, at the time, virtually the only people in Europe with any formal education) because Church authorities thought such practice interfered with their religious work.

Over the centuries, errors in copying led to profound changes in the works attributed to the ancients (see Figure 1.5). These works were changed even more when Church leaders, such as Thomas Aquinas, ed-

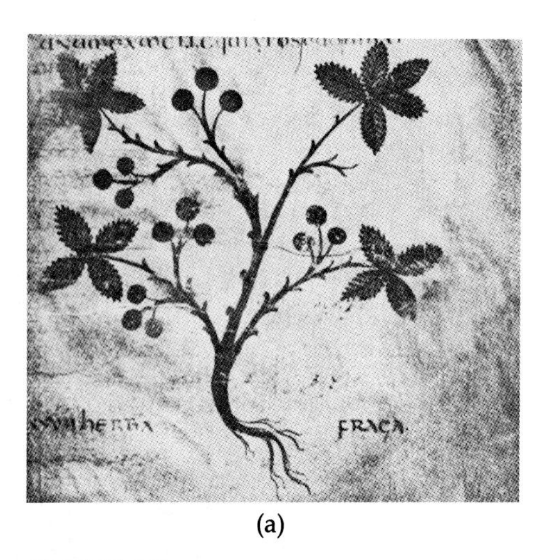
(a)

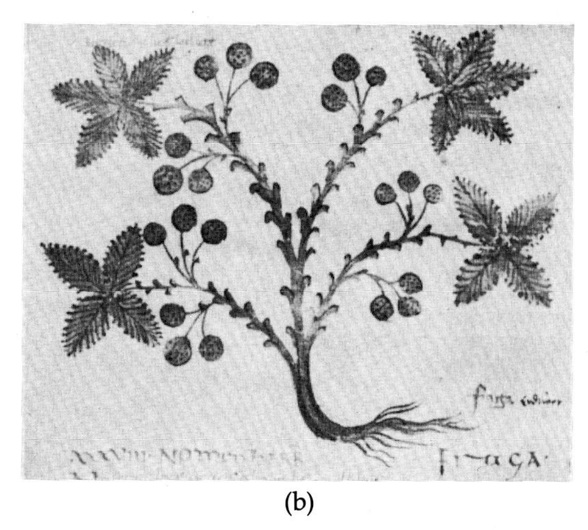
(b)

(c)

FIGURE 1.5

Herbal Illustrations

As these illustrations from ancient herbals show, depictions of actual plants became more and more stylized, to the point that they soon became useless as guides to plant identification. (a) A reasonably accurate copy, made c. 500 A.D., of a herbal from c. 50 B.C.. (b) An illustration from c. 1050 A.D.; do you think this drawing accurately depicts a strawberry plant? (c) By 1200 A.D., the illustrations were so stylized that they were useless for field identification, the purpose herbals were intended to serve.

ited out "discrepancies" with Church doctrine, thereby creating a strange melding of science and scripture. Anyone with the temerity to challenge the statements of Aristotle, Galen, or other early biologists ran the risk of being labeled a heretic and threatened with burning at the stake.

The West slept, dreaming of heaven and a life hereafter, while the scientific knowledge of the Greeks slid slowly back into the abyss of ignorance from which it had so laboriously been freed.

Arabic Science

When Plato's Academy was closed by the Christians in 529 A.D. (because Plato had been a heathen!), many of the exiled scholars (and their libraries) moved to Persia. In the century following the death of Mohammed in 632 A.D., his Muslim followers swept west across North Africa to Spain, and east to what is now Pakistan. Thus, with the conquest of Persia by the Arabs, much of the knowledge of the ancient Greek biologists fell into Arabic hands. Leadership in the sciences remained with the Arabs until the thirteenth century. However, this leadership was not so much the result of active research as it was the preservation of Greek knowledge, which was superior to the smatterings of science retained by the West.

The Crusades, and campaigns in Muslim-held Sicily and Spain, provided contact between Christians and Muslims from the eleventh to the fourteenth centuries, and the Western world became aware of the superiority of Arabic science. Works in Arabic, which had been translated by the Arabs from the Persian, were in turn translated into Latin by the Christians. Only much later were the original Greek manuscripts located and translated directly.

Universities and Scholasticism

In the twelfth and thirteenth centuries, universities began to be founded throughout Europe but especially in Italy. These institutions were very different from the universities of today in both design and function. However, their very existence was a dramatic departure from the past because their affiliation was generally to a city or region rather than to the Church. As such, they represented a break with tradition, and a reflection of the growing interest in learning.

In other respects, these early universities were very traditional institutions. The focus of attention was the study of ancient texts, not the conducting of new investigations. Universities sharpened wits rather than training the senses. This system of thought came to be known as **scholasticism,** and was characterized by an interest in words rather than things. Many of the university masters were, after all, clerics, and they overwhelmingly retained the traditions fostered for so many centuries by the monasteries.

Nonetheless, the blinders had been loosened, and over the next four centuries they were to fall completely. The great awakening was at hand.

RENAISSANCE BIOLOGY

The causes of the Renaissance have been the subject of extensive debate, and a full discussion of them is beyond the scope of this book. Contact with the East, because of the Crusades, trade, and the occasional explorations of a Marco Polo, all fostered an intense curiosity about life outside Europe. The medieval alchemists and physicians of Europe realized that the Arabs were far ahead of them (indeed, medicine was at that time known as "Arabic science"), and wanted to learn their secrets.

Interest in biology manifested itself on several fronts. An unexpected source of this interest was art. Art changed dramatically during the fifteenth century, as artists abandoned the two-dimensional, very stylized patterns of the past, and moved with great vigor into naturalism (see Figure 1.6). Their desire to portray the human figure naturalistically made them very interested in anatomy (see Figures 1.7 to 1.9), and some of them—most notably Michelangelo and Leonardo da Vinci—practiced grave robbing in order to perfect their knowledge.

Anatomists

Artists were not the only professionals with an interest in human anatomy. Lawyers were also interested in anatomy for what light it could shed on causes of death in instances where murder was suspected. In the fourteenth century, the medical school at the University of Bologna was subordinate to the law school, and so began legal human dissections—for the first time in 1500 years.

In 1543, the same year Copernicus published his heliocentric theories that were to rock the physical sciences, Andreas Vesalius (1515–1564), a young Flemish physician, published his findings in human anatomy and rocked the biological world (see Figure 1.10). In a magnificently illustrated treatise, he pointed out error after error in Galen's anatomy, and incurred the wrath and enmity of the more narrow-minded of his colleagues (who formed a sizable majority).

From our perspective, it is perhaps surprising that Galen's errors went uncorrected for as long as they did, and that much of a fuss occurred when they were finally identified. However, human dissection had

FIGURE 1.6

Botticelli's *Primavera*

In contrast to the herbal illustrations is Botticelli's depiction of various trees and flowers, drawn in 1487. It is so accurate that at least a dozen can be identified to the species level. His human figures are less successful.

FIGURE 1.7

Stylized Art

Margaritone's *Virgin and Child Enthroned,* from the late thirteenth century is two-dimensional and unrealistically proportioned.

FIGURE 1.8

Perspective in Art

Fra Angelico's *The Annunciation* (c. 1445) is one of the earliest works illustrating perspective. Note the efforts undertaken to suggest depth, in contrast to the two-dimensionality of Margaritone's work. The human figures, however, are decidedly unlifelike.

FIGURE 1.9

Naturalism

Leonardo da Vinci's *Madonna of the Rocks* (London version, c. 1506) illustrates mastery of both perspective and the human form.

FIGURE 1.10

Andreas Vesalius

Vesalius demonstrated mastery of both anatomy and perspective in his 1543 text. This is a self-portrait of the scientist and one of his subjects.

only recently been made legal. Moreover, it was performed in the medical schools by relatively ill-educated demonstrators while the professors read Galen aloud to the class. Finally, because Galen's work had been thoroughly infused with religious dogma, and because, until the seventeenth century, critics of these works were sometimes burned at the stake, it is understandable why the early Renaissance anatomists may have been reluctant to point out Galen's errors.

Some of Vesalius' contemporaries attempted to account for the differences Vesalius uncovered. For instance, the bent femur described by Galen "straightened" over the centuries because of the use of trousers. Renaissance man had only six bones in his sternum, not the seven found by Galen, because modern humans were degenerate. As ingenious as these arguments were, however, the damage had been done. The ancients were not infallible. They were not all-knowing. The dam had been breached, and a flood of biological research gushed forth.

Later in the sixteenth century, Bartolomeo Eustachi (1520–1574) rediscovered—and named—the eustachian tube (which had been first discovered by Alcmaeon almost 2200 years earlier but, like so much of Greek biology, long since been forgotten). Eustachi also demonstrated the existence of the sympathetic nervous system (see Chapter 14), and corrected a number of Vesalius' own errors in the circulatory and nervous systems.

Gabriele Fallopio (1523–1562), a pupil of Vesalius at the University of Padua, extended Vesalius' work with the nervous and reproductive systems, and gave his own name to the tubes that link the ovaries with the uterus. Fallopio was, in turn, succeeded by Hieronymo Fabrizzi (1537–1619) who taught at Padua for almost 50 years. Fabrizzi discovered the valves in veins, applied mechanical principles to the movement of muscles, and did pioneering work in embryology. With Fabrizzi, interest in anatomy began to shift from the purely descriptive to the functional. Physiology was about to come into its own.

Experimentalists and Empiricism

William Harvey (1578–1657), a graduate of Cambridge University, came to Padua in 1597 to study with Fabrizzi (see Figure 1.11). Five years later, he returned to England, and for many years he both lectured in London and served as physician to King James I. In 1628, at the age of 50, he published a small book on blood circulation that stood biology on its ear. Building on the work of the anatomists, Harvey laid bare the enormous flaws of Galen's physiology. These were not minor differences in anatomy, explainable by copyist errors or changes in the human condition since

FIGURE 1.11

William Harvey

Harvey was among the first scientists to use an experimental approach to the investigation of biological phenomena.

the time of Galen. Rather, they went to the very center of Galen's philosophy and rendered Galen's whole system of beliefs sterile and useless.

Vesalius had been content to expose Galen's anatomical errors, but for the most part he continued to accept Galen's physiology. Not so Harvey. Harvey estimated the volume of the heart, and using a very conservative figure for heartbeat, he calculated that the heart pumps at least 30 pounds of blood in just half an hour. (The correct figure is over 200 pounds.) How could so much blood be manufactured by the liver from food and then sweated off by the body, as Galen had concluded? Then Harvey severed an artery in an animal, and watched as the blood flow slowed and finally stopped. How could that observation be reconciled with Galen's notion that the blood ebbed and flowed through the blood vessels?

Harvey theorized that blood flow must be unidirectional, and demonstrated this point by holding his fingers at different points along a vein, thereby showing that the veins filled up from one direction only (see Figure 1.12). He concluded:

> The blood is carried in one direction by the arteries, in the other direction by the veins, in so great a quantity that it cannot possibly be supplied all at once from the food that is taken into the body. It is therefore necessary to conclude that the blood in animals is impelled in a circle, and is in a state of ceaseless movement.

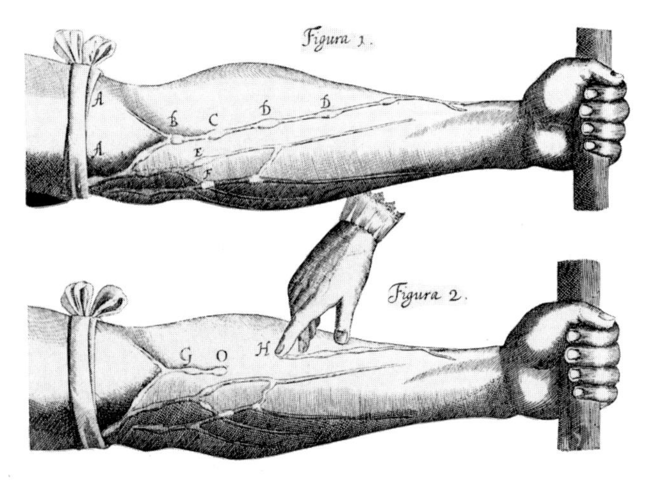

FIGURE 1.12

Harvey's Experiment

In these illustrations from Harvey's 1628 text on blood circulation, the simple but telling experimental design of Harvey's studies is readily apparent. Using a ligature to make the veins more obvious, Harvey demonstrated that pressing a finger against a vein caused it to drain on the side toward the heart. However, efforts to push the blood back down the vein were futile, because of the presence of one-way valves. Conclusion? Veins carry blood to the heart from the extremities in a one-way flow.

The age of experimentation—and the age of physiology—had begun (see Figure 1.13). Ironically, Harvey, by nature a very conservative man, did not see his work as issuing in a scientific revolution. To the contrary, he maintained he was simply returning to the traditions of Aristotle!

As it happens, Harvey's claim was not without merit, for both he and Aristotle had an empirical view of science. **Empiricism**—the reliance on direct observation and experimentation rather than on intuition and speculation—is the hallmark of modern science (see Chapter 2). Aristotle was not the pure empiricist that Harvey was, but he relied much more on observation than did the scholastics. Harvey carried the notion of observation to the level of experimentation (which we can think of as observations made under carefully controlled conditions). This approach was radically different from the scholastic school of thought, which still held sway in Harvey's time, but it was also a logical extension of Aristotle's methods. Thus, in linking his methodologies to those of Aristotle, who in the seventeenth century was a highly revered figure, Harvey neatly blunted much of the criticism that would otherwise have come his way.

Even as Harvey was initiating the experimental method, the philosophers were beginning to debate its merits. Francis Bacon (1561–1639) writing in the opening years of the seventeenth century, was a pure empiricist. He espoused the analytical, inductive methods to the exclusion of all else. He believed that by collecting all the facts, the results would automatically emerge. Above all, Bacon maintained that the scientist must not contaminate the purity of the process by consciously selecting the facts. This method is almost diametrically opposed to the methods of most of the later Greek and Roman naturalists, and the scholastics, all of whom used a deductive approach, and who were interested more in the logic of their arguments than in testing their thoughts experimentally.

Rene Descartes (1596–1650), writing a few years later, was (unlike Bacon) a practicing mathematician and scientist. (Indeed, he authored the first text on human physiology.) Because he stressed the importance of the scientist's own thought, albeit an objective and unbiased thought, Descartes differed sharply with Bacon. More importantly, unlike Bacon, he accepted the infinity of nature. One could never collect *all* the facts, as Bacon wished to do, because nature was infinite. Thus one had to espouse a method or process of ascertaining and selecting facts in a coherent manner.

Between them, Bacon and Descartes initiated the modern philosophy and methodology of science, even as Harvey was demonstrating what a successful methodology it was. Collectively, these three men were

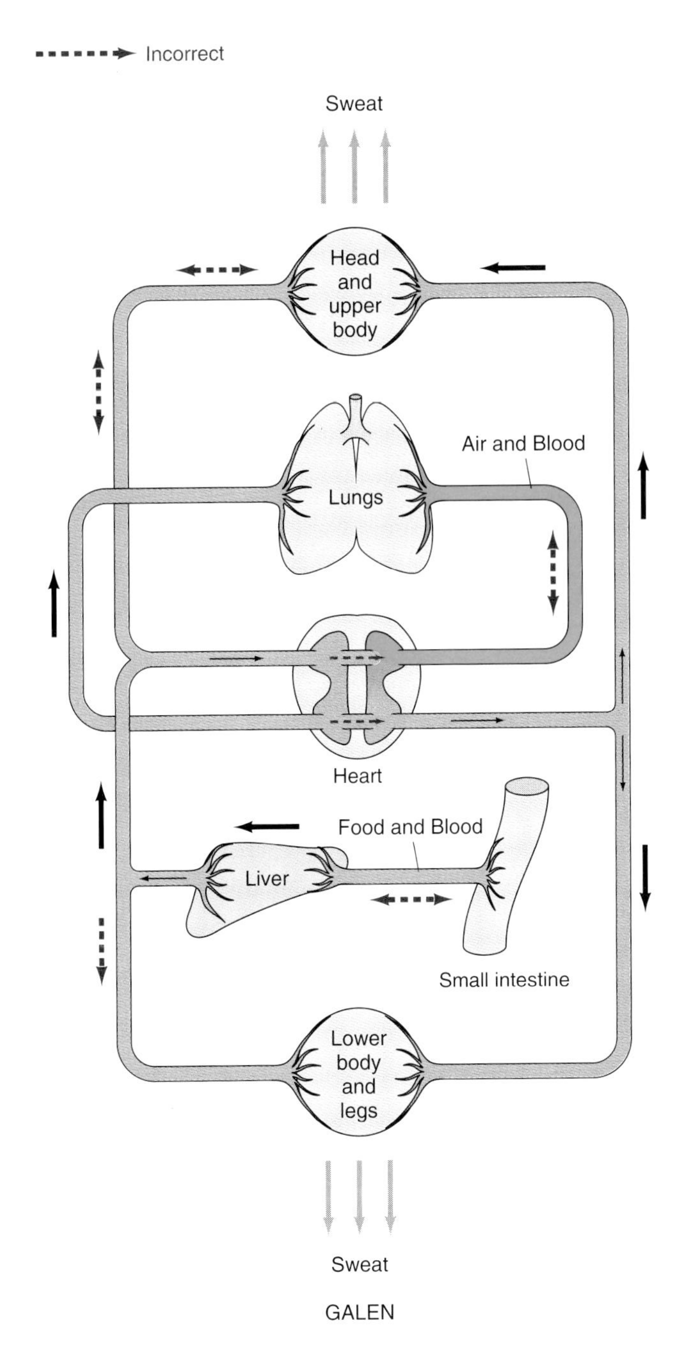

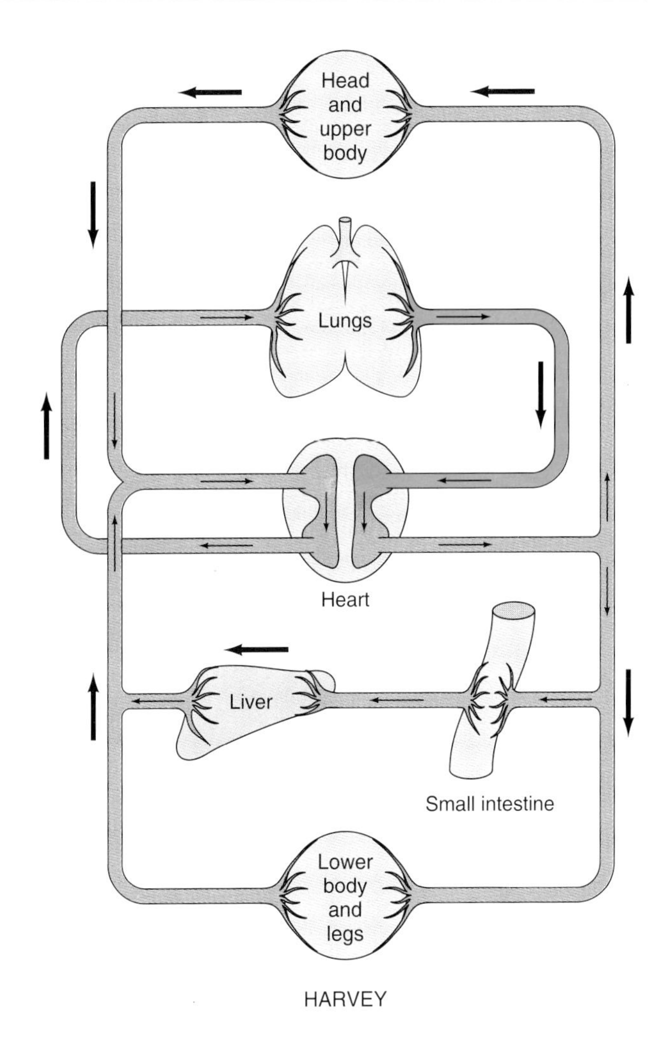

FIGURE 1.13

Galenic and Harveian Circulation

Juxtaposing Galen's and Harvey's theories of blood circulation illustrates their many differences. Note Galen's assumptions of two-way flow through the same vessel and movement of blood from one side of the heart to the other. His explanation of blood generation and loss (converted from food and lost as sweat) are not persuasive to the twentieth century mind.

responsible for much of the fervent scientific investigation that characterized the rest of the seventeenth century.

Humanism and Academies

In the early years of the seventeenth century, almost simultaneously in England, France, and Italy, wealthy patrons began to provide support and encouragement to many scientists. Not only did they provide funds, but their homes were often the sites where scientists (hitherto relatively solitary figures) were able to meet and confer (see Figure 1.14). The writings of men such as Bacon and Descartes, and the enormous impact of the scholarship of Harvey, encouraged these loose associations which, by the end of the seventeenth century, had blossomed into formal societies or academies.

The Royal Society received formal approval from Charles II of England in 1662, and the Academie des Sciences in France followed in 1668. Both societies soon began publishing reports from their members, and this constant dissemination of new scientific information provided a strong impetus for still more research.

Given the importance of universities in research today, why were such institutions not the site of the

FIGURE 1.14

Scientific Societies

Scientific societies became common after the middle of the seventeenth century. The French Academie des Sciences, shown here, had a royal patron and sumptuous meeting rooms at the royal library at Versailles.

explosion of biological research in the late Renaissance? At that time, most of the universities (Padua, Leyden, Oxford, and Cambridge excepted) were still mired in the intellectual morass of scholasticism. The academies, by contrast, were the product of **humanism,** the movement that began in the late fifteenth century as a repudiation of scholasticism. The humanists were interested in the study of humankind, not simply in the intellectually skimpy legacy of Latin misinformation so scrupulously studied by the scholastics. They turned to the original Greek writings and art for their models. Aided by the newly invented printing press, the humanists were able, by the beginning of the sixteenth century, to provide high quality translations of the original Greek to a broad audience hungry for information. Because the humanists were, for the most part, outside the universities, it is not surprising that the academies and societies, themselves outgrowths of the humanist spirit, also began outside the universities. Indeed, universities did not become major research centers until well into the nineteenth century.

Microscopists

The microscope was to Renaissance biology what the telescope was to Renaissance astronomy—both opened literally undreamed-of vistas for scientific exploration. The history of the early microscopists provides a fascinating example of the rapidity with which biological knowledge grew as a direct consequence of the availability of a particular instrument.

The earliest microscopes were constructed by Dutch spectacle makers late in the sixteenth century. However, it was a group of four men, born between 1628 and 1638, who, working entirely independently, achieved great and lasting success with the microscope. Indeed, so thoroughly did they capitalize on the microscope that it was not until the modern compound microscope was perfected near the end of the nineteenth century that any significant advances were made on the findings of these seventeenth century scientists.

Marcello Malpighi (1628–1694), working at the University of Bologna, put the finishing touches on Harvey's theory of blood circulation (see Figure 1.15). Harvey had speculated on the existence of a connection between arteries and veins, but it was Malpighi who actually witnessed blood moving through capillaries in the lung of a frog. In addition, he greatly extended Fabrizzi's work in embryology and provided a detailed internal anatomy of the silkworm at a time when it was widely believed that such simple creatures were devoid of internal organs.

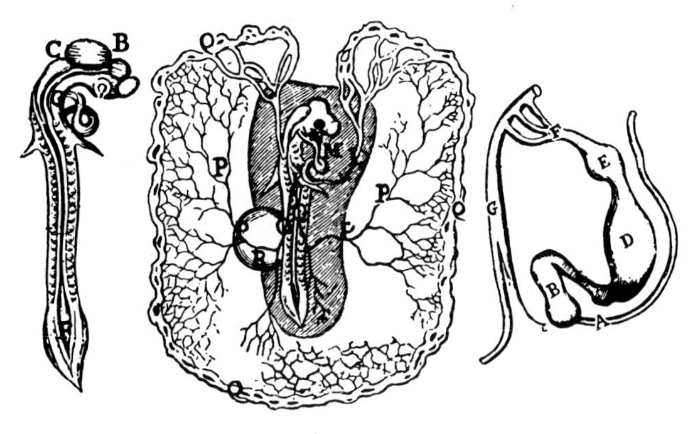

FIGURE 1.15

Marcello Malpighi

Malpighi was one of the early microscopists responsible for discovering a whole new realm of biology. His illustration shows the developing chick after 48 hours of incubation. The drawing on the right shows the developing heart.

Antony von Leeuwenhoek (1632–1723) owned a fabric shop in Holland, but spent all his spare time perfecting and using his simple home-made microscopes (see Figure 1.16). His work on animal tissues qualifies him as the father of histology, and he did remarkable work on insect development. However, his most spectacular studies were on protozoans and other microscopic denizens of ponds, in the course of which he observed bacteria, which were not rediscovered until the nineteenth century.

The Englishman Robert Hooke (1635–1703) was perhaps the most brilliant, ingenious, and versatile scientist of his day (see Figure 1.17). His observations with microscopes represented only a minor part of a most productive life, but his *Micrographia*, published when he was just 29, is a stunning work. In addition to his discovery of cells (see Box 1.1), he described

(a)

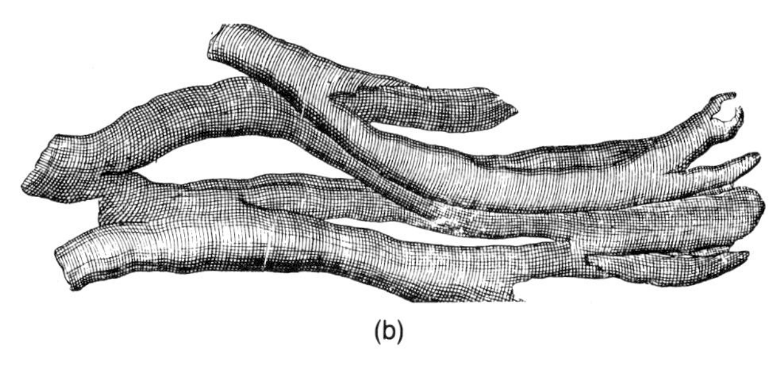

(b)

FIGURE 1.16

Antony van Leeuwenhoek

(a) Leeuwenhoek in his study, using his microscope and natural light. (b) Leeuwenhoek's illustration of cardiac muscle, showing the characteristic branched fibers. (Compare with Figure 12.9.)

(a)

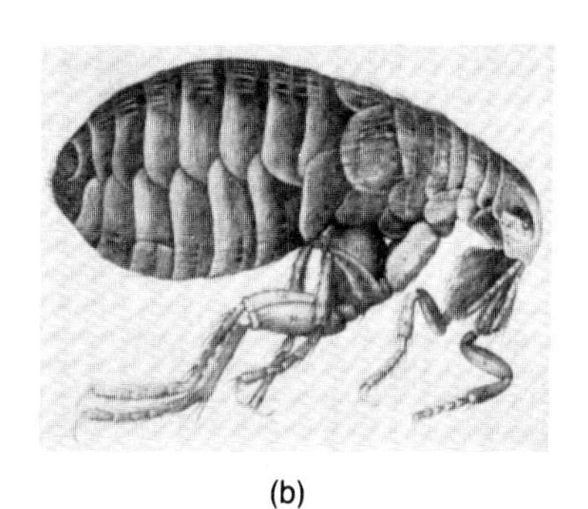

(b)

FIGURE 1.17

Robert Hooke

(a) Hooke's microscope, which was considerably more sophisticated than Leeuwenhoek's, included a lamp that could focus light on the specimen being observed. (b) Hooke's elegant drawing of a flea. The original drawing is almost half a meter long.

the structure and function of feathers, insect wings, sponges, and silk, all accompanied by masterful drawings.

Jan Swammerdam (1637–1680) received a medical degree in Holland, but his real interest was biology. He was the first true entomologist, and one of the pioneer comparative anatomists. Although he was only 43 when he died, his collected works, published as the *Biblia Naturae*, are nothing short of incredible in their detail and accuracy (see Figure 1.18). Some of his figures have never been surpassed, and the

BOX 1.1

Cell Theory

Unlike many of the other biological principles that emerged during the nineteenth century, the **cell theory** was not born in a cauldron of emotion and resistance, but instead emerged smoothly as the logical consequence of a long series of investigations.

In 1665, Robert Hooke first named and described cells. It is ironic, given the central position of cells in our understanding of life on earth, that Hooke was viewing slices of dead cork at the time. He had no appreciation of the central significance of cells to life, and was merely describing the cavities left behind as ghostly reminders of their former living occupants. What he called "cells" (after their appearance to the monastic cells of monks) were, in fact, merely cell walls. However, he did subsequently note the existence of cells when he described and illustrated the surface of living plants.

In 1805, Lorenz Oken, a prominent German naturalist, stated that all organisms "originate from and consist of cells." Oken did not support this early statement of the cell theory with evidence, however, offering it instead as a philosophical proposition. In 1831, an important piece was added to the puzzle when Scottish botanist Robert Brown identified the nucleus as a common feature of plant cells.

Matthias Schleiden, a German botanist who built on the work of Oken and Brown, concluded in 1838 that plants "are aggregates of fully individualized, independent, separate beings, namely the cells themselves," but he misidentified the nucleus as being the point from which new cells budded. The following year, Theodor Schwann, a zoologist colleague, extended Schleiden's thinking into the animal kingdom (where specialization of cells is more extreme and generalizations are therefore less obvious) by noting that animals are also composed of cells or cell products and that the cells are bounded by a membrane and contain a nucleus, just as they do in plants. Plants and animals now had a common link—the cell. Schwann also traced the development of animals back to the egg, which he correctly viewed as a single cell and the precursor for all the cells of the adult animal.

In 1858, Rudolf Virchow put the finishing touches on the cell theory when he stated:

> Where a cell arises, there must have been a cell before . . . There is no discontinuity, nor can any developed tissue be traced back to anything but a cell.

These thoughts were summarized in Virchow's famous aphorism, *Omnis cellula e cellula* ("Every cell from a cell"), which on the one hand extended Harvey's "All creatures come from an ovum" and on the other hand anticipated Pasteur's statement on spontaneous generation (see Box 2.1, Chapter 2).

Today, the cell theory is stated as three principles:

1. The cell is the basic organizational unit of living things (Schleiden).

2. All living things are composed of cells and cell products (Schwann).

3. Cells arise only from pre-existing cells (Virchow).

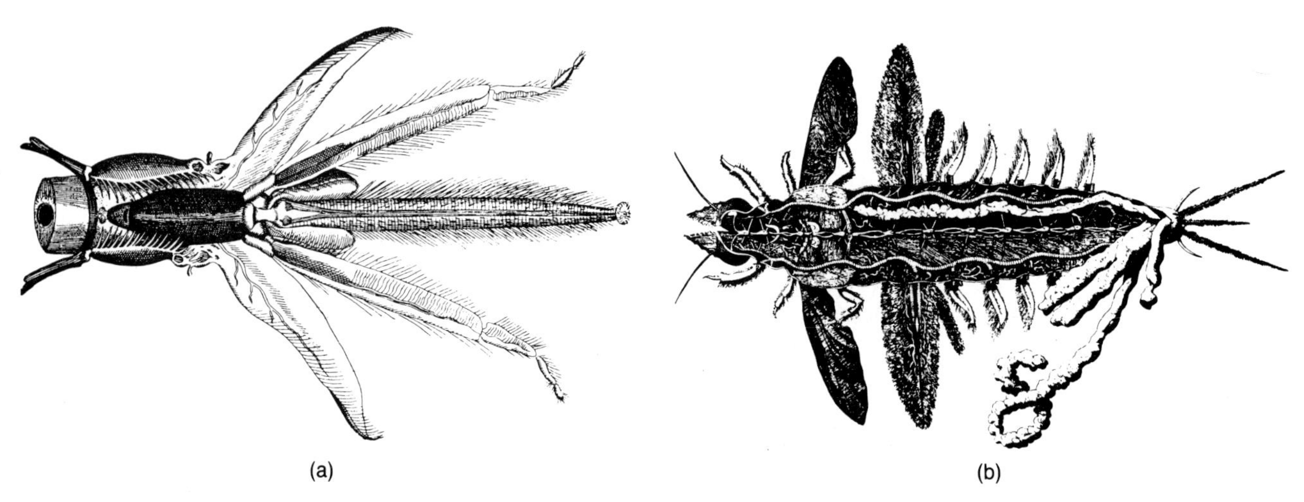

FIGURE 1.18

Jan Swammerdam
Swammerdam combined observation, dissection, and excellent artistic skills. (a) The mouthparts of a honeybee, correct in virtually every detail. (b) Swammerdam's dissection of a mayfly larva.

Biblia Naturae is still regarded as the finest collection of microscopial observations ever performed by one scientist.

Epigenesis Versus Preformationism

As the welter of scientific information began to accumulate during the seventeenth century, theoreticians scrambled to fit the data to a variety of hypotheses designed to explain one of life's oldest mysteries—the process by which complex organisms reproduce and develop.

The two major theories were **epigenesis** and **preformationism.** Epigenesis assumes an undifferentiated beginning, whereas preformationism, as the name implies, assumes that development consists of the growth of tiny, but preexisting, structures.

Today we know that epigenesis is the correct view (although preformation of a kind does exist in the form of genes that are essentially constant in their chemical composition, generation after generation). However, three centuries ago, things seemed much more complicated. First, those believing in epigenesis frequently also believed in **spontaneous generation.** Consequently, scientists opposed to spontaneous generation often found themselves branded as preformationists. Second, the early studies of Leeuwenhoek and Swammerdam on insects showed that many insects, from a very early stage of development, pass through a preformation-like sequence on their way to adulthood.

The two contrasting theories were first presented by Aristotle, who preferred epigenesis, although (in a classic example of chauvinism) he believed that it was the male who contributed "form" to the new individual, whereas the female merely contributed "substance." Harvey extended Aristotle's thoughts experimentally, and coined the phrase "ex ovo omnis"—all creatures come from an ovum. However, Harvey never saw a mammalian ovum, and erroneously interpreted the early embryo as the ovum.

Following the death of Harvey, preformationists became dominant, a logical outgrowth of the fact that the **mechanist** school of Descartes (in which the biological processes of living organisms were likened to the workings of the parts of a machine) came to dominate scientific thought during the latter part of the seventeenth century. Indeed, the microscope helped this shift because of the discovery of sperm—some of which were "seen" to have a tiny human inside (see Figure 1.19). Moreover, the appearance of structure in the developing hen's egg was almost magical (absent one day, present the next), and the subsequent growth of these structures seemed to imply that they had been present all along but were initially too small to be seen.

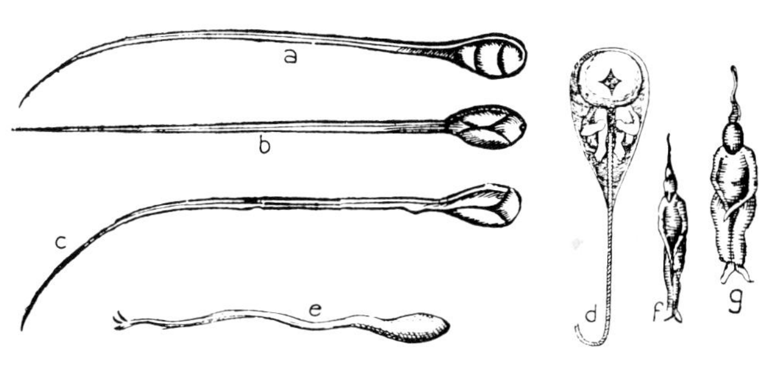

FIGURE 1.19

Sperm and Homunculi

Leeuwenhoek's drawings of sperm are shown on the left; a more creative illustration—showing a homunculus within, which is then dissected from out of the sperm—is on the right.

However, the preformationists ran into theoretical problems almost immediately. Which was the site of the preformation, the sperm or the ovum? "Spermists" and "ovists" debated that question endlessly. If the next generation was present in miniature in sperm or ova, what about subsequent generations? Some preformationists extended their argument to the point where they claimed that the extinction of a species occurred when the last of the line of these "boxes within boxes" was born. For humans, this was presumed to be at the time of resurrection predicted by Revelations.

The preformationist arguments were not laid to rest until the nineteenth century, when extraordinarily careful work by Karl von Baer on the development of eggs, and a correct understanding of the union of cells that characterizes fertilization, showed that epigenesis was the correct interpretation of development.

The handmaiden of epigenesis, spontaneous generation, was not disproved until Pasteur's work in the 1860s (see Box 2.1, Chapter 2).

Systematists

With the Age of Exploration, in the fifteenth and sixteenth centuries, came stories both real and fanciful of the enormous diversity of plants and animals in distant corners of the world. Renaissance biologists soon realized that not all of these had been described by Aristotle and Theophrastus, and so began the time (which continued until the end of the eighteenth century) when systematists dominated the biological landscape.

The basic problem facing systematists was devising an acceptable method of ordering and arranging the various types of plants and animals. In the fifteenth century, as new specimens arrived in great numbers from around the world, systematists were in a position analogous to a librarian blessed (or cursed) with daily deliveries of books and no shelves or classification system with which to group them. Of course, the problem of systematists was much greater, for books are inherently distinctive. Systematists had to decide when two specimens were sufficiently different to justify separate categories (and they often quarreled over those decisions); they needed to determine which criteria would be used in making placement decisions; they required the construction and adoption of a brief method of describing specimens; and most of all they needed an agreed-upon system of naming specimens so as to permit unambiguous identification by anyone using their publications.

The magnitude of these problems is revealed by a brief quote from a sixteenth century work on insects. The frustration of the author, as he describes his collection of grasshoppers and locusts, is palpable:

> Some are green, some black, some blue. Some fly with one pair of wings, others with more; those that have no wings they leap, those that cannot either fly or leap, they walk; some have longer shanks, some shorter. Some there are that sing, others are silent . . . Of the winged some are more common and ordinary, some are rare; of the common sort, we have seen six kindes all green, and the lesser of many colors.

None of that is terribly useful to the person wanting to identify the type of grasshoppers cavorting in a garden!

Because of their importance in medicine, plants were in particular need of classification. Several German systematists made substantial progress in plant classification in the early sixteenth century. Collectively, they traveled widely in Europe, grouped plants on the basis of the shape and position of leaves, stems, and flowers, described over 500 distinct types of plants (which they called **species,** the Latin equivalent of the Greek term used by Aristotle), and perhaps most importantly provided accurate illustrations of each species. At about the same time, Kaspar Bauhin (1560–1624), a Swiss botanist, began the system of identifying species by using just two words, rather than the dozen or more used by Aristotle. This **binomial** system was subsequently universally adopted.

The seventeenth century Englishman John Ray (1627–1705) ranks as one of the major founders of systematics. In a prolific career, Ray catalogued over 18,000 species of plants, and then turned his attention to animals, concentrating on mammals, reptiles, and insects. In addition to its broad scope, Ray's work was particularly significant because he attempted to group similar species into what he called "Natural Orders," something also attempted by Aristotle, although Ray was a good deal more successful.

However, to the extent that systematics may be said to have had a father, that distinction is universally afforded to Carl von Linné (1707–1778; see Figure 1.20), an eighteenth century Swedish naturalist. In some respects, this honor seems misplaced, for Linné was more a cataloguer than an innovator. However, he so successfully promoted the notion of the Latin binomial for the name of a species (to the point that he even Latinized his own name to Carolus Linnaeus!), and his use of higher categories (Order and Class) was so widely accepted that today's systematics literally begins with the publication of the tenth edition of his *Systema Naturae* in 1758. He gave Latin binomials, and descriptions, to about 7300 species of plants and 4000 species of animals; most of his binomials are still in use.

The genius of Linnaeus in using Latin binomials to label species was that the system avoided regionalism. Plants and animals often have common names that are different in different countries. What we call

FIGURE 1.20

Carl von Linné

Generally regarded as the father of modern systematics, Linné was so enraptured with Latin binomials that he referred to himself as "Carolus Linneaus."

BOX 1.2

Modern Systematics

The particulars of Linnaeus' system of classification have been repeatedly modified over the past 200 years, but the basic scheme he devised has persisted. Biologists have now described more than a million species of animals, and approximately 500,000 species of plants. Each species has been given a unique binomial, and categorized so as to indicate relationships with other species. Extinct species, including many known only from fossils, are classified in the same manner.

Although botanists and zoologists devised slightly different terminologies for their classification categories, the overall arrangements are very similar. In both, specific categories are called **taxa.** Consider how the human is classified by zoologists:

CATEGORY	TAXON
Kingdom	Animalia
Phylum	Chordata
Class	Mammalia
Order	Primates
Family	Hominidae
Genus	*Homo*
Specific epithet	*sapiens*

Let us examine this table from the bottom up. First, the specific epithet is the second half of the species name. Thus the species name for humans is *Homo sapiens,* the traditional Latin binomial, which consists of the genus (capitalized) and the specific epithet (lower case), with both genus and specific epithet italicized. A species, in the current convention, is defined as a population of organisms that is capable of interbreeding and producing viable (and fertile) offspring but that is reproductively isolated from other populations. It is the only natural biological category; all of the other categories are arbitrary distinctions, made for purposes of information retrieval.

As we move up the table, each category becomes more inclusive. As we have seen, the species name is, in every case, unique—there is only one *Homo sapiens*—but there can be other species within a given genus. In the genus *Homo,* for instance, there are no other living species—but there are two extinct species, *Homo habilis* and *Homo erectus,* both of which are thought to be our ancestors (see Chapter 23 for details).

The category Family is a still more inclusive unit. The Hominidae includes genera other than *Homo,* such as *Australopithecus* (see Chapter 23). The great apes (chimpanzee, gorilla, orangutan, and gibbon) are not placed within the Family Hominidae, but are within the next higher taxon, the Superfamily Hominoidea (intermediate categories are often created using the prefixes "super" or "sub").

The Order Primates includes not only humans and apes but also monkeys, baboons, and lemurs. However, the Primates are only one of more than a dozen Orders within the Class Mammalia, a taxon which includes all of the familiar haired, warm-blooded creatures from kangaroos to bats.

The mammals, along with the birds, reptiles, amphibians, fish and their relatives, constitute the Phylum Chordata, a taxon whose members are characterized by the presence of a nerve cord running the length of the back and by a parallel stiffening element, the **notochord,** which is the forerunner of the internal skeleton possessed by most of the chordates. There are almost two dozen phyla within the Kingdom Animalia (the actual number varies among different authorities—remember, these are arbitrary categories), ranging from sponges to the insects and their relatives, and from the starfish and their relatives to the chordates.

Although the overall pattern of classification now in use would be instantly recognizable by Linnaeus (he would, however, undoubtedly be most impressed by the sheer volume of described species!), the criteria used in classification have changed considerably since the time of Linnaeus. Linnaeus relied largely on obvious, external characteristics such as shape and color. Many times he gave species names to different color forms of what we now know to

(Continued on the following page)

BOX 1.2 (Continued)

be a single interbreeding species. The point is that the most obvious characteristics are not necessarily the best indicators of species differences.

At present, external anatomical characters are still used to distinguish species, but, in addition, internal anatomy, differences observed during embryonic development, behavioral and ecological differences, and even chromosomal differences may be used. Our current knowledge of chromosomes is such that chromosomes from two organisms can be compared at the molecular level, with the degree of similarity or dissimilarity then being used to classify the organisms (see Chapters 22 and 23). That approach would very quickly have had Linnaeus scratching his head!

the "English sparrow" is the "house sparrow" in England, for example—and the mallard duck has more than 30 common names within the United States alone! However, scientists use the scientific names *Passer domesticus* and *Anas platyrhynchos*, respectively, for these two species and avoid the ambiguity of different common names.

Systematists such as Linnaeus did not look beyond a perfected system of classification. However, their work led to a much larger question and, almost paradoxically, provided ammunition for both sides. Are species fixed and immutable, or are they capable of change? The very fact that all organisms can be placed in one or another species implies both uniformity within a species and distinct differences between species, and therefore argues for immutability. On the other hand, the fact that groups of species can be lumped into Orders and Classes suggests a biological relationship among them, implying evolution. The findings of the systematists of the sixteenth, seventeenth, and eighteenth centuries became the fodder for the evolutionist arguments of the nineteenth century.

Summary

Biology, the science of life, did not spring into being fully formed (as Athena purportedly did from the brow of Zeus), but instead was the consequence of an almost endless number of studies, frequently done in ignorance of other studies, conducted by philosophers and scientists over many centuries. The process continues today at an ever-accelerating pace.

From the standpoint of our culture, biology began with the ancient Greeks, and reached something of a pinnacle with Aristotle. The quality of their observations was often superb, but they frequently displayed a distressing tendency to speculate wildly on their findings, and did not subject their speculations to experimental testing.

From the time of the conquest of Greece by the Romans until the beginning of the Renaissance, not only were there few advances in biology, but the quality and quantity of the information passed on from generation to generation declined markedly. It was not until the work of Vesalius and Harvey, in the sixteenth and seventeenth centuries, that biological data surpassed that known by Aristotle to any significant degree, and it was not until the middle of the nineteenth century that some of Aristotle's findings were confirmed.

With the success of the experimental method by Harvey, and the philosophical writings of Bacon and Descartes, biology developed rapidly during the seventeenth century. Among the notable success stories were the microscopists, who opened new worlds for exploration; the academies, which sponsored scientific research in great quantity and quality; the developmentalists, who gradually worked their way through the traps of preformationism and who ultimately repudiated spontaneous generation; and the systematists, who paved the way for evolutionist studies in the nineteenth century.

Key Terms

scholasticism
empiricism
humanism
epigenesis
preformationism
spontaneous generation
mechanist
species
binomial

Box Terms

cell theory
taxa
notochord

Questions

1. Name three early Greek biologists. Characterize the type of biological investigation undertaken by the ancient Greeks.
2. Account for the relative disinterest in biology during Roman times and the actual decline in scientific understanding during the Middle Ages.
3. Distinguish scholasticism from humanism.
4. Explain the resurgence of interest in biology, especially anatomy, during the Renaissance. Why was the work of Vesalius significant?
5. How did Harvey's explanation of blood circulation differ from Galen's? Account for the fact that these two men described the same process so differently.
6. Why were the seventeenth century microscopists important in the history of biology?
7. Distinguish epigenesis from preformationism. Which theory is now viewed as correct? Why did it take so long for this controversy to be resolved?
8. What is the importance of systematics to biology?

CHAPTER

2

Biology Today

The concept of evolution is central to modern biological theory, yet it remains a controversial topic to the general public even today. Why is it so controversial? What exactly was the role of Charles Darwin in promoting evolution as a concept? How does evolution relate to genetics?

Our scientific understanding is currently advancing at unprecedented rates, thanks to the development of the scientific method. What *is* the scientific method? If it is so effective, why is there so much confusion over what is, or is not, safe to eat?

These are the kinds of questions we shall consider in Chapter 2.

THE INTEGRATION OF BIOLOGY

The growth of biological information that began in the Renaissance can be likened to the growth of a plant's root system; it begins as a handful of separate shoots but ultimately becomes a vast and interwoven network. The various disciplines of biology that started in the seventeenth, eighteenth, and nineteenth centuries as essentially independent sciences have increasingly crossed paths in the twentieth century so that the findings in one discipline now routinely, and often profoundly, affect the operation of another discipline.

The principal reason why biology has become one integrated science was the discovery of a single underlying theme that now dominates the biological landscape and gives direction and a sense of balance to biology as a whole. That theme is evolution—the "golden thread" that unites all of the biological disciplines. As we shall see in this chapter, and throughout this book, evolution is at the very core of modern biological research.

STABILITY VERSUS CHANGE

Since at least the time of Aristotle, theories about the two levels of organismal relationships—those between generations and those between species—have been central to biological thought. Are species stable, or do they change? If species are stable, how is stability maintained? If species change, how is change initiated? To the early biologists, it was obvious that "stability" was the norm, and "change" the exception. After all, cats give birth to cats, and humans to humans—and never the other way around.

However, it is also true that children often bear distinct resemblances to one or both parents, which indicates how important the parents are in determining the particular traits of their children. From that observation, it was an easy step for the ancients to explain oddities of various types as the result of a mating between members of different species. Thus the giraffe was thought to be the product of a camel having mated with a leopard. (Indeed, the scientific name for the giraffe is *Giraffa camelopardis.*) The ostrich was a cross between a camel and a sparrow. (The mind reels at the mechanical problems associated with such a mating!) The manatee was a cross between a fish and an Arab. (Bigotry was endemic in the tenth century.) Many other plants and animals, both real and mythical, were explained in the same way.

These explanations of change were novel and imaginative, and preserved intact the notion of stability within a species by limiting change to instances of hybridization between species. Heredity (or what we would now call "genetics") ensured species stability. Change (or what we would now term "evolution"), if it occurred at all, happened only in a quantum fashion, through the formation of hybrids. With the renewed interest in biology during and after the Renaissance, these two fields—genetics and evolution—were investigated separately. Only within the last 50 years has it finally been appreciated just how tightly the two are linked.

Early Evolutionary Theories

Although Aristotle's *scala naturae* has been called the first formal evolutionary theory, the fact is that Aristotle himself was no evolutionist, but instead believed in the fixity of species. An expanded version of his *scala naturae,* however, led to increased speculation about evolution during the eighteenth century. Even so, the emergence of evolutionary thought was far from automatic. Indeed, with the rise of systematics during the seventeenth and eighteenth centuries, evolution virtually disappeared from the biological lexicon. The fact that organisms could all be placed within one species or another suggested a uniformity within species and a discontinuity between species. Because all members of a species look very much alike (although not necessarily absolutely identical), and all are clearly distinguishable from members of any other species, there was no evidence for evolution—indeed, there was no need even to suggest it.

Another nail in the coffin of evolution came from the notion of the **type specimen.** A type specimen, which embodies all the characteristics of the species, is selected whenever a new species is first described. Other biologists are then able to compare specimens they have collected with the type specimen to determine if their specimens belong to the same species. As originally practiced, the type concept (**typology**) required stability in species—if a species changed with the passing generations, then the value of type specimens (and of systematics as a whole) would be greatly diminished. Typological thinking dominated systematics until the middle of the nineteenth century and provided a powerful retardant on theories of evolution.

Finally, most of the early systematists were devout Christians who saw their work as glorifying God, and who were generally prone to a literal interpretation of the Bible. John Ray noted that there was some flexibility of type within a species—that is, species were not absolutely fixed—but held that the number of species must be constant because the Bible said that God rested on the seventh day after He had created the plants and animals. Linnaeus was even stronger on this point:

> There are just as many species as the Infinite Being created in the beginning . . . At the beginning of the world there was created only a single sexual pair of every species of living thing.

However, even as Linnaeus was publishing his doctrinaire views, the winds of change were beginning to blow. Comte Georges-Louis Leclerc de Buffon (1707–1788), a wealthy French biologist who ranks with Linnaeus as the greatest biologist of the eighteenth century, compiled a massive collection of scientific information in his 44-volume *Histoire naturelle* (see Figure 2.1). He ridiculed Linnaean systematics as being oversimplified and arbitrary, and believed species and other categories to be artificial. Buffon was particularly troubled by rudimentary and vestigial organs (such as the muscles controlling movement in the human ear) that seemed to have no reason for being. At a time when the official Bible of the Church of England specified the date of the origin of the Earth to be 4004 B.C., Buffon claimed the Earth to be more than 100,000 years old. Indeed, some of his views were thought so heretical that, in 1751, the faculty of theology at the Sorbonne ordered Buffon to eliminate certain passages from his *Histoire naturelle.*

FIGURE 2.1

Comte de Buffon

Buffon was one of the first of the biologists of the post-Renaissance era to advance the notion of evolution, although he did so obliquely, presumably to avoid church censure.

FIGURE 2.2

Erasmus Darwin

Erasmus Darwin, the grandfather of Charles, championed the notion of evolution through the inheritance of acquired characteristics.

Although Buffon was a typologist, not an evolutionist, he discussed evolutionary problems at length and brought them to the attention of other biologists for the first time. Thus he paved the way for the evolutionists who followed. Erasmus Darwin (1731–1802; see Figure 2.2), the grandfather of Charles Darwin, was a country doctor with a strong interest in natural history. Much of his writing was speculative and untested by experimentation, but in several important respects he anticipated his grandson. His *Zoonomia*, published in 1794, was so controversial that it was placed in the Index of Prohibited Books. In this work, he espoused the notion that all life derived from a common "filament" that originated millions of years ago. (He was apparently referring to sperm cells, but was in essence anticipating the cell theory, which was not articulated until many years after his death.) Darwin thought that species changed over time, forming new species in the process. He pointed out various categories of change, such as metamorphosis, domestication, and climatically induced change, and concluded that these changes accumulate in individuals and are passed on to their offspring. This view came to be known as the **inheritance of acquired characteristics.** Darwin was not the first to articulate such a belief—indeed, it was widely accepted from the time of the ancients until the 1930s—but he articulated it at length and with great vigor.

FIGURE 2.3

Jean-Baptiste Lamarck
Lamarck extended the views of Buffon and Erasmus Darwin, and added his concept of the law of use and disuse to the basic idea of inheritance of acquired characteristics.

The French biologist Jean Baptiste Lamarck (1744–1829; see Figure 2.3) was one of the more tragic figures in all of science. He led a life of penury, was widowed four times, outlived most of his children, and when he died, blind and alone at 85, he was buried in a pauper's grave. His views were not well received during his lifetime, and he is often held up to ridicule even today. His "crime"? He misidentified the mechanism of evolution. Ironically, Lamarck was a good deal more correct than his contemporaries, many of whom have retained far more illustrious reputations.

In his *Philosophie zoologique,* published in 1809, Lamarck presented the first completely developed theory of evolution. Consequently, he deserves the credit for having replaced the notion of a static nature with a dynamic nature. Lamarck was particularly impressed by the profound effects of domestication, and

Ancestral giraffes had short necks
Successive generations kept stretching to reach leaves
Present-day giraffes have long necks

FIGURE 2.4

Use and Disuse
Lamarck explained the evolution of the long neck of the giraffe through use (or need) over a period of many generations. The constant stretching for still higher leaves favored such a development. In short, a changed environment prompted a species response.

noted that less dramatic variations also occur naturally. He attributed the source of these variations to the environment. A changing environment placed demands on organisms, with some structures no longer being needed and others requiring enhancement (see Figure 2.4). This **law and use and disuse** is inseparably identified with Lamarck. Unfortunately, in translating his works from the French, the word *besoin* was translated as "desire" rather than "need." Thus Lamarck has been wrongly quoted as saying that organisms consciously wished structures into existence, when he was, in fact, speaking of a need generated by a changed environment. Like Erasmus Darwin, Lamarck assumed that changes that developed during an organism's lifetime would be passed on to its offspring. Indeed, he is often credited with inventing the concept of inheritance of acquired characteristics (and then ridiculed for it), even though as an idea it predated him.

Lamarck was, in many respects, an excellent biologist. (In fact, he is responsible for the word "biology," a term he coined in 1802.) He separated both spiders and crabs and their relatives from the insects, brought order to Linnaeus' miscellaneous class of Vermes by distinguishing the true worms, and did important work on the starfish and their relatives. He also correctly held evolution to be a fact. However, he was given to rash speculation. More methodical scientists were easily able to destroy his more outlandish claims and, in so doing, discredit the entire notion of evolution as well. Consequently, advocates of evolution were a discouraged lot during the first half of the nineteenth century.

Charles Darwin and Natural Selection

Because he correctly and successfully enunciated the theory of evolution by natural selection, Charles Darwin (1809–1882; see Figure 2.5) has rightly been called the last of the natural historians, and the greatest. As a young man, he sailed aboard the scientific ship *Beagle* on a five-year collecting expedition (see Figure 2.6). His faith in the immutability of species appears to have been shaken by his observations on that voyage, but there was no particular sight or event that caused him to formulate his theory on the spot, as some authors have maintained. His notebooks do not reveal a first glimmering until 1838, almost two years after his return to England, and not until 1844 did he have his theory well in hand.

Darwin's epochal *On the Origin of Species by Means of Natural Selection* was not published until 1859. What prompted the long delay? The answer seems to be Darwin's full realization of the impact that his work would have, and his desire to anticipate, and rebut, all of the conceivable reactions against it. The memory of Lamarck's fate must certainly have been in his mind.

Indeed, Darwin was still several years from completing what he envisioned as a multivolume work when Alfred Russel Wallace, a young colleague, sent him a short paper for review (see Figure 2.7). Darwin was stunned to read that Wallace, while recovering from malaria in Malaya, had stumbled across the notion of natural selection as the engine of evolution. He reportedly thought briefly of destroying his own notes, so as not to appear to have stolen Wallace's ideas, but friends prevailed upon him to submit a companion paper to Wallace's. The two were published in 1858 (ironically, in the *Journal of the Linnaean Society*, a society dedicated to extending the work of Linnaeus, who was a vigorous foe of evolution). Darwin then quickly completed what he termed an abstract of his planned masterpiece (the "abstract" was 490 pages long!), and it was published the following year. It released a tidal wave of controversy that, more than a century later, has not yet fully subsided.

FIGURE 2.5

Charles Darwin

A meticulous biologist, Darwin did not publish his ideas on evolution by natural selection until at least 15 years after the idea had first come to him, and only after he had considered every possible alternative explanation for the data he had collected.

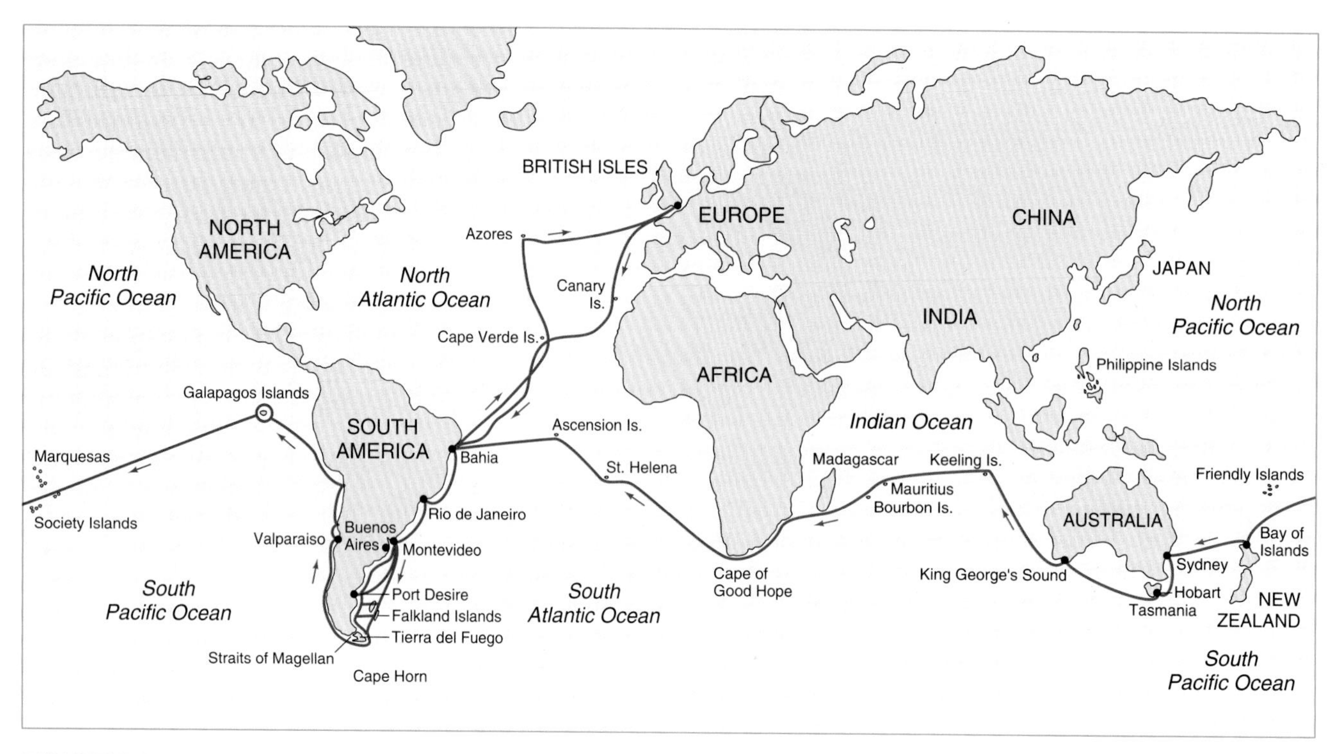

FIGURE 2.6

Voyage of the *Beagle*

As a young man, Darwin sailed on the *Beagle,* whose five-year mission was to boldly go where no mapping and collecting expedition had gone before. The route of the *Beagle* is shown here. Observations made during this voyage were to prove very important to Darwin as he began to consider the causative agent of evolution.

If we cut through the Victorian rhetoric, we can reduce the heart of Darwin's theory to a few lines:

1. Organisms produce more offspring than can possibly survive.

2. Within a species, particular characters vary from individual to individual (see Figure 2.8), and some of this variation is inherited.

3. Some of the variants are better adapted to their local environment than are others.

4. Offspring that are better adapted will, on the average, survive and reproduce with greater success than those that are less adapted. This process of differential reproductive success is known as **natural selection** (see Figure 2.9).

FIGURE 2.7

Alfred Russel Wallace

Wallace independently discovered evolution by natural selection and in so doing prompted Darwin to complete and publish his book.

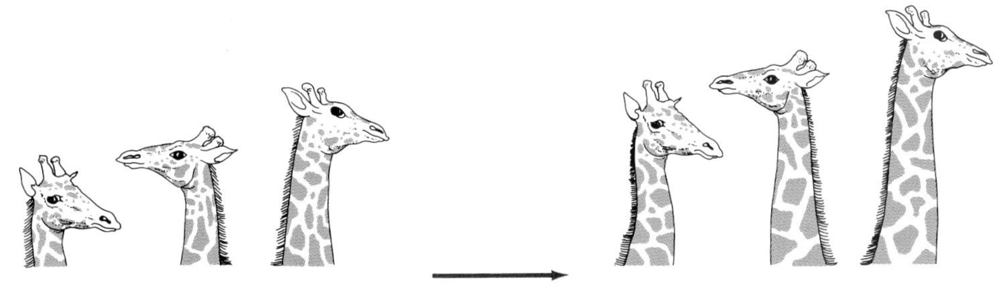

Ancestral giraffes had necks of varying lengths but were generally relatively short

Long necks are adaptive; natural selection eliminates those with the shortest necks

Present-day giraffes still show variation in neck length, but are generally very long

FIGURE 2.8

Giraffe's Neck Revisited

Darwin's explanation of the evolution of the giraffe's neck was that the giraffe population historically showed variation in neck length, and with passing generations those individuals with the longest necks were most favored by natural selection. In short, variation within the species existed first, rather than being induced by the environment, as Lamarck believed.

FIGURE 2.9

Natural Selection

The finches of the Galapagos Islands have undergone a spectacular adaptive radiation into a variety of habitats and food choices not utilized by mainland finches. Although Darwin captured many of these species in his visit to the Galapagos, their importance in the formulation of his theory is uncertain.

Note that natural selection theory holds that variation *precedes* changes in the environment rather than being *caused by* changes in the environment, as is required by both the use and disuse and the inheritance of acquired characteristics concepts.

Reaction to Darwin's Theory

Reaction of the establishment to the *Origin* was both swift and negative. Most of the scientific community and the Church of England promptly took the field against him. In the years following publication of the *Origin*, there were many more books of condemnation than of praise. Ministers inveighed against it from the pulpit, and there were numerous debates. At the same time, Darwin enjoyed substantial support from the public. The first printing of the *Origin* sold out on the first day of sale, and the book ultimately went through six editions. How can these apparently contradictory reactions be reconciled?

The public did not understand the subtleties of Darwin's thinking. They thought of evolution in the old *scale naturae* manner, as involving a perfecting principle with Man at the top of the pyramid, dominant over all life forms, and subservient only to God and the angels. The whole notion seemed to mesh well with the possibilities of social perfection, and the promise of an ideal hereafter—and here was scientific proof that evolution was at the heart of the universe! The ruling class justified their business practices as Social Darwinism, while at the same time Marx was so impressed with Darwin's views that he reportedly offered to dedicate the second volume of *Das Kapital* to him. (The Marx story may be apocryphal—but in any case, Darwin declined the honor.)

Reasons for opposition were more complex. The Church opposed Darwin's thinking because any theory of evolution requires an ancient origin of the Earth, and Church doctrine had established that the Earth was less than 6000 years old. Darwin's biological opponents were mainly typologists who believed in the fixity of species as a necessary outgrowth of the importance they ascribed to systematics. Beyond that, his theory of evolution by natural selection was completely different from any that had preceded it. Evolution by natural selection simply involved accommodation to the local environment. There was no direction, no purpose, no goal, no perfecting principle—it just happened.

In some respects, Darwin's timing was perfect. Biology was beginning to mature as a science, and the first bricks of its theoretical base were visible in such concepts as the cell theory. Younger biologists, unfettered by the trappings of the past, were ready to adopt new ideas. Most importantly, Darwin had amassed a mountain of evidence to support what was really a rather simple idea. Indeed, upon reading Darwin's book, Thomas Henry Huxley (see Figure 2.10), who was to become one of the more prominent biologists of the nineteenth century, was said to have muttered, "How extremely stupid of me not to have thought of that!" Many of these younger biologists eagerly trooped behind Darwin's banner, and the "Darwinists" became an important group in biology as the nineteenth century wound to a close.

Darwin and Variation

Although Darwin correctly identified natural selection as the driving force of evolution, he largely failed in his efforts to identify the source of the variation that is grist for natural selection's mill. As did virtually all the biologists of his day, Darwin completely misunderstood the mechanism of inheritance. His critics managed with some success to show the fallacies of Darwin's speculations on heredity and thus, by extrapolation, cast doubt on his theory of evolution by natural selection.

Darwin suggested three sources of variation in a population, and ranked them as follows:

1. Use and disuse (derived from Lamarck)

2. Environmentally induced variation that is then inherited (derived from Erasmus Darwin and Lamarck, among others)

3. Chance emergence of new characters in a spontaneous manner

The first two sources are variations of the now-defunct idea that characteristics acquired during an organism's lifetime can be inherited by the offspring. Chance emergence, however, is the equivalent of **mutation,** which biologists today recognize as the ultimate source of genetic variation. Darwin ranked chance emergence third in his scheme because of his problems in explaining inheritance. The prevailing view of the time was that inheritance involved the "blending," or averaging, of characters. How, then, could a single spontaneous variation be perpetuated in a population? It would be blended out of existence after the first generation.

Thus Darwin found himself on the horns of a dilemma. To avoid the problem of loss through blending, he would have to suggest that many identical variants must arise spontaneously. Yet this explanation would be suspiciously like some type of special creationism, the antithesis of natural selection.

Alternatively, he could accept the inheritance of acquired characteristics. This notion would allow for a number of identical variations having occurred at one time and would eliminate the need to invoke any sort of special creationism. However, the inheritance of acquired characteristics is also not compatible with

FIGURE 2.10

Thomas Henry Huxley
Huxley so enthusiastically agreed with Darwin's theory of evolution by natural selection that he soon became known as "Darwin's bulldog." Unlike Darwin, Huxley was a master debator and successfully defended his friend's theory in a sensational series of debates with Bishop Samuel Wilburforce.

evolution by natural selection, because it requires a response by organisms *to* the environment rather than the selection of existing variation *by* the environment. (Darwin himself pointed out the difficulties of using the inheritance of acquired characteristics to account for the origin of worker and soldier castes in ants, because these forms are sterile and therefore by definition have no descendants to inherit whatever characteristics they have acquired.)

Gregor Mendel and Genetics

The correct answer was known to Gregor Mendel (1822–1884), a monk born in what is now Czechoslovakia, who in 1866 discovered the mechanism of heredity. In marked contrast to Darwin's notion of evolution by natural selection, which was clearly an idea whose time had come (as evidenced by Wallace's independent findings), Mendel's findings were unappreciated and ignored for over 30 years. More than any of the other prominent biologists of the nineteenth century, Mendel was ahead of his time.

Yet neither did Mendel work in a vacuum. He was aware, for instance, of the pioneering hybridization experiments of the German botanists Joseph Koelreuter and Carl von Gaertner in the late eighteenth and early nineteenth centuries. Between them, they had created several hundred hybrids, and had found that the first generation hybrids were generally intermediate between the parents, whereas the second generation hybrids were very diverse. However, they offered no explanation for these interesting findings.

Mendel's genius was in concentrating on individual traits, rather than on whole plants, and in counting all of the hybrids (not just taking note of the most interesting ones) and then subjecting his data to statistical analysis. His good fortune was in choosing an assortment of traits of the pea plant, all but two of which happened to be genetically coded on a different chromosome, thus minimizing problems in genetic linkage (see Chapter 18 for details). However, his genius and good fortune were also his downfall, since no one else was interested in such trivial findings as the presence or absence of wrinkling in pea seeds. Other biologists, caught up in the excitement of evolution, were interested in major variations, such as size, strength, intelligence, or the type of variation shown by domestic animals. Mendel's paper, a masterpiece in design, execution, and significance, was published in 1866 but lay dormant until rediscovered in 1900 by three scientists working independently. Each of the three codiscoverers had independently conducted studies similar to Mendel's—indeed, one of them (Hugo de Vries) had data from more than 30 species of plants—before stumbling across Mendel's paper while reviewing the literature.

What had happened in the intervening 34 years? Why was the biological world ready for genetics in 1900 (as demonstrated by the virtually simultaneous experiments of three scientists working independently), but not ready in 1866? Certainly one important reason was the work of August Weismann (1834–1914) who, in 1885, proposed his **germ plasm** theory. Weismann maintained that the germ plasm (sperm and ova) was isolated from the rest of the cells of the body and therefore, despite being an ardent Darwinist, he rejected the inheritance of acquired characteristics that Darwin had somewhat unwillingly accepted. Moreover, Weismann also stated that "heredity is brought about by the transference from one generation to another of a substance with a definite chemical, and above all molecular, constitution."

Another factor in the acceptance of genetics was the improvement in microscopes, because interest once again developed in cells—the level of organization at which Weismann urged that the study of heredity be directed (rather than at the population level favored by Darwinists). Mitosis was observed in plant cells in 1875 (although the process was not named until 1880), and chromosomes were identified and named in 1888. Weismann predicted the existence of meiosis on theoretical grounds in 1887; it was actually seen for the first time the following year. (Mitosis and meiosis are discussed in detail in Chapter 17.) In 1892, Weismann concluded that:

> The complex mechanism for cell division exists practically for the sole purpose of dividing the chromosomes, and thus the latter are without doubt the most important part of the nucleus.

The stage was now set for the rediscovery of Mendel's paper.

The Modern Synthesis

We might have expected that the problem of heredity, Darwin's nemesis, would have been instantly laid to rest with the rediscovery of Mendel's paper and the overnight growth of genetics, but such was not the case. Indeed, it was not until the 1940s that the two fields—genetics and evolution—were united. Why was there such a long delay?

As paradoxical as it now sounds, the initial findings in genetics seemed to contradict the notion of natural selection. Hugo de Vries' early work on plants suggested that evolution took place by sudden jumps. (de Vries had plant varieties that differed on a wholesale basis, as a consequence of what he called a mutation. In actuality, the differences were caused by major chromosomal errors, as opposed to the much more common mutations of a single gene.) Thus the early geneticists believed that new species were formed by the occasional occurrence of a dramatic mutation, creating what they termed "a hopeful monster." Although the geneticists were firm believers in evolution, in their view natural selection played an insignificant role, serving only to eliminate unfit individuals.

The naturalists were a somewhat confused lot in the first two decades of this century. Although they accepted the *fact* of evolution, the *mechanism* of evolution was still a matter of active debate. Some followed the original Darwinian notion that variation arose both from genetic changes and from the inheritance of acquired characteristics. Others were neo-Darwinians, meaning that they agreed with Weismann that only natural selection was significant in causing evolutionary change. Still others were Lamarckians, believing only in the inheritance of acquired characteristics.

Perhaps most importantly, virtually the only time the geneticists and the naturalists spoke was to exchange disparaging remarks, and thus each tended to remain ignorant of progress in the field of the other. As a consequence, the naturalists were largely unaware that in 1910 the definition of "mutation" changed from the de Vriesian concept of a wholesale transformation creating a new species to the alteration of a single gene. The geneticists, in turn, were unaware that by the 1920s the naturalists were defining "species" as a population of interbreeding, but varying, individuals, as opposed to the older typological notion that denied variation within a species.

During the 1920s and 1930s, there were at least three major developments that served to bridge the gap between the geneticists and the naturalists. These were:

1. The showing by experimental geneticists that artificial selection in populations of laboratory animals and plants could profoundly change the form of the population;

2. The demonstration by mathematical geneticists that even a very small selective advantage could, if maintained over a sufficiently long period of time, lead to dramatic changes in the genetic constitution of a population; and

3. The development of population genetics, largely by naturalists, who showed that geographic variation within a population was both adaptive and genetically induced.

As naturalists and geneticists became aware of each other's work, it became clear to a number of prominent scientists in both fields that each had something to offer the other, and genetics became reconciled with evolution. Both groups now agree that mutations are the ultimate source of the variation among individuals, and that natural selection operates on this variation to increase adaptation to the environment. Since environments change, so, too, does the genetic constitution of a species, which is to say that species evolve.

Outgrowths of the Modern Synthesis

The problem of reconciling genetics and evolution provides a cogent example of the type of schism that has historically plagued biology. Until the beginning of the nineteenth century, biology was essentially either natural history, on the one hand, or medical anatomy and physiology, on the other, and each reflected an entirely different pattern of historical development and traditions. The factionalism increased dramatically during the nineteenth and twentieth centuries with the establishment of new disciplines: embryology (1828); cytology (1839); evolutionary biology (1859); ecology (1866); genetics (1900); animal behavior (1927); and molecular biology (1944). Most were fields unto themselves, focused on solving problems of immediate relevance to workers in that field. Certainly a major reason for the delay in the union of genetics and evolution was that there was such a division of interest between the two groups. The geneticists were experimentalists, and continued in the medical and physiological tradition. They were primarily interested in **proximate causes,** such as the mode of operation of genes in the development of the individual organism. The naturalists, on the other hand, were the spiritual descendants of the medieval herbalists, and their great interest was in describing and cataloging nature. However, Darwin, himself a naturalist, provided the first rational approach to the study of **ultimate causes,** which include such things as the adaptive role played by a particular structure or behavior in the life cycle of a given species.

The ramifications of the modern synthesis have been profound, for discussions of ultimate cause now pervade all of the biological disciplines. Many current studies in anatomy, for example, focus not on descriptive but on functional anatomy—why is a particular structure shaped the way it is? The term **adaptation** (see Chapter 22), once used exclusively by the Darwinists, is now common in anatomy, physiology, animal behavior, and ecology. Much of physiology is now devoted to the study of **homeostasis** (see Part 3), which refers to the ability of many organisms to maintain their internal environment essentially constant.

The modern synthesis has also dramatically affected systematics. Although the task of naming and classifying newly discovered species continues at an accelerating pace, much of the focus of systematics is now on constructing **phylogenies,** which are the biological equivalents of a family tree. Systematics has historically classified the diversity of life; the modern synthesis has provided the impetus to study the unity of life, because it stresses a common origin for all species.

The unity of life, a concept derived from evolutionary theory, is a theme underlying scientific investigations ranging from molecular biology to ecology. Studies in molecular biology have demonstrated that the chemical nature of the gene, the mechanics of protein formation, the composition of energy molecules, among many others, are common to all life forms—facts that strongly support, and are supported by, a single origin of life followed by evolution through natural selection. (These ideas are explored more fully in Chapters 3, 4, 18, and 22.)

Ecological studies (see Chapters 5, 24, and 25) focus on the transformation of energy, starting with sunlight and ending with the giant structural molecules of the cells of all organisms, thereby both demonstrating the interrelationship of all life forms, and illustrating the results of evolution in the form of the myriad kinds of adaptations developed by organisms in addressing their energy needs.

As we come to detailed descriptions of each of these areas later in the book, it is important that you remember the lesson biologists themselves were slow to learn: Biology is ultimately a unitary discipline because all of the components of all organisms are the reflection of a single force—evolution by natural selection.

THE SCIENTIFIC METHOD

As we saw in the introduction to Part 1, biology is not just facts and theories. It is also a process by which these facts are accumulated and theories are validated. Just as biological theories have evolved, so, too, has the process—indeed, changes in the process have largely been responsible for the discovery of new facts and the presentation of new theories.

The methods of biology today bear little resemblance to those used by the ancient Greeks. To Aristotle and his colleagues, Bacon's notion of "putting nature to the test" would have been an alien concept. Observations (when they were made at all) were meshed with philosophical beliefs, and the quality of logic used in reaching a conclusion was deemed far more valuable than the observation itself.

Deductive and Inductive Logic

The Greeks relied heavily on **deductive logic**—reasoning from a general premise to a specific conclusion. In particular, they used the **syllogism,** which Aristotle both developed and advocated as the major method of reaching a scientific conclusion. Syllogisms consist of a general premise, a specific premise, and a conclusion that is deduced from the premises. For example:

> All mammals give birth to live young.
> A dog is a mammal.
> Therefore, a dog gives birth to live young.

The dangers of this approach are manifold, but primarily involve the structure or the validity of the general premise. Suppose, for example, our choice of mammal in the syllogism given above was the duck-billed platypus. We would be forced to conclude that the platypus, because it is a mammal, necessarily gives birth to live young. In fact, the platypus lays eggs. Our conclusion is wrong because our general premise is very nearly, but not absolutely, correct.

Even worse problems arise if we invert the general premise. For example:

> All animals that give birth to live young are mammals.

If we begin with this premise, we must conclude that a whole host of animals that give birth to live young—from cockroaches to guppies—must be mammals when they indisputably are not.

Inductive logic (reasoning from a series of specific facts or observations to a general principle) was the approach advocated by Bacon and Descartes, and used so tellingly by Harvey. It is at the heart of the modern scientific method.

Modern Approaches

Modern science is an enormously diverse discipline, and so, too, are the methods of science. We would not expect an evolutionary theorist to employ the same methods as an investigator attempting to determine the potential health hazards of a new chemical, and indeed they do not. There is no single scientific method. There are, however, at least two elements that are shared by all modern methods of scientific investigation. These elements are *testability* and *repeatability*. Testability means that assumptions must be verified through experimentation or direct observation before they can be claimed to be valid. Repeatability means that others should be able to arrive at the same findings by following the same procedures.

Mendel's work serves as an elegant example of

the modern scientific method (see Figure 2.11). Mendel began by objectively and systematically winnowing a series of observations and reports on breeding experiments in plants and animals and then determining that some of these reports conflicted with the widely accepted blending notion of inheritance. For example, blending would predict that when a white ram was bred to a black ewe, the lambs would be grey. In fact, they are either white or black, but never grey.

From this misfit of data and notion, Mendel developed a **hypothesis** (a tentative explanation posed in such a way as to be testable). Mendel's hypothesis was that specific parental traits are not blended in the offspring, but rather that one trait dominates the other. He tested this hypothesis by crossing pea plants that differed in one or more traits, and found that the offspring were identical in that they all possessed just one of the parental traits. When he crossed the offspring, he found that most of their offspring had one of the original parental traits, a small number had the other original parental trait, but none was intermediate.

These results confirmed his hypothesis that dominance, and not blending, is the correct explanation of heredity. However, before stating his conclusion, Mendel further tested his hypothesis by repeating these experiments with a total of seven different traits in pea plants. Mendel's work was ultimately repeated on many other species of plants and animals, and his conclusions were found valid for all sexually reproducing organisms. Because his hypothesis has withstood repeated experimental testing, it now qualifies as a **theory.**

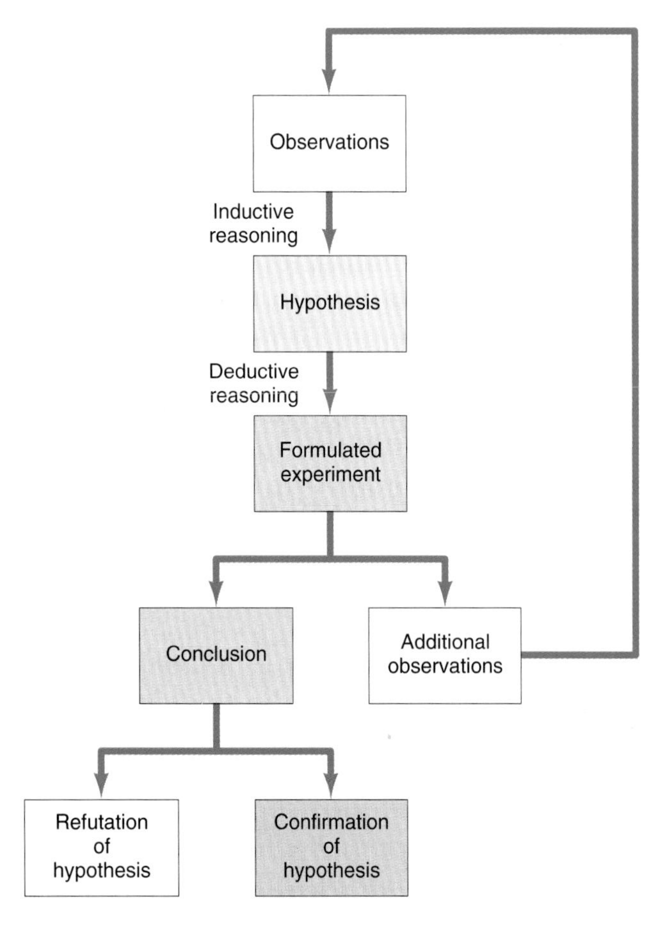

FIGURE 2.11

Scientific Method

Inductive reasoning from many observations yields a hypothesis, which is tested by a deductively formulated experiment. The hypothesis is always stated in the negative, and the experiment tests the validity of this negative (or null) hypothesis. (See Figure 2.12 for an example.) Results may support or refute the hypothesis or they may be inconclusive, requiring additional observations or a reformulation of the experiment.

Not all scientific findings are so clean-cut. Consider the hypothesis that saccharin may pose a threat to human health. Saccharin is obviously not a strong poison, for it has been widely used for over 80 years—but could it weaken the body in some way, and increase the likelihood of cancer? How do we design an experiment to test such a hypothesis?

First, we cannot experiment directly with humans, for obvious reasons, and therefore we must use another mammal and impute the findings to humans. Second, if we assume that the risk of saccarin are, at worst, marginal (since human saccharin users have not exactly been dropping like flies), we must have a very large population of our test mammal, and carefully examine each after feeding it saccharin. However, such an experiment would be prohibitively expensive. Therefore we must use relatively massive doses of saccharin on a much smaller population of test mammals in the hope that any negative results will occur at a rate high enough to be detected. Third, to be sure that each of our test mammals receives identical amounts of saccharin, we cannot rely on feeding, but must instead inject our animals with measured doses.

We now have three major design flaws in our experiment. Scientific experiments of this type are designed to isolate what is termed the **independent variable**—in this instance, the effect on humans of ingesting saccharin. We are looking for changes in **dependent variables**—growth of tumors, loss of weight, birth defects, and so on—that we can link to the independent variable. In order to forge this link, we must control all of the other potential independent variables. However, we have included three of them (different species; huge doses; injection rather than feeding) as part of our experimental design. We can treat a control population of our experimental species identically, housing them in adjacent cages, and injecting them with water instead of a saccharin solution—but that still will allow us only to make conclusions about the test species, not about humans (see Figure 2.12).

In fact, when rats were used in just this type of experiment, three rats out of 100 developed bladder tumors when injected with an amount of saccharin equivalent to a human dose of 800 bottles a day of saccharin-sweetened carbonated drinks. But saccharin is metabolized differently by the body if it is injected rather than being ingested; rats are not humans; and we do not drink 800 bottles of saccharin-sweetened carbonated drinks daily. Clearly, rats should stay away from hypodermic needles containing massive doses of saccharin—but is saccharin, as presently used, dangerous to humans? We cannot say conclusively. Science has provided us with information that may or may not be relevant. It has not provided us with final answers.

Has science failed? No, because final answers are not always possible. Science is a source of information about which society as a whole can react, but it does not and should not dictate policy. "Dose effects" of various substances, most of which must be tested on experimental animals, account for much of the controversy we see regularly in the newspapers today. The experiments are not the source of the disagreements—it is the conclusions that stem from these experiments. Is salt (sodium chloride) bad for people with high blood pressure? The accepted view is yes, and most individuals who are being treated for high blood pressure are placed on low sodium diets. Recently, however, one investigator concluded that it was the absence of calcium, not the presence of sodium, that was the real culprit.

These disagreements are annoying for people wanting to know what to do about their high blood pressure, but they exemplify how science works. There is no room for dogma in science. Scientists do not strive for ultimate truths. Rather, they search for conclusions that are the most compatible with the evidence. The "laws" of science are constantly being challenged, and occasionally they are overturned. For example, a basic law of physics for many years was the Law of the Conservation of Energy—"Energy cannot be created or destroyed." When Einstein formulated his famous equation, $E = mc^2$, that law had to be recast. As Einstein showed, energy can be created—from the destruction of matter, as atomic explosions so tellingly reveal. The new law is the Law of the Conservation of Matter and Energy—"Energy and matter can be interconverted, but the total of both is constant in the universe." Whether this law will still be valid 100 years from now is anyone's guess. What is certain is that at present it is the best explanation of thousands of experiments and observations dating back almost 200 years.

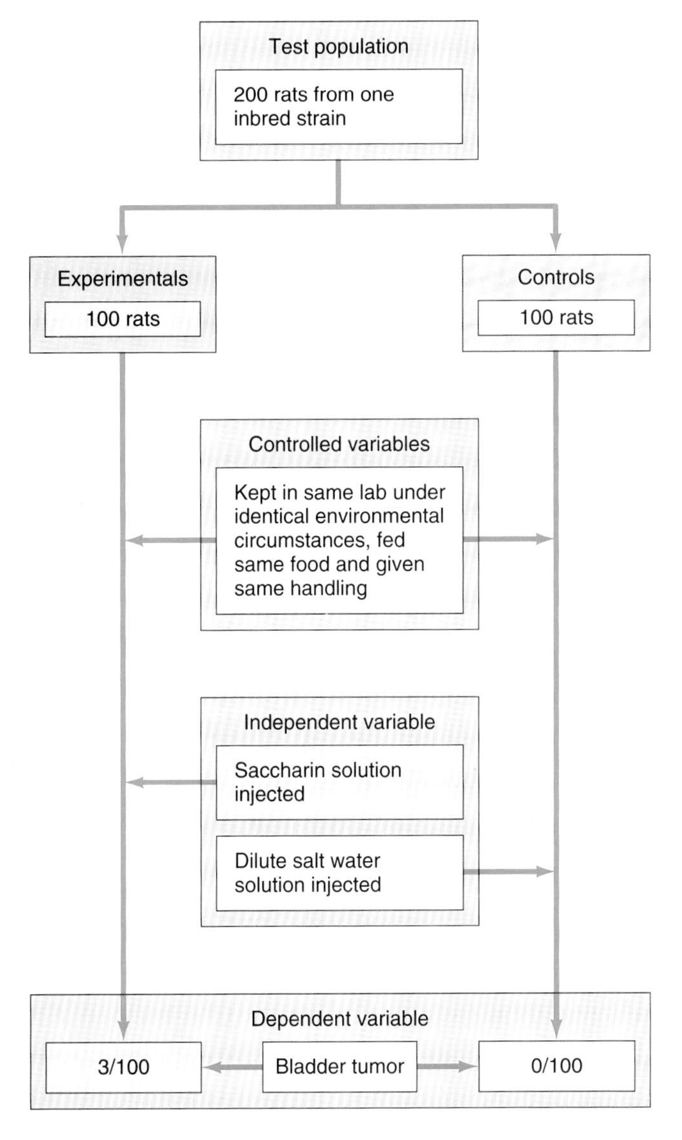

FIGURE 2.12

Experimental Design

Experiments are designed to isolate the test, or independent, variable—in this case, the injection of saccharin. The hypothesis is stated in the negative: "Rats injected with saccharin will develop no negative consequences." In this experiment, three of the experimentals, but none of the controls, developed bladder cancer. Is that ratio high enough to conclude that saccharin causes bladder cancer in rats? In other words, was the null hypothesis supported or refuted? Finally, what can we conclude about the potential dangers of saccharin to humans?

BOX 2.1

The Death of Spontaneous Generation

For thousands of years people believed that, given the proper environment, living organisms could arise directly from nonliving material, a belief known as **spontaneous generation.** Aristotle believed in spontaneous generation, at least for a variety of lower organisms. Jan Baptista van Helmont (1579–1644), the Flemish physician who showed that plants grew in size not principally because of materials taken from the soil but from water and air (a remarkably early presaging of the discovery of photosynthesis), nonetheless was a believer in spontaneous generation. In fact, he developed a recipe for mice: Stuff bran and old rags in a bottle, and leave in a darkened closet to produce mice! Even Isaac Newton (1642–1727), when not developing the calculus, speculated that plants arose from the attenuated tails of comets!

As ludicrous as those ideas seem to us today, it was not until the 1860s that Louis Pasteur (1822–1895), the great French chemist, tolled the death knell for spontaneous generation.

Why did this take such a long time? For one thing, it is impossible to disprove a universal negative, and thus the argument had to reach a point where the two sides could agree on what would constitute a definitive experiment. For another thing, the developments of science contributed ammunition to both sides for some period of time.

Some of the earliest work was done by Francesco Redi (1621–1697). Redi used what we would now term "controlled experiments." He placed meat in both open and stoppered flasks, and demonstrated that fly maggots developed only in the open flasks. To answer the criticism that he had deprived the meat of air, he then repeated his experiment using both open flasks and flasks covered with gauze—and again, only the open flasks developed maggots. However, in both cases, the meat putrefied, which left open the question of whether microorganisms might arise by spontaneous generation.

Microscopes enforced the notion that spontaneous generation might not make sense for large organisms but still made sense for small (and presumably simple) organisms. On the one hand, careful microscopic examination showed that plant galls did not arise spontaneously, but in response to the presence of tiny insect larvae. On the other hand, mixtures of water and hay quickly developed impressive arrays of protists, and these tiny organisms, new to science, quickly became the focus of spontaneous generation questions.

During the eighteenth century, as the scientific method continued to be improved, a classic encounter developed between John Needham (1713–1781) and Lazzaro Spallanzani (1729–1799). Needham boiled mutton broth and then placed the broth in sealed containers—after which the broth developed vast colonies of microorganisms. Spallanzani sealed the flasks first, and then boiled the broth for 30 to 45 minutes—and no organisms developed. Critics of Spallanzani's work maintained, however, that his harsh treatment destroyed the "vital force" (or, alternatively, that he had "spoiled" the air). Although Spallanzani had not convinced a skeptical world that spontaneous generation did not occur, he did attract the notice of one Nicolas Appert (1750–1841), a French chef, who, in 1810, used Spallanzani's technique to preserve food—and founded the canning industry!

Pasteur arrived at the question of spontaneous generation via a circuitous route. He had carried out a number of studies on fermentation, a subject near and dear to the hearts of French vintners, and he was convinced that fermentation was the result of the activities of living yeasts, rather than the then-accepted contrary view—that the yeasts arose spontaneously as a consequence of fermentation.

Fermentation had an interesting history, and in the middle of the nineteenth century, it occupied a critical position in the minds of the scientists of the day. Although fermentation as a process had been known for thousands of years, it was long associated with alchemy, not true science. After all, the addition of "ferments" (that is, what we now know to be yeast colonies) *transformed* (and that was the critical word to alchemists) passive substances into actively fermenting substances. Fermentation was also seen as being similar to decomposition, putrefaction, and infectious disease, and all of these processes were believed to develop spontaneously—and any microorganisms associated with these processes were thought to be the *product,* rather than the *cause,* of the process.

The point is that Pasteur's studies on fermentation, which led him to investigate the general question of spontaneous generation, had profound practical value. If Pasteur could satisfactorily dispose of the spontaneous generation question, then medical doctors (a singularly reactionary group in the nineteenth century) would have no recourse but to accept the notion that microorganisms cause disease, rather than the other way around.

Pasteur performed a number of ingenious experiments, but perhaps the classic experiment was one where he boiled a broth in a flask, and then immediately drew the neck of the flask out into an "S" shape—but left it open to the air. Bacterial spores in the air could not gain access to the broth because they settled out in the curve of the "S"—but

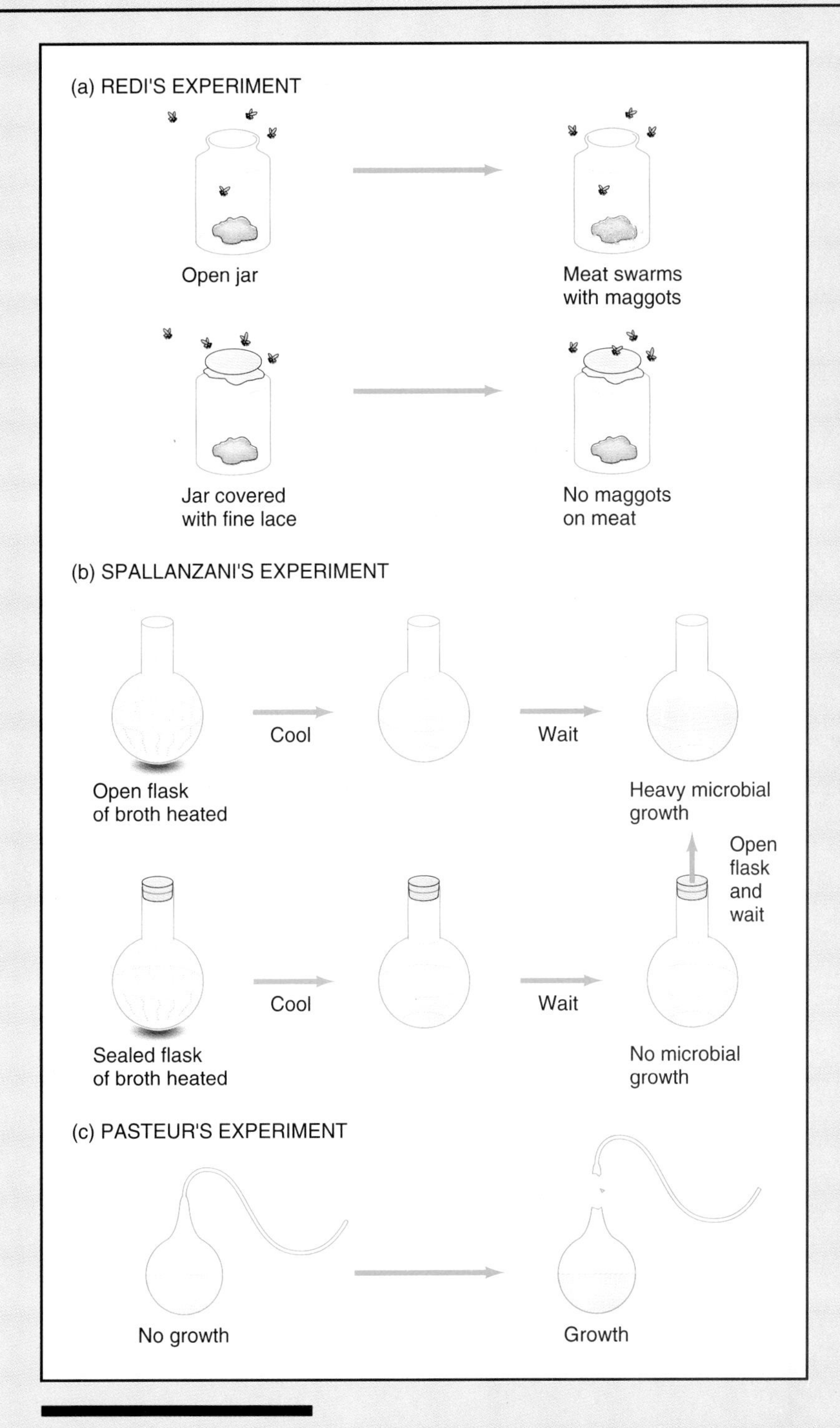

the criticism leveled against Spallanzani could not be made against Pasteur because the flask remained open to the air. The broth remained pure indefinitely—but when Pasteur broke the neck of the flask, the broth quickly spoiled. The only acceptable explanation was that the source of the spoilage was air-borne microorganisms that infected the broth, not that the microorganisms were being spontaneously generated by the broth. T.H. Huxley built upon this notion in coining the term **biogenesis** in 1870—the belief that living organisms always arise from preexisting living organisms.

Adherents of spontaneous generation continued to argue their case well into the twentieth century, but with increasingly fainter voices and smaller audiences. Pasteur extended his work from fermentation to diseases such as anthrax and rabies, and helped to establish the germ theory of disease (see Chapter 9 for more details). These seemingly arcane investigations into spontaneous generation were to set the stage for modern hygiene.

Spontaneous Generation

Redi's experiment (a) demonstrated that maggots did not form when flies could not reach the meat—but the meat spoiled, suggesting that spontaneous generation of microorganisms was possible. Spallanzani's experiment (b) showed no bacterial growth in the broth that was sealed from the air, but critics claimed that the experiment proved nothing since something vital had been destroyed when the air was heated. Pasteur's experiment (c) left the vessel open to air, but bacterial spores could not reach the broth until the neck of the flask was broken.

Observation and the Comparative Method

Not all branches of biology rely on the experimental method. Experimentation is frequently difficult in evolutionary biology, animal behavior, and ecology—just as it is in meteorology, paleontology, and astronomy. However, these fields are no less "scientific" merely because they do not employ the experimental method.

Biologists in these fields see themselves as observers of the ongoing "experiments" of nature. These are not controlled experiments, in the sense that an independent variable has been isolated, for there are usually many variables—and therein lies the challenge and the fascination. By making careful observations typically employing a comparative methodology (that is, making the same observations using a different species, or observing the same species in different habitats or at different times of the day or year), the adept biologist is able to discern which variables are the most important.

The comparative method has been in use for almost 200 years, and has resulted in some of the most impressive developments and discoveries in biology, including the growth of systematics, the development of ecology and biogeography, and the successful elucidation of the theory of evolution by natural selection.

Contemporary Applications

We all use the scientific method in our everyday lives. Decisions ranging from which laundry detergent to use to which variety of corn seed to plant typically employ the scientific method. It has become so commonplace in our culture that only with difficulty can we contemplate a society and a time when the scientific method did not exist.

Increasingly, our society will be called on to make judgments regarding possibly injurious materials, and the evidence will often be equivocal. Future problems are likely to involve low-incident effects or effects distantly related in time to the cause. For example, should chlorofluorocarbons have been banned as the propellant in aerosol cans because of the fear that they were causing a breakdown of the ozone layer in the upper atmosphere? Should sodium nitrite be prohibited as a meat preservative because it may cause cancer? (If so, what is the alternative for preserving meats?) Did Agent Orange, used extensively during the Vietnam War to defoliate trees, promote the incidence of cancer in soldiers unfortunate enough to be on the ground during the spraying? Answers to these kinds of questions depend on using and interpreting the scientific method, and these issues are only a few of the complex questions that have been debated in recent years. Many more will follow in the coming years. Only by having some knowledge of the workings of the scientific method can we make informed judgments about such controversies.

Summary

A central problem in biology almost since the first biological investigations has been to reconcile the seemingly irreconcilable notions of stability between generations (heredity) and the incredible diversity of life that is frequently patterned in such a way as to suggest the potential for change within a species (evolution). After a number of false starts by others, Charles Darwin correctly and convincingly presented a theory of evolution by natural selection that won wide, if at times grudging, support. The concept of natural selection embodies the notion that organisms are adapted to their environment, and changes that increase adaptiveness will tend to be retained in the population whereas changes that decrease adaptiveness will tend to disappear.

Darwin was unaware that a contemporary, Gregor Mendel, was almost simultaneously explaining the basic workings of genetics. The two fields—genetics and evolution—at first seemed as irreconcilable as the original notions of stability versus change, but by 1940 they were united in a grand evolutionary theory that today is undoubtedly the single most important theory in biology.

Although the history of biology dates back almost 2500 years, the great bulk of our biological knowledge has been discovered just within the last 200 years. This fact is directly related to the development of modern scientific methods, which emphasize inductive logic rather than deductive logic. Although different scientific problems call for somewhat different methods, all encompass the notions of testability and repeatability, and all are subject to review by other scientists who are free to determine independently that the findings initially reported are valid. The basic form of the scientific method has now been so completely integrated into our social fabric that we use it daily without recognizing it.

Key Terms

type specimen
typology
inheritance of acquired characteristics
law of use and disuse
natural selection
mutation
germ plasm
proximate cause
ultimate cause
adaptation
homeostasis

phylogeny
deductive logic
syllogism
inductive logic
hypothesis
theory
independent variable
dependent variable
independent variable

Box Term

spontaneous generation

Questions

1. Explain the relationship between genetics and evolution.
2. Distinguish between the theories of evolution through the inheritance of acquired characteristics and evolution by natural selection.
3. How is typological thinking in conflict with the theory of evolution by natural selection?
4. What is the law of use and disuse? Can it be reconciled with Darwin's theory of evolution by natural selection?
5. Outline the major elements of Darwin's theory.
6. Darwin's book on natural selection was highly controversial. Why did Mendel's discovery of the basic laws of genetics not create an equal controversy?
7. Why was it not until about 1940 that genetics and evolution were united?
8. What are the circumstances and processes that unite all living organisms?
9. Distinguish deductive from inductive logic.
10. When does a hypothesis become a theory?
11. Explain how independent variables, dependent variables, and controlled variables differ.

PART

II

The Cellular Basis of Life

If life, as we know it, has the cell as its fundamental unit, how and when did the first cell arise? How are cells organized and of what are they composed? What is it that cells *do,* anyway?

In the next three chapters, we shall consider these and other related questions. It is useful at this stage to keep a few points in mind as we undertake the exploration of life at the cellular level.

First, cells are composed of chemicals, just as is the nonliving world. While it is true that the chemicals found within cells are often very large and complex, and frequently are not found in the nonliving world, it is also true that the fundamental building blocks from which these chemicals are composed *are* found in the nonliving world. These building blocks—called **elements**—will be discussed in greater detail in Chapter 3. For the moment, however, it is important to realize that the living world utilizes elements found in the nonliving world to assemble the chemicals that characterize life.

Second, life has existed on the earth almost since the earth's beginnings, about 4.6 billion years ago. Because of the constant geological upheaval of the

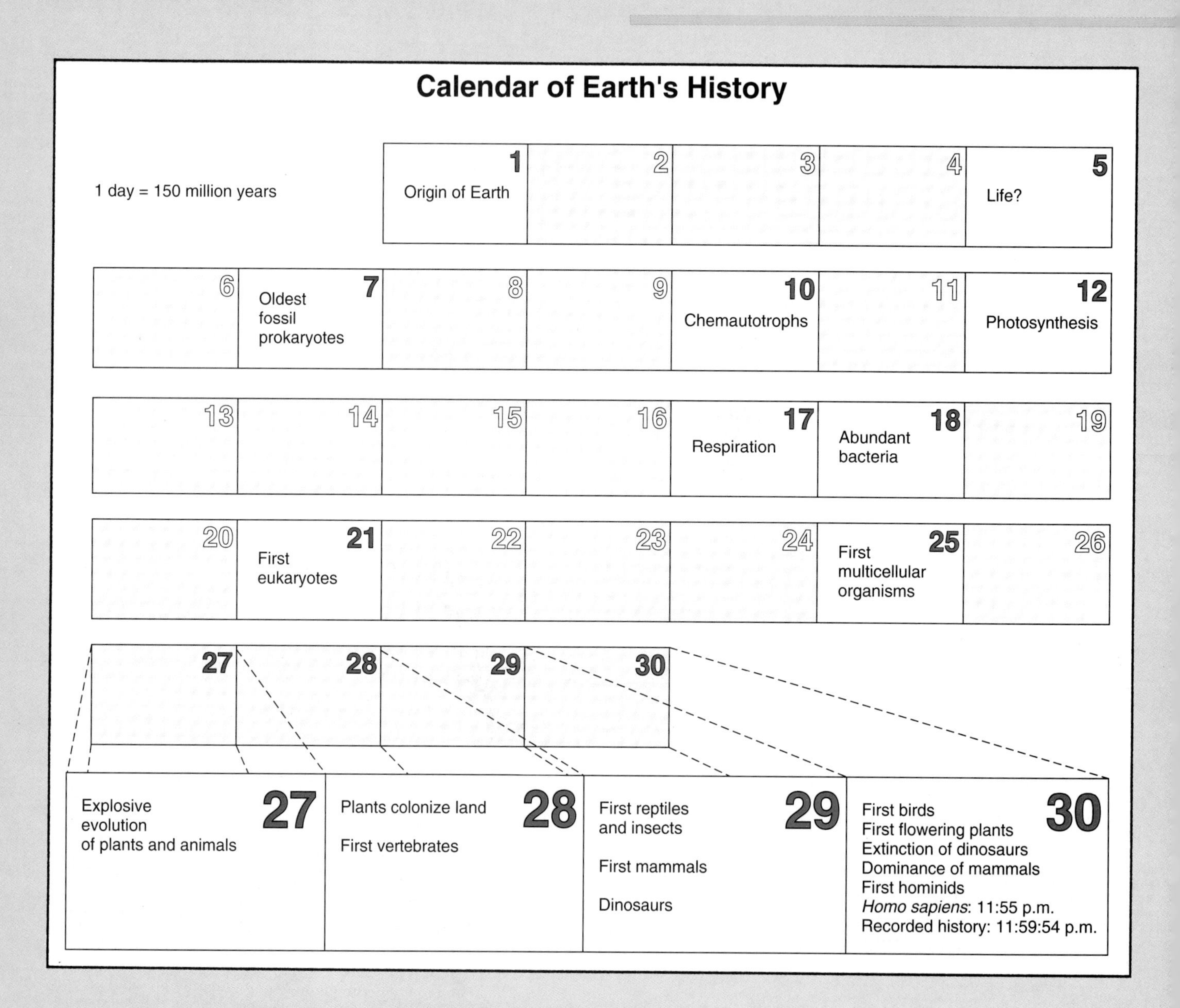

Calendar of Earth's History
1 day = 150 million years
1
Origin of Earth
2
3
4
5
Life?
6
7
Oldest fossil prokaryotes
8
9
10
Chemautotrophs
11
12
Photosynthesis
13
14
15
16
17
Respiration
18
Abundant bacteria
19
20
21
First eukaryotes
22
23
24
25
First multicellular organisms
26
27
28
29
30
27
Explosive evolution of plants and animals
28
Plants colonize land
First vertebrates
29
First reptiles and insects
First mammals
Dinosaurs
30
First birds
First flowering plants
Extinction of dinosaurs
Dominance of mammals
First hominids
Homo sapiens: 11:55 p.m.
Recorded history: 11:59:54 p.m.

earth's crust, we have no rocks that date to the very origin of the earth. However, the oldest rocks yet discovered, dating back almost 4 billion years, contain fossils of cells. These earliest cells are ancestral to all of the life forms on earth today, a fact seemingly belied by the incredible diversity of life. (Do we humans really share a common ancestor with a mushroom?) However, at the cellular level, a great many fundamental biological processes are shared by all organisms, testimony to a common ancestry.

Third, cells are organized. Under the typical Biology 101 microscope, cells look rather uniform in their contents—just a tiny bag of grainy jelly, generally with a central nucleus—but under the powerful electron microscope, the cell's true organization is stunningly revealed. During the past 40 years, it has become increasingly possible to link the structure, or anatomy, of the cells with the various chemical processes, or physiology, which characterize life.

Fourth, although we generally see cells dead and dyed on a microscope slide, these miniature mummies are as uncharacteristic of the living cell as a fossil dinosaur is of its once-proud self. Don't lose sight of the obvious. It isn't enough that cells simply *are*—they must *do* something. Life isn't static; it's dynamic. Cells are constantly engaged in interchange with the environment (which, for most human cells, is the internal environment of our bodies, but which might be a pond for various single-celled organisms). The process of living requires that each cell obtain certain chemicals from its environment (precisely which chemicals is in part a function of the particular cell); it also requires that each cell be able to dump its waste products back into its environment. These requirements have profound implications for the overall design of organisms, as we shall see in following sections. For the present, however, consider how a cell must be designed so as, on the one hand, to ensure that it stays intact, but, on the other hand, to permit it to exchange materials with the environment. Those needs dictate a great deal about the shape, size, and composition of cells.

Each day on this calendar is the equivalent of 150 million years. Note the historical dominance of prokaryotes, and the very recent appearance of humans.

CHAPTER

3

The Chemistry of Life

"What little interest I have in biology centers around knowing how my body works and understanding environmental problems. Why should I have to know anything about chemistry?"

If you feel that way, be of good cheer—there is only a modest amount of chemistry in the pages that follow. Yet without some knowledge of molecules and how they work, you would almost certainly be unable to answer certain questions: Why do you have the same amount of salt in your blood as your roommate, who smothers every morsel of food with salt (see Chapter 11)? Why is diabetes fatal if untreated (see Chapter 13)? Why are new cars required to use lead-free gasoline (see Chapter 25)?

Our world and our bodies are not insulated from atoms and molecules—to the contrary, they are composed of atoms and molecules. Before we know whether a bridge will stand or fall, we must first know something about the girders. Before we can understand how living things work, we must first be familiar with the basic substances of which they are composed.

IN THE BEGINNING . . .

Astronomers calculate the age of the universe to be about 12,000,000,000 years. "Twelve billion" is virtually meaningless when applied to measures of time. We are used to seeing numbers that large only in reference to money—the monthly trade deficit, for instance, or the annual revenue figures of major corporations. Let's put such numbers in some perspective: There are fewer than 1,000,000 individual letters in this book. It would take more than 13,000 copies of the book to include 12 billion letters—a stack of books taller than the Empire State Building! You would travel approximately 190,000,000 inches if you drove from New York to Los Angeles. A journey of 12 billion inches would take you more than halfway to the moon. In short, 12 billion is a huge number.

The origin of the universe remains shrouded in mystery. Most astronomers believe that the universe began with a "big bang"—an enormous explosion that propelled matter and energy outward with such force that, 12 billion years later, the pieces are still flying apart at incredible speeds. In recent years, with the observation that the stars and galaxies closest to the periphery have the highest velocity, the prevailing view has been that the universe had a single origin. However, some evidence suggests that the universe may be in a continuous cycle of expansion and contraction, meaning that there may have been any number of "big bangs"—with more to come at some unfathomable time in the future.

As the result of over 300 years of work by an international succession of chemists, we know that the earth is composed entirely of various combinations of 90[1] naturally occurring **elements** (see Figure 3.1). (More than a dozen additional elements have been created in the laboratory in recent years.) More importantly, many of these same elements have been detected in stars through the prismatic analysis of starlight, a finding that proves that the earth—and everything on it, including ourselves—is composed of the same "building blocks" as is the rest of the observable universe.

However, the proportion of these elements varies. Unlike the earth (or human beings—see Box 3.1), the sun and other stars are composed primarily of the two smallest elements—**hydrogen** and **helium.** Because of the intense heat and incredible gravitational pull in stars, **fusion reactions** occur, forming larger and more complex elements. (The same type of fusion reactions occur when a hydrogen bomb is detonated; atomic bombs, by contrast, use **fission reactions,** in which large elements are split.) Exploding stars dispersed these other elements throughout the universe.

Based on measurements of the age of rocks (by radioactive dating, a process discussed more fully in Chapter 22), scientists generally agree that the earth

[1]The number of naturally occurring elements is sometimes said to be 92. However, technetium (atomic number 43) and promethium (atomic number 61) are unstable, and are not found in nature. They were synthesized in the laboratory in 1937 and 1945, respectively.

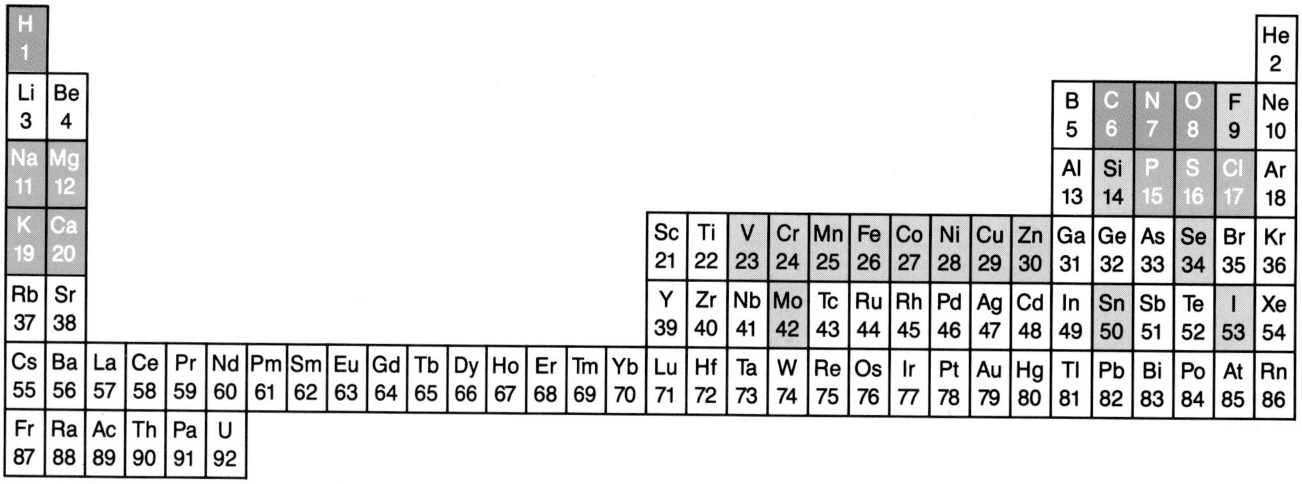

H 1																															He 2
Li 3	Be 4																									B 5	C 6	N 7	O 8	F 9	Ne 10
Na 11	Mg 12																									Al 13	Si 14	P 15	S 16	Cl 17	Ar 18
K 19	Ca 20															Sc 21	Ti 22	V 23	Cr 24	Mn 25	Fe 26	Co 27	Ni 28	Cu 29	Zn 30	Ga 31	Ge 32	As 33	Se 34	Br 35	Kr 36
Rb 37	Sr 38															Y 39	Zr 40	Nb 41	Mo 42	Tc 43	Ru 44	Rh 45	Pd 46	Ag 47	Cd 48	In 49	Sn 50	Sb 51	Te 52	I 53	Xe 54
Cs 55	Ba 56	La 57	Ce 58	Pr 59	Nd 60	Pm 61	Sm 62	Eu 63	Gd 64	Tb 65	Dy 66	Ho 67	Er 68	Tm 69	Yb 70	Lu 71	Hf 72	Ta 73	W 74	Re 75	Os 76	Ir 77	Pt 78	Au 79	Hg 80	Tl 81	Pb 82	Bi 83	Po 84	At 85	Rn 86
Fr 87	Ra 88	Ac 89	Th 90	Pa 91	U 92																										

FIGURE 3.1

Periodic Table of Elements

Positioning is based on the subatomic organization of the atoms of each element. Elements found in living organisms are shown in color. The elements found in greatest abundance in organisms are shown in darkest color; those in moderate amounts in medium color; and those in minute amounts in light color.

BOX 3.1

The Elements of Life

Although the matter of the universe is composed of 90 different elements, organisms on earth utilize only about two dozen. Some of the remaining elements are neutral, in a biological sense, in that they neither enhance nor detract from the vitality of organisms. Still others, such as arsenic, mercury, lead, and cadmium are highly poisonous to life. Why are there such differences among the elements?

With few exceptions, the elements of life share several of the following properties: relative abundance in the earth's crust, small atomic size, chemical reactivity, and solubility in water. The atoms of such elements would therefore have been abundant in the dissolved materials present in the primordial seas in which life presumably first arose.

The biologically neutral elements include a variety of gases present in minuscule amounts in the air we breathe. Helium, neon, argon, and xenon, among others, have no biological effect because they are chemically neutral. Their atoms have filled outer shells and they do not normally combine with other atoms. As such, they do not impinge on the welfare of organisms.

The deadly elements, such as mercury, lead, and arsenic, tend to be those possessing none of the properties listed for the elements of life. As life began and organisms evolved, they would rarely have encountered atoms of these elements and would therefore have developed no mechanisms for dealing with them. It has only been during the technological stage of human development that these elements have, on a regular basis, been freed from the earth's crust and become a part of our biological environment. Organisms have, for the most part, simply not had enough time to evolve defense mechanisms against them and for this reason exposure to these elements is generally deleterious and sometimes even fatal for the organism involved.

and the rest of our solar system originated approximately 4.6 billion years ago. The planets originated as separate coalescences of what we might euphemistically call "stardust" into bodies much smaller (and, therefore, much cooler) than the sun. All 90 elements were represented in this cooling and coalescing, although the smaller elements were generally more abundant.

As the earth gradually took shape, and the isolated bits of elements collected together, many of the elements began to react with one another, and to combine to form new and dramatic assemblages, called **compounds.** We have a particular interest in compounds, since we (and living things in general) are largely composed of compounds. However, before we are able to explain how compounds can form, we must first examine the composition of elements.

ELEMENTS AND COMPOUNDS

Each element is assigned a unique symbol, consisting of one to three letters, which are abbreviations of the name of the element in English (or, in a few instances, Latin). The symbols for hydrogen (**H**) and oxygen (**O**), for instance, derive from their English names, whereas symbols for **sodium** (Na, for *natrium*) and **potassium** (**K**, for *kalium*) come from their Latin names.

The **atom** is the smallest unit possessing all of the particular properties that characterize a given element. Except for isotopes (see below), the atoms of each element are identical, yet they are structurally different from the atoms of all other elements. Thus one speaks of an atom of oxygen or an atom of uranium.

Atomic Structure

Although atoms are considered to be the basic units of matter, each atom is nevertheless made up of still smaller particles. Many kinds of these **subatomic particles** have been identified, but only three are important in understanding the properties of elements.

An atom consists of a central core (**nucleus**), which generally contains two types of particles, **protons** and **neutrons** (see Figure 3.2). **Electrons,** a third type of particle, are found at a considerable distance from the nucleus. (The distance really is considerable. If the nucleus of a given atom were the size of the dot over a letter *i*, the nearest electron would be more than 100 feet away. This open space inside atoms is greatly collapsed in dwarf stars, a fact which explains why a teaspoonful of dwarf star would weigh several tons.)

Protons have an electrical charge that is designated positive, and electrons are assigned a negative charge. The aptly named neutrons have no detectable

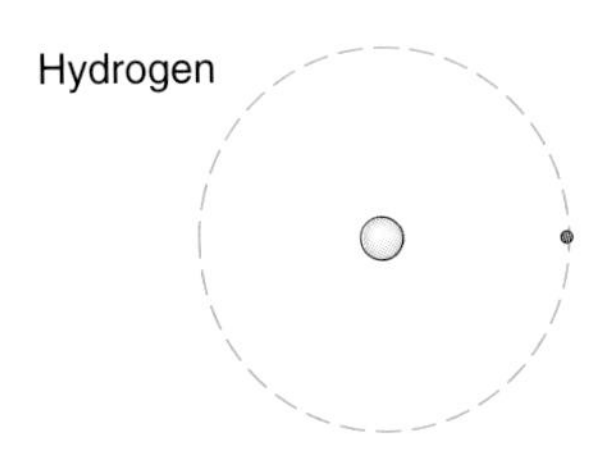

FIGURE 3.2

Atomic Structure of Hydrogen

For ease of illustration, atoms are usually represented by one or more circles in the nucleus, indicating the protons and neutrons, and an appropriate number of electrons in orbit around the nucleus. Hydrogen atoms have only one proton and one electron.

electrical charge. Protons and neutrons have virtually the same **mass,** some 1800 times as great as the mass of an electron. ("Mass" is often used as if it were a synonym for "weight." However, mass is independent of gravity, whereas weight is mass times acceleration due to gravity. On the surface of the earth, the mass and weight values of an object are interchangeable, since gravity's force is essentially constant everywhere on the earth's surface. On the moon, however, the mass of an object would be the same as it is on earth—but the weight of that object would be only one-sixth of the object's weight on earth, because the moon's gravitational force is only a fraction of the earth's gravitational force.)

Each atom of a given element has a specific number of protons in its nucleus, a number that is unique for each element. Because the number of electrons orbiting the nucleus of an atom is always equal to the number of protons within the nucleus, atoms are electrically neutral.

The number of protons of an element defines the element's **atomic number.** For example, because each atom of the element *carbon* has six protons in its nucleus, carbon's atomic number is 6. The **mass number** of an element is the sum of the number of protons *and* neutrons in the nucleus of each atom. For example, carbon atoms have six protons, and most also have six neutrons. Thus the mass number for the most common form of carbon is 12 (symbolized in the following way: ^{12}C).

Although each atom of an element always has the same number of protons, the number of neutrons may vary. These variant forms of a given element are called **isotopes.** For example, carbon 14 (^{14}C) is a carbon isotope with six protons and eight neutrons. Note that, because isotopes of a particular element have different numbers of neutrons, they also differ in their *mass* number. However, because they all have the same number of protons, they share the same *atomic* number.

Some isotopes are stable, but others are unstable, and they *decay.* (Used in this sense, "decay" means that they spontaneously emit subatomic particles, not that they smell bad.) Because of this emission, unstable isotopes are termed **radioactive.** We shall discuss the biological use of radioactive isotopes in Chapter 22.

Electrons and Shells

Because they are located at the periphery of atoms, electrons determine many atomic properties. We therefore need to examine electrons in greater detail.

To begin, because they move so rapidly around the nucleus (at about 1/100 the speed of light), it is not possible to locate an electron with precision at any given moment. In describing the position of electrons, scientists speak of "electron clouds" (see Figure 3.3). This does not mean there is literally a cloud of many electrons, but rather that an electron may be located anywhere within the area of the cloud at any particular instant.

However, the location of electrons is not entirely random. Depending on the amount of energy possessed by an electron, it will be found within one or another of the **shells** that ring the nucleus. The shell closest to the nucleus holds electrons with the lowest energy level, and each successive shell contains electrons possessing higher energy levels. The first shell can hold no more than two electrons, and the second can hold only eight. Subsequent shells are capable of holding more than eight electrons, but they are particularly stable when they contain eight. The relative chemical reactivity of atoms is determined by the number of electrons in the outermost shell, and it is the primary factor controlling the formation of associations between atoms. However, when an atom's outermost shell contains the maximum number of electrons, the atom is chemically nonreactive.

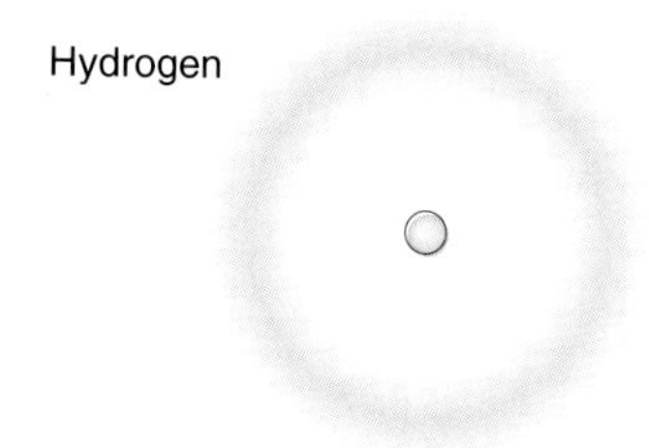

FIGURE 3.3

Electron Clouds

In a more accurate representation of atomic structure, the most probable location of the electron is represented by shading. However, even this drawing of a hydrogen atom is incorrect in its scale—with a nucleus drawn this large, the electron should be shown hundreds of meters distant.

CHEMICAL BONDING

As gravity gradually condensed the swirling gases that were destined to coalesce into the planet Earth, the atoms comprising the gases were brought into proximity. The very nature of atomic organization dictated interaction between these atoms. Free atoms are relatively rare in nature, at least in the earth's crust and in living organisms. Instead, atoms usually unite with other atoms to form compounds.

A compound is a substance that is formed from the combination of atoms of two or more elements in specific proportions. For example, water is a compound formed from the elements hydrogen and oxygen, in the proportion of two hydrogen atoms for every oxygen atom.

The **formula** of a compound contains information about the kinds and numbers of atoms that make up a **molecule,** which is the smallest unit that possesses all of the characteristics of the compound. (By convention, the term "molecule" is used in an even more expansive way, as the smallest stable unit of a compound or element. Thus a molecule of water, H_2O, consists of two atoms of hydrogen and one of oxygen; a molecule of oxygen gas, O_2, consists of two atoms of oxygen; and a molecule of the inert gas helium, He, consists of a single atom of helium. In the case of helium and other inert gases, "atom" and "molecule" are equivalent.)

A subscript indicates the number of atoms of each element comprising the molecule. For example, because each molecule of the sugar *glucose* contains six carbon atoms, 12 hydrogen atoms, and six oxygen atoms, its formula is $C_6H_{12}O_6$ (see Figure 3.4).

The forces that hold atoms together in molecules are called *chemical bonds.* A stable molecule may be formed if the electrons of the atoms can be shuffled so as to permit each atom to have a full outer shell of electrons. Three types of chemical bonding are of particular importance in organisms. In **ionic bonding,** molecules assume a stable configuration by the transfer of electrons from one atom to another. In **covalent bonding,** stability is achieved within the molecule by the sharing of electrons between atoms. The third kind, **hydrogen bonding,** is a special case always involving atoms of hydrogen.

Ionic Bonding

A molecule of table salt, *sodium chloride* (symbolized NaCl), is composed of one atom of sodium and one of chlorine (see Figure 3.5). Sodium has an atomic number of 11, which means that each sodium atom has 11 protons in the nucleus and 11 electrons in shells

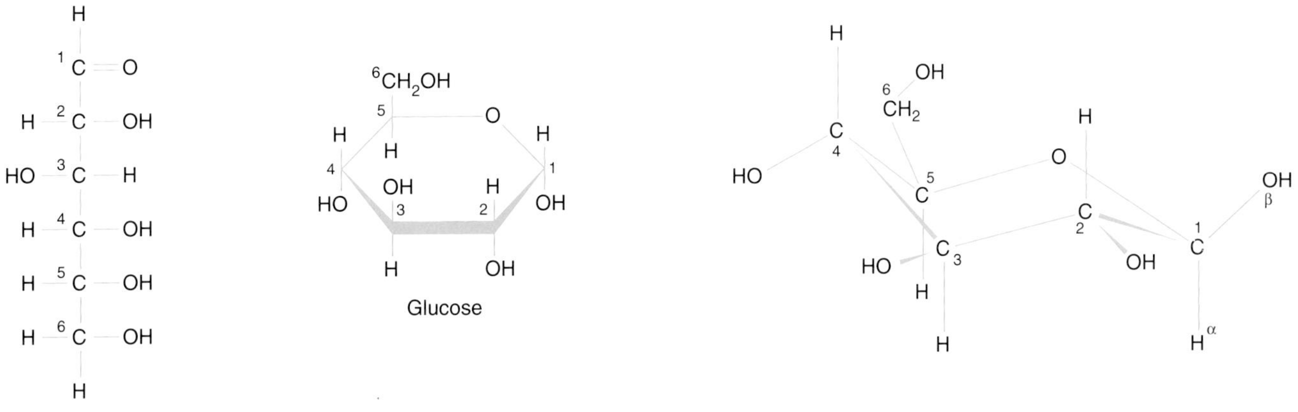

FIGURE 3.4

Chemical Formulas

Compounds can be represented by several different formulas, each correct in its own way. On the left the glucose molecule ($C_6H_{12}C_6$) is shown as a chain of carbon atoms, to which are joined hydrogen and oxygen atoms in a characteristic manner. However, glucose can also form a ring compound through the interaction of the first carbon with the oxygen of the fifth carbon (center). Depending on whether the oxygen attached to the first carbon is "down" or "up," the glucose may be designated as either alpha or beta. This seemingly trivial difference has profound implications in the formation of long chains of glucose molecules (see Figure 3.15). Finally, glucose rings are not two dimensional, as implied by the preceding formulas, but are twisted in three dimensions (right). In this form, the first carbon is bent down relative to the rest of the molecule in a "chaise lounge" shape. In another form, it may be bent up, in a "hammock" shape.

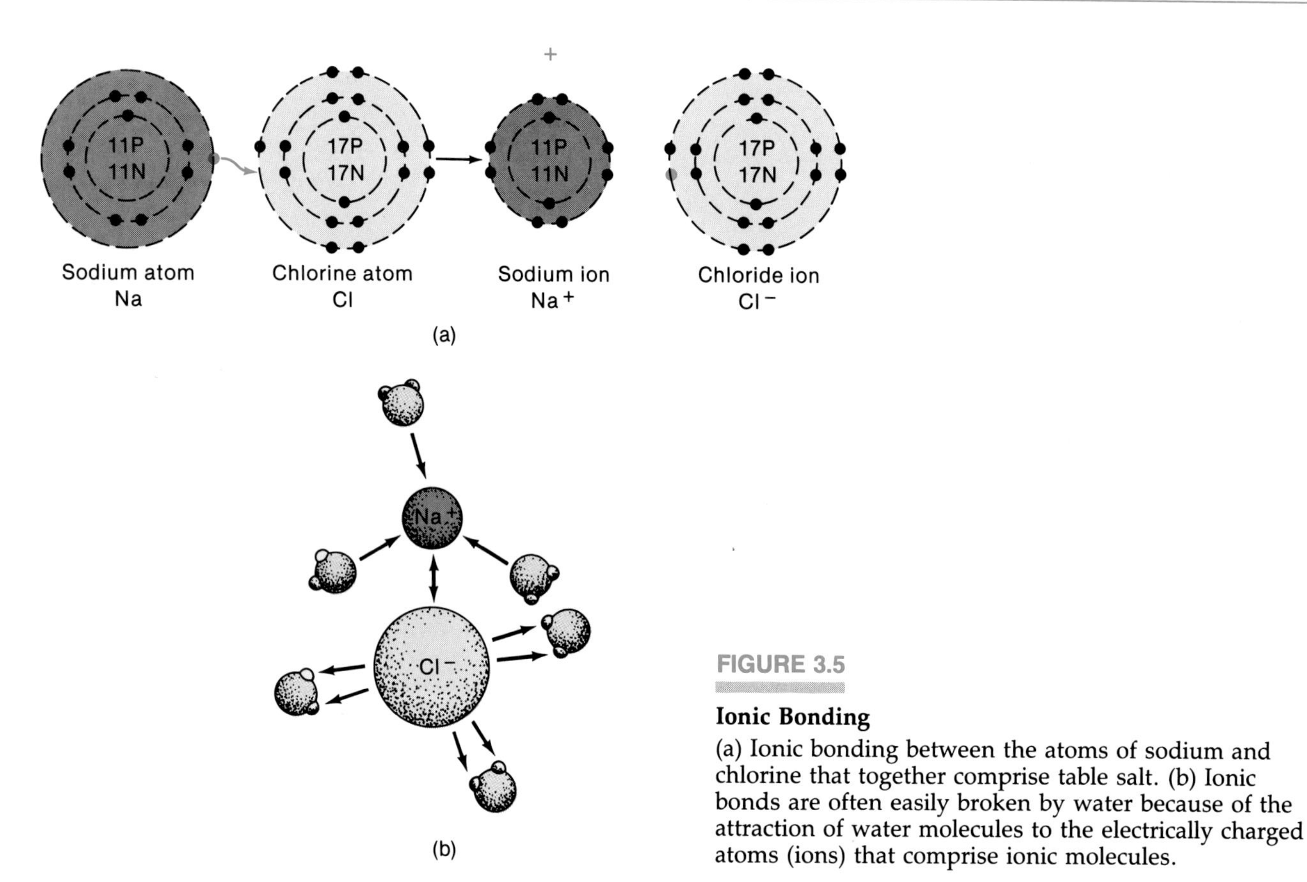

FIGURE 3.5

Ionic Bonding

(a) Ionic bonding between the atoms of sodium and chlorine that together comprise table salt. (b) Ionic bonds are often easily broken by water because of the attraction of water molecules to the electrically charged atoms (ions) that comprise ionic molecules.

around the nucleus. The first shell has two electrons, the second shell has eight electrons, but there is only one electron in the third shell (2 + 8 + 1 = 11). Atoms such as those of sodium, with only one electron in the outer shell, act as **electron donors.** Chlorine has an atomic number of 17, meaning that each chlorine atom has seven electrons in its outer shell (2 + 8 + 7 = 17). Atoms such as those of chlorine tend to be **electron acceptors.**

If a sodium atom collides with a chlorine atom, the lone outer electron of sodium is attracted into the orbit of the outermost shell of chlorine. Sodium attains stability, because it has its full second shell as its new "outer" shell. Chlorine also attains stability because its outermost (third) shell has eight electrons. However, the sodium atom, by losing an electron, has one more proton than it has electrons. Conversely, chlorine has one more electron than it has protons. As a consequence, sodium now has a net charge of +1 (symbolized Na^+), and chlorine now has a net charge of −1 (symbolized Cl^-).

Such electrically charged atoms are called **ions.** Ions of opposite charge are attracted to one another. This attraction results in ionic bonding. That is, an ionic compound such as NaCl is held intact by the attraction between positive and negative ions. When a solid ionic compound such as NaCl is dissolved in water, the ionic bond is broken and the solution contains separate Na^+ and Cl^- ions, each surrounded by water molecules. In organisms, ionic compounds are most commonly found in this *dissociated (ionized)* form, although the ions themselves may be bound to the electrically charged surfaces of other molecules. There are many biologically important ions, in addition to Na^+ and Cl^-. Some, such as K^+, are also formed by the loss of a single electron, but others, such as Ca^{2+} (*calcium*) and Mg^{2+} (*magnesium*), are formed by the loss of two electrons.

Covalent Bonding

A molecule of oxygen (O_2) does not form by ionic bonding. Because the outer shell of the oxygen atom has six electrons, a transfer of electrons cannot lead to a stable configuration for both atoms. However, the *sharing* of electrons by the two bonded atoms results in each attaining a stable configuration (see Figure 3.6).

Shared electrons are considered part of each atom involved in the sharing. In the case of oxygen, *two* pairs of electrons are shared. The covalent bond of the oxygen molecule is therefore said to be *double*. (In structural formulas this would be shown as O=O.) If there is only one pair shared, the bond is *single*; if three pairs are shared, it is a *triple* bond.

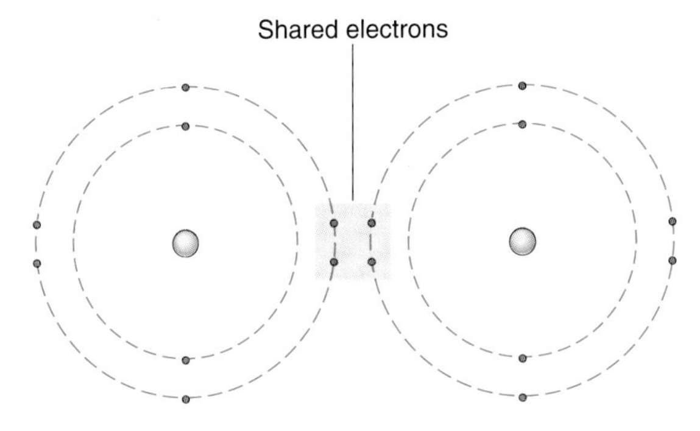

FIGURE 3.6

Covalent Bonding
The atoms of a molecule of oxygen share a total of four electrons, forming a very strong bond.

Hydrogen Bonding

In addition to ionic and covalent bonds, there are also weaker interactions among atoms and molecules. These are biologically very important, because they are responsible both for the three-dimensional shapes of large molecules and for some of their functional properties. They also explain some of the properties of particular compounds, such as water.

In some molecules formed by covalent bonds, the shared electrons tend to be attracted more toward one atom (generally, the larger atom, because the larger atom has more protons that attract the electrons). This unequal distribution of the electrons of a bond is called **polarity,** and it results in molecules with a small positive charge at one end and a small negative charge at the other. Both small and large molecules can show this type of charge distribution. Because of these slight charge differences, polar molecules are attracted to each other.

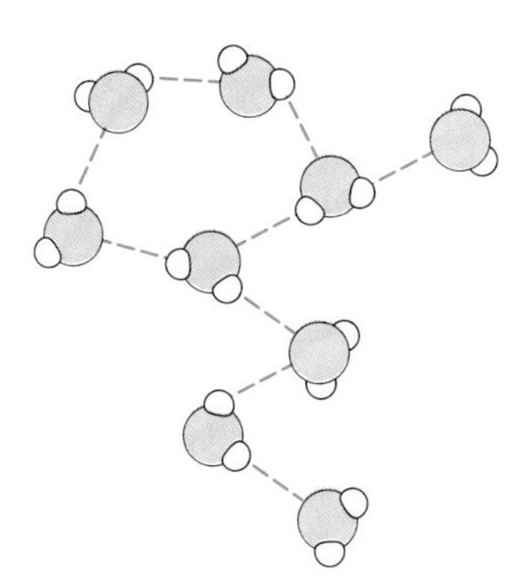

FIGURE 3.7

Hydrogen Bonding
Water molecules exemplify hydrogen bonding, a type of weak bonding between a hydrogen atom of one molecule, and an atom of oxygen or nitrogen in a second molecule. Hydrogen bonding in water explains many of its properties.

Hydrogen bonding is one type of attraction involving polar molecules. It occurs when a hydrogen atom covalently bonded to an atom of nitrogen, oxygen, or fluorine is attracted to a "lone pair" of electrons (that is, a pair of electrons not involved in bonding) on a *different* atom of nitrogen, oxygen, or fluorine (see Figure 3.7).

Hydrogen bonding can occur between molecules, or within a single large molecule. Hydrogen bonding between water molecules gives water many of its unique properties; hydrogen bonding within molecules of proteins and nucleic acids is of great importance in their organization, as we shall see in the next section.

WATER AND LIFE

Because of a variety of factors—the fact that the earth rotates relatively rapidly, that it has a mass sufficiently large to hold a gaseous atmosphere in place, that it is 93,000,000 miles from a sun of only modest size—the mean temperature of the earth is within the relatively narrow range in which water is a liquid (as opposed to being a gas or a solid). As a consequence, water was an available medium for the evolution of life—and it is so integrally a part of life that we are hard-pressed to imagine a form of life that is not water-based.

Water is the most abundant compound in organisms (60 to 95 percent by weight). Because virtually all of the chemical reactions that collectively comprise life occur in water, some familiarity with its properties is essential in our study of biology.

Water and Ice

Water molecules are bent, with an H-O-H bond angle of 104.5° (see Figure 3.8). Because water is a polar molecule, with oxygen forming the negative end and the hydrogens the positive end, hydrogen bonding can occur among water molecules. The significance of this hydrogen bonding is strikingly demonstrated by the structure of ice (see Figure 3.9). Each oxygen atom is surrounded by four hydrogen atoms, two of which are linked by covalent bonds, and two of which are the result of hydrogen bonds.

In ice, the hydrogen bonding produces a loose and dispersed organization. Consequently, the density of ice is low (that is, relatively few molecules in a given volume). Although solids are almost always more dense than the corresponding liquid, water is an exception. Thus ice floats on water, and lakes freeze from the top down, leaving liquid water below the ice. Obviously, the survival of many aquatic species

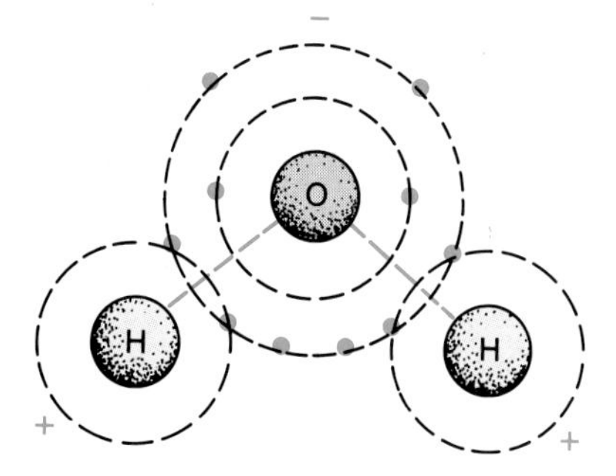

FIGURE 3.8

Water Molecule

The hydrogen and oxygen atoms are covalently bonded at an angle of 104.5°. This asymmetry permits hydrogen bonding between water molecules, because the oxygen "end" of the molecule is slightly negative relative to the hydrogen "ends."

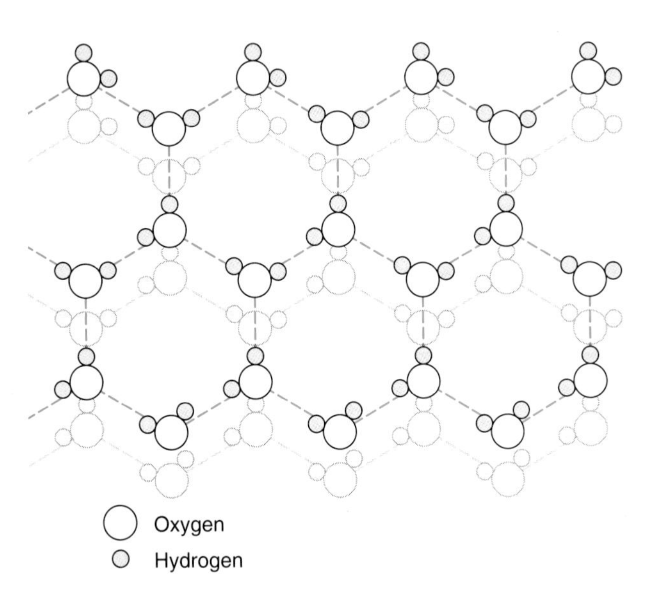

FIGURE 3.9

Molecular Arrangement in Ice

Hydrogen bonding accounts for the very regular and open arrangement of ice, since each water molecule forms a hydrogen bond with four others. (Liquid water is more dense than ice, because not every water molecule forms four hydrogen bonds, and the lattice-like arrangement seen in ice is somewhat collapsed.)

during the winter is predicated on this unusual property of water. On the other hand, since water expands as it solidifies into ice, the formation of ice crystals in living cells is disastrous, because it causes the cell membranes to rupture. This destruction helps to explain why severely frost-bitten fingers and toes do not "thaw out" well, and often need to be amputated.

Boiling and Freezing

Because of hydrogen bonding, water has significantly higher melting and boiling points than might be expected for a molecule of its size. (Ammonia, a nonpolar compound with virtually the same mass as water, freezes at −78° C and boils at −34° C.) Water also has an unusually high **heat capacity** (the amount of heat energy necessary to raise or lower its temperature). The high heat capacity of water minimizes sudden changes in temperature within our bodies and, on a larger scale, also stabilizes the temperature of ponds and lakes such that they are typically cooler in the summer and warmer in the winter than is the surrounding land.

Vaporization

The **heat of vaporization** is the amount of heat energy necessary to convert water from a liquid to a gaseous state—that is, to cause water to evaporate. The high heat of vaporization of water permits efficient cooling by evaporation (evaporation of one gram of water cools another 540 grams of water by 1° C). This high heat of vaporization is used by many organisms in regulating body temperature, in the form of sweating or panting.

Surface Tension

Hydrogen bonding also contributes to the high **surface tension** of water. Surface water molecules are attracted sideways and downward into the body of the liquid by hydrogen bonds. Consequently, the surface tension of water is three times as great as that of benzene, a nonpolar liquid with a mass more than four times that of water.

Surface tension causes liquids to form shapes with minimum surface area (if other forces, such as gravity, are negligible). The small droplets of water that form when water is spilled on a hard, nonpolar, surface show the effects of surface tension. Surface tension is used by such organisms as *water striders,* a group of insects that can skim across the surface of ponds (see Figure 3.10). Trees exploit surface tension in transporting water from the soil to its leaves.

Water as a Solvent

There is an apparent paradox in the origin of life. Although the first life forms originated in water and used water as the medium in which their chemical reactions occurred, they at the same time required mechanisms for staying intact in this unique substance, for in the absence of such mechanisms, they would fragment and dissolve.

FIGURE 3.10

Surface Tension

Water striders exploit the surface tension of water for locomotion. Hairs on the first and third pairs of legs prevent the insect from sinking, while the second pair of legs rows it across the water's surface.

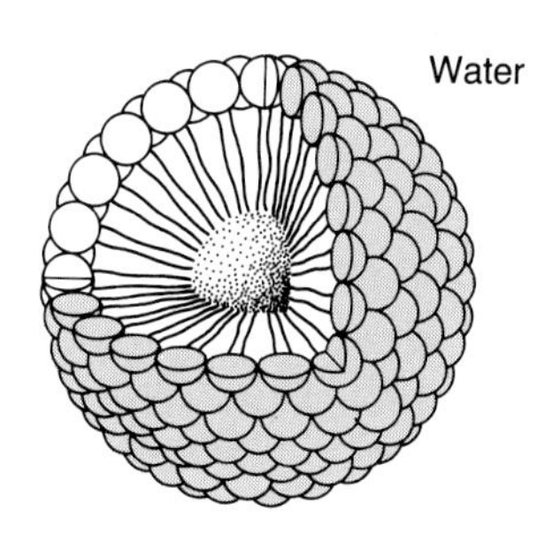

FIGURE 3.11

Micelle Formation

Molecules with hydrophobic and hydrophilic ends will form small clusters in a water solution with the hydrophilic ends facing out, rather like a herd of musk oxen facing danger.

On a simpler level, sugar, alcohol, and table salt are *soluble* (can be dissolved; L. *solvere*, "to loosen") in water, but gasoline, oxygen, and salad oil are essentially insoluble. Why?

Hydrophilic (Gr. *hydor*, "water" and *philos*, "loving") interactions are attractive forces between water and most other polar molecules or ions. These interactions disperse molecules of **solute** (the dissolved substance) among water molecules (the **solvent**, or dissolving substance).

By contrast, interactions between water and nonpolar molecules are **hydrophobic** (Gr. *phobos*, "fear"). Water molecules squeeze out hydrophobic molecules because the attractiveness of one water molecule for another generally exceeds that between water and nonpolar molecules. For example, because molecules of glucose and alcohol are polar, they are also hydrophilic and dissolve readily in water. Conversely, nonpolar molecules such as oxygen gas and salad oil are hydrophobic and are therefore relatively insoluble.

Some molecules have both hydrophilic and hydrophobic ends. When large molecules with both polar and nonpolar ends are mixed with water, they tend to form structures known as **micelles** (L. *micella*, "tiny grain"; see Figure 3.11). In the micelle, the nonpolar ends cluster together centrally, away from the water molecules, both because of positive interactions among the nonpolar ends and because hydrophilic attraction occurs between the polar ends and nearby water molecules. The phenomenon of micelle formation is closely related to the structure of biological membranes, as we shall see in Chapter 4, and it was the evolution of membranes that permitted cells to remain intact in water.

When an ionic compound is dissolved in water, the water molecules are oriented such that the negative O end of the H_2O is adjacent to the positive ion, and the positive H end of the H_2O is adjacent to the negative ion. Thus the ions have envelopes of water molecules surrounding them (see Figure 3.12). The attraction of water to ions (and vice versa) is sufficiently strong that most ionic compounds readily dissociate when placed in water. However, there is considerable variation among ionic compounds in their relative solubility, and in each case the water molecules must be much more numerous than the ions, which is why you can dissolve only so much table salt in a given volume of water before reaching saturation—at which point the Na^+ and the Cl^- are as likely to associate with each other as with a cloud of water molecules.

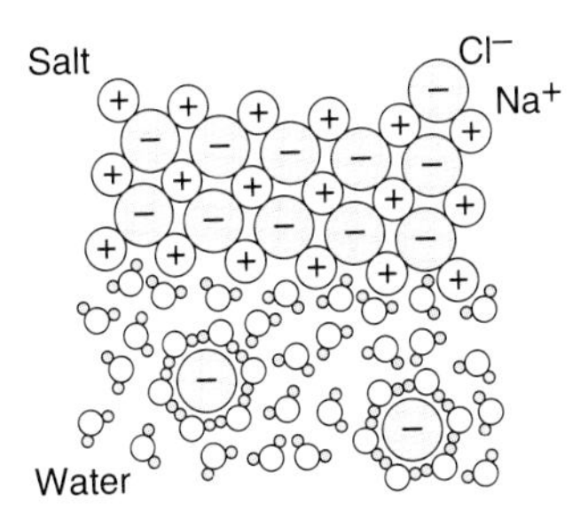

FIGURE 3.12

Hydration Spheres

The ions of ionic compounds dissolved in water are separated and surrounded by water molecules. Note that the orientation of the water molecules is different around Na^+ than it is around Cl^-. Can you explain why?

ACIDS, BASES, AND BUFFERS

In organisms one of the most important ions is H^+, the hydrogen ion. Its relative concentration affects how rapidly many chemical reactions occur, and influences the function and shape of protein molecules, among other effects.

A hydrogen atom consists of one proton in the nucleus, and one electron in a shell outside the nucleus. Thus a H^+ ion consists of a single proton. The H^+ ion is attracted by pairs of electrons not involved in bonding (lone pair), and tends to form a new chemical bond, as for example:

```
                 H                  H
                 |                  |
H+      +      :N—H    ———>      H—N—H+
                 |                  |
                 H                  H
Hydrogen ion   Ammonia        Ammonium ion
```

H^+ Concentration and pH

Since they are both mobile and electrically charged, ions in solution conduct electricity. For example, the mineral ions of tap water conduct electricity reasonably well, which is why hair dryers carry labels warning users not to use them in the bathtub. Because of the absence of mineral ions, you would be less subject to electrocution were you to bathe in distilled water. However, electrocution is still possible because of salts from your skin and because water does show a very slight tendency to ionize:

$$H_2O \rightarrow H^+ + OH^-$$

The standard way of designating the H^+ concentration in a solution is to utilize the **pH scale.** "pH" stands for "potential of hydrogen" and it is simply the logarithm of the H^+ concentration, expressed negatively so as to yield a positive number. On average, one water molecule in 10,000,000 (10^7) will be ionized at any given moment. Thus the H^+ concentration in distilled water is 1×10^{-7}, and the pH (negative logarithm) of distilled water is therefore 7.

The pH scale runs from 0 to 14 (see Figure 3.13). A solution is **acidic** if it has a higher H^+ concentration than OH^- concentration. If the H^+ concentration of a solution is lower than its OH^- concentration, it is said to be **basic** (or alkaline). Acidic solutions have pH values less than 7, and basic solutions pH values greater than 7. Distilled water (which has a pH of 7) is used as the reference standard for neutrality (neither basic nor acidic) since, when it ionizes, water of necessity forms equal numbers of H^+ and OH^- ions.

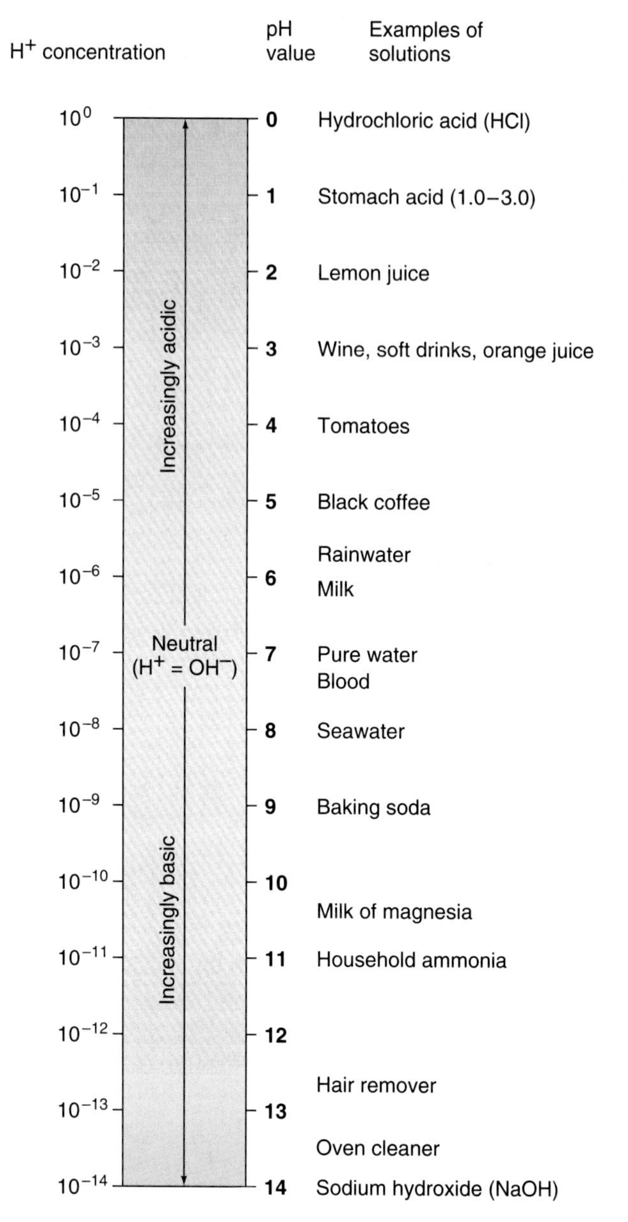

FIGURE 3.13

pH Values

Acids and bases are placed on a scale of 0 to 14, with 7 being neutral. Because it is a logarithmic scale, each difference in a whole number reflects a tenfold increase or decrease. Thus tomatoes are about ten times more acidic than coffee.

Because the scale is logarithmic, each increase or decrease in a whole number reflects a ten-fold change. The decibel scale for loudness of sounds and the Richter scale for tremors caused by earthquakes are also logarithmic scales.

Acids and Bases

Acids are defined as H^+ donors. The relative number of contributed H^+ ions determines the strength of the acid. Strong acids ionize completely; weak acids do not. Some examples of acids are H_2CO_3 (carbonic acid, a weak acid) and HCl (hydrochloric acid, a strong acid):

$$H_2CO_3 \longleftrightarrow H^+ + HCO_3^-$$
$$HCl \longrightarrow H^+ + Cl^-$$

Bases are OH^- donors (or H^+ acceptors). Some examples of bases are KOH (potassium hydroxide) and NaOH (sodium hydroxide):

$$KOH \longrightarrow K^+ + OH^-$$
$$NaOH \longrightarrow Na^+ + OH^-$$

In our daily lives, we routinely encounter acids and bases. The tingle that soft drinks give our tongue is caused by carbonic acid (which is produced when carbon dioxide is dissolved in water). Our stomach produces hydrochloric acid (see Chapter 6). Many drain cleaners contain sodium hydroxide.

Buffers

Even relatively minor changes in pH may profoundly affect the rate at which many biological reactions occur. Therefore the pH of the body must be maintained within narrow limits. Cells have an internal pH of approximately 6.8 (the actual value varies with cell type and level of activity) and, in humans, blood has a pH of 7.4. These values remain nearly constant even though many biological reactions either liberate or incorporate H^+ ions.

This constancy in pH is possible because of the presence of **buffers.** A buffer is a substance that prevents a change in pH by acquiring H^+ when the H^+ concentration of a solution is otherwise tending to increase and by releasing H^+ when the H^+ concentration is tending to decrease. The most important buffer system in human blood is *bicarbonate ion:*

$$\underset{\text{Bicarbonate ion}}{HCO_3^-} + H^+ \longleftrightarrow \underset{\text{Carbonic acid}}{H_2CO_3}$$

If H^+ is added to the blood, it combines with HCO_3^- to form H_2CO_3. If H^+ is removed from the blood, more H_2CO_3 ionizes and replaces the H^+ that was lost. (See also Chapter 10 for a discussion of buffers at work in the blood.)

A QUICK REPRISE

To review briefly, the earth is composed of the same elements as the rest of the universe. The structure of the atoms of these elements permits—indeed, mandates—the occurrence of certain fundamental chemical reactions, and the atoms of most elements combine with other atoms to form compounds.

In the next chapter, we shall discuss the steps that must have taken place on the road to life. If we are to understand the significance of each step, however, we must first know something about the particular class of chemicals that comprise living organisms.

ORGANIC MOLECULES

With the exception of water, almost all of the chemical compounds in organisms are **organic** compounds. Organic compounds, which derive their name from the erroneous belief of early chemists that they were found exclusively in living organisms, are chemical compounds that contain both carbon and hydrogen. The important characteristics of organic compounds derive from the properties of the element carbon.

Carbon has an atomic number of 6. That is, each carbon atom has six protons in its nucleus, and six electrons around its nucleus. Two electrons fill the first shell, and four are in the second shell. Because of this arrangement, a carbon atom can form covalent bonds with as many as four other atoms.

Although a carbon atom can bond with atoms of a variety of elements, it most commonly bonds with hydrogen, oxygen, nitrogen, or other carbon atoms. Carbon-to-carbon bonding makes possible organic molecules consisting of various lengths and shapes of carbon chains. The carbon chain establishes the skeleton of an organic molecule. Carbon atoms can form single, double, or triple covalent bonds with one another, and patterns of carbon bonding also contribute to the characteristics of a given organic compound.

Hundreds of thousands of different organic compounds have been isolated from organisms or synthesized in the laboratory. Why are carbon compounds so diverse? The principal reason for the existence of so many organic compounds is that carbon is a better chain-former than any other element. Although silicon chains as long as ten atoms have been described (silicon also has four electrons in its outermost shell see Box 3.2), organic compounds with more than 50 carbon-carbon bonded atoms in a chain are commonly found in organisms, and chains containing other elements such as nitrogen and oxygen along with carbon may consist of thousands of atoms.

These very large molecules are called **macromolecules.** Macromolecules are **polymers** (Gr. *poly,* "many" and *meros,* "parts")—large molecular chains composed of smaller organic molecules (**monomers**)

BOX 3.2

Must Life Be Carbon Based?

Carbon is capable of forming many different molecules—because it can form long chains with other carbon atoms and because one carbon atom can form chemical bonds with as many as four other atoms at one time. However, carbon is not the only element with these characteristics. Silicon, which occurs just below carbon in the periodic table, shares the same traits and is much more abundant than is carbon. Could silicon-based life exist somewhere in the universe?

A strong argument against such a possibility is that silicon atoms are considerably larger than carbon atoms. Not only are large atoms generally slower to engage in chemical reactions, but, because of their mass, compounds formed from large atoms tend to be solids, rather than liquids or gases. For example, one atom of carbon combines with two atoms of oxygen to form the gas carbon dioxide, the end product of cellular respiration and the primary waste gas of the body (see Chapters 5 and 10). Similarly, one atom of silicon can combine with two atoms of oxygen to form one molecule of silicon dioxide. However, silicon dioxide (also known as quartz) is a solid and is the primary component of sand. Imagine a living organism that produced quartz grains as a respiratory waste product. Plan to avoid being anywhere near it when it sneezes!

linked together. In living organisms, monomers are linked by **condensation** reactions (also known as *dehydration* reactions) in which the components of water molecules are removed (see Figure 3.14).

The converse of the condensation reaction is **hydrolysis** (Gr. *lysis*, "to split, loosen"), the reaction of a substance with water. In hydrolysis, macromolecules are split into their component monomers through the incorporation of water molecules, as occurs in digestion (see Chapter 6).

There are four principal types of macromolecules found in organisms: carbohydrates, lipids, proteins, and nucleic acids.

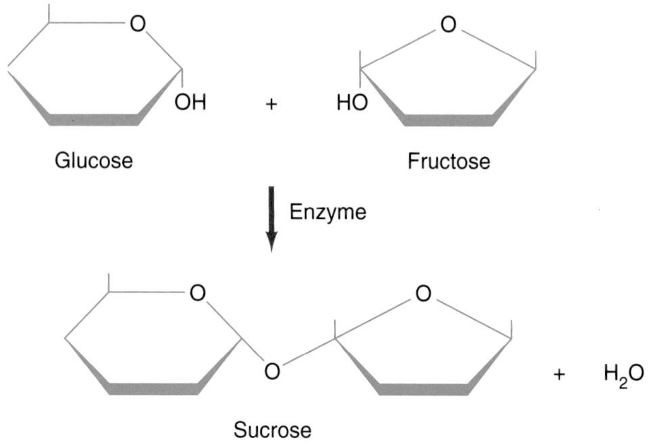

FIGURE 3.14

Condensation Reactions

Condensation reactions link two molecules by the enzymatic removal of a water molecule. Here, glucose and fructose are linked to form sucrose (table sugar) plus a molecule of water.

Carbohydrates

Carbohydrates (hydrates of carbon) are used as energy storage and structural compounds in organisms (see Figure 3.15). They have roughly one molecule of water for every atom of carbon. Thus a general formula for a carbohydrate is $(CH_2O)_n$.

The monomer units that make up carbohydrates are sugars containing three to eight carbon atoms. These simple sugars, best exemplified by glucose ($C_6H_{12}O_6$) are called **monosaccharides** (Gr. *monos*, "single" and *sakcharon*, "sugar"). Two of these monosaccharides may be joined together to form a **disaccharide**, of which **sucrose** (table sugar) is the most familiar. It is composed of one unit of glucose and one unit of **fructose** (fruit sugar), another six carbon monosaccharide. Chains of monosaccharides, called **polysaccharides** (Gr. *polys*, "many"), are exemplified by *starch* and *cellulose*, both of which are polymers of glucose, differing only in the way the glucose units are linked to each other (see Box 3.3).

Lipids

Another class of macromolecules, the **lipids** (Gr. *lipos*, "fat"), are structurally the most diverse class. "Lipid" is no more than a convenient term for compounds in the cell that are insoluble in water but soluble in hydrophobic solvents. Lipids resemble polysaccharides in having both structural and storage roles, and in being composed exclusively of carbon, hydrogen, and oxygen, but they differ from carbohydrates in that they contain relatively much less oxygen. The lipids include a number of subcategories, examples of which are shown in Figure 3.16.

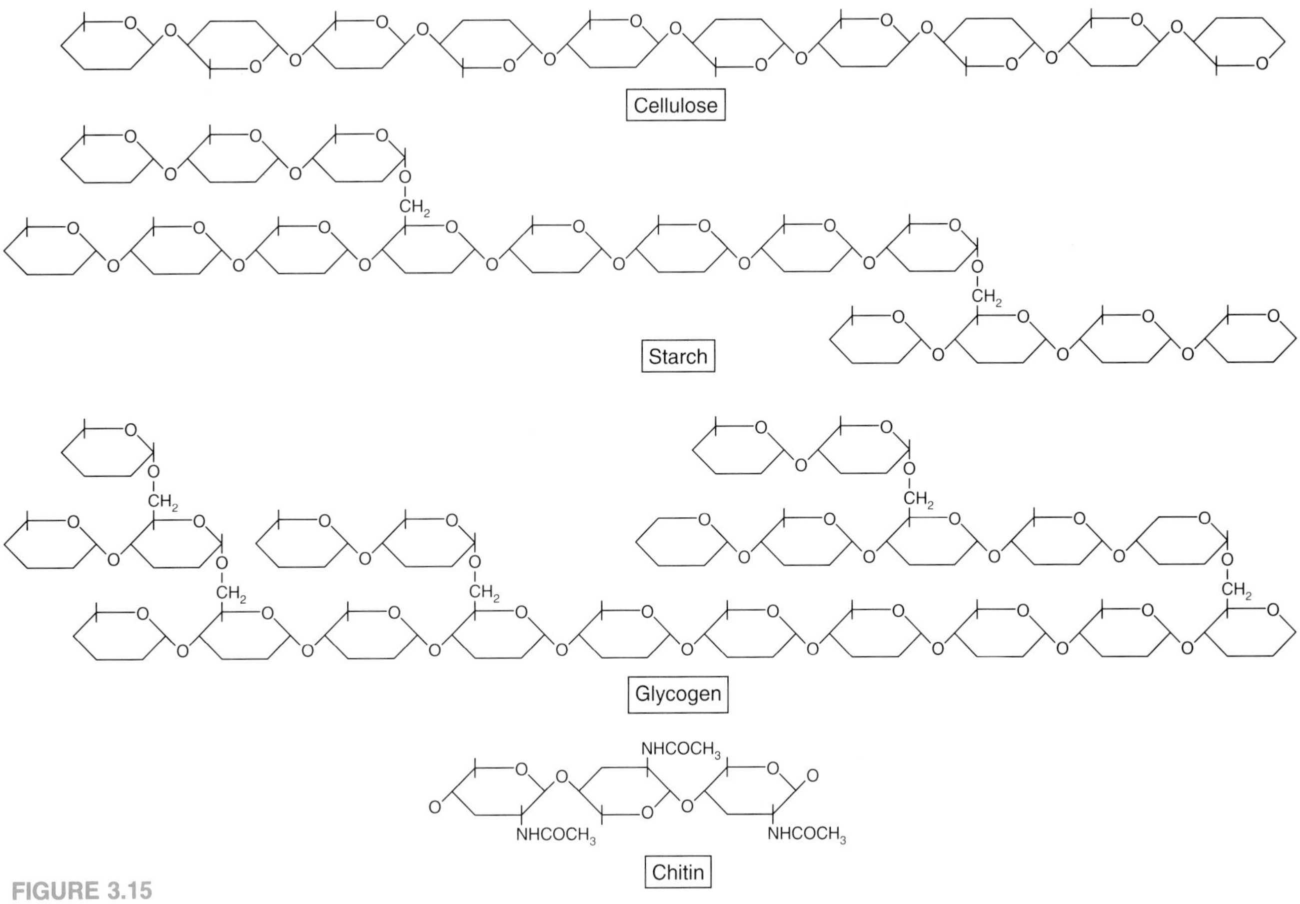

Cellulose

Starch

Glycogen

Chitin

FIGURE 3.15

Polysaccharides
Molecules of glucose may be linked in various ways. Cellulose is composed of beta-glucose units, which most animals have a very difficult time breaking. Starch and glycogen are both composed of alpha-glucose units, which animals can digest easily. Chitin, the dominant polysaccharide in the exoskeleton of insects, includes a nitrogen-containing side chain.

BOX 3.3

Deciphering the Code of the Chemists

We have already seen the problems associated with attempting to illustrate atomic structure in a way that is both accurate and easily perceived. These same problems arise in a more extreme form when we attempt to depict molecular structure, especially in the large organic macromolecules. Not only is it a problem to try to suggest three-dimensional structure on a two-dimensional page, but some molecules have the unfortunate habit of occurring in more than one form. Glucose, for example, can be found either as a chain or a ring (sugars of more than five carbons tend to form rings when dissolved in water), and it is not flat, but bent in one of two ways. If that weren't bad enough, glucose isn't the only sugar with the formula $C_6H_{12}O_6$—both fructose and galactose have the same formula, but differ in the positioning of a double bond and an —OH side chain. These differences are important—glucose and fructose combine to form sucrose; glucose and galactose combine to form **lactose** (milk sugar), a very different molecule.

What to do? Generally, the simplest possible method will be used to illustrate a given molecule. When you see simple rings, for instance, each corner of the ring is occupied by a carbon atom (unless some other symbol, such as O for oxygen or N for nitrogen, is specifically indicated). Examples are illustrated in this Box—but bear in mind, these are only shorthand methods of indicating what are really rather complex structures.

(Continued on the following page)

BOX 3.3 (Continued)

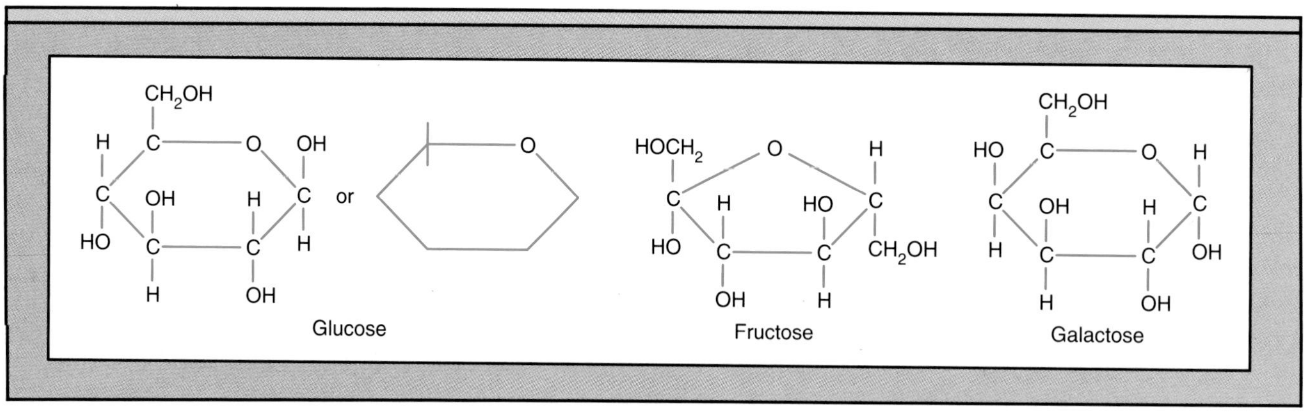

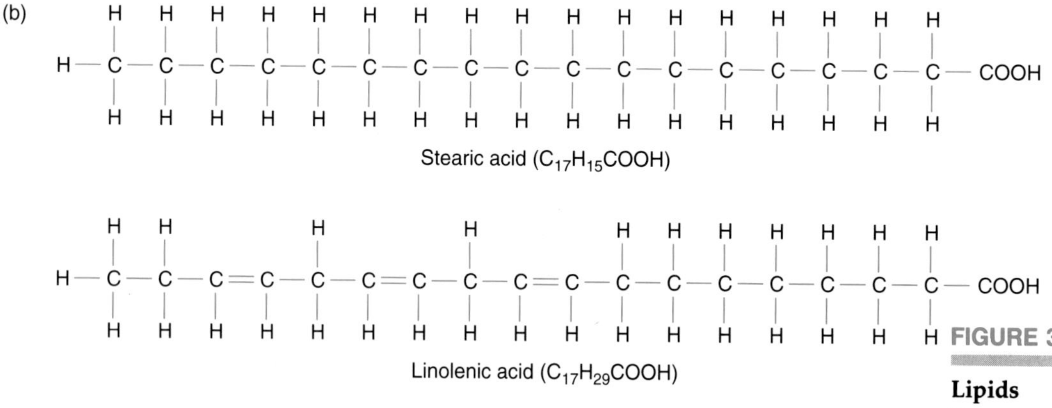

Glucose

Fructose

Galactose

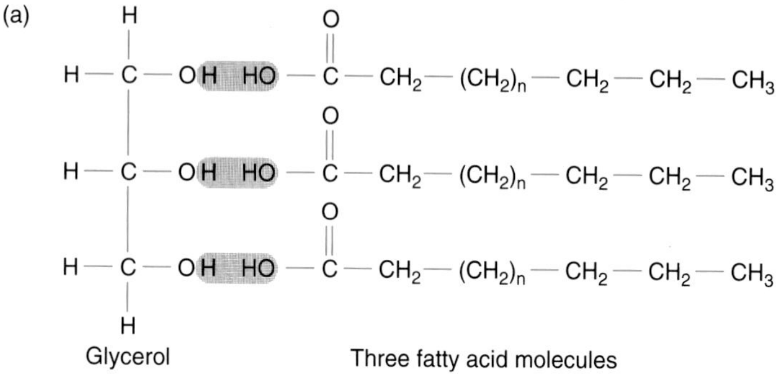

(a)

Glycerol

Three fatty acid molecules

(b)

Stearic acid ($C_{17}H_{15}COOH$)

Linolenic acid ($C_{17}H_{29}COOH$)

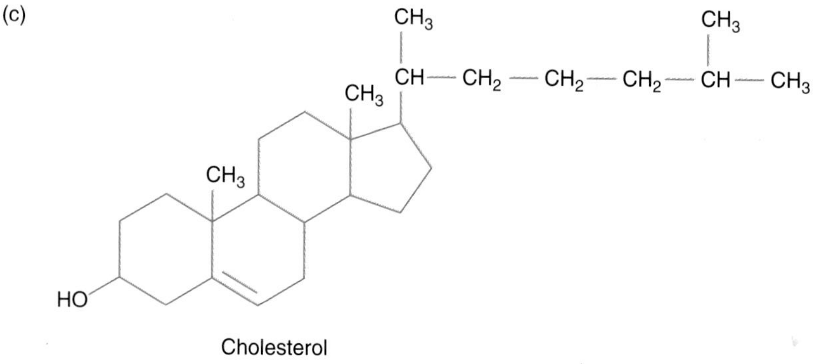

(c)

Cholesterol

FIGURE 3.16

Lipids

Fats are an important group of lipids. (a) They are composed of three fatty acid molecules linked by dehydration reactions to a molecule of glycerol, an alcohol. (b) Saturated fatty acids, found commonly in animals, have no double bonds (that is, they are "saturated" with hydrogen); unsaturated fatty acids, found commonly in plants, have at least one double bond. (c) Steroids, another important group of lipids, share a common skeleton of four linked carbon rings. Relatively minor changes in the side chains profoundly affect their biological activity. In humans, many hormones are steroids, and all are synthesized from cholesterol.

The **fats** are the most abundant and diverse of the subcategories of lipids. Fats consist of a molecule of the alcohol *glycerol* to which are attached three *fatty acids*. Fats are particularly important as long-term energy storage molecules. In a related group of molecules, the **phospholipids** (of critical importance in the structure of the cell membrane), one fatty acid is replaced by a phosphorus-containing **phosphate group.** The phospholipids will be discussed in greater detail in Chapter 4.

A second group of lipids, the **carotenoids,** are plant pigments that are responsible for the yellow and orange colors of many fruits and vegetables. *Carotene,* a carotenoid found in carrots and other vegetables, is split in our bodies to become two molecules of *vitamin A,* a substance needed for night vision (see Chapter 7).

Steroids are a third category of lipids, distinguished by being comprised of four linked rings, and differing from each other only in the presence or absence of certain side chains. Among the many steroids are the sex hormones *estrogen* and *testosterone* (the effects of which are discussed in Chapter 20); *cortisone,* the synthetic form of a hormone used extensively in the treatment of inflammation (see Chapter 13); and *cholesterol,* a substance implicated in cardiovascular disease (see Chapters 7 and 8).

Proteins

Protein molecules are composed of long chains of **amino acids**—organic compounds with both an *amino* group (-NH_2) and a *carboxyl* group (-COOH) linked to the same carbon atom (see Figure 3.17). Proteins form important parts of the structure of all organisms. Among other uses, some proteins function as **enzymes** (organic **catalysts** that accelerate the rate of specific chemical reactions without themselves being used up in the process; see Chapters 5 and 6); other proteins are contracting elements in muscle cells (see Chapters 4 and 12); still others help the body to resist infection by serving as **antibodies** (see Chapter 9). Like all carbohydrates and lipids, proteins are comprised of the elements carbon, hydrogen, and oxygen; unlike carbohydrates and lipids, proteins also contain nitrogen and sulphur.

There are 20 different amino acids that combine like the letters of the alphabet to make up the thousands of different proteins found in our bodies. The linkage between any two amino acids is a type of covalent bond, formed by a dehydration reaction, called a **peptide bond.** (At this point you may be asking yourself, "Why do biologists need another name for what is simply a covalent bond in a protein?" Actually, there is no need at all—the basis for the name comes from an early observation that proteins were split into small chains of amino acids by a stomach enzyme called **pepsin,** the activity of which we shall examine more completely in Chapter 6.)

Small chains of amino acids are often called *peptides* or *polypeptides;* the point at which a polypeptide chain becomes long enough to warrant being called a protein is somewhat arbitrary, though a chain of 100 or more amino acids will usually be called a protein.

Although for purposes of illustration proteins are often depicted as straight chains of amino acids, in reality no protein is so simple. The sequence of amino acids is called the **primary structure** of a protein (see Figure 3.18). However, all proteins also have a **secondary structure,** which consists either of a coil (or *helix,* as scientists would say, from the Greek word for "spiral"; *Helix* is also the name of a genus of snails), or what has been called a *pleated sheet,* with an appearance similar to the inner core of corrugated cardboard. Unlike the primary structure, which is maintained by covalent (peptide) bonds, the secondary structure is maintained by much weaker hydrogen bonds. Proteins having only a primary and a secondary structure are called **linear** (or *fibrous*) **proteins.** *Keratin,* the protein in hair and fingernails, is a linear protein.

Some proteins have, in addition, a **tertiary structure** wherein the linear spiral is bent back on itself in a variety of bizarre shapes (rather like what happened to your Slinky when your little brother got hold of it). The tertiary structure is also maintained largely by hydrogen bonds. Proteins with primary, secondary, and tertiary structures are called **globular proteins;** many of them function as enzymes. The specific shape of the globular protein is very important in its role as an enzyme. If the shape of the enzyme is changed such that it can no longer catalyze a chemical reaction that is vital to an organism, the reaction will not occur (or will occur too slowly) and the organism will die.

Despite the importance of enzyme shape, it is perhaps ironic that this shape is maintained by very weak hydrogen bonds that, once broken, often cannot be reestablished. Proteins that have lost their characteristic secondary and tertiary structures, usually because of exposure to high temperatures or to a changed pH, are said to be **denatured.** Egg whites, which are composed almost entirely of protein, become solid when cooked because of denaturation, an obviously irreversible process (try uncooking an egg sometime); bacteria in milk are killed during the pasteurization process (in which milk is heated to about 65° C) because their protein coats are denatured. To take a more dramatic example, high fevers and heat stroke can be fatal because the enzymes of our bodies begin to denature when the temperature of the body exceeds 40° C.

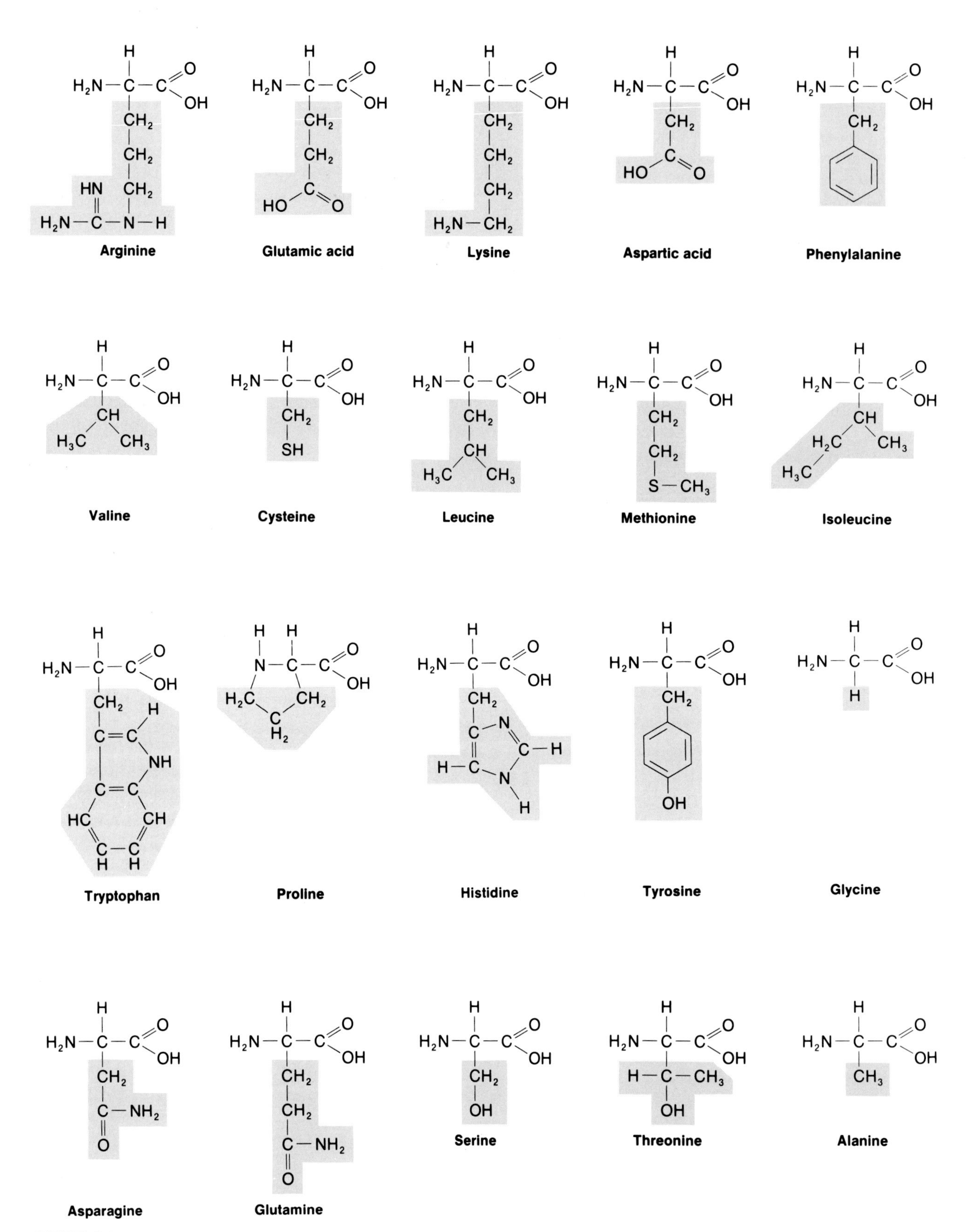

FIGURE 3.17

Amino Acids

The 20 amino acids found in proteins are shown here. Each shares a common carboxyl group (-COOH) and an amino group (-NH_2), linked by a carbon atom. To this carbon atom is joined a side chain that characterizes the particular amino acid. Dehydration reactions between amino acids form a particular type of covalent bond called a peptide bond.

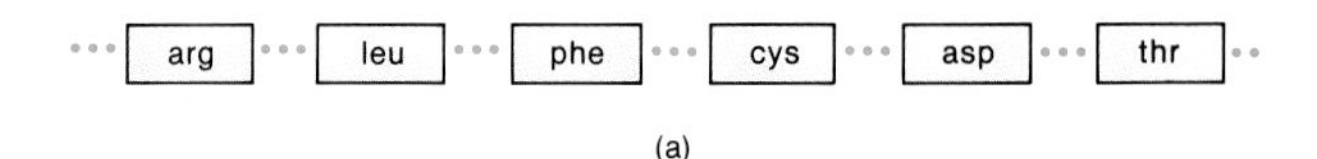

(a)

Alpha Helix

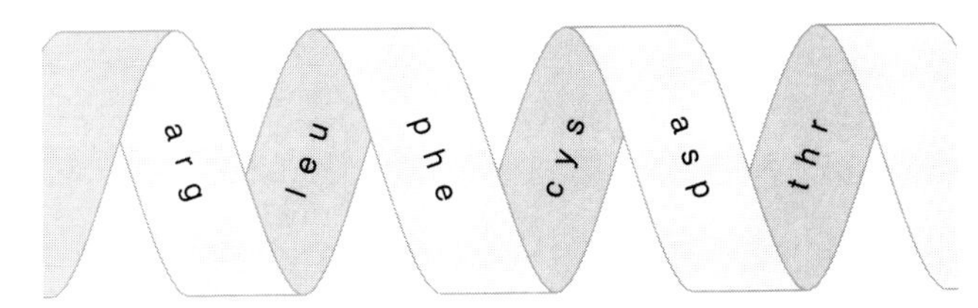

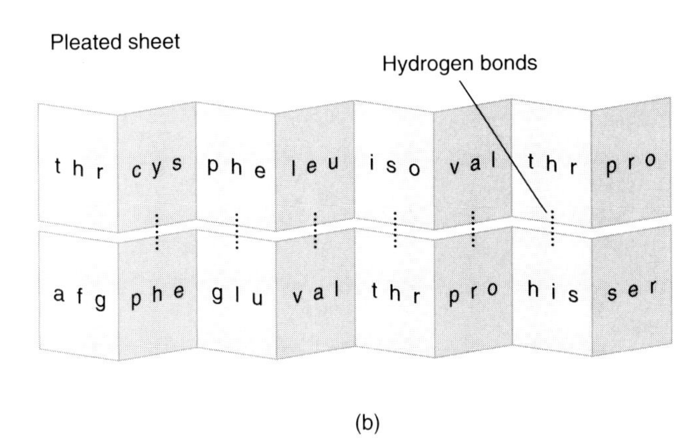

(b)

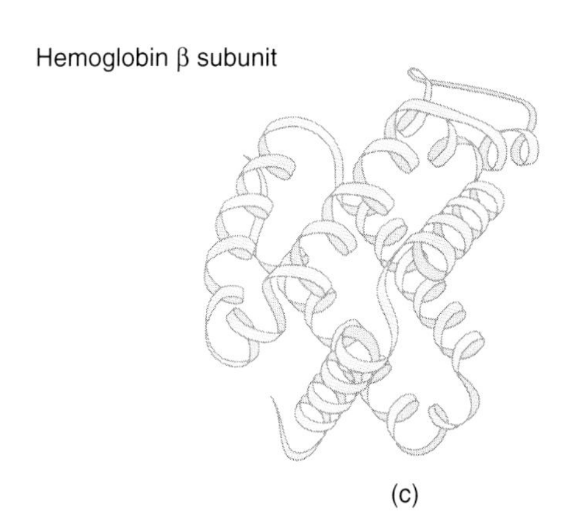

(c)

FIGURE 3.18

Structure of Proteins

The primary structure of proteins (a) is simply the order in which the amino acids occur in a particular protein. However, all proteins also have a secondary structure (b), which consists either of a tight spiral or a pleated sheet, both of which are maintained by hydrogen bonds. Globular proteins also have a tertiary structure (c), in which the protein is folded back on itself at particular points. The tertiary structure is also maintained by hydrogen bonds.

Nucleic Acids

Nucleic acids are macromolecules that contain genetic information. The principal nucleic acids in living organisms are **DNA** (deoxyribonucleic acid) and **RNA** (ribonucleic acid). Nucleic acids are linear polymers of **nucleotides.** A nucleotide is composed of a nitrogen-containing base, a five-carbon sugar, and phosphoric acid (see Figure 3.19). The genetic information

(a) Five-carbon sugar

(b) Phosphate group

(c) Nitrogenous bases

Pyrimidines (one ring): Cytosine, Thymine

Purines (two rings): Adenine, Guanine

(d) Three components linked

FIGURE 3.19

Nucleic Acids

All nucleic acids consist of a five-carbon sugar, a phosphate group, and one of five nitrogenous bases (four of which are shown here). The manner in which they are linked together is shown at the bottom of the figure.

contained in nucleic acids is encoded in the sequence of the nucleotides. The specific means by which genetic information is coded, translated, and expressed is discussed in Chapter 18. The elements from which nucleic acids are composed include carbon, hydrogen, oxygen, nitrogen, and phosphorus.

Summary

Although the universe had its origin perhaps 12 billion years ago, the age of the earth and the rest of the planets of the solar system is less than five billion years. As is everything else in the universe, the earth is composed of various combinations of 90 elements.

Elements are distinguished from one another by their atomic structures. An atom is the smallest unit possessing the properties and characteristics of a given element, and, except for isotopes, the atoms of each element are identical to each other but differ from the atoms of other elements. The differences arise because of variations in the way atoms are composed.

All atoms contain a variety of subatomic particles, of which electrons, protons, and neutrons are the most important. Electrons and protons are equal in number in every atom, but are opposite in electrical charge. Protons and neutrons give mass to the atom; electrons are virtually massless. The atoms of each element have a characteristic number of electrons and protons. The number of protons is the atomic number of the element. The number of protons plus neutrons is the mass number of the element.

Electrons occur in shells, spinning in orbits around the nucleus of the atom (the location of the protons and neutrons). Each shell is capable of holding at least eight electrons, with the exception of the innermost shell, which holds only two. Atoms with unfilled outer shells are prone to engage in bonding with other atoms, forming compounds. These bonds involve either a transfer of electrons (ionic bonds) or a sharing of electrons (covalent bonds).

Compounds containing carbon and hydrogen are called organic compounds, because they are typically found in, or are a product of, organisms. Carbohydrates, lipids, proteins, and nucleic acids are the four principal groups of organic compounds.

Key Terms

element
fusion reaction
fission reaction
compound
atom
subatomic particle
nucleus
proton
neutron
electron
mass
atomic number
mass number
isotope
radioactive
shell
formula
molecule
ionic bond
covalent bond
hydrogen bond
electron donor
electron acceptor
ion
polarity
heat capacity
heat of vaporization
surface tension
hydrophilic
solute
solvent
hydrophobic
micelle
pH
acid
base
buffer
organic
macromolecule
polymer
monomer
condensation
hydrolysis
carbohydrate
monosaccharide
disaccharide
polysaccharide
lipid
fats
phospholipid
carotenoid
steroid
protein
amino acid
enzyme
catalyst
antibody
peptide bond
primary structure
secondary structure
linear protein
tertiary structure
globular protein
denaturation
nucleic acid
DNA
RNA
nucleotide

Box Key Term

lactose

Questions

1. How old is the universe estimated to be? What is the estimated age of the earth?
2. Distinguish atoms from molecules and elements from compounds.
3. Approximately what fraction of the elements are essential for life?
4. Electrons are the smallest of the subatomic particles, yet they are the most important from the standpoint of chemical reactions. Why?
5. Distinguish between atomic number and mass number.

6. What is an isotope?
7. Distinguish ionic, covalent, and hydrogen bonding.
8. What is an ion?
9. List five properties of water that are important for life.
10. What is a micelle?
11. How does an acid differ from a base?
12. What is a buffer?
13. List the four classes of biologically important macromolecules.
14. Distinguish condensation reactions from hydrolysis.

CHAPTER

4

The Evolution and Organization of the Cell

If the cell is the basic unit of life, how (and when) did the first cell arise? Are cells fundamentally the same in all organisms? If they are, how can organisms such as oak trees and antelope use the same fundamental building block and yet be so different? If they are not, what is the nature of the differences among cells? Do cells have an internal organization, or are they merely small bags of biologically active chemicals?

These are the kinds of questions we shall address in Chapter 4.

THE ORIGIN OF LIFE

Pasteur's disproving of spontaneous generation in 1864 had the effect of discouraging any significant amount of speculation on the origin of life for more than 60 years. In the 1860s, both Darwin and Huxley informally discussed how life might have begun, but serious consideration of life's origins was not undertaken until the 1920s, and these first speculations were not tested experimentally until the 1940s and 1950s. Because DNA was not identified as the molecule of heredity until 1944, and its structure not elucidated until 1953, this late start in experimentation on life's possible origins is perhaps not surprising.

Since the 1950s, experimentation and speculation about the origin of life have continued at an accelerating pace, bolstered substantially by our greatly increased knowledge both of the biochemistry of cells and of the chemistry of the universe. As a consequence, it has become possible to construct what seems to be a reasonable scenario for the origin of life on earth.

MEANWHILE, 4.6 BILLION YEARS AGO . . .

As we saw in Chapter 3, it is relatively easy to imagine how a variety of simple inorganic compounds were formed during the origin of the earth, once we understand atomic organization and the effects of gravity that brought potentially reactive atoms together. However, it is a good deal more difficult, at least at first glance, to imagine the formation of the highly complex organic macromolecules that are the integral components of all life forms on earth today.

Part of the problem is resolved when we realize that the ancient earth was very different from the earth of today. For example, our present atmosphere consists largely of nitrogen (N_2) and oxygen (O_2). The original atmosphere of the earth probably consisted largely of ammonia, water vapor, methane, and possibly hydrogen cyanide. There are two reasons for making this assumption. First, these simple compounds are all formed of elements common in the universe. Hydrogen atoms comprise 93 percent of all atoms, and in a hydrogen-rich atmosphere it would combine readily with carbon to form methane (CH_4), with oxygen to form water vapor (H_2O), and with nitrogen to form ammonia (NH_3). Hydrogen cyanide (HCN) is formed from hydrogen, carbon, and nitrogen. A second reason for assuming that these compounds would have been present in the atmosphere of the early earth is that they are found in the atmosphere of Jupiter, Saturn, and the outer planets, all of which, because of their size and distance from the sun, are evolving more slowly than the earth.

The surface of the earth was also very different during its first few hundred million years. At this early date in the earth's history, our infant planet was still very much in its cooling stage, its outer layers only gradually becoming hardened rock. Water issued forth from geysers and volcanoes as superheated steam, rose into the cooler atmosphere, and condensed and fell as rain, only to boil back up again as it landed on semimolten rock. This pattern must have continued for thousands of years, as the surface of the earth slowly cooled to a temperature below the boiling point of water. Over the same period, the interminable rain steadily leached salts and other soluble materials from the rocks and, on the now-cooler earth, these dissolved salts were borne downward into depressions on the earth's surface to become the primordial oceans.

The oceans grew in size as the rain continued. They also increased in salinity, as more dissolved materials were carried into them by the rivers and streams running over bare rock. Thus, at a point more than four billion years ago, the ancient earth had oceans rich in dissolved minerals and gases, and an atmosphere of small organic and inorganic molecules. The stage was set for the creation of the molecules of life.

THE FORMATION OF SMALL ORGANIC MOLECULES

The old notion that organic compounds were inextricably linked to living organisms was refuted early in the nineteenth century when chemists synthesized an organic compound in the laboratory. Much more recently, the link between organic compounds and life has been further weakened with the discovery of organic molecules in meteorites and even in interstellar space. Sugars, fatty acids, amino acids, and nitrogenous bases have been found in meteorites; molecules containing as many as 11 carbon atoms have been detected in space. The logical conclusion is that atoms of the requisite elements, in the presence of an energy source, can combine to form relatively complex organic molecules **abiotically** (that is, in the absence of life).

What energy sources existed on the ancient earth that might have served to drive chemical reactions? A prime candidate is **ultraviolet radiation.** By one estimate, each square centimeter of the ancient earth's surface received almost 600 calories of energy in the form of ultraviolet radiation each year. In contrast, other sources of energy, such as volcanic eruptions, lightning, and radioactivity, have been estimated at less than ten calories annually per square centimeter.

Is it merely idle speculation to assume that the combination of large amounts of ultraviolet radiation, acting on an ammonia- and methane-rich atmosphere and a salt-laden ocean, could result in the formation of the monomers from which the macromolecules of life are assembled? Since 1953, the answer has been "no." In that year, Stanley Miller performed one of the more elegant experiments of modern science (see Figure 4.1). After several days of passing a spark (energy) through a chamber containing methane, ammonia, hydrogen, and water, Miller found that he had created four of the amino acids found in proteins, plus several of the fatty acids found in biologically important lipids. The scientific community was astounded. If Miller could achieve these results in a week, was it not reasonable to assume that ultraviolet radiation and lightning storms in the ancient earth, over a period of thousands or even millions of years, might have had the same effect?

Many other scientists, using various modifications of the Miller apparatus, have since duplicated and extended his results. Two of the nitrogenous bases found in nucleic acids have been created in a chamber containing ammonia and hydrogen cyanide. Formaldehyde (which in recent years has been discovered in interstellar space, and which therefore may have been present in the atmosphere of the ancient earth) is the simplest of the sugars (CH_2O). When very dilute formaldehyde solutions were passed over surfaces covered with clay (as may well have existed at the edges of ponds or oceans), both five and six carbon sugars were produced. Thus the basic building blocks of the carbohydrates, lipids, proteins, and nucleic acids have all been synthesized under conditions thought to duplicate those of the ancient earth, and in impressively short periods of time.

In addition to small organic molecules formed abiotically on earth, a significant amount of organic matter is believed to have arrived as a part of the asteroids and comets that bombarded the earth in large numbers during its earliest days (from 4.5 to 3.8 billion years ago). It is likely, however, that abiotic formation of organic molecules in the Earth's oceans was a far more important source of these materials.

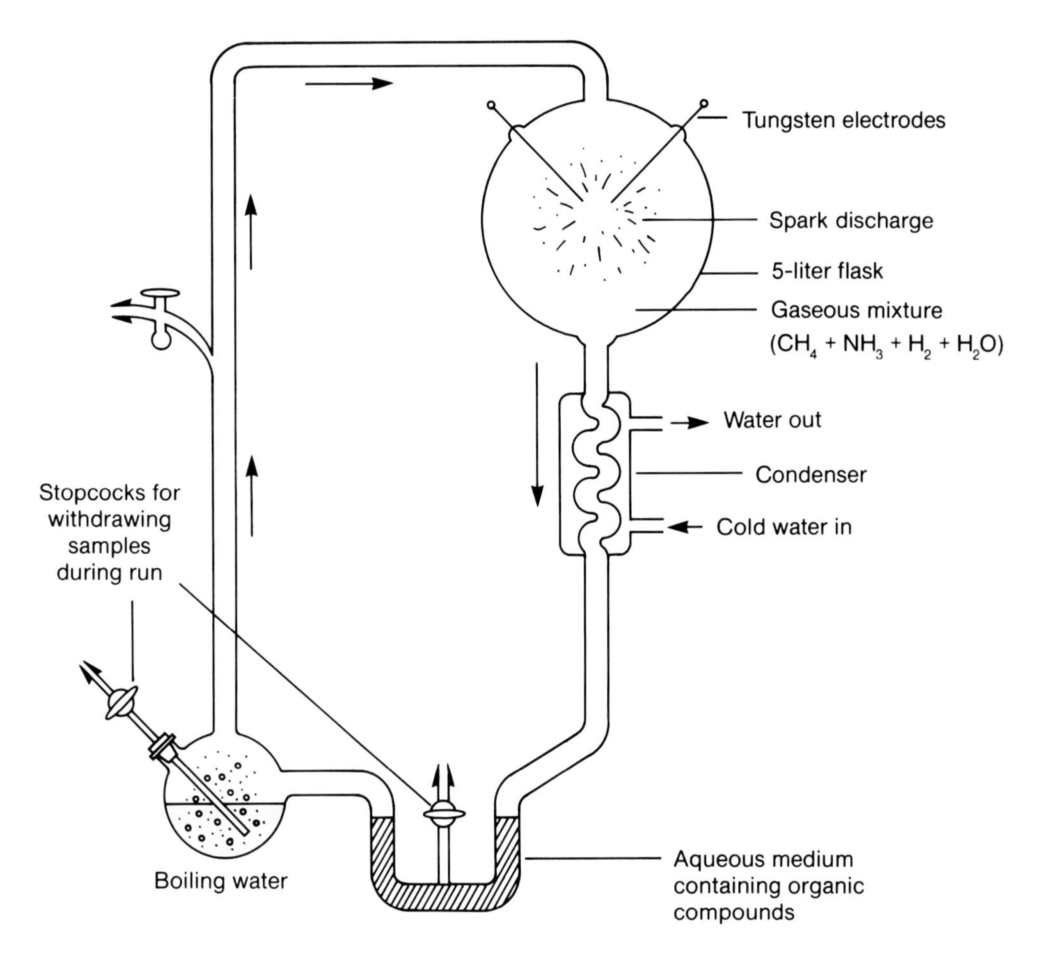

FIGURE 4.1

The Miller Apparatus

This is a diagrammatic representation of the apparatus used by Stanley Miller in his attempt to synthesize organic compounds under conditions that simulated those of the ancient earth.

THE FORMATION OF MACROMOLECULES

These monomers, of course, are only the components of macromolecules, not the macromolecules themselves. Is there evidence that these components could have combined into the macromolecules found in living organisms? (Would we ask such a question if the answer were "no"?)

1. In the presence of organic catalysts formed in Miller-style experiments, short proteins have been formed from dilute solutions of amino acids.

2. Larger proteins, called **proteinoids**, have been synthesized by heating amino acids in the absence of water, as have short chains of nucleic acids. (Just such a situation may have taken place at the bottom of lagoons on the ancient earth, which would have been subject to drying and exposure to ultraviolet radiation.) Some of these proteinoids have demonstrated enzymatic activity, thus again demonstrating the possibility that enzymes existed before there were living organisms.

3. Various combinations of sugar molecules have been formed when glucose, in a dilute formaldehyde solution, was exposed to ultraviolet light.

In short, simple laboratory experiments have demonstrated the distinct possibility that short chains of organic molecules formed spontaneously from a variety of small molecules presumed to be present in the ancient earth.

What about longer chains? Clay particles are known to adsorb various organic molecules, and may have served as a pattern, or template, on which proteins and/or RNA chains were assembled. These notions are admittedly speculative, and perhaps sound improbable—but given the enormous time spans involved (hundreds of millions of years), even highly improbable events become very possible indeed.

THE FORMATION OF PROTOCELLS

Although it seems likely that we shall never know precisely what the steps were that led from nonlife to life (see Box 4.1), certain possibilities command our

BOX 4.1

What Is Life?

Life does not exist in the abstract. It cannot be weighed or measured. It cannot be dissected out of a living organism, any more than the capacity for forward movement can be dissected out of an automobile. Life, in other words, is not an **inherent property** of some component of an organism. It is, instead, an **emergent property** of those very complex chemical assemblages that we know better as living organisms. As such, we do not define "life," as we might "gram" or "meter"; instead, we describe it, by looking for common denominators possessed by all living systems and not possessed by things we know not to be alive. (Note, incidentally, the use of the plural "denominators," there is no one trait which invariably separates life from nonlife.) In that sense, "life" is somewhat akin to the definition offered by a Supreme Court justice of "pornography"—"I know it when I see it."

There are many common traits, and lists of these traits are of various lengths. Some of the more important traits are summarized as follows:

1. *A high level of organization.* Living things are not mere bags of chemicals without form or clear function; rather, they are highly organized and chemically very complex.

2. *A capacity for growth, development, and reproduction.* Living things change over their lifetimes. They increase in size, mature, and finally they generally produce offspring that will constitute the next generation.

3. *An ability to interact with the environment.* Living things are partially, but not totally, isolated from the external environment. They must be able to acquire certain chemicals from the environment, and to release wastes back into the environment. However, since these activities are selective, organisms are not mere chemical subsets of the environment, but rather maintain higher or lower concentrations of particular chemicals.

4. *The capacity for metabolism.* Metabolism is the sum total of the many transformations of chemicals and energy that characterize any living organism.

5. *An inherited set of instructions that govern all of the above activities.* Life does not start over with every new generation. Rather, there is continuity from generation to generation through inherited instructions—that is, the genes—which are characteristic for each species in how they regulate and control growth, development, metabolism, environmental exchange, and reproduction.

attention. Two candidates in particular warrant a brief discussion.

Coacervate Droplets. Under appropriate aqueous conditions (proper pH, ion balance, and temperature), various lipids will spontaneously organize into coacervate droplets (L. *coacervare,* "to heap up") from 1 to 2 μm in diameter. They form a double-layered boundary with the surrounding water, and selectively admit lipid-soluble substances from the medium (see Figure 4.2). In so doing, they grow and form new droplets by budding or pinching in half. Molecules which gain access to a droplet tend not to be uniformly spread throughout its interior, but rather assume an orderly arrangement. Under the right conditions, coacervates can decompose glucose, a trait that (as we shall see in Chapter 5) is universally possessed by all living cells.

Microspheres. If the proteinoids discussed earlier are placed in hot water and allowed to cool, they form microspheres, which, under the microscope, closely (albeit superficially) resemble certain bacteria. These microspheres can swell or shrink, depending on environmental conditions, just as living cells will do. As do the coacervates, microspheres form a double-layered outer boundary and are capable of growing and budding off new microspheres. They also tend to cluster, as do many bacteria, and readily accumulate various organic molecules from the environment. Finally, unlike coacervate droplets, which tend to be relatively unstable, microspheres can remain intact for years.

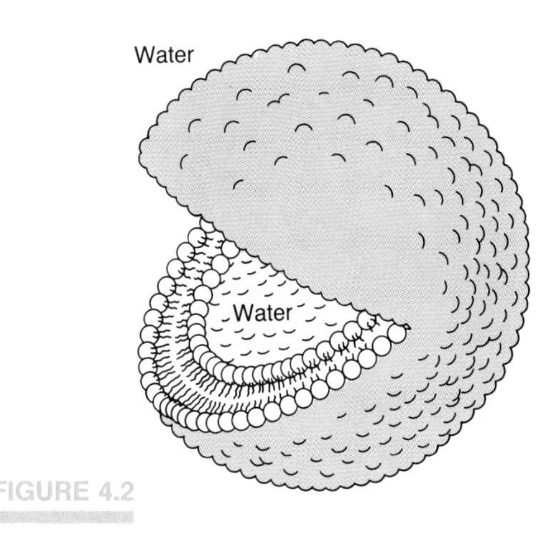

FIGURE 4.2

Coacervate Droplets

Coacervate droplets form as double-layered micelles when appropriately-sized lipids are added to a water medium.

Which was the intervening step between the development of macromolecules and a true living cell? We have no way of knowing. It may have been either coacervates or microspheres, or some combination of the two, or something entirely different (see Box 4.2). The point is that we can readily create, using conditions approximating those of the ancient earth, cell-like entities that could conceivably have served as shelter for the complex of chemical reactions we call life.

BOX 4.2

The Viruses

Are the viruses useful to study as an intermediate step in the evolution of life? Let's take a look.

By the beginning of this century, it was widely understood that many diseases were caused by microorganisms, and great effort was expended to isolate these microorganisms, and to culture them in suitable media. The term *virus,* as it was originally used, denoted any infectious disease agent. Subsequently, the term was restricted to the smallest of disease agents—those capable of passing through a fine porcelain filter. (We now know that they range in size from 20 to 300 nm.)

Although viruses were at first considered to be very small bacteria, they continued to puzzle biologists because they could not be cultured in any of the media normally used for bacteria. In 1935, scientists demonstrated that viruses were in fact remarkably different from bacteria or any known organism. In that year, the virus responsible for tobacco-mosaic disease (a condition wherein leaves of affected tobacco plants become mottled and wrinkled) was crystallized! Moreover, when these virus crystals were injected into healthy plants, they again caused the disease. In other words, the infectious agent was capable of becoming reproductively active even following crystallization. No organism possessed such a capacity.

So what are viruses? They are often represented as being "on the threshold of life." However, they display no independent metabolic activity and can reproduce only as parasites. They either enter or inject their nucleic acids into a host cell and reproduce by pirating its metabolic machinery. In reality, viruses are merely small sets of genetic instructions randomly "seeking" a compatible cellular environment in which to reproduce. In the absence of such conditions, many viruses crystallize and are as lively as a grain of sand.

(Continued on the following page)

BOX 4.2 (Continued)

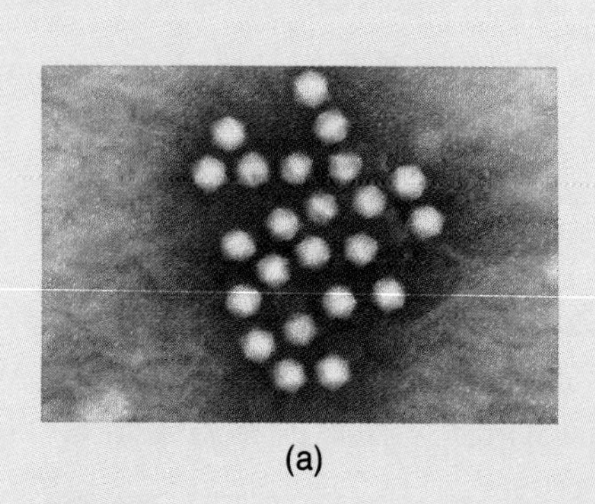

(a)

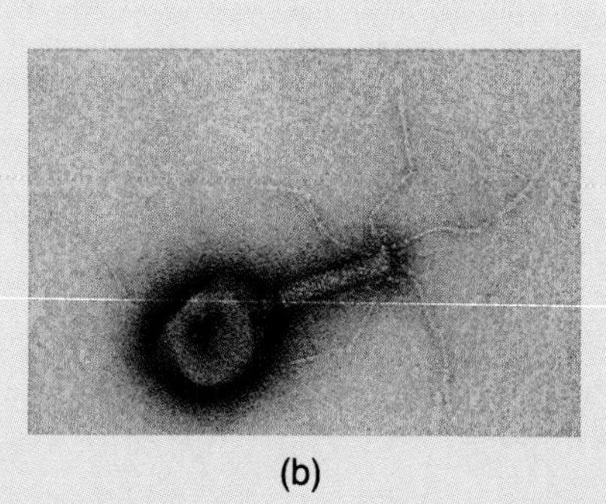

(b)

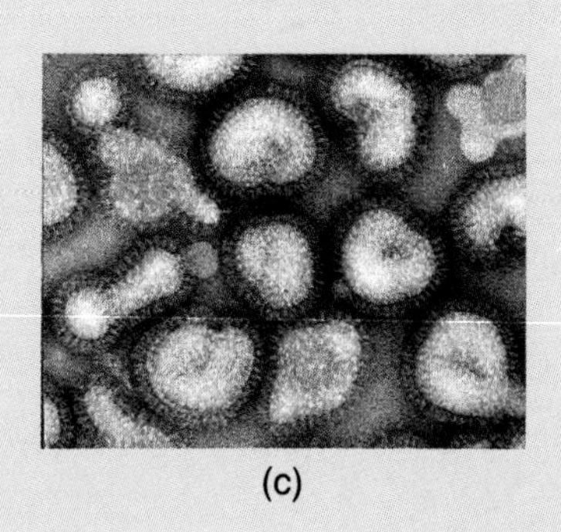

(c)

Viruses

Viruses are structurally very diverse. The adenovirus (a), which causes colds in humans, consists of 252 identical subunits composed of protein. Inside is a core of DNA. The influenza virus (b) has a core of RNA surrounded by a lipoprotein coat from which proteins protrude like tiny hairs. The bacteriophage (c) is a virus that attacks bacteria. It looks like a lunar lander, with spider-like "legs" protruding from a hexagonal "body" in which a DNA core resides.

Is a virus alive? Is it alive some of the time, and not alive at other times (that is, when crystallized)? The answers to these questions depend entirely on how we define the term "life"—and the experts do not always agree. A more important question is, "Do viruses represent a modern version of the several billion-year-old step that occurred as life arose, or are they merely degenerate descendants of much more complete life forms?" We cannot answer that question with complete assurance, but since viruses are parasites, at least in their present guise they must have evolved *after* the evolution of more sophisticated life forms. Thus, they are unlikely to be an evolutionary remnant of the step from nonlife to life.

Viruses are circles of DNA or RNA inside a protein coat. An even simpler disease agent are the **viroids**, which are small circles of DNA without a protein coat. They are known to be the agents of more than a dozen different plant diseases. Until very recently the **prions** were even more mysterious. Prions are proteins that were implicated in a disease of sheep and were also thought to be the agent of two rare human diseases. Because they contain no nucleic acid, how they could reproduce inside a cell was completely unknown. Now it is generally believed that these proteins are the *result*, not the *cause*, of the diseases with which they have been linked. They are probably natural proteins produced in excessive quantities and in erroneous arrangements, owing to genetic errors, or a "slow" virus (see Chapter 14).

GENETICALLY CONTROLLED METABOLISM AND REPRODUCTION

It is not enough, of course, that we simply account for a suitable membrane around an appropriate group of chemicals. The chemicals must *do* something, and that something must be to engage in reactions that we recognize as being fundamental to life.

On one level, it is not difficult to picture how the increasing sophistication of metabolic reactions might have arisen. If there were a particular substance (say, glucose) that was of great importance and value to the protocells (as indeed glucose is to organisms today), then glucose that had been formed abiotically would be quickly removed from the medium and utilized by the protocells. If, through what we might term **chemical evolution** (that is, natural selection in nonliving systems), a protocell acquired the capacity to convert another molecule common in the environment (say, galactose) to glucose, then clearly it would have an advantage over protocells without this capacity. Subsequently, a later protocell might have acquired the capacity to convert yet another substance in the medium to galactose (or directly to glucose) and in that way acquire a selective advantage over those protocells without such capacities. Gradually, a relatively sophisticated metabolism would have arisen.

However, the farther we go with our story, the more speculative it becomes. We have now apparently reached the ultimate in chicken-egg paradoxes. As living systems work today, there is the following sequence of events:

$$\begin{array}{l} \text{DNA} \xrightarrow{\text{(Enzyme)}} \text{RNA} \xrightarrow{\text{(Enzyme)}} \text{Protein (Enzyme)} \\ \downarrow \text{(Enzyme)} \\ \text{DNA} \end{array}$$

This sequence of reactions is actually much more complex than our simple diagram suggests. However, even at its simplest level the diagram indicates that each of the reactions listed (including the replication of DNA) requires enzymes to function. If enzymes are an end product of these reactions, how could DNA have come into being before there were enzymes to assemble it? How could DNA manufacture the first RNA without enzymes? How could RNA create enzymes without there first being enzymes to assist with the process? In other words, which came first—the DNA (chicken) or the enzymes (egg)?

Perhaps we can resolve our paradox. It is likely that the nucleic acids and the enzymes evolved together. RNA might have been the critical molecule. It can polymerize without enzymes and, in the presence of zinc ions, it can self-replicate. Moreover, it can catalyze certain reactions (this enzymatic ability of RNA was deemed significant enough that its discoverers were awarded the Nobel Prize in chemistry in 1989), and RNA may therefore have served as both "gene" and "enzyme" before either existed as we know each of them today. (This hypothesis is not as farfetched as it might at first seem. For example, certain recently discovered viruses, including the one that causes AIDS, have no DNA and instead use RNA as a template for synthesizing DNA.) Thus one possible scenario would be the initial evolution of several RNAs, the subsequent control of the RNAs over enzyme synthesis, and the ultimate development of DNA as a safe and stable way of providing a blueprint for all of the chemical instructions needed by the cell. These instructions would, of necessity, have included not only instructions regulating cell metabolism but also instructions for the reproduction of the cell. A haphazard reproduction, wherein the daughter cells might or might not receive a full set of metabolic instructions, would be strongly selected against.

The line between nonlife and life was not crossed in one fell swoop, to the rattle of drums and the skirl of bagpipes. Rather, there was a gradual accumulation of genetic information which both governed critical metabolic activities and regulated reproduction.

HETEROTROPHS AND AUTOTROPHS

Whenever it was that the line was crossed, and through whatever intermediate steps, it is clear that by 3.5 billion years ago, life existed on earth. That is, during the first billion years of the earth's history (though probably not much before 4.3 billion years ago) chemical evolution began, and it culminated in the development of life sometime before 3.5 billion years ago, by which time we see bacteria-like fossils in the oldest rocks available to us from earth (see Figure 4.3).

It seems likely that the first organisms were **heterotrophs** (Gr. *heteros,* "other" and *trophos,* "one that feeds"). Heterotrophs are organisms that require a preexisting source of organic molecules for both energy and assembly components. In the modern world, heterotrophs include many bacteria and single-celled organisms, and all fungi and animals. Heterotrophs are contrasted with **autotrophs** (Gr. *autos,* "self"), which are organisms capable of assembling organic molecules from inorganic precursors. Autotrophs include many bacteria and single-celled organisms, and virtually all plants. Most, but not all, autotrophs use sunlight as the necessary energy source and are therefore **phototrophs** (because they engage in photosynthesis).

The basis of the assumption that heterotrophs arose before autotrophs is that the simplest heterotrophs are less metabolically complex than are the simplest autotrophs. In addition, at the time of the first living organisms, the oceans would presumably

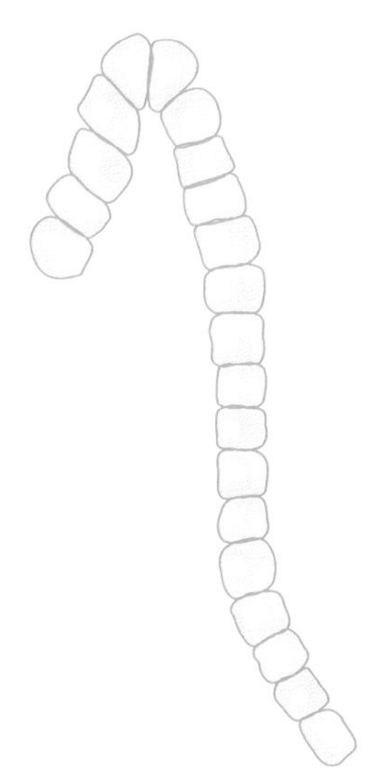

FIGURE 4.3

Fossil Bacteria

A diagrammatic representation of a fossil bacterium from 3.5 billion year old rock from Western Australia. There are several groups of modern bacteria which look very similar to this ancient fossil.

have been full of abiotically produced organic molecules ripe for the plucking. However, certain bacteria are **chemautotrophs**, meaning that they assemble organic molecules from inorganic molecules, using energy from various inorganic reactions rather than energy from the sun. It is likely that this is an ancient form of autotrophy and may even have existed before heterotrophy.

Regardless of whether the chemautotrophs originated before or after the heterotrophs, it is widely accepted that photosynthesis was a subsequent, albeit hugely successful, refinement. The basis of that statement is that photosynthesis produces oxygen gas as a waste product. O_2 is a highly reactive gas, and initially the O_2 produced by photosynthesis would immediately have been bound up with a host of potentially reactive molecules. However, ultimately the continued production of oxygen by photosynthesizers swamped the capacity of other molecules to bind to the oxygen, and gaseous oxygen began to accumulate in the atmosphere.

We have some rough notions of when these events occurred, based on the analysis of rocks and sediments from early in the earth's history. For example, about 2.5 billion years ago, iron suddenly showed up in ocean sediments, having become bound to oxygen and consequently precipitating out of the seawater. Based on the volume of oxygen necessary to precipitate such a large amount of iron, the source of the oxygen must have been photosynthesis. By 2.0 billion years ago, oxygen had accumulated to the level of 1 percent of the atmosphere (as compared to 21 percent today). Based on these data, it is thought that photosynthesis probably arose about 2.8 billion years ago, or well after the time of the earliest known fossils.

The impact of photosynthesis on the evolution of life was enormous. Its principal effects include the following:

1. *The proliferation of species.* This highly successful form of autotrophy provided a variety of species that served as valuable food sources for an increasing diversity of heterotrophs, including both active consumers and organisms that specialized in feeding on the remains of dead autotrophs. At the same time, many other species were evidently unable to overcome the problems of thriving in an oxygen-rich environment, and became extinct. Even today, there are species of bacteria that cannot live in the presence of oxygen, and survive only in oxygen-free environments (such as in decaying vegetation in marshes, for example). One of these is the species that causes botulism, a disease we shall consider in Chapter 14.

2. *The evolution of respiration.* As we shall see in Chapter 5, respiration (the cellular use of oxygen) is a highly efficient method of energy extraction—so much so that it is used by the overwhelming majority of organisms on earth today. Certainly, life as we know it would be impossible without respiration—and respiration became possible only because of the availability of atmospheric oxygen, itself an outgrowth of photosynthesis.

3. *Life on land.* Until about 1 billion years ago, life was restricted to aquatic environments. Why was there no life on land? The principal reason is that the ancient earth was constantly bombarded by high levels of ultraviolet radiation, and energy of that particular wavelength is very destructive to proteins and nucleic acids. Most organisms not shielded by a layer of water would be killed by such an onslaught. However, with the evolution of photosynthesis and the appearance of an increasing amount of O_2 in the atmosphere, an ultraviolet filter developed that screened out most of the ultraviolet radiation and permitted organisms to colonize the land. This filter is the **ozone** layer. When O_2 is bombarded by ultraviolet radiation, it absorbs much of the energy and in the process is converted to ozone (O_3). In Chapter 25, we shall have more to say about the ozone layer. (It is starting to develop holes but should survive at least until the end of the semester!)

4. *The end of spontaneous generation.* The presence of O_2 in the atmosphere created a situation wherein any new abiotically assembled organic molecules, if not ingested by rapacious heterotrophs, would certainly be destroyed by being chemically linked to oxygen. Thus, we come full circle. Just as the absence of atmospheric oxygen was necessary for the origin of life, the presence of atmospheric oxygen (while permitting an explosion of new species) ensured the end of any new and independent origins of life.

PROKARYOTES AND EUKARYOTES

When we examine the cellular structure of living organisms, we find that the cells of monerans differ sharply from the cells of all other kingdoms. Moneran cells are generally much smaller and have no defined nucleus. It is the absence of a nucleus that serves as the basis for the name given to moneran cells—**prokaryotic** (Gr. *pro,* "prior to" or "before" and *karyon,* "kernal"; see Figure 4.4).

The first organisms were prokaryotes, as were the first photosynthesizers. For over 2 billion years, prokaryotes ruled the earth. It was not until about 1.45 billion years ago that **eukaryotes** (Gr. *eu,* "true") evolved. All eukaryotes utilize respiration; they have a sophisticated form of cell division known as mitosis

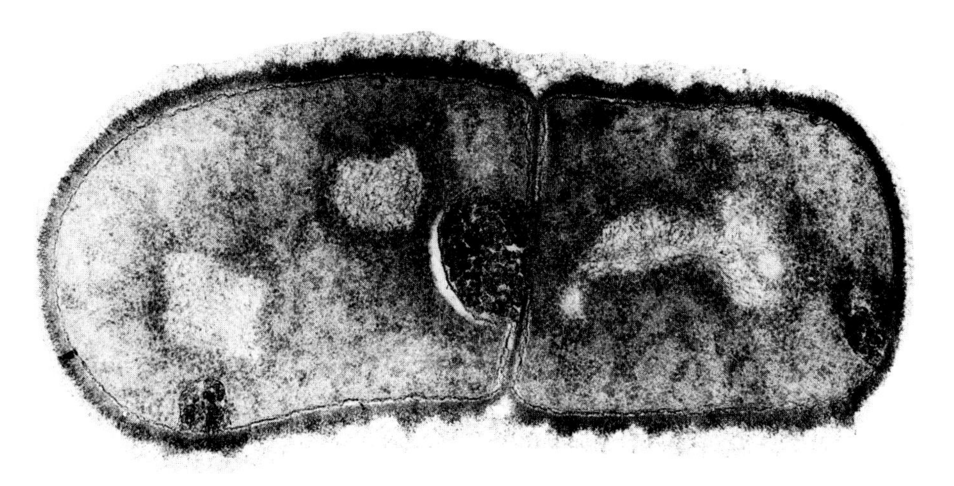

FIGURE 4.4

Prokaryote
Prokaryotic cells are found only in the monerans (bacteria and their relatives) and include a poorly differentiated nuclear region called the nucleoid; cytoplasmic organelles are limited to ribosomes.

(see Chapter 17); they reproduce sexually for the most part; they have chromosomes rather than the ring of DNA found in prokaryotes; and, as their name implies, they have a distinct nucleus.

The eukaryotes have been a remarkably successful and diverse group of organisms (see Box 4.3). Much of their success is based on their highly organized cell structure.

BOX 4.3

The Kingdoms of Life

Life was once simple for biologists. Living creatures were placed in one of two kingdoms—everything was either a plant or an animal, and therefore biologists were either botanists or zoologists. Mushrooms and their relatives were plants—and so were bacteria; some single-celled organisms were animals (protozoa), some were plants (algae), and some were gently fought over because they seemed intermediate. There weren't too many of the latter, however, and their existence was only a minor nuisance.

With our increased knowledge of cell structure, however, it became apparent that the bacteria were very different from plants. After all, the bacteria were prokaryotes—and fossils indicated they had been around for more than 2 billion years before the first indisputable plants arrived on the scene. Moreover, the algae contained one group with prokaryotic cells (the blue-green algae), which were really photosynthesizing bacteria. Thus they were renamed the **Cyanobacteria** and, with the other bacteria, comprised a newly recognized kingdom, the **Monera.**

At the same time, those single-celled plants and animals began to weigh on the conscience of scientists, and there was strong support for breaking them off as their own kingdom—the **Protista**—which are all the single-celled eukaryotes with a few multicellular algae thrown in just to keep things from being too pure.

At this point, it seemed to make no sense to have the plants include the mushrooms and their friends, so the **Fungi** came to be recognized as a valid kingdom.

Biologists relish controversy, however, so this five-kingdom scheme, although broadly accepted for the past 20 years, has continued to generate arguments. What is the relationship between the five kingdoms? In particular, did the three kingdoms of multicellular eukaryotes arise separately from the Protists, or from some more intermediate form? Perhaps more importantly, do each of the kingdoms include organisms that share a common ancestry, or have there been, for example, multiple times when the protists gave

(Continued on the following page)

BOX 4.3 *(Continued)*

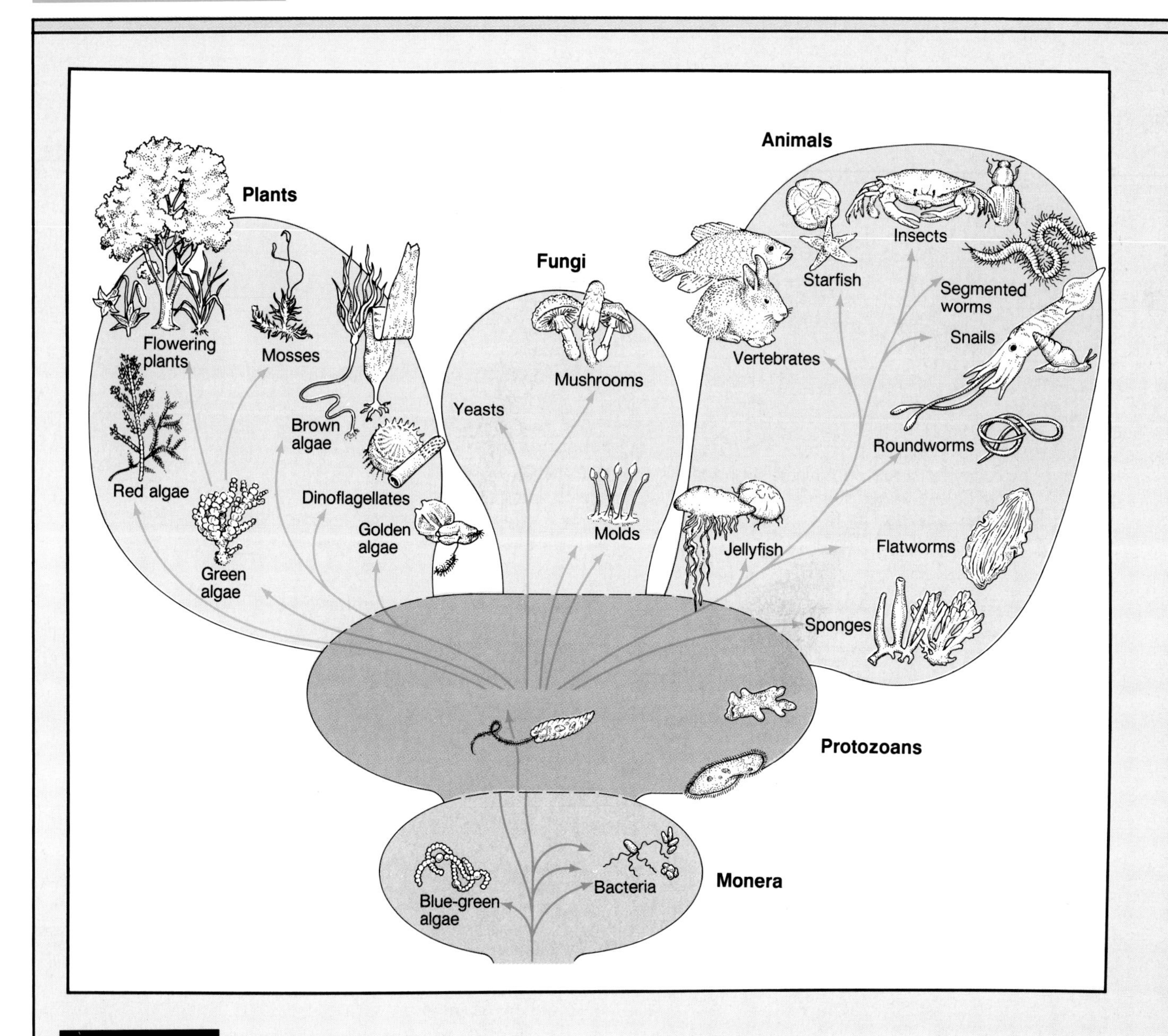

Kingdoms of Life

(Left) The traditional five-kingdom arrangement of life, first proposed by Robert Whittaker. (Right) A three kingdom arrangement, derived from a recent proposal by Carl Woese based on the degree of difference in the nucleic acids. The possible derivation of certain eukaryotic organelles from specific prokaryotes, forcefully argued by Lynn Margulis, is indicated with dashed lines.

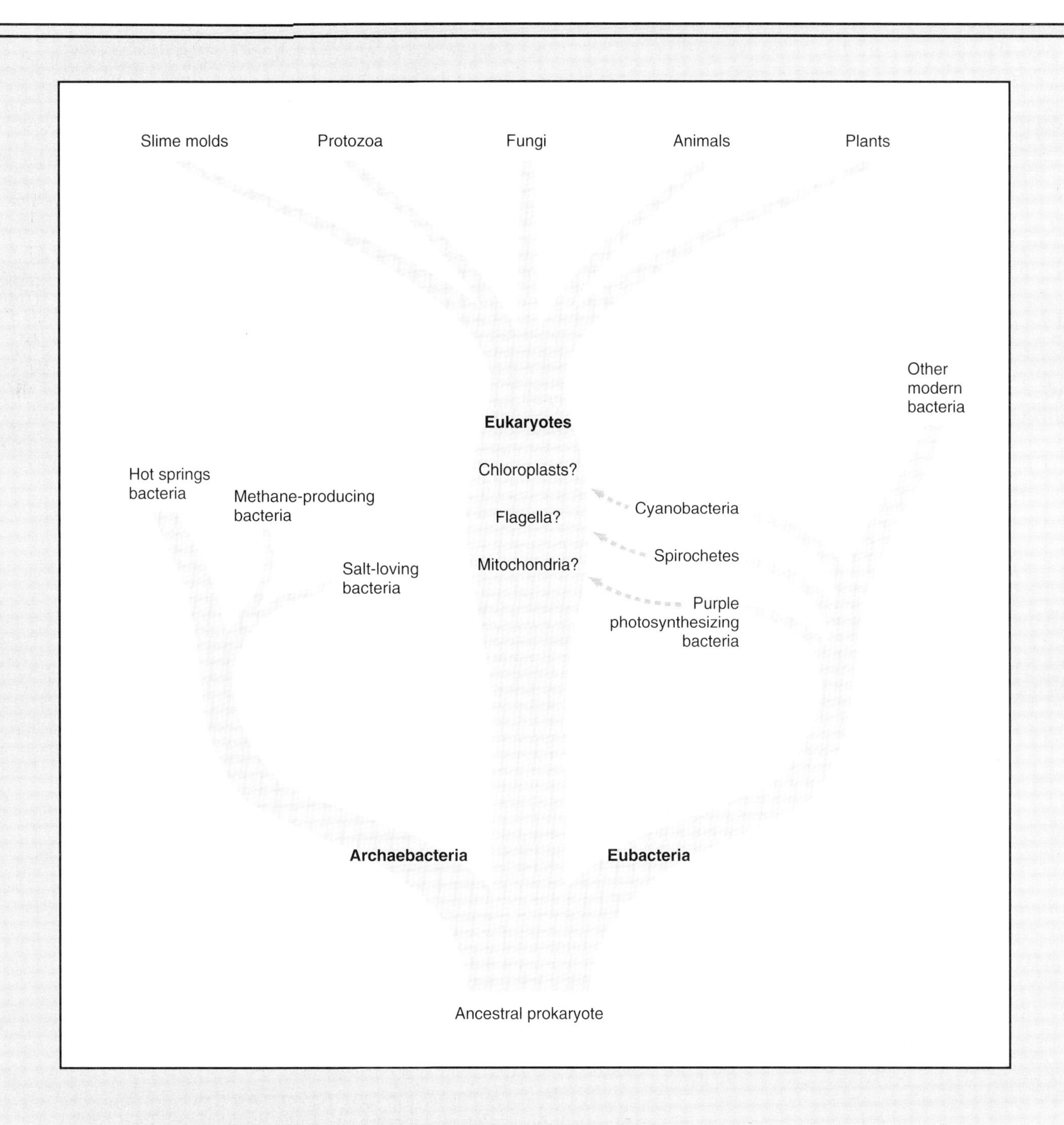

rise to organisms that we call "plants"? That is, are the kingdoms pure and biologically valid, or hybrid and mere terms of convenience?

Before those questions had been answered, a new scheme was proposed a few years ago that has become increasingly more widely accepted. The bacteria are now in two very different kingdoms. The overwhelming majority, including the Cyanobacteria, are included in the **Eubacteria** ("true" bacteria), but a few remaining groups of very ancient lineages are lumped as the **Archaebacteria** ("ancient" bacteria). A very troublesome group of organisms that have long seemed part protist and part fungi have been split off as the **slime molds,** and the eukaryote kingdoms in general all are presumed to have arisen from a common stem and then diverged.

One of the fascinations of biology is that so many of the really big questions remain unsettled!

THE EUKARYOTIC CELL

Under the light microscope, eukaryotic cells reveal little of their internal organization (see Figure 4.5). Without special fixatives and stains, the cell appears to consist of some outer boundary material (a cell wall in the case of plant cells, but just a highly flexible membrane for animal cells); a dark central nucleus; and a grainy amorphous soup called **cytoplasm** (Gr. *kytos,* "a hollow vessel" and *plasma,* "form" or "mold"). As its etymology would suggest, the function of the cytoplasm was initially unknown—but for the early cell biologists, this surely was the place where life resided.

TABLE 4.1 Units of Measures Used in Biology, on a Logarithmic Scale

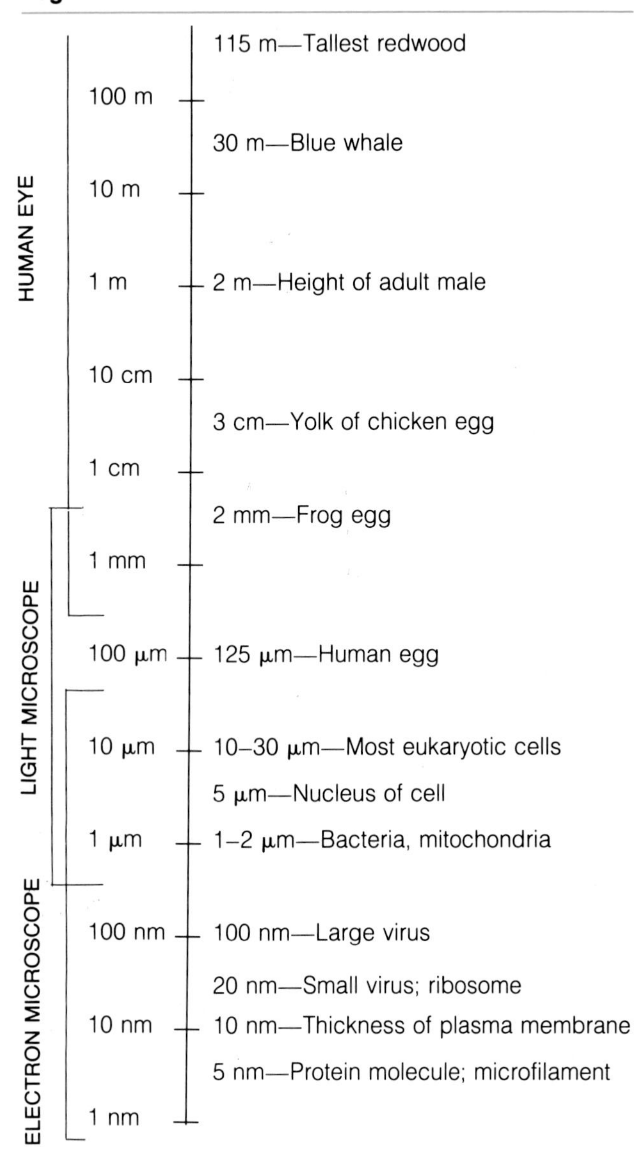

1 meter (m) = 10^2 centimeters (cm) = 10^3 millimeters (mm) = 10^6 micrometers (μm) = 10^9 nanometers (nm)

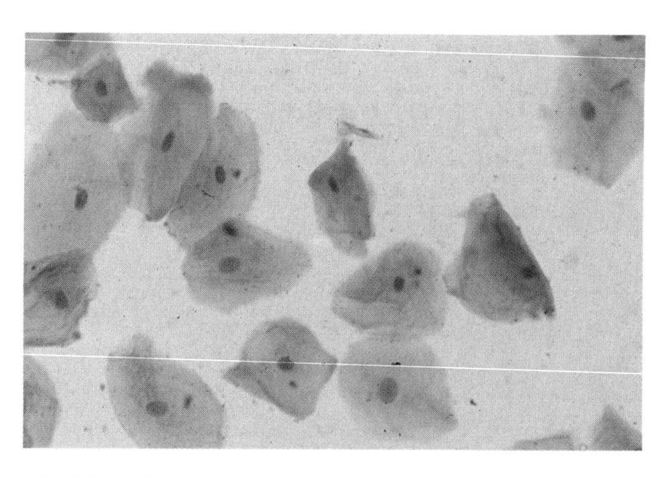

FIGURE 4.5

Eukaryote

Eukaryotic cells are found in all the other kingdoms of life. However, with the light microscope, it is not possible to see very much of their internal structure.

Because of its extraordinary powers of magnification, the electron microscope has completely revolutionized our understanding of cell structure and function (see Figure 4.6 and Table 4.1). We now see that the eukaryotic cell is organized so as both to permit many activities to occur simultaneously, and to ensure that others take place sequentially. Because the activities conducted by the cell are, fundamentally, chemical reactions, undertaking many chemical reactions either simultaneously or in an orderly sequence can be done only by compartmentalizing the cell. Thus we see that the eukaryotic cell consists of many specialized membranes, sacs, and chambers that segregate the various chemical reactions. These specialized regions of the cell are known as **organelles,** and different organelles are devoted to different activities of the cell, such as support and movement; the manufacture, secretion, and digestion of chemicals; energy transformations; and control and regulation.

Support and Movement

As organisms, we rely on our skin, bones, and muscles to provide support and movement. Cells have similar requirements, but resolve them in somewhat different ways.

Plasma Membrane. The plasma membrane is the envelope that encloses the cell and prevents its contents from spreading randomly into the environment. It is only about 10 nm thick, and consists largely of *phospholipids,* arranged in two layers. Phospholipids are molecules that resemble fats except that one of the fatty acids has been replaced by a phosphate-

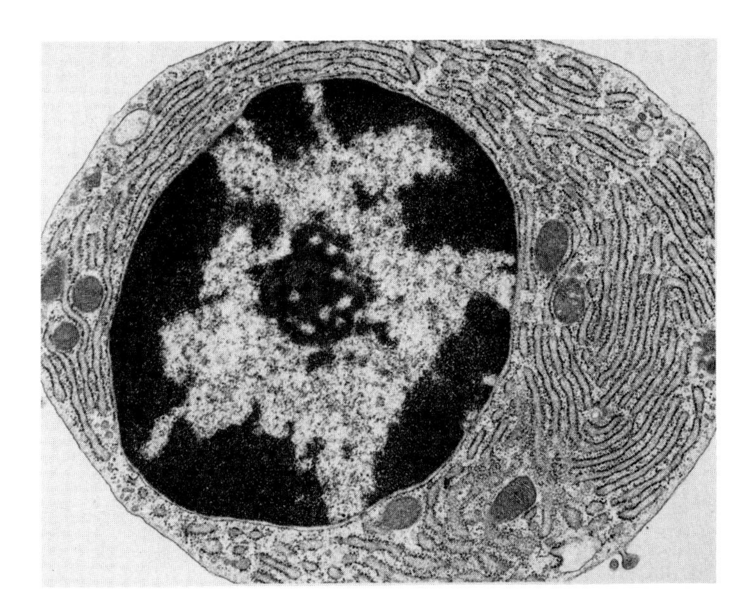

FIGURE 4.6

Eukaryote

In contrast to Figure 4.5, the electron microscope shows a variety of formed structures called organelles. These structures are graphically displayed in a highly stylized rendition, below. For purposes of clarity the organelles are not drawn to scale.

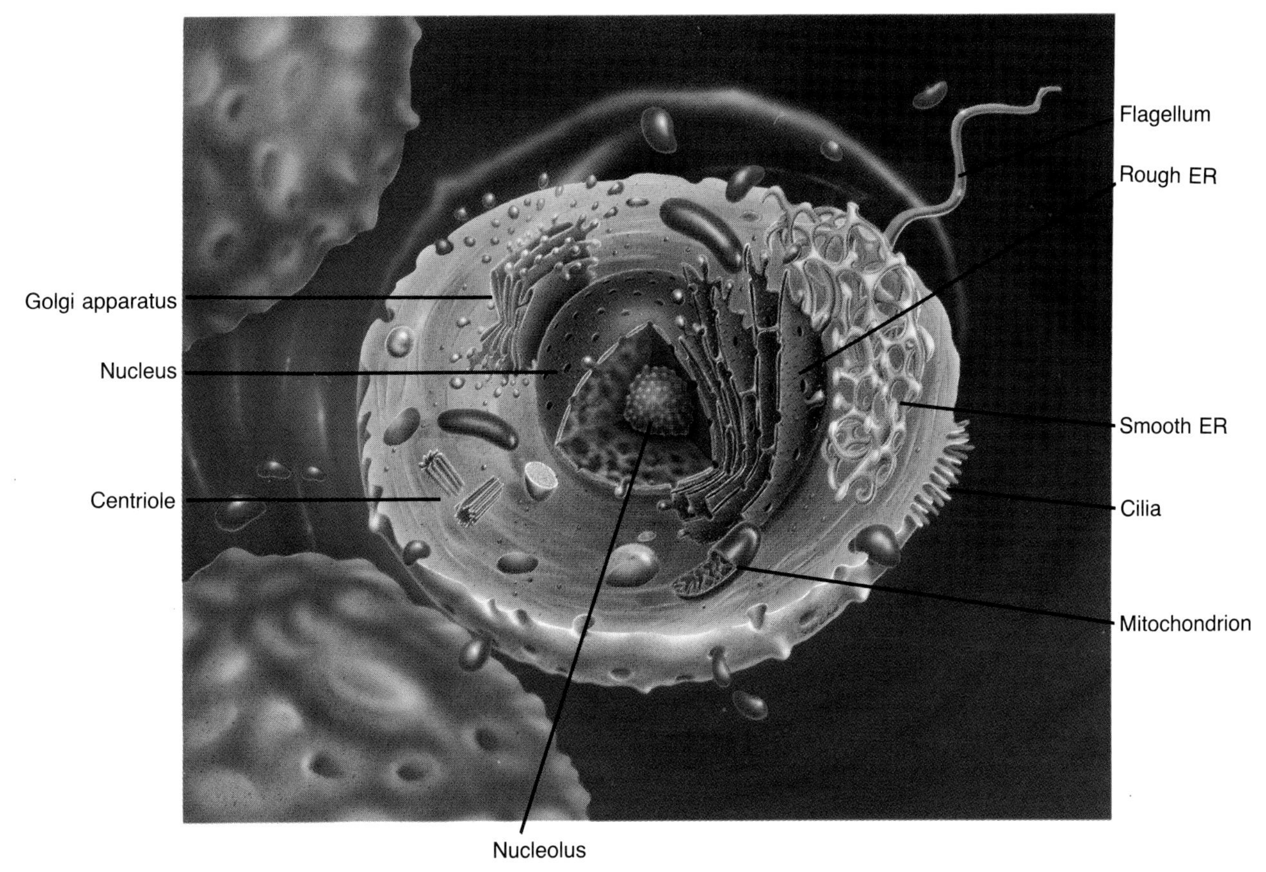

containing side chain (see Figure 4.7). The fatty acids are hydrophobic, but the phosphate-containing side chain is hydrophilic. As a consequence, the phospholipids orient such that the fatty acid ends of the molecules from each of the two layers of the membrane face each other, and the phosphate-containing side chains face the external environment and the internal cellular environment (both of which are water-based environments) (see Figure 4.8).

A primary function of the plasma membrane is to maintain the physical integrity of the cell—that is, the membrane must hold the cell's contents in place.

FIGURE 4.7

Phospholipids

Phospholipids are basically fat molecules, with one of the fatty acids removed and a phosphate group attached in its place. However, because the phosphate group is hydrophilic, it swings to the opposite end of the molecule from the hydrophobic fatty acids. Note that the presence of a double bond in one of the fatty acids causes the molecule to be kinked.

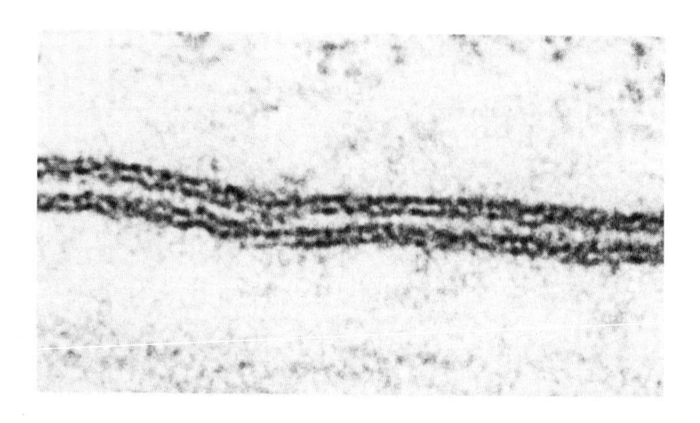

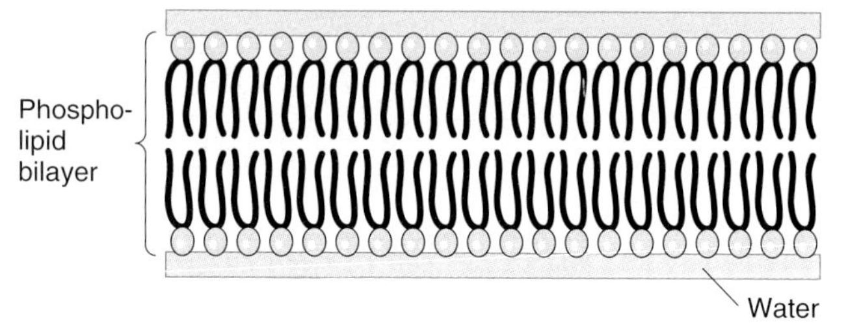

FIGURE 4.8

Plasma Membrane

Even with the electron microscope, the plasma membrane is seen only as a double line (top). The lines are actually the phosphate groups; the fatty acid portions of the molecules lie between the two rows of phosphate groups, as seen in the diagram at the bottom.

However, it cannot be an impermeable barrier for the simple reason that the cell must be able to take in nutrients from the environment, and to expel the wastes that it produces—and both nutrients and wastes must be capable of crossing the plasma membrane.

The need to be simultaneously a wall preventing leakage and a doorway permitting access puts seemingly impossible demands on the plasma membrane. Transport of materials across the membrane is discussed at length in Chapter 5, but for the moment it is important to note that the plasma membrane contains an extensive array of proteins imbedded in the phospholipid bilayer (see Figure 4.9). Some of these proteins are studded on the outer surface of the plasma membrane like warts on a toad, but others extend all the way through the membrane. Chemicals that are lipid-soluble (hydrophobic) can pass easily through the phospholipid bilayer, but hydrophilic substances must pass through channels in the proteins.

The plasma membrane also contains small carbohydrate chains, linked either to proteins or to phospholipids, which (along with certain of the proteins) characterize a cell as being of a particular type or being from a particular individual. It is the pattern of carbohydrates and proteins on the cell's surface, for example, that "tells" skin cells growing across a cut in your skin that they have reached the other side (that is, the other skin cells) and can stop dividing. Similarly, cells that are transplanted from another individual (as in a heart transplant, for example) will normally be rejected by the recipient because the presence of different types of carbohydrates and proteins on the cells of the transplanted organ characterize them as foreign. (The problem of tissue rejection is discussed in greater detail in Chapter 9.)

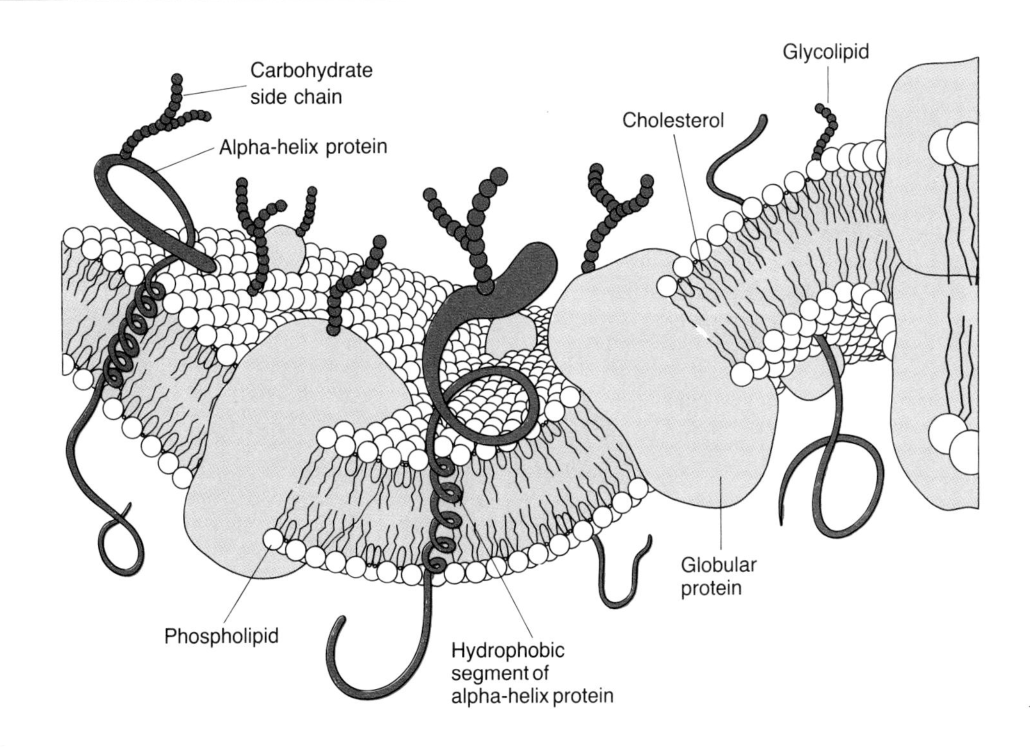

FIGURE 4.9

Plasma Membrane

The plasma membrane is dominated by phospholipids, but there are many other molecules in the membrane as well. These molecules are used for cell to cell recognition, or in the transport of substances into, or out of, the cell.

Cytoskeleton. The cytoskeleton is a constantly changing arrangement of linear proteins, strung together like Tinkertoys, which provides some amount of rigidity to the cell and prevents it from assuming the appearance of a tiny puddle of cytoplasm (see Figure 4.10). Its actual organization is currently the subject of intense investigation and speculation, but its major components are known to be **microfilaments** and **microtubules.** Many of the organelles appear to be secured to the cytoskeleton, and may move around the cell on "rails" formed by the microtubules, propelled by molecular "motors" of **dynein** and **kinesin.** Although these newly discovered molecules are known to consume energy as they work, the details of how they actually function are not yet clear. It is likely, however, that both function similarly to **myosin,** one of the contractile proteins in muscles (see Chapter 12 for the details of muscle contraction).

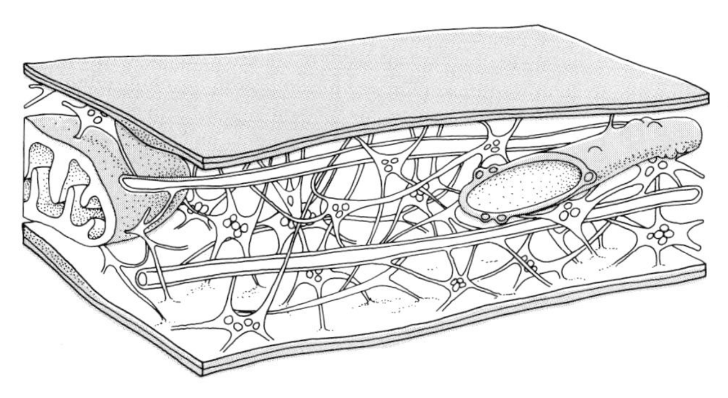

FIGURE 4.10

Cytoskeleton

The cytoskeleton is made up largely of microtubules and microfilaments that are constantly reorganized during the life of the cell. Above, a diagrammatic representation of the elements of the cytoskeleton.

Microfilaments. Microfilaments are composed of long but very thin (about 5 nm) strands of the globular protein **actin** (see Figure 4.11), which along with myosin is the principal contractile protein of muscle. These strands form spontaneously in the presence of elevated levels of Ca^{2+} and Mg^{2+} within the cell, and as a consequence of changes in the ion levels, microfilaments are constantly forming and breaking down.

Microfilaments are found just beneath the plasma membrane, and, in conjunction with dynein and kinesin, are responsible for a variety of effects including changing the shape of the cell, crawling movements, transport of small sacs of materials called **vesicles** to and from the plasma membrane, movement of organelles, and division of the cell.

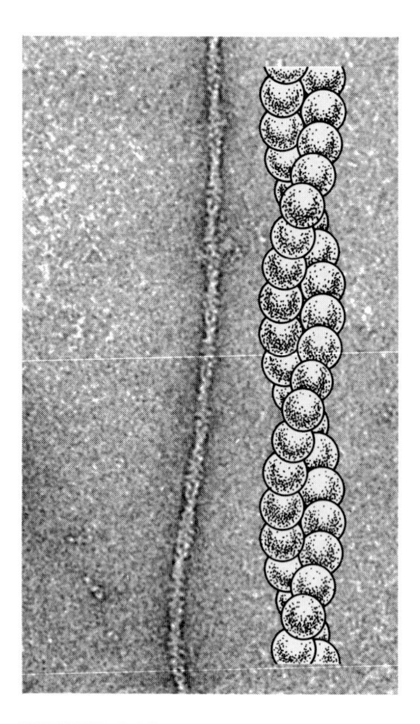

FIGURE 4.11

Microfilaments

Electron micrograph (a) and diagrammatic representation (b) of a microfilament. Note that microfilaments consist of two strands twisted around each other.

Microtubules. Microtubules are, as their name suggests, long, slender tubes about 25 nm in diameter (or five times thicker than microfilaments) and up to 25 μm in length (see Figure 4.12). They are composed of the globular protein **tubulin.** As was true of microfilaments, microtubules form spontaneously in the proper ionic environment, and disassemble with equal ease.

Microtubules are found in cilia and flagella, and also comprise the **spindle fibers** that separate chromosomes during cell division.

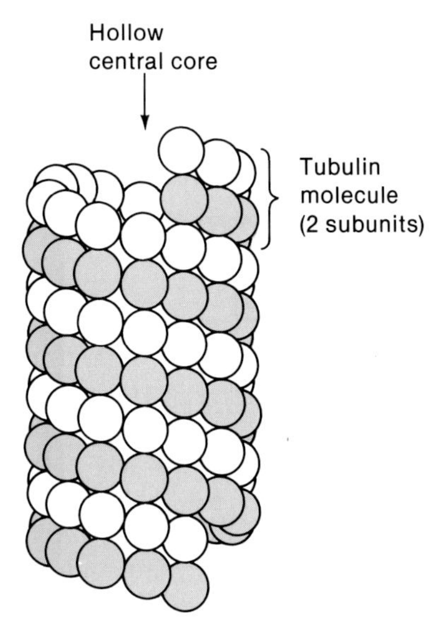

FIGURE 4.12

Microtubules

Microtubules are composed of tubulin molecules that are stacked so as to form a spiral wall around a hollow core.

Cilia and Flagella. Cilia (L. *cilium* "eyelash") and flagella (L. *flagellum,* "whip") are thin outpocketings of the cell covered by the plasma membrane. Under the light microscope, they are easily distinguished—flagella are typically only one or at most a few in number and are up to 200 μm long; cilia are always very numerous and are generally less than 25 μm in length. However, under the electron microscope, these superficial differences vanish, because their relatively complex internal anatomies are identical—nine pairs of microtubules arranged in a circle and surrounding two single microtubules, an arrangement known as 9 + 2 (see Figure 4.13). Interestingly, the anatomy of

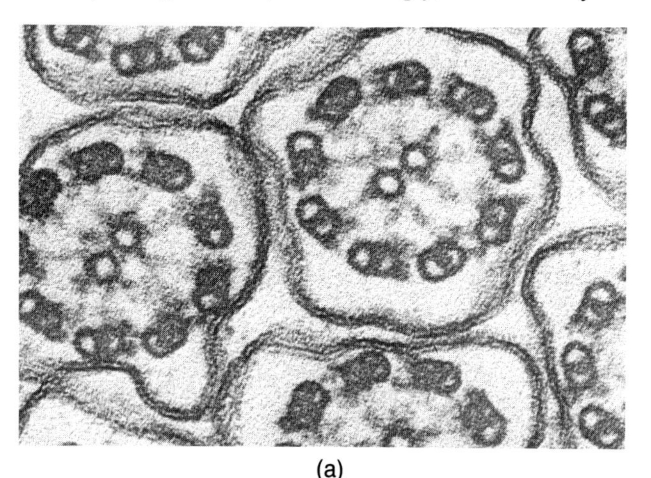

(a)

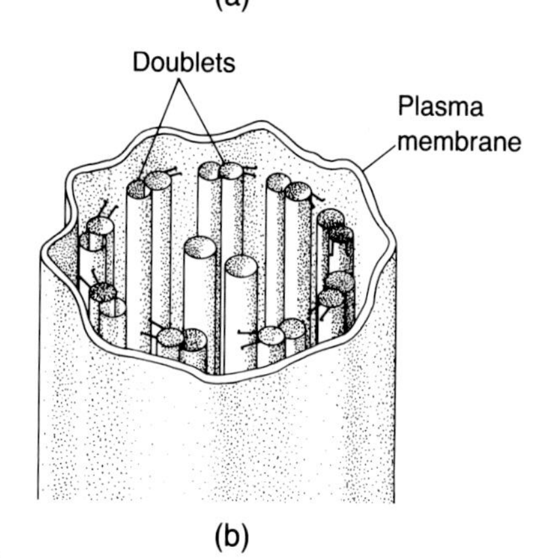

(b)

FIGURE 4.13

Internal Structure of Cilia and Flagella

Although they look quite different at first glance, cilia and flagella have the same internal structure: nine pairs of microtubules surrounding a central pair. These are revealed in the cross section of cilia in the electron micrograph (a) and in the diagrammatic representation (b).

cilia and flagella is identical throughout the diverse assemblage of organisms comprising the four kingdoms of eukaryotes. This remarkable similarity is one piece of evidence used to support the notion of a common ancestry for all eukaryotes.

Cilia and flagella function in cell movement. However, cilia row (forward stroke followed by a backward recovery), whereas flagella scull (the movement of a gondola oar, for those of you familiar with transport along the canals of Venice).

In the human body, ciliated cells are anchored, and rather than moving the cells, the cilia move materials across the surface of the cells. This activity is most readily appreciated in the respiratory system, where the cells lining the tubes of the system are covered with cilia (as many as 1 billion per square centimeter) that carry dust-laden mucus back up from the lungs to the throat where it can be spit out or swallowed. Nicotine tends to paralyze cilia, meaning that smokers have less efficient cleansing of the lungs. As a consequence, more mucus remains lodged in the respiratory tract, leading to "smoker's cough," among many other ailments.

Centriole and Basal Bodies. The centriole is a paired organelle, with each half lying at right angles to the other (see Figure 4.14). The two halves are composed of microtubules arranged in a ring of nine triplets, but without any central microtubules (9 + 0 arrangement). Centrioles are approximately 200 nm by 400 nm in size, and, in animal cells (they are not found in plant cells), function as an organizing center for the development of spindle fibers during cell division.

Basal bodies also have a 9 + 0 arrangement of microtubules and are found at the base of each cilium and flagellum. It is believed that basal bodies may arise from centrioles, and that they, in turn, generate the cilia and flagella.

Manufacture, Secretion, and Digestion

Much of what individual cells do relates to specific chemical transformations. Muscle cells manufacture contractile proteins; certain cells of the intestinal lining secrete digestive enzymes; some white blood cells ingest—and then digest—bacteria. These activities require the services of several different organelles.

Ribosomes. Ribosomes are the sites of protein formation. They are small (about 20 nm in diameter), exceedingly numerous (up to 500,000 per cell), and consist of RNA and protein in a roughly 2:1 ratio (see Figure 4.15). The mechanics of protein formation are discussed in detail in Chapter 18.

Endoplasmic Reticulum. Among the more interesting revelations of the electron microscope was the discovery of the endoplasmic reticulum (L., "a little net within the cell"). The ER, as it is commonly called, appears to be one continuous sheet complexly folded back on itself (see Figure 4.16). The result is a many-chambered sac enclosing as much as 10 percent of the volume of the cytoplasm.

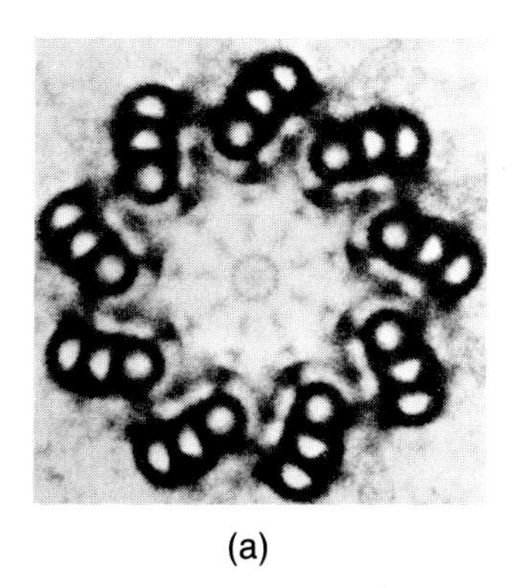

(a)

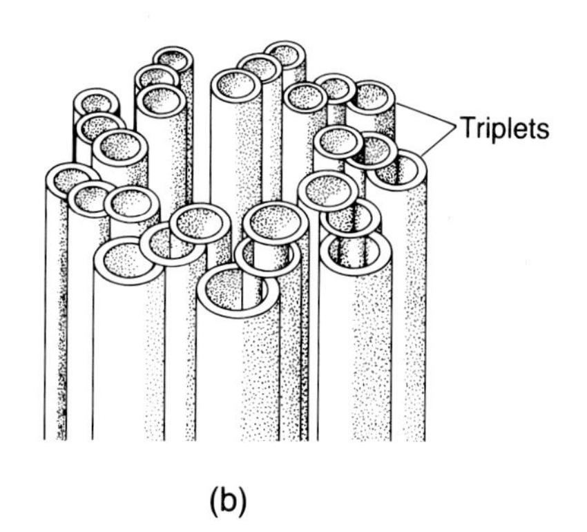

(b)

FIGURE 4.14

Centriole

Centrioles, which are also composed of microtubules, differ from cilia and flagella in two ways: First, they are internal structures, and therefore are not surrounded by the plasma membrane; second, their microtubules are arranged in nine triplets but without a central pair. This organization is shown in the electron micrograph (a) and in the diagrammatic representation (b).

A portion of the ER typically has ribosomes attached to it, giving it a saw-toothed appearance. This so-called **rough ER** is the site of protein formation for proteins that are going to be exported from the cell (such as digestive enzymes). Some ribosomes are also found away from the ER, attached to the cytoskeleton. These ribosomes are where proteins to be used within the cell are assembled.

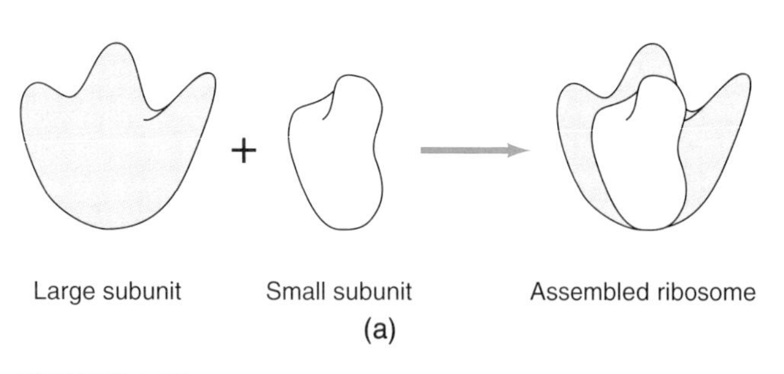

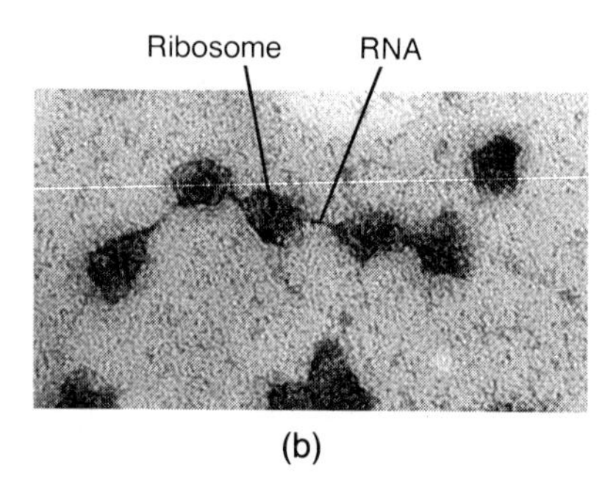

FIGURE 4.15

Ribosomes

The components of ribosomes (a) are synthesized in the nucleus but assembly takes place in the cytoplasm. In the electron micrograph (b), a series of ribosomes are seen attached to a molecule of RNA.

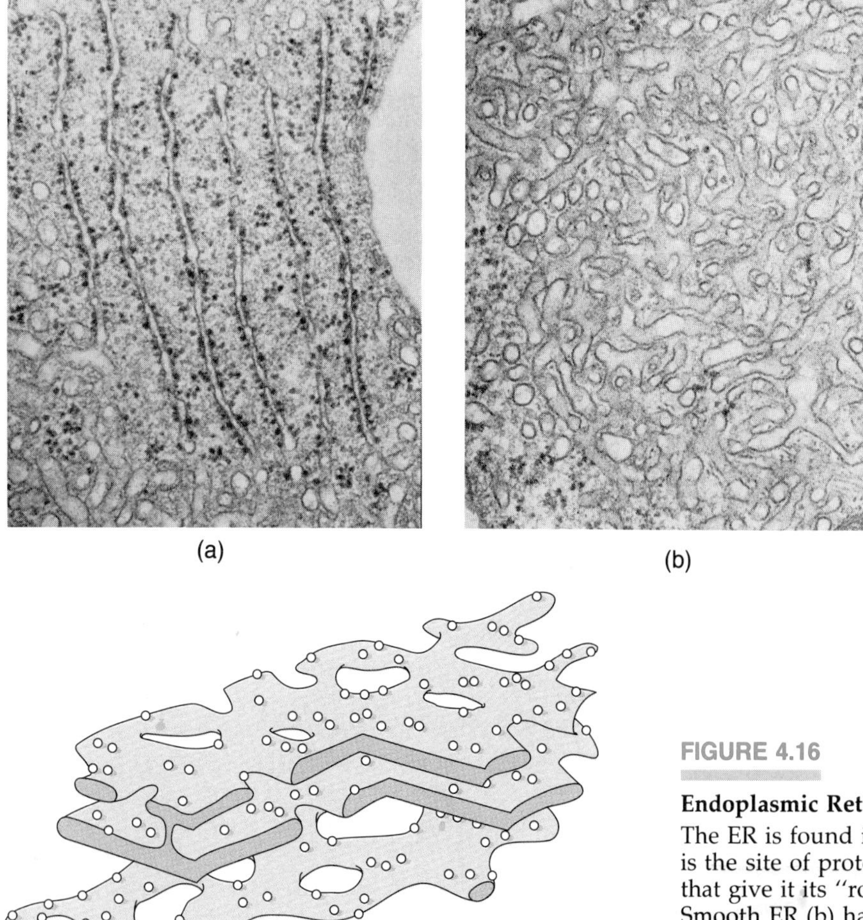

FIGURE 4.16

Endoplasmic Reticulum

The ER is found in two forms in the cell. Rough ER (a) is the site of protein formation; the tiny black specks that give it its "rough" appearance are ribosomes. Smooth ER (b) has no ribosomes; it is the production site of lipids. Diagrammatically, the rough ER is seen to be a series of interlinked folds (c).

Other portions of the ER are ribosome-free. These regions, called **smooth ER,** synthesize lipids. The proportion of rough to smooth ER varies among different types of cells, according to the relative amounts of protein and lipid they manufacture.

Far from being merely a passive structure on which ribosomes may be hung, we now see the ER as a highly dynamic organelle. It is capable of synthesizing more of itself, for example, and many of the proteins imbedded in the lipid bilayer making up the membrane of the ER are enzymes that catalyze various chemical reactions. In addition, proteins formed by the ribosomes of the rough ER are sequestered in vesicles and sent, via the smooth ER, to the Golgi apparatus for additional processing (see Figure 4.17).

Golgi Apparatus. The Golgi apparatus derives its name from its discoverer, Camillo Golgi, who in 1898 described a new organelle from cells treated with a special fixative and stain. Other investigators, unable to verify Golgi's claim when they used different fixatives and stains, suggested that his "organelle" was nothing more than some cellular debris resulting from his drastic methods of preparing the cell. Golgi died in 1926 with the issue of his organelle still very much in question. It was not until the electron microscope came into wide use in the late 1940s that he was vindicated—the Golgi apparatus does exist, and we now know just how important it is to the cell. There are usually anywhere from ten to several hundred Golgi apparatuses in every cell.

In appearance, the Golgi apparatus resembles a stack of ten to 20 hollow pancakes, with the smallest of the "pancakes" being closest to the plasma membrane of the cell (see Figure 4.18). The lowest pancake in the stack receives vesicles of proteins from the ER and passes the proteins from one pancake to another. The proteins are modified in various ways: To some, sugars are added whereas others are purified and rerouted to other organelles in the cell. The Golgi apparatus is also the site of polysaccharide synthesis, and these polysaccharides may be added either to lipids to create **glycolipids** or to proteins to create **glycoproteins.**

Ultimately, small vesicles containing the finished products are budded off the edges of the smallest pancake where they may

1. move to the plasma membrane to release protein enzymes to the outside;

2. move to the plasma membrane where the glycoproteins or glycolipids may be incorporated into the plasma membrane (in which case the phospholipid membrane surrounding the vesicle will also be incorporated into the plasma membrane); or

3. move to the lysosomes.

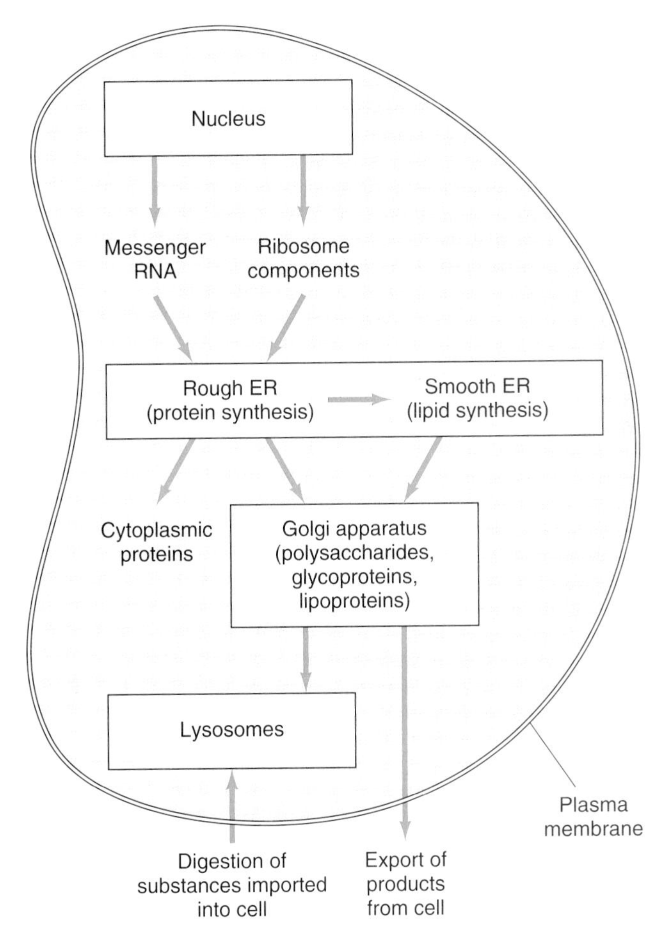

FIGURE 4.17

Functional Relationship of Organelles

A flow chart showing the interrelationship of the organelles involved in manufacture, secretion, and digestion, albeit in a highly simplified form.

Lysosomes. The lysosomes are known euphemistically as "suicide sacs" because they contain powerful digestive enzymes able to digest the cell if they were to escape the confines of the lysosomal membrane. They vary considerably in size but are commonly about 500 nm in diameter. Lysosomes contain as many as 40 different enzymes, all originally synthesized by ribosomes on the rough ER and modified and packaged into vesicles by the Golgi apparatus.

The manifold role of these enzymes includes the following:

1. In certain types of white blood cells, disease bacteria are engulfed by the cell and enclosed in a large vesicle. This vesicle is then fused with a lysosome, the enzymes destroy the bacteria, and any indigestible remains are transported back to the plasma

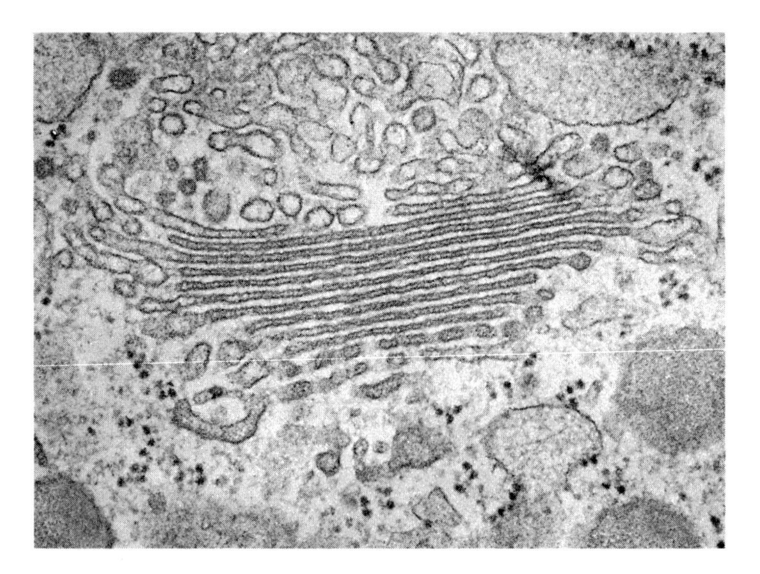

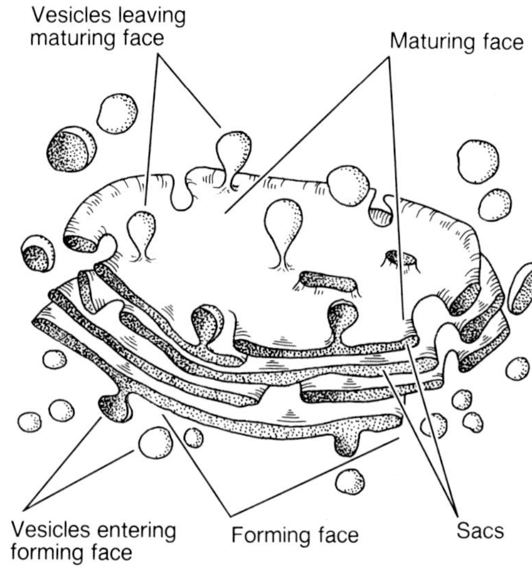

FIGURE 4.18

The Golgi Apparatus

An electron micrograph and a diagrammatic representation of the Golgi apparatus, which consists of flattened pancake-like chambers from which small sacs, called vesicles, leave and enter.

membrane in a vesicle and expelled. (Any useful molecules diffuse across the lysosomal membrane into the cytoplasm.)

2. Lysosomes also digest worn-out organelles, and recycle their components for fabrication into new organelles. The recycling of organelles is critical in maintaining the cell in a vigorous state. It is the failure to renew organelles that may underlie the aging process at the cellular level.

3. Cholesterol, an important molecule for a variety of reasons (as we shall see in Chapters 7, 8, and 13), is attached to a carrier molecule as it moves through the blood to individual cells. Lysosomes split the carrier molecule from cholesterol and permit the cell to use cholesterol for the synthesis of a variety of hormones.

4. In some instances, the lysosomes rupture and destroy the cell. As destructive as that sounds, it is often essential from the standpoint of development. Your hands and feet, for example, begin in the embryo as webbed pads. At an early stage, the cells of the webs between the fingers and toes are destroyed by lysosomal action, leaving the fingers and toes independent of each other. Organisms undergoing metamorphosis, such as tadpoles and insects in the pupal stage, rely heavily on lysosomal-mediated cellular destruction.

Under normal circumstances, the cell is able to synthesize new membrane around the lysosome as quickly as the enzymes destroy it from within. However, when the cell is deprived of oxygen even for relatively short periods of time, it cannot generate enough energy to maintain the membrane intact, and the enzymes escape. Despite the fact that the enzymes work best in the relatively acidic environment of the lysosome (as opposed to the more basic environment of the cell), they are nevertheless fully capable of destroying the cell—and cellular destruction by lysosomal enzymes is one reason why individuals who have been deprived of oxygen for more than a few minutes generally cannot be "brought back to life."

Finally, as we shall see in Chapter 19, genetic errors in the manufacture of lysosomal enzymes can be devastating. Tay-Sachs disease, which is relatively common in Eastern European Jews, results from the absence of an enzyme responsible for lipid metabolism. As a consequence, lipids accumulate, especially in the cells of the nervous system, with the result being retardation, blindness, and ultimately death in early childhood.

Peroxisomes. Peroxisomes are similar to lysosomes in being membrane-bound sacs of enzymes, but instead of the digestive enzymes found in lysosomes, peroxisomes contain detoxifying enzymes (see Figure 4.19). For example, peroxisome enzymes remove the amine group from amino acids and convert it to ammonia for disposal, with the balance of the amino acid being used for energy production. Alcohol, a potentially lethal chemical, is detoxified by peroxisomes in liver cells.

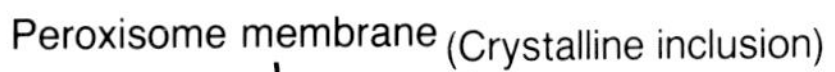

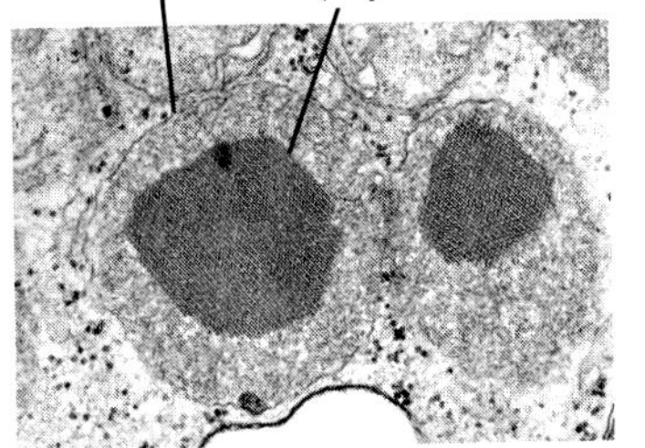

FIGURE 4.19

Peroxisome

Peroxisomes are small organelles that serve various functions in different cells but are primarily used in detoxification: the destruction, through enzymatic conversion, of chemicals that could injure or kill the cell.

Energy Transformation

Chloroplasts. Chloroplasts are the organelles in which photosynthesis occurs, a statement that should suggest to you that they are found only in photosynthesizing organisms, such as plants and some single-celled organisms. We shall discuss the details of photosynthesis in Chapter 5, but for now it is important to recognize the basic formula:

Carbon dioxide + Water
+ Energy (sunlight) → Glucose + Oxygen

This transformation requires the presence of chlorophyll, a molecule that absorbs certain wavelengths of light but reflects light at the wavelength we perceive as green. It is the chlorophyll of the chloroplasts that gives leaves their green color.

Chloroplasts are rather large organelles, easily seen under the light microscope (see Figure 4.20). They are roughly oval, measuring approximately 3×6 μm. Plant cells contain from one to several hundred chloroplasts. Each chloroplast is surrounded by a double membrane of lipoproteins, with the inner membrane deeply folded into stacks containing the chlorophyll.

Interestingly, chloroplasts contain their own ribosomes and a significant amount of DNA. They reproduce by dividing, as opposed to being assembled from components in the cytoplasm.

Mitochondria. Mitochondria are extremely important organelles in the production of energy. They are ubiquitous in eukaryotic cells. The production of energy is discussed in detail in Chapter 5, but for now remember this equation:

Glucose + Oxygen → Carbon dioxide
+ Water + Energy

(Those of you who immediately recognized this equation as the converse of the photosynthesis equation can now place a gold star beside your name. Plants

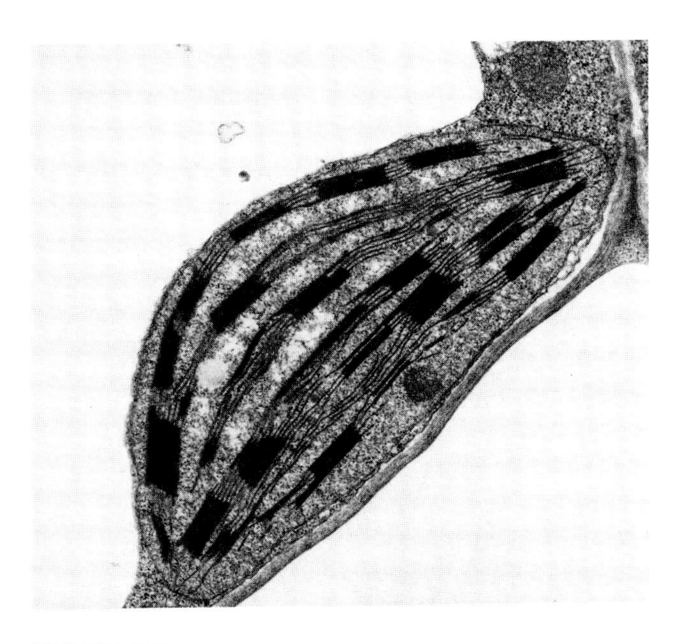

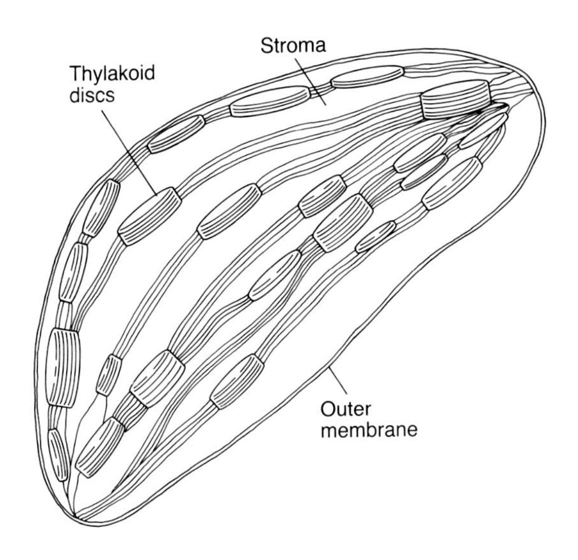

FIGURE 4.20

Chloroplast

Chloroplasts are double-membraned organelles found in organisms capable of photosynthesis. Their specific function is discussed in Chapter 5.

assemble glucose, the precursor of many macromolecules, from inorganic molecules, using the energy from sunlight. Animals extract the energy contained in glucose, and dispose of inorganic wastes—which the plants used to make more glucose. Tidy little arrangement, isn't it?)

This process of energy extraction is known as **cellular respiration.** It requires oxygen, and the source of this oxygen is the air we breathe. In fact, the *only* place where we utilize oxygen is in the mitochondria of our cells. We breathe, in essence, on their behalf.

Mitochondria are also large organelles, easily seen under the light microscope—although at about 1 × 4 μm they are usually somewhat smaller than chloroplasts (see Figure 4.21). Like chloroplasts, mitochondria have a double membrane, with the inner membrane deeply folded, and their own DNA and ribosomes (see Box 4.4). They reproduce by dividing in half—although the *rate* at which mitochondria divide is controlled by the nucleus of the cell. The presence of a double membrane creates two separate chambers—one between the two membranes and one inside the inner membrane—and energy extraction involves, in part, movement of ions from one chamber to another.

Control and Regulation

Nucleus. At five to seven μm, the nucleus is the largest of the organelles, and was the first to be described (by Robert Brown in 1831). Its name derives from the Latin word for "kernel" or "nut," an apparent reference to its spherical shape and dark coloration at the center of the cell. (Do not confuse the nucleus of the cell with the nucleus of atoms. Unfortunately, the same term has been used for these two very different entities.)

The nucleus is surrounded by a double membrane, with the outer membrane apparently continuous with the ER and possibly derived from it (see Figure 4.22). At various points, the two membranes fuse and create openings called **nuclear pores** through which materials may move between the interior of the nucleus (the **nucleoplasm**) and the cytoplasm. The pores are partially plugged with protein and RNA, and these molecules presumably regulate the passage of molecules through them. Since the nucleus contains no functional ribosomes, all of the proteins in the nucleus must be imported from the cytoplasm. Reciprocally, all of the RNA found in the cytoplasm has its origin in the nucleus.

The prominence of the nucleus bespeaks its importance to the cell, for it is responsible for regulating most of the cell's activities, especially in the areas of metabolism, differentiation, and cell division. A cell deprived of its nucleus is ultimately doomed, because it cannot maintain itself. Our red blood cells, for example, have no nucleus, and the lifespan of an individual red blood cell is limited to about four months. (The implications of this statement are discussed in Chapter 9.)

Chromosomes. The chromosomes have the unusual capacity of being invisible except when the cell is dividing—and then they are very readily observed even with a light microscope (see Figure 4.23). Chromosomes are composed of extremely long chains of DNA, associated with a protein core. During cell division, the DNA chains wind tightly around the core,

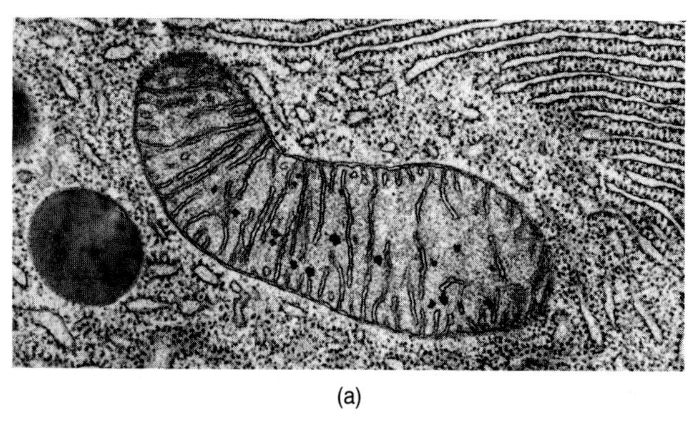

(a)

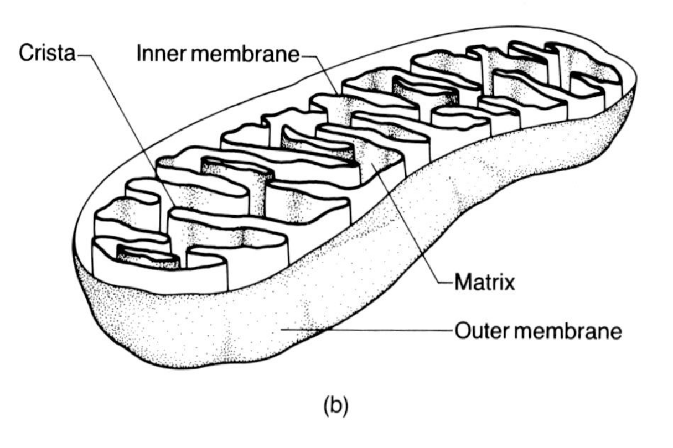

(b)

FIGURE 4.21

Mitochondrion

A mitochondrion is shown in longitudinal section in the electron micrograph (a) and in a diagrammatic representation (b). Mitochondria have both an outer and an inner membrane. The latter is complexly folded into ridges. As we shall see in Chapter 5, this double membrane structure is of vital importance in the extraction of energy from energy-rich molecules such as glucose.

BOX 4.4

The Origin of Organelles

As you can see, the differences between prokaryotic cells and eukaryotic cells are profound—and a troublesome question for many years has been, "Where did all those organelles come from?"

The initial answer was that organelles derived from infoldings of the plasma membrane. That answer was widely, if not enthusiastically, accepted, but as we came to know more about the structure of organelles, a plasma membrane origin was increasingly less satisfactory an answer. Some organelles, for instance, have no membrane (ribosomes); others have a double membrane (mitochondria and chloroplasts); still others just don't make sense as infoldings (cilia and flagella).

In recent years, an old and ignored idea has been dusted off and, in light of today's knowledge of organelles, has come to be broadly (but by no means completely) accepted. This notion declares eukaryotic cells to be assemblages of different types of prokaryotic cells—a sort of United Nations of bacteria, as it were.

What is the evidence?

1. Even the smallest of eukaryotic cells are always much larger than prokaryotic cells, an expected result if eukaryotic cells were comprised of a number of prokaryotic cells.

2. Some organelles have retained considerable independence. Both chloroplasts and mitochondria have DNA, RNA, and ribosomes of a size and type found in bacteria, and the chloroplasts in particular look a great deal like some free-living photosynthesizing bacteria.

3. There are numerous examples throughout the various kingdoms of life where one species has become an internal partner of another. An impressive variety of marine invertebrates, for example, have incorporated living algae into their bodies, and have become, in a functional sense, a photosynthesizing animal. Animals as diverse as cows and termites depend on intestinal bacteria and protists, respectively, to digest cellulose for use by the host. Indeed, one of the protists living inside termites' guts is covered with what were long thought to be cilia—but which turn out to be hundreds of spirochete ("spiral-toothed," a reference to the flagellum) bacteria, each imbedded in the protist with only the flagella pointing outward.

At present, the prevailing view is that chloroplasts, probably mitochondria, and possibly cilia and flagella are remnants of once free-living prokaryotes. Most of the rest of the organelles are believed ultimately to have arisen from the ER—but the origin of the ER is itself still fiercely debated, with the plasma membrane still being considered a possibility.

in preparation for distribution to the daughter cells, and in the process become dense enough to be easily seen. (The details of cell division are described in Chapter 17.)

DNA is critical to the survival of the cell because it determines the exact structure of all of the protein

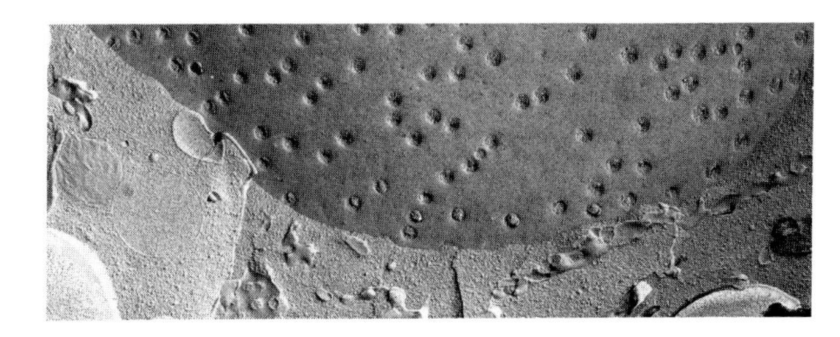

FIGURE 4.22

Nuclear envelope

The nuclear envelope is perforated by many small pores, which provide access between the nucleus and the cytoplasm.

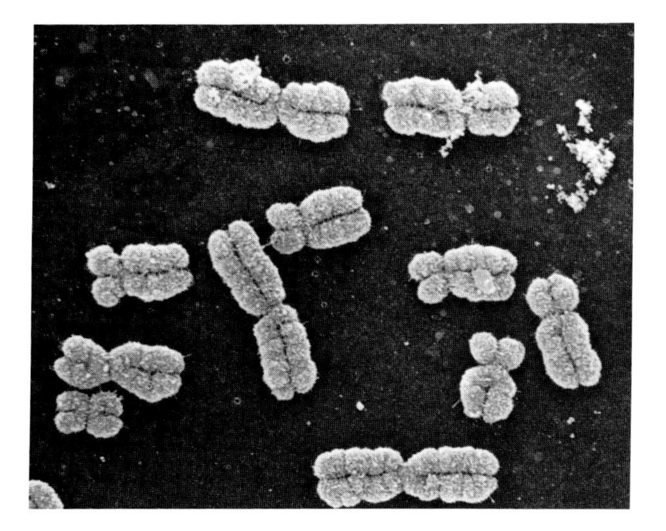

FIGURE 4.23

Chromosomes

Chromosomes are normally visible only during cell division, at which point they are, in terms of size, the dominant organelles in the cell.

molecules manufactured by ribosomes in the cytoplasm. Although the DNA remains in the nucleus, it nonetheless directs protein synthesis by sending short molecules of RNA (appropriately called **messenger RNA**) to the ribosomes via the nuclear pores. These molecules act as a template upon which amino acids assemble in an order characteristic of the particular protein. (The details of protein assembly are discussed in Chapter 18.)

Nucleolus. The nucleolus (L., "small kernel") is a dark-staining region of the nucleus. It may exist as a single organelle or in multiple form, with two being a common number. The nucleolus is simply a specialized part of one or more of the chromosomes, and is ultimately responsible for the manufacture of the RNA found in ribosomes, although the final assembly of the ribosomes occurs in the cytoplasm.

Summary

Life had its beginnings on earth shortly after the earth had *its* beginnings, some 4.6 billion years ago. Conditions on the earth were favorably disposed for the creation of life. The atmosphere was rich in various hydrogen, carbon, and nitrogen compounds, and there was no atmospheric oxygen. Energy, especially in the form of ultraviolet radiation, was very abundant.

Given these conditions, many small organic molecules presumably formed. (They are easily formed experimentally, under conditions that attempt to simulate those of the ancient earth.) The formation of macromolecules is somewhat more hypothetical, especially because the reactions that create them in the living cells of today require enzymes, which are themselves macromolecules. Nevertheless, there are a number of scenarios in which macromolecule formation can be explained.

Cells have a surrounding membrane, and so, in essence, do proteinoid microspheres and coacervate droplets, both of which can readily be formed in the laboratory under conditions simulating those of the ancient earth. If such a structure came to house the appropriate set of macromolecules, the stage would be set for crossing the threshold of life.

However, the cell is not just a bag of random macromolecules. Both the metabolic activities of the cell and its reproduction are genetically programmed, not haphazard events. The most difficult step in our scenario is accounting for the acquisition of this genetic program.

However it was done, fossils of bacteria-like organisms are known from rocks 3.5 billion years old. These earliest fossils were apparently heterotrophs, but the development of autotrophy—and subsequently of photosynthesis—followed in the next billion years. With the evolution of photosynthesis, gaseous oxygen began to collect in the atmosphere, and the way was paved for the evolution of more sophisticated organisms that were able to use oxygen for cellular respiration and able to colonize land.

These more sophisticated organisms had more sophisticated cells. Instead of the small, rather homogenous cells of bacteria, they possessed cells with many specialized regions, called organelles. Organelles allow the cell to perform many functions at the same time, and to sequence activities in a productive manner. Organelles serve in support and movement; in the manufacture, secretion, and digestion of chemicals; in transforming energy from one state to another; and in control and regulation of all the other functions.

Key Terms

abiotic
ultraviolet radiation
proteinoid
coacervate droplet
microsphere
chemical evolution
heterotroph
autotroph
phototroph
chemautotroph
ozone
prokaryote
eukaryote
cytoplasm
organelle
plasma membrane
cytoskeleton
microfilament
microtubule
dynein
kinesin
actin
vesicle
tubulin
spindle fiber
cilium
flagellum
centriole
basal body
ribosome
endoplasmic reticulum (ER)
Golgi apparatus
glycolipid
glycoprotein
lysosome
peroxisome
chloroplast
mitochondria
cellular respiration
nucleus
nuclear pore
nucleoplasm
chromosome
messenger RNA
nucleolus
myosin
rough and smooth ER

Box Terms

inherent property
emergent property
virus
viroid
prion
cyano bacteria
monera
protista
fungi
eubacteria
archaebacteria
slime mold

Questions

1. Briefly describe the kinds of evidence used to support the concept that life arose from abiotic beginnings early in the earth's history. Where is the evidence weakest?
2. Distinguish among heterotrophs, phototrophs, and chemautotrophs. Which presumably arose first in earth's history? Defend your choice.
3. What is the relationship between photosynthesis, the rapid evolution of multicellular organisms, and the beginnings of life on land?
4. What are prokaryotes? How do they differ from eukaryotes in terms of cell structure?
5. How are cilia and flagella different? How are they similar?
6. Explain the interrelationship of the endoplasmic reticulum, the Golgi apparatus, and lysosomes.
7. Relative to the other organelles, what is unusual about the structure of mitochondria and chloroplasts? What does that structure suggest about their possible origins?
8. If the nucleus is surrounded by a double-membraned "envelope," how does it communicate with the rest of the cell?
9. What are some essential features of life? Which of those features are possessed by viruses?
10. If you were to place all living organisms into kingdoms, how many kingdoms would you have? Defend your choice.

CHAPTER

5

Energetics

How the Cell Makes a Living

The body of a young man, recently washed up on the shores of San Francisco Bay, lies draped on a marble slab. Clancy, the pathologist assigned to this case, completes his autopsy. Looking up from his microscope, he announces to the police lieutenant standing nearby: "This was no accident or suicide. This man was murdered!" How does he know?

When yeast is added to a mixture of fruit or flour and sugar, it forms alcohol. How is this transformation accomplished? What happens when humans ingest sugar and fruit—do we synthesize alcohol?

Under certain extreme medical situations, patients may be nourished by the transfusion of glucose solutions directly into their blood. How does the body survive when the digestive system is bypassed in this way? In other words, how do the cells of the body utilize this glucose?

These and other questions will be answered in our discussion of cellular energetics.

THE NATURE OF THE PROBLEM

At its most fundamental level, life is a quest for energy. The fact that living organisms require a continual supply of energy is the answer to questions as diverse as why do lions stalk and kill zebras, why do most plants vie for the sunniest location—and why do we start thinking about what to eat for lunch only a few hours after breakfast.

In this chapter, we shall examine a number of aspects of energy acquisition and utilization at the cellular level. Recall that, in the last chapter, we learned *how* the cell is organized. Our examination of cellular energetics will not only help to explain *why* the cell is organized the way it is, but it will also provide the foundation for our discussion, in subsequent chapters, of the anatomy and physiology of human organ systems.

In order to see how the various components of our discussion of cellular energetics will be fitted together, consider the following questions and answers:

1. Why do cells need energy? (Because cells are subject to the laws of thermodynamics.)

2. That's not much of an answer, but I'll let it pass for the moment. OK, if cells need energy, where does it come from? (Cells obtain energy by importing energy-rich organic molecules from the environment, which means transporting them across the plasma membrane.)

3. All right, I'll buy that—but how does the cell actually *use* energy? (In the form of a special molecule called ATP, in which the energy released when the cell breaks down energy-rich organic molecules is stored for future use.)

4. You keep talking about these "energy-rich" molecules. How does the cell break them down? (Through the use of enzymes.)

5. Wait a minute. What's the relationship between ATP, enzymes, and the breakdown of organic molecules? (To answer that question, we need to take a look at something called "cellular respiration.")

6. So cellular respiration is the mechanism by which cells obtain energy? (Well, that's one of the ways—but some cells use another, more direct, method called photosynthesis.)

We shall consider each of these points in the order given.

THE CELL AND THERMODYNAMICS

Nowhere is the biological impact of the laws of physics more apparent than in the fact that the living cell needs a continual supply of energy. Obviously, cells need some external source of *matter* if they are to grow and divide; they cannot, after all, manufacture biochemicals out of nothing. But why do cells need a continual supply of *energy*? In constructing and disassembling complex molecules, why can't they simply recycle a one-time source of energy?

Both questions are answered by the laws of thermodynamics. Those imposing words mask two relatively straightforward principles. The **First Law of Thermodynamics** states:

> Energy cannot be created or destroyed, but it can be converted from one form to another.

Actually, this formulation of the law is not strictly correct; as Einstein projected, energy can be created from matter (as occurs when an atomic bomb explodes). However, since atomic explosions are somewhat beyond the capacity of individual living cells, the exception to the law has no pragmatic value to biological systems. Thus cells cannot create energy, but they can transform it (as occurs in the transformation of the energy present in the chemical bonds of a glucose molecule to the energy of movement produced by contracting muscle cells).

The **Second Law of Thermodynamics** states:

> When energy is changed from one form to another, the amount of usable energy decreases.

In other words, energy conversions are never 100 percent efficient; a certain fraction of the energy is always lost, typically as heat. For example, in burning gasoline, a car's engine converts less than 25 percent of the energy available in the gasoline to forward movement; the rest is lost as heat. The efficiency of chemical conversions within the cell is often higher, but, as the Second Law of Thermodynamics states, this efficiency never reaches 100 percent. Thus, because it cannot undertake chemical reactions without the loss of some energy as heat, the cell cannot endlessly recycle a one-time source of energy. To the contrary, it constantly needs to replenish its supply by importing energy from the environment, typically in the form of particular molecules that have large amounts of energy stored in their chemical bonds.

Obviously, if the cell needs energy (in the form of energy-rich molecules) from the environment, it also needs to have the capacity to move materials from the environment across its plasma membrane. However, this capacity poses potential perils. How does the cell prevent the same mechanisms that permit the uptake of materials from the environment from operating in reverse, causing the cell to lose essential materials back to the environment? If materials can flow in both directions across the plasma membrane, how does the cell avoid becoming, in a chemical sense, a mere subset of the surrounding environment?

TRANSPORT ACROSS THE CELL MEMBRANE

Cells have two primary methods by which they acquire (or lose) materials. One method takes advantage of basic laws of chemistry and physics and requires no expenditure of energy on the part of the cell. The second method attempts to thwart the basic laws of chemistry and physics and requires the expenditure of energy.

Transport Without Energy Expenditure

Diffusion is a basic law of physics that predicts the movement of molecules in fluids (that is, liquids and gases). The principle of diffusion states that molecules in a fluid will move from regions of high concentration to regions of low concentration (that is, *down* a concentration gradient) until the distribution of molecules is uniform. For example, a drop of food coloring placed in a beaker of water will gradually spread evenly throughout the beaker (see Figure 5.1). Diffusion is not based on some highly abstruse, theoretical model. Rather, it simply takes note of the fact that molecules in fluids are in constant motion and will, over time, assort themselves randomly.

Cells use the principle of diffusion to their advantage. Cells can acquire a substance by diffusion if

1. the substance is more concentrated outside the cell than inside the cell, and if

2. the substance is capable of passing across the cell's plasma membrane.

If the same criteria are satisfied in a reverse direction, cells can use diffusion to get rid of waste products. The acquisition of most salt ions and lipids by the cells of the small intestine and the elimination of the nitrogen-containing waste product urea (see Chapter 11) by the cells of the liver are accomplished by diffusion.

How are materials able to diffuse through the plasma membrane? There are fundamentally four methods:

1. Some substances, such as steroid hormones (see Chapter 13), are soluble in the lipids that are the primary constituents of the plasma membrane.

2. Some substances are small enough that they apparently simply slide between the lipid molecules of the membrane. Such substances are usually nonpolar (such as oxygen or carbon dioxide), or polar but uncharged (such as urea or ethanol).

3. Still other substances apparently move through channels created by proteins. For example, since the inside of the cell is negative relative to the outside, small positive ions are attracted from the environment and enter the cell through protein channels rather than through the lipid portion of the membrane (see Figure 5.2).

4. Finally, certain substances, such as glucose, are too large to pass between the lipid molecules of the membrane, and evidently cannot simply slide through a protein channel (because of size, chemical configuration, or electrical charge). In these instances, the substance binds with a specific membrane protein and, through a mechanism not yet completely understood (but perhaps involving rotation of the membrane protein), are deposited in the interior of the cell. This process is termed **facilitated diffusion,** because specific protein molecules in the plasma membrane *facilitate*, or make possible, the diffusion (see Figure 5.3).

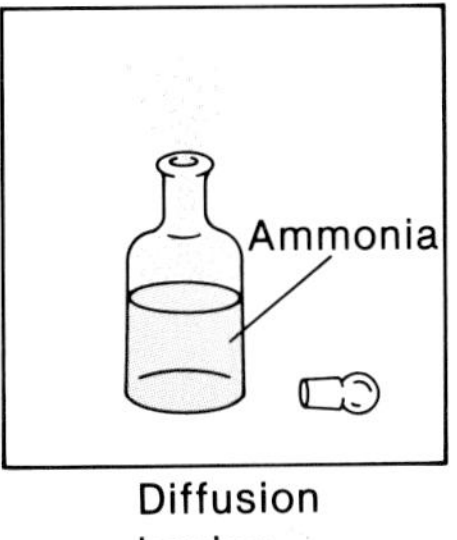

Diffusion begins.

Diffusion continues.

Equilibrium: no NET movement in any direction.

FIGURE 5.1

Diffusion
Molecules of fluids (liquids and gases) move freely and randomly from areas of high concentration down a concentration gradient to areas of low concentration until equilibrium is established. Here, a bottle of ammonia has been uncapped. Gradually, the ammonia molecules will disperse evenly throughout the room.

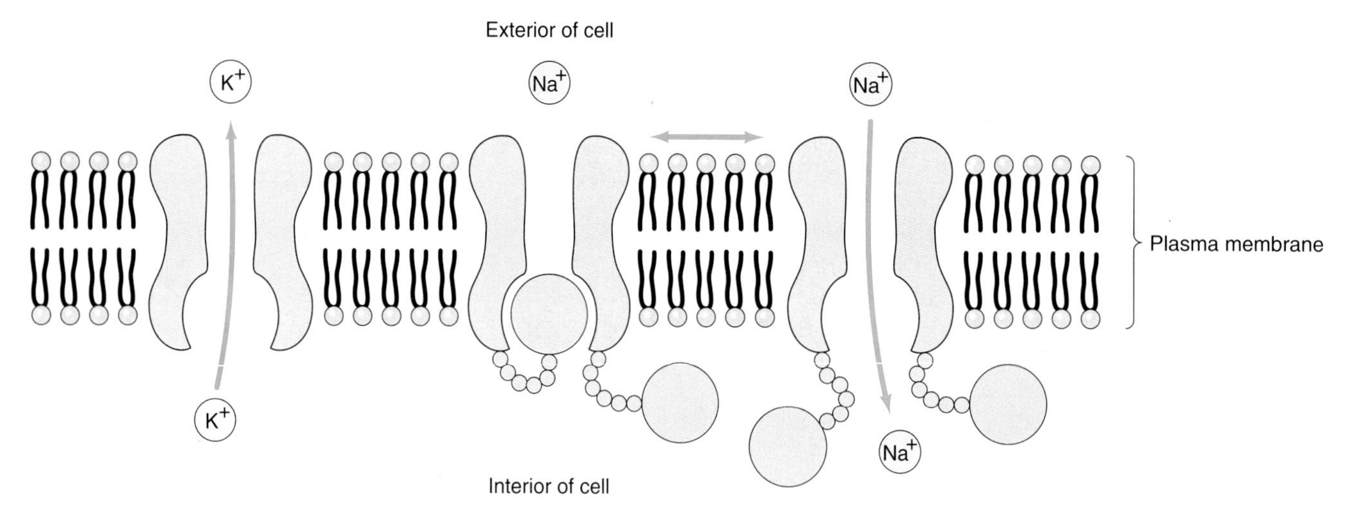

FIGURE 5.2

Gated and Ungated Protein Channels

Protein channels consist of a ring of protein molecules surrounding an opening through the plasma membrane. Most protein channels are characteristic for a particular ion. At least some of the potassium channels seem to be open more or less continuously. A majority of the sodium channels, on the other hand, are "gated" with a moveable protein plug. These gated channels become "ungated," or open, at critical points during the transmission of a nerve impulse along a nerve cell (see Chapter 14).

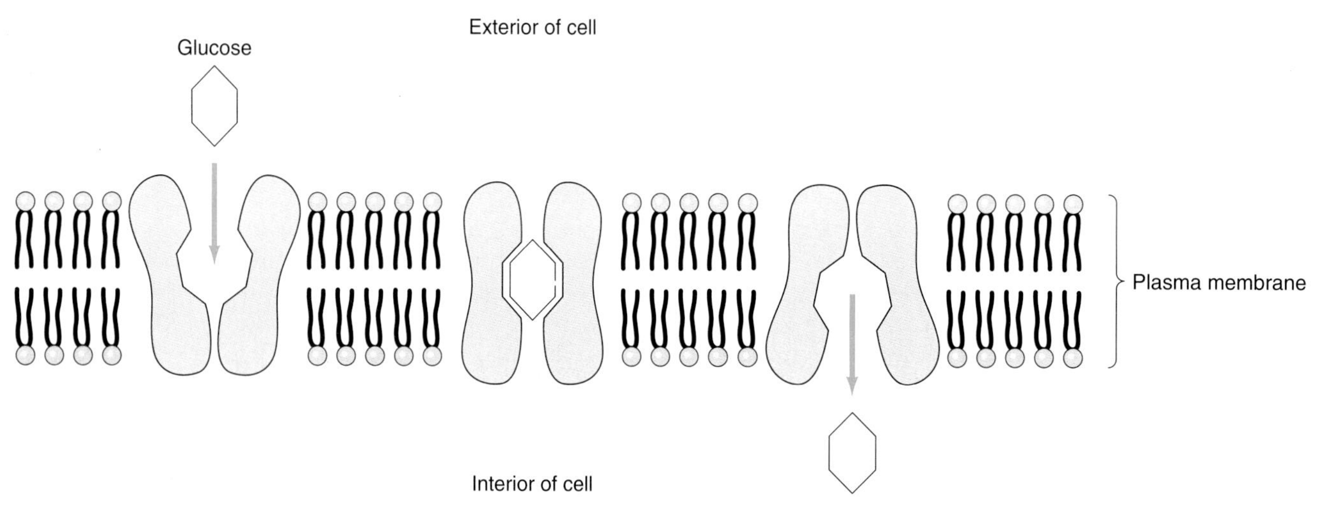

FIGURE 5.3

Facilitated Diffusion

Some molecules gain entry into the cell only by associating with particular proteins in the plasma membrane. In a non–energy-consuming manner, these protein molecules "facilitate" the movement of molecules (in this instance, glucose) into the cell, although the precise manner of their operation is still not well understood.

Interestingly, one molecule that moves freely back and forth across the plasma membrane is water, although precisely how this highly polar molecule moves across a largely lipid (and therefore hydrophobic) membrane is not presently known. Regardless, the diffusion of water across the plasma membrane is so important to life that the process is assigned its own name—**osmosis** (Gr. *osmos*, "impulse").

Because of osmosis, the concentration of molecules within a cell is frequently affected less by the movement of the molecules themselves than it is by the net movement of water into or out of the cell. For example, consider what happens when a cell is placed in a beaker of distilled water (see Figure 5.4). The contents of the cell may be thought of as a weak salt and protein solution; distilled water contains only water molecules. More water diffuses into the cell (through osmosis) than diffuses out because the water is more concentrated outside the cell than inside the cell (that is, movement is down the concentration gradient). Even if some salts diffuse out of the cell into the water of the beaker, the large molecules remaining inside the cell ensure a net movement of water into the cell until the cell finally bursts.

Now consider what happens when a cell is placed in a beaker of sea water. The salt concentration is higher in the sea water than it is in the cell. In other words, the water in the cell is more concentrated than the water in the beaker. Consequently, net movement of water is *from* the cell *to* the beaker (that is, from high concentration to low concentration), and the cell shrinks.

The movement of water across cell membranes is apparent in victims of drowning. A victim of drowning in fresh water typically shows some rupturing of the lung cells because inhaled water has moved by osmosis into the cells and has caused them to burst. A victim of drowning in salt water, on the other hand, shows no such damage to lung tissue because water has moved from the lung cells to the sea water in the lungs. Through microscopic examination of lung tissue, a pathologist can easily tell whether a victim drowned in salt water or in fresh water.

Transport Requiring Energy Expenditure

The two processes in which cells expend energy to move materials across the plasma membrane are distinguished primarily by scale:

1. In **active transport,** cells expend energy to move molecules or ions *up* a concentration gradient (see Figure 5.5). Moving molecules up a concentration gradient is akin to keeping a leaky bucket full of water— it can be done, but it requires lots of energy at the pump handle. Active transport is an important process in organisms because it is a principal means by which cells maintain internal concentrations of ions at levels different from those of the external environment.

The best-known example of active transport is the **sodium-potassium pump.** (Similar pumps exist for both calcium and hydrogen ions.) All cells have high internal concentrations of K^+ (potassium ion) and low internal concentrations of Na^+ (sodium ion) relative to their external environment, despite the fact that both ions can move through the protein channels in the plasma membrane. The cell in effect compensates for "leaks" by employing an active transport mechanism to pull K^+ back into the cell and to throw Na^+ out of the cell.

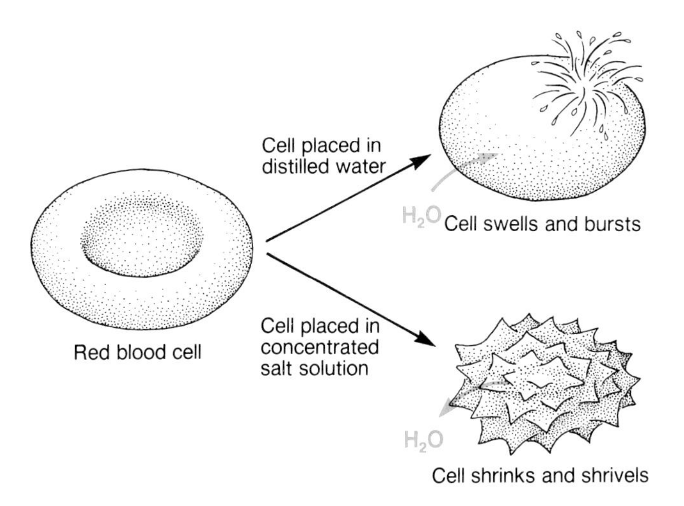

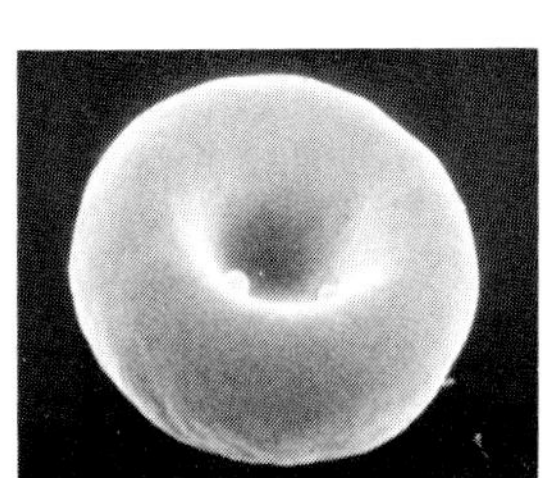

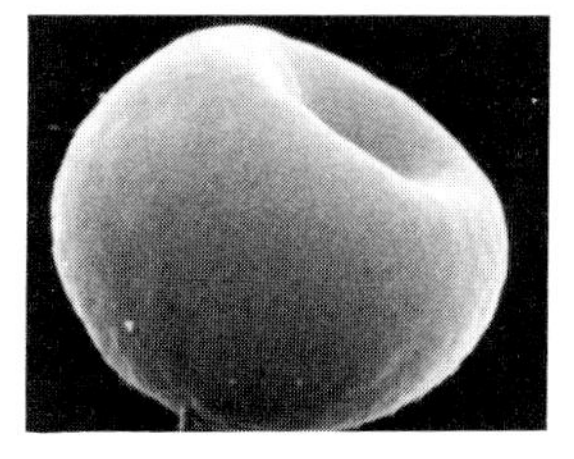

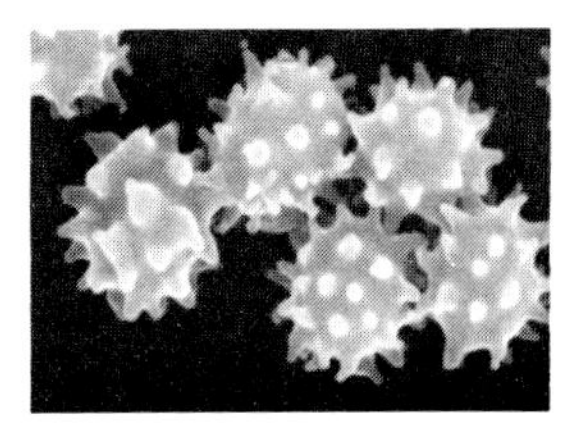

FIGURE 5.4

Effects of Osmosis on Red Blood Cells

Water moves from high concentration to low concentration by osmosis. If a cell is placed in distilled (that is, pure) water, water moves into the cell because water is more concentrated outside the cell than inside. However, if a cell is placed in sea water, water leaves the cell because water is more concentrated inside the cell than outside.

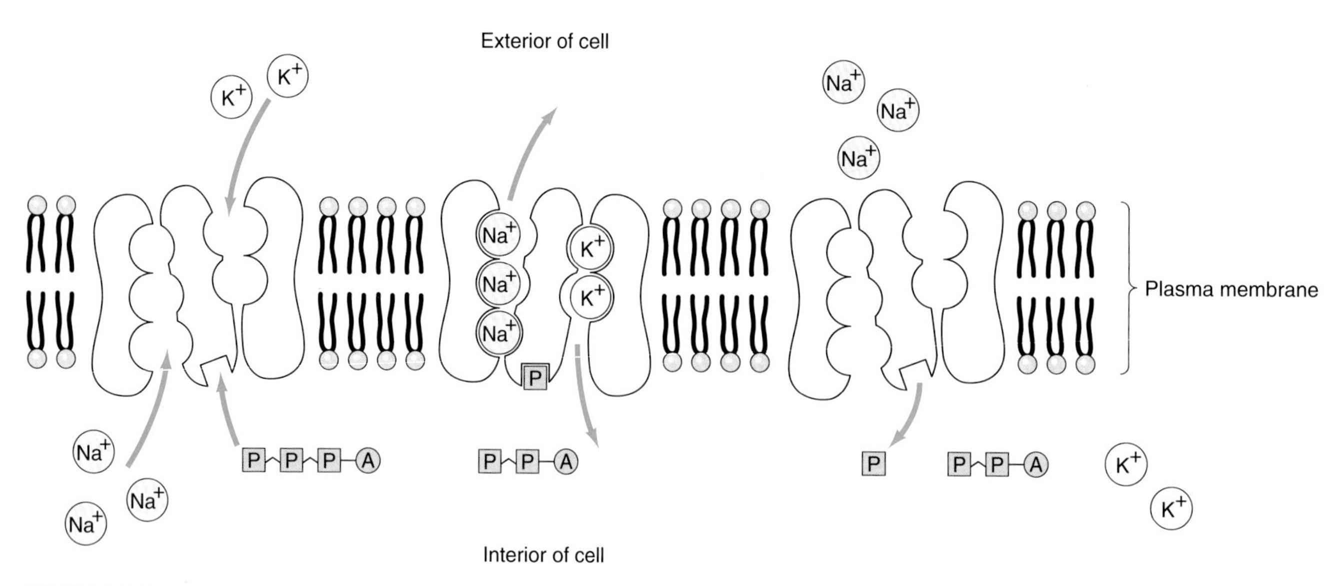

FIGURE 5.5

Active Transport: The Sodium-Potassium Pump
Although other protein channels for sodium and potassium exist, the regulation of sodium and potassium concentrations in the cell is largely achieved by the active transport of these two ions by an energy-consuming protein complex in the plasma membrane. For every molecule of ATP expended, three ions of sodium leave the cell and two ions of potassium enter the cell.

Precisely how the pump operates is not known; however, since cells whose energy-manufacturing machinery has been poisoned lose the ability to maintain the sodium differential, it is clear that the pump is indeed an energy-requiring mechanism. Moreover, we know that a plasma membrane protein is involved and that 2 K^+ are captured and 3 Na^+ are expelled for each energy molecule that is expended. (The fact that more positive ions are expelled than are captured accounts in part for the interior of the cell's being negative relative to the external environment.) We also know that each plasma membrane protein molecule involved in the pump can transport up to 300 Na^+ per second, and that the resting cell uses more than a third of all the energy it expends simply in running the sodium-potassium pump.

The sodium-potassium pump plays a particularly important role in the functioning of nerve cells (see Chapter 14).

2. **Endocytosis** (Gr., "within the cell") is an energy-requiring process that occurs when relatively large masses of material are passed into the cells through the formation of a **food vacuole,** a pinched-off piece of plasma membrane that surrounds the ingested material (see Figure 5.6). If the ingested material is large and visible under the microscope, the process is called **phagocytosis** (Gr., "cell eating"). Phagocytosis occurs, for example, when bacteria are engulfed by certain white blood cells (see Chapter 9). If the ingested material is in dissolved form, the process is termed **pinocytosis** (Gr., "cell drinking").

Pinocytosis is the only known mechanism whereby cells can take up certain very large molecules. Vitamin B_{12}, for example, is a molecule abundant in many types of food and it is essential for life (see Chapter 7), but it is too large to diffuse through the plasma membrane of the cells lining the small intestine. *Intrinsic factor,* a large protein produced by the stomach, binds with vitamin B_{12} and protects it from being destroyed by digestive enzymes. The combined molecules then attach to specific receptor sites on the cells of the intestinal lining and, once a sufficient number are present, a pinocytotic vesicle is formed and the molecules are taken into the cell (see Figure 5.7).

Individuals who have had large portions of their stomachs surgically removed because of cancer or ulcers frequently do not produce adequate amounts of intrinsic factor. Ultimately, they may show signs of vitamin B_{12} deficiency in the form of a type of anemia known as **pernicious anemia.** Fortunately, this condition is easily remedied by injection of vitamin B_{12}.

Exocytosis (Gr., "outside the cell") is an energy-requiring process that is the reverse of endocytosis. Through exocytosis, materials such as the indigestible

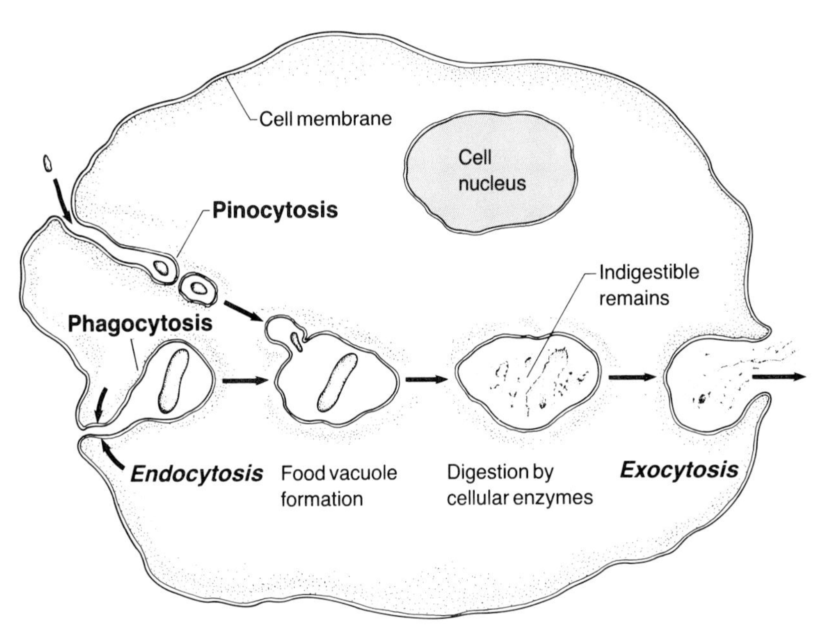

FIGURE 5.6

Endocytosis and Exocytosis

Endocytosis is a pinching in of a portion of the cell membrane, either as pinocytosis or phagocytosis, and the subsequent formation of a vesicle or food vacuole. The indigestible remains of a food vacuole are removed from the cell by exocytosis.

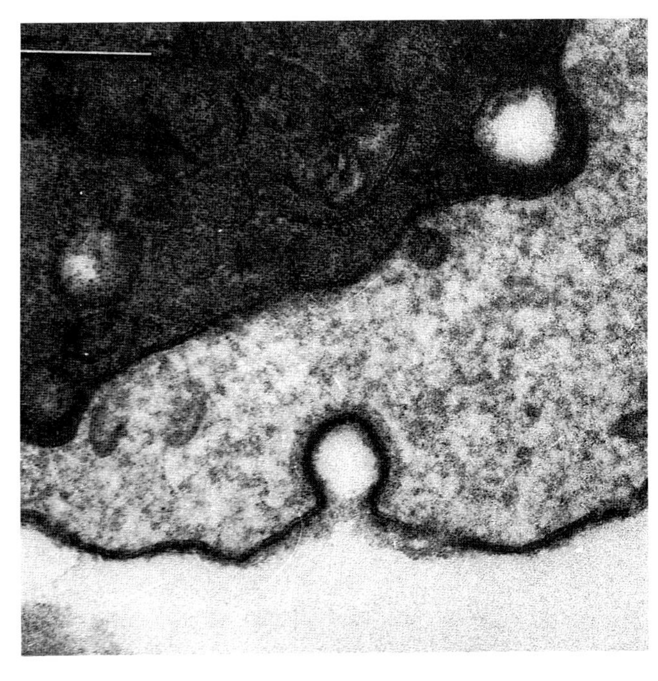

FIGURE 5.7

Pinocytosis

Pinocytosis begins with the movement of plasma membranes receptors for a particular molecule into an aggregation. As molecules attach to their receptors, the whole conglomeration of molecules begins to sink into the cell, a formation called a coated pit. The pit is subsequently pinched off to form a vesicle.

remains of food vacuoles are carried to the surface of the cell and expelled.

To summarize where we are at this point, we know that cells employ a variety of methods for moving substances across the plasma membrane, although ultimately these methods fall into two categories—those which require energy and those which do not. But what is meant by "energy-requiring"? What, precisely, is the cell doing when it "expends energy"—and where does the cell get this energy in the first place?

CAPTURING ENERGY: THE ROLE OF ATP

Cells store energy in the form of various organic molecules (certain carbohydrates and lipids being particularly favored for reasons we shall discuss in Chapter 7). It may seem odd—or even confusing—to talk of storing energy in the form of specific molecules, because that seems to imply that cells, in order to extract the energy, must somehow convert matter to energy. Of course, since cells possess no such thermonuclear capacity, there must be another explanation—and there is.

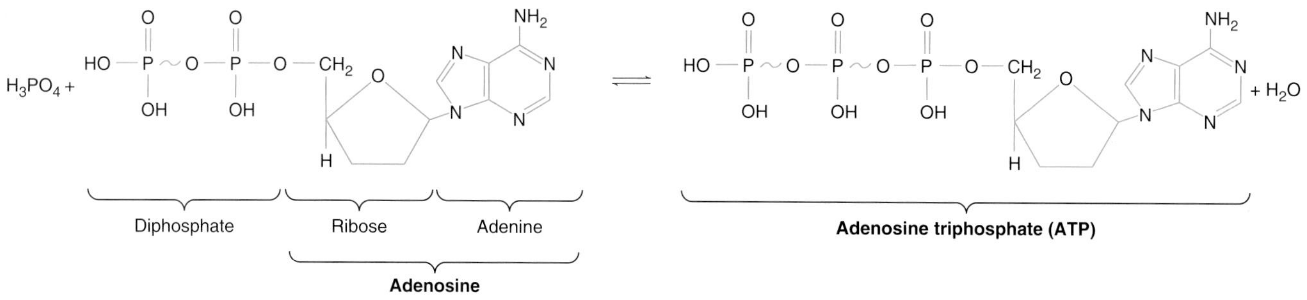

FIGURE 5.8

ATP

ATP is composed of a molecule of adenosine (ribose sugar plus adenine) and three phosphate groups. The last two phosphate groups are attached by so-called high energy bonds (wavy lines), and one or both may be detached during energy transfers to other molecules. Thus the cell expends energy when it converts ATP to ADP plus a phosphate group; it stores energy when it is able to attach a phosphate group to ADP and make ATP.

When the covalent bonds of a typical organic molecule are broken, and the resulting pieces are recombined into simple molecules such as carbon dioxide and water, a substantial amount of energy is released. We demonstrate this principle whenever we throw another log on the fire—the energy given off as heat by the burning log comes from the breaking of the covalent bonds of the cellulose molecules (the giant carbohydrates that comprise the bulk of the log).

Monosaccharides also possess considerable energy in their covalent bonds. For example, the "burning" (conversion to carbon dioxide and water) of 180 g of glucose yields 686 Calories of heat energy[1]. (A **Calorie** is the amount of energy needed to raise the temperature of 1000 g of water by 1° C—but you probably are more familiar with Calorie in such phrases as, "You must expend 7500 Calories in order to lose 1 kg of fat.")

Of course, cells don't "burn" glucose any more than they convert matter directly to energy through thermonuclear processes. Even if cells had the capacity to burn organic molecules, it would do them very little good, because they would have no way of directing the heat energy produced into processes that assemble, disassemble, or move specific molecules. Rather, what cells require is a way of focusing—or *transporting*—the energy derived from the destruction of glucose (or other organic molecule) to the molecules that are going to be used to create some required end product or function.

This process of transporting energy is solved by the use of the molecule **adenosine triphosphate** (more commonly known as **ATP**)—a molecule so essential to life that it is used by every type of living organism on earth.

ATP is very similar to a short piece of a nucleic acid (see Figure 5.8). It consists of a nitrogenous base, *adenine* (one of the five types of nitrogenous bases found in nucleic acids), to which is linked the same type of sugar as is found in RNA. When linked together, the adenine and the sugar become *adenosine*. Attached to the sugar are three phosphate groups (phosphates are also found in nucleic acids). Moreover, ATP is a molecule simple enough to have been synthesized in experiments simulating the early conditions on earth. As a result, we can reasonably assume that ATP was available for use by the first organisms.

One of the more difficult concepts we will encounter in the study of biology is the mechanics of ATP activity. It is one thing to learn that ATP is an energy transport and short-term energy storage molecule used in the construction of new, generally large, molecules by the cells of the body. It is quite another thing to visualize just how ATP works.

The three phosphates of ATP are bonded in series. The last, or terminal, phosphate is bonded by a covalent bond that is relatively easily broken. The terminal phosphate group is detached from the ATP molecule and attached to one of the molecules being used in a particular chemical reaction, at which point its energy is transferred to the chemical bonds of the forming molecule. Thus ATP supplies the *activation energy* (discussed in the next section) needed to power a reaction. Conversely, energy released in the breakdown of various organic molecules is used to attach a free phosphate group to a molecule of **ADP** (adenosine **di**phosphate), which is identical to ATP except that it has only two attached phosphate groups. When

[1] "Calorie" with a capital "C" is the equivalent of 1000 calories (with a lower case "c"). For that reason "Calorie" is also known as "kilocalorie." The "calories" of diet books are actually kilocalories.

the free phosphate group is attached to ADP, a molecule of ATP is formed. This results in a reversible reaction:

ADP + Ⓟ + Energy ↔ ATP

Remember, when we talk subsequently about "producing" or "using" ATP, we are not speaking of manufacturing the molecule out of thin air, or of destroying it by reducing it to a series of shell-shocked atoms. Rather, we are speaking of either joining Ⓟ to ADP, or of freeing Ⓟ from ATP.

By way of briefly summarizing our discussion, we now know that ATP is to a cell what electricity stored in your automobile battery is to your car—it's the "juice" that powers a host of different, and essential, activities. However, we also know that sugar molecules do not simply fall apart on their own, creating ATP in the process. (If you think otherwise, try watching a sugar cube for a week or two, just to see how much ATP it creates.) Since covalently bonded molecules are generally stable, how does the cell break them down for purposes of energy production?

Consider the role of enzymes.

ENZYMES

In order for two molecules to engage in a chemical reaction, the reactive portions of the molecules must come into close proximity, and they generally must do so with considerable velocity—a head-on collision, as it were. Because so many of the molecules of the body are extremely large, because their reactive portions are typically very small, and because they seldom possess the necessary velocity, the chance that the reactive sites will randomly come into proximity with sufficient energy to cause a chemical reaction is often very slight.

The energy needed to initiate a chemical reaction is called the **activation energy.** Because the motion of molecules increases with temperature, one way in which sufficient activation energy might be provided to start a given reaction would be to raise the temperature of the surrounding medium. (Indeed, the rate of most chemical reactions doubles for every increase of 10° C.) In making bread, for example, the dough is warmed to increase the rate at which yeast breaks down sugar, a process that releases the carbon dioxide responsible for causing the dough to rise. However, elevating the temperature cannot be done indefinitely—the yeast organisms die at temperatures above 50° C. As far as our own body chemistry is concerned, at 37° C we are already within a few degrees of our upper survivable limit—thus, increasing our body temperature to speed up the rate at which chemical reactions occur within our cells is simply not practical.

Yet at normal body temperatures, many vitally important chemical reactions simply do not occur rapidly enough—or, for all intents and purposes, do not occur at all. What, then is the alternative to raising our internal temperature?

The answer is to use **enzymes.** Enzymes are globular proteins that **catalyze** (increase the rate of) specific chemical reactions by reducing or eliminating the need for activation energy (see Figure 5.9). Like all catalysts, enzymes are not used up in the chemical reactions they promote; therefore only tiny amounts are needed to facilitate a particular reaction (see Box 5.1).

How do enzymes function? For one thing, shape is important. Molecules to be catalyzed by a given enzyme must be capable of fitting into the **active site** of the enzyme. Like the glass slipper and Cinderella's foot, each enzyme is highly specific for the reactant molecules of a particular reaction—which is why we have over 2000 different enzymes in our bodies!

Imagine two bowling balls, each with a white spot the size of a dime. The bowling balls represent large organic molecules; the white spots, their reactive sites. If you were to roll the bowling balls randomly around your living room, you might wait for weeks before the two white spots touched. But suppose you had a bowling ball holder that picked up the balls and oriented them so that the white spots touched each other. In a crude fashion, this is how enzymes work. By virtue of its specific shape, a given enzyme "recognizes" molecules of a matching configuration and

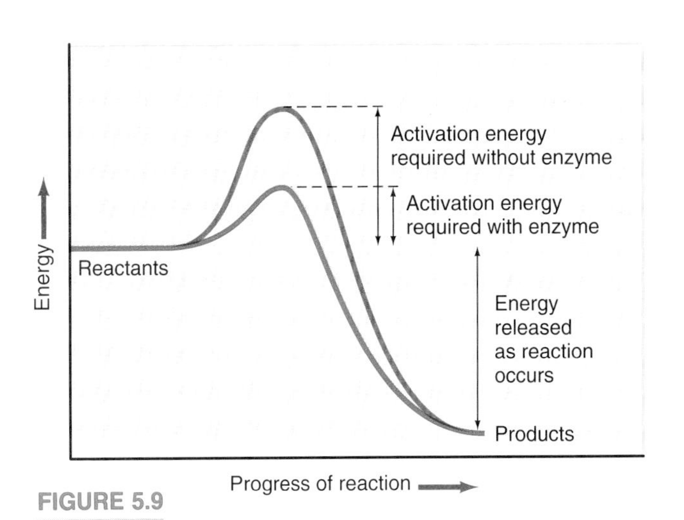

FIGURE 5.9

Enzymes and the Energy of Activation

Enzymes reduce the amount of activation energy necessary to power a reaction, thus making it much more likely to occur.

BOX 5.1

Enzymes and the Candy Man

How are liquid-center chocolate-covered cherries made? If your answer is, "By careful injection of a sugar solution using a hypodermic needle," you get points for ingenuity, but you are wrong. The chocolates are solid at the time of manufacture, and become liquid-centered only later.

Candy manufacturers add small amounts of the enzyme *invertase* to the fondant surrounding the cherry. Invertase catalyzes the conversion of sucrose (table sugar) to glucose and fructose. Because glucose and fructose have a high affinity for water, they gradually absorb water from the surrounding coating of chocolate until they go into solution—meaning that the original sugar coating of the cherry has been transformed into a thick syrup through the action of an enzyme.

either causes a molecule to be broken into two smaller pieces or brings two molecules together to form one large molecule.

Actually, it's a bit more complicated than that. In temporarily combining with **substrate** molecules (a substrate is a molecule destined to be chemically changed by enzymatic action), the enzyme distorts the configuration of the substrates, destabilizing them momentarily, and thereby allowing them to react chemically (see Figure 5.10). When we say "tempo-

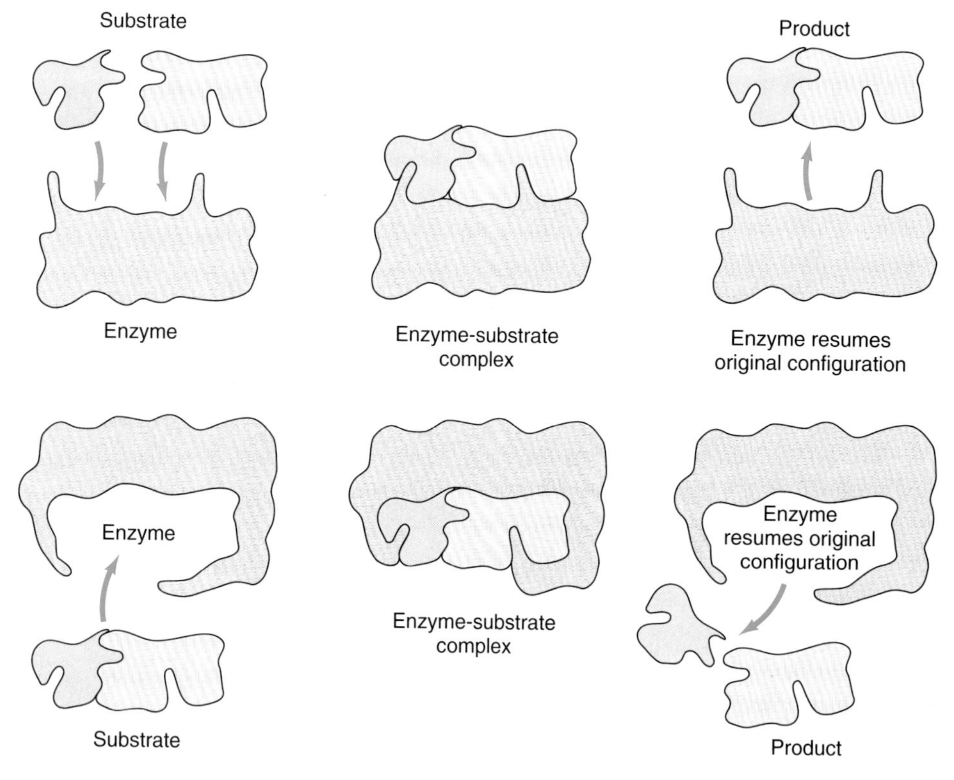

FIGURE 5.10

Enzymes and Substrates

Enzymes may either assemble or disassemble molecules. Generally, each enzyme can react with only one substrate, meaning that a different enzyme is required for each metabolic reaction. Note that as enzymes and substrates combine, there is a momentary change in the configuration of the enzyme that temporarily destabilizes the substrate, permitting it to become chemically reactive.

rarily" or "momentarily", we mean just that—for some enzymes, each molecule is capable of catalyzing over 100,000 reactions per second!

Certain enzymes do not function unless other molecules or ions are associated with their active site. When the assisting substance is an ion, such as Ca^{2+}, it is called a **cofactor;** when it is an organic molecule, it is called a **coenzyme.** Many *vitamins* (see Chapter 7) function as coenzymes—and now that you understand the importance of enzymes, you can readily see why it is that vitamins are essential to our well-being.

So where are we now? We know that the cell is capable of importing a variety of substances—including energy-rich molecules such as glucose—across its plasma membrane. We know that the cell uses enzymes to break apart these chemically stable energy-rich molecules in order to extract their energy. We also know that the cell stores this extracted energy in a special storage molecule called ATP. What we *don't* know at this point is *how* the cell choreographs these three steps.

GLUCOSE METABOLISM

Before examining the details of glucose metabolism, it will be useful to take a quick overview of the process (see Figures 5.11 and 5.12).

The first series of reactions in glucose metabolism is called **glycolysis** (Gr. *lysis,* "splitting" of glucose).

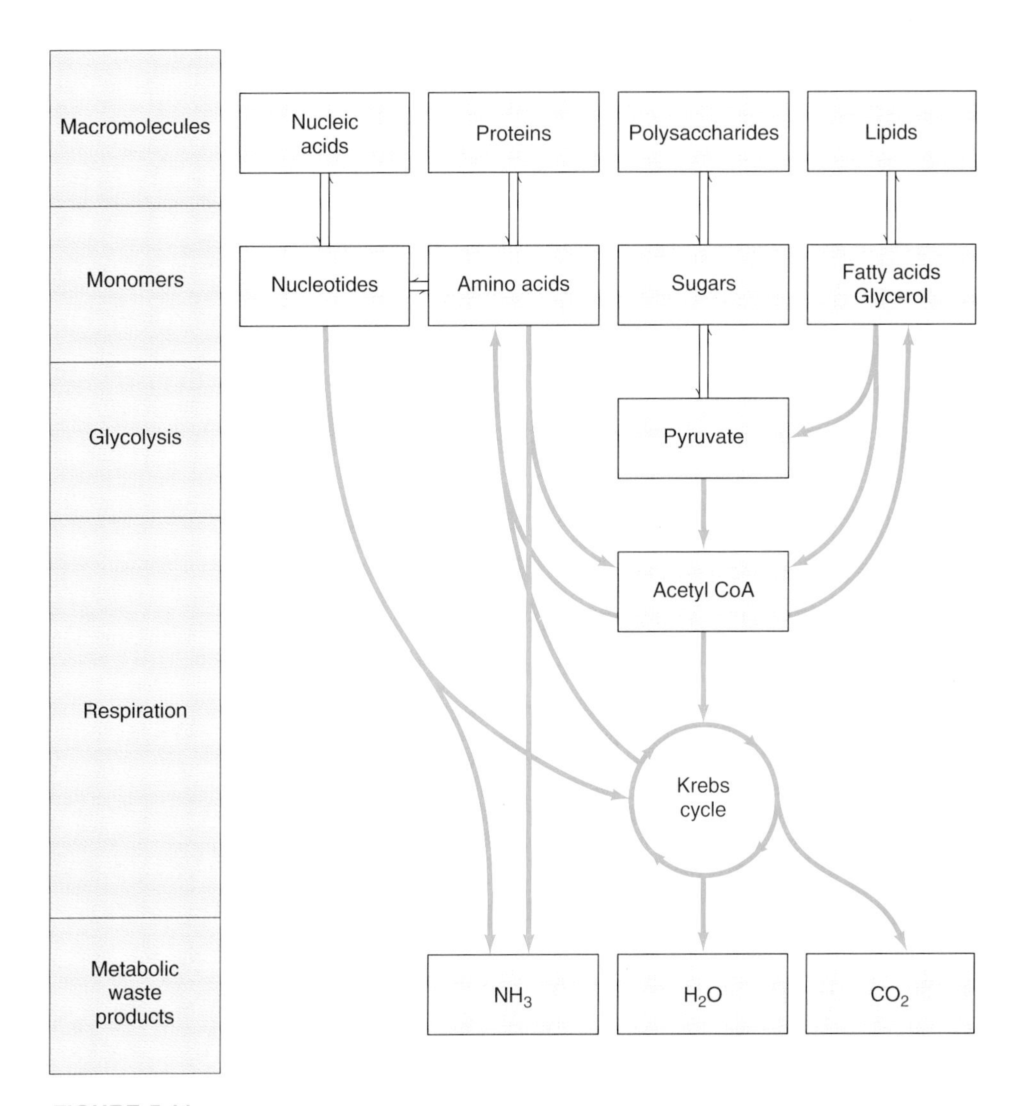

FIGURE 5.11

Glucose Metabolism

The overall conversion of polysaccharides for energy use is shown diagrammatically. Note, however, that nucleic acids, lipids, and proteins in excess structural need can also be transformed for energy use. In addition, a substantial amount of interconversion from one class of molecules to another is also possible.

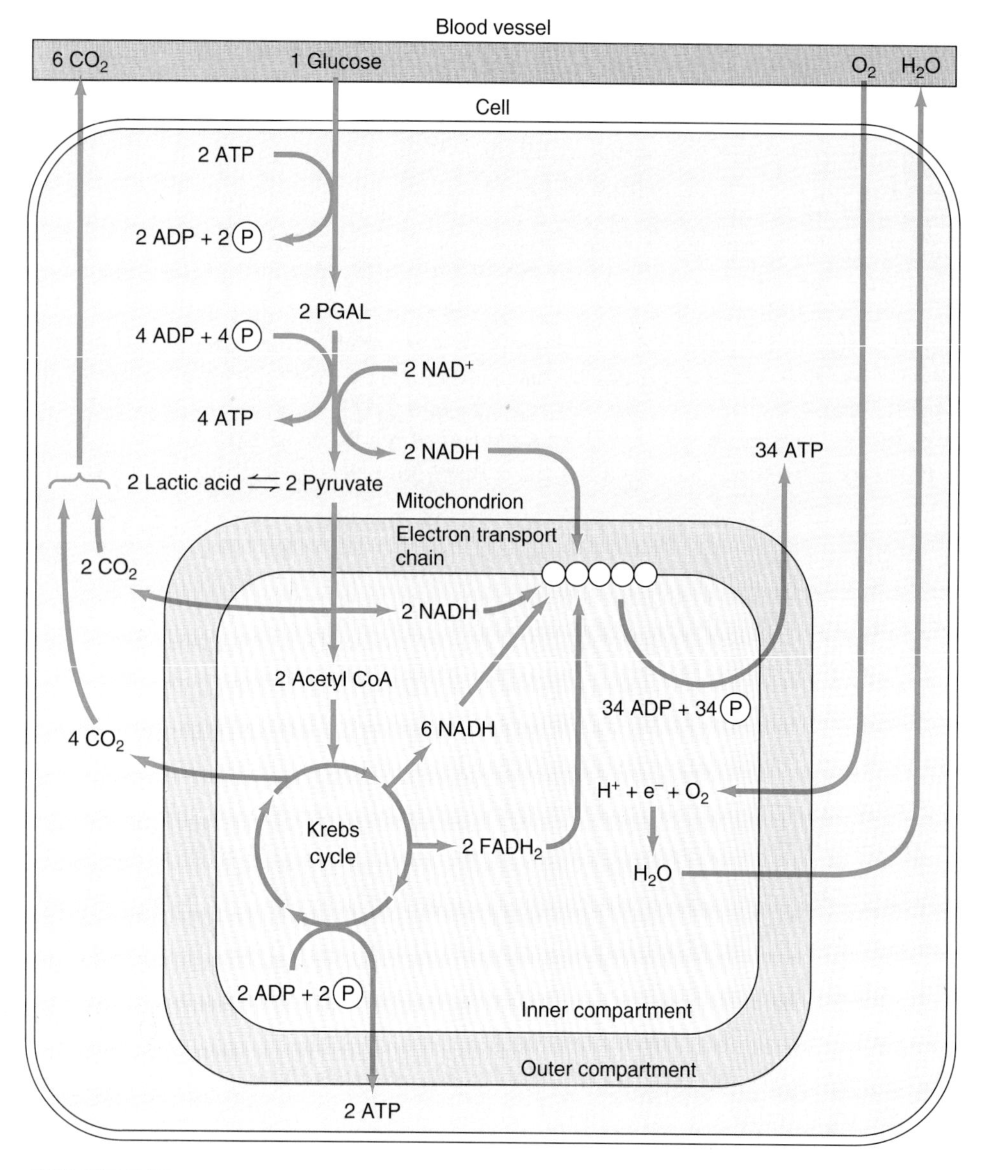

FIGURE 5.12

Summary of Glycolysis and Cellular Respiration

Glucose from the blood is taken up by a cell and converted by cytoplasmic enzymes to pyruvate, with a modest gain in ATP. Pyruvate is then passed into a mitochondrion where it enters the Krebs cycle, and the hydrogens from the pyruvate are transferred to the electron transport chain. As hydrogen ions move from the outer compartment of the mitochondrion into the inner compartment (where they combine with electron and with oxygen to form water), ATP is generated.

Glycolysis is a multi-step process that occurs in the cytoplasm of cells, and it does not require oxygen. The final step in glycolysis is the formation of two 3C molecules called *pyruvate*. Therefore the process does not decrease the number of carbon atoms still present in organic molecules ($C_6 \rightarrow 2C_3$). Pyruvate stands at an important branching point in carbohydrate metabolism, because different cells metabolize it in different ways.

Under **anaerobic** (no oxygen) conditions, pyruvate is **fermented.** In *alcoholic fermeniation,* which occurs in many simple organisms (for example, in brewer's yeast), pyruvate is converted to ethanol and carbon dioxide. In *lactic acid fermentation,* which occurs, among other places, in animal tissues with an insufficient supply of oxygen (for example, in muscle cells during extreme exercise), pyruvate is converted to *lactic acid.*

Under **aerobic** (oxygen present) conditions, pyruvate enters into a series of reactions occurring inside mitochondria. These reactions are known collectively as the **Krebs cycle,** named for Sir Hans Krebs, who in 1937 first elucidated their cyclical nature. (He was awarded the Nobel Prize for this work in 1953.) In the Krebs cycle, derivatives of pyruvate are further broken down by the removal of one carbon fragment in the form of carbon dioxide.

The metabolism of pyruvate in the Krebs cycle also provides electrons and hydrogen ions that are transferred to a succession of electron carriers (collectively termed the **electron transport system**) in the inner membrane of mitochondria. As the electrons cross the mitochondrial membrane (at which point they are reunited with their hydrogen ions and with atmospheric oxygen to form **metabolic water**), they release energy which is used to produce ATP. This use of oxygen, and the formation of carbon dioxide and metabolic water, are fundamental characteristics of the process known as **cellular respiration.**

Now let's examine the processes of glycolysis and cellular respiration in greater detail.

Glycolysis

The progressive breakdown of glucose in glycolysis begins with preparatory reactions that "activate" the molecule and make it possible for the succeeding reactions to take place. ATP molecules are used in these preparatory reactions, and their terminal phosphate groups are attached to the carbohydrate molecules in a process called **phosphorylation.**

The nine steps of glycolysis are as follows (see Figure 5.13):

1. A molecule of ATP is used to phosphorylate glucose to *glucose 6-phosphate* (the phosphate group is attached to the number 6 carbon of the glucose molecule).

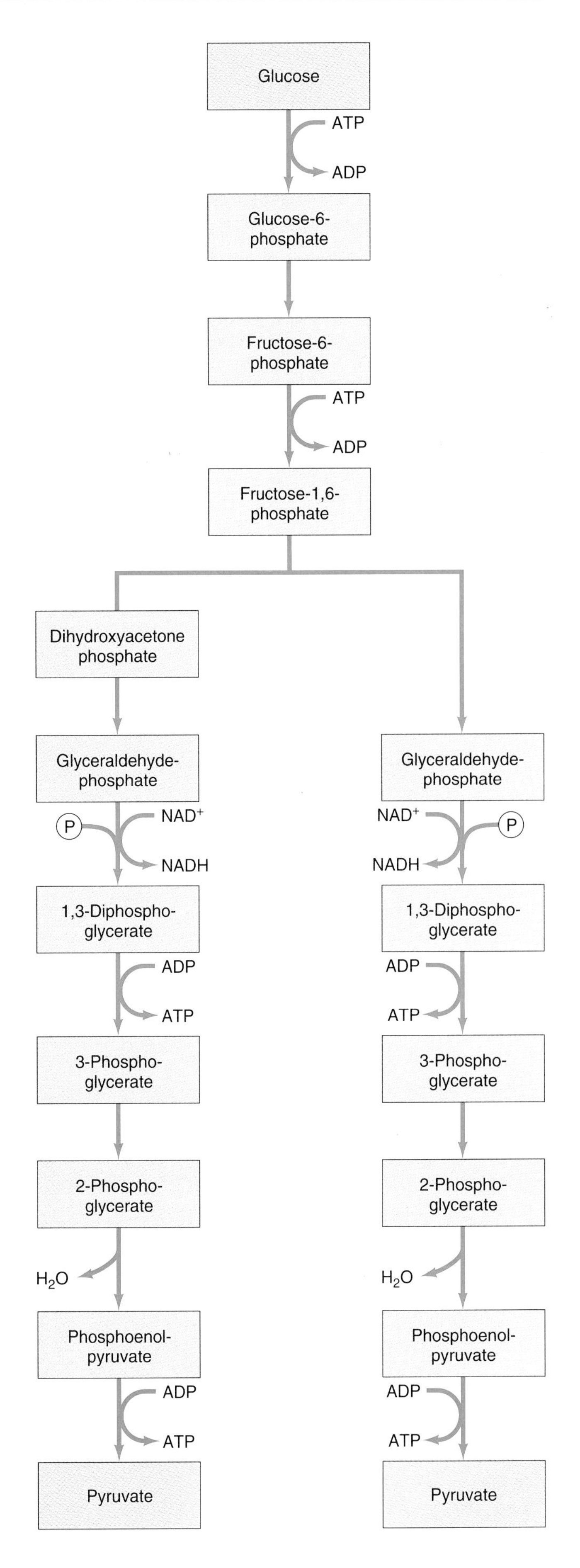

FIGURE 5.13

Glycolysis
Glycolysis consists of a series of nine transformations, as a single molecule of glucose is ultimately transformed to two molecules of pyruvate. These transformations occur in the cytoplasm, do not require oxygen, and produce a net gain of two ATP molecules for every glucose molecule transformed.

2. The molecule is reorganized to form *fructose 6-phosphate* (fructose is a 6 carbon sugar closely related to glucose).

3. A second molecule of ATP is used to phosphorylate fructose 6-phosphate to *fructose 1,6-diphosphate* (the phosphate group from the second ATP is added to the number 1 carbon).

4. Fructose 1,6-diphosphate is split into two interconvertible 3 C molecules, *dihydroxyacetone phosphate* and *glyceraldehyde phosphate* (or GAP); dihydroxyacetone phosphate is then converted to GAP.

5. In an energy-yielding reaction, each GAP gives up a hydrogen ion (that is, a proton) plus two electrons to an electron carrier molecule, *nicotinamide adenine dinucleotide* (or *NAD^+*) to form *NADH;* a second H^+ goes into solution in the cytoplasm; and a free phosphate group attaches to the number 3 carbon of the GAP molecules, thereby producing the unstable intermediate compound *1,3-diphosphoglycerate.*

6. This attached phosphate is released and combined with a molecule of ADP, to form ATP and *3-phosphoglycerate.*

7. The remaining phosphate is transferred to the number 2 carbon to form *2-phosphoglycerate.*

8. A molecule of water is removed from the 2-phosphoglycerate molecule to yield *phosphoenolpyruvate;* the energy yielded from the water loss is concentrated on the bond connecting the remaining phosphate.

9. The remaining phosphate is removed from phosphoenolpyruvate and combined with ADP to form ATP and *pyruvate.*

The net equation for glycolysis is therefore:

$$\text{Glucose} + 2\,NAD^+ + 2\,ATP + 4\,ADP + 2\,Ⓟ \rightarrow$$
$$2\,\text{Pyruvate} + 2\,NADH + 2\,H^+ + 2\,ADP + 4\,ATP + 2\,H_2O$$

Although the formation of pyruvate is technically the last step in glycolysis, it is not the last step in living organisms, since the pyruvate will either enter the Krebs cycle as a part of respiration or be fermented to lactic acid or ethanol (see Figure 5.14). Regardless of whether the end product is lactic acid or ethanol, the need for fermentation is the same: NAD^+ must be recovered in order to permit glycolysis of additional glucose molecules, and that means that NADH must be reconverted to NAD^+. NADH sheds a hydrogen ion and two electrons by donating them to pyruvate, thereby converting the pyruvate either to lactic acid or to ethanol plus carbon dioxide.

What has been the total energy gain for organisms undertaking glycolysis in anaerobic conditions? The answer is just 2 ATP per molecule of glucose (a total yield of 4 ATP minus the 2 ATP required for early phosphorylation steps). This amounts to little more

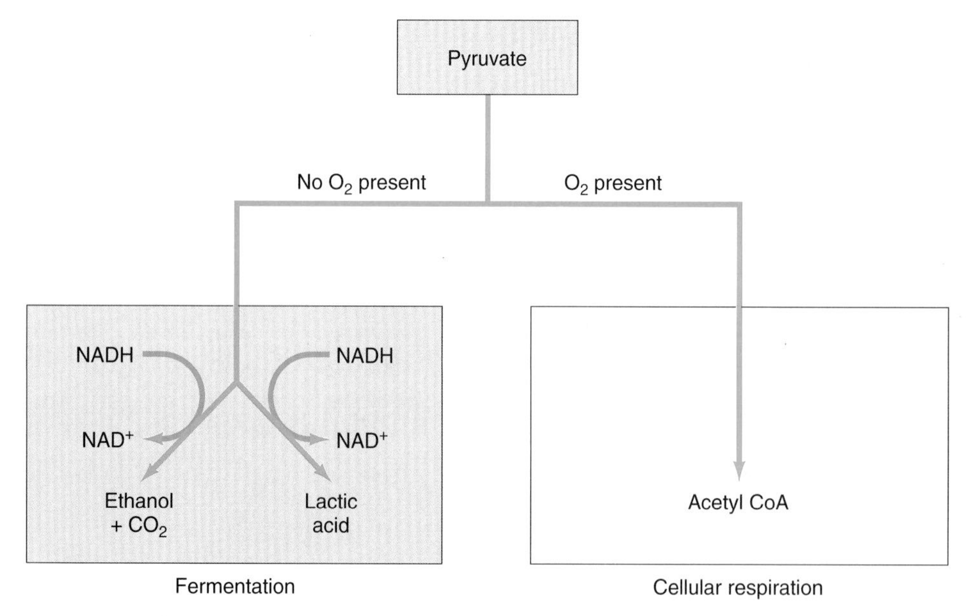

FIGURE 5.14

Fermentation

Pyruvate has two possible fates. In the presence of oxygen, it is respired by being converted to acetic acid and combined with coenzyme A to be fed into the Krebs cycle. In the absence of oxygen, it is fermented by being transformed either to ethanol plus carbon dioxide or lactic acid. Either of the fermentation transformations recovers the NAD^+ necessary for additional glucose breakdown.

than 2 percent of the total chemical energy bound up in a molecule of glucose. Most of the energy in the glucose molecule remains in the end products of fermentation—ethanol or lactic acid.

Cellular Respiration

Cells that are capable of using oxygen in respiration obtain a much greater energy yield from glucose metabolism. In respiration, pyruvate—the product of glycolysis—is not converted to lactic acid or ethanol, but instead is completely disassembled (see Figure 5.15).

As a first step, pyruvate crosses both the outer and inner membranes of the mitochondria from the cytoplasm, its point of origin. Before actually entering the Krebs cycle, a molecule of carbon dioxide is split off from the pyruvate. In addition, a hydrogen ion plus two electrons are attached to a molecule of NAD^+, forming NADH (a second H^+ goes into solution), and the remaining two-carbon fragment (*acetyl group*) attaches to a carrier molecule called *Coenzyme A.* This complex of the two-carbon fragment and Coenzyme A is referred to as *acetyl CoA,* and it is now ready to enter the Krebs cycle.

We will speak of the "turns" of the Krebs cycle as if it were a wheel, but clearly this is only an analogy. A "turn" of the Krebs cycle is illustrated in Figure 5.15. An acetyl group (2 C) from acetyl CoA enters the Krebs cycle and combines with a 4 C molecule (*oxaloacetic acid*) to produce a 6 C molecule (*citric acid*). Subsequent reactions of the Krebs cycle produce two molecules of carbon dioxide and reduce the length of the carbon chains from 6 C to 4 C. At this point, we are once again at the starting point, and the 4 C molecule is again available to combine with another acetyl group.

FIGURE 5.15

Cellular Respiration

Cellular respiration takes place in the mitochondria. The two-carbon acetyl fragment is joined with a four-carbon molecule to create a six-carbon molecule. In successive transformations, the two carbons from the acetyl are split off as carbon dioxide, and the hydrogens attached to carrier molecules that transport them to the electron transport chain. Further reconfigurations regenerate the original four-carbon molecule, which can then pick up another acetyl fragment and begin the cycle anew.

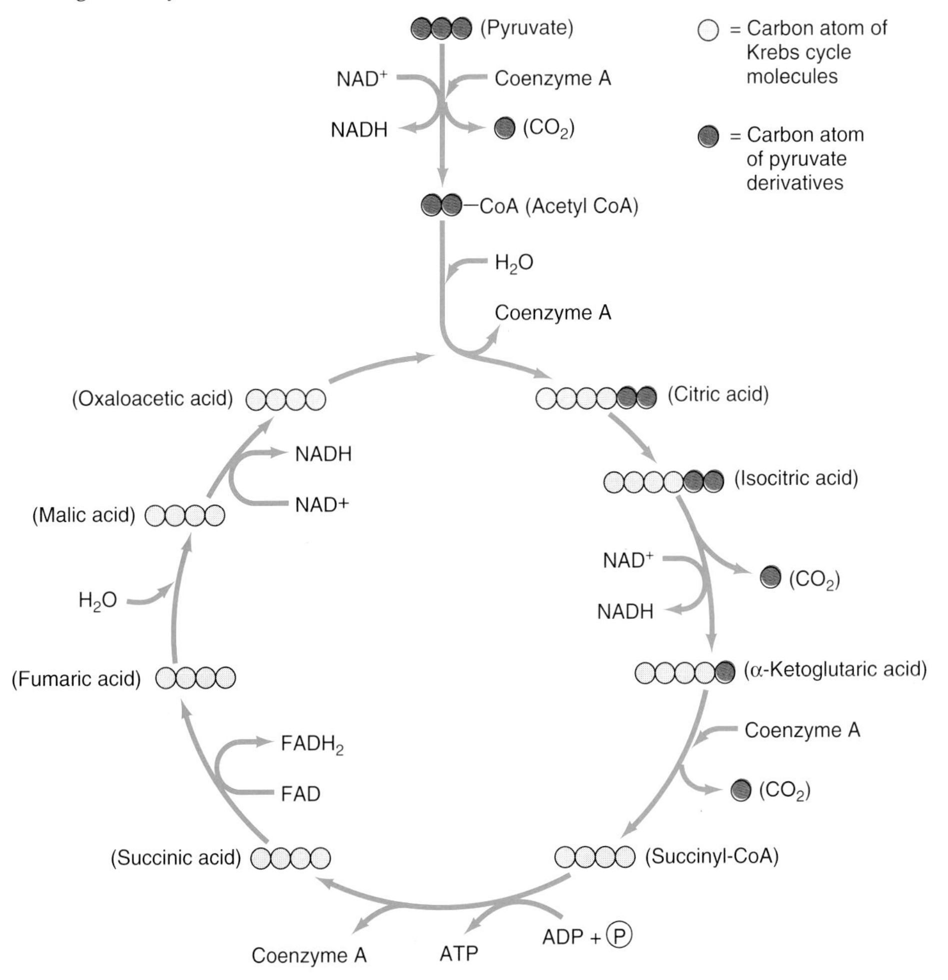

So, you might ask, is the point of this whole process merely to convert glucose to carbon dioxide, a waste product? Obviously not. It is not the *carbon* that is important in respiration; it is the *hydrogen*—both the protons (which are all an H^+ consists of) and the electrons, because it is the ultimate reunification of the protons and the electrons of hydrogen atoms that produces most of the ATP. How and where is this done?

At four distinct points in the Krebs cycle, pairs of electrons are transferred to electron carriers. In three instances, the compound serving as the electron carrier is NAD^+; in the fourth, the electron carrier is *flavin adenine dinucleotide* (or *FAD*). For every pair of electrons passed from NADH through the transport system, three molecules of ATP are formed; for electrons carried by FAD (as $FADH_2$), only two molecules of ATP are formed. In each "turn" of the Krebs cycle, one molecule of ATP also forms directly from the cycle. Adding all of these together, there is a total of 15 molecules of ATP formed for every molecule of pyruvate entering the Krebs cycle. The grand total, including the yield from glycolysis, is 36 ATP per starting molecule of glucose (see Figure 5.16). This is almost 40 percent of the chemical energy present in the bonds of the glucose molecule, a level of efficiency substantially exceeding that of such human inventions as the internal combustion engine. The remainder of the energy is released as heat energy.

Even here, however, apparent waste is put to use, for it is this "waste" heat from glycolysis and respiration that many organisms (including ourselves) use as a primary means of raising body temperature above that of the environment.

But *how* are the ATP molecules actually formed? The Krebs cycle, as we have seen, takes place in the inner compartment of mitochondria, and the electron transport molecules are located sequentially along the inner membrane (see Figure 5.17). As the electrons

FIGURE 5.16

Energy Release and ATP Yield During Glycolysis and Respiration

There is a total energy yield of 36 molecules of ATP for every molecule of glucose that passes through glycolysis and respiration.

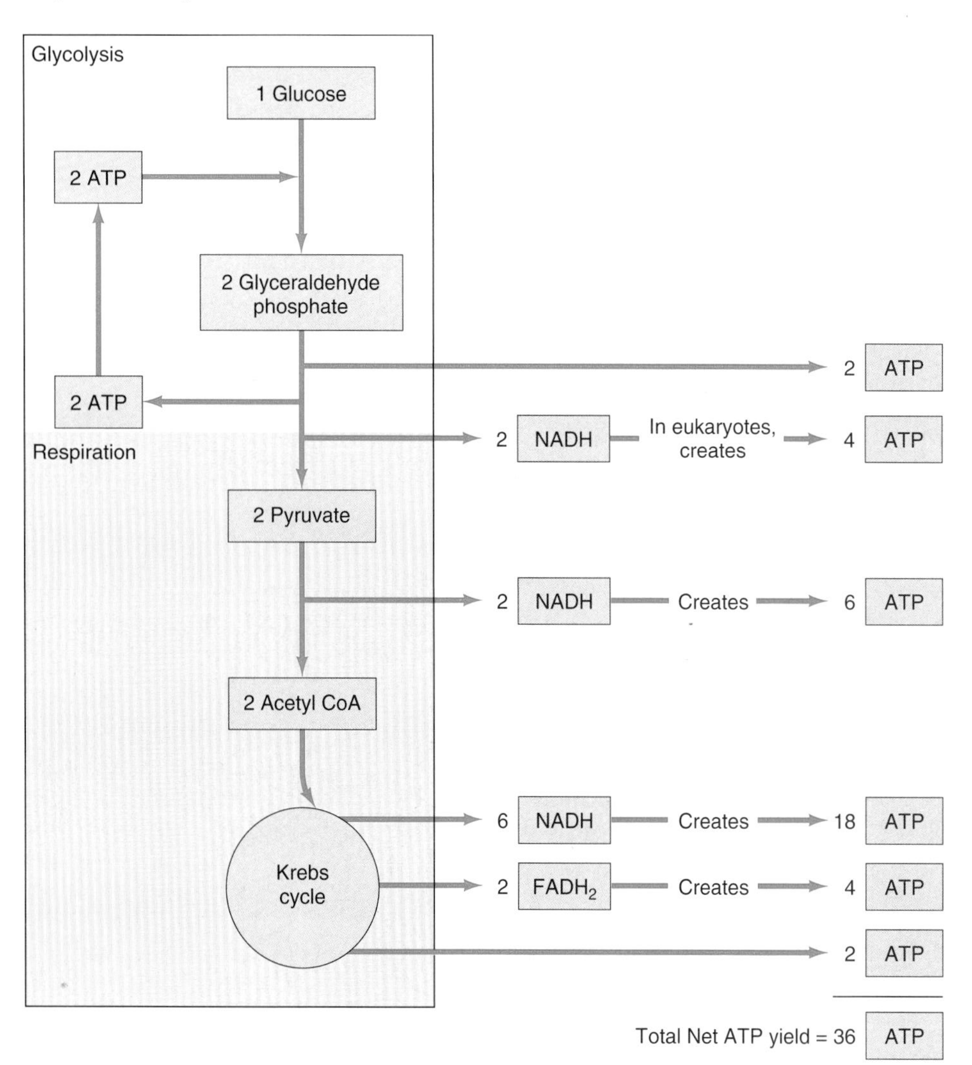

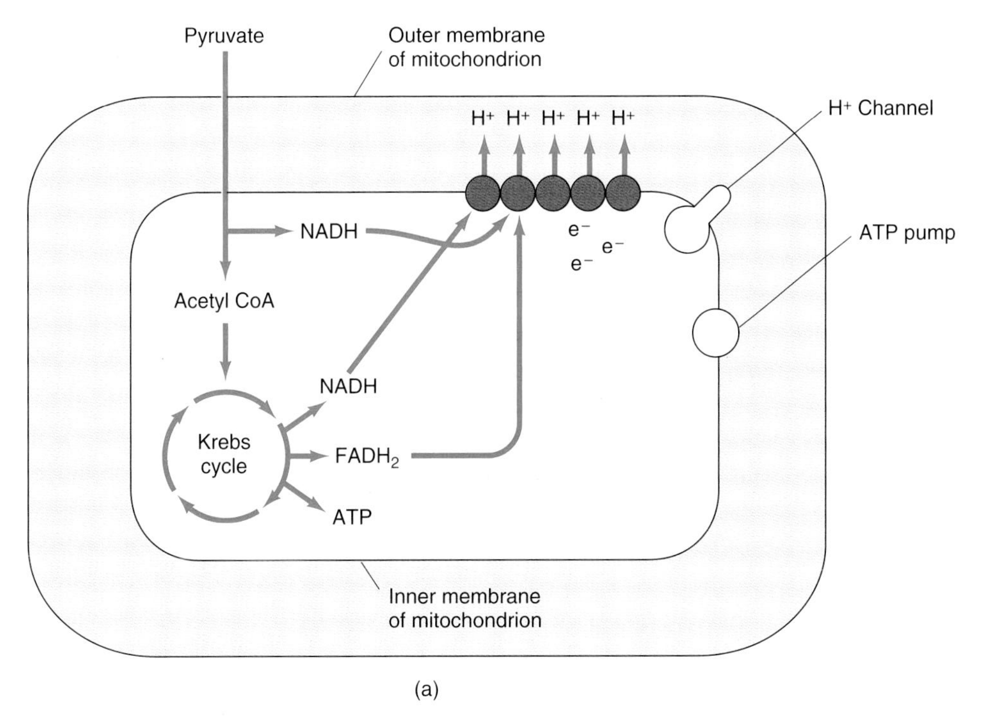

(a)

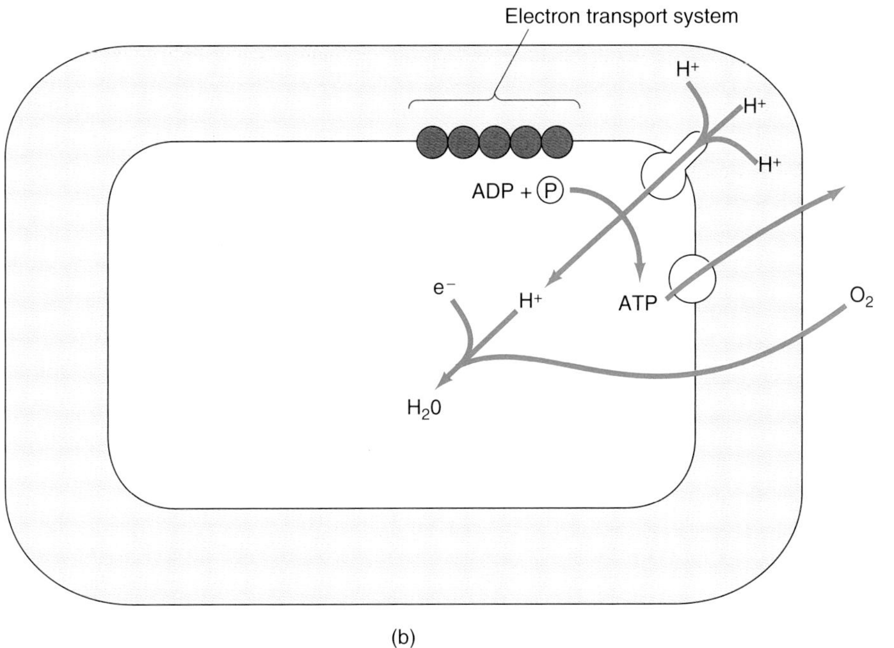

(b)

FIGURE 5.17

ATP Production in Mitochondria

(a) The carrier molecules of the electron transport chain are imbedded in the inner membrane of mitochondria. The Krebs cycle occurs in the inner compartment of mitochondria. As the electron transport chain receives electrons and hydrogen ions from NADH and $FADH_2$, it pumps the hydrogen ions from the inner compartment to the outer compartment. As a consequence, hydrogen ions become concentrated in the outer compartment. (b) Hydrogen ions diffuse back into the inner compartment through special channels. The energy released as they diffuse down the concentration gradient is used to attach a free phosphate group to ADP, making ATP. ATP is then transported from the mitochondria out into the cytoplasm of the cell. The electrons, hydrogen ions, and atmospheric oxygen are combined to form water.

are passed from NADH and $FADH_2$, and thence from one electron carrier to another in the electron transport system, the electrons and the H^+ are separated from each other. The electrons stay within the inner compartment, but the H^+ are pumped into the outer compartment. As a consequence, the outer compartment is rich in H^+ and is positively charged relative to the inner compartment. Mitochondria, with two areas of differing polarity, are therefore rather like batteries.

The inner membrane of mitochondria is impermeable to H^+ except through defined protein channels. Energy is produced as H^+ pours through these channels, very much as water produces energy as it pours across a mill wheel. Through processes not yet fully understood; this energy is used to create ATP by coupling Ⓟ to ADP.

The H^+, now in the inner compartment, are reunited with the electrons, and both are joined to oxygen (delivered to the mitochondria from the air we breathe) to form metabolic water.

Our discussion of glycolysis and cellular respiration has shown that cells solve their need for an external source of energy by transporting molecules of energy-rich organic compounds, such as glucose, across the plasma membrane from the external environment. Once the glucose molecule is inside the cell, specialized enzymes dismantle it sequentially, capturing the energy it contains by attaching free phosphate groups to ADP molecules, thereby creating ATP, a molecule that powers the cell's metabolic machinery. Carbon dioxide and water are spit out as the waste products of glucose metabolism. This process is constantly and endlessly repeated—and this is important—by virtually every cell of every multicellular organism on earth (and by a large majority of the single-celled organisms as well).

Fine and dandy—but where does all this glucose come from in the first place?

In the ancient earth, glucose and other complex carbohydrates were created spontaneously by the effects of volcanic and radiation energy on the "soup" of molecules and ions that comprised the ancient oceans. However, for the past three billion years or so, there has been a better, and more reliable, source of glucose—photosynthesis. Photosynthesis makes regular what was previously a random and haphazard process. That is, through sophisticated enzymatically controlled pathways, which evolved over hundreds of millions of years, the energy from a highly reliable source has been used to assemble glucose and other complex organic molecules from inorganic precursors—water and carbon dioxide. This "highly reliable" energy source is, of course, sunlight.

PHOTOSYNTHESIS

Before we examine the details of photosynthesis, we need to develop a clear picture of the overall process—including why it is that visible light, of all the sources of energy potentially available, became the form of energy used to power photosynthesis.

The **electromagnetic spectrum** (see Figure 5.18) includes radiation ranging from gamma and X-rays (short wavelength radiation) to radio waves (long wavelength radiation). All such radiation comes in discrete bundles called *quanta*. The amount of energy possessed by a given quantum of radiation is a function of its wavelength. Quanta with short wavelengths, such as ultraviolet radiation, have enough energy to break chemical bonds. This is why ultraviolet radiation is potentially dangerous to living cells. Infrared radiation has a substantially longer wavelength and therefore much less energy; quanta of infrared radiation are capable only of increasing the rate at which molecules vibrate (which is why infrared lights are often used to keep food warm in cafeterias). Visible light includes wavelengths of intermediate length; quanta of visible light possess sufficient energy to excite electrons in certain molecules without destroying the molecule in the process. We have such molecules in the retinas of our eyes (see Chapter 15); green plants and other photosynthesizers have such molecules in the form of **chlorophyll.** It is no accident that processes as seemingly different as vision and photosynthesis use radiation at the wavelength of visible light, because quanta at this point in the electromagnetic spectrum have just the right amount of energy to excite molecules without destroying them (see Figure 5.19).

Photosynthesis is a complex process involving many biochemical steps, not all of which are fully understood even today. However, it is useful, in attempting to understand photosynthesis, to think of it as two sets of chemical reactions (see Figures 5.20 and 5.21). The first set of reactions requires light, and it is here that (1) light energy is used to create the chemical bonds of high energy molecules required for the synthesis of glucose and other carbohydrates, and (2) water molecules are pulled apart, and oxygen gas is released as a waste product. These reactions, not surprisingly, are called the **light-dependent reactions.**

Reactions of the second set are collectively termed the **light-independent reactions,** because they do not require the presence of light energy. In the light-independent reactions, the high energy molecules obtained from the light-dependent reactions are used to couple carbon dioxide (from the air) and hydrogen (from the disarticulated water molecules) to form carbohydrates.

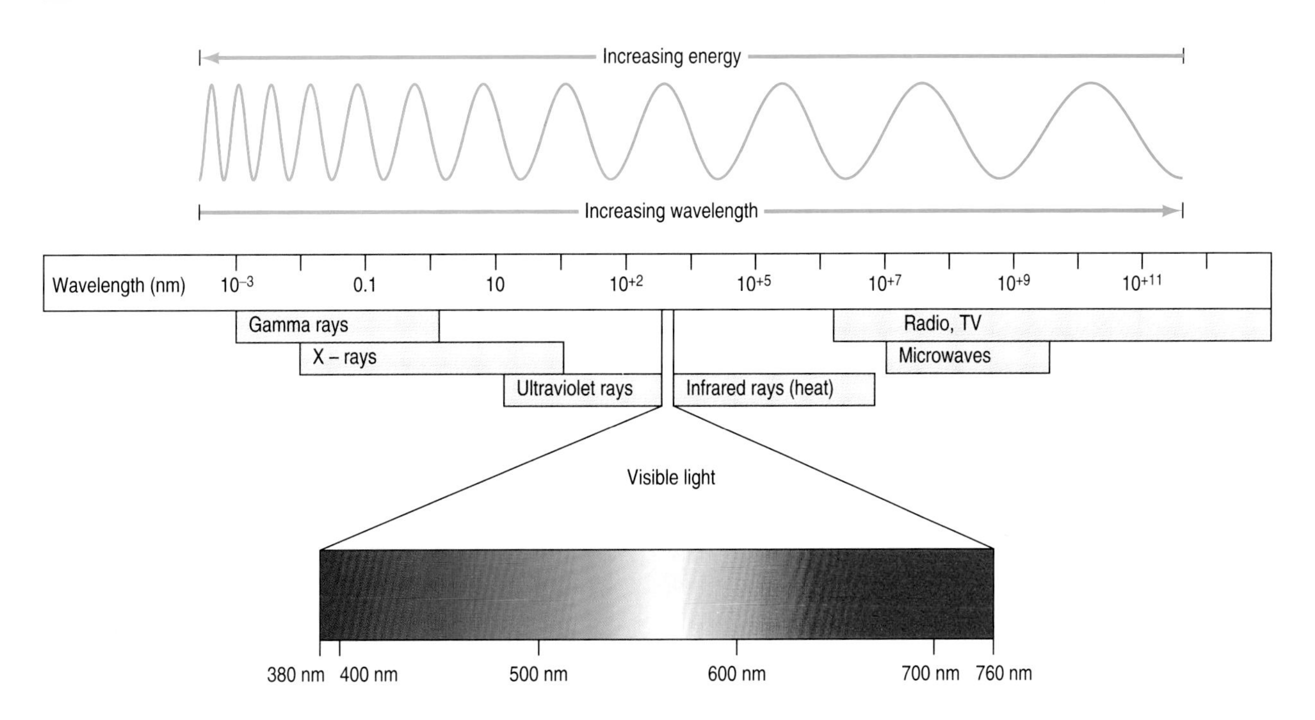

FIGURE 5.18

Electromagnetic Spectrum

The visible spectrum is a tiny slice of the electromagnetic spectrum. Electromagnetic radiation of shorter wavelength than the visible spectrum has increasingly higher energy, which quickly becomes destructive to biological molecules. Longer wavelength radiation has too little energy (except from major sources such as microwave ovens) to affect biological molecules. Energy of the visible spectrum has just enough energy to energize biological molecules without destroying them, which is why both vision and photosynthesis use energy of comparable wavelength.

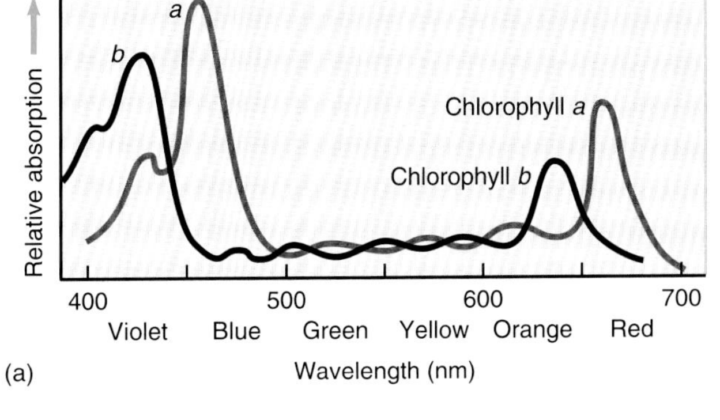

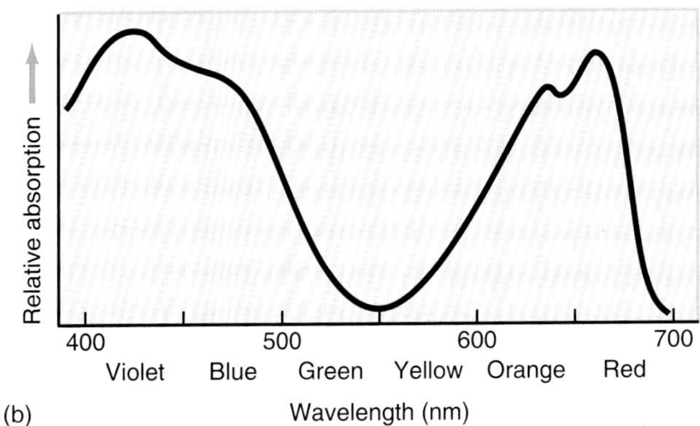

FIGURE 5.19

Absorption Spectra of Chlorophyll

(a) The absorption spectra of the two major types of chlorophyll. Both absorb light at either ends of the visible spectrum but reflect light in the middle of the spectrum. (b) When the absorption spectra of the two chlorophylls are combined with the carotenoids (accessory absorption pigments), most of the visible spectrum is absorbed, with the exception of wavelengths of intermediate value, which is why leaves generally appear green.

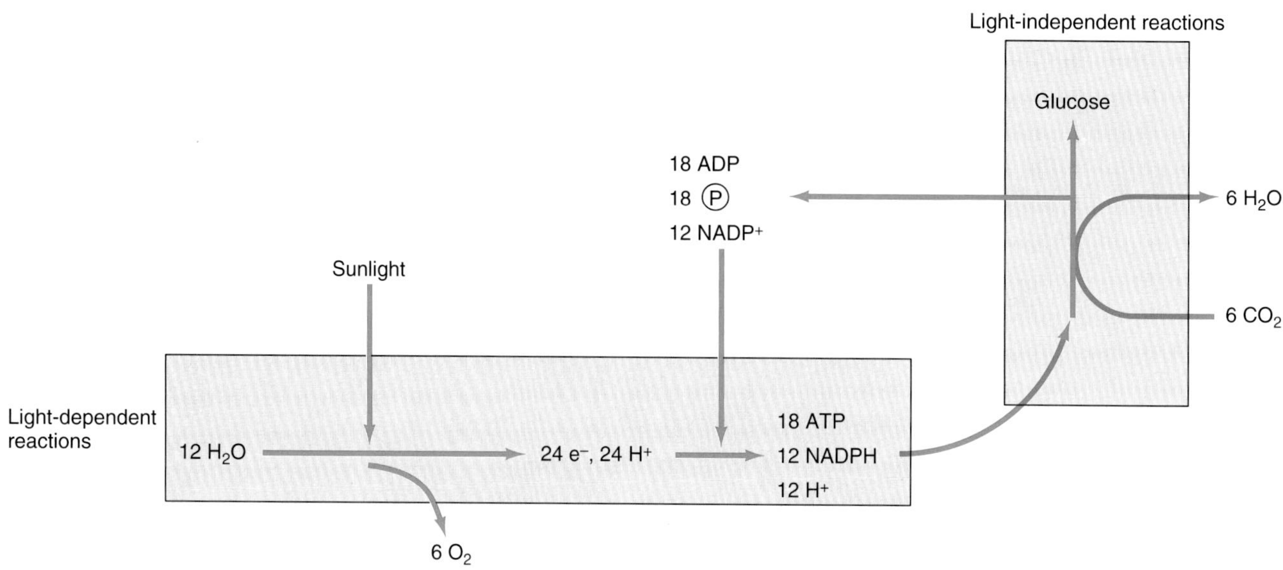

FIGURE 5.20

Overview of Photosynthesis
The light-dependent reactions are responsible for generating both ATP and NADPH, as well as releasing oxygen as a gas from water molecules. The light-independent reactions utilize both ATP and NADPH plus carbon dioxide to produce glucose and other carbohydrates.

Light-Dependent Reactions

Chlorophyll and other light-sensitive molecules are assembled in the chloroplast membrane into so-called **photosynthetic units,** each of which generally contains hundreds of molecules (see Figure 5.22). When a quantum of light is absorbed by any of the molecules in a photosynthetic unit, the light energy boosts an electron of the molecule to a higher energy level. Molecules with such energized electrons are said to be in the *excited state,* and excited molecules can transfer energy to other molecules having lower energy. This hand-off of energy moves eventually to the molecule with the lowest energy level within the unit—the *reaction center chlorophyll.*

An excited reaction center chlorophyll molecule loses an electron to a specialized receiver molecule with the imposing name of *nicotinamide adenine dinucleotide phosphate.* (For obvious reasons, this molecule is generally referred to by its acronym, *NADP*$^+$, the superscript indicating that, in the cell, it is typically found as a positively charged ion.)

After losing its electron, chlorophyll restores its normal complement of electrons by attracting an electron from a molecule of water. Since water is a lower energy molecule than is chlorophyll, its electrons are more attracted to chlorophyll than to water. In the process of losing its electrons, the water molecule is pulled apart. H^+ ions are destined to become a part of the glucose (or other carbohydrate) molecule, formed in a later stage of photosynthesis; oxygen atoms from destroyed water molecules associate and form oxygen molecules (O_2) that are released by the plant as a byproduct of photosynthesis:

$$2\,H_2O \rightarrow 4\,H^+ + 4\,e^- + O_2$$

In actuality, the photosynthetic system requires more energy than is provided by a single quantum of light in order to transfer an electron all the way from water to $NADP^+$. There are, in fact, two different types of reaction center chlorophylls, each with its own photosynthetic unit, and each must absorb a quantum of light (see Figure 5.23).

The reaction center chlorophyll of **Photosystem I** maximally absorbs light with a wavelength of roughly 700 nm and is therefore referred to as P_{700} chlorophyll. The reaction center chlorophyll of **Photosystem II** has an absorbance maximum at a wavelength of 680 nm (P_{680} chlorophyll). When a quantum of light activates a P_{700} molecule, the molecule contributes its excited electron to $NADP^+$. Simultaneously, the P_{700} chlorophyll picks up an excited electron from the P_{680} chlorophyll (which has also been activated by a quantum of light). Finally, the P_{680} chlorophyll restores its complement of electrons by attracting one away from a water molecule. It is important to note that the actual transfer of electrons from water to P_{680}, from P_{680} to

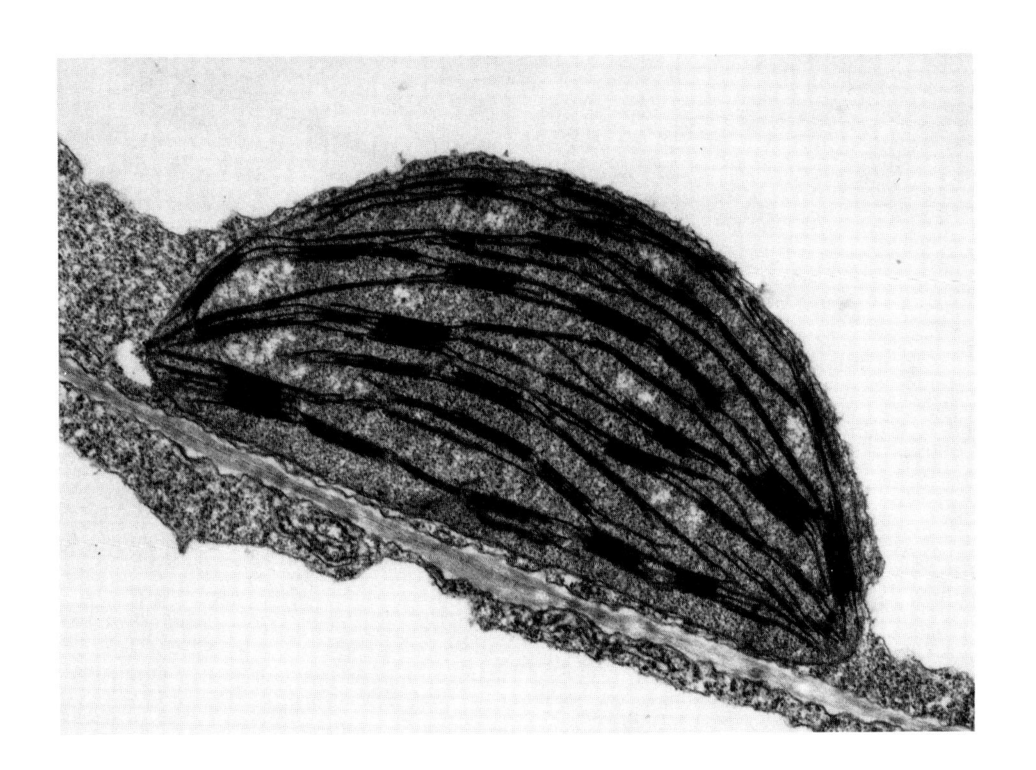

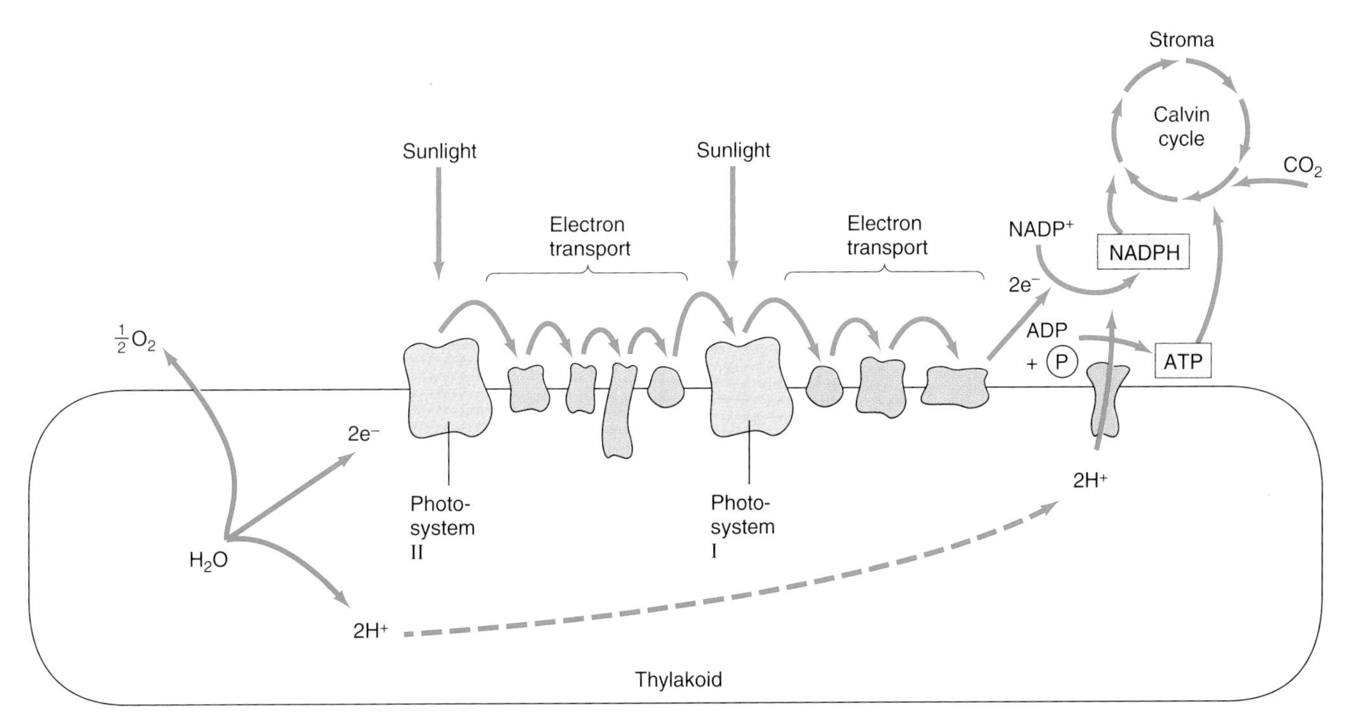

FIGURE 5.21

Functional Organization of Photosynthesis

Photosynthesis takes place in cells in the interior of leaves. These cells are rich in chloroplasts, a chlorophyll-containing organelle. Chloroplasts contain stacks of membrane called thylakoid discs, in a matrix called stroma. Hydrogen ions are stored in the interior of the thylakoids, and, in a manner reminiscent of mitochondria, the energy generated by the passage of hydrogen ions through special channels in the thylakoid membrane is used to create ATP. Chlorophyll molecules plus the molecules of the electron transport system are also located in the membrane of the thylakoids, whereas assembly of glucose molecules (the light-independent reactions) takes place in the stroma.

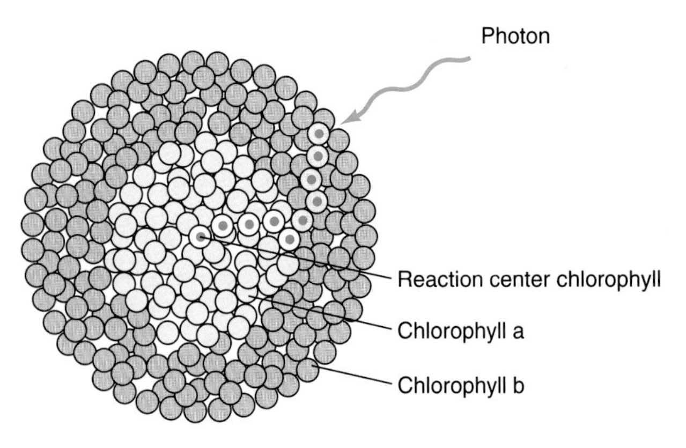

FIGURE 5.22

Photosynthetic Units

A number of chlorophyll molecules are grouped as a photosynthetic unit. Any of the chlorophyll molecules can react to being struck by a photon of light, but the excited electron is passed to chlorophyll molecules of lower energy (chlorophyll *a* molecules are of lower energy than chlorophyll *b*) until it reaches the reaction-center chlorophyll molecule—the chlorophyll with the lowest energy of all. Electron transport system molecules are associated only with the reaction-center chlorophyll molecule, but the other chlorophyll molecules can act as "antennae" to receive incoming photons of light.

P_{700}, and from P_{700} to $NADP^+$ is not done directly but rather through a complex of intervening molecules called *electron carriers;* collectively, these molecules are referred to as the electron transport chain. (It is also important to note that this electron transport chain is similar to, but not identical with, the electron transport chain found in cellular respiration.)

Not all of the energy absorbed from quanta of light is used simply to transport electrons. Some energy is also used to manufacture ATP (by attaching

FIGURE 5.23

Noncyclic Photophosphorylation (Photosystem I and Photosystem II)

When sunlight strikes the chlorophyll in Photosystem II, it causes an electron to be raised to a higher energy level. This electron is passed on to the molecules of the electron transport system before being handed off to the chlorophyll of Photosystem I. The electron replaces the electron lost by Photosystem I when it is struck by sunlight. That electron is also captured by the electron transport system and is ultimately passed to $NADP^+$, which picks up a hydrogen ion and become NADPH. The hydrogen ion, in turn, is produced by the destruction of water molecules, water being the source of the electrons used to replace the electrons lost from Photosystem II. For every 12 water molecules split, 24 hydrogen ions are released, and ultimately 24 ATP molecules formed. Of that number, 18 ATP are used in the synthesis of glucose; the rest are used for other metabolic activities.

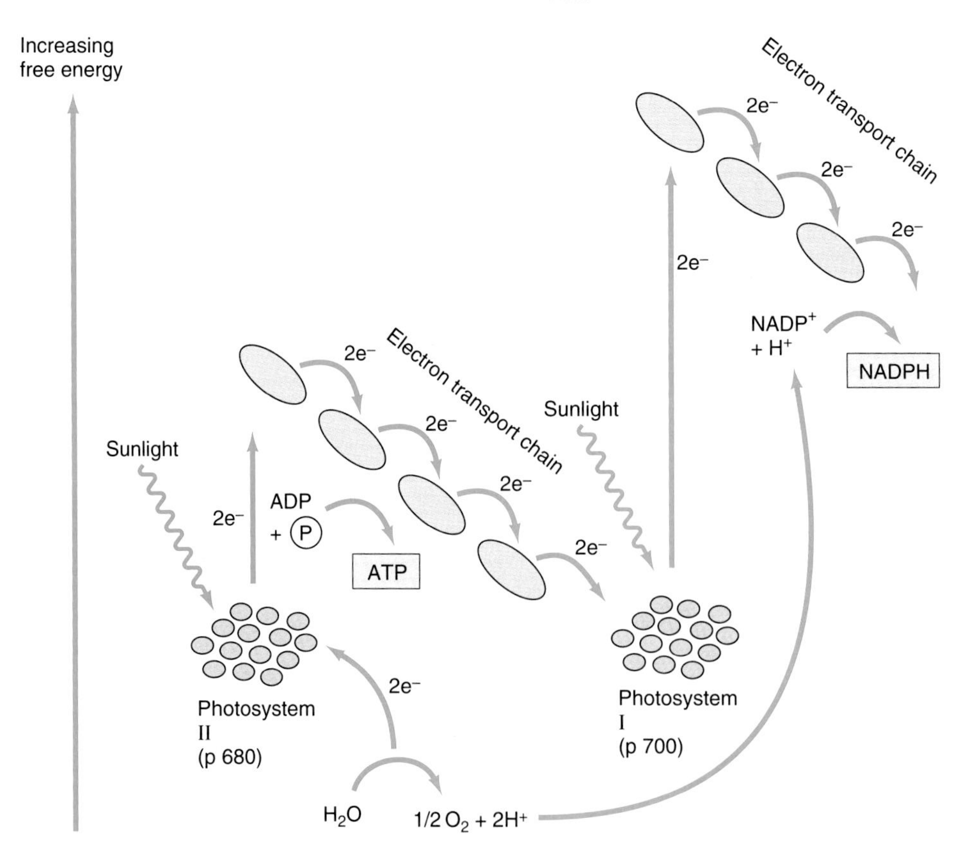

an existing phosphate group to a molecule of ADP); this process is called **photophosphorylation.** Photophosphorylation exists in two forms—*cyclic* and *noncyclic*.

Noncyclic photophosphorylation takes place as electrons move from water to $NADP^+$ (by way of Photosystems II and I, respectively). It is called noncyclic because the movement of electrons is one-way. Cyclic photophosphorylation is a different, and apparently more ancient, process (see Figure 5.24). Under specific conditions, electron movement in Photosystem I becomes cyclical. Chlorophyll continues to become excited, but there is no production of NADPH or destruction of water. Instead, electrons are permitted to return to the P_{700} molecule, via electron carrier molecules. This cyclical movement of electrons contains enough energy to permit the assembly of ATP.

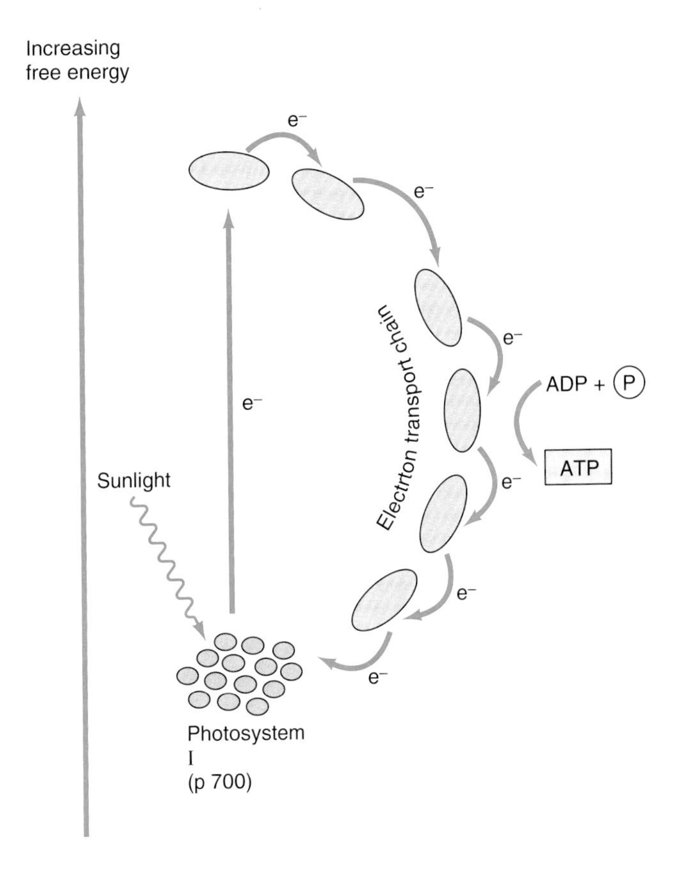

FIGURE 5.24

Cyclic Photophosphorylation

Plants can also undertake a cyclical form of photophosphorylation in which ATP is formed, but no NADPH. In this situation, light excites electrons in Photosystem I, the electrons are picked up by the electron transport system and used in the synthesis of ATP, but are ultimately returned to Photosystem I. Thus no water molecules are destroyed, and no hydrogen ions are produced.

Light-Independent Reactions

In the light-independent phase of photosynthesis, NADPH and ATP, produced in the light reactions, are used to incorporate carbon dioxide from the atmosphere and hydrogen from water molecules to create glucose and other carbohydrates (see Figure 5.25). This carbon dioxide incorporation occurs within the chloroplast. It is a multistep process, with each step requiring a specific enzyme. The overall process is called the **Calvin cycle,** after its discoverer, Melvin Calvin. (Calvin was awarded the Nobel Prize for this work in 1961.) It is also known as the *C_3 pathway*, because the first step involves the formation of a three-carbon compound, *phosphoglycerate* (or *PGA*).

It might be supposed that CO_2 would be bound to a two-carbon compound to make the three-carbon compound, but such is not the case. The immediate precursor of the C_3 compound is a C_5 compound, *ribulose bisphosphate* (or *RuBP*). The "carbon arithmetic" of the CO_2 incorporation reaction is:

$$\underset{(\text{RuBP})}{C_5} + \underset{(CO_2)}{C_1} \longrightarrow \underset{(\text{PGA})}{2\,C_3}$$

During the incorporation process, a six-carbon intermediate molecule is apparently formed, but it is present for such a short time before it splits into two three-carbon molecules that it has not yet been possible to isolate or identify it.

In order for the CO_2 incorporation process to continue, an adequate supply of the original C_5 molecule must be maintained. Thus five-sixths of the molecules produced are used to regenerate the supply of RuBP, thereby permitting the cycle to continue. Three molecules of ATP and two molecules of NADPH are required for each molecule of CO_2 incorporated into carbohydrate, for the continued regeneration of the cycle, and for the ultimate conversion of PGA to *glyceraldehyde phosphate* (or *GAP*), a molecule that is used to assemble a variety of organic molecules needed by the plant.

If incorporation of one CO_2 requires three ATP and two NADPH, incorporation of three CO_2 molecules will require three times as much. Thus nine ATP and six NADPH coming from the light-dependent reactions must be used in the light-independent reactions to produce a net gain of one GAP, which can be "withdrawn" from the Calvin cycle pool and used by the cell in the formation of carbohydrates.

While still inside the chloroplast, GAP is often converted (by running the glycolysis reactions in reverse) to *glucose phosphate,* and thence to starch, the fundamental carbohydrate storage molecule inside the chloroplast. During the day, when light is readily available, plants tend to form starch; at night, the starch is broken down to produce glucose.

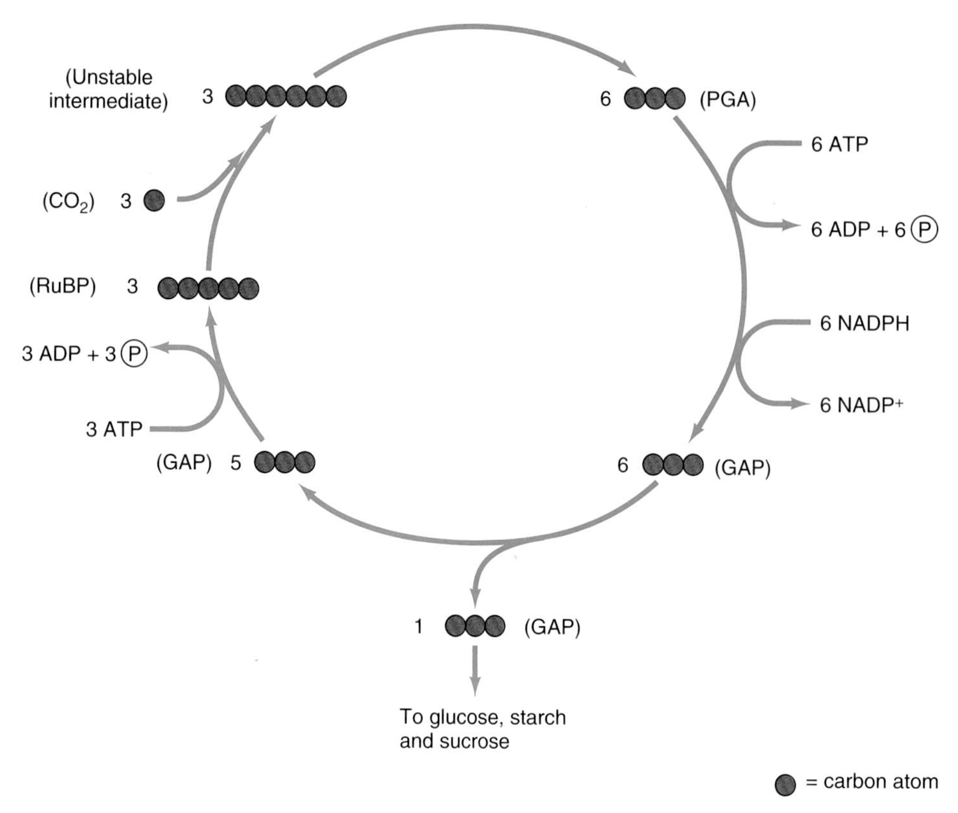

FIGURE 5.25

Calvin Cycle

With each "turn" of the Calvin cycle, one molecule of carbon dioxide is incorporated. Since glyceraldehyde phosphate (GAP) contains three carbons, it naturally follows that three "turns" of the cycle are necessary to produce one GAP molecule. This illustration summarizes the activities in the manufacture of one GAP molecule. Note the use of six NADPH molecules and nine ATP molecules to form one GAP molecule.

GAP can also be converted to *dihydroxyacetone phosphate* (or *DHAP*), which is the primary carbon molecule exported from the chloroplast to the cytoplasm. In the cytoplasm, DHAP is used to assemble glucose and fructose, which then are united to form *sucrose* (table sugar). Sucrose is the major export molecule from leaves to the rest of the plant, which is why commercial sugar production from sugar-rich plants such as sugar cane and sugar beet focuses on sucrose production. DHAP can also be converted to a variety of other molecules, including lipids and amino acids.

Summary

From the standpoint of energetics, the cell can be likened to a factory. Like any factory, the cell needs a regular supply of energy from an outside source, both because energy is needed to power essential chemical reactions and because the laws of thermodynamics predict an inability of any system to recycle energy without loss.

Many factories generate their own power. So, in essence, does the cell. By important energy-rich compounds from the environment across the plasma membrane, the cell gains possession of the "fuel" that can be "burned" for energy. Importation of chemicals is potentially a risky business, because the cell, by virtue of having a permeable membrane, must avoid the inadvertent acquisition of unnecessary or even dangerous substances while simultaneously protecting against the loss of essential substances. Sometimes the necessary outcome can be achieved only by the cell's expending energy.

Energy extracted from energy-rich molecules is stored in the form of ATP, a molecule that is universally used in living organisms to activate molecules and to power chemical reactions. Even with ATP present, however, many chemical reactions would occur at prohibitively slow rates at the relatively low internal temperatures required by organisms for their continued survival. Enzymes facilitate thousands of different reactions in cells by reducing the amount of activation energy needed for a given chemical reaction to occur.

Because most energy-rich compounds available to the cell are highly stable, the cell needs to use both ATP and enzymes to break the chemical bonds of their molecules. Some net gain of energy in the form of ATP can be achieved by metabolizing energy-rich compounds in the absence of oxygen, but most organisms use oxygen to complete the breakdown of glucose and other simple organic compounds to carbon dioxide and water. In the presence of oxygen, the cell can extract and capture almost 40 percent of the chemical energy of a glucose molecule.

Animals are, as a group, dependent on other organisms for their source of glucose. Most green plants, on the other hand, possess the capacity to assemble carbohydrates from water and carbon dioxide, using sunlight as an energy source. We call this process photosynthesis.

Key Terms

First Law of Thermodynamics
Second Law of Thermodynamics
diffusion
facilitated diffusion
osmosis
active transport
sodium-potassium pump
endocytosis
food vacuole
phagocytosis
pinocytosis
pernicious anemia
exocytosis
Calorie
adenosine triphosphate (ATP)
activation energy
substrate
cofactor
coenzyme
glycolysis
fermentation
Krebs cycle
electron transport system
metabolic water
cellular respiration
phosphorylation
photosynthesis
electromagnetic spectrum
chlorophyll
light-dependent reactions
light-independent reactions
Photosystem I
Photosystem II
photophosphorylation
Calvin cycle
anaerobic
aerobic
ADP
enzyme
catalyze
active site

Questions

1. What significance does the Second Law of Thermodynamics have to understanding the need of cells for an outside source of energy?
2. Distinguish between diffusion and active transport. To which category (if either) does osmosis belong?
3. What is ATP? Why is it important to cells? What is the cell's source of ATP?
4. How do enzymes function? Why do we have so many different enzymes in our bodies?
5. How does glycolysis differ from respiration? Where does fermentation fit into the picture?
6. Briefly explain the role of mitochondria in energy generation by the cell.
7. Explain the function of NAD and $FADH_2$. Where do they pick up electrons and where do they take them?
8. Comment on the similarities and differences between respiration and photosynthesis.
9. What happens in the light-dependent reactions of photosynthesis? What happens in the light-independent reactions?

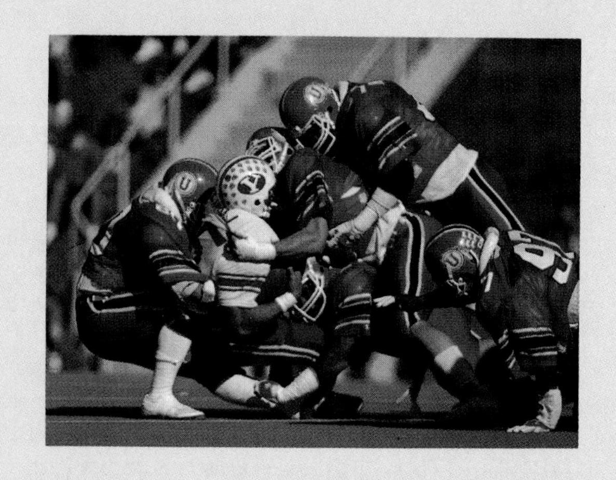

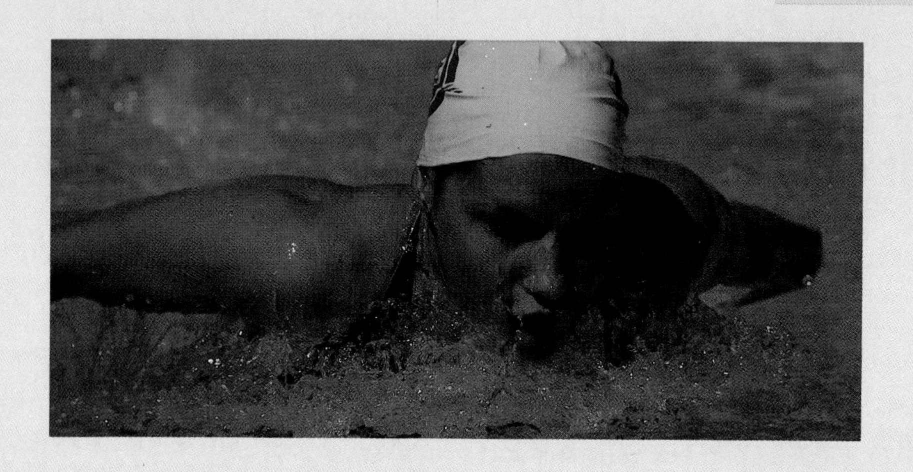

PART

III

Maintenance Systems

The seven chapters of Part 3 discuss the role of those organ systems that function to maintain the body in a balanced state (both anatomically and physiologically). However, before we begin an examination of how it is that they interrelate, let us first devote our attention to bridging the very large gap created in moving from a discussion of individual cells to a discussion of organ systems.

Two girders support our bridge. First, let us state the obvious: We humans are composed of many cells, and these cells are often specialized to perform particular functions. These specializations are described and categorized here. Second, we must examine the theoretical: Why is it that we are multicellular creatures? Why could we not be a single, very large cell? In other words, was multicellularity necessary, or simply a convenience?

Cells are inherently tiny entities. (There are a few exceptions, such as the egg of the ostrich, but even these are more apparent than real. Bird eggs begin as a single cell, but the development of the bird embryo is confined to a very small patch of activity, floating on a sea of yolk.) The reason why cells are never large—or, conversely, why large organisms are invariably multicellular—derives principally from surface area/volume relationships.

Imagine a brick-shaped cell 2 mm in length, and 1 mm both in thickness and in width. (Cells are rarely so regular in their dimensions, and this would be a very large cell—but the point at this stage is to demonstrate relationships, not to mirror reality.) The surface area of this cell, calculated on the basis of all six sides, would be 24 mm^2, and its volume would be 8 mm^3. All of the materials exchanged between the cell and its environment must pass across these 24 mm^2 of surface. Let us suppose, for the sake of argument, that the 8 mm^3 of volume requires all 24 mm^2 of surface area to meet its needs—that is, that the ratio of surface area to volume must always be 3:1 or greater.

Now suppose that we double the linear dimensions of our cell. It is now 4 mm in length, thickness, and width. Its surface area would now be 96 mm^2, and its volume would be 64 mm^3. The 3:1 ratio that we established as being mandatory to support the cell is now 1.5:1—which means that the cell cannot survive.

Expressed mathematically, what we are saying is that for every doubling in linear dimensions, there is a fourfold increase in surface area—but an eightfold increase in volume. Thus cells cannot increase in size indefi-

Problem

How does this cell, with a requirement of 3 mm^2 surface area for every 1 mm^3 of volume, increase in size?

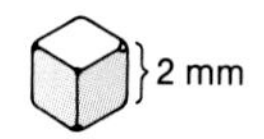

Surface area = 24 mm^2
Volume = 8 mm^3
Vol/s. a. ratio = 1:3

Alternative 1

Unacceptable to increase equally in all directions—ratio is insufficient.

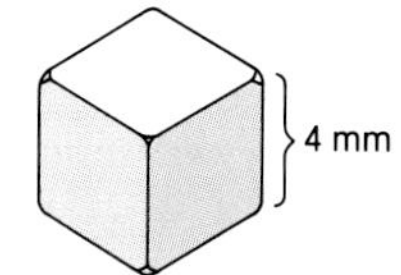

Surface area = 96 mm^2
Volume = 64 mm^3
Vol/s. a. ratio = 1:1.5

Alternative 2

Ratio is acceptable, but too much versatility is lost by linear shape.

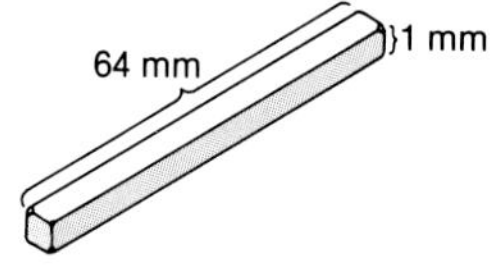

Surface area = 258 mm^2
Volume = 64 mm^3
Vol/s. a. ratio = 1:4

Alternative 3

Cell division preserves required ratio and offers the most potential for versatility in shape.

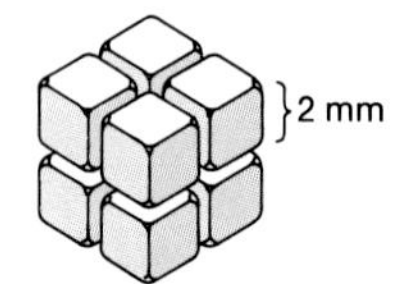

Surface area = 192 mm^2
Volume = 64 mm^3
Vol/s. a. ratio = 1:3

nitely without reaching the point where the surface area is no longer able to provide a large enough doorway to provide for the exchange needs the cell has with its environment.

To become large, organisms had to become multicellular. A few organisms are basically colonies of essentially independent cells, but in the overwhelming majority of multicellular species, groups of cells are specialized for particular functions needed by the organism as a whole. Groups of cells of one type, specialized for a particular purpose, are called **tissues.** Where various types of tissues are organized into a functional unit, that unit is called an **organ.** (Organs, in turn, may themselves be integrated into an **organ system.**)

As we shall see in the coming chapters, when we examine each organ system in turn, the role of an organ system is to contribute in a particular way to the overall welfare and functioning of the organism as a whole. Those organ systems that we characterize as "maintenance systems," which we shall discuss first, play a vital role in maintaining a stable internal environment. For example, the digestive, respiratory, and excretory systems all affect the composition of blood, either by adding, or removing, particular substances. These substances are either required by all of the cells of the body for their continued functioning, or they are waste products that must be removed.

Collectively, all of the organ systems contribute to **homeostasis**—the "steady state" or "dynamic equilibrium" that

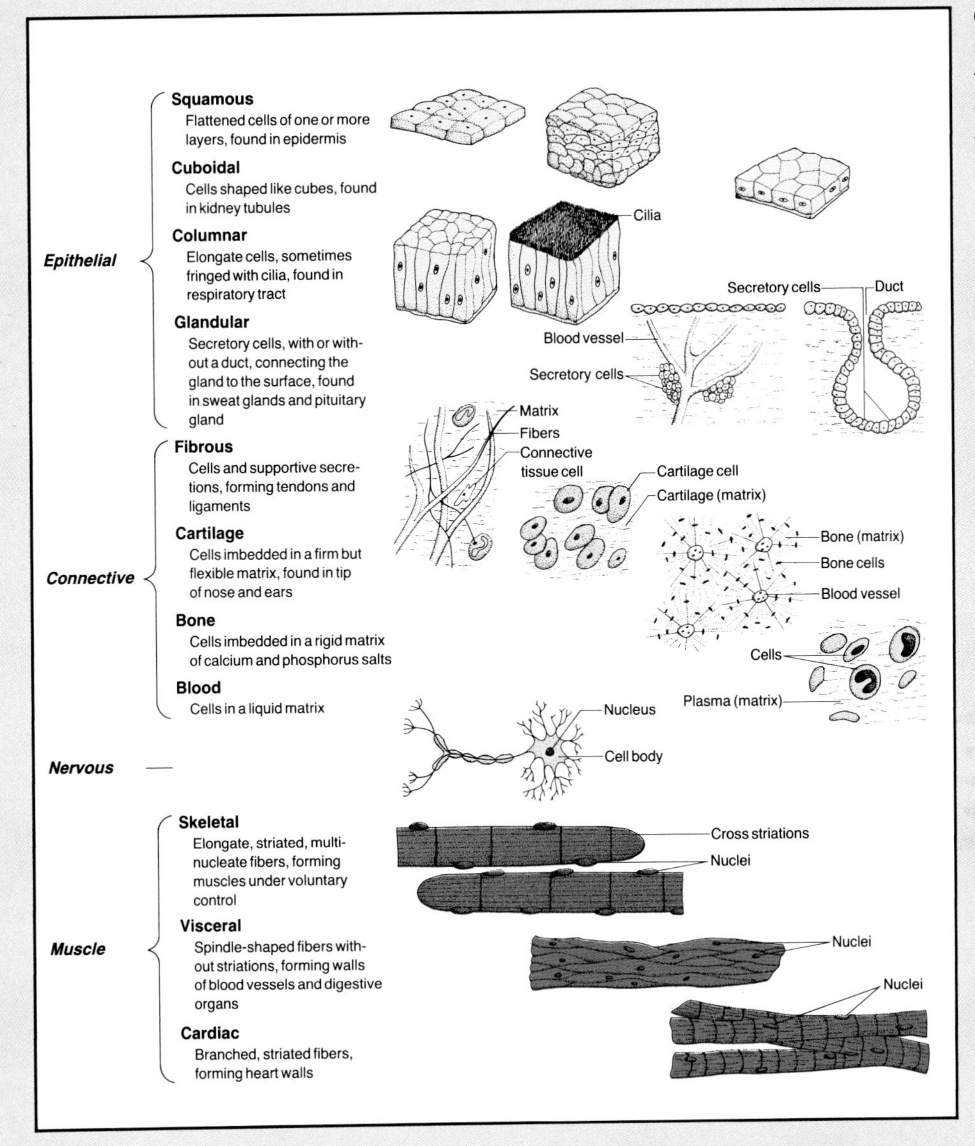

characterizes the internal environment of all organisms in the face of many potential variables. Consider just one example. Although the temperature of our environment varies widely, we maintain a constant internal temperature of approximately 37° C. How do we accomplish this feat? Temperature sensors in the brain respond to an elevated blood temperature by directing the sweat glands to produce perspiration. As the perspiration evaporates from the skin's surface, the skin is cooled. If the body temperature begins to fall below its normal level, the brain directs the skeletal muscles to contract rapidly (shivering), an action that metabolizes large amounts of glucose. Since about 60 percent of the energy contained in glucose passes off as heat, shivering creates a heat source that is capable of raising the temperature back to normal levels.

Finally, bear in mind that the organ systems in no way usurp the various cellular functions we discussed in Chapter 4 and 5; rather, they simply provide a setting in which the cells of the body can thrive.

CHAPTER

6

The Digestive System

Processing Nutrients

A snake handler can "milk" a poisonous snake and drink the venom with impunity. However, if the same snake were to bite him, the handler might die. Why is there a difference in result? Why can children eat garden soil without ill effects, but suffer infections if the same soil enters a cut or a scrape?

The digestive system is one of the most interesting yet most abused of the body's systems. It provides a doorway to the body far more substantial than its delicate structure would suggest, yet drug manufacturers make millions of dollars every year selling medicines for indigestion and constipation. Are these medicines really necessary?

These are the kinds of questions discussed in our chapter on the digestive system.

THE NATURE OF THE PROBLEM

Up to this point, our consideration of life processes has been at the level of the molecule and the cell. Thus, although we now have some understanding of certain processes—how the cell identifies metabolically useful molecules and transports them across its plasma membrane; how the cell recombines these molecules to form structural or storage molecules; or how the cell dismantles molecules for energy—all of that understanding still seems to leave us well short of knowing much about how a complex organism, such as a human, makes a biological living.

To take just one example of the problems of extrapolating from the single cell to the intact organism, speculate on the processes whereby "metabolically useful" molecules come to lie adjacent to the cells of the body. From your basic understanding of biology, you would no doubt assume that the circulatory system is responsible for transporting these molecules from the digestive system to all of the cells of the body (and you would be correct, although the details are somewhat more involved). However, that scenario trivializes the complex role of the digestive system itself.

Consider just a few of the problems faced by the digestive system:

1. Most of the molecules making up our food are much too large to cross a plasma membrane—yet in order to be useful to the body, they must pass across the intestinal lining and into the blood stream. Thus, one task of the digestive system is to produce the enzymes necessary to fragment macromolecules into small, transportable pieces. How does the digestive system produce these enzymes without digesting itself in the process?

2. We tend to eat relatively large meals at discrete (and not necessarily regular) intervals. How do we coordinate the production of digestive enzymes so as to coincide with the presence of food in successive organs of the system?

3. In order for digestion to function properly, the food must be exposed both to digestive enzymes and also to absorptive surfaces. In which organs of the system does digestion occur? Are the same organs involved in absorption? How is the food physically moved from one organ to another? What controls how long the food stays in any one region? What controls the elimination of indigestible materials from the digestive system?

4. Finally, our food is not necessarily sterile. How do we avoid becoming ill due to contamination of our food by bacteria and other disease-causing agents?

AN OVERVIEW OF THE PROCESS

The human digestive system is organized like an assembly line (or, more correctly, a disassembly line; see Figure 6.1). The teeth grind the food into small pieces, thereby facilitating enzymatic action; the salivary glands lubricate the food, thereby permitting easy swallowing; the stomach provides an acid bath, and the beginnings of protein digestion; the small intestine is the site of additional digestion and most of the absorption; secretions from the liver and pancreas further facilitate the digestive processes; the large intestine recaptures most of the water remaining in the food, and passes the residue out of the body.

It is important to recognize, at the outset, that the digestive system is a tunnel that runs through the body. The contents of the digestive tract are no more *inside* the body than a child floating in an inner tube is *inside* the inner tube. Before something can be said to be inside the body, it must first pass across the plasma membrane of a cell. Thus, until (and unless) the contents of the digestive system are transported across a plasma membrane, they remain outside the body.

This distinction is no mere exercise in semantics. It would be erroneous to assume, for instance, that we were nutritionally fit simply because we included all of the essential nutrients in our diet. These nutrients must be *absorbed* to do us any good. Otherwise, they will pass out of the body with the feces, without having provided us with any of the anticipated benefit. As we shall see in Chapter 7, nutrient absorption is a complex matter. Conversely, we can eat food contaminated with many kinds of bacteria (but not all!) without becoming ill—and a snake handler can "milk" the venom from a rattlesnake and drink it without ill effects—because in both cases the materials ingested are not, by our definition, inside the body. However, if those bacteria were injected directly into our blood—or if a snake handler were to drink rattlesnake venom when he had cracked lips or a bleeding ulcer, thereby ensuring direct entry into the blood—the effects would be much more dramatic.

THE ANATOMY OF DIGESTION

In humans the **digestive tract** consists of a series of specialized organs, all interconnected, stretching from mouth to anus. For the most part these organs function in one (or both) of two processes: **digestion** and **absorption**. Digestion is the fragmenting of food by mechanical and chemical means into sizes small enough to permit uptake by the cells lining the digestive tract. Absorption is the uptake of digested ma-

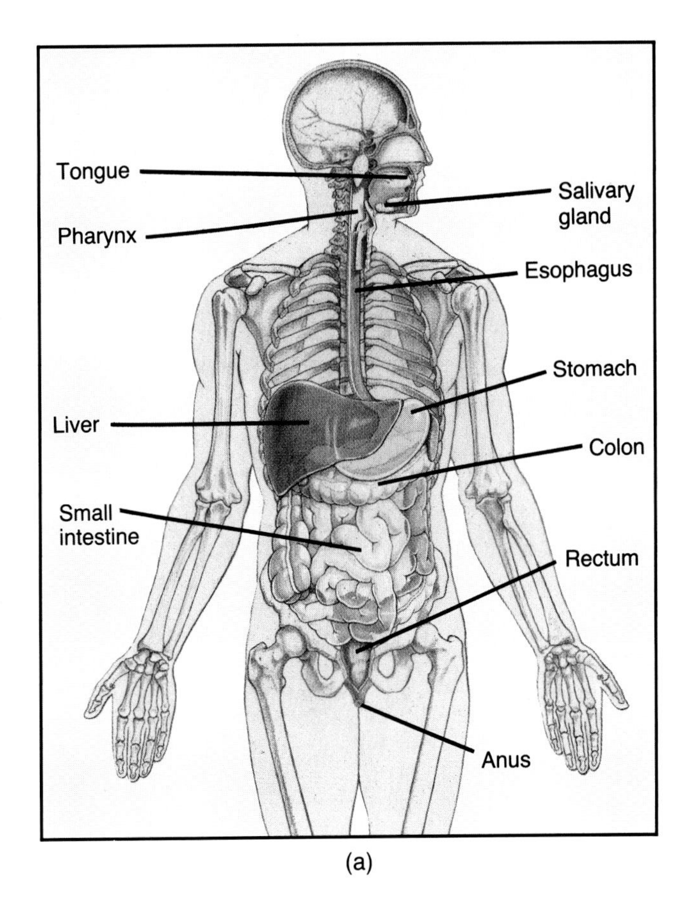

(a)

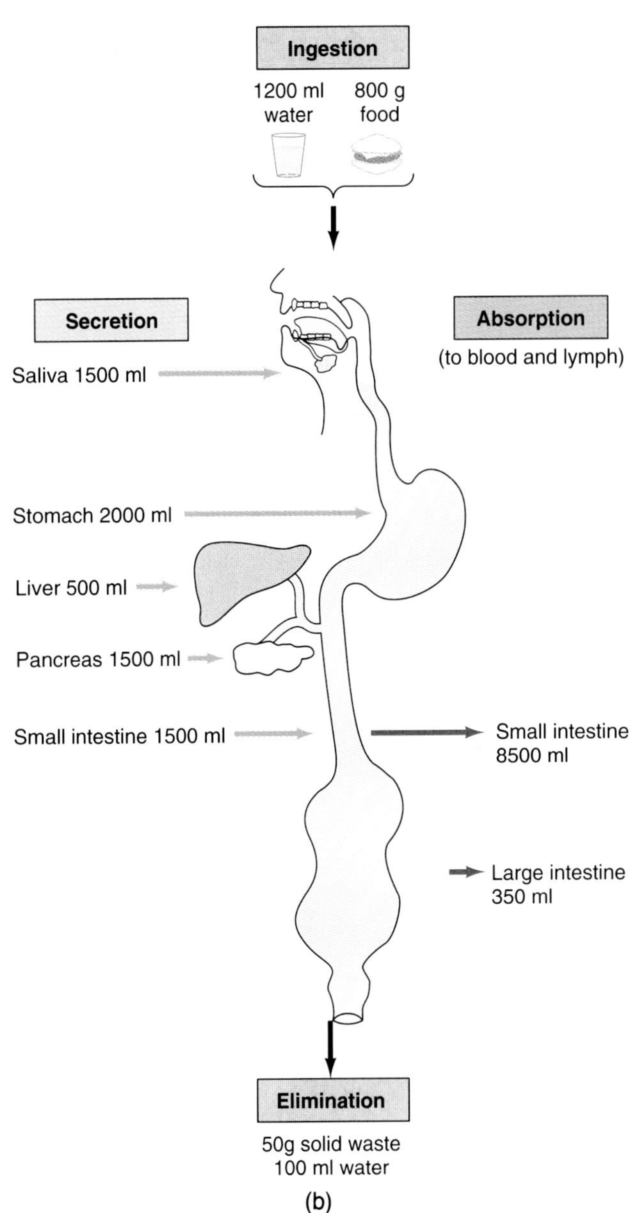

(b)

FIGURE 6.1

Human Digestive System

(a) An artist's rendering of the digestive system. It is fundamentally a tube running through the body that has become regionally specialized into a series of distinct organs. (b) A diagrammatic representation of the digestive system, showing secretion and absorption sites. Note the impressive volume of digestive secretions. Most of these secretions are reabsorbed along with digested food.

terials by these cells. The stomach, for example, functions primarily in digestion. In contrast, the large intestine functions almost exclusively in absorption. The small intestine has a part in both processes.

Let's consider the function of each component of the system individually.

The Mouth and Pharynx

As **omnivores** (eaters of both plants and animals; L. *omnis*, "all"), humans use the mouth in a variety of ways. A complex set of teeth, including incisors, canines, premolars, and molars, mirrors the complex human diet (see Figure 6.2). By contrast, **carnivores** (mean-eaters; L. *carnis*, "flesh") have little need for incisors; thus, their incisors are much reduced. Their canine teeth, by contrast, are markedly enlarged. **Herbivores** (plant-eaters), on the other hand, often have well-developed incisors and grinding molars, but generally have neither canines nor bicuspids. (Since canines are used in a dagger-like fashion to subdue prey, and since grass rarely puts up much of a struggle while being consumed, the absence of canines in herbivores is hardly surprising.)

Because as omnivores we have a mixed diet, our bodies must monitor whatever we ingest to ensure that it is edible. As one aid to monitoring, humans have a well-developed set of taste buds (see Chapter 15), not only on the tongue but on the roof of the mouth as well. (Dentures cover this latter set of taste

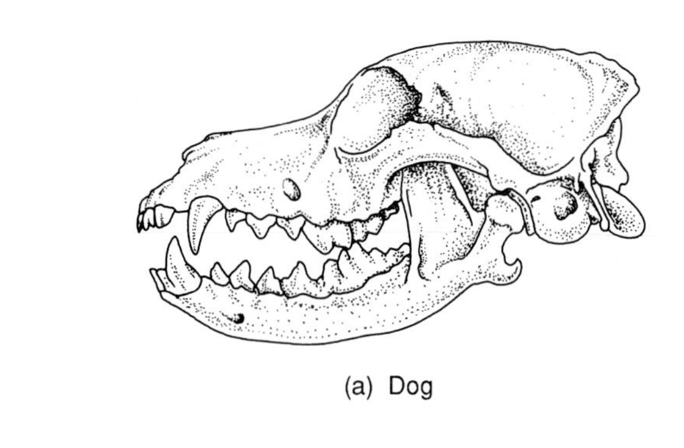

(a) Dog

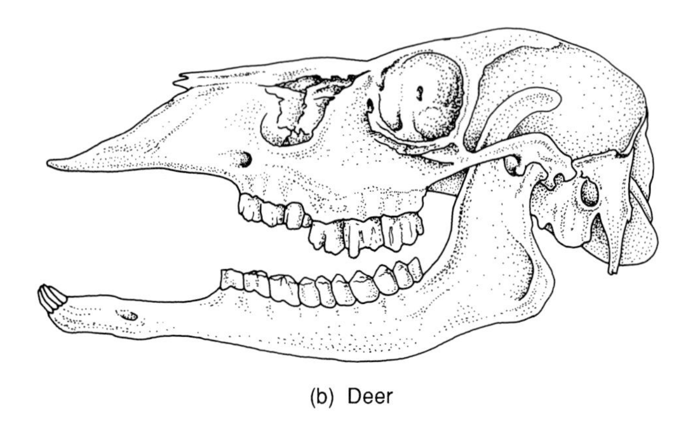

(b) Deer

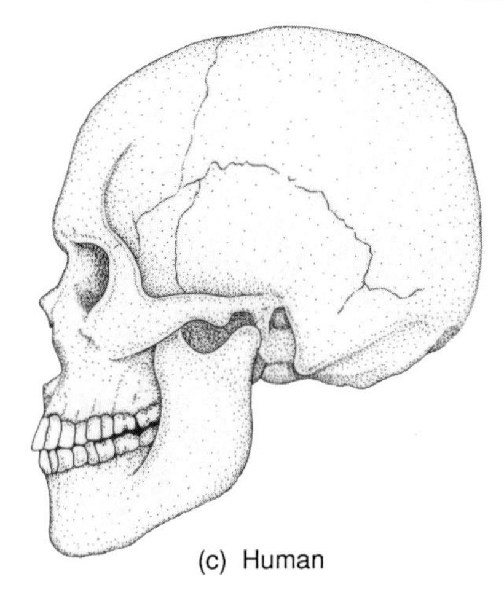

(c) Human

FIGURE 6.2

Teeth in Carnivores, Herbivores, and Omnivores
(a) As a carnivore, the dog has well-developed canine teeth, serrated molars, very small incisors. (b) By comparison, the herbivorous deer has an elongated set of flattened, grinding molars and premolars, no canine teeth, and lower incisors that nip off bits of vegetations. (c) Humans, as omnivores, have rather generalized dentition, with a complete set of teeth, and no particular specialization.

buds, which accounts for the frequent complaint from denture wearers: "Food just doesn't taste as good as it did when I had my own teeth.") In addition, humans tend to chew food for extended periods, during which time the food is not only ground up (thereby enhancing the digestive process) but also monitored by the nose as vapors from the food rise into the nasal cavity. (Thus hot food, with more vapors, generally "tastes" better than cold food, and when we have a stuffy nose, food isn't as tasty as when we are well and our nasal cavities are not congested.)

Carnivores tend to gulp their food, as anyone who has ever timed the family dog at his dinner knows. Because they have a more restricted diet, and because their food is so similar to their own body chemistry, carnivores require less monitoring and processing of their food.

By contrast, many herbivores seem never to stop chewing their food. This chewing is not a function of monitoring, since herbivores often have diets as monotonous as those of carnivores. Rather, extensive chewing increases digestive efficiency by fragmenting plant cell walls.

The mouth and the tongue are also important in swallowing—the movement of food from the mouth to the pharynx—a function aided by the liter of lubricating fluids produced by the **salivary glands** every day. Saliva contains secretions for attacking mouth bacteria that cause tooth decay. It also contains an enzyme called *salivary amylase* (L. *amylum*, "starch"). This enzyme begins the digestion of starch, the most common storage polysaccharide in plants.

The Esophagus

The **esophagus** (Gr. *oisophagus*, "passage for food") is a straight tube, approximately 25 cm in length and 2.5 cm in diameter. It links the pharynx with the stomach, but it has no digestive function of its own because it secretes no enzymes and absorbs no materials.

The lower portion of the esophagus is formed of visceral muscle (see Chapter 12), the same type of muscle found in the wall of the stomach and intestines. This section of the esophagus contracts rhythmically to form circles of contractions that move wavelike along the esophagus—a process called **peristalsis** (Gr. *peristellein*, "to surround"; see Figure 6.3). Peristalsis in the esophagus allows us to swallow even when we are standing on our head (although this method of eating is frowned upon in better restaurants).

Peristalsis also occurs in the stomach and intestines. In all of these organs, it has the effect of moving food farther along the digestive tract.

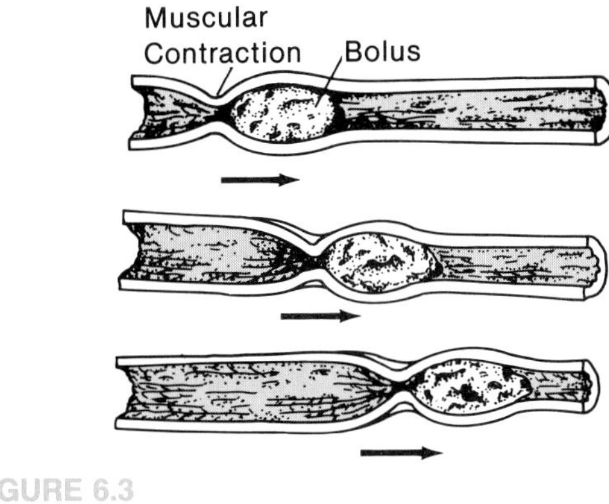

FIGURE 6.3

Peristalsis

A successive wave of contractions in the smooth muscles enveloping the digestive tract propels the bolus of semidigested food along the length of the system.

The Stomach

The **stomach** serves as a holding chamber for food. Depending on the size and composition of a meal and on the period of time since the previous meal, food may stay in the stomach for four hours or more. During that time the stomach gradually liquefies the food through a combination of churning movements, secretion of enzymes, and production of large amounts of mucus. **Sphincter muscles** (circular bands of muscle) seal off both ends of the stomach during this time to prevent leakage back into the esophagus or into the small intestine. In essence, the stomach acts as a rather sophisticated blender.

The stomach produces almost two liters of secretions daily. Mucus makes up the bulk of these secretions, but the cells of the stomach wall (see Figure 6.4) also produce HCl (hydrochloric acid) and the protein-digesting enzyme **pepsin.** Because pepsin is released in an inactive form, *pepsinogen,* the cells do not digest themselves during the production process. (Protein-digesting enzymes produced elsewhere in the digestive system are also secreted in an inactive form, for the same reason.) The independent production of HCl by other cells of the stomach wall creates a stomach pH as low as 1.5, and in this highly acidic environment pepsinogen is activated and converted to pepsin. A thick layer of mucus protects the walls of the stomach from the ravages of pepsin. As a further protection, the cells of the stomach wall are replaced every three days or so.

The Small Intestine

With a length of approximately six meters, the **small intestine** is actually the longest single segment of the digestive tract. The adjective *small* derives not from its length but from its width; at 3 cm, the small intestine is only about half the diameter of the large intestine.

The small intestine has rather arbitrarily been divided into three sections: the **duodenum,** the **jejunum,** and the **ileum** (not to be confused with the *ilium,* or hip bone). The arbitrariness of these subdivisions is illustrated by the name assigned to the first section: *duodenum* is derived from the Latin word for "twelve," a reference to the fact that the length of the duodenum was determined as equivalent to the width of twelve fingers (about 25 cm). The duodenum is of particular interest because it is the site where most of the intestinal enzymes are produced. It is also where ulcers are most likely to occur.

Once the stomach has finished liquefying a meal into what is euphemistically called **chyme** (Gr. *chymos,* "juice"), it does not simply dump this material into the small intestine, like a toilet being flushed. Rather, the chyme is allowed to seep through the **pyloric sphincter** (the boundary between the stomach and duodenum) a little at a time.

Movement of the chyme through the small intestine is not unidirectional, as one might assume. Bands of smooth muscle alternately contract and relax (a process called **segmentation**), sloshing the chyme back and forth, and ensuring both a thorough mixing of the chyme with the digestive enzymes and repeated contact of the chyme with the absorptive surfaces of the intestine.

More than any other section of the digestive tract, the small intestine is responsible for the breakdown (digestion) of large molecules and the absorption of the resulting small molecules. Digestion of these molecules occurs both through the secretion of digestive enzymes by the small intestine, and through the secretion of enzymes and other substances from adjacent organs that empty their secretions into the small intestine. See Table 6.1 for a description of digestive enzymes and their function.

The volume of secretions produced by the small intestine totals two liters per day, although most of this amount is in the form of mucus. However, the small intestine also produces a full complement of digestive enzymes that collectively can digest all four classes of macromolecules. The enzymes are contained in cells sloughed off into the cavity of the small intestine, and the rate of cell loss is such that the entire lining of the intestine must be replaced every 36 hours!

The human small intestine is designed for efficient absorption of materials. Not only is it long and narrow, but the area of its interior surface is increased by a factor of 10 by the presence of millions of tiny (0.5 to 1.5 mm), finger-like projections called **villi** (L. *villus,* "tuft of hair"; see Figure 6.5). Moreover, the cells that form the outer surface of each villus are

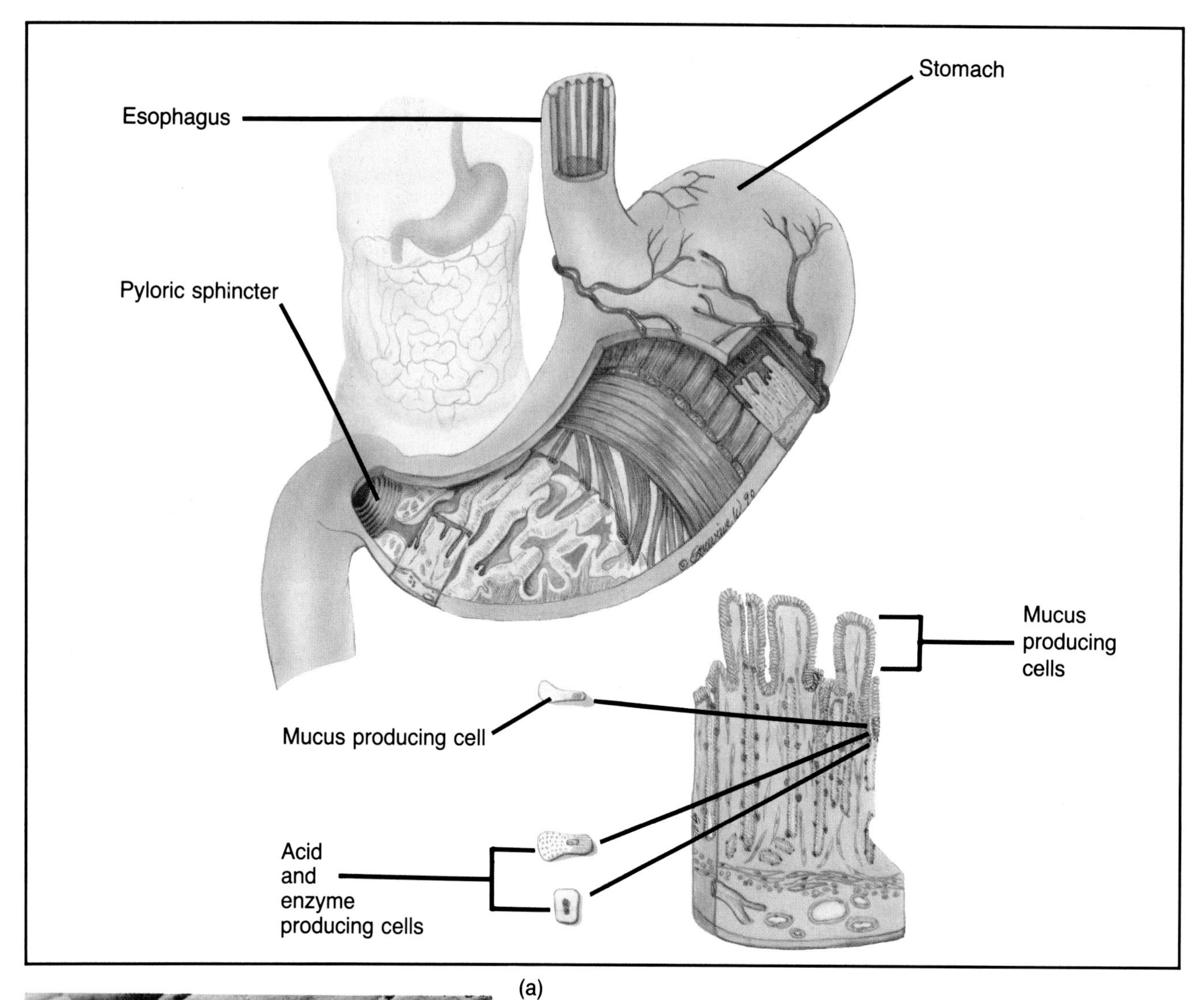

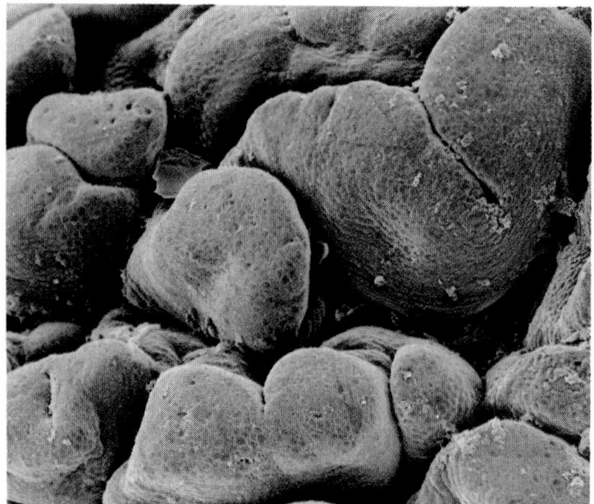

FIGURE 6.4

Stomach and Secretory Cells

(a) The various stomach secretions are produced independently by different cells, most of which are located in so-called gastric pits. (b) A scanning electron micrograph showing the ridged interior wall of the stomach very clearly.

covered by brush-like projections called **microvilli,** which further increases the surface area of the intestinal lining by twenty-fold. As a consequence of the presence of the villi and the microvilli, the absorptive surface of the small intestine is about 250 m^2, or roughly the area of a three bedroom house. If the small intestine were a smooth tube, lacking villi and microvilli, it would have to be 2 km long in order to have an equivalent surface area (and presumably we would have to transport such a monstrous organ in a wheelbarrow). Esthetically (and physiologically), villi and microvilli seem a better alternative.

TABLE 6.1 Digestive Enzymes and Function

LOCATION OF DIGESTIVE ACTION	SOURCE OF DIGESTIVE SECRETION	DIGESTIVE SECRETION	ENZYME	ENZYME ACTION
Mouth	Salivary glands	Saliva	Salivary amylase	Starch → maltose
Stomach	Stomach lining	Gastric juice	Pepsin	Proteins → peptides
		HCl	None	No enzymatic activity
Small intestine	Liver	Bile	None	No enzymatic activity
	Pancreas	Pancreatic juice	Trypsin, Chymotrypsin, Peptidases,	Peptides → amino acids
			Pancreatic amylase,	Starch and glycogen → maltose
			Pancreatic lipase,	Fats → fatty acids and glycerol
			Pancreatic nuclease,	Nucleic acids → nucleotides
	Small intestine	Intestinal juice	Peptidases	Peptides → amino acids
			Maltase	Maltose → glucose
			Sucrase	Sucrose → glucose and fructose
			Lactase	Lactose → glucose and galactose
			Lipase	Fats → fatty acids and glycerol
			Nucleases	Nucleotides → nucleic acids and sugars

The Accessory Digestive Organs

Three organs—the **pancreas,** the **liver,** and the **gallbladder**—play an important role in the digestive process, even though no food passes through them during digestion (see Figure 6.6).

Pancreas. The pancreas, an organ about 15 cm in length, lies at the curve of the duodenum. Its role as the source of the hormone insulin is widely known (Chapter 13); less widely known but equally important is its role in digestion. The pancreas produces not only a series of enzymes that can break down all of the major categories of food but also a large volume of *sodium bicarbonate,* a substance that neutralizes the acidic chyme entering the small intestine from the stomach. Neutralization of the chyme is important not only to protect the lining of the small intestine but also to create an environment in which the enzymes produced by the pancreas and small intestine can be activated. The digestive enzymes of the pancreas alone double the protein-digesting and triple the lipid-digesting capacities of the body.

Liver. The liver, at 1500 g the largest organ in the body, is a multilobed, roughly triangular organ lying largely opposite the stomach on the right side of the body. It is probably second only to the brain in organ complexity, and undertakes a myriad of functions, including many that have nothing to do with digestion (for example, production of red blood cells during embryonic development; production of various plasma proteins, including those associated with blood clotting; detoxification of alcohol and other substances—see Box 6.1). Its digestive functions include metabolism of most of the amino acids, sugars, and lipids absorbed by the small intestine, as well as the production of **bile.**

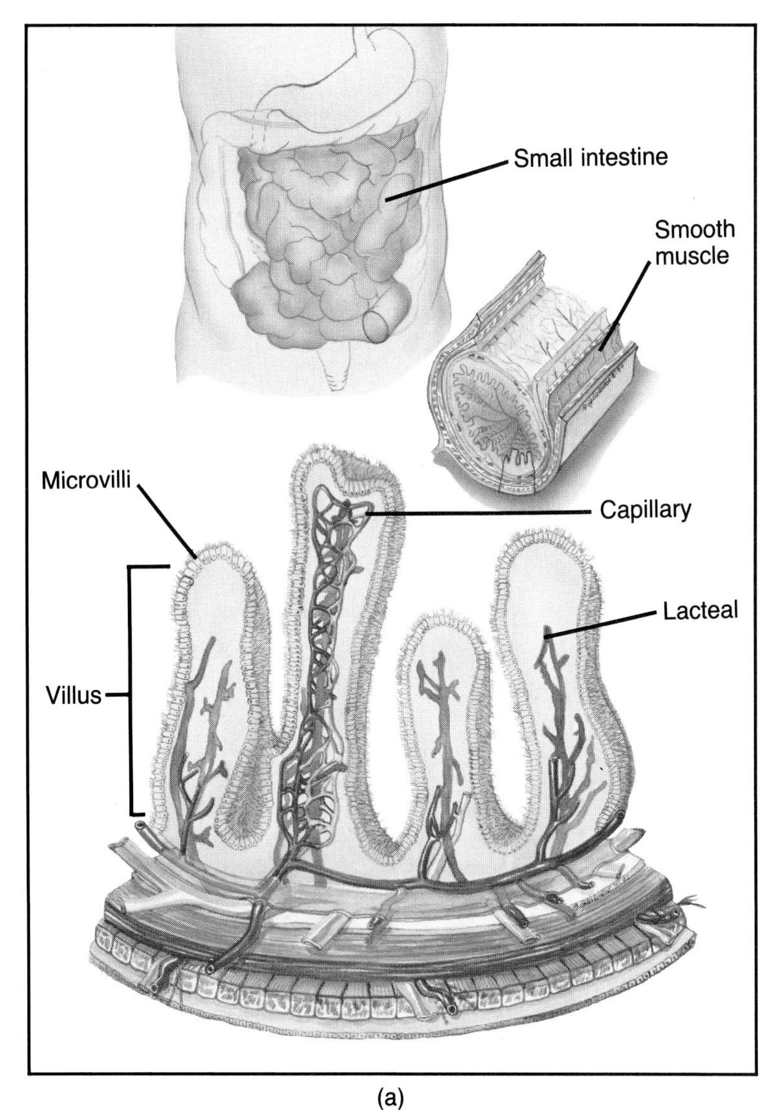

(a)

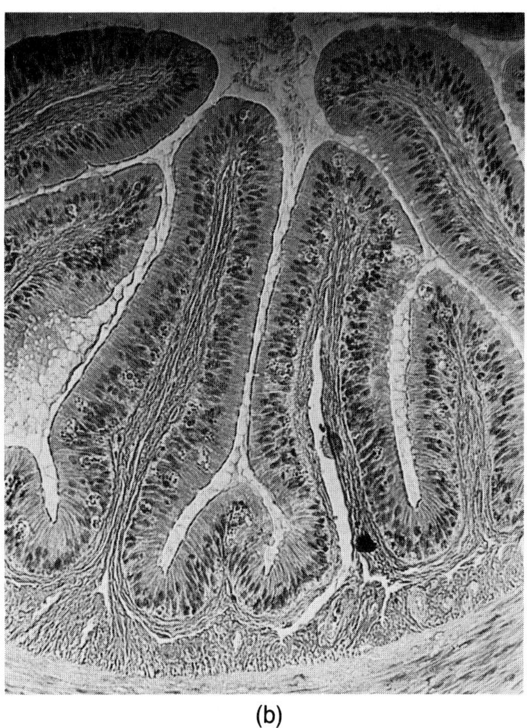

(b)

FIGURE 6.5

Anatomy of the Small Intestine

(a) An artist's rendering of the lining of the small intestine. (b) This micrograph of a section of the small intestine shows the prominent villi that characterize this region of the digestive tract.

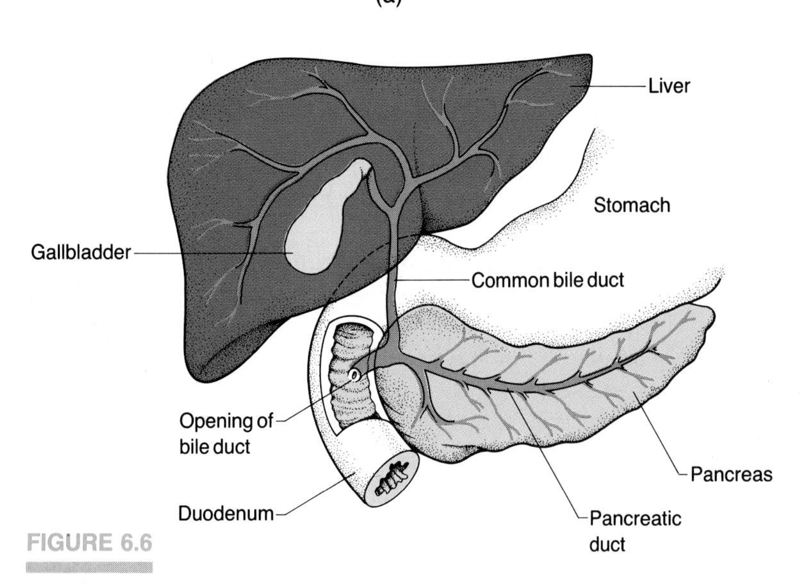

FIGURE 6.6

Interrelationship of Pancreas, Liver, Gall Bladder, and Duodenum

Note that the pancreatic duct joins the bile duct before entering the duodenum. Thus blockage of the bile duct by gallstones can present pancreatic secretions from reaching the small intestine.

BOX 6.1

Alcohol and the Liver

As you are no doubt aware (although certainly not from personal experience!), the intoxicating effects of grain alcohol (**ethanol**) diminish within a few hours of ingestion. The implication is that the body must have some mechanism for detoxifying what is actually a rather toxic substance—after all, alcohol is often used as a preservative. As is true for many toxic substances, the detoxification of ethanol occurs in that most versatile of organs, the liver.

Detoxification comes at a price. Liver cells convert ethanol to carbon dioxide and water by first stripping off two hydrogen atoms using NAD^+, forming NADH in the process. NADH, in turn, transports the hydrogen to the mitochondria, where they are passed along the electron transport chain, permitting the formation of ATP from ADP and free phosphates (see Chapter 5 for details). The fact that we can form ATP from the energy contained in the covalent bonds of the ethanol molecule means that alcohol has caloric value (unlike water, for example), which is why dieters are encouraged to reduce or eliminate their consumption of alcohol.

The problem with alcohol detoxification stems from the use of NAD^+, because now NAD^+ is not available for glucose metabolism. As a consequence, sugars and other energy molecules are not broken down by the liver cells, but instead are converted to fats. With prolonged and relatively intense alcohol use, the liver cells become swollen with fat and eventually die, to be replaced with scar tissue—a condition known as **cirrhosis** (Gr. *kirrhos*, "tawny," a reference to the yellow-orange appearance of a diseased liver). As more and more of the functional liver cells are replaced by scar tissue, the liver gradually becomes unable to perform its many vital functions. Cirrhosis and other liver diseases currently rank ninth among the various causes of death in the United States.

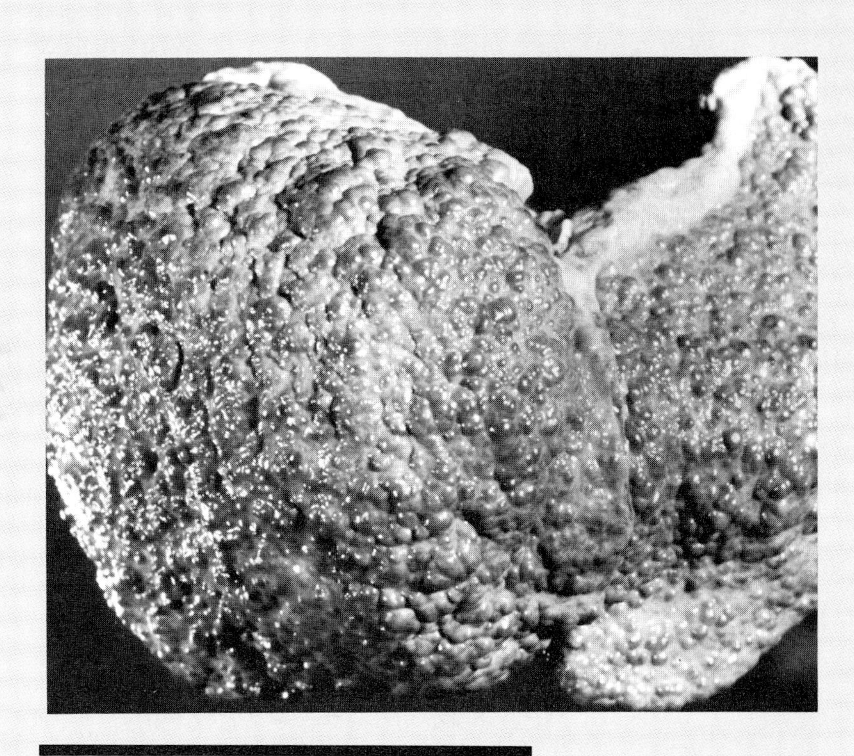

Surface view of a portion of a cirrhotic liver.

Bile is composed of *bile salts,* the lipids *cholesterol* and *lecithin,* and *bile pigments,* the last being the pigments that remain after the destruction of worn-out blood cells (see Chapter 9). Most of the salts and lipids are reabsorbed by the small intestine and are rapidly reconverted to bile. In contrast, most of the bile pigments pass out with the feces, to which they impart a characteristic brownish color. The rest of the pigments are reabsorbed and then passed to the kidneys to be excreted with the urine, to which they impart a characteristic yellowish color.

Gall Bladder. The liver produces bile at an essentially constant rate. However, bile is prevented from entering the small intestine by a sphincter muscle at the opening of the bile duct. When this sphincter is closed, the bile backs up and enters the gallbladder, a sac-like structure roughly 2.5 cm across and 6 cm long. The gallbladder not only stores bile but also concentrates it by absorbing much of the water.

The Large Intestine

The large intestine, or **colon,** consists of a tube 1.5 m in length and 6 cm in diameter, oriented in an inverted U-shape in the lower abdomen. The junction with the small intestine is in the lower right abdomen and is marked by the presence of a small blind sac called the **appendix** (see Figure 6.7).

Chyme moves slowly through the large intestine; it may take up to 24 hours to traverse the whole length. Peristaltic contractions of the large intestine are few; sometimes they occur as much as 30 minutes apart.

The large intestine makes no significant contribution to digestion because it produces no enzymes. Its principal function is to concentrate the 500 ml of chyme it receives daily by reabsorbing most of the remaining water—a process that converts the chyme to **feces** (L. *faeces*, "dregs"). In addition, cells of the intestinal wall release excess calcium and iron into the cavity of the large intestine, to be passed out with the feces.

Defecation

At periodic intervals—typically after a meal, as food moves from the stomach to the small intestine—the **gastrocolic reflex** triggers strong peristaltic contractions of the large intestine, and the residue of the undigested food moves to the **rectum,** the terminal 20 cm of the large intestine. The opening of the rectum, the **anus,** is guarded by two sphincter muscles. The internal sphincter is composed of visceral muscle over which the body can exercise no conscious control. The external sphincter, by contrast, is composed of skeletal muscle, and its control is a learned response, although one not generally mastered until the second year of life (as those of you who have diapered babies well know). Thus the urge to defecate can be resisted through control of the external sphincter (a fact that prevents no end of potentially embarrassing social blunders). However, because water continues to be absorbed from the feces while they remain in the rectum, prolonged control of the external sphincter can result in the feces becoming hard and painful to pass.

The frequency of defecation is highly variable, depending largely on the quantity and quality of food eaten. Young children may defecate several times daily; an elderly person with a very modest appetite,

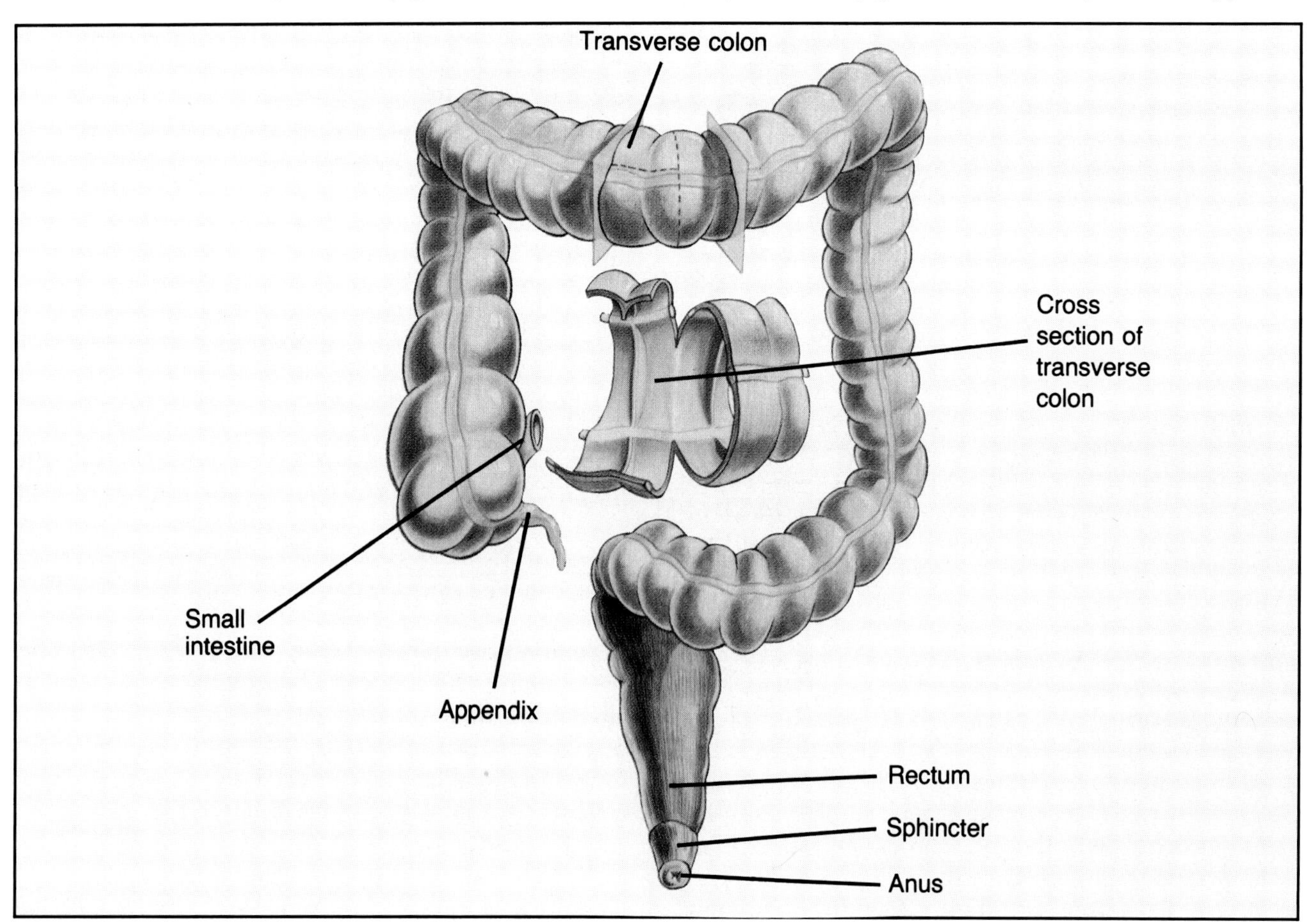

FIGURE 6.7

Large Intestine and Appendix

Note that the appendix is actually at the end of a short blind chamber of the large intestine. Blockage of the opening into the appendix can cause appendicitis.

as infrequently as once a week. There is no optimum frequency, despite laxative manufacturers' suggestions to the contrary. Moreover, the commonly held belief that feces are poisonous and their retention in the body is dangerous is simply not true. In fact, in certain pathological conditions, individuals have gone for weeks without defecating and have suffered no more serious consequences than substantial weight gain.

THE CHEMISTRY OF DIGESTION

As we learned in Chapter 5, most enzymes are very specific for particular classes of molecules. Thus, given the diversity of our diets, it should not surprise you to learn that we produce a large number of digestive enzymes. Before we examine how the production and release of these enzymes is coordinated, let us briefly follow the fate of each of the major classes of food molecules.

Carbohydrates

Carbohydrate digestion begins in the mouth, where salivary amylase splits some of the starch into disaccharides. The pancreas also produces amylase, and pancreatic amylase is functionally far more important in starch digestion that is salivary amylase. A series of intestinal enzymes splits the various disaccharides into monosaccharides, most of which are glucose.

Lactose (milk sugar, a disaccharide) requires a specific enzyme (*lactase*) for its digestion. Interestingly, although this enzyme is present in young children, it is scarce or absent in many adults. Moreover, the presence or absence of the enzyme varies among ethnic groups. For instance, most Asians and blacks produce very little lactase as adults, whereas only a small minority of Caucasians lack the enzyme. Individuals without the enzyme are said to be *lactose-intolerant;* large quantities of milk (though not cheese or yogurt where lactose has been broken down by bacterial action) cause cramps, diarrhea, and other signs of digestive distress, all created by the enthusiastic metabolism of lactose by bacteria in the large intestine.

Lipids

Lipid digestion begins in the stomach; however, pancreatic and small intestinal enzymes play the dominant role. Most of the dietary lipids are fats, and fats are relatively small molecules. Thus the role of the lipid-digesting enzymes is simply to split the fatty acids from glycerol.

The digestion of lipids is not terribly efficient, but that fact is not particularly significant, since fat molecules readily dissolve through the plasma membrane of the absorptive cells. Of greater importance is preventing the fat molecules from forming large clumps as they float along in the watery chyme. Clump prevention is the job of bile.

Bile functions to **emulsify** lipids (that is, to form a suspension) in a water solution. In other words, bile acts in the same way soap does when we wash our hands to remove grease or oil. The large globs of lipid in the digestive system are fragmented by the emulsifying power of bile, and their ultimate destruction by enzymes is thereby made much easier. In addition, bile aids in the absorption of lipids by the cells lining the intestine. The efficiency of lipid digestion and absorption is almost doubled by the presence of bile, even though bile itself does not function as an enzyme.

Proteins

Because of the size and structural complexity of proteins, their digestion is something of a challenge. Pepsin in the stomach breaks the bonds formed by two of the 20 amino acids, creating polypeptide fragments of various sizes in the process. Two protein-digesting enzymes from the pancreas split the bonds formed by five other amino acids, producing somewhat shorter polypeptides. Finally, a series of very specific enzymes from the small intestine (plus one from the pancreas) break off one amino acid at a time from the small polypeptides.

Nucleic Acids

Nucleic acids form only a tiny fraction of the total volume of food ingested, but they are still subject to enzymatic destruction. Most cells produce enzymes for breaking down both RNA and DNA within the cell proper, since these molecules are constantly being reorganized within each cell; cells of the digestive system release these enzymes into the intestinal cavity, where both DNA and RNA are broken down into their constituent parts.

ABSORPTION

The principal role of the stomach is digestion (especially of proteins), not absorption. However, the cells of the stomach wall do absorb some substances, notably water, alcohol, and weak acids such as aspirin. The capacity of the stomach to absorb alcohol explain why the effects of imbibing alcohol (especially on an empty stomach, with no food to provide a diluting effect) are so immediate.

Most absorption takes place in the small intestine, especially the jejunum and ileum. The villi of the small intestine are lined with a network of blood capillaries, and each villus also possesses an open-ended lymph vessel called a **lacteal.** Amino acids and sugars, along with other water-soluble nutrients, are collected by the blood capillaries and transported to the **hepatic portal vein,** which carries them to the liver to be metabolized (see Figure 6.8).

In contrast, fats and fat-soluble substances pass into the lacteals and are transported through the lymphatic system before finally entering the bloodstream near the heart. (For details of fat absorption, see Chapter 7). Although they are ultimately also metabolized by the liver, the fact that fats are dumped into the bloodstream in this fashion poses challenges for the body. (Recall that lipids are nonpolar, and tend to clump in water-based solutions, such as blood.) Indeed, it is widely believed that many of the more severe problems of the circulatory system result from failures—or overloads—of the lipid transport system (see Chapter 8).

The large intestine absorbs much of the water that we ingest or secrete and certain of the vitamins (most notably K and some of the B vitamins), which are byproducts of the large numbers of bacteria that live in the large intestine. (Indeed, these bacteria are sufficiently numerous that they comprise up to one-half of the dry weight of the feces.)

CONTROL AND INTEGRATION OF DIGESTIVE SECRETIONS

Digestion is obviously more efficient if the digestive enzymes are produced and released at the same time that food is ingested. Conversely, these enzymes should not be produced in abundance when no food is present. How is the production and secretion of enzymes coordinated with food ingestion?

The salivary glands have a basal secretion rate of about 0.5 ml/min, but that rate can increase by a factor of eight in the presence of food (especially acidic food that is neutralized by the somewhat basic saliva). This increase is mediated by nervous stimulation from the brain, in response to the detection of food in the mouth by taste and odor receptors.

The stomach continuously secretes mucus, HCl, and pepsinogen at a rate of about 0.5 ml/min, but the rate increases to about 3 ml/min shortly after a meal, as a consequence of the interplay among three factors: the digestive secretions just mentioned, the food in the stomach, and a hormone called **gastrin,** which is produced by specialized cells within the stomach itself, and which is responsible for increasing the rate at which the digestive secretions of the stomach wall are produced (see Table 6.2).

The production of gastrin is caused largely by the presence of food in the stomach, although not all foods are equally effective in stimulating production (see Figure 6.9). Protein is more effective than carbohydrates, for instance, and alcohol and caffeine are

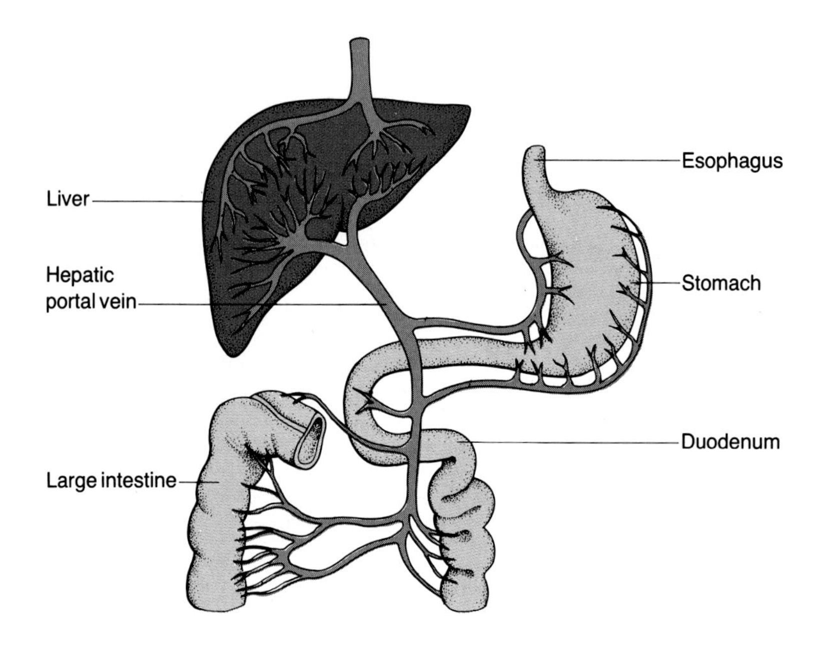

FIGURE 6.8

Hepatic Portal System
The hepatic portal vein drains the small intestine, and to a lesser degree the large intestine and stomach, and transports sugars and amino acids to the liver.

TABLE 6.2 Digestive Hormones

HORMONE	SOURCE	STIMULUS	TARGET ORGAN	FUNCTION
Gastrin	Stomach	Distension, protein, alcohol, caffeine	Esophageal sphincter	Closes
			Pyloric sphincter Stomach	Opens, Secretions, motility
Secretin	Small intestine	Acid, distension	Small intestine	Secretions
			Liver Pancreas	Bile secretion, Sodium bicarbonate secretion
			Small intestine/stomach	Inhibits motility
CCK	Small intestine	Fatty acids, amino acids	Gall bladder	Contraction
			Pancreas	Release of pancreatic enzymes,
			Stomach Brain	Inhibits motility, Satiation

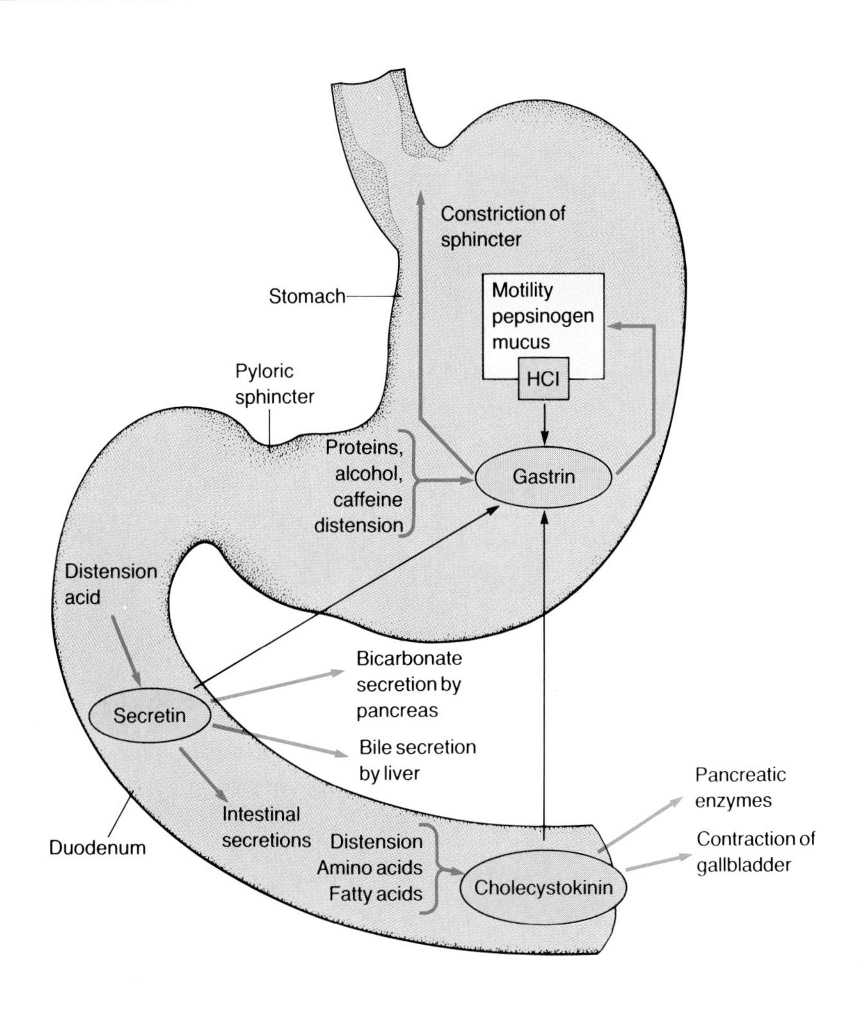

FIGURE 6.9

Hormonal Control of Digestive Secretions
Gastrin is produced by, and acts on, the stomach; secretin and CCK are produced by cells of the small intestine. Colored arrows indicate stimulation; black arrows indicate inhibition.

also powerful stimulants of gastrin release. Thus a soup course, a pre-dinner cocktail, or a cup of coffee with the meal all serve as aids to digestion. Conversely, alcohol or coffee taken on an empty stomach is counterproductive and even potentially harmful because stomach secretions will be produced in quantity when no food is present on which they can act. Therefore people with stomach ulcers are urged not to drink alcohol or caffeine-containing beverages.

Besides the type of food, the *volume* of food also influences the rate of gastrin production. Food volume is measured by the amount of distension, or stretching, of the stomach—the greater the distension, the more gastrin is produced. On the other hand, gastrin production is inhibited by a high acid concentration in the stomach, a condition reached only late in the digestive process, after large amounts of HCl have been produced and mixed with the food.

Gastrin controls the release of approximately 80 percent of the daily volume of stomach secretions. The remaining 20 percent is under the control of the nervous system. Mood and input from the senses play a role in controlling these secretions. An atmosphere conducive to eating—one that includes soft lighting, pleasant conversation, delectable odors—favors the release of stomach secretions. Conversely, an inimical atmosphere—one marked by garish cafeteria decor, arguments, and greasy food—tends to inhibit stomach secretions, and to favor the onset of indigestion.

The presence of chyme in the duodenum—or, more specifically, the presence of acid plus distension of the duodenum—trigger the release of the hormone **secretin,** which, in turn, causes the liberation of the intestinal secretions, promotes secretion of bile by the liver, and stimulates release of sodium bicarbonate by the pancreas.

If amino acids or fatty acids are present in the chyme, they trigger the release of a second intestinal hormone, **cholecystokinin** (Gr. *chole,* "bile"; *kytis,* "bladder"; and *kinin,* "to move") which is responsible for causing the gall bladder to contract, releasing bile, and for stimulating secretion of the pancreatic enzymes. In addition, cholecystokinin (or **CCK** as it is more commonly known) inhibits stomach contractions and thereby reduces the rate at which chyme enters the small intestine. Because CCK is triggered only by amino acids and fatty acids, meals high in protein and fat tend to remain longer in the stomach. The activities of CCK help explain why you are less likely to feel hungry at 10 A.M. if you breakfasted earlier on bacon and eggs than if you ate cereal and grapefruit.

CCK is also believed to act on the brain to indicate satiety. Thus some researchers believe that the morbidly obese are individuals who either produce too little CCK or whose brains are nonreceptive to this hormone.

PROBLEMS WITH THE DIGESTIVE SYSTEM

The digestive tract is the site of numerous problems and ailments of varying levels of significance, including the following.

Caries. Dental caries (tooth decay) is a problem for most people at some point in their lives. Bacteria feeding on sugars in the mouth produce acid, which etches increasingly larger cavities in the enamel and ultimately the dentin of the tooth. Efforts are underway to immunize against such bacteria, but until such efforts are successful, standard practice will remain the removal of carbohydrate-containing food particles through brushing, flossing, and regular visits to the dentist for the removal of tartar from the teeth.

Heartburn. Before it joints the stomach, the esophagus extends about 2.5 cm past the diaphragm. This small section of the esophagus acts as a sphincter and prevents food in the stomach from being regurgitated into the esophagus. The sphincter opens only during the act of swallowing, to allow food to pass into the stomach. We might assume that gravity alone would keep food in the stomach, but the abdominal cavity is at higher pressure than the chest cavity, and in the absence of a sphincter, stomach contents would tend to be pushed back into the esophagus. (In infants, the esophagus essentially ends at the diaphragm. Consequently, the sphincter is much less effective, which accounts for a much higher rate of regurgitation in babies than in adults, as any practiced baby burper can attest.)

Despite the presence of the sphincter, food is sometimes regurgitated into the esophagus. The resulting irritation of the sensitive esophageal tissues by the highly acidic stomach secretions causes heartburn—a totally erroneous term derived from the fact that the esophagus joins the stomach just below the heart.

The sphincter is constricted by the actions of the stomach hormone gastrin. Individuals who have had portions of their stomachs surgically removed produce less gastrin and consequently tend to suffer more frequently from heartburn.

Vomiting. Although generally regarded as one of the most revolting of bodily functions, vomiting is an important defense mechanism. Our sensory systems are not infallible in monitoring food quality, and occasionally we may inadvertently ingest toxic substances or *pathogens* (disease-causing organisms). Such materials might prove hazardous indeed if we had no way to get rid of them except to wait 24 hours or more for them to traverse the gut tube.

Vomiting involves relaxation of the esophageal sphincter and violent contraction of the diaphragm and the abdominal muscles. A variety of stimuli other than toxic food may induce vomiting—for example, gagging, excessive distension of the stomach, intense pain, dizziness, and a number of sights or smells. (Thus the odor of vomit or the sight of someone vomiting can give rise to a biological application of the "domino theory.")

Although vomiting is primarily a defense mechanism, excessive vomiting can be dangerous because the fluids lost are primarily the highly acidic stomach contents. Repeated acid loss can upset the delicate pH balance of the body, with potentially disastrous results.

Peptic Ulcers. Despite extensive safeguards, the lining of the gut may be attacked and digested by pepsin and HCl. Peptic ulcers are called *gastric ulcers* if they occur in the stomach (see Figure 6.10) and *duodenal ulcers* if they occur in the beginning of the small intestine. The latter are far more common, especially in the region of the duodenum between the stomach and the entry point of the common bile duct (which carries neutralizing bicarbonate solution from the pancreas). Overall, about 10 percent of adult Americans are afflicted with peptic ulcers at some point in their lives.

Just why some individuals develop ulcers and others do not is unclear; however, genetic predisposition, a stressful life-style, and poor eating habits all appear to be contributing factors. For people prone to ulcers, stomach irritants such as caffeine, alcohol, tobacco, and aspirin are off limits.

The danger with ulcers is that the gut lining may be eaten away until the underlying blood vessels hemorrhage (*bleeding ulcers*) or until a hole is formed in the gut wall (*perforated ulcers*). Under these conditions surgery may be necessary. Otherwise, drugs such as *cimetidine,* which blocks HCl secretion, are often effective in curing peptic ulcers.

Gallstones. When bile is concentrated in the gallbladder, the cholesterol present in the bile is virtually at saturated levels. In certain individuals (notably those who are one or more of the following: obese, diabetic, middle-aged, or female), the cholesterol may precipitate and form gallstones. Gallstones are very common, occurring in 20 percent of Americans. They are not a serious problem as long as they remain in the gallbladder, but if they move into the bile duct, they may block it and prevent the bile from flowing into the small intestine. The yellowish bile pigments are then forced back into the body and give a yellowish cast to the skin—a condition known as **jaundice** (Fr. *jaune,* "yellow"). Pancreatic secretions may also be blocked, since they share a portion of the bile duct. Fortunately, gallstones can often be treated without surgery. Patients can, in some instances, be treated with small doses of one of the bile salts over a period of several months; the bile salts cause the gallstones to dissolve gradually. Alternatively, the lithotripter, an instrument initially developed to fracture kidney stones by generating pressure waves (see Chapter 11), has been modified for use on gallstones. The applicability of these techniques depends on the urgency of the case, the type of gallstones (sand-like grains

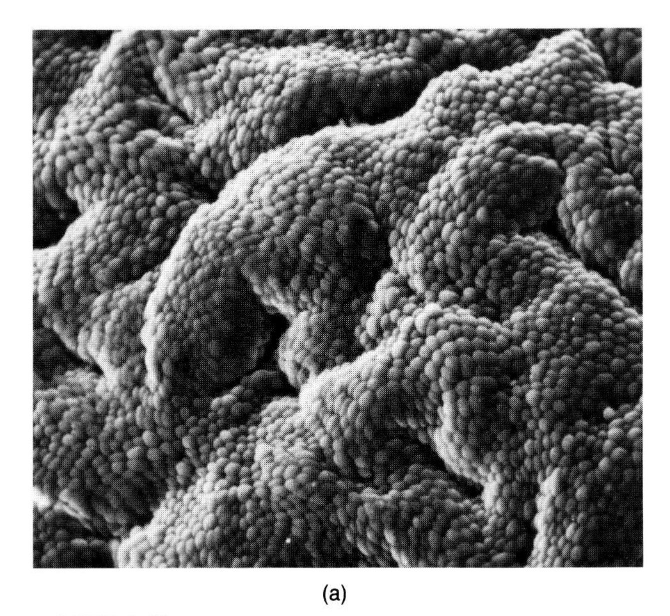

(a)

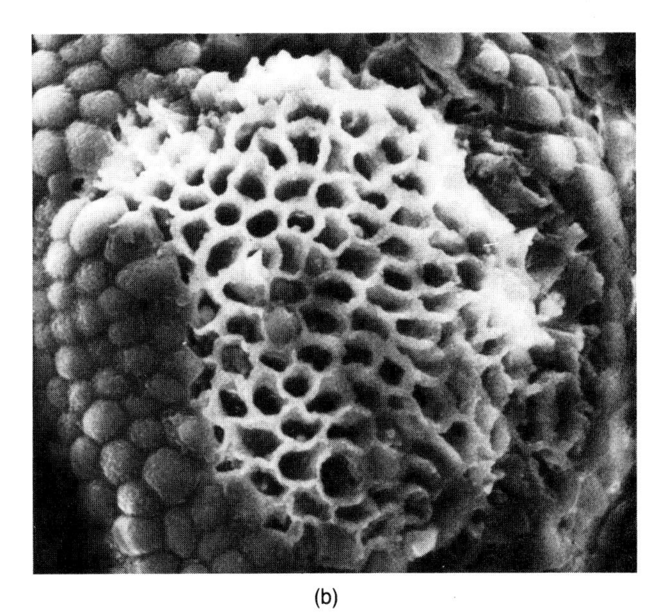

(b)

FIGURE 6.10

Peptic Ulcers
(a) The normal appearance of the stomach lining. (b) Ulcerated stomach lining.

versus gravelly stones), and the overall health of the patient.

Appendicitis. The appendix is a small blind sac near the junction of the small and large intestines. The appendix has no known function in humans, but it is subject to infection. In fact, appendicitis is so common that any severe pain in the right abdominal region—particularly if the pain is accompanied by tenderness and swelling—is immediately suspect as indicating appendicitis. If left untreated, the appendix may burst and cause the infection to spread from the gut to the lining of the abdominal cavity, a condition known as **peritonitis.** Peritonitis, which can also result from perforated ulcers, is an extremely serious disease and was frequently fatal until the discovery of antibiotics some 50 years ago. The progress of this infection illustrates the consequences of the development of a connection between the outside of the body (the digestive tract) and the inside of the body (the abdominal cavity itself).

Diarrhea and Constipation. Diarrhea (Gr. *diarrhein,* "to flow through") has many causes, but it is most frequently associated with intestinal infection. In essence it results from an increase in peristalsis that prevents the chyme from remaining in the large intestine long enough for most of the water to be reabsorbed. Diarrhea is the analog of vomiting; the two processes serve to empty the digestive tract and thereby to minimize the length of time the body's absorptive surface is exposed to disease organisms.

In conditions such as "intestinal flu," diarrhea is usually nothing more than a nuisance. However, in diseases such as **amebic dysentery** or **cholera,** death may result from loss of ions and from dehydration. (Recall that seven liters of fluid are secreted into the digestive tract daily, as compared with a blood volume of only about five liters. Obviously, most of these secretions must be reabsorbed for reuse.)

Constipation is the opposite of diarrhea, but it results more frequently from improper diet than from disease. Doctors are loath to prescribe laxatives for anything other than immediate relief; not only can the body become dependent on laxatives, but the problem can usually be corrected simply by an altered diet. Nonetheless, laxatives constitute one of the most common over-the-counter medications—their sales aided in large measure by television ads that suggest a rich and rewarding life is only a spoonful away.

Laxatives fall into four main groups:

1. Bulking laxatives, such as bran and fiber, which are not enzymatically digested and yet have the capacity to absorb water, leading in turn to an increased volume of soft fecal material.

2. Lubricants, such as mineral oil, which operate just as the name would imply. However, mineral oil interferes with the absorption of certain ions, such as potassium and calcium, and various fat-soluble vitamins. Thus the prolonged use of mineral oil is contraindicated.

3. Mineral salts, such as magnesium, which, because they are not absorbed, increase the osmotic pressure of the fecal material, resulting in less water uptake by the large intestine.

4. Irritants, such as castor oil, which speed up the rate of peristalsis, in an analogous fashion to certain pathogens. Again, the result is that the fecal material does not remain in the large intestine long enough for more than a fraction of the water to be removed. Certain natural foods, such as prunes (which are also rich in fiber), fall into this category.

Irritable Bowel Syndrome. Irritable bowel syndrome (also known as *spastic colitis*), is a condition marked by abdominal discomfort and irregular bowel function, frequently coupled with loud gurgling sounds. Constipation alternates with diarrhea, the latter characterized by excessive amounts of mucus. The condition is believed to be most commonly caused in response to emotional distress.

Inflammatory Bowel Disease. Inflammatory bowel disease, or chronic enteritis, is manifested as two related, although distinguishable, diseases, each of which affects about one million Americans. **Crohn's disease** (ileitis) primarily affects the small intestine; **ulcerative colitis** most commonly involves the colon and rectum. In both diseases, the walls of the intestine become ulcerated and inflamed, with the production of excessive amounts of mucus and, in some cases, blood in the stool.

The cause of these diseases is unknown. However, there is increasing evidence that they result from an autoimmune reaction, meaning that the white blood cells attack body tissue (see Chapter 9). What prompts the autoimmune reaction in the first place—pathogen or genetic predisposition—is still a mystery. The severity of these diseases tends to wax and wane, suggesting a possible emotional component, and existing drug therapy is only sometimes effective. Surgical removal of the affected regions is ultimately necessary in a significant proportion of patients. However, very recent findings of an overabundance of a particular neurotransmitter (see Chapter 14) and an abnormally high sensitivity to certain substances produced by white blood cells in sufferers from these diseases suggest new avenues for the development of more effective drugs in the near future.

Diverticulosis and Diverticulitis. The colon sometimes develops outpocketings, or *diverticula,* that may become infected. **Diverticulosis** (the presence of these outpocketings) is rather common in the elderly and is by itself not serious. It is generally treated with a diet containing increased amounts of leafy vegetables and grains to increase the rate of fecal formation—and such diets help avoid the problem in the first place. Occasionally, the diverticula become infected (**diverticulitis**), and may even perforate the colon. In such instances, surgical removal of the affected area is often mandated.

Summary

The digestive system is a tube through the body, designed to digest food and to absorb the resulting small molecules for use by the cells of the body. The anatomy and physiology of the system are closely intermeshed, such that release of digestive enzymes is correlated with the entry of food into the various specialized compartments of the system. Food that cannot be digested is voided in the form of feces.

Digestion begins almost as soon as the food enters the mouth because saliva contains a starch-digesting enzyme. Most digestion occurs in the stomach and small intestine, and is mediated by secretions from these two organs and from the liver and pancreas. By contrast, absorption is limited almost entirely to the small and large intestines. The absorbed ions and molecules are accumulated by the capillaries and lymph vessels of the intestinal lining. Most of the absorbed materials are then transported directly to the liver.

The digestive system is also an optimum habitat for many bacteria, some of which could pose a severe threat to the body as a whole, if they were to gain access through a leak in the system. Because of a series of dramatic changes in pH, coupled with the continued presence of various enzymes, most organic compounds, including the cell walls of many types of bacteria, are fragmented during their trip through the digestive tract.

Nevertheless, the system is delicate and rather easily disrupted. Vomiting and diarrhea are common responses to invasion by bacteria and viruses immune to digestion, but even more dangerous are breaks in the intestinal lining, for these openings can lead to the invasion of the body itself. Fortunately, antibiotics were developed not long after intestinal ulcers became common; thus, the advantages and disadvantages of modern society have been balanced rather evenly with respect to the digestive system.

Key Terms

digestive tract
digestion
absorption
omnivore
carnivore
herbivore
salivary glands
salivary amylase
esophagus
peristalsis
stomach
sphincter muscle
pepsin
pepsinogen
small intestine
duodenum
jejunum
ileum
chyme
pyloric sphincter
segmentation
villi
microvilli
pancreas
liver
gall bladder
bile
colon
appendix
feces
gastrocolic reflex
rectum
anus
emulsify
lacteal
hepatic portal vein
gastrin
secretin
cholecystokinin
caries
heartburn
vomiting
pathogen
peptic ulcer
gallstones
jaundice
appendicitis
peritonitis
amebic dysentery
cholera
irritable bowel syndrome
inflammatory bowel disease
Crohn's disease
ulcerative colitis
diverticulosis
diverticulitis

Box Terms

ethanol
cirrhosis

Questions

1. Distinguish digestion from absorption. Are all of the digestive organs equally involved in each?
2. How does the digestive system prevent being digested by its own enzymes?
3. Besides lubricating the food, what role does saliva play in the digestive process?
4. What are villi? What are microvilli? In which digestive organ are they found?
5. What is the role in digestion of the pancreas, liver, and gall bladder?
6. What is the function of the large intestine in digestion? What digestive enzymes does it produce?
7. Using appropriate examples, explain the role of hormones in the control of digestive secretions.
8. Where do the products of digestion go once they are absorbed?
9. What effect, if any, does alcohol have on the organs of digestion?

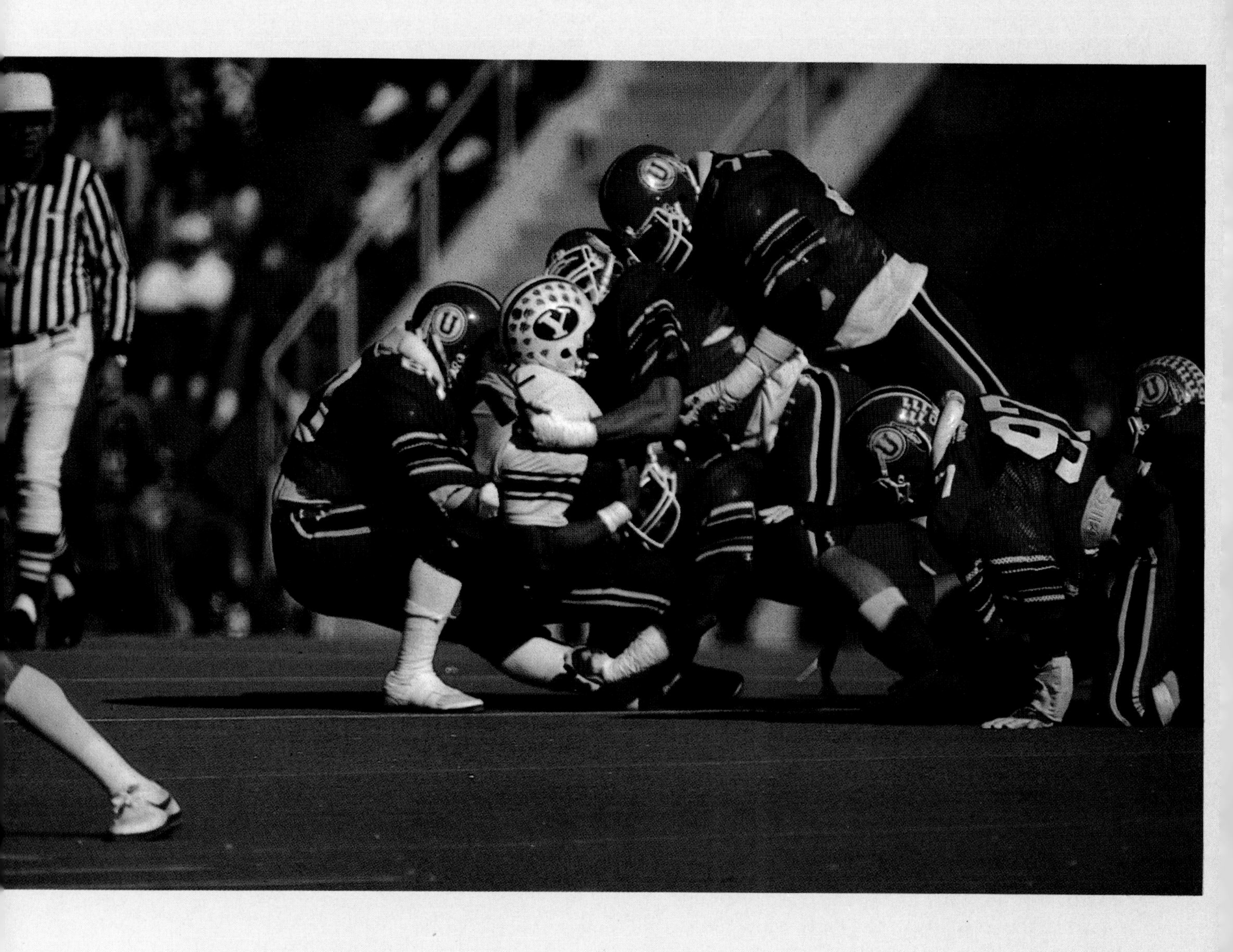

CHAPTER

7

Nutrition

Energy and Building Blocks

King Nebuchadnezzar II of Babylonia reportedly went mad and spent his days going about on all fours and eating grass. Had you been his nutritional advisor, would you have recommended a dietary supplement? Would vitamin pills have been enough? In short, could he have survived on grass alone?

Why do some people continue to add weight, even though they "eat like a sparrow," whereas other people cannot seem to gain weight no matter what they eat? What's the truth about oat bran, cholesterol, and vitamin supplements? Why can't government agencies agree on a single, ideal set of dietary recommendations?

These and other questions will be our focus as we discuss the subject of nutrition.

THE NATURE OF THE PROBLEM

In Chapter 4, we saw how (and where) cells assemble large molecules, such as proteins, from relatively simple molecules, such as amino acids. In Chapter 5, we discussed the capacity of individual cells to convert amino acids, glycerol, fatty acids, and various simple sugars into glucose or glucose derivatives to be fed into the Krebs cycle for energy derivation. Finally, in Chapter 6, we considered how the digestive system systematically fractures certain very large molecules into molecules that are small enough to cross a plasma membrane. Implicit in these discussions is something that we must now make explicit—the capacity of individual cells to convert one class of organic molecules to another class is extensive, but not without limits. The degree of that limitation varies from organism to organism; as a general rule, plants have greater capacities to interconvert molecules than do animals. Humans, for instance, have significant limitations on this transformational capability, meaning that our food must contain a particular diversity of molecules on a regular basis if life is to be sustained.

Missing in this discussion of the interconvertability of organic molecules is any consideration of the various ions we know to be essential to life, or of water (which we humans require in amounts greater than we can produce through the metabolism of glucose), or of those interesting coenzymes we call vitamins. These substances, too, must be in the food we eat. What, then, is the minimum diversity necessary in our food molecules? What are the consequences of falling short of this minimum? How can we be sure that the food we are eating contains the things that are essential—or that it doesn't contain too much of what is not essential? For that matter, are there any negative consequences of ingesting too much of these dietary essentials? Answering these questions leads us into a discussion of the outcome of digestive system functioning—nutrition.

THE ADEQUATE DIET

Few things in modern life seem more confusing than information and recommendations on what we should include in (or exclude from) our diets. At first glance, it seems odd that there should be confusion—after all, surely it is possible to approach the question scientifically, ascertain our nutritional needs, analyze various foods accordingly, and develop a single set of standards. In actuality, things are not that simple. Consider the following complicating factors:

1. *Individual differences:* Not only are people different genetically, but they lead very different life styles. A 40 kg retired woman bookkeeper has very different nutritional needs from her 100 kg lumberjack son—or his pregnant wife. No one standard will meet their diverse needs.

2. *In vitro* versus *in vivo: In vitro* refers to test tube science, whereas *in vivo* pertains to the living world. That is, it is one thing to analyze and determine the nutritional content of a particular food; it is quite another to know how well that food will be digested and absorbed by a human digestive system. Many ions, for example, are poorly absorbed, even though they may be abundant in the diet. Some nutrients interfere with other nutrients, or with medications, during absorption. Subtle biochemical differences among individuals may also lead to very different patterns of absorption. In short, nutrition is still an emerging discipline, and there is a great deal yet to be learned.

3. *Short-term* versus *long-term needs:* Many of the essential nutrients can be stored in the body for varying lengths of time; for others, we have much more limited storage capabilities. Thus, in attempting to define what the human body requires in terms of nutrients, various governmental agencies and other interested groups have wrestled (with only some success) with the problem of what time base to use in making recommendations. Most agencies use a daily time base that they quickly explain is intended only to represent an *average* set of nutritional needs. Many individuals, however, erroneously interpret these recommendations as being daily *requirements.*

4. *Conflicting data and objectives:* Most industrialized nations have a published set of recommended nutritional standards for their citizens (see Box 7.1). These recommendations often differ from country to country based on differences in interpretations of nutritional data, or assumptions about the availability of particular nutrients in particular foods. They also differ because not all purport to do the same thing: some list requirements, some list allowances, still others list standards. Great Britain, for instance, uses "standards," which are generally lower than the "allowances" used in the United States.

5. *Minima and maxima:* A great deal of confusion has resulted from misinterpretation of nutritional recommendations. Many people apparently believe that these recommendations represent absolute minima, below which life is immediately in peril. As a consequence, supplementation has become extremely widespread. As we shall see later in this chapter, many nutrients can be toxic when ingested in large quantities, and supplements of particular nutrients can often impair the absorption of other nutrients.

BOX 7.1

A History of Nutritional Recommendations

The first U.S. food guide was printed in 1916, prior to the discovery of all but one of the vitamins. As such, it emphasized a common sense approach to meal preparation, and categorized foods into five groups, with the assumption that regular consumption of representatives from these groups would ensure a balanced diet.

In 1943, the Department of Agriculture (USDA), mindful of the rationing underway during World War II, published a revised food guide that listed what it termed the Basic Seven food groups, with a recommendation of the number of servings of each per day. The Basic Seven approach was criticized for being unnecessarily complicated, and in 1956 the USDA printed another revision it termed the Basic Four, again with a recommended number of servings of each per day. Fats were dropped as a food group, in part because of growing concern about the perceived link between high fat diets and heart disease.

However, with the discovery of vitamins and essential minerals, the emphasis began to shift from diets emphasizing the organic materials needed for growth and repair to those that emphasized obtaining a sufficient supply of these newly discovered essentials. Thus, in 1941, the National Research Council (NRC) published the first recommendations on vitamins and minerals, setting what is called Minimum Daily Requirements (MDRs). In 1973, the NRC changed its terminology with respect to vitamins and minerals from Minimum Daily Requirements to Recommended Daily Allowances (RDAs), a term still in use today. The RDAs are intended to be high enough so as to prevent deficiencies in virtually all healthy people; they are also intended to represent averages over several days, rather than being a minimum that must be met every day. The RDAs are under continuous review and adjustment, with the most recent being the tenth edition of October 1989. This last edition lists intake suggestions for protein, 11 vitamins, and seven minerals.

In 1979, the USDA modified its Basic Four plan to include a fifth group consisting of fats, sweets and alcohol (all of which were labelled with "caution"). This new plan, entitled the Hassle-Free Food Guide (a name that could only have been used in the 1970s), also made note of the acceptability of peas and beans as a protein source.

Finally, in 1984, the American Red Cross, in conjunction with the USDA, promoted the idea of a six-category Food Wheel, with greater emphasis on whole grains, fruits, and vegetables. About the same time, the Senate Select Committee on Nutrition and Human Needs issued a report that emphasized the role of the diet in such diseases as diabetes, cancer, and cardiovascular diseases in general. This report made a number of recommendations focusing on reduced intake of salt, refined sugars, cholesterol, and fats in

Changed RDAs for Selected Vitamins and Minerals

	ADULT FEMALE		ADULT MALE	
NUTRIENT	1980	1989	1980	1989
Vitamin B_6 (pyridoxine)	2.0 mg	1.6 mg	2.2 mg	2.0 mg
Vitamin B_{12} (cobalamin)	3.0 μg	2.0 μg	3.0 μg	2.0 μg
Folic acid	400 μg	180 μg	400 μg	200 μg
Vitamin K	No RDA	65 μg	No RDA	80 μg
Sodium[1]	1100 mg	500 mg	1100 mg	500 mg
Calcium[2]	800 mg	1200 mg	800 mg	1200 mg
Magnesium	300 mg	280 mg	350 mg	350 mg
Zinc	15 mg	12 mg	15 mg	15 mg
Selenium	No RDA	55 μg	No RDA	70 μg
Iron	18 mg	15 mg	10 mg	10 mg

[1]Values for sodium are "estimated minimum requirement" not RDA
[2]Increased calcium values are for adults younger than 25 years of age

(Continued on the following page)

BOX 7.1 *(Continued)*

general, and increased intake of complex carbohydrates. Almost simultaneously, the National Institutes of Health issued similar (but not quite identical guidelines), and virtually the same recommendations were subsequently made by the American Heart Association, the National Cancer Institute, and the American Diabetes Association.

In sum, we can perceive three separate types of dietary recommendations, each of which has a different set of objectives. The first set, from the USDA, emphasizes a balanced diet. It states minimum intake levels. The second set, from the NRC, primarily addresses dietary essentials (that is, vitamins and minerals). The third set, from various sources, is focused on the role of the diet in causing (or preventing) diseases of excess. It sets maximum intake levels.

Given the multiplicity of lists, each with its own set of objectives, it is not surprising that the American public is so often confused regarding the "correct" diet.

THE ESSENTIAL NUTRIENTS

The essential nutrients include the following:

1. Organic molecules used for energy and for growth or repair
2. Vitamins
3. Minerals
4. Water

Let us consider each of these in turn.

ORGANIC MOLECULES

The organic molecules that are used either for energy or for structural purposes include the four groups introduced in Chapter 3: carbohydrates, lipids, proteins, and nucleic acids.

Carbohydrates

Dietary carbohydrates include both simple (glucose; fructose; sucrose) and complex (starch; glycogen; cellulose) molecules. The intestine is capable of absorbing only monosaccharides; thus the digestive processes involve splitting polysaccharides to their monosaccharide components (primarily glucose). We obtain carbohydrates almost exclusively from plant sources (milk being the only significant exception).

On average, American ingest about 300 g of carbohydrates per day, and carbohydrates provide roughly half of our total caloric intake. While it is true that we can synthesize carbohydrates from lipids and proteins, and therefore technically carbohydrates are not essential in the diet, their absence leads to problems similar to those associated with starvation—loss of sodium and water, followed by a loss of potassium, followed by muscle weakness; the use of protein for energy (either dietary protein or muscle protein); and poor fat metabolism leading to increased blood acidity. All of these problems can be avoided with as little as 60 g of carbohydrate in the diet, although the Food and Nutrition Board recommends 100 g as a minimum. In any event, the primary carbohydrate problem Americans have is not insufficiency but imbalance—the prevailing wisdom today being that we should double our intake of complex carbohydrates (to about 50 percent of our total caloric intake), and cut back correspondingly on our ingestion of simple sugars and fats. (Americans currently consume an average of almost 60 kg of sugar annually—and that amount continues to rise, despite the increased use of artificial sweeteners.)

Aside from their role in energy metabolism, carbohydrates are also important as the source of fiber in the diet (see Box 7.2). However, like anything else, there can be too much of a good thing. Excess amounts of fiber bind various essential minerals including calcium, zinc, and iron, and can be abrasive to the lining of the intestine. A diet that includes 15 to 20 g of fiber a day is recommended; presently, Americans average about 7 g per day.

Lipids

Most of the lipids in our diets are fats. In recent years, there has been considerable controversy regarding the significance of fats in the diet. Fat consumption in the United States increased from 34 percent of the diet in the 1930s to about 42 percent in the late 1950s through the mid-1960s. Since that time it has fallen steadily to about 36 percent in 1984. During that same period, the consumption of saturated and monounsaturated fats dropped from almost 20 percent to about 15 percent, whereas the consumption of polyunsaturated fats increased from about 2 percent to 7.5 percent.

The fact remains that our diet contains a higher proportion of fats than does the diet of any other developed country, and there is a correlation between high fat diets and disease, most notably certain forms of cancer (see Chapter 17) and cardiovascular disease. Diets that are rich in saturated fats (that is, fats having no double bonds; see Chapter 3) are correlated with high blood cholesterol levels, and high blood cholesterol levels are, in turn, correlated with the deposition of fatty deposits in the arteries (a condition known as atherosclerosis; see Chapter 8 for details). Atherosclerosis, in turn, is linked to high blood pressure, formation of blood clots, blockage of vessels, strokes, heart attacks, and aneurysms (ballooning of arterial walls). Since none of these conditions is desirable, the American public has been urged to reduce its collective blood cholesterol level by reducing its consumption of fats and cholesterol, especially saturated fats. Thus we have all witnessed the less-than-subtle advertisements of margarine and salad oil makers over the last few years, hammering away at our collective conscience by extolling both the virtues of polyunsaturated fats and the evils of butter. (This consciousness-raising has done some good—the recent drop in fat consumption is strongly correlated with a 46.3 percent decline in deaths from coronary heart disease since 1968.)

However, the "plant fats good/animals fats bad" litany is a gross oversimplification. Some plant fats are saturated—most notably palm oil and coconut

BOX 7.2

Fiber: How Much Oatmeal Can You Eat?

"Fiber" is one of those words that has a multiplicity of meanings and usages, leading to no end of confusion. Initially, the fiber content of foods was determined by subjecting various food items to harsh chemical treatment. The residue was known as "fiber," or more properly *crude fiber*. As a practical matter, it included most of the cellulose from plant cell walls, and little else.

In recent years, it has been recognized that there are several categories of macromolecules that resist digestion in humans, and therefore should be included as fiber. In addition to cellulose (particularly abundant in vegetables), there are other polysaccharides such as *pectin* (common in fruit), *hemicellulose* (abundant in grains), and *lignin* (which is not properly a carbohydrate at all, but is found in woody plants and is an important component of fiber). When all of these items are included as fiber, the term *dietary fiber* is used to distinguish it from crude fiber. Dietary fiber is generally three to four times more abundant than crude fiber, and in some foods may be ten times more abundant.

For purposes of examining the functions of fiber in the body, and as an aid in analyzing the various claims currently being made by cereal manufacturers and others, dietary fiber is best regarded as comprising two major types: *water-soluble fiber*, such as pectin, and *water-insoluble fiber*, such as cellulose.

Water-soluble fiber is important because it binds various substances, including bile salts and cholesterol, thereby reducing lipid uptake by the intestines. If no water-soluble fiber is present, the bile salts are reabsorbed by the small intestine. However, when bile salts are bound to water-soluble fiber, they are passed out in the feces, forcing the body to break down bodily cholesterol to form additional bile salts. Thus water-soluble fiber minimizes the absorption of cholesterol in the food, and reduces cholesterol produced by the body. Unfortunately, the trapped bile salts are converted by bacteria in the colon into substances that may promote cancer of the colon.

Water-insoluble fiber has a different set of functions:

1. It promotes peristalsis of the intestines and regular bowel activity because it is both indigestible and water-absorbent.

2. It promotes a feeling of satiety because of its effects on distending the digestive tract, a highly desirable outcome for anyone wishing to lose weight.

3. In large amounts, it correlates favorably with a reduced incidence of colon cancer, in part because it binds with the breakdown products of bile salts trapped by water-soluble fiber. A recent study also demonstrated that the consumption of high fiber cereals led to a sharp reduction in the production of abnormal cells in the colon of individuals with a history of colon cancer.

In short, a balance of water-soluble and water-insoluble fiber in the diet seems far better than excessive amounts of one, to the exclusion of the other. Thus, the oat bran (water-soluble fiber) now being touted as dietarily indispensable (at least by cereal manufacturers who sell oat bran) in the reduction of cholesterol levels should be balanced with wheat bran (water-insoluble), which seems to provide some protection from colon cancer.

(Continued on the following page)

BOX 7.2 (Continued)

Dietary Fiber in Selected Plant Foods

FOOD	AMOUNT	WEIGHT (g)	TOTAL DIETARY FIBER (g)	NONCELLULOSE POLYSACCHARIDES (g)	CELLULOSE (g)	LIGNIN (g)
Apple	1 med					
Flesh		138	1.96	1.29	0.66	0.01
Skin		100	3.71	2.21	1.01	0.49
Banana	1 small	119	2.08	1.33	0.44	0.31
Beans						
Baked	1 cup	255	18.53	14.45	3.59	0.48
Green, cooked	1 cup	125	4.19	2.31	1.61	0.26
Bread						
White	1 slice	25	0.68	0.50	0.18	Trace
Whole meal	1 slice	25	2.13	1.49	0.33	0.31
Broccoli, cooked	1 cup	155	6.36	4.53	1.78	0.05
Brussels sprouts, cooked	1 cup	155	4.43	3.08	1.24	0.11
Cabbage, cooked	1 cup	145	4.10	2.55	1.00	0.55
Carrots, cooked	1 cup	155	5.74	3.44	2.29	Trace
Cauliflower, cooked	1 cup	125	2.25	0.84	1.41	Trace
Cereals						
All-Bran	1 oz	30	8.01	5.35	1.80	0.86
Corn Flakes	1 cup	25	2.75	1.82	0.61	0.33
Grapenuts	¼ cup	30	2.10	1.54	0.38	0.17
Puffed Wheat	1 cup	15	2.31	1.55	0.39	0.37
Rice Krispies	1 cup	30	1.34	1.04	0.23	0.07
Shredded Wheat	1 biscuit	25	3.07	2.20	0.66	0.21
Special K	1 cup	30	1.64	1.10	0.22	0.32
Cherries	10 cherries	68	0.84	0.63	0.17	0.05
Cookies						
Ginger	4 snaps	28	0.56	0.41	0.08	0.07
Oatmeal	4 cookies	52	2.08	1.64	0.21	0.22
Plain	4 cookies	48	0.80	0.68	0.05	0.06
Corn	1 cup	165	7.82	7.11	0.51	0.20
Canned	1 cup	165	9.39	8.20	1.06	0.13
Flour						
Bran	1 cup	100	44.00	32.70	8.05	3.23
White	1 cup	115	3.62	2.90	0.69	0.03
Whole meal	1 cup	120	11.41	7.50	2.95	0.96
Grapefruit	½ cup	100	0.44	0.34	0.04	0.06

Adapted from Southgate, D.A.T., and others: A guide to calculating intakes of dietary fiber, J. Hum. Nutr. 30:303, 1976.

oil—and these are used extensively in, among other things, certain cereals (including some that are advertised as healthful because of their fiber content!) In addition, a recent study concluded that plant oils that have been "hardened" (by the addition of hydrogen and the corresponding elimination of some double bonds—an essential process to convert the oils to solids for use as margarines) promote increases in blood cholesterol levels that are at least as bad as those caused by saturated fats. Moreover, some animal fats are unsaturated (see Table 7.1), and there is growing (although not yet conclusive) evidence that certain monounsaturated fish oils may well block the deposition of cholesterol in the walls of blood vessels, thereby cancelling out some of the ills associated with saturated fats. Finally, although blood cholesterol levels are correlated with the amounts of saturated fats consumed, almost paradoxically there is no strong correlation between blood cholesterol levels and the amount of cholesterol consumed (see Box 7.3). Perhaps this fact is not as paradoxical as it first seems. We ingest about 500 mg of cholesterol daily (the actual

TABLE 7.1 Fat Content of Selected Foods

		FATTY ACIDS[1]		
			UNSATURATED	
FOOD	TOTAL FAT (%)	SATURATED (%)[2]	*Mono (%)*	*Poly (%)*
Salad and cooking oils				
Safflower	100	10	13	74
Sunflower	100	11	14	70
Corn	100	13	26	55
Cottonseed	100	23	17	54
Soybean	100	14	25	50
Sesame	100	14	38	42
Soybean, special processed	100	11	29	31
Peanut	100	18	47	29
Olive	100	11	76	7
Coconut	100	80	5	1
Margarine, first ingredient on label				
Safflower oil (liquid)—tub	80	11	18	48
Soybean oil (liquid)—tub	80	15	31	33
Butter	81	46	27	2
Animal fats				
Poultry	100	30	40	20
Beef, lamb, pork	100	45	44	2–6
Fish, raw				
Salmon	9	2	2	4
Mackerel	13	5	3	4
Herring, Pacific	13	4	2	3
Tuna	5	2	1	2
Nuts				
Walnuts, English	64	4	10	40
Walnuts, black	60	4	21	28
Brazil	67	13	32	17
Peanuts or peanut butter	51	9	25	14
Egg yolk	31	10	13	2
Avocado	16	3	7	2

Source: Adapted from *Fats in Food and Diet,* U.S. Department of Agriculture Information Bulletin No. 361.
[1]Total is not expected to equal total fat.
[2]Includes fatty acids with chains containing from 8 to 18 carbon atoms.
Note that all the vegetable oils contain some saturated fats, whereas all the animal fats contain some unsaturated fats.

BOX 7.3

Cholesterol: Up Close and Personal

Because the cholesterol-atherosclerosis connection figures so prominently in today's health-conscious society, let's examine the subject a little more fully. As usual, we will find that things are not nearly as simple as they first appear. First, we need to look at how cholesterol is transported in the blood.

When fatty acids and glycerol are absorbed by the cells of the intestinal lining (see Chapter 6), those cells recombine the fatty acids and glycerol to form fats (or, in nutritional terms, *triglycerides*) and then surround them with a protein coat before releasing them into the intercellular spaces where they are picked up by the lacteals. These tiny packages of protein and triglycerides are known as **chylomicrons** (see Box Figure).

The chylomicrons are transported through the lymphatic system and are dumped into the general blood circulation. As they move through the body, they are removed by fat tissue, and especially by the liver. The liver transforms them into **very low density lipoproteins (VLDLs),** which contain about 19 percent cholesterol by weight. As the VLDLs move back into the blood, the triglycerides are gradually removed by body cells, and the VLDLs are transformed into **low density lipoproteins (LDLs),** which are substantially richer in cholesterol (about 45 percent by weight). LDLs transport cholesterol through the blood for use by various body tissues—cholesterol being the forerunner of a variety of essential body chemicals, including vitamin D, various hormones, and bile salts, which are important in the digestion and absorption of lipids. Another group of lipoproteins, the **high density lipoproteins (HDLs),** transport

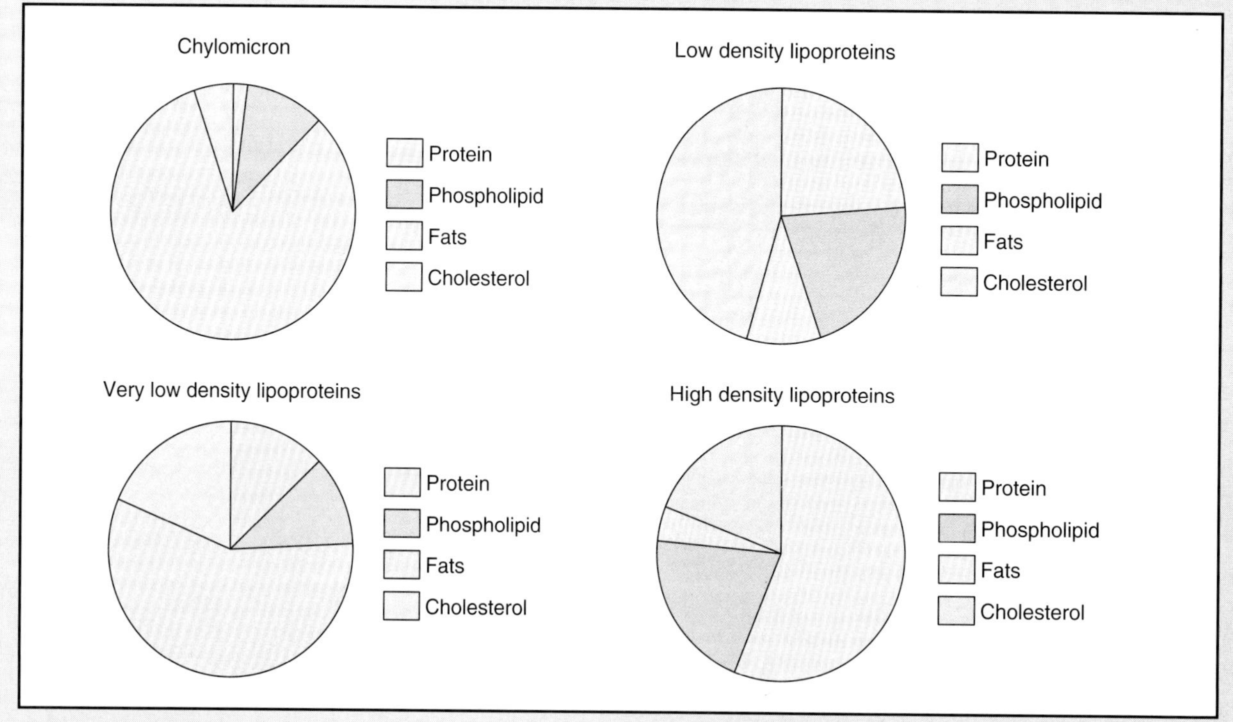

Fat Transport in the Body
The fat and cholesterol content varies significantly among chylomicrons, very low density lipoproteins, low density lipoproteins, and high density lipoproteins.

amount varies considerably depending on one's diet; remember, cholesterol is found exclusively in animal products, and therefore strict vegetarians have no dietary source of cholesterol). However, our bodies synthesize two to four times that amount from other dietary lipids.

Before we hastily vow to exclude all lipids from our diets, let us remember that at least two fatty acids

cholesterol from the cells back to the liver for conversion into bile, much of which is ultimately lost in the feces. HDLs contain only about 15 percent cholesterol by weight.

A relatively high proportion of HDL is considered good, because this is the group of lipoproteins that lowers blood cholesterol. Conversely, a relatively high proportion of LDL is troublesome, because they carry much larger loads of cholesterol, and, as we have seen, excess cholesterol in the blood leads to deposition of fatty deposits in the arteries.

To some extent, the ratio of LDL to HDL is genetically determined (which helps explain why some people can live seemingly forever on a high fat diet without ever developing cardiovascular disease); to some extent, it is a function of dietary habits and exercise. (Exercising, especially running, raises the level of HDLs and lowers blood cholesterol levels. Excessive coffee consumption apparently raises, and garlic apparently lowers, blood cholesterol levels. However, no one has yet determined the net effect of garlic-flavored coffee!)

Blood cholesterol readings below 200 mg per 100 ml of blood are considered good; levels between 200 and 240 mg per 100 ml of blood are considered problematic; and levels above 240 mg are grounds for intervention. However, what seems to be even more significant than total cholesterol readings as an indicator for possible cardiovascular problems in the future is the LDL/HDL ratio, which ideally should not exceed 4:1. As the American public becomes more sophisticated about such things, knowing one's LDL/HDL ratio will become more important than simply knowing one's total blood cholesterol reading—although whether either will ever become more essential as a conversation starter than knowing one's astrological sign remains uncertain at this point!

Cholesterol Content of Selected Foods

FOOD	AMOUNT	CHOLESTEROL (mg)
Milk, skim	1 c	5
Cottage cheese, uncreamed	½ c	7
Lard	1 T	12
Cream, light	1 fl oz	20
Cottage cheese, creamed	½ c	24
Cream, half and half	¼ c	26
Ice cream, regular	½ c	27
Cheese, cheddar	1 oz	28
Milk, whole	1 c	34
Butter	1 T	35
Oysters, salmon; cooked	3 oz	40
Clams, halibut, tuna; cooked	3 oz	55
Chicken, turkey; light meat, cooked	3 oz	67
Beef, pork, lobster; chicken, turkey, dark meat; cooked	3 oz	75
Lamb, veal, crab; cooked	3 oz	85
Shrimp, cooked	3 oz	130
Egg	1 yolk or 1 egg	250
Liver, cooked	3 oz	370
Kidney, cooked	3 oz	680

Source: Adapted from R. M. Feeley et al., "Cholesterol Content of Foods," *J. Amer. Dietet. Assoc.* 61 (1972): 134.

(*linolinic acid* and *linoleic acid,* both 18 carbons in length, and both found in many seeds, nuts, and animal products) are essential in the diet, although all of the other fatty acids can be synthesized. In the absence of these fatty acids, skin lesions and other problems may arise (see Figure 7.1). Deficiencies of these fatty acids are virtually unknown in adults, because we can store an extensive supply, but deficiencies can occur

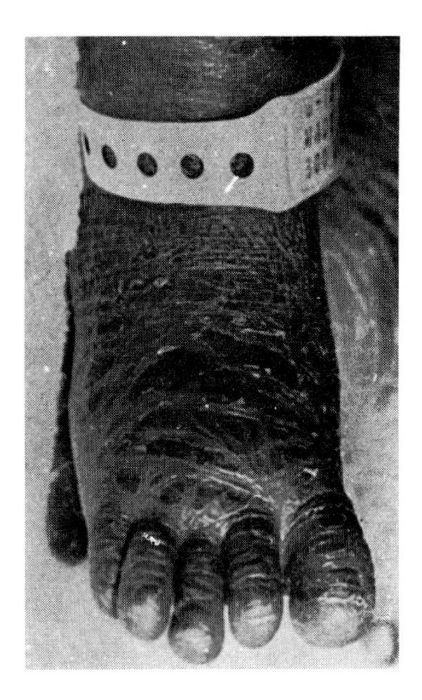

FIGURE 7.1

Essential Fatty Acid Deficiency Symptoms
Flaky skin on the foot of a patient given prolonged fat-free intravenous feeding.

TABLE 7.2 Essential and Nonessential Amino Acids

ESSENTIAL AMINO ACIDS	SEMIESSENTIAL AMINO ACIDS[1]	NONESSENTIAL AMINO ACIDS
Isoleucine	Arginine	Alanine
Leucine	Histidine	Asparagine
Lysine		Aspartic acid
Methionine		Cystine (cysteine)
Phenylalanine		Glutamic acid
Threonine		Glutamine
Tryptophan		Glycine
Valine		Proline
		Serine
		Tyrosine

[1]These are considered semiessential because the rate of synthesis in the body is inadequate to support growth; therefore these are essential for children. Recent studies indicate that some histidine may also be required by adults.

in infants fed a nonfat milk formula. (Since breast milk is from 6 to 9 percent linoleic acid, and since these fatty acids must constitute only 3 to 4 percent of the infant's diet to prevent deficiency, infants who are breast-fed never suffer from fatty acid deficiency.)

Proteins

It has been estimated that the human body contains more than 30,000 different proteins, in various quantities. Some 20 percent of the adult weight is protein: half this amount is in muscle tissue, and most of the rest in bone, cartilage, skin, and blood.

Fortunately, we do not require that those 30,000 different proteins be preformed in our diet. All of these proteins are composed of various combinations of 20 amino acids (see Chapter 3), and most of these can be synthesized from the eight to ten **essential amino acids** (EAAs) that must be present in the food we eat (see Table 7.2). (The reason for the apparent confusion regarding the actual number of EAAs is as follows: Eight are clearly essential at all times during our lives because we lack the capacity to synthesize them, and two others are required during childhood, because our ability to synthesize them is substantially less than our need for them at that point in our development. Moreover, there is some evidence that adults require one of these two in amounts beyond our ability to synthesize. Thus, you may see books listing the EAA number as eight, nine, or ten.)

We also have volume requirements for protein. The typical adult requires about 50 to 60 g per day of protein, and that number increases to about 80 g per day for pregnant and lactating women. This volume requirement is generally met in developed countries (Americans average over 100 grams of protein per day), but protein deficiency is a major health concern in many underdeveloped countries. **Kwashiorkor** (see Chapter 8) is a malnutrition disease resulting from protein insufficiency. The name is from the Ghanese word for "first-second," the concept being that a child receives sufficient protein by nursing until a second child is born—at which time it is abruptly weaned, and no protein source of comparable quality is available.

On the other hand, excess protein in our diets is not necessarily a good thing. Amino acids that are not needed for structural purposes are converted (depending on the particular amino acid) either to glucose or to fatty acids, following removal of the nitrogen-containing group by the liver. This nitrogen-containing group must then be excreted in the urine (see Chapter 11), and a substantial volume of water is required to maintain it in dilute solution (because high concentrations are poisonous). The long-term effects on the kidneys of dealing with the consequences of a high protein diet are thought to be deleterious, although this supposition has not been proved conclusively.

About 70 percent of the protein in American diets is from animal sources (that is, meat, eggs, and dairy products). Animal protein contains all of the EAAs and most are easily and efficiently digested. Plant sources, by comparison, are generally substantially less well digested, and most plants do not contain all of the EAAs (which means that vegetarians must eat a variety of plant types in order to ensure a diet that contains all of the EAAs; see Box 7.4).

Enzymes are necessary to interconvert amino acids. If a particular enzyme is missing because of a genetic error, toxic quantities of a given amino acid may accumulate. For example, a failure to manufacture the enzyme responsible for converting phenylalanine into tyrosine leads to very high levels of phenylalanine in the blood—with profound effects, including progressive deterioration of the central nervous system and mental retardation. This situation is considered at greater length in Chapter 19, but the point to be made at present is simply that in body chemistry, as in life in general, there are times when there can be too much of a good thing.

Nucleic Acids

Because most of the food we eat is cellular in nature (certain refined foods being an important exception), our diet regularly includes nucleic acids—DNA from the nucleus and RNA in the cytoplasm. These molecules are enzymatically broken into their components by the digestive system prior to absorption, and can be used by the body either for the assembly of more nucleic acids or for energy.

However, it is also true that the body can assemble nucleic acids from other organic molecules and minerals; thus, nucleic acids are not essential in the diet.

BOX 7.4

Vegetarianism

One of the most common alternative nutritional life-styles is vegetarianism. What, precisely, constitutes vegetarianism, and what are its advantages and disadvantages?

At least seven million Americans call themselves "vegetarians," although different people use the term in different ways. Consider the following continuum of dietary habits, and decide for yourself where the line should be drawn:

1. *Red meat abstainers* sharply reduce, or entirely eliminate, red meat (beef, pork, lamb, etc.) from their diets.

2. *Pollovegetarians* eliminate red meat entirely, but still allow poultry products.

3. *Pescovegetarians* eat fish but no red meat or poultry.

4. *Lactoovovegetarians* drink milk and eat cheese and eggs, but consume no meat as such. Subcategories include those who use milk and milk products, but not eggs (*lactovegetarians*) and those who have eliminated milk products but still eat eggs (*ovovegetarians*).

5. *Vegans* eat a wide variety of plant products but no animal products of any kind. They may or may not use nutritional supplements.

6. *Fruitarians*, as the name implies, eat only fresh or dried fruit, plus nuts, honey, and (for some) olive oil.

7. *Zen macrobiotics* restrict themselves to brown rice and herb tea. Less extreme diets allow whole grains and vegetables.

As you can see, vegetarianism means very different things to different people. The elimination, in whole or in part, of animal products eliminates all dietary cholesterol (since cholesterol is not found in plants), and generally sharply reduces the level of saturated fats. Because vegetarians tend to be nutrition-conscious, most also consume fewer refined foods, including simple sugars, and greater amounts of fiber. Their total caloric intake is generally less than that of meat eaters, and they are less inclined to be obese. All of those are generally favorable outcomes.

There is a negative side though. Vitamin B_{12} is found only in animal products, and therefore strict vegetarians must use a vitamin supplement. Iron deficiency can be a problem, since iron from vegetable products is much less well absorbed than is iron from animal sources. Pregnant women may have a more difficult time on a strict vegetarian diet, unless they use supplements extensively. However, vegetarians who use broad spectrum vitamin supplements extensively may suffer from vitamin A toxicity. Children who are on a strict vegetarian diet tend not to thrive. Vegans must also combine their plant materials carefully to ensure that they are receiving all of the EAAs. This means grains plus nuts plus peas or beans. Finally, the zen macrobiotic diet is deficient in many required nutrients, most notably protein, calcium, and especially vitamin C—which means that scurvy is usually the first deficiency disease to occur.

VITAMINS

Vitamins (L. *vita,* "life") are chemically diverse but share a common functional relationship—in their absence, debilitating diseases and even death may result. Although they are required only in very small amounts (since, as coenzymes, they are not used up in chemical reactions), they are truly required—they must be present and preformed in the food because, by definition, vitamins cannot be synthesized by the cells of the organism.

Vitamins are broadly classified as either fat-soluble or water-soluble. Fat-soluble vitamins are stored by the body; water-soluble vitamins are less well stored and excess amounts tend to be passed off rapidly in the urine.

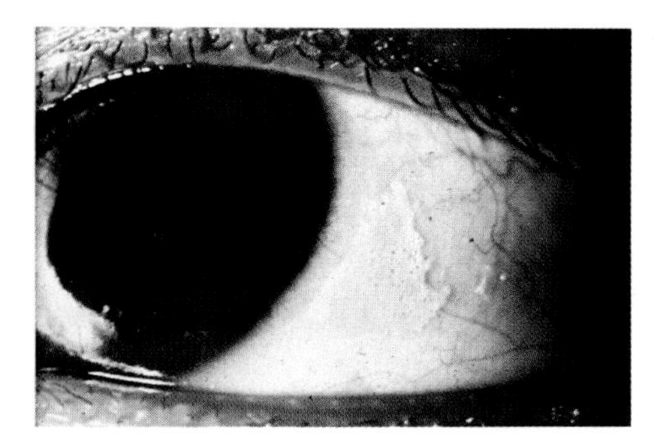

FIGURE 7.2

Vitamin A Deficiency in Children
An early sign of vitamin A deficiency is the development of a white patchy area, known as Bitot's spot, on the surface of the eye near the iris.

Fat-Soluble Vitamins

Vitamin A. Vitamin A functions in night vision, bone development, cell differentiation, and reproduction (see Table 7.3). We obtain most of our vitamin A from such animal sources as eggs and liver (and from milk, which is now generally fortified with vitamin A). However, a related compound, *β-carotene,* is abundant in various yellow fruits and vegetables, and our bodies can easily convert this substance to vitamin A. Thus, when your mother told you to eat your carrots because they were good for your eyes, it wasn't an old wives' tale (even if your mother was an old wife).

Because it is a lipid, vitamin A is absorbed in the same way as other lipids and is stored in the liver. However, vitamin A from animal sources is absorbed two to four times more efficiently than are carotenes from plants. Excess vitamin A (whether from animal sources or from vitamin supplements) can be toxic, leading to hair loss, the enlargement of the spleen and liver, and, in rare instances, death. No such toxicity exists from excess carotene in the diet because the body simply does not absorb the excess.

Vitamin A deficiency leads to respiratory infections and night blindness, followed ultimately by total blindness. Indeed, world-wide, vitamin A deficiency is the leading cause of blindness in children (250,000 cases annually, primarily in developing countries; see Figure 7.2). Vitamin A is also used topically in the treatment of acne, but in this form can cause severe birth defects, which means it cannot be used by pregnant women.

Vitamin D. Vitamin D exists in two forms, one from animal sources and one from plant sources. The body can also manufacture this vitamin, given sufficient exposure of the skin to ultraviolet radiation.

Vitamin D has a variety of effects in the body. It stimulates the synthesis of calcium- and phosphate-binding proteins in the intestinal lining, leading to a threefold increase in calcium uptake from the food. It interacts with the hormone parathormone (see Chapter 13) to increase the release of calcium into the blood. Finally, it increases the rate at which the kidneys reabsorb calcium and phosphate from the prospective urine. Deficiencies of this vitamin cause softening of the bones, manifested in children as **rickets** (see Figure 7.3).

FIGURE 7.3

Vitamin D Deficiency in Children
Rickets, the consequence of vitamin D deficiency, in a 2½-year-old child.

TABLE 7.3 The Fat-Soluble Vitamins

VITAMIN	DEFICIENCY SYMPTOMS	FUNCTIONS	SOURCES	DAILY REQUIREMENT	DISCOVERY	ISOLATION	SYNTHESIS
Vitamin A	Night blindness; dry, scaly skin; changes in lining of internal organs	Synthesis of visual pigment; development in bones and teeth	Carotene: dark green and yellow leafy vegetables; Vitamin A: liver and fortified dairy products	Infants: 375 RE[1] Children: 400–700 RE[1] Adults: 800–1300 RE[1]	1915	1937	1946
Vitamin D	Rickets (children); bone softening (adults)	Calcium metabolism in small intestine and bones	Fish; liver; fortified milk	Infants: 7.5–10 μg Children: 10 μg Adults: 5–10 μg	1918	1930	1936
Vitamin E	Uncommon in adults; destruction of red blood cells in infants	Neutralizes potentially destructive molecules	Vegetable oils, nuts, green leafy vegetables	Infants: 3–4 mg Children: 6–7 mg Adults: 8–12 mg	1922	1936	1937
Vitamin K	Hemorrhage; prolonged clotting times; delayed wound healing	Synthesis of prothrombin in liver	Dark green leafy vegetables	Infants: 5–10 mg Children: 15–30 mg Adults: 60–80 mg	1934	1939	1939

[1]RE = retinol equivalents
Daily requirement is from the National Research Council's Tenth Edition (1989) of Recommended Dietary Allowances (RDA), except where otherwise noted.

Our food is an erratic source of vitamin D, which is why so much of our prepared foods (flour; breakfast cereals) as well as milk have been fortified with this vitamin.

Vitamin E. Vitamin E is actually a mixture of at least eight naturally occurring chemicals, some of which are biologically much more active than are others. Because it has become known as the antisterility vitamin, it is the most common vitamin supplement in this country, with 12 million regular users. However, it appears that vitamin E's primary function is to neutralize agents, such as ozone, which can weaken plasma membranes, especially those of the red blood cells and cells in the lungs. Vitamin E also assists in cellular respiration and in the synthesis of hemoglobin.

Vitamin E is generally abundant in our diet, most notably in vegetable oils, and deficiencies appear to occur only in individuals who are not properly absorbing fats in the intestine. Conversely, although the long term effects of heavy supplementation are not yet known, excess vitamin E can interfere with the absorption of vitamins A and D.

Vitamin K. Vitamin K functions in blood coagulation (and since its discoverers were Danish, they used the symbol "K" from the Danish word "koagulation"). It is now known that vitamin K is essential in the formation of various clotting proteins, including prothrombin (see Chapter 9).

Vitamin K is actually a group of related molecules, some of which are common in green, leafy vegetables and some of which are formed by our intestinal bacteria. The presence of the latter ensures that deficiencies are uncommon, at least in adults, although supplementation is often given to nursing infants since vitamin K is scarce in breast milk. Toxicity, which can occur with large doses, is manifested as jaundice and anemia.

Water-Soluble Vitamins

Vitamin B_1. Vitamin B_1, now more generally known as **thiamin,** is a part of a coenzyme needed for carbohydrate metabolism (see Table 7.4). It is not well stored by the body, but it is abundant in many types of food. However, in countries that rely heavily

TABLE 7.4 The Water-Soluble Vitamins

VITAMIN	DEFICIENCY SYMPTOMS	FUNCTIONS	SOURCES	DAILY REQUIREMENT	DISCOVERY	ISOLATION	SYNTHESIS
Vitamin B_1 (thiamin)	Beriberi	Carbohydrate, lipid, and protein metabolism (coenzyme)	Liver, pork, nuts, enriched flour, milk, eggs	Infants: 0.3–0.4 mg Children: 0.7–1.0 mg Adults: 1.0–1.6 mg	1921	1926	1936
Vitamin B_2 (riboflavin)	Cracked lips, scaly skin, sore tongue, itching eyes	Carbohydrate, lipid, and protein metabolism (part of FAD)	Liver, meat, milk, eggs, cereals, green leafy vegetables	Infants: 0.4–0.5 mg Children: 0.8–1.2 mg Adults: 1.2–1.8 mg	1932	1933	1935
Vitamin B_3 (niacin)	Pellagra	Carbohydrate, lipid, and protein metabolism (part of NAD)	Red meat, fish, poultry, green leafy vegetables, enriched flour	Infants: 5–6 mg Children: 9–13 mg Adults: 13–20 mg	1936	1936	1867[2]
Vitamin B_5 (pantothenic acid)	Weight loss, intestinal disturbances, nervous disorders, irritability	Carbohydrate, lipid, and protein metabolism (coenzyme)	Widespread—red meat, fish, poultry, grains, various fruits and vegetables	Infants: 2–3 mg[1] Children: 3–7 mg[1] Adults: 4–7 mg[1]	1933	1938	1940
Vitamin B_6 (pyridoxine)	Irritability and convulsions in infants; anemia and skin lesions in adults	Protein metabolism (coenzyme)	Red meat, liver, eggs, bananas	Infants: 0.3–0.6 mg Children: 1.0–1.4 mg Adults: 1.6–2.2 mg	1934	1938	1939
Vitamin B_{12} (cobalamin)	Pernicious anemia; nerve cell degeneration	Synthesis of nucleic acids; formation of red blood cells	Red meat, poultry, fish, eggs	Infants: 0.3–0.5 μg Children: 0.7–1.4 μg Adults: 2.0–2.6 μg	1948	1948	1973
Vitamin C	Scurvy	Collagen formation, wound healing, general metabolism	Citrus fruits, berries, yellow and green vegetables	Infants: 30–35 mg Children: 40–45 mg Adults: 60–90 mg	1932	1932	1933
Biotin	Dermatitis in infants; nausea, anemia, dermatitis in adults	Carbohydrate, lipid, protein metabolism (coenzyme)	Liver, kidney, eggs, milk, cheese	Infants: 10–15 μg[1] Children: 20–30 μg[1] Adults: 30–100 μg[1]	1924	1935	1942
Folic acid	Anemia; spina bifida(?)	Synthesis of nucleic acids; formation of red blood cells	Pork, liver, nuts, green leafy vegetables	Infants: 25–35 μg Children: 50–100 μg Adults: 180–400 μg	1945	1945	1945

[1]Estimated safe and adequate daily dietary intakes
[2]Niacin was synthesized by organic chemists long before its role in human nutrition was discovered.
Daily requirement is the RDA from the National Research Council's Tenth Edition (1989), except where otherwise noted.

on "polished" rice (rice without the bran coat) deficiencies are common. Thiamin deficiency is very serious, because it leads to the development of **beriberi** (which translates to "I can't, I can't," a reference to the extreme lethargy and muscular weakness that characterizes the disease). Although the United States has mandated that flour be enriched with thiamin since 1941 (with the result that beriberi is virtually unknown in this country except among alcoholics who require very large doses of thiamin to metabolize the alcohol they are ingesting), beriberi is still a major disease throughout much of the world. It is, for example, the fourth leading cause of death in the Philippines. Infants are particularly susceptible, and death can come only hours after the first symptoms (see Figure 7.4).

Interestingly, some foods contain enzymes that destroy thiamin and can therefore lead to thiamin deficiency. These include raw fish and clams, and tea (although more than eight cups of tea a day are needed before there is significant destruction of thiamin). Conversely, there are no known toxic effects from high doses of this vitamin.

Vitamin B_2. Vitamin B_2, now more commonly known as **riboflavin,** is a part of several coenzymes including FAD (see Chapter 5). It is required for the formation of red blood cells and the synthesis of DNA, glycogen, and hormones of the adrenal cortex. Although riboflavin is widely distributed in foods, with milk being a particularly important source, deficiencies are relatively common since we can store only small amounts in the body (primarily in the liver and kidneys). To offset the possibilities of deficiencies, most flour is now enriched with riboflavin. Interestingly, riboflavin is rapidly broken down by sunlight, which is one reason why milk is generally packaged in opaque plastic or paper containers, rather than glass.

There are no specific diseases that are linked to riboflavin deficiency, but symptoms include depression, personality changes, and a general feeling of malaise. On the other side of the coin, there is no apparent toxicity problem, even in high doses.

Vitamin B_3. Vitamin B_3 (now more commonly known as **niacin**) is a part of NAD (see Chapter 5), a

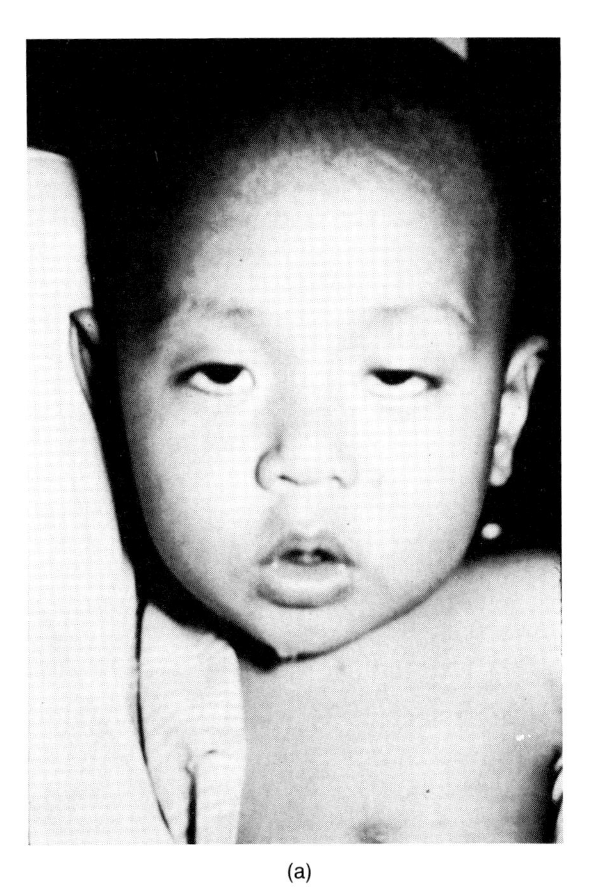

(a)

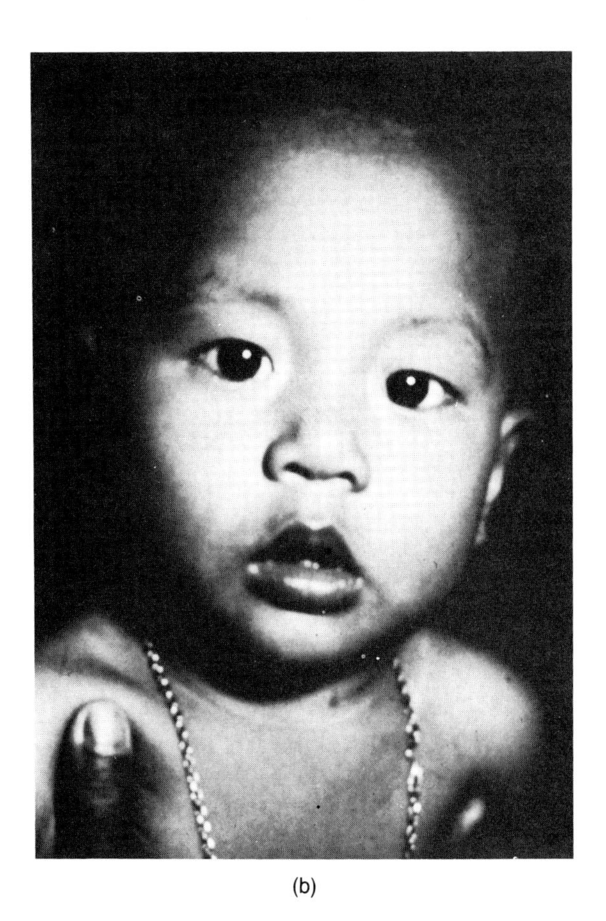

(b)

FIGURE 7.4

Thiamin Deficiency

(a) Child with thiamin deficiency. Note the characteristic drooping eyelids. (b) Same child after thiamin treatment.

hydrogen transport molecule used in cellular respiration. Niacin is relatively abundant in meat (especially organ meat), but since we can synthesize it from *tryptophan*, one of the essential amino acids, diets that are rich in tryptophan can make up for diets low in niacin. Tryptophan is abundant in peas and beans; both niacin and tryptophan are scarce in corn, however, and cultures where corn is a staple often experience the consequences of niacin deficiency.

These consequences are severe, manifesting themselves in the form of **pellagra,** a disease affecting the skin, the digestive system, and the central nervous system. Pellagra is characterized by the four D's—dermatitis, diarrhea, depression, and death—and it remains a serious problem in many developing countries. Indeed, as recently as the beginning of this century, pellagra was a major disease in the United States, with 100,000 cases annually, 10 percent of which were fatal (see Figure 7.5).

Toxicity is not a problem with niacin, except with extremely large doses. However, the currently popular practice of using tryptophan supplements is undergoing scrutiny because the use of such supplements has been linked to a painful, and sometimes fatal, blood disorder.

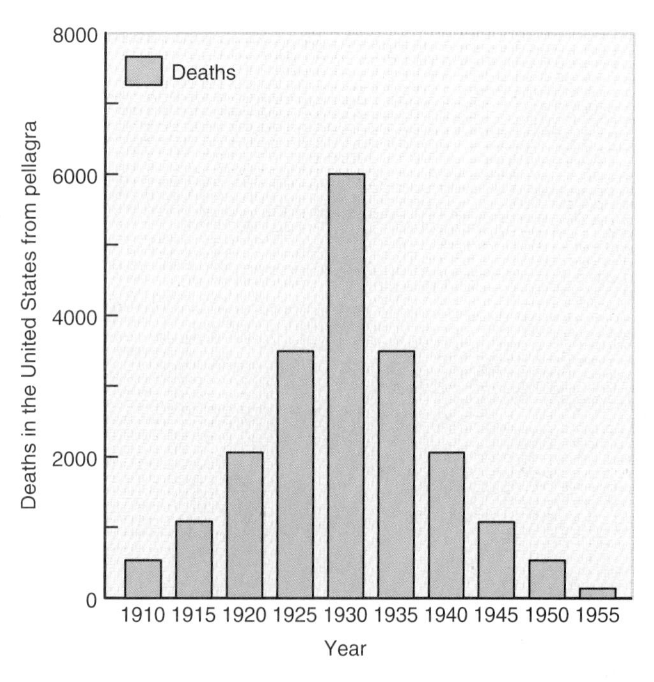

FIGURE 7.5

Pellagra in the United States

Although now virtually unknown as a cause of death in this country, pellagra was a serious disease, especially in times of economic distress (which hence led to poor diets), until niacin began to be added to bread in 1938.

Vitamin B_5. Vitamin B_5, now more commonly known by its chemical name, **pantothenic acid,** functions as a part of coenzyme A, a coenzyme essential in glucose metabolism (see Chapter 5). However, because of its widespread occurrence in many foods (hence, the name—Gr. *pan,* "all, universal"), deficiencies are very rare. Toxic effects from overdose are minor, with diarrhea being the most obvious effect.

Vitamin B_6. Vitamin B_6 is now generally called by its chemical name, **pyridoxine** (although it actually consists of three related chemicals). Pyridoxine is essential in protein metabolism, and in the healing of wounds. Only small amounts are needed and, because it is common in animal products, deficiencies are rare. However, when deficiencies occur, the results can be very serious, especially in infants where convulsive seizures and permanent brain damage can occur. Breast-fed babies obtain a sufficient supply of pyridoxine in the mother's milk, but formula-fed babies had some major problems in the 1950s before pyridoxine was added to nursing formulas. Interestingly, the use of oral contraceptives increases the need for pyridoxine. On the other hand, megadoses have been linked to nerve damage and paralysis.

Vitamin B_{12}. Vitamin B_{12}, or **cobalamin,** is required in the formation of red blood cells and nerve cells. Since we can store from five to seven years' supply in the liver, deficiencies are uncommon, but potentially very serious. Indeed, **pernicious anemia,** the disease that results from cobalamin deficiency, was inevitably fatal until 1926 when it was discovered that a diet high in liver could cure the condition. In most individuals, cobalamin deficiency stems not from an insufficient supply in the diet, but from an inability of the small intestine to absorb this vitamin (usually because of an insufficient supply of *intrinsic factor,* a substance produced by the stomach, and one which is essential in the formation of the pinocytotic vesicles needed for cobalamin uptake; see Chapter 5 for details).

Synthetic cobalamin is particularly important to vegetarians, since the natural product is found only in animal sources.

Vitamin C. Vitamin C, or **ascorbic acid,** is chemically very similar to glucose. Indeed, most species of animals (primates excepted) can convert glucose to ascorbic acid, and therefore do not require it preformed in the diet. Vitamin C has many functions in the body, including the formation and maintenance of connective tissues, teeth, and two of the chemicals used to relay information between nerve cells (see Chapter 14). It is also important in the absorption of iron and calcium.

With the exception of liver, our natural sources

of this vitamin are exclusively from plants. Although it is water-soluble, with excess amounts being quickly lost in the urine, we can store about a 90-day supply of this vitamin. Long-term deficiencies, however, lead to **scurvy,** the scourge of sailors until the discovery that fresh fruit prevented the disease. The British navy promptly began including barrels of limes with its shipboard stores, and British sailors just as promptly became known as "limeys."

We spend almost half a billion dollars annually on vitamin C supplements in this country, although (with the exception of alcoholics and formula-fed infants) deficiencies are very rare. Toxicity from overdoses is apparently not a major problem, although high supplementation is correlated with the formation of kidney stones.

Biotin. Biotin forms a part of at least four enzymes of critical importance in transfers of carbon dioxide during various metabolic processes, and in the formation of antibodies. It is common in many food sources, and in intestinal bacteria. As a consequence, deficiencies are very rare, although raw egg whites contain a substance that causes biotin to clump, hindering its absorption by the intestinal tract. (Fortunately, we would have to eat more than two dozen raw eggs a day to create a biotin deficiency!)

Folic acid. Folic acid functions in many metabolic reactions, including DNA and RNA synthesis. Although folic acid is common in many plant and animal products, and in intestinal bacteria, deficiencies of this vitamin are rather common—indeed, in the United States it is probably the vitamin that is most commonly deficient, despite the fact that we can store a four-month supply in the liver. In underdeveloped countries, as much as 30 percent of the population may be deficient in this vitamin, and folic acid deficiency may account for up to half of hospital admissions in those countries. Anemia is the primary symptom of deficiency. However, a recent study suggests that, in this country, *spina bifida,* a serious birth defect involving failure of the vertebrae to close around the spinal cord during embryonic development (see Chapter 21), could be reduced by 70 percent if pregnant women took folic acid supplements during the first six weeks of their pregnancy. As there is no known toxic effect from an overdose of folic acid, supplementation seems entirely reasonable, especially among pregnant and nursing women.

Although the range between a sufficient amount and a dangerous excess of a given vitamin is very broad, controversy still surrounds the question of necessary dosages. In 1971, Linus Pauling, twice a Nobel prize winner and one of the most distinguished scientists of the twentieth century, published a book claiming that very large doses of vitamin C were greatly effective in preventing colds. By inference, the standard minimum dose (based on an amount sufficient to prevent scurvy) was a below-minimum dose from the standpoint of colds prevention.

Several subsequent studies failed to confirm Pauling's findings, but the controversy continues. In more recent years, certain cancers and other diseases have also been treated with **megavitamins** (very large doses of a particular vitamin), again with controversial results. Pharmaceutical companies and advertising agencies have had a field day with vitamins, which are often advertised by means of a less-than subtle scare campaign (Are you *sure* you had a balanced diet today?") or by blatant appeal to children (for example, the marketing of fruit-flavored vitamin pills in the shape of cartoon characters). There is little doubt that most of the vitamin pills sold in the United States today—a $3 billion per year industry—are not needed by their buyers.

MINERALS

As we saw in Chapter 3, some 25 of the 90 naturally occurring elements are of biological importance in humans. We have already discussed four of these—hydrogen, nitrogen, carbon, and oxygen—at some length. The remaining 21 are categorized as **minerals,** and their presence in our diet is as essential to our well-being as is the presence of vitamins (see Figure 7.6).

For convenience, we can subdivide the minerals into two categories: the **macroelements,** which are found in relatively large amounts in our body, and the **microelements,** which are found in much smaller amounts. It is important to recognize at the outset, however, that despite their relative abundance in the body (or lack thereof) all of these elements are essential in the diet.

Macroelements

There are seven macroelements (see Table 7.5):

1. *Sodium:* Of the 60 to 100 g of sodium in the body, about one-third is in the bones, and two-thirds in the extracellular fluids. Sodium plays a number of major roles in the body, including neutralizing organic acids; assisting in the uptake of various nutrients (most notably, glucose) by the small intestine; and, as the primary positive ion in extracellular fluids, in water/salt balance. We ingest between 1 and 3 g per day in our food and, depending on personal proclivities, may add another 2 g by salting our food. For

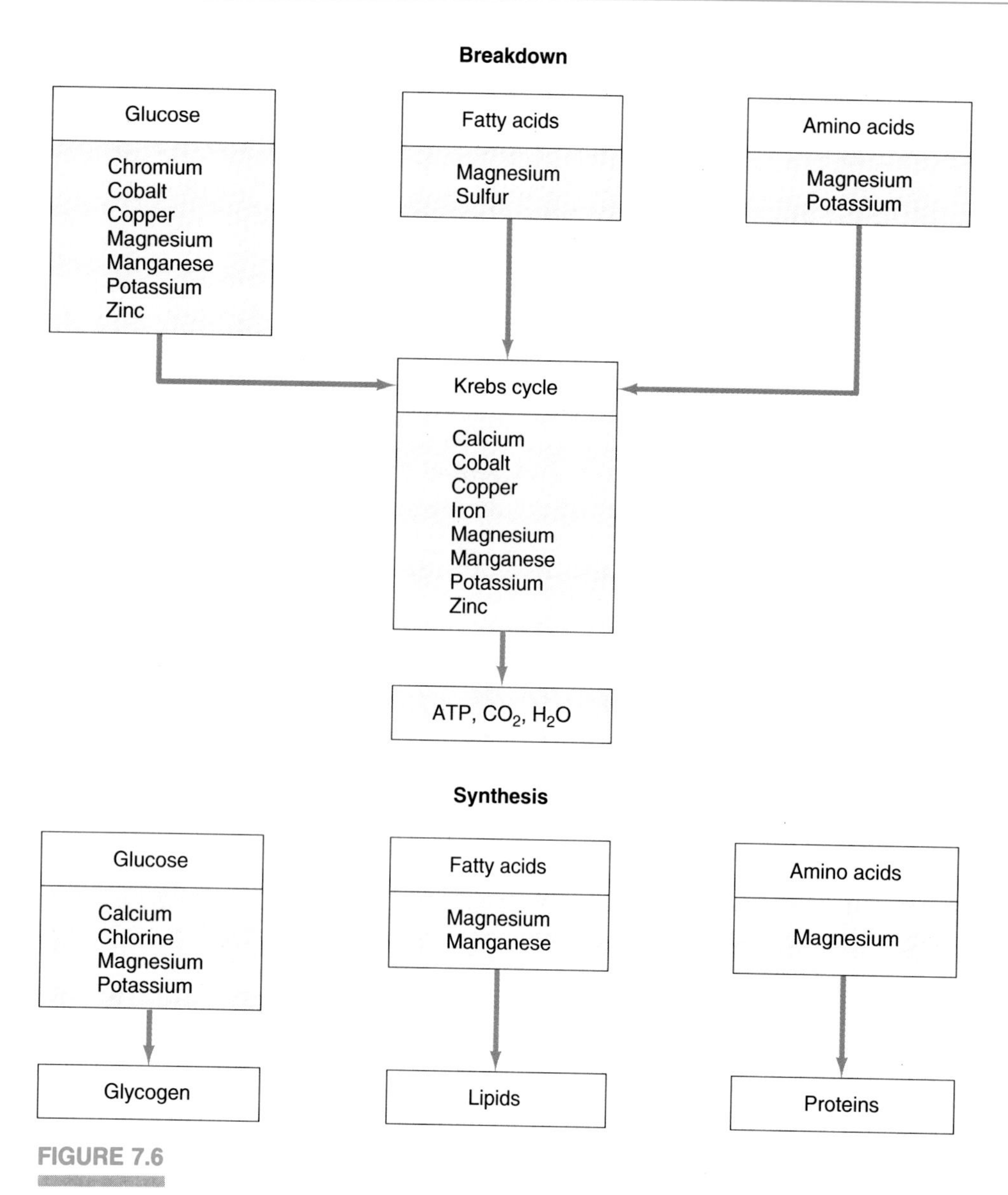

FIGURE 7.6

Minerals and Metabolism

Minerals catalyze certain metabolic reactions; they may also be a part of some metabolic enzymes or act as coenzymes in metabolism. Some minerals are essential in the conversion of organic molecules to the precursor molecules fed into the Krebs cycle; others are essential in the functioning of the Krebs cycle itself; still others are required for the assembly of large organic molecules from small ones.

replacement purposes, 500 mg per day would be sufficient; thus, our current rates of sodium ingestion are excessive and are thought by many doctors to play an important role in contributing to the high rates of hypertension in this country. Other authorities have suggested that it is actually the sodium/potassium balance that is important, and that presently we are ingesting too much sodium relative to our potassium intake. Data from one major study suggests that the real problem is a deficiency of calcium. In any event, prudence requires a reduction in our current rates of sodium consumption.

2. *Potassium:* We have about 200 g of potassium in our bodies. Unlike sodium, potassium is found primarily within cells where it plays important roles in cellular growth, as a catalyst for various reactions, including those for glycogen and protein synthesis, and, in conjunction with sodium, in the proper functioning of the nervous system (see Chapter 14). Potassium is widely distributed in our food, and deficiencies are uncommon with a normal diet, although significant amounts of potassium may be lost due to prolonged diarrhea and vomiting, or extensive use of certain diuretics.

TABLE 7.5 The Macroelements

MINERAL	FUNCTIONS	SOURCES	DAILY REQUIREMENTS
Sodium (Na)	Water balance, glucose absorption, nerve cell function	Table salt, milk, cheese, meats, eggs, vegetables	Infants: 120–200 mg[1] Children: 225–400 mg[1] Adults: 500 mg[1]
Potassium (K)	Water balance, protein synthesis, nerve cell function	Meat, fruit, vegetables, nuts, grains	Infants: 500–700 mg[1] Children: 1000–1600 mg[1] Adults: 2000 mg[1]
Chlorine (Cl)	Acid-base balance, growth, digestion	Table salt	Infants: 180–300 mg[1] Children: 350–600 mg[1] Adults: 750 mg[1]
Calcium (Ca)	Bones and teeth; blood clotting, nerve transmission, muscle activity, enzyme activation	Milk, cheese, eggs, green leafy vegetables, grains, nuts	Infants: 400–600 mg Children: 800 mg Adults: 800–1200 mg
Phosphorus (P)	Bones and teeth; ATP; plasma membrane; nucleic acids; glucose absorption	Milk, cheese, meat, eggs, grains, nuts	Infants: 300–500 mg Children: 800 mg Adults: 800–1200 mg
Magnesium (Mg)	Bones and teeth; coenzyme; smooth muscle activity	Milk, cheese, meat, seafood, nuts, grains, green leafy vegetables	Infants: 40–60 mg Children: 80–170 mg Adults: 280–355 mg
Sulfur (S)	Proteins; vitamins; blood clotting; energy transfer	Meat, milk, cheese, eggs, vegetables, nuts	No RDA; any diet adequate in protein contains sufficient sulfur

[1]Estimated minimum requirements
Daily requirement is the RDA from the National Research Council's Tenth Edition (1989), except where otherwise noted.

3. *Chlorine:* Chlorine is widely distributed in the body as a negative ion. It assists in acid-base balance, growth of body tissues, and, as hydrochloric acid, is essential in digestion. Chlorine is common in many of our foods (usually in the form of table salt), and deficiencies occur only in clinical conditions.

4. *Calcium:* Of the kilogram or more of calcium in our bodies, 99 percent is in our bones and teeth, where it provides strength and rigidity. The remaining 1 percent (10 g or so) is of vital importance, however, in such diverse functions as blood clotting, muscle contraction, the secretion of neurotransmitters, hormones, and digestive enzymes, plasma membrane permeability, glycogen metabolism in the liver, and enzyme activation. The NRC has recently recommended an increase in the recommended daily allowance of calcium in young adults—from 800 mg per day to 1200 mg per day—in recognition of the fact that bone density continues to increase until about age 30, after which there is a gradual decline. As a consequence, ingestion of additional calcium by young adults may provide some measure of protection against osteoporosis (see Figure 7.7), a degenerative bone condition of the middle-aged and elderly (see Chapter 12). The fact is, however, that most women and many men were not meeting the previous RDA, and it remains to be seen whether the new standards will promote increased ingestion of calcium.

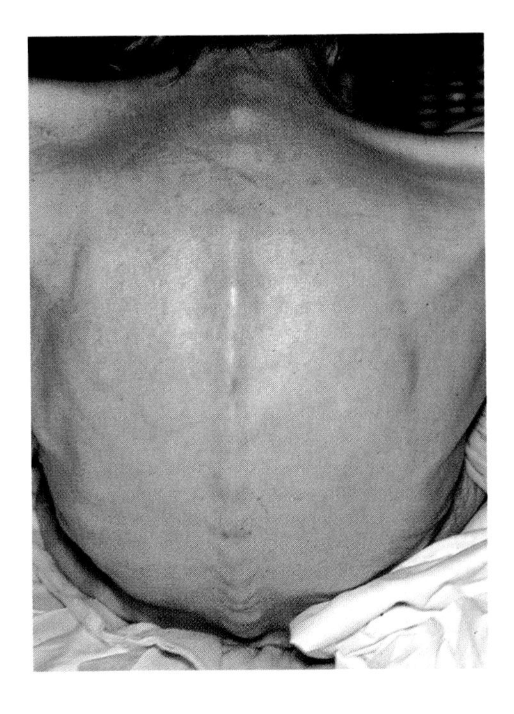

FIGURE 7.7

Osteoporosis

Loss of calcium from the bones, especially in post-menopausal women, can lead to a gradual weakening and collapse of bones. The "dowager's hump," a label that sounds quaint to today's ears, is a common example of osteoporosis.

5. *Phosphorus:* Between 85 and 90 percent of the phosphorus in the body (or about 750 g) is in the bones and teeth, and half of the remainder is in muscle tissue. The rest is scattered throughout the body, where it serves many key structural functions: The plasma membrane of cells, the energy molecule ATP, and the nucleic acids all contain phosphorus. It is also involved in the absorption of glucose and glycerol by the small intestine, in the transport of fatty acids in the blood, and as a buffer. Fortunately, because phosphorus is abundant in many foods, deficiencies are very uncommon, except in certain illnesses affecting phosphate absorption.

6. *Magnesium:* Adults have only about 30 g of magnesium in their bodies, most of which is in bone. The rest of the magnesium is concentrated within cells; extracellular magnesium levels are very low. Magnesium serves as a catalyst to several hundred reactions, mostly within mitochondria. It is also a coenzyme in protein synthesis, and affects the activity of smooth (i.e., intestinal) muscle. It may also help prevent the onset of atherosclerosis—which may be why people who drink soft water (low in magnesium and calcium) have a higher incidence of heart attacks and strokes than do people who drink hard water. Because it is a part of the chlorophyll molecule, magnesium is abundant in green, leafy vegetables. However, the typical American diet is only barely adequate in magnesium. Nevertheless, except in certain diseases or where diuretics are used extensively, magnesium deficiencies are rare.

7. *Sulfur:* The 350 g of sulfur in our bodies is widely dispersed, since sulfur is found in the cytoplasm of every cell. It is a part of three vitamins, and also functions in such diverse areas as blood clotting, energy transfer, and metabolism. It is also found in two of the 20 amino acids. Because it is widespread in our food, no deficiency symptoms are known.

Microelements

There are 14 microelements, all of which are found in much lower concentrations in the body than are the macroelements (see Table 7.6).

1. *Iron:* Of the 3.5 g of iron in the body, 70 percent is found in red blood cells, a large amount of the remainder is stored in the liver, spleen, and bone marrow, and the balance is found in enzymes involved in the electron transport system (a part of cellular respiration; see Chapter 5). We cannot excrete iron efficiently, and therefore iron balance in the body is regulated by limiting the amount of iron absorbed by the small intestine. The body is very efficient at recycling iron when red blood cells are broken down, but repeated blood loss, as occurs in menstruating women, necessitates dietary replacement of the lost iron. As a consequence, the RDA for women of reproductive age is roughly twice that for men. Iron is not particularly abundant in most foods. At present, fewer than 20 percent of women in this country (but more than 80 percent of the men) are receiving the RDA for iron.

2. *Zinc:* About 75 percent of the 2 g of zinc in our bodies is in bone; the rest is in some 40 enzymes associated with digestion and metabolism. Zinc also functions in the healing of wounds, the binding of certain hormones (including growth hormone) to plasma membrane receptors, and in the promotion of sexual maturity. This latter function—which has, in many people's minds, been equated with virility—accounts for the fact that zinc is among the most popular of over-the-counter nutritional supplements. In some respects, this supplementation is just as well—most Americans do not obtain the RDA for zinc in their diets. Finally, recent studies suggest that zinc deficiency may lead to or enhance PMS (premenstrual syndrome). Zinc helps regulate progesterone secretion (see Chapter 20), and low progesterone levels are correlated with PMS.

3. *Fluorine:* Fluorine plays an important role in strengthening teeth and bones. Where fluorine is a part of the drinking water, tooth decay is reduced by as much as 60 percent as compared with populations drinking fluorine-free water (but for another view, see Box 7.5). Fluorine also promotes calcium uptake by bone and is regularly used in the prevention and treatment of osteoporosis.

4. *Iodine:* About 50 percent of the 20 mg of iodine in the body is in muscle tissue, 20 percent in the thyroid gland (since iodine is a part of two hormones produced by this gland), and the rest is scattered throughout the body. Iodine is abundant in seafood and in plants grown in soils once covered by the ocean, but deficiencies were historically common in areas such as the Great Lakes region and Switzerland where iodine is not present in the soil. With the addition of iodine to salt, the incidence of iodine deficiency in this country has dropped dramatically—from 47 percent of Michigan school girls being deficient in 1921, to 1 percent in 1951. The role of iodine is discussed more completely in Chapter 13.

5. *Selenium:* The amount of selenium in the soil varies enormously from place to place, with the result that both deficiencies and toxicities of this element can occur (although toxic doses are far more common in livestock than in humans—excluding humans using selenium supplements). Selenium's role in the

TABLE 7.6 The Microelements

ELEMENT	FUNCTIONS	SOURCES	DAILY REQUIREMENT
Iron (Fe)	Hemoglobin; electron transport system enzymes	Meat, eggs, enriched flour, dark green vegetables, nuts	Infants: 6–10 mg Children: 10 mg Adults: 10–30 mg
Zinc (Zn)	Bone; metabolic and digestive enzymes; wound healing; hormone function	Seafood, liver, red meat, milk, cheese, eggs, grains	Infants: 5 mg Children: 10 mg Adults: 12–19 mg
Fluorine (F)	Teeth and bones; calcium metabolism	Fish, drinking water	Infants: 0.1–1.0 mg[1] Children: 0.5–2.5 mg[1] Adults: 1.5–4.0 mg[1]
Iodine (I)	Thyroid gland	Iodized salt, seafood	Infants: 40–50 μg Children: 70–120 μg Adults: 150–200 μg
Selenium (Se)	Teeth; functions in conjunction with Vitamin E	Seafood, vegetables, grains	Infants: 10–15 μg Children: 20–30 μg Adults: 55–85 μg
Manganese (Mn)	Connective tissues, bone; metabolic enzymes	Cereals, peas, beans, leafy vegetables	Infants: 0.3–1.0 mg[1] Children: 1–5 mg[1] Adults: 2.5–5 mg[1]
Copper (Cu)	Blood, connective tissue, nervous tissue, enzymes	Liver, red meat, seafood, grains, nuts, peas, beans	Infants: 0.4–0.6 mg[1] Children: 0.7–2.5 mg[1] Adults: 1.5–3 mg[1]
Molybdenum (Mo)	Two enzymes	Beans, grains, milk, leafy vegetables	Infants: 15–40 μg[1] Children: 25–250 μg[1] Adults: 75–250 μg[1]
Chromium (Cr)	Glucose metabolism	Grains, cereals, meat	Infants: 10–60 μg[1] Children: 20–200 μg[1] Adults: 50–200 μg[1]
Cobalt (Co)	Vitamin B_{12}	Meat and dairy	No RDA established
Nickel (Ni)	Nucleic acids; iron metabolism	Fruits and vegetables	No RDA established
Vanadium (V)	Bone development; iron and lipid metabolism; growth	Widespread	No RDA established
Silicon (Si)	Connective tissue development; growth	Widespread	No RDA established
Tin (Sn)	Growth	Widespread	No RDA established

[1]Estimated safe and adequate daily dietary intakes
Daily requirement is the RDA from the National Research Council's Tenth Edition (1989), except where otherwise noted.

body overlaps that of vitamin E, with the result that abundant vitamin E in the diet largely compensates for deficiencies of selenium. Other functions of selenium are not presently well understood, although selenium deficiencies have proved a significant health problem in China, where they have been linked to a degenerative heart disease.

6. *Manganese:* Manganese is essential in the development of connective tissues and the skeleton, and serves as a part of several enzymes, including those used in urea formation, cholesterol synthesis, and mitochondrion structure. Our bodies contain about 15 mg of manganese, most of which is in the pancreas, liver, kidneys, and bone. Manganese is relatively abundant in fruits and vegetables, and particularly abundant in tea. Deficiencies of this mineral are unknown in humans, except in clinical conditions such as diabetes or kwashiorkor.

7. *Copper:* Copper is essential in the formation of red blood cells and in the synthesis of various connective tissue proteins, the skin pigment melanin, and the myelin sheath around nerve cells. We have about 100 mg of copper in our bodies, half of which is in bone and muscle. Because copper is widely distributed in foods, deficiencies are rare.

8. *Molybdenum:* The 9 mg of molybdenum in our bodies is concentrated in the liver, kidneys, adrenal glands, and red blood cells. It is an essential part of

BOX 7.5

Fluoride Revisited

As this text goes to press, the long-simmering fluoride debate has bubbled over once again. There are two issues in the current controversy:

1. *Does fluoride reduce the incidence of tooth decay?* Rates of tooth decay in U.S. children have dropped by more than 50 percent in the last 20 years. How much of this drop is due to the addition of fluoride to the drinking water? Ironically, the current popularity of fluoride makes this question difficult to answer. Over 50 percent of the U.S. population now drinks fluoridated water, but almost everyone uses fluoridated toothpaste or mouthwash, or drinks beverages made with fluoridated water. Thus, finding suitable control populations (see Chapter 2) is very difficult. Nevertheless, the National Institute for Dental Research states that the difference in decay rates for children in fluoridated versus nonfluoridated areas has dropped from more than 50 percent in the 1950s to 25 percent in 1987, as a consequence of the overall improvement in nutrition and dental care, the implication being that fluoridated water has contributed as much as all other factors combined in reducing the rate of tooth decay. Antifluoridationists vigorously contend that the value of fluoride in preventing tooth decay is greatly overestimated.

2. *Is fluoride a carcinogen?* The National Toxicology Program recently released data from a two-year experiment, during which time rats and mice drank water with different levels of sodium fluoride. No animals drinking fluoride-free water, or water with a fluoride concentration of 11 parts per million (ppm), developed cancer. However, in the population drinking water containing 45 ppm of fluoride, one of the 50 male rats developed a form of bone cancer known as *osteosarcoma*. In the population of rats drinking water containing 79 ppm of fluoride, four of 80 male rats developed osteosarcoma. No mice or female rats developed osteosarcoma at any of the tested fluoride levels.

What do these findings mean for humans? The Environmental Protection Agency's maximum allowable concentration of fluoride in drinking water is 4 ppm, although the amount generally added to municipal drinking water is 1 ppm. There has been no apparent increase in the incidence of osteosarcoma in humans that could be correlated with the addition of fluoride to drinking water over the past 40 years, nor is there any evident clustering of this disease in areas where the water is fluoridated. Indeed, there are only about 750 cases of this cancer in the U.S. annually, from all causes.

How would you assess the risks and benefits of adding fluoride to our drinking water? Do you place greater value on the evidence for a widespread, but arguably modest benefit (a reduction in tooth decay rates), or on the presently much weaker evidence for a restricted, but very serious risk (osteosarcoma)? Tough call, isn't it?

two enzymes. Molybdenum deficiencies in humans are unknown.

9. *Chromium:* We have less than 2 mg of chromium in our bodies, although in the Far East this value may be quadrupled. Chromium is essential in the utilization of glucose, apparently by increasing the number of plasma membrane receptors for insulin. It is relatively abundant in most grain and cereal products. Deficiencies are correlated with glucose intolerance and an increased incidence of Type II diabetes (see Chapter 13). Interestingly, a diet high in simple sugars causes the body to excrete chromium. Thus a diet low in simple sugars is a good preventive measure for the onset of Type II diabetes—something which has long been recommended, but now we finally understand the underlying reason.

10. *Cobalt:* The only known role for cobalt is as a part of vitamin B_{12}, the role of which was discussed earlier. However, it is suspected that cobalt is also required in the synthesis of several enzymes. Cobalt is abundant in foods from animal sources, and deficiencies of the element (although not the vitamin) are unknown.

11. *Nickel:* Nickel enhances the utilization of iron and is also associated with the nucleic acids. This element is relatively abundant in fruits and vegetables, and deficiencies are rare.

12. *Vanadium:* Because vanadium is known to be essential for many species of animals in such varied functions as bone development, reproduction, the metabolism of iron and lipids, and growth, and because vanadium is found in human tissues, it is assumed to be an essential element for humans. However, at present little is known of its role in the human body.

13. *Silicon:* In mice and rats, silicon stimulates growth and promotes connective tissue development, but little is known about the particulars in humans. However, because it is so widespread in the environment, deficiencies are highly unlikely.

14. *Tin:* Tin is believed to be required for growth; deficiencies result in loss of hair, skin infections, and poor growth. The average diet supplies adequate amounts of tin, one important source being the tin leached from opened cans of acidic materials such as tomato juice stored in the refrigerator!

In addition to these elements, aluminum, arsenic, boron, and cadmium—all known to be toxic above a certain level—are currently listed as "probably essential" in minute amounts.

WATER

Water constitutes about 60 percent of the weight of our bodies. We lose about 6 percent of our bodily water each day—an amount which, after about three days, is fatal, something known to anyone who has ever watched any movie involving deserts and cowboys. In short, water is an essential part of our diet.

In the body, water has so many roles that, as we saw in Chapter 3, it is difficult to imagine any life form that is independent of water. It is the solvent in which our cellular chemistry functions; it is necessary in hydrolysis reactions; it functions as a lubricant in joints, and, in the form of perspiration, as a temperature regulator.

The details of water regulation in the body are discussed in Chapter 11.

SPECIAL PROBLEMS

The subject of nutrition is, as you perhaps can now see, complex. It represents entire academic departments at many universities—and we cannot possibly cover all the details in one chapter. However, just in case things still seem a bit too simple, let us consider three categories of problems, specifically:

1. In what ways can our health or our choice of diet interfere with absorption of essential nutrients?

2. Do our needs for essential nutrients change or differ as a function of our age or sex?

3. What special dietary problems face the athlete—or the obese individual—and how are they best resolved?

Problems in Nutrient Absorption

As we noted earlier in the chapter, the presence of a given nutrient in the food in no way ensures that it will be absorbed by the body. It may, at first glance, seem odd that anything we label a "nutrient" would be allowed to pass through the digestive tract without being absorbed, but there are good reasons why such a situation occurs. For example:

1. Several minerals are poorly absorbed by the body. Less than half of the magnesium in the diet is absorbed, for instance, and as little as 10 percent of the calcium and iron may be absorbed.

2. Some nutrients enhance the absorption of other nutrients. Vitamin D, for example, enhances the absorption of both calcium and zinc; protein in the diet enhances the absorption of both calcium and copper.

3. Some nutrients interfere with the absorption of other nutrients. Zinc, copper, and chromium interfere with each other; supplementation of one may therefore cause a deficiency of the others. High levels of phosphate in the diet can interfere with calcium and iron absorption. Fiber can capture iron, zinc, and copper, and reduce their rate of absorption.

4. The source of a nutrient is often important. A substantial fraction of the iron from animal tissue, for example, is in the same form as that found in blood; this form of iron is readily absorbed. Iron from plants, by comparison is very poorly absorbed. Thus, even though spinach is relatively rich in iron, it is not easily assimilated, Popeye's claims to the contrary notwithstanding.

Gender and Age Differences in Nutrient Needs

One of the difficulties about establishing RDAs is in allowing for individual differences based on age, sex, and body size, among other potential variables. It stands to reason that large, active, growing individuals will require more nutrients than small, sedentary, mature individuals, and indeed such is the case—although the increase is not proportional for every nutrient. Women of reproductive age require fewer nutrients than men of the same age (because, as a rule, they are smaller in size), but require more iron, for instance, because of blood loss during menstruation. Similarly, pregnant women require additional nutrients, but not proportionally more of all of the nutrients.

The Care and Feeding of Athletes

A maintenance diet, measured in kilocalories, is about 2500 kcal per day for young men and about 2000 kcal per day for young women (see Table 7.7). This amount is insufficient, however, to provide for the needs of athletes in training (or lumberjacks, coal miners, and other very physically active individuals, for that matter). Runners and swimmers may expend 1000 kcal

TABLE 7.7 Your Daily Energy Requirements

1. Determine energy needed to maintain a basal metabolic rate (BMR):

   Women: 0.9 kcal/kg body weight/hr
   Men: 1.0 kcal/kg body weight/hr

   (For example, the energy needed to maintain a BMR in a 50 kg woman: 0.9 × 50 × 24 [hours in a day] = 1080 kcal)
2. Add energy cost associated with general activity level as a fraction of BMR:

   Sedentary = 20% of BMR
   Light = 30% of BMR
   Moderate = 40% of BMR
   Heavy = 50% of BMR

   For example, if our 50 kg woman is a student, her activity level is light, and the energy costs associated with that level of activity are 30% of 1080 kcal (BMR) = 324 kcal.
3. Add the energy costs associated with eating and digesting the food you consume.
   (These energy costs are approximately 10% of the kcal contained in the food eaten each day. Using a calorie chart our 50 kg woman finds that she is consuming 1400 kcal on average each day; the energy costs associated with eating and digesting this food is therefore 10% of 1400 kcal = 140 kcal)
4. Add the values in lines 1, 2, and 3.
   (Our 50 kg woman student is expending 1080 + 324 + 140 = 1584 kcal each day. However, since she is consuming only 1400 kcal a day, she will lose weight at the rate of 184 kcal [1584 − 1400] a day. There are approximately 7500 kcal in 1 kg of fat. Therefore, if she maintains an equivalent level of activity and energy consumption, in about 40 days [7500 ÷ 184 = 40] she will lose about 1 kg in weight.)

per hour, and must ingest equivalent energy to maintain their weight (see Table 7.8).

Higher caloric intake for athletes, however, is not the area of controversy. Rather, it is what *form* this energy should take (that is, should it be largely protein, since the athlete is building muscle?), and what other nutrients are needed in amounts exceeding the needs of the general population.

The idea that athletes need a higher proportion of protein in their diets because they are building muscle is not generally valid (although younger athletes who are still growing, and athletes involved in contact sports where tissue damage is common may need slightly more protein than the average individual). The primary reason for this is that the American diet is already rich in protein—and excess protein requires greater water intake to metabolize, increases calcium loss in the urine, and forces greater activity by the kidneys to maintain the body's pH within normal limits.

The balance between carbohydrates and fats is somewhat more difficult to characterize, since different types of activity affect the metabolism of these two classes of compounds differently. During low-intensity exercise, much of the muscles' supply of energy comes from fat metabolism. However, during high-intensity exercise, the muscles are much more dependent on glucose and glycogen. In addition to the intensity and duration of the exercise, the balance struck between fats and carbohydrates also depends on such variables as level of conditioning and the amount of oxygen available to the muscles. For instance, training increases the efficiency of fatty acid use and diminishes the demand on glycogen.

Because they are performing at such a high level, even very well-trained marathon runners must rely on stored glycogen for a significant fraction of their energy needs. Although even the leanest marathoner has enough stored fat to run several marathons, most individuals are capable of storing no more than about a two-hour supply of glycogen—an amount insufficient for the duration of a marathon—and marathoners regularly speak of "hitting the wall" (a feeling of sudden and enormous exhaustion) late in a race as they deplete their glycogen stores.

Some athletes practice glycogen loading as a solution. A week before a major race, the athlete trains heavily to exhaust his glycogen reserves, and then eats a low carbohydrate–high fat and protein diet for several days. Two days before the race, the athlete shifts to high carbohydrate diet, and exercises very little, during which time the muscles store large amounts of glycogen. Several studies have indicated that performance in distance races is significantly better in glycogen loaders. However, the athlete tends to retain water in such a regimen, resulting in a stiff or heavy feeling in the muscles. There is also concern about increased blood acidity, increased blood cholesterol, and possible heart problems as a consequence of these dramatic dietary shifts. Accordingly, some experts recommend simply eliminating the low carbohydrate phase, and increasing the high carbohydrate phase to three days, which (when coupled with low activity) leads to increased glycogen loading. In any case, glycogen loading probably should not be done more than three times a year.

Some runners have simply gone on very low fat diets as a way of emphasizing carbohydrate importance. However, the heart depends heavily on fatty acids for its energy needs—and very low fat diets have actually led to heart attacks in some athletes.

Water needs to be ingested regularly during heavy exercise to replace water lost by perspiration. Salt replacement is less critical, since proportionately less salt than water is lost, and since our diet is normally high in salt in any case. Salt tablets may be needed, however, if the water loss regularly exceeds 3 kg—which can happen within 90 minutes in high temperature and humidity conditions.

Contrary to popular opinion, athletes have no special vitamin and mineral needs. As we have seen, these substances are not used up in the production of energy, and therefore their need does not increase as a function of activity level.

Weight Reduction and Maintenance

With the recent interest in nutrition and exercise, the issue of weight reduction and maintenance has taken on heroic proportions. Diet books regularly appear on the list of best sellers; weight clinics pop up like mushrooms; newspapers of the type most commonly seen at supermarket check-out registers greet each New Year (the time of resolutions, remember?) with quick-fix weight loss techniques of the stars. What should we make of all this?

First, let us recognize that, as a society, we currently have a fixation on slimness—especially for women. Insurance company weight tables suggest only a modest differential between men and women of the same height and build—but society seems to want women to be about half the weight of men (women should all be 5′7″ and weigh 110 lbs; men should all be 6′1″ and weigh 190 lbs).

Second, the fact is that obesity (15 percent or more over ideal body weight, where the excess is fat, not muscle) is very common in this country, and it poses significant health risks, including problems with the cardiovascular system, respiration, diabetes, arthritis, and chronic illnesses in general. The National Center for Health Statistics estimates that almost 20 percent of American men and almost 30 percent of American women are overweight—almost 40 million people in all! However, fewer than four million are moderately or severely overweight (that is, more than 40 percent over average weights, see Figure 7.8).

There is no question that people differ greatly in their individual metabolisms. Some people utilize food more efficiently than others, and will tend to gain weight when others are merely maintaining weight. People differ with respect to the number of fat cells they have, or with respect to certain hormonal levels.

TABLE 7.8 Energy Costs for Different Activities in a 70 kg Adult Male

ACTIVITY	KCAL/MIN
Basketball	9.0–10.0
Boxing	9.0–10.0
Cleaning	4.0– 4.5
Coal mining	6.0– 8.0
Cooking	3.0– 3.5
Dancing	3.5–12.5
Eating	1.0– 2.0
Fishing	4.0– 5.0
Gardening	3.5– 9.0
Horse riding	3.0–10.0
Painting	2.0– 6.0
Piano playing	2.5– 3.0
Running	9.0–21.0
Scrubbing floors	7.0– 8.0
Standing	1.5– 2.0
Swimming	4.0–12.0
Typing, electric	1.5– 2.0
Walking	1.5– 6.0
Writing	2.0– 2.5

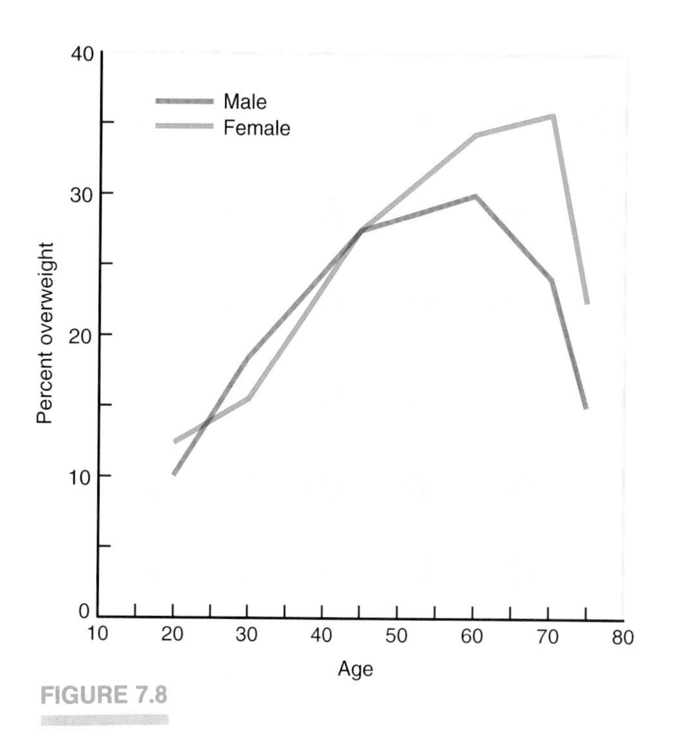

FIGURE 7.8

Obesity by Age

The incidence of obesity rises with age in both men and women, but is more rapid in women. The percentage of obese men in the population begins to fall when men reach their 60s, but does not drop in women until they reach their 70s. One reason for the decrease in percentage of obesity with age is that obese individuals tend to die earlier, and obese men in particular are at risk.

There are cultural differences as well that contribute to weight problems—and many obese adults were obese as children, often because their parents encouraged them to "clean their plates" at dinner.

One recent study concluded that caloric intake per unit of body weight was not the primary difference between the diets of lean and obese individuals. Rather, it was that obese people ate more fat and fewer carbohydrates than did lean people—the implication being that high fat diets tend to promote obesity, irrespective of total caloric intake.

The bottom line, however, is that 1 kg of fat is worth about 7500 kcal. Ingesting 7500 more kcal than are needed to operate the body will, all things being equal, result in the acquisition of 1 kg of body fat. Reducing caloric intake, or increasing activity levels, by 7500 kcal will result in the loss of 1 kg of fat.

The problem is that 7500 kcal is an enormous amount of energy. A person weighing 70 kg would have to run 15 km every hour for eight hours to expend 7500 kcal! Similarly, the average individual would have to go without food entirely for three or four days to eliminate 7500 kcal from the diet—or, looking at it another way, eliminating 7500 kcal would mean reducing food intake by 40 to 50 percent for a week. The point is that weight loss is not easily or quickly accomplished, despite claims to the contrary (but see Box 7.6).

Weight gain tends to be an insidious process. Suppose a woman who was maintaining her weight were to add one doughnut (125 kcal) at mid-morning every day to her diet. Barring no other changes, at the end of one year she would have gained 6 kg (about 13 lb) in weight. Unfortunately, what goes on slowly usually must be lost the same way.

Because our bodies are so efficient, the best method of weight loss combines a modest reduction in food intake with a regular exercise program. Moderate exercise, almost paradoxically, often results in a reduction of appetite, making reductions in food intake more easily managed. Slow, steady weight reduction—perhaps 0.5 kg per week—is a more reasonable target than very rapid weight loss. Programs promising rapid weight loss—2 kg per week or more—are either based on water loss (which is quickly regained), or are extremely spartan (meaning that persistence in the program is difficult), or are simply fantasies.

BOX 7.6

Eating Disorders: Anorexia and Bulimia

Anorexia nervosa is a very serious eating disorder affecting approximately 0.5 percent of girls between the ages of 12 and 18. (Older women and males have a much lower incidence.) It is manifested in the willful suppression of appetite, based on the perception (which may or may not be valid) by the individual that she is overweight or is inclined to become overweight. Sustained periods of near-starvation lead to shrinkage of the brain and reduced mental acuity, significant changes in endocrine function, and deficiencies in such minerals as iron, copper, and zinc, all of which tend to reinforce the behavioral manifestations of anorexia. In other words, anorectics tend to spiral downward, sometimes with fatal consequences (usually because of mineral imbalances leading to cardiac arrest).

The cause of the disorder is not well understood. Some authorities feel that it results from a problem with the area of the brain associated with eating; others conclude that it is a psychological disorder, often the result of a deteriorating mother-daughter relationship. Treatment includes psychotherapy, behavioral modification, use of appropriate medications, and hospital feedings. Recovery may take a year or more, and is not always successful—only 50 percent remain free of symptoms, and another 25 percent have reduced symptoms. If untreated, about 5 percent of anorectics die.

Bulimia, or "binge-purge" behavior, is characterized by periodic eating binges, followed by induced vomiting and extensive use of laxatives. Victims of this disorder are also usually young women, and although they are often of normal weight, their method of weight control can lead to serious problems, such as loss of teeth enamel from the acidic vomit, mineral imbalance from laxative use and from vomiting, rectal bleeding, and liver, kidney, and heart damage. Some researchers think that this condition results from abnormalities in CCK production or reception (see Chapter 6); others feel that it is primarily psychological, resulting from attempts to obtain approval from unresponsive, perfectionist parents, and societal pressure to be thin.

The incidence of this condition in the United States is unknown, but as many as 20 percent of college women are estimated to have had some experience with bulimia. Bulimia is often more hidden than anorexia (since the victim is often within normal weight limits), and bulimics frequently suffer from the disease for six or seven years before seeking help. Antidepressant drugs, coupled with psychotherapy and behavioral therapy, are generally effective.

Summary

The science of nutrition is still in its infancy. Analyzing the content of food is easy; knowing how the individual deals with this food in his or her digestive system is much more difficult.

Nevertheless, we now know that everyone requires a variety of organic molecules for energy and for growth and repair; small amounts of organic molecules that function as coenzymes and are called vitamins; a variety of minerals (although the precise number and amounts of each are not yet fully elucidated); and water.

There is a considerable difference between what nutritionists think we should be eating and what actually constitutes our diet. Americans tend to eat more protein, more sugars, and considerably more fats and cholesterol than is currently recommended, and only about half as much complex carbohydrates and fiber as is currently recommended. Our caloric intake is generally more than sufficient for our activity level; unless we eat less, exercise more, or both, most of us tend to put on weight as adults.

Although we have enormous capacities for the metabolic conversion of molecular types, there are certain minima that we must have in our diet. At least eight of the 20 amino acids, for instance, must be in our diets, as well as two particular fatty acids, since we cannot create any of these molecules from other molecules. These molecules are known as essential amino acids and essential fatty acids, respectively.

Vitamins are chemically diverse, but share a common functional relationship. There are four fat-soluble vitamins, and nine water-soluble vitamins. The typical American diet contains adequate amounts of these vitamins, especially because flour, milk, and cereals are so often fortified with vitamins, and except in particular circumstances (pregnant and nursing women), vitamin supplements are not generally necessary.

Minerals are a different story. Many Americans have marginal deficiencies of minerals such as iron, calcium, and zinc, and supplementation can be problematic since many minerals compete with each other at absorption points in the small intestine.

There are considerable differences in nutritional needs based on age, gender, body size, and activity levels, although the single biggest variable is level of caloric intake. Vitamin and mineral needs vary much less, primarily because they are not used as energy molecules.

Key Terms

essential amino acids
kwashiorkor
vitamin
vitamin A
vitamin D
rickets
vitamin E
vitamin K
thiamin
beriberi
riboflavin
niacin
pellagra
pantothenic acid
pyridoxine
cobalamin
pernicious anemia
ascorbic acid
scurvy
biotin
folic acid
megavitamins
minerals
macroelements
microelements

Box Terms

chylomicron
very low density lipoprotein (VLDL)
low density lipoprotein (LDL)
high density lipoprotein (HDL)
anorexia nervosa
bulimia

Questions

1. How would you respond to the question, "What constitutes an adequate diet?"
2. What is your opinion of the average diet of Americans today? What changes would you suggest, if any?
3. What are the major issues surrounding fiber intake?
4. What are the major issues regarding the consumption of cholesterol?
5. What is the role of vitamins in the body? Describe the effects of deficiencies in any two vitamins.
6. What is meant by the terms "essential fatty acids" and "essential amino acids"?
7. What is the role of minerals in the body? Describe the body's specific use of any two minerals.
8. Discuss both obesity and eating disorders as problems in America today.
9. What are the dangers, if any, of using special vitamin or mineral supplements?

CHAPTER

8

The Circulatory System

An Internal Sea

In the summer of 1974, both before and after he resigned the presidency, Richard Nixon suffered from *phlebitis*, an inflammation of the veins of his legs. This was a life-threatening condition because of the possibility of blood clots in his lungs. What does a leg disease have to do with lung problems?

Four of the top 15 causes of death in the United States are diseases of the circulatory system (heart disease; stroke; hardening of the arteries; high blood pressure). Heart disease, the number one killer, is responsible for almost 550,000 deaths annually—a figure that exceeds the total number of Americans killed in World Wars I and II! What factors of our life-style contribute so significantly to the deterioration and malfunction of our circulatory system?

These and other questions are considered in our discussion of the circulatory system.

THE NATURE OF THE PROBLEM

We humans take the existence of a circulatory system very much for granted. However, it is instructive to think about how and why circulatory systems first arose. Unicellular organisms obviously have no need for one, because they can rely on diffusion and active transport to meet their requirements for food, oxygen, and the voiding of waste products. Initially in their evolution, multicellular organisms solved their problems in the same way. The cells of multicellular organisms are not, for the most part, mortared in place, like bricks in a wall, but instead are stacked at odd angles, like rocks in a pile. For that reason, in primitive marine organisms such as sponges and jellyfish, sea water can percolate through the intercellular spaces, and the individual cells are able to rely largely on diffusion to meet their needs.

Although more sophisticated organisms have tended to develop an outer skin or shell that is largely impermeable to the environment (this is especially true for terrestrial forms that must avoid desiccation), they have nevertheless retained the basic exchange methods of the sponges and jellyfish. The fluid percolating between the cells, now called **interstitial fluid** (L. *interstitium,* "small space or crevice"), provides

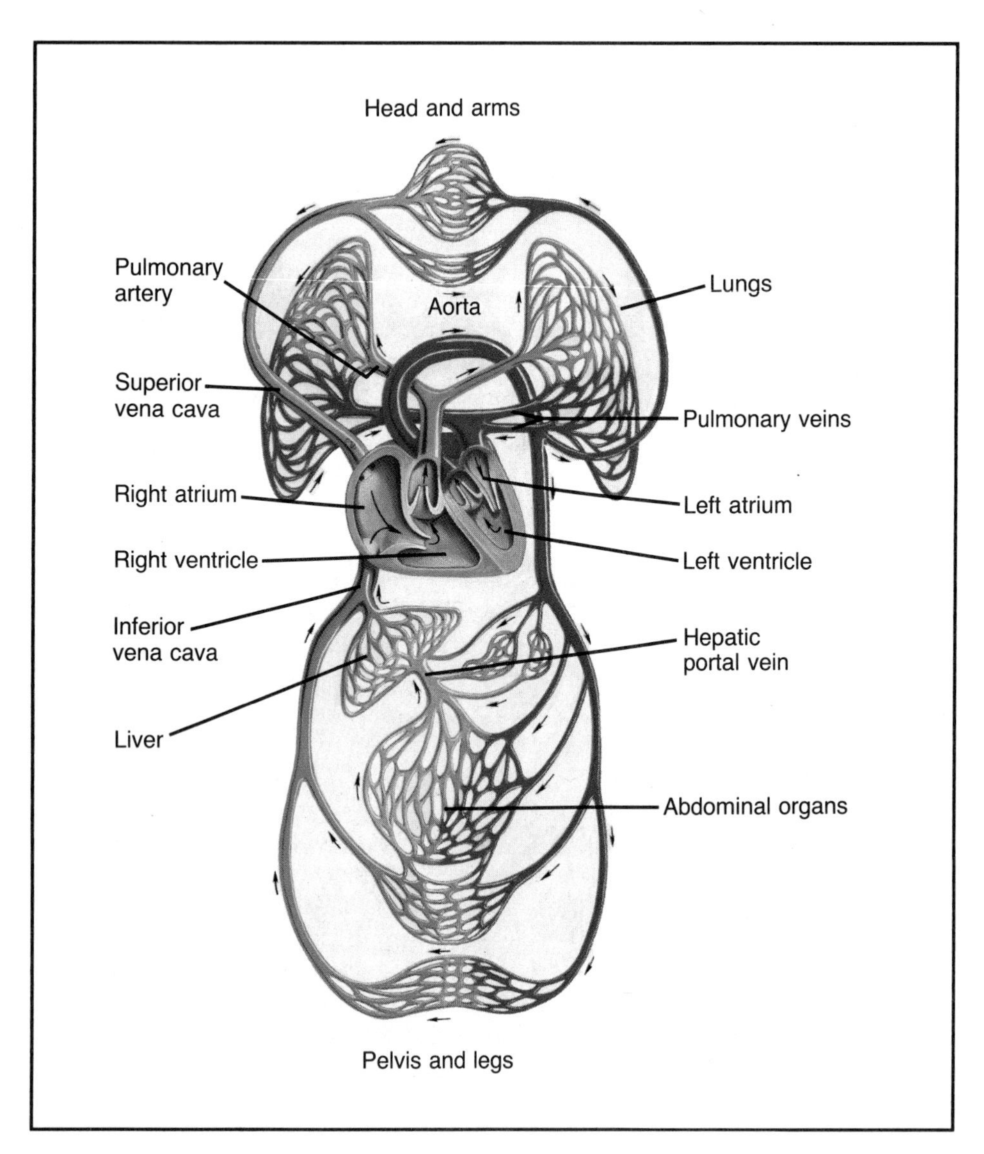

FIGURE 8.1

Human Blood Circuits

In all mammals, blood leaves the heart in vessels called arteries, and returns to the heart in vessels called veins. Exchange of substances between the blood and the cells of the body can occur only in the capillary beds that link arteries and veins.

oxygen and glucose to cells and receives carbon dioxide and nitrogenous wastes from cells. Interstitial fluid is therefore the functional equivalent of the primordial sea in which life first arose.

Of course, the volume of interstitial fluid is relatively small, especially when compared to the infinite oceans, and if it were not constantly replenished, the cells would quickly exhaust its store of oxygen and glucose, and saturate it with carbon dioxide and nitrogenous wastes. Replenishment and purification of the interstitial fluid is the role of the digestive, respiratory, and excretory systems.

However, the specialized systems that replenish and purify the interstitial fluid are not spread uniformly throughout the body, but instead are anatomically localized. At first glance, this arrangement would seem to pose significant problems to the welfare of the organism. How does the organism ensure that the composition of the interstitial fluid is essentially uniform throughout the body, rather than being rich in oxygen only near the lungs, for instance, but poor elsewhere?

One answer is to circulate the interstitial fluid past or through the various organs that affect its composition, thereby ensuring a more or less uniform mixing of the substances dissolved in the fluid. Many animals, ranging from insects to lobsters to clams, use an arrangement of this type. Open-ended vessels accumulate interstitial fluid and pass it on to one or more pumping organs ("hearts"), which circulate the fluid throughout the body. Circulatory systems of this type are called *open*.

Open systems are relatively inefficient, however, because circulation is necessarily rather slow. Most large animals, and all vertebrates, use a *closed circulatory system*, and the fluid flowing through the closed network of vessels is called *blood*. The problem with a closed system is in providing a way for substances carried by the blood to be exchanged with the interstitial fluid, on the one hand, and with the organs of respiration, digestion, and excretion, on the other. We shall examine the solutions to this problem in this chapter.

The human system is representative of mammalian circulation systems in general. It consists of the following (see Figure 8.1):

1. A four-chambered **heart,** which actually serves as a pair of independent but physically linked pumps.

2. A network of intake, outflow, and connecting vessels, each type of which is structurally and functionally distinct.

3. Blood, which is responsible for the following:

(a) The transport of oxygen, nutrients, and wastes.

(b) The transport of chemical messengers, called *hormones* (see Chapter 13), from their site of production to their site of activity.

(c) The transport of cells and chemicals involved in fighting disease organisms.

(d) The regulation and maintenance of a constant internal body temperature.

(e) The sealing of leaks in vessels, rather like the operation of a puncture-proof tire.

What began as an apparently simple problem—recycling the fluid used as the external environment of the body's cells—has ended up with a very complex solution indeed. We will consider the plumbing of the circulatory system—the heart and vessels—in this chapter, and the function of the blood in the next chapter.

THE HEART AND CIRCULATION

As we have seen, the mere presence of blood is insufficient to meet the needs of the individual cells of the body, because the cells would quickly utilize all of the available oxygen and nutrients in the blood. To be effective, blood must be circulated, so as to allow replenishment of its oxygen and nutrients and removal of the waste products of the cells by those systems of the body that specialize in each of the above-mentioned functions. Circulation, in turn, demands a heart.

The Design of the Heart

The heart of birds and mammals represents the pinnacle of vertebrate heart evolution. The original four-chambered heart of fish, in which each chamber passes the blood forward to the next chamber, has been drastically modified into two side-by-side, two-chambered pumps. Why was it necessary to tamper with fish heart design? After all, it works for the more than 20,000 species of fish.

Circulation in fish is relatively simple (see Figure 8.2). Blood leaving the heart is sent immediately to the gills to be enriched with oxygen; this **oxygenated blood** then proceeds **anteriorly** (forward) to the head or **posteriorly** (backward) to the body. It is then collected by both anterior and posterior veins and routed again to the heart.

However, the circulatory system in fish is, of necessity, a low pressure system. The heart cannot beat too powerfully, or the thin-walled capillaries of the gills would rupture. Blood leaving the gills is at still lower pressure, and blood therefore circulates rather slowly through the rest of the body.

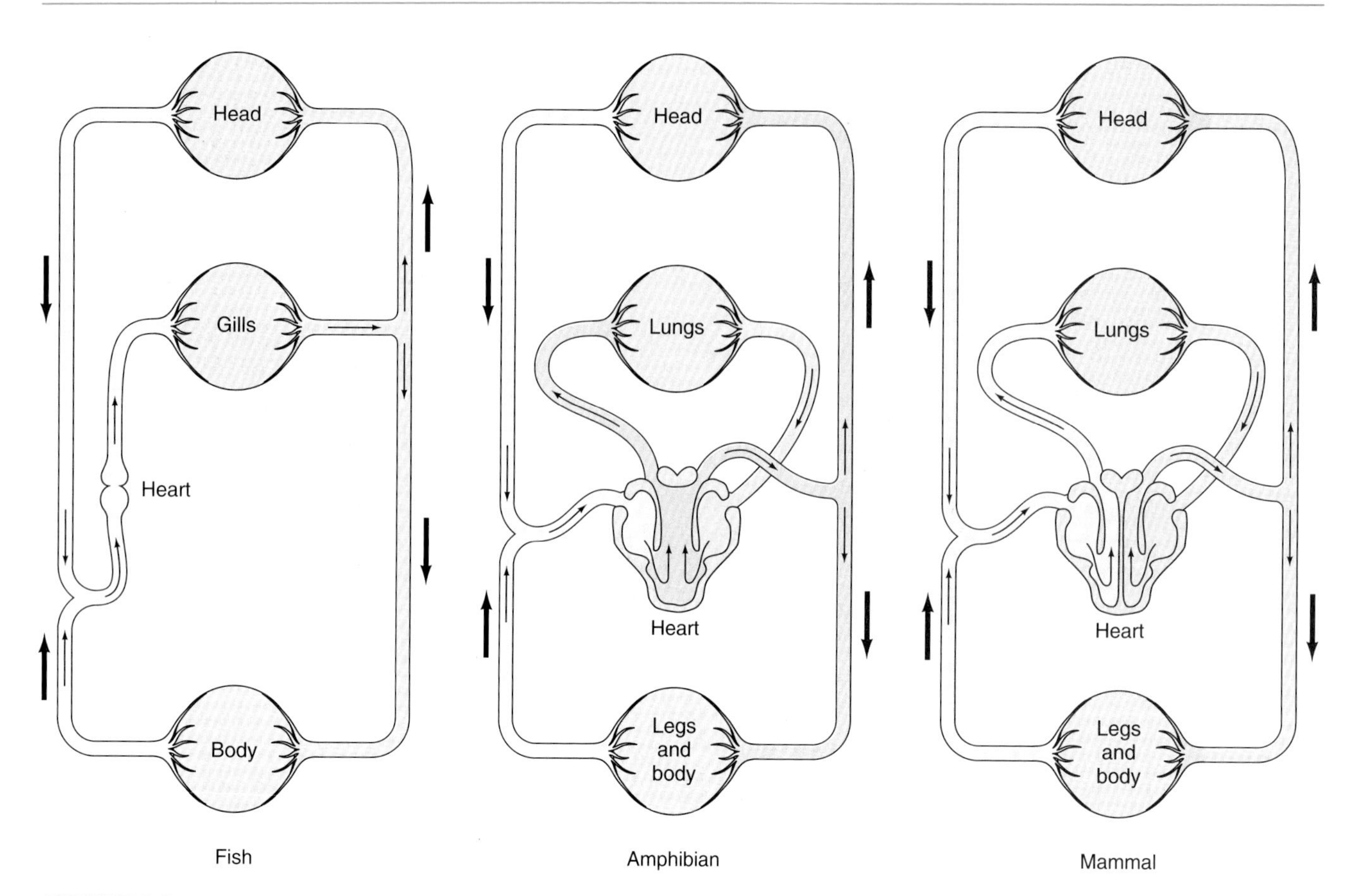

FIGURE 8.2

Blood Circuits in Fish, Amphibians, and Mammals
In fish (a), the blood does not return to the heart immediately following gas exchange in the gills. Therefore, because the blood must pass through a capillary bed before it travels to the body, blood pressure must be low and thus blood circulates rather slowly in fish. In amphibians (b), blood returns to the heart following gas exchange, but since there is only one ventricle, oxygen-rich blood mixes with oxygen-poor blood, and the efficiency of oxygen transport to the body is therefore impaired. In mammals (c), blood returns to the heart following gas exchange in the lungs, but because there are two atria and two ventricles, there is no mixture of oxygen-rich with oxygen-poor blood. Moreover, blood pressure is high, meaning that blood circulates rapidly.

The need of both birds and mammals to maintain a constant body temperature necessitates a more efficient system. Rather than the single pump of fish, birds and mammals employ a double pump, housed in a single organ. The right half of the heart receives blood from the body and pumps it, at relatively low pressure, to the lungs for oxygenation. The left half of the heart receives oxygenated blood from the lungs and pumps it, at high pressure, to the organs of the head and body. In terms of their circulatory patterns, the two pumps are so independent that they could, in theory, be located in different parts of the body. However, it is much more efficient, in terms of control and coordination, to have them linked as a single organ.

BLOOD FLOW THROUGH THE HEART

As a first step in understanding how the heart functions, let us consider the path followed by a single drop of blood as it moves through the heart (see Figure 8.3). If our hypothetical drop is returning from the head, having given up much of its oxygen to the cells of the brain, it passes through the **superior vena cava** (L., "hollow vein"), one of the two major collecting vessels of the body. (Had our drop been returning from the legs, it would have traversed the other vessel, the **inferior vena cava,** which collects blood from all parts of the body below the heart.) The two vena cavae fuse just at the level of the heart, and

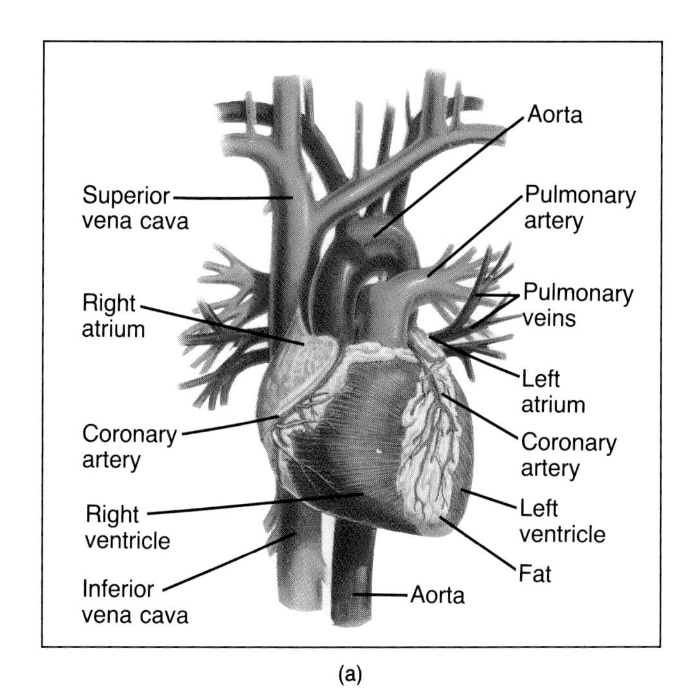

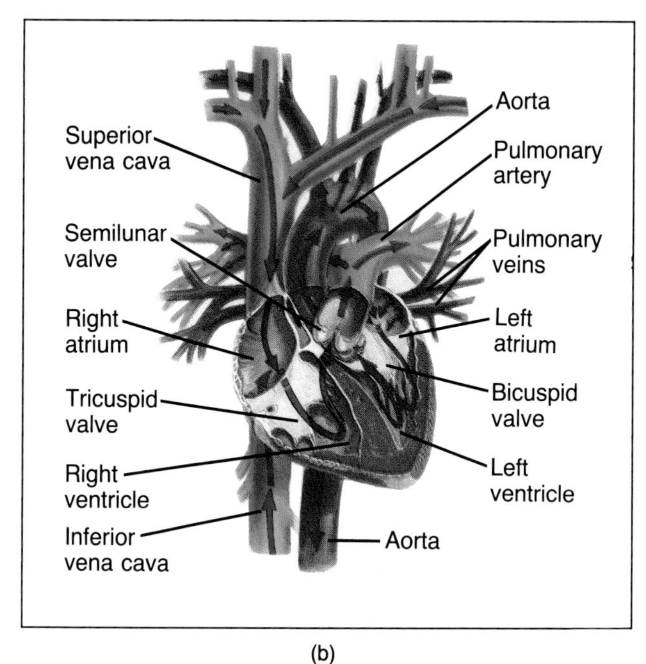

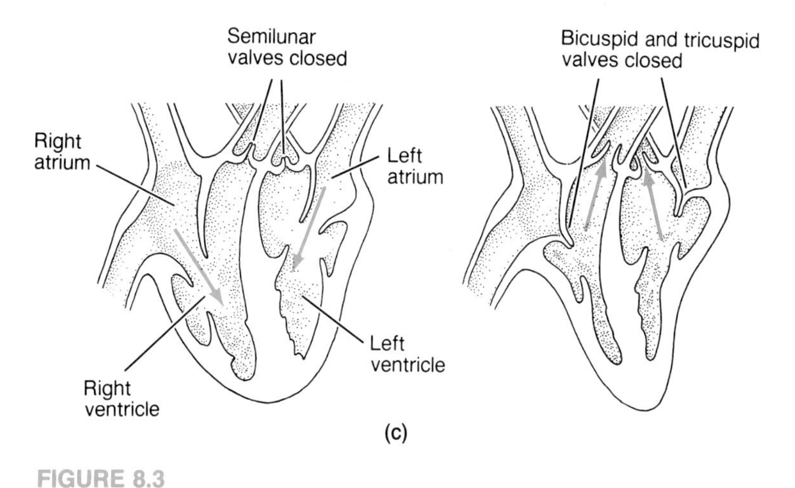

FIGURE 8.3

Human Heart

(a) Superficial view of the heart and associated blood vessels. (b) Interior view of the heart, showing the pathway of blood flow. Valves ensure that blood flow is unidirectional. The tricuspid valve is on the right side of the heart, the bicuspid on the left. Semilunar valves guard the entry of the aorta and pulmonary arteries. (c) The heart during contraction. The atria contract simultaneously, as the bicuspid and tricuspid valves open. A fraction of a second later, the valves snap shut as the ventricles begin their powerful contraction.

empty into the **right atrium** (L. *atrium,* "entrance hall").

From the right atrium, our drop of blood is pumped into the **right ventricle** (L. *ventriculus,* "a stomach"). (In reality, only about 20 percent of the blood in the atria is actually *pumped* into the ventricles. The other 80 percent enters passively, when the valves between the atria and ventricles are opened.) As the right ventricle contracts, the three-part **tricuspid valve,** located between the right atrium and ventricle, snaps shut, thereby preventing our drop of blood from returning to the atrium. Instead, it is propelled into the **pulmonary artery** (L. *pulmonis,* "lung"), which quickly divides into two vessels, one supplying each lung. The pulmonary artery trunk contains a **semilunar valve** (named for the half-moon shape of its flaps).

The flaps collapse against the sides of the vessel as blood from the ventricle enters, but between beats they fill with blood and bulge out like a little boy's back pocket into the cavity of the vessel, thereby closing it and preventing our drop of blood from seeping back into the ventricle as it relaxes between beats.

Once our drop of blood has been oxygenated by the lungs, it returns to the heart via one of four **pulmonary veins** (two from each lung), all of which enter the **left atrium.** From the left atrium, our drop is pumped into the left ventricle which, as it contracts, causes the **bicuspid valve**[1] between the left atrium and ventricle to snap shut. Our blood drop is therefore obliged to leave the ventricle by way of the **aorta,** the major artery of the body, and from there will move up to the head or down to the body. The aorta also contains a semilunar valve that prevents blood from seeping back into the left ventricle (but see Box 8.1).

[1]Because of a fanciful similarity to the two-cusped hat, or mitre, worn by bishops, the bicuspid valve is sometimes called the **mitral valve.** T. H. Huxley, the nineteenth century British biologist who was the primary defender of Darwin, the grandfather of Aldous and Julian Huxley, and the coiner of the word "agnostic," provided his anatomy students with a method of remembering that the mitral valve is located on the left side of the heart: Huxley pointed out that, at least in his opinion, bishops were never right.

BOX 8.1

Heart Diseases

Angina pectoris: An angina (Gr. *anchone,* "a strangling") attack consists of a feeling of breathlessness or choking, typically following exertion or emotional distress. It is *not* a heart attack, in that it does not result in tissue damage or pose an immediately life-threatening situation. Instead, it is a signal that the muscular walls of the ventricles are not receiving sufficient oxygen.

The muscular ventricular walls do not obtain oxygen directly from the blood they pump. Rather, oxygen is supplied by a pair of **coronary arteries** (L. *corona,* "crown", a reference to the position of the coronary arteries atop the heart). Because of the amount of work done by the heart, and the relatively small size of the coronary arteries, there is little margin for error. Thus, if the coronary arteries become hardened or constricted, they cannot expand to supply the heart muscle with the additional volume of oxygen-laden blood it needs when stressed or exerted. The result? Angina

Myocardial infarction: An infarction is an area of tissue death. A myocardial infarction is death of an area of *myocardium*—the heart muscle. You may know this term under its more colloquial name—*heart attack.* Heart attacks remain the single leading cause of death in this country—540,000 deaths each year, and 1.5 million heart attacks annually.

Heart tissue death results from a blockage of one of the coronary arteries, typically by the formation of a clot or **coronary thrombosis** (Gr. *thrombos,* "a lump or clot"). It is almost always preceded by vessel disease (see Box 8.2, "Problems with the Blood Vessels), and frequently, but by no means always, occurs in individuals who have had a history of cardiac pain (including angina).

Death of the heart muscle results from its being deprived of oxygen. Depending on the size of the area of muscle involved, the severity of the heart attack may range from virtually symptomless to death. Because it works harder than the other chambers of the heart, the left ventricle is more vulnerable to interruption of its blood supply than are the other chambers. As a consequence, myocardial infarctions are much more common in the left ventricle than in any other area of the heart.

The principal problem arising from a myocardial infarction does not stem from the fact that an area of the heart muscle no longer functions, for such dead tissue is ultimately replaced with scar tissue, but from the decay products of the dead cells. These substances may interfere with the passage of impulses from the SA and AV nodes, often leading to the development of rapid twitching of the ventricles, a condition known as **ventricular fibrillation.** This is an immediately life-threatening situation, calling for instant stimulation of the heart, either by pounding on the chest or by the use of electrical "paddles" to "jump start" the heart into a regular beat. Ventricular fibrillation is most common in the first few hours after a heart attack, which is why heart attack victims are usually initially placed in an intensive care ward.

Other problems stemming from a myocardial infarction include the development of blood clots inside the ventricle; rupture of the heart wall (because of the strain of the beating heart on the area of dead tissue); damage to the muscles controlling the bicuspid valve; and infection or inflammation of the heart, due to the effects of tissue breakdown.

Heart block: If a myocardial infarction damages the SA-AV conductive system, the heartbeat may become irregular (a condition known as heart block or **arrhythmia**). In such instances, an **artificial pacemaker** may be installed. A wire is threaded through a vein in the upper arm, down the superior vena cava, and through the right atrium to the right ventricle. A battery-powered unit sends electrical impulses at regular intervals, and effectively substitutes for the natural pacemaker. Some modern pacemakers have a variable frequency, because they pick up and amplify the frequency produced by the SA node. However, this type of pacemaker can be used only when the SA node has not itself been damaged by heart attack.

Valvular heart disease: The most common cause of valve disease is **rheumatic fever,** a once-common outgrowth of tonsillitis and strep throat, which today is usually controlled by modern antibiotics. The bicuspid and tricuspid valves are most commonly affected; they develop nodules that prevent them from closing completely. Consequently, there is some leakage back into the atria when the ventricles contract, a result easily detected by anyone familiar with the normal heart sounds. (Through a stethoscope, the heart is heard to produce two sounds. The first, usually characterized as *lupp,* is the sound of the bicuspid and tricuspid valves closing, at the end of the atrial contraction. The second, *dub,* is the sound of the semilunar valves closing, at the end of the ventricular contraction. Damaged valves create a *ssss* noise, as blood escapes back into the atria from the ventricles, or into the ventricles from the pulmonary arteries or aorta.)

In severe cases, **congestive heart failure** may result. Typically, the bicuspid valve is more susceptible to failure, and, as the left ventricle contracts, it forces blood back through the failed valve into the left atrium and lungs, causing the individual literally to drown in his own fluids. Fortunately, modern surgical techniques now permit replacement of faulty valves where conditions warrant.

Heart **murmurs** are any irregularities in normal blood flow, as revealed by peculiarities in heart sounds. They are commonly present from birth, and are not necessarily serious.

CONTROLLING THE HEARTBEAT

The muscle cells that comprise the bulk of the heart's tissues differ from muscle cells found elsewhere in the body in a number of important respects. Among the more significant differences is that **cardiac** (Gr. *kardia,* "heart") muscle cells contract spontaneously, even in isolation from other cells. That is, if you were to view living individual cardiac muscle cells in a water suspension under the microscope, you would see them randomly contracting every second or two.

Although spontaneous contraction of individual cells is an important prerequisite for an organ that must operate as a pump, obviously the heart cannot function simply as a mass of randomly contracting muscle cells. How are the individual muscle cells coordinated? As it happens, because cardiac muscle cells are inherently excitable, they will, as a group, contract at the rate established by the area of the heart with the fastest autonomous rhythm. In the mammalian heart, this group of cells, called the **pacemaker,** is located in the wall of the right atrium (see Figure 8.4). It is actually all that remains of a once-separate chamber of the heart (still found in fishes) called the **sinus venosus.** For that reason, the proper name for the pacemaker is the **sinoatrial (SA) node.**

Impulses from the SA node spread rapidly across the muscle tissue comprising the atria, causing these excitable cells to contract virtually as a unit. Thus the left and right atria contract simultaneously. However, impulses are prevented from reaching the ventricles by a ring of fibrous tissue that acts as an insulator.

This ring of tissue is broken in only one place, the location of a second cluster of excitable cells called the **atrioventricular (AV) node.** The AV node transmits impulses from the SA node a fraction of a second after the atria contract. This brief delay ensures that all four heart chambers do not contract simultaneously, a situation that would result in immediate heart failure, because the powerful ventricles would force blood back into the atria through the open bicuspid and tricuspid valves.

Radiating from the AV node is a specialized strand of conducting tissue known as the **bundle of His,** which ultimately divides into a series of individual fibers known as **Purkinje fibers.** The impulse from the AV node is carried first to the bottom (apex) of the heart, and then back along the sides, thereby ensuring that ventricular contraction begins at the apex, and moves back and toward the base. Thus the blood is pumped in the proper direction.

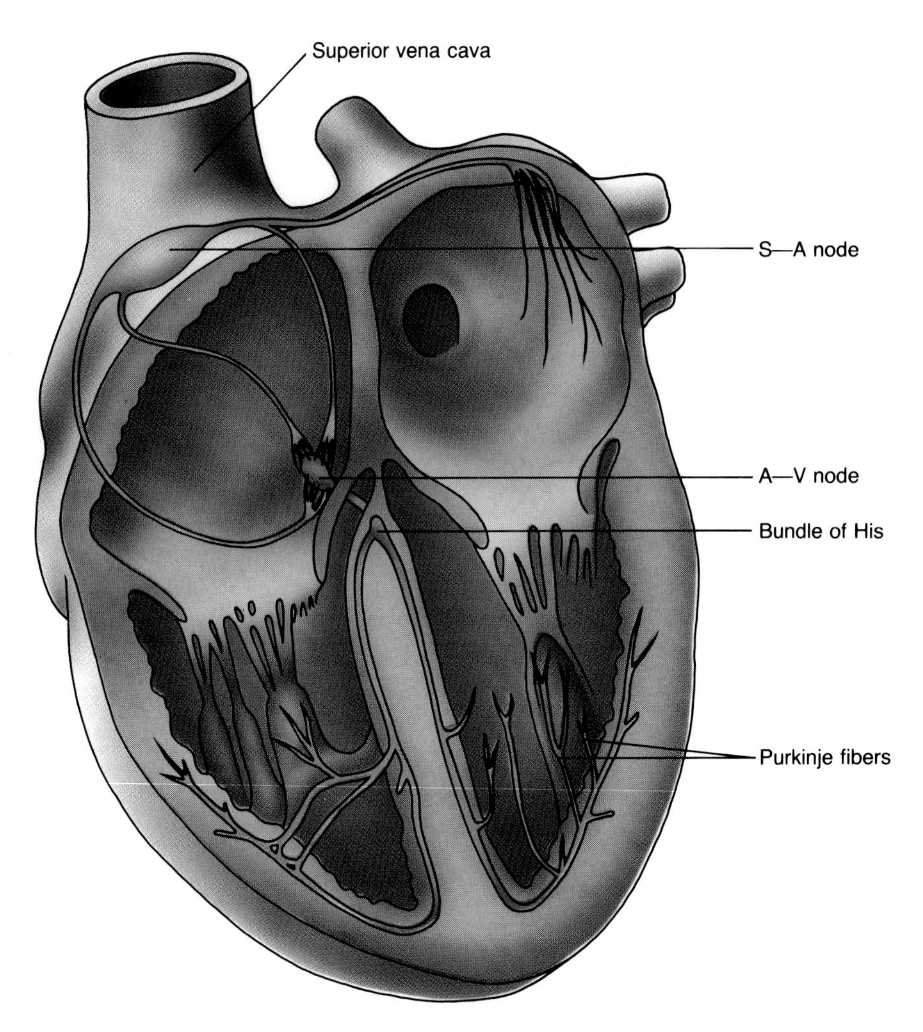

FIGURE 8.4

Pacemaker

The initiating stimulus for a heart beat occurs in the SA node, high in the right atrium. The atria contract, the stimulus is relayed by the AV node to the ventricles, by way of the bundle of His, and the ventricles contract.

BLOOD PRESSURE

It is essential that the ventricles pump with sufficient force to move the blood through the whole system. (This is particularly true of the left ventricle, which must propel the blood through the entire body. For that reason, the left ventricle is more powerfully constructed and contracts with greater force than does the right ventricle, which must deliver blood only a relatively short distance to the lungs.) However, the ventricles cannot pump with too much vigor lest the capacity of the blood vessels be exceeded and rupture, much as a garden hose would burst if linked to a fire hydrant.

In an individual at rest, the left ventricle produces a blood pressure of about 140 mm of mercury. That is, it pumps with a force sufficient to raise a column of mercury 1 mm in width to a height of 140 mm. Values appreciably higher than this level pose the very real dangers of burst vessels; values much lower may cause dizziness and fainting spells, for the pressure may be insufficient to push an adequate amount of blood "uphill" to the brain.

In the **brachial artery** (the main artery to the arm, which is where blood pressure is normally measured), the blood pressure is about 120 mm of mercury (see Figure 8.5). Actual blood pressure readings are expressed as a fraction (for example, 120/80). The first number is the **systolic pressure** (Gr. *systole*, "to contract or shorten"), measured during ventricular contraction. The second number is the **diastolic pressure** (Gr. *diastole*, "to expand or dilate"), measured between contractions.

The pressure of the blood drops as the blood moves into progressively narrower, but much more numerous, vessels (see Figure 8.6). By the time the blood returns to the heart via the vena cavae, the blood pressure has dropped to less than 10 mm of mercury.

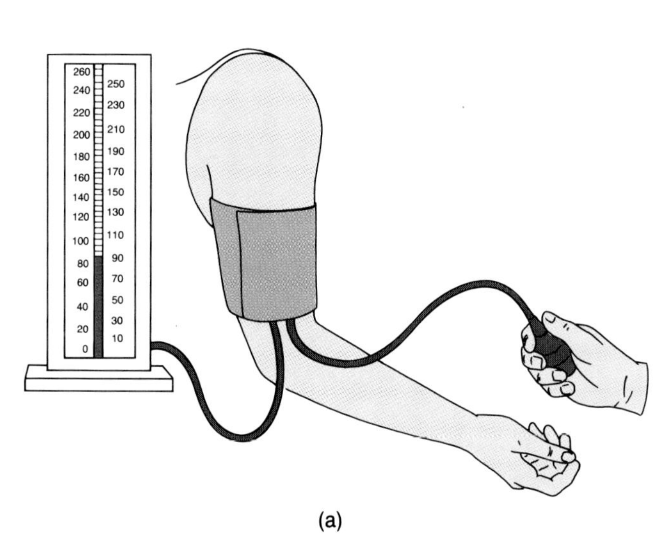

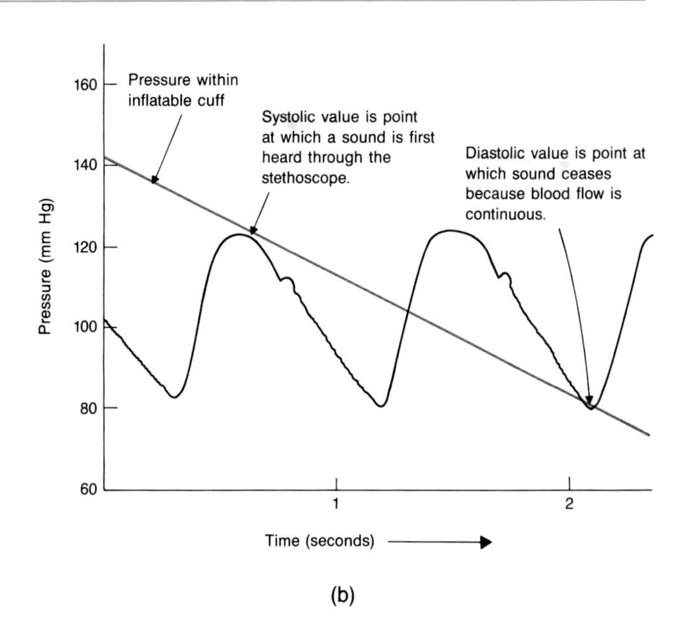

FIGURE 8.5

Measuring Blood Pressure

(a) Blood pressure is measured by a sphygmomanometer (Gr. *sphygmos*, "pulse"), an instrument that includes both an inflatable cuff and a pressure scale. (b) Initially, the pressure in the cuff cuts off all arterial flow. As pressure is relieved, arterial flow begins in spurts (that is, as the heart contracts) and then becomes continuous. The points at which these two events occur are the systolic and diastolic pressures, respectively.

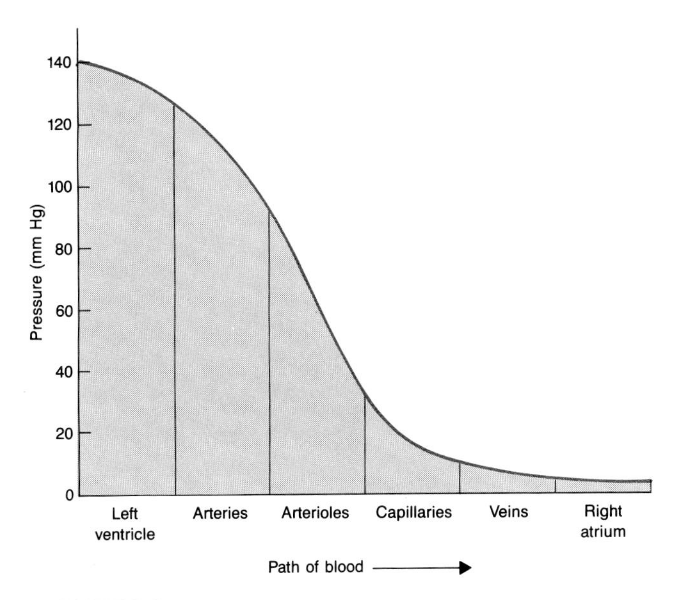

FIGURE 8.6

Changes in Blood Pressure

From an initial pressure of 140 mm in the left ventricle, blood pressure drops quickly as the blood moves through the narrow, but elastic, arterioles. Blood pressure is near zero by the time the blood returns to the heart.

CARDIAC OUTPUT

In an individual at rest, the heart beats approximately 70 times per minute, and with each beat the left and right ventricles each pump 80 cm^3 of blood into the aorta and pulmonary arteries, respectively. This amount is referred to as the **stroke volume,** and stroke volume times heart rate equals **cardiac output.** Thus the cardiac output of an individual at rest is (70×80 cm^3 =) 5.6 liters/minute. In an individual exercising heavily, the heartbeat rate may exceed 180/minute, and the stroke volume may reach 160 cm^3, for a cardiac output of almost 30 liters/minute. How is this fivefold range of cardiac output regulated?

Although the SA node acts as the internal pacemaker of the heart, it is itself regulated by stimulatory and inhibitory centers in a portion of the brain known as the **medulla** (see Chapter 14 for a discussion of brain structure and function). Normally, the inhibitory center dominates, and the inherent capacity of the SA node to stimulate the heart at the rate of 100 times per minute is modulated to a resting rate of 70 beats per minute. Inhibitory stimuli from the brain travel down the **vagus nerve,** the principal nerve of the **parasympathetic nervous system,** which is one of two branches of the nervous system that supply the internal organs and regulate activities that we do not consciously control, such as heartbeat and digestive activity (see Figure 8.7).

The other branch, the **sympathetic nervous system,** operates antagonistically to the parasympathetic system, and stimulates heart activity. This is the system that predominates when the individual becomes active. Not only is the heartbeat rate increased, but blood is preferentially shunted to the muscles that are being exercised.

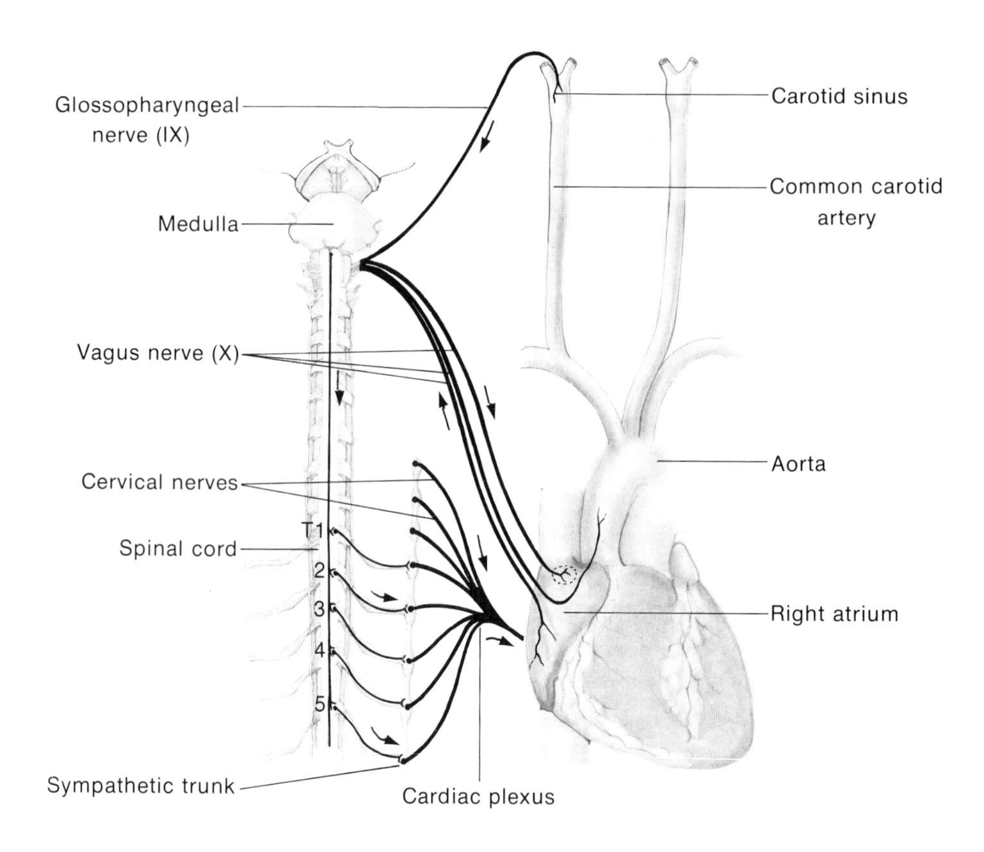

FIGURE 8.7

Regulation of Heart Beat and Stroke Volume

The frequency and strength of the heartbeat are regulated by nerves and hormones. Pressure sensors in the carotid sinus, aorta, and the atria relay information to the medulla. Stimulation occurs through the sympathetic nervous system and through the release of hormones from the adrenal gland; inhibition occurs through the action of the vagus nerve of the parasympathetic nervous system.

However, exercise is not the only situation where heartbeat rate increases. All of us can remember moments of absolute terror when, as young children, watching some particularly scary movie on television, we suddenly became aware that our heart was beating furiously, even though we were riveted in place. This type of situation (which is explained more fully in Chapter 13) exemplifies the action of **adrenal gland,** which produces the hormone **adrenaline** (now more commonly known as *epinephrine*). This hormone prepares the body for immediate action ("fight or flight"), a part of which involves increasing heart activity.

As we saw above, increasing the heartbeat rate is only one part of increasing cardiac output. How are changes in stroke volume achieved?

Within rather broad limits, the strength of ventricular contraction is directly related to how much ventricular muscle is stretched. During vigorous physical activity, the volume of blood returned from the muscles to the heart increases, causing the ventricles to stretch and, as a consequence, to beat more vigorously, increasing the stroke volume.

THE STRUCTURE AND FUNCTION OF THE VESSELS

The vessels of the circulatory system are highly specialized to perform the very different tasks of transporting blood at high pressure from the heart to the tissues and organs of the body; exchanging gases, nutrients, and wastes with the cells of the body; and carrying waste-laden and oxygen-depleted blood back to the heart (see Box 8.2). The vessels responsible for each of these tasks are, respectively, the **arteries** and **arterioles;** the **capillaries;** and the **veins** and **venules.**

Structurally, the vessels show important similarities. The lining of all of the vessels is composed of

BOX 8.2

Problems with the Blood Vessels

Hypertension: Hypertension, or high blood pressure, is both common (one-third of all adult Americans are hypertensive—58 million people!) and potentially dangerous, for a variety of reasons discussed below. Blood pressures are considered elevated if they measure more than 140/90.

Hypertension results from excessive constriction of the arterioles. In a small minority of cases (**organic hypertension**), the cause is known to be a tumor of the adrenal glands (see Chapter 13), or kidney disease. However, in most cases, the cause is unknown. This form of high blood pressure is called **essential hypertension.**

Candidates for the cause of essential hypertension are beginning to emerge. First, defective **ATPase** (the transport molecule of the sodium/potassium pump) has been identified as the cause of hypertension in rats, a finding which may have implications for humans. Second, very recently a hormone that increases the strength of the heartbeat and constricts the arteries has been identified in humans. It is chemically identical to the East African arrow poison *ouabain* (which natives extract from the seeds of a tropical vine), and its source in humans is believed to be the adrenal glands (see Chapter 13). Elevated levels of this hormone would almost certainly cause hypertension.

If untreated, hypertension causes a strain on the heart (because of its having to pump blood through narrowed vessels). Frequently, the heart becomes enlarged, and is prone to failure. The blood vessels are susceptible to excessive strain and wear, and may weaken and rupture, or become scarred and thickened, further reducing their internal diameter. Finally, kidney failure is a common consequence of hypertension, because the kidneys are particularly sensitive to reduced blood flow and elevated blood pressure.

Atherosclerosis: Atherosclerosis is the major type of hardening of the arteries. It is characterized by the development of fatty plaques (which often become calcified) on the inside walls of arteries. It is the primary cause of angina pectoris, and it predisposes the individual to myocardial infarctions. Atherosclerosis also contributes to the formation of aneurysms and strokes (see below), and, when it occurs in the arteries of the hands and feet, to the development of gangrene. It is therefore a serious and much-researched disease.

Although atherosclerosis is correlated with hypertension, the reasons for its inception are still very much a matter of debate. We discussed the relationship between high blood cholesterol levels (especially high LDL levels) and atherosclerosis in Chapter 7. Now let us examine a current scenario: Excessive LDL is scavenged by a type of white blood cell called **macrophages** (see Chapter 9). As a consequence of ingesting too much LDL, macrophages convert to **foam cells,** which accumulate along arterial walls. These cells produce a substance called **PDGF** (platelet derived growth factor), which stimulates the growth and proliferation of smooth muscle cells in the adjacent arterial walls. These

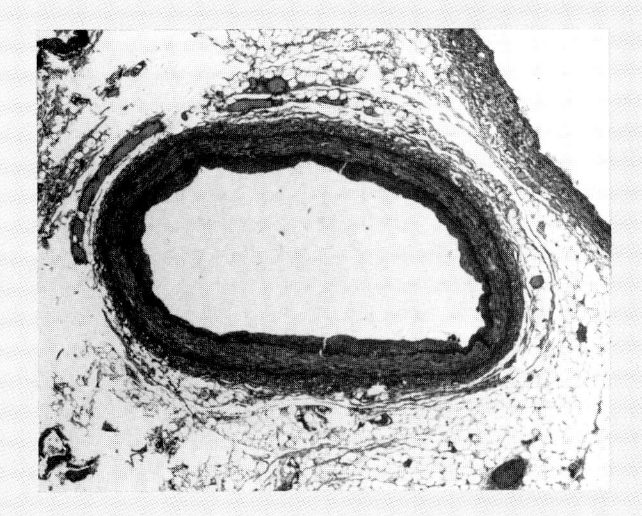

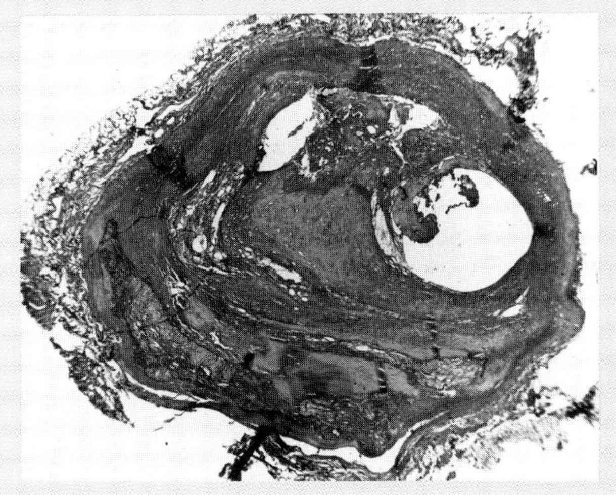

Atherosclerosis
In this cross section of the coronary artery, the vessel's diameter has been markedly reduced by the development of atherosclerotic plaques along the vessel walls.

(Continued on the following page)

BOX 8.2 (Continued)

nodules of smooth muscle are further increased in size by being surrounded by connective tissue, and their irregular surface tends to attract fatty deposits, especially if LDL levels remain high. The result is a narrowing of the arteries, and the development of sites for the formation of blood clots.

Interestingly, monounsaturated fatty acids (see Chapter 7) sharply reduce the likelihood of LDLs causing macrophages to convert to foam cells, which is why there is so much current focus on increasing the proportion of monounsaturates in our diet.

Although a diet low in cholesterol does not necessarily lead to a drop in blood cholesterol levels, the incidence of atherosclerosis (and, therefore, of heart attacks) is lower in individuals who are not obese, have a diet relatively low in cholesterol, exercise regularly, are free from diabetes, and are not living highly stressful lives. Premenopausal women show markedly less atherosclerosis than do men of comparable age, because the female hormone **estrogen** acts to lower blood cholesterol levels. Finally, one recent study suggests that β-carotene (the plant pigment that our bodies can convert to vitamin A; see Chapter 7) may help prevent atherosclerosis. The study found that men with a history of cardiovascular disease who took a 50 mg carotene pill every other day had half as many strokes and heart attacks over a six-year period as men taking placebos.

Aneurysms: An aneurysm is a weakened area of an artery that balloons out and ultimately ruptures, typically with disastrous results. Such weakened vessels may be a product of prolonged hypertension, especially when coupled with atherosclerosis, or may be present from birth as a congenital weakness. Very often the sudden death of a young athlete, outwardly the healthiest of individuals, is found to be due to the rupture of a congenital aneurysm.

Recently, a defective gene that codes for a form of **collagen** (a protein of connective tissues; see Chapter 12) has been linked to the development of aortic aneurysms. A simple saliva test can determine if a person has this mutation, and a positive test would be followed by ultrasound scanning on a regular basis to identify aneurysms before they rupture.

Stroke: Stroke (now known as *cerebrovascular accident* and, in earlier times, called *apoplexy*) results from the injury or death of brain tissue due to a failure of the brain's blood supply. Typically, the cause is a clot in one of the arteries supplying the brain (**cerebral thrombosis**). About 20 percent of strokes are caused by the rupture of an artery (**cerebral hemorrhage**), either because of atherosclerosis, hypertension, or both.

The consequences of a stroke depend on the location of the stroke and the amount of brain tissue involved. An immediate partial paralysis is common, because of disruptions in the nerve tracts leading to the spinal cord, although recovery generally occurs if the area affected is not large. Unfortunately, especially in strokes caused by cerebral hemorrhage, the damage to the brain is often extensive, and stroke remains a leading cause of death in the United States. Indeed, 400,000 Americans suffer strokes each year, and more than 100,000 die—but the number of deaths from strokes is now only half what it was

endothelium, the same type of tissue that lines many internal organs. Indeed, the capillaries are composed exclusively of endothelium. All of the other vessels also have an outer covering of connective tissue. Between these two layers, in the arteries, arterioles, and veins (but not the venules) is found a layer of smooth muscle in which elastic fibers are imbedded. The thickness of this layer varies greatly with the type of vessel (see Figure 8.8).

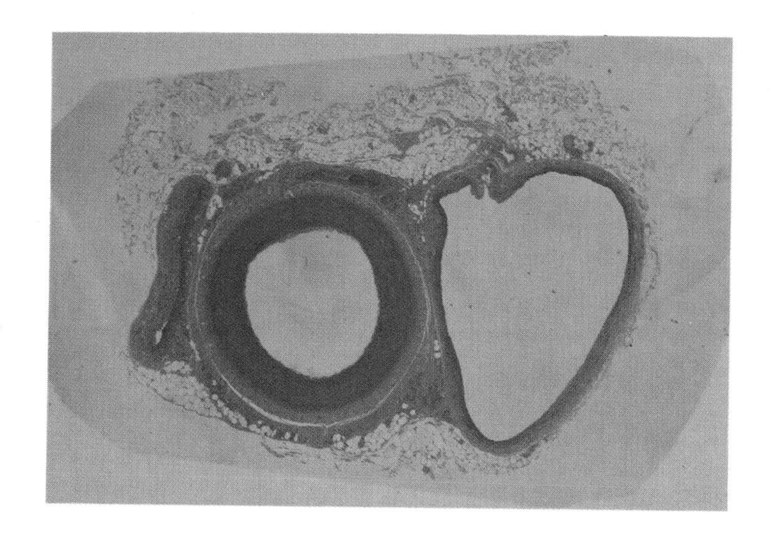

FIGURE 8.8

Structure of Arteries and Veins

A vein and artery as revealed by the scanning electron microscope. In this instance, the inside diameter of the vein (left) is greater, but the wall is substantially thinner than is the wall of the artery (right).

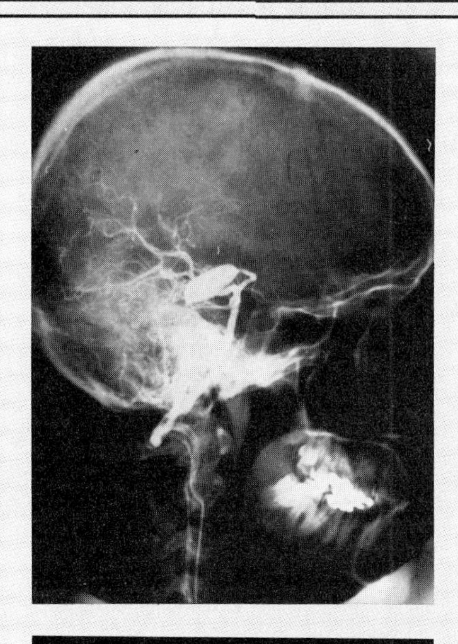

Aneurysm of a Blood Vessel in the Brain

just a decade ago, thanks largely to improved treatment techniques (see Box 8.3).

Phlebitis: Phlebitis (Gr. *phleb,* "vein") is an inflammation of the veins, most commonly in the legs. Although the disease is serious in its own right, much more serious is the likelihood that blood clots may form at the site of inflammation, break loose, and be carried back to the heart. Since the veins increase in diameter as they near the heart, the clots generally do not lodge until they pass through the right atrium and ventricle and are pumped to the lungs. As the pulmonary arteries narrow, the clots sooner or later arrive at a vessel too small to pass through, and plug it. A lodged clot is called a **pulmonary embolus** (Gr. *embolos,* "anything put in") and, when large or numerous, is a life-threatening situation because of the risk that pulmonary exchange will simply cease.

Blood clots are also more likely to form following prolonged bed rest. For this reason, post-surgical patients and women who have just had babies are urged to begin to walk again as soon as possible, a practice that has sharply decreased the incidence of deaths due to pulmonary emboli.

Varicose veins: An important mechanism in the return of blood to the heart is the massaging effect of muscles on the veins. Some of the veins, most notably those in the legs, are not located within or between muscles, but are between a muscle and the skin. Individuals who are overweight or who must stand a great deal may develop varicose (L. *varicosus,* "enlarged vein") veins as the skin gradually loses its elasticity, and the venous valves begin to fail. Prominent veins in the arms, so beloved by body builders in particular (and teenage boys in general) are not properly termed varicose, because the valves have not failed and the blood continues to move through the veins efficiently.

Varicose veins frequently develop in pregnant women who not only are carrying around substantially more weight than usual, but who also suffer cramping of the intestinal veins because of the growth of the fetus. Generally, these diminish or disappear shortly after childbirth. The problem of varicose veins is more severe in the elderly, because they are generally less active, and the poor circulation through the varicose veins frequently leads to phlebitis, ulceration, or hemorrhage.

Hemorrhoids are varicosities of the hemorrhoidal veins that ring the anus (external hemorrhoids), or of veins associated with the rectum (internal hemorrhoids).

ARTERIES AND ARTERIOLES

The arteries and arterioles constitute the delivery component of the circulatory system (see Figure 8.9), for together they are responsible for conveying blood at relatively high pressure to the body's tissues. Given their function, it is not surprising that they have very thick walls, composed largely of smooth muscle and elastic fibers.

The inside diameter of these vessels ranges from 25 mm in the aorta to 30/1000 mm in the smallest arterioles. A well-established principle in physics, and one which is pertinent to our discussion of blood flow, is that the resistance of a tube through which a fluid is flowing increases by a factor of 16 for every halving of the radius of the tube. Because of the great decrease in vessel diameter, the blood pressure drops from about 140 mm of mercury in the aorta to less than 50 mm in the arterioles.

The reason that the blood pressure does not drop even more is that the system does not, of course, consist of one vessel with a gradually narrowing diameter, but instead is a multibranching system. Thus there is an interesting relationship between the speed (velocity) of blood flow, and the total cross-sectional area of various components of the circulatory system.

By way of illustrating this point, consider that the aorta receives 80 cm^3 of blood with each beat of the heart. Since the heart beats 70 times a minute, about 100 cm^3 of blood enter the aorta each second in an

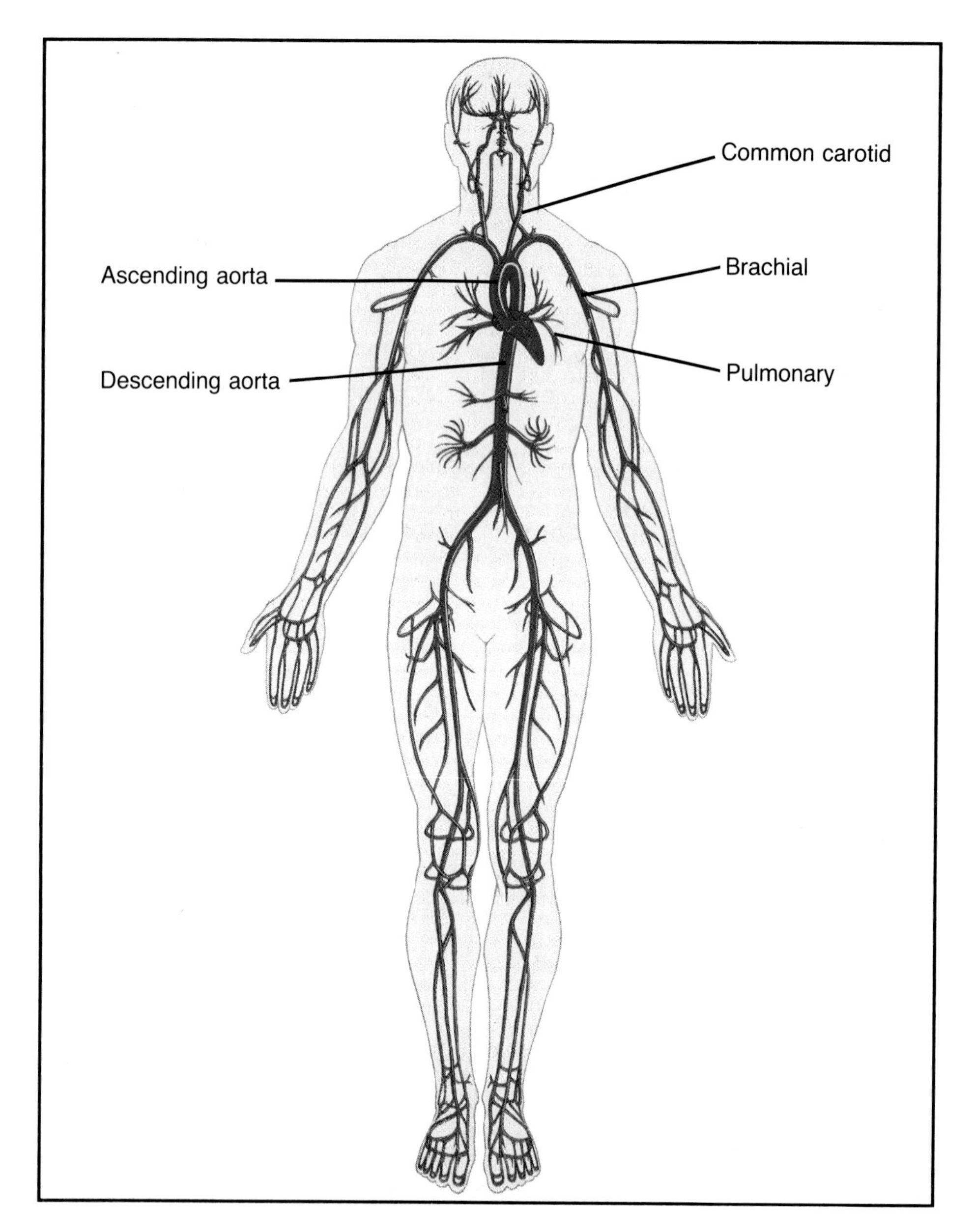

FIGURE 8.9

Human Arterial System

Although there is considerable individual variation in the precise arrangement of arteries, this illustration is a good general representation of the human arterial system.

individual at rest. The cross-sectional area of the aorta is approximately 4 cm^2, and therefore the velocity of the blood in the aorta is (100 cm^3/4 cm^2) = 25 cm/sec. By comparison, the total cross-sectional area of the arterioles is about 40 cm^2 (despite their individually narrow diameter), which means that the velocity of the blood in the arterioles is (100 cm^3/40 cm^2) = 2.5 cm/sec. Therefore, by the time the blood reaches the end of the arterioles, it is moving much more slowly, and at much lower pressure. As we shall see shortly, both these facts are of vital importance to the effective operation of the capillary system. (See also Box 8.3).

The arterioles also play a vital role in permitting, or adjusting to, changes from the resting condition. Although all organs of the body are well supplied with blood vessels, not all receive equivalent amounts of blood. In the individual at rest, for example, the kidneys receive a disproportionate amount of blood, relative to their mass. During exercise, however, the muscles receive a massive increase in blood flow (as does the skin, for purposes of temperature regulation), while the internal organs (except the heart) receive correspondingly less. The brain receives approximately the same amount of blood, irrespective of the activity of the individual. How are these changes brought about? Consider the following examples:

1. You decide, after six months of concerted lethargy, to get back into shape—in one afternoon. Donning your newest jogging outfit, you take to the track and run the mile in 5½ minutes—the last three laps in increasing oxygen debt. You finish the mile making sounds like a leaky steam engine and collapse on the grass. One of your friends notices that your face

BOX 8.3

Treatment of Heart and Vessel Diseases

In the United States, fully one-half of all deaths from all causes is caused by diseases of the heart and blood vessels. Even so, during the past decade, there has been a 20 percent drop in the number of such deaths (to slightly under one million annually), in large measure because of the kinds of treatments listed below.

Aspirin: Aspirin inhibits the "stickiness" of platelets (see Chapters 9 and 13), and reduces the likelihood of spontaneous blood clot formation. A standard dose aspirin (300 mg) every other day is recommended for individuals who are at risk for strokes, or who have angina, vessel disease, or *claudication* (L. *claudicare,* "to limp," leg pain caused by inadequate blood flow to the calf or thigh muscles).

Nitroglycerin: Nitroglycerin has been used for many years in the treatment of angina. It relaxes the muscles in the walls of both arteries and veins, thereby expanding the diameter of the coronary arteries themselves, permitting more blood to reach the heart muscle. Dilated vessels lead to a reduction in the volume and pressure of blood returning to the heart, thus reducing the workload of the heart.

Beta blockers: Beta blockers, introduced about 20 years ago, are used primarily by patients with angina, hypertension, and survivors of a first heart attack (in whom these drugs have been responsible for a reduction of 30 percent in mortality). Beta blockers interfere with certain activities of the sympathetic nervous system, and counter tendencies for increased heart activity and constriction of arterial walls.

Calcium channel blockers: Calcium channel blockers are an even newer group of drugs now used extensively by angina patients. By blocking channels in the plasma membrane of muscle cells of the heart and blood vessels (through which calcium enters during nerve and muscle activity; see Chapters 12 and 14), these drugs prevent spasms of the coronary arteries, reduce overall heart activity, and lower blood pressure. Cardiac muscle, unlike skeletal muscle, employs an active transport system in the plasma membrane—the sodium-calcium exchange—to regulate calcium levels inside the cells. Since calcium channel blockers interfere with this transport mechanism, they function effectively on cardiac muscle but do not interfere with the activity of skeletal muscles.

Antihypertensives: Medications used to reduce blood pressure include **diuretics** (drugs that increase urine output and reduce total blood volume, thereby reducing blood pressure), and beta blockers and calcium channel blockers, discussed above. More recently, two new substances have come into use. **ACE inhibitors** (angiotensin converting enzyme), as the name implies, inhibit the production of **angiotensin,** a hormone that raises blood pressure by "instructing" the kidney to recapture more sodium and water from the forming urine. **ANF, (atrial-natriuretic factor)**, is a hormone produced by the atria, which increases the rate of sodium ("natrium" in Latin) excretion by the kidneys by inhibiting the rate of renin, angiotension, and aldosterone secretion (see Chapters 11 and 13 for details of these hormones). Because water accompanies the excreted sodium, the blood volume is reduced, and the blood pressure falls.

Interestingly, there is also some recent evidence that 1 g per day supplementation with vitamin C may reduce blood pressure in borderline hypertensive individuals.

Clot dissolvers: Heart attack victims who are still alive upon reaching the hospital are now commonly given medication to dissolve the clot in the coronary arteries, enabling blood flow to be restored to the entire heart muscle. These medications have reduced the death rate of hospitalized heart attack victims by as much as 40 percent. The medications include **streptokinase,** an enzyme that dissolves blood clots, and **tissue plasminogen activator (TPA),** an anticoagulant produced in small quantities by the body but available in quantity today through the miracles of genetic engineering (see Chapter 19). Although these medications are still very expensive, they often permit the length (and, hence, the cost) of a hospital stay to be reduced by several days.

Ribose administration: A recent study has demonstrated that one reason why heart attack patients are at such high risk during the first ten days after a heart attack is that cardiac muscle is very slow to resynthesize normal levels of ATP, and the heart is therefore in a weakened condition. If ribose is given for the first five days, however, ATP levels return almost immediately to normal levels. Extensive field testing of ribose treatment is now underway.

TGF treatment: A new study in rats has pointed the way to new treatments in humans (although these have not yet been undertaken). Damage to heart muscle (as occurs following an infarction) causes an increase in the number of **neutrophils** (a type of white blood cell; see Chapter 9) which adhere to the blood vessels in the area of damage. Neutrophils produce a substance called **tumor necrosis factor (TNF),** which causes further damage to cardiac muscle cells. At the same time, there is a drop in the production of TNF's antagonist, a chemical called **TGF-β (transforming growth factor),** which is normally produced by heart muscle cells. Damage to cardiac muscle in rats following an infarction was dramatically reduced if TGF-β was administered—and this finding may have profound implications in the treatment of heart attack patients.

(Continued on the following page)

BOX 8.3 (Continued)

Angioplasty: Individuals who are diagnosed as having atherosclerosis in their coronary arteries (and who are therefore at risk for a heart attack) are more commonly being given an *angioplasty.* (Some 250,000 angioplasties are now performed annually in the United States.) This procedure involves threading a thin, deflated balloon into a vein in the thigh, tracking its movement through the heart and into a coronary artery, and then inflating the balloon at the site of the atherosclerosis. The balloon pushes the cholesterol plaque into the wall of the artery, and restores the vessel's diameter to something approaching normal. This technique is not without some risk, and it has been criticized as effecting only a temporary cure, since cholesterol deposits tend to reaccumulate—but it is much less invasive (and costly) than is open heart surgery.

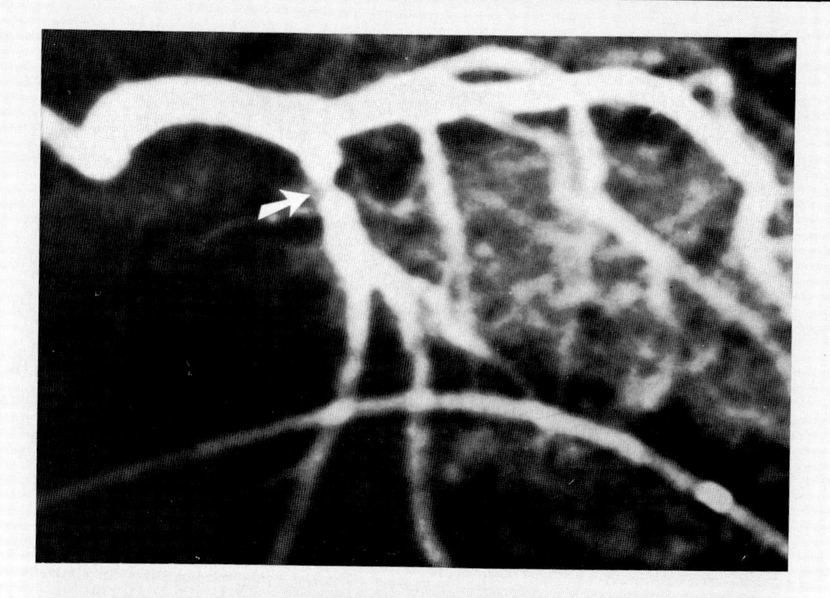

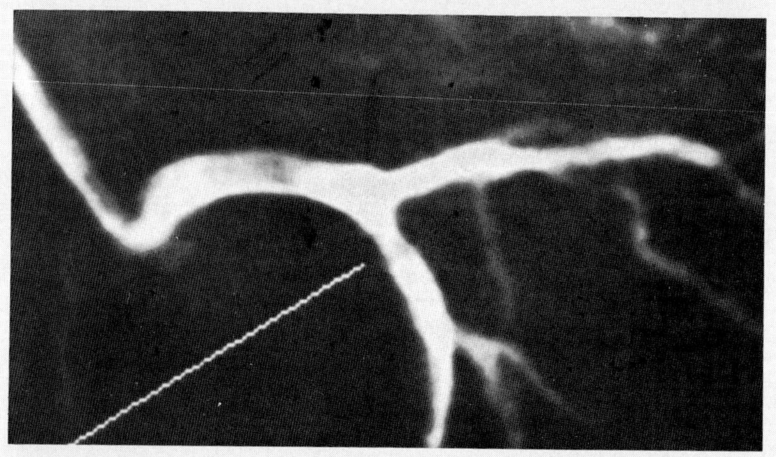

Angioplasty
(top) An X-ray view of a severely narrowed coronary artery (arrow). (bottom) A view of the same vessel following angioplasty.

Lasers: Lasers are now used for so many procedures that many surgeons are predicting the end of scalpel surgery within 20 years. Lasers are presently used to remove cholesterol plaque from clogged leg arteries, and soon will be used on the much smaller (and constantly moving) coronary arteries. Since the plaques are fragmented into tiny pieces, there is no significant risk of embolisms, and the vessel walls are rendered much smoother than is possible with angioplasty.

Coronary bypass: Some 290,000 coronary bypass operations were performed in 1986, and since its introduction in 1970, more than 2 million Americans have had this procedure. Indeed, it has become so common that a recent case-by-case review found that 14 percent of the operations in the study were "inappropriate," and another 30 percent were "of questionable necessity." Coronary bypass involves splicing one or more alternate pathways around the area of narrowing in the coronary arteries. The number of new pathways varies from case to case ("triple bypass," "quadruple bypass," etc.). Grafting materials come from a vein removed from the patient's leg (other vessels take over for the removed vein), and, in about 10,000 cases a year, from umbilical

matches the school colors (crimson and gray) and that you are sweating profusely (the grass beneath you is beginning to die). What has happened?

Control of blood delivery to the various organs is largely accomplished by the arterioles. During exercise, the muscles quickly exhaust their available oxygen, and produce large amounts of carbon dioxide, nitrogenous wastes, and heat, all of which tend to dilate the arterioles, allowing more blood to flow through them. Rising body temperature also dilates the arterioles supplying the skin, causing the skin to

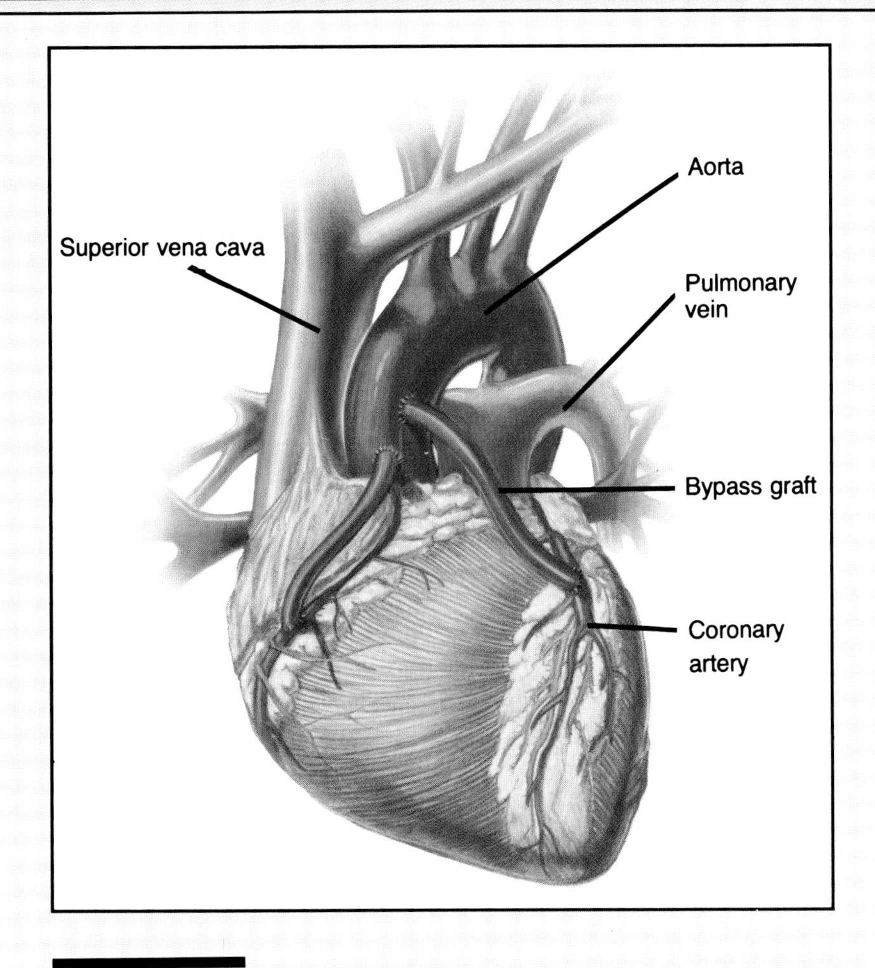

Coronary Bypass

Through the use of natural or synthetic vessel grafts, surgeons can create one or more bypasses around a narrowed region of a coronary artery.

veins salvaged from umbilical cords. Coronary bypass operations are serious surgeries, with all of the risk associated with any open-heart procedure, and the grafted vessels themselves often become blocked with cholesterol deposits, necessitating additional surgery. Nevertheless, there is no question it has often been a life-saving operation.

Coronary transplants: The most dramatic procedure of all is the coronary transplant. Heart transplants came into vogue in the early 1970s, but, although the surgeries were successful, the patients almost invariably died because the transplant was rejected (see Chapter 9). Accordingly, mechanical hearts began to be developed and were first used in 1982. The first patient survived 182 days, and a second lived 620 days—but problems with spontaneous clots and other difficulties with the pump dampened some of the enthusiasm for this treatment. In 1984, a surgeon attempted to transplant a baboon heart into a human infant with a defective heart—but Baby Fae lived only 21 days. However, beginning in 1979 the immunosuppressant drug **cyclosporine,** extracted from a newly discovered fungus, began to be used in transplants, and the success of the transplants increased markedly. Heart transplants are more common than ever (about 1700 annually in the United States), and the survival time has greatly increased.

Interestingly, one problem now facing heart transplant patients is atherosclerosis. The endothelial cells of the coronary arteries of the transplanted heart are damaged by an unusual immune reaction that is not thwarted by cyclosporine. Endothelial cells normally produce both a blood vessel dilator and an enzyme that dissolves blood clots; damaged cells produce neither. Moreover, damage to endothelial cells promotes the initiation of atherosclerosis. These facts not only help explain the problem for heart transplant patients (and why they sometimes require a second transplant); they also explain why 25 percent of patients receiving angioplasties (a procedure that commonly causes endothelial damage) require a second angioplasty within six months of the first one. Finally, the reason why smoking increases the risk of atherosclerosis and heart attacks is that smoking damages the endothelium of blood vessels.

become flushed, and increasing the rate at which heat is dissipated from the body. (The same principle operates in the application of liniments or a heating pad to sore muscles. The irritants in the liniment, and the heat from the heating pad, cause the arterioles supplying the muscle to expand, and blood flow to the muscle increases, often leading to an uncramping of the muscle.)

2. You are 12 years old and the prettiest girl (or handsomest boy) in your sixth grade class announces, in front of the entire class, that she (or he) thinks

you're "cute." Sixty-four eyes, accompanied by thirty-two smirks, suddenly swing in your direction. Your face feels warm and red—you are blushing. Why?

The arterioles are supplied by branches of both the parasympathetic and the sympathetic nervous systems. During emotional crises, such as being an embarrassed 12 year old, the sympathetic system is inhibited, and, under the unrestrained stimulation of the parasympathetic system, the arterioles in the skin expand, causing the emotional flushing we call a "blush."

3. You are four years old and, in the process of exploring a garbage can, manage to spill a half-empty ketchup bottle all over yourself. You begin to cry and, covered in scarlet, you walk into your house. Your mother, convinced that you have just been hit by a bus, collapses to the floor in a heap. What causes such a dramatic reaction?

Fainting is a more extreme reaction to severe emotional stress. When the sympathetic nervous system is strongly inhibited, all of the blood vessels expand, and the blood pressure drops rapidly, causing the brain to be deprived temporarily of blood, which results in fainting.

The role of the arterioles in the maintenance of blood pressure is vital. An otherwise healthy individual can lose up to 1.5 liters of the five to six liter blood supply and still maintain an essentially normal blood pressure. Pressure detectors in the aorta and **carotid sinus**[2] relay blood pressure information to the brain, and the brain, in turn, increases or decreases the activity of the sympathetic nervous system. If the blood pressure has fallen (as would occur with substantial blood loss or after fainting), the activity of the sympathetic nervous system increases, and the arterioles are constricted, lowering the pressure of the blood entering the capillaries. (The significance of this reduction in capillary blood pressure will be discussed in the next section.) Increased sympathetic activity also causes the heart to beat more powerfully and more rapidly, and the veins to become constricted. The net result is that the diminished supply of blood is passing through a narrower network of vessels, which causes the blood pressure to rise to almost normal levels.

[2]The carotid sinuses (one on each side of the neck) are located at the point where the common carotid artery, which carries blood to the head, divides into internal and external branches. Anatomically, the carotid sinus is just behind and below the angle of the jaw. The reason why a karate chop to this area is so dangerous is that the blow mechanically stretches the pressure detectors, causing profound inhibition of the sympathetic system and stimulation of the parasympathetic system. As a consequence, the heartbeat rate decreases, the blood vessels dilate, and the blood pressure drops very sharply, causing unconsciousness and, in severe instances, death.

Capillaries

To say that the capillaries link the arteries and the veins is akin to saying that the brain separates the scalp from the tongue—both statements are true, but in both instances they obscure the more important role of these vital structures. In a real sense, the rest of the circulatory system simply plays a support role to the capillaries, for it is in the capillaries—and only in the capillaries—that exchange between the blood and the interstitial fluid occurs.

No cell is more than a fraction of a millimeter away from a capillary, and the total surface area of the capillaries is on the order of 700 m^2, the area of a very large house. Moreover, although most capillaries are less than 1 mm in length, they are so numerous that their combined length in the human body has been estimated to exceed 90,000 km. Thus, even though capillaries have a diameter of only 8/1000 mm (just 1/1000 mm wider than the diameter of a red blood cell), they are so abundant that their combined cross-sectional area is 2800 cm^2, which means that the velocity of blood flow is ($100\ cm^3/2800\ cm^2 =$) 0.3 mm/sec. This low velocity facilitates capillary exchange (see Figure 8.10).

Capillary exchange is a complex process. Movement of materials from the capillaries to the interstitial fluid occurs through diffusion, bulk flow, and pinocytosis. Small molecules, such as O_2 and CO_2, diffuse readily through the membrane of the cells that comprise the capillary wall. Water-soluble materials move through **capillary pores,** the spaces between the en-

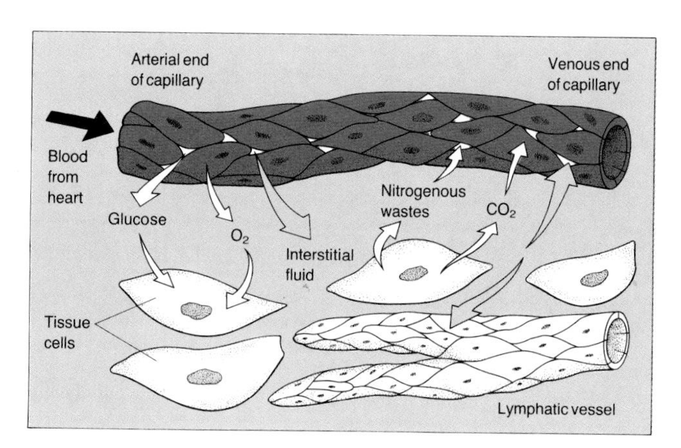

FIGURE 8.10

Capillary Exchange

At the arterial end of capillary beds, some of the noncellular portion of the blood (including nutrients) leak out of the capillary and into the interstitial fluid. At the venous end of capillary beds, some of the interstitial fluid (including waste products) re-enters the capillary. Excess interstitial fluid is accumulated by lymphatic vessels.

dothelial cells that constitute the capillary walls. The pores are very narrow and effectively exclude the transit of blood proteins and blood cells. However, blood pressure forces smaller materials out through the pores, a process known as **bulk flow,** just as water pressure forces water out of a perforated garden hose. Proteins and other large molecules are moved by pinocytosis (see Chapter 5).

Two primary factors influence the rate, and direction, of the movement of materials between the capillaries and the interstitial fluid:

1. Blood pressure in the capillary
2. Osmotic pressure of the blood.

The **osmotic pressure** of the blood is a measure of the blood's tendency to retain water, and its value is essentially constant. Blood pressure values, however, differ significantly between the arterial and venous ends of the capillary. Thus:

	ARTERIAL END	VENOUS END
Outward blood pressure	32 mm Hg	12 mm Hg
Inward osmotic pressure	25 mm Hg	25 mm Hg
Net outward pressure	7 mm Hg	
Net inward pressure		13 mm Hg

Thus, at the beginning (arterial end) of the capillary bed, net movement of materials is *from* the capillary *to* the interstitial fluid. Because of the drop in blood pressure as the blood moves through the capillary bed, net movement of materials at the venous end is *from* the interstitial fluid *to* the capillary.

The result is that there is a turnover in interstitial fluid, "new" fluid being produced at the arterial end of the bed, and "old" fluid being taken up by the capillary at the venous end of the bed. Thus there is a constant replenishment of nutrients for the surrounding cells, and a constant removal of cellular wastes that are borne along as the water medium is returned to the capillary at its venous end.

Blood pressure is the primary determinant in establishing the direction of capillary exchange. In the capillary beds of active muscle, arteriolar expansion ensures a relatively high blood pressure in the capillary, which enhances outward flow. Reciprocally, the arteriolar contraction that occurs following a major blood loss decreases the blood pressure of the blood entering the capillaries, which enhances the inward flow from the interstitial fluid. Indeed, the blood volume returns essentially to normal following a major blood loss because of the uptake of a like volume of interstitial fluid. (Replacing the lost blood cells takes several days or even weeks.)

Only about 25 percent of the capillary beds of the body are in active use at any given moment (though of course there is a constant turnover in the capillaries that constitute the active fraction). Precapillary sphincters operate in concert with the arteriolar sphincters, permitting or preventing blood flow through the bed. In nonactive capillary beds, **through-flow channels** are used to shunt blood from arterioles to venules (see Figure 8.11).

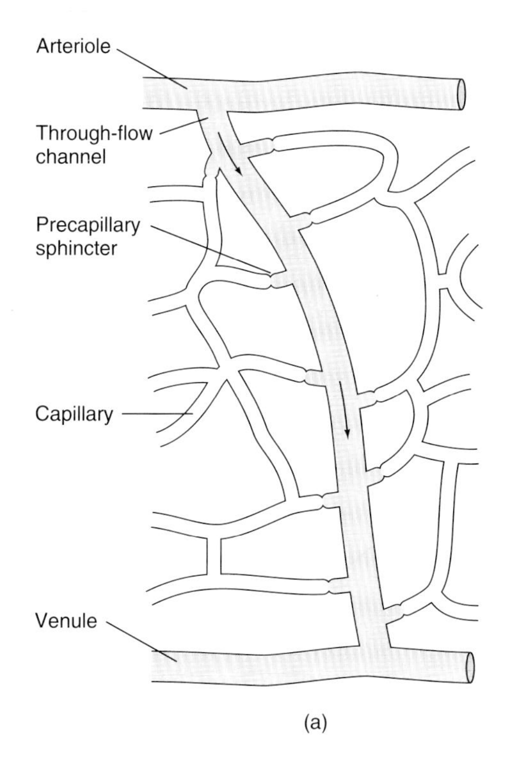

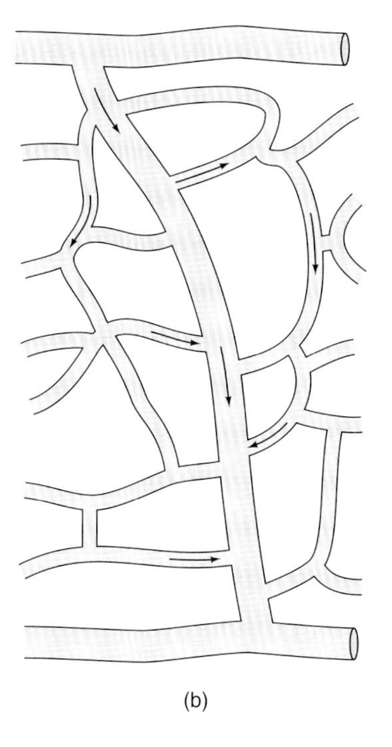

FIGURE 8.11

Precapillary Sphincters
Sphincter muscles in the smallest arterioles may be open (permitting blood into the capillaries) or closed (routing blood into a through-flow channel)

Venules and Veins

The venules and veins are responsible for the return of blood to the heart (see Figure 8.12). Because the blood is at relatively low pressure, the venules and veins do not need to be nearly as robust as the arteries and arterioles. Indeed, the venules possess no muscular layer at all (being composed entirely of an inner endothelial layer and an outer connective tissue coat), and the muscular layer in the veins is much thinner than in the arteries.

How can venous blood, with a blood pressure of less than 15 mm of mercury at the venous end of the capillary bed, be returned to the heart? Why does gravity not cause all of the blood to collect in the feet?

Part of the answer is that the veins are equipped with a series of one-way valves that permit the blood to flow only toward the heart (see Figure 8.13). In a sense, these valves operate like a bucket brigade, with the blood being "passed on" from one valve to the next. Another part of the answer is that the muscles of the body, which frequently surround the veins, "massage" the veins as they contract, squeezing the blood along, rather as toothpaste is squeezed from a toothpaste tube. Finally, as the blood moves through the inferior vena cava, the mechanics of breathing, which creates a partial vacuum (see Chapter 10), aids in blood movement.

Normally, the veins are rather flaccid. However, just as we saw for the arterioles, the veins undergo constriction (leading to higher blood pressure) following injury and blood loss, for they, too, are innervated by the sympathetic nervous system.

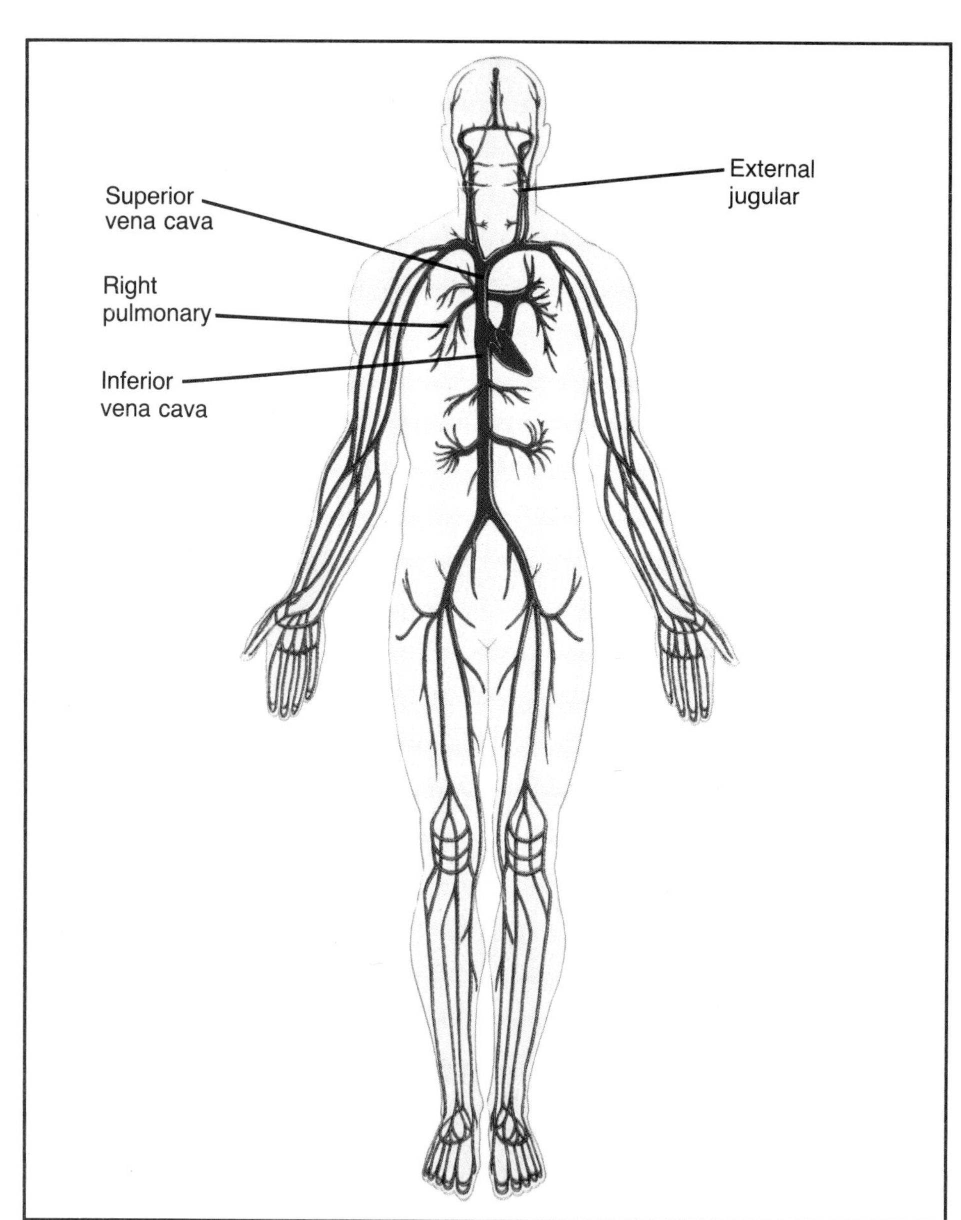

FIGURE 8.12

Human Venous System

As was the case with the arterial system, the venous system varies somewhat from individual to individual with respect to the precise points of vessel branching, but this artist's representation is a good general view of the human venous system.

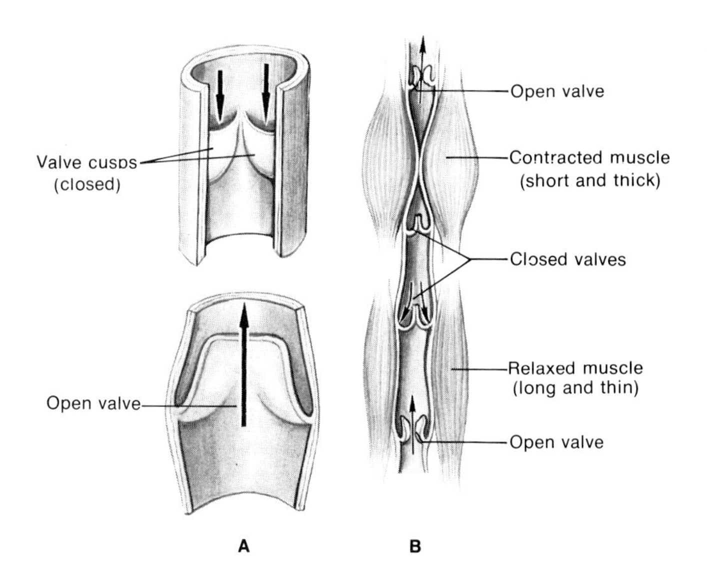

FIGURE 8.13

Venous Return
Skeletal muscles surround most of the major veins in the arms and legs. When these muscles contract, blood is forced upwards toward the heart. One-way valves prevent the blood from flowing back down the veins when the skeletal muscles are relaxed.

THE LYMPHATIC SYSTEM

The exchange of materials between the blood and interstitial fluid depends on a delicate balance between blood and osmotic pressures. In theory, the volume of material leaving the blood and entering the interstitial fluid at the arterial end of the bed must be precisely equaled by the volume of interstitial fluid reentering the blood at the venous end of the capillary bed. However, because blood completes a circuit through the body approximately once every minute, if the exchange process is not absolutely 100 percent efficient, there will be a rapid and progressive loss from the blood to the interstitial fluid. As it happens, of the 20 liters of interstitial fluid formed daily, only 17 to 18 liters reenter the blood directly. The remaining two to three liters are reabsorbed by the **lymphatic system** (see Figure 8.14).

The lymphatic system consists of a series of blind-ended vessels that link virtually all parts of the body and ultimately connect to the venous system through one-way valves just before the superior vena cava empties into the right atrium. (Blood pressure is lowest at this point in the system, and the chance of blood being inadvertently pumped into the lymphatic vessels is therefore minimal.)

The lymph vessels are particularly permeable to protein, and any protein that escapes from blood is rapidly picked up by the lymph vessels. The lymph also transports lipids from the digestive system (see Chapter 6). Together, these protein and lipid molecules exert a sufficient osmotic pressure to accumulate any excess interstitial fluid, and thereby to ensure that the interstitial fluid level remains essentially constant.

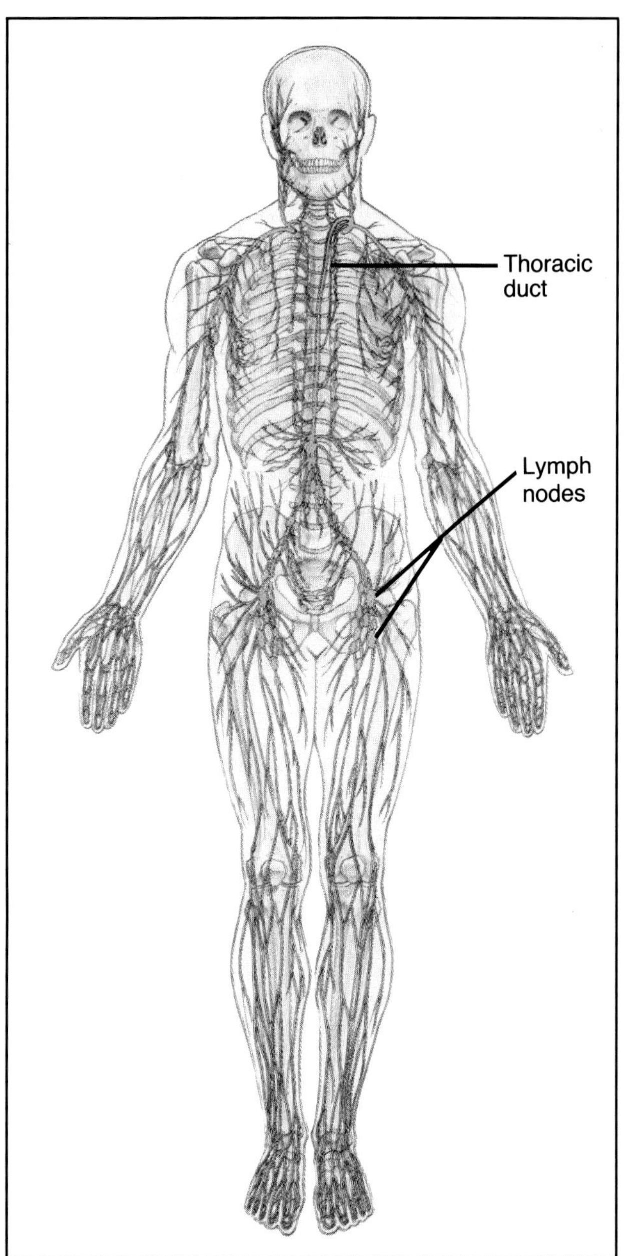

FIGURE 8.14

Human Lymphatic System
A network of vessels collects excess interstitial fluid and returns it to the blood just prior to the point where venous blood enters the right atrium (that is, at the point where blood pressure is lowest). Lymph nodes, important in filtering and monitoring the fluid in the lymphatic system, are shown as small swellings.

Movement of lymph through the system is similar to the movement of blood in veins. That is, the random movements of muscles and organs massage the lymph along, and a series of one-way valves prevents backflow.

The lymphatic system also contains a series of filtering chambers called **lymph nodes,** and a number of small organs, including the **tonsils** and **thymus gland.** These organs play an important role in the immune response, discussed in Chapter 9.

Summary

The human circulatory system is much more than a set of irrigation pipes that transport nutrients to the other tissues of the body. The blood is in intimate and dynamic association with the interstitial fluid, and together they bathe the cells of the body, just as the sea bathes unicellular and simple multicellular organisms. Providing for the needs of all of the cells of the body with the small volume of liquid represented by the blood and interstitial fluid requires exquisite maintenance and control.

Blood circulation in humans and other mammals involves a double circuit. The right atrium and left ventricle, respectively, receive blood from, and deliver blood to, the body. The left atrium and right ventricle, respectively, receive blood from, and deliver blood to, the lungs.

Valves prevent backflow and ensure one-way circulation. Heartbeat is regulated by the pacemaker, a remnant of an ancient heart chamber still found in fish.

Blood pressure must be maintained within narrow limits to ensure safe and effective transport of nutrients. Exchange between capillaries and interstitial fluid is particularly dependent on suitable blood pressure in the capillaries.

The lymphatic system is an adjunct of the circulatory system proper and is essential for the return of excess interstitial fluid to the blood.

Key Terms

interstitial fluid
heart
hormones
oxygenated blood
anterior
posterior
superior vena cava
inferior vena cava
atrium
ventricle
tricuspid valve
pulmonary artery
semilunar valve
pulmonary vein
bicuspid valve
aorta
cardiac muscle
pacemaker
sinus venosus
sinoatrial (SA) node
atrioventricular (AV) node
bundle of His
Purkinje fibers
brachial artery
systolic pressure
diastolic pressure
stroke volume
cardiac output
medulla
vagus nerve
parasympathetic nervous system
sympathetic nervous system
adrenal gland
adrenalin
artery
arteriole
capillary
vein
venule
endothelium
carotid sinus
capillary pores
bulk flow
osmotic pressure
through-flow channels
lymphatic system
lymph nodes
tonsils
thymus gland

Box Terms

angina pectoris
coronary artery
myocardial infarction
coronary thrombosis
ventricular fibrillation
heart block (arrhythmia)
artificial pacemaker
valvular heart disease
rheumatic fever
congestive heart failure
heart murmurs
organic hypertension
essential hypertension
ATPase
atherosclerosis
macrophage
foam cell
PDGF
estrogen
aneurysm
collagen
stroke
cerebral hemorrhage
phlebitis
pulmonary embolus
varicose vein
hemorrhoid
beta blocker
calcium channel blocker
diuretic
ACE inhibitor
angiotensin
atrial natriuretic factor (ANF)
streptokinase
tissue plasminogen activator (TPA)
neutrophil
tumor necrosis factor (TNF)
transforming growth factor (TGF)
angioplasty
coronary bypass
coronary transplant
cyclosporine

Questions

1. What is the relationship between interstitial fluid and blood?
2. Describe the path of blood as it flows through the heart.
3. Explain the functioning of the pacemaker. What prevents all four chambers of the heart from contracting simultaneously?
4. Why is the diastolic pressure usually significantly lower than the systolic pressure?
5. How is cardiac output regulated?

6. Why does blood pressure drop as the blood travels through the system? Are some regions more significant in this regard than others? Why?
7. Briefly explain capillary exchange. Why is there a net flow of materials out of the capillary into the interstitial fluid at the arterial end and a net flow into the capillary at the venous end?
8. If venous blood is at such low pressure, how does blood in our feet get back to the heart?
9. What is the role of the lymphatic system regarding the maintenance of blood volume?

CHAPTER

9

Blood

Transport and Immunity

Your best friend, who has a bad case of the flu, has just sneezed on you. Like clockwork, three days later you begin to develop the first symptoms of the disease—and three days after that you are reasonably sure that you are going to die. Yet within another few days you begin to recover, and within two weeks of initial exposure, the flu is only a memory. How does the body respond to disease? Why do we become sicker for a while, and then slowly recover? And what about vaccinations—what's going on there?

Millions of people suffer from various *autoimmune* diseases—diseases in which the defense system of the body attacks the body itself. What accounts for such bizarre behavior—and why do most of us escape having such diseases?

A major problem with organ transplants is *rejection.* Why would our bodies fight organ transplants that could save our lives?

These are the kinds of questions we shall address in our chapter on the blood.

THE NATURE OF THE PROBLEM

In Chapter 8 we saw how the circulatory system was constructed, how the blood moves through the vessels, and how exchange occurs between the blood and the interstitial fluid. However, there are still many unanswered questions. Since oxygen dissolves very poorly in water, how does the blood transport the enormous quantity of oxygen required by the hundreds of trillions of cells that collectively constitute our body? As a liquid tissue under pressure, loss of blood can be expected to result whenever a blood vessel is damaged—but blood is too valuable a tissue to lose through routine injuries. How is the system designed to seal off vessels that are damaged, so as to minimize blood loss? Finally, because the blood vessels penetrate every portion of the body, the blood can inadvertently serve as a delivery system for disease agents, which can rapidly be spread throughout the body from their point of entry. How does the body overcome invasion by potentially dangerous microorganisms?

BLOOD

Like all tissues, blood is composed of several types of cells. Unlike other tissues, however, blood has a liquid matrix, and the various types of cells are therefore free to move about and intermingle.

In addition to cells and cell fragments, blood contains a large variety of proteins—some of which are involved in clotting—and many small molecules and ions, all in a water solution. Blood with all of these components is called **whole blood.** Blood minus the blood cells is called **plasma.** Blood minus both the cells and the clotting proteins is called **serum.** About 45 percent of whole blood consists of cells.

The cellular portion of the blood is responsible for defense against disease, the transport of oxygen and some carbon dioxide, and the initiation of blood clotting (see Figure 9.1). The plasma transports glucose, salts, nitrogenous wastes, hormones, and some carbon dioxide, and it aids in clotting.

The two major classes of blood cells are the **erythrocytes** (Gr. *erythros,* "red"), or red blood cells, and the **leukocytes** (Gr. *leukos,* "white"), or white blood cells. A third class, the **platelets,** are more correctly characterized as cell fragments. All are descended from **stem cells,** which are located in the blood-producing regions of the body (see Figure 9.2). (Differentiation of stem cells is discussed in Chapter 21.)

Blood cells are actually produced at several sites in the body. During embryonic and fetal development, they are produced in the liver and spleen. In children, blood cells are manufactured in the marrow of the long bones of the arms and legs. Adults produce blood cells in the cavities of the ribs, sternum, vertebrae, and pelvis.

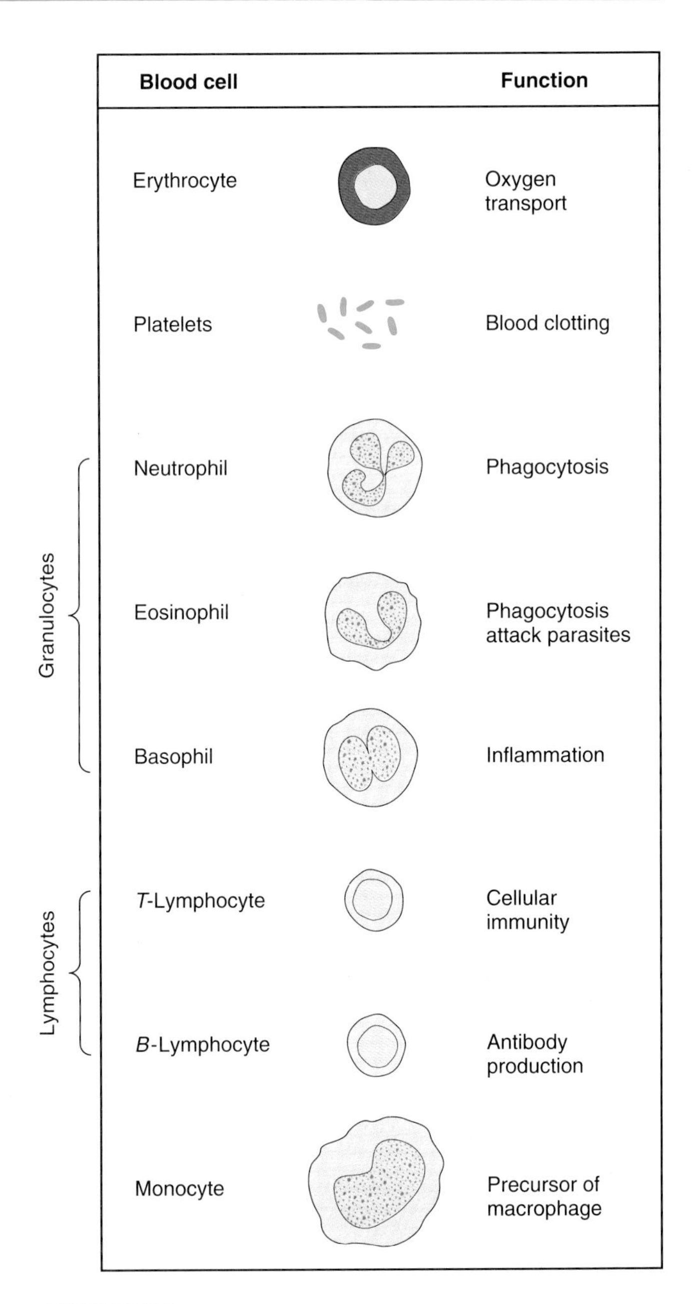

FIGURE 9.1

Human Blood Cells

The principal types of cells found in circulating blood are shown here. The white blood cells readily leave the blood and enter tissues of the body; hence, their proportion in the circulating blood is not indicative of their overall numbers in the body.

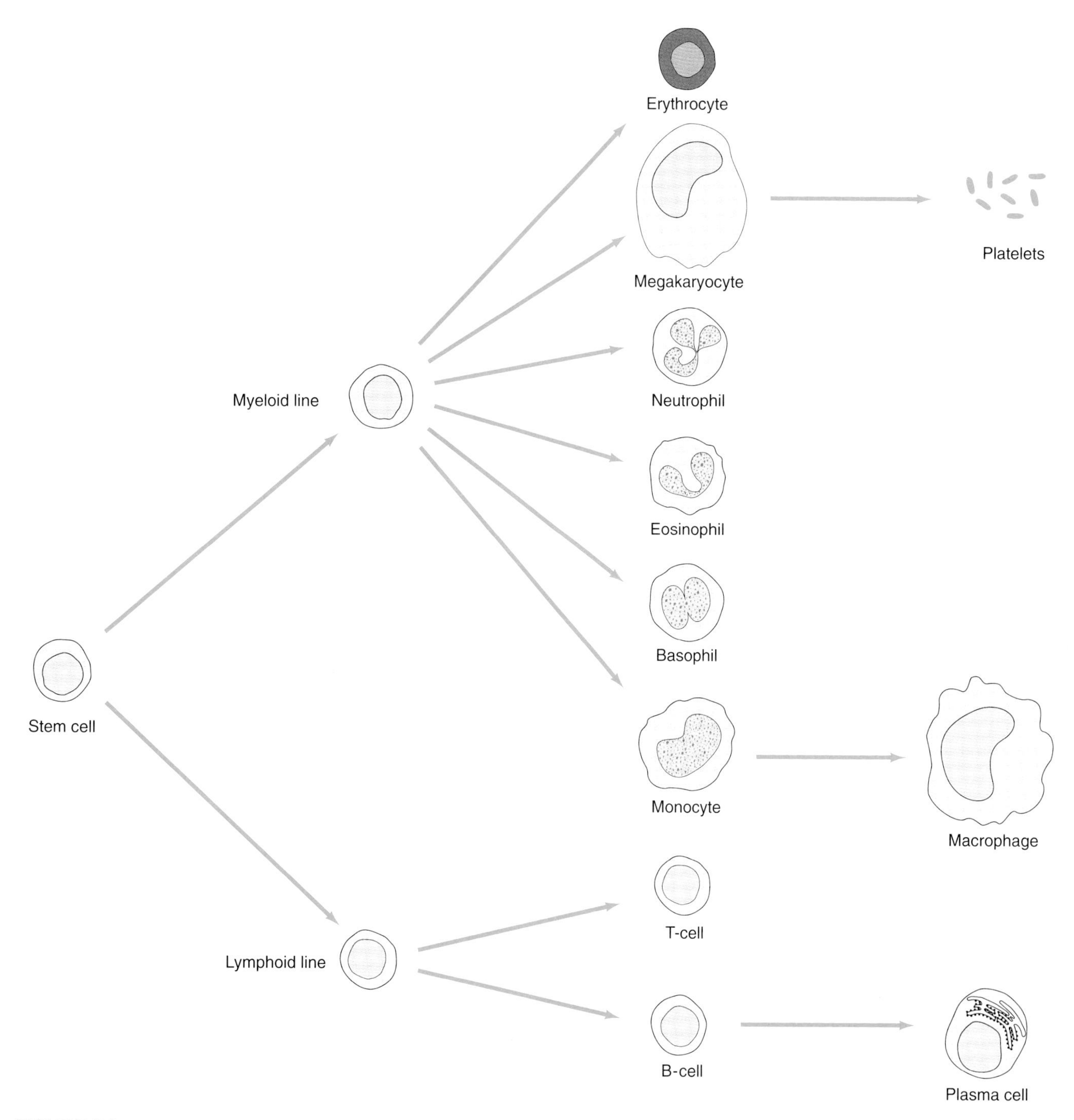

FIGURE 9.2

Derivation of Blood Cells
All blood cells are derived from a common precursor cell known as a stem cell. Most blood cells are a part of the myeloid line; lymphocytes derive from the lymphoid line. Intermediate steps in differentiation are not shown.

ERYTHROCYTES

With a diameter of only 7/1000 mm, erythrocytes are the smallest cells in the body. They are also among the most numerous, with about 5 million/mm^3 of blood. Because the erythrocytes of mammals lack a nucleus, they possess a characteristic biconcave shape (see Figure 9.3b). Since a cell's nucleus is responsible for the maintenance and repair of the cell, and since red blood cells are constantly buffeted about, bouncing off other cells and being squeezed through the tight bends of capillaries, it is not surprising that the normal lifetime of an erythrocyte is only three or four months—during which time it travels almost a thousand miles! This short lifetime means that we must replace our 25 trillion erythrocytes at the rate of three million every second!

Damaged erythrocytes are destroyed by cells in the spleen and liver, and the iron they contain is transported to the bone marrow for use in the synthesis of new erythrocytes. Excess iron is stored in our liver, as it is in all mammals, which is why your mother was always trying to get you to eat liver as a child.

Erythrocytes are highly specialized for the transport of blood gases (O_2 and CO_2). Each erythrocyte contains approximately 250 million iron-containing molecules of **hemoglobin,** which are arrayed like stacks of miniature poker chips around the periphery of the cell. Because every hemoglobin molecule can transport up to four molecules of oxygen, every erythrocyte is theoretically capable of carrying one billion oxygen molecules.

The importance of hemoglobin to the body is readily evidenced. One liter of plasma can carry only 2 ml of O_2 and 3 ml of CO_2 in solution. By contrast, one liter of whole blood can transport 200 ml of O_2 and 300 ml to 600 ml of CO_2, and at the same time act as a buffer for potential changes in the pH of the circulatory system. (The details of gas transport and exchange are discussed in Chapter 10).

Since erythrocytes serve exclusively as hemoglobin containers, why are the cells necessary at all? Why is the hemoglobin not simply floating free in the plasma, as are so many other protein molecules of the blood? The problem with such an arrangement is that the blood would be as thick as molasses. Erythrocytes in effect take hemoglobin out of solution, thereby sharply reducing the amount of dissolved substances in the plasma, which permits more efficient movement of the blood.

PLATELETS

An obvious hazard to a system that relies on the movement of a liquid tissue at high pressure is the potential for damage or destruction to a vessel and the resulting loss of the contents. Given the enormous importance of the blood in maintaining homeostasis, it is not surprising that a protective mechanism has evolved to prevent (except in extreme instances) the loss of substantial amounts of blood due to damage to the vessels. This mechanism is referred to as *clotting*.

Integral in the formation of blood clots are the *platelets*, which are fragments of large cells, called **megakaryocytes** (Gr., "big nucleus cells"), found in bone marrow. Platelets number about 250,000/mm^3 of blood.

The mechanics of blood clotting is very complex, but the essentials are as follows: Tissues that are damaged (as by a sliver, cut, or bullet wound) attract platelets that adhere to the walls of the damaged blood vessels and, within seconds of the injury, form a **platelet plug.** At the same time, muscles in the walls of the blood vessels contract (a process augmented by the release of chemicals from the platelets forming the plug), narrowing the diameter of the vessel. This process is known as **vasoconstriction.**

Even as these processes are occurring, the initial steps of clot formation begin to occur (see Figure 9.3). Substances released by the platelets combine with plasma factors (in the presence of calcium ions) to produce **thromboplastin.** Thromboplastin is also produced from substances released by injured tissue. Sharp cuts, as from a razor blade, damage relatively few cells, and only a small amount of tissue thromboplastin is produced. Consequently, the clotting of such cuts is slower than when tissue damage is more extensive (see also Box 9.1).

Thromboplastin acts as an enzyme to convert the plasma protein **prothrombin** into its active form, **thrombin.** In turn, thrombin enzymatically removes some small branches of the very large, abundant, and soluble plasma protein **fibrinogen,** thereby converting it to the insoluble protein **fibrin.** Fibrin molecules are highly attracted to one another, and join end to end and side to side to form a meshwork of strands around the wound, trapping blood cells in the process, and forming a **blood clot.**

The final step, **clot retraction,** is also under the control of the platelets. Microfilaments (see Chapter 4) in the platelets adhering to the fibrin clot pull the fibrin strands closer together, using the same type of contractile protein found in muscle cells (see Chapter 12 for a discussion of muscle contraction). The clot becomes more compact, as the liquid portion (serum) is squeezed out, and the edges of the wound are brought closer together. If the clot is exposed to air, as in a skin cut, it dries out and becomes a *scab*.

It is important to note that in order for clotting to be effective, it must be a localized process. If clotting were to occur throughout the body, all the blood would be transformed into one giant clot, rendering it as useless as if it had all escaped from the original cut. Widespread clotting does not occur because the enzymes necessary to convert the plasma proteins prothrombin and fibrinogen are neutralized almost as quickly as they form, and therefore their effects are limited to the immediate area of the cut.

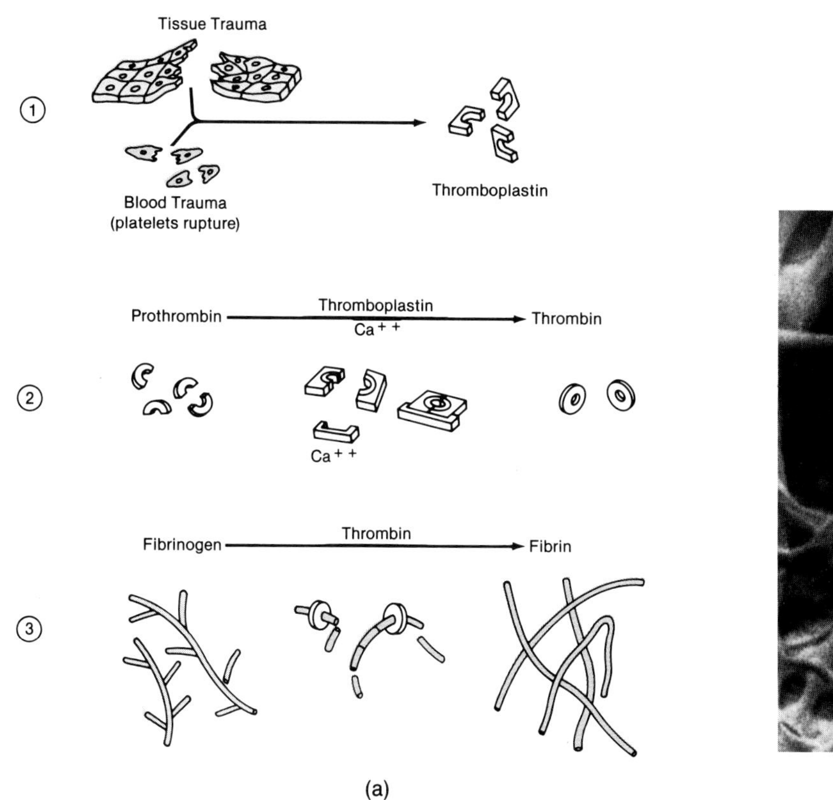

FIGURE 9.3

Blood Clotting

(a) Damage to tissues or blood vessels causes a cascade of reactions that are summarized here by the development of an enzyme called thromboplastin (see also Box 9.1). Thromboplastin catalyzes the conversion of the blood protein prothrombin into the enzyme thrombin. Thrombin, in turn, catalyzes the conversion of the inactive blood protein fibrinogen into the highly interactive molecule fibrin, which forms a mesh that ensnares blood cells (b) and creates a clot.

BOX 9.1

Problems with the Blood and Lymph

Anemia: Anemia (Gr. *anaimia,* "without blood") is a general term used for any condition in which the blood lacks either an adequate number of erythrocytes or an adequate amount of hemoglobin. Blood loss caused by injuries is one of the common causes of anemia. Because the **spleen** (an organ located near the liver) holds a certain amount of blood in reserve, the loss of half a liter of blood (as in a blood donor clinic) poses no significant problems, but losses over a liter may prove serious.

Diets that are iron-poor may lead to anemia because of the inability of the body to synthesize sufficient hemoglobin. **Iron-deficiency anemia** is the most common type of anemia, especially in infants and adolescents who are growing rapidly and who therefore are synthesizing relatively large numbers of erythrocytes. In adults, the condition usually arises because of some type of chronic blood loss, rather than because of insufficient iron in the diet (television commercials for iron supplements notwithstanding), although as we saw in Chapter 7, women of reproductive age commonly do not include sufficient iron in their diets.

Pernicious anemia is due to the failure of the body to take up a sufficient supply of vitamin B_{12}, a vitamin essential in the formation of red blood cells (see also the discussion of pinocytosis in Chapter 5). **Aplastic anemia,** in which the bone marrow fails and the production of red blood cells simply ceases, is a common result of radiation poisoning and is

(Continued on the following page)

BOX 9.1 (Continued)

one of several likely causes of death in individuals who have been exposed to high doses of radiation.

Yet another form of anemia, **sickle-cell anemia,** is limited almost exclusively to blacks (although very similar diseases also occur among certain white populations). In sickle-cell anemia, the erythrocytes contain an aberrant form of hemoglobin, and as oxygen is given up in the capillary beds, this unusual form of hemoglobin causes the erythrocytes to assume a sickle or holly-leaf shape. Cells of this configuration are prone to lodge or rupture in the capillary beds, and they also cause progressive damage to various organs, most notably the kidneys and heart. Sickle-cell anemia is a genetic condition for which some 10 percent of black Americans are carriers. (The genetics and evolution of this disease are discussed more fully in Chapters 18, 19, and 22.)

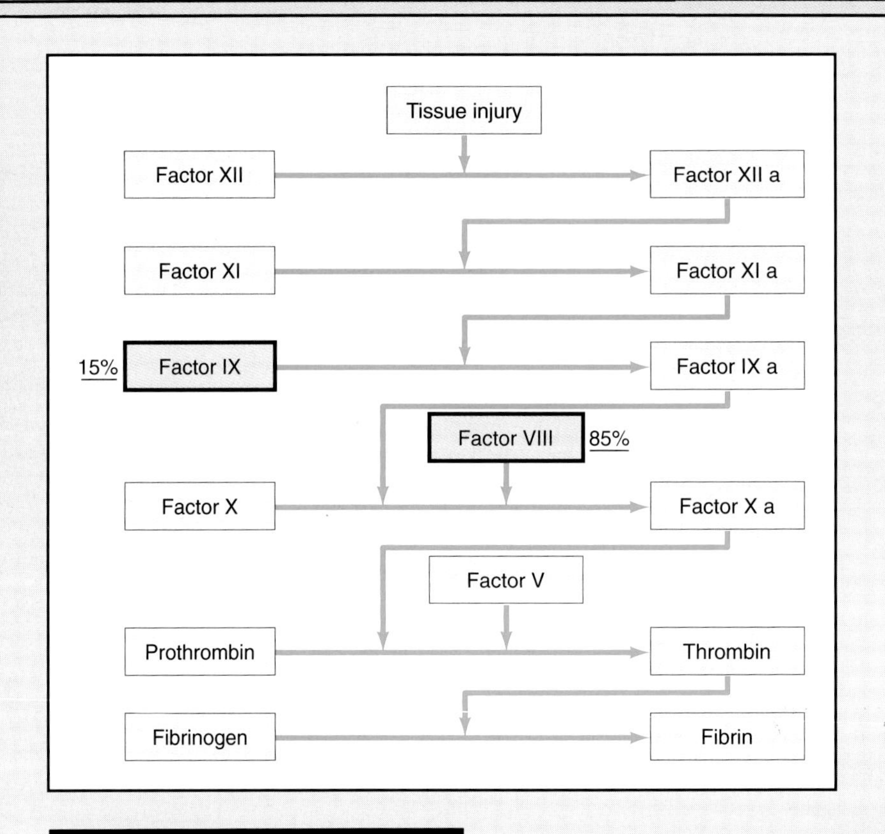

Some Details of the Clotting Cascade

Although by no means complete, this illustration shows where hemophilia has its site of action. Some 85 percent of hemophiliacs lack Factor VIII; the remaining 15 percent lack Factor IX. In either case, the result is a failure of the clotting cascade, and the failure of fibrin to develop in sufficient quantity to permit normal blood clotting.

Leukemia: Leukemia is the uncontrolled replication of white blood cells. It is a form of cancer, but, because blood is a fluid tissue, it is a cancer in which there is often no tumor. **Chronic leukemia** is a condition in which elevated white blood cell counts may be present continuously for many years without any significant threat to the individual's health, although the metabolic drain of manufacturing these huge numbers of cells (up to 100 times the normal level) is enormous and may eventually prove fatal. **Acute leukemia,** by comparison, is characterized by the invasion of lymphatic tissues by abnormal white blood cells and the subsequent development of tumors in these tissues. The irony in acute leukemia is that the white blood cells, though very numerous, are immature and therefore do not function normally. Thus the individual is highly susceptible to other diseases. Anemia and excessive bleeding (due to diminished numbers of erythrocytes and platelets, respectively), are common in acute leukemia.

Hemophilia: There are several different kinds of problems associated with blood clotting. Because vitamin K is essential for the synthesis of prothrombin, a deficiency of this vitamin can cause clotting failure. In phlebitis, the tendency to form "unwanted" clots is countered by the administration of **anticoagulants,** substances that interfere with the normal clotting process. Their mode of action is generally either to compete with vitamin K (thereby blocking the synthesis of prothrombin), or to block the action of thrombin.

Anticoagulants are used in several commercial forms of rat and mouse poisons, where they cause uncontrolled internal bleeding and death. However, they are also produced naturally by such blood-sucking animals as leeches and mosquitoes. (It is easy to imagine the havoc that would result were the blood to clot halfway up Ms. mosquito's proboscis!)

Of all the clotting problems, none has received the attention and fame of hemophilia, because this genetic disease, virtually limited to males but transmitted by females, profoundly

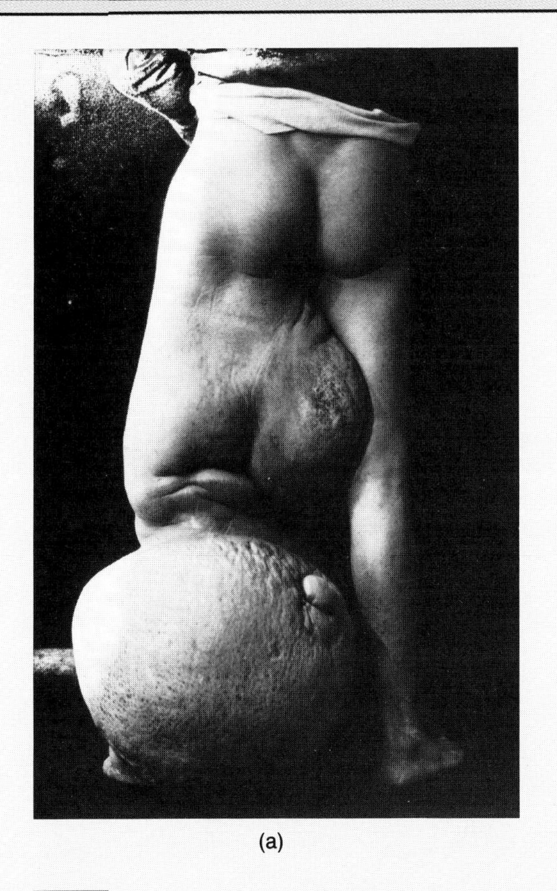

(a)

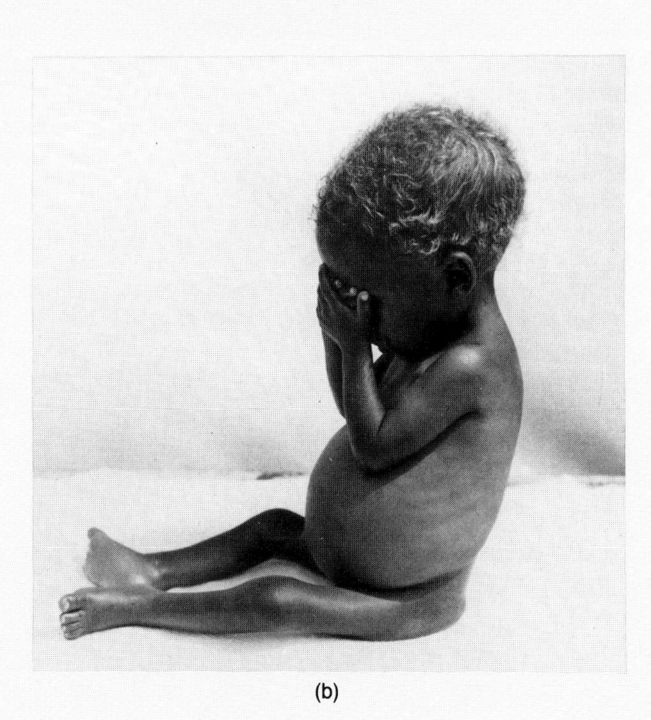

(b)

Edema
(a) Elephantiasis, a form of edema caused by blockage of the lymphatic vessels by parasitic worms. (b) Kwashiorkor, a form of edema caused by insufficient protein intake.

affected the royal houses of Europe (see Chapter 19 for details). Hemophiliacs lack factor VIII, one of the factors that are essential for the activation of platelet thromboplastin. Clotting can still occur, if tissue thromboplastin is present, but it is very much slower and may not be sufficient to prevent extensive blood loss even from relatively minor cuts or scrapes. Hemophiliacs must therefore receive transfusions or injections of factor VIII at regular intervals.

Edema: Edema (Gr. *oeidema,* "a swelling") is the accumulation of abnormally high levels of interstitial fluid. There are a variety of causes, but two are especially graphic. **Elephantiasis** is a condition resulting from the blockage of lymphatic channels by a parasitic worm. Typically, this condition affects an arm or leg, but other areas of the body may also be involved (for example, the scrotum). Because return of lymph is thwarted, the affected area of the body becomes massively swollen.

Kwashiorkor results from starvation. After fat and carbohydrate reserves are metabolized, the protein is sacrificed. The muscles atrophy, and the proteins normally present in the blood and lymph diminish sharply. As a consequence, the osmotic pressure of the blood drops, and the interstitial fluid begins to accumulate, rather than being returned to the circulatory system proper. Typically, this accumulation is in the abdominal region, which becomes bloated and provides a stark contrast to matchstick-like arms and legs.

LEUKOCYTES

With a diameter varying from 12/1000 to 20/1000 mm, leukocytes (which are actually clear, rather than white) are two or three times the size of erythrocytes, but from 500 to 1000 times less numerous (5000 to 10,000/mm^3 of blood). However, unlike erythrocytes, leukocytes are not restricted to the circulatory system proper but instead can move freely through most of the body's tissues. Thus they are actually more abundant in the body than their prevalence in the blood would suggest.

Five kinds of leukocytes are usually distinguished (see Figure 9.4). Three of them are often lumped under the general term of **granulocyte,** because of their granular appearance; they form two-thirds of the leukocytes of the blood. **Lymphocytes** constitute another 30 percent, and **monocytes** contribute 5 percent.

Understanding leukocytes can be accomplished only by considering the role they play as a part of the **immune system**—the complex set of mechanisms responsible for waging the continual battle a body must fight to repel or destroy hostile microorganisms.

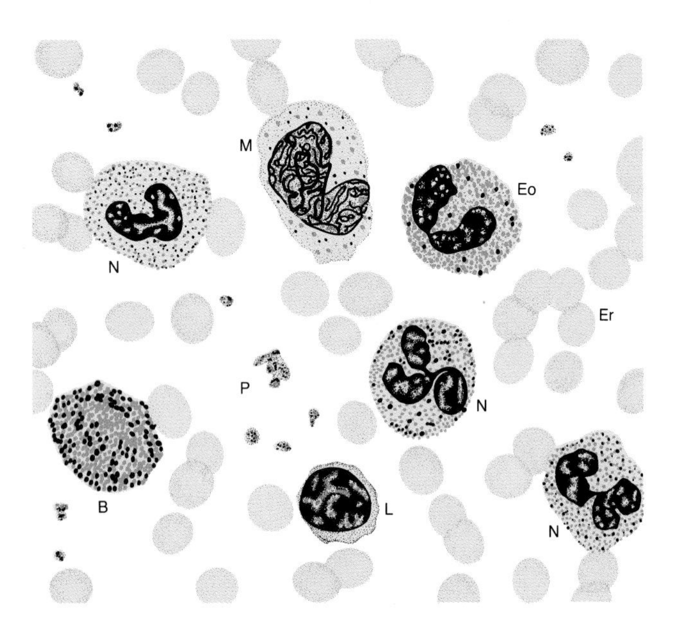

FIGURE 9.4

Leukocytes

The major categories of leukocytes found in circulating blood are shown here. The neutrophils (N), eosinophils (Eo), and basophils (B) are all called granulocytes, because of their granular inclusions. They are sometimes called *polymorphonuclear,* because of their characteristically irregular nucleus. Monocytes (M) transform into macrophages in the tissues, and lymphocytes (L) are involved in specific immune reactions. Platelets (P) and erythrocytes (Er) are also shown.

THE IMMUNE SYSTEM

In 1898, H. G. Wells wrote one of the classics of science fiction—*The War of the Worlds.* In this book, Wells described the invasion of our planet by technologically superior Martians, who quickly overran our defense efforts. All seemed lost—and then, quite suddenly, the Martians began to die. It seems they had no resistance to our earthly bacteria.

The colonization of the New World by Europeans in the sixteenth through the nineteenth centuries was marked by repeated epidemics among native populations who were exposed to certain European diseases for the first time. Even comparatively mild diseases such as measles caused tens of thousands of fatalities.

These two stories—one fictional, the other very real—tell us a good deal about our immune systems. Bacteria have been around much longer than have humans. During our entire evolution as a species, we have been exposed to many kinds of bacteria and other types of **pathogens** (Gr. *pathein,* "to suffer" and *genes,* "something that produces"—i.e., something that is disease-causing). Obviously, at least some of our ancestors survived these encounters (or our presence today would be difficult to explain!). Over time, many presumably once-virulent bacteria became relatively innocuous, meaning either that we destroy them quickly when we are exposed to them today, or that we tolerate their presence in our bodies because of some mutual benefit (such as the vitamin-producing bacteria that colonize our large intestine). It is easy to understand that it is not generally in the best interests of pathogens to kill their hosts. By the same token, it is also not in their best interests to be destroyed by the host's immune system. A kind of evolutionary arms race has ensued, as host and pathogen strive to gain the upper hand. New types of pathogens, such as the AIDS virus, tend to be virulent, as do new strains of rapidly mutating viruses such as those causing influenza. We have one set of mechanisms for dealing with routine exposure to the common pathogens, and a distinctly different set of mechanisms for novel infections. These two mechanisms are, respectively, the *generalized defense mechanisms* and the *specific defense mechanisms.*

GENERALIZED DEFENSE MECHANISMS

We can identify four rather different categories of mechanisms used by the body to repel or resist all manner of potential invaders:

1. Barriers

2. Inflammation
3. Attack chemicals
4. Attack cells

Barriers

The principal barrier to invading organisms is the *skin*. Except for a few openings, the skin (see Chapter 12) covers the entire body and is virtually impenetrable to most invaders. To be sure, some organisms, such as mites, can burrow into the skin, and others, ranging from viruses to parasitic worms, can be injected by biting insects. However, despite these prospective breaches through the wall, the skin is an extremely important and effective barrier—a point best illustrated by a consideration of the consequences occasioned by a major skin loss, as occurs in severe abrasions or burns. Infection in such areas is virtually a given, and indeed death by infection is a major risk for severe burn victims.

The air passageway to the lungs creates a very large opening through the skin. All manner of potentially toxic substances and microorganisms are inhaled as we breathe. What prevents massive infections in the warm, moist environment of our lungs? One answer is that much of this material is trapped by the *mucus* secreted by the cells lining the air passages. Many of these cells are also *ciliated*, and armies of beating cilia transport material-laden mucus out of the respiratory system to the throat where it is swallowed.

Not all of the barriers are mechanical. Consider another major opening in the skin—the mouth. How do we avoid infections from eating food that is often contaminated with potential disease agents? Cells in our mouth secrete the enzyme **lysozyme,** which splits the bonds of a compound found in the cell walls of bacteria. The stomach, of course, produces an acid bath in the form of *HCl*, which few bacteria can survive—and those that do are immediately thrust into the alkaline environment of the small intestine. Various benign organisms thrive in our large intestine, and in other openings such as the *vagina*, and the competition they provide inhibits successful invasion by more toxic organisms.

Inflammation

Physical and chemical barriers can never be entirely effective—especially for those of us of the clumsy persuasion who regularly find novel ways of scuffing, cutting, or in other ways providing access points through the skin.

Inflammation is the general response to such breaches of the skin (see Figure 9.5), and its magnitude can range from the small red sores we know as *pimples* (infected skin pores) to the swelling of an entire limb due to extensive infection. Inflammation helps to direct the body's resources against invading organisms by the following methods:

1. It increases the blood supply to the site of the breach (the area becomes red).

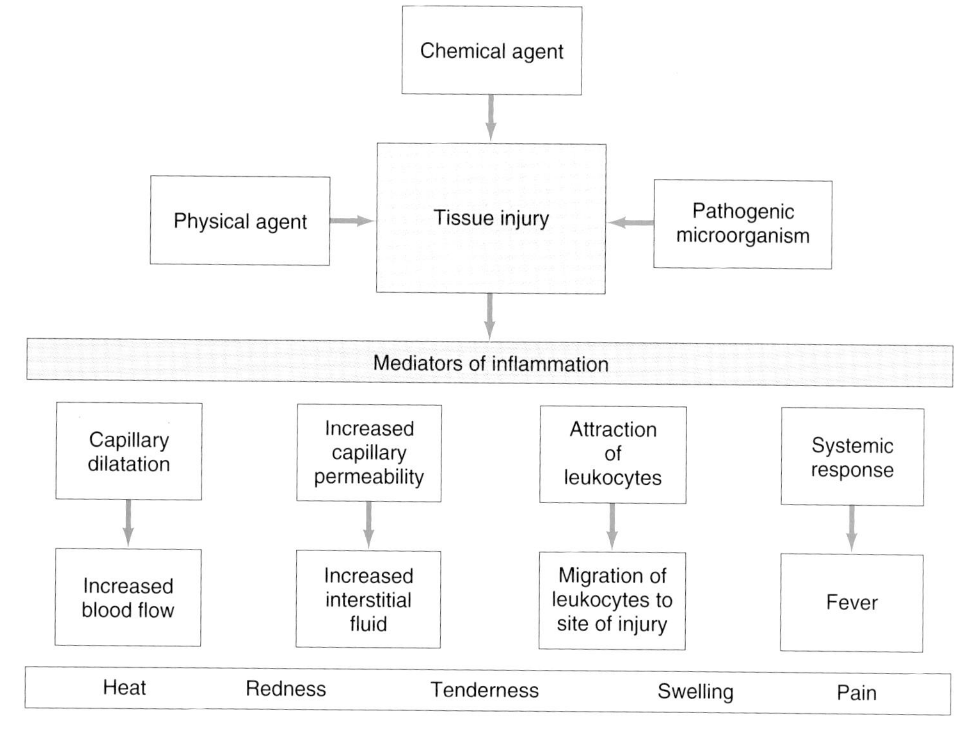

FIGURE 9.5

Inflammation

Inflammation begins with a tissue injury from any of several sources. Several different cells and chemicals ("mediators of inflammation") interact to produce a series of effects, all of which ultimately result in the red, tender, painful swelling we know as inflammation.

2. It increases capillary permeability which permits more extensive movement of leukocytes and specific chemicals into the interstitial fluid (the area becomes swollen).

3. It increases the sensitivity of the area thereby reminding us not to injure the area further (the area becomes tender or even painful).

Attack Chemicals

The principal attack chemicals used in generalized defense are the chemicals of the **complement system** (see Figure 9.6). The complement chemicals include about 20 plasma proteins that are normally in an inactive form but that, in a manner very reminiscent of the clotting mechanism, can be activated in a sequential cascade upon exposure to invading organisms, most notably bacteria.

The complement proteins operate in three ways:

1. Some of the complement proteins attract and activate attack cells, and the attack cells then destroy the invaders.

2. Other complement proteins attach directly to the bacterial wall where their presence greatly increases the likelihood that the bacterial cell will be recognized as foreign and engulfed or split open by attack cells.

3. Still other complement proteins attack the bacterial plasma membrane, causing it to rupture. These complement proteins are particularly effective when combined with the effects of lysozyme.

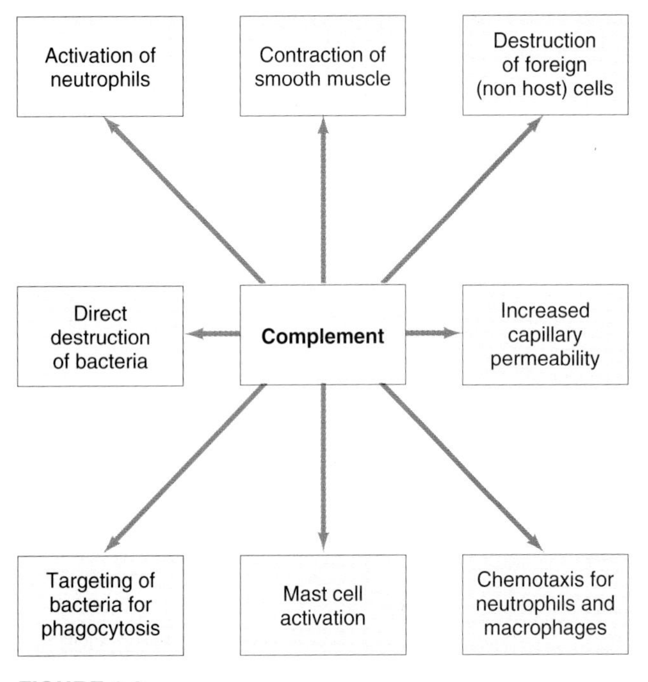

FIGURE 9.6

The Complement System

The proteins of the complement system collectively have a broad range of effects, some of which are shown here. As is true throughout the immune system, there is a great deal of two-way interaction. Thus what is not shown in this diagram are the many agents that stimulate a more complete expression of the complement system.

Attack Cells

There are two primary classes of attack cells:

1. **Natural Killer (NK) cells,** which include several cell types. The most common is a class of lymphocytes that attach to tumor cells or cells infected with viruses and secrete chemicals that destroy the attacked cell.

2. **Phagocytes,** which are leukocytes that destroy their victims by engulfing them.

How do attack cells distinguish tumor cells, or cells infected with a virus, from normal body cells? That question was recently answered when scientists discovered that bits and pieces of all of the proteins manufactured by a cell are displayed on the cell's surface, in a type of on-going quality control test. Tumor cells produce abnormal amounts of a few proteins; cells infected with a virus are of course manufacturing viral proteins that are very different from the usual type of protein produced by a cell. Thus attack cells can readily distinguish both tumor and virally infected cells from normal body cells.

NK cells are also aided by the production of **interferon,** a chemical produced by virally infected cells, which increases the resistance of uninfected cells and which also attracts NK cells. In essence, the infected cell is committing cellular suicide when it attracts NK cells, but since the viruses within the infected cell are also destroyed, the infected cell is acting for the greater good of the body as a whole.

Phagocytes, as their name implies, use a very different approach in the destruction of invading organisms or damaged body cells—they engulf their victim in a vacuole, and fuse the vacuole to a lysosome, digesting their prey in the process.

Phagocytes include a variety of cells. Most of the granular leukocytes are phagocytic, including the **neutrophils** (which form about 90 percent of the granular leukocyte population) and the **eosinophils** (which constitute most of the remainder). The neutrophils (so-named because they can be stained for microscopic viewing with a neutral stain) act primarily against bacteria; the eosinophils (named for their capacity to bind with eosin, an acidic stain) are thought to function primarily against parasitic worms.

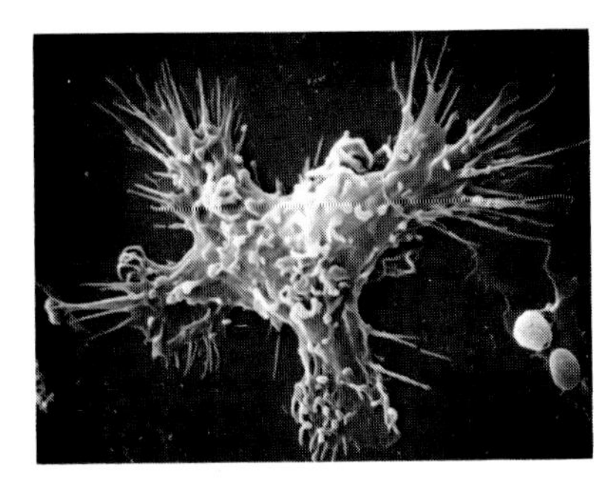

FIGURE 9.7

Macrophage
Macrophages develop from monocytes in the tissues of the body where they engage in many actions critical to the immune system, including phagocytosis of foreign cells.

Another group of phagocytes emerges from the monocyte lineage. Although monocytes are not particularly abundant among the various types of leukocytes found in the circulating blood, they transform in the tissues of the body into **macrophages** ("big eaters"), which are killing machines up to five times the size of monocytes (see Figure 9.7). Macrophages may be free-ranging in the interstitial fluid, or they may be localized in particular organs, including the liver, kidney, spleen, lung alveoli, brain, and lymph nodes, where they are responsible for engulfing any foreign materials that pass through or invade those organs.

MOVEMENT TO INFECTION SITES

Phagocytes and NK cells show much of their activity at sites of infection. How do they locate these sites? Four interrelated mechanisms are involved (see Figure 9.8).

1. **Margination:** The term used to describe the tendency for leukocytes to stick to capillary walls. Between 60 and 75 percent of the granulocytes and monocytes are attached in this fashion at any one time. The products of infection greatly increase the "stickiness" of capillary walls near the site of infection. Thus, by increasing the rate of margination, infection sites cause the build-up of a large local population of leukocytes.

2. **Diapedesis:** The capacity of leukocytes to squeeze through pores in the capillary walls. These pores are dilated near infection sites, permitting diapedesis to occur more readily.

3. **Amoeboid Movement:** In the interstitial fluid, the movement of granulocytes and macrophages in an amoeboid fashion—sometimes as rapid as three times their own length in a minute.

4. **Chemotaxis:** Movement in response to chemicals. Different chemicals either attract or repel phagocytes and NK cells. It is by chemotaxis that the amoeboid movement of leukocytes is directed.

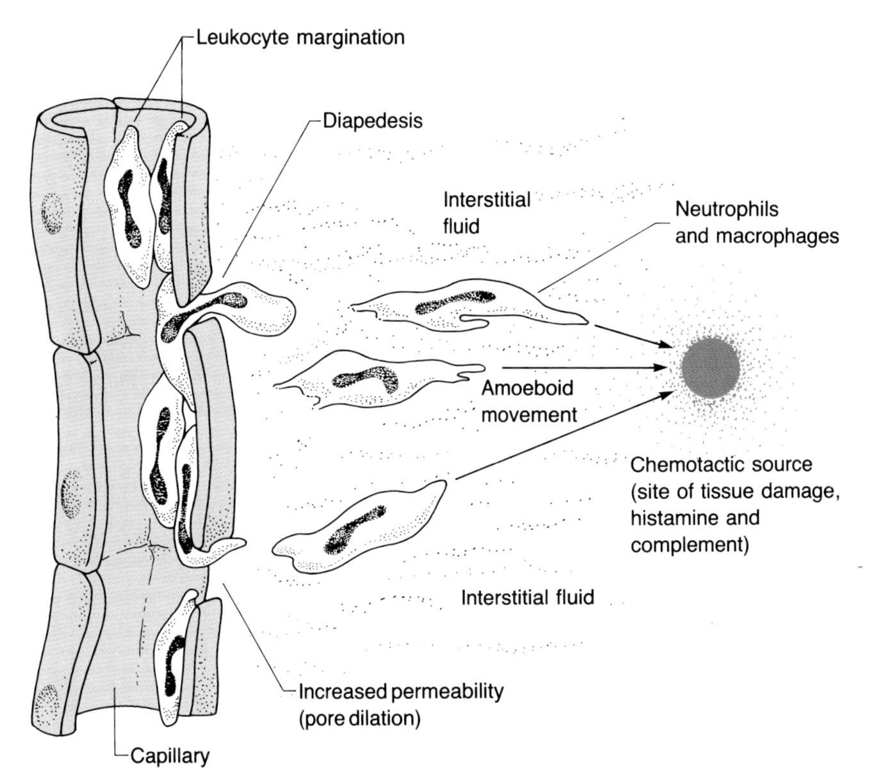

FIGURE 9.8

Initial Response to Injury or Infection
Leukocytes (primarily neutrophils) adhere to capillary walls near the sites of infection (a process known as margination). Spaces between the cells making up capillary walls increase in size, permitting neutrophils to leave the circulatory system (a process known as diapedesis). Histamine and complement proteins provide a chemical gradient up which neutrophils and macrophages migrate (by a process known as chemotaxis).

Thus it is through the interaction of margination, diapedesis, amoeboid movement, and chemotaxis that large local populations of phagocytes and NK cells build up at infection sites. Moreover, within a few hours of the onset of an infection, the number of granulocytes will increase by as much as five times, owing to the stimulating effects of chemicals released by other leukocytes. Increased numbers of monocytes are also formed, and the rate at which monocytes transform into macrophages increases.

THE MECHANICS OF PHAGOCYTOSIS

How do the granulocytes and macrophages correctly identify the appropriate subjects for phagocytosis? Three factors are particularly important:

1. **Electrical charge.** The outer surfaces of most body cells and tissues have negative electrical charges on their surface. These repel the similarly charged granulocytes and macrophages. Foreign substances and dead tissues, however, often bear positive electrical charges.

2. **Surface texture.** Objects with rough surfaces are more likely to be phagocytized than are objects with smooth surfaces.

3. **Immune recognition.** Foreign cells are marked as appropriate subjects for phagocytosis through the actions of the complement system and the immune reaction (discussed in the next section). Both granulocytes and macrophages, for example, have receptors that bind specifically to one of the complement proteins once it has attached to a pathogen.

The granulocytes are the kamikaze warriors of the body, in the sense that they die as a consequence of phagocytizing foreign materials. Granulocytes can ingest from five to 25 bacteria before the toxins released by the bacteria (or the digestive enzymes released by the lysosomes within the granulocytes), cause the granulocyte to die. Granulocytes live only two or three days in any case—and their short life span accounts for the fact that we must produce new granulocytes at the rate of 80 million each minute!

Macrophages are more resilient. They are much larger cells and can ingest as many as 100 bacteria. Moreover, they phagocytize a larger series of materials than do the granulocytes, including dead tissue and dead granulocytes. Finally, they are also capable of releasing their digestive end-products by exocytosis, and may survive for months or even years (although many do succumb following extensive phagocytosis). The dead macrophages and granulocytes, together with dead tissues, form **pus.**

SPECIFIC DEFENSE MECHANISMS

Specific defense mechanisms include those used with microorganisms that, for various reasons, do not activate the complement system or provide binding sites for phagocytes. At the heart of this component of the immune system are the lymphocytes. (Table 9.1)

An Overview of the Lymphocytes

The lymphocytes can be divided into two different populations of cells: **B cells,** which mature in the bone marrow of adults (and the liver of fetuses), and **T cells,** so-named because they migrate from the bone marrow and mature in the **thymus gland** (located just under the *sternum,* or breastbone). B and T cells function in very different ways. B cells attack bacteria and toxic molecules by producing and releasing **antibodies**—proteins that bind to the surface of the invader and thereby mark it for destruction. T cells, by comparison, primarily attack virally infected cells and destroy these cells by secreting a protein called **perforin** (see Figure 9.9). As its name implies, perforin creates large openings in the plasma membrane, causing the cell to rupture, and thereby preventing replication of the virus. (NK cells and granulocytes, which we discussed earlier, also use perforin, and a very similar protein is also a part of the complement system.)

How Lymphocytes Identify Their Victims

The B and T cells are, taken collectively, a very potent defense mechanism. How do they identify invaders? More precisely, how do they distinguish invaders (or invaded cells, in the case of viral infections) from the body's own, healthy tissue?

In order to function effectively, the immune system must distinguish between "self" and "non-self," and then preserve the former and destroy the latter. Recognition of "self" occurs because cells of the immune system—especially T cells—have receptors that precisely fit identifying markers found on all of the cells of one's body. These markers are genetically coded by genes that, in all mammals, form the **major histocompatibility (MHC) complex.** (In humans, the MHC complex is often referred to as the **human leukocyte antigen [HLA] complex.**) There are several hundred genes in this complex (all on chromosome 6), and the combinations that are possible ensure that unrelated individuals rarely possess the same precise combination of genes, hence rarely precisely the same cellular markers.

During maturation in the thymus gland, T cells that react strongly to MHC proteins in conjunction with other surface proteins are destroyed, although

TABLE 9.1 Leukocytes and Defense

CELL TYPE	FUNCTION	MODE OF OPERATION
Lymphocytes:		
1. Cytotoxic T cell	Attack cell	Destroys virally infected cells by using perforin to create holes in the cell membrane
2. Helper T cell	Facilitates activity of T and B cells, and other attack cells	Responds to antigen-presenting cells and stimulates development of T and B cell clones by producing cytokines
3. Suppressor T cell	Modulates activity of T and B cells	Unknown
4. B cell	Once activated, B cell descendants produce antibodies	Transformed into plasma cell following activation by antigen, antigen-presenting cell, and/or helper T cell
5. Plasma cell	Production of antibodies	Antibodies attach to antigens, marking them for destruction by phagocytic cells
6. Memory cell	Retained portion of an activated B or T cell clone, capable of rapid response to reinvasion by same pathogen	Same as cytotoxic T cells, helper T cells, and plasma cells
Other leukocytes:		
7. Neutrophil	Phagocytic cell	Attacks bacteria; prominent in inflammation response
8. Eosinophil	Phagocytic cell	Attacks parasitic worms and complexes of antigen and antibodies
9. Basophil	Inflammation response	Releases histamine, causing dilation of capillaries
10. Macrophage	Phagocytic and antigen-presenting cell	Attacks pathogens and damaged body cells; helps activate B and T cells
11. Natural killer	Attack cell	Destroys tumor or virally infected cells by using perforin

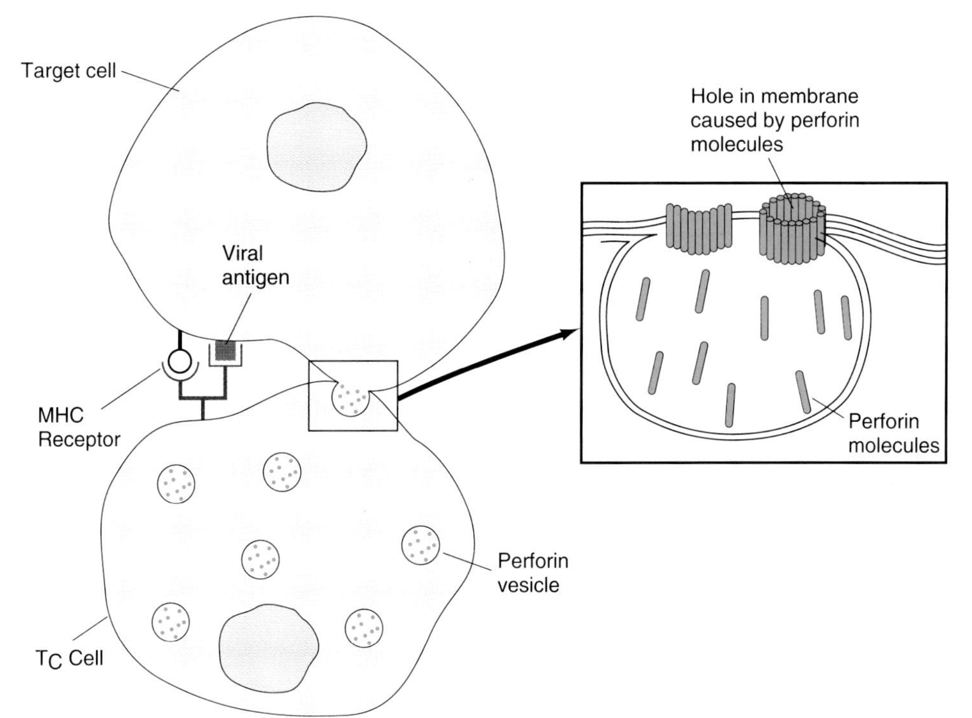

FIGURE 9.9

Mode of Action of Cytotoxic T Cells

Through processes described later but involving identification of the target cell (in this case a cell that has been invaded by a virus), T_C cells merge perforin-containing vesicles with the plasma membrane and release perforin immediately adjacent to the target cell. The long perforin molecules then line up, rather like the staves of a barrel, forming a series of large openings in the plasma membrane of the target cell. Additional toxic agents may then be introduced into the now-doomed target cell. NK cells and some of the granulocytes also use perforin in their attacks.

the mechanism by which this destruction occurs is not yet known. Autoimmunity (discussed later in this chapter) may occur when these overreactive cells escape destruction. Mature T cells eventually leave the thymus and migrate to lymphoid organs such as the spleen, tonsils, and lymph nodes.

B and T cells also possess receptors for specific molecules that are "non-self" in origin. (Stem cells lack "non-self" receptors; they develop only during the maturation process of B and T cells.) Each B and T cell has only one type of "non-self" receptor, but each has many copies of that receptor. During the maturation of B and T cells, new receptors are randomly produced by the interaction of at least five highly variable genes. The consequence is that the body has millions of T cells with different "non-self" receptors, and probably hundreds of millions of B cells with different "non-self" receptors. That is not to say that each cell is unique. The total number of B and T cells is so large—and one billion new B and T cells are produced each day!—that inevitably some cells of each type have the same receptors. Cells with the same receptors are called **clones.**

Despite their similarity in possessing "non-self" receptors, B and T cells differ greatly in the form of these receptors, and they usually react to different portions of the "non-self" molecule. The T cell receptors are less well known, but consist of two groups of highly variable proteins embedded in the plasma membrane of the cell. B cell receptors, by comparison, are relatively well understood, because free-floating versions of their receptors are released into the blood as antibodies.

Antibodies are a particular type of **glycoprotein** (proteins with attached sugars) called **immunoglobulins.** There are five classes of immunoglobulins, each assigned a letter (see Figure 9.10). In descending order of abundance they are IgG, IgA, IgM, IgD, and IgE. All consist of two large (or *heavy*) protein chains, and two small (or *light*) chains, and the ends of all four chains are highly variable. Although each class of immunoglobulin has a characteristic molecular shape, most vaguely resemble a Y, with the two arms of the Y containing the variable ends of the four chains. This is the portion of the antibody that attaches to a particular surface protein or carbohydrate of the invader (see Figure 9.11). (These surface molecules are called **antigens,** because their presence causes *anti*body *gen*eration.) The base of the Y, on the other hand, is a stable region that precisely fits surface proteins on phagocytic cells. Thus, once antibodies attach to an invading cell, phagocytic cells are able to "recognize" it as an invader by attaching to the other end of the antibody and then destroying the invader.

As just described, the activities of B and T cells are somewhat misleading, since, in reality, the various B and T clones are inactive until they encounter the appropriate "non-self" molecule. The process of activation of the system is rather complex, since there are many interactions taking place more or less at the same time. Let us try to thread our way through the activation of the lymphocytes.

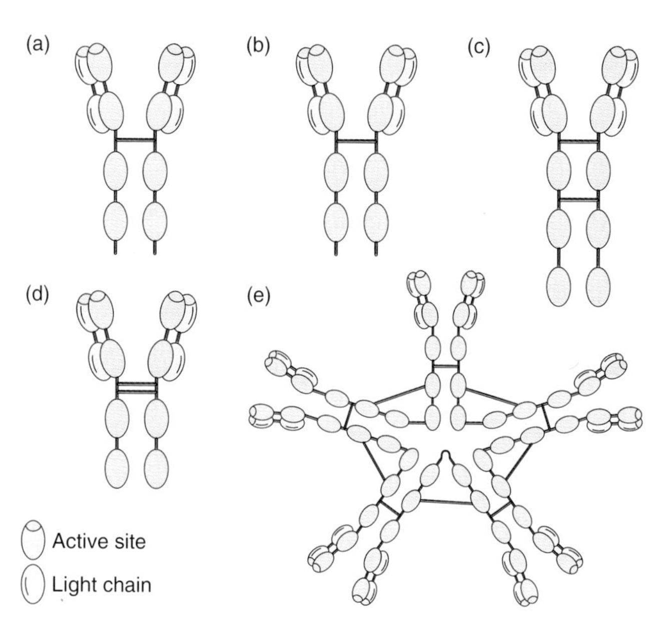

FIGURE 9.10

Immunoglobulins

Antibodies formed by plasma cells are glycoproteins called immunoglobulins. Humans have five classes of immunoglobulins, each of which consists of two "heavy" chains (each containing four or five regions), linked by one or more bonds, and two "light" chains (each consisting of just two regions), all arranged in a vaguely Y-shape. The tips of the two arms of the "Y" are highly variable in their configuration, and these are the active sites by which the antibody attaches to the antigen. When antibodies attach to antigens on the surface of cells, those cells are marked as targets and are so recognized because the base of the stem of the "Y" fits receptors on several classes of killer or phagocytic cells. (a) IgA, (b) IgD, (c) IgE, (d) IgG, (e) and IgM. IgA and IgD have characteristic tailpieces at the ends of their long chains; they differ from each other in the amount of carbohydrate (not shown) associated with each. IgE has five (not four) regions in its heavy chains. IgM is generally found in groups of five molecules.

Activating the Lymphocytes

The major steps in lymphocyte activation are as follows (see Figure 9.12):

1. When a pathogen (or some other "non-self" entity) to which a person has not previously been

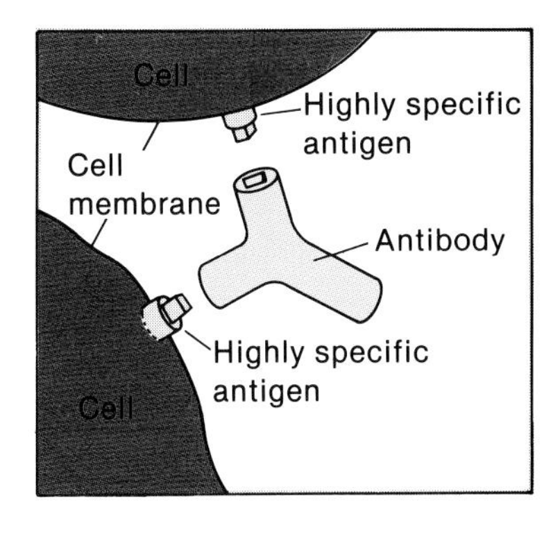

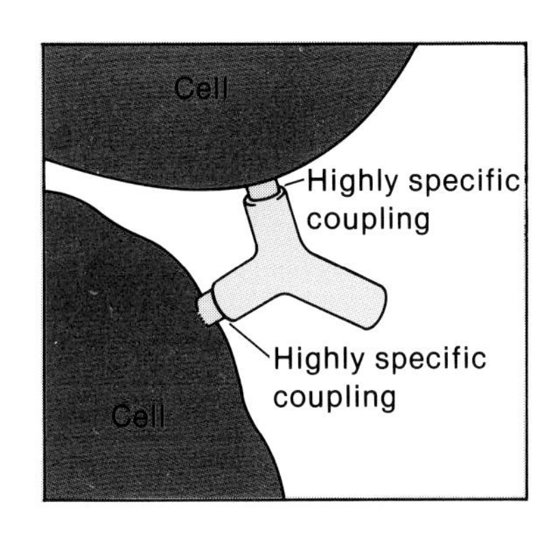

(a)

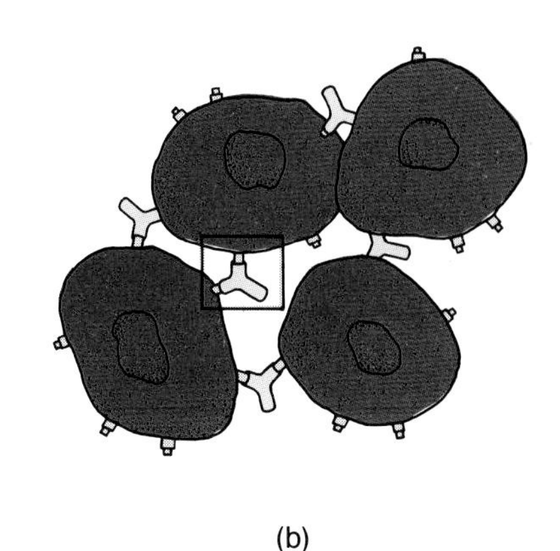

(b)

FIGURE 9.11

Cell Clumping by Antibodies

(a) Antibodies may attach to free-floating antigens, or to antigens on the surface of cells. (b) In the latter instance, clumps of cells may be linked together by antibodies and earmarked for destruction.

exposed enters the body, it will eventually be ingested by a phagocyte such as a macrophage (although the pathogen may have multiplied by this time, and illness could be the consequence). The antigen is fragmented in the lysosomes, and pieces of ten to 20 amino acids in length are delivered to the plasma membrane and displayed next to MHC protein.

2. The macrophage then "presents" these foreign proteins to a class of T cells known as **T helper cells** (T_H cells). When macrophages function in this way they are known as **antigen presenting cells** (APCs), because these foreign proteins function as antigens. (Some other body cells can also function as APCs, including cells infected with virus, but the macrophages are particularly important in this regard). Only the clone of T_H cells that has receptors that bind well with both the MHC protein and the antigen will become activated. (Activation also involves the production and release of chemical signals, the details of which are discussed in the next section of this chapter.)

3. Activated T_H cells subsequently undergo a series of divisions, and then interact with and activate the clone of B cells that possesses surface receptors capable of binding with the antigen. (B cells can respond directly to pathogens, but their full development requires interaction with T_H cells.)

4. The activated B cells undergo a series of cell divisions, and most of the daughter cells mature into **plasma cells.**

5. Plasma cells produce antibodies at very high rates—with each cell producing as many as 2000 antibody molecules each second! The antibodies attach to the surface antigens of the pathogen and mark it for destruction by various leukocytes.

6. In addition to activating B cells, the T_H cells also activate members of a class of T cells known as **killer T cells** (sometimes known as *cytotoxic T cells*, or T_C cells). Again, the T_H cells must locate the clone of T_C cells having surface receptors that exactly match the antigen (see Figure 9.13).

7. The activated T_C cells undergo a series of divisions, and then bind directly to pathogens, secreting chemicals that destroy the pathogen.

8. T_H cells also activate such attack cells as granulocytes, macrophages, and NK cells. These cells then attack the pathogen with greater vigor than occurs if they are not stimulated by T_H cells.

9. A third class of T cells, known as **suppressor T cells (T_s cells)** are responsible for turning B and T_H cells off once the pathogen is under control. The shutting down of the system is further aided by the gradual drop in antigen, which results in reduced stimulation of the T and B cell clones, and they stop dividing.

10. The remaining B and T cells from the activated clone remain as **memory cells,** and can be activated very promptly upon reinfection by the same pathogen.

The process as outlined takes several days to complete. During much of that time, the pathogen often multiplies rapidly—and we become ill. We start to recover only once our immune system begins to gain the upper hand.

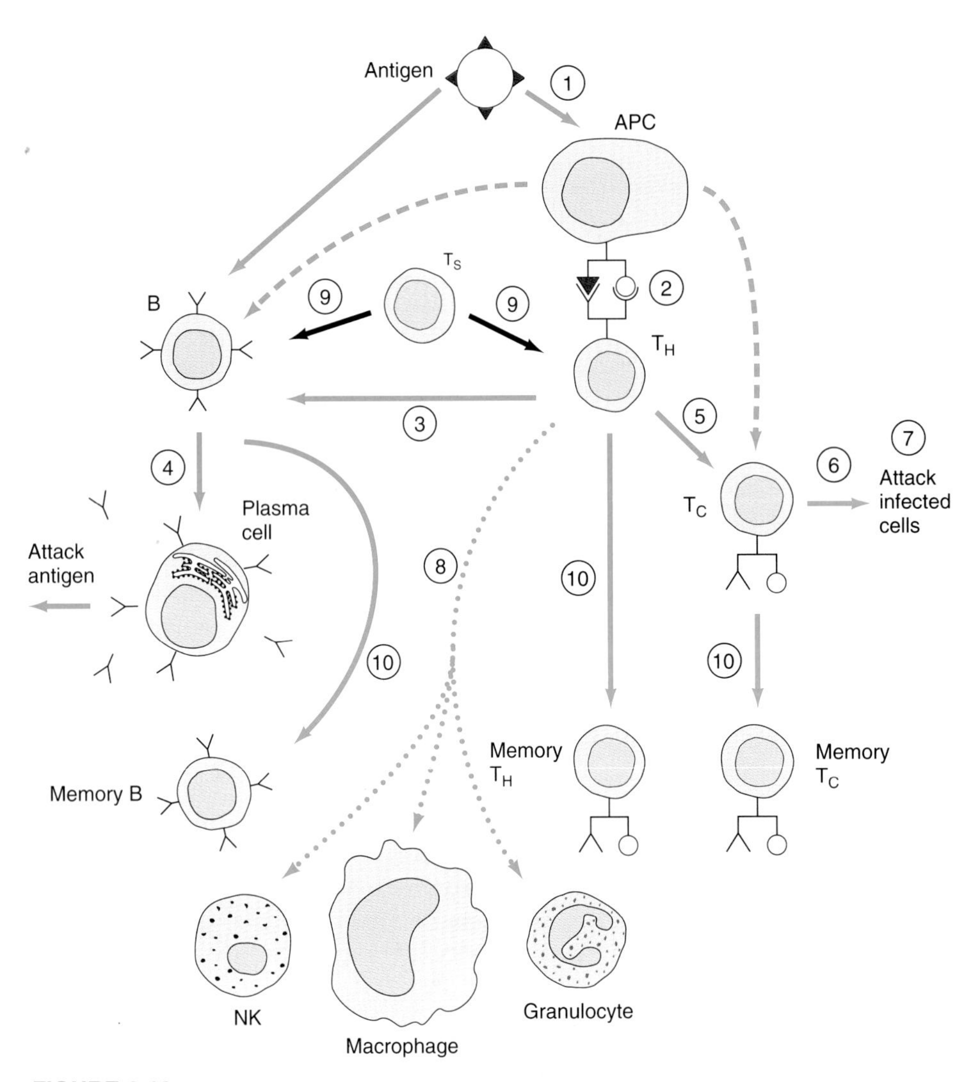

FIGURE 9.12

Details of Lymphocyte Activation

(1) Foreign cell is phagocytized by a macrophage. (2) Acting as an antigen-presenting cell (APC), the macrophage displays portions of the foreign cell next to its MHC proteins. A helper T cell with the appropriate configuration of plasma receptors reacts to the MHC-antigen complex. (APC cells can also stimulate B cells and cytotoxic T cells directly, as shown by the dotted lines.) (3) The T_H cell activates a B cell having receptors that correspond to the antigen. (4) A B cell clone forms, leading to the development of plasma cells that produce antibodies to attack the antigen. (5) The T_H cell also activates a T_C cell having receptors that correspond to the antigen. (6) A clone of T_C cells forms, and (7) they attack cells possessing the antigen. (8) By secreting cytokines, T_H cells stimulate the activity of NK cells, macrophages, and granulocytes. (9) Suppressor T cells modulate the response of B and T_H cells. (10) Memory B, T_H, and T_C cells form and are retained indefinitely.

Chemical Signals and the Immune System

A complicating factor in the priming of the immune system is the array of chemical signals used within the system. These chemical signals, or **cytokines,** are polypeptides, and they function in a manner very similar to hormones (see Chapter 13). Those cytokines that are produced by lymphocytes are called **lymphokines.** The various classes of cytokines and lymphokines at work in the immune response include the following, (see Table 9.2):

1. **Interleukins:** There are at least eight different interleukins (IL-1 through IL-8), most of which are produced by T cells. They function primarily to stim-

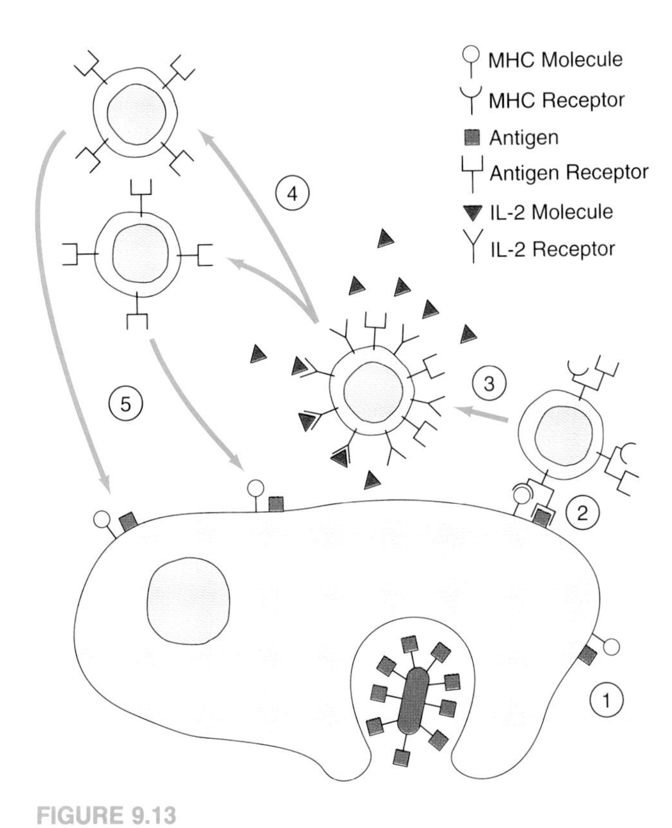

FIGURE 9.13

Details of Clone Formation

(1) Macrophage engulfs a bacterium. (2) Bacterial antigens are displayed in conjunction with MHC proteins to a T cell having surface receptors that match the antigen. (3) The now-activated T cell produces both IL-2 and IL-2 receptors, thereby stimulating its own growth and division. (4) A clone of T cells forms, members of which can themselves be activated by APC cells.

ulate the growth and development of both T and B cells (see Figure 9.14).

2. **Tumor Necrosis Factor:** Two TNFs have been identified, one produced by activated macrophages and the other by activated lymphocytes. Their name derives from their cytotoxic effect on cancer cells, but they also attack virally infected cells, and they stimulate both the production of IL-1 and such effector cells as granulocytes and NK cells. TNF receptors occur on virtually all cells.

3. **Interferon:** There are at least three interferons, which collectively stimulate macrophage and NK cell activity, and increase viral resistance in uninfected cells. They are produced by T cells, and by cells that are infected with a virus.

4. **Growth Factors:** Various leukocytes produce growth factors that stimulate the development of stem cells along particular lines. These growth factors are considered in detail in Chapter 21.

What makes things even more confusing is that the activities of these cytokines broadly overlap. Without going into excessive detail, however, it is possible to discern some basic trends.

When a macrophage presents an antigen to a T_H cell, the macrophage also releases IL-1, which stimulates the T_H cell to produce both IL-2 and IL-2 receptors. IL-1 also activates NK cells and further stimulates the macrophage. IL-2 binds to the IL-2 receptors and stimulates a cascade of cell divisions of cells of the T_H clone. IL-2 also activates both monocytes and NK cells, and promotes the development of a clone of B cells (see Figure 9.15). Activated T and NK cells produce interferon, which further stimulates the macrophage.

TABLE 9.2 Selected Cytokines: Their Source, Targets, and Effects

CYTOKINE	SOURCE	TARGETS	EFFECTS
IL-1	Macrophages, B cells	T and B cells, macrophages, granulocytes, NK cells	Activation and simulation
IL-2	T cells	T and B cells, monocytes, NK cells	Growth factor for T cells; activates and stimulates others
IL-4	T cells	T and B cells	B cells growth factor
IL-5	T cells	B cells	Growth and differentiation
IL-6	T and B cells, macrophages	B cells	Late growth and differentiation
TNF	Macrophages	T and B cells, granulocytes, various other tissues	Activation and stimulation of much of the immune system
IFN	T cells	Macrophage, B cells, NK cells, cytotoxic T cells	Stimulates and activates broad spectrum immune response (especially antiviral)

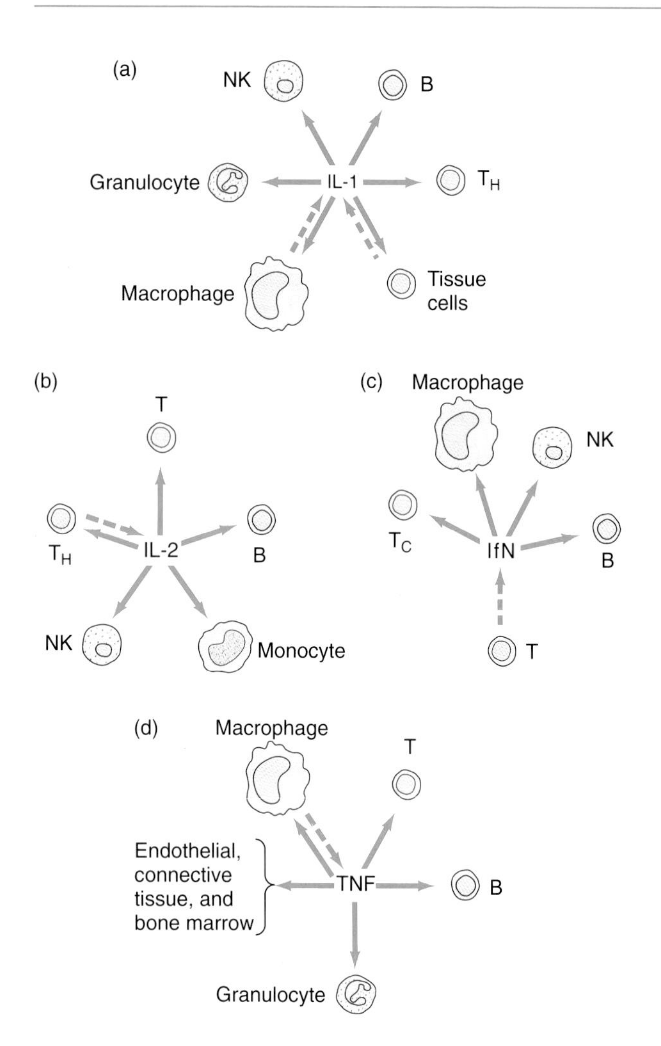

FIGURE 9.14

Some Selected Cytokines

(a) IL-1, produced by macrophages and some tissue cells, stimulates the activity of granulocytes, NK cells, B cells, and T cells, as well as macrophages. (b) IL-2, produced by T_H cells, stimulates T and B cells, monocytes, and NK cells. (c) Interferon, produced by T cells, stimulates T_C and B cells, macrophages, and NK cells. (d) Tumor necrosis factor, produced by macrophages, stimulates B and T cells, granulocytes, and a variety of body tissues.

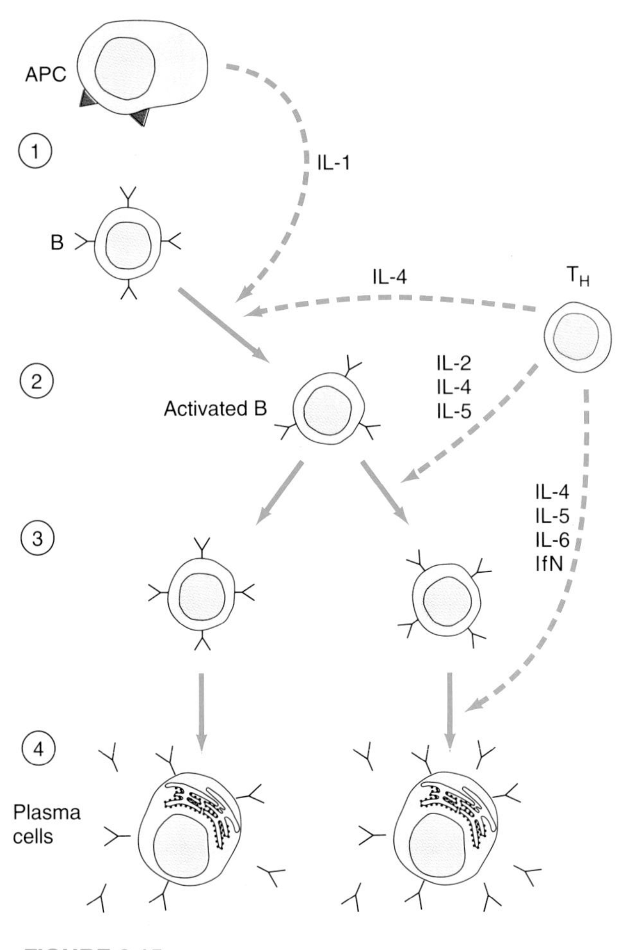

FIGURE 9.15

Details of T_H Activity

(1) APC presents an antigen to B cell. (2) IL-1, produced by macrophage, and IL-4, produced by T_H cell, stimulate activation of B cell. (3) IL-2, IL-4, and IL-5, produced by T_H cell, stimulate the growth and division of a B cell clone. (4) IL-4, IL-5, IL-6, and interferon, produced by T_H cell, stimulate the development of plasma cells.

When a macrophage encounters and engulfs bacteria coated with antibody, it produces TNF and interleukins. These cytokines stimulate activity not only by B and T lymphocytes, but also by various other tissues in the body needed for repair of tissue damage—and the bone marrow itself, where stem cells are directed to differentiate along particular lines so as to ensure an increased production of leukocytes. TNF also causes many of the symptoms associated with illness, including fever (because it adjusts the body's thermostat—located in the brain—upward).

All of the cells activated by cytokines from the macrophage produce their own cytokines; some of them further stimulate the macrophage, while others enhance the activities of the macrophage by directing more growth factors to the bone marrow, and still others enhance the potency of the granulocytes and help direct them to the areas of infection.

The similarity of hormones and cytokines is further evidenced by the fact that lymphocytes have receptors for many hormones and certain neurotransmitters (chemical messengers of the nervous system; see Chapter 14). In addition, the thymus gland responds both to hormones and messengers from the nervous system (see Figure 9.16). It has long been recognized that stress, acting through both the nervous and hormonal systems, inhibits the immune re-

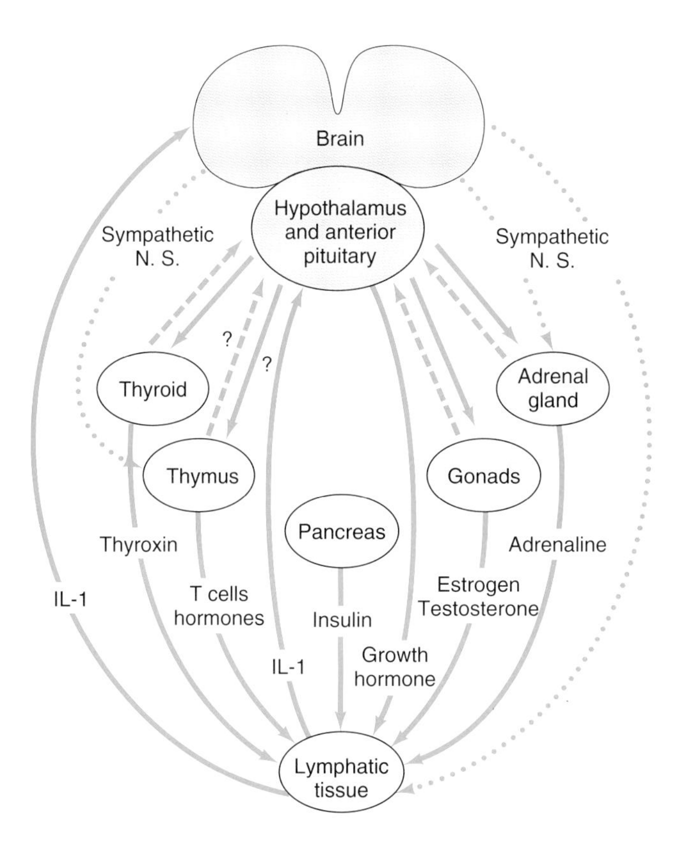

FIGURE 9.16

Hormones, the Brain, and the Immune System
The role of the brain in the functioning of the endocrine system is now well known (see Chapter 13). However, the role of the brain and the hormones in the functioning of the immune system is only now beginning to be understood. There are many examples of hormones affecting the lymphatic tissue of the immune system, as well as examples of direct nervous stimulation of lymphatic tissue by the brain. A hormonal relationship between the thymus and the hypothalamus and anterior pituitary is hypothesized but has not yet been conclusively demonstrated.

sponse—people under stress simply become ill more often. For example, cancer patients about to undertake chemotherapy often have the nausea and diminished immune functioning associated with chemotherapy even before actually receiving the therapy. Now that we know that the cells of the immune system can respond directly to hormones and neurotransmitters, we have an explanation for how it is that the nervous and endocrine systems exert their effects. The long-range therapeutic implications of this understanding are enormous. Indeed, it has spawned an entirely new (and still very controversial) discipline—psychoneuroimmunology, the study of the interaction of the conscious brain with the immune system.

In summary, the full picture is not yet known, but what we do know makes sense functionally: The various components of the immune system are operating not as independent agents but in a highly coordinated manner—a coordination that is accomplished through the use of messenger molecules.

The Immune System—A Reprise

Separating our defense mechanisms into "generalized" and "specific" is useful in attempting to sort out the various ways in which our bodies deal with pathogens (see Table 9.3), but it is somewhat misleading insofar as there is abundant cooperation among the various mechanisms. Antibodies, for example, are much more effective in combination with complement proteins—and, indeed, the manufacture of antibodies is one way in which complement proteins are activated in the first place. Secretions of the various lymphocytes stimulate phagocytosis, and the macrophages play a key role in the activation of the lymphocytes. As is so often the case, in the immune system the whole is much more than just the sum of the parts.

IMMUNITY AND VACCINATION

Most diseases provoke a sufficiently strong response that the memory cells remaining from the activated clones of B and T cells provide a lasting immunity to subsequent exposure to the disease agent. But does that mean that immunity to a particular disease comes only *after* one survives having had the disease in the first place? Fortunately, for an increasingly longer list of diseases, the answer is "no."

In 1796, an English physician named Edward Jenner undertook a very dangerous experiment. For many years it had been recognized that *smallpox,* a disease that was annually responsible for the deaths of tens of thousands of children, could be caught only once. However, Jenner noted that dairymaids who had had *cowpox,* a much less serious disease, never seemed to contract smallpox. Could he induce cowpox in someone who had never had either disease, and create immunity to smallpox?

An eight-year-old boy, James Phipps, was his subject. Jenner removed pus from a cowpox sore on the hand of Sarah Nelmes, a dairymaid, and inserted it into two cuts he had made on the boy's arm. The boy subsequently caught cowpox. Six weeks later, Jenner introduced pus from a smallpox sore into James's arm, but the boy did not contract smallpox. The first **vaccination** (L. *vacca,* "cow") was successful.

Today, vaccinations are used for a variety of diseases. They take three forms:

1. **Dead organisms:** Some disease organisms still induce the formation of antibodies even though the

TABLE 9.3 Generalized and Specific Defense Mechanisms of the Body

	GENERALIZED DEFENSE MECHANISMS	
Category	*Subcategory*	*Mechanism*
Barriers	Skin	Physically prevents invasion of most pathogens
	Mucus and ciliated cells	Traps particles entering respiratory system
	Enzymes	Attack pathogens entering mouth and digestive system
Inflammation		Increases blood flow to injured area
	Basophils	Release histamine, particularly in response to viral infections; closely related to mast cells
	Mast cells	Release histamine and other inflammatory chemicals
Attack chemicals	Complement system	Puncture bacterial membrane; mark bacteria for destruction by other cells; attract phagocytes
Attack cells	Natural Killer (NK) cells	Attack and destroy certain tumor and virally infected cells
	Macrophages	Phagocytosis of pathogens and cell debris
	Neutrophils	General phagocytes, especially of bacteria
	Eosinophils	Phagocytosis; attack parasitic worms

	SPECIFIC DEFENSE MECHANISMS	
Category	*Subcategory*	*Mechanism*
B lymphocytes		Clonal response to specific antigens
	Plasma cells	Antibody factories for specific antigens
T lymphocytes	Helper T cells	Stimulate B and cytotoxic T cells
	Cytotoxic T cells	Clonal response to specific membrane-bound antigens; attack "non-self" cells and virally infected cells
	Suppressor T cells	Modulate response of T and B cells
Memory cells		Remnants of activated B and T cell clones that permit immediate attack on pathogens to which there has been a previous exposure
Macrophages		Antigen-presenting cell in the activation of T and B cells (especially helper T cells)
Killer cells		Bind to antibodies attached to antigens on cell surfaces and destroy the cell

organisms are dead. Many bacterial vaccinations, including those for *typhoid fever*, *diphtheria*, and *pertussis* (whooping cough), utilize dead organisms.

2. **Attenuated organisms:** Attenuated organisms are living, but have been rendered much less potent by a variety of culture techniques. Many viral vaccinations, including those for *polio*, *measles*, *smallpox*, and *yellow fever*, utilize attenuated organisms.

3. **Toxins:** Some diseases are caused not directly by a disease organism, but indirectly by the waste products, called *toxins*, that the organisms produce. Immunity to the disease can therefore be imparted by vaccinating the individual with the toxins themselves (which have been treated so as to render them harmless). This method is used for such diseases as *botulism* and *tetanus*.

Immunity to most of these diseases lasts for years or even for life. However, a succession of shots, or occasional "booster" shots, must generally be given to ensure a high level of antibody formation. The world-wide campaign against smallpox has been so effective that several years ago the World Health Organization declared the disease to be eradicated—the first known instance where a disease has been completely wiped out. (Fortunately, smallpox, unlike many diseases, occurs only in humans, and there is therefore no population of alternate hosts—as there is in rabies, for instance—that can perpetuate the disease.)

Vaccinations for other diseases have also had dramatic results. In 1921, there were 200,000 new cases of diphtheria in the United States; in 1986, there were no new cases. In 1952, there were 57,000 new cases of polio in this country; in 1986, there were two new cases. Other diseases still await the development of extensive vaccination programs. For example, vaccines now exist for hepatitis B (500,000 to 1,000,000 current infections in the United States), and for pneumococcal pneumonia (20,000 to 40,000 deaths annually), but they are not yet in widespread use.

Some disease agents, such as those for influenza or the common cold, mutate rapidly, rendering useless antibodies that were formed against their previous genetic structure. For this reason, cold and flu vaccinations are only partially effective and then only for short periods of time. Consequently, only individuals at high risk normally are vaccinated against these diseases. Nevertheless, an extensive vaccination program for influenza could sharply reduce the number of deaths from this disease (10,000 to 40,000 annually in the United States, but more than 50,000 in the 1984-1985 epidemic).

Other diseases are sufficiently rare that there is no general vaccination policy. Veterinarians and other handlers of animals are vaccinated against rabies, in the same way that your dogs and cats are, but it is so rare in the general population that routine vaccinations are not given.

What happens if you are bitten by a rabid animal, or by a venomous snake? **Passive immunity** involves the injection of preformed antibodies for the disease organism or toxin directly into your body. This form of immunity is short-lived, because you are producing no antibodies yourself, nor are memory cells being created, both of which occur in **active immunity** (immunity resulting from exposure to a disease organism or vaccine). However, it is entirely effective in the short run.

Of course, as we know full well, there are other mechanisms besides vaccinations that may be used in the treatment of disease. Foremost among these are **antibiotics.**

The use of antibiotics is now so extensive in the treatment of disease that it is difficult to imagine doing without them. However, they are entirely a product of the twentieth century. *Sulfa drugs* were discovered in 1908, but were not used medically until the 1930s. *Penicillin* was discovered in 1928, but was not used in the treatment of disease until 1940. Other drugs, such as *Streptomycin, Aureomycin,* and *tetracycline* are of even more recent vintage. The development of antibiotics represents one of the great achievements of medicine, and their use over the last 50 years has saved millions of lives.

Antibiotics are used primarily in the treatment of diseases caused by bacteria. They generally work in one of four ways:

1. Interference with the synthesis of the bacterial cell wall (see Figure 9.17);

2. Inhibition of the functioning of the bacterial plasma membrane;

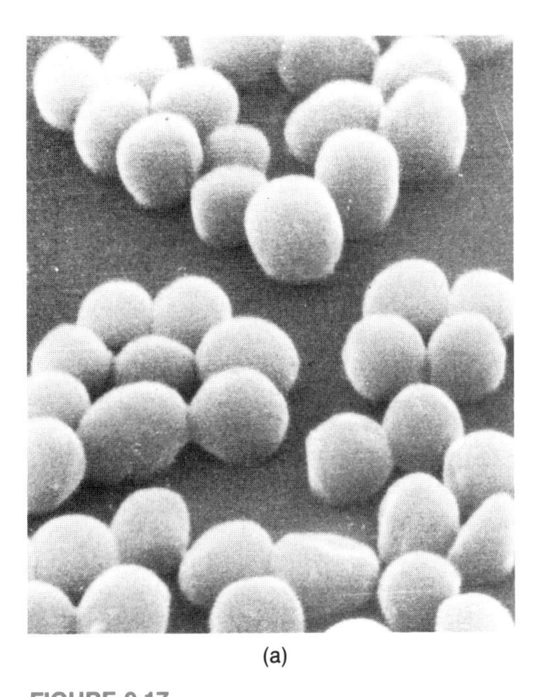

(a)

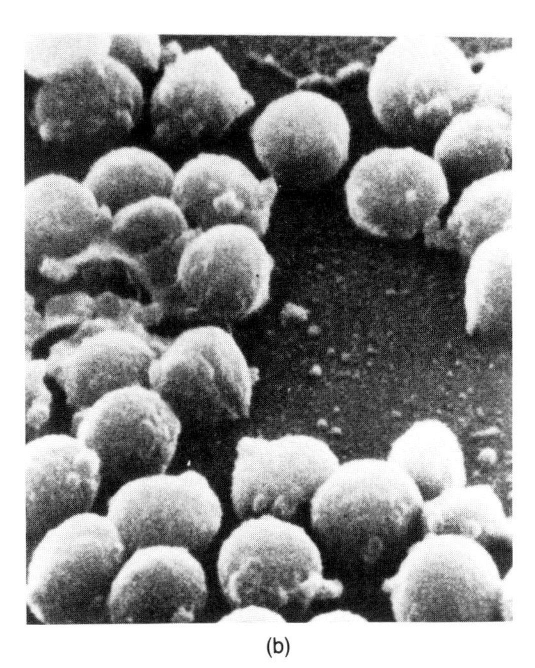

(b)

FIGURE 9.17

Effect of Antibiotics

(a) The bacterium *Staphylococcus aureus.* (b) The same species when treated with penicillin. Note the evident changes in the cell wall.

3. Interference with metabolic functions unique to bacteria; and

4. Competitive inhibition—the antibiotic resembles some essential compound needed by the bacteria, and when the bacteria mistakes the antibiotic for the essential compound, its metabolic machinery is impaired.

All of these mechanisms depend on the fact that bacterial cells are different from human cells in many ways. Thus drugs that are lethal to bacteria often have no effect whatsoever on human cells.

Viruses are a different matter. As we saw in Chapter 4, viruses function by pirating the metabolic machinery of our cells (the infected cells being those that have membrane receptors that are compatible with the attachment proteins on the shell of a particular virus). The consequence is that there are only a few points in the life cycle of a virus where the virus can be attacked by drugs without simultaneously jeopardizing the welfare of our cells. Thus, while there are many antibacterial drugs, there are presently only a handful of antiviral drugs.

PROBLEMS WITH THE IMMUNE SYSTEM

The primary problems of the immune system result either from overactivity (hypersensitivity) or underactivity (deficiency). Under the former, we shall consider allergy, autoimmunity, and the rejection of tissue transplants (see Figure 9.18); under the latter, we shall consider immune deficiency syndromes, both inherited and acquired.

Allergies

As we saw earlier, IgE is normally the least abundant class of immunoglobulins. However, individuals who lack a dominant allele (see Chapter 18) develop greatly elevated levels of IgE. (They may also lack an appropriate number of T_S cells, which would lead to an overreactive immune system.) In any case, following the usual antigen presenting cell-T_H cell-B cell sequence, a clone of B cells begins producing large amounts of IgE that is specific for the antigen. (In the case of allergies, the antigen is known as an **allergen.**) The stable ends of the IgE molecules bind to a type of granulocyte known as the **basophils** (the rarest of the granulocytes), and to **mast cells,** that are very much like basophils but are embedded in tissues throughout the body. When the variable end of the IgE binds to an allergen, the mast cells release large quantities of **histamine,** which causes local swelling of tissues, especially in mucous membranes where allergens often first encounter the immune system. This entire sequence is of value in fighting parasitic worm infections (something that still occurs in one-third of the world's population), but is obviously counterproductive in allergic reactions. Let's look at some examples of this type of allergy.

Hay Fever. The pollens of various plants, most notably the grasses and ragweed, often act as allergens, and the allergic reaction occurs in the nose. The release of histamine causes severe swelling and redness in the tissues of the nose and eyes, along with accompanying discomfort. Various drugs that act as **antihistamines** substantially diminish the severity of the reaction.

Asthma. *Asthma* is a more serious allergy, because the allergic reaction occurs in the bronchioles of the lungs, causing muscular spasms in the bronchiolar walls and great difficulty in breathing. Antihistamines are of less use in asthma, because the problem is caused by a different group of chemicals from ruptured cells.

Anaphylactic Shock. *Anaphylactic shock* is the most serious of the allergies, because the effects are bodywide, rather than localized. Typically, it results from the injection of an allergen such as bee stings or penicillin directly into the blood. (For this reason, your physician normally insists that you remain in his office for a few minutes following a penicillin shot, to ensure that you do not go into anaphylactic shock.)

The effects of histamine release throughout the bloodstream cause widespread dilation of peripheral blood vessels, which leads to a rapid drop in blood pressure (shock), and possible death. *Noradrenaline* is used to counter the effects of histamine, thereby minimizing shock.

A very different type of hypersensitivity occurs in such conditions as poison ivy or other forms of **contact dermatitis** (inflammation of the skin). The reaction to skin contact with allergens reaches its maximum about 48 hours after contact—much longer than the reaction time for hay fever, asthma, or anaphylactic shock.

The reason for the difference is that the reactive compounds of poison ivy or poison oak, for example, are not themselves antigenic, but only become so after penetrating the skin and binding with body proteins. The antigenic response occurs to the allergen-protein combination, although not all individuals become sensitized. The key cell is the **Langerhans cell**—an antigen-presenting cell derived from stem cells and found in the epidermis. Langerhans cells induce the migration into the epidermis of lymphocytes, followed by macrophages, over the next 48 hours, by which time considerable swelling and reddening of the skin has occurred.

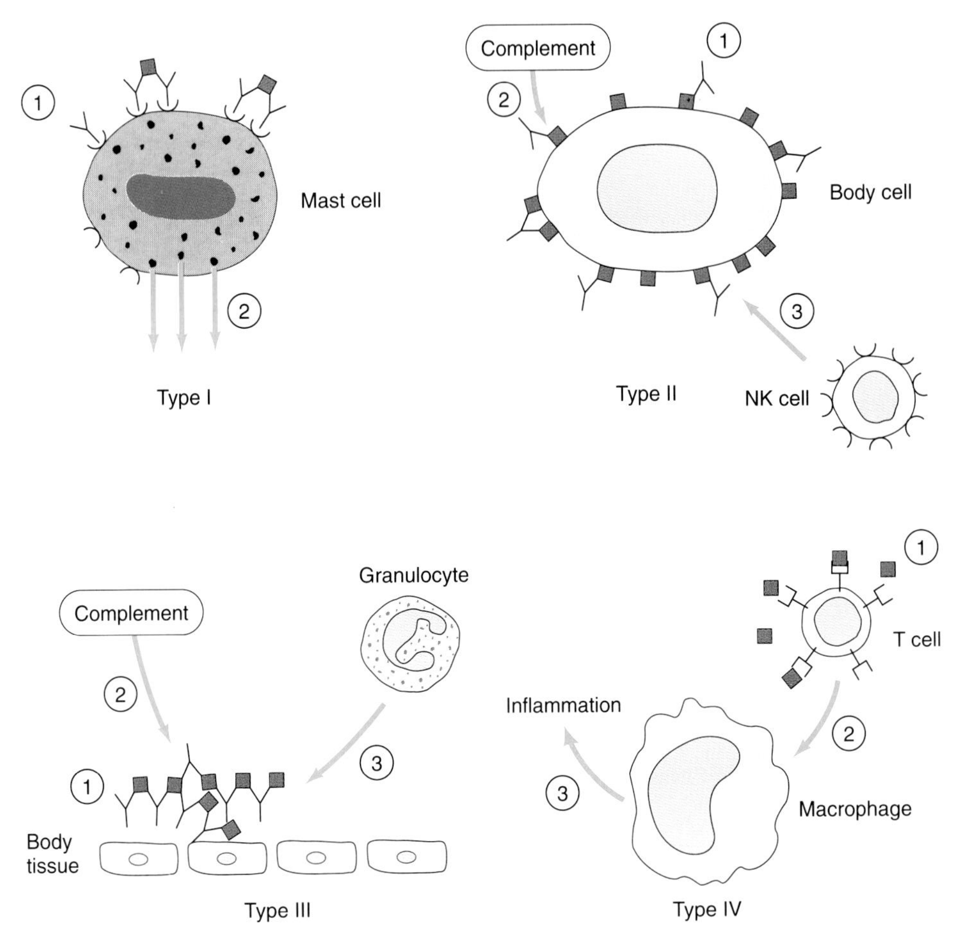

FIGURE 9.18

Hypersensitivity Responses

Type I: Mast cells bind excessive amounts of IgE (1) and degranulate, releasing histamine (2). This is the classic allergic response, as occurs in hay fever. *Type II:* Antibodies form to body's own tissues (1) and these tissues are then attacked by complement proteins (2) or killer cells (3). This is the classic autoimmune response. *Type III:* Complexes of antigens and antibodies are deposited in body tissue (1), where they are attacked by complement proteins (2) or killer cells (3). Immune complexes form in many diseases, including both autoimmune (rheumatoid arthritis, lupus) and pathogenic (malaria, leprosy). *Type IV:* Sensitized T cells release cytokines when they encounter antigen (1), activating macrophages (2) and causing inflammation (3). This is classic delayed hypersensitivity (since it requires from one to 14 days following exposure for symptoms to develop), and it is exemplified by dermatitis caused by poison oak or poison ivy.

Autoimmunity

Recognition of "self" by B and T cells develops gradually during embryonic and fetal development. Even foreign materials injected into a fetus will be regarded as "self," and exposure to these same materials after birth will not cause the development of an immune response.

However, for a variety of reasons (aging; the destruction of tissues normally sheltered from the immune system; the effects of certain disease agents; the resemblance between a particular foreign molecule and a molecule of "self"; and other factors that have not yet been identified), recognition of "self" may break down, and the immune mechanisms begin to act against the body's own tissues (see Figure 9.19). The fact that possession of particular MHC antigens is correlated with increased susceptibility to certain diseases, such as *juvenile diabetes, rheumatoid arthritis, multiple sclerosis, ankylosing spondylitis, myasthenia gravis,* and *Addison's* and *Graves' diseases*—some or all of which are autoimmune diseases—suggests that one important cause of autoimmunity may be the failure of T cells to recognize certain MHC antigens as "self" (see Table 9.4).

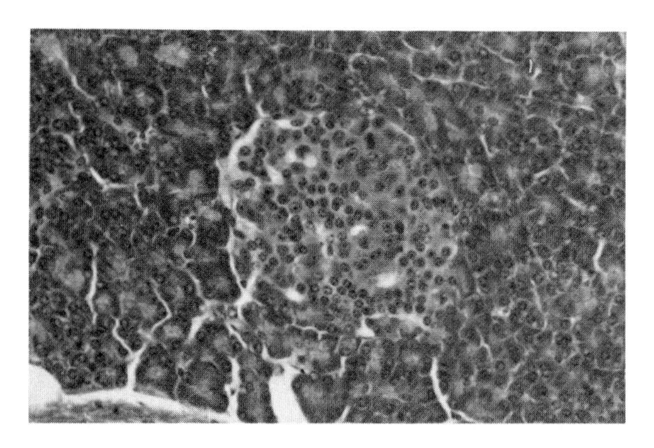

FIGURE 9.19

Juvenile Diabetes

Juvenile diabetes results from an autoimmune attack on the insulin-producing cells of the pancreas. This photomicrograph shows normal human insulin-producing tissue.

However, autoimmunity does not arise solely because of overactive T cells. The complement system can also play an important role. As we saw earlier, some of the complement proteins attack the plasma membrane; others stimulate inflammation and attract neutrophils that may attack connective tissue. An overactive complement system has been implicated in such diseases as *myasthenia gravis, hemolytic anemia,* and *rheumatoid arthritis* (all of which are discussed elsewhere in this book in greater detail). Drugs that interfere with the complement system are now being tested on people with these autoimmune diseases.

A variety of diseases are known, or suspected to be, autoimmune in nature. They include the following:

Glomerulonephritis. *Glomerulonephritis* commonly develops as a consequence of strep throat or other streptococcal disease (that is, a disease caused by *streptococcal bacteria.*) Antigen-antibody complexes lodge in the glomeruli of the kidney, where they and the adjacent kidney tissues are attacked by phagocytes. The resulting inflammation may cause blockage of the glomeruli and kidney failure, although recovery is usually good.

Rheumatic Fever. *Rheumatic fever* is another disease that may follow streptococcal infections. Proteins released from the streptococci promote antibody formation, but for reasons that are not entirely clear, these antibodies also attack body tissues, most notably the heart valves (see also Chapter 8).

Myasthenia Gravis. In *myasthenia gravis,* muscle fibers fail to become depolarized following the transmission of an action potential from a neuron (see Chapter 14). Antibodies are formed to the protein that functions as the receptor for the neurotransmitter *acetylcholine.* As a consequence, the antibodies prevent receptor-antibody binding, and the muscles fail to contract properly.

Systemic Lupus Erythematosus. *Lupus* is a very severe autoimmune disease occurring most frequently in young women. The immune system attacks many different body tissues at one time, causing extensive damage, anemia, and, rather frequently, death (usually due to kidney failure).

Autoimmunity may also be a factor in *schizophrenia.* However, some diseases are only secondarily autoimmune, in the sense that they are initially triggered by infective agents. The development of effective antibiotics has, for instance, all but eliminated rheumatic fever in this country, because the streptococci are killed before they can promote heart valve damage.

Human trials are now underway in the treatment of rheumatoid arthritis and multiple sclerosis, in which patients are given oral doses of collagen and myelin, respectively. (These are the two proteins that are attacked in the immune disorders.) The idea is that people seldom become sensitized to proteins that are

TABLE 9.4 MHC Alleles and Autoimmune Disease

DISEASE	MHC ANTIGEN	FREQUENCY AMONG PATIENTS	FREQUENCY IN GENERAL POPULATION	INCREASED RISK FACTOR
Ankylosing spondylitis	B27	71–100%	3–12%	90
Myasthenia gravis	B8	38–65	18–31	4.5
Juvenile diabetes	DR3	23–36	14–19	2.3
	B8	19–55	2–29	2.4
Multiple sclerosis	DR3	50	21	3.8
	DR4	42	19	3.5
	B7	12–46	14–30	1.7
Rheumatoid arthritis	DR2	47–70	15–31	4.3
	DR4	38–65	18–31	4.4

ingested. Such a regimen greatly reduced the symptoms of the two diseases in rats, and the surmise is that human sufferers will therefore also benefit.

Another treatment for autoimmune diseases is to vaccinate patients with an inactivated form of their own T cells for the particular disease. The patient then creates antibodies that destroy all of the T cells in that particular clone. Alternatively, antibodies to the specific disease can be linked to a lethal molecule and then injected into the body. Both approaches have been successfully accomplished in mice, and the latter approach is now being tested in small groups of patients with multiple sclerosis or rheumatoid arthritis.

Finally, a promising treatment for all autoimmune and tissue transplant problems has recently been modeled in rats. In the treatment of juvenile diabetes (see Chapter 13), insulin-producing cells were transplanted to the thymus gland, where they were seen as "self." Subsequent transplants from the same donor, transplanted elsewhere in the body, were not attacked or rejected. In theory, the same technique could be used for any organ about to be transplanted, thereby avoiding the need for immunosuppressive drugs.

Transplantation and Rejection

The first, and still the most common, tissue transplant was blood transfusion. Early efforts at blood transfusion were not always successful, and not infrequently ended with the death of the recipient. Not until 1900 was it finally realized that blood possessed antigenic properties, such that antibodies in the plasma of the recipient often react with antigens on the erythrocytes of the donor.

At least 20 different blood groups have been identified in human erythrocytes, but two are of particular importance because they are strongly antigenic. These are the **ABO system** and the **Rh system.**

The ABO system involves two related antigens—*type A* and *type B*—and people may possess one, both, or neither. The system of plasma antibodies is just the reverse, which means that individuals with neither antigen possess both antibodies (see Table 9.5). Proper blood typing eliminates the dangers in transfusions, as long as the transfusion is of compatible blood (see Figure 9.20).

The Rh system, which was not discovered until 1940, differs in that Rh antibodies, rarely present initially, appear only after a previous exposure. Individuals are designated either "positive" or "negative" with respect to the class of antigens they possess. A first addition of positive blood to an Rh^- person will trigger antibody formation, and a second addition will cause a massive immune reaction.

This was once a serious problem in Rh^- women married to Rh^+ men because during pregnancy, some fetal cells often leak back into the mother's circulation at birth (see Figure 9.21). If the baby is Rh^+, the mother will form antibodies to Rh^+ erythrocytes and in subsequent pregnancies the fetal erythrocytes are destroyed by antibodies from the mother. However, at birth massive transfusions of Rh^- blood, which does not promote an immune reaction, permits sufficient oxygen-carrying capacity to be maintained. In the meantime, the maternal antibodies are removed by normal metabolic processes. Currently, Rh^- women giving birth to Rh^+ babies are injected with anti-Rh antibodies immediately after delivery. This antibody eliminates Rh^+ erythrocytes and prevents the development of Rh^+ antibodies by the mother.

Other types of tissues and organs that have been transplanted include skin, liver, heart, lungs, and kidneys. The principal problem in each instance is not surgical implantation but the prevention of rejection—the development of an immune response to the transplanted tissues.

T cells are particularly important in tissue rejection, because it is the T cells which function in infections of host cells (rather than with pathogens floating freely in the blood, the task of B cells). Thus T cells are particularly attuned to distinguish between "self" and "non-self."

Transplants between identical twins, or (as in the case of skin grafts) from one part of the body to another, rarely pose rejection problems, because the transplant is recognized as "self." However, transplants from other individuals are an entirely different matter. Tissue typing between a proposed donor and

TABLE 9.5 Blood Types and Transfusability

BLOOD GROUP	ANTIGENS ON RED BLOOD CELLS	ANTIBODIES IN PLASMA	TRANSFUSIONS CAN BE ACCEPTED FROM:	TRANSFUSIONS CAN BE GIVEN TO:
O	None	Anti-A, anti-B	O	O, A, B, AB
A	A	Anti-B	A, O	A, AB
B	B	Anti-A	B, O	B, AB
AB	A, B	None	A, B, AB, O	AB

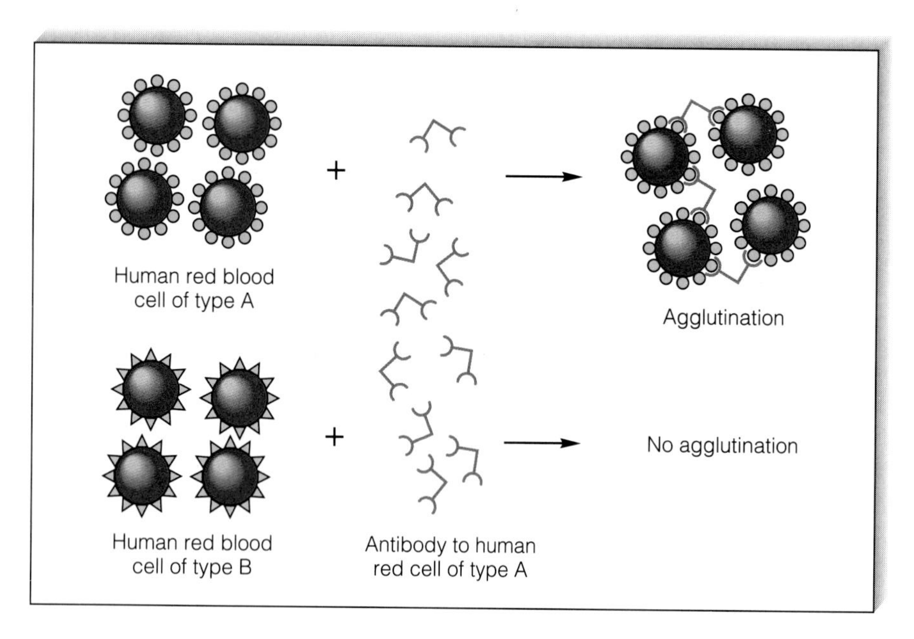

FIGURE 9.20

Agglutination Reaction

If Type B blood is transfused into a Type B individual (bottom) there is no agglutination, since the cell antigen is B and the plasma antibody is anti-A. However, if Type B blood is transfused into a Type A individual (top), agglutination occurs because the host has plasma anti-B antibodies that adhere to the transfused red blood cells. (In addition, the anti-A antibodies of the transfused blood will adhere to the host's red blood cells, thus compounding the problem.)

a recipient involves a determination of how many antigens are held in common. With paired organs, such as kidneys, relatives of the recipient are generally tested first because they are likely to be antigenically more similar than nonrelatives.

Even with a good match, drugs that suppress the immune response (such as glucocorticoids, ACTH, or cyclosporine) must be given to reduce the likelihood of rejection. Unfortunately, these drugs also reduce the recipient's capacity to resist disease and infection, necessitating the striking of a careful balance between rejection of the organ and illness and death due to disease. Cyclosporine, for instance, inhibits the production of IL-2, an interleukin that we saw previously to be vital in promoting the development of B and T cell clones. The absence of a T cell clone for the "non-self" molecules of the grafted tissue will permit it to remain functional in the body—but with no IL-2, new B and T cell clones for subsequent pathogens will also fail to develop. Moreover, cyclosporine also depresses the activity of all of the lymphokines, and can cause kidney damage.

Even so, cyclosporine has made an enormous difference in organ transplants. In 1981, before the introduction of cyclosporine, only 26 liver transplants and 62 heart transplants were performed in the United States, simply because the rejection rate was so high. In 1987, almost 1200 liver transplants and over 1500 heart transplants were performed, with success rates of 70 percent and 82 percent, respectively. In addition, more than 9000 kidney transplants are now performed annually, with a success rate for this procedure of 95 percent.

Tissues that are largely noncellular in nature—for example, those in the cornea of the eye and in bones and tendons—present few rejection problems, and can be grafted or transplanted with great success. Corneal transplants, for example, which were first performed in 1905, currently have a 95 percent success rate among the 35,000 procedures performed annually in the United States.

Immune Deficiency Disorders

Immune deficiencies take several forms. One of the rarest, and most dramatic, is **SCID** (*severe combined immunity disorder*), in which the number of lymphocytes is so low that the individual has virtually no immunity to any pathogen. David, the famous "boy in the bubble," was isolated from the outside world inside a plastic chamber from the time of his birth in 1971 until a transplant of stem cells from his sister was performed in 1984. Unfortunately, that transplant contained a latent virus (Epstein-Barr) that induced cancer in David, and he died shortly there-

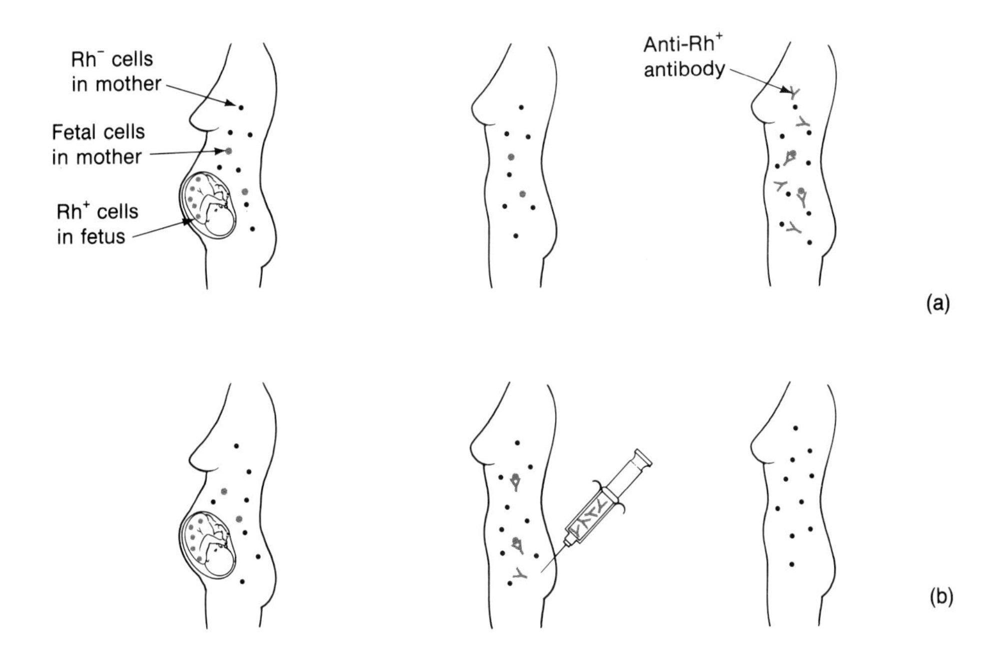

FIGURE 9.21

Rh Incompatibility

(a) During a first pregnancy of an Rh negative woman married to an Rh positive man, a few red blood cells from the Rh positive fetus may enter the circulatory system of the mother, where they will stimulate production of antibodies against the Rh positive cells. (b) In a second pregnancy, antibodies may cross the placental barrier and attack fetal red blood cells. This problem can be avoided if, following the first pregnancy, the mother is injected with anti-Rh positive antibody that destroys any fetal blood cells in the mother's circulation without provoking her immune system into developing her own anti-Rh positive antibodies. (In essence, this represents passive immunity to Rh positive antigen.)

after. (SCID is discussed in greater detail in Chapter 19.)

During the past decade, great attention has been focused on a disorder called **AIDS** (*acquired immune deficiency syndrome*). As its name implies, this disease creates a deficiency in the immune response, and the individual then frequently succumbs to infections and diseases that a person with a fully functioning immune system would quickly eliminate.

A great deal has been learned about AIDS in the past few years, and this knowledge has profoundly clarified our understanding of the immune system in general. The causative agent is a virus (see Figure 9.22) and, as we saw earlier, viruses invade whichever tissue possesses a plasma membrane receptor that the virus can recognize. Unfortunately, in the case of the AIDS virus, the most susceptible cells are the central cells of the immune system itself—the macrophages and T_H cells. The T_H cells are killed by the infection, but the macrophages remain alive and continue to be a source of new generations of the virus.

Because macrophages normally function as antigen-presenting cells, and T_H cells normally coordinate the activities of both B and T_C cells, anything that

FIGURE 9.22

AIDS Virus

The tiny size of viruses is apparent from this scanning electron micrograph of a T cell covered in AIDS virus ready to escape and infect other cells of the body.

cripples macrophages and T_H cells necessarily cripples the immune system as a whole. The rest of the immune system remains intact, but nonfunctional, like an army that has lost its generals and is no longer capable of a coordinated battle plan. Moreover, other body tissues may secondarily become infected with the AIDS virus. The nervous system is particularly susceptible, and paralysis and profound brain deterioration often are manifested late in the course of the disease. (The virus does not actually invade nerve cells, but instead infects macrophages and specialized immune cells in the central nervous system. When these cells die, they release enzymes that damage or destroy the surrounding nerve cells.)

Fortunately, the AIDS virus is not very contagious because, unlike the cold virus, for example, it can be transmitted only through the direct exchange of body fluids (most notably, blood and semen). However, the incubation time—the time from initial infection until the disease becomes manifest—may be five years or more, during which time the infected, but seemingly healthy, individual may infect other people. Thus, from its initial identification as a disease in 1981 to the present time, perhaps 10 million individuals worldwide are believed to have been infected, and more than 100,000 people have died of the disease in the United States alone.

Although no vaccine has been developed as of this writing, there are several drug therapies now available. The earliest of these was the drug **AZT,** which prevents the assembly of more viral DNA by selectively competing with a DNA component. Although AZT does not significantly interfere with normal DNA synthesis by uninfected cells, over long periods of use, the drug can affect tissues that are dividing rapidly (such as blood). Anemia is therefore a common side effect of AZT use. Moreover, AZT does not seem to work as well on infected macrophages as it does with infect T_H cells—and there is concern that a reservoir of the virus may remain in the macrophages indefinitely, even with continued AZT treatment.

Summary

Blood is a highly specialized fluid tissue, with a wide range of responsibilities. Erythrocytes transport oxygen from the lungs to the interstitial fluid, and erythrocytes plus the plasma transport carbon dioxide from the interstitial fluid to the lungs. Platelets assist in blood clotting, thereby preventing loss of this vital tissue. Leukocytes are involved in defense against infection and disease, and the complexity of the mechanisms involved is staggering.

All of the blood cells are derived from stem cells located in the bone marrow, with the ratio of each type of cell determined by growth factors secreted by a variety of tissues. Certain leukocytes, such as the granulocytes and the macrophages, destroy invading organisms or damaged or infected body cells by phagocytosis. Other leukocytes, such as the natural killer cells and the cytotoxic T lymphocytes, kill virally infected cells or cancer cells by producing chemicals that rupture the plasma membrane of the attacked cells. Still other leukocytes, notably the B lymphocytes, produce antibodies that adhere to foreign proteins (whether free-floating toxins or proteins which are part of the cell structure of an invading cell), and these attached antibodies then greatly increase the likelihood of phagocytosis by granulocytes and macrophages.

Some leukocytes—such as the natural killer cells, the granulocytes, and to a lesser extent, the macrophages—are capable of independent and immediate action against foreign cells, but other leukocytes, most notably the T and B lymphocytes, require activation by antigen-presenting cells before they are fully functional. Chemical messages are constantly being sent back and forth among the various leukocytes, to ensure coordinated attacks against pathogens and other threats.

Novel pathogens must first be identified before B and T cells can mount an attack, and mounting an attack may take several days (while colonies of antibody-producing B cells and cytotoxic T cells are being developed), during which time we may become ill. Modern medicine has thus been concerned with the development of vaccines that can activate these cells without inducing the disease, thereby providing immunity to the disease organisms. Vaccines now exist for many diseases, and vaccinations are now a part of every young child's life.

The immune system is no more free from problems than is any other system. Hypersensitivity can lead to the development of allergies (wherein nonpathogenic substances, such as pollen, are treated as disease agents) or autoimmunity (wherein the body's own cells are treated as foreign by the immune system and are attacked). Tissue transplants are also recognized as foreign, and unless immunosuppressive drugs are used, such transplants are ultimately rejected by the body.

Finally, the immune system can itself fall victim to certain disease agents, of which the AIDS virus is currently the most serious, since it opens the door for invasion by other pathogens that are normally easily defeated.

Key Terms

whole blood
plasma
serum
erythrocyte
platelet
leukocyte
stem cell
hemoglobin
megakaryocyte
platelet plug
vasoconstriction
thromboplastin
prothrombin
thrombin
fibrinogen
fibrin
blood clot
clot retraction
granulocyte
lymphocyte
monocyte
immune system
pathogen
lysozyme
complement system
Natural Killer (NK) cell
phagocyte
neutrophil
eosinophil
macrophage
margination
diapedesis
chemotaxis
B cell
T cell
thymus gland
perforin
Major Histocompatibility (MHC) complex
Human Leukocyte Antigen (HLA) complex
clone
glycoprotein
immunoglobulin
antigen
T helper (T_H) cell
antigen presenting cell (APC)
plasma cell
killer T (T_C) cell
suppressor T (T_S) cell
memory cell
cytokine
lymphokine
interleukin
tumor necrosis factor
interferon
growth factor
vaccination
toxin
passive immunity
active immunity
antibiotic
allergy
allergen
basophil
mast cell
histamine
antihistamine
hay fever
asthma
anaphylactic shock
contact dermatitis
Langerhans cell
autoimmunity
glomerulonephritis
rheumatic fever
myasthenia gravis
lupus
ABO system
Rh system
Severe Combined Immunity Disorder (SCID)
Acquired Immune Deficiency Syndrome (AIDS)
AZT

Box Terms

anemia
spleen
iron-deficiency anemia
pernicious anemia
aplastic anemia
sickle-cell anemia
leukemia
chronic leukemia
acute leukemia
hemophilia
anticoagulants
edema
elephantiasis
kwashiorkor

Questions

1. If erythrocytes have no nucleus, how do they divide? That is, what is the source of new red blood cells?
2. Describe the general steps involved in blood clotting. Why is clotting a problem in people with hemophilia?
3. Inflammation is often painful. Why, then, is inflammation beneficial when we have an infection?
4. What is the complement system? What does it have to do with immunity and body defenses?
5. Do NK cells and neutrophils attack their targets in the same manner? If not, how do they differ?
6. How do attack cells find sites of infections?
7. How do B cells and T cells identify, and respond to, the presence of foreign cells or pathogens? If their functions broadly overlap, why do we have both B and T cells? Wouldn't one type of cell be enough?
8. Briefly describe antibodies. How do they function?
9. What are cytokines? What role do they play in the immune system?
10. How do vaccinations work? Are they the same as antibiotics?
11. What is the relationship between allergies and the immune system? What is autoimmunity?
12. Why has AIDS proved to be such a deadly disease? Why can't the immune system deal more successfully with it?

CHAPTER

10

The Respiratory System

Gas Exchange

In 1900 the two leading causes of death in the United States were tuberculosis and pneumonia, both of which are lung diseases. By 1990, tuberculosis was of minimal significance as a cause of death in this country, but emphysema, a lung disease unknown in 1900, had become the tenth leading cause of death. Why has there been such a shift—and why are the lungs so vulnerable a pair of organs?

Every summer a number of swimmers drown after breathing deeply several times before diving into the water. Every winter a number of young children are pulled from icy ponds and lakes after being under water for many minutes—and survive. What's going on here?

We generally think of hiccups as representing nothing more than an occasional nuisance, but in the years before his death, Pope Pius XII (1876–1957) suffered repeatedly from serious attacks of hiccups that lasted for weeks at a time and kept him from sleeping. What are hiccups, and what do they have to do with the respiratory system?

These are some of the questions we shall consider in our discussion of how the respiratory system functions.

THE NATURE OF THE PROBLEM

As we saw in Chapter 5, the cells of all multicellular organisms (and the majority of unicellular organisms) utilize oxygen at the end of the electron transport system as the ultimate acceptor of hydrogen ions and electrons. Without oxygen, the metabolism of glucose ceases at the level of fermentation, and the amount of energy extracted from glucose therefore drops by almost 95 percent. Most organisms (including ourselves) are so dependent on the level of energy produced by cellular respiration that they cannot survive without it for more than a very limited time. That is to say, they cannot survive without oxygen. For human beings, the "very limited time" is generally less than five minutes, after which the level of cell death in the brain and other critical tissues reaches a point where permanent damage and ultimately organismal death ensues.

Given their dependence on oxygen, it is not surprising to learn that multicellular organisms have evolved various strategies for securing a supply sufficient for the needs of their cells. In large measure, the habitat of the organism dictates the solution to the problem, because in order to pass across a cell membrane, oxygen and carbon dioxide (the gases involved in gas exchange) must be dissolved in water. That means that the structure or organ used for gas exchange must be both moist and thin enough to permit rapid diffusion of gases.

Unlike terrestrial organisms, aquatic organisms face no dangers from desiccation of their moist gas exchange structures, and therefore these structures can be external. However, because oxygen dissolves so poorly in water (about 0.5 ml/liter, depending on temperature and salinity, as compared with an abundance of 210 ml/liter in air), acquiring enough oxygen becomes a significant problem. To compensate for the relative scarcity of dissolved oxygen in water, aquatic organisms frequently have gas exchange organs with a very large surface area, and they often take advantage of existing water currents, or create their own currents, in order to maximize the efficiency of gas exchange. (Most fish, for example, alternately open and close their mouths and gill covers to create a current across the gills. Sharks, which do not have gill covers, swim constantly, with their mouths open.)

Terrestrial organisms, on the other hand, have an

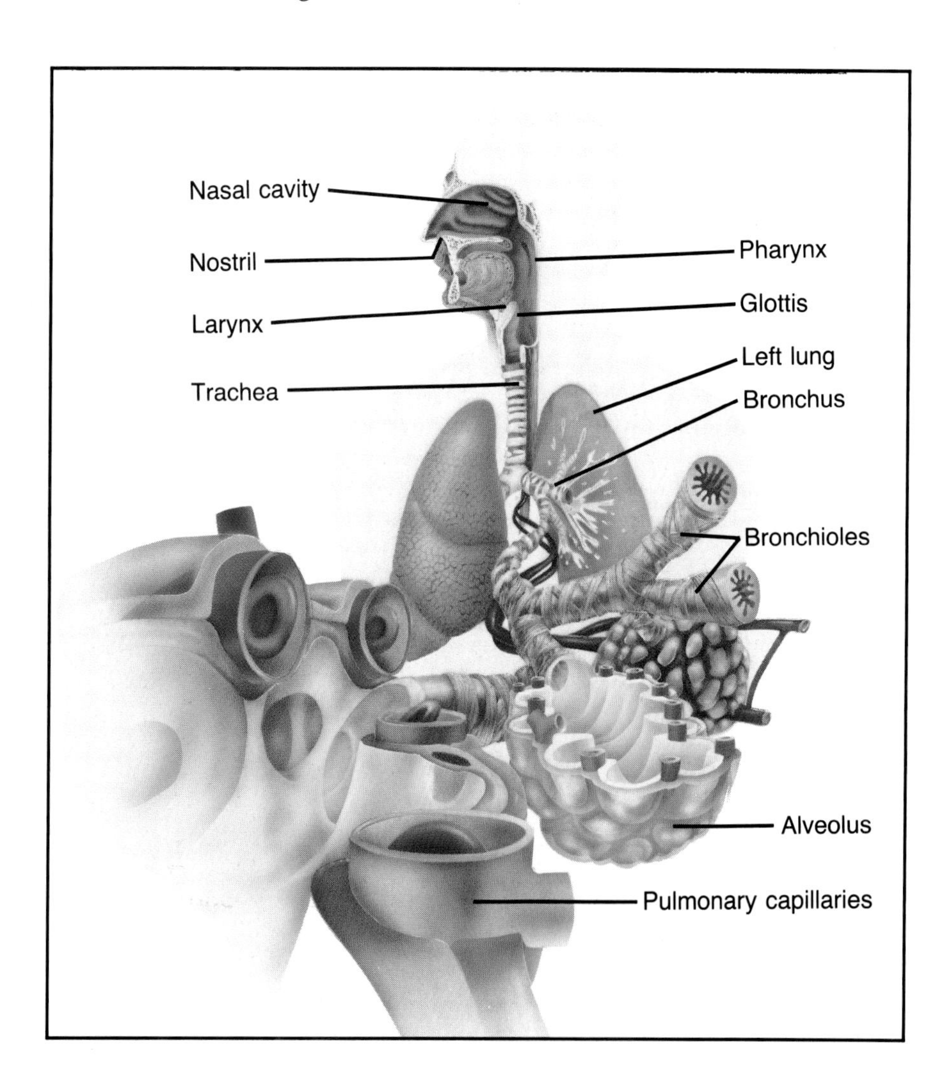

FIGURE 10.1

Human Respiratory System

When the diaphragm contracts, air pressure forces air through the nose or mouth, down the pharynx, trachea, and bronchi, and into the gas exchange areas of the lungs. When the diaphragm relaxes, air is pushed from the lungs, through the bronchi, trachea, and pharynx to the mouth and nose, where it leaves the body.

abundant supply of oxygen but must avoid desiccation. If gas exchange occurs across the moist skin of the body, as happens in organisms such as salamanders and earthworms, then those organisms must live in damp environments, protected from the sun. (That is why you do not see salamanders wandering through the desert, or why earthworms so quickly go to the earthworm farm in the sky when they are stranded on sidewalks after a rain.) Organisms living in drier environments cannot use their skin for gas exchange and must therefore have internal gas exchange surfaces. In humans and other land vertebrates, we call these structures **lungs.**

It is important to recognize at the outset that the lungs are not the sites of oxygen *utilization.* The lungs do not "use" oxygen any more than the digestive system "uses" food. As we saw at the beginning of Part 3, organ systems regulate the cellular environment; they do not, as a rule, take over cellular functions. Thus, in order to understand the functioning of the respiratory system, it is essential to understand that the role of the lungs is limited to supplying the body's delivery system—the blood—with oxygen, and to removing carbon dioxide from the blood. (In theory, the lungs could consist of a network of passages and chambers, providing each cell of the body with oxygen directly. That is essentially the situation in insects, for example. In practice, the lungs of all vertebrates are merely blind sacs, but they are abundantly invested with blood vessels.)

Before looking at the details of gas exchange, however, let us first examine the components of the human respiratory system (see Figure 10.1).

THE AIR PATHWAY

The typical entry points of air into the air pathways are the **nostrils,** which in all terrestrial vertebrates are paired openings separated by a largely cartilaginous **nasal septum.** The nostrils open into paired **nasal cavities,** which extend from the **palate** (a horizontal plate of bone and soft tissue that separates the nasal cavities from the mouth) upward to the floor of the braincase. Within each naval cavity are three bony ridges covered with a mucus-secreting membrane that is richly endowed with blood vessels. Particulate material in the air that gets past the coarse hair of the nostrils is trapped in the mucus, and the cilia that cover the cells sweep the dust-laden mucus to the rear of the nasal cavity where it is swallowed—at the rate of half a liter a day! (Cold viruses or substances causing allergies irritate this tissue to the point where the ciliary system is overwhelmed—and we develop a "runny nose".)

In addition to being cleaned as it passes through the nasal cavities, the air is also warmed and moistened. The efficiency of this process is remarkable. Even very dry air is virtually saturated with water by the time it reaches the rear of the nasal cavities.

Immediately behind and below the nasal cavities is a large chamber called the **pharynx** (Gr., "throat"). Not only do the food and air pathways share the pharynx, they in fact cross over at this point, a situation which provides the opportunity for food to enter the air pathway with potentially deadly consequences. Why are the two pathways not completely separated? The answer is that the hazards of choking are presumably outweighed by the benefits of being able to use the mouth as an alternate entry point to the air pathway. Although air is not cleaned or moistened nearly as well when it enters through the mouth, greater volumes of air are possible (since the mouth is a larger opening), which is why mouth breathing is typical during periods of exertion. Moreover, in the absence of this alternate air pathway, nasal congestion from a cold or allergy might well prove fatal!

As air leaves the pharynx, it passes through an opening, the **glottis** (Gr. *glossa,* "tongue"), into the **larynx** (also called the voice box or Adam's apple). (See Box 10.1). When a person swallows, the larynx rises, causing the glottis to be covered by a flap of tissue called the **epiglottis,** thus preventing food from entering the lungs (see Figure 10.2).

Contained within the larynx are the **vocal cords,** which not only function in speech but also act both as a physical shield against the entry of anything other than air, and as a valve that allows air pressure to build up in the lungs before it is suddenly expelled as a *cough.*

Once past the larynx, the air enters the **trachea** (or windpipe), a tube roughly 2.5 cm in diameter and 15 cm in length. The trachea is reinforced by a series of U-shaped pieces of cartilage, the open ends of which face the esophagus, thus allowing food to pass easily from the mouth to the stomach. (Food would be hung up by the tracheal rings if they completely encircled the tube.) If the upper respiratory tract becomes blocked, the trachea is sometimes cut open in a dramatic surgical procedure known as a **tracheotomy.**

At the end of the trachea, air passes into the **bronchi** (Gr., "windpipe"), one of which passes to each lung. Structurally, the bronchi are like the trachea in that they are lined with cartilaginous rings. However, as the bronchi split into successively smaller branches, the cartilage reinforcements are gradually lost.

The bronchi and their subdivisions—the **bronchioles**—are lined with a mucous membrane covered with cilia. The mucus captures most of the remaining particles in the air and the cilia transport them back up the bronchi to the larynx, where they are coughed out or swallowed.

Clustered like grapes around the end of each of the bronchioles are tiny blind chambers called **alveoli**

BOX 10.1

The Larynx

A classic example of the tendency for biological systems to adapt in particularly resourceful ways is the use by mammals and birds (and a few other vertebrates) of exhaled air to produce sounds used in communication. In humans, these sounds are called *speech*.

Speech is produced by the vibration of the vocal cords as air passes across them. The vocal cords are housed within the larynx, a largely cartilaginous box derived from the gill supports of fish, our distant ancestors.

The vocal cords are normally folded to the side, but they can be engaged if their owner wishes to speak. Within broad limits, we can control the frequency of sounds (that is, produce both high- and low-pitched sounds) by altering the tension on the vocal cords—just as the note produced by a guitar string may vary, depending on how tight or loose the string is.

Overall length of the vocal cords is also important in determining frequency, just as the length of a guitar string determines the pitch of the notes played. By pinching (and thus shortening) a guitar string between your finger and a fret, you can play a higher note. By the same token, because women's vocal cords are shorter, women generally have higher voices than men. Longer vocal cords require a larger larynx, which is why the larynx is more prominent in men (and has hence been named an Adam's apple rather than an Eve's apple).

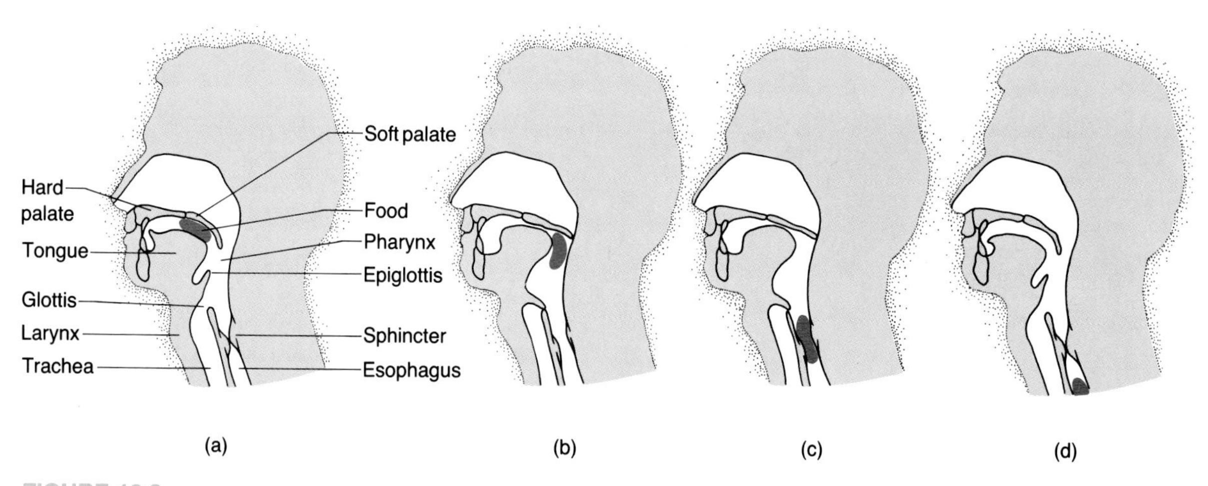

FIGURE 10.2

Swallowing

(a) Normally, the opening from the pharynx to the trachea is open and the opening from the pharynx to the esophagus is closed, since breathing occurs much more frequently than swallowing. (b) When food is swallowed, the tongue rises, forcing the food backward and simultaneously forcing the soft palate up, thus preventing food from moving up into the nasal passages. At the same time, the larynx rises, causing the epiglottis to be deflected over the glottis, which thus prevents food from entering the trachea. (c) The sphincter to the esophagus opens, permitting food to enter the esophagus. (d) As the food moves down the esophagus, the tongue and larynx resume their normal positions, and the epiglottis rises, allowing air once again to enter the trachea through the glottis.

(L. *alveus,* "a cavity"). The lungs contain hundreds of millions of alveoli, which are the actual sites of gas exchange. (The rest of the air pathway serves simply to conduct the air to the alveoli.) Functionally, the alveoli are analogous to the villi of the digestive system in that they vastly increase the surface area of the lungs. Each human lung has a surface area of more than 30 m^2, or roughly the area of a good-sized living room!

So much for the structure of the air pathway. How does it function?

INHALATION AND EXHALATION

The respiratory system of land vertebrates is two-way—that is, air must enter and exit through the same opening. Thus we cannot breathe merely by running about with our mouths open (although, as we saw, a similar strategy works well for sharks which, like all fish, have a one-way system). Nor can we rely on simple diffusion, as can organisms such as earthworms, which breathe through their skin. Instead, land vertebrates employ one of two diametrically-opposed techniques for both **inhalation** (breathing in) and **exhalation** (breathing out).

Frogs use **positive pressure breathing** (the piston approach). During inhalation, air in the mouth is pushed into the lungs as the floor of the mouth contracts. (The nostrils are closed at this point to prevent the air from exiting.) During exhalation, the nostrils open and the abdominal muscles contract, squeezing the lungs. In contrast, all mammals, including ourselves, use **negative pressure breathing** (the bellows approach). The volume of the chest cavity is increased when the chest muscles raise the rib cage and the **diaphragm** (Gr. *dia*, "through" and *phragma*, "a fence")—a horizontal muscle separating the chest cavity from the abdominal organs—contracts and flattens (see Figure 10.3). Together, the actions of these muscles increase the volume of the chest cavity and

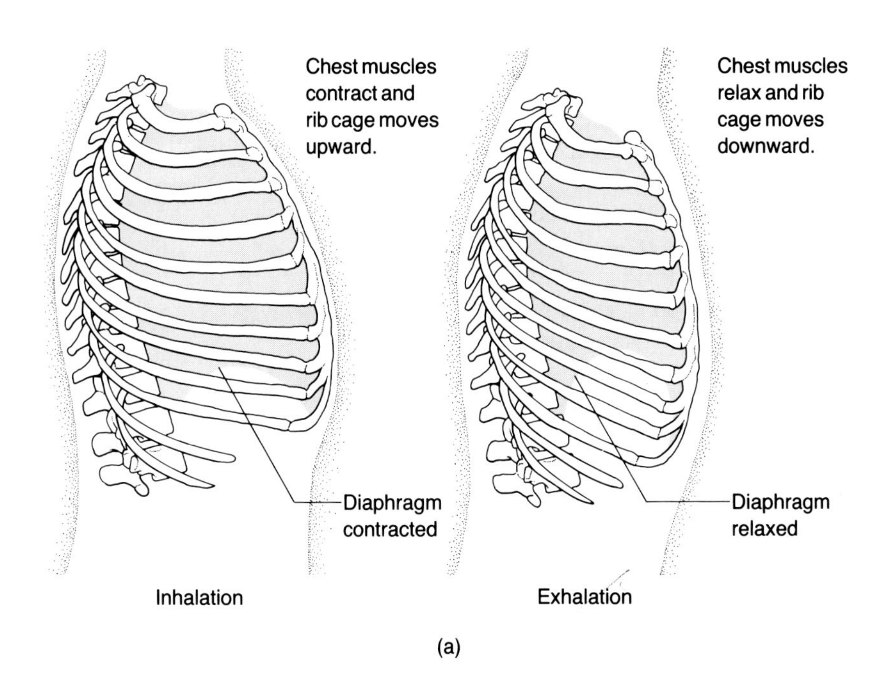

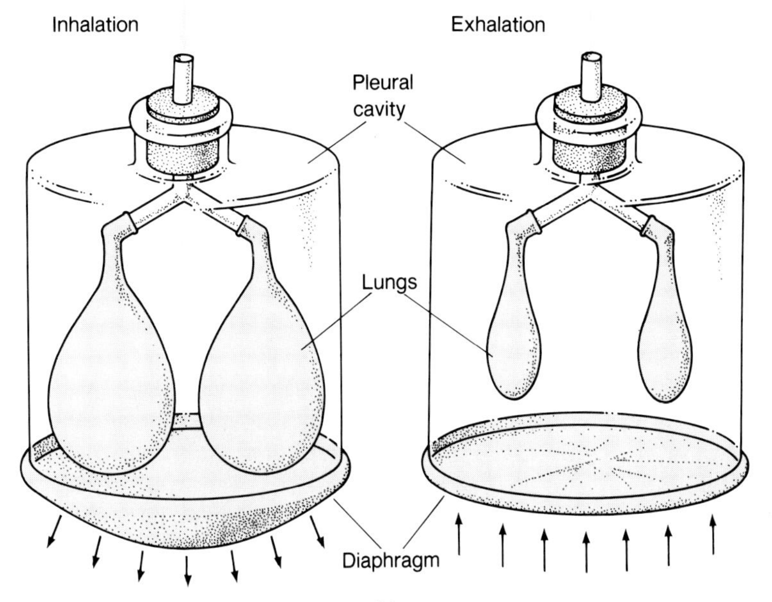

FIGURE 10.3

Inhalation and Exhalation

(a) Diagrammatic representation of inhalation and exhalation, showing changes in the size of the chest cavity because of actions of the chest muscles and diaphragm. (The colored areas represent the relative size of the lungs.) (b) A model showing how inflation of the lungs occurs when the diaphragm is contracted, creating negative pressure in the chest cavity.

reduce the atmospheric pressure in the lungs by 1 to 2 mm of mercury. (Normal atmospheric pressure at sea level is 760 mm of mercury—not much of a vacuum is formed!) However, even this small differential is sufficient to cause about 500 ml of air to rush into the air pathway. Of this amount, about 350 ml enters the lungs; the other 150 ml occupies the upper portions of the air pathway, where no gas exchange occurs.

Exhalation is very similar to squeezing a bellows, except that exhalation is a passive process. The volume of the chest cavity diminishes when the chest muscles and diaphragm are relaxed, and return to their resting positions, and about 500 ml of air is forced out of the air pathway. At rest, we perform the actions of inhalation and exhalation approximately 12 times per minute.

Integral in the proper functioning of the lungs is their containment in the chest cavity. The chest cavity is virtually surrounded by a protective armor composed of ribs and muscles. Covering both the inside of the chest cavity and the outside of the lungs is a shiny, slippery tissue called the **pleura** (Gr., "rib" or "side"). As the lungs alternately fill and empty, they are protected from friction and wear by a thin film of lubricating fluid produced by the pleura—much as the wheel bearings of a car are protected by a thin film of grease.

A midline septum divides the chest cavity into right and left halves, with the result that each lung is completely enclosed in its own chamber. Thus a breach in the chest cavity, as from a wound or injury, may cause the collapse of one lung, but the other normally will stay inflated and functional. In short, regarding our lungs, our eggs are in two baskets.

GAS EXCHANGE AND TRANSPORT

Gas exchange occurs between the alveoli of the lungs and the capillaries that surround the alveoli. The walls of the alveoli are extremely thin, and the air in the alveoli is separated from the blood in the capillaries by a distance of only 0.5 μm, which is 1/15 the diameter of a red blood cell, or 1/200 the diameter of a human hair! Oxygen dissolves in the watery film on the surface of the alveoli, and even though the diffusion of dissolved gases is notoriously slow, the distance is so small that oxygen moves quickly from the alveoli to the capillaries.

How much oxygen does the body use? The answer: not nearly as much as is available. At a rate of 12 breaths per minute, and with 350 ml of air (only 21 percent of which is oxygen) entering the lungs with each breath, roughly 800 ml of oxygen ($12 \times 350 \times .21 = 882$) enters the lungs each minute. Of this amount, only about 200 ml is actually picked up by the five liters of blood passing through the lungs every minute; the remaining 600 ml of oxygen passes back out with the exhaled air (see Figure 10.4). Finally, only about 20 percent of the oxygen being transported by the blood is used by the cells, and less than 10 percent of the carbon dioxide in the blood is exhaled through the lungs.

Why is the system so seemingly inefficient? At rest, the cells of the body utilize 200 ml of oxygen per minute and produce 200 ml of carbon dioxide per minute. Therefore these are the volumes that must be exchanged in the lungs. However, since we are not always at rest, it is essential that we have a reserve capacity. Highly active tissues can draw on the additional 800 ml of oxygen per minute being transported by the blood but not needed by cells at rest. If this utilization occurs, the blood is easily able to accumulate additional oxygen in the lungs from that fraction of the oxygen that is normally exhaled.

A large amount of carbon dioxide stays in the blood because of its importance as a buffer. In any event, the volume of carbon dioxide lost through the lungs is always essentially equal to the volume of oxygen acquired by the blood, irrespective of the level of activity of the body.

Oxygen Transport

As we saw earlier, oxygen dissolves very poorly in water. If oxygen were transported in solution, the oxygen needs of the cells could never be met by the body's relatively small volume of blood. In fact, one liter of blood can carry at most 3 ml of oxygen in solution. Thus in one minute the five liters of blood in the body could transport no more than 15 ml of oxygen in solution to the cells—far short of the 200 ml the cells need even at rest.

Most of the oxygen is transported to the cells by being bound to the blood pigment **hemoglobin**—a complex molecule consisting of a protein portion (*globin*) connected to four nonprotein rings (*heme*), each of which contains an ion of iron. One molecule of oxygen combines with each ion of iron; thus, each hemoglobin molecule has the potential for transporting four molecules of oxygen.

The acquisition and release of oxygen by hemoglobin is a function of three variables: (1) partial pressure of oxygen, (2) blood pH, and (3) temperature (see Figure 10.5).

Partial Pressure. The concept of *partial pressure* refers to the contribution a given gas makes to the pressure of the atmosphere or of any other environment. For example, at sea level the pressure of the atmosphere is 760 mm of mercury. Since the air is 21

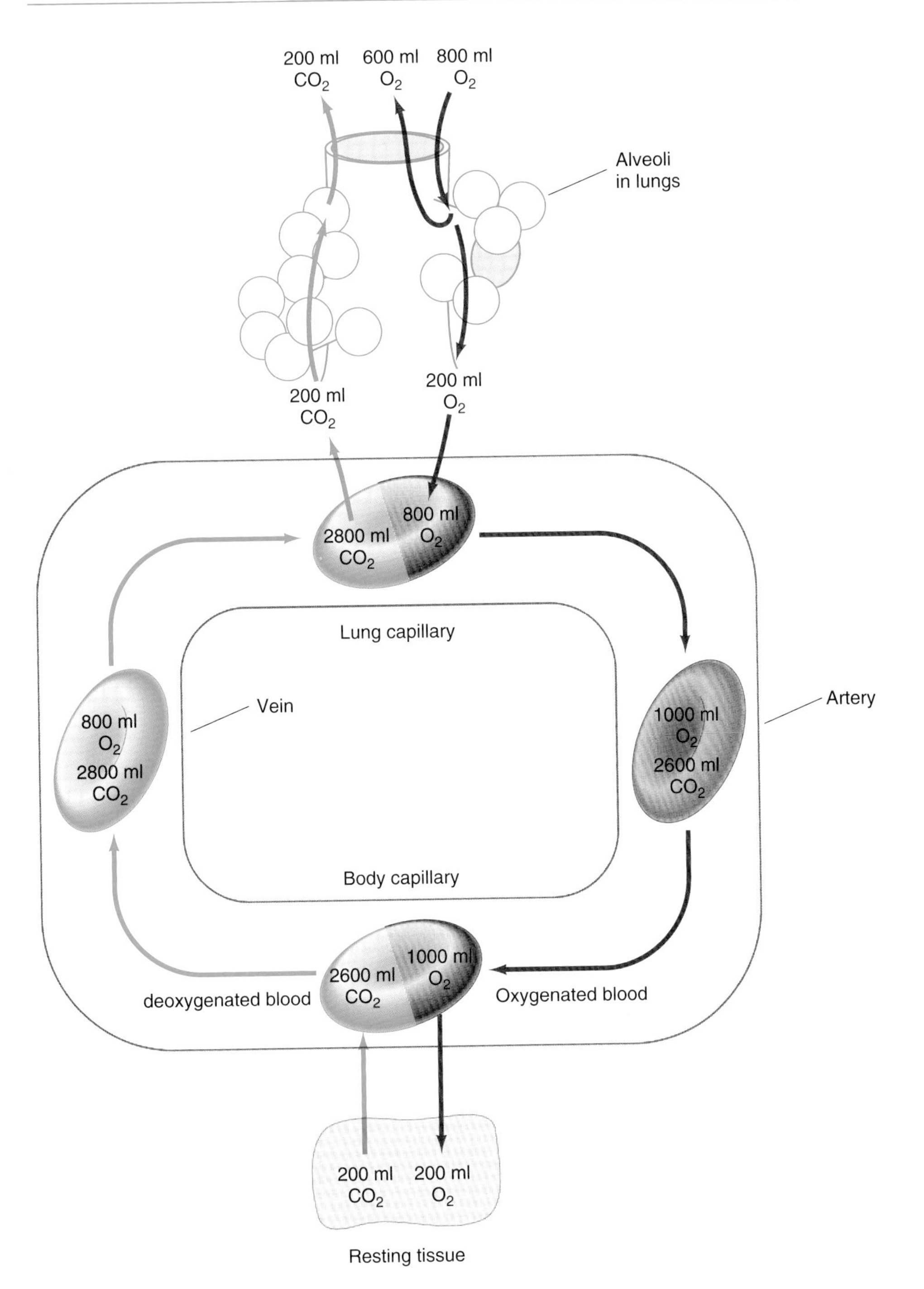

FIGURE 10.4

Gas Exchange Between Atmosphere and Tissues

Each minute 800 ml of oxygen and 5 liters of blood enter the lungs—but the blood acquires only 200 ml of the available oxygen. Because the 5 liters of blood entering the lungs already have 800 ml of oxygen, adding 200 ml in addition means that the 5 liters of blood leaving the lungs each minute are transporting 1000 ml of oxygen to the cells of the body. At rest, the tissues use only about 200 ml, which accounts for the presence of 800 ml still in the blood when it returns to the lungs. Resting tissues produce about 200 ml of carbon dioxide each minute, an amount added to the 2600 ml of carbon dioxide normally borne by our 5 liters of blood. Of the new total of 2800 ml of carbon dioxide arriving at the lungs each minute, only about 200 ml are given up for discharge into the air.

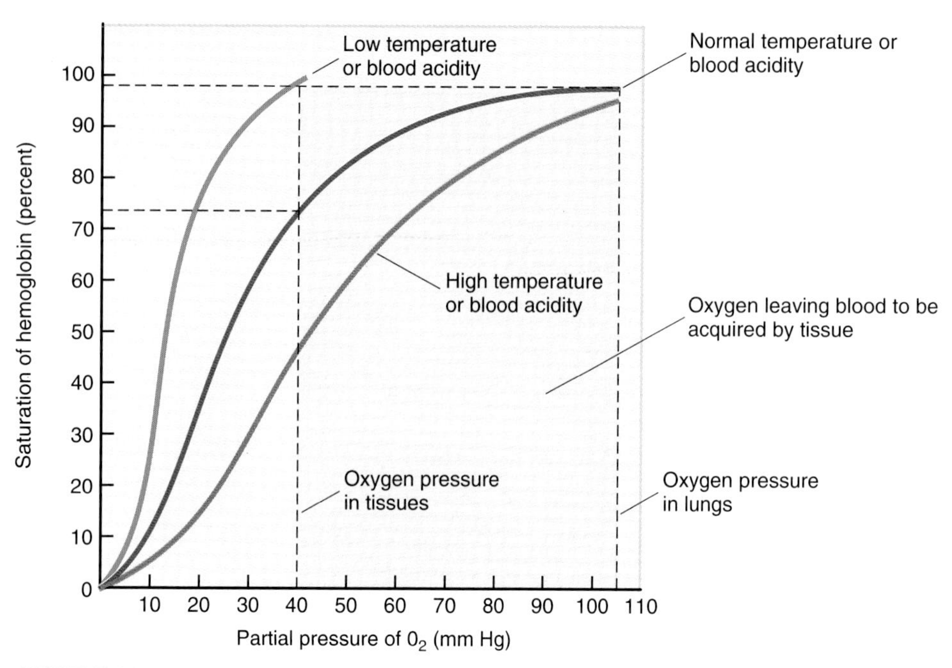

FIGURE 10.5

Dissociation Curves of Hemoglobin
The middle line shows the curve at normal temperature and pH values. The area between the broken lines represents the oxygen that dissociates from hemoglobin and enters the tissues. In tissues that are metabolically active because of acid conditions or high temperature, the dissociation curve moves to the right, meaning that a larger fraction of the oxygen will dissociate from the hemoglobin and enter the tissues. Conversely, in metabolically inactive tissues, the dissociation curve moves to the left, meaning that less oxygen will dissociate from hemoglobin and enter the tissues.

percent oxygen, the partial pressure of oxygen at sea level is 160 mm of mercury (.21 × 760 = 160). (See Box 10.2).

Because of the mixing of "new" and "old" air in the lungs, the partial pressure of oxygen in the alveolar air is only about 105 mm of mercury. However, the uptake of oxygen by hemoglobin is not strictly linear; instead it forms a lazy S-shaped *dissociation curve*. Because of the shape of the curve, about 90 percent of the hemoglobin is saturated at partial pressures of oxygen less than 60 mm of mercury; the remaining 10 percent of saturation is accomplished only gradually, even at relatively high partial pressures. Thus, at 105 mm of mercury, the partial pressure of oxygen in alveolar air is sufficient to ensure more than a 90 percent saturation of hemoglobin.

How does gas exchange take place between the blood and the cells of the body? The partial pressure of oxygen in the cells is only about 40 mm of mercury—an amount sufficient to maintain only about 75 percent hemoglobin saturation. Thus, as the blood passes by the cells, about 15 percent of the oxygen being transported by the hemoglobin is released and diffuses to the cells. Under resting conditions, however, most of the oxygen remains bound to the hemoglobin—a fact that accounts for the observation that blood returning to the lungs is still relatively rich in oxygen.

Blood pH. The oxygen/hemoglobin dissociation curve is steep at high pH (alkaline conditions) and shallow at low pH (acidic conditions). A shallower curve means that more oxygen is given up by the blood at a particular partial pressure value. Cells that are metabolizing rapidly, such as muscle cells during exercise, produce relatively large amounts of carbon dioxide that, in solution, produces a weak acid. As blood enters this zone of acidity it gives up more of its oxygen, which is exactly what these cells need in order to continue their rapid metabolic rate.

Temperature. The effect of temperature on the acquisition and release of oxygen by hemoglobin is similar to the effect of blood acidity, and for the same

BOX 10.2

Life at High Altitudes

For the 1968 Olympic Games in Mexico City, most athletes either arrived earlier than usual, or undertook weeks of training at high altitudes. Mexico City is more than 2000 meters above sea level—which means that if athletes in most events came directly from sea level to such a high altitude, they would have been at a significant disadvantage. Why?

Atmospheric pressure at sea level is 760 mm of mercury. At higher altitudes, the atmosphere is thinner and consequently weighs less. Therefore the partial pressure of oxygen is less as well, and oxygen uptake by hemoglobin is impaired.

After just a few weeks at high altitude, the body adjusts by manufacturing more red blood cells than usual. Even though oxygen uptake in the lungs may be less efficient, the total volume of oxygen being transported by the blood remains roughly constant because the body has more red blood cells to aid in transport. Thus the Mexico City olympians needed an extended period of training at high altitudes to give their bodies time to increase the production of red blood cells.

Does the reverse hold true? Is an athlete who has trained at high altitudes at an advantage when competing at sea level with individuals who have not had high altitude training? The evidence is mixed, but many athletes believe it to be the case. Some have even eliminated the middle step by simply having a liter or so of their own blood (which they had placed in storage several weeks earlier, permitting the body time to manufacture replacement cells) transfused back into their bodies just before an athletic event. This process, called **blood doping,** is now prohibited by most athletic associations.

Of course, the body's capacity to adjust to high altitudes is not infinite. There are few permanent settlements anywhere in the world situated above 4000 meters, simply because the body has to labor too hard just to obtain sufficient oxygen at such altitudes. Mountain climbers going past 6000 meters almost always wear oxygen masks because oxygen deficiency quickly leads to mental confusion—clearly a dangerous condition for a mountain climber.

reason. Elevated temperatures are a consequence of increased metabolic rate. Because dissociation increases at elevated temperatures, more oxygen will be available to supply the needs of metabolically active cells.

Carbon Dioxide Transport

Carbon dioxide is substantially more soluble in water than is oxygen, and its transport by blood reflects this fact. However, less than 10 percent of the carbon dioxide is transported as dissolved gas; the overwhelming portion (about 70 percent) is transported as ions, as shown in the following reversible reaction:

$$CO_2 + H_2O \leftrightarrow \underset{\text{(Carbonic acid)}}{H_2CO_3} \leftrightarrow \underset{\text{(Biocarbonate ion)}}{HCO_3{}_-} + H+$$

The conversion of molecular carbon dioxide to bicarbonate ions permits a diffusion gradient to be maintained between the cells and the blood, and ensures that most of the carbon dioxide produced as cellular waste is taken up by the blood.

The remainder of the carbon dioxide (roughly 20 percent) is carried bound to hemoglobin, though not bound at the same site as oxygen; thus, a molecule of hemoglobin can transport both oxygen and carbon dioxide at the same time.

The partial pressure of carbon dioxide in the air is very low (less than 1 mm of mercury). Thus, when the blood returns to the lungs, molecular carbon dioxide is readily given off, to be exhaled. The loss of the molecular carbon dioxide also shifts the equilibrium of the reaction given above from right to left, meaning that some of the hydrogen and bicarbonate ions combine to form carbonic acid, and some of the carbonic acid breaks down into water and carbon dioxide gas. As a consequence of the decreased number of hydrogen ions, the blood becomes less acidic (that is, the pH rises).

As the blood circulates through the body, it picks up carbon dioxide from the cells. As carbon dioxide accumulates, the equation gradually shifts from left to right: Some of the carbon dioxide combines with water to form carbonic acid, and some of the carbonic acid breaks down into bicarbonate and hydrogen ions. As a consequence of the increased number of hydrogen ions, the blood becomes more acidic (that is, the pH falls).

In summary, the pH of the blood rises as a function of the rate of exhalation of carbon dioxide by the lungs, and falls as a function of the rate of production of carbon dioxide by the cells. However, because maintaining a stable blood pH is vital to the welfare of the body, it stands to reason that these two variables—exhalation of carbon dioxide and metabolic production of carbon dioxide—must be tightly linked. With that as a premise, consider how the rate of gas exchange is regulated in the body.

REGULATION OF GAS EXCHANGE

As we have seen, when a person is at rest roughly 500 ml of air are exchanged with each breath. This volume is called the **tidal volume,** and it represents just a fraction of total lung capacity (see Figure 10.6). Together, the lungs have a capacity of almost 6000 ml, but that amount can never be inhaled or exhaled at one time because a **residual volume** of about 1500 ml remains in the lungs and prevents them from collapsing. The 4500 ml that potentially can be exchanged with maximum effort is called the **vital capacity.** By varying the volume of air inhaled between the extremes of 500 ml and 4500 ml per breath, and by varying the rate of breathing, the amount of gas actually exchanged can also vary widely.

Consider the following example. At a resting level of 500 ml per breath (150 ml of which occupies the nasal cavities and pharynx and is therefore not involved in gas exchange in the alveoli), and a breathing rate of 12 breaths per minute, 4000 ml of air enter and leave the lungs every minute. By contrast, at maximum capacity (4500 − 150 = 4350 ml/breath) and with a breathing rate that can reach 50 breaths per minute, more than 200,000 ml of air theoretically could enter and leave the lungs each minute. As a practical matter, this theoretical limit can rarely be maintained for more than a few seconds (because of the problems of exhaling and inhaling 4500 ml of air every second), but the example illustrates the capacity of the respiratory system to adjust to changing demands.

How does the body initiate a change in the rate of gas exchange? Increased physical activity requires the cells to produce additional energy. The cells need more oxygen to produce this energy; consequently, the amount of oxygen in the blood drops, and the carbon dioxide level in the blood rises. Thus the initial impact of increased activity is a change in both blood oxygen and blood carbon dioxide levels.

Theoretically, the body could use changes in the blood levels of either gas as a cue to alter breathing rates. In fact, the body does monitor both gases, but changes in carbon dioxide levels are apparently more important.

Where are the gases in the blood monitored? Receptors for blood oxygen and carbon dioxide are located in the aorta and in the **carotid arteries,** which carry blood to the head (see Figure 10.7). The brain stem also monitors carbon dioxide levels and, through nerves running to the diaphragm and the muscles of the rib cage, causes changes in breathing rate. Thus high levels of carbon dioxide in the blood ultimately cause an increased breathing rate. The increased rate causes more carbon dioxide to be passed off through the lungs, and simultaneously more oxygen is taken up by the blood. As the blood carbon dioxide level gradually falls to normal values, the breathing rate also declines (but see Box 10.3).

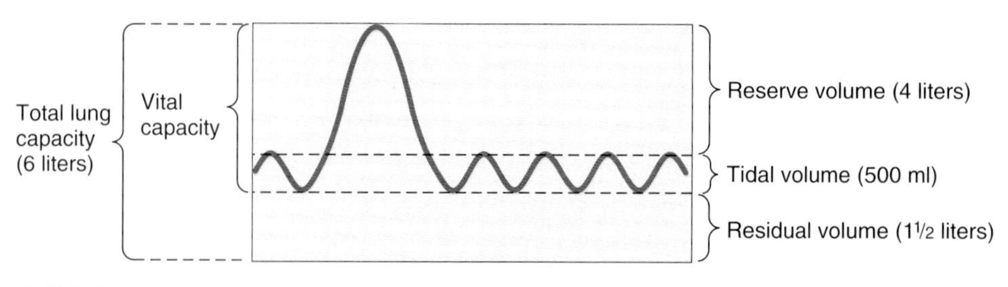

FIGURE 10.6

Lung Capacity

Total lung capacity in adult males may exceed six liters, but in normal breathing only about 0.5 liters are exchanged with each breath (tidal volume). Even with maximal effort, only about 4.5 liters can be exchanged (the vital capacity), because 1.5 liters will remain to keep the lungs inflated (residual volume).

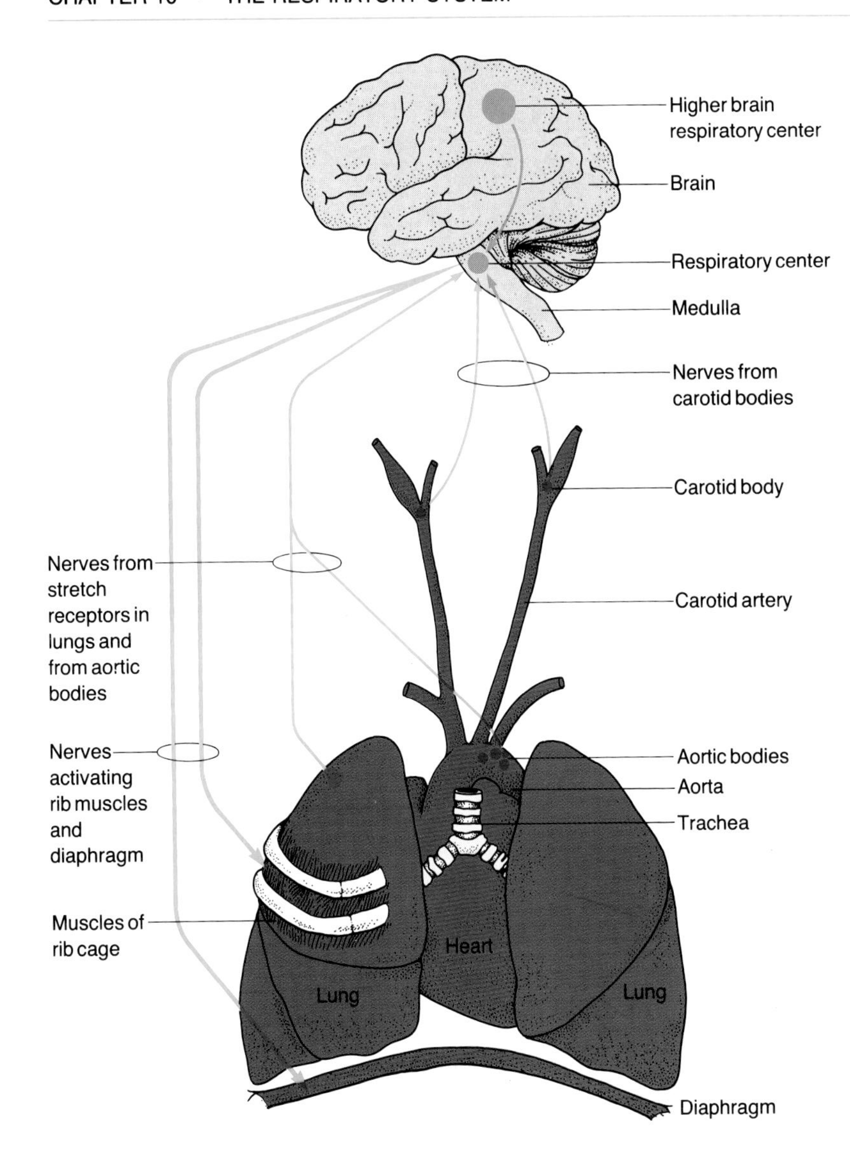

FIGURE 10.7

Breathing Control System

Breathing rates are regulated by a respiratory center in the medulla of the brain, but they can be influenced by conscious control from the cerebral lobes of the brain. The respiratory center responds directly to changes in hydrogen ion concentration in the blood and to stimuli from nerves running from the carotid and aortic bodies, which measure blood oxygen and carbon dioxide levels. Nerves also run to the respiratory center from stretch receptors in the lungs. These receptors monitor how deeply breathing is occurring. The rate and intensity of breathing is controlled by messages sent from the medulla to the diaphragm and the muscles of the rib cage.

BOX 10.3

Hyperventilation and Hypoventilation

Changes in the rate and in the volume of our breathing occur as an automatic response to changes in the blood concentrations of oxygen and carbon dioxide. However, we also have considerable conscious control over our breathing patterns. By exercising this control, we can alter blood gas concentrations—sometimes with unfortunate results.

Hyperventilation is a voluntary increase in breathing rate, as opposed to one which occurs subconsciously in response to changes in blood gas concentrations. Before competing, athletes frequently hyperventilate to boost blood oxygen levels, a vain effort since the blood is already saturated with oxygen at normal breathing levels. Hyperventilation does cause more carbon dioxide to be exhaled, however, and as a consequence the pH of the blood rises. The respiratory center in the medulla is therefore "tricked" into dramatically slowing the breathing rate. The consequence may be temporary unconsciousness—a dangerous consequence if the athlete happens to be a swimmer. Every year a number of swimmers drown because they hyperventilated excessively prior to swimming underwater.

(Continued on the following page)

BOX 10.3 (Continued)

The reverse of hyperventilation—**hypoventilation** (holding one's breath)—is much less dangerous. Little children may achieve an impressive level of purple coloration in the midst of a full-blown temper tantrum, but they can never do serious injury to themselves by holding their breath. Even if hypoventilation were to cause a momentary loss of consciousness, the brainstem, free of the conscious override, would reinitiate breathing immediately.

An interesting variant of hypoventilation occurs when people—especially young children—fall into icy water. Almost every winter we read of some child who has plunged into an icy pond or lake, is pulled from the water many minutes later showing no vital signs. The next day the child is sitting up in a hospital bed, asking for a second dish of chocolate ice cream. What is going on here—and why is this recovery phenomenon so closely linked with winter conditions, rather than with summer swimming pool mishaps; which so often end tragically?

The answer is that the "dive reflex" is at work. Cold water on the face causes the heartbeat to drop and the blood to be preferentially shunted to the brain and heart, and away from the muscles. As the body temperature drops—which happens quickly in a child in icy water—the heart rate falls even more, to about two beats per minute at a body temperature of 32°C. Below 25°C, the heart stops and death is imminent—but at intervening temperatures, life continues, albeit weakly. The greatly reduced metabolic rate of individuals in this situation often permits them to survive without oxygen for 20 minutes or more, well beyond the time normally thought possible.

The lesson is that resuscitation of drowning victims—especially of children pulled from very cold water—should be attempted even when no life signs are present. Many authorities believe that a significant fraction of the 8000 deaths by drowning each year in this country (most of which are children) could be prevented if these actions were taken.

RESPIRATORY PROBLEMS AND DISEASES

Because the respiratory system is constantly exposed to the atmosphere, and because the warm, moist conditions of the system are ideal for many pathogenic organisms, the respiratory system is especially prone to disease.

Upper Respiratory Diseases. Viral diseases such as *colds* and *influenza* ("flu") are the most common infectious diseases. Colds are caused by any of more than 100 closely related (and apparently constantly mutating) viruses. In stable populations people quickly develop immunity to the local viruses, but changes in locale or in the composition of the population (for example, at the start of a new school year) bring exposure to different viral strains, and colds are frequently the result.

Influenza is typically a more serious disease and is frequently accompanied by fever. Again, the viruses are quick to mutate; thus last year's flu shot is generally ineffective against this year's flu.

Strep throat is a particularly severe sore throat caused by streptococcal bacteria. The toxins produced by these bacteria may invade various parts of the body and cause rheumatic fever and possible heart valve damage or inflammation of the kidneys. Fortunately, strep throat usually responds well to a variety of antibiotics if its diagnosis is timely (although development of particularly virulent and resistant strains in recent years—such as the one that killed Jim Henson, creator of Sesame Street and of the Muppets—is causing great concern in the medical community).

Sneezing and *coughing* are not diseases but symptoms. Both result from irritation of the mucous membranes of the respiratory tract, although the cause may vary from dust or pepper to excess mucus produced by cells infected with cold viruses. Sneezing and coughing both involve an explosive release of air, which generally transports the offending materials up and out of the tract.

Bronchitis. *Bronchitis* is an inflammation of the bronchi or the bronchioles. It frequently becomes a chronic condition in smokers and in individuals who have suffered repeated respiratory infections. Its presence increases the likelihood of additional respiratory problems, including pneumonia and emphysema.

Asthma. *Asthma* (Gr., "to breath hard") has a variety of causes, with allergic reactions (see Chapter 9) blamed for 50 percent of adult cases and 90 percent of children's cases. Proportionately, about twice as many children (especially black urban boys) contract the disease as adults, but half the children show diminishment of their symptoms as they mature. There is also a hereditary component to the disease—it tends to run in families.

In asthma, the bronchi becomes swollen, and the muscles lining the smaller bronchioles contract in spasms. As a consequence, exhalation becomes difficult, and the individual begins to wheeze. Fortunately, modern medications (if they are on hand during an attack) generally provide almost instant relief. However, asthma is not a trivial ailment, and it seems to be becoming more severe. The death rate from asthma doubled during the 1980s (to about 4000 per year), and hospitalizations rose 67 percent for adults and 225 percent for children during that same decade. Most authorities blame increasing air pollution for these grim statistics.

Moderate exercise—especially swimming, with the attendant warm, moist air—is generally beneficial to asthmatics (contrary to the belief that they should avoid exertion at all costs).

Hiccups. *Hiccups* occur because of spasms of the diaphragm that cause a sudden burst of air to rush into the larynx. As that happens, the glottis closes, and the impeded air movement produces the "hiccup." Probably no human ailment has occasioned more bizarre "remedies." One remedy that generally works well, however, is to swallow a teaspoon of granulated sugar. The mechanical effect of the granulated sugar moving down the esophagus apparently interferes momentarily with the impulses to the diaphragm being carried by the *phrenic nerve,* which lies immediately behind the esophagus. Once the train of impulses has been broken, the diaphragm resumes its normal activity, and the hiccups are just a memory.

Pleurisy. Sometimes the pleural membranes lining the chest cavity and the outer surface of the lungs become inflamed, and breathing becomes difficult and painful as the sore surfaces rub against each other. This condition is called *pleurisy,* and although it generally responds well to antibiotics, occasionally the pleural membranes fuse permanently to the wall of the chest as *adhesions.*

Pneumothorax. *Pneumothorax* is the medical term for a collapsed lung. This condition most commonly results from a chest injury that breaches the outer wall and permits air to enter the chest cavity, but it may also result from the internal rupture of the lung itself or from diseases such as pneumonia or pleurisy. There is increasing evidence that a variety of air pollutants may greatly increase the frequency of pneumothorax. Fortunately, the lung generally heals rapidly, and typically reinflates itself without specialized medical procedures.

Hypoxia. *Hypoxia* is the failure of sufficient oxygen to reach the cells of the body and has a variety of causes, including choking. One common cause of hypoxia is *carbon monoxide poisoning.* Carbon monoxide (CO) results from the burning of carbon compounds in a low-oxygen environment. The most common source of carbon monoxide is the internal combustion engine. If a car's exhaust system has no leaks, or if the car is moving, causing the carbon monoxide to be blown away, there is no danger. However, keeping a car engine running in a closed space or running an engine with a leaky exhaust system permits a buildup of carbon monoxide that can be fatal even at low levels (for example, three hours' exposure to an atmosphere with 0.07 percent CO).

Carbon monoxide competes with oxygen for the same binding site on hemoglobin, but it binds much more tightly than does oxygen. Gradually, most of the binding sites are taken over by carbon monoxide, and the cells of the body begin to die because of oxygen deprivation. Since carbon monoxide is both colorless and odorless, it is a particularly insidious poison.

Emphysema. *Emphysema* (Gr., "inflation") is a serious and disabling disease in which the alveoli gradually break down, thereby sharply reducing the area of gas exchange surface in the lungs. The primary cause of emphysema is cigarette smoking, although air pollutants may also contribute substantially and may cause similar kinds of diseases (such as, *black lung disease* in coal miners, or *asbestosis* in asbestos workers). Unfortunately, emphysema is often not diagnosed until substantial damage has been done to the lung. Once present, the damage cannot be repaired. Victims are chronically short of breath and very limited in physical abilities. Frequently, victims must use an oxygen mask during some or all of their waking moments.

Pneumonia. *Pneumonia* (Gr. *pneumon,* "lung") is an inflammation of the lungs and is caused by bacteria, viruses, fungi, or a variety of chemicals (for example, if a child swallows gasoline, the gasoline may be regurgitated and find its way into the lungs, where it can cause pneumonia). Because of antibiotics, bacterial pneumonia is no longer the scourge it was as recently as 50 years ago. Viral pneumonia is still serious, but for a person in general good health who is receiving adequate care, it is normally not life-threatening.

Tuberculosis. *Tuberculosis* is a bacterial disease that may occur in virtually any part of the body, although the lungs are the most common site. It was formerly referred to as *consumption,* because the body was gradually consumed by the disease. Sufferers literally coughed their lives away. Keats, Chopin, and Thomas Wolfe, along with millions of others who lived (and died) in the nineteenth and early twentieth centuries, were victims of this disease.

Tuberculosis responds well to modern medical treatment if it is diagnosed early, for the disease organism is susceptible to many antibiotics. Because of the highly contagious nature of tuberculosis, chest X-rays were routinely administered to all school-aged children as recently as 25 years ago. Today, the X-ray has been replaced by the tuberculin skin test as a routine screen. Anyone who shows a positive reaction (which indicates exposure to the disease organism) is a logical candidate for an X-ray, but since positive reactions are becoming increasingly rare as the disease itself becomes rare, the need for routine X-rays is in most part of the United States a thing of the past. The disease itself, however, is on the increase, in part because it commonly occurs in conjunction with AIDS. On a worldwide basis, it may be the most serious of all contagious diseases.

Lung Cancer. The evidence has become increasingly stronger that the primary cause of lung cancer is cigarette smoking. Although many smokers never get cancer—and nonsmokers are by no means immune—the risk of lung cancer is almost ten times greater for people who regularly smoke a pack or more a day than for people who have never smoked. Cigarette consumption in the United States now exceeds 600 billion cigarettes a year, and lung cancer deaths approach 100,000 a year. Historically, the increase in cigarette consumption has been mirrored by a steady increase in lung cancer deaths. Because cigarette smoking is also linked to increases in a variety of other diseases—most notably, heart and circulatory diseases—cigarette smokers have twice the overall mortality rate of nonsmokers.

Summary

The term *respiration* has a dual meaning in biology. It refers both to the metabolic utilization of oxygen (cellular respiration) and to the organ system that permits the exchange of gases between the circulatory system and the environment. Respiratory organs are commonly internal in terrestrial organisms, a location that minimizes the dangers of desiccation. They are frequently external in aquatic organisms, because of the relative scarcity of dissolved oxygen in water.

The respiratory system in humans consists of a series of tubes and passages linking the nose and the mouth with the lungs. Breathing is made possible by muscular movements that change the volume of the chest cavity; filling and emptying of the lungs is a consequence of this changing volume.

Almost all oxygen in the blood is transported by being attached to hemoglobin, the pigment of red blood cells. Some carbon dioxide is transported in the same manner, but most is either dissolved in the plasma or transported as bicarbonate ions.

By varying both the volume of air inspired with each breath and the number of breaths per minute, the body can vary the volume of exchanged gases over an enormous range. We can exercise some conscious control over breathing, but our breathing rate is primarily controlled by stimuli from the brain stem in response to the level of blood gases (especially the concentration of hydrogen ions).

The respiratory system is prone to many diseases, in large measure because it provides a warm, moist, and relatively passive environment for pathogenic organisms. In addition, various air pollutants cause a variety of respiratory problems, of which lung cancer and emphysema are the most significant.

Key Terms

lungs
nostrils
nasal septum
nasal cavity
palate
pharynx
glottis
larynx
epiglottis
vocal cords
trachea
tracheotomy
bronchi
bronchioles
alveoli
inhalation
exhalation
positive pressure breathing
negative pressure breathing
diaphragm
pleura
hemoglobin
tidal volume
residual volume
vital capacity
carotid artery
bronchitis
asthma
hiccups
pleurisy
pneumothorax
anoxia
emphysema
pneumonia
tuberculosis
lung cancer

Box Terms

blood doping
hyperventilation
hypoventilation

Questions

1. Explain how air and food channels cross over in the pharynx. How do we normally avoid getting food into our respiratory system?
2. Discuss the mechanics of breathing. What is the role of the rib muscles and the diaphragm?
3. Why is so-called deoxygenated blood still so rich in oxygen? Why is so much of the waste gas, carbon dioxide, retained by the blood rather than being exhaled through the lungs?

4. What is the nature of the association between hemoglobin and oxygen that hemoglobin can acquire oxygen in the lungs but give it up to the tissues?
5. How does the body ensure that additional oxygen will be supplied preferentially to metabolically active tissues?
6. Why can't we increase our lung efficiency by exhaling all the air in our lungs between inhalations?
7. How does the body regulate breathing rates? What factors are monitored?

CHAPTER

11

The Excretory System

Regulating the Internal Sea

It's a July night in a southwestern ranch town, and a bunch of good ol' boys are tossing back a few at the local tavern. Within a short time their digestive systems will absorb the bulk of their intake of beer and add it to the blood. How does the body respond to a dilution of the blood? For that matter, how does the body even know that the blood has been diluted?

Some people salt virtually every morsel of food they eat. Much of this salt is absorbed by the small intestine. What prevents the blood from becoming so salty that it draws water from surrounding tissues by osmosis? How—and where—is excess salt removed from the blood?

If you contract bacterial pneumonia, your doctor may give you a shot of penicillin. Does the penicillin remain in your blood forever? If not, how does the body "recognize" this, or any of the other modern drugs, and remove it?

In 1976 Howard Hughes, along with 50,000 other Americans, died of kidney failure. What are the physiological consequences of kidney failure? Why are these consequences severe enough to cause death?

These questions are considered in our discussion of the excretory system.

THE NATURE OF THE PROBLEM

Imagine that you had to obtain all of your nutritional needs—indeed, everything you needed to survive—from a tank of liquid only half your size. Imagine also that you had to dispose of all of your waste products into this same tank. Leaving the obvious aesthetic concerns aside, you would clearly have to create a very efficient filtering and disposal system to avoid polluting yourself with your own wastes.

As it happens, these are very much the problems actually faced by the cells that collectively constitute your body. As we saw in Chapters 8 and 9, our body's cells depend on the blood and interstitial fluid for their survival. All of the nutrients required by our cells, including oxygen, are delivered by the blood and interstitial fluid, and these two bodies of fluid, which together constitute less than half the volume of fluid of the cells themselves, are also the site into which cellular wastes (primarily carbon dioxide and nitrogenous wastes) are dumped. We have seen, in Chapters 6 and 10, how the digestive and respiratory systems ensure a continuous supply of nutrients and oxygen, and how carbon dioxide is removed. Now let us examine how cellular wastes, and excess water and minerals, among other things, are removed from the blood and interstitial fluid.

FLUID COMPARTMENTS OF THE BODY

Our bodies are about 60 percent water by weight, allowing for some variation from individual to individual based on the relative amount of body fat. About two-thirds of this water is within cells (*intracellular fluid*) and one-third is outside the cells (*extracellular fluid*). The extracellular fluid is further subdivided into various discrete reservoirs, including the fluid inside the eye that maintains the shape of the eyeball, and the cerebrospinal fluid surrounding and cushioning the brain and spinal cord (see Chapter 14). However, in terms of volume, the two major categories of extracellular fluid are the *interstitial fluid* (about 16 percent of body weight) and the *plasma* of the blood (about 4 percent of body weight).

As we saw in Chapter 8, there are two dynamic exchanges constantly underway between major fluid compartments of the body. The first is the exchange between the cells and the interstitial fluid; the second is the exchange between the interstitial fluid and the plasma. The waste products of the cells move from the cells to the interstitial fluid and finally to the plasma. The plasma, in turn, is freed of these wastes by various organs of the body of which the kidneys are the most important. It is remarkable testimony to the design efficiency of the human body that a fluid representing less than 3 liters in volume (the plasma) can receive wastes from the cells (which represent 23 liters of fluid volume) and that the plasma can then be cleansed by a pair of organs—the kidneys—which together weigh only about 300 g.

THREATS TO FLUID HOMEOSTASIS

Fluid homeostasis refers to the need for the extracellular fluid of the body to be maintained in a balanced equilibrium with respect to such variables as water and salt balance, pH, metabolites, and various miscellaneous substances that are purposely or accidentally introduced into the body. For organisms such as jellyfish, which use sea water as their extracellular fluid, the idea of fluid homeostasis is nonsensical, because the sea is, for all intents and purposes, constant in its composition (and certainly no effort by a jellyfish is likely to change that composition). However, for most terrestrial organisms, which rely on a self-contained "sea" (the blood and interstitial fluid), fluid homeostasis is essential for life, because even modest variations from the norm in any of the variables listed above are life-threatening.

Yet it is not enough that we remove the waste products of cells. We must also account for a number of variables: in the quantity and the quality of the food we eat, the amount of acid produced by the metabolism of our food and by our inclination either to exercise (acid-generating) or to sit quietly, the abundance or scarcity of particular salts in our food, and the volume of water that we drink or that we lose by sweating profusely (or not at all). For the most part, adjusting for these variables and thereby maintaining fluid homeostasis is the task of the kidneys. However, before we examine how these small organs are able to perform such seemingly miraculous feats, let us examine how they are put together.

EXCRETORY OR URINARY?

Excretion is the process by which metabolic wastes, including carbon dioxide, nitrogenous wastes, and metabolic water, are separated from the blood and eliminated from the body. Although the kidneys control the elimination of nitrogenous wastes and water, carbon dioxide leaves the body by way of the lungs, organs we generally place in the respiratory system, not the excretory system. Conversely, the kidneys also remove nonmetabolites in addition to the metabolites just mentioned. Thus, in the interests of accuracy, the kidneys and associated structures should really be called the *urinary system* rather than the more traditional *excretory system*.

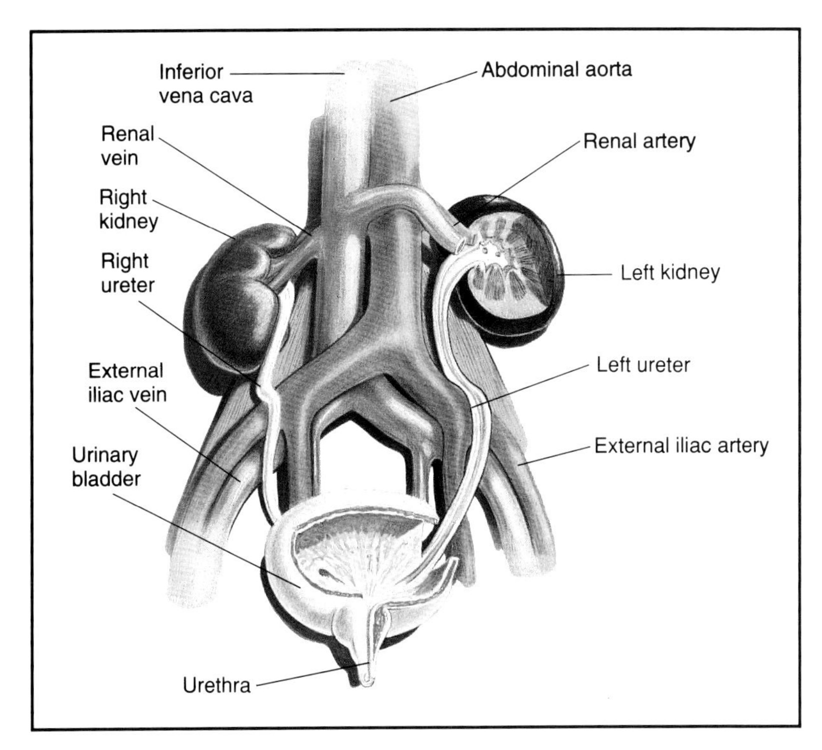

FIGURE 11.1

Human Urinary System
The human urinary system consists of two kidneys, two ureters, a single urinary bladder, and a single urethra.

THE URINARY SYSTEM

The urinary system consists of two **kidneys,** a pair of **ureters**—tubes that lead from each kidney to a single **urinary bladder**—and the **urethra,** a tube leading from the bladder to the outside (see Figure 11.1).

Microanatomy of the Kidney

The kidneys are a pair of bean-shaped organs[1] located on either side of the vertebral column at roughly the level of the last rib. The right kidney is somewhat lower than the left. In cross-section the kidney reveals two distinct layers—an outer **cortex** (L., "bark") and an inner **medulla** (L., "marrow") (see Figure 11.2). Although the kidneys are small organs (about 10 cm by 6 cm), together they contain more than 60 km of microscopic tubes in the form of some 2.5 million 4 cm-long **nephrons** (Gr. *nephros,* "kidney")—the functional units of the kidneys.

The nephron is made up of a number of components. The first is **Bowman's capsule,** a cuplike structure that forms the filtering portion of the nephron. Collectively, the capsules of the 2.5 million nephrons provide the body with about 1 m^2 of filtration surface.

From the capsule, which is located in the cortex, the nephron continues as the *convoluted tubule,* which itself is in three parts: the first portion, adjacent to Bowman's capsule, is called the **proximal convoluted tubule;** the intermediate section, which, in many nephrons, descends down into the medulla and then back up into the cortex in a hairpin-like fashion is called the **loop of Henle** (with **descending** and **ascending arms**); and the final portion is the **distal convoluted tubule.** Ultimately, the convoluted tubules of a series of nephrons fuse to become a **collecting duct,** and the ducts fuse to form the ureter.

Imbedded in each Bowman's capsule, rather like a baseball in a glove, is a knot of arterial capillaries called the **glomerulus** (L., "a small ball or round knot") (see Figure 11.3). (They are called arterial capillaries because no gas exchange occurs here, as it does in typical capillary beds.) The glomeruli supply blood to the capsules; the convoluted tubules, in a complex series of processes, form urine from the filtered blood.

Urine Formation

Three processes, discussed in the following subsections, are involved in transforming blood into urine (see Figure 11.4).

Filtration. The kidneys receive about one-fifth of the blood in the body at any given moment—that is, approximately 1 liter every minute, or roughly 1500 liters each day. Blood enters the glomeruli at relatively high pressure (about 60 mm Hg), but the glomeruli are thick-walled and resistant to rupture. Pores in the glomeruli, which are both large and abundant, permit free movement of all the dissolved materials in the

[1]As an illustration of the difference between a zoologist and a botanist, to a botanist a bean is kidney-shaped.

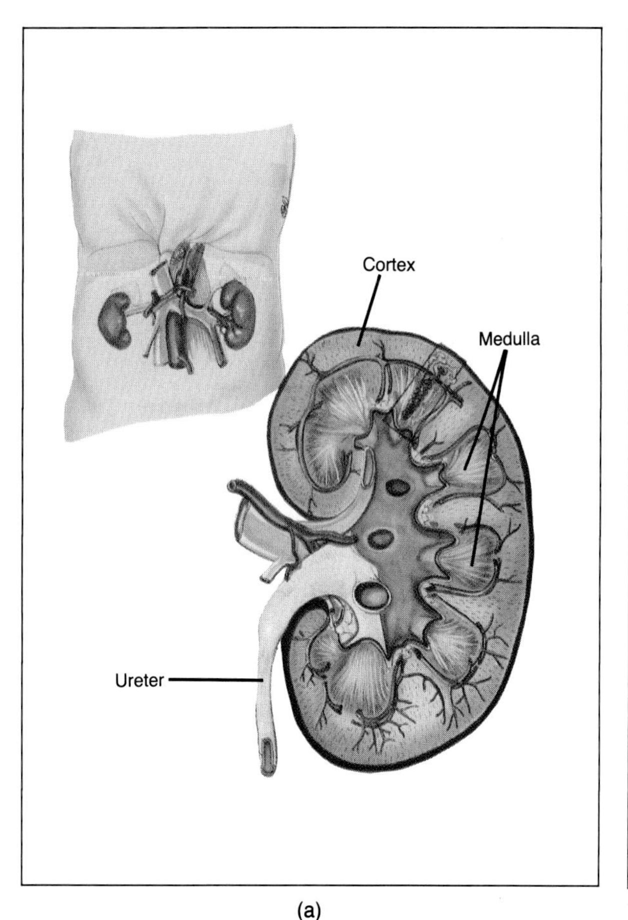

(a)

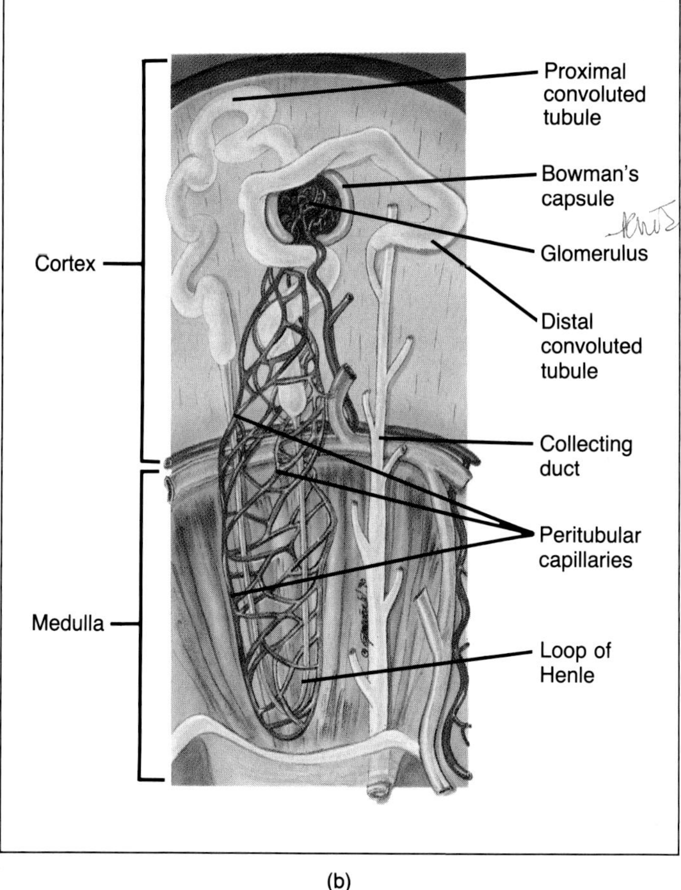

(b)

FIGURE 11.2

Kidney and Nephron

(a) The kidney in longitudinal section. Note the distinct separation between the outer cortex and the inner medulla. (b) A single nephron. About 80 percent of the nephrons in the human kidney are confined to the cortex where they function primarily in the excretion of nitrogenous wastes. The other 20 percent plunge deeply into the medulla. These nephrons are of vital importance in maintaining the appropriate salt/water balance.

blood except the large proteins (see Table 11.1). Thus about 20 percent of the blood plasma (whole blood minus cells) that enters the glomeruli is filtered through the pores and across the wall of Bowman's capsule into the cavity of the nephron. Logically enough, this material is called **glomerular filtrate,** and we produce about 180 liters of filtrate each day.

Reabsorption. Since we do not produce 180 liters of urine each day, it is obvious that not all of the glomerular filtrate becomes urine. Instead, the cells of the nephron recapture essential molecules and ions from the filtrate and return them to the blood, leaving behind nonessential substances (and enough water to keep them in solution) as urine.

Reabsorption is not a perfect process because it is not limitless. If, as happens in diabetes mellitus (see Chapter 13), the amount of glucose in the blood is such that the amount entering the glomerular filtrate exceeds 225 mg per minute (about twice the normal level), the reabsorptive capacity of the tubular cells is exceeded, and glucose begins to be lost in the urine, even though glucose could never be considered a waste product.

Secretion. Secretion is a backup mechanism in which the final constitution of urine can be altered through the active transport of various materials from the cells of the tubule walls into the cavity of the tubules. Most of the small molecules and ions in the plasma enter the tubule during filtration, and subsequently a select few are reabsorbed. However, H^+ and much of the K^+ enter the tubule by secretion, as do such "foreign" chemicals as penicillin and certain other drugs—in part because their large size precludes effective filtration.

(a)

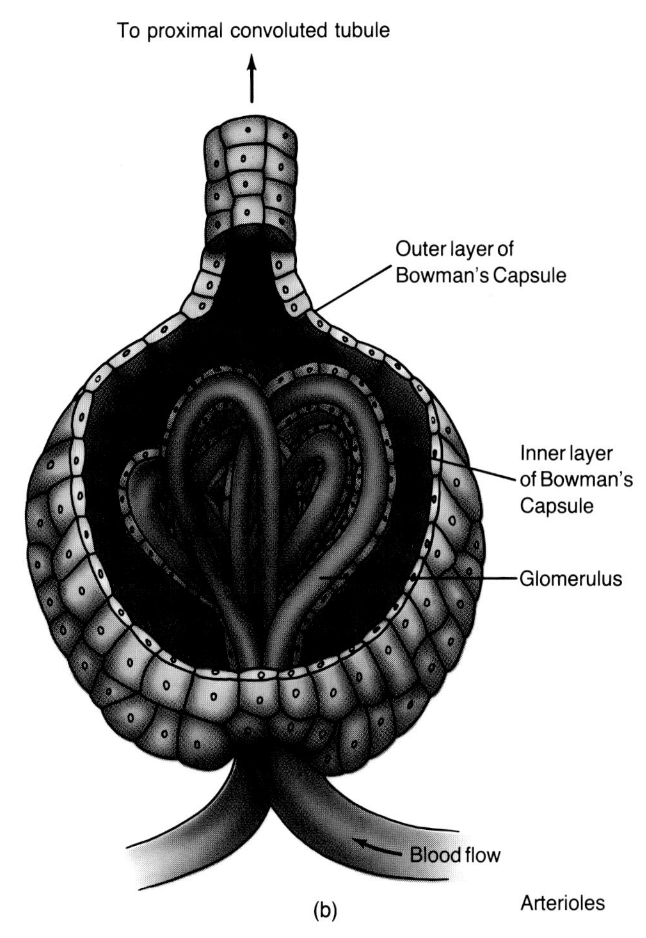

(b)

FIGURE 11.3

Functional Anatomy of the Glomerulus

(a) A scanning electron micrograph of the surface of a glomerulus. Part of Bowman's capsule has been cut away to reveal the glomerulus, but the wall of the capsule is still visible at the periphery of the photograph. (b) Artist's rendering of the relationship between a glomerulus and a capsule. Note that the inner wall of the capsule is pressed tightly against the capillaries of the glomerulus. Substances are therefore simultaneously filtered by the wall of the glomerulus and the inner wall of the capsule and are then in the cavity of the nephron.

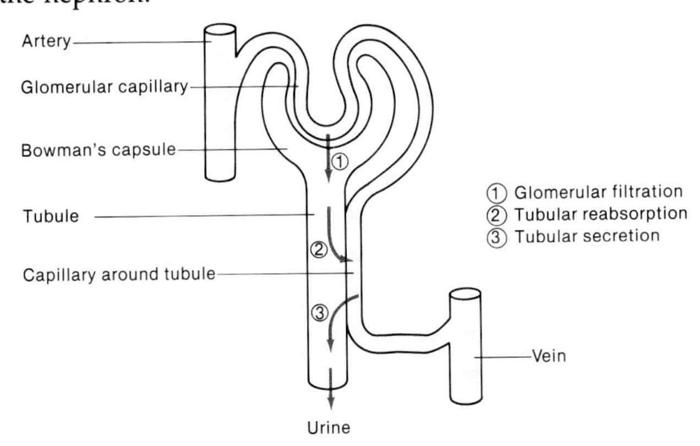

FIGURE 11.4

Physiology of Urine Formation

(1) About 20 percent of the blood plasma passes through the glomerular pores and into the cavity of the nephron as glomerular filtrate. (2) Many ions and molecules are subsequently selectively reabsorbed by the cells of the tubule walls. (3) Other ions and molecules are secreted into the filtrate, primarily by the portions of the nephron located in the cortex. The end result of these activities is urine.

TABLE 11.1 Water Intake and Loss

INTAKE	VOLUME IN ML	LOSS	VOLUME IN ML
Drinking	1400	Diffusion through skin and exhalation	700
Eating	700	Perspiration	100
Production of metabolic water (cellular respiration)	200	Defecation	100
		Urination	1400
Total intake	2300	Total loss	2300

BOX 11.1

The Artificial Kidney

Although the kidneys are complex organs, much of their function is based on the physical processes of filtration and diffusion—processes that are relatively easily duplicated in the laboratory. Thus the artificial kidney was among the earliest of the artificial organs to be developed.

The artificial kidney consists of a long tube of cellophane with pores roughly the diameter of those in the glomerulus. Surrounding the cellophane is a bath warmed to body temperature and containing a fluid of special composition. Blood is passed through the cellophane tube, and small ions and molecules diffuse from the blood to the surrounding fluid if they are more concentrated in the blood than in the fluid. (Diffusion of materials through a semipermeable membrane is called **dialysis** (Gr., "separation"), hence the term *dialyzing fluid* for the surrounding fluid.) In this way waste products such as urea and excess salts can be removed from the blood.

The artificial kidney, although a lifesaver for thousands, is a poor substitute for the real thing. Dialysis is physically and psychologically wearing; it may require as much as six hours a session three times a week and it presents a continuing risk of infection and blood clots. Kidney transplants (see Chapter 9) are therefore a better alternative, but the demand for kidneys continues to outstrip the supply of organ donors. Over 7000 new patients begin dialysis each year in the United States, and there are presently approximately 60,000 people now dependent on dialysis in this country. By comparison, fewer than 10,000 kidney transplants are performed annually in the United States.

Dialysis Machine

Arterial blood is pumped slowly through dialyzing (that is, semipermeable) tubing that is surrounded by dialysis fluid. Small molecules and ions diffuse into the dialyzing fluid and are discarded. The blood is then returned to the body, after first passing through a trap to remove any bubbles or clots that many have formed.

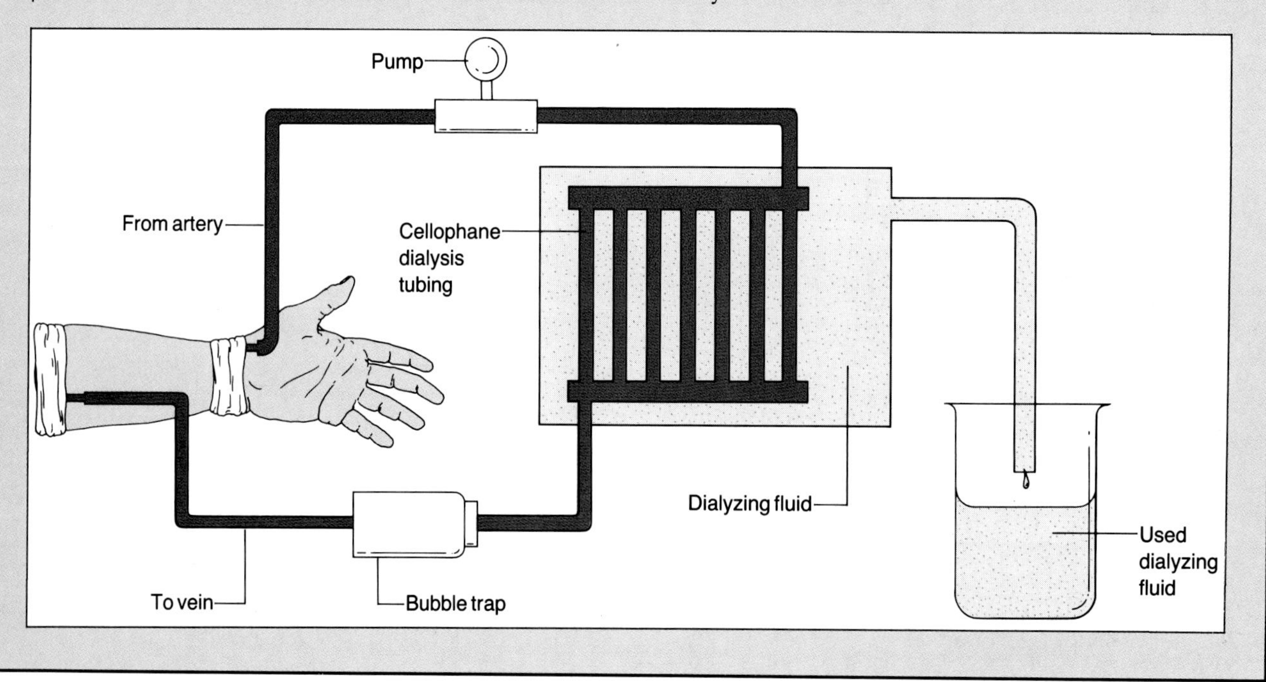

Concentration of the Urine

Now that we have considered the microanatomy of the nephron and the various steps by which urine is produced, let us put anatomy and physiology together to understand how urine is concentrated.

The volume of the filtrate is reduced in each area of the nephron, but the manner by which this reduction is accomplished varies widely from area to area (see Table 11.2). In part these differences result from differences in permeability to water or in the ability of particular cells to engage in active transport. In part they result from the fact that the kidney is comprised of an outer cortex and an inner medulla

TABLE 11.2 Filtration and Excretion Levels of Key Substances

SUBSTANCE	AMOUNT FILTERED PER DAY	AMOUNT EXCRETED PER DAY	PERCENT REABSORBED	CONCENTRATION URINE/ CONCENTRATION IN PLASMA
Water (liters)	180	1.8	99.0	Varies
Sodium (grams)	630	3.2	99.5	0.9
Potassium (grams)	22	2.1	90.1	12
Glucose (grams)	180	0	100	0
Urea (grams)	475	262	54.1	70
Uric acid (grams)	5.5	0.6	88.8	14

(tissues with very different properties), and many of the nephrons are partly within the cortex and partly within the medulla.

Proximal Tubule. The proximal tubule is located entirely in the cortex. The cells that form the proximal tubule are metabolically very active, with high concentrations of mitochondria. Moreover, on the side facing the cavity of the tubule, they are covered with *microvilli* (see Figure 11.5). The presence of the microvilli increases the surface area of the interior of the proximal tubule by a factor of 20. Proteins in the filtrate are removed by *pinocytosis* (see Chapter 5), and glucose, amino acids, and Na^+ are removed by active transport. Cl^- passively follows the active transport of Na^+. As a consequence of all of this reabsorption, water follows passively by osmosis, and the volume of filtrate is more than halved by the time it reaches the loop of Henle (see Figure 11.6). At the same time, K^+, H^+, and ammonia are actively secreted into the filtrate by proximal tubule cells.

Loop of Henle. The loop of Henle passes deeply into the medulla, a tissue in which (for reasons to be explained shortly) the interstitial fluid becomes increasingly rich in urea (the primary nitrogenous waste product, discussed at length later in this chapter) and dissolved ions (especially Na^+ and Cl^-). This very

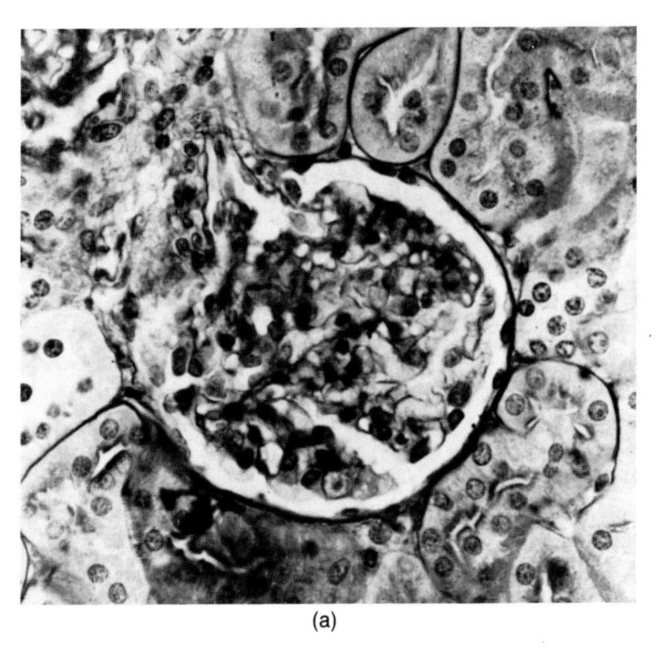

(a)

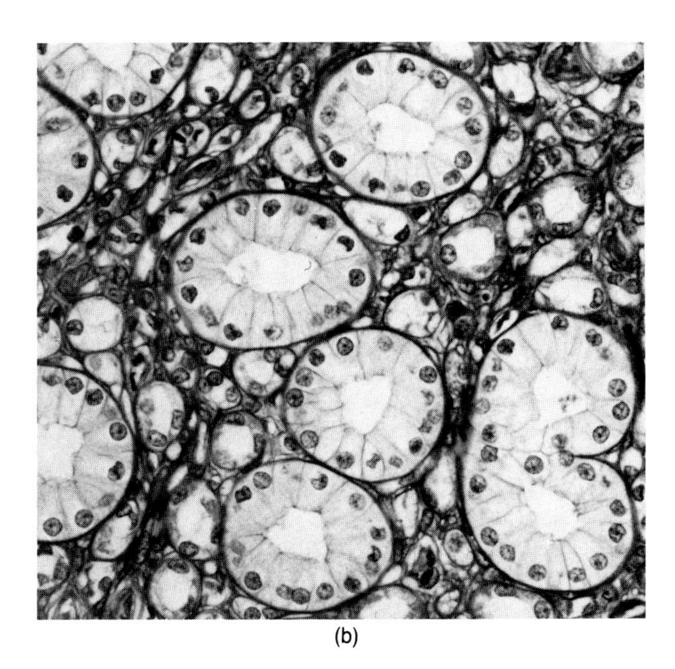

(b)

FIGURE 11.5

Cross Sections of Nephrons

In this scanning electron micrograph the microvilli lining the proximal convoluted tubule and the immediate juxtaposition of the peritubular capillary to the nephron are very evident. (a) Light micrograph of Bowman's capsule and glomerulus; (b) light micrograph of proximal convoluted tubules. Note that the cells have microvilli on their interior surface.

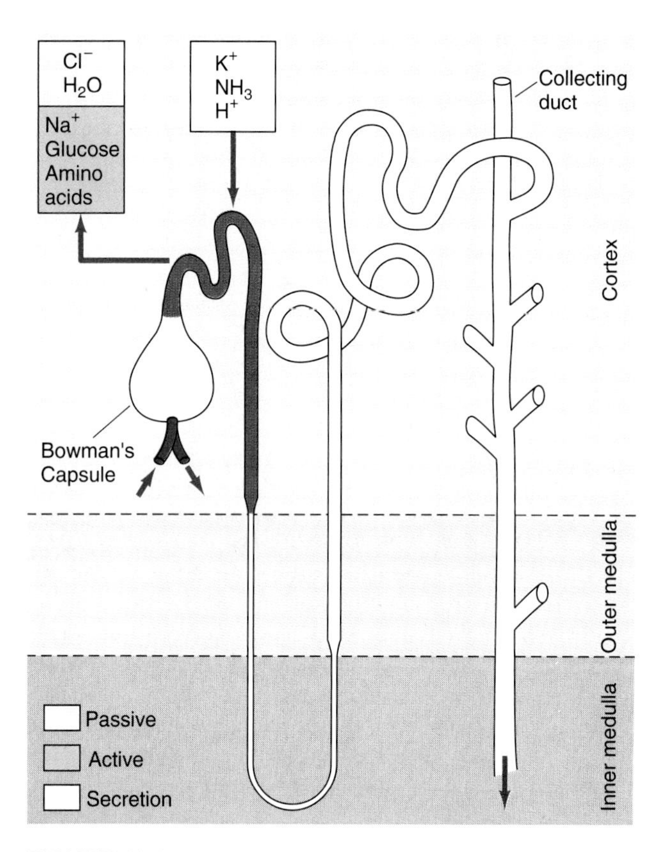

FIGURE 11.6

Events in the Proximal Convoluted Tubule

In the proximal tubule the volume of the filtrate is sharply reduced by the active transport of sodium ion, glucose, and amino acids, and by the passive diffusion of chloride ion and water. In addition, ammonia, hydrogen ion, and potassium ion are secreted into the filtrate.

salty fluid therefore exerts a powerful osmotic pressure on the filtrate. The cells that make up the descending arm of the loop of Henle are thin walled, and highly permeable to water. (However, they are relatively impermeable to urea, Na^+, and Cl^-.) As a consequence, additional amounts of water are drawn out of the filtrate as it moves down the loop, and only half as much filtrate leaves the loop as entered it.

The cells of the ascending arm of the loop of Henle are, by contrast, relatively impermeable to water and urea, but are highly permeable to Na^+ and Cl^-. Because of the loss of water in the descending arm, the filtrate is now very concentrated. As it rises in the ascending arm, it passes through zones of decreasing concentration of interstitial fluid, and as a consequence Na^+ and Cl^- diffuse from the filtrate into the interstitial fluid (see Figure 11.7).

Distal Tubule. The portion of the distal tubule in the outer medulla is characterized by cells that are largely impermeable to the passage of water and urea. However, these same cells actively transport Cl^- out of the filtrate, and various positive ions (primarily Na^+, K^+, and Ca^{2+}) follow passively. The cells of the cortical portion of the distal tubule, on the other hand, actively transport Na^+ out of the filtrate, and Cl^- and HCO_3^- (bicarbonate ion) follow passively (see Figure 11.8). Because of the reabsorption of ions, the filtrate is even more dilute than the interstitial fluid at this point. Therefore, since the cells of the cortical portion of the distal tubule are, unlike the cells of the medullary portion, permeable to water, water leaves the filtrate and enters the interstitial fluid by osmosis. Finally, just as occurred in the proximal tubule, the cells of the distal tubule actively secrete K^+, H^+, and ammonia into the filtrate.

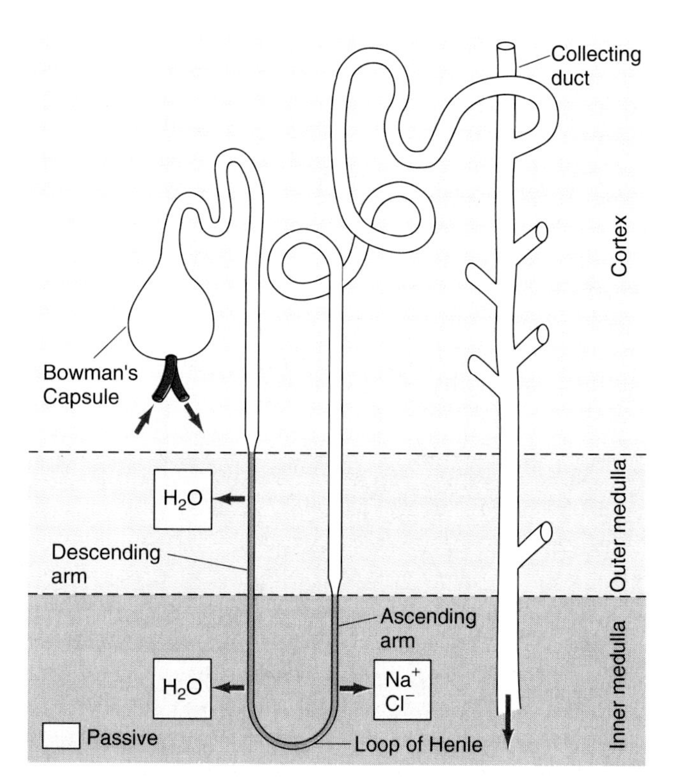

FIGURE 11.7

Events in the Loop of Henle

Water flows passively out of the descending loop of Henle as the loop passes through an increasingly concentrated interstitial fluid. The cells of the ascending arm are impermeable to water, but sodium and chloride ions diffuse passively into the interstitial fluid (which is increasingly less concentrated as the filtrate moves back up the loop).

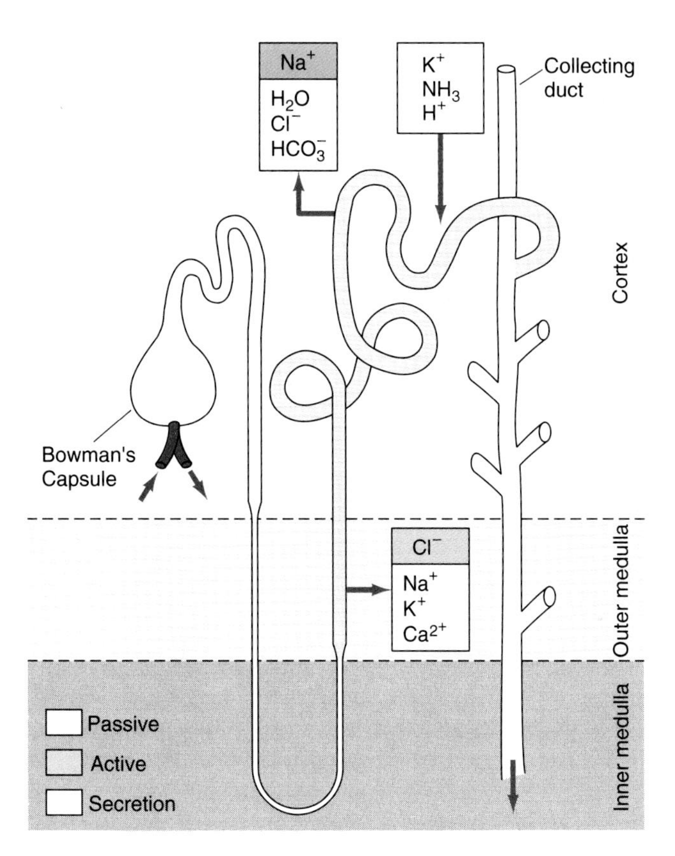

FIGURE 11.8

Events in the Distal Convoluted Tubule

In the medullary portion of the distal tubule, chloride ion is actively transported out of the filtrate, and sodium, potassium, and calcium ions follow passively. In the cortical portion of the distal tubule, sodium ion is active transported out of the filtrate, and chloride ion, bicarbonate ion, and water follow passively. Potassium and hydrogen ions, and ammonia, are secreted into the filtrate.

Collecting Duct. Because the collecting duct passes through the medulla, and because the interstitial fluid becomes progressively more concentrated the deeper one goes in the medulla, the filtrate is subjected to increasingly more powerful osmotic pressure as it moves down the collecting duct. However, the amount of water that leaves the filtrate is a function of the permeability of the cells of the collecting duct to water, and the permeability of these cells is under the control of a hormone. (The details of how this hormone functions are discussed in the next section of this chapter.)

The cells comprising that portion of the collecting duct in the cortex are almost impermeable to the passage of urea. However, the medullary portion of the collecting duct is somewhat permeable to urea, especially when the cells are rendered permeable to water by the presence of the hormone mentioned above. Since essentially no urea has been reabsorbed to this point, the concentration of urea in the filtrate is very high, and it readily diffuses into the interstitial fluid of the inner medulla (see Figure 11.9).

Putting It All Together. Now that we have examined the specific role of each section of the nephron, let's try putting it all together (see Figure 11.10). The profound reduction in volume which occurs in the nephrons, whereby 180 liters of glomerular filtrate is reduced to two liters of urine, is achieved through the interplay of osmosis, diffusion, and active transport, with considerable regionalization of these activities along the length of the nephron. There are also important differences between the cortex and the medulla of the kidney. In particular, the interstitial fluid of the medulla is much more concentrated (that is, saltier) than is the interstitial fluid of any other area

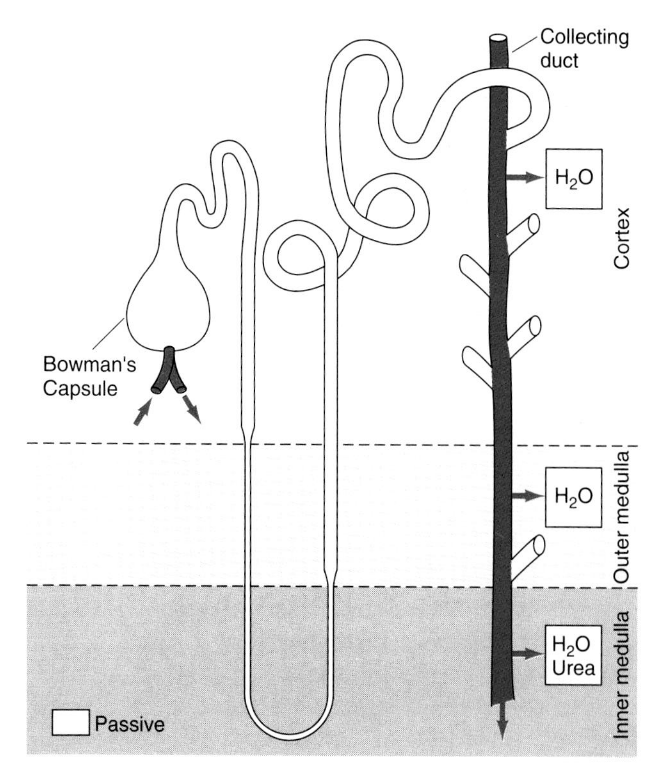

FIGURE 11.9

Events in the Collecting Duct

In the collecting duct, water diffuses passively out of the filtrate as the duct passes through increasingly concentrated interstitial fluid. In addition, urea passively diffuses out of the filtrate in the inner medulla.

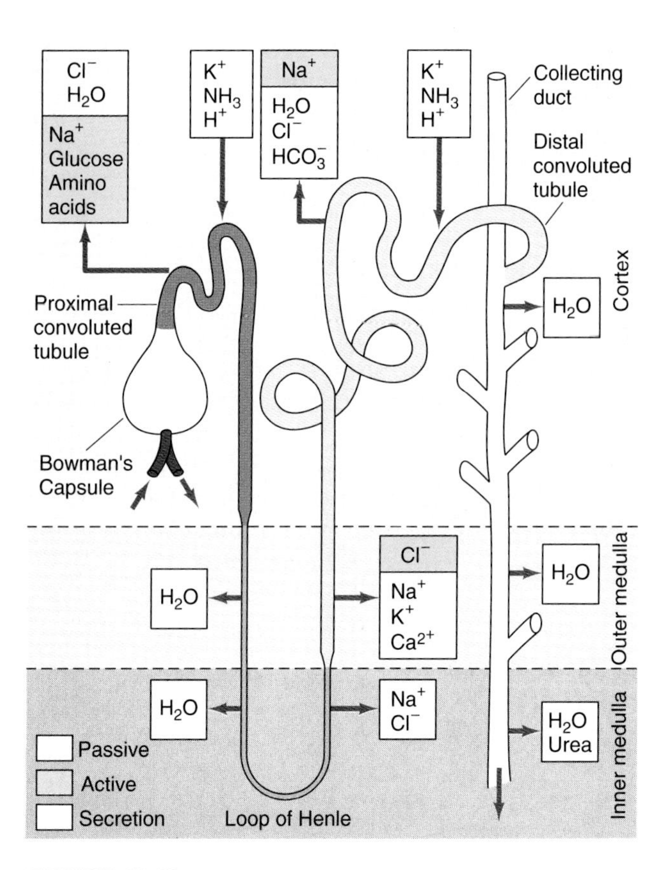

FIGURE 11.10

An Overview of Nephron Function

This figure combines the information in Figures 11.6–11.9.

of the body. It is essential that this be the case, in order for large amounts of water to be reabsorbed from the loops of Henle and the collecting ducts—but *how* is the medullary interstitial fluid concentration achieved and maintained? Why do the reabsorbed water and salts not immediately enter the capillaries surrounding the nephron, to be reincorporated into the blood (which is what happens to the water and salts reabsorbed in those regions of the nephron located in the cortex)?

These questions have two answers. First, blood flow in the medulla is very sluggish. The medulla receives less than 2 percent of the kidney's blood supply. Quite simply, urea and salts move into the interstitial fluid of the medulla faster than they can be removed by the surrounding blood capillaries. Second, and more importantly, the blood vessels in the medulla use a counter-current mechanism.

Counter-Current Mechanism. Counter-current mechanisms, which involve exchange between closely adjacent fluids moving in opposite directions, are common in biological systems. For example, the arterial blood supply of the feet of Arctic birds and mammals is side-by-side with the venous return. Heat from the outwardly flowing arterial blood continuously warms the incoming venous blood such that the temperature of the body is consistently greater than the temperature of the feet. The peritubular capillaries surrounding the nephron are parallel to the loop of Henle and therefore have the same hairpin shape. These capillaries are highly permeable to water and ions. As the blood moves deeper into the medulla, water passes out from the capillary into the increasingly more concentrated interstitial fluid, and Na^+, Cl^-, and urea diffuse into the capillary from the interstitial fluid. However, as the blood heads back toward the cortex, movement of these materials takes place in the opposite direction. The end result is that the blood leaves the medulla with essentially the same concentration of materials as it entered, and there is therefore very little net movement of salts and urea out of the medullary interstitial fluid (see Figure 11.11).

On the other hand, strictly speaking the nephron itself does not function as a counter-current system because there is no (or very little) exchange between closely-adjacent fluids moving in opposite directions. (Some urea does reenter the ascending arm when the amount being reabsorbed by the collecting duct is large, a feature which accounts for the fact that the concentration of urea in the body remains essentially constant despite variations in the volume of urine.) It was at one time believed that salts recycled between the ascending and descending arms, but current evidence does not support that contention.

Rather, concentration of the filtrate occurs because of regional specialization in the cells of the nephron and in the kidney itself. The only region of the nephron which is relatively permeable to urea is the inner medullary portion of the collecting duct. As a consequence, roughly half of the urea diffuses from the filtrate into the interstitial fluid of the inner medulla, thereby increasing the osmotic pressure of the fluid. Water, but not salt ions, is drawn out of the filtrate in the descending arm because of the osmotic pressure of the interstitial fluid, but it cannot reenter the ascending arm because the cells of the ascending arm are impermeable to water. Salt ions, but not water, diffuse out of the filtrate in the ascending arm, thereby increasing the osmotic pressure of the interstitial fluid. The net result is an interstitial fluid which is increasingly concentrated in dissolved particles as one moves more deeply into the medulla.

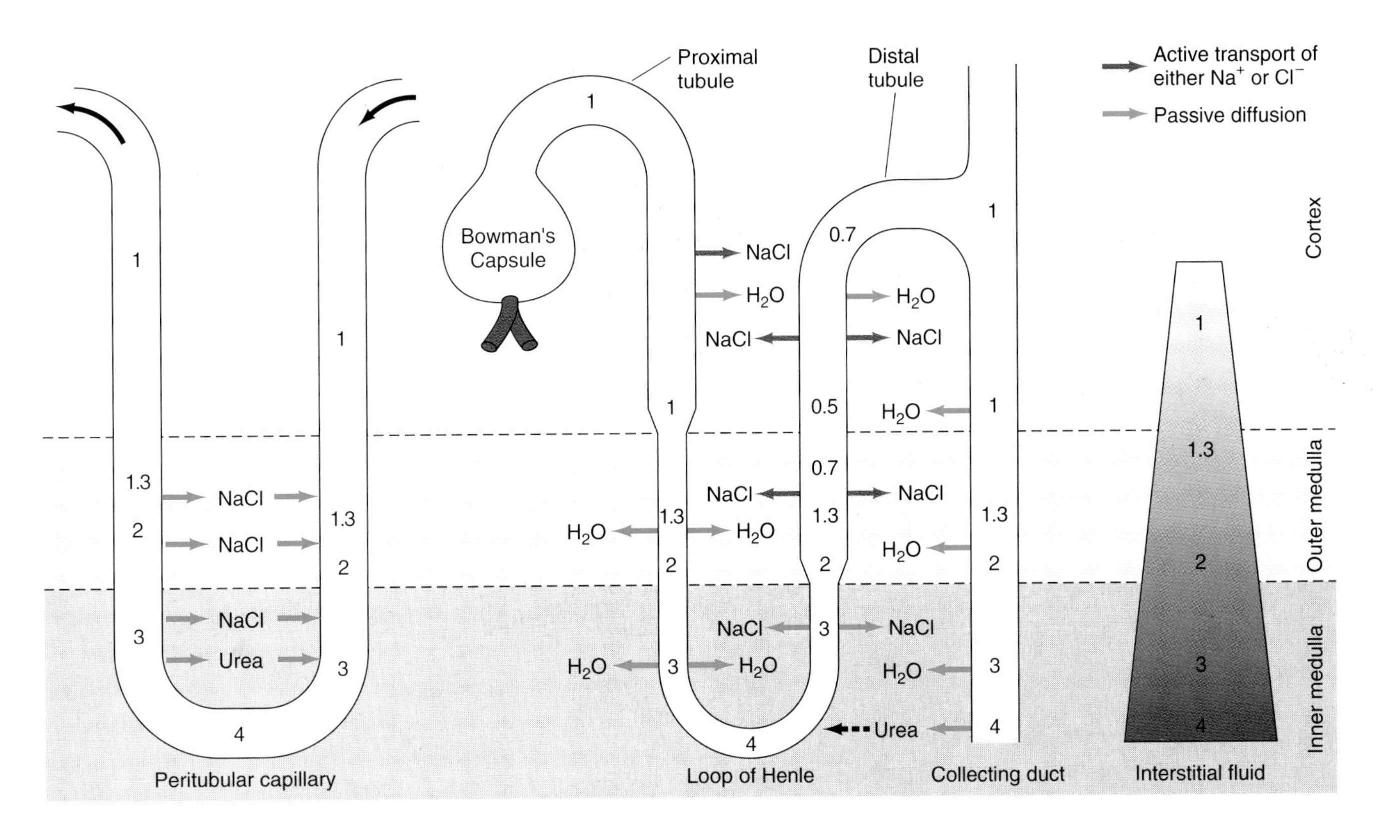

FIGURE 11.11

The Counter-Current Mechanism

The counter-current mechanism is not involved directly in concentrating the filtrate. In terms of overall concentration, the filtrate is in equilibrium with the interstitial fluid from the proximal tubule to the beginning of the distal tubule. (Numbers represent the relative concentration of dissolved particles, with blood equal to 1.0.) Active transport of Na^+ and Cl^- in the cortical and medullary portions of the distal tubule, respectively, lowers the concentration of the filtrate to a level more dilute than blood. Because of the uptake of water, in the collecting duct the filtrate is once again at the same concentration as the interstitial fluid. Urea diffuses out of the filtrate at the base of the collecting duct (a small amount re-enters the loop of Henle), helping to create a very concentrated interstitial fluid. Passive diffusion of Na^+ and Cl^- from the ascending arm of the loop of Henle further concentrates the interstitial fluid.

The counter-current mechanism does function in the peritubular capillaries that directly overlie the nephron. (They are drawn to one side in this illustration for the sake of clarity.) Na^+, Cl^-, and urea diffuse into the capillary as the blood descends into the medulla, but diffuse right back out again as the blood rises towards the cortex. In essence, there is a direct exchange between the two arms of the capillary, and there is very little net movement of Na^+, Cl^-, or urea out of the interstitial fluid of the medulla. Thus the counter-current mechanism is not directly responsible for concentrating filtrate, but it does preserve the highly concentrated medullary interstitial fluid without which concentration of the filtrate could not occur.

REGULATION OF KIDNEY FUNCTION

The kidneys have the task of maintaining the blood not only reasonably free of nitrogenous wastes and miscellaneous chemicals but also reasonably balanced with respect both to salt and water ratios and to pH. These are very different kinds of challenges. It is one thing for the kidneys to be able to "recognize" a waste molecule and then to dispose of it; it is quite another for them to be capable of discarding precisely the correct amount of salts, water, and hydrogen ions, regardless of our level of activity or our food and water intake on any given day. How is this regulation achieved?

Regulation of Water Level

The body can do little to regulate the volume of water lost in the feces or in exhaled air. In addition, water lost in perspiration is primarily a function of the need to maintain a stable body temperature rather than of the need to maintain a constant proportion of water in the blood. On the intake side, the body has little control over the amount of water produced metabolically or ingested as part of our food. Thus regulation of water level is primarily achieved by the balancing of two variables: the amount of water drunk in response to thirst and the amount of water voided in the urine (see Table 11.3).

When the body is dehydrated, blood volume is reduced, and consequently blood pressure drops. The drop in blood pressure is monitored by the brain, which both initiates the thirst reflex and increases production of **antidiuretic hormone (ADH).** (A *diuretic* is any substance that promotes urine production; conversely, an *anti*diuretic is any substance that inhibits urine production.) High levels of ADH reduce the volume of water leaving the body as urine by increasing the diameter of the pores of the cells of the collecting duct of the nephrons. Thus, in the presence of ADH, less water is lost in the urine, more water is retained by the body, and blood volume ultimately increases as a consequence, leading to higher blood pressure. Conversely, if the body has an excess of water, very little ADH is produced, the walls of the collecting duct remain largely impermeable to water, and a urine as much as five times as dilute as plasma is produced in copious amounts.

Two common molecules interfere with the action of ADH. Alcohol reduces the amount of ADH produced by the brain, a situation that leads inevitably to increased urine output. Thus rest rooms in bars are frequented not so much because of the *volume* of liquid being imbibed but because of the *components* of that liquid. Caffeine (in coffee, tea, or cola) also interferes with ADH, but does so not by interfering with ADH production but with its effect on the cells of the collecting duct. In a manner not completely understood, caffeine impairs the capacity of ADH to increase the size of the membrane pores, and, as a consequence, less water is reabsorbed and more leaves the body as urine. The similar consequences of both alcohol and caffeine suggest that, to the extent that a hangover is caused by dehydration, ingesting a second diuretic (coffee) to "sober up" may be counterproductive in trying to get rid of a headache.

Regulation of Salt Level

The body generally has no difficulty in obtaining sufficient salts, because they are abundant in any balanced diet (see Chapter 7). Regulation of salt balance is therefore primarily achieved on the "loss" side of the ledger. There are several sites of salt loss, although they vary widely in significance. For example, the amount of salt lost in perspiration and tears is small and is not well regulated. At least some of the salt lost in the feces is regulated, but primary responsibility for the regulation of salt balance in the body falls to the kidneys.

TABLE 11.3 Reabsorption Rates of Different Areas of the Nephron

REGION OF NEPHRON	ENTERING VOLUME (ML/MIN)	REABSORPTION (%)
Proximal tubule	125	65
Loop of Henle	45	15
Distal tubule	25	10
Collecting duct	12	10
Urine entering bladder	1	—

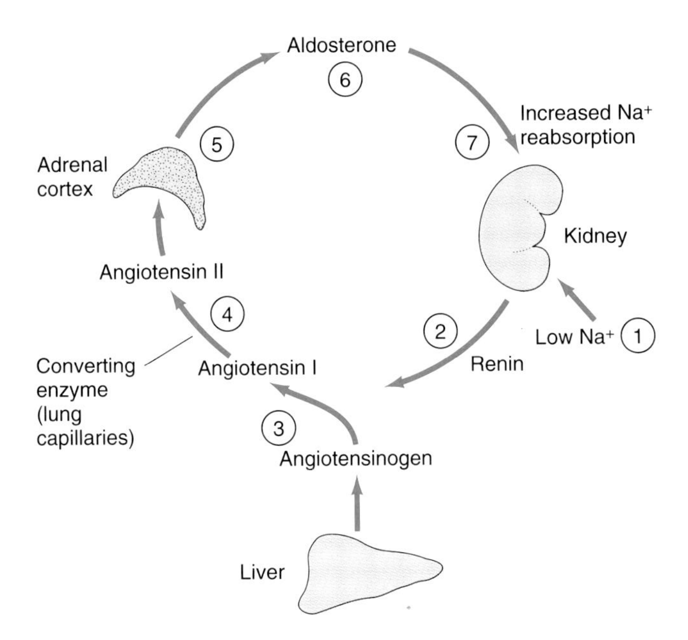

FIGURE 11.12

Control of Blood Sodium

When blood sodium levels drop (1), cells in the kidney respond by secreting renin (2). Renin catalyzes the transformation of angiotensinogen (3), a blood protein produced by the liver, into angiotensin I. Converting enzyme, produced by cells in the capillaries of the lungs, converts angiotensin I into angiotensin I(4), which stimulates the adrenal cortex (5) to produce aldosterone (6). Aldosterone acts on the kidney to increase reabsorption of sodium ion (7).

The principal salt ion in the extracellular fluid (blood and interstitial fluid) is Na^+. Thus the problem of salt balance is reduced to a determination of what proportion of the Na^+ present in the glomerular filtrate will be reabsorbed by tubule cells and what proportion will remain in the filtrate to be lost as urine. This is not a trivial problem: Variations in Na^+ levels of more than 20 percent from its mean concentration are life-threatening. So precise are the controls, however, that, as a practical matter, Na^+ levels seldom change by more than 1 percent during the course of a day.

The kidneys control Na^+ levels in the body in an intricate fashion (see Figure 11.12). If Na^+ levels in the blood begin to drop, the nerves of the body become excitable (for reasons explained in Chapter 14), and the kidneys produce a protein called **renin.** Renin acts as an enzyme to convert the plasma protein **angiotensinogin** (produced by the liver) to **angiotensin I.** An enzyme called **converting enzyme,** located on the walls of capillaries in the lungs, converts angiotensin I into **angiotensin II.** Angiotensin II, in turn, stimulates the adrenal cortex to produce **aldosterone** (see Chapter 13), and aldosterone "instructs" the kidneys to reabsorb more Na^+ and thus diminishes the rate at which it is lost in the urine. Conversely, if blood Na^+ levels are high, less aldosterone is produced, and larger amounts of Na^+ are lost in the urine.

Aldosterone has its effect primarily on the cells of the distal tubule and the collecting duct. By the time the filtrate has reached these areas of the nephron, about 90 percent of the Na^+ has already been reabsorbed, but absorption of the final 10 percent is a function of aldosterone levels. The kidneys filter approximately 600 g of Na^+ each day (about ten times the amount actually present in the body). When aldosterone levels are high, as little as 0.1 g of Na^+ is lost. However, if aldosterone levels are low, as much as 40 g of Na^+ may be lost. Indeed, when ADH levels are high and aldosterone levels are low (meaning that little water but much Na^+ are lost in the urine), the urine may be as much as 3.5 times as concentrated as blood, in terms of the total percentage of all dissolved materials (most of which is Na^+).

Partially offsetting the effects of aldosterone is the hormone **atrial natriuretic factor (ANF).** Produced by the atria, it inhibits Na^+ reabsorption directly, by acting on the nephrons, and indirectly, by inhibiting the secretion of both renin and aldosterone. In the normal scheme of things, ANF is secondary in Na^+ regulation to aldosterone. ANF is now used in the treatment of hypertension (see Chapter 8).

Regulation of Blood Acidity Level

The regulation of blood pH is a complex process involving not only the kidneys but also the respiratory system and a series of buffers in the blood (see Chapter 10). However, only the kidneys have the capacity to excrete H^+. The pH of the urine can vary from 4.5 to 8.0, depending on the amount of H^+ being excreted. Of course, there is a limit to how acidic the urine can become before it begins to damage the delicate cells of the kidney. Nevertheless, a substantial amount of H^+ can be passed out with the urine without an increase in the acidity of the urine through the simple expedient of combining it with **ammonia** (NH_3), and forming the ammonium ion (NH_4^+).

Regulation of Nitrogenous Wastes

Ammonia, chemically the simplest of the nitrogenous wastes, is a powerful poison. To prevent the concentration of ammonia in the blood from reaching lethal levels, we would have to produce huge volumes of very dilute urine. This method of excretion is eminently practicable for organisms living in fresh water (which must produce copious urine in any event just to rid themselves of the water they acquire by osmosis), but copious urine production is a poor choice for most terrestrial organisms.

Some more-or-less terrestrial organisms nevertheless do utilize ammonia as a waste product and thus do produce volumes of urine that are very large relative to body weight. For example, the daily volume of urine in frogs is equal to 25 percent of their body weight, and in earthworms the urine volume is 60 percent of their body weight.

Most terrestrial organisms employ another solution for excreting nitrogenous wastes: They convert ammonia into a less lethal substance. Humans and other mammals produce **urea,** which consists of two ammonia molecules linked by a molecule of carbon dioxide. Because urea is not very toxic, daily urine volume in humans is only 2 percent of body weight. (Humans also produced some ammonia, as anyone who has strayed near a diaper pail can attest—although most of that ammonia smell results from the action of bacteria in breaking urea down into its ammonia components.)

An even more efficient method of excreting nitrogenous wastes, used by birds and reptiles, is the production of **uric acid,** which is so nontoxic that it can be excreted as a paste. Thus the organism can conserve huge amounts of water. Humans also produce small amounts of uric acid. Perhaps the reason we do not produce it more extensively is that we are constructed very differently from birds and reptiles. In those organisms, the pastelike uric acid passes from the bladder (or, in birds, directly from the kidneys) into a common chamber with the rectum. In humans (and other mammals) the contents of the bladder pass through a narrow urethra, and trying to pass a paste through the urethra would be both difficult and painful. Freud's statement "anatomy is destiny" seems to have special significance here.

It is interesting to note that only about half of the urea formed by the body is actually lost in the urine.

BOX 11.2

Diseases of the Urinary System

The urinary system is somewhat less prone to disease and other problems than are many of the other systems. Among the more common are the following.

Nephritis

Nephritis, an inflammation of the kidney, is a relatively common disease (three million cases are reported annually in the United States). Nephritis often stems from strep throat. Nephrons may become blocked and thus permit only a small amount of urea to be excreted, or the glomeruli may be damaged and thus permit blood proteins—or even blood cells—to enter the tubule, where they are passed out with the urine. (Runners occasionally pass some blood in the urine after a strenuous workout, particularly if they are dehydrated. In these instances, the presence of blood in the urine is not a significant problem and should not be confused with nephritis.) The seriousness of nephritis depends on the degree of impairment of the kidneys.

Nephrosclerosis

Nephrosclerosis is a complication resulting from severe hypertension (see Box 8.2, Chapter 8) or from diabetes mellitus (see Chapter 13). Hypertension typically affects the arterioles supplying the glomeruli. Not only is normal kidney function impaired, but the narrow sclerotic arteries may not permit a supply of blood adequate to nourish the cells of the kidneys, in which case the kidneys will gradually degenerate. Diabetes tends to affect the glomeruli directly, although the arterioles are also generally involved. Proteins leak through the damaged glomeruli and are lost in the urine, the consequence sometimes being massive disturbances of the osmotic relationship between the blood and the interstitial fluid.

Uremia

Uremia (or *renal failure*) is a symptom, not a disease. It refers to the presence of excessive amounts of urea in the blood and indicates malfunctioning (or nonfunctioning) kidneys. Most cases result from chronic nephritis. It is not the excessive urea levels in the blood that pose the problem (urea is relatively nontoxic); it is that high urea levels indicate kidney failure, and kidney failure means potentially fatal changes in salt con-

The other half is reabsorbed, primarily from the collecting duct as it passes through the medulla. The urea tends to remain in the interstitial fluid of the medulla, where it plays an important role in contributing to the high osmotic pressure of this fluid. When ADH is present and, as a consequence, more water is reabsorbed from the collecting duct, more urea is also reabsorbed, because ADH also increases the permeability of the cells of the collecting duct to urea. Thus the concentration of dissolved materials in the interstitial fluid of the medulla is kept both high and balanced.

Urination

By the time the glomerular filtrate reaches the end of the collecting duct, all transformations are complete and the fluid has become **urine.** Movement of urine down the ureters to the bladder is accomplished by peristaltic contractions of the ureters, which are lined with muscle.

The urinary bladder has a capacity of from 500 to 1000 ml. However, whenever roughly 300 ml of urine are present, stretch receptors in the walls of the bladder transmit a stimulus to the spinal cord which, in turn, stimulates contraction of the bladder. This contraction opens the urethra at its bladder end (an area known as the *internal sphincter*), and urination occurs automatically.

In children past the age of about two, an override system develops whereby urination can be consciously delayed through the contraction of muscle at the external end of the urethra (an area called the *external sphincter*). At a somewhat later age the ability to urinate at will—even in the absence of stimuli from a filled bladder—is developed. The fact that control of urination matures in two distinct steps accounts for situations in which children who claim they "don't have to go to the bathroom" suddenly need to do so in the worst way only 15 minutes later—just as the family car enters the freeway.

centration or in pH level of the blood. Dialysis or kidney transplant are often the only solutions.

Cystitis

Cystitis, an inflammation of the bladder, is caused by invasion of bacteria up the urethra. Because the urethra is so much shorter in women, cystitis is much more common in women than in men. Cystitis is usually easily treated with appropriate antibiotics.

Gout

Gout was traditionally blamed on too much rich food and wine. It actually results from the precipitation of excess uric acid in the joints. (Precipitation may also occur in the kidney and produce kidney stones.) Gout is a very painful condition, though drugs that lower the uric acid level in the blood can now alleviate the discomfort.

Kidney Stones and Other Mechanical Obstructions

Kidney stones are formed primarily by the precipitation of uric acid, although why they occur in some individuals and not others is unclear. Problems occur when the stones move through the ureter. Not only do they cause excruciating pain, but they may scar the ureter or even lodge, in which case surgery may be necessary.

A recent development in the treatment of kidney stones is the lithotriptor, a device that focuses ultrasonic waves on the stones and shatters them without the necessity of surgery. Lithotripsy is now being done on an outpatient basis, where formerly kidney stone surgery required extensive hospitalization. The lithotriptor has recently been modified for use on gallstones.

A variety of other agents may produce mechanical obstructions that negatively affect kidney performance. For example, heavy metals such as lead, mercury, and cadmium are particularly lethal to kidney cells. Also, mismatched blood transfused into the body forms clumps of red blood cells that rupture and release hemoglobin, which, in turn, blocks the glomeruli. Indeed, the degree of danger posed by a mismatched transfusion depends on how much kidney damage results.

Summary

Most terrestrial animals, including humans, face the essential problem of ensuring that the fluid that bathes the cells of the body (namely, the interstitial fluid) is kept relatively constant in composition. A major component of this task is to control such potential variables as salt and water balance, pH, and to remove metabolites. In humans, most of these activities are performed by the kidneys, which are the principal organs of the excretory system.

The majority of our organ systems exhibit a close interrelationship between anatomical entity and physiological activity (for example, the digestive tract and digestion). Such is not the case with excretion and the excretory system. Excretion is, strictly speaking, the removal of metabolites (carbon dioxide, metabolic water, and ammonia), and the excretory system actually controls only the latter two. However, because the "excretory" system also controls salt and pH balance, it would be better termed the "urinary" system, since that term implies only that urine is formed but says nothing about its composition. Similarly, a better term for the physiological activities commonly called excretion would be "fluid homeostasis," which includes all of the physiological events involved in regulating the constitution of interstitial fluid and blood.

The functional unit of the kidney is the nephron. Large amounts of glomerular filtrate are formed from blood plasma, but the nephron reabsorbs the great bulk of this filtrate before it leaves the body as urine. In addition to filtration and reabsorption, the nephron also engages in secretion, and chemicals ranging from antibiotics to hydrogen and potassium ions enter the prospective urine in this manner.

The volume of water and amount of sodium ions reabsorbed are largely under the control of two hormones: antidiuretic hormone and aldosterone. These hormones function by affecting the permeability of the cells of the nephron.

In terms of dissolved materials, human urine may be more than three times as concentrated as blood. This high concentration is achieved because many of the nephrons loop deeply into the medulla of the kidney, and the interstitial fluid of the medulla is extremely rich in ions and urea. As a consequence, water is drawn from the filtrate and the filtrate becomes very concentrated. Once concentrated, the filtrate is referred to as urine.

Urination occurs as a reflex that begins with activation of the stretch receptors of the bladder when it is approximately half full. By the age of two, most children have mastered the socially necessary art of overriding this reflex by conscious concentration of the external sphincter, a band of muscle at the exterior end of the urethra.

Key Terms

fluid homeostasis
excretion
kidney
ureter
urinary bladder
urethra
cortex
medulla
nephron
Bowman's capsule
proximal convoluted tubule
loop of Henle
distal convoluted tubule
collecting duct
glomerulus
filtration
glomerular filtrate
reabsorption
secretion
counter-current mechanism
antidiuretic hormone
renin
angiotensinogin
angiotensin I
converting enzyme
angiotensin II
aldosterone
atrial natriuretic factor
ammonia
urea
uric acid
urine

Box Terms

dialysis
nephritis
nephrosclerosis
uremia
cystitis
gout
kidney stone

Questions

1. In general terms, what is the function of the kidneys with respect to the maintenance of homeostasis in the body?
2. Describe the various parts of a nephron.
3. What are the three processes involved in creating urine? In what regions of the nephron do they occur?
4. What is the counter-current mechanism? What does it have to do with urine formation?
5. What is antidiuretic hormone? What part of the nephron does it affect, and what are the consequences of its presence or absence?

6. What is the role of the kidney in regulating blood sodium levels?
7. What are the main nitrogenous waste products? Why do we have more than one?
8. Explain the functioning of the artificial kidney.

5
25
25

CHAPTER

12

Skin, Bone, and Muscle

Support and Locomotion

An 83-year-old woman slips on the edge of a sidewalk, falls, and breaks her hip. She may never walk again. Her five-year-old great-grandson falls from the window of his third-story bedroom, suffers a greenstick fracture of his leg, and is running again in four weeks. Why is there such a difference?

A scientist discovers the thighbone of an extinct dinosaur and is able to reconstruct a model of the whole animal from this one bone. Is it pure speculation, or is there validity to his methods?

Most of us could throw a baseball more than 60 miles an hour or lift 75 pounds off the floor. If we are that strong, why do we have so much trouble opening a stuck window?

Because of their prominence, we are quick to recognize the roles played by the muscular and skeletal systems, but we often fail to appreciate either the remarkable degree of interrelationship between these two systems or their limitations on our activities. However, muscles and bones are subject to the same physical laws that are all organs, and these laws limit how our bones and muscles serve us, as we shall see in the following chapter.

THE NATURE OF THE PROBLEM

Imagine for a moment that the calendar has been turned back almost one billion years, and that you are a unicellular organism living in an ancient ocean. Even though you have no brain, you are contemplating (remember, this is imagination!) what it would be like to live on land as a multicellular animal. Think of the problems you would have to overcome. The air is dry: you would have to avoid desiccation. Because the cells of animals have no walls, they are very pliant: You would need some form of supporting structure to hold up your body. You would no longer be able to rely on cilia or on the random action of waves to move you about: You would need some very different mechanism for locomotion.

Of course, you are not a unicellular organism living one billion years ago. You are a land-dwelling, multicellular animal whose ancestors long ago solved these problems and who therefore can now take these solutions very much for granted. Our dry skin retards water loss and prevents desiccation; the bones of our skeletal system keep us from looking like a beached jellyfish; our well-developed muscular system gives us the ability to move about readily.

Before lapsing totally back into complacency, however, think a little more about these systems. The skin is not completely water-tight: we perspire. A skeletal system is a good solution for the support problem—but how can our bones be capable of growing, as indeed they must during our childhood years, and yet still be solid enough to provide the support we need? We move because our muscles have the ability to contract—but *how* do they contract? What is happening to muscles at the cellular level?

Let us take a brief look at each of these organ systems.

SKIN AND SUPPORT

Although the skin is the heaviest organ of the body, it is never more than 6 mm thick—yet it is deceptively complex. The skin consists of two distinct layers (see Figure 12.1). The external **epidermis** (Gr. *dermis*, "skin") is composed of many layers of **epithelium** (see introduction to Part 3), the outermost layers of which are dead, flattened cells that together form a water-impermeable sheet which keeps you from inflating like a balloon when you climb into the bathtub. The innermost layer of epithelial cells is the **stratum germinativum** (L., "germinating layer"), which provides replacement cells for the epidermis as it is gradually worn away. The stratum germinativum is also the location of the **melanocytes** (Gr. *melos*, "black"), which are cells containing the pigment **melanin.** The amount of melanin present determines how fair or dark our skin appears.

Below the stratum germinativum is the **dermis,** composed of loosely organized connective tissue. The dermis supports the epidermis and binds it to the underlying **subcutaneous** (L. *cutis*, "skin") **tissue.** The subcutaneous tissue, in turn, binds the skin loosely to the inner tissues, allowing the skin to slide about freely during movement. Fat cells are located in the subcutaneous tissue, providing insulation and protection for the underlying tissues. Some of us have more "insulation and protection" than others, as you may have noticed.

The epidermal cells contain large amounts of the linear protein **keratin** (Gr. *keratos*, "horn"), which is also found in hair and nails. Keratin gives the epidermis its characteristic toughness. Nevertheless, the cells of the epidermis are constantly being sloughed off in small amounts (or in sheets, if you happen to have had a sunburn). During our lifetime, we shed about 25 kg of skin in this manner. The skin is particularly subject to changing conditions, and a sudden decision by its possessor to become a manual laborer is reflected by the production of extra layers of epidermis in high wear areas, a manifestation known as *callouses*.

The epidermis is laminated onto the dermis in a manner similar to the way that the outer layer of a piece of plywood is attached to the core. Too many sets of tennis on the first warm day in spring can cause this lamination to work loose in areas of stress, permitting fluid-filled spaces to form between the layers. We call such structures *blisters*. The fact that the dermis is alive and well-endowed with nerve endings is amply demonstrated by our reaction to the seepage of perspiration into a broken blister.

In addition to nerve endings, the dermis is richly blessed with capillaries, hair follicles, and a variety of glands. **Sweat glands** are responsible for releasing water and salts onto the surface of the skin, where evaporative cooling brings about a reduction of body temperature. The volume of perspiration produced is largely a function of the degree of overheating of the body, although perspiration may also be produced by emotional stimulation (the "cold sweat" of fear, the "sweaty palms" of a nervous individual). **Sebaceous glands** are associated with hair follicles, and produce an oily secretion that helps to prevent drying of the skin.

The hair follicles are actually infoldings of the epidermis, and as the cells at the base of the follicle divide, they are incorporated into the shaft of the hair. As do all good epidermal cells, these hair shaft cells produce abundant amounts of keratin before they die, which is why the organic structure of the outer epidermal cells and the hair is so similar. (Finger- and

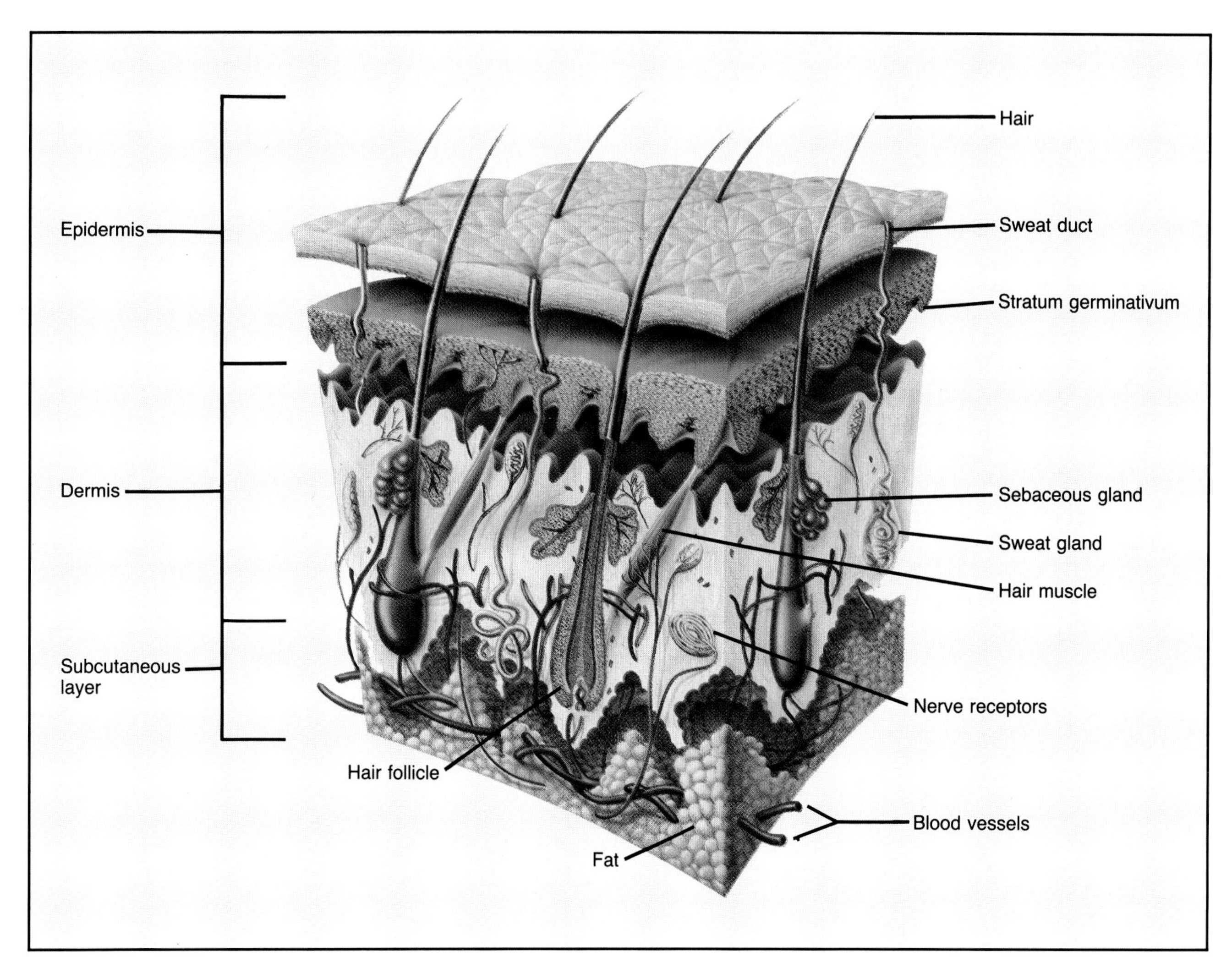

FIGURE 12.1

Skin
This artist's rendition of the skin presents some diagrammatic evidence of the complexity of the body's heaviest organ.

toenails are produced in a similar manner, and again both are composed largely of keratin.)

Two types of proteins in the dermis are of great importance in its role as an organ of support. **Collagen,** a protein also found in tendons and ligaments, acts as a kind of organic baling wire, holding everything in place. It is the toughness of collagen that keeps our skin from becoming baggy at the knees and ankles, like a cheap pair of pantyhose. However, nothing lasts forever, and the gradual stretching of collagen is responsible for the development of wrinkles later in life. **Elastin,** the second protein, gives skin its unique property of snapping back into place when stretched, since, as its name implies, elastin has a rubberlike quality and will return to its original length after having been stretched.

In sum, the skin performs a number of vital functions: Since its outermost layers are dead, it provides protection for the underlying living cells from bacterial attack, mechanical damage, desiccation, and (by producing more melanin) from ultraviolet radiation (see Box 12.1). The epidermis and dermis also provide structural support for the superficial blood vessels and nerves, and for the fat cells. Finally, the sweat glands are an essential part of temperature regulation.

BOX 12.1

Growing Skin, Removing Wrinkles, and Stimulating Hair

The importance of the skin to the overall health of the body is most dramatically demonstrated in the case of severe burns. Historically, there has always been a high correlation between the amount of skin lost and the likelihood of survival. For example, in people under the age of 30, death is usually inevitable if the skin loss is greater than 75 percent, and death is most unlikely if the loss is less than 25 percent. (These percentages drop rapidly as the age of the individual increases.) Death usually results from a combination of fluid loss and infection. Skin grafts are routinely performed in cases of severe burns, but because the skin is living tissue, grafts must normally be from the burn victim himself, because grafts from others will be rejected. Obviously, a primary problem in severe burn cases is that there is not enough unburned skin available to transplant.

Within the past decade, several new procedures have become available. One is a clear film of polyurethane plastic that is laid directly over the burn. It is porous to oxygen, which allows white blood cells to keep the burned area free of bacteria, but is largely impermeable to water, which means that skin cells from the periphery of the burn can migrate into the burned area. However, it is not as useful where the burn is severe, and the tissue has been destroyed to some depth.

A second type of material, composed of silicone rubber, nylon, and collagen, can be used in severe burns, although after about two months it must be removed and replaced with skin grafts from the patient.

A third grafting material is even more promising. It has an "epidermis" of silicone and a "dermis" of proteins derived from cowhide and sharkskin. The artificial dermis stimulates growth of the patient's own collagen, and ultimately is replaced by regenerated tissue. The artificial epidermis is removed after a month, and replaced with small grafts of skin from the patient. Dozens of patients with burns ranging from 50 to 85 percent of the body have recovered with this graft.

Finally, and most promising of all, is the promise of growing new skin in culture. In one instance, two young children with burns over 80 percent of their bodies were saved by skin cultured from small pieces taken from their armpits. These small pieces were cultured and ultimately grew into 1 m^2 of new skin.

THE SKELETAL SYSTEM

The two major types of tissue found in the skeletal system are **bone** and **cartilage.**

Bone is a much more complex tissue than is generally realized. We think of bone as being very strong, because it must be capable of retaining its form during the contraction of powerful muscles. However, bone must also be capable of repair and of three dimensional growth, which means that it cannot simply be an inert, dead substance, as is tooth enamel, for example.

Moreover, in order to permit movement, the bones of the skeleton cannot be fused into a solid framework, but must, in places, be jointed (or **articulated**), like a suit of armor. Finally, since the entire body must be moved during locomotion, the skeleton must not be burdensomely heavy. All of these needs dictate certain anatomical outcomes. Before dealing with the anatomy of bone, however, let us briefly review the organization of the skeleton.

The Human Skeleton

As every schoolchild knows, the bones of our body are not scattered randomly about the body, but rather are organized into a framework we call the **skeleton** (Gr. *skeletos,* "dried up" see Figure 12.2). The organization of the skeleton is both highly species-specific and also highly predictive of the overall appearance of the body. Thus paleontologists are capable not only of "fleshing out" the actual appearance of an organism from its skeleton but frequently of predicting with considerable accuracy the overall structure of the organism from just a handful of bones (because the size and shape of a given bone is very closely correlated with its function.)

For purposes of convenience, the skeleton is divided into two major categories: the **axial skeleton** (which is aligned along the primary axis of the body) and the **appendicular skeleton** (which pertains to the appendages—the arms and legs).

Axial Skeleton. The axial skeleton is comprised of the *skull,* the *vertebral column,* and the *ribs.*

The skull, which surrounds and protects the brain, consists of a total of 29 bones. The braincase, or **cranium,** consists of eight bones that entirely encase the brain, except for small entry and exit points for nerves and blood vessels. (By far the largest of these is the opening through which the spinal cord passes.)

The *facial bones* include seven pairs of bones com-

Perhaps the most exciting aspect of this type of treatment is that it offers the promise of scar-free healing. Normally, extensive skin damage results in the formation of scar tissue, which is characterized by the presence of excessive amount of connective tissue invasion from below the skin, the formation of an abundance of irregularly arranged collagen strands, the absence of elastin (scar tissue does not stretch easily), and a randomized pattern of blood vessels.

Cultured epidermis, on the other hand, promotes a more normal development progression: Collagen fibers form in layers, elastin is formed, and the growth of new blood vessels is orderly, not random. The reason for the difference is apparently due to the production of large quantities of growth factors by the cultured epidermis—the same factors that are responsible for growth and development in the fetus (see Chapter 21). Scar-free healing is not universal with the use of cultured epidermis, but the fact that it can occur at all gives great hope for the treatment of future burn victims.

In comparison to the tragedy of severe burns, the wrinkling of skin with age seems a mundane problem. However, a derivative of vitamin A, **retinoic acid,** long used in the treatment of severe acne, has been shown to reduce the fine facial wrinkles and blotchiness caused by overexposure to the sun, although skin irritation is a common side effect. Even though this substance appears to have only minimal impact on the normal wrinkling associated with age, our culture's fascination with looking young generated very great interest in this product. As you might have guessed, following the initial announcement of the first experimental trials of retinoic acid, the value of the stock of the manufacturer climbed eight points in two days!

And while we are discussing the effects of aging, let us not forget the subjects of baldness. Male pattern baldness is caused by a gene on the X chromosome (see Chapter 19) and is a common problem for many middle-aged men in our society. A drug developed a decade ago for hypertension has been recently found to stimulate hair growth in many men when rubbed directly on the scalp. The hair growth in the formerly bald areas is somewhat sparse, and the treatment seems to work best where the baldness is limited and relatively recent in origin. Nevertheless, stock prices in the parent company jumped $13 a share overnight, giving some indication of how seriously our culture views baldness.

prising the nose, the orbits of the eyes, and the cheekbones, plus a single bone separating the nostrils and a single lower jaw, or **mandible.** Unlike the other bones of the skull, the mandible has a moveable articulation, an obvious necessity that permits us to bite and chew our food.

Finally, the skull also includes the *hyoid* bone in the larynx and three pairs of bones in the middle ear. These are the smallest bones in the body. Their function is described in Chapter 15.

The vertebral column is comprised of 33 bones called *vertebrae* (L., "joint"). The seven **cervical vertebrae** of the neck, the 12 **thoracic vertebrae** of the chest, and the five **lumbar vertebrae** of the abdomen are, indeed, jointed, but the five vertebrae of the **sacrum** are fused, thereby providing support for the bones of the hips. The four vertebrae of the tailbone (**coccyx**) are mere vestiges of the tails once proudly borne by our ancient primate forebears.

Although there are regional differences in their structure, all vertebrae consist of a series of *processes,* to which muscles are attached; an *opening* through which the spinal cord passes; and a large, flattened *body.* Between the bodies of adjacent vertebrae are the **intervertebral discs,** which act as shock absorbers and friction pads. These discs become less flexible with age and are prone to rupture, which can result in painful pressure on spinal nerves.

The 12 pairs of ribs, which protect the lungs and serve as attachment sites for the muscles used in breathing (see Chapter 10), are tightly articulated with the 12 thoracic vertebrae in back, and the first 10 pairs are joined in front to the breastbone, or **sternum.** The last two pairs of ribs are called the "floating ribs," because they are not joined to the sternum; instead, they end in muscle tissue at the sides of the body.

Appendicular Skeleton. The appendicular skeleton consists of the **pelvic** (hip) **girdle,** the **pectoral** (shoulder) **girdle,** and the bones of the legs and arms.

The pelvic girdle consists of three pairs of bones that in humans are tightly fused. These bones are very tightly articulated in back with the sacrum, but in front they are linked together by cartilage. The pelvic girdle thus forms a bowl, with a central opening through which pass the digestive, urinary, and reproductive outlets. This is therefore the opening through which a baby must pass during delivery. Not surprisingly, the opening is larger and flatter in women than in men. Moreover, hormones weaken the cartilage holding the halves of the girdle together, thereby

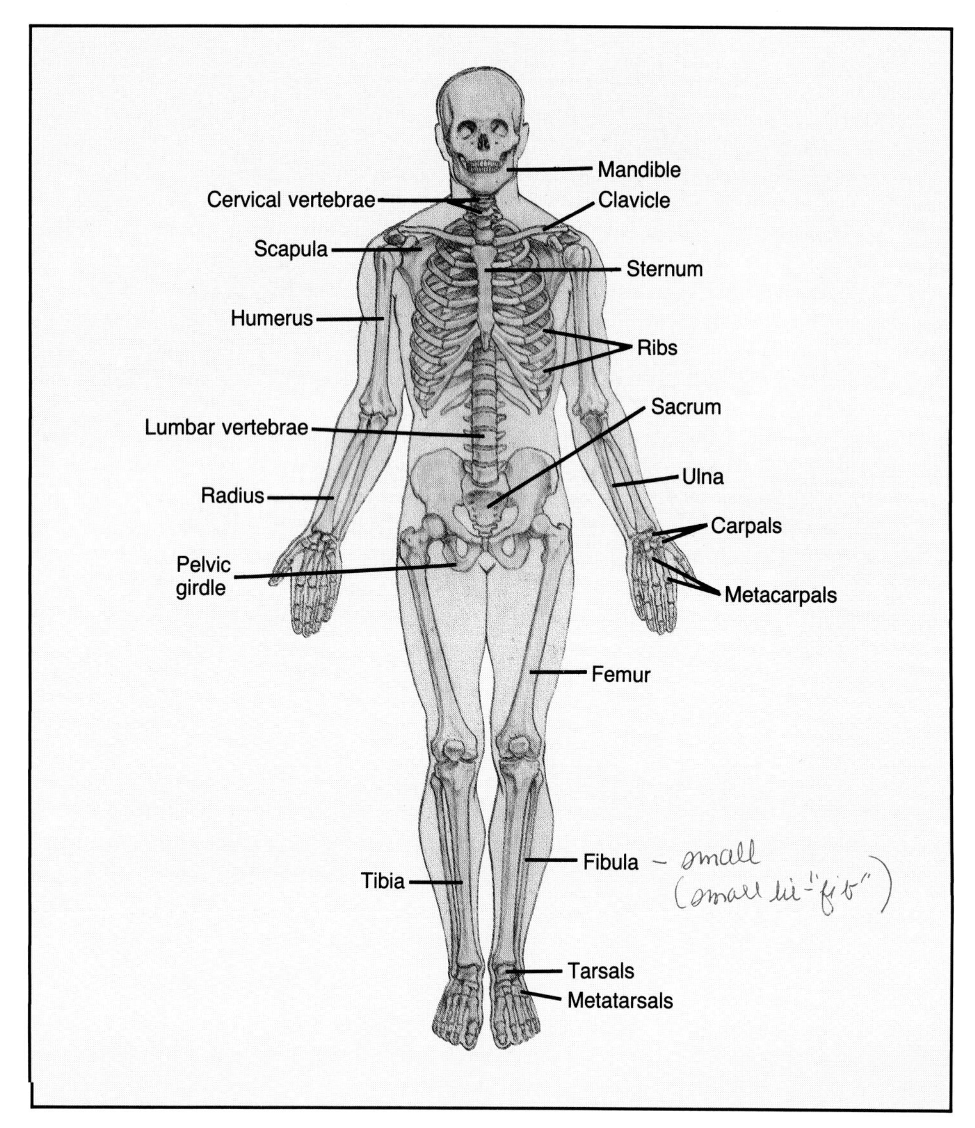

FIGURE 12.2

Human Skeletal System

The skeletal system gives us form and shape, as well as a convenient site to attach and protect various organs.

allowing it to stretch during delivery, creating a somewhat larger opening for the baby to pass (see Chapter 20).

A deep opening in the side of the pelvic girdle provides the socket in which the head of the **femur** (thighbone) rests. The femur is the largest single bone in the body. Below the knee, the leg is composed of two bones, the shinbone or **tibia,** and the slender **fibula,** both of which articulate with one of the seven **tarsals,** the bones that collectively form the ankle and heel. The foot consists of five **metatarsals,** and five toes, four of which are comprised of three bones, and one (the big toe) of two bones.

The pectoral girdle consists of a pair of shoulder blades (**scapula**) in back and a pair of collar bones (**clavicle**) in front. The rod-like clavicles connect the scapulae to the sternum; the clavicles are tightly articulated at both ends. The shield-shaped scapulae, however, are held in place only by muscles and are therefore reasonably mobile, permitting much greater mobility to the arms than the pelvic girdle permits to the legs. (You can turn your hand through 360°—palm up to palm up—but your foot only through 180°.)

A shallow depression in the scapula forms a rather loose articulation point for the bone of the upper arm (the **humerus**)—which is why shoulder dislocations are so much more common than are hip dislocations. The elbow joint is formed by the articulation of the humerus with one of the bones of the forearm—the

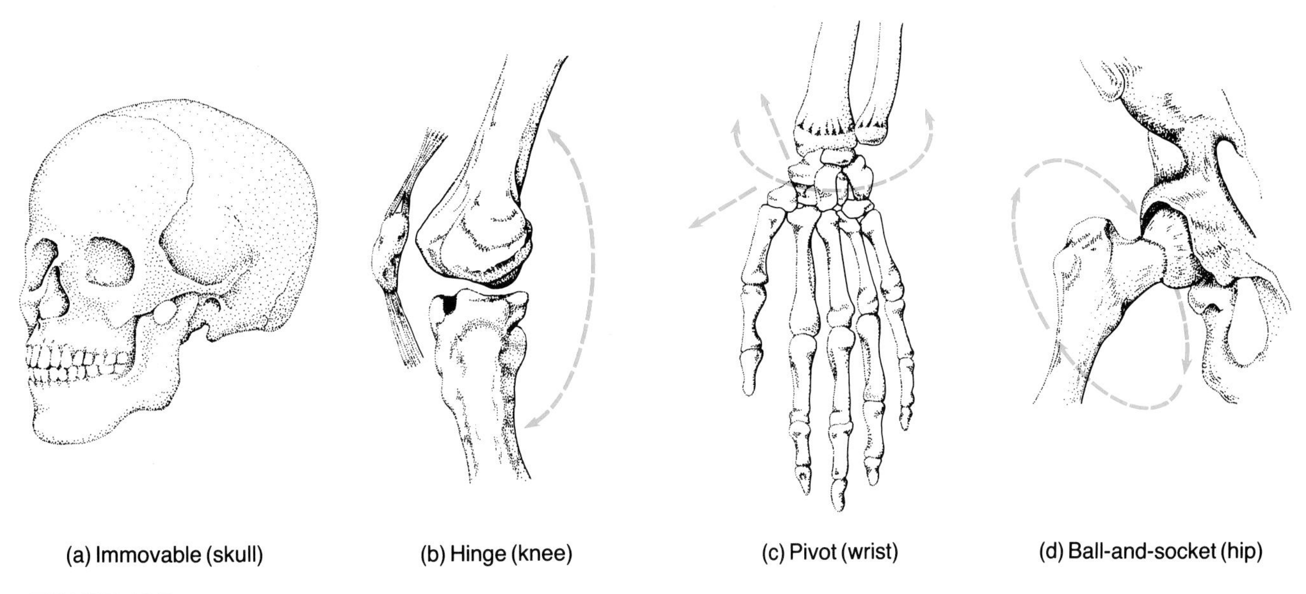

FIGURE 12.3

Types of Joints
We generally think of joints as being moveable, but some are not—and others can move in one or more planes of movement.

ulna. (The second bone of the forearm, the **radius,** articulates with the ulna, not the humerus, which permits us to turn the hand palm up or palm down without moving the upper arm.)

The eight bones of the wrist (the **carpals**) are in two rows of four bones each. The bones of the second row articulate with the five **metacarpals** which form the back of the hand. Each finger articulates with one of the metacarpals, and each contains three bones (except the thumb which, like the big toe, has only two).

Joints

As we have seen, the junction between two bones is called a joint, or articulation. Fundamentally, there are four types of articulations (see Figure 12.3):

a. **Immovable articulations,** such as those found between the bones comprising the adult skull;

b. **Hinge articulations** (one plane of movement), such as the knee or elbow;

c. **Pivot articulations** (two planes of movement), such as the ankle or wrist;

d. **Ball-and-socket articulations** (three planes of movement), such as the shoulder or hip.

Unlike immovable articulations, movable articulations must be constructed so as to avoid friction. Not only are the ends of the bones covered with a smooth layer of cartilage, but surrounding the joint is a **synovial membrane** that produces a lubricating fluid. Thus the bones never touch, but slide over one another, separated by a thin film of lubricating fluid, in the same way that oil prevents the bearings in a bicycle wheel from coming in direct contact with a metal surface (see Figure 12.4).

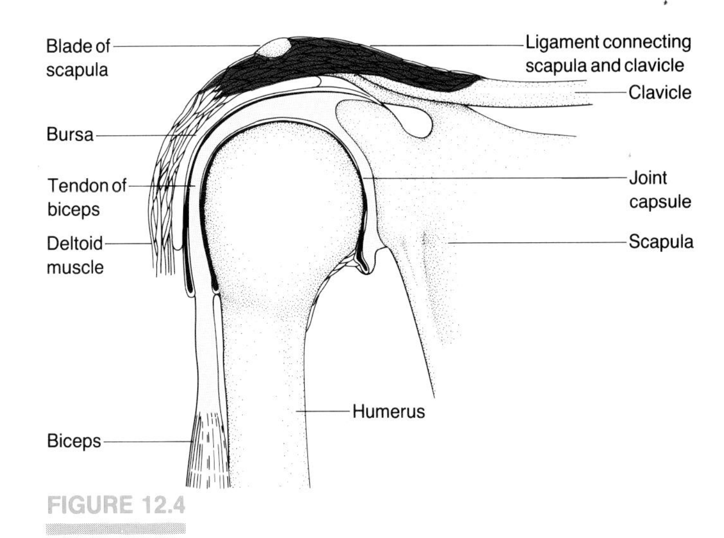

FIGURE 12.4

Human Shoulder Joint
In order to show the articulation between the humerus and the scapula (note how limited an area it is), one set of ligaments that bind the clavicle to the scapula has been omitted from this diagram.

In addition to the need to avoid friction, joints must also permit unhindered movement within a certain range while at the same time ensuring that movement does not exceed this range. For example, you can bend your lower leg backward at the knee, but (hopefully!) not forward. There are three anatomical adaptations ensuring that movement at a given joint does not exceed certain limits: the shape of the articulating surfaces, as in the elbow; the use of muscles, as in the scapula; and the use of ligaments.

Ligaments (L. *ligare,* "to tie or bind") are specialized straps of connective tissue that join bone to bone more tightly than do muscles, but not so tightly as to form an immovable articulation. Although tendons lack the flexible response that characterizes muscles, they represent a considerable gain in terms of weight and energy conservation. For example, although the hip is a ball-and-socket joint, theoretically capable of extensive movement in three dimensions, in fact we can bend forward, but not backward, at the hip while keeping our legs straight (see Figure 12.5). Ligaments prevent this unnecessary movement (falling over backward at the hip) and thus use much less energy than would be needed by a large set of muscles.

The Structure, Growth, and Repair of Bone

Bones are actually a complex mixture of inorganic calcium and phosphate salts, interwoven by strands of the protein collagen (which we also saw in the dermis), in roughly a 2:1 ratio (see Figure 12.6). Together, the salts and protein form the **bone matrix.** During embryonic development, the collagen and other organic molecules of the matrix are synthesized by bone cells called **osteoblasts** (Gr. *osteon,* "bone"). Once the matrix has been formed, the osteoblasts mature to become **osteocytes** that live in cavities, or **lacunae,** of the bone. The lacunae are linked by tiny canals, called **caniculi,** and are arrayed as a series of concentric rings around a central cavity called the **Haversian canal** (named for the English anatomist Clopton Havers who discovered them in the latter part of the seventeenth century), through which run blood vessels and nerve fibers. The entire system looks rather like a three-dimensional spider web.

The second type of bone cell, the **osteoclast,** is a giant cell that dissolves bone by producing acids and enzymes. Osteoclasts operate in conjunction with the osteocytes to permit growth and repair, the activity of both types of cells being in rough balance during adult life. The rate of breakdown and regeneration of bone diminishes with age, which is the reason why broken bones in the elderly are much more serious, and require much more time to heal, than do fractures in infants and young children.

Bones, especially the long bones of the limbs are hollow tubes, not solid bars. The outer layer (which is surrounded in life by a connective tissue sheath called the **periosteum**) is called **compact** (or solid) **bone.** The inner layer, which is spongy, is **cancellous** (L. *cancellare,* "to make lattice-like") **bone.** Within the cancellous bone, the arrangement of the salts and collagen is along lines of stress, the net effect being a structure that is relatively light in weight but exceptionally strong. **Red marrow,** a specialized blood-producing tissue, is found in the cancellous portions of many bones (see Chapter 9).

Maintenance of bone strength is largely a function of the dynamics of osteoclast and osteocyte activity in response to the continual action of muscles pulling against the bone. Paralysis, prolonged bed rest, or space flights (no gravity, hence much reduced muscle activity) all serve to reduce the activity of the osteocytes and, consequently, the thickness and strength of bone (see Box 12.2).

Although bone is the dominant tissue in the adult skeleton, the importance of cartilage is difficult to overstate. Cartilage forms the entire skeletal system of sharks and some other fish, but it is much more limited in terrestrial vertebrates, presumably because it is not as *rigid* as bone. In humans, cartilage forms the tip of the nose, the ears, "pads" in the knees, much of the intervertebral discs, and the ends of bones at the joints.

Perhaps its most important role, however, is in

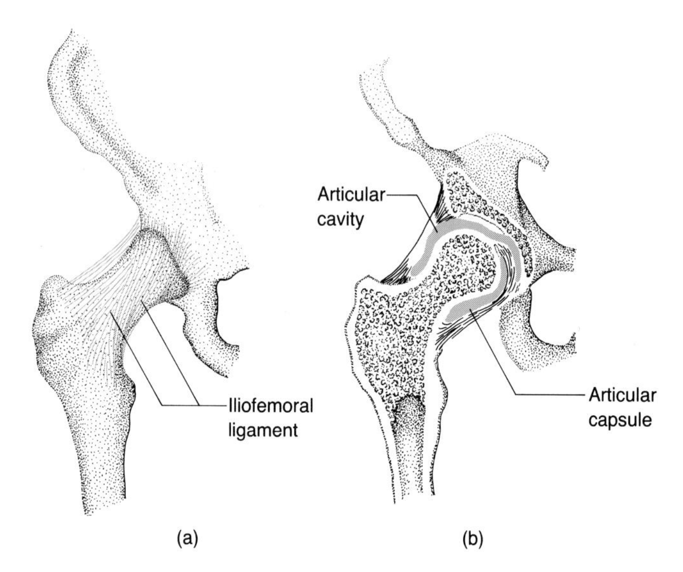

FIGURE 12.5

Human Hip Joint

(a) Anterior view, showing the main ligaments responsible for limiting backward movement at the hip. (b) Longitudinal section, showing the articular cavity and capsule.

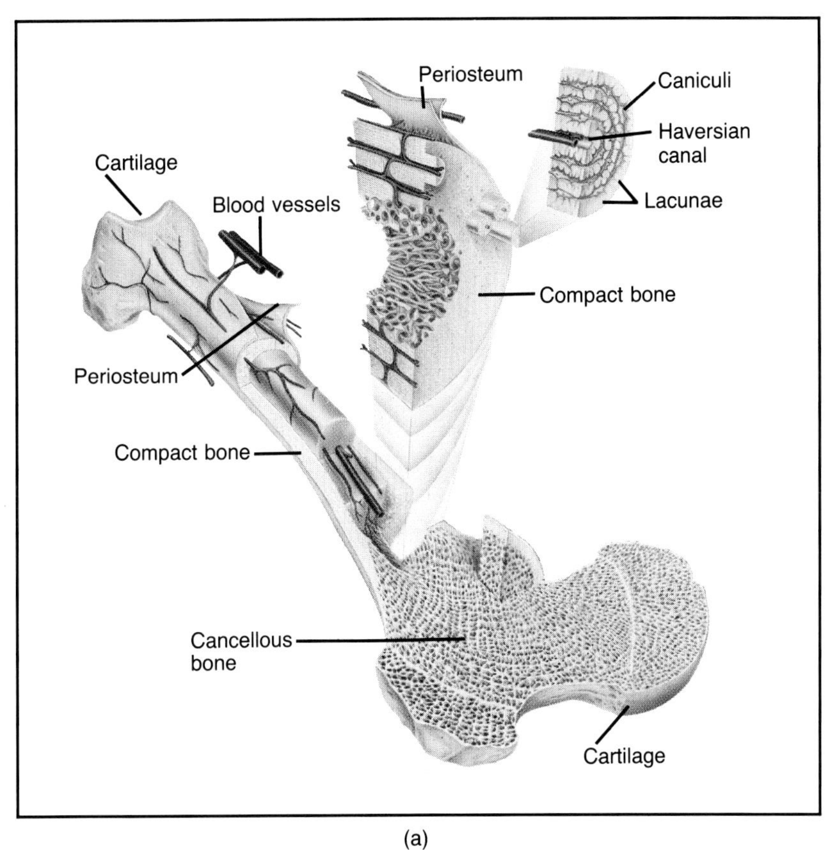

(a)

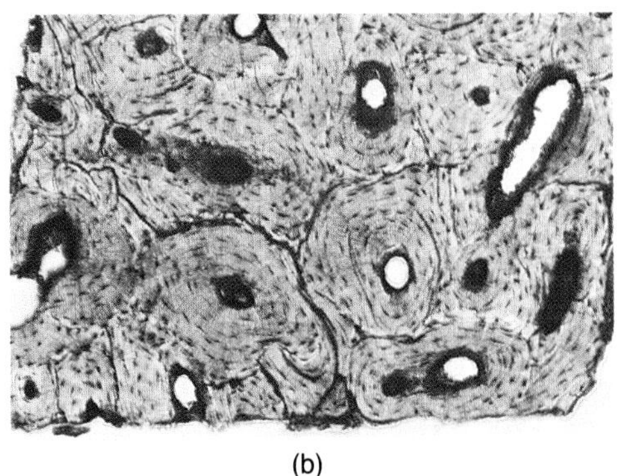

(b)

FIGURE 12.6

Structure of Bone

(a) Diagrammatic view of a section through a long bone. Blood vessels run through the Haversian canals, carrying nutrients to the osteocytes scattered about the canal. (b) A light micrograph of Haversian systems. In life, osteocytes would be present in the small cavities surrounding the canals.

BOX 12.2

New Bones from Old

There are a host of conditions—ranging from bone tumors to developmental abnormalities to surgery following accidents—where bone trauma is so extensive that bone from the ribs or pelvis must be grafted onto the damaged bone. Obviously, such procedures are both painful and risky, and the amount of bone that can be transplanted is limited. Grafts from fresh or cadaver donors face the problems of rejection (see Chapter 9).

However, a new treatment appears to be revolutionizing bone grafting. Bone from cadavers (either human or animal) is pulverized and bathed in solvents to remove most of the calcium and phosphorus. The remaining granules are then combined with water to form a paste which can be directly implanted into a graft site. This new material transforms adjacent cells into osteoblasts, the osteoblasts form cartilage, and, over a period of several months, bone replaces cartilage, just as it normally does during fetal and early childhood development. In one instance, a six-year-old boy with a chronic condition in which the skull fails to grow had his entire skull replaced with a new one formed completely from this grafted tissue!

permitting growth of the long bones during infancy and childhood (see Figure 12.7). During embryonic and fetal development, the long bones (but not the flat bones of the skull) are performed in cartilage, and only gradually during childhood does bone replace the cartilage. Even as this replacement process is underway, however, the cartilage continues to grow near the ends of the bones, permitting an increase in length. At maturity, these cartilaginous zones of growth are finally replaced by bone, and bone growth ceases.

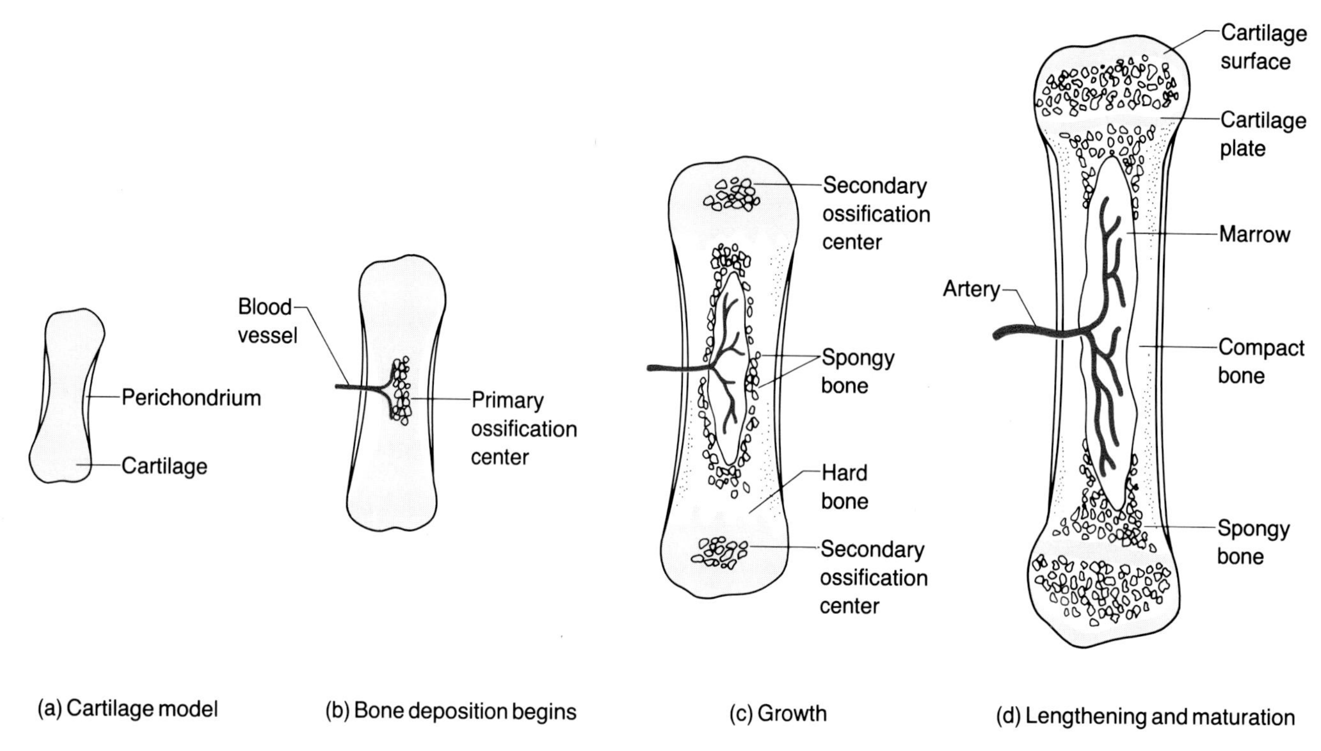

FIGURE 12.7

Growth and Development of a Long Bone

(a) Long bones begin as cartilage models during embryonic development. (b) Bone deposition begins even before birth. (c) During infancy and childhood the shaft of the bone becomes more ossified, with growth occurring because of continued development of the cartilage plates. Secondary ossification sites occur near the ends of the bone. (d) During adolescence, growth continues, but gradually the cartilage plates are replaced by bone. Once this occurs, no further increase in bone length is possible.

MUSCLES

Muscles are used in such varied ways as the support and movement of the skeleton (see Figure 12.8), the movement of food along the digestive tract, and the propulsion of blood through the blood vessels. As it happens, each of these functions employs a structurally distinct form of muscle (see Figure 12.9).

Muscles associated with bone are called **skeletal muscles.** They form by far the largest mass of muscle in the body. Under a microscope, skeletal muscle appears *striated,* because of the alignment of certain types of proteins involved in muscle contraction.

Skeletal muscle is composed of bundles of very long (up to 30 cm), multinucleated, cylindrical cells called **muscle fibers.** Skeletal muscles generally taper at their ends, forming **tendons** (Gr. *tenein,* "to stretch"), which are strap-like bands composed primarily of collagen that connect the muscle to bone. Skeletal muscles are used in contractions that are rapid and powerful, but of short duration.

The muscles of the internal organs, which are rarely found as distinct muscle masses, are termed **visceral muscles** (L. *viscus,* "inner parts of the body"). The cells in some visceral muscles—for example, in the muscle that controls the diameter of the pupil of the eye—are easily separated one from another, but generally they are loosely fused into broad sheets or bundles. Because they do not possess the characteristic striations of skeletal muscle cells, visceral muscles are sometimes called *smooth muscles.* Unlike skeletal muscles, the contractions of visceral muscles are generally slow and relatively weak but are frequently of long duration.

The third type of muscle is restricted to the heart. **Cardiac muscle** is something of a hybrid of skeletal and visceral types. It appears striated under the microscope, but unlike skeletal muscles cardiac muscle fibers are branched. Moreover, cardiac muscle fibers are bound to their neighbors by means of **intercalated discs,** structures that are found only in cardiac muscle. The result is that cardiac muscle operates as a

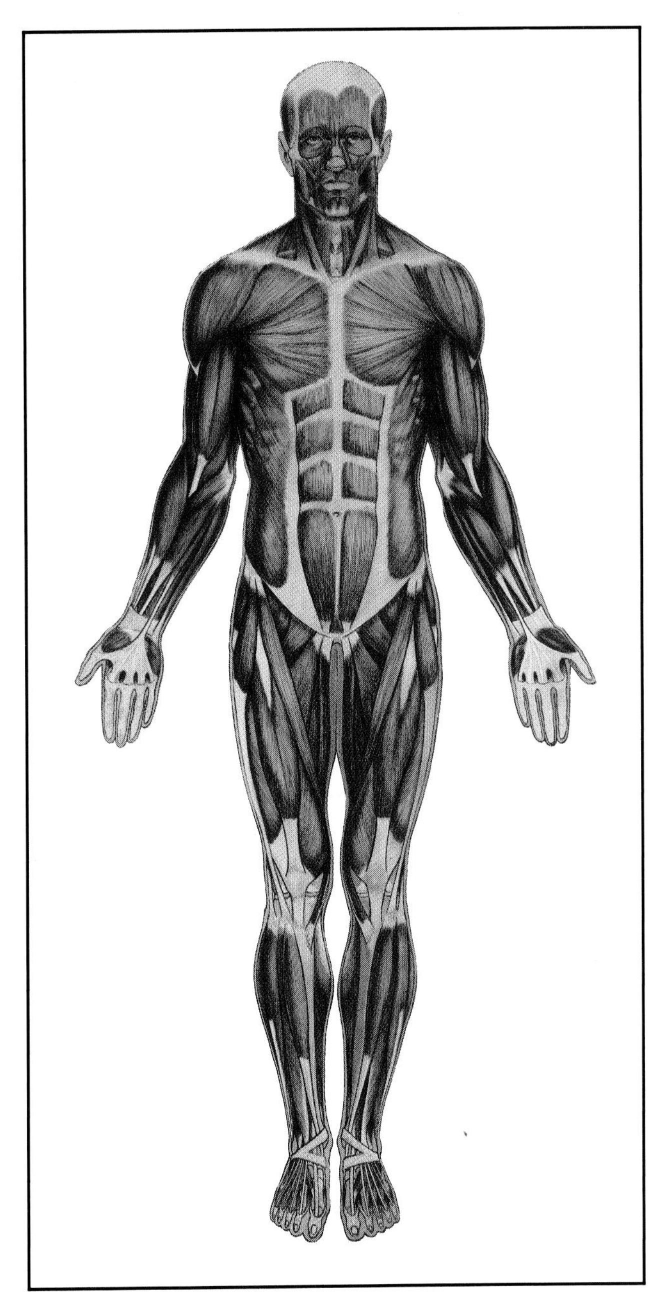

FIGURE 12.8

Human Muscular System
Muscle literally fleshes out the skeletal system, and permits the movements and expressions that characterize our species. The muscles shown here are all skeletal muscles.

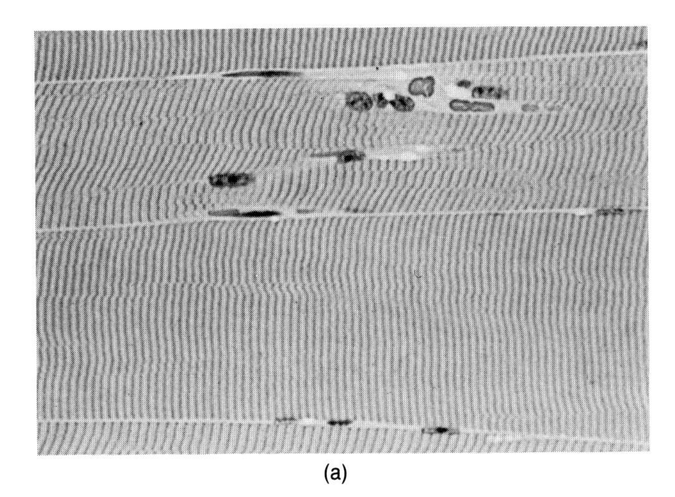

(a)

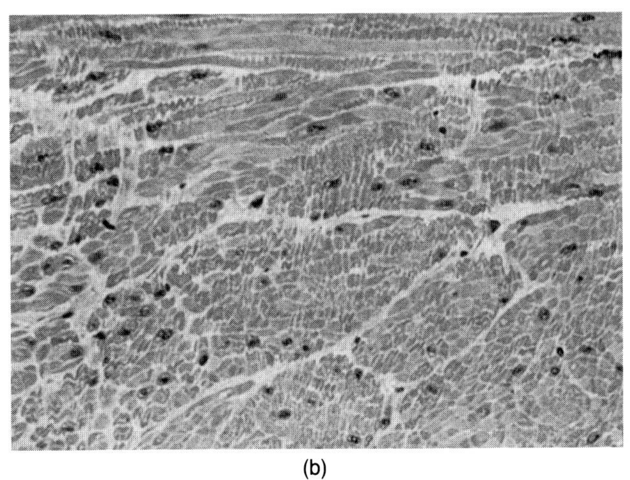

(b)

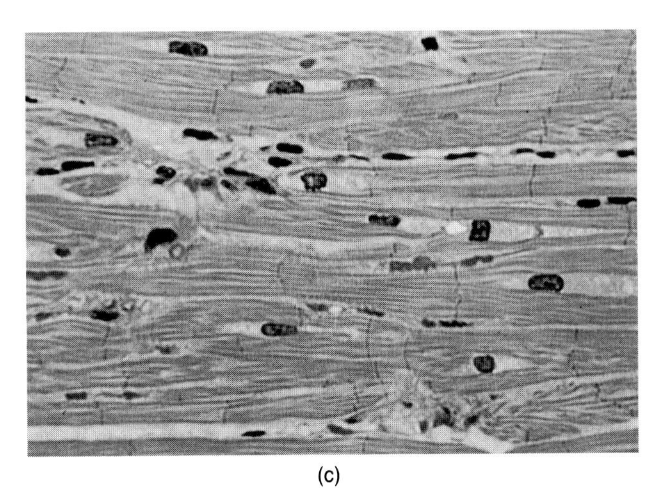

(c)

FIGURE 12.9

Types of Muscle Tissues
There are three primary types of muscle tissue. (a) Skeletal muscle is characterized by very elongate, multinucleate cells with characteristic striations. (b) Smooth muscles have spindle-shaped cells that are usually organized into sheets of tissue, and are without striations (hence, smooth). (c) Cardiac muscles are striated, but unlike skeletal muscles the cells are branched and are fused to neighboring cardiac cells in a unique manner.

syncytium (a cluster of cells that functions essentially as a single cell—an arrangement that helps to explain how it is that the cells of the heart contract together during a heart beat). As you presumably know from personal experience, cardiac muscle is capable of very frequent, powerful, rhythmic contractions.

Contraction of Muscle Fibers

In order to understand how a large muscle such as the calf muscle is able to shorten, thereby allowing you to stand on tip-toe, it is necessary to look at what is happening inside a single muscle fiber. Muscle fibers are organized very differently from the "typical" cell we discussed in Chapter 4, and the specialized anatomy of the muscle fiber is reflected by the specialized terminology used in labeling it.

Organization of Muscle Fibers. As we have just seen, the muscle fibers of skeletal muscle are extremely long, thin cells, running parallel to each other along the length of the muscle. Aside from the many nuclei and mitochondria found in each muscle fiber (can you guess why mitochondria are so abundant?), the most prominent feature inside the fiber is a specialized structure called the **myofibril,** hundreds or even thousands of which are found in each muscle fiber. Like the muscle fibers themselves, the myofibrils are long, parallel structures oriented along the length of the muscle (see Figure 12.10).

Muscle fibers also differ from regular cells in the structure of their plasma membrane and endoplasmic reticulum. The plasma membrane, which, in muscle fibers, is called the **sarcolemma** (Gr. *sarkos,* "flesh"), has many regularly spaced openings that are, in fact, the entryways to tubules penetrating deeply into the muscle fiber. Associated with the tubules are special sacs of the endoplasmic reticulum (called the **sarcoplasmic reticulum** in muscle fibers), which contains Ca^{2+}. The Ca^{2+} is released into the myofibrils when a nerve impulse is transmitted to the muscle and, as we shall see shortly, it is the presence of Ca^{2+} within the myofibril itself that permits the muscle to contract.

When skeletal muscle was first examined under the electron microscope, in the hope that knowing more about its ultrastructure would yield clues to how it functions, scientists were struck by the highly organized pattern of lines and stripes seen in each myofibril. Moreover, because the myofibrils are parallel to each other, these lines and stripes continued across the entire muscle fiber, which explained the striations characteristic of skeletal muscles when viewed under the light microscope.

Much of the early descriptive work on the ultrastructure of myofibrils was done by German scientists who, oddly enough, chose to name the various lines and stripes in German, rather than in English. These various lines and stripes are now known by the first letters of their original German names (see Figure 12.11).

Each myofibril consists of many **sarcomeres,—** repeating units, arranged end-to-end like a line of dominoes, and running the length of the myofibril. The sarcomere is bounded at each end by the **Z line,** which also happens to be the point at which the tubules from the sarcolemma penetrate into the muscle fiber.

Inside each sarcomere are two types of protein filaments. It is the distribution of these filaments that accounts for the banded appearance of the sarcomere. **Thin filaments,** which are anchored at one end to a Z line, are composed largely of molecules of **actin** (the same molecule found in the cellular organelles known as microfilaments, which we examined in Chapter 4). **Thick filaments,** which are found in the central region of the sarcomere (the so-called **A band** region), are composed of **myosin** molecules. The ends of the two types of filaments broadly overlap.

In electron micrographs of contracted muscle, the appearance of the sarcomeres is somewhat different than it is in relaxed muscle. The **H zone** (the space between the ends of the thin filaments) is virtually obliterated, and the **I band** (the space between the ends of the thick filaments) is much narrower, with the result that the sarcomere itself is noticeably shorter than is a sarcomere from relaxed muscle.

At this point, it is possible to visualize a muscle contraction as resulting from the action of thousands of sarcomeres, arranged end-to-end in each myofibril, all shortening at the same time. Fine and dandy— but *how* is this shortening within a sarcomere accomplished?

Contraction of the Sarcomere. To understand how a sarcomere can shorten, we have to increase the magnification still further and examine the details of the structure of the thick and thin filaments (see Figure 12.12).

The thick filaments are, as we have seen, composed of myosin molecules. Myosin molecules are shaped something like a golf club—they are essentially linear molecules with a bulbous portion, or head, at one end. The myosin molecules are arranged in a thick filament such that the heads radiate out along both ends of the filament. Each head has both an ATP binding site and an actin binding site. In the presence of ATP, the heads actually pivot, allowing alternate attachment and detachment in a progressive fashion along the length of the actin molecules found in the thin filaments. Thus the thin filaments are pulled toward each other by the bonds that are formed, broken, and formed again by the myosin molecules of the thick filaments.

Since we can presume that all muscles are generally rich in ATP, why is myosin not constantly bonded to actin, causing the muscle to be in a permanent state of contraction? The answer is found in the structure of the thin filament.

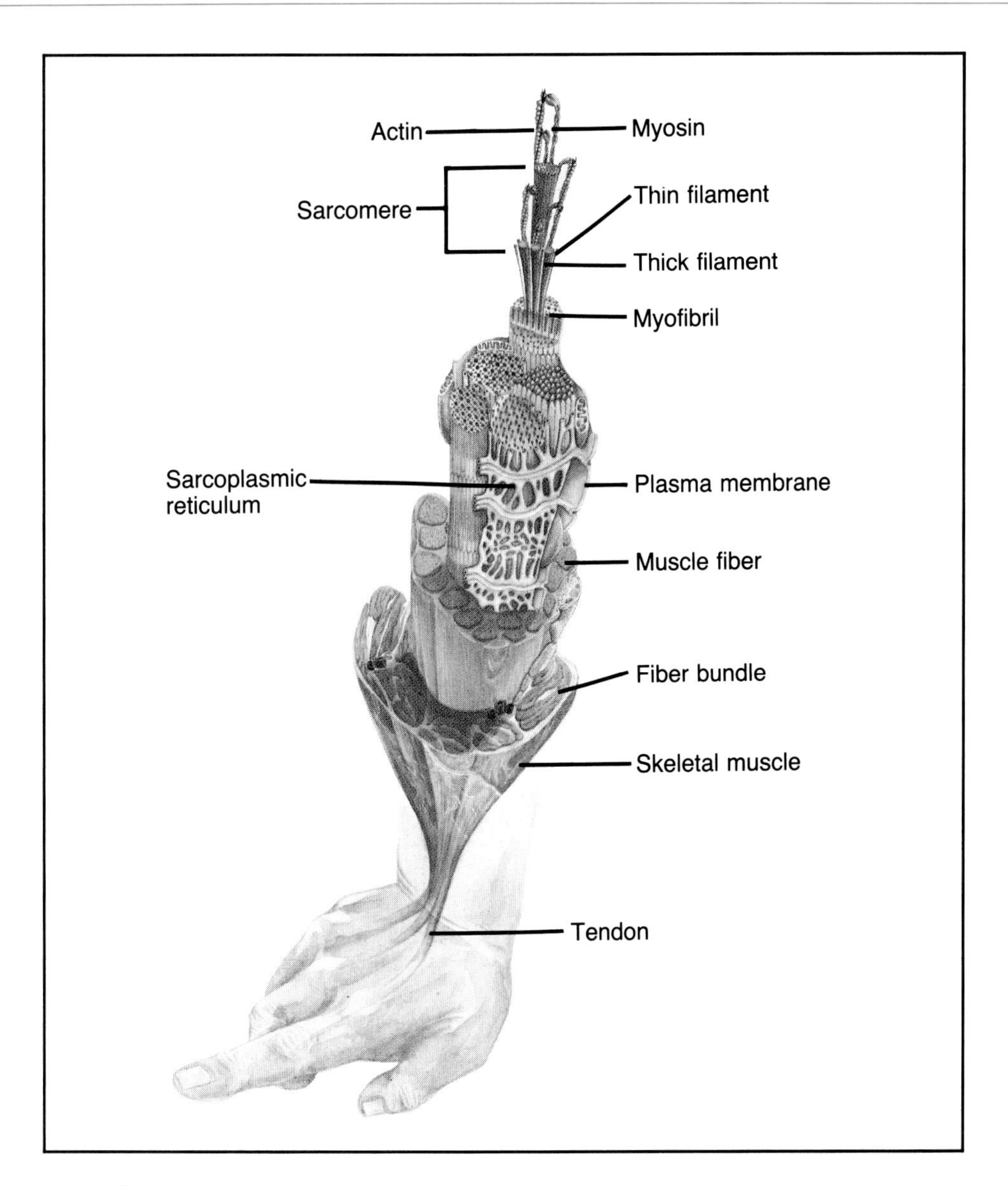

FIGURE 12.10

Structure of Muscle

Individual muscles are formed of many extremely long, thin muscle cells, each of which contains large numbers of myofibrils that are themselves made up of individual contractile units, called sarcomeres, arranged end-to-end. Within the sarcomeres are thin filaments of actin, and thick filaments of myosin. Movement of the thick filaments along the thin filaments permits the shortening we associate with contracting muscle.

The thin filaments are composed of two twisted chains of globular actin molecules. Wrapped around these two chains are a pair of twisted strands of another protein called **tropomyosin.** In the resting muscle, the tropomyosin strands cover the active sites on the actin molecules where myosin would normally bind. The tropomyosin strands, in turn, are held in place by the third protein found in the thin filaments, the globular protein **troponin.** When a nerve impulse stimulates a muscle (see Chapter 14), Ca^{2+} is released from sacs of the sarcoplasmic reticulum and enters the sarcomeres (see Figure 12.13). In the presence of Ca^{2+}, the configuration of the troponin molecules changes, which in turn permits a shifting of the tropomyosin strands and exposure of the active sites on the actin molecules. The myosin molecules from the

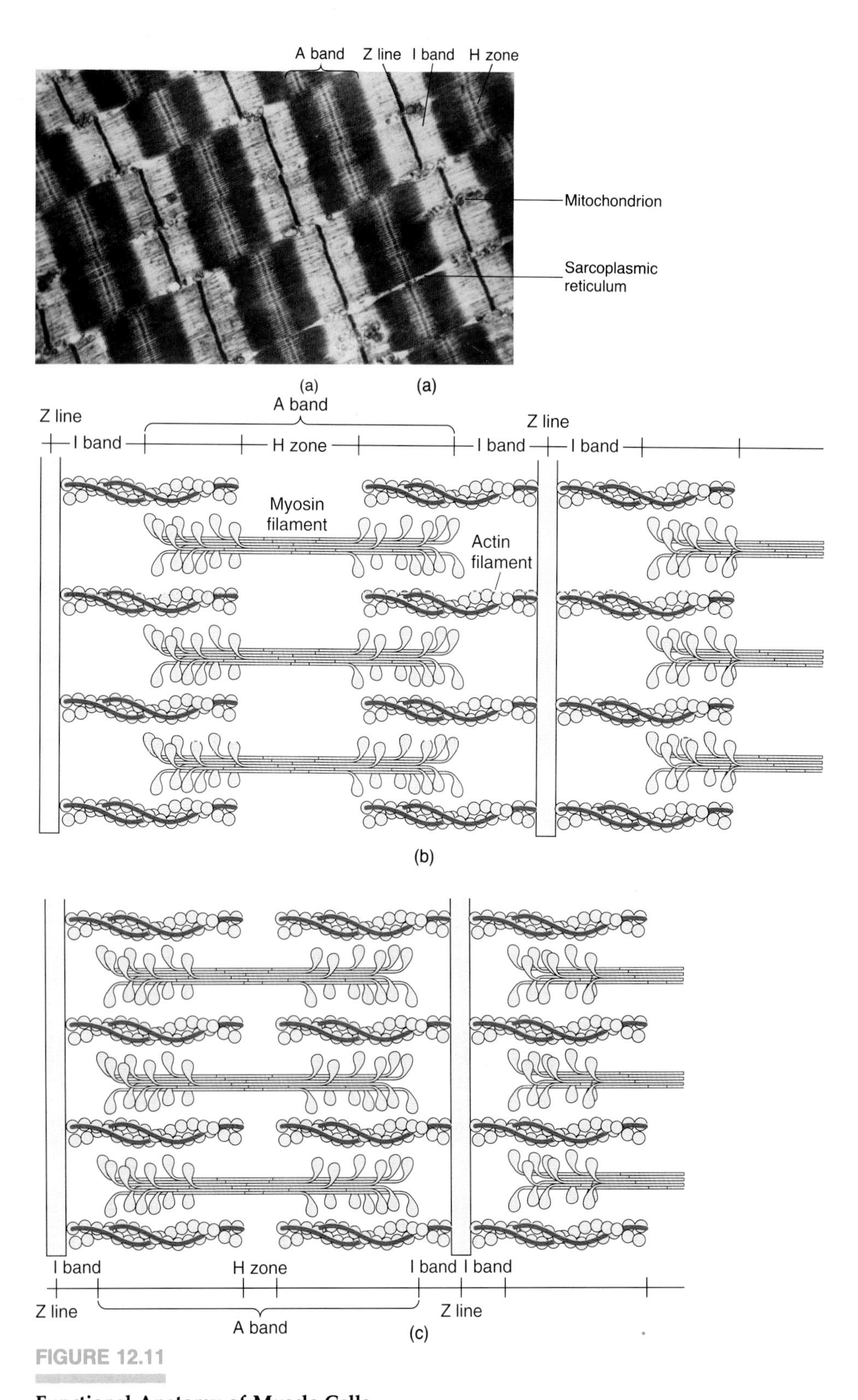

FIGURE 12.11

Functional Anatomy of Muscle Cells

(a) Electron micrograph of several sarcomeres. Note the characteristic pattern of lines and bands. (b) An artist's rendering of a single relaxed sarcomere, showing both lines and bands as well as the thin and thick filaments. (c) A contracted sarcomere. Note which bands and lines change in size—and why.

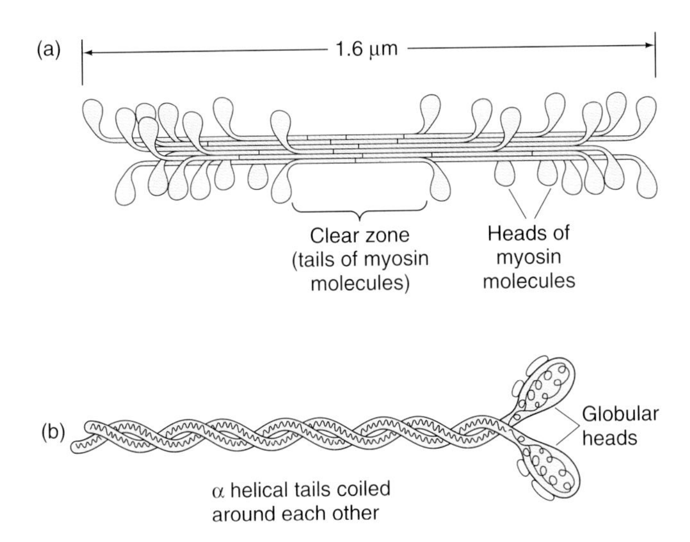

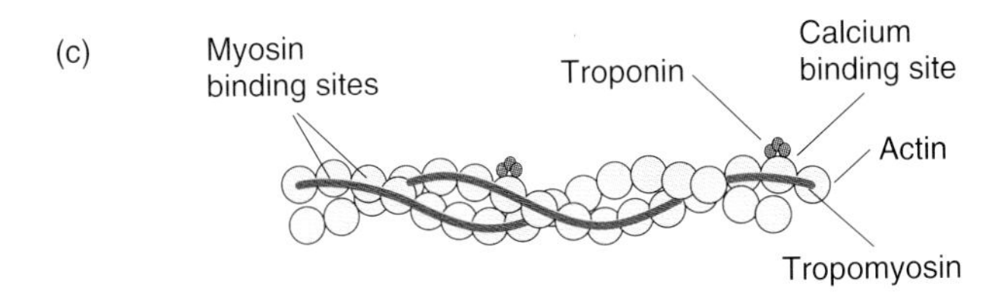

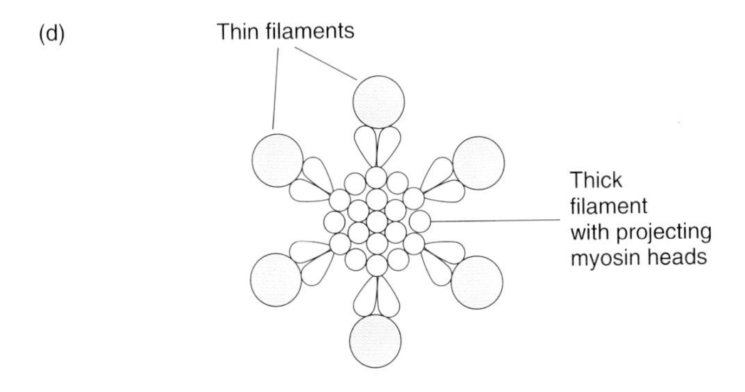

FIGURE 12.12

Structure of Contractile Proteins of Muscle

(a) Overall structure of a thick filament. Note the numerous myosin molecules, each of which has a club-like end. (b) Detailed view of a single myosin molecule. (c) The structure of a thin filament. Note the presence of both tropomyosin and troponin, proteins that are vital in regulating muscle contraction. (d) We sometimes lose sight of the fact that the sarcomere has three dimensions. When viewed end-on, each thick filament is seen to be surrounded by six thin filaments, not just two as Figure 12.11 might suggest.

thick filaments immediately bind with the actin molecules, and the sarcomere contracts. If there are no additional nerve impulses, the Ca^{2+} is immediately pumped back into the sacs of the sarcoplasmic reticulum, which causes troponin to regain its normal configuration, forcing the tropomyosin strands back across the active sites of the actin molecules. When actin-myosin binding ceases, the sarcomere relaxes.

How Muscles Function

Now that we understand the contraction of muscles at the microscopic level, let us examine how the entire muscle actually functions in the body. At the gross level, muscles function by shortening from one-third to one-half their relaxed length. Thus the larger skeletal muscles (the tendons of which, by necessity, always cross over a joint) function by pulling one bone toward, or away from, another bone.

A corollary to that statement is that muscles can function only by contracting, not by expanding. That is, muscles pull, never push. For example, a primary function of the **biceps** muscle is to pull the lower arm towards the upper arm. (When you stand with your arms at your sides, palms facing forwards, and bend your arms at the elbows in order to touch your fingertips to your shoulders, you are using your biceps.) However, the biceps are not responsible for straightening the arm out again—that is, they do not push the lower arm away from the upper arm by expanding. Rather, muscles are arranged in *antagonistic pairs*, to allow for both flexing and extending (see Figure 12.14). Thus, when the arm is straightened, it is the **triceps** muscle, running along the back of the upper arm, that contracts while the biceps relaxes.

Muscles must also be capable of contractions of different intensities. For example, you would use a less intense contraction to hold an egg between your thumb and forefinger than you would use if you were bending bottle caps—and so much for the egg if you get the two mixed up!

Muscles are composed of a great many individual muscle bundles, each containing many muscle fibers. Although each bundle is either contracted or relaxed, the strength of the contraction of the entire muscle is a function of the number of muscle bundles that are contracting. Interestingly, even in sleep, all muscles are in a state of partial contraction known as **tonus.** (Total relaxation is achieved only in death—and even then, the muscles become rigid within a few hours of death, a condition known to all mystery fans as **rigor mortis.**) A state of tonus is necessary to maintain posture and form. The specific bundles that are in a contracted state are constantly changed over time, to avoid fatigue.

At the other extreme, a muscle in a prolonged state of complete contraction may go into **tetany** (Gr. *tetanos*, "spasm"). This situation occurs when a muscle cramps, as may occur after a period of unusual exercise, and a return to normal tonus levels may take anything from a few seconds to hours or even days (see Box 12.3).

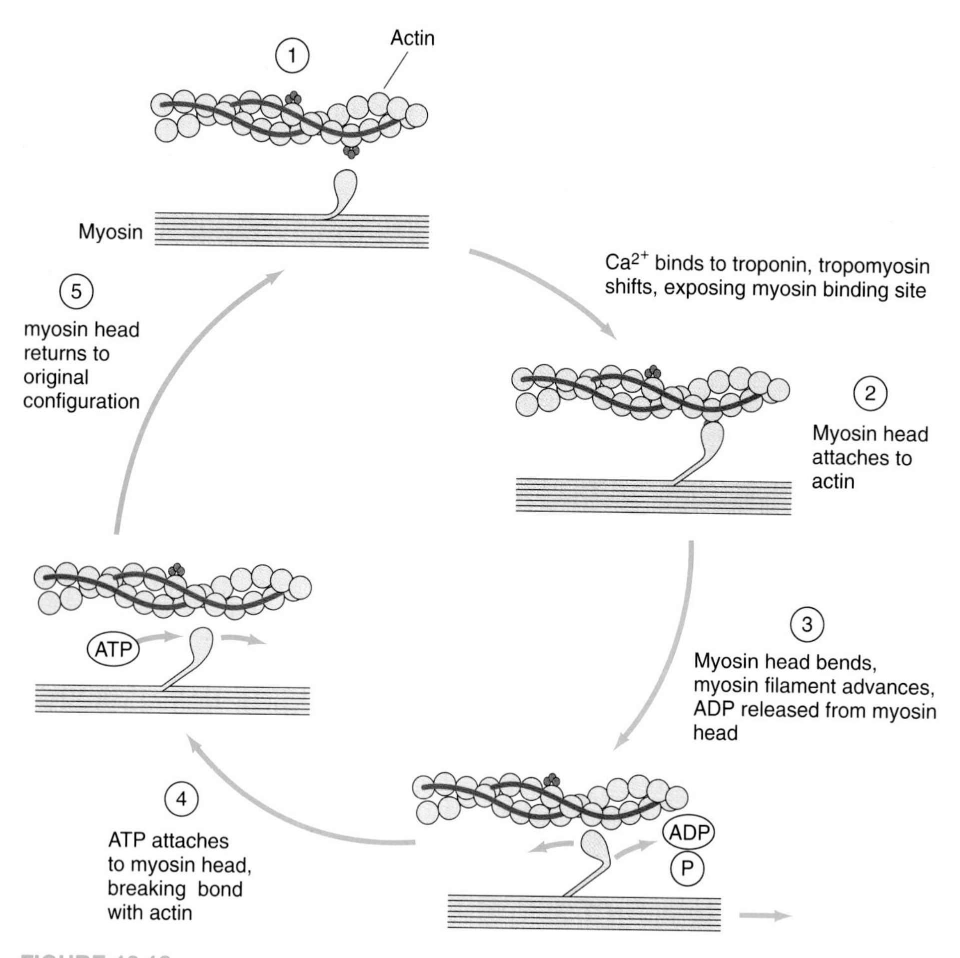

FIGURE 12.13

Interrelationship of Contractile Proteins

(1) At rest, troponin holds tropomyosin over the myosin binding sites on actin, and there is therefore no binding between the thin and thick filaments. (2) With the release of Ca^{2+} following a nerve impulse, and its binding with troponin, tropomyosin shifts, exposing the myosin binding site, and a bond forms between a myosin head and an actin molecule. (3) At this point the myosin head actually flexes, shifting the myosin filament to the right with respect to the actin filament. ADP plus a phosphate group, which have been attached to the myosin head to this point, are released. (4) A molecule of ATP attaches to the myosin head, and the energy of ATP is used to break the bond between myosin and actin. (5) The myosin head returns to its initial configuration and is now ready to form another bond with actin. This cycle of activity will continue as long as Ca^{2+} is present (or until the ATP is exhausted).

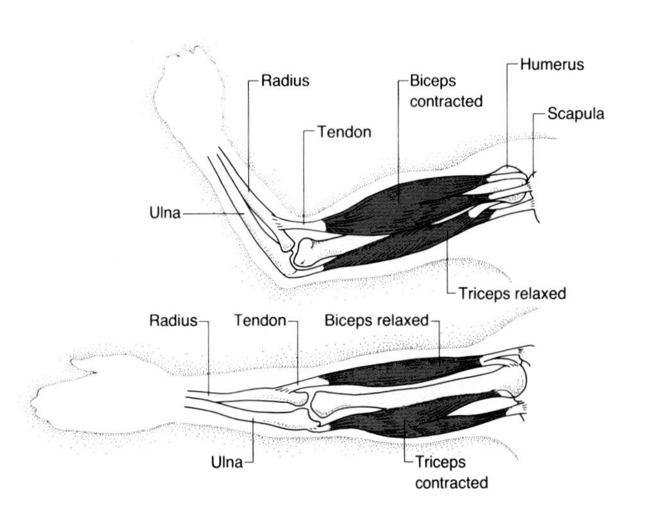

FIGURE 12.14

Antagonistic Muscles

Skeletal muscles are often arranged in antagonistic pairs, with one muscle responsible for closing a joint, and the other for opening it. Here, as the biceps contracts, it closes the elbow joint; the elbow joint opens when the triceps contracts.

BOX 12.3

Problems with Muscles and Bones

Most ailments of the skeletal and muscular systems are injuries, rather than diseases, but because of the importance of these two systems to our mobility, even minor injuries can be debilitating. Among the more common problems are the following.

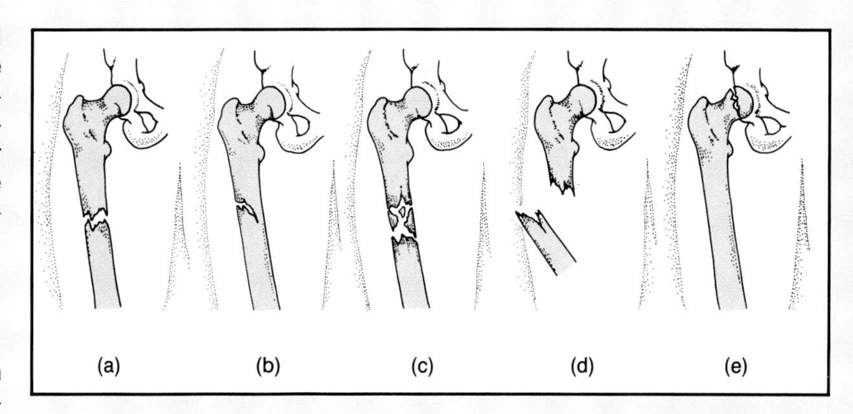

Types of Bone Fractures
(a) Simple; (b) greenstick (incomplete break); (c) comminuted (multiple breaks); (d) compound (skin pierced); and (e) articular.

Cramps

Cramps are muscle spasms wherein the muscle remains in a continuously contracted state. They may arise from a variety of sources, but most commonly they are due to a combination of fatigue and excessive use. Minor cramps are often quickly relieved by stretching the affected muscle; more serious cramps (for example, a "charley horse" of the thigh muscles) may disappear only after several days.

Most serious muscle diseases are actually problems of the nervous system or the neuromuscular junction and are discussed in Chapter 14. *Muscular dystrophy,* a genetic disease, is discussed in Chapter 19.

Sprains

A *sprain* is an injury to a joint and may involve either the joint capsule, associated tendons, or both. Severe sprains may actually be more serious than fractures, because they tend to give rise to chronic weakness in the affected joint. Treatment for sprains involves chilling for the first 24 hours, to reduce swelling, followed by gentle application of heat, and immobilization of the joint. *Dislocations* are severe sprains in which the bone is displaced out of the joint capsule.

Fractures

Fractures range from minor cracks in the bone to multiple or compound fractures involving severe damage to the bone and surrounding tissues. The age of the affected individual is also important in assessing the seriousness of the injury, for bones in older people tend to be more brittle and to heal more slowly.

Bursitis

Bursitis is an inflammation of a **bursa** (L., "purse"), which is a fluid-filled sac that surrounds tendons at the joints. Bursitis occurs in such conditions as "tennis elbow" and "housemaid's knee." Though painful, it normally responds rapidly to rest and heat.

Arthritis

Arthritis is an inflammation of a joint, and may arise from a variety of causes. Indeed, there are more than 100 different conditions that can lead to arthritis. It is a disease of enormous significance in this country. Some 37 million individuals in the United States suffer from one form of arthritis or another, with a total cost to the economy of some $8.6 billion each year—$4.4 billion of which is for hospital and nursing home services. Over one million new cases are identified each year. Major subcategories include:

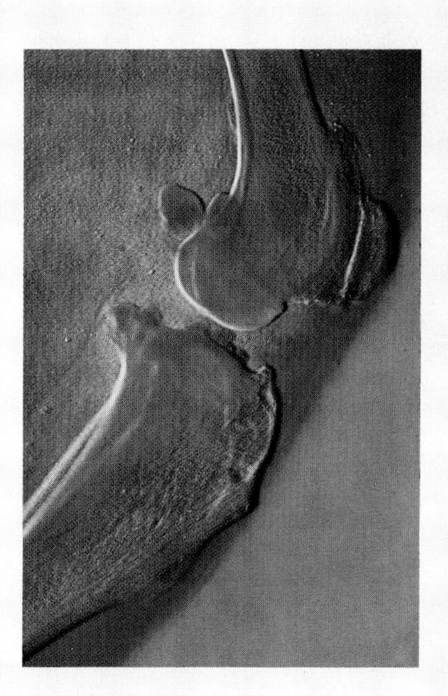

Osteoarthritis
Osteoarthritis of the knee.

1. **Osteoarthritis** is the most common form of arthritis, with 16 million cases in this country. It is found primarily, but not exclusively, in the el-

(Continued on the following page)

BOX 12.3 (Continued)

derly, where a wearing down of the smooth cartilage surfaces of a joint, and their replacement by bone spurs, causes painful movement of the affected joint. It does not result simply from overuse, but apparently because of an imbalance between the breakdown and regeneration rates of cartilage in the joints. There is a genetic component to this disease, especially in arthritis of the hands. The gene controlling the formation of collagen has recently been identified. A mutated form of this gene produces defective collagen, which wears out at the joints more quickly than normal collagen. As many as six million osteoarthritics may possess this mutation. For osteoarthritis in general, women sufferers outnumber men 3:1.

2. **Rheumatoid arthritis** is a chronic condition affecting more than two million adults and almost 75,000 children under the age of 18. The precise cause is unknown, although there are indications of both an infective agent and an autoimmune reaction (see Chapter 9). It is frequently devastating in its effects, causing

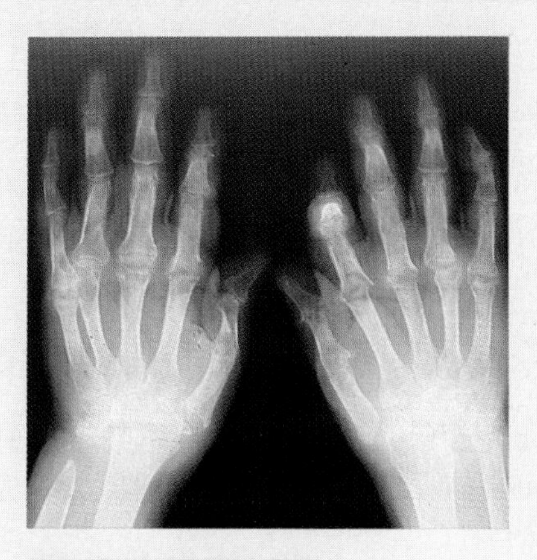

Rheumatoid Arthritis
Rheumatoid arthritis of the hand.

crippling malformations of joints throughout the body. Again, women sufferers outnumber men 2:1 in the adult form of the disease, and 6:1 in the juvenile form.

3. **Gout,** which we briefly considered in Chapter 11, affects more than one million individuals in this country, 80 percent of whom are men. The ball of the foot and big toe are the most commonly affected areas.

4. **Ankylosing spondylitis** is a form of arthritis that affects the vertebral column, and greatly restricts movement as a consequence. It can also lead to heart and lung problems. It afflicts more than 300,000 Americans, 70 percent of whom are women.

5. **Lyme disease** is the newest form of arthritis, having first been identified in Old Lyme, Connecticut, in 1975. In this instance, the cause of the disease is known—a spirochete bacterium that is transmitted by several species of ticks. Some 14,000 cases were reported in the United States between 1980 and 1988, although 5000 of those were from 1988 alone. Vaccines are now in use for dogs and cats, with a human vaccine (where standards are higher) expected by 1995. In addition to stiffness in the joints, Lyme disease can cause a wide array of symptoms, including dizziness, heart arrhythmia, facial paralysis, and memory loss.

MUSCLES AND BONES AS LEVER SYSTEMS

In order to understand the functional design of bones and muscles, it is useful to think of them as lever systems. (In this sense, muscles and bones are providing yet another example of how it is that biological systems are obliged to obey the laws of physics.) There are three principal classes of levers, and specific muscles and bones in the human body illustrate each of them (see Figure 12.15).

In Class I lever systems, the mechanical advantage achieved is a function of the ratio of the distance between the weight and the fulcrum (the **weight arm**) and the fulcrum and the point at which the force is applied (the **power arm**). Archimedes illustrated the mechanical advantage of Class I levers when he said that he would be able to move the world if only he had a place to stand. (He would also need a fulcrum, of course, to say nothing of a rather extraordinary pole!)

Because of the mechanical efficiency of this class of lever, it might be assumed that all the muscles and bones of the body would be of this type. In actuality, this class of levers is poorly represented in the body, although one example is provided by the muscles running up the back of the neck that hold the head upright, with the neck vertebrae serving as the fulcrum.

In Class II lever systems, the weight is between the fulcrum and the point of application of the power. Therefore the weight arm is entirely subsumed by the power arm. The ratio of the two arms determines the mechanical advantage. Nutcrackers and wheelbarrows are commonplace examples of Class II lever systems.

There are relatively few Class II lever systems in the human body. One example is provided by the foot and **gastrocnemius** (calf) muscle. This muscle inserts on the heel, through the Achilles tendon. When the muscle is contracted, the heel is raised upward, causing the whole weight of the body to be lifted—one is standing on one's toes.

Treatments vary with each type of arthritis, but most involve efforts to reduce inflammation of the affected joints, and thereby to permit greater freedom of movement. More than nine million sufferers take high doses of aspirin or aspirin-like medications every day—but 10,000 of those die each year from gastrointestinal complications, such as ulcers, brought on by these medications. Cortisone (a synthetic corticosteroid hormone; see Chapter 13) is more effective but has serious side effects with prolonged use. Several new classes of drugs offer great hope, although all have some unpleasant side effects. In recent years, many individuals have had artificial joints installed, replacing joints damaged beyond repair. However, for the immediate future, it would appear that arthritis will continue to represent a major medical problem to a society growing collectively older.

Osteoporosis

Osteoporosis is a weakening of the bones with age, due primarily to calcium loss. It is the major factor in hip fractures in the elderly. In extreme cases, bones are so brittle that a simple cough may lead to broken ribs. The familiar "widow's hump" is an outward manifestation of the disease. Osteoporosis affects 25 million Americans, more than 80 percent of whom are women. Indeed, 90 percent of all women over the age of 75 have some signs of the disease. There are several reasons why women are more affected than men: their skeletons are less dense to begin with; by age 85, they will lose fully one-third their skeletal mass (men lose only one-tenth by the same age); and post-menopausal women incorporate new calcium into their bones very poorly. Osteoporosis now costs this country $12 billion annually, a cost that will rise rapidly with our aging population.

There are some answers, but they apply primarily to preventing or slowing the onset of the disease, not to curing the disease once it is fully developed. Premenopausal women —especially women in their teens and early 20's—should increase their calcium intake so as to meet the current RDA (see Chapter 7). They should also have a regular exercise program—running is particularly good for building skeletal strength. They should limit their consumption of caffeine, which interferes with calcium absorption by the small intestine. In one study, elderly persons with a history of drinking more than two cups of coffee (or more than five cups of tea) a day were found to be 69 percent more likely to develop osteoporosis than were caffeine abstainers. Finally, women should consider estrogen therapy following menopause, since it is the absence of estrogen that leads to calcium loss from the skeleton and poor calcium absorption by the small intestine. Unfortunately, estrogen therapy increases the risk of breast cancer. However, a drug now being tested has been found to increase bone density in post-menopausal women by about 5 percent in two years. It functions by interfering with osteoclast activity.

Most of the muscles and bones of the body exemplify Class III lever systems in which no mechanical advantage can occur because the weight arm is always longer than the power arm. At first glance, it seems paradoxical that so many of the muscles and bones are constructed so as to sacrifice mechanical advantage, since that term has such a desirable ring to it. However, there are two relatively simple reasons.

First, because muscles cannot contract by more than one-half of their overall length, even the longest muscle can shorten by only a few cm. However, in order to gain mechanical advantage, the power arm must move through a longer arc than the weight arm. Thus, in Class I and II lever systems, the bone being moved will move an even shorter distance than the few centimeters represented by the contracting muscle. Therefore such a system would work very well if there were any real advantage in a series of powerful twitches, but it would not work well to produce extensive movement. Class III lever systems, however, are designed to produce extensive movement. When you touch your fingertips to your shoulder, your hand moves through an arc of perhaps one meter, and this movement is achieved by a contraction of the biceps of about 5 cm.

Second, both the power arm and the weight arm of any lever must move through their respective arcs in the same period of time. Thus, in the same time it takes for the biceps to contract 5 cm, the fingertips move one meter. Most mammals are characterized by speed, either because they must catch their prey, or because they must avoid becoming prey. Speed, in turn, requires the sacrifice of mechanical advantage for the ability to move the limbs rapidly through large arcs with relatively short (but powerful) muscle contractions. As a consequence, we might not be able to pick up an anvil with one hand, or open a stuck window, but we can throw a baseball 60 miles an hour or more.

Reliance on Class III lever systems limits the potential range of body design, because to maximize efficiency the major limb muscles must be located near

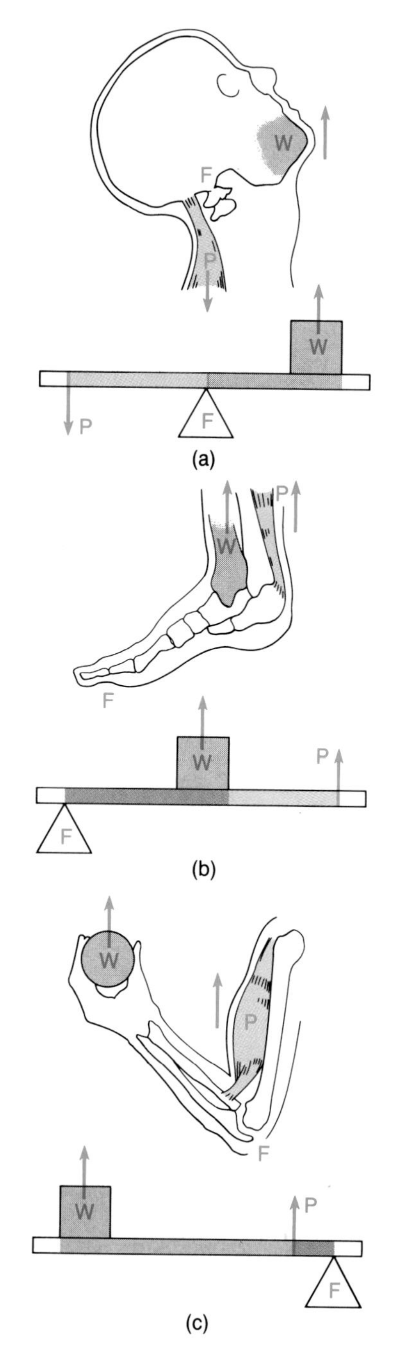

FIGURE 12.15

Muscles and Bones as Lever Systems

(a) In Class I levers the power arm is generally longer than the weight arm. (b) In Class II levers the power arm is always longer than (and includes) the weight arm. (c) In Class III levers the power arm is always shorter than (and is included by) the weight arm.

the center of the body, lest too much effort be expended simply in moving a muscle-burdened leg or arm. Consider the very slender legs of deer or horses—or the functioning of our own hands. The muscles that move the fingers are not located in our hands. If they were, our hands would look like baseball gloves. Not only would they be very heavy to move, but we would lose enormous dexterity (can you imagine trying to type, wearing baseball gloves?) Instead, these muscles are located in the forearm, as you can readily demonstrate by placing the fingers of your left hand around your right forearm and wiggling the fingers of your right hand.

MODIFICATIONS FOR BIPEDALITY

One of our most characteristic features is our bipedal method of locomotion, which distinguishes us from virtually all other species of mammals. Many species can stand or even shuffle about on their hind legs, but only humans use bipedal walking and running as the normal method of locomotion. (The bipedal gait of kangaroos and jumping mice is a very different method, because they jump rather than walk.) It is not surprising, therefore, that our bones and muscles have been modified to permit this method of movement. A brief review of these modifications will not only demonstrate the nature of evolutionary change but will also exemplify the relationship between structure and function.

The Vertebral Column

The vertebral column of apes and other primates is slightly bowed, as it is in most mammals (see Figure 12.16). The vertebral column of humans, however, is shaped like a drawn-out S. Although this arrangement limits our ability to flex our back (try duplicating your dog's ability to catch its tail or to lick the inside of its hind leg some time when you have access to a good chiropractor), the "S" shape enhances our abilities to walk erect by positioning the feet, pelvis, and skull in a vertical line. Thus the act of remaining erect becomes, for humans, less of the balancing act it is for the apes, and the amount of muscular energy needed to remain erect is also considerably less. Moreover, our skull is centered over our vertebral column, rather than being thrust forward as in apes and in most other mammals. (In humans, the brain case is ballooned backwards, and the jaws are compressed into a face, rather than being lengthened into a muzzle.) Thus we save on the muscular effort needed to keep our head erect (compare the huge necks of animals such as cows to our own).

The Pelvis

There are also considerable differences in the shape of our pelvis as compared with that of a gorilla or chimpanzee. Our shorter, dish-shaped pelvis accom-

BOX 12.4

Shape and Function of the Human Jaw

Place your fingertips on the back part of your cheek near the jaw angle, and clench your teeth tightly. You should be able to feel the sudden bulging of the contracting **masseter** muscles. Repeat the procedure with your fingertips on your temples. This muscle is the **temporalis.** Why do we have two different muscles simply to close the jaw?

The type of food eaten by a particular species, the structure of the jaw, and differential reliance on either the masseter or temporalis muscles are all closely correlated. For example, a major task of the jaw of carnivores is to withstand the struggle of their prey as they try to escape. For these animals, the temporalis muscle is vital, because it contracts at 180° to the direction of the struggling prey. Many carnivores, including the large dog breeds, have a crest running along the midline of the skull that increases the area of attachment of the temporalis muscles.

In contrast, the jaws of herbivores are designed for maximum reliance on the masseter muscles (since plants rarely struggle as they are eaten!), which do a more efficient job of effecting the grinding action of the molars than do the temporalis muscles.

As omnivores, we rely on both sets of muscles for the different tasks necessitated by our varied diet. Thus, our jaw, like our diet, is a hybrid of carnivore and herbivore.

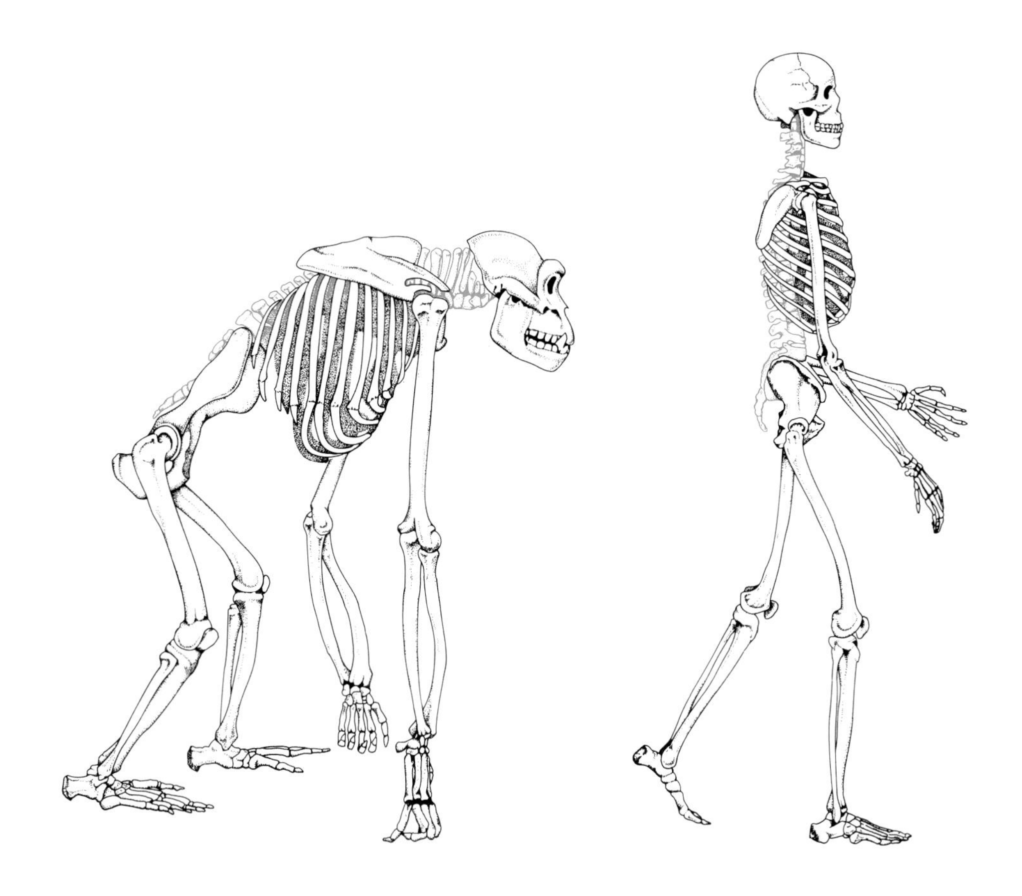

FIGURE 12.16

Gorilla and Human Skeletons

Note relative limb length, shape, and orientation of the limb girdles, the "S" configuration of the human vertebral column (as opposed to the more "C" configuration in the gorilla), and the position of the head relative to the top of the vertebral column.

modates a very large, but compact, **gluteus maximus** muscle, a muscle that is relatively small and strap-like in the gorilla. The gluteus maximus is used in running, or climbing stairs. Gorillas are incapable of doing either well, at least in bipedal form.

The Foot

The human foot is very different from the feet of all other primates, and the differences relate almost entirely to our bipedal form of locomotion (see Figure 12.17). The three most significant differences are as follows:

1. The development of the ball of the foot, which really amounts to an alignment of the toes, with a corresponding loss of opposability of the big toe (the big toe in apes is as opposable as the thumb).

2. The development of a foot arch, which allows a rotation forward on the ball of the foot during walking or running.

3. The increased development and refinement of a heel, which provides a site for insertion of the greatly enlarged gastrocnemius muscle. This muscle lifts the body up and forward, causing the weight of the body to be shifted forward onto the ball of the foot, as occurs in walking and running.

Structurally, the human foot lies between the largely unmodified mammalian foot of apes and the highly specialized feet of such animals as deer and horses. Unlike an ape's foot, the human foot permits an increase in the functional length of the entire limb, because the ankle joint is used as fully as the hip or knee joint, permitting a longer stride. However, the proportions of the human hind limb are still well short of those of the horse and other highly modified species (see Figure 12.18). In humans, the ratio of the distance from the ball of the foot to the ankle relative to the total length of the hind limb is about 1:6, whereas the ratio is 1:3 in a horse. It is partially because of the rather primitive structure of our hind limb that we are relatively slow-moving creatures.

The Shoulder

The human shoulder is very different from the shoulders of other primates. Although this difference is not essential for bipedal locomotion, it is an outgrowth of such locomotion. Once the forelimbs were no longer directly involved in locomotion, they became free for other functions.

The human clavicle is well developed. This bone assists in keeping the shoulders out to the side. It is lost or very reduced in many of the rapidly running mammals, because its presence would interfere with

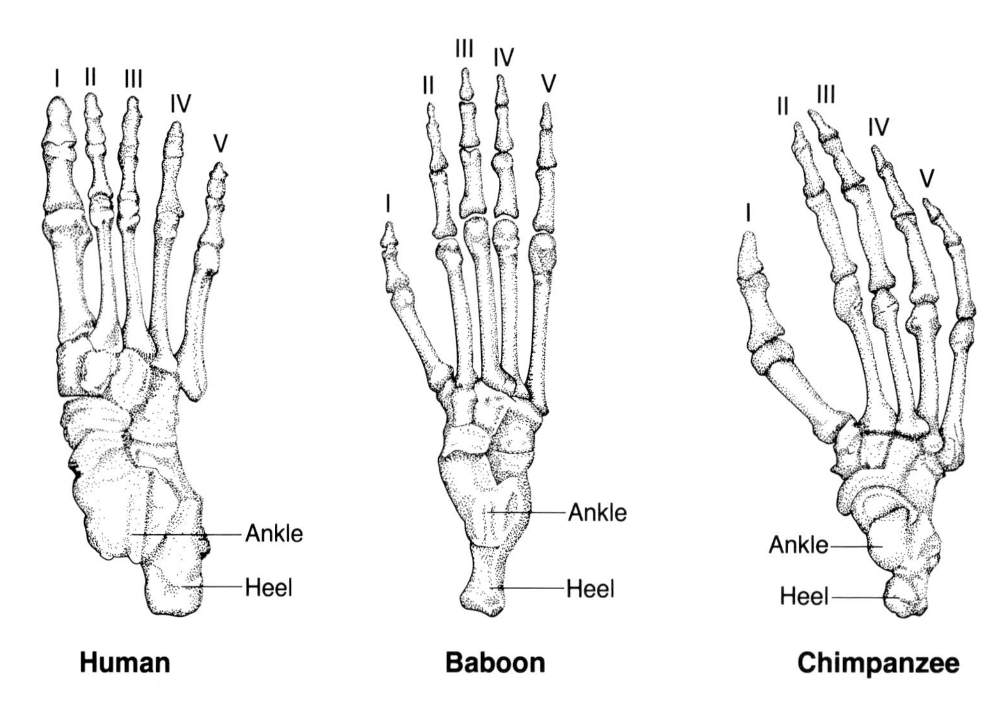

FIGURE 12.17

Human, Baboon, and Chimpanzee Feet
Note the loss of opposability of the big toe in humans, and the hand-like appearance of the foot of the chimpanzee.

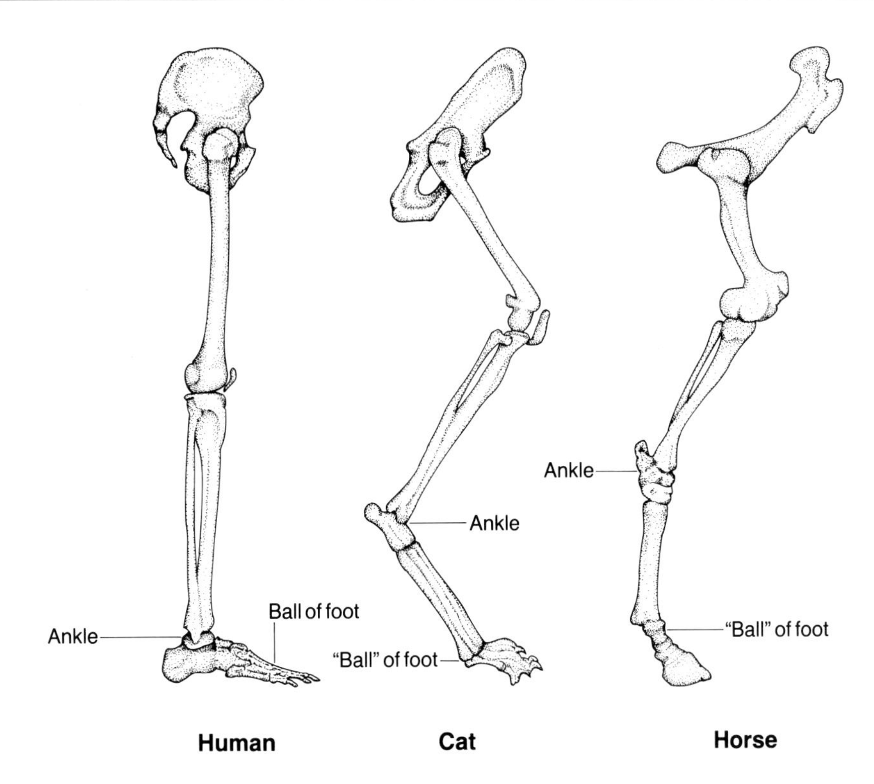

FIGURE 12.18

Hind Limbs of Human, Cat, and Horse
Note the length of the feet (toe to ankle) relative to the overall length of the leg.

their locomotion. Bone is too rigid to be an effective shock absorber, as anyone who has had a broken collarbone knows full well. (There is a tendency to thrust out the arms to break a fall, but it is the collarbone, not the fall, that is commonly broken.) Rapidly running mammals, such as antelope or lions, catch their entire weight on their forelimbs as they run; the presence of a clavicle would simply increase the likelihood that it would be broken. Instead, these animals use contracted muscle as a shock absorber, and this is one reason why they have heavily muscled shoulders.

The human scapulae lie almost flat along the back, the edges pointed toward each other. By contrast, in dogs, cats, and most primates, the scapulae are at the side of the rib cage, the edges pointing almost straight up. This arrangement permits more effective shock absorption by providing a point of attachment for the shock absorbing muscles.

BOX 12.5

Shock Absorption

Imagine standing on a chair and jumping off. Unless you are totally inept, you would land with bent knees and on the balls of your feet, taking advantage of the enormous strength and size—and, hence the shock-absorbing capacities—of your calf and thigh muscles. If you were to do the same thing stiff-legged, the shock of impact would pass largely undiminished up your skeleton, and the chances are very good that the impact of even this slight drop would be sufficient to create a first-class headache and perhaps shatter a few of your teeth.

The point is simply that providing the force necessary for movement is only the beginning of the role of our muscular system, which must also function to absorb the shock of landing as we jump or run. The skeletal system is virtually incompressible, and serves essentially no role in shock absorption, as our example above illustrates.

In humans, scapular rotation has permitted much greater lateral movement of the arms. Stand up and raise your arms in front of you, elbows straight, thumbs up, with the palms of the hands facing each other. With the arms at shoulder level, swing them apart as far as they will go. Assuming that you are in reasonably good shape, you will find that the angle formed by your arms exceeds 180°. Should you try that same movement on your pet dog or cat, you will hear a snapping sound once the angle exceeds about 140°. The point is simply that the rotation of the scapulae in humans has greatly increased the mobility of the arm, which, of course, helps maximize the manipulative abilities of the hand, and this scapular rotation has been undertaken to a much greater degree in humans than in any of the other primates.

Summary

There are more than 200 bones and more than 600 separate muscles in the human body. In size and weight, they form the dominant systems of the body. Indeed, along with the skin, the bones and muscles *are* the body, at least as we generally think of it, because it is through these systems that our body is given form and movement.

The skin plays an important role in the support of the outer tissues, and in maintaining body form. However, most support and movement is provided by the muscles and bones. The bones are organized into a skeletal system, which is subdivided into axial and appendicular skeletons.

Bone is a complex tissue, composed of both living and nonliving components. The living components, in the form of various types of bone cells, are essential in permitting bone growth and repair. The long bones of the body are preformed in cartilage, which is only gradually replaced by bone during the childhood years. Some cartilage remains in the adult, most notably in the ears, nose, and intervertebral discs.

The three types of muscles (skeletal, visceral, and cardiac) are isolated from each other in the body. They perform significantly different functions and are visibly distinct from one another when viewed under the microscope. Skeletal muscle is striated, and an examination of this tissue under the electron microscope shows in great detail the alternate pattern of thick and thin filaments which, by sliding across each other, are ultimately responsible for muscle contraction.

Bones and muscles together form lever systems, primarily of the Class III type which sacrifices mechanical advantage for speed and amount of movement. This arrangement permits very rapid, long strides of the legs from relatively slow and short contractions of muscles.

The relationship between form and function is elegantly revealed by examining those modifications of the basic primate skeletal patterns which, in humans, have permitted the effective development of bipedal locomotion.

Key Terms

epidermis
epithelium
stratum germinativum
melanocytes
melanin
dermis
subcutaneous tissue
keratin
sweat gland
sebaceous gland
collagen
elastin
bone
cartilage
articulation
skeleton
axial skeleton
appendicular skeleton
cranium
mandible
cervical vertebrae
thoracic vertebrae
lumbar vertebrae
sacrum
coccyx
intervertebral disc
sternum
pelvic girdle
pectoral girdle
femur
tibia
fibula
tarsal
metatarsal
scapula
clavicle
humerus
ulna
radius
carpal
metacarpal
synovial membrane
ligament
bone matrix
osteoblast
osteocyte
lacunae
caniculi
Haversian canal
osteoclast
periosteum
compact bone
cancellous bone
red marrow
skeletal muscle
muscle fiber
tendon
visceral muscle
cardiac muscle
intercalated disc
syncytium
myofibril
sarcolemma
sarcoplasmic reticulum
sarcomere
Z line
thin filament
actin
thick filament
A band
myosin
H zone
I band
tropomyosin
troponin
biceps
triceps
tonus
rigor mortis
tetany
weight arm
power arm
gastrocnemius
gluteus maximus

Box Terms

retinoic acid
cramp
sprain
fracture
bursitis
bursa
arthritis
osteoarthritis
rheumatoid arthritis
gout
ankylosing spondylitis
Lyme disease
osteoporosis
masseter
temporalis

Questions

1. Briefly describe the structure of the skin.
2. What is the axial skeleton? How does it differ from the appendicular skeleton?
3. Are all joints the same? If not, how do they differ?
4. Briefly describe the structure of bone. How do nutrients reach the living cells of the bone?
5. How do bones grow during childhood? Why do we no longer increase in height at maturity?
6. Distinguish among the three types of muscle tissues.
7. At the molecular level, how does a skeletal muscle shorten during contraction?
8. Describe muscles and bones as lever systems. Discuss the significance of mechanical advantage (strength) versus range and speed of movement.
9. List three skeletal differences that distinguish humans from other primates.
10. What is arthritis? Is it a single condition?

UTAH

PART

IV

Integration

As we all know, our cells are specialized into tissues, organs, and organ systems, each with a vital role to play in the operation of our bodies. However, how does each organ know what the other is doing? How does the body as a whole respond to changes in the external environment? How, in short, are our physiological functions coordinated for best effect?

The fact that cells are capable of communicating with each other is not a recent discovery. The production of electrical signals by nerve cells has been known for more than a century; the existence of chemical signals, called **hormones** (Gr. *hormaein,* "to stimulate"), which are produced by certain tissues and glands and delivered by the blood to distant **target organs,** has been known since 1902; the ability of certain cells to induce particular modes of differentiation in adjacent tissues during embryonic development has been known since 1912 (although the definitive paper was not published until 1924).

Initially, these categories of communication were seen as being independent of each other. In their zeal for categorization, biologists were quick to label those cells employing electrical changes as components of the **nervous system.** Cells that produce hormones were placed in an anatomically dispersed **endocrine system** (Gr. *endon,* "within" and *kreinen,* "to separate," a reference to the fact that hormones are produced by glands without ducts, meaning that their products must be secreted directly into the bloodstream, as distinct from glands with ducts, such as the sweat glands of the skin.) Cells that produce inducers during embryonic development were conveniently distinguished from the nervous and endocrine systems on a temporal basis—that is, they are active early in embryonic development, before the nervous and endocrine systems are fully developed and are thereafter silent. Thus we have three convenient pigeonholes for three distinct processes.

Unfortunately, biological phenomena tend to resist such convenient categorization, and the boundaries of these three categories have become increasingly fuzzy in recent years. To give just a few illustrations:

First, the nervous system does indeed use electrical changes in communication—but only along the length of individual cells. Communication *between* nerve cells requires the use of chemicals, and in several instances these chemicals are also hormones (in the sense that identical chemicals are produced by cells outside the nervous system proper and are released into the blood where they ultimately affect specific target organs). Thus the use of chemicals in communication is common to both the nervous and endocrine systems.

Second, certain brain cells produce hormones that regulate the activity of specific endocrine glands. Thus the notion that the nervous and endocrine systems are truly independent systems of communication within the body is simply not true—the nervous system in fact regulates much of the activity of the endocrine system.

Third, so-called "local" hormones have recently been discovered. These include chemicals that affect the activity of adjacent cells without being transmitted through the body by the circulatory system. In other words, these are substances that have aspects of both hormones and inducers, and their existence muddies the waters that were once seen as clearly separating the endocrine system from embryonic development.

Fourth, we now realize that "development" is not a process that is limited to embryonic and fetal life. The constant regeneration of all of the various blood cells from undifferentiated tissue in the bone marrow is but one of many examples of development that occur throughout the life of the individual—and the specific development of each type of blood cell is under the control of chemicals that are usually called "hormones," although they could as easily be labeled "inducers."

In a paradoxical manner, all of this blurring of distinctions between and among the three once clearly separated processes clarifies, rather than obscures, the true picture of what is happening inside the living organism. The cells of the body must communicate with each other if the organism is to function as a single entity, rather than exist as a colony of isolated and independent cells and tissues. The nature of this communication between cells is, in every instance, chemical (remember, electrical changes are used only *within* cells of the nervous system). There are many kinds of communication necessary within the body, and it should not be surprising to us that there are, as a consequence, many kinds of responses to chemical signals. For example, consider the following responses to an appropriate chemical signal:

1. Muscle cells immediately contract.
2. Nerve cells transmit electrical impulses along their often-impressive lengths extremely quickly.
3. Certain cells in the digestive system produce digestive enzymes.
4. Bone cells absorb or release Ca^{2+}.
5. Cells in the testes or ovaries develop into sperm or eggs.
6. Precursor cells in the bone marrow differentiate along specific developmental pathways into particular types of blood cells.
7. Cells in the brain subtly change the nature of their association with other brain cells and store memories.
8. Cells of the immune system are activated.

There are many more examples that could be listed.

If we look at the kinds of reactions that particular cells can have to chemical signals, we are frequently able to separate them into discrete categories—a task that is often worthwhile, since it permits us to focus on a particular type of cells or tissue and to consider what underlies the commonality of the response. Therefore we will continue to follow tradition and discuss the endocrine system (Chapter 13) separately from the nervous system (Chapter 14), and talk about developmental processes in yet another chapter (Chapter 21). In doing so, however, it is important that we not forget the underlying commonality that all these processes share: the use of chemicals as the mode of communication between cells.

Brunswick
2000

CHAPTER

13

The Endocrine System

Long-Term Coordination

Two individuals undergo surgery for enlarged thyroid glands, and equivalent amounts of glandular tissue are removed from each. One person recovers quickly and is soon leading a normal life once again; the other dies of convulsions a few days after the operation. Why?

Why does too little iodine in the diet of young children cause dwarfism and mental retardation? Why is diabetes a leading cause of blindness? Why does a coma result from both too much and too little insulin?

In each case, the proper functioning of one of the endocrine glands—the subjects of this chapter—is impaired.

THE NATURE OF THE PROBLEM

The fact that both our internal and our external environments are constantly subject to change creates a never-ending challenge for our homeostatic control mechanisms. If the body is to meet this challenge, our tissues and organs must be capable of communicating with each other. For the most part, this communication relies heavily on the circulatory system. This reliance is not particularly surprising since the blood and interstitial fluid constitute our internal environment, and, as we have already seen, many body tissues are sensitive to (and react to) particular changes in blood chemistry. Because these changes are generally the result of the metabolic activity of various organs, it is not a particularly large step to arrive at the point where certain tissues secrete very specific chemical messages into the blood that are reacted to by particular tissues and organs often located in distant parts of the body.

However, as we have described it, this scenario gives rise to many questions. What kinds of messages are sent, and what processes do they regulate? How is the message made specific for just one particular organ and no other? If an organ does receive a message, how does it react? That is, precisely what is it that takes place inside the cells of the receiving organ as a consequence of having received a message? Finally, if receiving a message causes a tissue or organ to change in some way, how does it change back (or does it)? That is, if we liken a message to a doorbell, what is the counterpart to taking our finger off the doorbell? How, in effect, does the organ know when the message ends?

In the hope that you are by now sufficiently intrigued, let us examine the endocrine system.

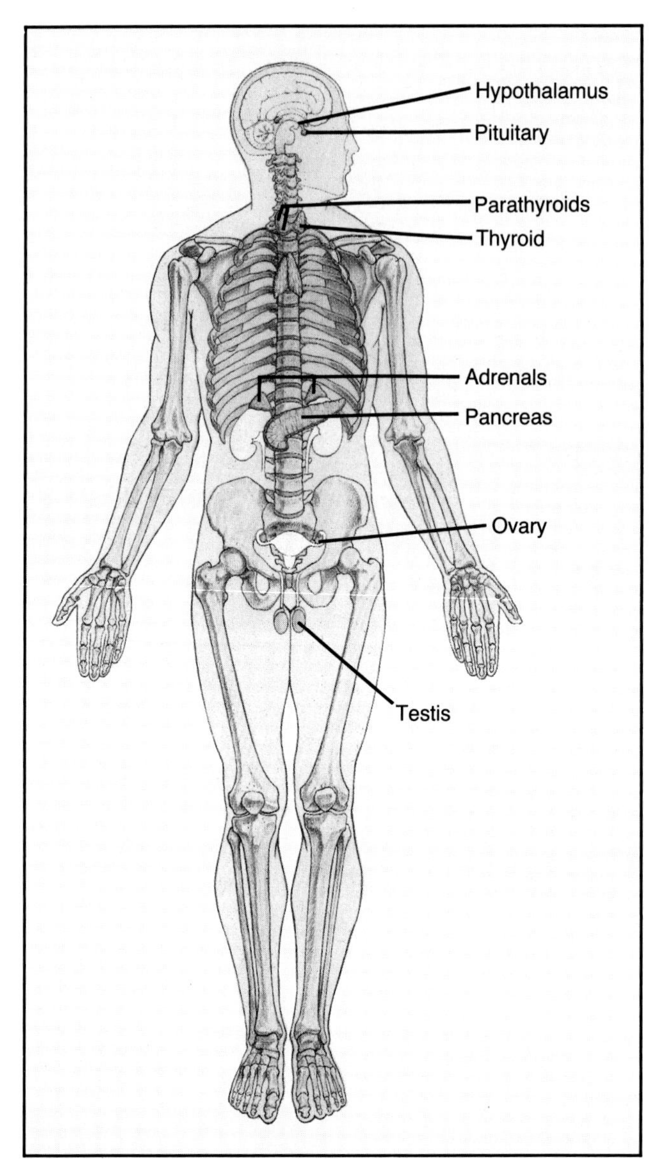

FIGURE 13.1

Major Glands in the Endocrine System

Note that, unlike the other systems, there is no anatomical continuity among these glands. They are a system because of similar physiological functions, not because they share a common evolutionary and embryological heritage.

THE ENDOCRINE SYSTEM

Unlike the other organ systems, the endocrine system is not anatomically continuous (see Figure 13.1). In fact, it is not really a system at all. Rather, it consists of those glands and tissues scattered throughout the body that share the capacity to secrete hormones. **Hormones** are chemical messengers that are secreted into the bloodstream. They travel to various parts of the body, ultimately influencing the metabolic activity of one or more **target organs,** typically either by affecting the rate at which they produce particular enzymes, or by directly influencing the activity of preexisting enzymes (see Table 13.1).

The fact that hormones travel via the bloodstream is important definitionally (see Box 13.1). **Endocrine** (hormone-producing) **glands** are ductless (since their products are secreted directly into the circulatory system); in contrast, **exocrine glands**—which include salivary, sweat, sebaceous, and digestive glands, among many others—have ducts, and their products are released onto body surfaces (such as the skin or the interior lining of the digestive system).

TABLE 13.1 Major Endocrine Glands and Their Hormones

SOURCE TISSUE OR ORGAN	HORMONE	TARGET TISSUE OR ORGAN	PRIMARY ACTIONS
Hypothalamus	Prolactin release inhibiting hormone (PIH)	Anterior pituitary	Inhibits production of prolactin
	Growth hormone releasing hormone (GHRH)	Anterior pituitary	Stimulates production of growth hormone
	Somatostatin	Anterior pituitary	Inhibits production of growth hormone
	Thyroid stimulating hormone releasing hormone (TRH)	Anterior pituitary	Stimulates production of thyroid releasing hormone (TSH)
	Gonadotropin releasing hormone (GnRH)	Anterior pituitary	Stimulates production of follicle stimulating hormone (FSH) and luteinizing hormone (LH)
	Corticotropic releasing hormone (CRH)	Anterior pituitary	Stimulates production of adrenocorticotropic hormone (ACTH)
	Oxytocin[1]	Uterus	Stimulates uterine contractions
		Mammary glands	Stimulates release of milk
	Antidiuretic hormone (ADH)[1]	Kidney	Increases rate of water reabsorption from glomerular filtrate
Anterior pituitary	Follicle stimulating hormone (FSH)	Ovaries	Stimulates maturation of egg follicle
		Testes	Stimulates maturation of sperm cells
	Luteinizing hormone (LH)	Ovaries	Stimulates ovulation
		Testes	Stimulates testosterone production
	Prolactin	Mammary glands	Stimulates milk production
	Growth hormone (GH)	Body-wide effects	Stimulates growth through diverse means
		Liver	Stimulates production of somatomedin
	Thyroid stimulating hormone (TSH)	Thyroid gland	Stimulates production of thyroxin
	Adrenocorticotropic hormone (ACTH)	Adrenal cortex	Stimulates production of adrenal steroid hormones
Thyroid	Thyroxin, triiodothyronine	Body-wide effects	Increases metabolic rate
	Calcitonin	Bone	Lowers blood calcium levels by promoting uptake of calcium by bone
Parathyroids	Parathormone	Bone, intestine, kidneys	Raises blood calcium levels by promoting liberation of calcium from bone, uptake of calcium by intestine, and reabsorption of calcium by kidney
Pancreas	Insulin	Many tissues	Lowers blood glucose levels by promoting glucose uptake by cells; increases protein synthesis
	Glucagon	Liver	Raises blood glucose levels by converting glycogen to glucose and by promoting gluconeogenesis
Adrenal cortex	Mineralocorticoids	Kidney	Promotes sodium reabsorption and potassium excretion
	Glucocorticoids	Many tissues	Promotes gluconeogenesis; inhibits inflammation and immune function
	Adrenal sex hormones	Sex organs	Early development of sex organs (especially in the male)
Adrenal medulla	Adrenaline (epinephrine)	Liver, fat tissue, heart	Promotes glycogen breakdown and gluconeogenesis; increases cardiac output

(Continued on the following page)

TABLE 13.1 ***(continued)***

SOURCE TISSUE OR ORGAN	HORMONE	TARGET TISSUE OR ORGAN	PRIMARY ACTIONS
	Noradrenaline (norepinephrine)	Blood vessels	Constricts blood vessels and raises blood pressure
Testes	Testosterone	Many tissues	Promotes secondary sexual characteristics and development of sperm cells
Ovaries	Estrogen	Many tissues	Promotes secondary sexual characteristics and growth of uterine lining
	Progesterone	Uterus	Prepares uterine lining for implantation
Stomach and intestine	Gastrin, secretin, cholecystokinin	Organs of digestive tract	Regulates activity of stomach, intestine, and associated organs
Liver	Somatomedin	Many tissues	Stimulates growth
Heart	Atrial natriuretic factor (ANF)	Kidney, blood vessels	Stimulates sodium excretion; lowers blood pressure

[1]Oxytocin and ADH are produced by the hypothalamus but stored in, and released from the posterior pituitary.

BOX 13.1

Local Hormones

As we noted earlier, the line between "classic" hormones produced by distinct endocrine glands and substances that are more properly called "inducers" has become very fuzzy. Consider, for example, some categories of what might be termed "local" hormones.

Prostaglandins

Almost 20 different *prostaglandins* have been identified, all of which are apparently synthesized from phospholipids in the plasma membrane of cells. Rather than being produced by distinct glands, they are instead synthesized rather generally by many different types of tissues, and they have a host of functions, including the following:

1. Stimulation or relaxation of visceral muscle
2. Dilation or constriction of blood vessels
3. Dilation or constriction of the bronchioles of the lungs
4. Stimulation of the peristaltic rate of the small intestine
5. Stimulation of the inflammation response
6. Enhancement of pain reception
7. Modulation of transmission of nerve impulses
8. Promotion or inhibition of blood clotting

While some of these are undoubtedly related (the first four all involve visceral muscle action), the list is still impressive for its diversity.

Prostaglandins are used medically to induce labor in women who are late in delivering, and to initiate abortion. However, in other instances, the goal has been to block their effects. For example, prostaglandins are thought to be primarily responsible for the often painful cramping that many women experience during menstruation. Interestingly, the role of aspirin, long a mystery, has been clarified with the discovery of the prostaglandins, because aspirin appears to block the effects of many prostaglandins—in particular, those responsible for the enhancement of inflammation, pain, and blood clotting. Since spontaneous blood clots are a primary cause of heart attacks, many people prone to heart attacks are now taking an aspirin every other day to control the blood clotting enhancing prostaglandin. (Fortunately, aspirin does not affect the prostaglandin that inhibits blood clotting.)

Endorphins and Enkephalins

The *endorphins* are small proteins of about 32 amino acids in length that mimic the effects of morphine and opium. They are released in times of stress, and inhibit nerve cell activity in certain regions of the brain, reducing pain and enhancing the feeling of well-being. (The so-called "runner's high" may result from the release of endorphins.) They apparently play other roles as well: They have been implicated in the buildup of cholesterol in the blood of individuals under stress, and in slowing the rate of cell division (their capacity to inhibit the growth of certain cancer tumors is now being tested).

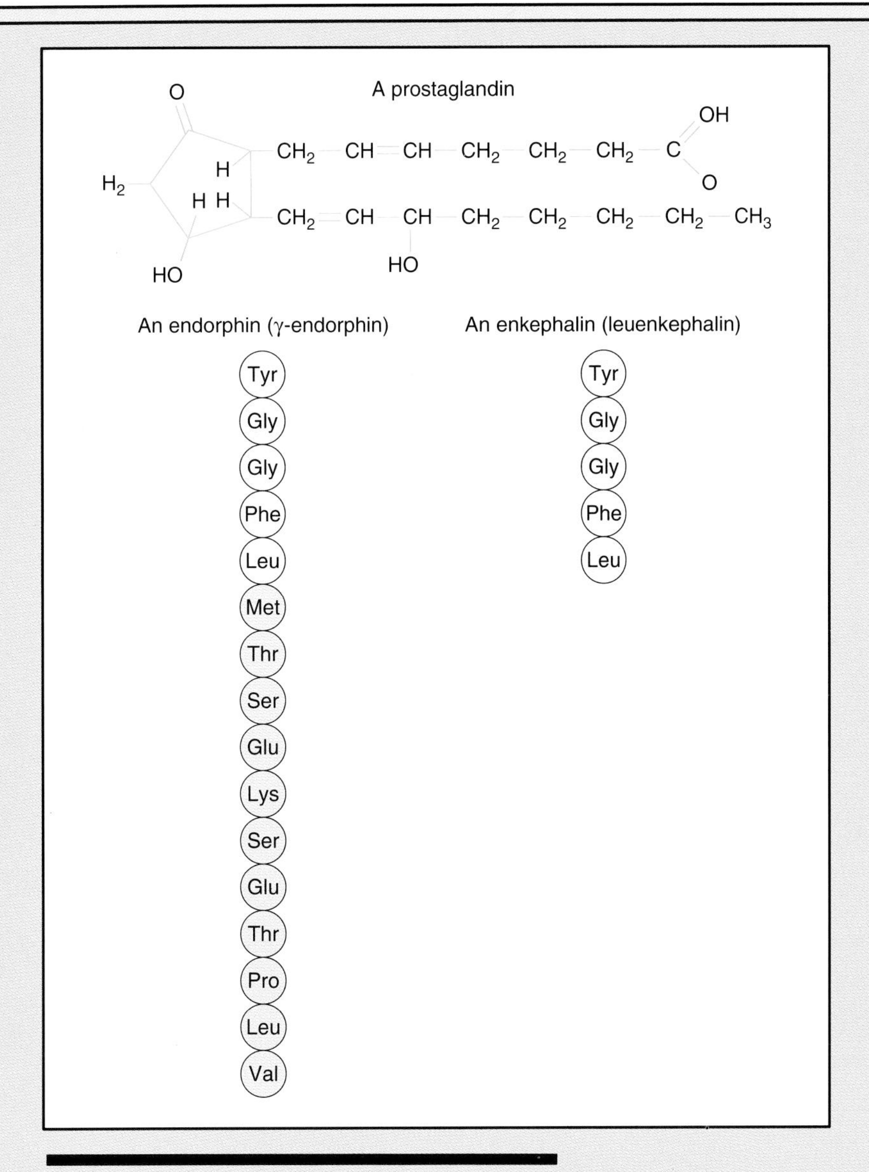

Prostaglandins, Endorphins, and Enkephalins

Structural formulas for a prostaglandin, an endorphin, and an enkephalin.

The *enkephalins* are much smaller molecules (only five amino acids in length), and they apparently enhance the perception of pain by integrating impulses from our pain receptors.

Growth Factors

In recent years, a great many "growth factors"—substances that stimulate cell division and maturation in specific tissues—have been discovered. (Indeed, the 1988 Nobel Prize in physiology or medicine was awarded to two individuals who were at the forefront of this research.) Several of these growth factors are discussed at length in Chapter 21. However, from the standpoint of categorization, they appear to straddle traditional boundaries. To the extent that they are secretions from one cell that affect another cell, they may be considered hormones. However, to the extent that they cause differentiation in tissue, they operate more like inducers. To some extent, of course, this is simply an exercise in semantics. What is exciting is learning of their existence, because these are the substances necessary for stimulating the healing process in skin, for example—and in the future they may be the substances used to induce regeneration in such tissues as nerve and muscle, which traditionally have been regarded as essentially incapable of regeneration. Their medical role in the treatment of cancer (which is basically uncontrolled cell division) is, as yet, uncertain, but potentially very exciting.

THE NATURE OF HORMONES

Before we discuss the functions of specific hormones, it is necessary to generalize on the mechanisms and chemistry of hormone functioning.

Negative Feedback

If hormones are to serve a regulatory function, they obviously must be produced at the proper time and in the proper amounts. These requirements imply a control system governing both the initiation and the cessation of hormone production. Typically, hormone population is regulated by **negative feedback.** The concept of negative feedback is best introduced by analogy. Regulation of your home's internal temperature through the interaction of the thermostat and the furnace is an example of negative feedback. As the temperature of the house falls, a circuit in the thermostat closes and the furnace is activated. The product of the furnace (heat) is ultimately responsible for causing the thermostat circuit to open again, at which point the furnace shuts down and stops producing heat. The repeated on/off cycle of the furnace, regulated by the thermostat, maintains the temperature of the house at a very nearly constant level.

Hormones acts in a similar manner (see Figure 13.2). As the product they are responsible for initiating, or controlling, increases or decreases, hormone production falls or rises. Thus the hormone *stimulates* the cells of the target organ; the product of the target organ *inhibits* the rate at which the hormone is produced. *How* these activities are initiated and maintained for particular hormones is the focus of much of the rest of this chapter. First, however, we need to understand the chemical composition of hormones and how hormones interact with their target organs.

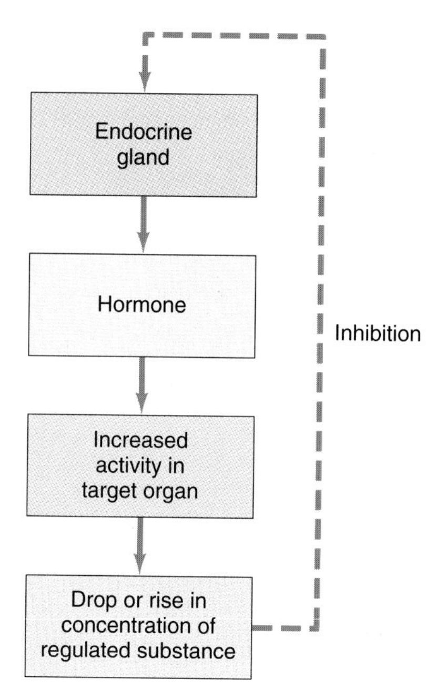

FIGURE 13.2

Negative Feedback
The endocrine system is largely self-regulating because of the use of negative feedback. The output of an endocrine gland (the hormone) triggers a response in the form of a changed concentration of some substance. That change in turn inhibits continued functioning of the endocrine gland, which means that the hormone stops being released. The cessation of hormone production in turn results in a reversal of the concentration of the regulated substance. This leads to a removal of the inhibition—and hormone production starts all over again.

The Chemistry and Action of Hormones

Chemically, hormones are basically of two types. (1) small strings of amino acids (some long enough to qualify as proteins) or derivatives of individual amino acids, and (2) steroids (one of the classes of lipids; see Chapter 3). Hormones of the pituitary, pancreas, and thyroid are in the former group; those of the adrenal cortex and gonads (that is, the testes and ovaries) are in the latter.

The mode of action of the two classes of hormones also differs (see Figure 13.3). Most hormones that are protein or amino acid derivatives (thryoxin and insulin are exceptions) do not pass directly into the cells of their target organs. Rather, they bind with specific receptor sites in the plasma membrane of target cells, where they trigger the conversion of ATP into cyclic adenosine monophosphate (**cyclic AMP**). Cyclic AMP is the **second messenger,** the substance that actually activates the appropriate physiological activity of the target cell (which is generally a cascading series of enzymatic reactions resulting in the production and release of a particular end product, such as glucose).

If many different protein hormones use the same second messenger, why is the end product not identical in each case? That is, how can different target organs respond differently to the same messenger? The answer is that the initial binding of these hormones to the plasma membrane of the cells of their respective target organ is highly specific. Thus the particular chemical configuration of an individual hormone molecule allows it to bind only to the receptor molecule imbedded in the plasma membrane of the cells of its target organ. Other hormones (with different molecular configurations) cannot bind to these receptor molecules and therefore the cells of the target organ become active only in the presence of one par-

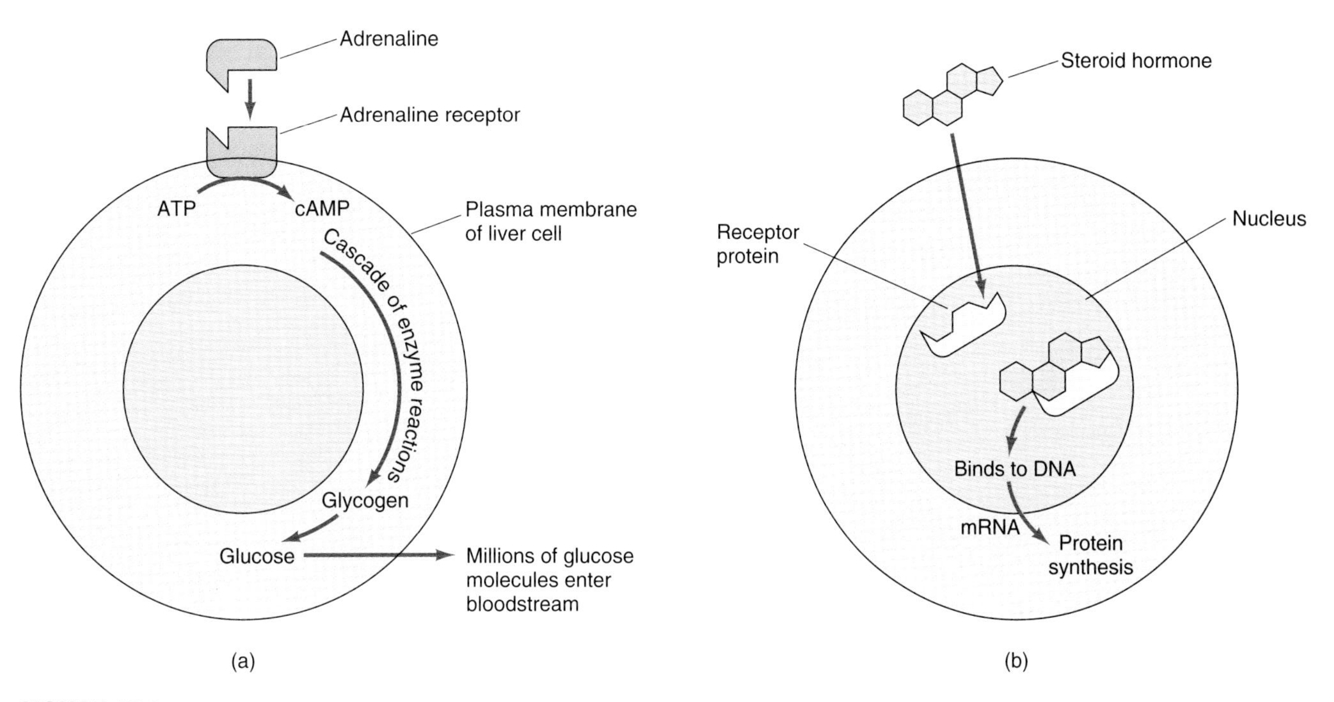

FIGURE 13.3

Modes of Action of Hormones at the Cellular Level

(a) Most proteinaceous hormones do not pass through the plasma membrane but instead bind with a specialized receptor, thereby triggering the production of cyclic AMP, which in turn catalyzes a specific set of biochemical reactions. This diagram illustrates the mode of action of adrenaline, which has been particularly well studied. (b) Steroid hormones readily pass through the plasma membrane and bind with a receptor in the nucleus. Together, the hormone-receptor complex activates a particular gene, thereby triggering synthesis of a specific protein.

ticular hormone. Therefore a common middle step—the use of cyclic AMP as the second messenger—is not inconsistent with the fact that the cells of the body respond differently to the various protein hormones. Moreover, the cascade of enzymatic reactions that is initiated by the production of cyclic AMP is not the same in every target organ, for the simple reason that the cells of different target organs contain different enzymes. Thus, just as our cells use a common energy storage molecule (ATP) to power a host of different chemical reactions, target organs can use a common second messenger (cyclic AMP) to initiate a host of different enzymatic reactions.

The steroid hormones operate much more directly, because they are lipid-soluble and readily pass through a cell's plasma membrane. However, in order to have an effect, a given steroid hormone must first bind with a specific receptor protein, located in the nucleus of the cells of the target organ. The now-united hormone and receptor protein then activate specific genes that control the synthesis of one or more proteins.

Although there are significant differences among particular hormones, generally within hours of the time they are produced both protein and steroid hormones are inactivated, either by enzymatic destruction or by excretion in the urine. Thus, to answer our earlier question, the message stops when the messenger has been destroyed.

THE HYPOTHALAMUS AND THE PITUITARY

With a weight of just four grams, the **pituitary gland** (L. *pituitosis*, "discharging mucus") is only the size of a lima bean, yet for many years it was known as the "master gland" of the body, both because it produces so many different hormones and because many of its hormones direct the activity of other endocrine glands. However, some years ago it was realized that the *posterior* portion (towards the back of the head) of the

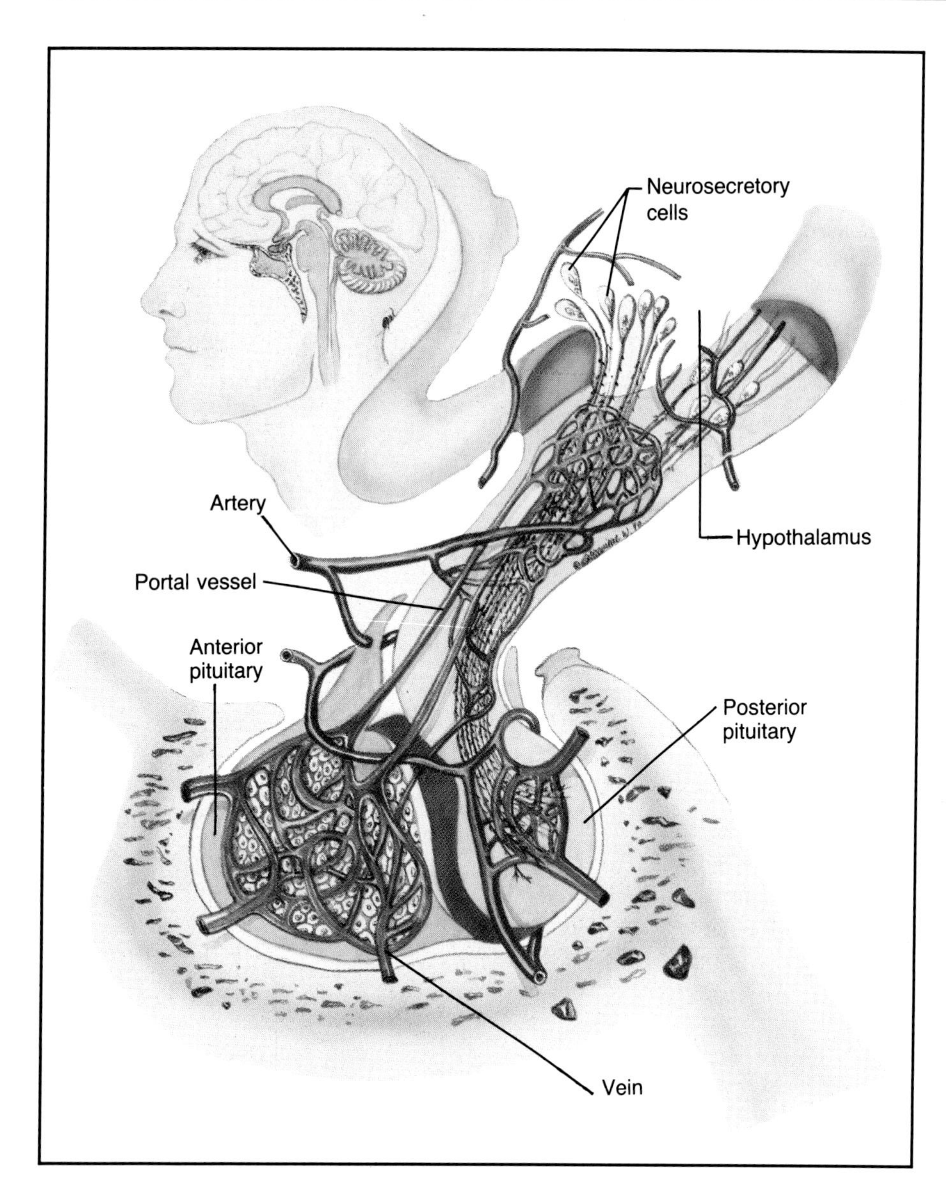

FIGURE 13.4

Pituitary Gland

The pituitary gland consists largely of two lobes. The posterior lobe is physically a part of the hypothalamus, which secretes hormones that are then stored in the posterior lobe. The anterior lobe is derived embryologically from the tissue of the roof of the mouth, and is linked to the hypothalamus only by blood vessels. Releasing and inhibiting hormones from the hypothalamus travel down these blood vessels and thereby determine the level of production and release of anterior pituitary hormones.

pituitary did not produce its own hormones. The actual production of the so-called "posterior pituitary hormones" occurs in a portion of the brain called the **hypothalamus,** which is located just above the pituitary gland (see Figure 13.4). Embryologically, the posterior pituitary arises from, and remains connected to, the hypothalamus. Hormones secreted by the hypothalamus are attached to carrier proteins and are transported down specialized nerve cells to the posterior pituitary where they are stored.

The more forward, or *anterior,* portion of the pituitary does not develop from the brain, but instead originates during embryonic development as an outgrowth from the roof of the mouth. It is adjacent to both the posterior pituitary and hypothalamus, but it is not intimately linked to either. Nevertheless, a few years ago it was discovered that, although the anterior pituitary does produce its own hormones, it does so only under the direction of the hypothalamus. The hypothalamus produces both releasing and inhibiting factors (these "factors" are actually themselves hormones) for virtually all of the hormones produced by the anterior pituitary, and depending on which factor is produced, the secretion of a particular hormone from the anterior pituitary is either stimulated or inhibited.

The one exception is the anterior pituitary hormone **prolactin,** which is apparently automatically produced except when held in check by a hypothalamic inhibiting factor (called, appropriately enough, **prolactin release inhibiting hormone,** or **PIH**). Normally, this hormone is in constant production, and no prolactin is produced. However, nervous stimuli from the breasts, caused by the sucking actions of a nursing infant (see Chapter 20), inhibit the production of the inhibiting factor, permitting the anterior pituitary to produce prolactin. Prolactin, in turn, stimulates the mammary glands to produce milk, and this production continues as long as the infant nurses regularly—an elegant example of **positive feedback.**

The finding that the brain is the master switch for the master gland establishes a close union between the two coordinating systems of the body. Previously, the concept of two coordinating systems operating independently had been vaguely disquieting—how were the inevitable conflicts resolved? Now we have an answer. The brain takes priority.

This finding also explains how the level of hormone production can be regulated. If the hormones of the anterior pituitary were regulated exclusively by the activities of the target organs in negative feedback loops, there would be no conceivable way in which the overall production level of either the pituitary hormones or the target organs could be altered. To refer to an earlier analogy, it is as if you could never raise or lower the setting on your furnace thermostat once the initial setting was made. As we shall see, altered output of certain pituitary hormones, because of changed circumstances in the body, is absolutely critical for normal functioning. Hence, the interposition of the hypothalamus in the negative feedback loop is analogous to a finger on the thermostat dial.

Posterior Pituitary

The posterior pituitary is responsible for the storage and release of two hormones. **Oxytocin** stimulates contraction of the uterus of the pregnant female at the time of birth and also promotes milk release in the nursing mother. It is discussed in more detail in Chapter 20. **Antidiuretic hormone** (**ADH** or vasopressin) increases the rate at which water is reabsorbed by the kidneys (see Chapter 11). The two hormones are chemically very similar. Each is just nine amino acids in length, and seven of the nine are identical in both hormones.

Release of these hormones by the posterior pituitary is controlled by nerve impulses from the hypothalamus. In the case of oxytocin, the hypothalamus receives nervous stimuli from the stretched uterus of the pregnant female, and from the nipples of the breasts (when the infant is nursing). For ADH, the hypothalamus monitors the concentration of salt in the blood. If the salt concentration rises, the hypothalamus stimulates the release of ADH by the posterior pituitary.

Failure of the hypothalamus to produce ADH (most commonly because of a tumor that destroys the cells responsible for its production) results in the condition known as **diabetes insipidus.** This is a totally different disease from **diabetes mellitus,** which results from a lack of insulin (see below). In both cases, the volume of urine increases, but in diabetes mellitus the presence of sugar in the urine gives it a sweet taste (L. *mel,* "honey"), whereas in diabetes insipidus the urine is not sweet (hence, insipid). The name of the brave soul who first detected the difference in taste has, unfortunately, been lost to science.

The volume of urine produced in the complete absence of ADH is staggering, sometimes reaching 15 liters a day. Not only does this loss in urine require an equivalent ingestion of water (plus the salts that are inevitably lost in this massive urine output), but an enormous amount of energy is required to warm this water up to body temperature. Fortunately, the disease is easily treated by the administration of ADH.

Anterior Pituitary

The anterior pituitary produces six hormones:

1. **Follicle stimulating hormone (FSH),** which causes maturation of egg follicles in the ovaries of women and maturation of sperm cells in the testes of men.

2. **Luteinizing hormone (LH),** which causes rupture of the mature egg follicle in women and the production of the hormone testosterone in the testes of men.

3. **Prolactin,** which causes milk production in nursing mothers.

4. **Growth hormone (GH),** which promotes growth by affecting a host of different metabolic activities, most notably the formation of proteins.

5. **Thyroid stimulating hormone (TSH),** which stimulates production of certain hormones of the thyroid gland.

6. **Adrenocorticotropic hormone (ACTH),** which stimulates the production of a series of hormones by the cortex of the adrenal gland.

The three anterior pituitary hormones involved in reproduction (FSH, LH, and prolactin) are discussed in greater detail in Chapter 20. The other three hormones are discussed below.

GROWTH HORMONE

Growth hormone is a small protein consisting of 191 amino acids. Unlike most hormones, it does not have a specific target organ, but instead exerts a body-wide effect in influencing growth. Specifically, it does the following:

1. Increases cell size;
2. Increases the rate of cell division;
3. Increases the rate of protein synthesis;
4. Promotes the mobilization of fatty acids;
5. Decreases the rate of carbohydrate use; and
6. Indirectly stimulates the growth of cartilage and bone by causing the liver to form several related proteins (known collectively as **somatomedin**), which directly stimulate cartilage and bone growth.

Although GH has its most dramatic effect on children, especially in promoting growth of the long bones, GH levels decline only marginally in adults. GH formation is governed by both stimulating and inhibiting factors produced by the hypothalamus. (These are, respectively, **growth hormone releasing hormone,** or **GHRH,** and **somatostatin**). The rate of production of these factors, in turn, appears to be a function of the level of protein in the blood. Low blood protein levels, as occurs in individuals who are starving, or in children suffering from the protein deficiency disease kwashiorkor (see Chapter 8), leads to massive increases in GH production (as much as ten times normal levels). Low blood glucose levels also promote GH secretion. Finally, GH secretion generally also increases during the early hours of sleep.

Problems with Growth Hormone

Excesses or insufficiencies of most hormones play havoc with normal body functioning, and GH is no exception. Underproduction of GH in children leads to very slow body growth, which may result in **pituitary dwarfism.** Most pituitary dwarfs produce none of the anterior pituitary hormones. As a consequence, they never enter puberty (because of the absence of LH and FSH—but see Box 13.2).

Overproduction of GH, usually the result of a tumor of the anterior pituitary, leads to gigantism if this condition develops in a child. If untreated, the tumor ultimately destroys the anterior pituitary, and all hormone production ceases from this gland, usu-

BOX 13.2

Abusing Hormones

In 1985, recombinant DNA techniques (discussed in detail in Chapter 19) were used to manufacture growth hormone (previously available only from cadavers and at great expense). Growth hormone thus became available in quantity, and at a more affordable price—from $6000 to $20,000 for a year's supply. Its availability has been a godsend to the 15,000 children in the United States who are pituitary dwarfs. Within months of FDA approval of the hormone, however, doctors began to be approached by the parents of normal-sized children who wanted their children treated in order to increase their chances for athletic success. In one instance, the father of a 6½ foot 16-year-old boy obtained a supply illegally, and then asked the family physician what dose he should be giving his son!

The long-term effects of excess GH in otherwise normal children is not well understood but presumably would involve the risk of acromegaly. However, even if there were no adverse long-term effects, what are the ethical concerns of administering a hormone to children for the purpose of giving them a possible advantage in athletics? On an international level, what should the policy of, say, the Olympic committee be and what is the likelihood that any policy of restriction could be enforced? Detection might well be impossible because GH is broken down quickly by the body.

The use of **anabolic steroids** is, of course, well known, especially after the disqualifications that marred the 1988 Olympic Games. Anabolic steroids are testosterone analogs and cause an increase in muscle mass and strength in both men and women. They also cause liver damage, stunted growth (in adolescents), infertility, and a host of psychological problems. Distressingly, the use of these artificial hormones is no longer restricted to athletes. A recent nationwide survey indicated that almost 7 percent of male high school seniors have used anabolic steroids—and fully one-quarter of them did so only to improve the appearance of their bodies!

It is ironic that the more we learn about how our bodies function, the more opportunities we find for misusing the information.

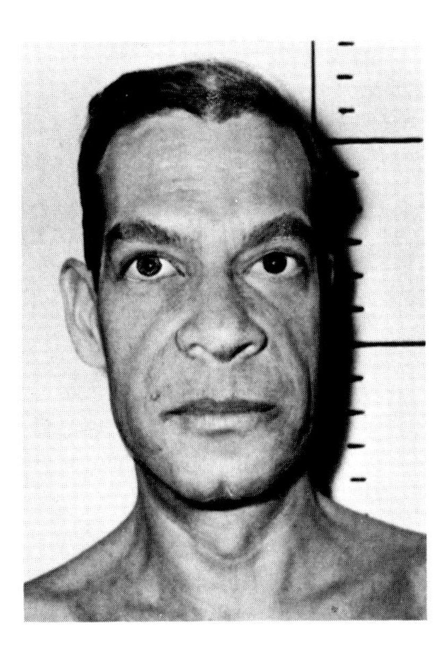

FIGURE 13.5

Acromegaly
Excess production of GH in adulthood can cause acromegaly, a condition characterized by growth of cartilage and a corresponding coarsening of facial features.

ally with fatal results. Historically, most giants have died at relatively young ages, probably because of anterior pituitary deficiency (although one famous exception reputedly met his end from a rock hurled by a sling). A significant number of giants also develop diabetes, because initially the tumor causes overproduction of the anterior pituitary hormones, and large amounts of GH and ACTH cause elevated blood sugar.

Development of a tumor in adults does not lead to a sudden spurt in height, because GH promotes growth in the long bones only until the cartilage growth zones are completely replaced by bone, an event that generally occurs in the late teens. However, remaining cartilaginous areas continue to be affected by GH, as do some of the soft tissues, most notably the tongue, liver, and kidneys. Cartilage in the nose, chin, and joints of the hands and feet continue to grow, and the facial features become coarsened (see Figure 13.5). This condition is known as **acromegaly** (Gr. *akros,* "tip, point" and *megas,* "large").

THE THYROID GLAND

The **thyroid gland** is located just below, and on either side of, the larynx (Adam's apple) (see Figure 13.6). At 25 grams, it is the largest of the endocrine glands. It produces three hormones: **thyroxin, triiodothyronine,** and **calcitonin.** The first two are closely related chemically (they are derivates of the amino acid **tyrosine**), and affect metabolic rate by stimulating the synthesis of enzymes involved in cellular respiration and by accelerating mitochondrial reproduction. They also regulate the synthesis of protein and the distribution of particular proteins within the body. Calcitonin is involved in regulating the level of blood calcium.

Thyroxin and Triiodothyronine

At the time of production, about 90 percent of thyroid output is thyroxin, and only 10 percent is triiodothyronine. However, thyroxin slowly converts to triiodothyronine as it moves through the blood; the result is only about 60 percent of the hormone reaching the cells of the body is thyroxin. Triiodothyronine is about four times more effective than thyroxin, but it breaks down about four times more quickly. Because both hormones promptly bind with plasma proteins and are released only very slowly, there is a lag of two or three days before an injection or an increase in the rate of production of thyroxin has any noticeable effect.

Production of thyroxin and triiodothyronine is under the control of the hypothalamus and the anterior pituitary (see Figure 13.7). The hypothalamus produces a releasing factor consisting of just three amino acids, but bearing the imposing name of **thyroid stimulating hormone releasing hormone,** or **TRH.** TRH stimulates the anterior pituitary to produce TSH. TSH stimulates growth of the thyroid gland and because so much of the gland is composed of secretory cells, high levels of TSH result in the production of large amounts of thyroxin.

Under most circumstances, the negative feedback loop is between thyroxin and TSH. However, when there is a need to increase the overall level of thyroxin, as occurs in a move to a cold climate, the hypothalamus produces additional amounts of TRH.

Problems with the Thyroid Gland

Both thyroxin and triiodothyronine contain the element *iodine.* About 50 mg of iodine are required annually for the production of these two hormones. Iodine is relatively abundant in the ocean (hence, in marine organisms as well), and also in plants growing in soils that were at one time part of the sea bed. However, in certain parts of the world, including Central Europe and the Great Lakes region of North America, iodine is essentially absent from the soil.

Without an adequate supply of iodine, thyroxin cannot be manufactured in sufficient quantities. Consequently, the thyroxin level never reaches the point where it effectively inhibits TSH production. The unchecked high levels of TSH cause the cells of the thyroid gland to continue to grow and divide, which

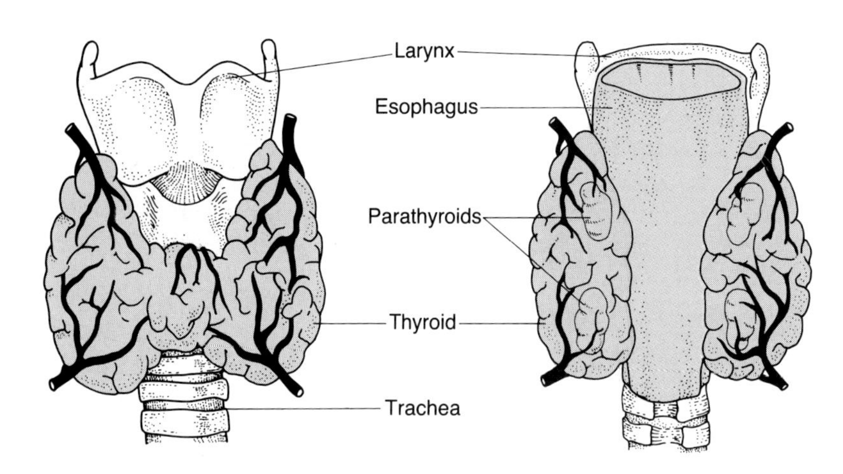

FIGURE 13.6

Structure and Location of Thyroid and Parathyroid Glands

The thyroid gland is located just below the larynx; the four parathyroid glands are buried toward the rear of the thyroid gland. The gland receives a generous blood supply—five times its weight in blood every minute.

would normally lead to increased production of thyroxin, but this cannot happen if there is an insufficient supply of iodine in the diet. The net result is a greatly enlarged thyroid gland, called a **goiter** (L. *guttur,* "throat").

When it was discovered that goiters resulted from iodine deficiency, this essential element was added to table salt (but only after violent debate, much like the more recent controversy regarding the fluoridation of drinking water). To this day, much of the salt used in this country is "iodized."

In addition to a goiter, the underproduction of thyroxin also depresses the metabolic rate. When thyroxin levels are very low, adults develop **myxedema** (Gr. *myxa,* "mucus" and *oidema,* "swelling"), a condition in which there is massive swelling of body tissue due to increased volumes of interstitial fluid. In addition, blood cholesterol increases, and, consequently, so does the incidence of atherosclerosis.

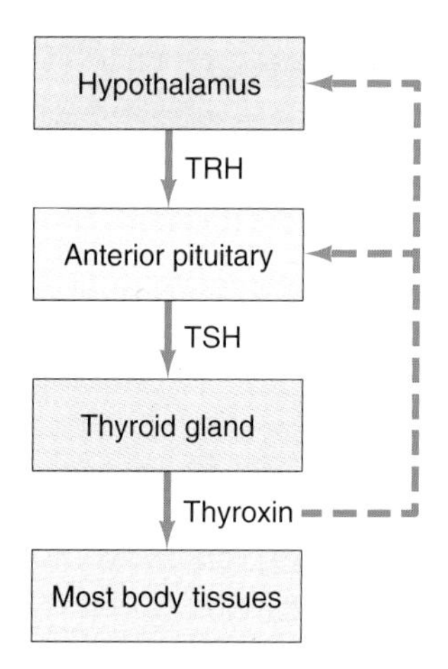

FIGURE 13.7

Negative Feedback of Thyroxin and Its Chemical Formula

Thyroxin has a negative feedback loop with both the hypothalamus and the anterior pituitary.

Thyroxin insufficiency is even more serious in infants. The lowered metabolic rate inhibits growth, and the individual becomes a hypothyroid dwarf (see Box 13.3). Moreover, because metabolic activity is normally very high in nerve tissue during the first year of life, inhibition of this activity leads to an improper maturation of brain cells and consequent retardation, a condition known as **cretinism.**

Fortunately, thyroxin treatments are highly effective, and the hormone can be administered orally (since, unlike most protein hormones, it is not destroyed by digestive enzymes). However, these treatments must be undertaken soon after birth, because once the brain cells have failed to mature, cretinism is irreversible.

Overproduction of thyroxin is less common, and results generally either from thyroid tumors or an autoimmune condition known as **Graves' disease,** in which there is substantial growth of the thyroid gland due to a biochemical failure in the negative feedback loop. (Autoantibodies destroy the TSH receptors in the plasma membrane of thyroid gland cells.) Because this failure does not stem from an iodine insufficiency, excess amounts of thyroxin are produced, causing irritability, weight loss, and hyperactivity. Treatment typically involves either surgery or the injection of carefully measured amounts of radioactive iodine, which destroys portions of the thyroid gland as it is taken up by thyroid cells. Both George and Barbara Bush suffer from this disease.

The medical use of radioactive iodine to destroy portions of the thyroid gland has an unfortunate counterpart in our nuclear age. One of the effects of the meltdown in 1986 of the nuclear reactor in Chernobyl in the Soviet Union was the production and release into the atmosphere of radioactive iodine. Throughout much of Europe, people rushed to pharmacies for dietary iodine. The idea was to overload the body with normal iodine, to prevent the uptake by the body of any radioactive iodine present in the

BOX 13.3

Midgets, Dwarfs, and Pygmies

There are a number of causes of extremely short adult stature. Pituitary dwarfs (GH deficiency) have normal body/limb proportions, and are sometimes called **midgets** (in recent years, this term has sometimes been regarded as pejorative).

Relatively more common are dwarfs. **Achondroplastic dwarfs,** who have a normal-sized head and body but very short limbs, possess a recessive gene that interferes with normal growth of the long bones (see Chapter 19). **Hypothyroid dwarfs,** or *cretins,* have short stature due to the absence or insufficiency of thyroxin and are frequently retarded. Because insulin operates synergistically with GH, Type I diabetics may show some reduction in body stature, but not to the point of dwarfism (since a level of insulin low enough to stunt body growth to that point would almost surely be fatal).

Pygmies have normal body/limb proportions but are considerably larger than pituitary dwarfs. They produce normal amounts of GH but are deficient in the plasma membrane receptor molecule for GH.

food and water, and the consequent destruction of the thyroid gland. Although this idea was sound in principle, execution was something else. There were many instances of people attempting to drink tincture of iodine, which is commonly used to treat superficial injuries to the skin—and which is highly irritating to the skin, let alone to the mouth and digestive system. Thus, while it may be true that "a rose is a rose is a rose," it is *not* true (at least for dietary purposes) that, "iodine is iodine is iodine."

Calcitonin

The third thyroid hormone, calcitonin, is not involved in general metabolic rate, nor is its synthesis directed by TSH or other pituitary hormones. Instead, in a very simple negative feedback loop, the production of calcitonin increases with a rise in blood calcium levels, and falls as the blood calcium level drops.

Calcitonin is a small protein of 32 amino acids. It lowers blood calcium levels by promoting the incorporation of calcium into bone through activation of the osteocytes and suppression of the osteoclasts (see Chapter 12). It is particularly active immediately after meals rich in calcium.

THE PARATHYROID GLANDS

Surgical removal of excess thyroid gland tissue has long been an accepted method of treatment for goiter. However, in the early days of this surgical procedure, some patients died from convulsions shortly after surgery. Examination of the removed tissue ultimately revealed the presence of a distinct type of glandular tissue imbedded in the thyroid gland itself. These glands (there are four of them) came to be known as the **parathyroid glands;** each is about 5 mm in diameter.

The parathyroid glands produce the hormone **parathormone,** which operates antagonistically to calcitonin by causing blood calcium levels to rise. Parathormone, a protein of 84 amino acids, activates the osteoclasts (causing them to release calcium and phosphate into the blood), promotes the absorption of calcium by the intestine (see Box 13.4), and stimulates the kidneys to reabsorb calcium and to excrete phosphate.

Parathormone operates more slowly, but much more powerfully, than calcitonin. Like calcitonin, parathormone is in a feedback loop directly with blood calcium, rather than with a pituitary or hypothalamic hormone. The regulation of blood calcium is critical for the well-being of the individual, because of the importance of calcium in nervous transmission and muscle contraction (see Chapter 12). Normal blood calcium levels are between 9 and 10 mg/100 ml of blood. A drop to 4 mg, or a rise to 17 mg, is fatal.

Problems with the Parathyroid Glands

In the absence of treatment (which usually consists of large doses of vitamin D or injection with parathormone), death results when the parathyroid glands are removed because of the continual loss of calcium through the kidneys. At levels below 4 mg/100 ml of blood, the nerves begin firing spontaneously, causing the muscles to contract in an uncoordinated fashion. Ultimately, the muscles of the chest fail to function adequately in breathing, and the individual suffocates.

Excess production of parathormone usually is caused by a tumor of one or more of the parathyroid

BOX 13.4

Calcium and Vitamin D

Vitamin D could be classified both as a vitamin and as a hormone. Our skin produces vitamin D under the influence of ultraviolet radiation from the sun. However, under certain conditions—for example, during northern winters—not enough sunlight reaches the skin, and vitamin D must be ingested, just like other vitamins. Synthetic vitamin D is now widely available, and it is routinely added to a variety of food, most notably milk. However, until well into the 1950s, American children were regularly dosed during the winter with cod liver oil, because the livers of several types of marine fish, including cod, are especially rich in this vitamin.

Under the influence of parathormone, vitamin D is altered by the kidney to its active form, in which state it causes the formation of a calcium-binding protein in the cells of the intestine. Because vitamin D has its effects at some distance from its site of formation, it is acting in a hormone-like manner.

In the absence of either parathormone or vitamin D, little calcium is absorbed from the intestine, and, in growing children, this lack leads to weakened bones that may bend under the weight of the child, a condition known as **rickets** (see Chapter 7). Fortunately, with the addition of vitamin D to milk, rickets is no longer a common problem in this country.

glands. Calcium is liberated from bone, causing bone weakness and spontaneous fractures, and the high level of calcium in the blood increases the likelihood both of kidney stone formation and of irregularity in the contraction of cardiac muscle.

THE PANCREAS

In addition to its critical role in the production of digestive enzymes, the pancreas is also an endocrine gland, producing two major hormones, **insulin** and **glucagon,** which operate antagonistically in regulating blood glucose levels. Let us first examine how insulin, a small protein of 51 amino acids, functions.

When glucose in food is absorbed by the cells of the small intestine, it is carried to the liver by the blood. The fate of glucose in the liver is determined, in large measure, by the level of glucose already in the circulating blood. As the blood glucose level rises past 120 mg/100 ml of blood the pancreas secretes increasing amounts of insulin (see Figure 13.8). Insulin has the following effects:

1. It promotes the conversion of glucose in the liver to the short-term storage molecule *glycogen.* (Normally, about 60 percent of the glucose in a meal is stored as glycogen).

2. It stimulates the uptake of glucose from the blood by muscle and fat cells.

3. It inhibits the metabolic breakdown of both glycogen and fats.

4. It increase the rate both of amino acid uptake by cells and of protein synthesis.

The first three functions all lower blood glucose levels. So, indirectly, does the fourth—by reducing the amount of available amino acids in the blood, insulin forces cells to rely more heavily on glucose for their energy needs.

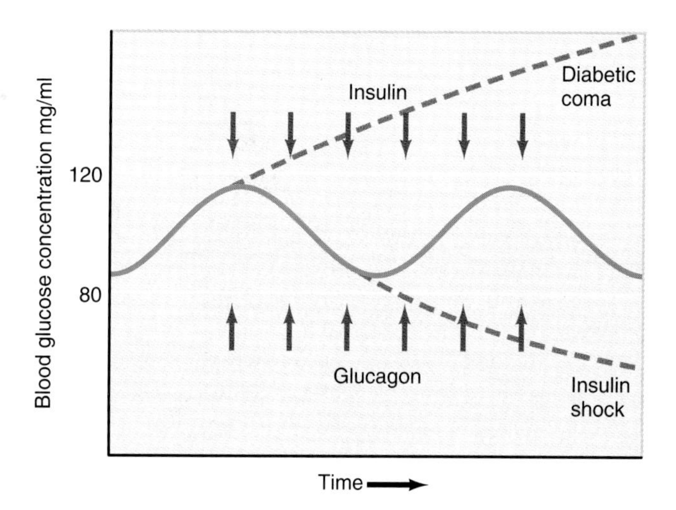

FIGURE 13.8

Hormonal Regulation of Blood Glucose Levels

As blood glucose levels rise to about 120 mg/ml of blood, the rate of insulin secretion increases, and blood glucose is taken up by the body's tissues more readily. As blood glucose levels fall to about 80 mg/ml of blood, the rate of glucagon secretion increases, and glucose is released into the blood from storage sites. Failure of insulin to act leads initially to high blood glucose levels and diabetic coma. Failure of glucagon to act (or insulin levels that are too high) leads to hypoglycemia (or insulin shock).

When blood glucose falls below about 80 mg/100 ml of blood, another group of cells in the pancreas begins to produce glucagon. Glucagon, a small protein of 29 amino acids, raises blood glucose levels by stimulating the conversion of glycogen to glucose and by promoting **gluconeogenesis**—the formation of "new" glucose from the conversion of amino acids.

Insulin production alone can maintain blood glucose levels within fairly narrow parameters, as long as the individual is relatively sedentary and is eating regularly. However, in fasting or exercising individuals, glucagon secretion is particularly important, because at that point the body needs both glycogen breakdown and gluconeogenesis in order to maintain an adequate blood glucose level.

Problems with the Pancreas

Several hormones operate in concert with glucagon to increase blood glucose levels. However, only insulin acts to lower them. Thus it is a pivotal link in the endocrine system, and failures of this link are relatively common. These failures lead to elevated blood sugar levels, or **hyperglycemia** (Gr. *hyper,* "above, excessive" and *glukus,* "sweet"). The most common cause of hyperglycemia is *diabetes mellitus* (the modifier is necessary to distinguish this disease from the much rarer diabetes insipidus, discussed earlier).

Diabetes mellitus exists in two forms. One form (Type I, or juvenile-onset diabetes) usually arises during infancy or childhood. It apparently develops in genetically predisposed individuals as a consequence of the destruction of the insulin-producing cells in the pancreas by autoimmunity (see Chapter 9). The antibody believed to be responsible for the destruction of these cells was identified in 1988, and the current hope is that it can be detected and neutralized in prediabetic individuals, thereby sparing them from the ravages of the disease. As it stands, Type I diabetics generally produce little or no insulin, and as a consequence they must take insulin by injection every day. Over one million people in the United States suffer from this disease.

The second form (Type II, or adult-onset diabetes) develops during adulthood. Again there is a heritable component (the incidence of diabetes tends to run in families, and in some populations such as the Pima Indians, the incidence of Type II diabetes may exceed 50 percent of the adult population). However, it is simply the *tendency* for the disease, rather than the disease itself, that is inherited. That is, if the body is placed under sufficient stress (obesity, serious diseases or injuries, pregnancies in rapid succession, advanced age), the disease may become manifest. However, it is usually less severe than the Type I form, and may often be managed by dietary restrictions alone, or by a combination of diet and oral medication that stimulates the production of insulin by the pancreas. More than eleven million people in the United States have Type II diabetes.

The two forms of the disease are actually very different. Whereas Type I diabetes results from insulin deficiency, Type II may be due to an insufficiency of the insulin receptor protein in the plasma membrane of cells. Insulin or insulin-stimulating therapy operates to swamp the system, ensuring that a sufficient amount of insulin attaches to the plasma membrane, even though the amount of receptor protein is low. Significantly, a deficiency of receptor protein can be detected biochemically long before diabetes itself develops. Thus susceptible individuals can then take precautions to avoid the onset of diabetes.

Very recently, a new model for the onset of Type II diabetes has emerged. The same cells in the pancreas that produce insulin also produce a small protein called *amylin,* a substance that is believed to fine-tune the insulin response. Because amylin injections in rats produce symptoms akin to Type II diabetes in humans, the model suggests that Type II diabetics may be producing too much amylin. If that is true, medications that neutralize amylin could be used to prevent the onset of Type II diabetes.

Untreated diabetes leads to a marked rise in blood glucose. At very high levels of blood glucose, the kidneys are incapable of reabsorbing all of the glucose present in the glomerular filtrate, and it begins to be lost in the urine. In order to keep the glucose in solution, more water must also be passed in the urine, and untreated diabetics therefore produce large volumes of urine and have increased thirst. Moreover, because fats and proteins are metabolized for energy production in the absence of insulin, the untreated diabetic tends to lose weight, even though his appetite may be sharply increased. The increase in fat breakdown also leads to an increasingly acidic blood, which in turn causes profound disruptions in metabolic activities. These acids combine with Na^+ during the formation of urine, and this loss of Na^+ is devastating, with coma and death the inevitable result.

With the isolation and first clinical use of insulin in 1922 by the Canadian scientists, Frederick Banting and Charles Best, diabetes quickly changed from a death sentence to a manageable disease. However, insulin therapy in no sense duplicates the normal condition, and diabetics remain a high risk population for a variety of other disease conditions, such as atherosclerosis, that result from high levels of blood cholesterol. Accumulation of cholesterol in the small arteries of the eye and kidney accounts for a high incidence of blindness and kidney failure in diabetics.

In fact, diabetes ranks third among the causes of death in the United States.

Unlike thyroxin, which last for days in the bloodstream, insulin is rapidly destroyed by the cells of the liver (in as little as ten minutes in the case of naturally produced insulin). Moreover, the level of insulin secretion varies widely according to the time of day, and the size and quality of the meal. Diabetics on insulin therapy frequently must on a daily basis inject a long-action insulin (made by uniting it to a large protein to decrease solubility) in order to increase overall carbohydrate metabolism, and then supplement this injection with additional injections of short-term insulin at mealtime. Moreover, because the amount of insulin required is very much a function of the quality of food ingested, the diet must be carefully regulated. In the event of illness or injury, which result in changed metabolic rates, insulin doses must be adjusted accordingly. In short, insulin therapy prevents death, but in no sense represents a cure.

Much more rarely pancreatic tumors may cause an overproduction of insulin, a condition known as **hypoglycemia.** Ironically, the symptoms are similar to hyperglycemia, because again the cells are not obtaining a sufficient supply of glucose—except in this instance, it is due to the conversion of too much blood glucose to glycogen. Since the cells of the nervous system are entirely dependent on blood glucose for their energy supply, fainting spells and blackouts are common in hypoglycemics. Treatment consists of glucose injections in emergency cases, followed by surgical removal of the tumor. Interestingly, a comatose hypoglycemic will, if given an injection of glucose, regain consciousness in less than one minute, and the return to normalcy is essentially absolute (although temporary). This startlingly abrupt return to consciousness is one of the most dramatic events in medicine.

Problems may also arise with insulin therapy. Too little insulin fails to curb completely the symptoms of diabetes; too much insulin causes **insulin shock,** a form of hypoglycemia, with fainting, blackouts, and even death a distinct possibility.

THE ADRENAL GLANDS

The adrenal glands are a pair of glands that, in humans, are located just above the kidneys (see Figure 13.9). They consist of two distinct layers of tissue, each with very different embryonic origins and each functionally distinct from the other. However, both layers secrete hormones. The outer **adrenal cortex** secretes a series of steroid hormones; the inner **adrenal medulla,** which is derived from nervous tissue, secretes hormones that are derivatives of the amino acid *tyrosine* (as is the thyroid hormone thyroxin).

The Adrenal Cortex

The adrenal cortex produces three major groups of hormones: **mineralocorticoids, glucocorticoids,** and **adrenal sex hormones.** Collectively, these affect "salt, sugar, stress, and sex".

Mineralocorticoids. The mineralocorticoids are a group of steroids that regulate the balance of ions in the blood and interstitial fluid. Of these, by far the most important is **aldosterone** (see also Chapter 11). Aldosterone stimulates the cells of the kidney tubule to reabsorb Na^+ from the glomerular filtrate, and to secrete K^+ into the filtrate.

Both Na^+ and K^+ are essential in the transmission of nerve impulses (see Chapter 14 for details), and therefore the concentration of each within the body is critical. Interestingly, the concentration of Na^+ in the blood and interstitial fluid varies only slightly, regardless of the amount of aldosterone present, because water tends to "follow" Na^+ (by osmosis), whether the Na^+ is retained or lost. Thus, in the absence of aldosterone, excessive amounts of both Na^+ and water are lost, and the volume of blood and interstitial fluid may drop as much as 25 percent. If excess aldosterone is secreted, Na^+ and water are retained in quantity, and there is a correspondingly large increase in the volume of blood and interstitial fluid.

K^+ does not follow the same pattern. High levels of aldosterone cause the K^+ concentration in the blood and interstitial fluid to drop. At approximately half the normal concentration, muscle weakness and paralysis develop. Low levels of aldosterone lead to an increase in K^+ concentration. At approximately twice the normal concentration, heart function is seriously impaired (because of arrhythmia and weak contractions). In the total absence of aldosterone, and without appropriate therapy, death results in less than a week.

The role of the renin-angiotensin system in governing aldosterone secretion was briefly discussed in Chapter 11. However, this is not the only, nor even the most important, factor affecting aldosterone production. Even more important is the concentration of K^+ in the blood and interstitial fluid. Na^+ levels, and ACTH production, also play significant roles in affecting aldosterone production, although the role of ACTH is largely passive. That is, ACTH is required for healthy cortical tissue, and, in its absence, the tissue responsible for aldosterone production begins to degenerate. However, ACTH does not play an important stimulatory role as it does with the glucocorticoids.

Glucocorticoids. Like the mineralocorticoids, the glucocorticoids include several steroid hormones, but one is strongly dominant. It is the steroid **cortisol.**

Although the mineralocorticoids have a more dramatic short-term effect on the well-being of the individual, the glucocorticoids are very important for long-term good health. Without the glucocorticoids, basic metabolic processes are upset, and the individual is unable to respond to even mildly stressful situations. As a consequence, otherwise minor illnesses may be fatal.

Cortisol has two important categories of effects: "sugar" and "stress." First, it is responsible for affecting a variety of metabolic functions. It stimulates gluconeogenesis (hence the name of this group of hormones). In so doing, blood glucose levels may rise by as much as 50 percent. It also mobilizes both amino acids from stored proteins, and fatty acids from stored fats.

Second, when stressed, the body postpones activities not of immediate benefit, including immune functioning and the inflammation response. Inhibition of these functions is caused by cortisol which, in a manner described below, the body produces in response to stress. Cortisol prevents the development of inflammation by stabilizing the membranes of lysosomes, the cellular organelles known colloquially as "suicide sacs" because they contain powerful enzymes that can destroy the cell if ruptured. Just such rupturing normally takes place in inflammation, resulting in the release of the protein *histamine,* which causes tissue swelling because of its effects on local blood circulation (see Chapter 9 for details of the effects of histamine).

Cortisol also reduces fever and suppresses the overall activities of the immune system. Because it reduces inflammation, a synthetic form of cortisol called *cortisone* is often administered to athletes, for whom inflammation may mean a long ride on the bench. However, because it suppresses the immune system, cortisone is not generally administered over an extended period of time, nor is it given to individuals with active infections (such as ulcers, for example). Similarly, although cortisone is used to treat arthritis, which causes swelling at the joints, pain, and reduced mobility, it must be used sparingly. Continued and extensive use not only lowers the body's ability to fight off other diseases and infection but also increases the blood glucose level, forcing the pancreas to produce more insulin, which increases the risk of developing diabetes.

Unlike the mineralocorticoids, where ACTH plays only a permissive role, ACTH is vitally linked to glucocorticoid production. Stress of any type (physical or emotional) is perceived by the hypothalamus (which receives information from other parts of the brain and nervous system), causing it to secrete a releasing factor called **corticotropic releasing hormone,** or **CRH.** CRH, in turn, stimulates the production of ACTH by the anterior pituitary, and ACTH stimulates the conversion of cholesterol by the adrenal cortex to all three classes of cortical hormones, though with cortisol far and away the most abundant. Increases in circulating cortisol inhibit both the hypothalamus and the anterior pituitary, shutting down both CRH and ACTH production, in a classic negative feedback loop.

Adrenal Sex Hormones. The adrenal cortex produces both **androgens** (male sex hormones; Gr. *andros,* "male") and **estrogens** (female sex hormones; Gr. *oestros,* "frenzy"). Estrogens are produced only in very small amounts, and the volume of both androgens and estrogens is not large in comparison to the volume produced by the mature glands. Nevertheless, the androgens produced by the adrenal cortex probably play a role in the early development of the male sex organs.

The Adrenal Medulla

The adrenal medulla, which constitutes only about 15 percent of the weight of the entire gland, is not under the control of ACTH, nor does it produce steroid hormones. Despite its anatomical proximity to the cortex, it has a very different embryonic origin, and, in direct response to nervous stimulation during times of stress, it produces two hormones, **adrenaline** (epinephrine) and **noradrenaline** (norepinephrine). The two hormones have broadly overlapping effects on the circulatory system (stimulation of heart activity) and on metabolic rate (promotion of gluconeogenesis; elevation of metabolic rate by as much as 100 percent).

The adrenal medulla is considered more fully in the discussion of the sympathetic nervous system (Chapter 14).

Problems with the Adrenal Glands. The principal problems associated with the adrenal glands involve the adrenal cortex, not the medulla. They can be briefly described as too little, too much, and wrong kind.

Addison's disease is the name given to the failure of the adrenal cortex to produce significant volumes of hormones, usually as the result of the effects of disease (tuberculosis; cancer). The severity of this condition depends on the degree of the deficiency, but because adrenal hormones are essential for life, it is potentially fatal.

Individuals suffering from Addison's disease have limited resistance to stress, impaired kidney function, muscle weakness, lowered blood pressure, and an increase in skin pigment, giving a peculiar bronze color to the skin. It is by no means a rare condition, and probably contributes significantly to the death rate, but the deaths are more commonly attributed to the effects of some stressful condition that a person with fully functional adrenals would have been able to resist.

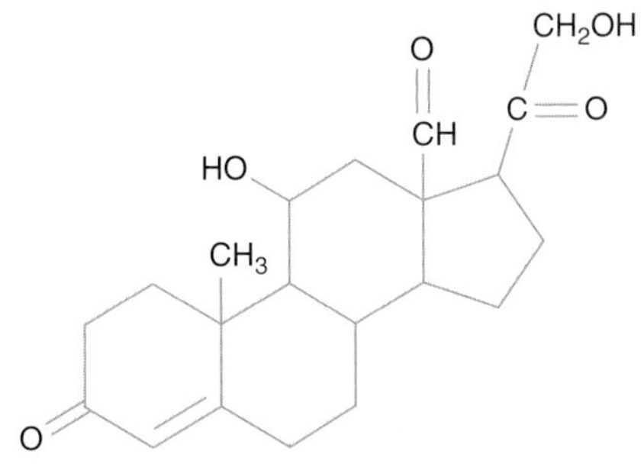

Aldosterone

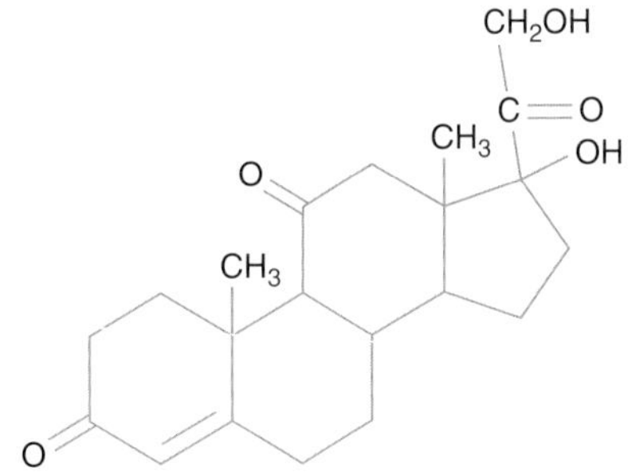

Cortisone

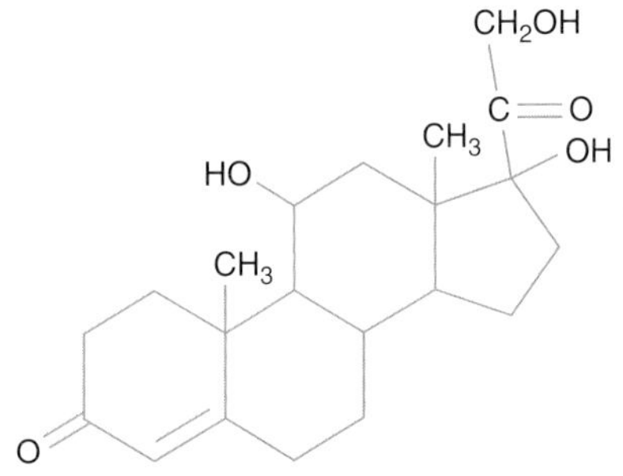

Cortisol

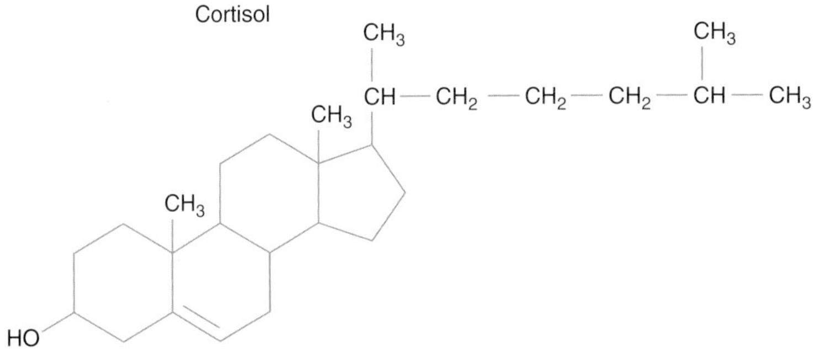

Cholesterol

(a)

FIGURE 13.9

Adrenal Gland

(a) The structural formulas for three corticosteroids, all derived from cholesterol (above).
(b) The position of the adrenal glands relative to the kidney, and a longitudinal section of an adrenal gland showing the outer cortex and inner medulla. (c) Negative feedback loop of the corticosteroids. Note the singular importance of cortisol in this loop, and the involvement of the cerebral cortex (on facing page).

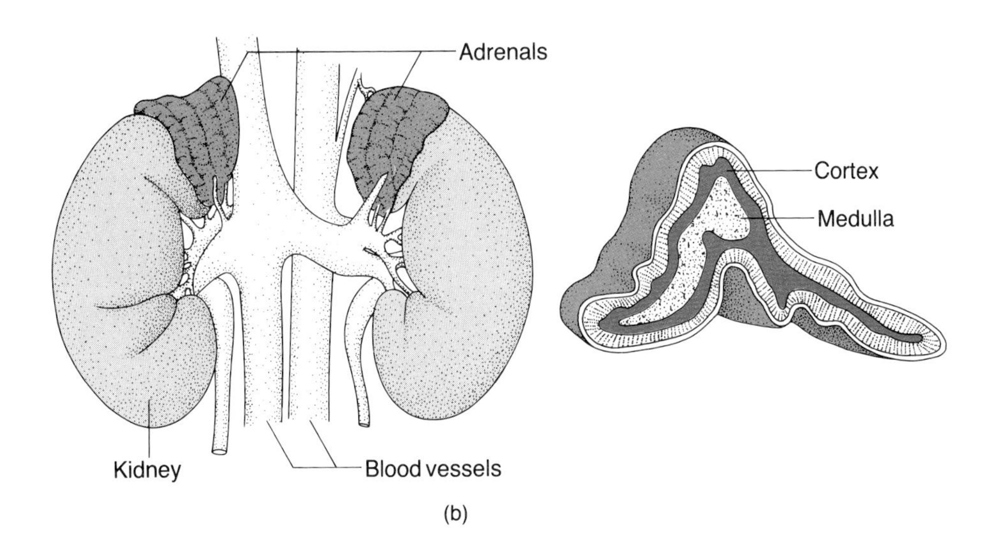

(b)

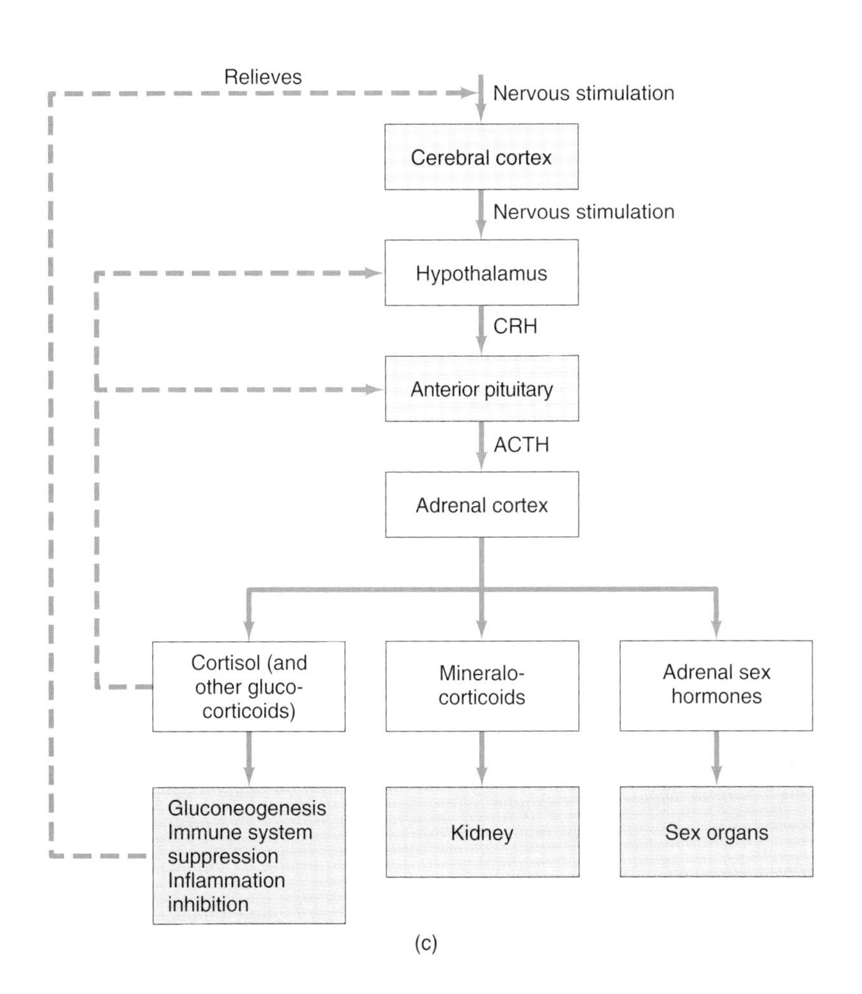

(c)

John F. Kennedy suffered from a mild case of this disease, and received treatment for it over a period of years. However, even under treatment, individuals with Addison's disease are highly susceptible to the effects of stress, because they cannot increase their output of glucocorticoids. It is obviously not a condition conducive to meeting the demands of the presidency!

Cushing's disease results from the overproduction of adrenal hormones, usually because a tumor in either the pituitary or the adrenal gland itself. Most of the effects stem from an overproduction of cortisol, and include breakdown of protein (with resulting muscle and bone weakness), elevated blood pressure and blood glucose levels (with diabetes a not-uncommon result), a swollen and often hairy face, and a tendency for bruising of the skin. Treatment involves removal of the tumor; otherwise, the condition is generally fatal.

Adrenogenital syndrome is a condition in which excessive amounts of adrenal androgens are produced. This condition is essentially symptomless in adult males, in whom the volume of testosterone from the testes greatly exceeds the volume of androgens from the adrenal glands. However, in women and children, adrenogenital syndrome has a masculinizing effect, leading to growth of the genitals in children, and to the production of facial hair in women. (The "bearded ladies" of circus fame, when authentic, were women suffering from adrenogenital syndrome.)

This condition results either from a tumor of the adrenals or from a genetic condition where the enzyme for synthesizing cortisol is absent, and large volumes of androgens are produced in its place. The absence of cortisol leads to increased production of ACTH, which further stimulates the production of androgens. Treatment is either the surgical removal of the tumor, or the direct administration of cortisol to "turn off" ACTH production.

BLOOD GLUCOSE LEVELS—A REPRISE

Blood glucose levels are elevated by glucagon, the glucocorticoids, adrenaline, growth hormone, and the hunger drive (which leads to the ingestion of food). Blood glucose levels are lowered only by insulin. When viewed in this way, it is not difficult to understand why hyperglycemia is both common and severe—there are at least five forces opposing insulin, and there is no backup system.

Normally, blood glucose levels are maintained on a day-to-day basis by the combined effects of eating and glucagon on the one hand, and insulin on the other. Both adrenaline and the glucocorticoids are produced in times of stress, and in that sense there is some overlap between them. However, adrenaline is more important in times of stress of short duration. For instance, when you were in grade school and the class bully told you that he was going to "get you" during recess, your adrenal medulla undoubtedly became active. Conversely, the glucocorticoids are more important in long-term stress—as for instance when you are in the hospital recovering from your recess meeting with the bully.

Summary

A study of the endocrine system has value perhaps as much for the elegance of the negative feedback loops as for the knowledge that comes from understanding the effects of each individual hormone. Certainly no other system exemplifies the workings of homeostatic control mechanisms—and by inference the importance of homeostasis itself—as well as the endocrine system.

There was a time, not many years ago, when the endocrine system was seen as being an autonomous system largely under the governance of the pituitary gland. We now know that reality is far more complex. Not only are the actions of the pituitary largely governed by the hypothalamus, but the pituitary is involved in the control of only some of the other hormones, most notably those in which the levels of hormone output fluctuate broadly. The hormones that control highly regulated processes (blood sugar and ion levels, for example) are more commonly found in very simple negative feedback loops with the substance they control.

It is also increasingly apparent that many of the endocrine glands (thyroid, pancreas, adrenals, and the pituitary itself) are amalgams of several different tissue types, which very often have nothing more in common than a shared location.

Key Terms

hormone
target organ
endocrine gland
exocrine gland
negative feedback
cyclic AMP
second messenger
pituitary gland
hypothalamus
prolactin
Prolactin release inhibiting hormone (PIH)
positive feedback
oxytocin
Antidiuretic hormone (ADH)
diabetes insipidus
diabetes mellitus
Follicle stimulating hormone (FSH)
Luteinizing hormone (LH)
Growth hormone (GH)
Thyroid stimulating hormone (TSH)
Adrenocorticotropic hormone (ACTH)
somatomedin
Growth hormone releasing hormone (GHRH)
somatostatin
pituitary dwarf
acromegaly
calcitonin
Thyroid stimulating hormone releasing hormone (TRH)
goiter
myxedema
cretinism
Graves' disease
parathyroid gland
parathormone
insulin
glucagon
gluconeogenesis
hyperglycemia
hypoglycemia
insulin shock
adrenal cortex
adrenal medulla
mineralocorticoid
glucocorticoid

adrenal sex hormone
aldosterone
cortisol
Corticotropic releasing hormone (CRH)
androgen
estrogen
adrenaline
noradrenaline
thyroid gland
thyroxin
triiodothyronine
Addison's disease
Cushing's disease
adrenogenital syndrome

Box Terms

prostaglandins
endorphins
enkephalins
growth factors
midget
achondroplastic dwarf
hypothyroid dwarf
pygmy
rickets
anabolic steroids

Questions

1. Explain, with an example, the concept of negative feedback.
2. What is the nature of the difference in mode of action between steroid hormones and most protein or amino acid–derived hormones?
3. What is the relationship between the anterior pituitary, the posterior pituitary, and the hypothalamus?
4. Distinguish diabetes mellitus from diabetes insipidus. Which hormones are involved in each?
5. What hormones are involved in regulation of blood calcium levels? What is their source tissue or organ?
6. Why do you think thyroxin deficiencies have an impact that is so much more dramatic on babies and young children than on adults?
7. Discuss the regulation of blood glucose. Which hormones are involved on a daily basis, and which on an as-needed basis?
8. Distinguish Type I and Type II diabetes. Which one routinely requires insulin injections?
9. How do the hormones of the adrenal cortex and the adrenal medulla differ? Are they chemically similar? Are they all regulated by hormones from the pituitary? Do they have overlapping functions?

CHAPTER

14

The Nervous System

Short-Term Coordination

Late one Saturday night, as his babysitter is engrossed in watching a horror movie on television, ten-year-old Lester Bratt sneaks into the living room and drops a rubber spider into the babysitter's lap. To his great delight, she screams and leaps to her feet, overturning a TV table full of munchies in the process. Why do we humans, with our magnificently analytical brains, still show reflexive behavior of the type just described?

On a June morning, near some pea fields in New Jersey, the pilot of a spray plane is rolling a drum of the insecticide *parathion* into his plane, prior to spraying the fields for pea aphids. Suddenly, the top of the drum works loose, and the pilot is drenched in the insecticide. Panicked, he turns and runs towards his car, parked a few hundred yards away, but, before he reaches it, he drops to the ground and dies of convulsions. What was the physiological cause of the convulsions?

The activity of many of our internal organs is regulated by the nervous system. How does an organ "know" whether to increase or decrease its activity when it receives a stimulus from a nerve?

In each case, the answer relates to the functioning of the nervous system, the subject of this chapter.

THE NATURE OF THE PROBLEM

Although hormones do an exemplary job of coordinating and regulating a variety of body processes, they are for the most part involved in problems of homeostasis and growth, where speed of response is not at issue. Imagine a situation where, out of the corner of your eye, you catch a glimpse of a rock whistling toward your head. If your sense organs communicated with your muscles by hormonal secretion, perhaps 30 seconds after you had been knocked unconscious your muscles would receive the message to duck.

A less dramatic example is provided by the simple task of writing. The hand-eye coordination involved in this seemingly mundane task is actually very complicated. For instance, you require constant feedback from the muscles controlling your hand and fingers in order to move your hand as you write and to grip the pen or pencil with the right pressure. You also require continuous feedback from your eyes to keep your hand movement smooth and regular. (Even though your brain "remembers" how to direct your hand to form letters, the fine-tuning of writing depends on feedback from your eyes. To demonstrate the importance of vision to good writing, try doing a little with your eyes closed.) Perhaps most obviously, you also must have extensive involvement of the brain to ensure that what you are writing makes sense, is grammatically correct, and so forth. It would take a very active imagination to visualize how all these activities could be coordinated using only hormones.

As we saw in Chapter 12, skeletal muscle is characterized by speed and strength of contraction. In order to maximize its capabilities, we require virtually instantaneous coordination and feedback, capabilities that are beyond the scope of hormones. The solution, as we all know, is provided by the nervous system. But how does the nervous system work?

THE ANATOMY OF NERVES

Before answering that question, let us take a quick look at how the nervous system is organized. If we ignore the brain and spinal cord for the moment, the most obvious manifestation of the nervous system in the body is the presence of **nerves.** Nerves vary in size, but they are commonly about the diameter of veins, arteries, and ligaments. (Indeed, nerves and arteries are easily confused by students doing their first anatomical dissections, and more than one surgical resident, in the pressure of the emergency ward, has mistakenly stitched nerve to artery while attending to the slashed wrists of would-be suicide patients.)

Nerves are a little like telephone cables (see Figure 14.1). They have an outer connective tissue sheath but are principally composed of hundreds of individual nerve cells called **neurons,** which run the length of the nerve. Each neuron is isolated from all other neurons by a layer of fatty insulation called **myelin** (Gr. *myelos,* "marrow"), which is derived from a type of associated cell called **Schwann cells.** (Other associated cells, called **glial cells,** are found in the brain and spinal cord.) It is myelin that gives nerves their white color. Neurons on the surface of the brain have no myelin and appear gray in color—hence, the term "gray matter" as another name for the brain. Because the Schwann cells are very much shorter than are most neurons, there are gaps, or **nodes,** in the myelin sheath between adjoining Schwann cells. These nodes are of great importance in the functioning of neurons, as we shall see shortly.

The Neuron

Although it can possess a bewildering array of shapes (see Figure 14.2), a neuron is not likely to be mistaken for any other type of cell. In every instance, it consists of a small *cell body* with at least one elongated process. These processes tend to be of two types. The shorter, more numerous branches fancifully resemble the branches of a tree. They are called **dendrites** (L. *dendron,* "tree"). The longer processes, called **axons** (Gr., "axis"), often number only one to a neuron.

Neurons can be very long cells, although in humans they are always exceedingly thin. The cell bodies of the neurons responsible for controlling the muscles you use to wiggle your toes are located in your spinal cord, as much as a meter or more away from their site of action!

COMMUNICATION WITHIN A NEURON

The basis of neuronal functioning is the capacity of neurons to undertake electrical changes. As is true of all the cells in the body, the concentration of ions within the neuron is different from the concentration of ions in the surrounding interstitial fluid. Specifically, the concentration of Na^+ is much higher outside the neuron, whereas the concentration of K^+ is much higher inside the neuron.

How is the distribution of these ions achieved and maintained?

The Resting Potential

In a neuron at rest, Na^+ and K^+ are maintained in their respective positions by active transport. The sodium-potassium pump, located in the plasma membrane of the neuron, exchanges Na^+ and K^+ (see

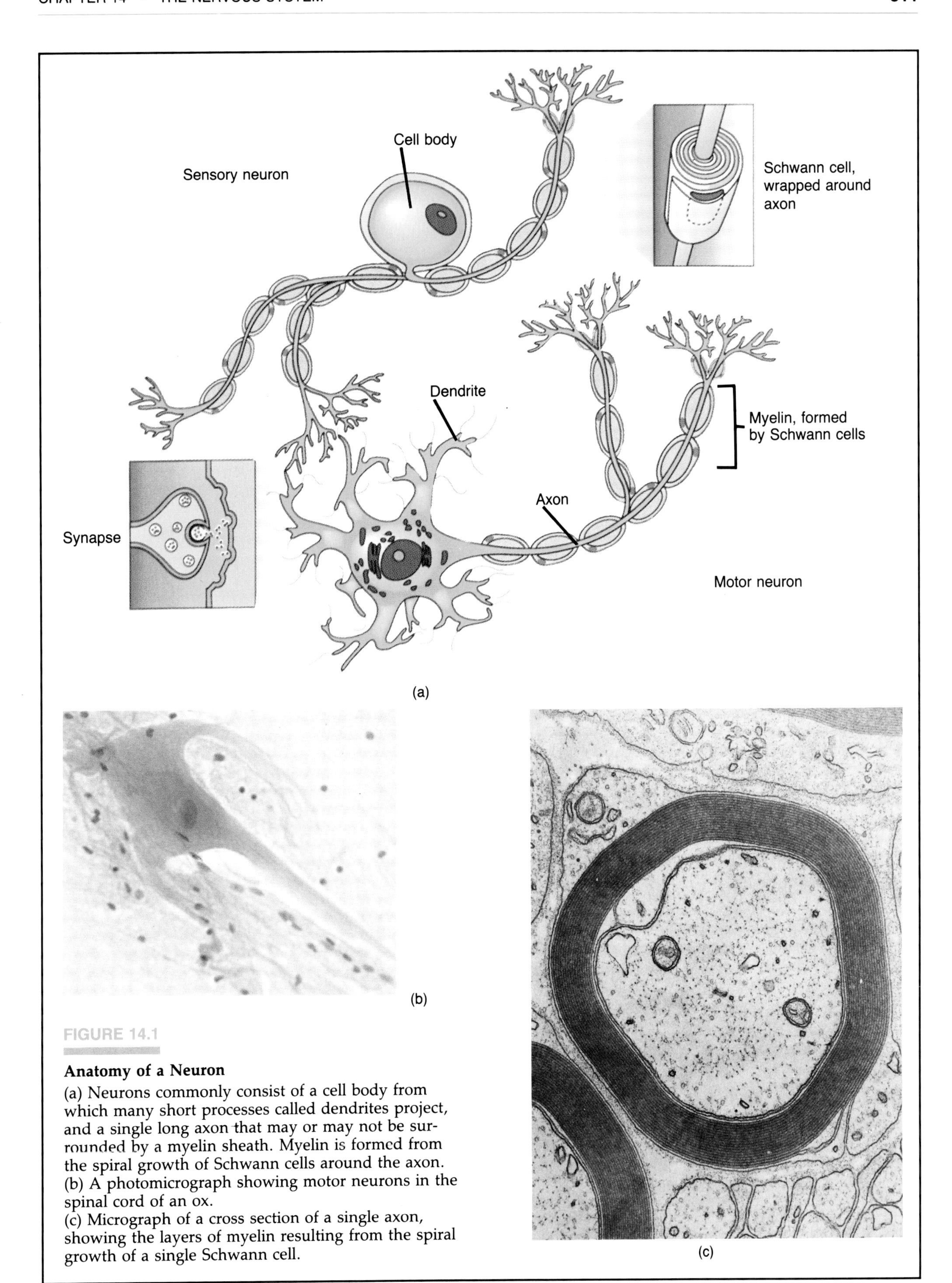

FIGURE 14.1

Anatomy of a Neuron

(a) Neurons commonly consist of a cell body from which many short processes called dendrites project, and a single long axon that may or may not be surrounded by a myelin sheath. Myelin is formed from the spiral growth of Schwann cells around the axon.
(b) A photomicrograph showing motor neurons in the spinal cord of an ox.
(c) Micrograph of a cross section of a single axon, showing the layers of myelin resulting from the spiral growth of a single Schwann cell.

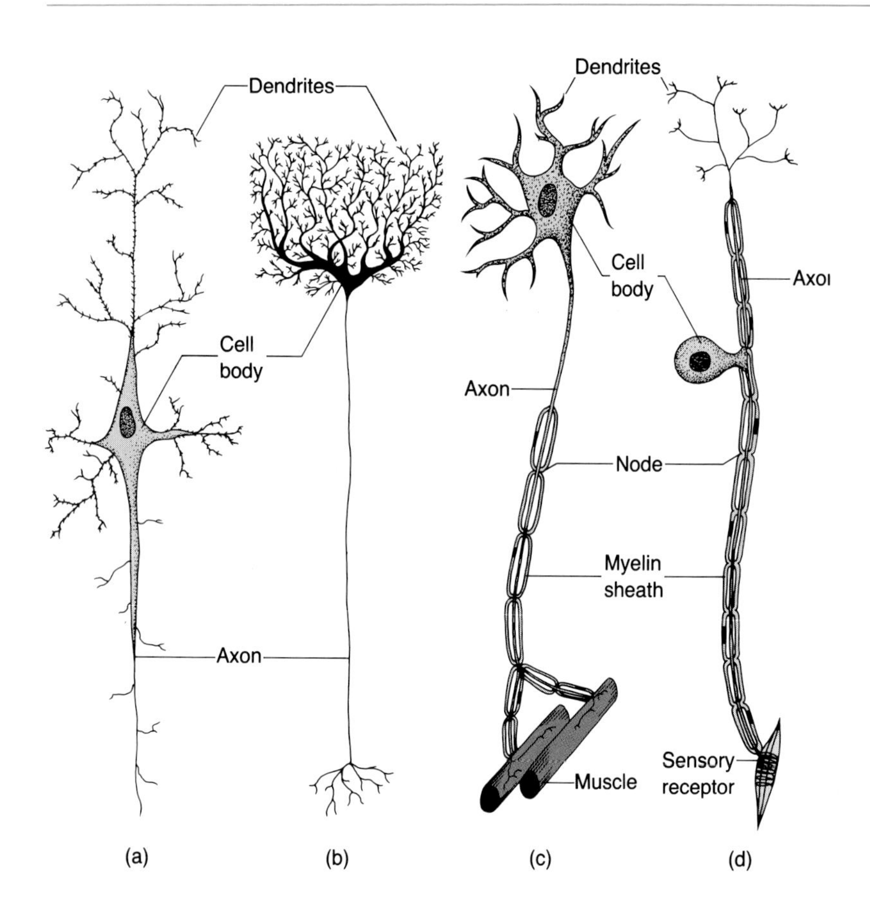

FIGURE 14.2

A Variety of Neurons

Neurons come in many shapes and sizes. (a) A pyramid cell from the cerebral cortex. (b) A Purkinje cell from the cerebellum. (c) A motor neuron. (d) A sensory neuron.

Chapter 5 for details). Making life more difficult for the pump, however, is the tendency for K^+ to move back into the interstitial fluid by diffusing very rapidly through open ion channels in the plasma membrane.

In contrast, most of the ion channels for Na^+ are apparently closed, or *gated*, for there is virtually no inward diffusion of Na^+ from the interstitial fluid. Incidentally, specific channels exist for each of the major ions (K^+, Na^+, Ca^{2+}, and Cl^-), and the channels of the positive ions are structurally very similar (see Figure 14.3). They consist of four proteins, each of which runs from one side of the plasma membrane to the other. On the inside of the membrane each of these four proteins has a linear "chain" of several amino acids that terminates in a 19 amino acid "ball." An ion channel is closed when one or more of the balls plugs the opening of the channel. (Some of you—those who bear close watching—might be interested in knowing that the venom of certain scorpions is potentially lethal because it contains a chemical that binds to the outer side of K^+ channels in neurons, and in so doing effectively shuts down the nervous system.)

Let us turn our attention once again to K^+. Although many of the K^+ channels are open, the potential outward movement of K^+ is offset not only by the actions of the sodium-potassium pump but also by the many large negative organic ions within the neuron that cannot pass through the plasma membrane. These negative ions attract K^+ and tend to keep it within the neuron.

The net result is an excess of negative charges inside the neuron. Na^+ is effectively excluded by active transport, and K^+, though relatively abundant within the neuron, is not present in sufficient quantity to offset all the large negatively charged organic ions.

The difference in electrical charge between the inside and outside of the cell can actually be measured. It amounts to some −70 mV (millivolts, that is 70/1000 of a volt). The negative sign indicates that the interior of the neuron is negative relative to the outside. This is the **resting potential** of the neuron, and it is constant, being maintained by the balance of forces represented by active transport, diffusion gradients, and electrical attraction and repulsion.

The Action Potential

The fact that neurons maintain an electrical gradient across their plasma membranes tells us nothing about the hallmark of the nervous system—its capacity to send messages. As you might expect, however, knowing about the resting potential is essential for understanding the neuronal code, because that code employs a momentarily disrupted resting potential

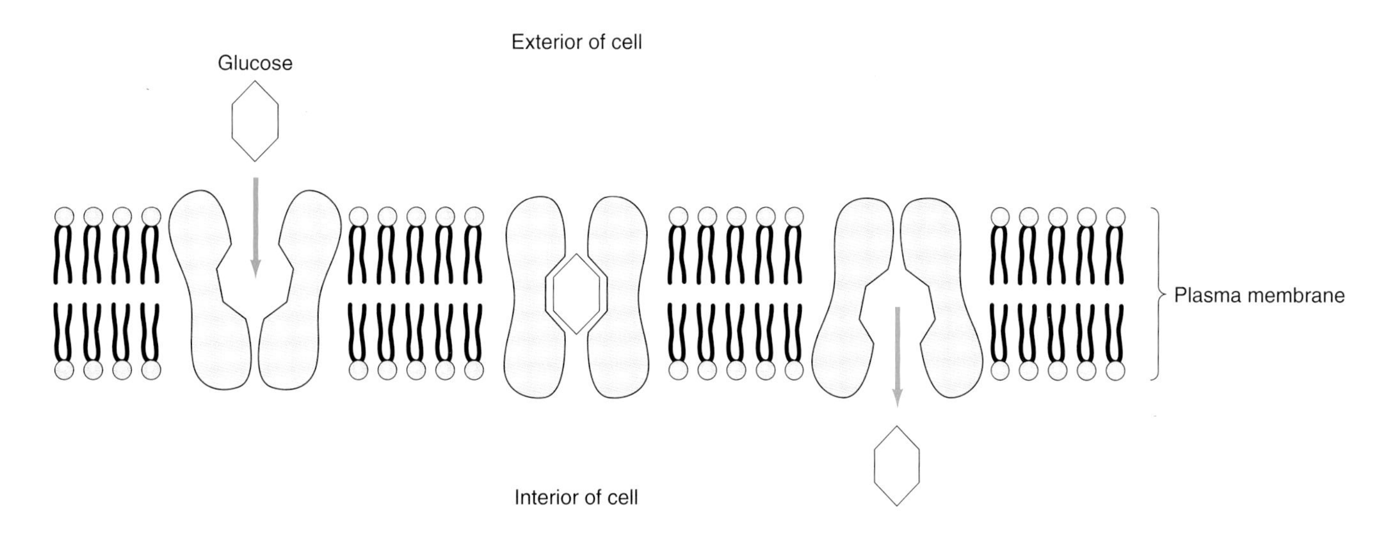

FIGURE 14.3

Gated and Ungated Ion Channels
Most sodium channels are gated, meaning that entry of sodium into the neuron is precluded by the presence of a "plug" of amino acids attached to one of the four protein molecules that make up the channel. When a neuron is stimulated, the plug falls to one side, permitting sodium ions to diffuse rapidly into the neuron.

called an **action potential,** which occurs whenever a stimulus (pressure, pH change, certain chemicals, etc.) of sufficient strength is applied to a neuron.

The action potential consists of two distinct phases (see Figure 14.4). The first phase involves **depolarization** of the plasma membrane, which occurs because the stimulus causes gated Na^+ channels in the plasma membrane to open and the membrane suddenly becomes 5000 times more permeable to Na^+ than it is normally. Na^+ diffuses rapidly into the neuron in such abundance that the interior becomes momentarily *depolarized* (that is, positively charged, with a value of roughly +30 mV).

The second phase involves **repolarization** of the plasma membrane. This step occurs when gated K^+ channels in the plasma membrane open and the membrane momentarily becomes 50 times more permeable to K^+ than it is at rest. Simultaneously, the gates close on the Na^+ channels, and the neuron becomes essentially impermeable to Na^+ diffusion once again. The consequence of these two events is an immediate cessation of the diffusion of Na^+ into the neuron and a rapid escalation in the diffusion of K^+ out of the neuron. This exodus of positively charged ions repolarizes the neuron and restores the normal resting potential value of −70 mV (see Figure 14.5). The sodium-potassium pump, which was temporarily overwhelmed by the momentary change in plasma membrane permeability to the two ions, quickly substitutes K^+ for Na^+, and restores the characteristic resting potential distribution of ions. The entire action potential lasts less than 2 msec (that is, 2/1000 of a second)!

In the formation of an action potential, not all of the axon is depolarized at one time. Rather, there is a wave of depolarization, which moves along the axon at speeds up to 100 m/sec, depending on the size and character of the neuron. As quickly as one portion of the axon becomes depolarized, the section immediately behind it becomes repolarized. Thus, because some neurons may be a meter or more in length, at any given moment a particular neuron may be transmitting several action potentials, meaning that there are several separate waves of depolarization moving along the neuron at different points.

Think of a neuron at rest as being analogous to a row of dominoes (see Figure 14.6). Initiation of an action potential is analogous to knocking over one domino—and both action potential and toppling dominoes then move as a wave. Now imagine that it were possible to restore the dominoes to their upright position almost as quickly as they were knocked over. That activity would be equivalent to the repolarization of the neuron and the reestablishment of the resting potential. In such a magical row of dominoes, one could imagine several points along the row where dominoes were toppling and then promptly being set up again, only to topple once more. This situation is analogous to several action potentials sweeping along a neuron at one time.

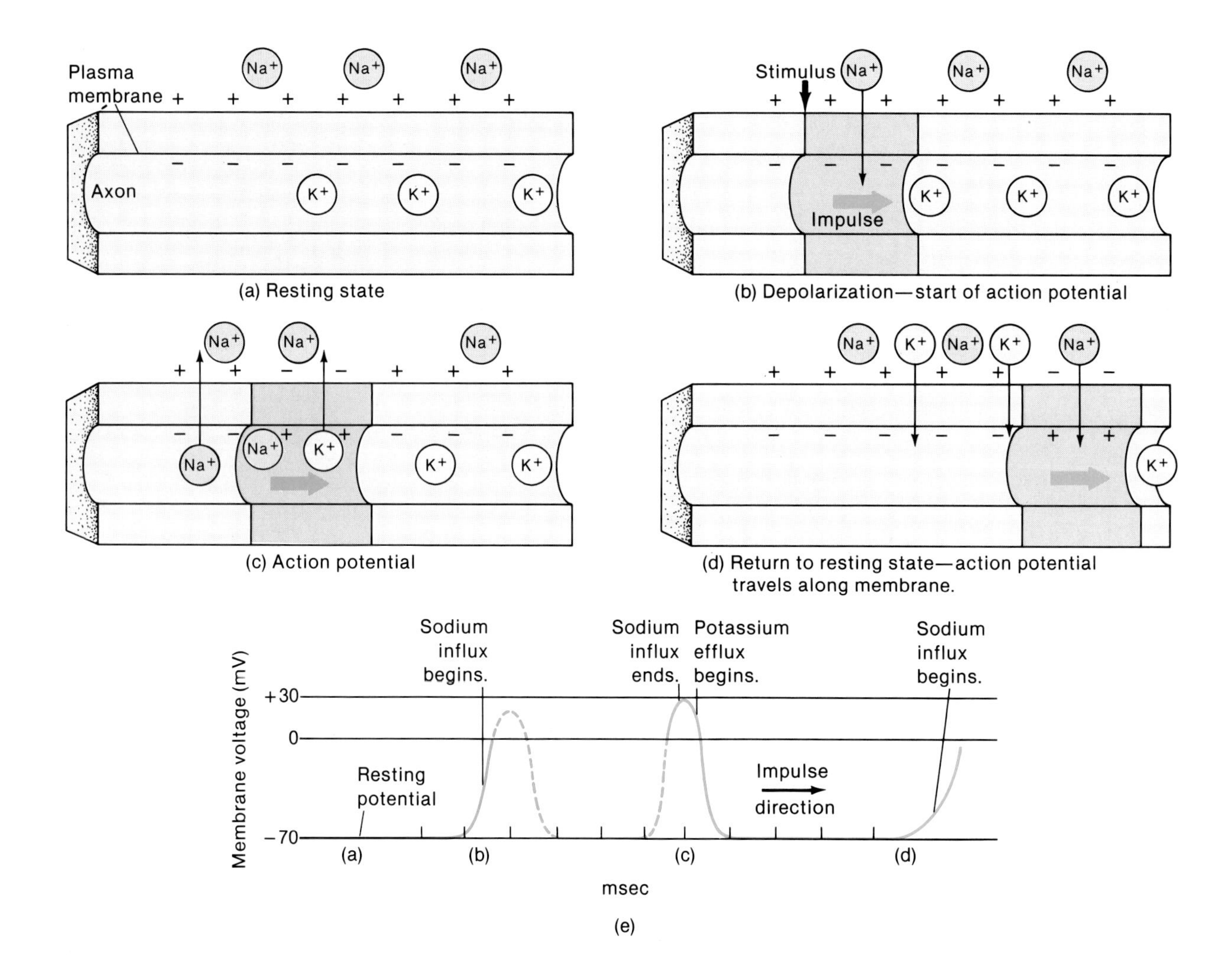

FIGURE 14.4

Action Potential

(a) A small piece of the axon of an inactive neuron. Note the characteristic distribution of charge (positive on the outside), and the distribution of positive ions (sodium on the outside). (b) With the application of a stimulus, the resting potential is disturbed and an action potential begins. At the region of the axon that has been depolarized, gated sodium channels have opened, and sodium ions diffuse into the axon, changing the membrane voltage from −70 mV to a peak figure of +30 mV. (c) The action potential continues, as potassium channels open and potassium ions diffuse rapidly out of the neuron, moving the membrane potential back toward its resting value. At the same time, the nerve impulse continues to move down the axon; in the area just depolarized, sodium ions are actively transported out of the neuron. (d) The nerve impulse continues to move down the axon, in the form of a wave of depolarization that ultimately will sweep the entire length of the axon. In the areas of the axon previously depolarized, potassium ions are actively transported back into the axon and sodium ions continue to be evicted. (e) If we look at membrane voltage levels as the action potential is being generated, we can link particular points on the curve with the four steps we have just described.

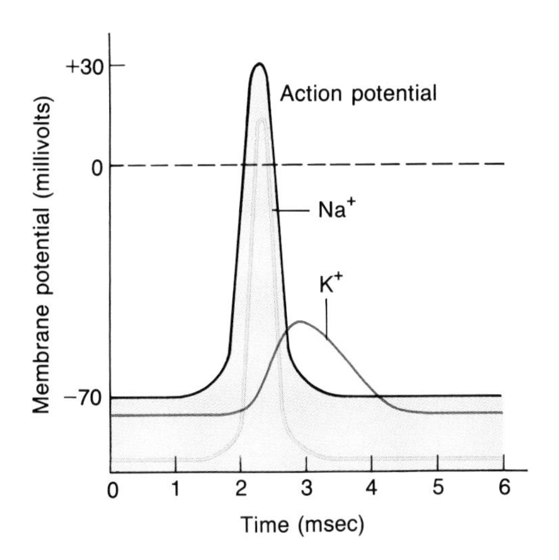

FIGURE 14.5

Relationship of Action Potential to Ion Movement

By superimposing the action potential on top of graphs showing the movement of sodium and potassium ions, it is possible to see that the spike of the action potential is due to the inflow of sodium ions, and the restoration of resting potential values is due to the outflow of potassium ions a fraction of a second later. Note that the entire action potential lasts only about 2 msec.

There are several things about action potentials that are unusual:

1. Unlike electrical current in a wire, the value of an action potential does not diminish as it moves along the neuron. This is because the action potential is constantly being reinitiated by depolarization of adjacent plasma membrane—it is *not* a passive phenomenon, as is the movement of electrical current in a wire.

2. In neurons with myelin sheaths, the action potential jumps from node to node, because it is only at the nodes that the cell can exchange ions with the

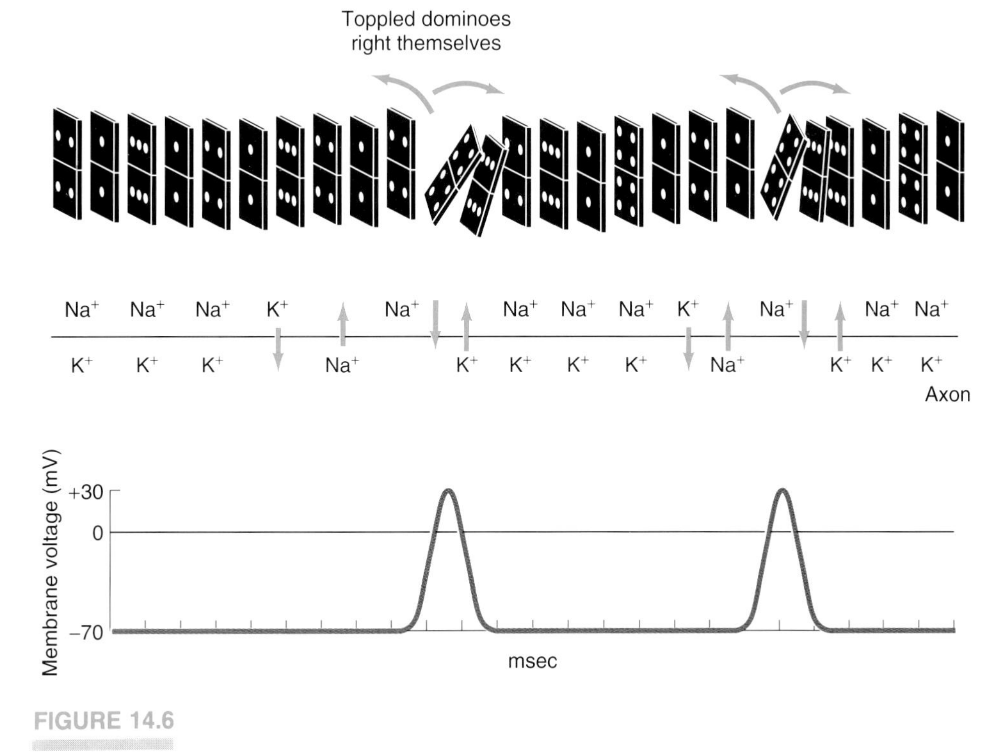

FIGURE 14.6

Toppling Dominoes and Action Potentials

One way to visualize the manner by which an action potential moves along a neuron is to think of it as akin to a row of toppling dominoes. The action potential cannot move backward because the dominoes have already toppled. Therefore action potentials move only away from areas that have already been depolarized. On the other hand, a neuron is obviously not permanently put out of commission once it has conveyed an action potential. To the contrary, it can convey many action potentials at one time, but all spaced by the equivalent of one or two msec. Thus our dominoes must be capable of righting themselves almost immediately.

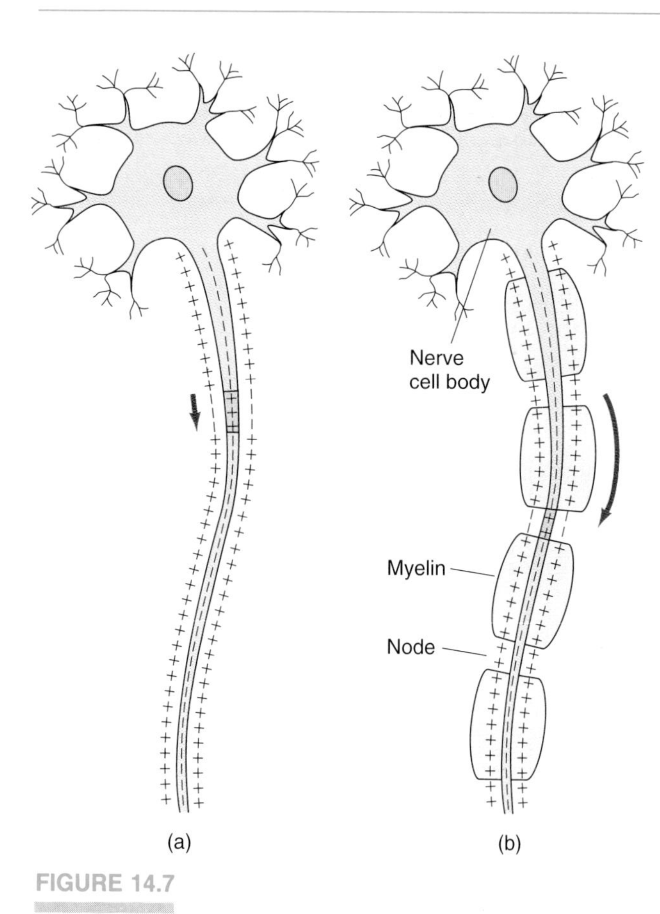

FIGURE 14.7

Saltatory Conduction

(a) In neurons without a myelin sheath, every portion of the neuron is successively depolarized. (b) In neurons with a myelin sheath, the wave of depolarization skips from node to node. Myelin sheaths not only insulate neurons, but also permit much more rapid transmission of nerve impulses. It is not surprising to find that many neurons that have very long axons also have a myelin sheath, because it is in long axons that speed of transmission is significant.

interstitial fluid. This **saltatory conduction** (L. *saltare,* "to leap") greatly speeds up the rate at which action potentials move along a neuron (see Figure 14.7). It also reduces the amount of ATP needed to power the sodium-potassium pumps, since only those pumps at the nodes need to function.

3. In any given neuron, a stimulus will either be strong enough to initiate an action potential or it will not. Once initiated, an action potential invariably sweeps along the entire length of the neuron. The concept that all of the neuron ultimately undergoes depolarization is known as the **all-or-none rule.** Finally, all action potentials—regardless of whether the initiating stimulus is only barely adequate or much stronger than adequate—have the same value.

If action potentials are always of the same magnitude, how is information about the strength of the stimulus transmitted? In graphic terms, how is your brain informed that you just drove a nail through your foot, as opposed to pricking your skin with a pin?

There are two answers. First, a massive insult to the nervous system (e.g., a nail through your foot) will cause many neurons to be depolarized. Your brain will receive information from many sources and interpret these multiple messages as indicating a serious injury (the nervous system's equivalent of a multiple alarm fire). Second, the stimulus may cause repeated depolarizations of the same neuron. Whether from the same or different neurons, the brain uses the frequency (number per unit time) of signals it receives as the primary index of the strength of the stimulus.

COMMUNICATION BETWEEN NEURONS

As we have just seen, communication along a neuron is accomplished by electrical changes. However, even though individual neurons may be exceedingly long, eventually the action potential reaches the end of the axon. What happens at that point? How does one neuron pass its message on to a second neuron? Does a "spark" jump across the gap separating the two neurons?

The Synapse

The junction between the axon of one neuron and the dendrites or cell body of a second neuron is called the **synapse** (Gr. *synapsis,* "a union, joining") (see Figure 14.8). The **synaptic cleft** is the actual space, about 10 to 20 nm wide, between the two neurons. In order to transmit an action potential from the first neuron (**presynaptic neuron**) to the second (**postsynaptic neuron**), chemical transmission is substituted for electrical transmission.

When an action potential reaches the tip of the axon of the presynaptic neuron, it causes the plasma membrane to become highly permeable to Ca^{2+}. Because Ca^{2+} is much more abundant in the interstitial fluid than in the cytoplasm of the neuron, it rapidly diffuses into the neuron where it causes tiny sacs in the axon to fuse with the plasma membrane and to release their contents into the synaptic cleft. Each of these sacs contains up to 10,000 molecules of a class of chemicals known as **neurotransmitters.** If we exclude neurotransmitters in the brain, the most commonly used neurotransmitter in the body is **acetylcholine.** Let's consider its mode of action.

Acetylcholine diffuses across the cleft and attaches to receptor molecules on the membrane of the postsynaptic neuron, causing ion channels to open.

If enough receptor sites are occupied (and typically this requires as many as 50 action potentials arriving at the postsynaptic neuron almost simultaneously), the membrane of the neuron is disturbed sufficiently as to create a new action potential.

Once released, what prevents these acetylcholine molecules from causing repeated action potentials in the postsynaptic neuron? The answer is that a second molecule lies in wait in the synaptic cleft. This molecule is **cholinesterase,** an enzyme that splits the acetylcholine molecules, thereby destroying their effectiveness. Because one cholinesterase molecule can destroy 20 million acetylcholine molecules every minute, not much cholinesterase is needed to destroy the acetylcholine released by an action potential.

In essence, the acetylcholine molecules swamp the cholinesterase destructors temporarily, running the gauntlet as it were, but even if enough of them survive the trip intact to depolarize the membrane of the postsynaptic neuron, they are promptly destroyed by cholinesterase, thus preventing an endless series of action potentials from being generated. The remnants of the acetylcholine molecules are then reabsorbed by the presynaptic neuron, and reconstituted for subsequent reuse.

FIGURE 14.8

Structure of the Synapse

(a) A scanning electron micrograph showing the tips of several dendrites forming synapses with another neuron. (b) The synapse in diagrammatic form. (1) As the action potential reaches the end of the presynaptic neuron, it causes calcium channels to open, and calcium ions diffuse rapidly into the neuron. (2) Calcium activates enzymes that cause the sacs of neurotransmitter (acetylcholine in this instance) to fuse with the plasma membrane of the neuron.
(3) Acetylcholine molecules are released into the synaptic cleft and attach to receptors on the surface of the postsynaptic neuron. If enough receptors are occupied, the postsynaptic neuron will be depolarized, and an action potential generated. (4) Cholinesterase molecules deactivate the acetylcholine molecules by splitting them in half. The deactivated acetylcholine molecules are recycled by being reabsorbed by the presynaptic neuron and reassembled.

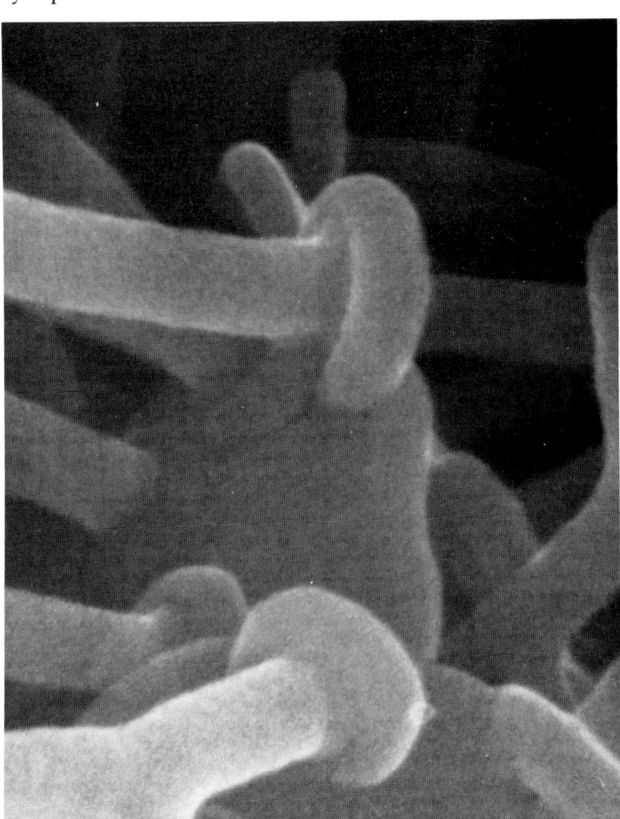

(a)

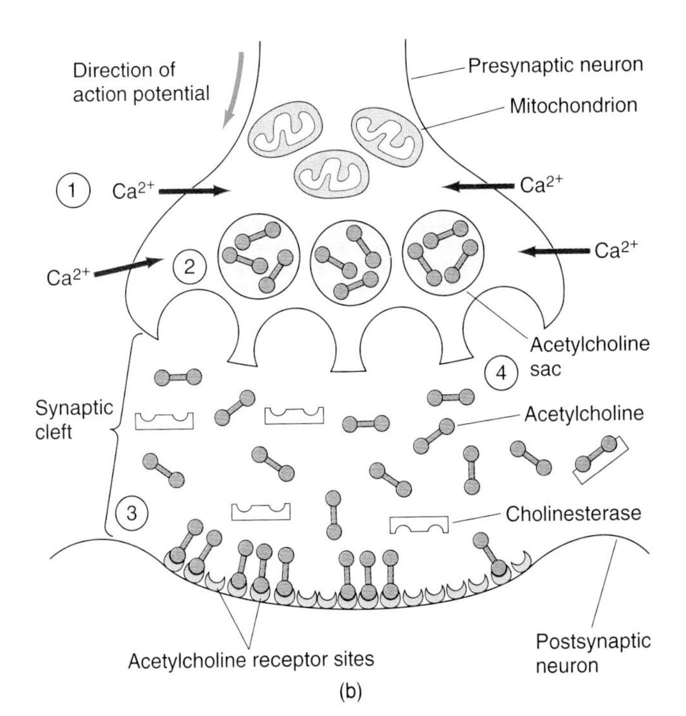

(b)

Neurotransmitters

Why do we have synapses? Before we answer that question, it is important to recognize that synaptic transmission is somewhat more complicated than our discussion might suggest. We humans use over 60 different neurotransmitters, each of which appears to serve a distinct function. In addition, one neuron may produce more than one type of neurotransmitter, and individual neurons may have as many as 10,000 synapses. Thus there is ample opportunity for a variety of messages to be sent and received.

For example, not all neurotransmitters are used to generate an action potential in the postsynaptic neuron, and not all neurotransmitters function in the same way throughout the body. Acetylcholine is usually excitatory in nature, meaning that it tends to depolarize the membrane of the following neuron, but it is inhibitory in cardiac muscle. Inhibition is accomplished either by opening Cl^- channels, thereby allowing Cl^- to diffuse into the cell, or by opening K^+ channels, allowing K^+ to diffuse out of the cell. In either case, the cell (be it postsynaptic neuron or muscle fiber) becomes **hyperpolarized**, meaning that the value of its resting potential is -90 or -100 mV. Under such conditions, many more excitatory impulses from other neurons will be required before an action potential will be initiated in the cell.

The picture is, of course, by no means restricted to acetylcholine. Other neurotransmitters include *glutamic acid*, an excitatory neurotransmitter in the brain; *gamma amino butyric acid (GABA), adrenaline, noradrenaline, dopamine, glycine,* and *serotonin*, all of which are generally inhibitory neurotransmitters; and the *endorphins*, which are *modulators* (substances that do not directly affect impulse transmission, but which sensitize the postsynaptic neuron to other neurotransmitters).

Let us return to the question, "Why do we have synapses?" Perhaps the answer is by now obvious. If the nervous system were "hard-wired" throughout, so as to permit the propagation of action potentials by direct electrical changes, there would be no opportunity to modify the message. Synapses are decision points. It is there that all the reinforcing and conflicting messages are integrated, and a "decision" is made whether or not to generate a new action potential. These kinds of "decisions" are, not surprisingly, of particular importance in the brain.

The Neuromuscular Junction

The **neuromuscular junction,** or motor end-plate, is the area of contact between a neuron and a muscle fiber (see Figure 14.9). Although it is structurally similar to a synapse, physiologically it functions in a somewhat different manner (see Box 14.1).

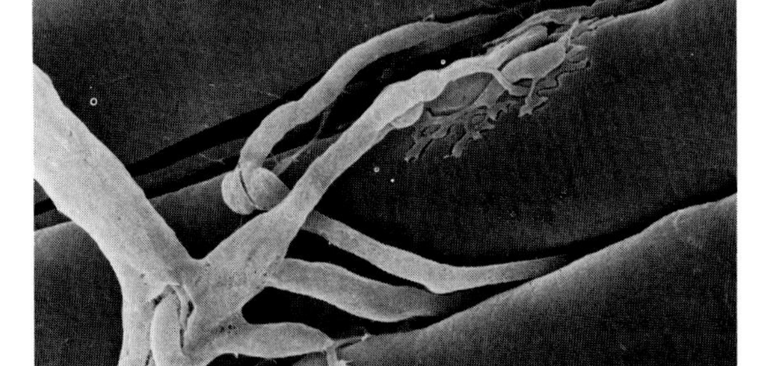

(a)

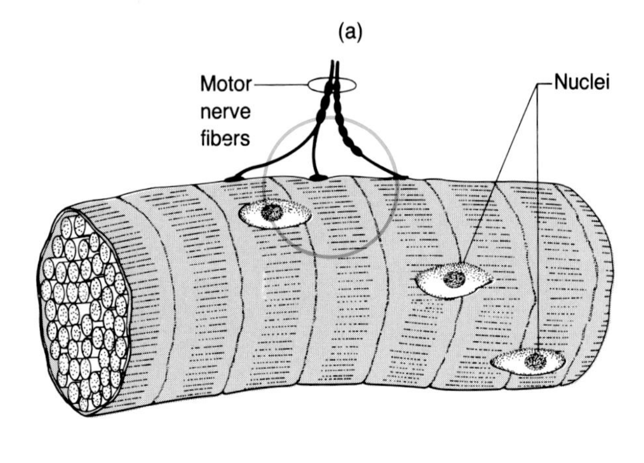

(b)

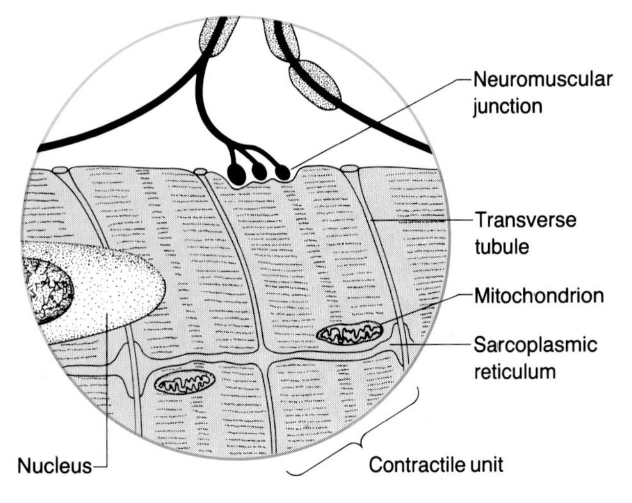

(c)

FIGURE 14.9

Neuromuscular Junction

(a) Scanning electron micrograph of neurons innervating muscle fibers. (b) Motor neurons and a myofibril. (c) A neuromuscular junction.

Acetylcholine is the neurotransmitter used in all neuromuscular junctions involving skeletal muscle. Acetylcholine binds to receptors on the membrane of the muscle fiber, opening both Na^+ and K^+ gates, and causing a depolarization of the membrane. As the wave of depolarization sweeps across the membrane, Ca^{2+} is released into the cytoplasm of the muscle fiber, and the presence of Ca^{2+} initiates contraction of the muscle (see Chapter 12 for details).

Within 1/30 of a second, the Ca^{2+} is pumped back out of the cytoplasm, and contraction of the muscle fiber ceases. Contractions of longer duration therefore require successive releases of Ca^{2+}.

Because every neuron innervates several muscle fibers, a single action potential can cause contraction in a number of muscle fibers simultaneously. The exact number is a reflection of the degree of fineness of control. Small muscles, such as those of the eye, may have as few as six muscle fibers innervated by a single neuron, whereas in the large muscles of the thigh and calf a single neuron may innervate hundreds or even thousands of muscle fibers. The number of neurons activated at any one time determines the strength of the muscle contraction as a whole.

FIGURE 14.10

Human Nervous System

Artist's rendering of the human nervous system. The dominance of the central nervous system is immediately apparent.

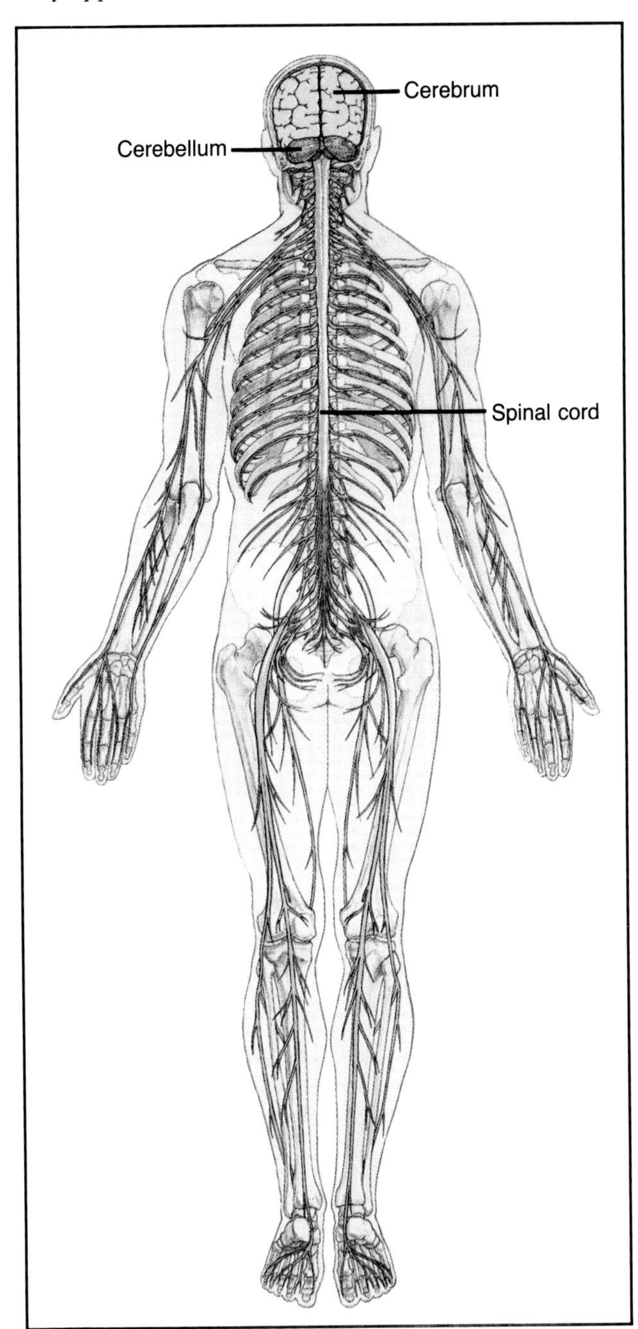

ORGANIZATION OF THE HUMAN NERVOUS SYSTEM

To this point, we have considered only the functioning of single neurons. Understanding how individual neurons do their job is obviously indispensable to our understanding of the functioning of the nervous system as a whole. However, unlike the situation with muscle fibers, we cannot blithely extrapolate from the activities of a single neuron to the operation of the entire nervous system. The brain is a bit more than a collection of 100 billion individual neurons, in the same way that life is a bit more than a mixture of specific chemicals. Both represent synergistic systems demonstrating emergent properties—properties that cannot be predicted simply from a knowledge of the components.

Before we begin discussing the brain, however, let us spend some time understanding the overall organization of the human nervous system (see Figure 14.10). As we do, keep in mind that individual neurons are either **sensory neurons** (those carrying information *from* a sense receptor *to* the brain or spinal cord), **motor neurons** (those carrying information *from* the brain or spinal cord *to* an effector organ such as a muscle or gland), or **interneurons** (those found entirely within the brain or spinal cord, where they function in relaying information from sensory to motor neurons as well as to other interneurons).

The overall organization of the human nervous system can be summarized as follows (see Figure 14.11):

I. **Peripheral nervous service:** Nerves running to and from the brain and spinal cord;

Problems with Nerves and Neuromuscular Junctions

Because the nervous system has such a critical role in coordinating body functions, problems with this system are often devastating. Some areas of the system are more prone to disruption than are others. For instance, the neuromuscular junction is a particularly susceptible area. Problems include the following:

Blockage of acetylcholine release. **Botulism** is a disease in which a toxin produced by the bacterium *Clostridium botulinum* (occasionally found in improperly preserved canned goods) blocks the release of acetylcholine. It is so potent that less than 1/1,000,000 g may be fatal. Recovery is generally good if antitoxin is administered quickly, although it may be weeks or even months before full use of the muscles is regained, as the toxin is notoriously long-lasting. Recently, the toxin has been used therapeutically in the treatment of certain muscle diseases characterized by extreme contraction of the muscle. Minute doses of the toxin are injected directly into the affected muscle, allowing it to relax. The treatment is repeated every few months as necessary.

Blockage of muscle membrane receptor sites. As in the blockage of acetylcholine release, blockage of the receptor sites results in paralysis. This result occurs in **curare** poisoning, curare being the poison used by the blowgun hunters of the Amazon basin. Curare has a shape similar to acetylcholine, and it occupies receptor sites on the muscle membrane. However, curare does not trigger an action potential, nor is it broken down by cholinesterase. As long as curare occupies the receptor sites, acetylcholine molecules are destroyed by cholinesterase before they can stimulate the muscle. Treatment for curare poisoning involves immediate administration of a cholinesterase inhibitor, which prevents the premature destruction of the acetylcholine molecules.

Perhaps the most interesting medicinal use of the paralytic drugs was the use of curare as an anaesthetic. Only rather recently have the traditional anaesthetics (chloroform and ether) been replaced, despite the fact that both frequently have significant after-effects, most notably nausea.

A number of years ago, curare was tested as a potentially useful anaesthetic for certain operations, both because it was considered relatively safe (although a respirator had to be used), and because its effects could be almost instantly terminated by the administration of a cholinesterase inhibitor. However, patients experienced an unexpected reaction. They claimed to have been fully conscious during the operation, able to hear everything that was said, and able to feel the scalpel as it cut into their skin, but unable to cry out or speak because of the paralysis produced by the drug.

Surgeons at first attributed these sensations to hallucinations induced by the drug, but after it was used in an operation on a physician, who promptly confirmed the statements of the earlier patients, the effects of the drug were investigated further. Ultimately, it was realized that the drug interfered only with neuromuscular junctions, and not with the synapses of the sensory system. The patients had indeed been fully conscious and unimpaired in their senses during the operations!

Curare is still used in surgery as a muscle relaxant, but (you will be pleased to know) only in conjunctions with a true anaesthetic.

The autoimmune disease **myasthenia gravis,** which presently affects half a million Americans, is characterized by the destruction of the acetylcholine receptors of muscle fibers and consequent weak or erratic muscular contractions. Treatment with cholinesterase inhibitors, such as *physostigmine* and *neostigmine* (both of which are chemically related to the organophosphate insecticides—see below) provides substantial relief. They compensate for reduced acetylcholine activity by marginally interfering with cholinesterase function, thereby allowing the acetylcholine sufficient time to depolarize the muscle fiber.

Impairment of cholinesterase activity. If cholinesterase activity is impaired, the acetylcholine molecules remain active, causing repeated depolarizations of the muscle membrane and (ultimately) convulsions. Most nerve gases and insecticides operate in this manner, by irreversibly binding to cholinesterase. (It is no accident that these two groups of poisons are similar in their mode of action. Some insecticides used today, such as the organophosphate **parathion,** were originally developed as nerve gases and, on a dose-per-body-weight basis, are about as toxic to humans as to insects.) Treatment consists of the immediate administration of **atropine,** a drug that blocks acetylcholine and artificially recreates the normal situation at the neuromuscular junction. Ironically, atropine, a curare mimic, is itself potentially lethal. In this instance, one poison is being used to counter another.

Atropine has been used for centuries, although not always as an aid to health. It is the active compound of the deadly nightshade plant, so-called because it was a favorite poison in the Middle Ages and early Renaissance. The plant is also known as *belladonna,* because, at this same time in history, women who wished to be perceived as beautiful ("bella donna") used eyedrops containing plant extract. The atropine interfered with the acetylcholine receptors in the iris of the eyes, with the result that the pupil remained dilated (the desired sign of beauty). Modern ophthalmologists use a related compound in examining the eyes, because examination of the interior of the eye is facilitated when the pupil is dilated.

Not all nerve problems are focused on the neuromuscular junction. For example, **multiple sclerosis** is a complex of related autoimmune diseases wherein the insulating sheath of myelin surrounding the nerve cells is attacked by T lymphocytes (see Chapter 9) and replaced by scar tissue, thereby disrupting the normal transmission of nerve impulses. It is a disease of young adults, generally first diagnosed in patients between the ages of 15 and 40.

MS often has an up-and-down course. Apparently, myelin can be repaired by certain antibodies, but eventually succumbs to the continued attack of T cells and macrophages. Experimentally, mice with an MS-like disease showed improved walking ability when given specific (IgG) antibodies. Whether these antibodies interfered directly with the T cells, or simply stimulated the manufacture of myelin, could not be determined. However, the potential for antibody therapy in human sufferers of MS is obvious.

A related disease is **amyotrophic lateral sclerosis** (Lou Gehrig's disease). It is characterized by similar scarring, but it has a different developmental pattern. It occurs in people aged 40 to 70, and 5000 Americans are stricken each year. Like MS, ALS is a progressive disease, with death ultimately resulting from pneumonia or respiratory failure. One promising experimental treatment uses large doses of the hormone TRH in an attempt to slow the progress of the disease.

Polio (poliomyelitis) is, in this country, virtually a relic disease because of the enormous effectiveness of the Salk and Sabin vaccines that have been employed since the mid-1950s. However, as recently as 1952, there were 60,000 new cases in the United States alone. The virus that causes the disease gains access to the body through the digestive system, and then invades the gray matter of the spinal cord (hence the name of the disease—Gr. *polios*, "gray"). The virus destroys the cell bodies of motor neurons, and the ultimate consequence depends entirely on how many neurons cells were destroyed. Partial or complete paralysis is common, and because nerve cells cannot divide, this paralysis is generally permanent.

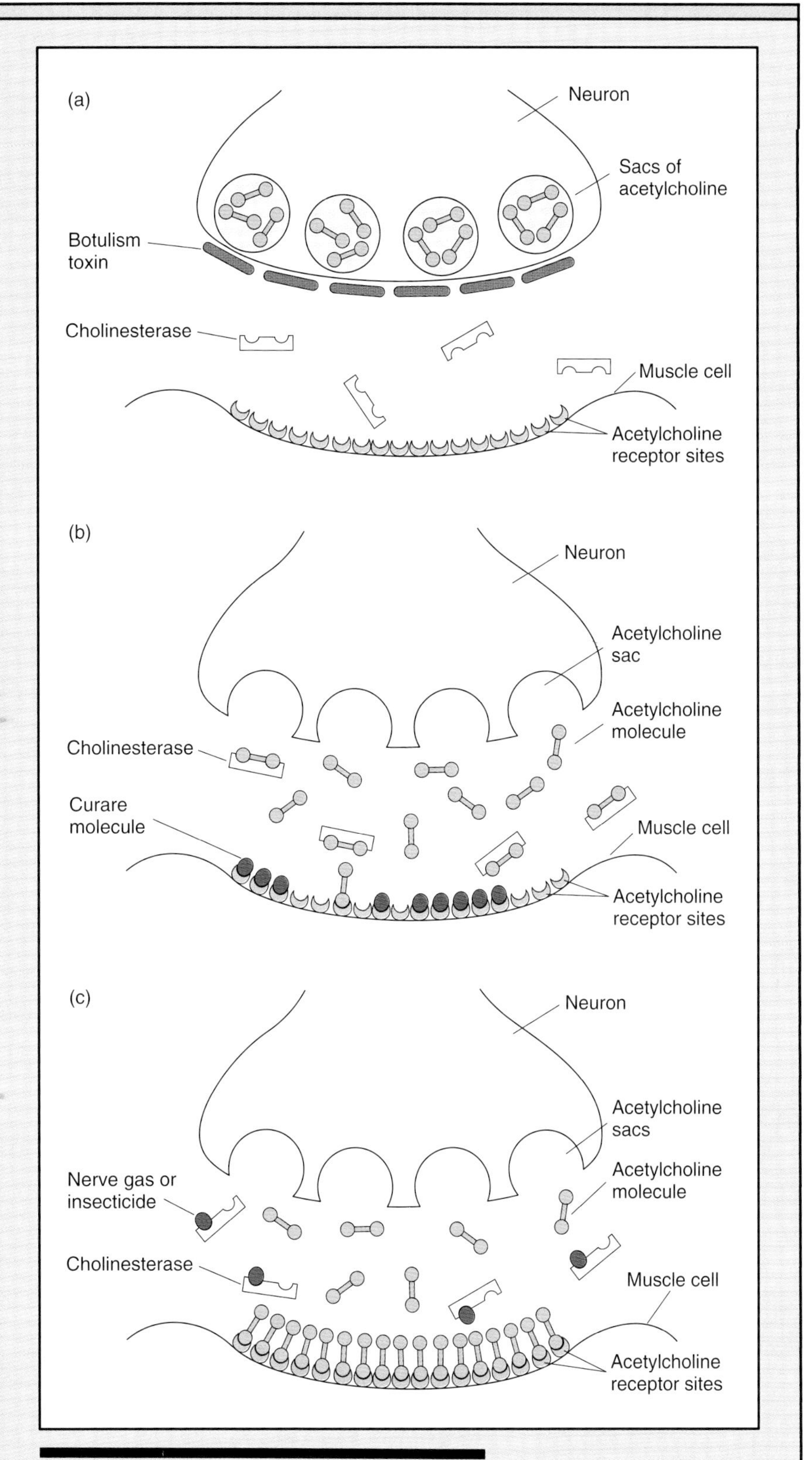

Problems with the Neuromuscular Junction

Various chemicals or disease agents can interfere with (a) acetylcholine release, (b) acetylcholine binding sites, or (c) destruction of acetylcholine by cholinesterase, any of which creates potentially devastating effects.

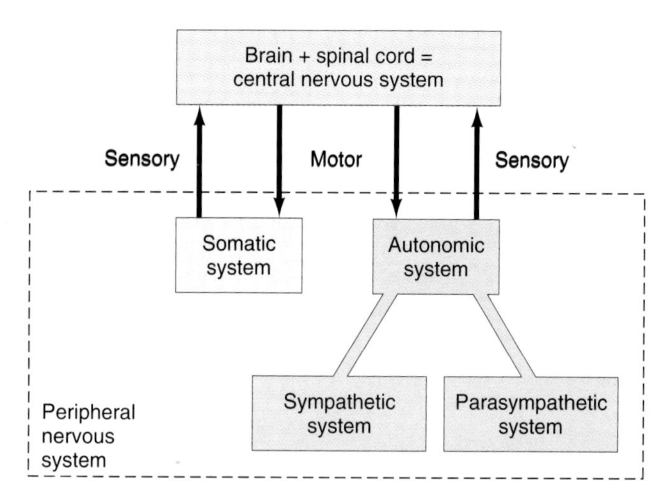

FIGURE 14.11

Divisions of the Nervous System
Note the two-way flow of information between the peripheral nervous system and the central nervous system. Both receptors and effectors tend to be peripheral, but communication between them inevitably involves the central nervous system.

A. **Somatic system:** Nerves supplying the skeletal muscles and skin;

B. **Autonomic system:** Nerves supplying the internal organs;

1. **Sympathetic system:** Nerves that stimulate the heart and inhibit most other internal organs;

2. **Parasympathetic system:** Nerves that stimulate most internal organs, except the heart;

II. **Central nervous system:** The brain and spinal cord.

THE PERIPHERAL NERVOUS SYSTEM

In humans and in other mammals, the nerves of the peripheral nervous system comprise 12 pairs of **cranial nerves,** originating in the brain, and 31 pairs of **spinal nerves,** originating in the spinal cord. Each of the 12 pairs of cranial nerves is assigned an appropriate Roman numeral. Most of the cranial nerves are highly specialized, but the spinal nerves are much more regular (see Figure 14.12).

Each spinal nerve consists of a **dorsal root** and a **ventral root,** which almost immediately fuse into a single nerve. Sensory neurons enter the spinal cord through the dorsal root, whereas motor neurons exit the spinal cord via the ventral root. The cell bodies of the motor neurons are within the spinal cord; however, the cell bodies of the sensory neurons are in a swelling of the dorsal root called a **ganglion** (Gr., "tumor").

The two roots no sooner fuse into a single spinal nerve than each nerve splits into three branches. The first branch passes to the skin and muscles of the front of the body; the second branch leads to the skin and muscles of the back; and the third branch innervates the internal organs. The first two branches collectively constitute the **somatic system,** which regulates voluntary actions. The third branch forms the **autonomic system,** which governs actions that are largely involuntary.

The Somatic System

The somatic system consists of both motor and sensory neurons. These neurons are integrated by interneurons with varying degrees of complexity. Even humans, however, with large and highly intricate brains, have some exceedingly simple sensory/motor pathways, sometimes involving only one sensory and one motor neuron. These very simple circuits control a type of behavior known as **reflexes.**

The Reflex Arc. Although the movement of action potentials along a neuron is very rapid, transmission at synapses is, by comparison, relatively slow. Thus, where an immediate response is essential, the ideal circuit will have a minimum number of synapses. Such circuits are called **reflex arcs.**

A sliver of glass flying up from a dropped bottle will trigger the blink reflex as soon as the glass touches an eyelash. Under those conditions at least, blinking is not the kind of reaction about which you would want to spend much time thinking. Similarly, if you accidentally touch a hot stove, you pull your hand away immediately, in a withdrawal reflex. Theoretically, you could wait until your sense of smell confirmed that your hand was, indeed, melting, but you are clearly much better off relying exclusively on the pain receptors in your fingertips, because the damage to your hand will be minimized.

One of the best-known reflexes is the knee-jerk reflex, which is generally tested by your physician during a physical examination. While you are sitting in a relaxed manner, with your legs dangling over the edge of the examining table, the physician strikes your leg just below the knee with a little rubber hammer, and your leg immediately straightens to some degree. What is the mechanism whereby the leg is straightened—and what function does this leg straightening have other than serving the interests of doctors?

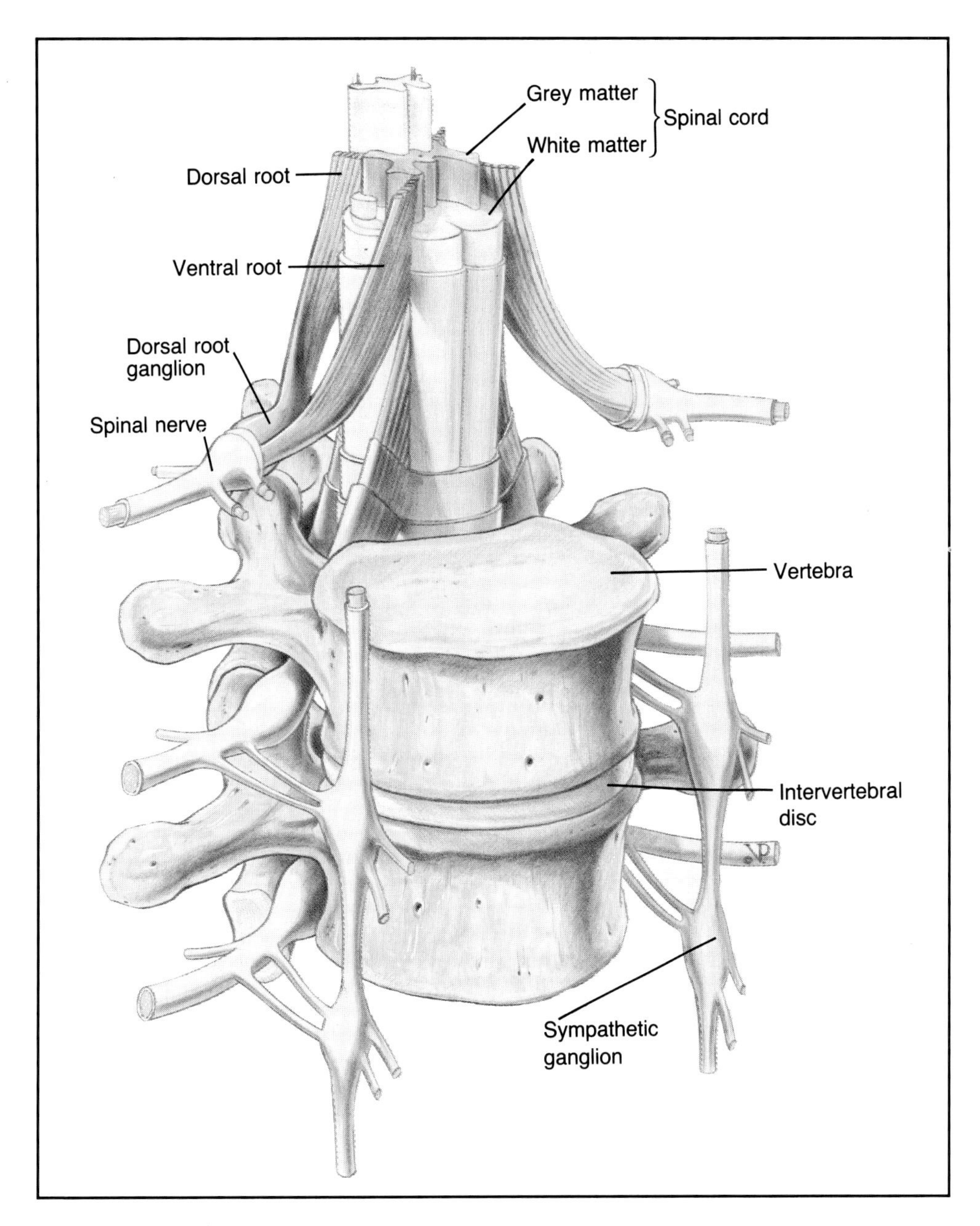

FIGURE 14.12

Spinal Nerves

A pair (left and right) of spinal nerves emerges from between each of the vertebrae as dorsal and ventral roots, which then fuse to form a single nerve that quickly branches in three.

When the tendon that runs from your thigh muscles, through the kneecap, and inserts on the front of the shin is stretched by the impact of the hammer, specialized muscle fibers in the thigh muscle are also stretched. They produce action potentials that are transmitted up sensory neurons, without synaptic interruption, to the spinal cord. These action potentials are then transmitted, through a synapse, to motor neurons that run directly to the thigh muscles. In response to the action potentials, the thigh muscles contract, and the leg straightens.

What is the purpose of this reflex? You may accept it on faith that it does not exist simply for the physician's convenience. The stretch receptors are also activated if your leg suddenly buckles; immediate contraction of the thigh muscles prevents you from collapsing. This reflex is also at work when you are running, although it takes a while to master, as anyone watching a ten-month-old infant practicing how to walk can attest.

But, you may protest, I can willfully prevent my leg from straightening; I can keep my hand on the stove; I can stop myself from blinking when something touches my eyelash. How does all that fit in with the notion of a simple two-neuron reflex?

First, through the use of interneurons and by the receipt of messages from other sense receptors, the brain is ultimately "informed" both of the factors surrounding the reflex and of the occurrence of the reflex response. For example, long after you have withdrawn your hand from the hot stove, the pain receptors in your hand continue to generate action potentials, and the brain registers this information—as well as the fact that your hand is no longer in contact with the stove. This relaying of information in no way contradicts the notion of a reflex, however, because none of it is necessary to the action of the reflex itself.

Second, you can prevent a reflex only when you can anticipate it. If you *know* the physician is about to rap your leg with his hammer, you can contract the muscles that are antagonistic to the thigh muscles, and freeze the leg in place. That ability does not nullify the existence of reflexes in humans, but it does demonstrate the level of control our brains have achieved over reflexes.

Such control is a peculiarly mammalian trait. We train our children and our dogs and cats to resist the urinary reflex, for example, until they are in a socially acceptable situation, but any attempt to housebreak your pet turtle will be in vain. Reflexes are much more a part of the behavioral repertoire of lower vertebrates than they are in ourselves, but it should be apparent that, even in humans, these simplest of neuronal integrations are of vital importance.

The Autonomic System

The autonomic system consists of sensory and motor neurons that innervate the internal organs (see Figure 14.13 and Box 14.2). It is a more complex system than is the somatic system for the simple reason that internal organs are not arranged in antagonistic pairs, as are the skeletal muscles. It is easy to understand why the arm bends if the biceps is stimulated, or why it straightens if the triceps is stimulated. However, what happens when the small intestine is stimulated by the autonomic system? Does peristaltic activity rise or fall? Does absorption increase or decrease? Do the glands secrete more or less? The situation is simply not analogous with the contraction of skeletal muscles.

As it happens, each internal organ is innervated by two sets of neurons. One set, comprising the **parasympathetic system,** generally increases the activity of the internal organs (except the heart). The other set, comprising the **sympathetic system,** usually decreases the activity of the internal organs. This dual system allows more fine-tuning of organ activity than would be possible if activity were purely a function of the amount of stimulation from a single system.

When an action potential arrives at some internal organ, how does the organ "know" which system is sending the message? At the synapse with the internal organ, the sympathetic system uses noradrenaline as the neurotransmitter, whereas the parasympathetic system uses acetylcholine. Different neurotransmitters initiate different responses by a given internal organ.

The sympathetic and parasympathetic systems are anatomically segregated. Parasympathetic neurons are carried by cranial nerves III, VII, IX, and X, and by three spinal nerves from the sacral region. Cranial nerves III, VII, and IX serve the head and neck exclusively, but cranial nerve X, the **vagus** (L. "wandering"), innervates organs in the neck, chest, and most of the abdomen. Sympathetic neurons, in contrast, are carried by many of the spinal nerves in the thoracic (chest) and lumbar (abdominal) regions. No cranial or spinal nerve carries both sympathetic and parasympathetic neurons.

Functionally, the sympathetic and parasympathetic systems gear the body for very different types of contingencies. The parasympthetic system prepares the body for relaxed, vegetative, functions. It stimulates the activity of the internal organs largely by causing expansion of the abdominal blood vessels.

By contrast, the sympathetic system overrides the internal organs by causing constriction of the internal blood vessels, allowing the blood to be sent to the muscles instead. If you imagine yourself in a state of complete panic, you will be able to visualize the effects of sympathetic stimulation. Your mouth feels dry (salivary activity is impaired), the hair on the back of your neck rises, you have goosebumps and sweaty palms, and you may feel a sudden urge to go to the bathroom. If someone were suddenly to tap you on the shoulder, you might leap a great deal higher into the air than you thought yourself capable. You are ready for immediate action (see Box 14.3).

THE CENTRAL NERVOUS SYSTEM

The central nervous system, or CNS, consists of the spinal cord and the brain.

The Spinal Cord

The spinal cord is a cable of neurons that runs from the base of the skull to the tip of the vertebral column. In cross section, it has the appearance of a gray but-

BOX 14.2

Biofeedback

For many years, scientists distinguished the autonomic system (Gr. *autos,* "self" + *nomos,* "law") from the somatic system by noting that the latter was under conscious control whereas the former was not. This statement is not entirely true. Recently, it has been discovered that many people (perhaps all) have the capacity to influence autonomic activity consciously.

This finding has been put to good use in such disorders as hypertension (high blood pressure). Individuals with hypertension can often be trained to lower their blood pressure or heart rate consciously, or to perform other activities long thought beyond the reach of the conscious brain. This finding once again demonstrates how much our brains have evolved over those of our vertebrate ancestors in which autonomic activity is, indeed, autonomous.

Our "discovery" of the conscious control of the autonomic system presumably was greeted with great relief by the Indian yogi who have practiced such methods for generations!

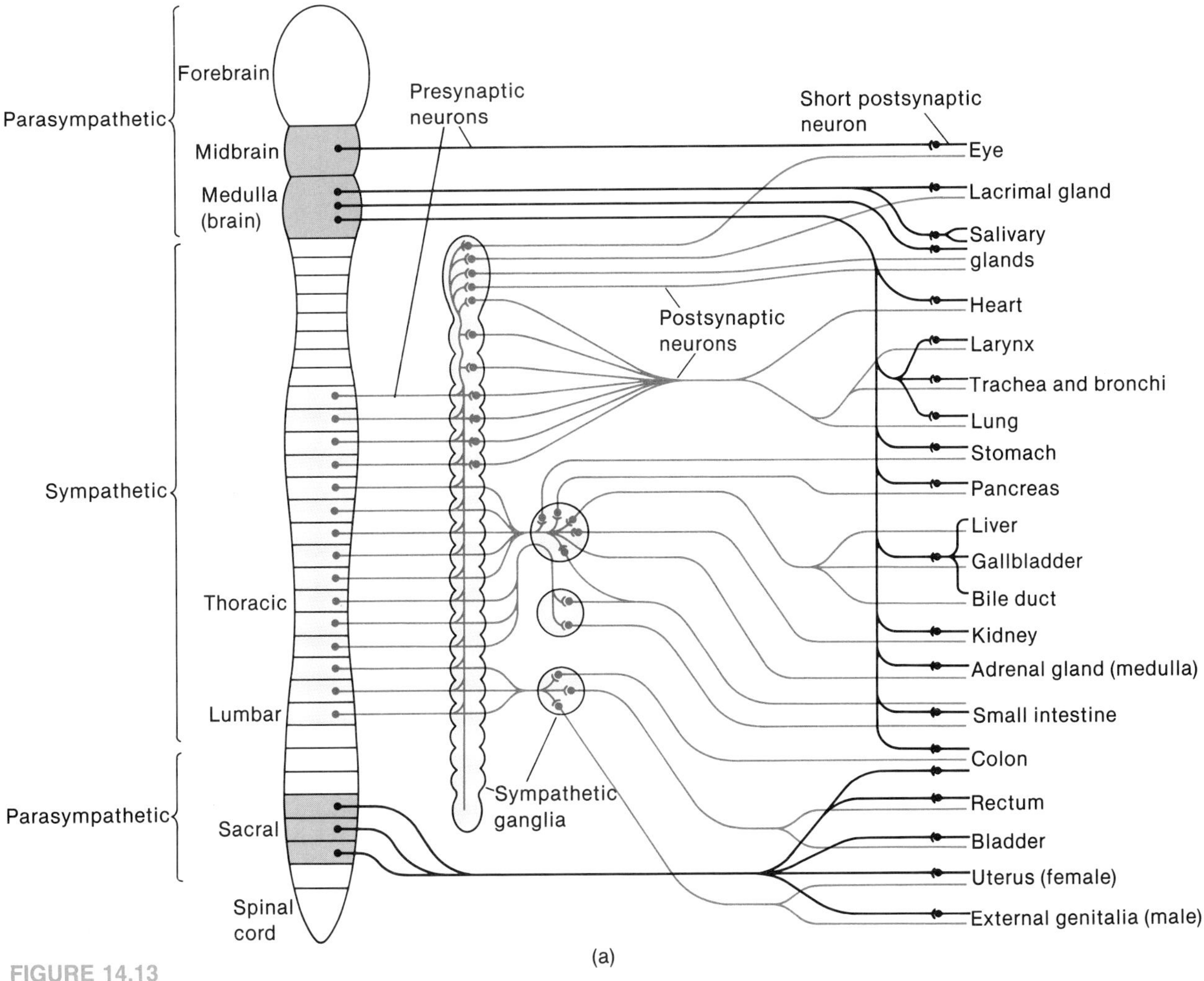

FIGURE 14.13

Autonomic Nervous System

The autonomic nervous system consists of the sympathetic system, which employs spinal nerves in the chest and abdominal regions, and the parasympathetic system, which uses four cranial and three sacral nerves. The two systems have diametrically opposite effects on the organs they innervate.

BOX 14.3

The Autonomic Nervous System and the Adrenal Medulla

The sympathetic nerve fibers utilize noradrenaline as the neurotransmitter at the synapse with the affected organ. Noradrenaline is also produced by the adrenal medulla as a hormone (Chapter 13), along with much larger volumes of the chemically very similar hormone adrenaline. Does the sympathetic nervous system merely duplicate the activities of the adrenal medulla?

The duplication is more apparent than real, because the sympathetic system functions on an immediate basis, whereas the adrenal gland functions over a period of hours or even days. Suppose that you had no sympathetic system and had to rely exclusively on the adrenal medulla in order to maximize blood flow to the muscles. Suppose, too, that you were suddenly accosted by a knife-wielding maniac. Roughly a minute after your adrenal medulla began producing adrenaline, all of your muscles would begin receiving increased blood flow. Of course, during this minute-long interval, you might become somewhat the worse for wear, following the attentions lavished on you by the knife-wielder.

In a real situation, your sense organs convey information about this dangerous intruder to your brain, spinal cord, and sympathetic nervous system, all within one second. The adrenal medulla is also activated and remains active for some time following the initial shock. Thus the sympathetic nervous system and the adrenal medulla operate in concert to ensure both an immediate and prolonged response to danger or other crises.

Is it simply a coincidence that the nervous and endocrine systems are both using the same chemical? (Would we ask such a question if the answer were, "yes"?) Notice in Figure 14.16 that innervation of the internal organs by both sympathetic and parasympathetic systems involves a two-neuron sequence. In the case of the sympathetic system, the synapse between the two neurons is in a ganglion near the spinal cord, whereas in the parasympathetic system the synapse is generally in the wall of the organ itself (meaning that the postsynaptic neuron is very short). Presynaptic neurons in both systems use acetylcholine as the neurotransmitter; only the postsynaptic neurons of the sympathetic system use noradrenaline (see Box Table).

The one exception to all of this is the adrenal medulla. Not only is it not innervated by the parasympathetic system, but there is no two-neuron sequence in the sympathetic system—the neurons run all the way from the spinal cord to the adrenal medulla without interruption. Where, then, are the postsynaptic neurons? Perhaps the answer is by now obvious to you. The postsynaptic neurons have, in this instance, been modified to form the adrenal medulla itself. Rather than stimulating any organ, these postsynaptic neurons release their neurotransmitter directly into the bloodstream, at which point the substance is no longer called a neurotransmitter. It is called a hormone.

Neurotransmitters of the Autonomic Nervous System

DIVISION OF ANS	PRESYNAPTIC NEURON	POSTSYNAPTIC NEURON
Parasympathetic	Acetylcholine	Acetylcholine
Sympathetic	Acetylcholine	Noradrenaline (norepinephrine)

terfly on a field of white. The gray matter is composed of millions of nerve cell bodies; the white matter is made up of axons that are covered in myelin.

The Brain

What we call the brain is actually just the highly modified anterior region of the spinal cord. The point at which the spinal cord ends and the brain begins (the opening of the skull) is entirely arbitrary, for within the nerve tracts that constitute the spinal cord and brain there is no dramatic change in appearance at this point (see Box 14.4).

For purposes of convenience, the human brain can be subdivided into three principal portions, the **brainstem,** the **cerebellum,** and the **cerebrum** (see Figure 14.14).

Brainstem. The brainstem is the region in the floor of the brain that leads directly from the spinal cord. The brainstem begins with the **medulla oblongata,** from which emanate eight of the 12 cranial nerves.

Since the medulla is responsible for monitoring such vital functions as breathing, heart rate, and blood pressure, damage to this portion of the brain is often fatal. For example, the reason that hanging is such an effective mode of execution is that it causes the neck to be broken at the junction of the vertebral column and the skull. As a consequence, the medulla or adjacent region of the spinal cord is either severed

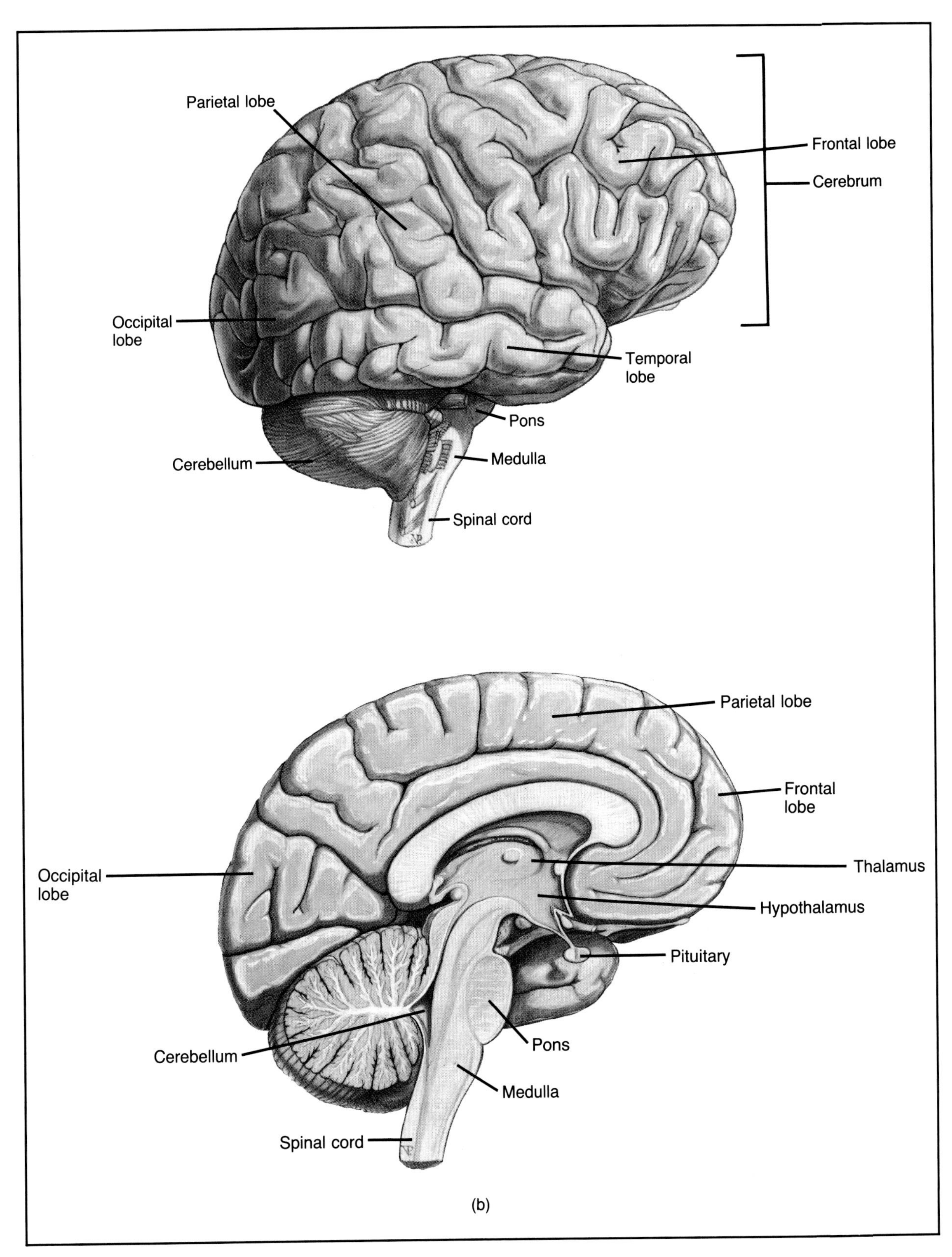

FIGURE 14.14

Surface and Midline Section of the Brain

(a) The external view of the brain shows the physical dominance of the cerebral hemispheres. (b) The midline section shows the continuation of the brainstem into the center of the brain, and some of the structures associated with it.

BOX 14.4

Protecting the Brain

The brain and spinal cord are very soft and delicate structures. How are they protected? The answer is in two very different ways. The skull and spinal cord provide mechanical protection; a complex blood supply system provides chemical protection. Let us consider the mechanisms at work in each instance.

Although it is obvious that the skull and vertebral column prevent routine bumps from becoming fatal injuries, what stops the brain and spinal cord from being rattled around in their protective chambers, with potentially disastrous consequences? The answer is that they float in a pool of **cerebrospinal fluid (CSF),** which prevents direct contact of nervous tissue and bone. A series of filmy tissues collectively known as the **meninges** (Gr., "membranes"), lying just under the skull, surrounds the brain and spinal cord and holds the fluid in place.

The CSF is secreted by a specialized tissue, called the **choroid plexus,** located at the top of the brainstem. It is derived from blood, and it resembles plasma in that it is free of cells but differs from plasma in its ionic concentration. The entire volume of CSF is only about 150 ml; it is replaced every three or four hours, and the excess fluid drains back into the venous circulation.

Injury to, or infection of, the brain or spinal cord frequently may be confirmed by a **spinal tap**—a drawing off of a small amount of cerebrospinal fluid by a needle inserted at the base of the vertebral column. The presence of blood cells or pathogens in the fluid indicates the nature of the problem.

A very different kind of protection is provided by what is termed the **blood-brain barrier.** The brain and spinal cord have a greater need for homeostasis than does the rest of the body. After all, many circulating hormones are also neurotransmitters, and minor changes in the blood concentration of such key ions as K^+ could be devastating to brain function. How can the brain be insulated from these chemicals while at the same time be linked to the circulatory system proper for purposes of receiving nutrients and releasing wastes?

The blood capillaries that supply the brain are very different from capillaries elsewhere in the body. In the rest of the body, the cells that comprise the capillary walls are loosely fitted together, and a great deal of exchange between the blood and interstitial fluid takes place through the openings between the cells. In the capillaries of the brain, the cells of the capillary walls are actually fused together in what is called a **tight junction.** (Tight junctions are also found in the cells of the choroid plexus, in cells that line the intestine, and in other tissues where leaks between cells would be undesirable.) Therefore the only way materials can leave or enter the interstitial fluid in the brain is by specific ion channels, facilitated diffusion, or active transport.

Large volume nutrients, such as glucose, amino acids, and so forth, move from the capillaries to the interstitial fluid of the brain by facilitated diffusion. Micronutrients are handled by the choroid plexus. The choroid plexus has been called "the kidney of the brain." It actively regulates the molecular concentration of the CSF both by transporting such substances as vitamins and certain ions out of the blood and into the CSF, and by actively transporting small organic ions, breakdown products of neurotransmitters, and various antibiotics out of the CSF and into the blood. Since the cell layer separating the CSF from the brain has no tight junctions, substances can move freely between the interstitial fluid and the CSF.

With so guarded an entry to the brain, how does the brain monitor the level of circulating hormones for purposes of negative feedback (see Chapter 13)? The answer is that the blood-brain barrier does not completely surround the brain. The pituitary and hypothalamus are supplied by regular capillaries, and hormones can move freely between the blood and interstitial fluid in this area of the brain.

The blood-brain barrier is not complete in another respect. Lipid-soluble substances will diffuse rapidly through the plasma membrane of the cells comprising the capillary walls and enter the interstitial fluid in the brain. This fact has two important medical consequences:

1. Medications that are not lipid-soluble (such as penicillin) tend not to cross the blood-brain barrier. In some instances, infections of the brain must therefore be treated by injections directly into the cerebrospinal fluid;

2. Most of the drugs of abuse are abused precisely because they *are* lipid-soluble and rapidly enter the brain. These include ethanol, caffeine, nicotine, and heroin, among many others. Interestingly, even though heroin is quickly converted by the brain to morphine (in which form it has its effects), morphine itself is substantially less lipid-soluble than is heroin. Thus injecting morphine does not give the "rush" that heroin does, which may be why morphine is less commonly abused than is heroin.

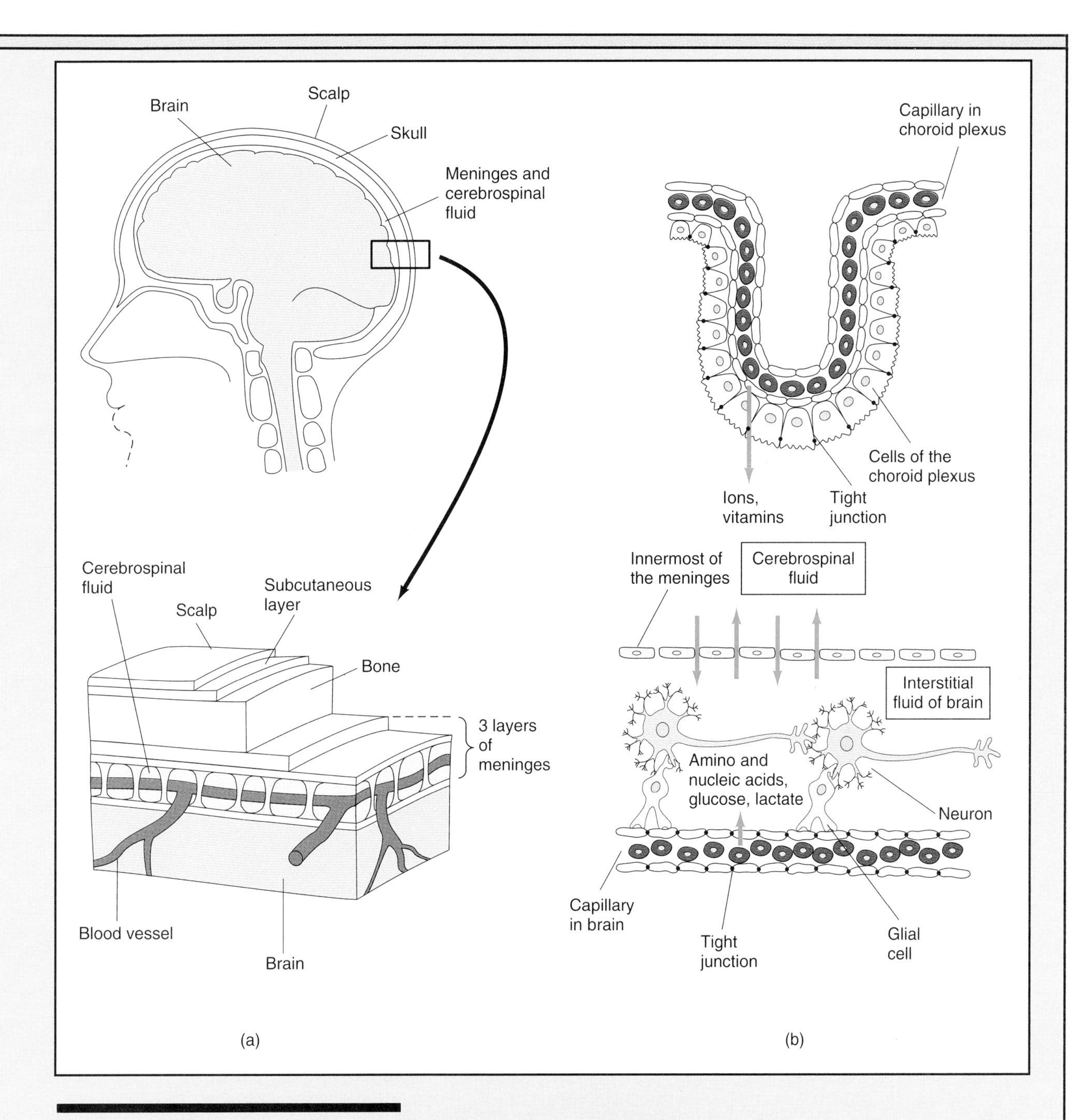

Meninges and the Blood-Brain Barrier

(a) The relationship of the meninges to the scalp and skull above, and the brain below, is readily apparent. (b) The workings of the blood-brain and blood-cerebrospinal fluid barriers are illustrated in this drawing. Note the reliance on facilitated diffusion and active transport because of the extensive nature of tight junctions in the brain capillaries and the capillaries of the choroid plexus. The cells of the choroid plexus take up various ions and vitamins from the capillaries of the plexus and transport them to the cerebrospinal fluid. The interstitial fluid in the brain receives nutrients such as glucose, lactate, and amino and nucleic acids by facilitated diffusion through the endothelial cells of the brain capillaries. Free flow between the interstitial fluid of the brain and the cerebrospinal fluid allows nutrients to move from one to the other.

or seriously damaged. Breathing ceases, and the individual dies of suffocation. For the same reason, an accident victim with a broken neck must be handled with extreme care, lest would-be rescuers further damage the spinal cord. (It should be noted that paralysis due to a damaged spinal cord results not only from tissue trauma but from the release of chemicals from the grey matter that interfere with nervous signal transmission in the white matter. Recently, certain drugs have been identified that, if administered within a few hours of the injury, minimize the damage caused by the release of these grey matter chemicals and preserve a greater amount of functioning.)

Immediately above and in front of the medulla is the **pons** (L., "bridge"), which serves not only to initiate certain facial movements and to receive input from the ears (both via cranial nerves), but also to transfer information between the medulla and the parts of the brain lying more anteriorly.

Anterior to the pons is the **thalamus** (Gr. *thalamos*, "inner chamber") which runs along the sides of the brainstem. The thalamus receives sensory impulses from all of the major sense organs except the nose and relays them to higher brain centers for processing.

Running through portions of the medulla, pons, and thalamus is a formation called the **reticular activating system** (RAS) (see Figure 14.15). This system receives information from all of the sensory and motor tracts running up and down the brainstem. The RAS serves as a kind of filter, wherein our attention is focused only on changes in the environment that are particularly important. (This is how we are able to read in a noisy airline terminal, for example, but still be attuned to announcements that might be important to us.) When the RAS is relatively inactive, we sleep. Barbiturates depress RAS activity even further, and if it is damaged, we may lapse into a coma (see Box 14.5).

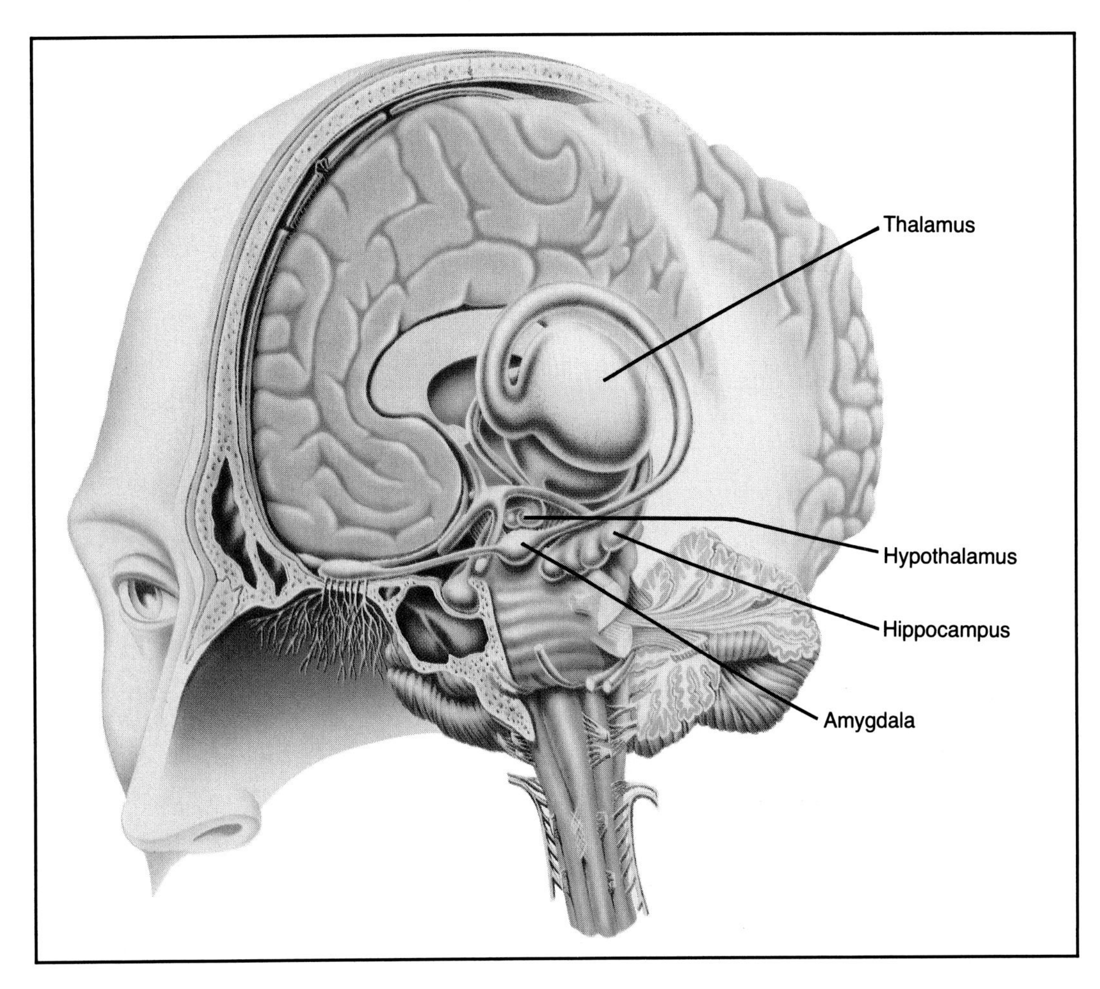

FIGURE 14.15

Brain with some of the components of the RAS and Limbic System

The reticular activating system (RAS) controls the overall degree of brain activity, including sleep. The limbic system controls emotion and motivation, and is also important in the formation and retrieval of memories.

BOX 14.5

Drugs and the Brain

The central nervous system is affected by many drugs, most of which are pharmaceutical agents of enormous importance in the practice of modern medicine. However, because of the extraordinary complexity of the CNS, the precise functioning of these drugs has not been readily understood. For example, although *aspirin* has been marketed as a drug for 80 years, its pharmacology is still only partially determined.

Other drugs are drugs of abuse, in the sense that they are taken not for therapeutic purposes but for their mood-altering effects. For purposes of convenience, we can distinguish five categories:

Stimulants. Stimulants include such common drugs as **caffeine** and **nicotine,** as well as more esoteric drugs such as the **amphetamines.**

Cola drinks, tea, and especially coffee all contain caffeine. Caffeine functions by enhancing synaptic transmission in the CNS, and by blocking the breakdown of cyclic AMP. It also contributes to infertility in women. Caffeine's very ubiquitousness causes most people to think of it as innocuous, but some physicians regard it as more powerful (hence, potentially more dangerous) than many prescription drugs, and people with ulcers, heart disease, or kidney problems are urged to eliminate it from their diet. The body rather quickly becomes tolerant of caffeine, forcing the individual to ingest more to achieve the desired effect. Correspondingly, a sudden reduction in caffeine intake has a depressant effect, as most inveterate coffee-drinkers can attest if they have to forgo the morning cup.

The problems potentially caused by nicotine are perhaps best illustrated by noting that tobacco plants remain virtually insect-free because almost no insects can ingest nicotine and live. Indeed, nicotine was widely used as an insecticide in the latter part of the nineteenth century, before synthetic insecticides became available. (No, you cannot give yourself immunity to mosquitoes by smoking so many cigarettes that an insecticide-level dose of nicotine is built up in your blood!) Nicotine stimulates the sympathetic nervous system, causing an increase in heart rate and a constriction of small blood vessels. It also mimics the effects of acetylcholine at some receptors scattered throughout the body. Tolerance to nicotine occurs quickly. Soon, the stimulation passes from a "lift" above normal to a lift only from the trough of depression that occurs during periods when the nicotine level in the blood drops.

Amphetamines include the so-called "uppers," most notably "speed" (methamphetamine). Chemically, the amphetamines resemble adrenaline, and they stimulate the release of noradrenaline and dopamine by the brain. They are sometimes used in weight reduction, because, for a few days, they are effective appetite suppressants. However, they are potentially dangerous drugs. The individual may feel euphoric at first but this euphoria is soon replaced by a letdown. Continued use can lead to amphetamine psychosis, which mirrors the symptoms of paranoid schizophrenia, and to the increased likelihood of strokes. Tolerance develops quickly, necessitating increasingly larger doses to achieve the desired effect.

Depressants. Depressants are so named because they cause a depression of CNS activity (specifically, the reticular activating system), not because they create a feeling of being depressed. The two principal depressants are **alcohol** and **barbiturates.**

Alcohol has been extensively used by many societies for thousands of years. The depressant effects of the drug are well illustrated by the rapid decline in motor skills (due to enhanced GABA activity, which inhibits synaptic transmission by hyperpolarizing the neuron) and, at higher doses, from the impairment of such basic functions as breathing, the usual cause of death in overdoses. Alcohol is largely detoxified by the liver, but liver cells are destroyed in the process, and prolonged destruction leads to *cirrhosis* (see Chapter 6). Other effects of alcohol may include irreversible brain damage, epilepsy, psychosis, and sexual impotence. Moreover, there is a strong physical addiction, and withdrawal is severe.

Barbiturates include many of the prescription sleeping pills. They act by reducing CNS activity in general, probably by interfering with Na^+ and K^+ movement. The effects may be so long-lasting that a person may need amphetamines in order to feel wakeful in the morning—and then require barbiturates to go to sleep again the next night. When accompanied by alcohol, barbiturates may be fatal, because of the combined depression of the CNS. Most of the movie and rock stars who have died of drug overdoses were using what proved to be a lethal combination of these two depressants. Barbiturates also lead to a muting of mood control, and individuals who take them may suddenly become violent. Ironically, rapid withdrawal may be more dangerous than addiction, because as the CNS strives to regain normal levels of activity, it may overcompensate, leading to convulsions and sometimes to death.

Tranquilizers. Tranquilizers reduce anxiety and tension without the general depressant activity of the barbiturates. The major tranquilizers enhance the activity of the parasympathetic system. In the last 35 years, they have been of enormous impor-

(Continued on the following page)

***BOX 14.5** (Continued)*

tance in the treatment of psychiatric patients, and more than any other single factor have led to a near-elimination of the padded cells and straitjackets of the asylums of old. They tend not to be drugs of abuse. Not so the minor tranquilizers, such as Valium and Librium, which enhance the action of GABA, lead to the opening of Cl^- channels, the hyperpolarization of neurons, and the inhibition of synaptic transmission. Tolerance and physical dependence to the minor tranquilizers is rapid, and this fact, when combined with their extraordinarily broad usage, makes them major drugs of abuse.

Hallucinogens. The hallucinogens include a variety of natural plant products as well as such synthetics as **LSD** (lysergic acid diethylamide). LSD chemically resembles serotonin, and it binds to serotonin receptors, especially in the reticular activating system. Side effects range from temporary motor impairment to spontaneous hallucinating and schizophrenia. The plasma membrane receptor for marijuana in the brain was recently identified, and the gene that codes for the receptor has also been identified and cloned. When marijuana binds to this receptor, it inhibits the enzyme responsible for synthesizing cyclic AMP, a substance that, as we saw in Chapter 13, is essential in cell responsiveness.

The long-term medical consequences of marijuana use are still being actively debated. It is known to reduce motor ability, however, and can cause a variety of respiratory ailments. It is suspected of contributing to lung cancer as well, but medical proof on that point is not well established at present.

Narcotics. From the medical standpoint, narcotics are opium-like compounds that simulate endorphin production, leading to a relief of pain and the inducing of sleep. The narcotics include both synthetic and natural compounds, ranging in strength from Demerol (meperidine) to **morphine.** The euphoria produced by the stronger narcotics, such as **heroin,** is far more pleasurable and powerful than that produced by the hallucinogenic drugs, but the effects steadily fade with extended use. As the periods between drug use gradually become more and more unpleasant, the drug soon is used only to maintain something approximating a normal state. There is a tremendous psychological addiction because withdrawal is so unpleasant, although unlike the barbiturates, withdrawal is seldom fatal because convulsions do not occur.

Legally, the narcotics also include cocaine, a stimulant that is chemically far removed from the opium-like drugs. Cocaine blocks Na^+ channels and inhibits the uptake of dopamine, allowing continued stimulation of pleasure centers. The consequences, however, are that the number of dopamine receptors on the postsynaptic neurons are reduced, which causes craving for additional cocaine.

As has been repeatedly and tragically demonstrated, fatal overdoses (as a result of heart and respiratory failure) are common. Indeed, recent studies indicate that doses as small as 150 mg (about one-sixth the dose normally taken by abusers of the drug) can cause constriction of the coronary arteries and a significant reduction in the amount of blood flowing to the ventricular walls. In addition, cocaine binds to the left ventricle and aorta, where it interferes both with the movement of Na^+ across plasma membranes and with nerve signals from the brain to the heart.

Beneath the thalamus is the **hypothalamus,** which occupies the floor of the brainstem. In addition to its role in directing the pituitary gland (Chapter 13), the hypothalamus also governs such basic behavior as hunger, thirst, and sex, emotions such as fear and anger, and general body functions such as temperature regulation.

The hypothalamus is not the only region of the brain associated with emotion. Rather, it is part of a ring-like system of brain centers known collectively as the **limbic system,** which connects the hypothalamus to higher brain centers by way of the thalamus. The limbic system apparently controls various emotional responses, such as laughter and crying, and presumably governs our rage and fear reactions to ensure (generally) a socially acceptable response.

An even more important function of the limbic system is memory storage and retrieval. Although the anatomical equivalent of a memory has not yet been identified, long-term memory is thought to involve subtle changes in the synapses of neurons throughout the cerebrum. However, we do know which brain structures are involved in at least some memories. These structures include two components of the limbic system that lie deep in the cerebrum. They were named for their anatomical appearance long before anything was known of their function—hence, the names **hippocampus** (Gr., "seahorse") and **amygdala** (Gr., "almond"). These structures route messages to the storage centers of the cerebrum by way of the thalamus and hypothalamus.

The hippocampus is needed for the storage of recent memories. However, its role in the formation and recall of specific memories diminishes with the passage of time since the event, suggesting that permanent memory forms elsewhere, presumably in the cortex.

The amygdala also functions in memory associations. We have all experienced the sensation of seeing a ripe apple and remembering instantly how apples taste, or of hearing a familiar voice on the telephone and simultaneously recalling the person's face. Those kinds of memories are stored in different areas of the cerebrum, but they are linked by the amygdala. When the amygdala is damaged—by a stroke, for example—these associations are no longer made—and the individual may "forget" how a particular food they are looking at tastes, or "not remember" what a person looks like when told his name or when hearing his voice on the telephone.

There is an interesting tie-in between memory and emotion. Very often, if a person is in a highly emotional state, the memories formed are particularly strong. The amygdala, which as we have seen is linked to both the hypothalamus (emotion center) and the cerebrum (memory center), has many neurons that produce endorphins (see Chapter 13). The endorphins not only act as modulators at synapses but also function to decrease the perception of pain and to enhance the feelings of pleasure. It is believed that these neurons are particularly active when stimulated by neurons from the emotion center of the hypothalamus. Thus the connection between memory and emotion—both of which are served by the limbic system—is not as odd as it might first have appeared.

Cerebellum. The cerebellum is a baseball-sized portion of the brain immediately above the pons. We owe our great ability for fast, coordinated, fine-tuned movements to the computer-like integrating capacities of the cerebellum. The cerebellum also receives information from the balance centers of the inner ear and helps us maintain proper posture and spatial orientation. A standard component of most physical examinations is an exercise where the patient is asked to close his eyes, stretch his arms out to the side, and touch his fingertips together. A slow, wobbly, or inaccurate performance may indicate a damaged cerebellum.

Cerebrum. The cerebrum consists of left and right hemispheres, linked by a major nerve tract, the **corpus callosum.** The cerebral hemispheres are so large in humans that, when viewed from above, they totally obscure the other portions of the brain. The cerebrum is largely an integrative center, where information from a variety of sources is received, assessed, and acted upon through the initiation of signals to appropriate motor centers (see Figure 14.16). Each of the hemispheres is subdivided by deep grooves into the *frontal, temporal, parietal,* and *occipital lobes,* the names deriving from the bones of the skull that roughly overlie each lobe.

The two hemispheres are identical in gross anatomy but differ functionally. The left hemisphere is normally dominant to the right, and, because the left hemisphere controls the right side of the body (due to nerve tracts from the hemispheres crossing over in the corpus callosum), most people are right-handed. Where the right hemisphere dominates, the individual is left-handed. Speech and memory centers are also regional. For instance, in right-handed individuals, the speech center is in the left hemisphere, which is why a stroke in the left hemisphere can interfere with a person's speaking ability, whereas a stroke in the right hemisphere will have no such effect.

The outer layer of the cerebrum is called the **cortex** (a general term for any outer layer, used also in the kidney and adrenal glands, among others). This is where most of the action takes place. Although the cortex is only a few millimeters thick, it contains fully 10 percent of all the neurons in the brain. (In total, the brain is estimated to contain 100 billion neurons, and almost one trillion glial cells.) The cortex is gray in color because of the presence of millions of nerve cell bodies, and it corresponds to the central region of the spinal cord. The interior of the cerebrum is white because of the presence of myelinated axons, and it corresponds to the outer region of the spinal cord.

Relative to the spinal cord, the gray and white matter have switched places in the cerebrum (the same is also true of the cerebellum). Why? Presumably, it is because positioning the cell bodies on the outside allows for the presence of more cell bodies. So, too, does the folding so characteristic of the cerebrum and cerebellum, another device for expanding surface area while holding volume stable (just as in the lungs, intestine, and so forth). The surface area of the cerebrum is three times that of the inside surface of the skull. It is the total area of the cerebral cortex that allows the behavioral complexity that serves, more than any other single factor, to distinguish us from other species of vertebrates.

Details of brain anatomy and function are well beyond the scope of any introductory text (at the moment, they are, for the most part, well beyond the scope of researchers, too!). However, two broad generalizations are worth mentioning. First, specific areas of the cerebral hemispheres control specific functions, and therefore damage to these areas, by a stroke or injury, may result in blindness, deafness, loss of speech, paralysis on one side, and so on (see Box 14.6). Second, some general functions are apparently much less localized. Intelligence, for instance, appears to be a widespread function, which means that damage to specific areas does not cause a sudden drop in I.Q.

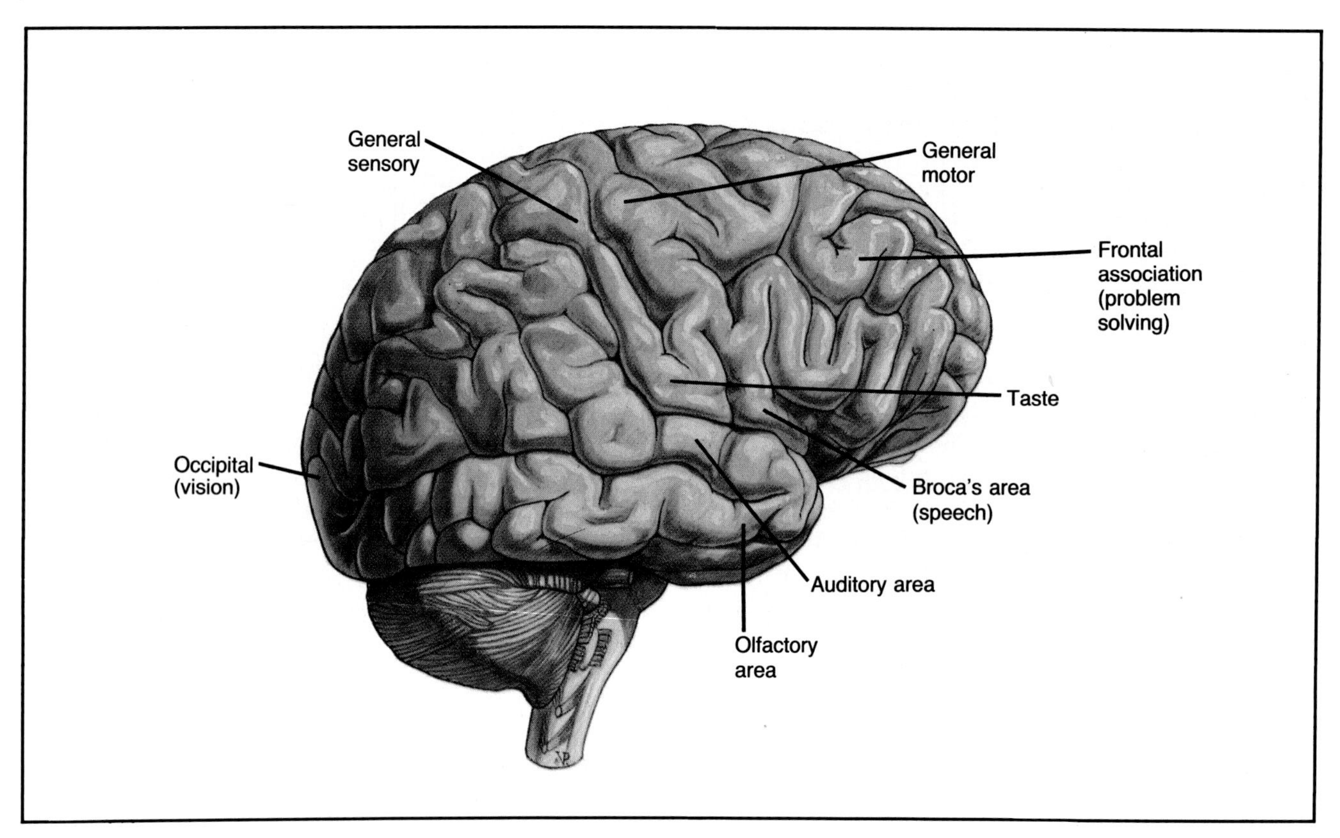

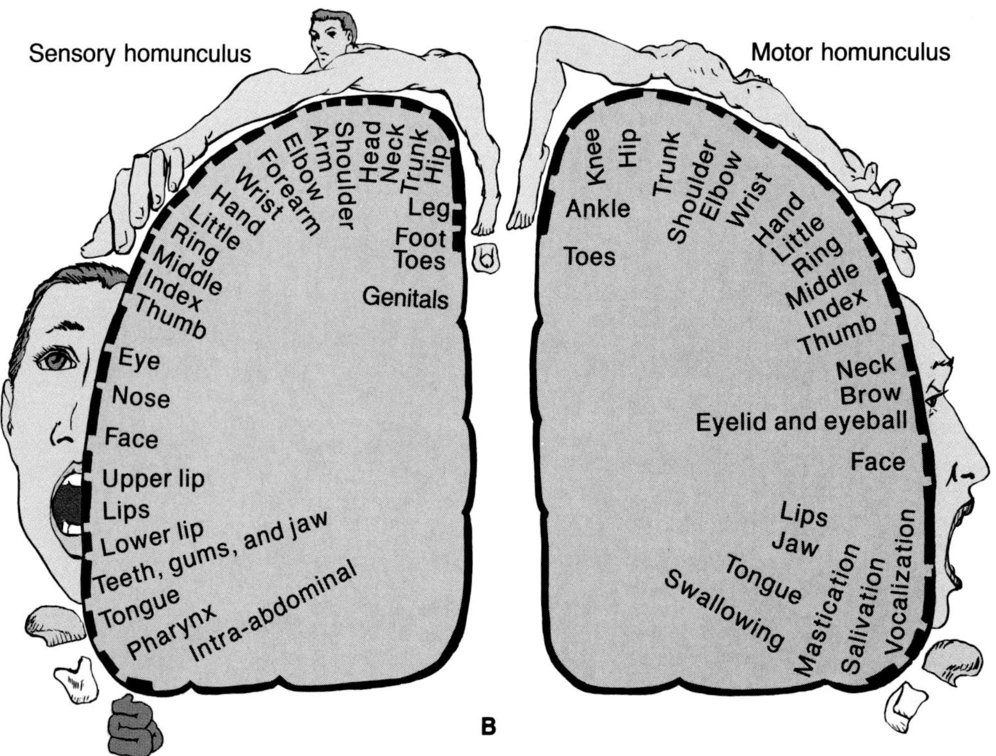

FIGURE 14.16

Functional Map of the Cerebral Cortex

The relative area of the cerebral cortex devoted to motor and sensory integration of particular regions of the body generally bears no relation to the relative size of those body regions. The hands and the lips and mouth, for example, have a disproportionately large share of both sensory and motor regions of the cerebral cortex.

BOX 14.6

Problems with the Brain

Because of the brain's complexity and enormous importance to normal functioning, problems affecting it are often particularly severe. By far the most common problem is stroke, considered in Chapter 8. Other brain problems include the following.

Meningitis. Meningitis is an inflammation of the meninges, the membranes that cover the brain and enclose the cerebrospinal fluid. It can be caused by viruses or bacteria. Viral meningitis is not generally a severe disease, but bacterial meningitis can be extremely serious. Bacteria normally cannot cross the blood-brain barrier, but occasionally some gain access, either by injury (e.g., a severe skull fracture), or by sinus or middle ear infection.

Viral Encephalitis. Viral encephalitis is a serious disease, often causing permanent damage and occasionally death. Several related viruses can cause encephalitis; most are transmitted by certain species of mosquitoes. Horses are susceptible to the same viruses, and equine encephalitis is a serious problem for horse breeders. In humans, *Herpes simplex* (see Chapter 20) can also cause encephalitis.

Cerebral Palsy. Cerebral palsy results from the destruction of portions of the motor centers of the cerebrum, generally because of an injury, either before or after birth. It is manifested by varying levels of impairment of motor functions, including speech. However, there is no interference with intelligence, contrary to what many uninformed people think.

Alzheimer's Disease. Alzheimer's disease is a progressive degenerative disease of the brain, with a course of between six and 20 years, which affects 10 percent of Americans over the age of 65, and 47 percent of those over the age of 85. Presently, there are four million Americans with the disease, and it is the fourth leading cause of death in adults.

Between 10 percent and 30 percent of the cases are caused by an inherited mutation; the rest are assumed either to be a response to an as yet unidentified virus, or to some environmental toxin. Alzheimer's victims show marked increase in the amount of aluminum in the brain, a finding which has caused many people to trade in their aluminum cookware. However, aluminum is very abundant in the earth's crust, and we are regularly exposed to large amounts of this element without harm. Thus the significance of its abundance in the brains of Alzheimer's patients is simply not yet known.

It is also known that, in people with Alzheimer's disease, the production of acetylcholine in the brain drops dramatically, to as little as 10 percent of normal levels. The acetylcholine-producing neurons are clustered in the basal forebrain, a part of the limbic system. They change dramatically, forming plaques and tangles of a type also seen in people with Down syndrome (see Chapter 19).

These plaques are composed of fragments of a protein normally found in the plasma membrane. For reasons not yet understood, a large number of fragments (about 40 amino acids in length), called **amyloid protein,** break off and accumulate in the extracellular spaces, crowding and ultimately destroying the adjacent neurons. The gene that codes for the protein is located on chromosome 21, the chromosome that (as we shall see in Chapter 19) is present in triplicate in people with Down syndrome. At present, there are no effective treatments for the disease, but the possible role of a hyperactive amyloid protein gene (which could conceivably be turned off before full-blown Alzheimer's disease develops) has researchers very excited about future treatment possibilities.

Parkinson's Disease. Parkinson's disease, which is characterized by rigidity and tremor, results from the destruction of certain motor centers in the brain stem by causes that are not well understood. Because of these losses, the secretion of dopamine is severely limited, and the unaffected acetylcholine-secreting neurons become overly active. Some 350,000 Americans suffer from the disease at present.

Until 1989, drug treatments were largely focused on slowing the rate of progress of the disease. However, a new drug introduced that year offers great promise in delaying the development of the disease indefinitely. At the same time, experimental surgery, involving the transplantation of dopamine-secreting cells from the adrenal cortex (or, in a highly controversial procedure, from aborted fetuses) directly into the brain appears, in some instances, to provide substantial relief.

Epilepsy. Epilepsy is characterized by an excess of activity of a particular portion of the brain, often because of an injury. An epileptic seizure occurs when the activity of the damaged area exceeds a critical threshold, and, in a very loose analogy, the brain temporarily "short circuits." Both *grand mal* and *petit mal* seizures apparently involve the reticular activating system, but differ because grand mal attacks are much more intense and involve much more disturbed brain wave patterns. Both types of attacks can generally be effectively controlled by medication.

Slow Viruses. In Chapter 4, we briefly discussed disorders once believed to be caused by a type of protein called *prions* but are now believed to be caused by a *slow virus* (i.e., a virus that has its effects sev-

(Continued on the following page)

BOX 14.6 (Continued)

eral years after initial exposure), although the actual agent has yet to be identified. There are apparently two slow virus diseases that occur in humans. They are quite rare but are very similar to a common disease of sheep called *scrapie*, and a disease of cattle called *mad cow disease*. One of these diseases, **kuru**, is known only from a few New Guinean tribes who eat the brains of their dead, a method of transmission which we would immediately conclude poses no threat for Americans, among whom such behavior is presumably rare. However, the second disease, **Creutzfeldt-Jakob disease (CJD)**, is found in this country and at least 14 cases are known to have resulted from the injection of growth hormone or gonadotropins (see Chapters 13 and 20) extracted from the pituitary glands of cadavers. Other cases have resulted from unsterile instruments used in neurosurgery, corneal grafts of the eyes, and even from root canal procedures.

There is already some evidence that these diseases can be transmitted between species (and all of them may ultimately prove to be variants of the same disease). An epidemic of mad cow disease in Britain has resulted in the destruction of 15,000 cows since 1986—and that epidemic occurred after scrapie-infected sheep carcasses were turned into protein supplements fed to cows.

Can CJD be transmitted from animals to humans? CJD is a very rare (albeit fatal) disease. However, there has recently been an outbreak of CJD in rural Czechoslovakia—an area of intensive sheep farming. The incidence of CJD in this area is several hundred times the normal incidence of one in 1,000,000. Some experts see this as the beginning of a major, world-wide, epidemic.

Mental Illness. In recent years, many mental illnesses have been shown to be caused by chemical imbalances in the brain. **Manic depression** and **schizophrenia** both appear to involve such imbalances, although the actual onset of these diseases may well be triggered by specific stressful events in an individual's life. Schizophrenia, which in at least some cases appears to result from the overproduction of dopamine, has an incidence of about one in 100 adults world-wide. It is apparently not one disease but a family of diseases, with one or more of several genes contributing in an additive fashion, any or all of which can be influenced by environmental circumstances.

Depression (which affects almost ten million Americans) is characterized by low serotonin levels, and a deficiency of serotonin binding sites.

Obsessive-compulsive disorder, a disabling disease affecting almost four million Americans, and characterized by prolonged bouts of bizarre behavior (e.g., washing one's hands over and over again), has also been linked to abnormal serotonin levels. Chemotherapy is extensively used in treating these diseases.

Summary

The obvious structural and functional complexity of the human brain is belied by the relative simplicity of the neurons that compose the brain. As in a computer, a neuron is limited to a binary system. It is either "on" or "off" because, in the presence of a stimulus, a wave of depolarization, known as an action potential, is either generated or it is not. Complexity results from the massing together of many billions of these individual units, although the results are much more impressive than in any computer.

The transformation from resting potential to action potential involves the sudden but very temporary shift of ions from one side of the plasma membrane of the neuron to the other. However, chemical transmission is required to convey information between neurons, and a variety of neurotransmitters function in this role, acetylcholine being the most common outside the brain.

Associations between neurons and muscle fibers are called neuromuscular junctions. Although they are very similar to synapses, neuromuscular junctions are far more susceptible to interference by a variety of potentially lethal agents.

The nervous system is subdivided into peripheral and central portions. The peripheral portions include both somatic and visceral nerves, the latter (autonomic nervous system) being further divisible into sympathetic and parasympathetic components, with diametrically opposite effects on the various internal organs. The central nervous system includes the spinal cord and brain.

Key Terms

nerve
neuron
myelin
Schwann cell
glial cell
node
dendrite
axon
resting potential
action potential
depolarization
repolarization
refractory period
saltatory conduction
all-or-none rule
synapse

synaptic cleft
presynaptic neuron
postsynaptic neuron
neurotransmitter
acetylcholine
cholinesterase
hyperpolarize
neuromuscular junction
sensory neuron
motor neuron
interneuron
peripheral nervous system
somatic system
autonomic system
sympathetic system
parasympathetic system
central nervous system
cranial nerve
spinal nerve
dorsal root
ventral root
ganglion
reflex
reflex arc
spinal cord
brainstem
cerebellum
cerebrum
medulla oblongata
pons
thalamus
reticular activating system
hypothalamus
limbic system
hippocampus
amygdala
corpus callosum

Box Terms

botulism
curare
myasthenia gravis
parathion
atropine
multiple sclerosis
amyotrophic lateral sclerosis
polio
cerebrospinal fluid (CSF)
meninges
choroid plexus
spinal tap
blood-brain barrier
tight junction
caffeine
nicotine
amphetamines
alcohol
barbiturates
tranquilizers
hallucinogens
LSD
narcotics
morphine
heroin
cocaine
meningitis
viral encephalitis
cerebral palsy
Alzheimer's disease
amyloid protein
Parkinson's disease
epilepsy
slow virus
kuru
Creutzfeldt-Jacob disease (CJD)
manic depression
schizophrenia
obsessive-compulsive disorder

Questions

1. Distinguish nerves from neurons. What is myelin?
2. Discuss the formation of an action potential. Can one neuron transmit more than one action potential at one time?
3. What is a synapse? What happens to action potentials at synapses?
4. Explain the effects of nerve gas or insecticides in terms of the functioning of the neuromuscular junction.
5. What is the autonomic nervous system? What does it do, and how does it differ from the somatic nervous system?
6. Discuss the operation of a reflex arc. Why do you suppose that we sophisticated organisms retain reflexes (as opposed to having everything decided by the brain).
7. What is the function of the reticular activating system and the limbic system?
8. What is the function of the cerebellum?
9. Describe the ways in which the brain is protected both mechanically and chemically. What is the blood-brain barrier?
10. What are "slow viruses" and what do they have to do with the brain?

CHAPTER

15

The Sense Organs

Windows on the World

You are a kindergarten teacher. Jack and Fred, the two most rambunctious children in your class, get into a tussle on the playground during recess. Jack manages to land a right cross to Fred's eye. Fred runs to you, not to complain, but to ask why he sees stars when it is the middle of the morning. How do you explain it?

Linda and Wendy are performing a biology experiment. Linda stands and closes her eyes. Wendy extends Linda's arms so that the right arm is slightly higher than the left. She then asks Linda to ascertain, with her eyes still closed, the relative position of her arms. Linda answers promptly and correctly. Which of the five senses did Linda use to determine the position of her arms?

Why do most foods taste better hot than cold? Why do people with false teeth complain that food is no longer tasty? Why does food so often taste bland during a severe head cold?

"Two-Ton Tony" McGurk, an enforcer for the local mob, is attacked one night by Guido Libido, an out-of-town rival. In the course of their scuffle, Tony slips and hits the back of his head on the curb. Rushed to the hospital with a skull fracture, he nonetheless quickly recovers—but is permanently blind. How could a blow to the back of the head induce blindness?

In this chapter, which deals with the sense organs, we address these and other questions.

THE NATURE OF THE PROBLEM

In Part 4 we have been concerned about communication between and among the many specialized tissues and organs that make up our bodies. Communication, as we have seen, is necessary both for purposes of coordination of the metabolic activities of different tissues (that is, the maintenance of homeostasis), and for ensuring the appropriateness of a response (such as withdrawing your hand when it touches a hot stove). We have investigated the intricate feedback loops by which hormonal secretion is regulated. We have also examined in some detail how messages are conveyed in the nervous system, and (in much less detail) how the brain is organized to analyze input and to determine output.

At this point, however, there is still a major gap in our understanding of communication within the body—namely, the input side. By comparison, it is relatively easy to understand motor output. Action potentials initiated in the CNS direct the contraction of muscles. The particular group of neurons that transmit the action potentials determines which muscle it is that contracts. The strength of the contraction is a function both of the number of neurons carrying action potentials and of the number of action potentials received by the muscle per unit time. Finally, contraction of the muscle is a direct consequence of the action potentials causing the opening of Ca^{2+} channels and the consequent removal of a chemical inhibition on the contractile proteins within the muscle cell.

However, the sensory, or input, side of the system is much less clear. Which environmental changes do we recognize as stimuli? For that matter, what *are* "stimuli"—and how is it that some environmental changes (but not others) result in action potentials in sensory neurons? Since all of the various sense organs send messages to the brain by using the same kind of action potentials, how can the brain distinguish one message from another and respond appropriately? For that matter, how does the brain even know which organ is sending the message? Until we answer these questions, our understanding of communication and integration within the body will be very incomplete.

SENSE ORGANS AND STIMULI

We generally speak of five senses—sight, hearing, smell, taste, and touch—largely because of the prominent sense organs associated with all these abilities save the last. In reality, however, we possess many more than just five senses. We also respond to things like temperature, orientation of the body, position of the limbs, pain, and various changes in the composition of the blood.

Moreover, human sensory abilities by no means exhaust all of the possibilities. Many snakes detect their warm-blooded prey by the heat radiated by the prey animal. Some species of fish orient by means of electrical fields. Many insects can detect polarized and ultraviolet light, as well as changes in atmospheric pressure. Some birds navigate using the earth's magnetic field. Undoubtedly, other examples exist of which we are presently unaware.

Types of Receptors

We can categorize our senses in various ways. We can distinguish between senses that monitor our internal environment and senses that monitor our external environment—although some senses, such as pain, cross over this boundary. We can distinguish complex sense organs from raw nerve endings—although a number of senses utilize something of an intermediate between these two extremes. Perhaps the easiest method of categorization is based on the nature of the stimuli to which we respond. (Used in this way, "stimulus" simply means any chemical, mechanical, or energy change in the environment that the body can perceive through the use of its sensory receptors.) Thus, based on commonality of stimulus type, we can distinguish five types of sense organs:

1. **Mechanoreceptors,** which respond to deformation of the receptor itself or of surrounding cells;

2. **Pain receptors,** which respond to physical or chemical damage to tissues;

3. **Chemoreceptors,** which respond to the presence of various types of chemicals in air, food, or internal body fluids;

4. **Thermoreceptors,** which respond either to heat or to cold; and

5. **Light receptors,** which respond to electromagnetic radiation of particular wavelengths in the so-called visible spectrum.

Properties of Receptors

Before beginning a discussion of individual sense organs, it is useful to delineate the characteristics that sense organs have in common. First, sense organs function as **transducers**—which means they are capable of converting particular kinds of environmental changes into a neural response (that is, an action potential). The basis of this transduction is that the particular environmental change opens or closes gated ion channels in the plasma membrane of the cells of the receptor, causing either depolarization or hyperpolarization of the sensory neuron (which either functions as the receptor cell or is synapsed with the receptor cell, depending on the particular sense organ).

Second, the various types of receptor organs are particularly sensitive to one group of stimuli and largely insensitive to all other stimuli. However, this insensitivity is not absolute. The rods and cones of the retina of the eye, for example, will initiate action potentials as a result of pressure (such as a blow to the eye); consequently, a person who is hit in the eye may "see stars."

Third, the action potentials produced by receptor organs are no different from any other type of action potentials. The brain is able to distinguish among the action potentials of the various sense organs because each sense organ transmits its action potentials to a particular region of the brain. Thus any stimulus to the eye—whether a view of the Alps or the impact of a fist—will be transmitted to the same region of the brain, and the brain will interpret the stimulus as light. Hence, a person "sees stars" after being hit in the eye because the brain interprets action potentials from the sensory cells of the eyes as light, regardless of the actual source of the stimulus.

Fourth, receptors *adapt* to stimuli after a period of time. That is, when receptors are exposed to continuous stimulation, they ultimately cease to respond. However, the period of activity before a response ceases may vary from less than one second to more than two days, depending on the type of receptor. For instance, certain touch receptors adapt very quickly. Their utility is therefore limited to registering sudden changes in pressure on the skin. In contrast, stretch receptors in the muscles and tendons adapt very slowly and are constantly sending messages regarding the degree of flexion of a joint or contraction of a muscle.

Differences in the rate of adaptation are useful. For example, the constant touch of clothing against skin is not something that requires continuous transmission to the brain. Indeed, if you sit still and try to "feel" the touch of clothing on your skin, you will probably be unable to do so. However, if you move your arm quickly, chances are you will "feel" the movement of your shirt or blouse against your skin.

Conversely, you are constantly being informed of the position of your arm, for example; even if you are not looking at it, you know just how it is oriented. However, if your arm or leg "falls asleep," you are, for a brief time, only marginally conscious of the limb and have difficulty controlling its movement. The continuous relay of information from muscles and tendons is indispensable for coordinated movement.

Adaptation has two sources. First, the structure of the receptor may be affected by the initial stimulus and rendered incapable of further stimulation almost immediately. This situation is common in many rapidly adapting sense organs. Second, the nerve fiber leading from the sense organ may become **accommodated**—that is, even direct stimulation of the nerve will fail to elicit a response. Accommodation is a much slower mechanism.

MECHANORECEPTORS

Mechanoreceptors include the receptors of the various tactile senses, the position sense, hearing, and equilibrium (see Figure 15.1).

Tactile Senses

The tactile senses include touch, pressure, vibration, itch, and tickle. The first three are variants on the same theme, and the same types of receptors respond to all three. *Touch* is restricted to superficial stimulation of the skin; *pressure* is defined as deformation of the deeper tissues; *vibration* is a rapidly oscillating touch or pressure. These receptors are not distributed equally: They are very abundant in the lips, somewhat less abundant in the fingers, and rather scattered in the back. The area of the brain devoted to processing sensory information from each of these areas is, not surprisingly, directly related to the number of receptors each area has.

The receptors used for touch, pressure, and vibration include the following:

1. **Free nerve endings,** which are common in many parts of the body (e.g., in the "white" of the eye). They respond to a variety of stimuli, including touch and pressure.

2. **Hair end-organs,** which are nerve fibers wrapped around the base of every hair. Deflection of the hairs triggers an action potential from the hair end-organs, which are important touch receptors. They adapt rapidly but remain effective detectors of movement on the skin—for example, the movement of a crawling insect.

3. **Pacinian corpuscles,** which consist of onion-like, concentric layers of tissue surrounding a nerve ending. Deformation of the corpuscle causes the nerve ending to initiate an action potential. These receptors function largely in detecting vibration because they adapt very quickly to continuous contact.

4. **Meissner's corpuscles,** which consist of a greatly expanded series of terminal nerve filaments surrounded by a capsule. They are common in the nonhairy areas of the skin—most notably, the fingertips and lips—where they are responsible for fine-grained touch and texture reception.

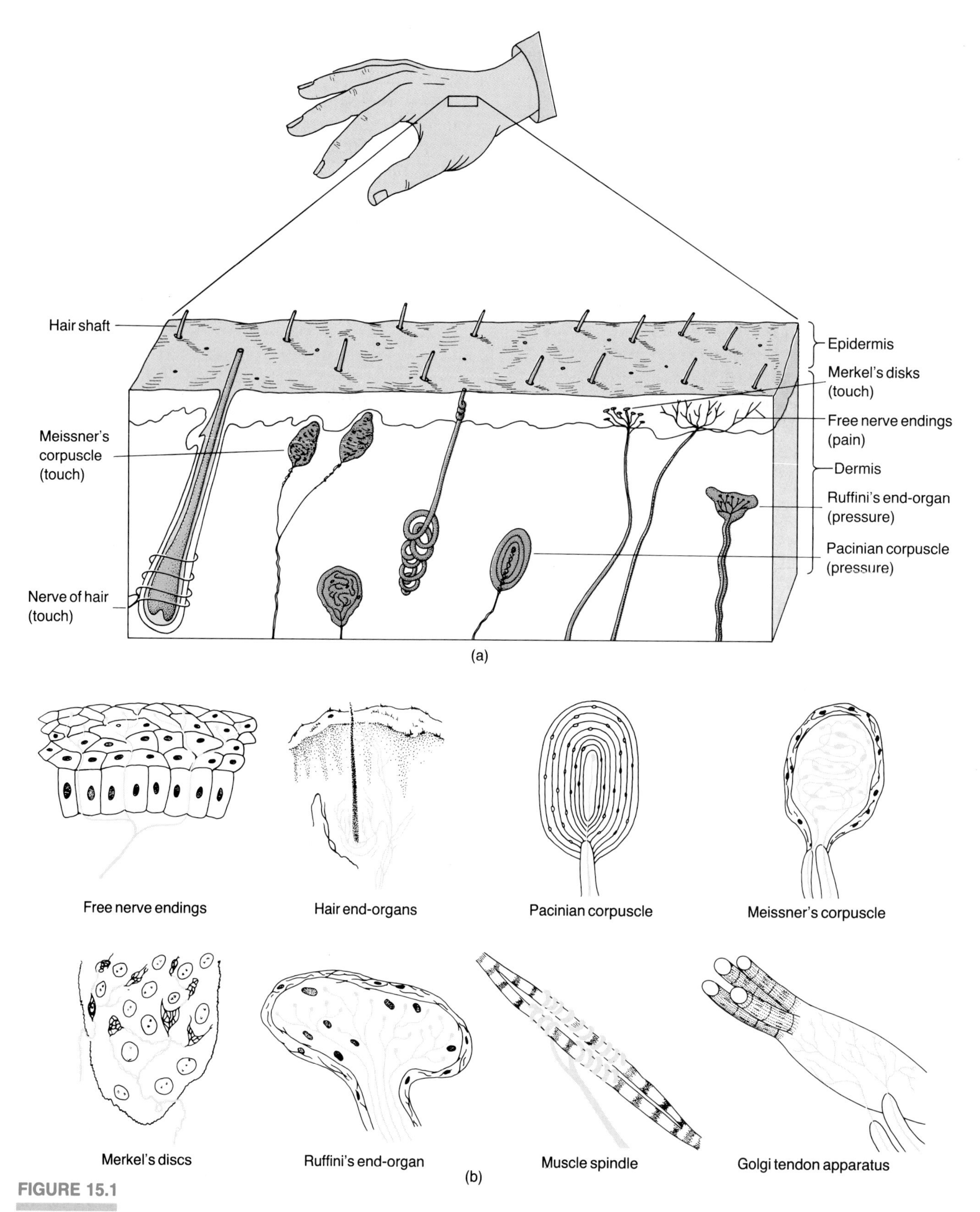

FIGURE 15.1

Mechanoreceptors

(a) A section of skin, showing the approximate locations of many of the mechanoreceptors. (b) A close-up view of the mechanoreceptors used in tactile and position senses.

5. **Merkel's discs,** which are another type of encapsulated expanded nerve ending. They are common in both hairy and nonhairy parts of the skin, and are responsible for detecting more prolonged skin contact than is detected by Meissner's corpuscles, because the discs adapt much more slowly.

6. **Ruffini's end-organs,** which are multibranched nerve endings located in the deeper portions of the skin and other tissues. They are very slow to adapt, and they therefore detect continuous, deep pressure or touch.

7. **Baroreceptors,** which consist of a multibranched network of nerve endings in the carotid artery that monitor blood pressure (see Chapter 8). Higher pressure increases the rate of action potentials being sent from the baroreceptors to the medulla oblongata, whereas lower blood pressure reduces the rate of action potentials.

The other two tactile senses, *itch* and *tickle,* are detected by rapidly adapting free nerve endings in the most superficial portions of the skin. The itch sensation triggers the scratch reflex; scratching either dislodges the irritant or, when repeated enough, causes pain that masks the itch.

Position Sense

Position sense includes both static and dynamic components. That is, the brain receives information from all parts of the body regarding the relative orientation of body parts when at rest (static) and when moving (dynamic). Position sense receptors include the following:

1. **Muscle spindles,** which are tiny, specialized muscle fibers that lack contractile proteins in the center. A nerve fiber is wrapped around the central portion, and the whole spindle is attached to regular muscle fibers. When a muscle is stretched, the central portion of the muscle spindle changes shape, and the nerve fiber is stimulated. These are the receptors involved in the knee-jerk reflex (Chapter 14).

2. **Golgi tendon organs,** which are multibranched nerve fibers located in tendons just beyond the end of the muscle fibers. Unlike the muscle spindles, which measure relative length of the muscle, Golgi tendon organs measure the amount of muscle tension.

Both muscle spindles and Golgi tendon organs react to static and dynamic conditions. In static conditions, a regular but relatively low frequency pattern of action potentials emanates from these organs. However, when muscle length or tension changes suddenly, there is an explosive increase in the number of action potentials. Thus the brain is kept fully informed about the level of muscle activity.

Pacinian corpuscles and particularly Ruffini end-organs are abundant in joint capsules, and they play an important role in monitoring motion at a joint. Because they adapt slowly, the Ruffini end-organs are also important in monitoring the static position of a joint.

How does the brain distinguish variations in position? For example, how does the brain determine whether the elbow joint is almost fully extended or half-closed? Different groups of receptors become active, depending on how open or closed a particular joint happens to be. Since each group of receptors supplies a discrete area of the brain, the brain determines the actual position of a limb, even in static situations, according to the region of the brain that is receiving signals.

Hearing

Any vibrating object—from the reed of a clarinet to the vocal cords of a tenor—causes alternating bands of compression and rarefaction of molecules in the air. If the object is vibrating at a frequency of between 20 and 20,000 cycles per second, and if the vibration is of sufficient magnitude, we are able to detect the vibration as *sound*—that is, we can *hear* it.

Anatomy of the Ear. The ear has three main divisions (see Figure 15.2):

1. The **outer ear,** consisting of the two flaps of tissue that decorate the sides of your head, plus the **auditory canal,** which leads to the **eardrum.**

2. **The middle ear,** consisting of three small, articulated bones. Going from outside to inside, these are the **malleus,** the **incus,** and the **stapes** (L, "hammer," "anvil," and "stirrup," respectively, based on their shapes). The malleus is tightly attached to the eardrum, whereas the stapes abuts the inner ear at the so-called **oval window.** The **Eustachian tube** leads from the middle ear to the throat and provides an air passage that permits equivalent air pressure on both sides of the eardrum.

3. The **inner ear,** which includes the **cochlea** (Gr. *kochlias,* "snail"), and the organs associated with the maintenance of balance and posture.

Physiology of the Ear. The eardrum vibrates in response to sound. These vibrations are transmitted via the middle ear bones to the cochlea. Because the surface area of the eardrum is so much larger than the surface area of the oval window (and, to a lesser degree, because the bones of the middle ear act as a lever system), the pressure exerted on the cochlea is

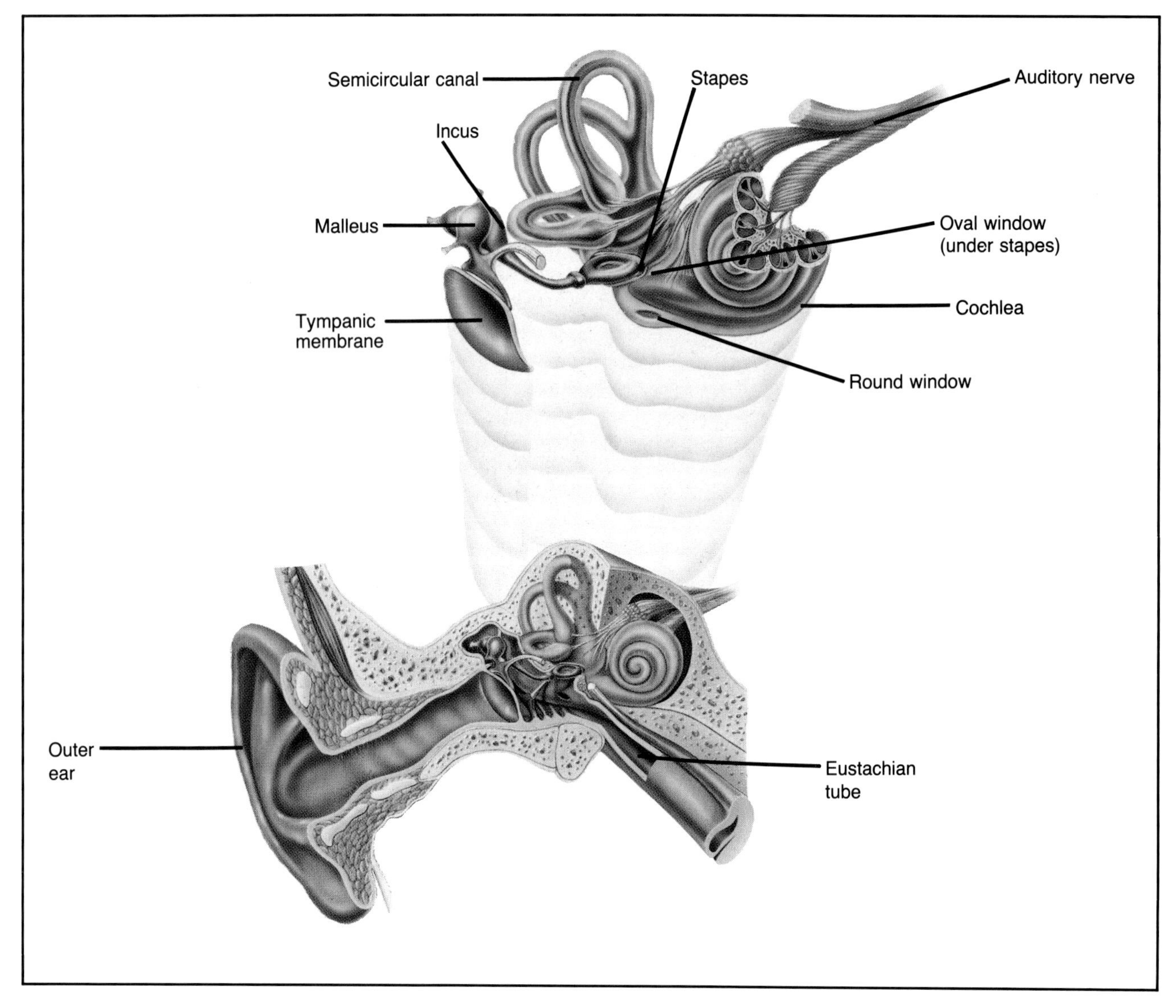

FIGURE 15.2

Anatomy of the Ear

An artist's rendering of the human ear showing the anatomical relationship of the various structures used in hearing and equilibrium.

more than 20 times greater than the pressure of the sound waves on the eardrum. This increase in pressure is essential, because the cochlea is filled with fluid, and substantial pressure is needed to set up vibrations in fluid.

The cochlea converts vibrations to nerve impulses. Anatomically, the cochlea is spiraled, rather like the snail that gives it its name. For ease of analysis, however, we may pretend it is shaped like a straight cone (see Figure 15.3).

At first glance, the cochlea would appear to have a significant design failure. How do vibrations of the eardrum, transmitted by the middle ear bones to the cochlea, set up vibrations of cochlear fluid? Liquids, unlike gases, strongly resist compression. Therefore the bands of compression and rarefaction that sound creates in air cannot easily occur in an enclosed liquid—and the cochlea cannot expand, because it is encased in bone.

Cochlear anatomy is further complicated by the presence of a pair of membranes running almost the full length of the cochlea and effectively dividing it into three parallel chambers (see Figure 15.4). The upper membrane is the **vestibular membrane** and it walls off an upper **vestibular canal.** The lower membrane is the **basilar membrane,** and it walls off the **tympanic canal.** Between the two canals, and isolated from the fluid in the canals, is the **cochlear duct.**

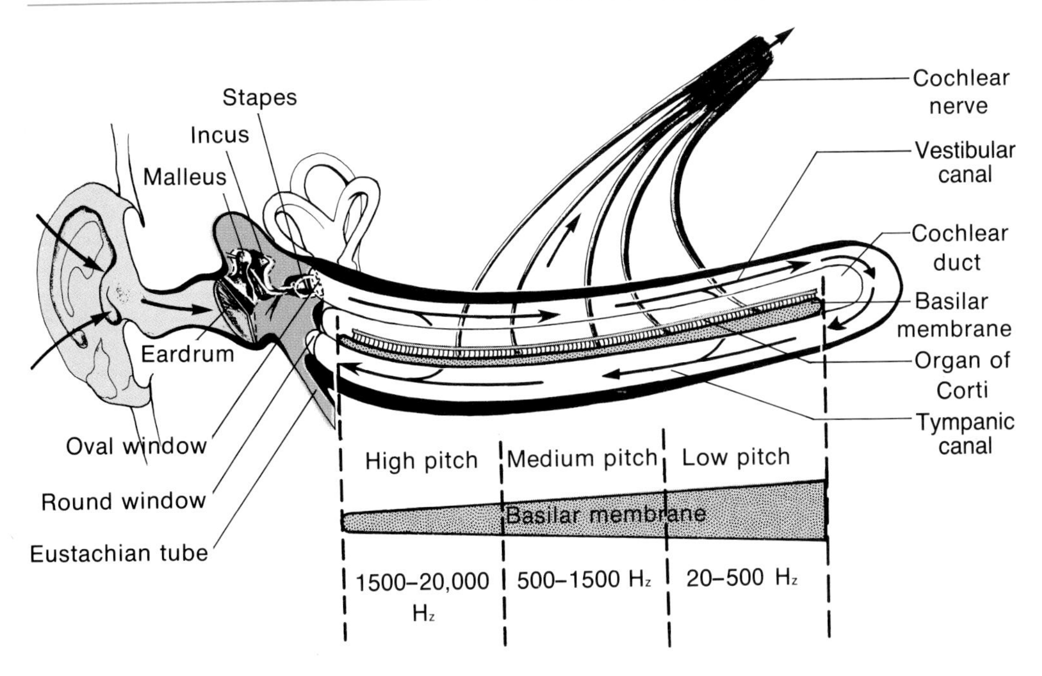

FIGURE 15.3

Diagrammatic Representation of the Cochlea
In this "uncoiled" cochlea, pressure waves can be seen moving away from the oval window along the length of the vestibular canal, and then returning to the round window by way of the tympanic canal. The basilar membrane is seen within the cochlear duct that lies between the vestibular and tympanic canals. Note that the width of the basilar membrane increases toward the end of the cochlea.

As sound vibrations are transmitted from the eardrum, through the bones of the middle ear, and to the oval window of the cochlea, the vibrating oval window creates vibrations in the cochlear fluid, which pass along the vestibular canal, around the end of the cochlear duct, and back along the tympanic canal. At the end of the tympanic canal is another membrane called the **round window,** which bulges outward in absorbing the vibrational energy in the cochlear fluid. Thus it is the presence of the round window that permits the oval window to vibrate in the first place. Note that the vibrational frequency of all the components of the system—the eardrum, the oval and round windows, and the cochlear fluid itself—is the same.

Perception of Sound. The basilar membrane contains the **organ of Corti,** a structure composed of hair cells that synapse with sensory neurons leading to the brain. The organ of Corti is the transducer that converts vibrational energy into action potentials. Understanding the structure of the organ of Corti will help us to comprehend how it is that we are able to perceive sound.

Along the length of the cochlea and forming an essential part of the basilar membrane are some 20,000 to 30,000 **basilar fibers.** The basilar fibers are imbedded in bone at one end but are free at the other. The fibers are by no means uniform; at the base of the cochlea they are short and stiff, but at the tip they are long and relatively flexible. Because they are free at one end, the fibers can vibrate like a series of tiny tuning forks. Because they differ in length and flexibility, different groups of fibers vibrate at different sound frequencies. Sounds of high frequency (short wavelength) cause the fibers near the base of the cochlea to vibrate, whereas sounds of low frequency (long wavelength) cause vibration of the fibers near the tip of the cochlea.

The hair cells of the organ of Corti are attached in rather elaborate fashion to the basilar fibers. Immediately above the hair cells, and just touching their cilia, is the **tectorial membrane.** When sound of a particular wavelength causes vibration of a particular group of basilar fibers, the hair cells attached to those vibrating basilar fibers bounce up and down, but the tectorial membrane remains stationary. Consequently, the cilia of the hair cells are deflected, and the deflection initiates action potentials that are relayed to the brain. Presumably, the brain distinguishes sounds of different frequencies based on the particular region of the brain that is receiving action potentials, but this assumption has not yet been verified in humans.

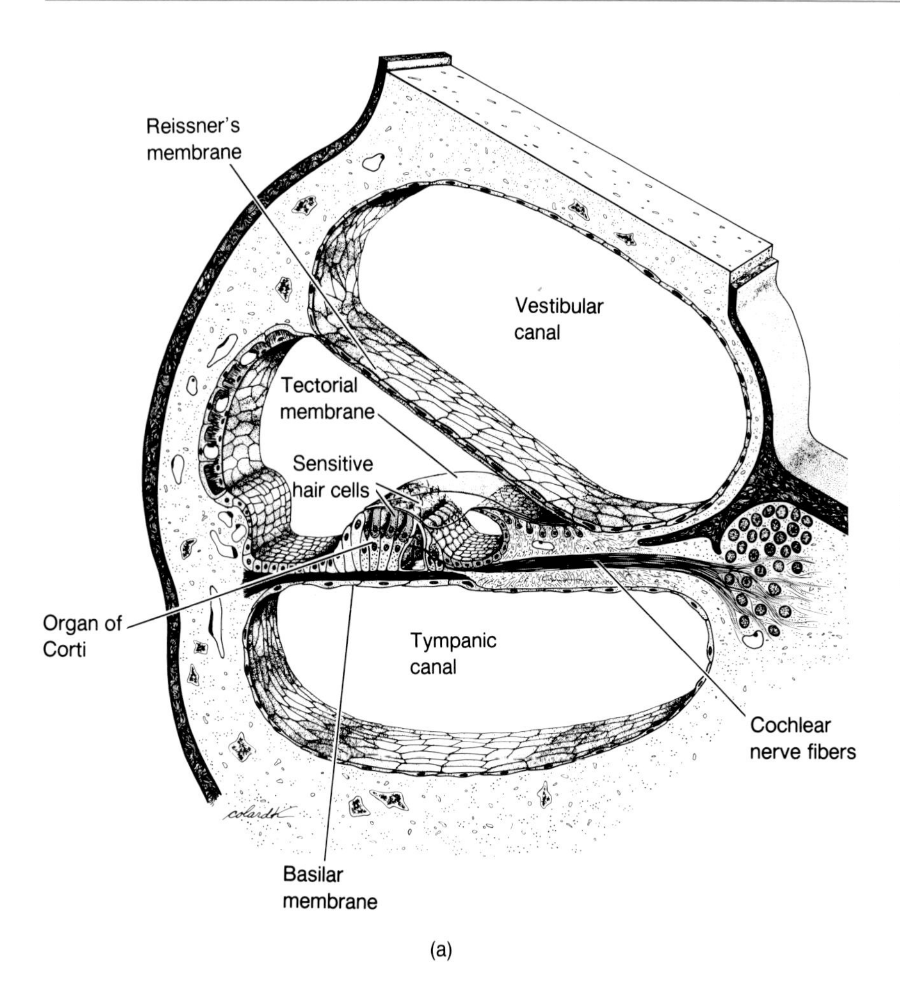

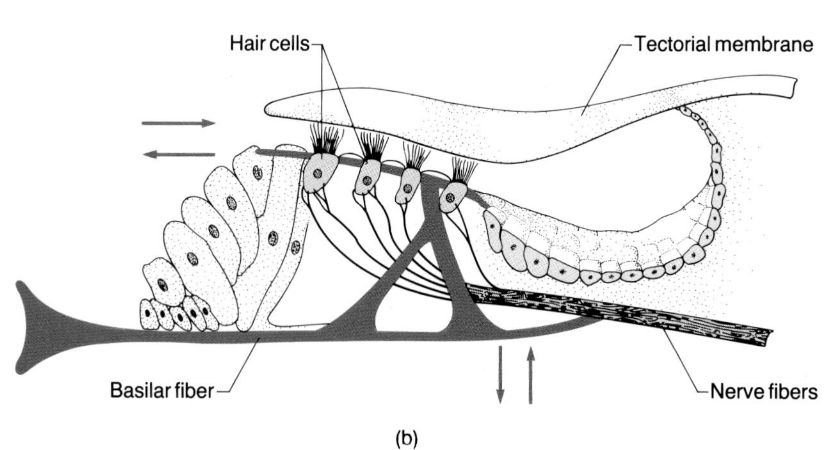

FIGURE 15.4

Cross Section Through the Cochlea and the Organ of Corti

(a) The relationship among the three passageways within the cochlea—the vestibular and tympanic canals and the cochlear duct—is evident in this cross-sectional view of the cochlea. (b) In a close-up view of the organ of Corti, it is apparent that if the basilar fiber vibrates in harmony with the pressure waves moving through the cochlear fluid, the cilia of the hair cells will be deflected as they move back and forth against the tectorial membrane. If action potentials result as a consequence of this deflection, a nerve impulse will pass to the brain where it will be interpreted as sound.

Perception of Loudness. We have established that sounds of different frequencies stimulate different regions of the basilar membrane, and that this regional response permits us to distinguish a range of sound frequencies. So far, so good—but how are differences in the *loudness* of sounds perceived?

The intensity of sound is measured by the **decibel scale,** which is a logarithmic scale. That is, intensity increases by a factor of ten with every increase of one bel (10 decibels). Thus a sound with the intensity of six bels (60 decibels) is ten times louder than a sound of five bels (50 decibels). Intensity of sound is normally measured in decibels rather than in bels because the human ear can distinguish sounds that differ in intensity by as little as one decibel. (Unfortunately, using decibels as the common unit of measure tends to obscure the origin of the term, which was named for Alexander Graham Bell.)

Loudness and frequency are interrelated in human hearing, because we can perceive sounds of intermediate frequency (500 to 5000 cycles per second) at much lower intensities than we can perceive sounds of very low or very high frequency. The hearing threshold for a sound of 100 cycles per second, for example, is 10,000 times greater than the threshold for a sound of 2000 cycles per second.

The ear registers loudness of sound by using one or more of three mechanisms:

1. Amplitude of vibration: Loud sounds cause greater deflection of the eardrum than do soft sounds, and a greater deflection of the eardrum produces a correspondingly greater amplitude of vibration of the basilar membrane. Consequently, the hair cells are more disturbed, and they produce action potentials at higher rates.

2. Summation: Because of an increase in the amplitude of vibration, loud sounds cause adjacent portions of the basilar membrane to vibrate and thus to stimulate adjacent hair cells. The result is an increase in the number of hair cells that are activated in response to a sound.

3. Variation in threshold: Some hair cells are apparently stimulated only by loud sounds, and their stimulation is a cue to the brain that the sound is of high intensity.

We can respond to sounds that range in loudness from one decibel to more than 120 decibels. Because the decibel scale is logarithmic, a sound of one decibel is less than one trillionth (1×10^{-12}) as loud as a sound of 120 decibels. One reason we can respond to this enormous range is that we can dampen very loud sounds by contracting muscles in the middle ear. Contraction of these muscles effectively locks the middle ear bones in place and permits reduction in intensity of up to 40 decibels. Even so, very loud sounds may, over a period of time, permanently reduce auditory acuity (see Box 15.1).

Contraction of the middle ear muscles also occurs when we speak, because the vocal center of the brain coordinates both the vocal cords and the middle ear muscles. Thus, we do not deafen ourselves when we speak or shout but we do lose the ability to monitor our speech as accurately as we can monitor the speech of others. Consequently, we recognize the tape-recorded voices of our friends much more readily than we recognize our own.

Equilibrium

Gravity plays an important role in our lives, in ways both overt and subtle. If we lean too far in one direction or another, the effects of gravity take over. We "lose our balance" and fall. If loss of balance can occur at rest, imagine how easily it can occur in someone running at top speed. Yet under normal conditions people rarely suffer loss of balance. Apparently we must be able to detect gravity in some way and thereby compensate for any tendency to lose our balance during movement.

Organ of Equilibrium. The primary organ of equilibrium is a closed system of fluid-filled sacs and canals called the **vestibular apparatus** (see Figure 15.5). It consists of three **semicircular canals,** each oriented at 90° to the others, and two chambers oriented perpendicularly to each other: the **utricle** (L. *utriculus,* "little bag") and the **saccule** (L. *sacculus,* "small sac"). The vestibular apparatus is contiguous with the cochlea, although it has a very different function.

Special patches of hair cells are located in the utricle, the saccule, and in swellings at the base of each of the semicircular canals. Movement of the head causes fluid in at least one of the semicircular canals to flow (because the semicircular canals are located in three planes, no movement of the head can go undetected), and the flow of fluid initiates action potentials by the hair cells.

This process is easy to visualize if you imagine what happens to a full cup of coffee that you pick up and move suddenly. As the cup begins to move, the coffee tends to lag behind, sloshing over the back side of the cup. When you promptly stop moving the cup, in response to the hot coffee dripping over your fingers, the coffee tends to continue to move, sloshing over the front side of the cup. If you had sensory hair cells around the rim of the cup, you would be able to detect, without looking, which direction the cup was moving by the response of particular hair cells to shifts in the coffee. Detection of shifts in the fluid, of course, is precisely what the hair cells in the semicircular canals are doing, and that is why they are so important in registering sudden changes in rate of rotation (that is, acceleration and deceleration) of the head.

In the utricle and the saccule, the hair cells are covered by a gelatinous mass containing crystals of calcium carbonate [$Ca(CO_3)_2$]. For each conceivable position of the head, gravity will cause a particular area of the calcium carbonate layer to deflect particular hair cells and thus to trigger action potentials. The utricle and the saccule therefore function primarily as gravity detectors. In addition to informing the brain about the orientation of the head at any given moment, they also register minor changes (as small as 0.5°) in head position and assist in allowing the body to maintain equilibrium and a normal posture.

Other Organs Aiding Equilibrium. The vestibular apparatus senses movement and orientation of the head only. Receptors in the joints of the neck verte-

BOX 15.1

Problems with the Ears

The primary problem affecting the ears is deafness, of which there are two basic types. More than two million Americans suffer significantly from one or the other.

Conduction Deafness. In conduction deafness, the bones of the middle ear become "frozen" because of calcification or other problems, and transmission of sounds from the eardrum to the cochlea is impaired. Modern surgical techniques are frequently very successful in curing this form of deafness. In mild cases, a hearing aid, which amplifies sounds, is often very useful. About 10 percent of deaf people suffer from conduction deafness.

Nerve Deafness. In nerve deafness, the conducting system of the ear is functional, but the cochlea or the auditory nerve is impaired. As they grow older, most people lose some auditory acuity to high-frequency sounds. Deafness to low-frequency sounds often results from prolonged exposure to very loud noises. General deafness to all frequencies is often due to damage to the organ of Corti by disease or by drugs such as streptomycin.

Conduction deafness and nerve deafness are easily distinguished. A vibrating tuning fork is held near the ear. As the vibrations in the tuning fork gradually diminish in amplitude, a point is reached where the patient can no longer hear it. The butt of the tuning fork is then immediately touched to the bone at the base of the ear (the *mastoid process*). In patients with conduction deafness, bone conduction through the mastoid process will permit the individual to hear the tuning fork; individuals with nerve deafness will not be able to hear the tuning fork.

In the last few years, cochlear implants have been tried with some success for nerve-deaf individuals having a problem in the cochlea, not in the auditory nerve itself. Although still very imperfect, the most recent implants permit a majority of users (who were totally deaf) to understand familiar voices and to have telephone conversations. These versions employ a series of more than 20 tiny electrodes, each responsive to a small set of sound frequencies, running from an implanted receiver to the auditory nerve.

Cochlear Damage

(a) Normal hair cell structure in the cochlea of a guinea pig. (b) Same view after a 24-hour exposure to rock music at 120 decibels.

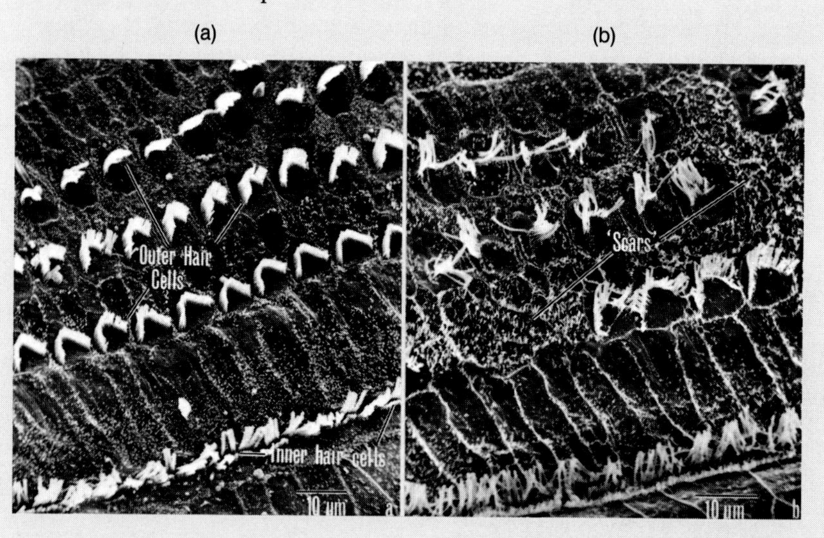

brae record postural relationships between the head and the body. The eyes are also important in maintaining equilibrium, as you can easily demonstrate by trying to balance on a narrow beam with your eyes closed (it is much easier with your eyes open).

The relationship between the vestibular apparatus and the eyes is apparent in another context as well. Using information received from the vestibular apparatus, the brain directs the muscles that move the eyes. Thus you can move your head and still keep your eyes focused on the same point. This ability may not seem remarkable, but if you have ever watched a movie filmed with a hand-held camera, you know how jerky the picture appears. The camera, unlike the human eye, makes no automatic compensation for movement. In contrast, you have no trouble keeping your eyes focused on a given object, even when you are running.

Of course, the vestibular apparatus merely transmits information about head position and movement. The brain must react to this information by sending the appropriate motor messages to ensure compensatory muscle activity that results in maintenance of equilibrium.

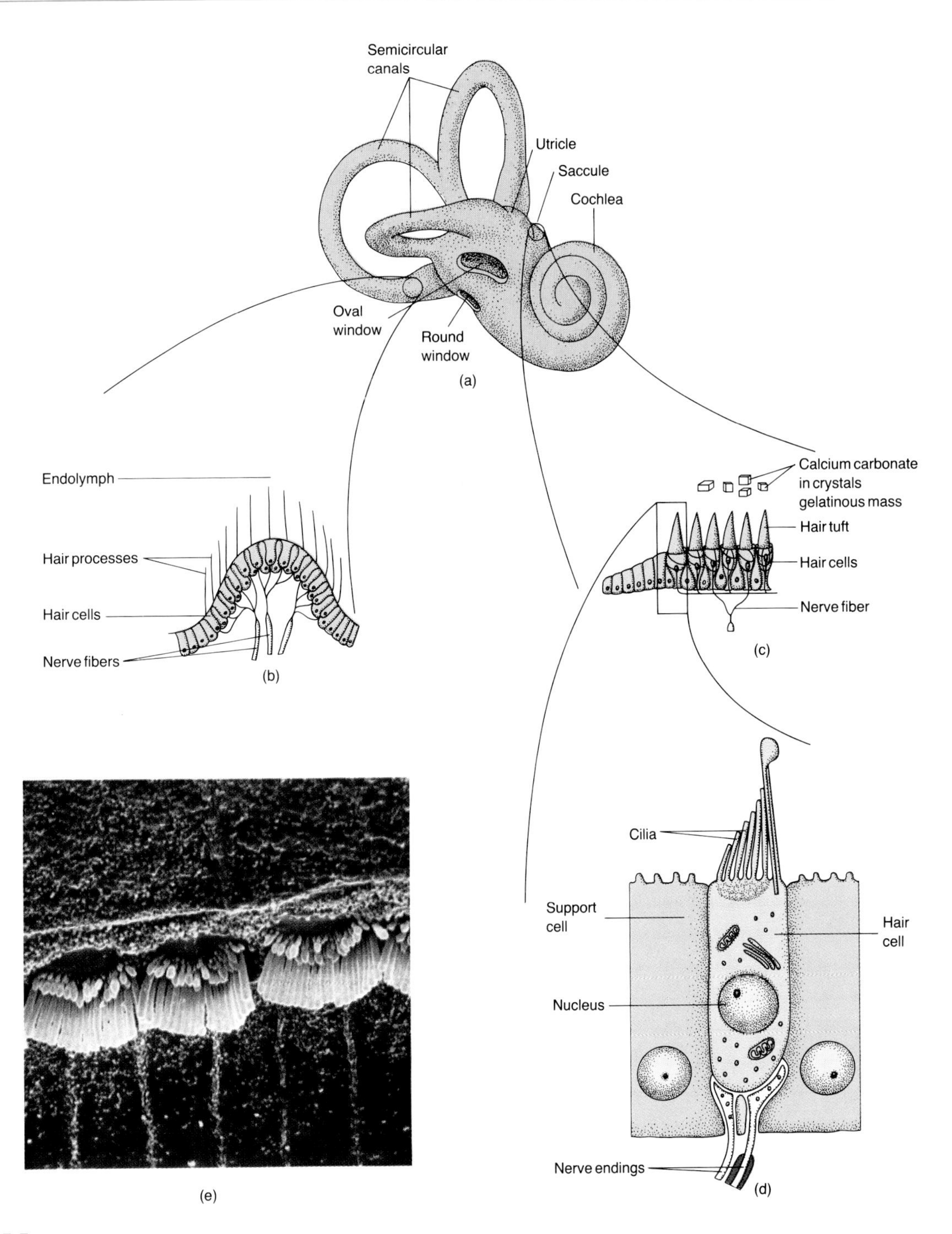

FIGURE 15.5

Vestibular Apparatus

(a) The semicircular canals and the cochlea. (b) The interior of the swelling at the base of each semicircular canal. Movement of the fluid within the canal will deflect the hair cells, generating an action potential. (c) Hair cells within the saccule. Note that, in the saccule (and utricle) the hair cells are covered with a gelantinous mass containing calcium carbonate crystals. As this mass moves, in response to changes in the orientation of the head, certain hair cells will be deflected, again generating an action potential. (d) A close-up view of a hair cell from the saccule. (e) A scanning electron micrograph of the cilia of a hair cell.

PAIN RECEPTORS

No specialized organs exist for the perception of pain. Rather, pain is caused by stimulation of free nerve endings. Stimulation may result from mechanical damage, from extreme heat or cold, or from the presence of certain chemicals. Pain can be classified into three types:

1. Pricking pain is superficial pain most commonly caused by a cut or other wound to the skin.

2. Aching pain is pain felt deep within the body, as opposed to being felt on the surface, and it may vary in intensity.

3. Burning pain is pain induced by burns.

Important differences exist in the nerves that transmit the various types of pain. Aching and burning pain are transmitted by small, unmyelinated fibers at very low rates of speed (as low as 0.5 m/sec). Pricking pain is transmitted rapidly, by myelinated fibers, at rates up to 30 m/sec. Thus pricking pain is perceived first, though it is frequently followed by one of the other types. Unlike most receptors, pain receptors adapt very little—which is why your finger continues to hurt long after you have stopped hitting it with a hammer, for example.

The Physiology of Pain Perception

When tissues are damaged, as by a cut or burn, damaged cells release their contents into the adjacent interstitial fluid, causing the fluid to become locally more acidic (see Figure 15.6). Greater acidity activates an interstitial fluid enzyme called **kallikrein,** which splits a large inactive protein also found in the interstitial fluid into small pieces, each nine amino acids in length. These small molecules are called **bradykinin,** the most potent of all known pain-producing molecules. (The venom of many wasps and snakes contain bradykinin, which explains why wasp stings are so painful.) At the site of injury, bradykinin initiates a cascade of events that include the following:

1. Bradykinin opens the junctions between the cells that constitute capillary walls, permitting blood plasma and white blood cells to flow into the injured area. The white blood cells phagocytize damaged cells and any foreign organisms that may have gained access at the injury site; plasma leakage causes the local swelling of tissue so characteristic of injuries.

2. Bradykinin also binds to receptors on nerve endings, causing them to initiate the action potentials that we perceive as pain. In addition, the nerve endings release a small molecule called **substance P** (for "pain") which binds to a type of cell common in connective tissues called a **mast cell.** Substance P also shuts down the K^+ channels of the neurons, exciting them and making them more responsive. Bradykinin also binds to the mast cells, and in response the mast cells release **histamine,** which further widens cellular junctions in the capillary, allowing more of the molecules that are kallikrein and bradykinin precursors to enter the injured area and to continue the pain-producing process.

3. Finally, bradykinin also binds randomly to cell membranes, initiating the release of **prostaglandins** (see Chapter 13), which bind to nerve endings and send additional pain messages to the brain.

Antihistamines reduce swelling by inhibiting histamine; aspirin, acetaminophen (e.g., Tylenol), and ibuprofen (e.g., Advil) all diminish pain by inhibiting prostaglandin production and activity; the opiate drugs, such as codeine and morphine (as well as the body's own endorphins and enkephalins), act directly on pain centers in the brain. The identification and production of antagonists for bradykinin are presently (for obvious reasons) the subject of intense research by pharmaceutical companies.

CHEMORECEPTORS

The body has many chemoreceptors—even if the term is restricted to the detection of chemicals for which there is a relay to the brain and a corresponding response. The relative acidity of the blood, its gas and glucose levels, and its concentrations of many hormones are all monitored, in whole or in part, by the brain. Most of these phenomena have been considered elsewhere in this text. Two sets of chemoreceptors, however, deserve special attention: those that provide the sense of taste and the sense of smell.

Taste

Taste receptors are in the form of **taste buds**—small clusters of modified skin cells only a fraction of a millimeter across (see Figure 15.7). Most of the more than 10,000 taste buds in the mouth cavity are located on the tongue. (The raised bumps you see when you look at your tongue in the mirror are *not* the taste buds. They are the **papillae,** which contain many of the taste buds. The taste buds themselves are housed in pockets, and open to the outside by way of a small pore.) Action potentials are generated when appropriate chemicals (dissolved in the watery environment of the mouth) bind to the microvilli of the taste bud cells. These cells are not themselves neurons—they are specialized skin cells that synapse with neurons.

Taste buds begin to degenerate after mid-life—a fact that accounts for the general complaint among older people that food just doesn't taste as good as it used to. Some taste buds are located on the roof of the mouth. These are effectively shielded by dentures, and denture wearers also commonly complain about how tasteless their food has become.

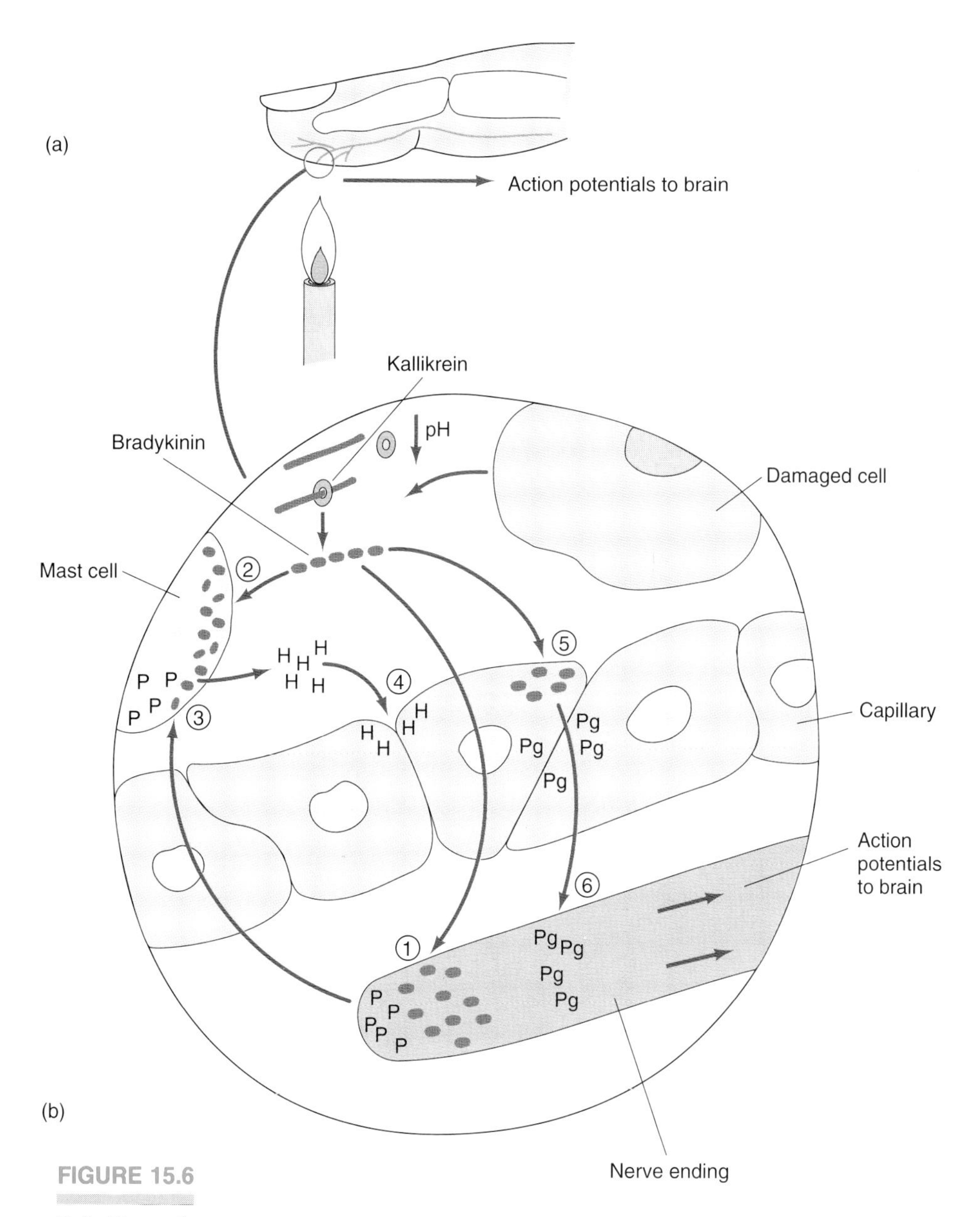

FIGURE 15.6

Pain Reception

(a) Pain is initially perceived by action potentials passing to the brain from nerve endings that serve as pain receptors. (b) Increased acidity in the interstitial fluid (as a result of the release of the contents of damaged cells) activate the enzyme kallikrein, which splits an interstitial protein into small pieces called bradykinin. Bradykinin (1) binds to nerve endings, generating more pain impulses and (2) binds to mast cells, causing the release of histamine. (3) Nerve endings release cells, accelerating histamine release. (4) Histamine widens the gaps in the endothelium of capillaries, causing swelling as the interstitial fluid increases in volume in the affected area. (5) Bradykinin also binds to capillary walls, initiating the production of prostaglandins. (6) Prostaglandins bind to nerve endings, generating additional nerve impulses to the brain.

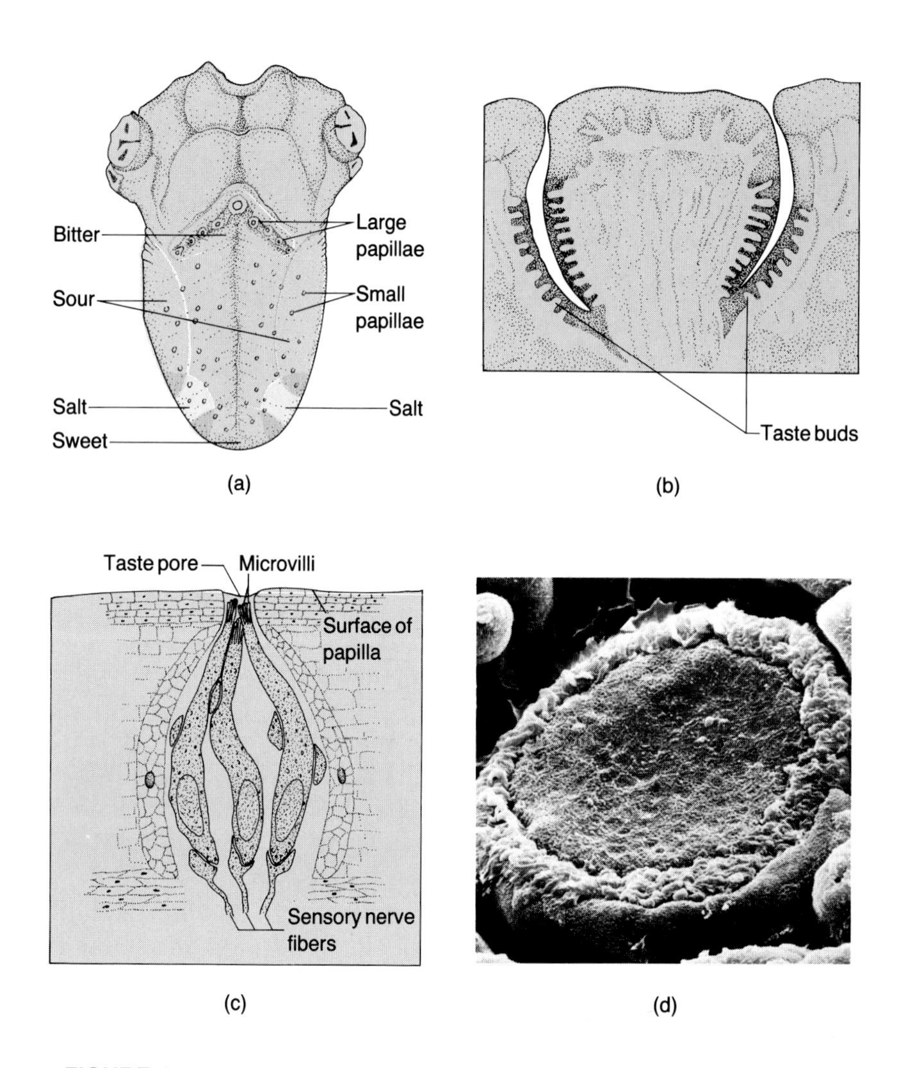

FIGURE 15.7

Tongue and Taste Buds

(a) The tongue, showing the approximate distribution of the four types of taste buds. (b) The relationship of papillae to the taste buds. (c) A longitudinal section through a taste bud. (d) A scanning electron micrograph of the taste pore of a taste bud. The center of the pore is filled with microvilli from the ends of the cells within the taste bud.

The sense of taste may be divided into four basic taste categories, although it is not clear that we have four corresponding types of taste buds. (In fact, there is growing evidence to rebut the old notion that a given taste bud will respond to only one of the four primary tastes.) The categories of taste include the following:

1. *Sweet taste,* a complex taste, is caused by many chemicals, not just sugars (*saccharin,* for example is 600 times sweeter than sucrose but is chemically very distinct from the sugars). We respond to sugar solutions as dilute as 1:250.

2. *Salty taste* is caused by the ions of salts, although some salts "taste" saltier than others. We respond to salt solutions as dilute as 1:500 (salt to water).

3. *Sour taste* is caused by acids. The strength of the reaction is proportional to the strength of the acid (i.e., the hydrogen ion concentration). We respond to sour solutions as dilute as 1:135,000.

4. *Bitter taste,* more complex than the salty or sour tastes, is caused by a broad variety of nitrogen-containing plant products such as nicotine, caffeine, quinine, and morphine. (Bitter food is generally rejected, which is just as well, because many of the bitter chemicals are poisonous.) We respond to bitter solutions as dilute as 1:2,000,000.

Most foods contain mixtures of the four tastes, and we distinguish among foods by the relative strengths of the primary tastes as well as by the sense of smell. Taste, of course, is ultimately highly subjective. For example, some people relish food that others find sickeningly sweet. Saccharin tastes bitter, or metallic, to many people. Precisely why this is so is not known.

Smell

The danger in all introductory biology texts is that, in the interests of presenting biological facts and principles in a simplified way, the reader may conclude that, except for a few loose ends, most of the important biological details have already been discovered. Such a conclusion would be very wrong—we have only scratched the surface. As regards our understanding of the sense of smell, we have not even accomplished that much. About all we can do is to describe the anatomy of olfaction.

The sense of smell is restricted to two small patches of **olfactory membrane** located in each nostril (see Figure 15.8). Within a total of 5 cm^2 of tissue are more than 10 million olfactory cells, each tipped with six to eight cilia. These olfactory cells are themselves neurons, and the cilia are branches of dendrites. (Incidentally, some breeds of dogs have more than 200 million olfactory cells, which helps to explain their greater sensitivity to odors.) The cilia presumably bear receptors that are sensitive to molecules of particular shapes. We do not know what these receptors are, nor can we explain how it is that we can distinguish so many different odors. We do not even know how many primary odors we can detect. One hypothesis claims that humans can detect seven primary odors, but it is known that insects can identify many more primary odors than just seven—so it is certainly possible that humans can, too.

It is clear that, to have an effective odor, a substance must be volatile, water soluble (in order to be able to dissolve through the mucus that protects the delicate olfactory membrane), and somewhat lipid soluble (in order to be able to penetrate the plasma membrane of the olfactory cells). Because much of what we "taste" are actually the volatile odors detected by the olfactory membrane from the food in the mouth (hot foods give off such odors more than cold foods), we find a tasty, hot meal disappointingly bland when we are "plugged up" with a head cold.

Almost everyone is "blind" to one or more smells, but at least some of this "smell blindness" can be overcome by repeated exposure to the smell. This finding suggests similarities between the olfactory and the immune system—in both, repeated exposure strengthens the response. (In the case of the olfactory cells, the receptors for a particular smell apparently proliferate.) Moreover, unlike other neurons, olfactory neurons can replicate.

On the other side of the coin, adaptation in olfactory cells is relatively rapid. In other words, we quickly become "used to" particular smells—a critical ability when we find ourselves trapped on an elevator next to a person with an overabundance of after-shave or perfume.

THERMORECEPTORS

We not only distinguish between feeling hot or cold, we are also conscious of a series of gradations between the two extremes. The receptors that permit this range of sensations include warmth receptors, cold receptors, and heat-pain and cold-pain receptors. The actual perception depends on how many of each type of receptor are initiating action potentials.

Scientists have discovered no distinct organ for warmth reception; warmth receptors may simply be

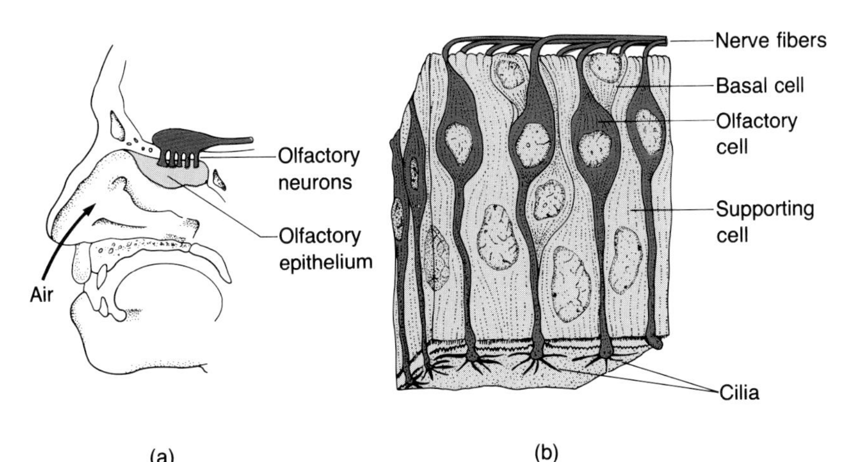

FIGURE 15.8

Nose and Olfactory Membrane

(a) The position of the olfactory epithelium within the nose. (b) Arrangement of cells within the olfactory epithelium.

free nerve endings. However, the body has special cold receptors in the form of multibranched, myelinated fibers that innervate the bottom of the epidermis. Both types of receptors are most common in the lips and least common across the back; however, cold receptors are ten times more abundant than warmth receptors. (By comparison, pain receptors are 30 times more abundant than cold receptors.) Thermoreceptors are most abundant in the skin, for obvious reasons, although there are also temperature-sensitive neurons in the hypothalamus that monitor the temperature of the blood.

Thermoreceptors apparently function because the proteins that constitute the Na^+ channels experience structural changes as a consequence of a change in temperature. Thermoreceptors adapt quickly but are very receptive to sudden changes in temperature. We feel colder or warmer when the temperature is changing than we do once it has stabilized. Thus, for example, a warm bath or a shower always feels hottest when you first enter.

LIGHT RECEPTORS

Although all cells show a vague light sense, the eyes are the specialized organs of light reception. In many respects, vision is the most intriguing of the senses, if only because transducing light energy to an action potential seems so different from simply detecting some mechanical change or the presence of a chemical. Moreover, our sense of vision goes well beyond the mere detection of light—we actually form a moving picture of our environment.

Anatomy of the Eye

The human eye is a complex organ (see Figure 15.9). At the tissue level, it consists of three layers, with some supporting structures. The outermost layer is the **sclera** (Gr. *skleros*, "hard"), a tough, thick tissue to which are attached the six muscles responsible for movement of the eyeball. Beneath the sclera is the

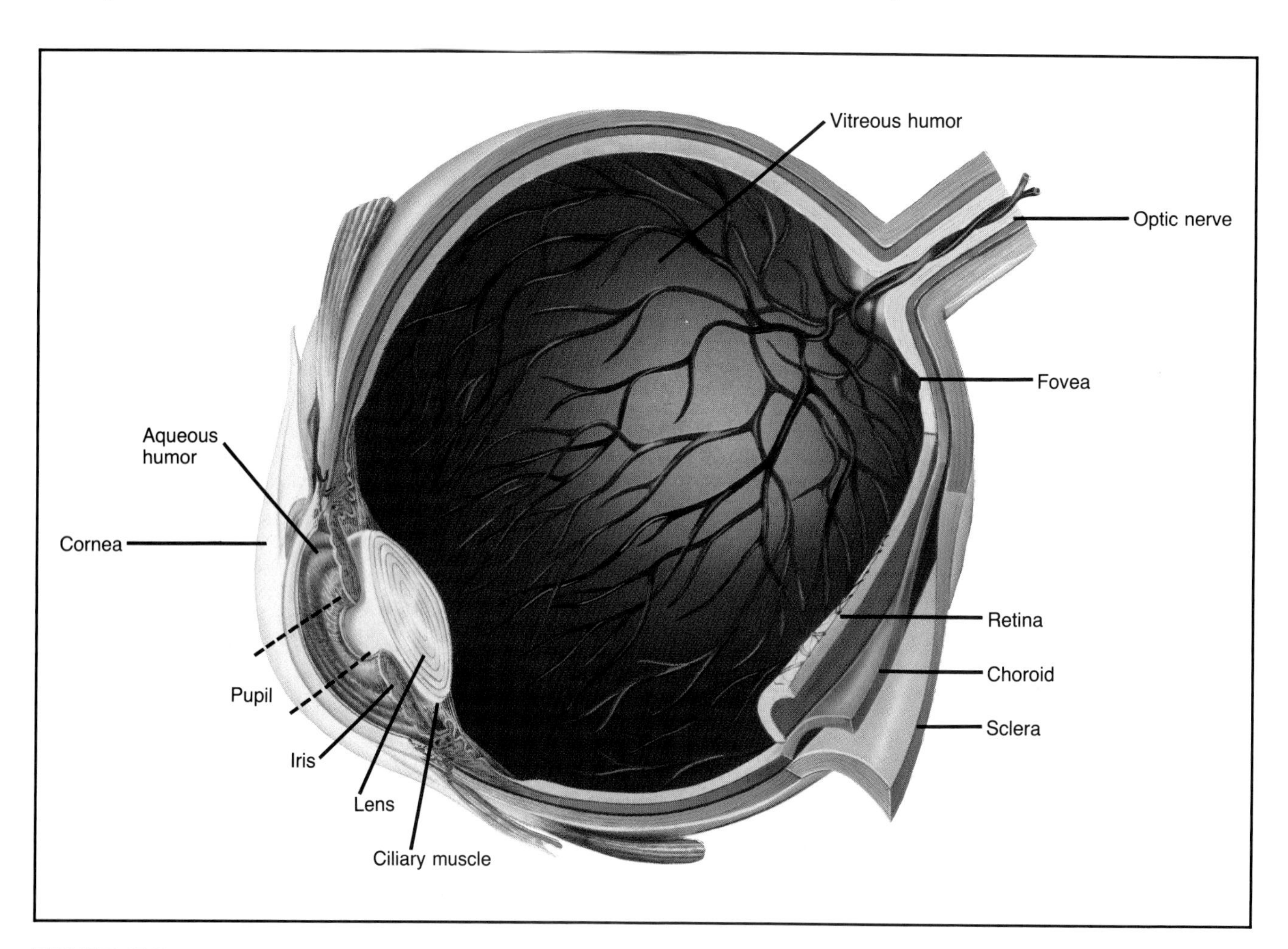

FIGURE 15.9

Anatomy of the Eye

A partial cutaway view of the human eye, showing the major anatomical features associated with vision.

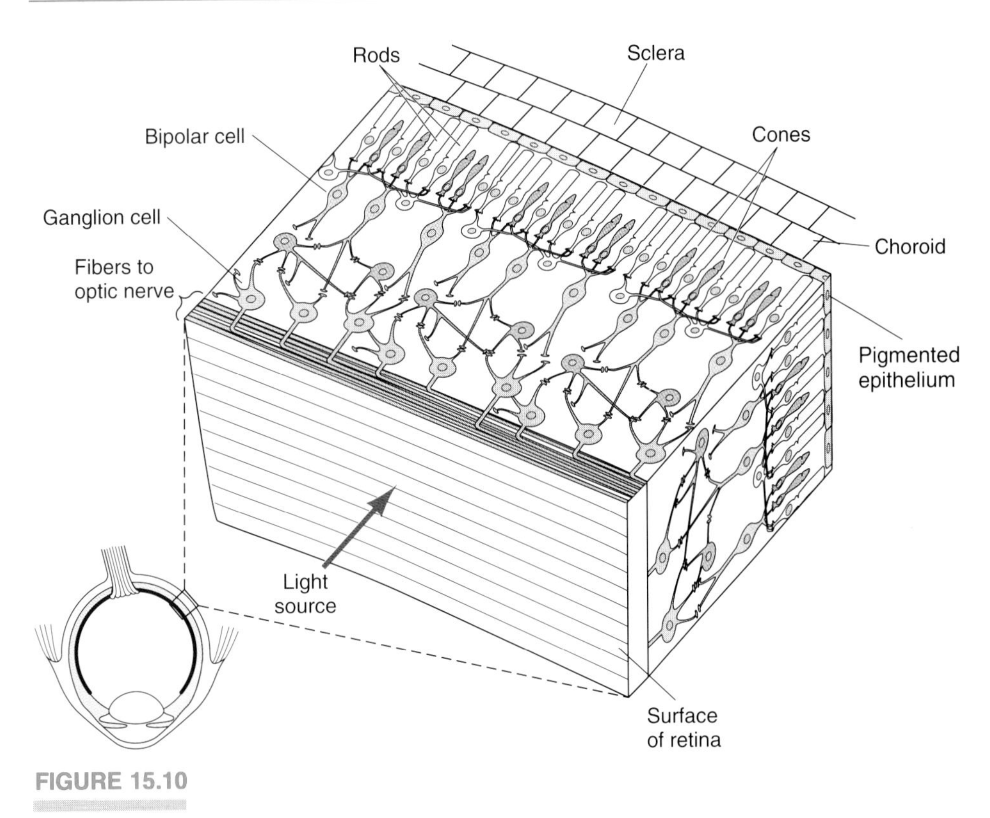

FIGURE 15.10

Organization of the Retina
The retina is organized precisely opposite to what might be expected. Rather than being pointed toward the lens, the light-sensitive cells (rods and cones) are pointed toward the back of the eyeball where they respond to light reflected from a layer of pigmented cells at the back of the retina, adjacent to the choroid.

choroid (Gr. *chorion,* "skin"), a heavily pigmented layer, well supplied with blood vessels, that absorbs stray light rays. (In nocturnal mammals, pigment in the choroid cells is withdrawn at night, exposing light-reflecting crystals which increase visual acuity. This is why your cat's eyes seem to "glow" at night.) The innermost layer is the **retina** (L. *retis,* "a net") which is the light-sensitive region of the eye.

Each of these layers is interrupted in the front of the eye. The sclera is present as the white of the eye, but a portion of the sclera is transparent, so as to permit the passage of light rays. This region is known as the **cornea** (L. *cornu,* "horn"). The cornea bulges out to some degree because of an underlying chamber filled with a watery fluid called the **aqueous humor.**

The choroid is not complete, either. In the front of the eye a portion of the choroid is visible through the cornea as a colored ring called the **iris,** which contains an opening of variable size, the **pupil.** The **lens** is located immediately behind the pupil. It separates the aqueous humor from the more gel-like **vitreous humor** which gives the eyeball its shape.

The retina is the least complete layer, being totally absent from the front portion of the eye. It is subdivided into three layers of cells, which are oriented precisely opposite to what one would expect (see Figure 15.10). The layer adjacent to the vitreous humor is composed of **ganglion cells,** which merge to become the **optic nerve,** which runs to the brain. Beneath the ganglion cells are the **bipolar cells,** which transmit action potentials to the ganglion cells. The layer adjacent to the choroid contains the sensory cells of the eye—three million color-sensitive **cones,** which are responsible for day vision, and 100 million **rods,** which are responsible for black and white night vision, and for detecting movement.

Physiology of the Eye

The physiology of the eye can be separated into two components: changes in the lens and activation of the rods and cones.

Changes in the Lens. The lens is not a hard, unyielding object like a camera lens; rather, it is soft and elastic. It is held in place by a series of ligaments that can be tightened or loosened by the contraction of the **ciliary muscles.** When the ciliary muscles are contracted, the ligaments relax and the lens is free to assume a spherical shape in which it can focus light on the retina reflected from objects near at hand (see Figure 15.11). In contrast, when the ciliary muscle is relaxed, the ligaments tighten, flattening the lens and focusing the light from more distant objects on the retina (but see Box 15.2). The curved cornea is also responsible for some light focusing, especially for distant objects.

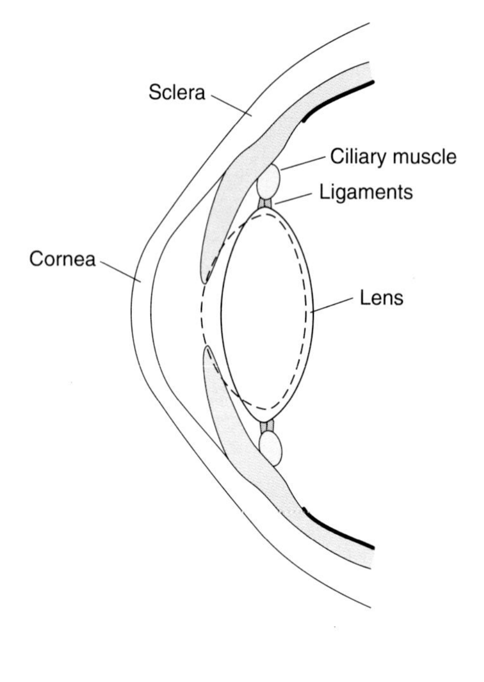

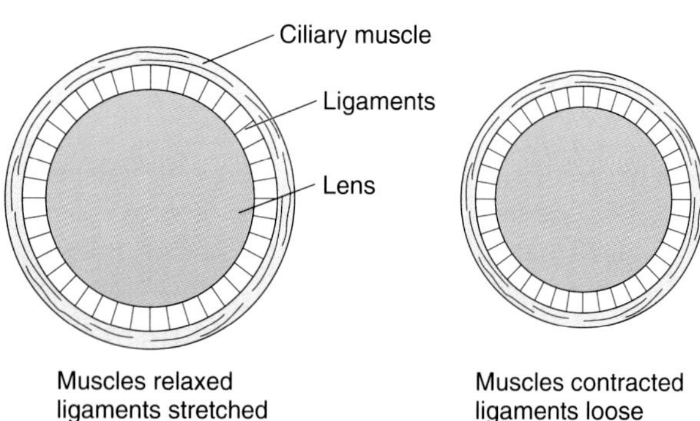

FIGURE 15.11

Changes in the Lens

The lens is primarily responsible for focusing light on the retina, but unlike a camera lens the lens of the eye is flexible. Ligaments link the edges of the lens to the ciliary muscle that surrounds the lens. When the muscle is relaxed, it is at its greatest diameter, and the ligaments pull against the lens flattening it for distance viewing. When the muscle contracts, its diameter shrinks, tension on the ligaments diminishes, and the lens assumes a more convex appearance, permitting focus on objects near at hand. Note that it is the front of the lens that increases its curvature, pushing the iris out into the aqueous humor; the back of the lens changes very little.

The lens switches from flat to spherical very rapidly (in less than one second, as you can demonstrate by changing your focus from this page to an object on the wall and then back again). The precise mechanism that mediates changes of focus in unknown, but apparently there is a negative feedback loop involving the sharpness of focus on the retina.

In addition to changes in the focal power of the lens, the size of the pupil determines how much light will be admitted to the eye. Changes in the size of the pupil are controlled by the **pupillary reflex,** wherein the amount of light striking the retina triggers a motor response from the brain that instructs the iris to open or close. (Doctors routinely test accident victims for possible brain damage by flashing light in their eyes to determine the presence or absence of the pupillary reflex.) The size of the pupil can range from 1.5 mm to as much as 8 mm, depending on light conditions.

Activation of Rods and Cones. The rods and cones point toward the back of the eye rather than toward the lens, and the light rays must therefore pass through the ganglion and bipolar layers before reaching the rods and cones. However, in the **fovea** (the region of greatest visual acuity, made up almost entirely of very small cones and measuring less than 1 mm in diameter), the ganglion and bipolar layers of the retina are displaced, and light can reach the cones directly.

Structurally, rods and cones are very similar, except that the terminal portion of the cell is rodlike in the rods and conical in the cones (see Figure 15.12). Both rods and cones contain light-sensitive chemicals that absorb photons (quanta) of light, and, in the process, initiate an action potential. This process has been particularly well studied in the rods (see Figure 15.13).

A rod cell is composed of an outer segment and an inner segment, which are linked by a narrow isthmus that forms from a modified cilium. The sodium-potassium pump is in full operation in the inner segment, meaning that Na^+ is constantly being moved from the cell to the interstitial fluid. In the outer segment, however, molecules of cyclic guanosine monophosphate (**cyclic GMP**) maintain the Na^+ channels open, permitting Na^+ to flow into the outer segment from the interstitial fluid. (Cyclic GMP is almost identical to cyclic AMP, which we encountered in Chapter 13, except that guanine replaces adenine.) Na^+ flows from the outer segment through the modified cilium to the inner segment. Because, at any given moment, there is more Na^+ in the rod than is the case in a typical neuron, the resting potential of the rod is re-

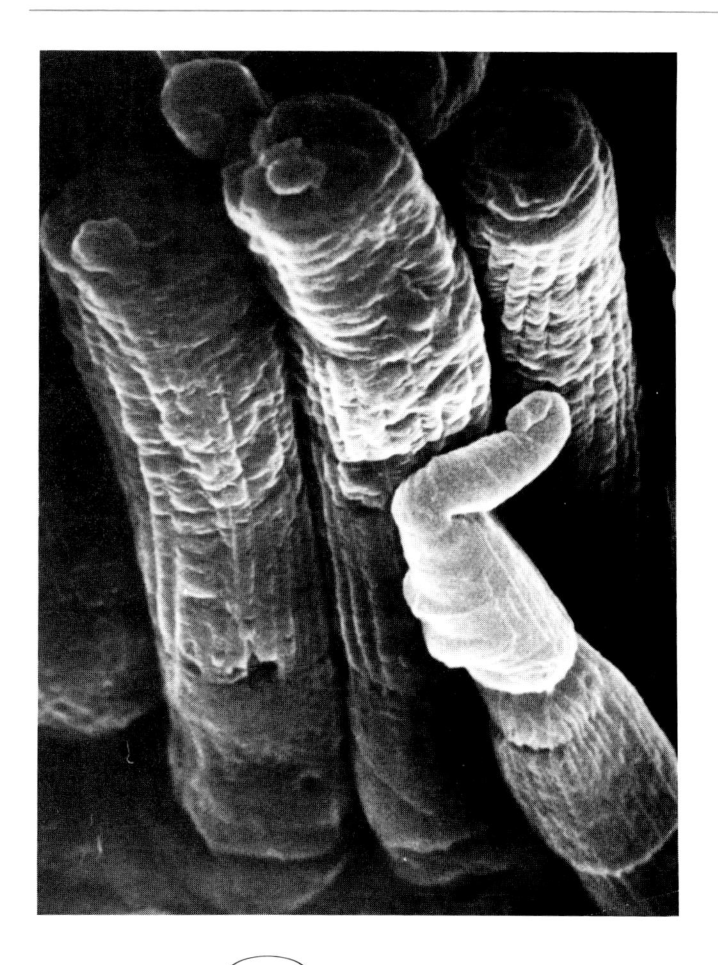

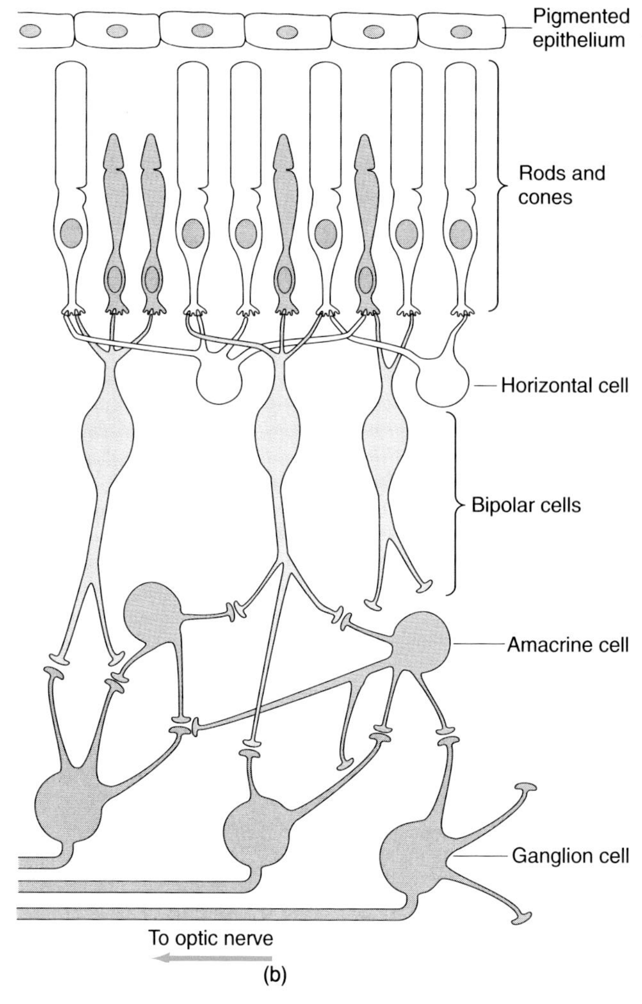

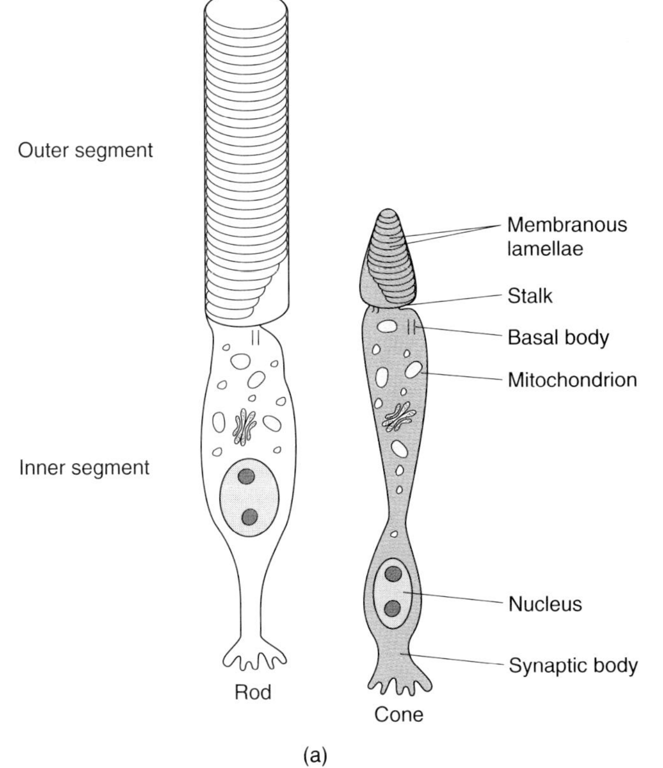

FIGURE 15.12

Rods and Cones

(a) A scanning electron micrograph and a drawing of rod and cone cells. (b) Artist's rendering of the human retina. Action potentials are passed from the rods and cones to the brain by means of a complex pathway, involving horizontal cells, bipolar cells, amacrine cells, and ganglion cells. Our understanding of the full range of functioning of these cells is still very incomplete.

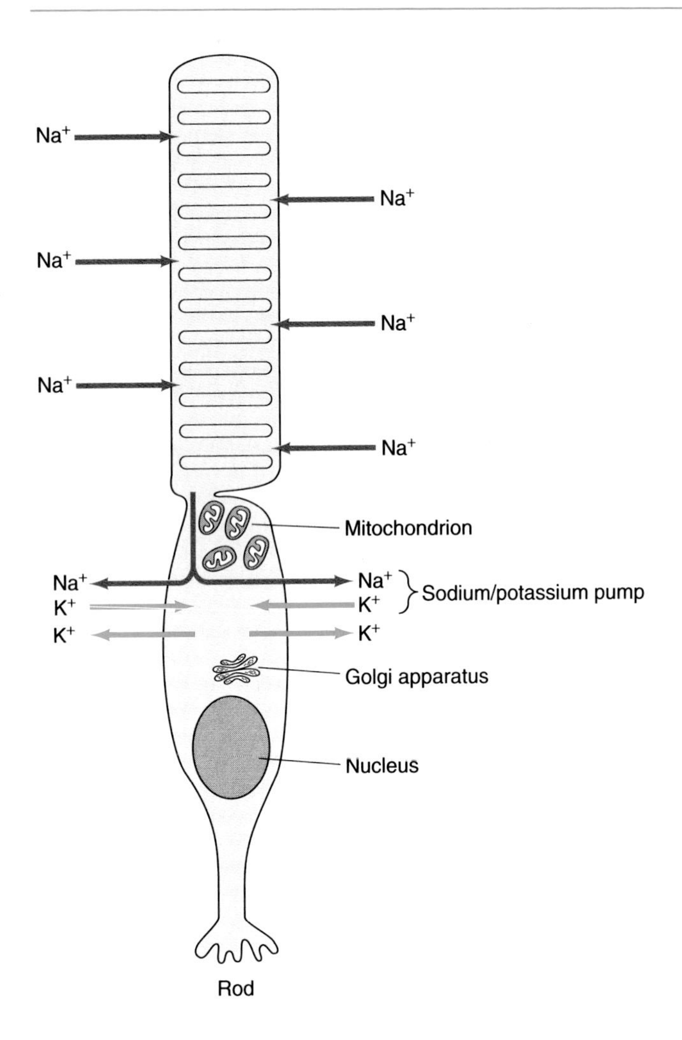

FIGURE 15.13

Resting Potential in a Rod

The outer and inner segments of a rod are physiologically distinct and are anatomically united only by a narrow isthmus of cytoplasm. Many of the sodium ion channels in the outer segment are ungated, the gate being held open by molecules of cyclic GMP. Consequently, sodium ions leak into the outer segment, lowering the value of the resting potential. As sodium flows into the inner segment, however, the sodium-potassium pump moves sodium back out of the rod, and retrieves the potassium that tends to leak out of the inner segment.

duced from the typical value of −70mV to approximately −40 mV.

In each outer segment are some 2000 flattened disks, stacked like a roll of coins. These discs are covered with a total of about 100 million molecules of **rhodopsin** (see Figure 15.14). Rhodopsin consists of a multilooped protein of 348 amino acids, called **opsin,** which is imbedded in the membrane of the disks, and a small molecule derived from vitamin A, called **retinal.** The retinal molecule, which is bent in the middle, sits in the center of the opsin molecule.

Retinal, like chlorophyll (see Chapter 5), is capable of absorbing electromagnetic energy having a wavelength of between 400 and 750 nm (which, of course, corresponds to the visual spectrum), although its peak absorption is at 505 nm. In absorbing light energy, retinal does not lose an electron, as we saw chlorophyll does. Rather, what happens is that the retinal molecule straightens.

When retinal straightens, it changes the shape of the surrounding opsin molecule. In its changed shape, opsin acts as an enzyme, and it sets off a small series of reactions that end in the conversion of several thousand molecules of cyclic GMP to noncyclic GMP. In its noncyclic form, GMP cannot maintain the Na^+ channels open in the outer segment, and more than one million Na^+ are denied access to the outer segment (see Figure 15.15).

As a consequence of the closed Na^+ channels in the outer segment, but with continued action of the sodium-potassium pump in the inner segment, the rod cell as a whole becomes hyperpolarized and assumes an inside potential of some −80mV. This change in

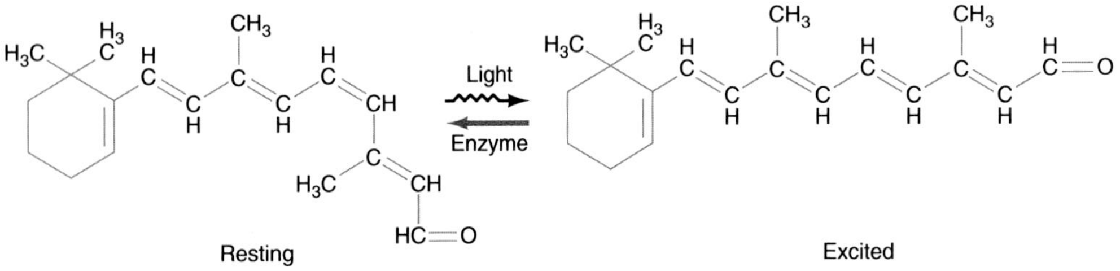

FIGURE 15.14

Retinal Molecule

Retinal is to the eye what chlorophyll is to green plants—the molecule that reacts when struck by photons of light at wavelengths between 400 and 650 nm—the visible spectrum. Retinal responds not by losing an electron but by changing its configuration from "bent" to "straight."

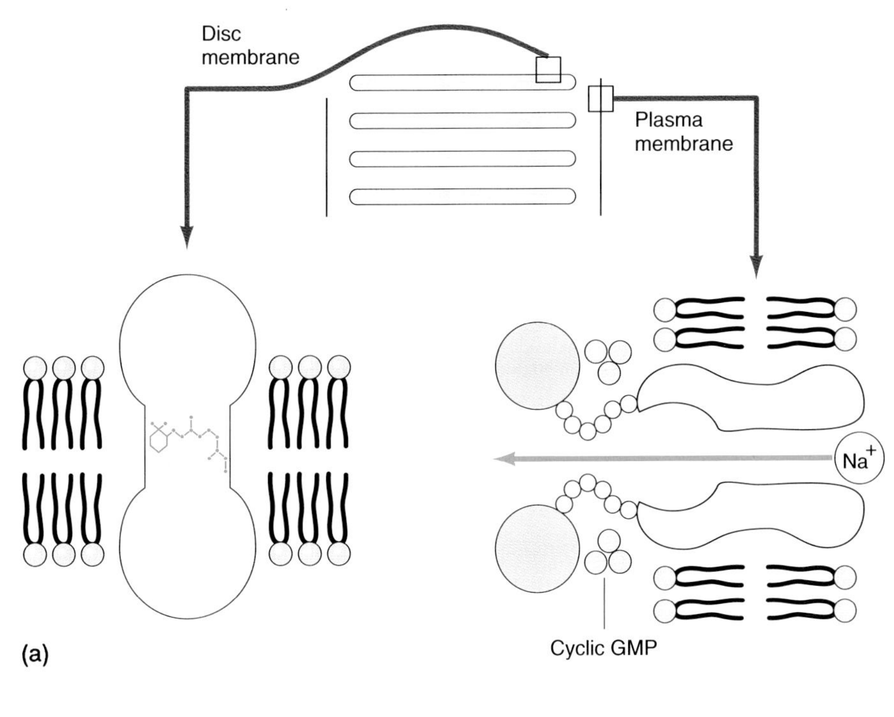

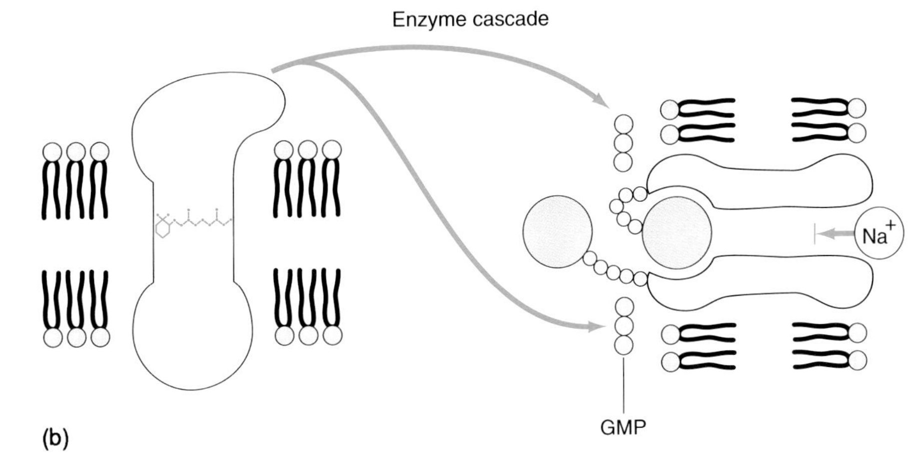

FIGURE 15.15

Action Potential in a Rod

(a) Retinal is located inside a multilooped protein called opsin. Together, retinal and opsin constitute the molecule rhodopsin. Each of the 2000 or so discs in the outer segment of a rod cell has about 50,000 rhodopsin molecules imbedded in the disc membrane. (b) When a photon of light strikes a retinal molecule, the consequent straightening of the retinal forces a configurational change in opsin, activating its enzymatic capabilities. A small cascade of enzymatic reactions then ensues, resulting in the conversion of cyclic GMP to noncyclic GMP. In its noncyclic form, GMP cannot maintain the sodium channels open, but sodium continues to be pumped from the inner segment. As a result of fewer positively charged ions inside the rod, the membrane potential of the rod drops to −80mV, and an action potential is initiated.

potential is conveyed as an action potential to the bipolar cells; the bipolar cells, in turn, transmit it to the ganglion cells and hence to the brain. Enzymes quickly reverse the process by restoring rhodopsin to its original configuration—but the system is so sensitive that a single photon of light, striking just one of the 100 million rhodopsin molecules in a rod, can initiate an action potential.

The rod cells function primarily in dim light, where retinal is abundant. In bright light much of the retinal is temporarily converted to vitamin A, and the sensitivity of the eye is correspondingly reduced. The transformation either of retinal to vitamin A or of vitamin A to retinal is largely accomplished within a minute or two (recall, for instance, the experience of entering or leaving a darkened theater on a sunny day). By means of these transformations, the eye can alter its light sensitivity over a 500,000-fold range.

In contrast to the rods, the cones are responsible for bright light vision and for the perception of color. They have the same overall structure as the rod cells, except their outer segment is smaller and conical. They are approximately 100 times less sensitive to light than are the rod cells—but they respond four times more quickly. Thus our visual acuity is substantially better in bright light.

Understanding color vision is somewhat more difficult. There are three types of cone cells, which respond maximally to light of 430, 535, and 575 nm (for which reason they are called blue cones, green cones,

BOX 15.2

Problems with the Eyes

Because the eyes are very complex organs, it is hardly surprising that they are subject to a number of malfunctions, among which are the following.

Color Blindness. *Color blindness* is due to one of several recessive alleles of genes on the X chromosome (see Chapter 19). Because males have only one X chromosome, the recessive alleles are expressed, and color blindness is therefore much more common in men than in women. Blue color blindness is rare, but 6 percent of men lack the gene controlling the production of green cones, and 2 percent lack the gene for red cones. Both groups, therefore, have difficulty distinguishing red from green.

Myopia, Hyperopia, and Presbyopia. *Myopia* (nearsightedness) and *hyperopia* (farsightedness) generally result because an eyeball is, respectively, too long or too short. In myopia, the light rays reflected by an object are generally focused on a point in front of the retina, although objects near the eye can be focused on the retina because of the accommodating powers of the lens. In hyperopia, distant objects can be seen clearly, but, because of the shortness of the eyeball, the lens cannot accommodate sufficiently to focus objects near at hand. *Presbyopia* is a condition that affects many middle-aged people. They become farsighted because the lens loses the ability to accommodate and thus cannot focus on objects near at hand.

Astigmatism and Cataracts. *Astigmatism* results from a misshapen cornea or lens that causes the focal point to be changed depending on the angle of the light. Consequently, there is no one point of sharp focus. A *cataract* is a clouding or opacity of the lens that frequently develops in older people. Modern surgical techniques permit the removal of a defective lens and (with the aid of very powerful glasses and/or a lens implant) the restoration of reasonable visual acuity.

Glaucoma. The shape of the eyeball is largely maintained by the presence of humors in the eye that are maintained at a pressure of 12 to 20 mm Hg. For reasons not well understood, the pressure of these fluids can sometimes rise to as much as 70 mm Hg. Unfortunately, at pressures above about 30 mm Hg the artery that supplies the retina collapses, and the retinal cells die. This condition is called *glaucoma,* and it is a leading cause of blindness.

Retinitis Pigmentosa. In *retinitis pigmentosa* the retina gradually degenerates from the periphery inward with a corresponding loss of vision. Retinal cells are replaced by cells containing an abundance of the pigment **melanin.** The cause of this condition is presently under intense investigation, and a cure has yet to be found.

Macular Degeneration. *Macular degeneration* is the most common cause of blindness in adults over the age of 65. It results from the proliferation of blood vessels in the **macula** (L., "spot"), the yellowish area surrounding the fovea in the center of the retina. These blood vessels crowd out the rods and cones, sharply reducing visual acuity. Recently, the use of tiny laser "hits" in a grid pattern across the macula was found to destroy excess blood vessels without damaging excessive numbers of retinal cells.

and red cones, respectively; see Figure 15.16). They all contain retinal, but the protein in which the retinal sits is somewhat different in each case, which accounts for the difference in wavelength responsivity. Our perception of color is based, it is presumed, on the relative degree of stimulation of each of the three types of cones that are stimulated by a given image, and the interpretation of this information by the visual cortex of the brain.

From Eye to Brain. Although the chemistry of vision has in recent years been rather well elucidated, the neurophysiology of vision is much less well understood. We have already noted that each eye possesses about 3 million cones and 100 million rods. Since fewer than 1 million optic nerve fibers leave the retina, each nerve fiber must serve an average of more than 100 rods and cones. Such an arrangement seems odd, because it suggests a reduction in visual acuity far below the theoretical maximum. What is the advantage of having more rods and cones than optic nerve fibers?

In the fovea, which is populated almost exclusively by cones, the ratio of cones to optic nerve fibers is almost 1:1—an arrangement that accounts for the fovea's very high visual acuity. However, as one moves toward the periphery of the eye, the rods and cones become larger, and the number of cells served by a

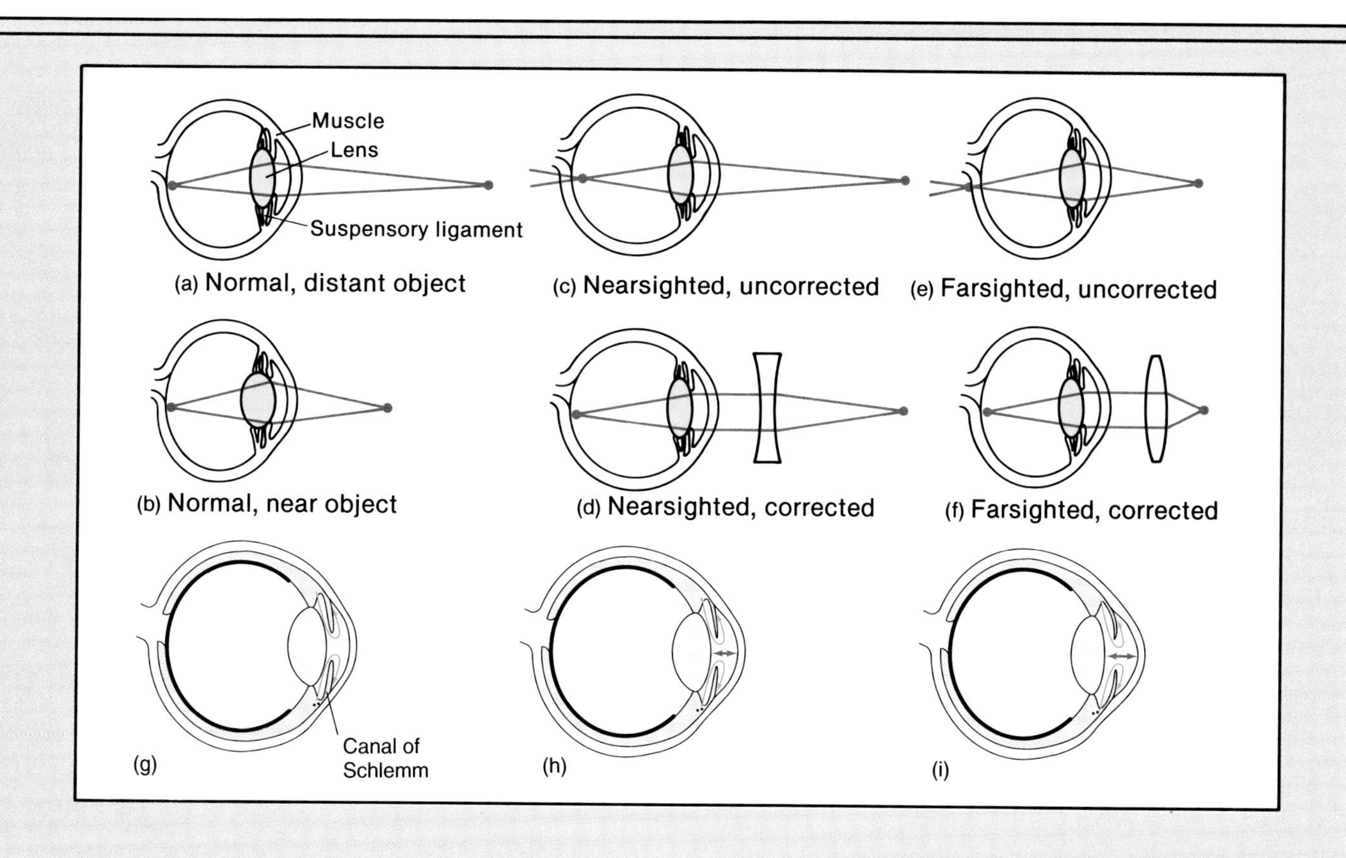

Eye Problems

Diagrams (a) and (b) illustrate accomodation of the lens of the normal eye in focusing on objects both near at hand and distant. In (c) the eyeball is too long for distant objects to be accommodated; a biconcave lens (d) is needed to augment the lens of the eye. In (e) the eyeball is too short for near objects to be accommodated; a biconvex lens (f) is needed to augment the lens of the eye. The aqueous humor drains through a canal and ultimately returns to the blood (g). If the canal is blocked, pressure can increase in the chambers of the eye, a condition known as glaucoma. Some cases of glaucoma respond well to drugs (h); more severe cases of blockage may require surgery (i).

single optic nerve fiber increases. Therefore visual acuity decreases toward the edges of the visual field. However, these regions are very light sensitive because hundreds of rods may be served by a single nerve fiber. The action potentials of these rods quickly surpass the threshold needed to initiate an action potential in the nerve fiber, whereas a much stronger signal would be needed from a cell served by a single nerve fiber.

As just described, the function of the rods and cones in providing the stimuli that the brain interprets as visual images can be generally appreciated—but that description significantly understates the complexity of the retina (to say nothing of the complexity of the visual cortex). Two other types of cells provide cross-linkage in the retina: **Horizontal cells,** which link various rods and cones together before synapsing with bipolar cells; and **amacrine cells,** which synapse with many bipolar and ganglion cells. More than 30 structurally distinct forms of amacrine cells have been described, and in total the amacrine cells use virtually all of the neurotransmitters found in the brain. It is known that some amacrine cells stimulate and others inhibit, and that they are involved in permitting us to detect movement, and to determine the direction of movement. Nevertheless, a great deal of work is yet to be done on the functional architecture of the retina.

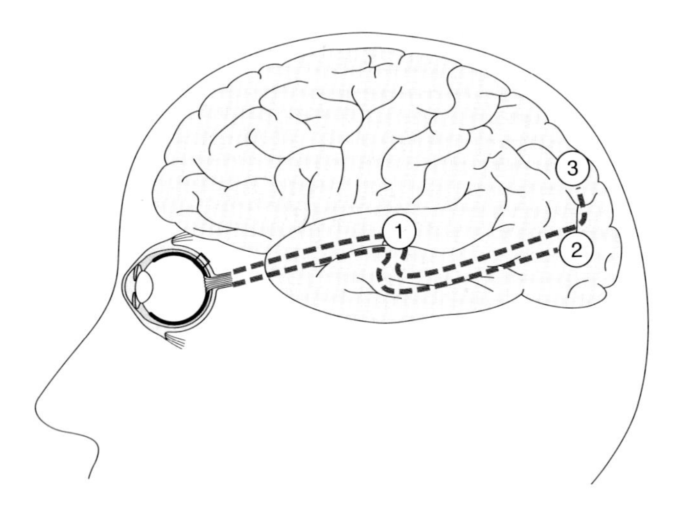

FIGURE 15.16

From Eye to Brain

The optic nerve runs from the eye to a region of the thalamus (1), then to the primary visual cortex at the very back of the cerebral lobes (2), and finally to the secondary visual areas, located toward the top of the brain (3). It is here that the action potentials are interpreted as vision.

Stimuli from the optic nerve of each eye are ultimately relayed to both sides of the brain, which permits us depth of vision. The principal area responsible for interpretation of the signals as visual pictures is a portion of the cerebral cortex located at the back of the head in the occipital lobes of both hemispheres. Thus damage to the visual cortex—for example, in an accident or because of a stroke—can cause blindness, even though the eyes and optic nerve may be unaffected.

Summary

To paraphrase Aldous Huxley, our sense organs are our doors of perception, and relative to most animal species, we are very well endowed. Nevertheless, many species have much greater acuity in some senses than we have, and some species can perceive stimuli of a type we cannot perceive at all.

The sense organs can be divided into mechanoreceptors (for the tactile senses, position sense, hearing, and equilibrium), pain receptors, chemoreceptors (for taste and smell), thermoreceptors (both hot and cold), and light receptors (for vision). Sense organs range from very simple (e.g., the nerve endings of pain reception) to very complex (e.g., the cochlea of the ear or the retina of the eye). In every case stimuli from the sense organs are ultimately conveyed to the brain, where distinct regions of the cerebral cortex identify, process, and interpret the stimuli.

A complex nervous system is a prerequisite for the development of diverse and sophisticated sense organs because the huge number of stimuli produced every second by the human sense organs must be assimilated and used. Thus the price we pay for being able to see and enjoy a beautiful sunset is being obliged to smell the air pollution the next morning.

Key Terms

mechanoreceptor
pain receptor
chemoreceptor
thermoreceptor
light receptor
transducer
accommodation
hair end-organ
Pacinian corpuscle
Meissner's corpuscle
Merkel's disc
Ruffini's end-organ
baroreceptor
muscle spindle
Golgi tendon organ
outer ear
auditory canal
eardrum
middle ear
malleus
incus
stapes
oval window
eustachian tube
inner ear
cochlea
vestibular membrane
vestibular canal
basilar membrane
tympanic canal
cochlear duct
round window
organ of Corti
basilar fiber
tectorial membrane
decibel scale
vestibular apparatus
semicircular canal
utricle
saccule
kallikrein
bradykinin
substance P
mast cell
histamine
prostaglandin
taste bud
papillae
olfactory membrane
sclera
choroid
retina
cornea
aqueous humor
iris
pupil
lens
vitreous humor
ganglion cell
optic nerve
bipolar cell
cone
rod
ciliary muscle
pupillary reflex
fovea
cyclic GMP
rhodopsin
opsin
retinal
horizontal cell
amacrine cell

Box Terms

conduction deafness
nerve deafness
color blindness
myopia
hyperopia
presbyopia
astigmatism
cataract
glaucoma
retinitis pigmentosa
macular degeneration

Questions

1. What are the five classes of receptors found in humans that receive sensory stimuli from the surrounding environment?
2. What is transduction? How does it pertain to the functioning of sense organs?
3. What is adaptation, as used with reference to sense organs? Give an example of sensory adaptation.
4. Why might a blow to the side of the head cause ringing in the ears, even though there was no object in the environment that was ringing?
5. Give three examples of mechanoreceptors. What specific function does each play in the body?
6. Briefly explain how the organ of Corti converts vibrations in the cochlear fluid into action potentials that are sent to the brain.
7. To which of the four major categories of taste are we most responsive? Can you guess why?
8. Briefly describe how action potentials are initiated by rod cells in response to light.
9. Distinguish between conduction deafness and nerve deafness. For which is a hearing aid most beneficial?
10. What changes occur in the eye that lead to the need for reading glasses in so many adults of middle age?

UTAH

CHAPTER

16

Our Behavioral Heritage

Ethology

Why do blind children smile?

What role, if any, does heredity play in human behavior?

Why do humans kill each other with an abandon not shared by any other species?

Why isn't sexual receptivity in women seasonal as it is in the females of virtually all other species?

These are the kinds of questions addressed by one of the newest branches of biology—ethology, the study of the biological bases of behavior.

THE NATURE OF THE PROBLEM

In the last three chapters, we have examined how chemicals called hormones regulate the workings of the body; we have explored communication within and between nerve cells and looked at the ultimate in neuronal complexity, the human brain; and we have considered the various ways we use to detect changes in both our internal and external environments. Underlying all of these discussions has been the biological necessity of regulating and coordinating our various organ systems so as to ensure the maintenance of homeostasis. However, we are much more than finely tuned homeostatic machines. We share with all living organisms the need to *interact* with our environment, not merely to perceive it. For most of our existence as a species, we have needed to find food, water, shelter, and mates, and to avoid becoming the prey of other organisms. We have needed to assimilate sensory data and to react to it. In short, we have needed to *behave.* Thus it is fair to say that behavior and homeostasis are the two outcomes of all this coordination and integration we have been talking about for the last three chapters.

When we consider behavior from a biological standpoint, we are asking questions such as the following:

1. To what extent is our behavior a logical outgrowth of our evolutionary, genetic, and developmental heritage, as opposed to being the consequence of our environment and upbringing?

2. Can we understand anything of human behavior by studying the behavior of animals, or is our nervous system so sophisticated that such studies are essentially meaningless in understanding why we behave the way we do?

3. Are we creatures of free will, and therefore in total control of our behavior and actions, or are we prisoners of our genes, helpless marionettes controlled by the hand of fate and the strings of DNA in our cells?

In discussing these questions, it is important to recognize at the outset that, in western culture at least, we have a long history of overestimating our place in nature. Only 500 years ago, we still faithfully believed that the earth was the center of the universe—the sun, the planets, and the stars all revolved around us. Straightening out this fiction cost a few astronomers their lives at the stake—human society was not anxious to give up its special place in the universe. Then, during the last century, along came a fellow named Darwin who told us that we are an integral part of the biological world, and, like all creatures, are the product of evolution. Rather than there being a "Man and the animals" dichotomy, Darwin told us that we share a biological heritage with all other life forms. Although we have grudgingly acknowledged our true position in the universe, even today some people find the notion of evolution too tough to swallow. That is understandable (albeit intellectually inexcusable)—after all, having once believed that we are special creatures, different in very fundamental ways from all other living things, it is difficult to accept evolution.

Even if we believe in evolution, however, we often still claim independence from the laws that affect all other organisms, an independence, that of course, accounts for much of our lifestyle—overeating, underexercising, misusing drugs, overpopulating the earth, contaminating the environment, and causing the extinction of other species. Only in the last few years have we begun to recognize that we are, in fact, subject to the same rules that dictate how the rest of the biological world must operate.

Yet there is still one area where we can sit smugly and complacently. We are, after all, intelligent and reasoning creatures—and we are, in large measure, responsible for our actions and in control of our behavior. (The modifier "in large measure" is necessary for two very different reasons. First, we have generally acknowledged the work of Freud and his followers who have attempted to show how early environmental experiences can shape—in some extreme cases, misshape—our adult behavior. Second, even when our behavior as individuals is not an expression of our free will, the exceptions occur only because of the intervention of another human being.)

Unfortunately, a few cracks have appeared in our facade of complacency. Many of the mental illnesses that were once thought to be environmentally induced (and which, therefore, were treated by psychotherapy), have been found to have a physiological (that is, genetic) basis, and are now treated by medication. Emotional and mood changes associated with premenstrual syndrome (see Chapter 20), once dismissed as hypochondria, are now thought to have a hormonal basis. The question we therefore need to resolve is not whether genetics affects our behavior; rather it is how *pervasive* is genetic influence on our behavior. To reject the possibility of genetic involvement is to be doctrinaire, and that is not good science. What, then, is the evidence for a genetic impact on human behavior?

ETHOLOGY

Although its roots are very old, **ethology** (the study of the biological bases of behavior) is a relatively new branch of biology. The foundations of this discipline originated with studies of animal species that are much simpler than our own. However, ethology's potential in the understanding of human behavior was dramatically illustrated with the awarding of the 1973 Nobel Prize in physiology or medicine to the three men who are generally acknowledged to be the founders of the discipline—Konrad Lorenz, Niko Tinbergen, and Karl von Frisch. As this potential becomes realized, ethology will play an increasingly more central role in biology in the future.

Ethology and Comparative Psychology

Ethology originated in Europe, and it remains an important force there. In the United States, however, psychology reigns supreme. How do these disciplines differ? Ethologists are primarily interested in four facets of behavior:

1. The adaptive value of behavior
2. The evolution of behavior
3. The development of behavior in the individual
4. The physiological basis of behavior

Psychologists, on the other hand, are much more interested in the last two than in the first two. Moreover, because their interests differ, so too do their theories, terminology, and, most notably, their methods. The methodological differences between psychologists and ethologists can best be summarized by the often-quoted saying that psychologists place an animal into a box and watch it, whereas ethologists climb into the box themselves and look out at the free-ranging animal. Put another way, psychologists generally use a laboratory approach, whereas ethologists generally use observation of animals in their natural habitat, or in a setting which simulates natural conditions.

Basic Ethological Theory

Ethologists consider behavior to be the outcome of the interaction of four factors. These factors, which we shall consider individually, are releasers, fixed action patterns, motivation, and (in many, but by no means all, instances) learning. Collectively, these factors are governed by activities of the nervous and endocrine systems.

Releasers (sometimes called *sign stimuli*) are specific signals that prompt the performance of a specific behavioral response. These signals commonly involve either communication between members of a species, or the detection of prey or the avoidance of predators.

For example, in mating season, male European robins will attack any small reddish object (which usually is the chest plumage of other male robins, but can be a headless ball of feathers positioned by an ethologist). Toads will attempt to swallow any small, linear object which is moving horizontally at a low rate of speed (which usually is a worm or slug, but can be a small stick being pulled by an ethologist). Young chicks will scatter for cover whenever an appropriately-sized shadow moves across the chicken yard (which usually is a hawk, but can be a cardboard model under the control of an ethologist). Male mosquitoes will attempt to mate with any object having the right "feel" and producing a sound of appropriate frequency (which usually is a female with a characteristic wing beat frequency, but can be a tuning fork covered with cheesecloth being held by an ethologist).

The point is that releasers are generally very simple signals, and represent a small subset of the total number of such signals which potentially might be used. Normally, they are very reliable, but our examples demonstrate how readily many animals can be deceived. Why *are* these animals so easy to fool? Why aren't they more analytical?

The answer is that most species of animals, especially nonmammalian species, have relatively simple nervous systems with little (or nothing) in the way of an associative cerebral cortex. Thus the opportunity to be analytical simply is not there—the nervous system is too small and too simple. Instead, in performing a particular behavior, these species are obliged to rely on one or two highly dependable stimuli, and to ignore all of the other possible stimuli.

Very often, releasers and specific behaviors are arranged sequentially, giving rise to a behavioral package which, at first glance, looks impressively complex, but which is really just a series of individually simple behaviors. Consider how the bee fly distinguishes its prey, the honeybee, from all of the other insects flying around. Any flying object about the size of a honeybee releases pursuit behavior by the bee fly. The odor of nectar or pollen emanating from the flying object releases attack behavior by the bee fly. The feel of branched, rather than straight, hairs on the flying object (only bees have branched hairs, an apparent adaptation to the collection of pollen) releases biting behavior by the bee fly. Three sensory modalities (vision, olfaction, and touch) are used to detect three different releasers, releasing three linked behaviors, which collectively result in the identification of the proper prey.

FAPs and IRMs

The type of behavior initiated by a releaser is called a **fixed action pattern,** or **FAP.** (An alternative name in current use is *motor program*). FAPs are characterized by being highly stereotyped behaviors that vary little from performance to performance among individuals of the same species. They are, in essence, an all-or-none response to a particular releaser, and, once initiated, require no feedback during their performance. For example, the movements by which insects groom themselves are FAPs. If, during wing grooming, a fly's wings are carefully cut off, the wing grooming movements continue even though the wings are no longer there. (A less dramatic method of demonstrating the same thing is to observe the behavior of flies which, due to a genetic mutation, are born wingless. They still perform wing grooming movements, despite the absence of wings!)

To our eyes, such behaviors seem pointless—even ridiculous—but ethological theory explains these features of FAPs by assuming the presence of particular neural circuits in the brain called **innate releasing mechanisms,** or **IRMs.** An IRM is activated upon the perception of the appropriate stimulus (i.e., the releaser) and, once activated, the IRM issues the motor signals to the muscles of the body responsible for performing the FAP, whether or not feedback occurs.

Implicit in this discussion of FAPs and IRMs is the notion that these behaviors (or, more correctly, the underlying neural circuitry) are genetically determined and inherited. Thus we see terms such as *instinct* or *innate behavior* being used as synonyms. Unfortunately, these are value-laden terms that mean different things to different people. We shall explore the relationship between genetics and environment and between inherited versus learned behavior shortly. First, however, we must consider some other important features of FAPs.

Motivation and Drive

It would be wrong to conclude, based on our discussion to this point, that animals are automatons, blindly reacting to a series of releasers that randomly present themselves as the animal mindlessly stumbles around the environment. It is important to recognize that, at any given moment, some behaviors have a greater priority than others, and that these priorities can change over time. Escape behavior, for obvious reasons, usually has a very high priority, especially in species subject to extensive predation. Feeding behavior is periodic in most animals, and its priority is a function of the nutritional state of the animal. Mating behavior is generally highly seasonal.

We explain the changing hierarchy of behavior in a given species by invoking the concept of **drive** (i.e., *motivation*). For example, in vertebrates, the hunger center is in the hypothalamus. In a hungry toad, action potentials from the hypothalamus to the appropriate centers in the cortex initiate greater responsiveness to a prey moving in its visual field. Put another way, whether or not a toad attacks a prey animal depends on how hungry it is. Thus a slug will release attack behavior in a hungry toad but will be ignored by a satiated toad. In short, motivation affects the performance of a given FAP.

In birds, an increase in daylight causes a decrease in the level of a hormone that inhibits the production of gonadal hormones. Thus, as the days grow longer in the spring, the levels of gonadal hormones increase, and these hormones are responsible for priming the bird to perform all of the complex behavior associated with the mating season—territorial defense, singing, courtship, nest building, and so forth—most of which are absent during the rest of the year, even though releasers are generally present year round.

Let's examine a particular case, to see the interplay of releasers and motivation. Consider the behavior of a male dove when he meets a female dove in the early spring. The female acts as a releaser for the series of FAPs in the male, which we collectively call the mating dance. It is important that the male's performance be highly stereotyped and typical of the species, for it is through his mating dance that the female recognizes the male as one of her own species and consequently accepts his advances. This is not the time for improvisation. The male who improvises dances alone.

Now suppose that we keep the male caged and alone during the mating season. We will find that the normally high standards of the male begin to deteriorate. After a time, the male will court a stuffed dove placed in his cage. Later, a pigeon will suffice. Still later, the male will respond to a ball of rags. Ultimately, the male (by now certifiably psychotic) will perform the mating dance to an empty corner of the cage. Such actions, in the absence of a releaser, are called **vacuum activities.** (You might have your own name for such desperate behavior.) The performance of vacuum activities, after the prolonged absence of the appropriate releaser, is a characteristic feature of many FAPs. We explain such behavior by suggesting that, where motivation or drive is very high, the IRM can eventually trigger the performance of FAPs even in the absence of a proper releaser.

Consider some additional examples. You are walking through the rain forests of eastern Zaire when suddenly you encounter a group of mountain gorillas feeding on some shrubs. The big silverback male is offended by your intrusion, and he rips a small tree out of the ground and dashes it to bits on the rocks. Then he stands erect and thumps his chest. Even if you had never seen a gorilla before, you would probably conclude that he was not performing in this way because of a sudden hatred of small trees. This is an example of **redirection activity,** in which an FAP (attack behavior), prompted by an appropriate releaser (intruding human), is not directed at the releaser but instead at another object (the small tree). We do the same thing when we kick the wall or punch a pillow out of frustration with a low grade.

Watch a pair of bull elk squaring off for control of a harem of females. Between jousts, they frequently stop to browse on some shrubbery. Are they hungry because of their exertions? On the contrary, such behavior exemplifies **displacement activity,** in which an inappropriate FAP suddenly appears in an out-of-place context. In an analogous fashion, a third grader, asked a question to which he does not know the answer, may respond by scratching his head.

Both redirection and displacement activities frequently occur in conflict situations—instances in which the motivation to attack is evenly balanced with the motivation to flee. The jousting elk, torn between attack and retreat, may do neither for a time but instead engage in a totally unrelated behavior—in this instance, grazing.

In the gorilla example, had you taken a step forward while the gorilla was bashing the tree, you would have forced the issue. You might have had the pleasure of watching this enormous creature flee from your advance. You might also have had the misfortune of having him attack and dismember you. One way or the other, he would have forgotten about the tree, because either flight or fight would have become dominant, and the redirection activity would have disappeared.

FAPs and Reflexes

Ethologists talk about FAPs, whereas psychologists speak of reflexes. How do these two terms differ? Actually, there are many similarities, but there are also differences, and these differences go beyond the fact that the two terms had their origins in different disciplines. To summarize some of the differences:

1. Reflexes occur only in response to a stimulus; FAPs may occur as a vacuum activity.

2. FAPs may appear in other contexts (i.e., as displacement and redirection activities); reflexes never do.

3. Reflexes tend to be simple, often involving only a portion of the body; FAPs are generally complex, and often involve the entire body.

Learning

The last of the factors that interact to produce behavior is **learning.** Learning is much more important in some species than in others and, except for the most rudimentary types of learning, it is dependent on the availability of an abundance of associative neurons, a circumstance best met by the mammalian cerebral cortex.

It is both difficult and arbitrary to establish a satisfactory set of learning categories, for the simple reason that we have only the vaguest sense of what is happening at the neuronal level during learning. We can, however, identify and characterize what appear to be some different types of learning.

Classical Conditiong. Classical conditioning is perhaps best exemplified by Ivan Pavlov's famous experiments on dogs. Pavlov noticed that his dogs began to salivate upon seeing their food. By ringing a bell as he gave the dogs their food, Pavlov conditioned the dogs to salivate in response to the bell, even when no food was served. In short, the dogs learned to substitute one stimulus for another. It was initially believed that classical conditioning was the basis of most learning, but it is now recognized that other learning methods are of greater importance in most species.

Operant Conditioning. As expounded by Harvard psychologist B.F. Skinner, operant conditioning (i.e., trial and error learning) underlies a great deal of learned behavior. A rat learns that, by pressing a particular lever, or by running a maze, food becomes available (see Figure 16.1). In nature, many animals learn how to find, select, and eat a particular type of food by trial and error. Hunting behavior or seed-eating behavior is inherited, to be sure, but perfection of the techniques required by each is learned.

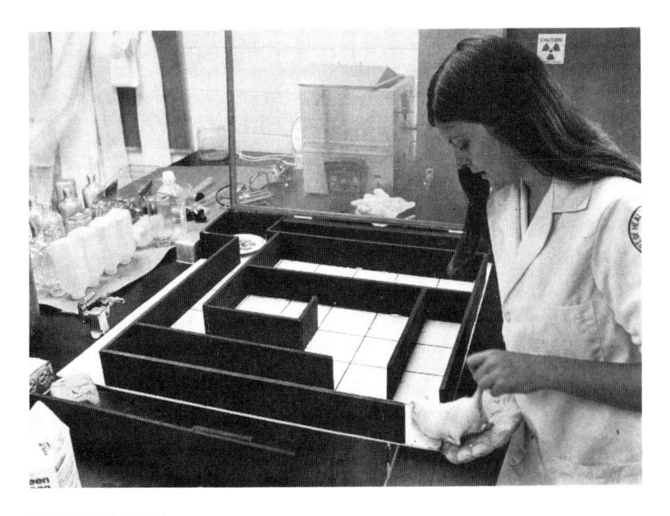

FIGURE 16.1

Operant Conditioning
A rat has learned to associate running a maze with receipt of food.

Habituation and Sensitization. Habituation and sensitization are mirror opposites, but both derive from repeated exposure to a particular stimulus. Habituation is a "screening out" of stimuli that are deemed unimportant in a particular setting. If, for example, you drop a folding chair on a cement floor immediately behind your unsuspecting dog, the odds are very good your dog will reach warp speed before the echoes have died out. However, your dog can *learn* to ignore such stimuli. (Indeed, this particular exercise is routinely conducted with dogs who are being trained in obedience for dog shows. Ignoring sudden, loud noises is essential for successful performance in the noisy environment of a dog show.) There is no pain or punishment that follows the loud noise; therefore, it can be ignored. Conversely, when reward or punishment follows an otherwise innocuous stimulus, an animal may become hypersensitive to the stimulus and respond very vigorously.

Imprinting. There are some behaviors that look a great deal like FAPs—they are species-specific, involve simple releasers, are highly stereotyped in their performance, and are not easily modified—yet they cannot easily be explained as inborn. For example, how does a newborn lamb identify its mother from the rest of the flock (and vice versa)? How does a spawning salmon find the particular place in the particular stream where it hatched from an egg many years earlier? How does a homing pigeon know where "home" is?

These behaviors all exemplify a special form of learning known as **imprinting.** Imprinting takes place during a brief period of time in an organism's life (sometimes only minutes in length) called the **critical period.** It does not require reward or reinforcement and, once imprinted, the behavior is permanent and irreversible. Young geese, for example, follow the first moving object they see (especially if it makes a noise) during a ten-hour period beginning 18 hours after hatching. Normally, this object is the mother, but it can be an ethologist (see Figure 16.2)—or even a box containing a ticking clock pulled around by a string. At maturity, imprinted geese use the object of their imprinting as a model for mate selection—and if the object is an ethologist, the results can be embarrassing for both the ethologist and the geese!

This kind of learning is by no means limited to geese. Many species of birds and mammals learn many behaviors by imprinting—including species-specific songs and calls, the location of desirable nest sites, or the characteristics of suitable mates. Consider the development of singing in the white-crowned sparrow. The young bird must hear the species-specific song during a critical period in the first two months of life. Several months later, it must be able to hear itself perfect its song. If the bird does not hear its species song during the critical period, or if it is deafened before it can hear itself sing, it never perfects the species song. However, if it is deafened after it perfects its song, it continues to be capable of singing perfectly.

FIGURE 16.2

Imprinting

Under natural conditions, young ducks and geese faithfully follow their parent. However, under experimental conditions, they may imprint on the first moving object they see in their first day of life. Here, the object of their affection is Konrad Lorenz, one of the founders of ethology.

Cultural Learning. Cultural learning involves the transmission of specific information from generation to generation. As you might surmise, this kind of learning is largely restricted to species where extended parental care is the norm. Young chicks learn from their mothers which seed-like objects are edible. The young of many species of birds learn to recognize enemies by relating the appearance of a novel animal to the utterance of a warning call by older birds.

Sometimes cultural learning can be seen evolving. For instance, in Britain during the 1940s a particularly brainy member of a species of bird with the intriguing name of "great tit" learned how to pry the caps off milk bottles left on doorsteps and thus to gain access to the milk (see Figure 16.3). Within a few years, milk theft had become endemic, as flocks of great tits, their milk bottle cap flipping behavior honed to perfection, descended on unsuspecting British doorsteps. (Some even began following the delivery trucks!) Obviously, this learned behavior had been taught to successive generations of great tits by the parents of the younger birds.

Cultural learning is particularly prevalent in social primates, with its ultimate development being reached in humans (as exemplified by your presence in this biology course).

Inductive Reasoning. Inductive reasoning is so sophisticated a type of learning that it was long considered to be restricted to humans. It involves insightful behavior—that is, the relationship of abstract concepts—in novel situations. The manufacture of spear points and arrow heads from flakes of rock is an obvious example, as is tool making in general. Unfortunately, our collective ego was dealt a cruel blow with the discovery of tool use by a variety of other species. Sea otters crack open clams by bashing them against rocks that they position on their chests; sea gulls achieve the same end by dropping clams onto rocks from the air; a species of finch uses thorns to extract insect larvae burrowing in trees; chimpanzees strip the leaves off small, straight branches and use the branch to impale termites in a shish kebab fashion (see Figure 16.4); chimpanzees also use dry leaves to sponge up water. While it is true that no species comes close to humans in the extent of its inductive reasoning, we cannot claim sole ownership of this type of behavior. Thus we must distinguish ourselves behaviorally from other species on quantitative, not qualitative, grounds.

FIGURE 16.4

Inductive Reasoning
A chimpanzee uses a leaf as a drinking vessel.

FIGURE 16.3

Cultural Learning
Raccoons have learned to open garbage cans to obtain food.

HUMAN BEHAVIOR

Now that we have considered the various components of behavior from an ethological standpoint, it is time to attempt to relate ethological theory to human behavior. However, because this is a highly controversial subject, we must proceed methodically.

Nature Versus Nurture

A major debate that has only recently begun to settle down has pitted the ethologists against the psychologists (with some defections from both camps) on the question of whether all human behavior is learned (sometimes phrased as "environmentally determined") or whether some is genetically programmed and inherited.

There are two arguments that have been used to dismiss the importance of genetically determined behavior in humans. First, there is the observation that genes function not in a vacuum but rather in a variable environment. It therefore follows that any characteristic, whether structural or behavioral, must always be the result of an interaction between genetics and environment. Second, it is pointed out that genes code for proteins, not behavior (see Chapter 18), and therefore we cannot properly speak of a gene for a given behavioral trait. Let's consider each of these objections in turn.

Of course, it is true that genes and environment interact during the developmental process, with both contributing to a particular end product. However, that is not the same as saying that the relative contribution of each is equivalent in every instance. For example, such anatomical characteristics as weight, height, and number of limbs differ in the relative importance of genetic and environmental influences, and we should expect nothing less from behavioral traits. Of the three characteristics just mentioned, we accept as a matter of course that adult weight is largely determined by environmental factors, whereas we acknowledge virtually total genetic control over the development of limbs. The fact that drugs such as *thalidomide* (see Chapter 19) can alter the environment (in this case, the intrauterine environment) so severely as to prevent normal limb development does not alter the fundamental genetic control of this trait.

A more useful way of looking at the interaction of heredity and environment is to note that genes set developmental limits within which environmental modification is possible. Sometimes these limits are very broad—we can use our network of associative neurons in all manner of novel ways when we engage in inductive reasoning. Sometimes these limits are narrower, as in imprinting where the influence of the environment is restricted to a brief period. Finally, these limits are sometimes very narrow indeed, as in FAPs where only a very aberrant environment during development will alter the behavioral program.

The second argument is really an attempt to trivialize the distinction between genetically determined behavior and learning. We readily talk about the gene for blue eyes, or the gene for sickle-cell anemia as a convenient and useful shorthand. We *know* that genes code for proteins, and the end product (blue eyes or sickle-cell anemia) is many developmental steps removed from these proteins—but we also know that these developmental steps are themselves highly ordered. A genetic error in the manufacture of hemoglobin is not *casually* linked to sickle-cell anemia; it is *causally* linked, in a direct and inevitable manner.

Certainly it is true that behavior has a temporal component—unlike blue eyes it is present one minute and gone the next—but if every time it occurs, it is precisely the same (i.e., an FAP) then it can be said to be as constant as an anatomical trait, and as clearly linked to a genetically controlled developmental process.

A recently reported ten-year study of twins separated at infancy and reared apart sheds light on the general question of the role of heredity in human behavior. The study compared the variance in behaviors between identical and fraternal twins. About 70 percent of the variance in IQ was found to be genetic, and on multiple measures of personality and temperament, occupational and leisure time interests, and social attitudes, separated identical twins were about as similar as identical twins reared together, suggesting a strong genetic component to such complex personality traits.

Do Humans Have FAPs?

The knee-jerk reaction to this question is an emphatic no. We do not want to acknowledge anything less than total cerebral, voluntary control over our behavior—any more than our great-great-grandparents wanted to acknowledge evolution or our ancestors in the sixteenth and seventeenth centuries wanted to acknowledge that the earth was not the center of the universe.

However, the answer is very important to us, for at least two reasons. First, we humans can always use a little humility. If we constantly think in anthropocentric (human centered) terms, we will never be willing to acknowledge the constraints that physical and biological laws place on us, nor will we gain the perspective necessary to understand our place in nature. Second, we must be willing to recognize and to compensate for whatever genetic heritage we have that does not serve us well in a modern age. Many ethologists hold that humans are innately aggressive, because we are so prone to kill others of our species. If this is true, we must explore ways of channeling such energies into something more productive than wars.

The notion of FAPs in humans is gradually becoming more widely accepted, although there is still great resistance to the idea by many social scientists as well as by lay people. Ethologists marshal three types of evidence to defend their position.

Different Populations. Until relatively recently in our history as a species, most human populations were very sedentary, and had only limited contact

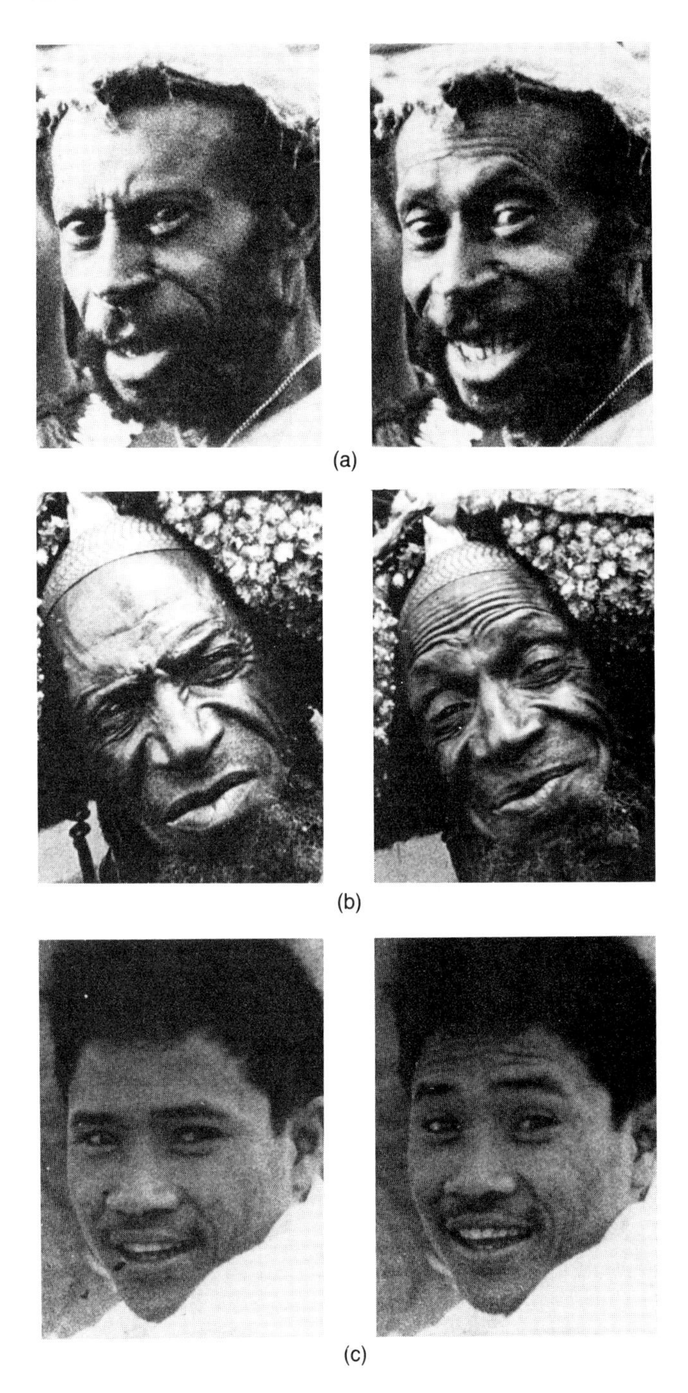

FIGURE 16.5

Eyebrow Flash

A quick up-and-down movement of the eyebrows as a greeting response is shown in (a) a Balinese from the island of Nusa Panida; (b) a member of the Huri tribe of New Guinea; and (c) a member of the Woitapmin tribe of New Guinea. In all likelihood you use the same gesture yourself.

with other populations. In the absence of such contacts, the presence of common behavioral traits used for the same purpose by widely separated populations suggests a genetic basis for such traits. In other words, they are FAPs.

Several such traits have been described, of which the eyebrow flash may be the most interesting (see Figure 16.5). You have probably used this behavior yourself in situations when you are engaged in conversation with a friend and an acquaintance walks by. Since you may not want to interrupt your conversation or to snub your acquaintance, you may well acknowledge recognition of your friend by flipping your eyebrows up and then down quickly. (If the person is someone you don't like, you may flip something else, but that signal is not universal—yet.)

This use of the eyebrow flash as an affirmative greeting is used widely, though not universally, by many human societies, most of which have had little contact with each other. A similar argument can be made for the embarrassment response, smiling, frowning, pouting, and so forth.

Different Species. FAPs are species-specific, meaning that the particular behavior is performed by virtually all members of the same species. However, species specificity does not mean species exclusivity. Much of the fascination of the monkey cage in zoos stems from the fact that the expressions and behaviors of monkeys commonly mirror our own. These similarities are so striking that we can frequently identify the context of a particular behavioral interaction from photographs of primate facial expressions (see Figures 16.6 and 16.7).

How did this similarity among different species arise? Did we learn expressions from monkeys, or did they learn them from us? Neither alternative is tenable. A more reasonable explanation is that these behaviors were present in our common ancestors and some have been retained in both monkeys and humans. This conclusion naturally suggests genetically determined behavior—that is, the presence of FAPs in humans.

Human Development. The newborn infant exhibits many highly stereotyped behaviors, including searching movements for the mother's breast and clinging (see Figure 16.8). Walking and swimming movements can also be demonstrated, although these behaviors soon disappear, because the infant's overall behavior regresses during the first three months. Prematurely born children may be even more adept at such behavior. Regardless of whether these behaviors are sophisticated reflexes or FAPs, they are unquestionably not learned.

Deaf-blind children are isolated from the sensory input of two of the primary senses and therefore many problems are involved in teaching them. (The education of Helen Keller provides a notable example.) Despite these learning problems, children who are both deaf and blind from birth nevertheless develop

(a)

(b)

FIGURE 16.6

Commonality of Behavior in Primates
Tense mouth face (used as a confident threat expression) in (a) a gorilla and (b) a politician.

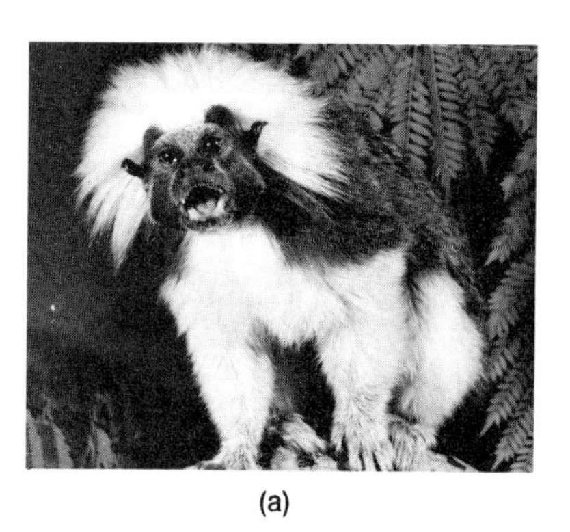

(a)

(b)

(c)

FIGURE 16.7

More Commonality of Behavior
Open mouth threat in (a) a cotton-top marmoset; (b) a red uakari; and (c) a politician.

the normal complement of facial expressions for the major emotions, including smiling, laughing, frowning, pouting, and so forth (see Figure 16.9). Many of these expressions subsequently regress, presumably because of the lack of feedback. Because regression occurs due to the *absence* of environmental stimulation, it is difficult to explain the acquisition of these behaviors in the first place as being due to the *presence* of environmental stimulation (i.e., learning). On the contrary, the ethologist would say that these are genetically determined FAPs that became manifest at the appropriate developmental stage.

Another type of developmental evidence can be found in comparisons of the developmental sequences of behaviors in humans and other primates (see Figure 16.10). Many of the same behavioral traits occur in the same sequence in different species, even though the age of first appearance may differ among species.

In balance, three independent lines of evidence support the presumption of genetically determined FAPs in humans. The fact that FAPs have a genetic basis does not, of course, mean that we *must* perform them once the proper releaser is presented. We have conscious control of our emotions, to at least some degree, and we can control FAPs such as those for facial expressions equally well, although some of us are better at it than others. Those exceptionally adept individuals who can feign various emotional expressions at the drop of a hat go on to become great actors—or great con artists!

SOCIAL ORGANIZATION

So much for the behavior of individuals. What does ethology tell us about interactions among groups of individuals of the same species (see Box 16.1)—especially when that species is our own?

Because of inferences that can be drawn regarding human social organization, it is only logical that we should be interested in the social organization of other

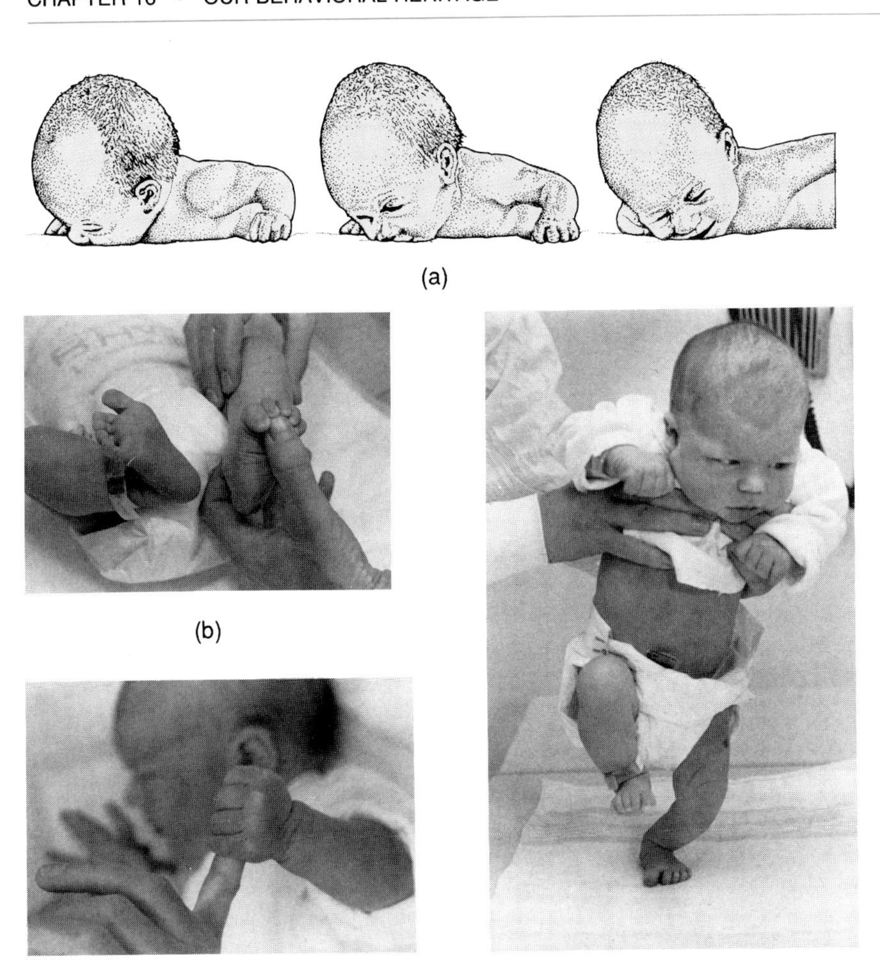

FIGURE 16.8

Innate Behavior in Newborns

(a) Rhythmic searching for the mother's breast. (b) The plantar reflex (curling of the toes). (c) Grasping when presented a finger. (d) Walking movements.

primates. In addition to our interest in the organization of these societies, we are also interested in how the societies arose and are maintained. Foremost among the mechanisms that maintain societies is communication, which operates as a social adhesive.

Communication

Communication may involve the use of the following sensory modalities.

Olfaction. Odor is a very common method of communication in many species, as you are no doubt aware if you have ever tried to walk a male dog past more than three consecutive vertical objects. Chemical communication has the advantage of being long-lasting and unambiguous, but it is not an effective method of conveying a succession of messages quickly. (Imagine trying to carry on a "conversation" by wafting odors back and forth!) Nevertheless, this modality is the one of choice in many mammals because most species use only a small number of messages.

Audition. Except for a few species of insects, sound communication is essentially a vertebrate invention. In order to be effective, sound communication requires not only a structure with which sound can be perceived (an "ear" of some sort), but also requires some apparatus that is capable of producing a variety of sounds. As it happens, most species have a very limited "vocabulary." For example, frogs can produce only three or four different sounds, birds generally do not exceed ten, and even apes have fewer than 30 distinct sounds.

Sound has the advantage of being instantly perceivable over a broad area, and it is highly effective even in wooded areas, where visual communication is limited. Typically, it has a "here I am" function, which serves to intimidate others of the same sex, and to lure members of the opposite sex. In primates, sound is commonly used to reinforce visual communication.

Vision. Visual cues disappear instantly, and are effective only over a short range (one has to be seen). Moreover, they are easily disrupted in a forest setting.

FIGURE 16.9

Innate Behavior in a Deaf-Blind Child
Laughter is a behavior that this girl could not easily have learned and which is therefore presumably innate.

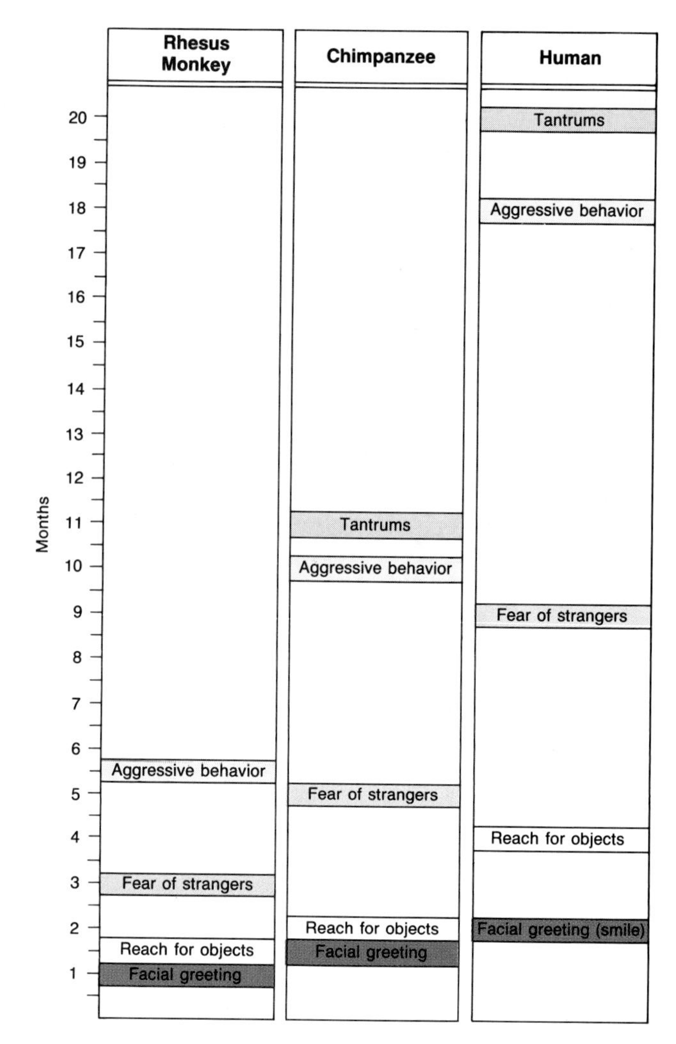

FIGURE 16.10

Development of Behavior in Primates
A comparison of the age at which particular behaviors first are manifested in rhesus monkeys, chimpanzees, and humans. It is apparent that many of the behaviors are similar, and they occur in the same sequence, but they do so much more slowly in humans than in other primates.

Nevertheless, they are the most important method of communication in most primate societies. As we move through the vertebrates, to mammals, primates, and ultimately to humans, we find that the facial musculature is increasingly more refined, allowing increasingly greater numbers of different facial expressions. Over 30 such expressions are known in humans, for example, and they collectively communicate a whole range of feelings and emotions.

Ritualization and Aggression

With increasing complexity, the number of interactions between members of the society necessarily increases. Some of these interactions are encounters over a limited resource such as food, resting places, or members of the opposite sex. Many mammals are equipped with lethal weapons of offense or defense, such as elongated canine teeth, claws, antlers, and hooves. It is obvious that many encounters could rather easily result in bloodshed and death, results that are undesirable if a species is to survive beyond a single generation.

However, many of these interactions have become *pro forma,* and the amount of physical damage is consequently limited. Over time, various displacement activities and **intention movements** (so-called because they anticipate an action, as when a bull paws the ground before charging) have evolved new communicative purposes. Such behavior is called **ritualization,** a highly exaggerated stereotyped behavior that serves a purely communicative role, very frequently in courtship, aggression, and, in social species, in the establishment of dominance hierarchies (see Box 16.2).

The effectiveness of ritualized behavior in communication, both in terms of avoiding ambiguity and of reducing the likelihood of injury, is impressive. If you were facing that gorilla who was bashing a tree around, you would know enough not to assume that he was using it to shoo mosquitoes. His message is very clear, not only to you, but also to other gorillas.

BOX 16.1

Sociobiology

"Sociobiology" is the title of both an emerging field of behavior and a seminal book by Edward O. Wilson of Harvard University. The book gave rise to the field, and both are extraordinarily controversial.

Wilson advanced the notion that the social structures of both humans and other animals are at least partially determined by genetics, an idea which, by itself, is far from heretical. However, as happened with Darwin, interest groups began using this idea to further their own purposes, and twisting it to mean something that it did not. Some people found racial overtones in Wilson's thesis, because a few individuals who have argued that whites are more intelligent than blacks used Wilson's notion of a genetic basis for some human behaviors as "evidence" that trying to teach blacks to learn as much as whites was fruitless. Marxists, who believe that the human social structure can be perfected, saw Wilson's thesis as an impediment to the acceptance of their beliefs, and attacked him accordingly. Still others saw a resurgence of the Social Darwinism of the last century, itself a regrettable corruption of Darwin's theory.

Despite all the controversy, sociobiology is coming of age. It will continue to be controversial, as it confronts dearly held notions, including the idea that men and women would be behaviorally identical but for the role enforcement dictated by an oppressive modern society. (The jury is still out on that question, but many people have already formed unshakeable conclusions.)

A dog or wolf will thwart an attack by another of its species by exposing its neck or belly (the two most vulnerable areas) to its aggressor. Male bighorn sheep always "attack" each other horn to horn, rather than broadside.

What about ritualized behavior in humans? Some authorities maintain that humans have developed very little ritualized behavior. Yet we all recognize a clenched fist, a scowl, or a finger wagged under the nose as signs of anger and hostility, and to some extent, these behaviors are mitigated by appeasing gestures such as an open hand, a smile, or looking away.

The fact is, however, that, as a species, we have only recently perfected our capacity for lethal behavior. It is rather difficult to beat someone of your own size to death with your fists, but it takes nothing more than a twitch of your finger to put a bullet into his brain. Therefore a more correct view of ritualized behavior in humans would be that we have never evolved the level of ritualized aggression and appeasement gestures necessary to deal with our recent technological advances. Moreover, because of our ability to override genetically based behavior by conscious thought, what little ritualized behavior we have is all too often subverted by appeals to "the fatherland," or "the flag," or by the swish of new uniforms.

We can hardly evolve effective ritualized appeasement gestures overnight, but by comparing our actions with those of lethally equipped animals, we can (and should) recognize our shortcomings and try to compensate for them.

Altruism

Altruism is defined as an action that does not serve an individual's immediate self interest—such as risking your life to save a drowning stranger. It seems to be the very antithesis of territoriality and aggression, both of which characterize selfish behavior. Yet altruism is common in many social species, including our own. Leaving aside moral judgments for the moment, selfish behavior makes sense evolutionarily, since selfish behavior can be expected to result in increased numbers of offspring. Self*less* behavior, on the other hand, would tend to reduce reproductive success (to the extent that one risks one's life), and therefore, if altruistic behavior has a genetic basis, we would conclude that, over time, it would gradually be eliminated from a population. Since that has not happened, does it mean that altruism is a learned behavior?

The answer to that question is no—at least in many species. Most honeybees are sterile workers who give their lives in defense of the colony. This behavior is certainly not learned. It has a genetic basis and is explained by the concept of **inclusive fitness,** which points out that enhancing the reproductive success of close relatives can result in many of the individual's own genes being passed on to the next generation. Thus if at the cost of my own life I save the lives of my six brothers and sisters, with each of whom I share 50 percent of my genes (on an average), I am leaving behind three times as many of my own genes (6 × 50 percent) as I would if I had saved myself at the cost of their lives.

BOX 16.2

Ritualized Behavior and the Cuteness Response

A component of ritualized behavior that deserves a special note is that of the nature of interactions between adults and the young. Adults generally do not attack the young of their own species, especially in the more social species. How are the young distinguished from the adults, other than on the basis of size?

The answer is that the juvenile's head and body shape is markedly different from that of the adult in most species of birds and mammals. The juvenile profile evokes what has been called the **cuteness response** in humans, and it is arguable that some similar response is operating in other animals as well.

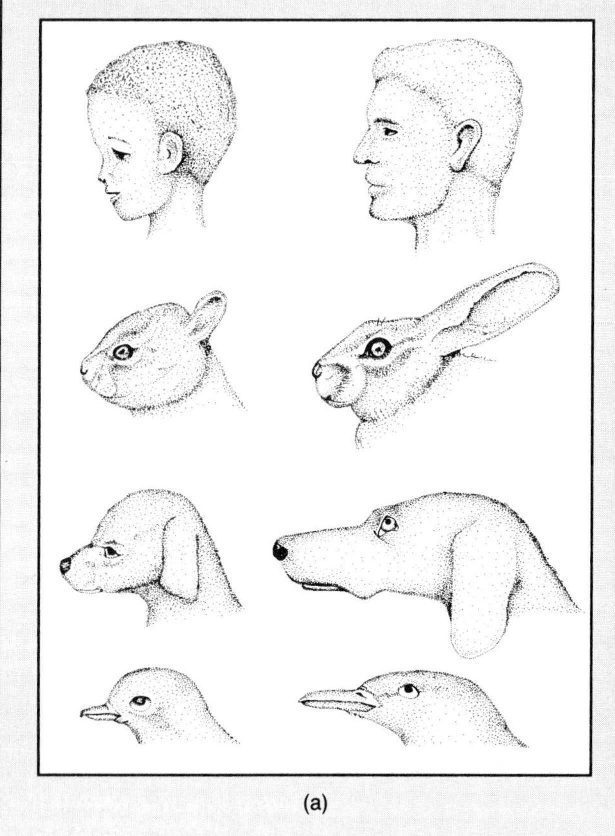

(a)

(b)

Cuteness Response

(a) Contrast in juvenile and adult profiles. (b) Mother and infant chimpanzee. Cute, isn't it?

Workers in a honeybee colony are sisters. If the reproductive success of their mother, the queen, is great enough (in terms of the number of new queens she produces, all of which are sisters to the workers), then it makes evolutionary sense for other sisters to forgo their own chance of reproductive success and instead to behave altruistically.

What about societies, including our own, where individuals are not closely related? First, we should note that our highly diverse populations are a recent development. Only a few thousand years ago, human populations were organized into genetically homogenous tribes—and altruism was presumably far more common within a tribe than between tribes. Thus altruism in modern humans may be a carryover from a time when being altruistic benefited kin.

Second, we must also note that altruism can be reciprocal. Grooming behavior is common in many social primates, but it is a mutual activity—quite literally, a case of, "I'll scratch your back if you scratch mine." Thus altruism, despite the cachet we tend to give it, ultimately may be a selfish activity, offered only in the expectation of future benefit. The genetic basis of this behavior in humans is, understandably, a matter of hot debate at present.

The Evolution of Human Social Organization

There are many dangers in extrapolating from the behavior of our fellow primates to the behavior of humans. However, if we proceed cautiously, we can make some educated guesses about the evolution of human social behavior over the past several million years. It is generally accepted that three changes must have occurred, each of which represents a dramatic departure from anything observed in modern nonhuman primates. These include the construction of many kinds of tools, the development of a sophisticated language, and the formation of the pair bond.

Use of Tools. Tool use in prehumans can be traced back more than one million years, when the long bones of prey animals were split and the sharp edges used as scrapers. Stone tools date back several hundred thousand years, and they show increasing sophistication as we move toward the present.

Tools were important to the evolution of human social structure because they ultimately allowed a division of labor within the society. Such an arrangement is unlike any of the nonhuman primate societies, wherein, despite the appearance of sophisticated interactions, in the final analysis it is every ape for himself. With the exception of nursing infants, all individuals must be capable of foraging for themselves or face starvation. In short, there is no social security for elderly apes.

Presumably, prehumans underwent this same stage, but the development of tools meant that an individual too old to hunt, for example, could still be useful to the society by making weapons and other tools. Because of this utility, other members of the group would presumably be willing to provide him or her with food.

Communication. Apes and monkeys can and do use sounds for communication, but it would be erroneous to conclude that they have developed a language. No species save our own possesses all of the design features of language.

The time at which language development began in prehumans is unknown, but the argument has repeatedly been made that the advantage of language was so great that it provided the selective pressure for the evolution of the large brain that characterizes humans.

Human evolution is discussed in Chapter 23, but it is interesting to note that a major difference between modern humans and Neanderthals is the shape of the skull. It has been suggested that the shape of the Neanderthal skull, especially the formation of the mouth region, was such that little verbal refinement would have been possible. If true, they would have been at a considerable disadvantage to Cro-Magnons (our immediate ancestors) who were presumably much more articulate. This disadvantage may explain why the Neanderthals disappeared so suddenly in the fossil record. (This argument has recently been challenged with the discovery, in a Neanderthal fossil, of the tiny bones used to support the larynx, suggesting that their laryngeal structure was similar to our own. Perhaps Neanderthals had verbal facility after all.)

Recent research suggests a strong genetic component to the acquisition of language. Shortly after birth, infants can distinguish among 40 consonants, including many not present in their native language. This ability is lost later in childhood, as the child begins to use the consonants representative of the native language. Deaf children begin to babble about the same age as children who are not hearing impaired, suggesting that the beginnings of speech occur as a result of a genetically programmed developmental sequence, not because of encouragement from doting parents. Without auditory feedback, children who are born deaf do not learn to speak. However, if deafness occurs only after the development of speech, the individual retains the ability to speak even without auditory feedback. This pattern mirrors the acquisition of bird song, discussed earlier.

Many of the neural circuits responsible for language facility have been identified in the human cerebral cortex. One area of the brain, known as **Broca's area,** is responsible for grammatical refinement. If this area is damaged by a stroke or through injury, the individual is able to speak, but in a grammatically incorrect manner. Damage to **Wernicke's area,** which is responsible for the storage of both written and spoken language, results in the individual's speaking grammatically correct nonsense. Still other centers of the brain are responsible for imbuing speech with emotional overtones. Damage to these areas results in speech that is correct in form and meaning but spoken in an emotion-free monotone. The point is that the human brain is genetically pre-wired (in terms of neuronal circuitry) for the acquisition of language. Facility in a particular language is a learned behavior that is built into these circuits, in a way very analogous to the manner in which many birds acquire their species-specific song.

Pair Bond. Although many species of birds and mammals, including some primates, mate for life, pair bonding does not occur in nonhuman primates living in colonies larger than family groupings. Moreover, the absence of the pair bond in the other social primates seems to hinge on the relative absence of sexual receptivity by the female. As in most animals, sexual receptivity in the overwhelming majority of nonhuman primate females is confined to brief periods just prior to ovulation. The dominant male in a baboon troop or a group of gorillas consorts with any female who is ovulating and who is, as a consequence, sexually receptive.

In contrast, the human female is, at least in theory, continuously sexually responsive. Consequently, a permanent association between a given male and female might well have arisen because of this constant sexual receptivity. Pair bonding would have had a profound effect on the social structure of prehumans, for it would have permitted increased division of labor. It is likely that the females stayed in camp, cared for the children, and gathered food found growing nearby, while the males cooperatively hunted prey too large for any one male to hunt alone. Food would then be shared, depending on whether the males or the females were the more successful.

With the females leading a relatively stationary existence, as opposed to being constantly on the move as in nonhuman primate societies, more prolonged child care became possible. Ultimately, the infant became increasingly more dependent on the mother, in the sense that adulthood was not achieved for many years. The pattern was thus established by which extended periods of learning could take place between generations.

In sum, the development of tools, language, and pair bonding were presumably the events responsible for the evolution of a highly integrated and cooperative society, with a degree of sharing and a division of labor among the members totally unknown in nonhuman primates. The evolution of this complex society in turn gave rise to culture—the transmission of information and beliefs between generations, without which we would be compelled to rediscover fire and the wheel with every generation.

Summary

Ethology is the study of the biological basis of behavior. Because of its biological basis, ethology focuses particular attention on the adaptation and evolution of behavior, in addition to its physiology and development.

Ethologists pay particular attention to a type of genetically determined behavior called fixed action patterns, and they have amassed considerable evidence suggesting that FAPs occur in humans.

Ethologists have also been very interested in the role of behavior in the evolution of human societies. Tool use, communication, and the pair bond all are deemed of critical importance in the evolution of human social organization. Unlike many species, however, we show relatively little ritualized behavior, which may account for our propensity toward violence.

As much as any branch of biology, ethology tells us that we differ in degree, but not in kind, from other animals. This revelation should be greeted not with dismay, but instead with an acceptance of our own shortcomings and a recognition of the need to compensate for them. Thus, when we become iconoclastic, we must take care that the beliefs we shatter are not those that make us what we are but instead are those that hold us back from what we might become.

Key Terms

ethology
releaser
fixed action pattern (FAP)
innate releasing mechanism (IRM)
drive
vacuum activity
redirection activity
displacement activity
learning
classical conditioning
operant conditioning
habituation
sensitization
imprinting
critical period
cultural learning
inductive reasoning
intention movement
ritualization
altruism
inclusive fitness
Broca's area
Wernicke's area

Box Terms

sociobiology
cuteness response

Questions

1. To what extent do you believe our behavior to be a logical outgrowth of our evolutionary, genetic, and developmental heritage, as opposed to being the consequence of our environment and upbringing? Defend your answer.
2. What are releasers? Give an example.
3. Describe the relationship between FAPs and IRMs.
4. Describe the relationship between drive and vacuum activities.
5. Distinguish between redirection activities and displacement activities.
6. Distinguish between classical conditioning and imprinting.

7. Some authorities distinguish humans from other species based on our use of cultural learning. Is this a valid criterion? Why or why not? Is the use of inductive reasoning a better criterion? Why or why not?
8. What evidence exists to support the notion that humans have FAPs?
9. What is ritualized behavior? Give an example.
10. What is altruism? How does it relate to human behavior?

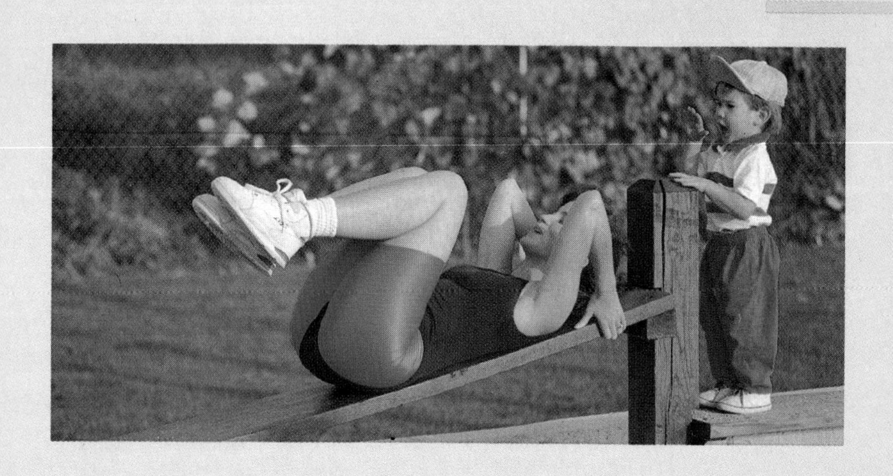

PART

V

From Generation to Generation

Up to this point we have considered the operation of the cell, and the manner in which the various organ systems of the body function both to provide an environment appropriate for the survival of cells and to achieve a whole that, when integrated, exceeds the sum of its parts.

Our focus has thus moved from the cell through the organ systems to the organism, and has considered both anatomy and physiology—but it has remained directed at a single generation of cells or organisms. Left unaddressed at this point is the manner in which cells replace themselves (although not all cells do), as well as the manner in which organisms replace themselves. Nor have we yet considered the process by which a single cell (the fertilized egg) becomes, through various stages, an embryo, a fetus, a child, and an adult, and finally, through the aging process, how it dies.

These questions are important both biologically and philosophically. We need to keep in mind, for instance, the distinction between life at the cellular level and life at the organism level. Organisms—at least highly sophisticated multicellular organisms such as ourselves—die. For them, as organisms, life eventually ends. At the cellular level, however, life never ends. Cells share a heritage unbroken by death going back almost 4 billion years. Life does not begin anew with each generation for the simple reason that it never stops. Rather, it carries on, in the form of the construction of an individual from a living blueprint—a blueprint carried in code within each fertilized egg.

Thus we come to the ultimate problem. How is the evolutionary legacy of a species transmitted between generations? That is, how is it that a fertilized human egg transforms into an adult human, rather than a pig, an oak tree, or something entirely new to the world? Finally, how are we designed, as organisms, to facilitate the formation of the next generation?

We shall consider these questions as we examine how cells divide, how hereditary information is passed from generation to generation, how our bodies function in the process of reproduction, and how individuals develop from single cells.

CHAPTER

17

Cell Division

Mitosis and Meiosis

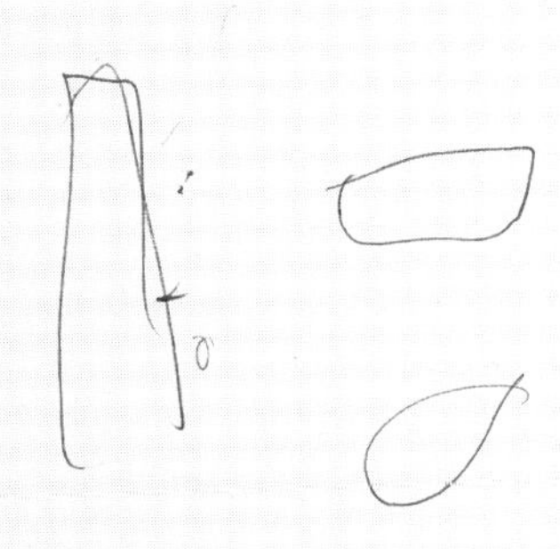

What happens to the various organelles when a cell divides? What happens to them when the sperm and egg fuse at the time of fertilization? How do cell division and fusion avoid the gain or loss of genetic material—or do they?

What causes cancer? Why does treatment of cancer so often include both surgery and chemotherapy? Why can't we be vaccinated against cancer as we are against so many other serious diseases?

These are the kinds of questions we shall consider as we examine the processes by which cells divide.

THE NATURE OF THE PROBLEM

Multicellular organisms become multicellular as a consequence of the repeated divisions of a single fertilized egg. Even if we leave aside (until Chapter 21) the problems of differentiation of these cells into specific tissues, the implications of that opening sentence are staggering. For example:

1. How does a cell accomplish the feat of dividing into two precisely equivalent daughter cells—or are the daughter cells *not*, in fact, precisely equivalent?

2. What special problems are posed in the formation of **gametes** (Gr. *gamos*, "marriage"), which are cells (such as sperm and eggs) designed to fuse together into a single cell during the process of fertilization—or are there, in actuality, no special problems?

3. Cells of some tissues—skin, intestinal lining, blood—are constantly being replaced, although various activities that we undertake—a cut to the skin, donation of a half liter of blood—demand very intensive, but temporary, replacement activity. How do cells "know" when to divide and when not to divide—and what are the implications of an error in the signals used to regulate cell division?

As a first step in answering these questions, we must look at the life cycle of a typical cell.

THE CELL CYCLE

Not all cells are capable of dividing. Specialized cells, such as nerve and muscle cells, do not divide, and they regenerate poorly, at best, if they are damaged or destroyed. (A severed spinal cord is a devastating injury not because of the physical amount of the damage, but because the damaged nerve cells have very limited capacity to repair themselves. An injury to the skin of similar magnitude might well heal without even leaving a scar.) Similarly, mammalian erythrocytes have no nucleus, and therefore are incapable of dividing. The basic rule seems to be that the more specialized a cell is, the less likely it is to divide.

The processes involved in cell division were hot topics during the last half of the nineteenth century. In 1880, the German biologist Walther Flemming reported seeing separation of what appeared to be a tangle of threads in the nucleus of an animal cell. (Similar, but less precise, observations had been made on plant cells five years earlier.) The "threads" turned out to be chromosomes, the organelles we now know as the carriers of the hereditary information that is transmitted from cell to cell and from generation to generation. However, the role of chromosomes in inheritance was not fully elucidated until 1902 (by the American biologist, Walter Sutton). At the time of Flemming's work, they were just "threads"—and Flemming named the process of their separation **mitosis** (Gr. *mitos*, "thread").

For many years, the attention of scientists was drawn to the events occurring in mitosis, which were after all readily observable under the microscope. When a cell was not undergoing mitosis, it was said to be in the "resting" stage, since no activity could be seen. However, we now recognize mitosis as being simply the culmination of a great deal of activity that occurs during the wrongly named "resting" stage. This activity is at the molecular level and is therefore not directly observable under the microscope, but the fact that we cannot see the activity makes it no less real.

Thus instead of a "mitosis—no mitosis" pattern of cellular activity, we now talk about a **cell cycle** (see Figure 17.1), which includes the following stages (each of which must be completed before the next stage can commence):

1. *G_1 stage* (*gap* stage, sometimes called the "growth" stage): The cell grows rapidly, producing large amounts of enzymes and structural proteins.

2. *S stage* (*synthesis* stage): The cell produces a duplicate set of DNA and chromosomal proteins, thus ensuring the availability of a sufficient supply of these substances for subsequent cell division.

3. *G_2 stage* (second gap, or growth, stage): The cell prepares for division by synthesizing the specialized proteins responsible for partitioning the chromosomes as the cell divides.

4. *M stage* (*mitosis* stage): The nucleus divides to form a pair of daughter nuclei, each possessing the same number of chromosomes as did the parent cell. Following the division of the nucleus and still within the M stage, the cytoplasm generally (but not always) divides in a process known as **cytokinesis,** isolating each daughter nucleus in its own cell with an appropriate volume of cytoplasm and associated organelles. Thus **cell division** requires both mitosis and cytokinesis (see Figure 17.2).

Cells that are actively dividing may be in any of these four stages at a given moment. Specialized cells that do not divide are usually in the G_1 stage.

MITOSIS AND CYTOKINESIS

Mitosis is believed to be initiated by the following events: During interphase a protein called **cyclin** accumulates in the cytoplasm of the cell (see Figure 17.3). Cyclin combines with a protein produced by a

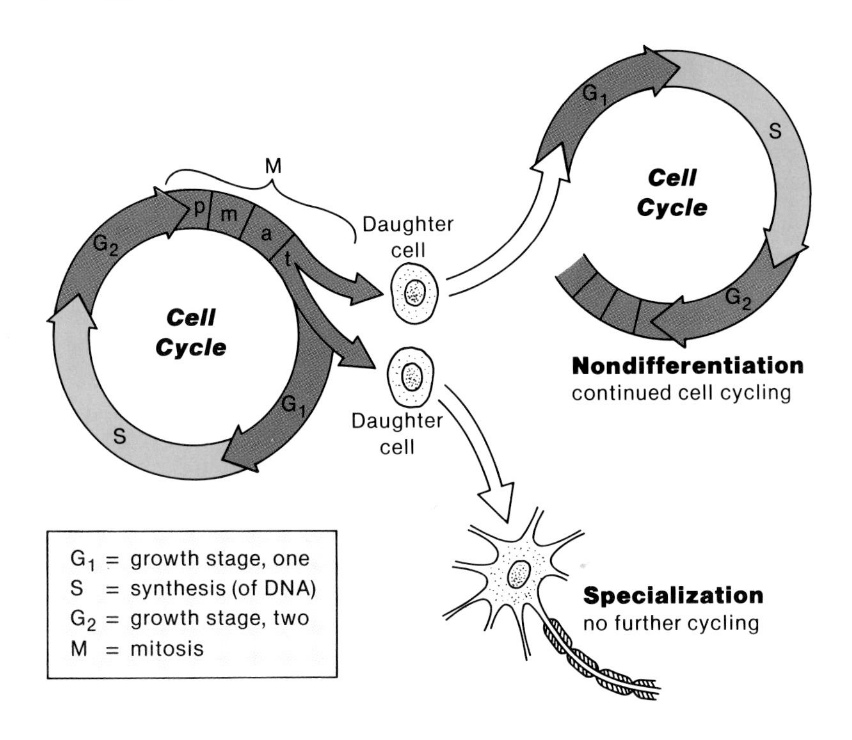

FIGURE 17.1

The Cell Cycle

Cells that are actively dividing move through the four stages of the cycle, ending with mitosis and cytokinesis. Some specialized cells do not divide and stay permanently in the G_1 stage.

FIGURE 17.2

Mitosis and Cytokinesis

Scanning electron micrograph of an animal cell in the process of dividing into two daughter cells.

FIGURE 17.3

Regulation of Mitosis

In cells that are regularly undergoing mitosis, a protein called cyclin is synthesized and accumulates between mitotic divisions (1), and binds to a protein manufactured by a gene labelled *cdc2* (2). Together, these two proteins become an inactive form of a substance called MPF (3). Enzymes modify the inactive molecule, converting it to active MPF (4), a substance that not only initiates mitosis (5), but also initiates a negative feedback loop by activating an enzyme that breaks down cyclin (6). Loss of the cyclin component of the MPF molecule inactivates MPF, leaving only the *cdc2* protein (7). Without the presence of MPF, the enzymes responsible for cyclin breakdown are inactivated, and the synthesis of cyclin begins once again (1).

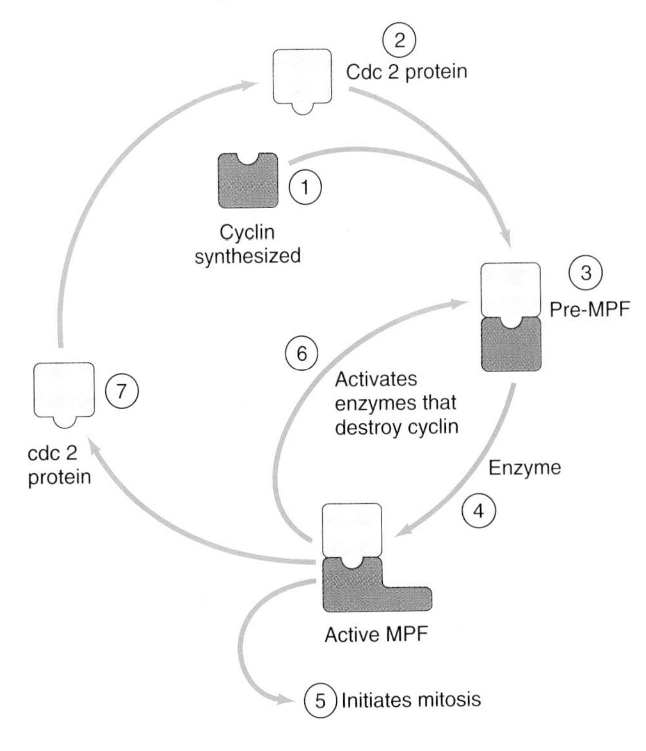

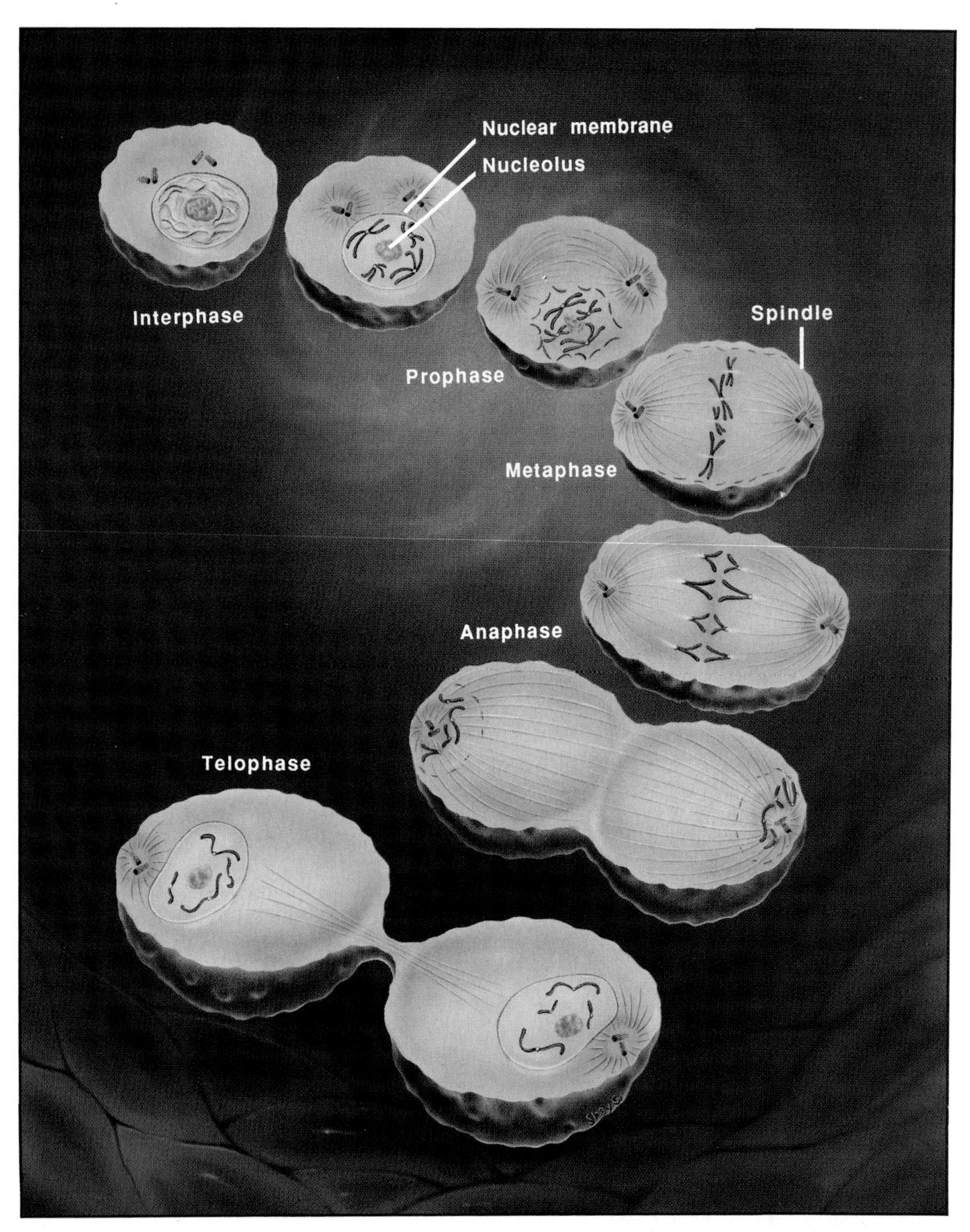

(a)

FIGURE 17.4

Mitosis—An Artist's Conception and the Real Thing

(a) For convenience, mitosis is regarded as a four-step sequence. In prophase, the nuclear membrane disappears, and the chromosomes condense and begin to become visible as distinct structures. Metaphase is characterized by the lining up of chromosomes. In anaphase the chromatid pairs separate and move to opposite sides of the cell. In telophase the nuclear membrane begins to redevelop, the chromosomes begin to elongate and disappear from view, and cytokinesis normally follows promptly. These micrographs of mitosis show (b) early prophase; (c) late prophase; (d) metaphase; and (e) anaphase.

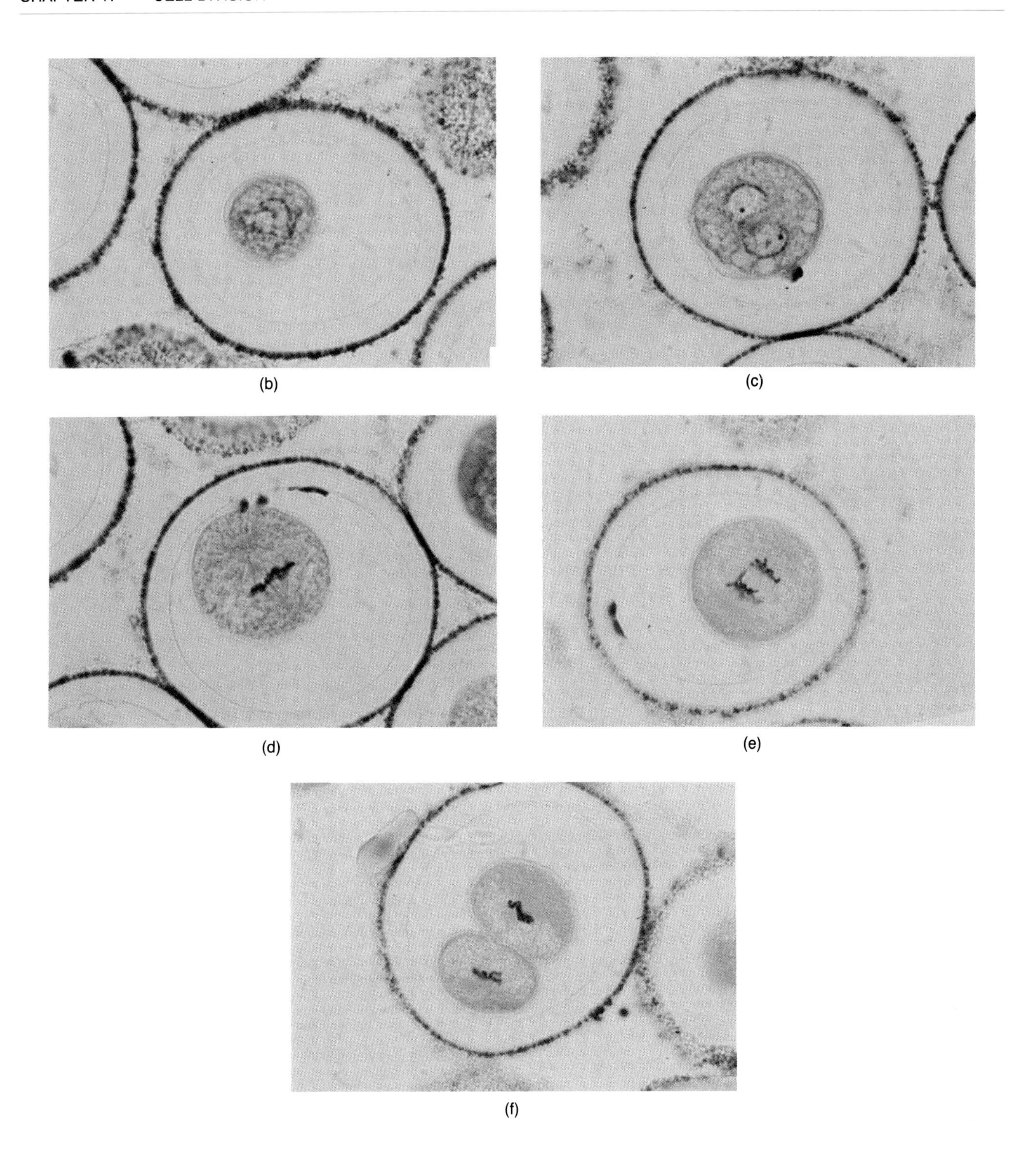

gene known as *cdc2* to form a molecule called *pre-MPF* (for **maturation promoting factor**). Pre-MPF is then modified by enzymes into MPF, and this substance both triggers mitosis and activates enzymes which degrade cyclin (thus avoiding an endless succession of mitotic divisions). This entire process is apparently evolutionarily very conservative in eukaryotes—MPF is found in organisms as evolutionarily distant as yeast, starfish, frogs, and mammals!

Mitosis is divided, for convenience, into four phases, although, from the cell's perspective, it is a continuous event (see Figure 17.4). In a typical animal cell, these four phases include the following activities:

Prophase. Generally, chromosomes cannot be seen as distinct entities under the light microscope, because they are strung out as thin, very elongate, threads. However, in mitosis the chromosomes become visible because they shorten by coiling into a series of distinct, sausage-like objects. Precisely why this occurs is unknown, but it has been postulated that separation and movement of chromosomes (both of which occur in later stages of mitosis) are facilitated by this shortening. In any event, the initiation of *prophase* (Gr., "before"; the first stage of mitosis) begins when the shortening, thickening chromosomes first become visible. Because the chromosomal materials have already doubled (during S stage), each chromosome now consists of two strands (called **chromatids**), linked by a button-like **centromere.**

The **spindle apparatus** and the **asters** also begin to form during prophase. These structures are composed of microtubules and are organized by the centrioles, which divide before mitosis begins and move to opposite sides of the cell early in prophase. (See Chapter 4 for details on the structure of microtubules and centrioles). The microtubules of the asters radiate away from the centrioles and attach to the plasma membrane of the cell. The spindle apparatus lies between the two centrioles, and some of the microtubules of the spindle apparatus attach to the centromeres of the chromosomes.

Even as the chromosomes are shortening and the spindle apparatus is beginning to assemble, the nuclear membrane disappears, as does the nucleolus. The Golgi apparatus disperses as small vesicles; RNA synthesis and pinocytosis cease; protein synthesis drops by a factor of four; and the microtubules and microfilaments of the cytoskeleton disassemble and rearrange, destabilizing the organization of the cytoplasm. As the chromosomes begin to move toward the midline of the cell, prophase comes to an end.

Metaphase. Metaphase (Gr., "between") begins where prophase left off. The chromosomes are lined up along the central axis of the cell by a shortening of the microtubules which are connected to the centromeres. Metaphase ends as the centromeres begin to split as a consequence of the continued shortening of the microtubules.

Anaphase. Anaphase (Gr., "again") is the shortest, and probably the most dramatic, of the stages of mitosis, for it is during anaphase that all of the centromeres simultaneously split and the formerly joined chromatids separate as *daughter chromosomes*. Because of the orientation of the chromosomes during metaphase, when the centromeres split the daughter chromosomes move to opposite sides of the cell as complete sets, thus ensuring that both daughter cells possess a full complement of chromosomes.

Precisely how the spindle apparatus functions during anaphase is still a matter of intense study. It appears to involve two activities. First, the spindle microtubules that attach to the centromeres shorten (by being gradually disassembled at the spindle end, not by actual contraction), causing the daughter chromosomes to be pulled toward opposite poles of the cell. Second, other spindle microtubules do not attach to the centromeres, but merely overlap their counterparts. These microtubules slide over one another, pushing the two ends of the spindle farther apart. Anaphase ends with the movement of the daughter chromosomes to opposite poles of the cell.

Telophase. Telophase (Gr., "end, completion") marks the end of mitosis, and in many respects it is the converse of prophase. The chromosomes begin to uncoil and gradually disappear as distinct bodies; the nucleoli and nuclear membranes (there are now two nuclei, of course) begin to reform, and the spindle microtubules begin to disappear as the tubulin subunits break off one by one.

In sum, mitosis is a process whereby a single nucleus containing one set of duplicated chromosomes is transformed into two nuclei, each of which contains one set of nonduplicated chromosomes.

Cytokinesis begins during telophase (sometimes, during late anaphase) with the formation of a **cleavage furrow,** which makes the cell look as if it is being constricted by an invisible rubber band. In actuality, the furrow is created by the action of microfilaments that are contracting just below the plasma membrane. This contraction continues until the cell is completely pinched in two, each half containing one of the daughter nuclei produced by mitosis.

It is not essential that all of the organelles in the cytoplasm be equally divided between the daughter cells. As long as each daughter nucleus contains a full complement of chromosomes, the genetic instructions for the assembly of additional organelles will be present, and the cell will restore its organelle numbers during the G_1 stage that follows mitosis. Mitochondria arise from preexisting mitochondria, however, and it is therefore essential that each daughter cell possesses at least one mitochondrion. In most cells, the number of mitochondria is sufficiently large, and their distribution sufficiently uniform, that virtually any equivalent division of cytoplasm will ensure that mitochondria will be found in both daughter cells.

MEIOSIS

In some organisms, including many species of protozoa, cell division (that is, mitosis + cytokinesis) is the principal method not only of growth but of reproduction, the two often being the same thing in unicellular organisms. However, cell division simply will not work as a means of reproduction in any but the simplest of multicellular organisms. You can imagine the problems you would face if you wished to duplicate yourself in some giant cellular division. Not only would all of your cells have to divide at essentially the same time (an impossibility for certain specialized cells, as we have already seen), but they would also have to migrate so as to allow a separation into a new organism. All in all, it would seem to be an impractical method of reproduction, to say nothing of the problems of deciding which of you had the rights to your nondividing tennis racket or rock albums.

Obviously, we need another process—and we have one. It is a form of nuclear division called **meiosis.** However, meiosis does more than merely solve the problems of reproduction in multicellular organisms. It also provides a mechanism wherein the genetic information possessed by two individuals can be combined in creating a third individual. We recognize this combining as *sexual reproduction.*

Meiosis—The Principle

The process of meiosis is integrally involved in the process of sexual reproduction. It may seem strange that a form of cell (more correctly, nuclear) division could have very much to do—let alone be "integrally involved"—with sexual reproduction. However, meiosis makes sense only if its connection with sexual reproduction is first understood and appreciated. Reciprocally, sexual reproduction is utterly dependent on the existence of meiosis.

Let us examine the significance of those statements. Suppose, for the moment, that we are living at a time in life's history before the "invention" of sexual reproduction (that is, at some point a billion or more years in the past) and that we have been assigned the task of designing the details of sexual reproduction, paying particular attention to potential pitfalls.

To begin, we recognize that sexual reproduction consists, at the cellular level, of the fusion of two gametes from two different individuals. (As you know, there is a little more to it at the level of the individual, but that discussion must wait until Chapter 20.) Right at the outset we encounter our first pitfall. If two gametes combine, the number of chromosomes in the fused cell will be twice what it was in the gametes. We cannot have the chromosome number doubling with every generation—it wouldn't be long before every cell was (metaphorically speaking) crammed to the gills with chromosomes. Obviously, we must avoid this pitfall by reducing the chromosome number in the gametes to precisely one-half the number possessed by the other cells in the organism. That way, when two gametes fuse, the chromosome number characteristic of the species will be restored (see Table 17.1). Because we humans, for example, have 46 chromosomes in the cells of our bodies, 46 is the chromosome number characteristic of our species (see Figure 17.5). Thus, following the logic we just developed, our gametes should contain only 23 chromosomes. This reduction in chromosome number accounts for the meaning of the term "meiosis" (Gr. *meioun,* "to make smaller").

But just a minute. That solution avoids one pitfall, but it creates another. How can we be sure that each gamete will possess the *right* 23 chromosomes? Suppose that we identify each chromosome and number them from one through 46 (see Figure 17.6). Assume

TABLE 17.1 Chromosome Number in Various Species

PLANTS		ANIMALS	
Organism	Chromosome Number	Organism	Chromosome Number
Lily	12	Scorpion	4
Garden pea	14	Housefly	12
Corn	20	Opossum	22
Giant sequoia	22	Salamander	24
Pasta wheat	28	Mouse	40
Bread wheat	42	Human	46
Tobacco	48	Gorilla	48
Sugar cane	80	Cow	60
Horsetail	216	Chicken	78
Adder's tongue fern	1262	Geometrid moth	224

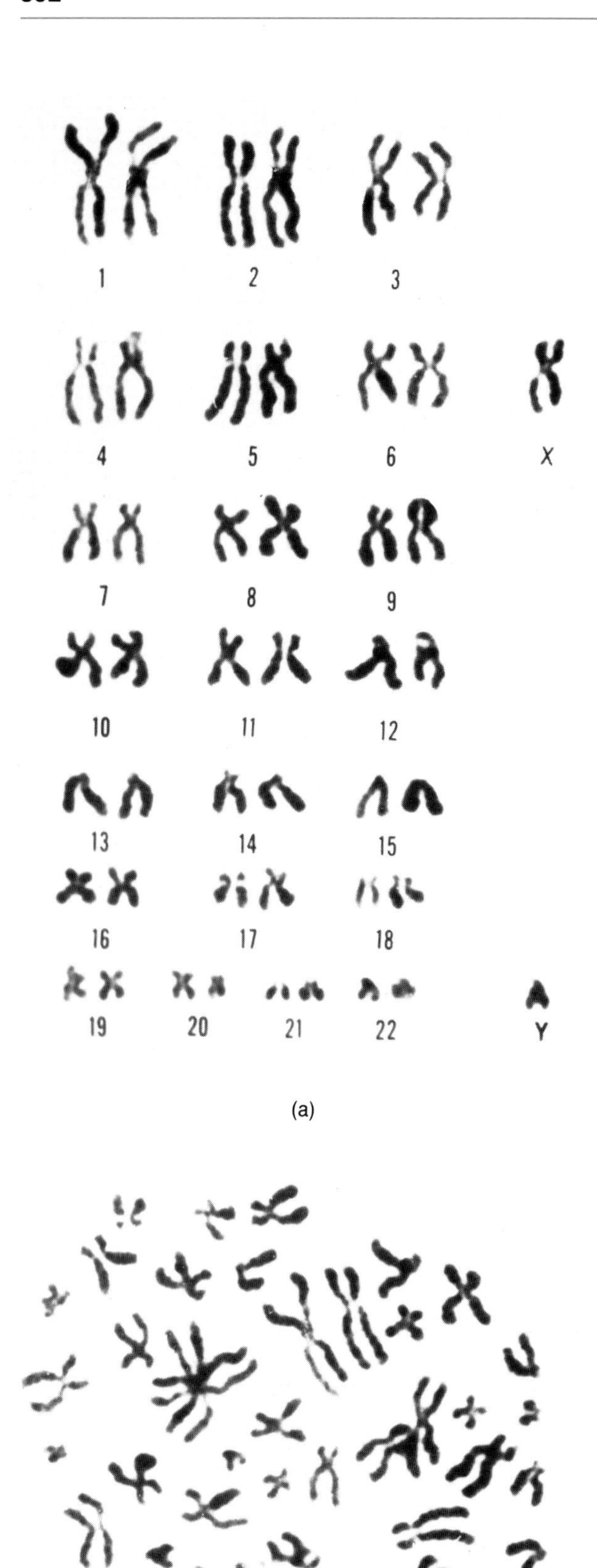

FIGURE 17.5

Human Chromosomes

(a) A chromosome spread of a cell from a human male at metaphase. (b) The same chromosomes arranged by pairs. Note the presence of 22 pairs of chromosomes plus a single Y chromosome at the end of the last row, and a single X chromosome at the end of the second row. The presence of both an X and Y is characteristic of cells from males.

that a given sperm cell (which, as we have just concluded, can have only 23 chromosomes) possesses chromosomes numbered one through 23. If that sperm fuses with an egg which also possesses chromosomes numbered one through 23 (or, for that matter, any 23 chromosomes other than those numbered 24 through 46), some of the 46 chromosomes that characterize our species will not be present in the resulting individual, a consequence that would at best be very detrimental and, at worst, fatal.

There is a solution, and it is obvious once you think about it for a moment. (Go ahead, take a moment now. Got it?) Of course—you build in chromosomal redundancy. We really have only 23 different chromosomes, but in our body cells each chromosome has a counterpart version of itself called a **homologue.** When we reduce our chromosome number to create a gamete, we must do it so as to ensure that each gamete has one complete set of chromosomes—that is, one member of each homologous pair. Thus, when the gametes fuse, they create a cell with two sets of chromosomes—one set from each parent—thereby restoring the chromosome number characteristic of the species (46 in the case of humans). When the fertilized egg becomes a sexually mature individual and begins to produce gametes, it does not matter whether its gametes contain the 23 chromosomes that came from the mother, the 23 chromosomes that came from the father, or (as would seem most likely) some combination of the two sets—just so long as each gamete contains a full set of 23 different chromosomes.

In short, we can avoid the pitfall of loss of essential genetic information during the halving of the chromosome number—which we have determined to be essential in meiosis—simply by devising a means of assorting the chromosomes by complete sets. That is a significant, but by no means insurmountable, problem.

There is one other major pitfall we must avoid if meiosis is to play a useful role in achieving the logical consequence of sexual reproduction—the enhancement of genetic diversity within the species. We know from Chapter 2 (with more details to follow in Chapter 22) that evolution by natural selection mandates a diverse population (otherwise, a sudden environmental change could eliminate the entire species). We

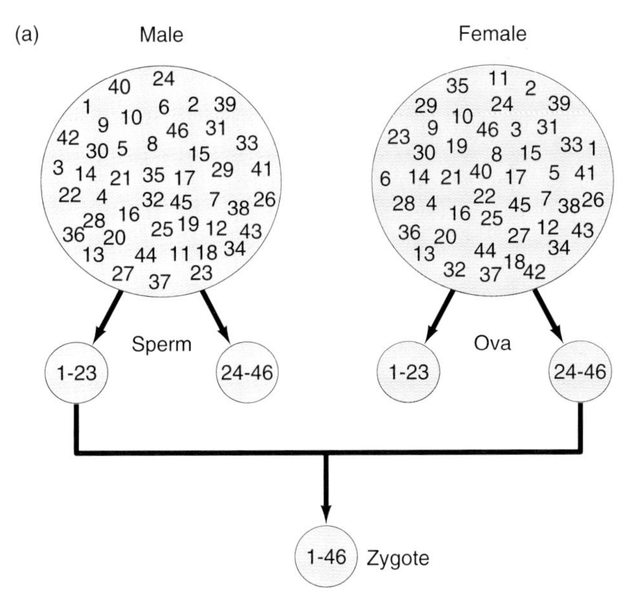

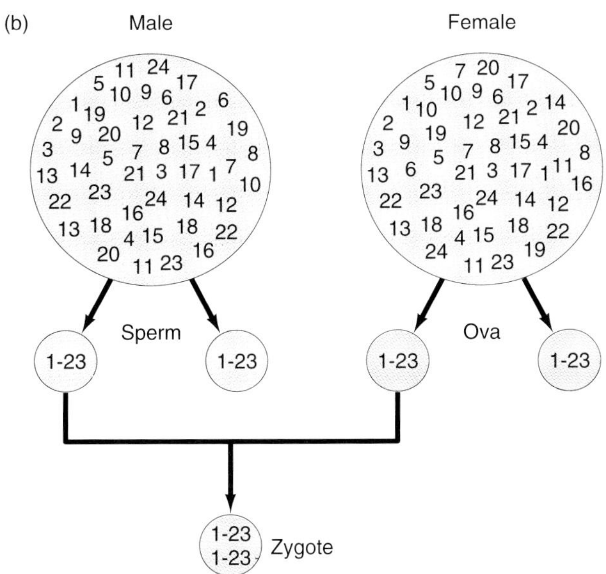

FIGURE 17.6

Problems with Reduction in Chromosome Number
If we assume that, in the formation of gametes, cells must halve their chromosome number (since fertilization involves equal contribution of genetic material from two cells), humans would have problems if each of their 46 chromosomes were different. What are the chances that a zygote would form from the fusion of a sperm carrying 23 chromosomes, and an egg carrying 23 *different* chromosomes? We see such a fusion in (a), but obviously the likelihood of a sperm with one-half of the chromosomes fusing with an egg carrying the other half (with no duplications or omissions) is very remote. However, if every chromosome had a counterpart—that is, if chromosomes occurred in pairs—then any mechanism that separated members of pairs of chromosomes would produce sperm and eggs all of which would have one complete set of chromosomes. Fusion of such gametes would produce a gamete with two complete sets of chromosomes, and the possibility of error (that is, of missing or duplicated chromosomes) would be sharply reduced. That is, of course, precisely what happens (b).

also know from Chapter 2 (with more details to follow in Chapter 18) that much of the diversity in a population is genetic and that the ultimate source of genetic diversity is mutation. Finally, we know from Chapter 3 and from our study of mitosis (with more detail to follow in Chapter 18) that genes are specific sections of chromosomal DNA. Thus we want to avoid the pitfall of mitosis—where daughter cells are genetically identical—and devise ways to enhance genetic diversity in the gametes that form in meiosis. Genetically diverse gametes will, in turn, result in a genetically diverse population of individuals.

In summary, our design for meiosis must include the following characteristics:

1. A chromosomal division that reduces the chromosome number by half—from the two sets possessed by body cells to a single set in the gametes; and

2. Mechanisms that maximize genetic diversity among the gametes without jeopardizing the need for each gamete to contain one full set of chromosomes.

As you might guess, these design features are precisely what we find in meiosis (see Figure 17.7). Let's examine the particulars.

Meiosis—The Process

Meiosis is a very restricted event within the individual organism because it is used only in the creation of gametes. Thus, in ourselves, meiosis is confined to the testes and ovaries.

Meiosis involves two successive cell divisions, which result in the formation of four cells, each with half the number of chromosomes present in the original cell. The first division (*meiosis I*) is a **reduction division,** because it is here that the number of chromosomes in each cell is reduced by one-half. By contrast, *meiosis II* is an **equational division,** because the number of chromosomes per cell remains constant.

The stages of meiosis are given the same names as those in mitosis, but there is a very significant difference in prophase I of meiosis as contrasted with prophase in mitosis. As the chromosomes begin to become visible (due to their shortening and thickening), it is apparent that the tips of the chromosomes are attached to the nuclear membrane at specific points. Moreover, and this is very significant, homologous chromosomes are lined up side-by-side, held together by a matrix of RNA and protein such that each chromosome precisely matches its homologous partner. This pairing of homologous chromosomes is

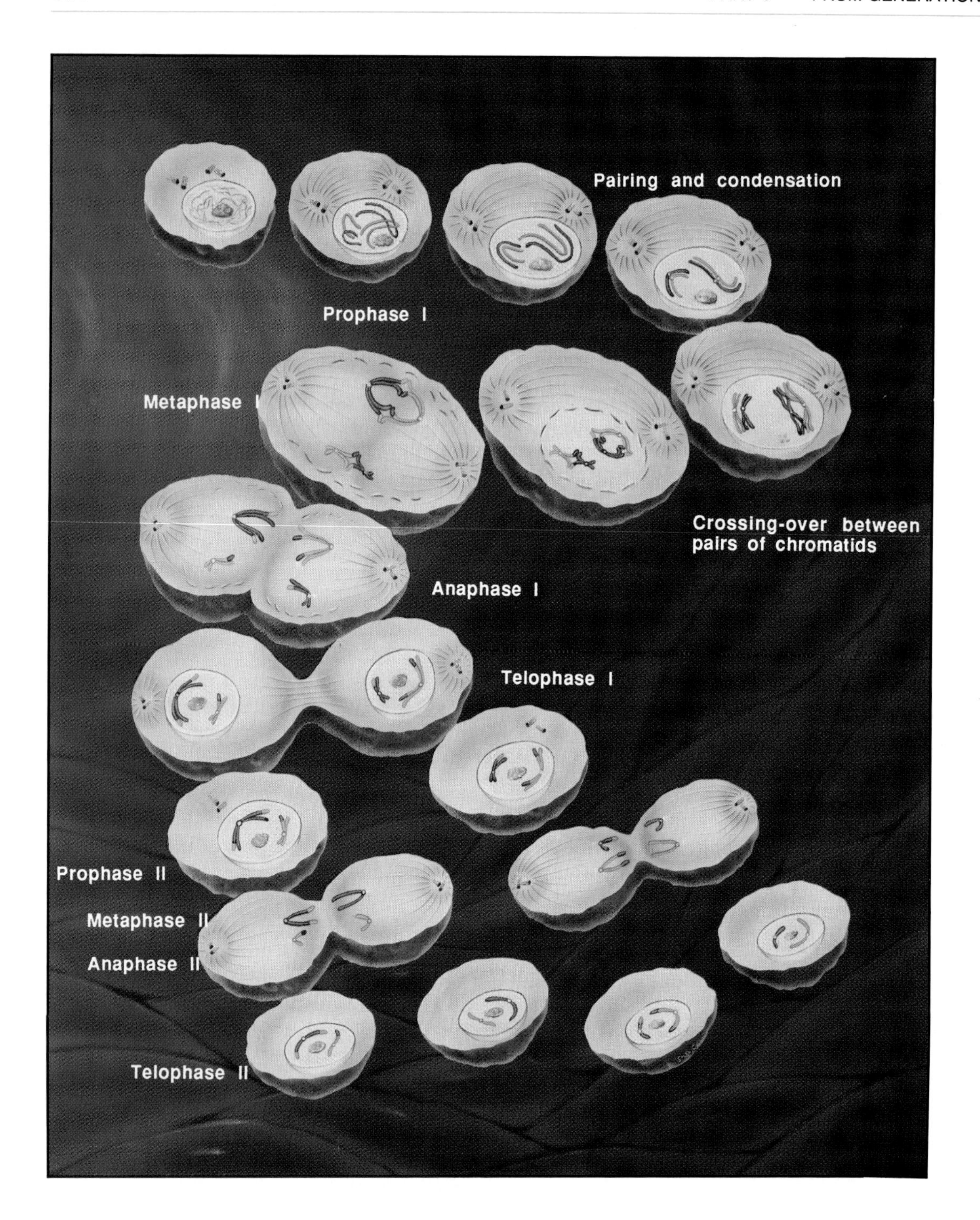

FIGURE 17.7

Meiosis

Meiosis looks very much like two successive mitotic divisions. However, in the first division, pairs of homologous chromosomes are separated, thereby reducing the chromosome number in each daughter cell to half that in the parent cell. In the second meiotic division, chromatids separate in a division that is much more mitotic in form. The end result: four daughter cells, each with half the number of chromosomes characteristic of the species.

called **synapsis** (Gr., "connection, junction"; see Figure 17.8), and it has no counterpart in mitosis.

Synapsis is not the only oddity of prophase I. Recall that each chromosome in prophase consists of two chromatids, linked by a centromere. Thus during synapsis a total of four chromatids (two from each chromosome of the homologous pair) are lined up in parallel. At this point one of the most dramatic events in cellular biology takes place. Nonsister chromatids

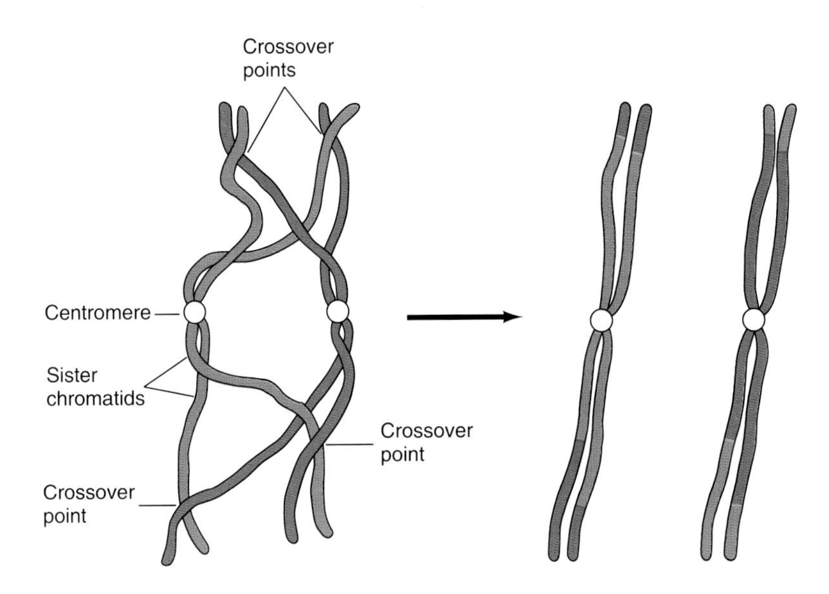

FIGURE 17.8

Crossing Over and Chiasmata

In late prophase I of meiosis, chromatids of pairs of homologous chromosomes entangle in a process called crossing over. The crossover points are called chiasmata; exchange between chromatids of equivalent portions of genetic material commonly occurs at this stage.

overlap each other at various points in a process called (explicitly, if not very creatively) **crossing over.** The crossover points themselves are called **chiasmata** (Gr., "crosspiece," from the Greek letter *chi*) and at these points the chromatids actually break and trade sections! (Protein complexes determine the precise points of the breakage and ensure that the break points are in equivalent locations in both chromatids.) There may be several chiasmata in one set of chromatids. The consequence is that the sister chromatids are now no longer genetically identical, and we have therefore set the stage for our goal of increasing genetic diversity among the daughter cells in meiosis.

Crossing over has other ramifications as well. In the absence of crossing over, organisms with very large numbers of chromosomes would be advantaged, because variability in genetic constitution could only result from the shuffling of whole chromosomes (and you can do more shuffling if you have more chromosomes). However, because of crossing over, there is no inherent advantage in a species having a large number of chromosomes. Thus the existence of crossing over accounts for the fact that the chromosome number characteristic of a species varies enormously in eukaryotes.

As in mitosis, a spindle apparatus forms between the two poles of the cell during prophase I, and some of the microtubules attach to the centromeres. However, because the homologous pairs of chromosomes are still linked together by chiasmata, as the chromosomes are being lined up to begin metaphase I, they line up in homologous pairs, not singly as in mitosis.

Something that furthers our goal of increasing the genetic diversity of the gametes is the fact that the lining up of homologous pairs of chromosomes during metaphase I is random (see Figure 17.9). That is, there is no mechanism that ensures that every chromosome on the left, for example, is from the father. Because there are 2^{23} (over eight million) possible ways of lining up the maternal and paternal chromosomes, each cell undergoing meiosis may be presumed to have a different alignment of chromosomes. It therefore follows that each gamete is, for all intents and purposes, genetically unique, in that each possesses its own particular blend of maternal and paternal chromosomes. Thus the chance of two individuals being genetically identical (i.e., the chance of genetically identical sperm fusing with genetically identical eggs), even assuming common parents, is less than one in 64 quadrillion (64×10^{12})—and that is *before* we take into consideration the effects of crossing over. Despite their frequency on television dramas, unrelated but identical individuals would seem a virtual impossibility.

During anaphase I, the shortening of the spindle microtubules causes the chiasmata, *not* the centromeres, to break. The result is that one chromosome (still

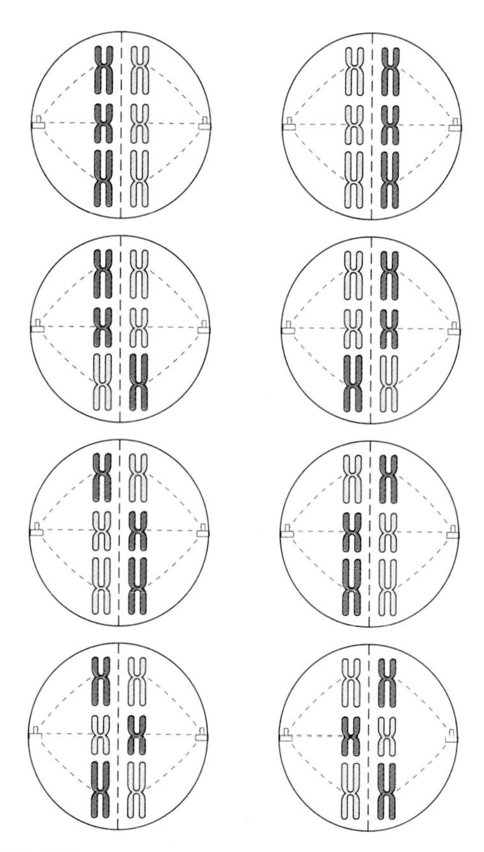

FIGURE 17.9

Randomness at Metaphase I

Although the chromosomes line up in homologous pairs at metaphase I, there is no "rule" that says all chromosomes inherited from the father are on the left, and all from the mother on the right. To the contrary, whichever member of any pair is on the left or the right occurs entirely by chance. In humans, with 23 pairs of chromosomes, the likelihood that a given gamete possesses only the chromosomes inherited from one parent is remote—indeed it is two chances in 2^{23}. In this hypothetical organism with only three chromosome pairs, there are still eight (2^3) different combinations—and that is before we factor in the effects of crossing over.

consisting of two chromatids) from each homologous pair moves to opposite poles of the cell. Thus the daughter cell that forms in telophase I contains only 23 chromosomes.

The second meiotic division is a pure mitotic division, differing from a regular mitosis only in that there are just 23 chromosomes (rather than the usual 46) available to line up along the cell midline. Theoretically, this second division would not even be necessary, but meiosis presumably evolved from mitosis, and therefore the chromosomal material was duplicated during the S stage that preceded meiosis I. Thus meiosis II is necessary to separate the chromatids and to split the centromere. However, because of the crossing over that took place in prophase I, the sister chromatids are no longer genetically identical. The consequence is that the four cells formed at the end of telophase II (and following cytokinesis) contain varying mixtures of genes and chromosomes from both parents, and each cell is therefore genetically distinct from the others.

As we have already emphasized, meiosis is a specialized form of nuclear division used only in the production of gametes. In human males, gamete production begins at puberty and is under the control of hormones (see Chapter 20). In females, on the other hand, meiosis I actually occurs during fetal development, but the prospective eggs are held in a kind of suspended animation until puberty begins. It is interesting to note that all of the cells destined to be released as eggs during a woman's entire reproductive life are already present at the time of her birth.

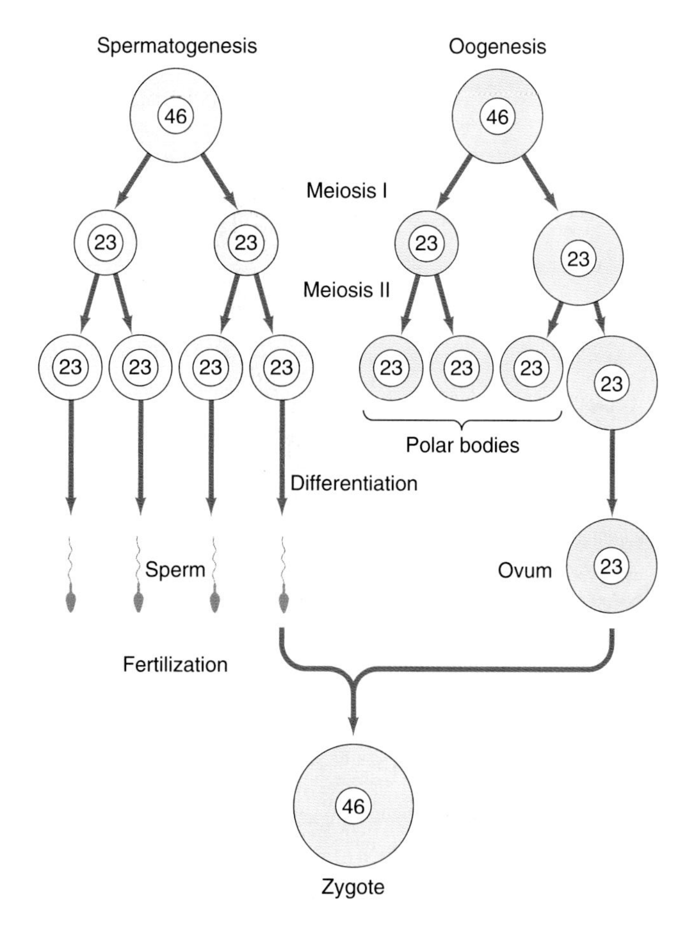

FIGURE 17.10

Formation of Gametes

In humans, gametes are sperm and eggs. Four functional sperm normally form from one parental cell, but only one functional egg forms, since one cell receives virtually all the cytoplasm. The other daughter cells are nonfunctional polar bodies that do not engage in zygote formation.

There is one significant difference between meiosis in males and meiosis in females. All four daughter cells in males become functional sperm cells (which are essentially flagellated nuclei with a few mitochondria to power the flagellum). However, in females, only one cell becomes a functional egg (see Figure 17.10). All of the cytoplasm which supports the fertilized egg during its early development comes from the female, the male contribution being nuclear only. Consequently, the egg is a relatively large cell. Cytokinesis during telophase I and II is unequal, with one daughter cell retaining virtually all of the cytoplasm. The other daughter cells are basically just nuclei surrounded by a thin layer of cytoplasm. They are called **polar bodies,** and they play no role in fertilization.

CANCER

Although the mechanics of cell division are now relatively well known, we are still largely ignorant of how a cell regulates its division. Not all cells divide. Those that do divide do so at different rates. Why? The same type of cell may divide readily in culture, but rarely in place in the body. Why not? If we suffer a cut or abrasion to the skin, cell division is promoted. How? Once the cut has healed, and new skin tissue has replaced the damaged tissue, cell division stops (or returns to normal levels). Why?

These are not esoteric questions. Failures in the processes regulating cell division are relatively common, and often have serious consequences. One of these consequences is called **cancer** (L., "crab"). One in three Americans will contract cancer at some point during his or her life, and cancer will kill one in five (see Table 17.2). To put those ratios into real numbers, in 1987, 965,000 Americans developed cancer, and 483,000 died from this disease (see Figure 17.11). Cancer ranks second in the causes of death in this country (following only failures of the cardiovascular system).

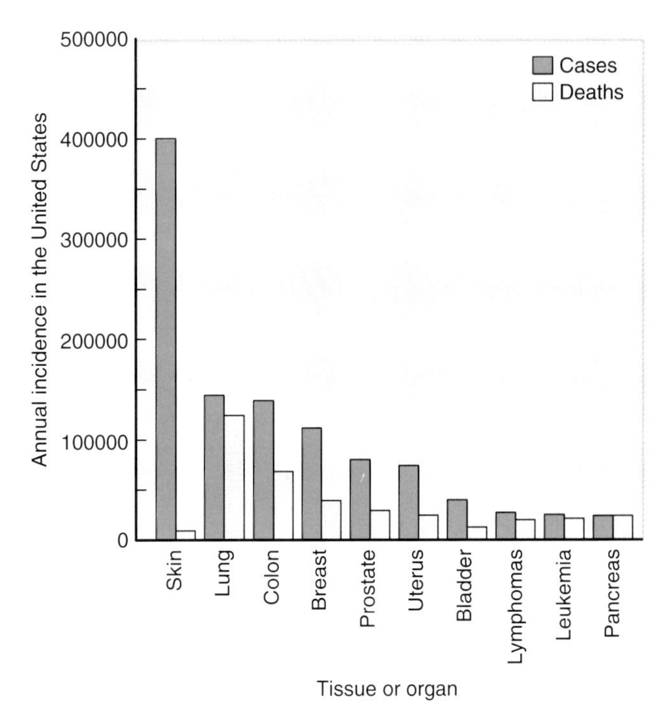

FIGURE 17.11

Types of Cancer

The incidence of cancer is not evenly distributed among the organs of the body, and the cure rate at present varies enormously.

TABLE 17.2 Five-Year Survival Rates for Various Cancers

TYPE OF CANCER	FIVE-YEAR SURVIVAL RATE	TYPE OF CANCER	FIVE-YEAR SURVIVAL RATE
Thyroid	83%	Kidney	42%
Melanoma	68%	Non-Hodgkin's lymphoma	35%
Uterus	66%	Ovary	34%
Breast	65%	Multiple myeloma	17%
Bladder	62%	Leukemia	14%
Larynx	61%	Stomach	13%
Prostate	57%	Lung	10%
Hodgkin's disease	55%	Esophagus	5%
Colon-rectum	45%	Pancreas	1%

Survival for all stages, adjusted for normal life expectancy. Average survival for both sexes is used when the neoplasm occurs in both sexes. (*Source: Cancer patient survival,* Report #5, DHEW Publication No. (NIH) 77–992.)

Death from cancer occurs because of several different factors. In order of significance, these are as follows:

1. Infection (cancer suppresses the immune system)

2. Interference with vital organs (due to the growth of tumors)

3. Starvation (the body wastes away as vital resources are directed at nourishing the cancer)

4. Hemorrhage (internal bleeding due to a tumor's interference with major blood vessels)

If we better understood the signals by which cells turned cell division on and off, we would presumably be able to prevent and cure most cancers.

Characterization of Cancer

There are over 100 different kinds of cancer, each distinct from the others in terms of rate of growth, age of onset of the disease, type of tissue involved, tendency to spread, and lethality. Uncontrolled cell division also occurs in noncancerous diseases such as *psoriasis* and *warts,* but the critical difference is that cancer is capable of spreading throughout the body and invading other tissues, a process known as **metastasis** (Gr., "to change").

In most cases, the dividing cells are clumped together as a **tumor,** although in some cancers, such as *leukemia,* the tissue involved (leukocytes) by its very nature precludes the development of a distinct tumor. However, not all tumors are cancerous. They are characterized as either **benign** or **malignant,** and benign tumors are not cancerous because they cannot metastasize. Nevertheless, they may still be dangerous and even life-threatening, as when they crowd vital organs such as the brain. Moreover, because benign tumors sometimes become malignant, they are generally removed once discovered.

Malignant tumors, by definition, do metastasize. Many of the tiny clumps of cancerous cells that break off the tumor and enter the blood or lymphatic system are destroyed by the leukocytes, but any that escape are capable of initiating a new tumor elsewhere in the body. Growth of new tumors may take months or even years after the initial onset of cancer, which explains why continued vigilance is necessary even after a malignant tumor has been surgically removed or destroyed.

Not all cancers show the same propensity for metastasis. Some types of skin and brain cancers rarely metastasize, whereas breast and lung cancers do so readily. Moreover, no one cancer can invade all tissues. For example, lung cancers often invade the brain, whereas prostate cancer commonly invades bone. These facts provide further evidence of the earlier statement that "cancer" is a hodgepodge of conditions sharing only the common trait of metastasis.

There are certain traits of cancer cells that distinguish them from normal cells.

1. *No density-dependent inhibition of cell division.* (Normal cells cease dividing once they are in contact with other cells. Cancer cells continue to divide).

2. *No anchorage dependency.* (Normal cells must be in contact with some substrate in order to divide. Cancer cells have no such requirement and will readily divide even when they are maintained in a liquid suspension in a test tube).

3. *No requirement for serum growth factors.* (Growth factors—see Chapter 21—play a major role in regulating the division of normal cells. Cancer cells divide in the absence of growth factors from other cells by producing their own).

4. *Failure to differentiate.* (Normal cells mature into functional cells. Cancer cells stay immature and therefore perform no useful role in the body.)

These differences are important in understanding the basis of cancer, and in treating the disease. However, the threat to life is not primarily from these cellular features. As a tumor grows, it eventually becomes large enough to be detectable, and careful surgery can remove it in most instances. The threat to life is that the tumor will have metastasized before being detected, in which case surgical removal of the tumor will not effect a cure, since small colonies of cancer cells will be left behind, often in regions of the body remote from the tumor site.

What is involved in metastasis?

1. Cells must break loose from the tumor.

2. These cells must then gain access to the circulatory system.

3. The cells must escape being destroyed by leukocytes.

4. The cells must leave the circulatory system and locate in some tissue of the body where they are capable of growing and reproducing.

5. The incipient tumor must be capable of inducing the growth of new blood vessels to nourish the growth of the tumor.

Each of these steps represents a potential point of intervention by clinicians seeking to prevent or cure the disease.

Causes of Cancer

There have been few controversies in medicine that have been longer, or more bitter, than the arguments about the causes of cancer. One sign of just how far this controversy has ranged was the awarding of the 1926 Nobel Prize to a Danish microbiologist, Johannes Fibiger, "for discovering a parasite that causes cancer." Fibiger fed cockroaches containing nematodes (a small parasitic worm) to rats. When he autopsied the rats, he found that they had stomach cancer. Therefore, he concluded that the nematodes caused the cancer! There has been no independent verification of this finding, but eating cockroaches laced with nematodes is still not to be encouraged, for esthetic reasons if no other.

What has sometimes been missing from discussions of the cause, or causes, of cancer is the realization that environmental agents or pathogens (the two most favored classes of suspects) can never be more than the penultimate cause. The ultimate cause must lie in the genetic machinery of the cell itself.

Genetically, cancer can arise because of the following:

1. A gene mutation that causes uncontrolled growth and cell division. In mutated form, these genes are called **oncogenes** (Gr. *onkos*, "mass, bulk," a reference to the development of a tumor), and they arise from normal genes called **proto-oncogenes** (a reference to their ability to transform into oncogenes). Proto-oncogenes are of two types:

    (a) Genes that produce a protein responsible for promoting or inhibiting cell division.

    (b) Regulator genes that control the activity of genes in category (a). (As we shall see in Chapter 18, regulator genes are normally located adjacent to the genes they control.)

2. The mutation of a gene that normally suppresses the development of cancer cells and tumors. Tumor suppression may occur either within a cell, or between cells of different types. Thus tumor suppressor genes include genes which:

    (a) Affect the production of hormone, growth factors, or their receptors.

    (b) Control the proper functioning of the immune system.

    (c) Control chromosome repair or stability.

    (d) Control cell differentiation or cell death.

    (e) Affect mitosis.

    (f) Influence the ability of cancer cells to invade new tissues or to promote the growth of blood vessels to nourish new tumors.

How do these errors or mutations arise in the first place? During chromosomal duplication and cell division, three types of errors may occur that can give rise to cancer:

1. A structural error in the sequence of bases in either proto-oncogenes or suppressor genes;

2. Separation (because of crossing over) of the regulatory gene from the gene it controls;

3. Loss of a suppressor gene, a regulatory gene, or a gene that normally inhibits cell division.

Mutations and Cancer

More than 60 oncogenes have now been identified, and their role in specific cancers has been elucidated. In some instances, a single mutation leads directly to cancer formation. More commonly, however, there is a stepwise accumulation of mutations over time. For example, in colon cancer, the loss of a suppressor gene on chromosome 5 leads to increased cell growth. When a proto-oncogene known as *ras* (which also has been implicated in cancers of the breast, pancreas and lung) mutates, and when a suppressor gene located on chromosome 18 is lost, **adenomas** (Gr. *aden*, "gland"; benign tumors with a gland-like appearance) develop. Malignancy develops with the loss from chromosome 17 of a suppressor gene called *p53* (which also occurs in some breast and lung cancers), and other chromosome losses lead to metastasis. (The actual sequence of mutations may vary.) Accumulating this number of mutations takes time, which is why lung, breast, and colon cancers tend to be diseases of middle age or later. Cancers in children and young adults presumably require fewer mutations.

For example, consider the onset of **retinoblastoma** (cancer of the retina). Approximately one child in 20,000 develops this cancer by the age of four. These are children who inherit a chromosome 13 with a missing suppressor gene. If, in any retinal cell, the corresponding gene on the homologous chromosome should happen to mutate, the affected cell will continue to divide, forming a retinal tumor. Interestingly, mutation of this same gene in lung tissue is responsible for more than half of all lung cancers. Deletion of a suppressor gene also occurs in *Wilms' tumor* (a kidney tumor occurring in one in 10,000 children under the age of five, and involving chromosome 11).

The genetics of most of the other cancers is presently less well known. Changes in chromosome 3 (the actual sites have not yet been determined) occur in many lung cancers, and changes in chromosome 1, 6, 7, and 9 have been reported in **melanoma** (a very

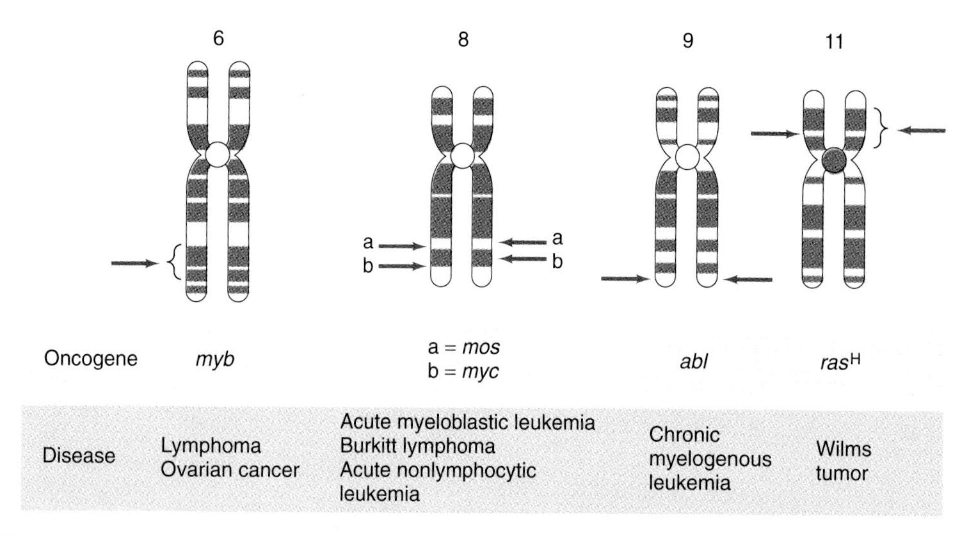

FIGURE 17.12

Oncogenes

The position of five oncogenes on four chromosomes is shown, and some of the cancers in which they have been implicated are listed.

serious form of skin cancer). In total, ten different sites on eight different chromosomes have been implicated in various tumors (see Figure 17.12).

Virally Transmitted Oncogenes

In 1910, a young researcher named Peyton Rous discovered that a form of cancer in chickens was transmitted by a pathogen. That discovery was alternately ignored and disputed, but Rous was ultimately vindicated and received the Nobel Prize in Physiology or Medicine in 1966. The infectious agent proved to be an unusual type of virus known as a **retrovirus.** That name derives from its method of reproducing. Retroviruses contain RNA, but not DNA. Once the RNA is in the host cell, it initiates the formation of a DNA copy which is then inserted into the host's DNA. (The AIDS virus, for example, is a retrovirus.)

This newly formed DNA contains only a few genes (after all, viruses are structurally rather simple), but in the case of Rous's retrovirus, one of the genes turned out to be not a virus gene at all, but a chicken gene that presumably had been inadvertently acquired from some ancient host. (Another strain of this same virus contains no such gene—and does not cause cancer.) As it happens, the chicken gene in question promotes cell division and is therefore a proto-oncogene. The gene has mutated slightly in the virus copy and is now a full-blown oncogene.

The Onset of Cancer

Despite the role of a virus in causing cancer in chickens, very few human cancers are caused by viruses. What, then, accounts for the transformation of a proto-oncogene into an oncogene?

The kinds of transformations we have been talking about are all various forms of mutations (see Figure 17.13). Some mutations involve errors in the sequence of nitrogenous bases within the gene itself (*point mutations*). The others involve changes in the sequence of genes along the chromosome (*chromosomal mutations;* see Chapter 18 for details). Substances that induce mutations are called **mutagens.** The mutagens responsible for the onset of cancer in humans are primarily chemical, although radiation is also a powerful mutagenic agent (see Table 17.3).

So if we avoid mutagens, we can avoid cancer, right? Well, it's not that simple. The problem is that any chemical that causes the death of cells creates a situation where cell replacement is necessary—and dividing cells are much more prone to mutate than are nondividing cells. When rats and mice were fed chemicals at the maximum tolerated dose for extended periods, 212 of 350 synthetic chemicals caused cancer. However, naturally occurring chemicals produced by food plants were also tested in the same fashion—and 27 of 52 caused cancer. (These 27 chemicals are found in 57 different plants that we eat as food!)

Of course, the doses of these chemicals given to rats and mice were much greater than would be consumed by humans—but the point is that we are constantly exposed to low levels of chemicals, both natural and synthetic, that can give rise to cancer and that are impossible to avoid completely.

The same situation is true regarding exposure to radiation. The National Academy of Sciences pub-

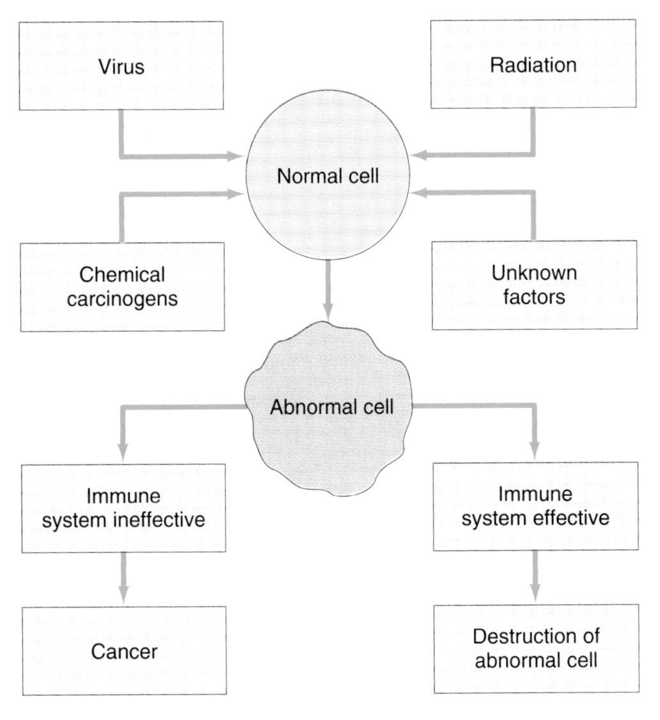

FIGURE 17.13

Onset of Cancer

Normal cells may become abnormal (precancerous) because of a variety of causes. However, cancer does not develop unless these cells escape destruction by the immune system. Why the immune system is sometimes unsuccessful in destroying abnormal cells is far from clear at present.

lished its fifth committee report on the biological effects of ionizing radiation in 1990, and that report concludes that the dangers of low-level radiation are considerably greater than had previously been estimated. Interestingly, less than 20 percent of the radiation to which Americans are exposed comes from human sources (primarily X-rays and nuclear medicine); by far the greatest amount (over 50 percent) comes from exposure to the element **radon.** Radon occurs in nature as an unstable gas that seeps out of the ground and releases radiation as it breaks down into other elements ("radon daughters"). Exposure to radon and radon daughters is estimated to be responsible for up to 20,000 cases of lung cancer in the United States annually. Better ventilation of basements in those areas of the country where radon is abundant in the soil could reduce the risk considerably.

Even if the figure of 20,000 cases a year is correct, radon would still be responsible for less than 15 percent of all cases of lung cancer. The primary villain in this disease, as presumably everyone now knows, is cigarette smoking (see Figure 17.14). Cigarette smoke contains more than 6000 different organic molecules, at least 30 of which induce cancer in laboratory animals (see Table 17.4). Why, then, is there any controversy over the link between cigarette smoking and lung cancer? There are two answers to that question:

1. *Time lag.* Lung cancer is primarily a disease of the middle-aged or elderly long-term smoker. The time required for the necessary accumulation of several mutations is 15 or 20 years, which is why there is such a lag between an increase in the smoking rate in the population and the incidence of lung cancer. Because of this temporal separation of triggering event (smoking) and consequence (lung cancer), many people are prone to dismiss any link between the two.

TABLE 17.3 Occupations and Cancer

OCCUPATION	CARCINOGEN IDENTIFIED	LOCATION OF CANCER
Chimneysweeps; manufacturers of coal gas	Hydrocarbons in soot, tar, oil	Scrotum; skin; bronchus
Chemical workers; rubber workers; manufacturers of coal gas	2-Naphthylamine; 1-naphthylamine	Bladder
Chemical workers	Benzidine; 4-aminobiphenyl	Bladder
Asbestos workers; shipyard and insulation workers	Asbestos	Bronchus; peritoneum
Sheep dip manufacturers; gold miners; some vineyard workers and ore smelters	Arsenic	Skin; bronchus
Makers of ion-exchange resins	Bis (chloromethyl) ether	Bronchus
Workers with glues, varnishes, etc.	Benzene	Bone marrow (leukemia)
Poison gas makers	Mustard gas	Bronchus; larynx; nasal sinuses
PVC manufacturers	Vinyl chloride	Liver
Chromate manufacturers	Chrome ores	Bronchus
Nickel refiners	Nickel ore	Bronchus; nasal sinuses
Isopropylene manufacturers	Isopropyl oil	Nasal sinuses

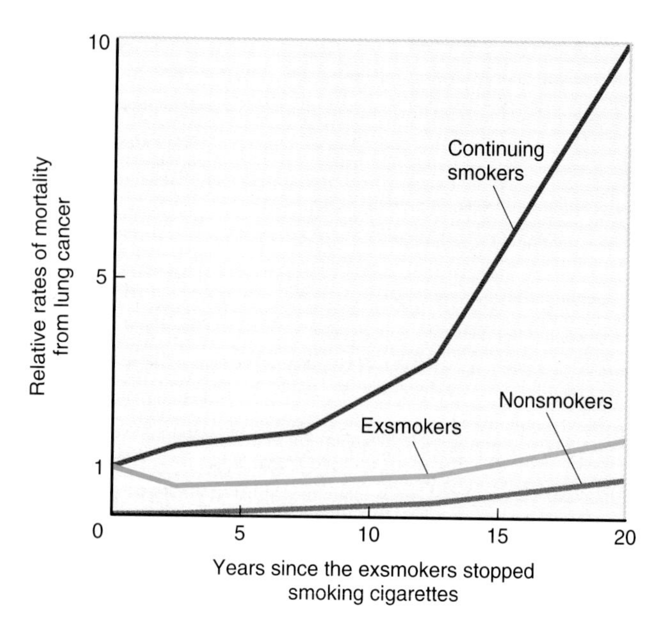

FIGURE 17.14

Cancer and Smoking I

Relative rates of lung cancer mortality among nonsmokers, ex-smokers, and continuing smokers of cigarettes over a 20-year period. The incidence of fatal lung cancer rises sharply after about 10 years of smoking, and continues to rise even more sharply in succeeding years.

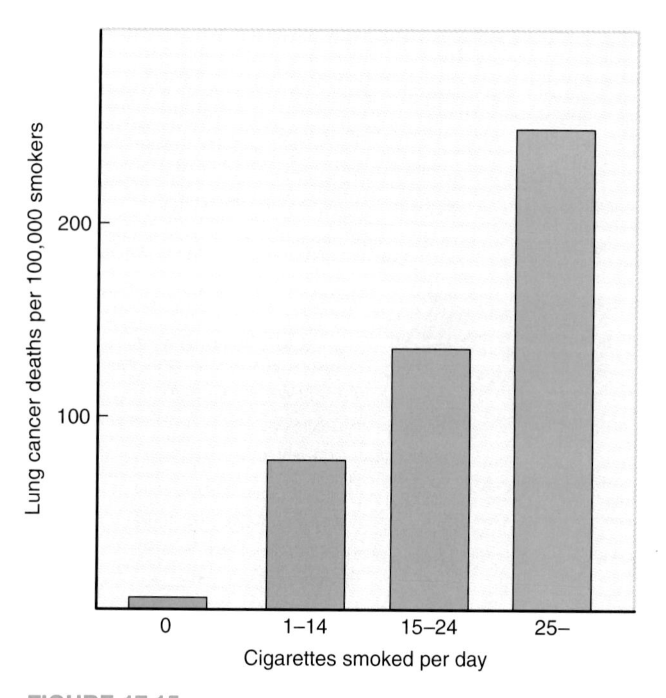

FIGURE 17.15

Cancer and Smoking II

The frequency of deaths from lung cancer in relation to the number of cigarettes smoked.

TABLE 17.4 A Sample of Chemicals from Cigarette Smoke

Acetaldehyde	Fluorene
Acetone	Formaldehyde
Acetylene	Hexane
Acrolein	Hydrazine
Aminostilbene*	Indeno[1,2,3-*cd*]pyrene*
Ammonia	Indole
Arsenic*	Isoprene
Benz[*a*]anthracene*	Methane
Benz[*a*]pyrene*	Methanol
Benzene*	Methylcarbazole
Benzo[*b*]fluoranthene*	5-Methylchrysene*
Benzo[*c*]phenanthrene*	Methylfluoranthene*
Benzo[*j*]fluoranthene	Methylindole
Cadmium*	*β*-Naphthylamine*
Carbazole	Nickel compounds*
Carbon dioxide	Nicotine
Carbon monoxide	Nitric oxide
Chrysene*	Nitrobenzene
Cresols	Nitroethane
Crotonaldehyde	Nitromethane
Cyanide	*N*-Nitrosodimethylamine*
DDT	*N*-Nitrosomethylethylamine*
Dibenz[*a,c*]anthracene*	*N*-Nitrosodiethylamine*
Dibenzo[*a,e*]fluoranthene*	Nitrosonornicotine*
Dibenz[*a,b*]acridine*	*N*-Nitrosonanabasine*
Dibenz[*a,j*]acridine*	*N*-Nitrosopiperdine*
Dibenzo[*c,g*]carbazone*	*N*-Nitrosopyrrolidine*
N-Dibutylnitrosamine*	Phenol
Dichlorostilbene	Polonium-210*
2,3-Dimethylchrysene*	Propene
Dimethylphenol	Pyridine
Ethane	Sulfur dioxide
Ethanol	Toluene
Ethylphenol	Vinyl acetate
Fluoranthene	

*Indicates proven carcinogen.

2. *Uneven results*. Not all smokers develop lung cancer; some nonsmokers do. That fact alone is enough for some smokers to be persuaded that the link between cigarette smoking and lung cancer is overstated. (Of course, they must ignore the fact that nonsmokers constitute more than half the general population, but only 4 percent of the lung cancer victims.) Nevertheless, with all the mutagens in cigarette smoke, why is there not an even stronger correlation between smoking and lung cancer?

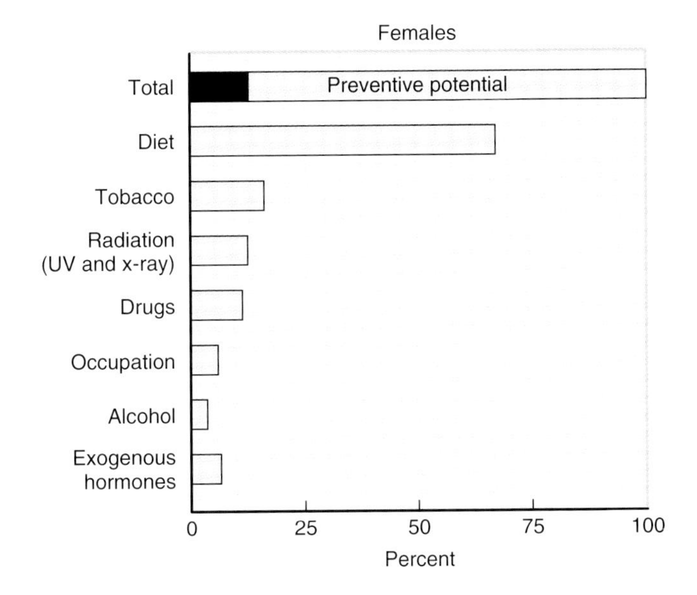

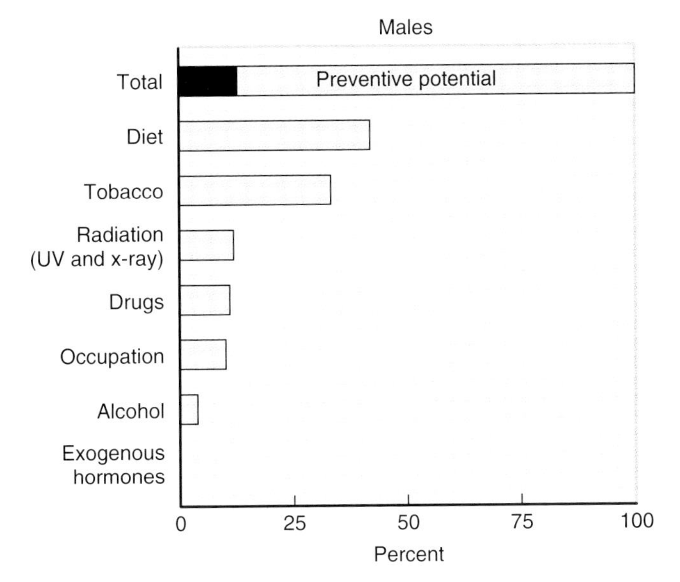

FIGURE 17.16

Environmental Factors and Cancer

The incidence of cancer could be dramatically lowered by reducing or eliminating exposure to a variety of environmental factors known or suspected to promote the development of cancer.

The answer depends on an understanding of the onset of cancer. Mutagens can affect any gene in a cell—but there are at least 50,000 genes in human cells, only about 60 of which are known to be proto-oncogenes. Thus the likelihood of a mutation in a proto-oncogene from a single event is extremely small. Moreover, several such mutations may have to occur before a full-blown cancer develops. However, you can readily see that the more events there are (that is, the more cigarettes smoked), the higher the probability of a mutation in a proto-oncogene. Thus two-pack-a-day smokers ultimately have a cancer rate that is 40 times the rate of the nonsmoking population (see Figure 17.15).

In the case of other cancers, modest dietary changes may be of considerable value (see Figure 17.16). High fiber diets are thought to provide some protection against colon cancer. Conversely, high fat diets are correlated with several cancers, most notably breast cancer (see Figure 17.17). Diets that are high in fat tend to lead to obesity, and estrogen levels rise in response to obesity. (Since high fat diets are also correlated with cardiovascular disease—see Chapters 7 and 8—we have two good reasons to change our dietary habits.)

In addition to reducing our intake of fat, we could increase our consumption of *cruciferous* vegetables—broccoli, cabbage, Brussels sprouts, and so forth—because they contain *indole-3-carbinol*, a chemical that helps convert estrogen in the blood into an inactive form. Asian women, who eat much more cruciferous vegetables than do Westerners, have a much lower incidence of breast cancer (see Table 17.5).

Wait a minute—what does estrogen (a hormone produced by the ovaries; see Chapter 20) have to do with breast cancer? As it happens, many breast tumors are initially estrogen-dependent. Estrogen receptors on the cancer cells capture estrogen molecules circulating in the blood, and the estrogen promotes the gradual growth of the tumor. (This role of estrogen helps explain why breast cancer is so much more common in women than in men.)

However, not all breast tumors are estrogen-dependent, and some that start out requiring estrogen later lose their estrogen receptors and thrive without this hormone. The post-surgical treatment of women with breast cancer now centers on determining if their tumor cells are estrogen-dependent. If they are, these women can be given *tamoxifen*, an estrogen mimic, which will be picked up by any remaining cancer cells, but which will not promote growth of a tumor. Women whose tumor cells are not estrogen-dependent, on the other hand, may require more extensive chemotherapy and/or radiation to eliminate any remaining cancer cells. Interestingly, soybeans contain a tamoxifen-like molecule, which may be another reason why breast cancer is less common in Asian women—for whom soybeans are a dietary staple—than in Westerners. Vitamin C is also thought to provide some protection against breast cancers. One group of researchers estimates that if all American women consumed the RDA for vitamin C every day, new cases of estrogen-dependent breast cancer would drop by 16 percent.

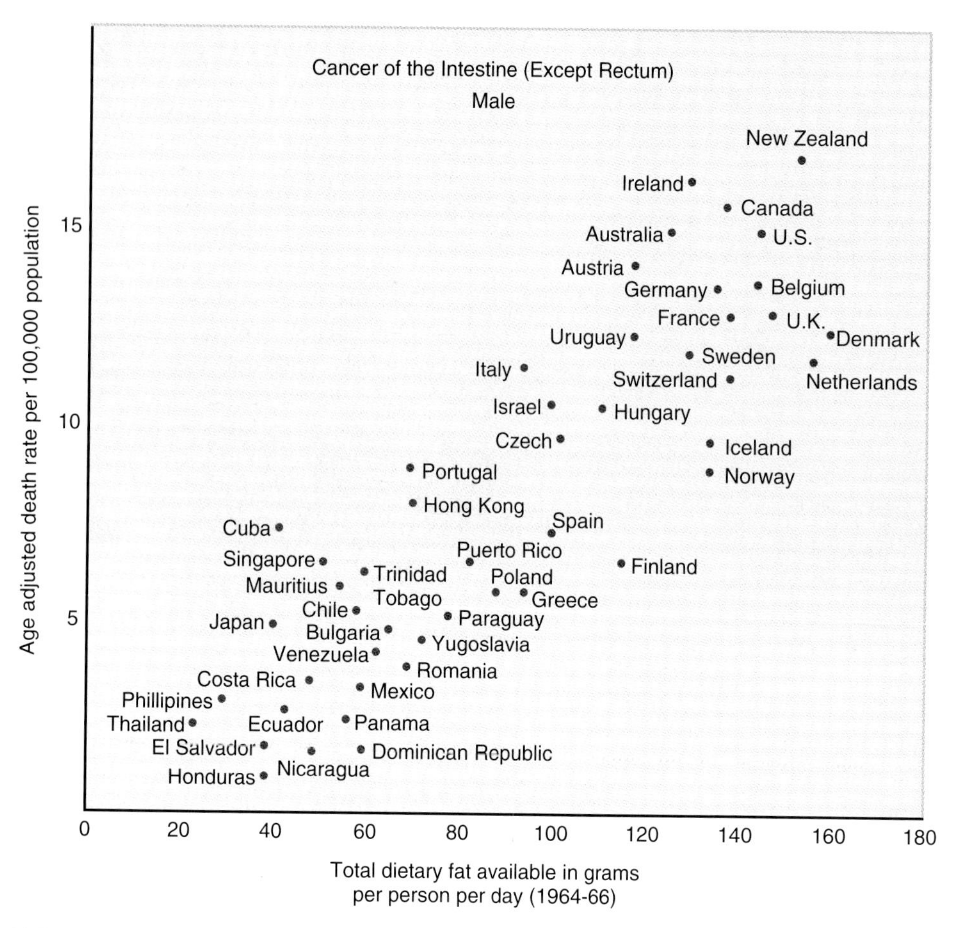

FIGURE 17.17

Intestinal Cancer and Diet

Data from this study show a clear correlation between the mortality rate from intestinal cancer and the abundance of fat in the diet.

TABLE 17.5 Relative Cancer Rates Among Native Japanese, the Descendants of Japanese Immigrants, and U.S. Whites

CANCER LOCATION	RELATIVE CANCER MORTALITY RATES		
	Japanese	*Offspring of Migrants*	*U.S. Whites*
Stomach	100	38	17
Colon	100	288	489
Pancreas	100	167	274
Lung	100	166	316
Leukemia	100	146	265

Source: W. Haenszel and M. Kurihara, "Studies of Japanese Migrants. I. Mortality from Cancer and Other Diseases Among Japanese in the United States." *Journal of the National Cancer Institute* 40 (1968), 43–68.

The Treatment of Cancer

There are several major methods of cancer treatment, which are often combined in individual cases:

Surgery. By scalpel, electric needle, or freezing, tumors can be destroyed, and surgery remains a favored treatment for many types of cancers. Often it is combined with other treatment methods to increase the likelihood that there will be no recurrence of the cancer.

Radiation. Radiation, either by radioactive drugs or by linear accelerators or other "guns," is increasingly effective for certain types of cancers, notably those of the throat and the lymph nodes (Hodgkin's disease). It is based on the finding that certain cancers are more susceptible to destruction by radiation than are the surrounding tissues. Radiation and **TNF** (*tu-*

mor necrosis factor; see Chapter 9) complement each other very well and are now commonly used in conjunction.

Chemotherapy. The treatment of cancers with drugs began 40 years ago with the discovery that estrogen slowed the growth of cancers of the prostate gland, a gland in men associated with the reproductive tract (Chapter 20). Since that time, there has been a rapid increase in the use of drugs in cancer treatment. Some drugs interfere with DNA replication. Since most cancer cells are dividing rapidly, their activity is impaired. Other drugs are antimetabolites that slow or impede the rate of growth of individual cells. Still other drugs interfere with RNA synthesis or protein formation. Often, these drugs are used in combination, in part because some tumors become resistant to drugs, and in part because not all the cells of a tumor are functioning the same way.

One recent technique, used especially in small tumors of the respiratory tract, uses dyes that are preferentially absorbed by cancer cells. Laser light is then used to alter these dyes chemically, causing them to kill the cancer cells.

Eyedrops that contain substances inhibiting the formation of new blood vessels are now being used on corneal tumors. Not only does the denial of new blood vessels starve the tumors, but it prevents the blindness that can sometimes develop simply because of the proliferation of blood vessels in the cornea. Similarly, a substance produced by blood platelets called **PF4 (platelet factor-4)** can prevent the growth of new blood vessels, and has successfully prevented cancer cells from forming tumors in mice. Human trials are expected to begin shortly.

Immunization. Because cancer cells are not foreign to the body in the same way that a bacterial cell is foreign, it was long assumed that standard immunization techniques would be unavailing in cancer treatment. Recently, however, **monoclonal antibodies** have come into use. In this technique, mice are injected with human cancer cells in response to which they manufacture antibodies. Much of the antibody production takes place in the spleen. The mice spleens are then removed and the spleen cells are fused with cancer cells from another mouse. These combined cells reproduce rapidly, and large amounts of antibodies are produced. Antibodies that react with proteins found only in cancer cells are then selected and are injected back into the patient where they attack the tumor, or are tagged with radioactive materials that are transported directly to the tumor cells.

The problem is that, because these antibodies are produced by mouse cells, they often trigger an immune reaction. Monoclonal antibodies obtained exclusively from human cells are still very expensive and difficult to obtain—but in the first clinical use of human monoclonal antibodies, the tumors of malignant melanomas disappeared within days following injection with carefully matched antibodies.

Immune System Activation. Vaccine-like injections of fragmented melanoma cells have also been used on patients with advanced melanomas, and the consequent activation of the immune response caused dramatic shrinkage of tumors, and in some cases caused the tumors to disappear.

Invigorating the body's own ability to fight cancer directly is now being attempted on several fronts. As we saw in Chapter 9, the body has several mechanisms for destroying cancer cells, and it is reasonable to assume that incipient cancer cells are regularly being destroyed in all of us. In that sense, the development of a tumor is the consequence of the failure of the immune system to respond adequately at the outset—and the spontaneous disappearance of cancer (although one of the rarest events in medicine) presumably results from a re-activation of the immune system.

In one study, lymphocytes were removed from a patient, cultured with interleukin-2 for several days, and then transfused back into the patient (along with the IL-2), as **LAKs (lymphokine-activated killer cells)**. LAKs caused complete or partial regression of tumors in 25 percent of patients suffering from a variety of cancers, and were particularly successful with kidney cancer and melanoma. **TILs (tumor-infiltrating lymphocytes),** a type of cytotoxic T cell first discovered in 1986, can be extracted directly from tumors, cultured with IL-2, and then injected back into the patient. Preliminary tests showed a 55 percent regression of tumors.

Another technique that combines immuno- and chemotherapy involves attacking the products of certain mutated genes, the presence of which distinguishes cancer cells from normal cells. Genes that control the production of growth factors or growth factor receptors are particularly good candidates, since these function at the level of the plasma membrane, rather than deep within the cell. Cancer cells are known to produce many types of growth factors—breast cancer cells, for example, produce at least five—and these growth factors bind with plasma membrane receptors and stimulate growth of the cell, and possibly metastasis.

One candidate gene is called *neu;* it encodes a protein that appears to be a growth factor receptor. Women whose breast tumor cells contain extra copies of this gene are more likely to have a recurrence of cancer than women whose breast cancer cells contain only a single copy of the gene. (The implication is that more growth factor receptors lead to more aggressive growth of cancer cells.) The same situation obtains for ovarian cancer. Antibodies to the protein produced by *neu* are now available, and clinical trials on patients with advanced breast tumors are currently underway. Two of the growth factors that bind to the *neu* receptor protein have also been identified, and chemicals that can bind to these growth factors (thereby preventing them from binding to the receptor protein) have been produced and are also now being used in clinical trials.

In summary, the techniques now being used to treat cancer are beyond the wildest imaginings of science just a few years ago, and are a direct result of the enormous increase of our understanding of the biology of the cell.

Summary

Mitosis and cytokinesis were among the first of the cellular events to be observed by the early light microscopists. Today, we tend to take them for granted, as virtually everyone is at least vaguely aware that cells have the capacity to divide. However, cell division is an enormously important phenomenon, because it underlies our capacity for growth, repair, and tissue specialization—and, where performed erroneously, for cancer.

Mitosis and cytokinesis do not result in an increase in genetic variability, and therefore are not valuable mechanisms for reproduction in times of environmental change. Meiosis, a specialized form of division that involves a halving of the chromosome number and a parallel reshuffling of the genetic material, is used in the formation of sperm and egg cells, the cells that fuse at the time of fertilization in sexually reproducing animals. Because of the random alignment of chromosomes during meiosis, and the process of crossing over, every sperm and every egg is genetically distinct, and this genetic variation is of critical importance in evolution (Chapter 22).

Cancer can be described as mitosis running amok. It is the result of mutations of genes critical to the cell cycle, which in some instances are inherited but which more commonly result from exposure to chemical agents or radiation. New techniques for treating cancer hold great promise for a general cure for this condition in the next few years, but in the meantime it remains the second leading cause of death in the industrialized world.

Key Terms

gamete
mitosis
cell cycle
cytokinesis
cell division
cyclin
maturation promoting factor (MPF)
prophase
chromatid
centromere
spindle apparatus
aster
metaphase
anaphase
telophase
cleavage furrow
meiosis
homologue
reduction division
equational division
synapsis
crossing over
chiasma
polar body
cancer
metastasis
tumor
benign
malignant
oncogene
proto-oncogene
adenoma
retinoblastoma
melanoma
retrovirus
mutagen
radon
tumor necrosis factor (TNF)
platelet factor-4 (PF4)
monoclonal antibody
lymphokine-activated killer cell (LAK)
tumor-infiltrating lymphocyte (TIL)

Questions

1. Briefly explain the cell cycle. How is entry into mitosis regulated?
2. Outline the four phases of mitosis, and briefly describe the primary events of each.
3. If humans have 46 chromosomes, and if a fertilized egg is the product of the union of the genetic material of a

sperm and egg, does that mean that a fertilized egg has 92 chromosomes? If it does not, explain why.

4. In what fundamental ways does meiosis differ from mitosis?
5. Does meiosis occur in all cells? If it does not, to which group of cells is it limited?
6. What is crossing over? Why is it important? When does it occur?
7. Distinguish between benign and malignant tumors. Are benign tumors really benign, in the sense of being harmless?
8. How do cancer cells differ from noncancerous cells at the level of their genes?
9. If smoking causes lung cancer, why does the cancer take so long to develop? Why don't all smokers contract lung cancer?

CHAPTER

18

Genetics

The Inheritance of Information in Code

What happens when two genes are on the same chromosome? Are they always or only sometimes inherited as a unit? Why are some inherited conditions limited to one sex—and why is that sex usually the male? If genes are on chromosomes, how do they control the synthesis of proteins that takes place on ribosomes in the cytoplasm? For that matter, how does the gene "instruct" the cell to assemble particular proteins consisting of particular sequences of amino acids?

These questions, and many others, will be answered in our discussion of genetics, the topic of this chapter.

THE NATURE OF THE PROBLEM

It was one of the earliest, and most debated, questions in biology: "How are parental traits inherited by the offspring?" Although easily posed, the question was not accurately answered until 1866. And even then the world seemed unprepared for the answer because it remained ignored until 1900!

We discussed the fundamentals of this question in Chapter 2, in conjunction with the integration of genetics and natural selection to form a unified theory of evolution. In Chapter 17 we discussed the cellular events associated with the duplication and separation of hereditary material during cell division, and the special problems that arise in the formation of reproductive cells (gametes). In light of these discussions, it is now time for us to consider some of the major questions that have concerned biologists in the twentieth century:

"What is the chemical nature of the hereditary material?"

"How do genes direct cellular metabolism?"

"How are proteins synthesized?"

"How are genes turned on and off?"

On an even broader scale we can ask, "How are the processes of life controlled and regulated?"

Although there are still many gaps in our understanding, we now have answers to most of these questions. These answers all began with Mendel.

FIGURE 18.1

Gregor Mendel
Mendel, who lived from 1822 to 1884, is generally regarded as the father of genetics because of his work on inheritance in the garden pea.

MENDELIAN GENETICS

In honor of Gregor Mendel (see Figure 18.1), the Augustinian monk who first formulated an accurate explanation of inheritance, the branch of genetics that deals with the inheritance of traits as units has been termed **Mendelian genetics.** By contrast, **molecular genetics,** discussed later in this chapter, is the study of the biochemical activity of the molecules involved in inheritance.

Mendel, who lived a quiet, monastic life during the middle of the last century, was by no means the first individual to investigate the nature of inheritance. However, unlike his predecessors, Mendel had a strong interest in mathematics and a powerful scientific mind. Equally important, as we shall see shortly, he had the good fortune to select a relatively simple genetic system—a series of anatomical traits of the common garden pea.

Mendel obtained strains of pea plants that were inbred for each of seven different characteristics: seed texture; seed color; flower color; flower position; pod shape; pod color; and plant height (see Figure 18.2). He had two distinctly different types for each of these characteristics. When Mendel planted his first seeds, he began a historic eight year study that culminated in an oral report in 1865 to the Brunn (Czechoslovakia) Society for the Study of Natural Science, followed by an article in 1866.

The Law of Segregation

Mendel first attempted to discover what happened when inbred strains differing in only one characteristic (such as flower color) were cross-fertilized. Garden peas generally self-pollinate, and therefore Mendel was obliged to hand-pollinate the flowers. For each of the contrasting seven traits, Mendel found that only one of the parental traits (*P generation*) appeared in the *F_1 generation* ("first filial"; L. *filius*, "son").

However, when Mendel allowed the F_1 plants to self-pollinate, the F_2 plants that resulted showed *both*

Trait	Dominant vs Recessive		F_2 Generation Results Dominant form	Recessive form	Ratio
Seed shape	 round	 wrinkled	5474	1850	2.96:1
Seed color	 yellow	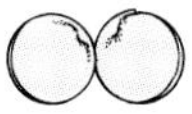 green	6022	2001	3.01:1
Flower color	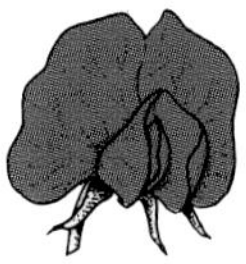 purple	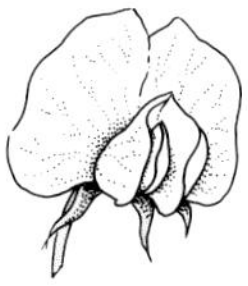 white	705	224	3.15:1
Pod shape	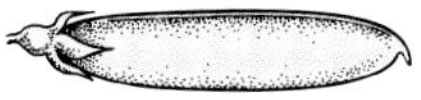 inflated	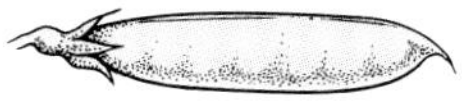 constricted	882	299	2.95:1
Pod color	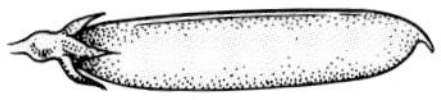 green	 yellow	428	152	2.82:1
Flower and pod position	 axial (along stem)	 terminal (top of stem)	651	207	3.14:1
Plant height	 Tall ×	 Dwarf	787	277	2.84:1

FIGURE 18.2

Mendel's Experiments

Mendel crossed inbred strains of peas that differed in a total of seven traits. In the second generation of the crosses he obtained data that demonstrated, for each trait, that one strain dominated the other in approximately a 3 to 1 ratio.

BOX 18.1

Pangenesis

It is interesting that only two years after Mendel published his work, Darwin, who continued to be thwarted by his inability to explain how variation arose in a population, resurrected the notion of **pangenesis,** an idea that dated back to 1651. In his book *The Variation of Animals and Plants Under Domestication,* Darwin suggested that heredity was accomplished by "gemmules," which are produced by all of the parents' tissues and which collectively constitute the hereditary component of the gametes. Following fertilization, the gemmules spread out and develop into the appropriate tissues of the offspring.

As Darwin explained it, not all of the gemmules would necessarily be expressed, and unused gemmules could appear in later generations. Thus characters could skip a generation. Moreover, gemmules would account for the *use and disuse* notion, because few or no gemmules would be produced by inactive tissues, whereas many gemmules

Pangenesis and Germ Plasm Theories of Gamete Formation

In pangenesis, all body parts contribute genetic material to the sex cells. In the germ plasm theory, only the gonads contribute genetic material to the sex cells. Thus changes to the other organs of the body (such as amputation, for instance) are not passed on to the sex cells for inheritance by the offspring.

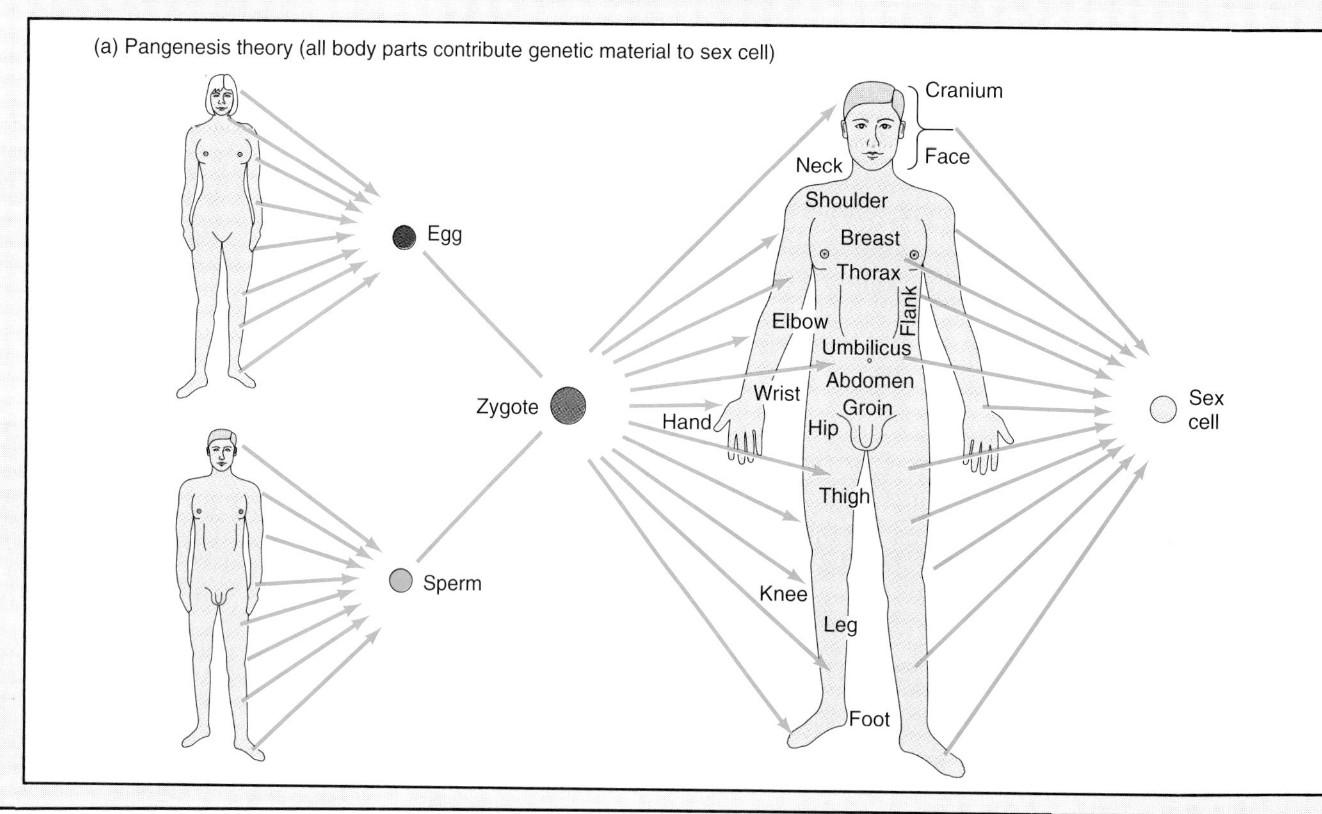

of the P generation traits. This finding was totally at odds with the most prominent theory of inheritance of the time, which held that parental traits were blended in the offspring, never to reappear in any subsequent generation (see Box 18.1). Mendel termed the trait that appeared in the F_1 **dominant,** and the trait that was absent from the F_1 **recessive.**

Remember that neither Mendel nor anyone else of his time knew anything about chromosomes, genes, or even mitosis and meiosis. Nevertheless, Mendel correctly deduced that each characteristic must be coded for by two inherited particles. The basis for this conclusion stemmed from his interest in mathematics.

First, Mendel noted that in the F_2 of each of his crosses there were approximately three times more dominants than recessives (see Box 18.2). Second, he used algebraic notations for dominance and recessiveness, assigning a letter in upper case for domi-

would be produced by active tissues. Limb or tail gemmules circulating in the blood would explain how salamanders and lizards can regenerate limbs and tails, respectively—and malformations were simply the results of wrong gemmules being expressed at a particular site. Finally, gemmules could "mutate," which would explain not only the source of variation, but also why variation was not blended out.

Thus, with this complicated but ingenious interpretation of pangenesis, Darwin managed to escape the enormous problems his theory of natural selection faced from the "blending" theory of inheritance. However, he was hard-pressed to explain sexual dimorphism (that is, how it is that males and females of a particular species often look very different, despite being the product of the same parental types)—and, of course, in the final analysis his endorsement of pangenesis was simply wrong.

The notion of pangenesis was ultimately replaced by the **germ plasm** theory of August Weismann, who cut off the tails of 22 generations of mice and showed that there was no decrease in tail length between the first and the twenty-third generation. His point was that changes occurring in nonreproductive tissue—which is segregated from reproductive tissue early in embryonic life—is not transmitted to the next generation.

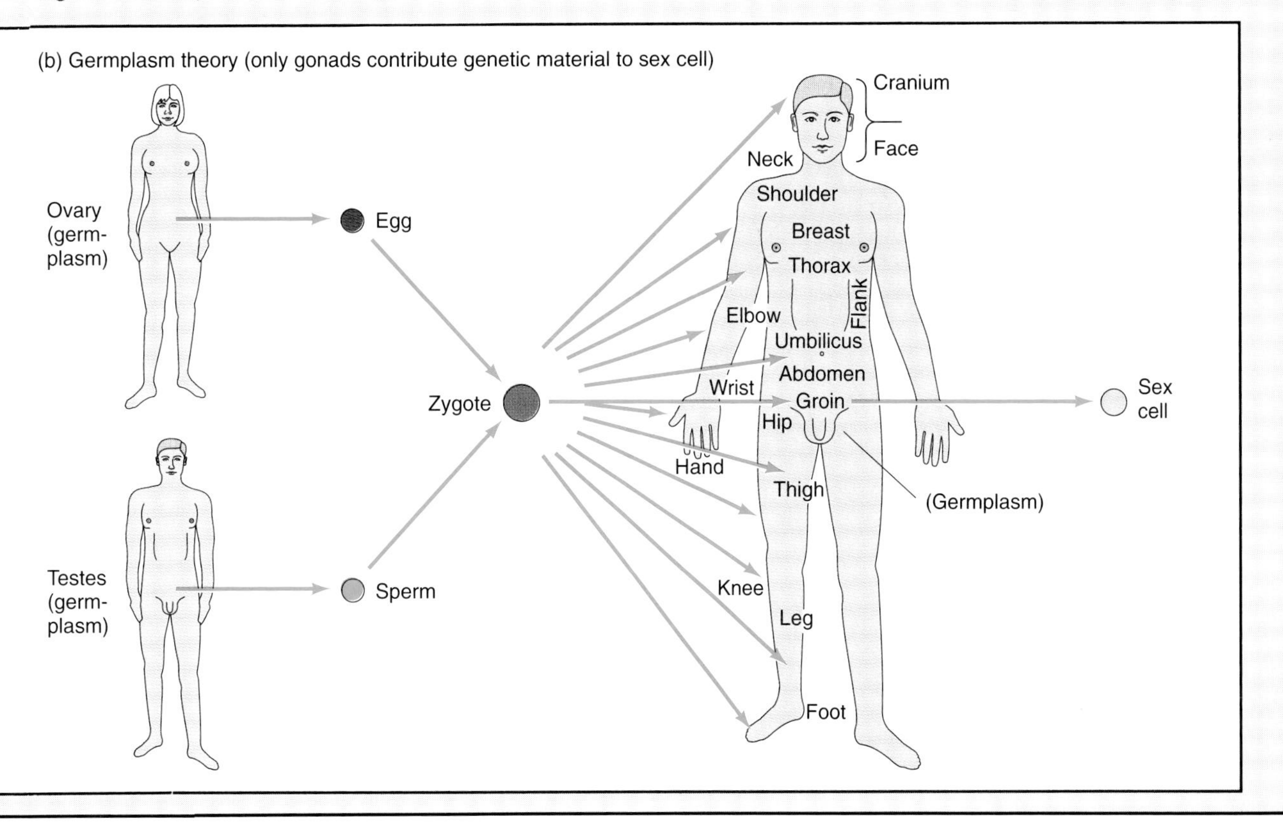

nance and the same letter in lower case for recessiveness (see Figure 18.3). Since both dominant and recessive traits reappeared in the F_2, both must somehow have been present in the F_1, even though only the dominant trait was manifested. Thus, if *Y* symbolizes the dominant yellow seed color, and *y* symbolizes the recessive green, the F_1 plants must be *Yy*. Since the F_1 plants have two inherited particles, it stands to reason that the P and F_2 plants must have two as well. The parental plants were inbred strains; therefore they must have been *YY* and *yy*, respectively. Finally, the F_2 must have consisted of a mixture of *YY*, *Yy*, and *yy*. Because one in four (3:1 ratio, remember?) was green (*yy*), logically one in four must be *YY* and two in four *Yy*. Mendel tested this last conclusion by planting just the yellow seeds and found that only one-third bred true (that is, were *YY*), indicating that his conclusion was valid.

BOX 18.2

The Punnett Square

Shortly after the turn of the century, an English geneticist named R.C. Punnett devised a simple method of mapping crosses in a grid, thereby providing a picture of Mendel's mathematical ratios. Beginning students have been in his debt ever since.

Suppose we are crossing plants with round seeds and plants with wrinkled seeds. We know that round is dominant to wrinkled, and we therefore designate it as *R*. Conversely, the wrinkled trait is designated as *r*. If the cross is between plants from inbred strains, all of the gametes of the round strain will contain the dominant *R* and all of the gametes of the wrinkled strain will possess the recessive *r*. Thus a grid will look like this:

All of the F_1 will possess the configuration *Rr*. If these are now crossed among themselves, on the average, 50 percent of the gametes will contain *R* and 50 percent will contain *r*. Hence the following grid:

Because *RR* and *Rr* look the same (the presence of a single dominant *R* is sufficient to create the round shape), there will be an average of three round seeds for every wrinkled seed—a ratio of 3:1, as Mendel had found.

Mendel performed the same types of crosses with plants differing in the other characteristics. From all these findings, Mendel developed what has come to be known as Mendel's First Law, or the **Law of Segregation.** In brief, it states:

> Two hereditary particles control each trait, and these are segregated into different gametes in equal proportions.

Thus, in a small series of carefully designed experiments, Mendel effectively put an end to the blending theory of inheritance.

The Law of Independent Assortment

Mendel's initial experiments all involved differences in a single trait—what we now call a **monohybrid**

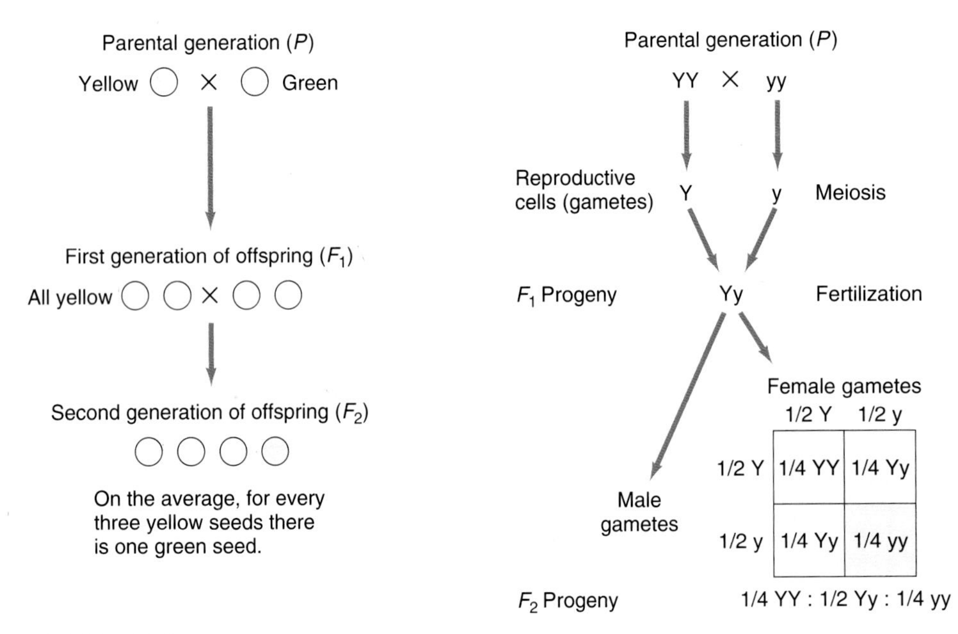

FIGURE 18.3

Law of Segregation

(Left) When Mendel crossed a pea with yellow seeds and a pea with green seeds, the offspring all were plants with yellow seeds. However, when these offspring were crossed, their offspring (that is, the second generation) averaged three plants with yellow seeds for every plant with green seeds. (Right) We can explain these results by assigning letters to each of the traits. The yellow seed parents are *YY*; the green seed parents are *yy*. The first generation are all *Yy*, meaning that they have yellow seeds. The second generation, however, has one plant in four with *yy*, meaning that there are three plants with yellow seeds for every plant with green seeds.

cross. Having formulated his first law, he was curious to learn if different traits interfered with each other during gamete formation. He therefore attempted a number of **dihybrid crosses**—crosses between inbred plants that differed in two traits.

One of Mendel's crosses was between plants with round, yellow seeds and plants with wrinkled, green seeds. From his monohybrid studies, he knew that round was dominant to wrinkled, and yellow dominant to green. Because he was crossing inbred lines, the appropriate symbols are

$$RRYY \text{ vs. } rryy.$$

Mendel was not surprised to see that the F_1 consisted entirely of round, yellow seeds, all of which would have been

$$RrYy.$$

However, when Mendel allowed the F_1 to self-pollinate, he found not just the two parental types in the F_2, but four distinct types (see Figure 18.4). They included not only round/yellow and wrinkled/green, but round/green and wrinkled/yellow. Not only that,

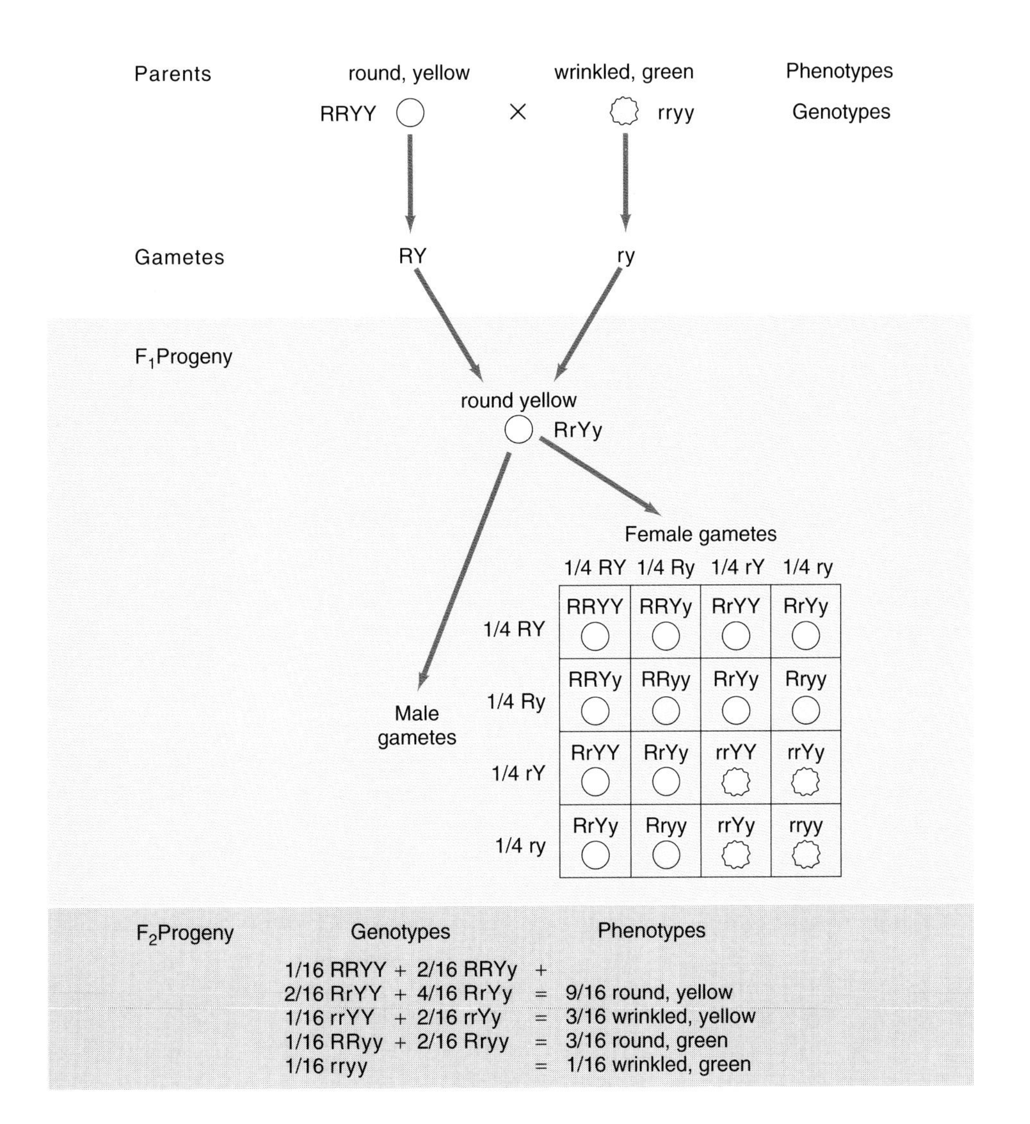

FIGURE 18.4

Law of Independent Assortment

When Mendel crossed pea plants differing in two traits (plants with round, yellow seeds and plants with wrinkled, green seeds), the first generation plants all had round, yellow seeds. However, the second generation had a ratio of nine plants with round, yellow seeds, to three plants with round, green seeds, to three plants with wrinkled, yellow seeds, to one plant with wrinkled, green seeds. That is, these two traits assorted independently of each other, and the ratio was a product of the expected single trait ratio (that is, 3:1 × 3:1 = 9:3:3:1).

but they were in a 9:3:3:1 ratio, with round/yellow the most abundant, and wrinkled green the least abundant.

Again, the mathematician in Mendel made him realize that this 9:3:3:1 ratio was, in fact, $(3:1)^2$. That meant that the two traits assorted themselves completely independently of each other. From this finding, confirmed with other dihybrid crosses, came Mendel's Second Law, the **Law of Independent Assortment,** which states:

> In the formation of gametes, the two hereditary particles for a given trait are distributed independently from those of any other trait.

Mendel even tried trihybrid crosses and found they, too, obeyed this Second Law.

These and other findings were presented both orally and in writing, but they attracted little attention. As if to rub salt in the wounds of the obscurity in which he labored, Mendel had the misfortune of being advised by Carl Naegli, a prominent botanist who doubted Mendel's findings, to repeat his experiments on another plant species. Naegli suggested the hawkweed, and Mendel labored unsuccessfully for five years attempting to verify his findings on this second species.

Failing eyesight, increased administrative duties in the monastery, and a total lack of success forced him to give up, never knowing whether the genetic model he had devised was valid only for peas, or if it was valid for all organisms except for the hawkweed. It is probably unnecessary to note that the second alternative was true, although it was not confirmed until years after Mendel's death in 1884. The hawkweed, although it produces seeds, reproduces asexually most of the time, a fact unknown to either Mendel or Naegli. It seems a pathetic finale for one of the great minds in nineteenth century biology.

POST-MENDELIAN GENETICS

Unlike most scientific luminaries, who are usually just slightly ahead of their time, Mendel was years in front of other biologists, and, for that reason, he was simply ignored. He was not a prominent scientist; his findings were published in a very modest journal, and his mathematical orientation was just not understood by most biologists. However, when Mendel's paper was "rediscovered" in 1900 (see Chapter 2), scientists realized that Mendel's notion of pairs of heritable particles that separate during gamete formation meshed very well with the observation, made a few years earlier, that chromosomes occur in pairs that separate in the formation of gametes. Were Mendel's particles arrayed on chromosomes?

The beginnings of the answer to this question were provided by Thomas Hunt Morgan and his students at Columbia University during the first two decades of this century. Morgan wanted to experiment with a species that had a generation time shorter than that of the pea, and he selected a small species of fly, *Drosophila melanogaster,* known familiarly as the fruit fly. These flies were easily reared on the yeast that grows on decaying fruit, and their life cycle is less than two weeks long, a decided improvement on the garden pea, which manage only two or three generations a year in temperate climates.

However, unlike the pea, there were no inbred strains in *Drosophila*. Eventually, a white-eyed male appeared in one of the cultures (normally, *Drosophila* has red eyes), and Morgan bred this male to a red-eyed female. The F_1 were all red-eyed, but in the F_2 roughly one-third of the flies had white eyes. Those results demonstrated not only that the white eye trait was inherited, but that the inheritance pattern followed Mendelian ratios. There was just one small problem—the only flies that had white eyes were male! All of the females and half the males had red eyes. Why weren't there any white-eyed females?

Sex Linkage

Even as Morgan and others labored to unravel the mysteries of inheritance, other scientists endeavored to understand the significance of mitosis and meiosis. Mitosis was observed in large numbers of species, and it was noted that, in some species of plants and in most species of animals, the number or arrangement of chromosomes differed between males and females. Commonly, but not always, the male possessed either a mismatched pair or was missing one chromosome, whereas the female possessed a true pair. Because it was presumed that these chromosomes were involved in sex determination, they came to be known as the **sex chromosomes** (as distinct from the **autosomes**—the other chromosomes that exist in matched pairs in both males and females). Males of species possessing a mismatched pair were designated as XY; those possessing only a single sex chromosome were designated as XO; and females with a matched pair of sex chromosomes were XX.

Morgan's choice of *Drosophila* as an experimental species was indeed fortuitous, because it proved the ideal species not only for breeding experiments but also for microscopic examination. The cells—and their chromosomes—in the salivary glands of the larvae are huge, and *Drosophila* was found to have only three pairs of autosomes and one pair (XY male; XX female) of sex chromosomes.

It occurred to Morgan that his F_2 findings in the white eye cross could best be explained by assuming that the trait for white eyes is transmitted on the X chromosome. Since the trait is recessive, it will be manifested only when there is no dominant trait blocking it.

To test this hypothesis, Morgan crossed a white-eyed male to an F_1 female. If white eye is transmitted on the X chromosome, then both males and females would show white eyes, since the F_1 female would carry traits for both normal and white eyes. Morgan found that half the males and half the females were white-eyed, and correctly concluded that white eyes were transmitted on the X chromosome. Traits transmitted on the sex chromosomes are said to be **sex linked.**

Linkage and Chromosomal Mapping

Morgan and his students soon collected a large number of true-breeding strains of *Drosophila,* and performed large numbers of dihybrid crosses. Almost immediately, however, their results failed to follow Mendel's predicted 9:3:3:1 ratio. Frequently, only the parental types were represented in the F_2. Sometimes, the recombinant pair were present in the F_2, but at frequencies much lower than expected. What was happening?

Recall that *Drosophila* possesses only four pairs of chromosomes, but, of course, has thousands of genes. It is evident, therefore, that in many of the dihybrid crosses, the genes for the two hybrid traits would have been on the same chromosome. However, the transfer of equivalent parts of the chromosome during crossing over (see Chapter 17) will tend to separate genes on the same chromosome, with the likelihood of separation being a function of how close they are together. Thus dihybrid crosses in which the F_2 showed an almost 9:3:3:1 ratio must have traits widely separated on the chromosome. Those in which there was strong deviation from the 9:3:3:1 ratio must be very close together on the chromosome. Using the frequency of separation as an index, Morgan and his co-workers were able to map the chromosomes, showing the relative location of the traits they were studying (see Figure 18.5).

The Chromosomal Theory of Inheritance

From the studies of Morgan and other early twentieth century geneticists (all of whom built on Mendel's solid foundation) and from the findings of the microscopists who so thoroughly detailed the happenings of the living cell arose the *chromosomal theory of inheritance.* Using modern terminology, it states the following:

1. The hereditary units (now called **genes**) are found on the chromosomes.

2. Except for gametes, each cell possesses two sets of chromosomes, one set from the father, and the other from the mother; cells with two sets of chromosomes are said to be **diploid** (symbolized as $2n$).

3. Gametes, which have undergone meiosis, have only one set of chromosomes and are called **haploid** (symbolized as $1n$).

4. Each gene is found at a characteristic point, or *locus,* on a chromosome.

5. Genes may exist in various chemical states, or **alleles,** and generally one allele is dominant to other, recessive, alleles, meaning that it is expressed whereas recessive alleles are masked.

6. Individuals that possess the same allele on each of their sets of chromosomes are said to be **homozygous** with respect to that trait; individuals with different alleles on each of their sets of chromosomes are **heterozygous** (or *hybrid*).

7. The genetic constitution of an organism is its **genotype;** its outward appearance is its **phenotype.**

8. The genes on a particular chromosome tend to be inherited together, but they may be separated by crossing over during meiosis, the likelihood of separation being a function of the distance between the two genes.

9. The chemical structure of genes may be altered during the construction of new chromosomes; such altered genes are called **mutations.**

Other Post-Mendelian Findings

Over the years, the simple Mendelian Laws have been modified by a variety of findings. These include the following:

Incomplete Dominance. Alleles are not always in a dominant-recessive relationship. Flower color in many plant species is very akin to the blending theory of inheritance that Mendel's results displaced. For example, it is fortunate that Mendel did not choose snapdragons for his experiments (see Figure 18.6), because plants with red flowers crossed with white flowers produce plants with pink flowers!

Multiple Allelic Systems. Mendel worked with systems in which there were only two alleles. However, for some traits, there are a large number of potential alleles, and the dominance relationships may be complex. For instance, in the ABO blood groups (Chapter 9), type A and type B are **codominant** (both

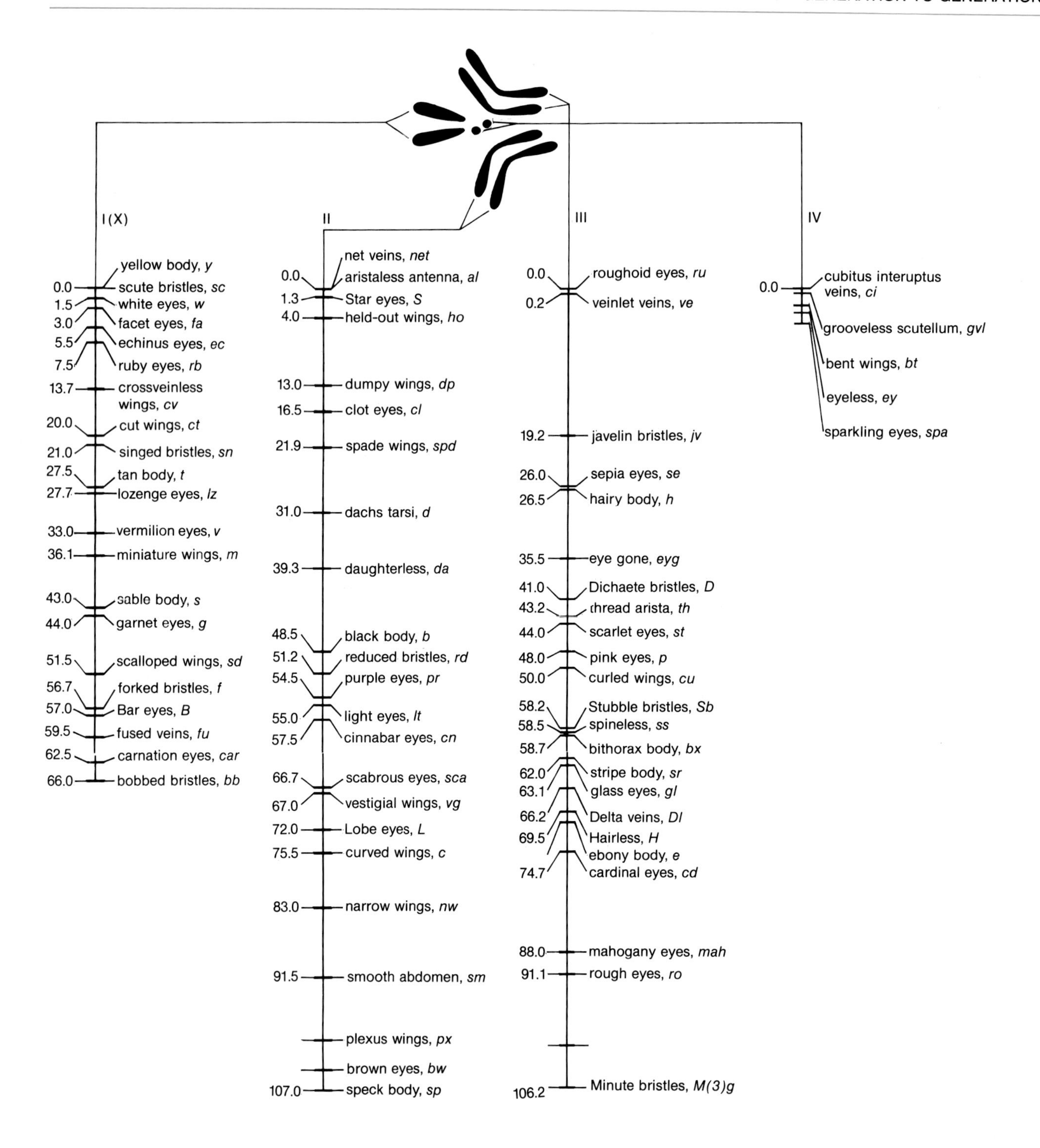

FIGURE 18.5

Chromosomal Mapping

Morgan also pioneered the development of chromosome maps as a way of positioning particular genes relative to other genes. Here are just a few of the genes identified in *Drosophila,* with their relative location on the four chromosomes found in this species.

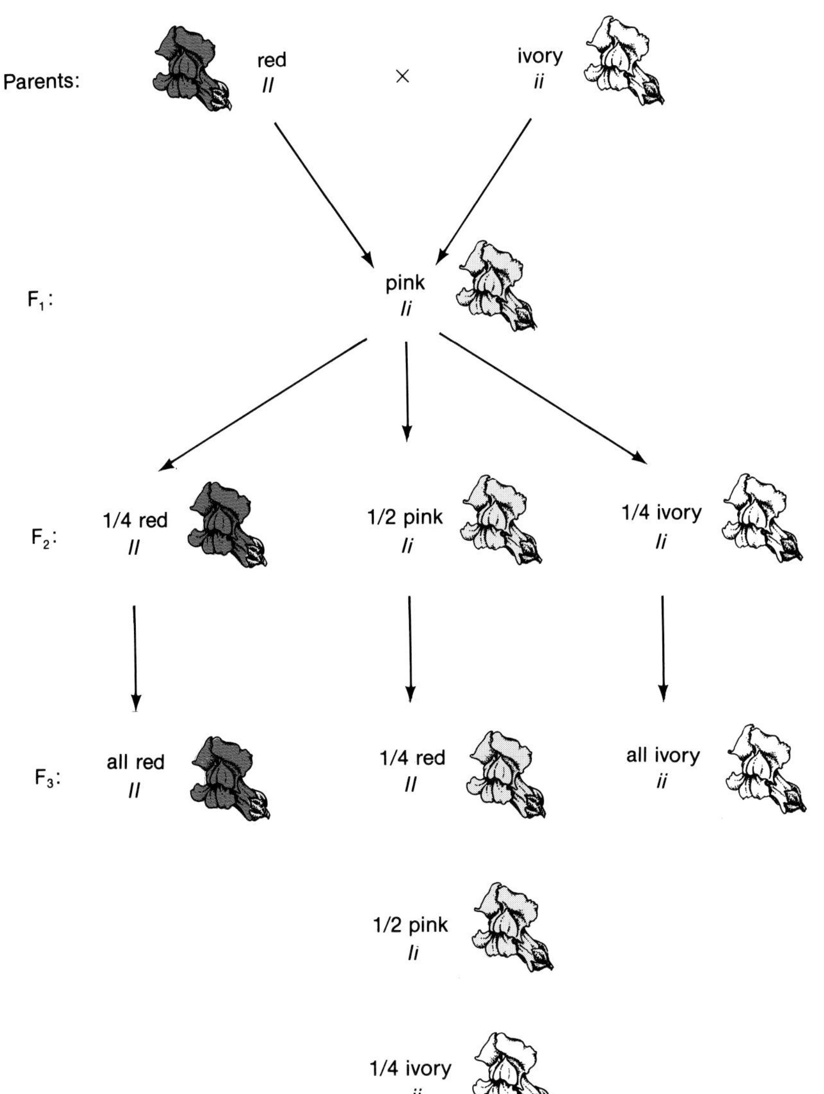

FIGURE 18.6

Incomplete Dominance

Fortunately, Mendel chose peas, not snapdragons, for his work. The offspring of a cross between a plant with red flowers and a plant with white flowers have pink flowers, a finding that would have supported the blending notion of inheritance that was prevalent in Mendel's time.

antigens are produced in an AB individual), but both type A and type B are dominant to type O. Thus it would be possible for a heterozygous type B (*Bo* genotype) mother and a heterozygous type A (*Ao* genotype) father to produce a type O (*oo* genotype) child, a finding that would have had Mendel doing some extensive head-scratching.

Polygenic Inheritance. Some traits, such as skin color or height, are controlled by a series of unlinked genes, each with at least two alleles. These traits are therefore not "either/or" situations, but instead exemplify graded responses, along a continuous gradient (see Figure 18.7).

Epistasis. One gene may mask the effects of other genes, a phenomenon called **epistasis.** It is not the same as dominant/recessive relationships, because it involves different genes, not alleles of the same gene. For example, a dominant allele is responsible for synthesizing the skin pigment *melanin;* other genes control the deposition of melanin and, hence, the overall skin color. *Albinism,* the failure to produce melanin, is a homozygous recessive condition. Thus, in albinos, none of the other genes that normally affect skin color can function, since all require melanin as a first step.

Limited Penetrance. Sometimes the presence or absence of a particular trait is due to a constellation of interactions between a gene and its neighbors or the environment. The tendency to contract various diseases such as tuberculosis and diabetes is inherited, but whether or not the individual will actually become tubercular or diabetic depends on a variety of other factors.

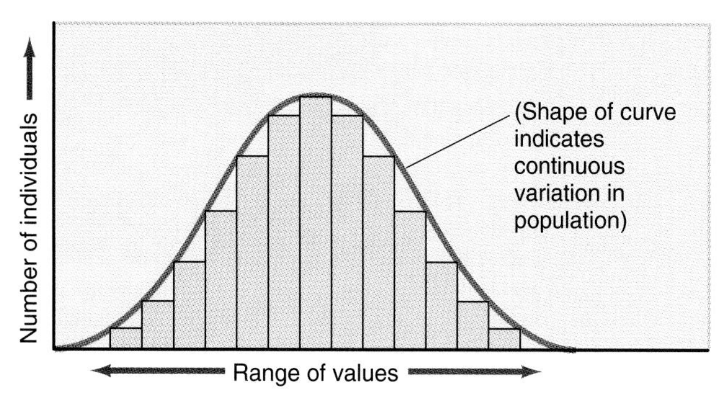

FIGURE 18.7

Polygenic Inheritance

Traits such as height in humans are controlled by the interaction of several genes. Variation is therefore continuous, not discrete. This graph illustrates a classic bell curve.

Obviously, the total picture is a good deal more complex than what Mendel first described, but to him we owe that most important first step.

MOLECULAR GENETICS

The chromosomal theory of inheritance united breeding experiments with the cellular events observed under the microscope, and thereby provided a physical explanation for inheritance. Genes exist on chromosomes, and chromosomes can be seen moving about during mitosis and meiosis. But what *are* genes? What are they composed of? How do they function? The chromosomal theory of inheritance answered none of these questions, but many scientists from the 1920s to the present day have spent their careers providing answers.

As the details of cellular chemistry gradually became understood, it was obvious to most scientists that genes must be proteins. Of all the large molecules of the cell, only proteins seemed variable enough to constitute the thousands of different genes in every cell.

Nucleic acids had been discovered in cells in 1868, but they did not elicit much interest. However, in 1944, a group of scientists found that they could alter the predictable outcome of experiments on the pathogenicity of certain bacteria by adding an enzyme that destroyed DNA but was harmless to protein. Was DNA the molecule of heredity? If it was, sniffed most scientists, it was true only for bacteria—proteins, they were sure, must hold the key for eukaryotes.

Shortly thereafter, another group of scientists who were experimenting with viruses that attack bacteria found that only the virus' internal DNA, and not its external protein coat, was injected into the bacterial cell. Since new viruses were manufactured by the pirated metabolic machinery of the bacteria, obviously the viral DNA must be the molecule responsible.

Viruses are even simpler than bacteria, and many scientists remain unconvinced that DNA had any significance to more complex organisms. However, the fact that it was the molecule of heredity in both viruses and bacteria was sufficiently intriguing to kindle renewed interest in DNA in general and its structure in particular.

The Structure of DNA

Chemical analysis had previously demonstrated that DNA consisted of a chain of *deoxyribose* molecules (a 5-carbon sugar) linked by phosphate groups. It was also known that one of four different nitrogen-containing bases was attached to each deoxyribose molecule. Two of these bases, *cytosine* and *thymine*, are single-ring *pyrimidines*. The other two, *adenine* and *guanine*, are double-ring *purines* (see Figure 18.8). It was initially assumed that all four bases were present at roughly equal frequencies, and indeed that is true for some species. However, in 1950, when a variety of species were analyzed, it was found that the base ratios varied from species to species, although the ratios from different tissues or individuals within a given species were always consistent. More importantly, it was noted that the amount of adenine always equaled the amount of thymine (A = T), and the

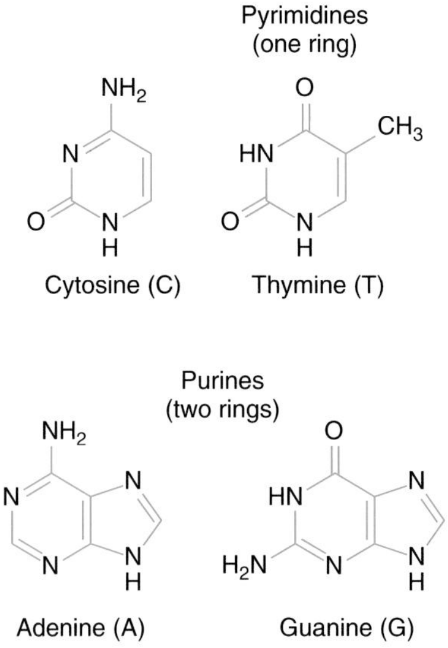

Pyrimidines (one ring)

Cytosine (C) Thymine (T)

Purines (two rings)

Adenine (A) Guanine (G)

FIGURE 18.8

Nitrogenous Bases in DNA

The four nitrogenous bases found in DNA include the pyrimidines cytosine and thymine, and the purines adenine and guanine.

FIGURE 18.9

Watson and Crick
In 1953 two young investigators at Cambridge University, James Watson and Francis Crick, first deduced the structure of DNA. For this work, they were awarded the Nobel Prize in 1962.

amount of guanine always equaled the amount of cytosine (G = C).

At about the same time, other investigators were analyzing DNA crystals by X-ray. They found that the molecule was long and thin, it had a consistent width and a repetitive structure, and it was coiled like the springs of an innerspring mattress. The door was open for someone to put all the pieces together, and to formulate a working model of the molecule. There were many who aspired, but the first ones through the doorway were James Watson and Francis Crick (see Figure 18.9).

The Watson-Crick Model. From the analytical and X-ray studies mentioned above, Watson and Crick knew that A = T and C = G, and that the width of the DNA molecule was constant. However, adenine and guanine are larger molecules than thymine and cytosine. Why didn't the DNA molecule bulge and thin with the alternate appearance of the purines and pyrimidines?

The answer, as Watson and Crick deduced from experimenting with models of the molecules, was that DNA exists as two complementary strands, with weak hydrogen bonds between A - T and C - G holding the two strands together in a double helix (see Figure 18.10). This model explained why A = T and C = G, and also accounted for the regular width of the molecule—with each "rung" of the spiraled ladder consisting of both a purine and a pyrimidine (see Figure 18.11).

With the publication of their model in 1953, Watson and Crick changed the focus of molecular genetics from how DNA was composed to how it functions.

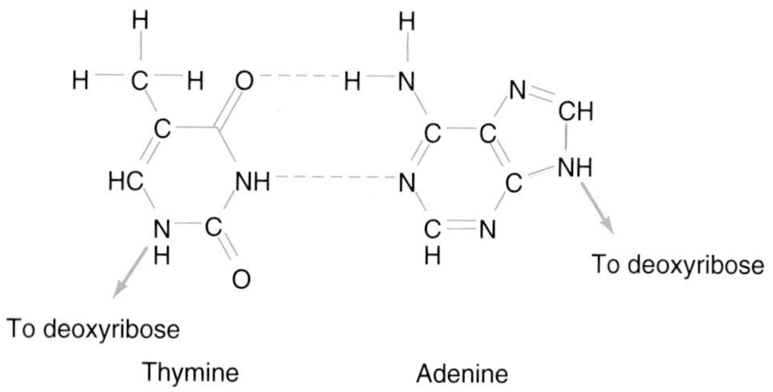

To deoxyribose

Thymine Adenine

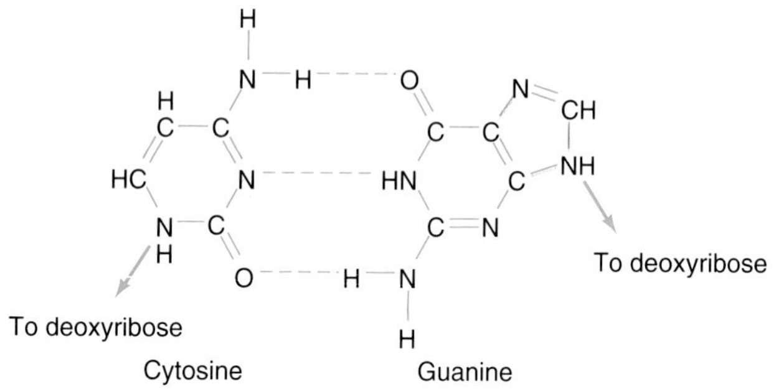

To deoxyribose

Cytosine Guanine

FIGURE 18.10

Hydrogen Bonding Between Purines and Pyrimidines
Thymine always bonds with adenine, and cytosine with guanine. Therefore the frequency of thymine is always the same as adenine, and cytosine equals guanine.

DNA Replication

One beauty of the Watson-Crick model was that it permitted easy speculation regarding the way in which DNA replicated itself before cell division. Hydrogen bonds are chemically very weak (Chapter 3). If the two DNA strands were to separate, each could serve as a template, or mold, for a new strand, and the net result would be two separate double strands.

In fact, this is exactly how replication occurs, as was proved in experiments reported in 1958. Enzymes cause the double helix to "unzip," creating a **replication fork** as the hydrogen bonds are broken (see Figure 18.12). Sugars, phosphates, and bases present in the nucleus are then assembled in such a way that each base is positioned opposite its complementary base. In this way, two new chains of DNA are formed, each complementary to one of the original chains. The end result is two complete and identical DNA double chains.

The error rate in DNA replication is extraordinarily low—perhaps only one in 10 billion bases. When DNA replicates in the absence of enzymes, however, the error rate is one in 100 bases. How do enzymes function to reduce the error rate?

Enzymes function in three ways: selection of bases, proofreading, and mismatch repair. **DNA polymerase** is the enzyme that selects the bases; following base selection, the error rate is one in 100,000 bases. An enzyme that is a part of or at least associated with DNA polymerase serves as a proofreader, eliminating wrong bases. Following proofreading, the error rate is one in 10 million. Mismatch repair is accomplished by a complex of enzymes not directly connected with DNA polymerase, and it further reduces the error rate to one in 10 billion.

RNA Transcription

From experiments on bread molds during the 1940s, scientists understood that mutated genes caused enzyme deficiencies. It was assumed, therefore, that each gene controlled the synthesis of one enzyme, a notion that came to be known as the *one gene–one enzyme concept.*

However, DNA and DNA replication occurs exclusively in the nucleus, whereas enzyme and other protein synthesis takes place in the cytoplasm. How does DNA direct an event that occurs in the cytoplasm?

Enter RNA, the other nucleic acid. The presence of RNA in cells had been known for some time, and the possibility that it served as a go-between, linking the nucleus and the cytoplasm, began to attract serious attention (see Figure 18.13).

RNA Structure

Although both are nucleic acids, RNA differs from DNA in a number of important ways:

1. The sugar in RNA is ribose; the sugar in DNA is deoxyribose.

2. In RNA, the nitrogenous base *uracil* substitutes for thymine.

3. RNA almost always exists as a single strand; DNA almost always is double-stranded.

4. RNA is very much shorter than DNA.

RNA Synthesis

In a process known as **transcription,** RNA is synthesized in much the same way that DNA duplicates (see Figure 18.14). A portion of the DNA double strand separates, and one of the strands serves as a template for the construction of a molecule of RNA, under the influence of the enzyme **RNA polymerase.** The length of separated DNA may be thought of as corresponding to a gene. Subsequently, the newly formed RNA molecule migrates through the pores of the nuclear envelope and enters the cytoplasm. (In actual practice, many copies of RNA—often thousands—are successively transcribed from a single gene.)

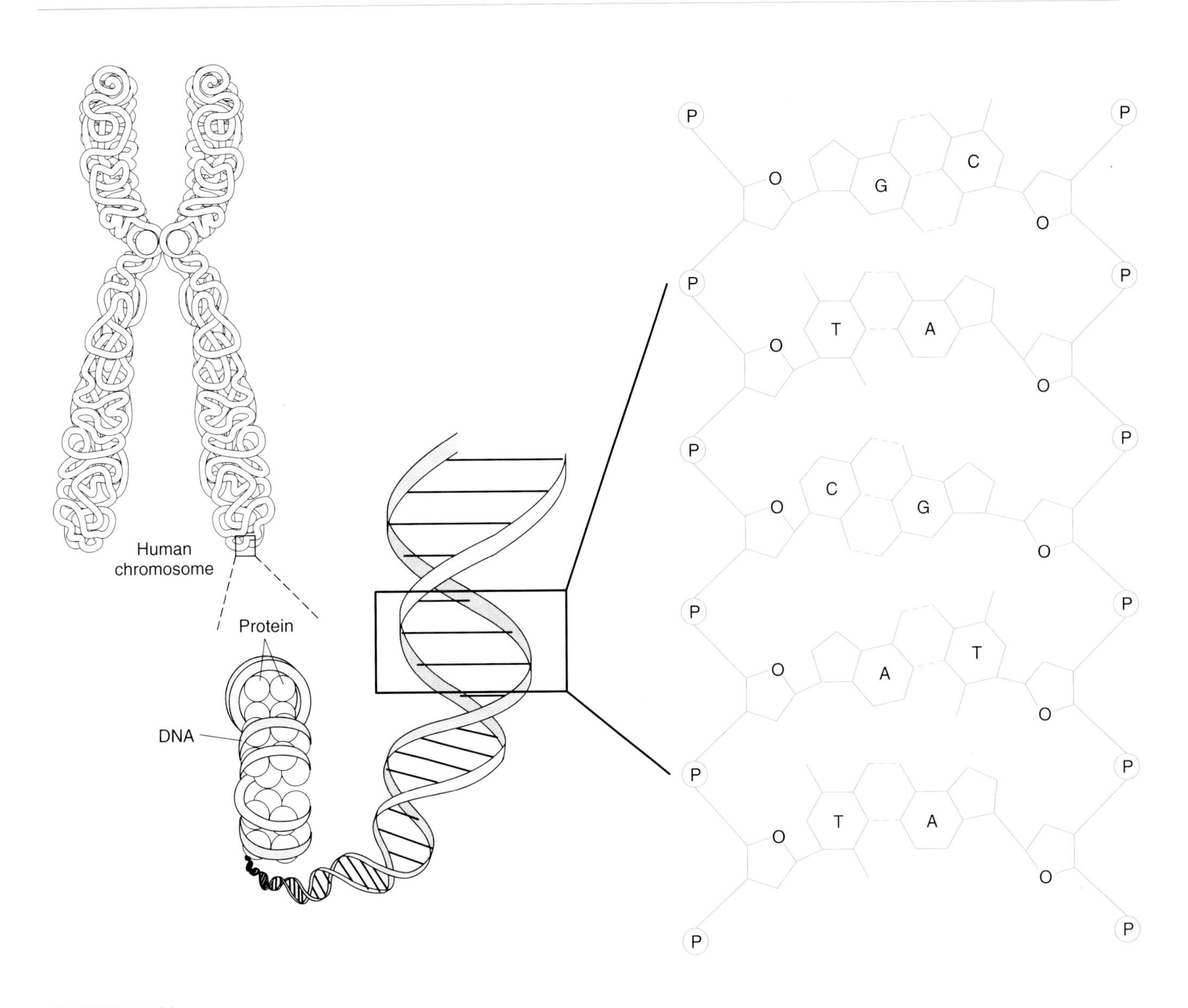

FIGURE 18.11

Organization of Chromosomes

During mitosis, eukaryotic chromosomes are seen as DNA strands tightly wound around proteins. The DNA strand is actually a double strand, the two strands being held together by hydrogen bonds between purines and pyrimidines. There is a backbone along each strand of alternating sugars (deoxyribose) and phosphates; a purine or pyrimidine is attached to each sugar.

PROTEIN SYNTHESIS

The sequence of bases along the DNA molecule can be thought of as a code, although a code limited to four symbols. Nonetheless, with a long enough string of these four symbols, unique packets of information can be coded. These code packets are then transcribed in RNA molecules. So far, so good. But how is the sequence of amino acids in a protein directed by the sequence of bases in the nucleic acids? That is, how does the cell crack the DNA code?

These questions formed the basis of much of the research in molecular biology during the 1960s and 1970s. Although there is still a good deal that we do not know, the basic steps are now rather well understood.

The Triplet Code

Cracking the DNA code was done in the same way that most codes are cracked—by logic and deduction. Proteins are composed of 20 different amino acids.

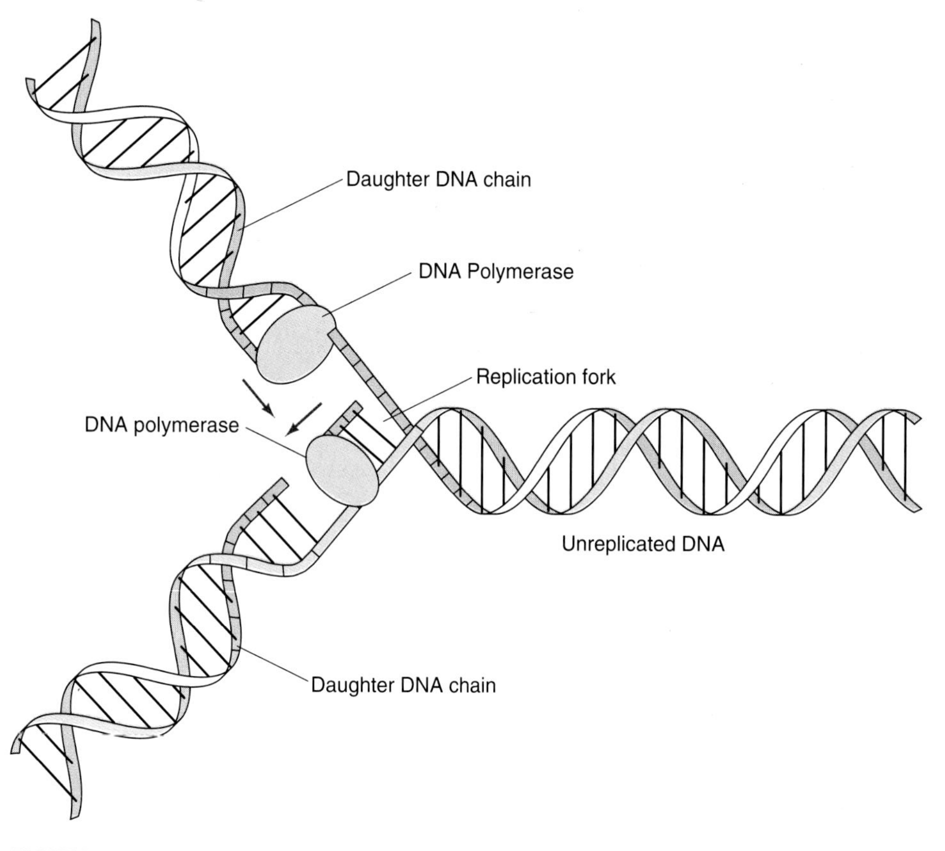

FIGURE 18.12

DNA Replication

Enzymes "unzip" the DNA molecule, and each half of the molecule acts as a template for the assembly of a new counterpart strand.

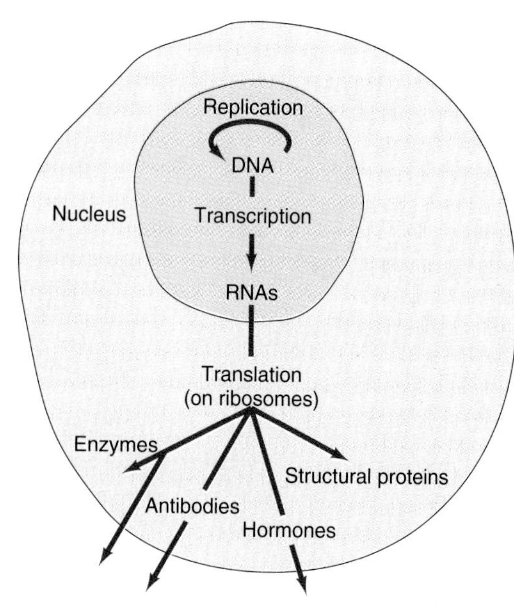

FIGURE 18.13

Replication, Transcription, and Translation

Replication and transcription occur in the nucleus; translation occurs in the cytoplasm.

However, there are only four different nitrogenous bases in DNA. It is therefore impossible for one base to code for each amino acid. Furthermore, even if a pair of bases coded for each amino acid, there would still be only (4 × 4 =) 16 possible pairs. It would require at least a triplet of bases to provide enough different combinations to permit a unique identifier for each of the 20 amino acids, although a triplet code provides (4 × 4 × 4=) 64 different combinations—and a good deal of redundancy.

Based on direct experimental work, all 64 triplets (called **codons**) have been determined to have a particular function in code (see Figure 18.15). Sixty-one triplets code for particular amino acids (some amino acids are coded for by as many as six different triplets, some as few as one), and three triplets signal a "stop" in the construction of a protein. Moreover, since every protein begins with the amino acid *methionine*, the codon for methionine also serves as the "start" signal. Thus every one of the 64 possible codons is, in fact, meaningful in protein synthesis.

(a)

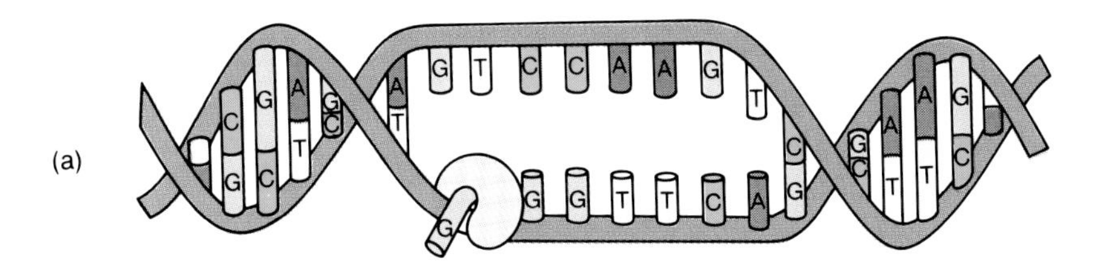

(b)

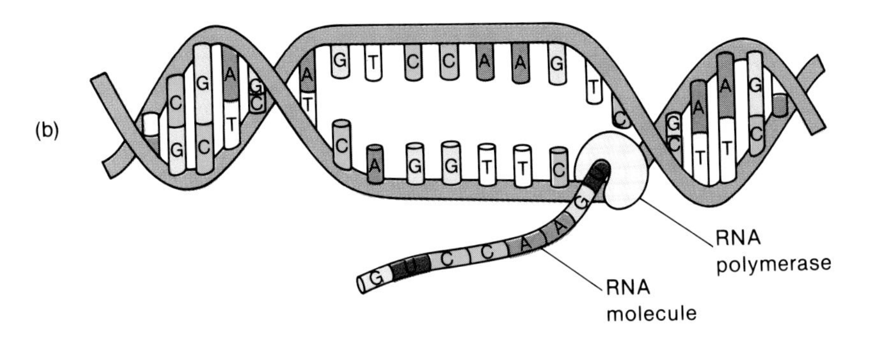

(c)

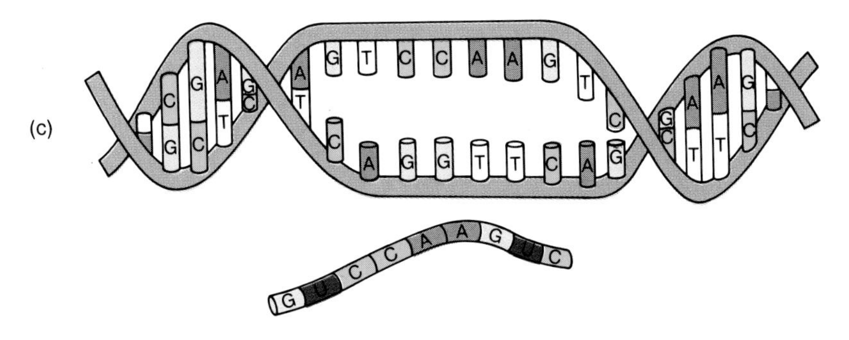

(d)

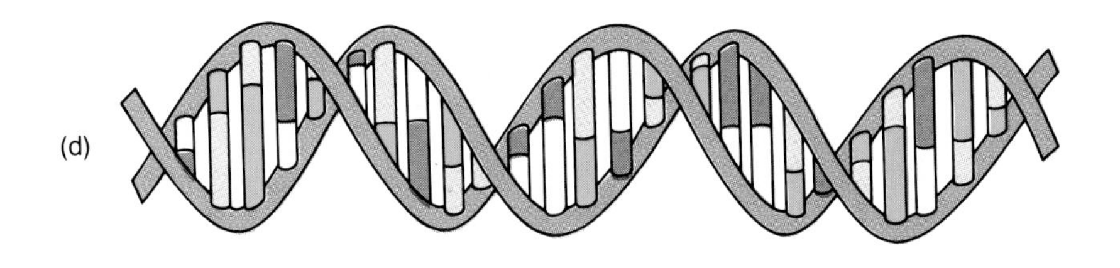

FIGURE 18.14

RNA Transcription

The enzyme RNA polymerase initiates the construction of a molecule of messenger RNA. In this process, only one of the DNA strands serves as a template. DNA unwinds as the RNA is transcribed, and rewinds as the RNA moves on. Once complete, mRNA moves through pores in the nuclear envelope and into the cytoplasm.

First Letter	Second Letter: U	Second Letter: C	Second Letter: A	Second Letter: G	Third Letter
U	UUU Phe UUC Phe UUA Leu UUG Leu	UCU Ser UCC Ser UCA Ser UCG Ser	UAU Tyr UAC Tyr UAA Stop UAG Stop	UGU Cys UGC Cys UGA Stop UGG Trp	U C A G
C	CUU Leu CUC Leu CUA Leu CUG Leu	CCU Pro CCC Pro CCA Pro CCG Pro	CAU His CAC His CAA Gln CAG Gln	CGU Arg CGC Arg CGA Arg CGG Arg	U C A G
A	AUU Ile AUC Ile AUA Ile AUG Met	ACU Thr ACC Thr ACA Thr ACG Thr	AAU Asn AAC Asn AAA Lys AAG Lys	AGU Ser AGC Ser AGA Arg AGG Arg	U C A G
G	GUU Val GUC Val GUA Val GUG Val	GCU Ala GCC Ala GCA Ala GCG Ala	GAU Asp GAC Asp GAA Glu GAG Glu	GGU Gly GGC Gly GGA Gly GGG Gly	U C A G

FIGURE 18.15

Triplet Code

Each group of three letters, one from each of the four bases in the code (here, the bases in mRNA are used) represents a specific amino acid or instruction.

A Trio of RNA Molecules

The cell has the following three types of RNA molecules, each with a different composition and function (see Table 18.1):

Messenger RNA. Messenger RNA, or **mRNA,** is the name given to the RNA that carries the coded message for protein assembly from the DNA out to the cytoplasm of the cell. It is a straight chain molecule, generally composed of at least 300 bases (which would permit a protein of 100 amino acids).

Transfer RNA. Transfer RNA, or **tRNA,** which is also transcribed from DNA, consists of about 90 bases coiled in a variety of ways such that an active site of three bases is exposed (see Figure 18.16). This site is called the **anticodon,** and it pairs with a specific codon on the messenger RNA. At the other end of the tRNA an amino acid is attached. Cells have 20 different types of tRNA, each of which is specific for a particular amino acid.

Ribosomal RNA. Ribosomal RNA, or **rRNA,** is the primary constituent of the ribosomes, the small cytoplasmic organelles on which protein assembly takes place (see Chapter 4). Like the other RNA molecules, rRNA is also transcribed from DNA.

Translation

Protein synthesis, or **translation,** occurs through the interaction of the three types of RNA. A molecule of mRNA leaves the nucleus, carrying instructions for the synthesis of a particular protein. One or more ribosomes bind to the mRNA at a special binding site on the ribosome (see Figure 18.17). Each ribosome also has two binding sites for tRNA. A molecule of tRNA that has the anticodon to match the "start" codon of the mRNA (i.e., the tRNA carrying methionine) occupies the first of the binding sites. Subsequently, a tRNA with the anticodon to match the next codon of the mRNA occupies the second ribosomal binding site.

Both tRNA molecules have an attached amino acid. As they occupy the binding sites, the amino acids are brought into proximity, and peptide bonds are formed between them. The first tRNA molecule, now freed of its amino acid passenger, then falls away from its ribosomal binding site, and, as the ribosome moves along the mRNA molecule, the second tRNA molecule moves over to take the place vacated by the first tRNA molecule. A third tRNA molecule slides into the site vacated by the second tRNA. In this way, enzymes and other proteins are formed, one amino acid at a time (see Figure 18.18).

How does the cell "know" how much protein to form? Once the last ribosome has moved along the mRNA molecule, the mRNA is destroyed by enzymatic action, thereby signalling an end to the formation of that particular protein until more mRNA molecules specific for the protein are synthesized in the nucleus. However, the lifespan of different mRNA molecules can vary enormously, as will be discussed in the next section.

GENE EXPRESSION

Before concluding this discussion of the function of genes, let's briefly consider two other aspects: an unexpected event in the formation of mRNA, and the way in which gene expression is regulated.

Introns and Exons

The description of genes and the formation of mRNA at this point in our discussion is overly simplistic. One oversimplification that we must now address is the fact that the mRNA molecule that leaves the nucleus is rarely a precise complement to the strand of DNA upon which it was transcribed. Prior to leaving the nucleus, sections of the mRNA molecule are enzy-

TABLE 18.1 The Types of RNA

TYPE OF RNA	APPROXIMATE SIZE (NUMBER OF BASES)	FUNCTION
Transfer (tRNA)	90	Carries each amino acid to its correct position in a growing polypeptide chain
Ribosomal (rRNA)	Three different kinds: 100, 1,500, and 3,000	Provides the framework for the construction of ribosomes
Messenger (mRNA)	Varies between 300 and 5,000	Carries the genetic information in a sequence of codons that determine the sequence of amino acids in proteins

FIGURE 18.16

Transfer RNA

(a) Diagrammatic representation of a tRNA molecule. An amino acid attaches at one end; the anticodon is at the opposite end. (b) A skeletal model of tRNA; (c) a schematic diagram of the three-dimensional structure of tRNA. From (b) and (c), it is obvious that (a) is a highly stylized representation, although it is the one most commonly used.

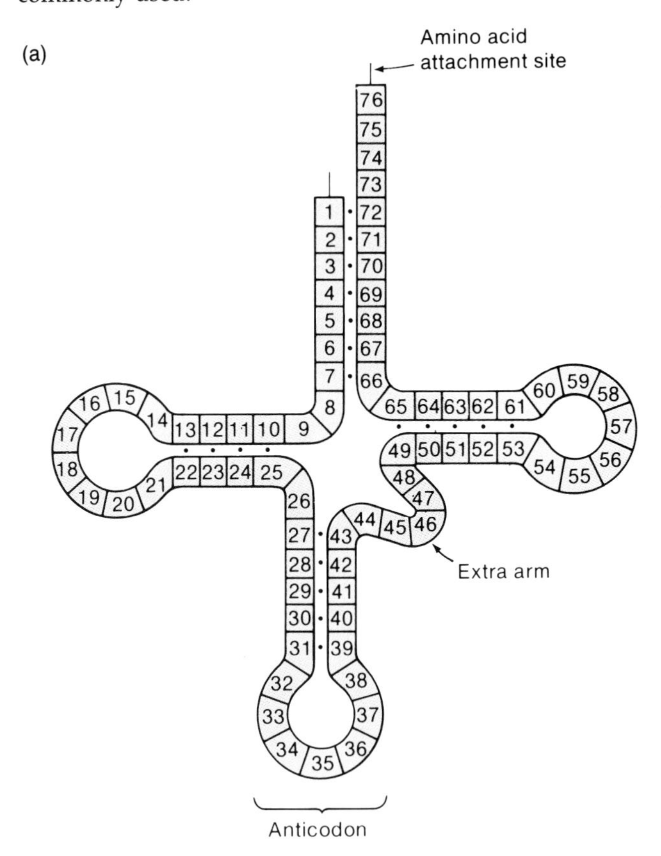

matically snipped out. The DNA that corresponded to the deleted sections are called **introns** because they are *intervening* sections. The DNA that corresponds to included sections of mRNA are called **exons** because they are *expressed.*

Introns may range in size from 65 to 100,000 bases, and one gene may have as many as 50 introns. In such genes the mRNA complements of the exons may be combined in various ways. Indeed, it is this great capacity for various combinations that underlies the enormous potential diversity of antibodies (see Chapter 9).

Regulation of Gene Expression

It is easy to speculate on the various places where gene regulation could occur: at the point of transcription; at the point of mRNA editing; at the point of exit from the nucleus; at the point of translation on the ribosome; or at the point where the chain of amino acids is converted to a functioning protein. However, at least in eukaryotes, very little is known about how gene expression is, in fact, regulated.

Gene expression in prokaryotes, by comparison, is far better understood, and similar mechanisms may

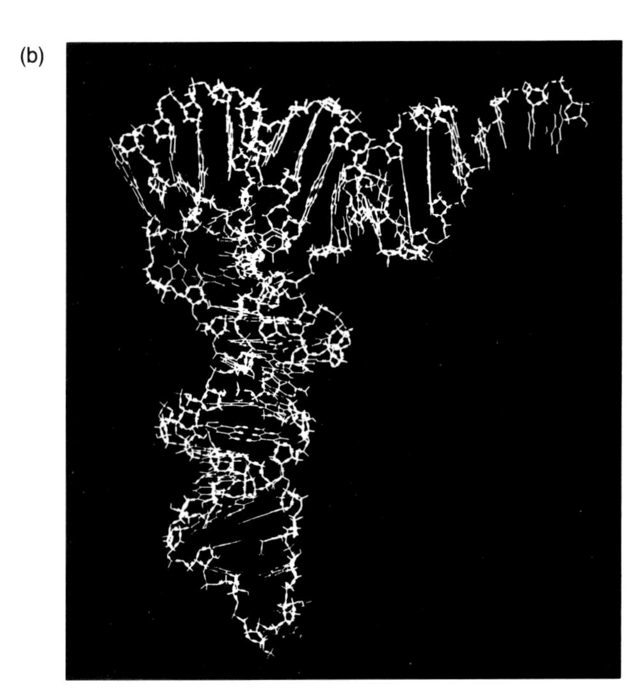

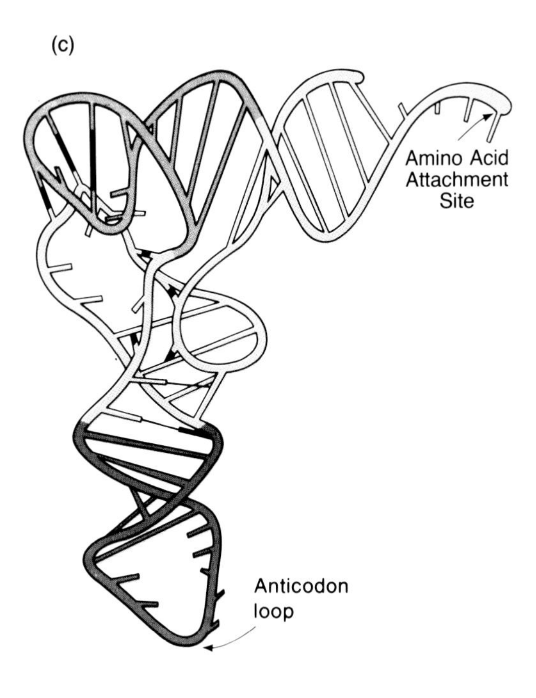

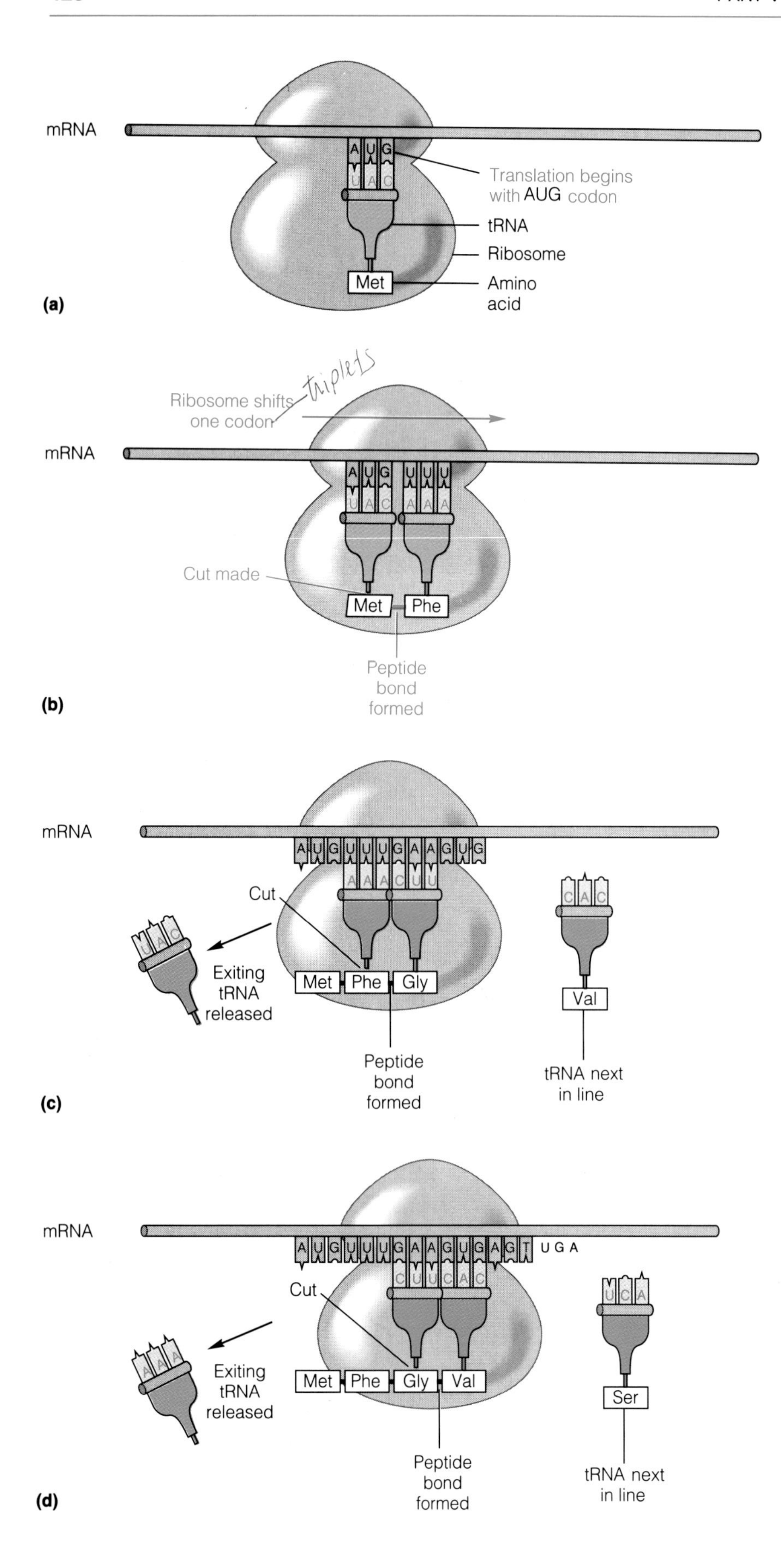

FIGURE 18.17

Translation

(a) A molecule of mRNA attaches to a ribosome. Protein synthesis is initiated with the AUG codon and the tRNA for methionine. (b) The ribosome shifts one codon, and a second tRNA brings a second amino acid; a peptide bond forms between the two amino acids. (c) The ribosome shifts another codon, permitting a third tRNA to bring in a third amino acid; the first tRNA is released back into the cytoplasm, where it may pick up another amino acid. (d) The process continues, until the ribosome reaches a "stop" codon, at which point protein assembly ceases, and the protein is released. Other ribosomes will continue to move down the same mRNA molecule, assembling more copies of the protein, until enzymes degrade the mRNA.

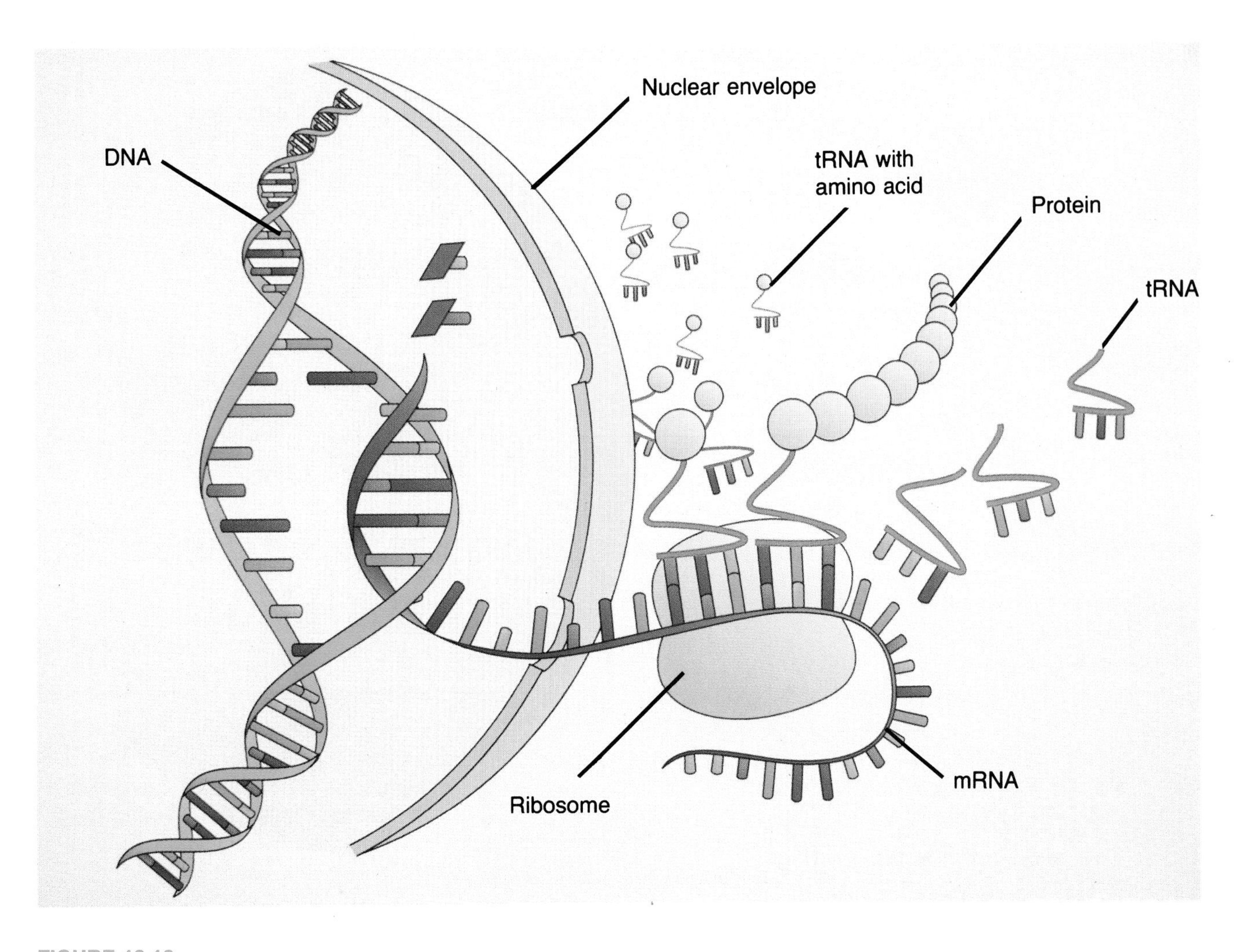

FIGURE 18.18

DNA, Transcription, and Translation
An artist's view of the events of transcription and translation.

well be at work in eukaryotes. In bacteria (and almost certainly in eukaryotes), some genes are *structural,* in that they code for proteins used by the cell in a variety of fashions; other genes are *regulatory,* in that they code for proteins that regulate the expression of structural genes (see Figure 18.19). For example, the bacteria *Escherichia coli* (popularly known as *E. coli*) is generally a benign denizen of our intestines. It has the capacity to produce the enzyme *lactase,* an enzyme that splits *lactose* (milk sugar, a disaccharide consisting of one molecule of glucose, and one molecule of another 6-carbon sugar, *galactose*). If its host has milk in the diet, *E. coli* produces lactase; if the host has no milk in the diet, *E. coli* produces no lactase.

How does *E. coli*—a very simple, brainless unicellular organism—know when to produce, or not produce, lactase? The answer lies in the manner in which the gene responsible for lactase production is regulated (see Figure 18.20).

E. coli actually has three structural genes involving lactose lying side-by-side. The first of these codes for the enzyme that splits lactose; the second codes for a protein that facilitates the transport of lactose across the cell membrane; the third codes for an enzyme that assists in the metabolism of galactose. Since none of these enzymes or proteins is needed in the absence of lactose, it is logical that all three would be regulated as a unit.

Immediately adjacent to the three structural genes are two regions of DNA known as the **operator** and the **promotor.** If an mRNA molecule is to be transcribed from the three structural genes, RNA polymerase must first attach to the operator. However, adjacent to the operator is a regulator gene that codes

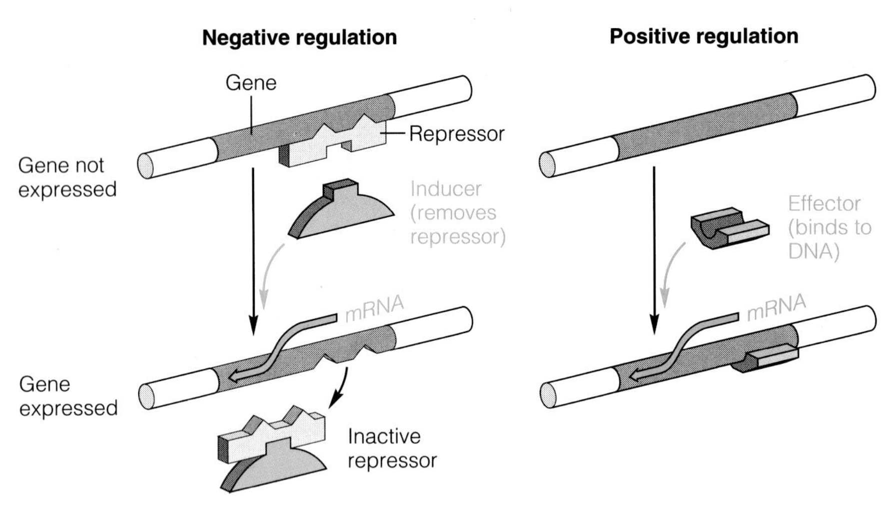

FIGURE 18.19

Gene Regulation

Gene regulation can be either negative or positive. In negative regulation (left) the regulator is an inducer that removes a repressor and permits gene expression. In positive regulation (right) the regulator is an effector that binds to DNA, thereby stimulating gene expression.

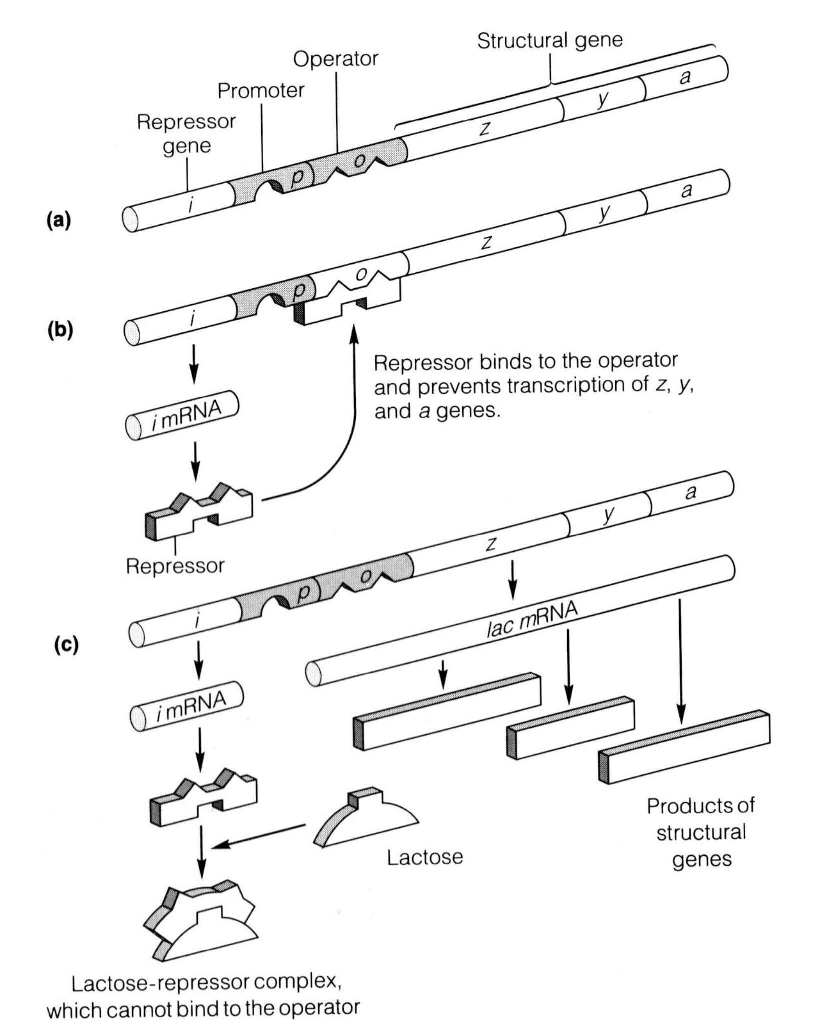

FIGURE 18.20

Regulation in the Lac Operon

The lac operon (a) consists of a repressor gene, a promotor and operator site, and three structural genes. In the repressed state (b), the product of the repressor gene (the repressor protein) overlaps the operator and promotor sites, thereby preventing expression of the structural genes. In the induced state (c), lactose combines with the repressor protein, and the three structural genes are expressed.

for a protein known as a **repressor.** If there is no lactose present, the repressor protein binds to the operator in such a way as to overlap the promotor and prevent RNA polymerase from attaching. Thus, in the absence of lactose, the repressor ensures that the structural genes are not activated.

However, if lactose is present, lactose molecules combine with the repressor in such a way that the repressor cannot attach to the operator. With the repressor otherwise occupied, RNA polymerase can bind to the promotor, mRNA transcription from the structural genes begins, and lactase begins to be synthesized. As the level of lactose in the cell falls (because of the activity of lactase), the repressor molecules are ultimately relieved of their lactose burden, and once again repress the transcription of mRNA from the structural genes.

Thus the answer to the question we posed earlier is simply that lactose regulates the production of lactase. If lactose is present, lactase is produced; if lactose is absent, no lactase is produced. What looks to be very clever behavior by a bacterium is, in reality, a simple negative feedback mechanism.

Regulation in eukaryotic cells is more complex. The RNA polymerases are very different in eukaryotes; the DNA is less accessible because it is within a nuclear envelope and is wound around spools of protein, and the regulator binding sites are often some distance removed from the structural genes. All of this suggest that eukaryotic regulatory mechanisms must be different from prokaryotic mechanisms. However, what evidence is now available—and it is still not much since eukaryotic cells are very difficult to work with—suggests that the regulatory mechanisms of prokaryotes and eukaryotes are really rather similar.

One regulatory mechanism that is apparently universal in eukaryotes (although the overall importance of this mechanism is still not clear) is the manufacture of **antisense mRNA.** Recall that we said that mRNA is transcribed from only one of the DNA strands. If mRNA is transcribed from the other DNA strand, it is antisense RNA, and it will bind with the usual, or *sense* form of mRNA, rendering it useless. By varying the ratio of sense and antisense mRNA, a cell could theoretically regulate the rate of protein synthesis. Antisense mRNA is also now being examined as a possible therapeutic tool in genetic and viral disorders, including AIDS.

Another regulatory mechanism in eukaryotes is the alteration of the lifespan of mRNA molecules. Some mRNA molecules are enzymatically degraded in minutes; others survive for months. Obviously, those surviving longer can produce more protein—and this is just what happens in certain cells. About 95 percent of the protein in erythrocytes is hemoglobin, and during the maturation process of erythrocytes, the lifespan of the mRNA for hemoglobin is greatly extended, whereas other mRNA molecules are degraded very quickly. Precisely how these changes are accomplished is not yet clear, although it is known that some hormones function by binding with particular mRNA molecules, thereby preventing their destruction by enzymes. These mRNA molecules are then able to synthesize larger amounts of the protein for which they code (see Chapter 13).

MUTATIONS

Point mutations consist of errors in one or two bases. Although this may seem, at first glance, to be a minor problem, in reality point mutations are frequently very serious. Point mutations are of two types.

Base Substitution. The most trivial type of error would seem to be the erroneous substitution of one base for another during DNA replication. In some cases, the "new" codon that is formed codes for the same amino acid, and the mutation is undetectable. In other cases, a different amino acid is coded for, and the outcome can be devastating.

The codon for the amino acid *glutamic acid* is GAA. The codon for the amino acid *valine* is GUA. In fact, the hemoglobin of sickle cell anemia differs from normal hemoglobin because of this difference alone. The substituted amino acid distorts the shape of the hemoglobin molecule. Because erythrocytes are composed very largely of hemoglobin, the altered shape of the molecule leads to an altered shape of the blood cell. This difference in cell shape, in turn, promotes numerous circulatory problems, including blood clots and kidney failure. Recently, the oncogene that causes bladder cancer was found to differ from the normal allele by a substitution of thymine for guanine in a critical codon (see Chapter 17).

Insertions and Deletions. Suppose that an extra base is inserted during DNA replication, or that a base is accidentally omitted. Since the code is read in clusters of three, whenever a base is added or dropped, the rest of the code will be misread. Depending on where they occur, insertions and deletions may have virtually no effect or be lethal.

Not all mutations result from errors. **Transposons** are particular sequences of DNA that can move from chromosome to chromosome. For that reason, they are sometimes called "jumping genes." Because position and interaction of genes is often critical for their proper functioning, random insertion of transposons frequently leads to an increase or decrease in the expression of a particular gene. Some scientists believe that transposons become especially active during times of environmental stress, when increased genetic variation might be assumed to be most beneficial to the population (see Chapter 22).

Summary

Although Gregor Mendel is known as the "father of genetics" (despite his status as a monk!), he worked in such obscurity that his "child" might well have been considered stillborn. Because he was not a professional scientist, because he published in an obscure journal, and especially because of his strong mathematical orientation, his successful elucidation of the basic laws of heredity went largely unnoticed and unrecognized for 34 years.

Mendel correctly deduced that organisms had a pair of hereditary factors (what we now call "genes") for many traits, that these factors were separated in the formation of gametes, that the factors were not always identical (because they might be what we now call "alleles"), and that, in such instances, one factor dominated the other. He also determined that factors for different traits were inherited independently of each other.

Since the "rediscovery" of Mendel's work in 1900, the field of genetics has developed very rapidly. Exceptions to Mendel's basic laws have been discovered—linkage, sex linkage, epistasis, codominance—but the basic principles remain just as Mendel stated them.

With the realization that Mendel's hereditary factors were located on chromosomes, it was only natural that the next quest would be to identify the chemistry of the chromosome and the mechanism of action of genes. Watson and Crick successfully interpreted the structure of DNA as a double helix, and subsequent work explained how DNA was replicated, and how, by integrating the chemical activity of three types of RNA, it directed the synthesis of proteins.

We are now at a very exciting stage, for not only is it becoming possible to create bacterial "factories" to synthesize otherwise very costly biochemicals, but it soon will be possible to transplant genes directly into the cells of individuals who have genetic defects and in that way to cure genetic diseases, not merely treat them symptomatically. Less than 50 years ago most scientists thought that genes were composed of proteins. Few areas of science have progressed as rapidly in as short a period of time.

Key Terms

Mendelian genetics
molecular genetics
dominant
recessive
Law of Segregation
monohybrid cross
dihybrid cross
Law of Independent Assortment
sex chromosome
autosome
sex linkage
gene
diploid
haploid
allele
homozygous
heterozygous
genotype
phenotype
mutation
incomplete dominance
codominance
polygenic inheritance
epistasis
limited penetrance
replication fork
DNA polymerase
transcription
RNA polymerase
codon
messenger RNA
transfer RNA
anticodon
ribosomal RNA
translation
intron
exon
operator
promotor
repressor
antisense mRNA
point mutation
base substitution
insertion
deletion
transposon

Box Terms

pangenesis
germ plasm

Questions

1. What is the Law of Segregation? How does it differ from the Law of Independent Assortment? Give examples of each.
2. What are sex-linked traits? Why are they limited to (or at least much more common in) one sex than another?
3. Explain how it might be possible for a man with blood type A, and a woman with blood type B, to conceive a child with blood type O.
4. Imagine that Gregor Mendel had a twin brother, Angus, who carried on plant breeding experiments with snapdragons at the same time that his more illustrious brother was crossing pea plants. Angus finds that when he crosses a red-flowered plant with a white-flowered plant he gets plants with pink flowers. How can he reconcile his findings with what Gregor has learned from pea plants?
5. Briefly explain DNA replication.
6. Briefly explain RNA transcription. Does RNA have the same nitrogenous bases as DNA? If not, what are the differences?
7. What are the major types of RNA and what are their purposes? Explain how they interact in the assembly of a protein.
8. What is the triplet code? What does it have to do with protein formation?
9. *E. coli*, a bacterial denizen of our large intestine, manufactures the enzyme lactase only when lactose (milk sugar) is available. How does *E. coli* regulate the production of lactase?
10. Distinguish between pangenesis and the germ plasm theory of inheritance.

CHAPTER

19

Human Genetics

Blueprints and Problems

Queen Victoria ruled England for more than 60 years. She had nine children, and outlived her husband by 40 years. Despite her impressive health record, one of her sons died of hemophilia, and the thrones of Russia and Spain fell in part because this disease also afflicted Victoria's grandchildren and great-grandchildren. Why did Victoria herself show no signs of the disease?

Anne and Mary, identical twins, were married in a double wedding on March 20, 1964. They lived in the same neighborhood, and saw each other almost daily. In September 1964, Anne and Mary both developed a case of *rubella* (German measles). On January 29, Anne gave birth to a healthy baby boy. On April 13, Mary gave birth to a boy who was blind, deaf, and mentally retarded. What accounts for the difference?

Beth Fletcher loved children, and her growing family was a constant delight. However, she was miserable during the early months of each of her pregnancies. When she became pregnant for the fourth time, early in 1961, her cousin in England sent her some *thalidomide,* a new drug that greatly lessened the severity of morning sickness. The drug worked well, for Beth had virtually no problems during her pregnancy. However, her daughter was born with no arms, and only a pair of incompletely formed hands projecting from her shoulders. What happened? If her daughter marries and has children, will they share their mother's deformity?

These questions involve an understanding of human genetics, the subject of this chapter.

THE NATURE OF THE PROBLEM

As is true for all organisms, much of the observable variation in humans is due to genetic differences among individuals. It is also true that a significant amount of human variation is environmentally caused. We examined some of the "nature versus nurture" controversy in Chapter 16; now we must revisit that subject, albeit from a slightly different perspective.

To state the obvious, the development of an individual from a fertilized egg to a sexually mature adult involves the interaction of genes with their environment. However, not all traits involve the same balance of interaction between genes and environment. Some traits are merely circumscribed by genes, but their actual manifestation is primarily due to the effects of environment. Other traits are normally almost entirely under genetic control, but unusual environmental circumstances may have a significant impact on the usual function of the genes. Still other traits are very tightly controlled by genes, and the role of the environment is not to alter the expression of the genes, but simply to permit such expression to occur. Understanding human genetics means, among other things, distinguishing among these three classes of traits.

One way of exploring the range of genetic expression in humans is by examining traits controlled by single genes, by several genes acting in concert, and by wholesale chromosomal changes.

SINGLE GENE EFFECTS

The number of genes on the chromosomes of our cells is estimated to be between 50,000 and 100,000 (see Box 19.1). Each of these genes is responsible for coding the amino acid sequence of a particular protein, and the proteins, in turn, are ultimately responsible for virtually every activity associated with the development of the individual from the fertilized egg to the growth, repair, and maintenance of cells and tissues. Perhaps one day, with a complete knowledge of the base sequence of all of the genes, and the corresponding identity of the proteins, it will be possible to ascertain the ways in which the proteins interact in sustaining life. Alternatively, we may find that the various permutations and combinations of interactions of between 50,000 and 100,000 proteins are simply beyond the range of human comprehension.

In any event, we are still decades away from such knowledge. The 1990 edition of *Mendelian Inheritance in Man* lists 4937 inherited characteristics, with more being reported virtually every week. In addition, as of September 10, 1990, the loci of 1884 genes had been determined. At present, much of our knowledge of gene functioning has come from an awareness of what happens if a mutation occurs. That is, just as our understanding of organ physiology has been greatly aided by our assessment of disease, so, too, is our understanding of genetics enormously assisted by identifying the impact of a mutation. Thus, when we attempt to list those human traits that we know to be under genetic control, we find that most of them are pathological conditions caused by a mutation of the normal gene, because in most instances it is only by linking pathology to mutation that we can deduce the role of the normal gene.

In this portion of the chapter, we shall examine some representative examples of mutations that cause abnormalities—some very profound—at the level of the single gene. Subsequently, we shall explore multiple gene effects, and wholesale abnormalities in the chromosomes.

Autosomal Dominant Disorders

Approximately 800 human traits have been linked to particular dominant alleles of the **autosomes** (chromosomes other than the sex chromosomes). Some of these are relatively innocuous. For example, the so-called *piebald trait*, which is manifested by a white forelock of the hair, or *wooly hair*, a condition in Caucasians where the hair superficially resembles that of blacks (although it is much more brittle), are hardly life-threatening conditions. *Cleft chin*, another trait controlled by a dominant allele, may even be preferred to the normal condition—it certainly has been an asset to Kirk Douglas!

Some dominant traits, however, are somewhat more serious. *Myopia* (nearsightedness; see Chapter 15), for example, is relatively easily corrected by the use of eyeglasses—but eyeglasses have been in use for only a few hundred years. For most of human history, myopia was uncorrected, and those suffering from it were presumably enormously disadvantaged.

There are many other human traits controlled by dominant autosomal alleles.

Achondroplastic Dwarfism. In the United States, approximately one child in every 10,000 live births is an achondroplastic dwarf. Mutations of the involved gene are common—most achondroplastic dwarfs are born to parents of normal height. The mutant allele affects the production of one form of collagen, a key ingredient of cartilage, and the cartilage growth plates of the long bones of the arms and legs grow very slowly, if at all (see Chapter 12). As a consequence, the individual has very short arms and legs. Achondroplastic dwarfs are invariably heterozygotes, as the homozygous dominant condition is lethal.

Brachydactyly and Polydactyly. There are several different genes that can cause **brachydactyly** (short fingers and toes) and **polydactyly** (extra fingers and toes; see Figure 19.1). In the case of polydactyly, the number of extra fingers and toes, and whether or not both hands and feet are affected, varies considerably from individual to individual.

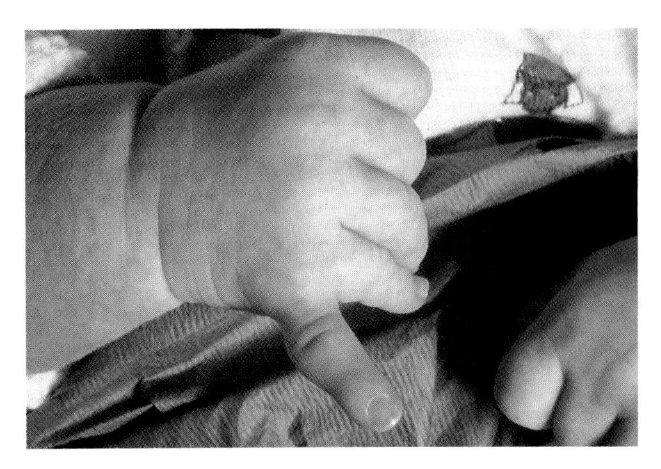

FIGURE 19.1

Polydactyly

Caused by a dominant allele, polydactyly is characterized by extra fingers and toes.

Familial Hypercholesterolemia (FH). With an incidence of one in 500, FH is one of the most common and serious of genetic diseases. The dominant allele causes a malformation of the LDL receptor of liver cells (see Chapter 7), with the result that blood cholesterol levels rise to twice the normal level. Myocardial infarctions (heart attacks) are common in such individuals as early as age 35. People who are homozygous for the condition (roughly one in one million) have extremely high blood cholesterol levels and generally suffer myocardial infarctions while still in their teens.

Huntington Disease. Huntington disease is relatively rare, affecting only one individual in 18,000 in this country, but has become rather well known because it caused the death of famed folk singer Woody Guthrie. It is a particularly nasty disease, since it results in the deterioration of nerve cells, with a corresponding loss of motor control and memory, the onset of periods of rage, and finally—after ten to 15 years—death. Moreover, it usually strikes about the age of 40, by which time the individual has generally started a family—and each child then is obliged to live with the fact that there is a 50 percent probability that he or she will develop the condition as well.

The gene responsible has been traced to chromosome 4, but as of this writing, there is still no effective treatment.

Marfan Syndrome. The gene for Marfan syndrome has been traced to chromosome 15. This gene apparently controls the formation of the connective tissue protein *fibrillin*. Defective fibrillin causes weakened connective tissue, leading to spontaneous ruptures in major blood vessels.

Marfan syndrome affects one in 10,000 Americans. People with this condition tend to be tall, lanky, and long-fingered—and many die before the age of 40. Several basketball and volleyball stars who have collapsed and died on the court were subsequently found to have had Marfan syndrome. The famed violinist Paganini, who died when he was 58, almost certainly had the disease—and quite possibly so did Abraham Lincoln, who was showing many signs of impending cardiovascular failure when he was assassinated, at the age of 54.

Neurofibromatosis (NF). Neurofibromatosis, which is a highly variable condition characterized by the growth of multiple small tumors of the central and peripheral nervous systems, is caused by dominant alleles of two different genes (see Figure 19.2). NF is relatively common, afflicting 100,000 Americans, and has become well known because of the case of Joseph Merrick (who has generally been misidentified as John Merrick), the so-called "elephant man" of Victorian England, and the subject of both a play and a movie in recent years. Ironically, Merrick, who was grotesquely deformed (his skull had a circum-

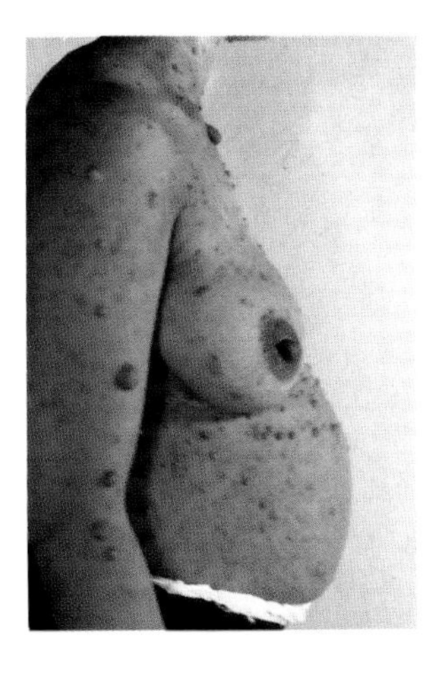

FIGURE 19.2

Neurofibromatosis

A dominant condition characterized by the appearance of many small tumors (neurofibromas) on the skin.

BOX 19.1

The Human Genome Project

A **genome** can be defined as the total amount of genetic information contained in an organism. The Human Genome Project is attempting to determine the sequence of nitrogenous bases in all of the chromosomes of a human—an estimated three billion bases!

A project of this magnitude was inconceivable only a few years ago, when sequencing the bases of even a single gene was beyond our reach. However, current techniques in recombinant DNA technology (discussed in some detail later in this chapter) permit the sequencing of up to 500 bases in a single experiment. The complete base sequence has now been determined in mitochondria (16,569 base pairs on a single, continuous double-stranded loop), and in several types of viruses containing up to one million bases.

The current cost of sequencing bases averages between $3 and $5 per base. Thus the cost of determining the human genome would be between $9 and $15 billion, although improved techniques will almost surely reduce this cost by at least one-half in the coming years.

The Human Genome Project has been under active discussion since 1985. Following a series of blue-ribbon meetings and panel reports, some $17 million was earmarked for the Human Genome Project in the 1988 federal budget, although scientists estimate that $200 million per year will be needed for about 15 successive years in order to complete the project. In October 1989, the Secretary of the Department of Health and Human Services created the National Center for Human Genome Research (NCHGR), with James Watson (of Watson and Crick fame) as its director. Its 1990 budget was some $60 million, most of which was awarded as research grants.

At the same time the Department of Energy embarked on a parallel project, although it has had less than half the funds available to the NCHGR. A Memorandum of Understanding was signed in 1988 to ensure coordination of these two projects. The DOE is primarily interested in mapping all of the expressed genes (exons), but it is not interested in *introns* (see Chapter 18). NCHGR is interested in both exons and introns. In addition, several foreign governments have announced intentions to participate, although coordinating these efforts will prove a formidable task.

Is spending perhaps $5 billion over 15 years to determine the precise sequence of all of the bases comprising the human genome worthwhile, in terms of money and time? Will the information that results be used in unintended, and perhaps harmful, ways? Obviously, it is impossible to say at this juncture. Nevertheless, therapeutic intervention into genetic diseases cannot

ference of nearly one meter), almost certainly did not suffer from NF at all, but from the much rarer, and far more serious, condition known as *Proteus syndrome.* In an interesting twist of fate, the play and movie led to an outpouring of public support for research into NF, and this support, in turn, led to the identification of one of the genes responsible for the disease.

This gene was isolated in July 1990 and was found to be located on chromosome 17. The protein formed by the NF gene is very similar to a protein called *GAP,* which deactivates the protein formed by the *ras* gene, an oncogene involved in perhaps 25 percent of human cancers (see Chapter 17). In the mutated form, the NF protein does not deactivate the *ras* protein, leading to the benign tumors of Schwann cells, called *neurofibromas,* which characterize the disease.

Penetrance and Expressivity

The preceding discussion noted, with respect to both polydactyly and neurofibromatosis, that considerable variation exists from individual to individual regarding the extent of the manifestation of the condition. At first glance, this statement seems odd—a person either has the mutated allele or he or she does not. Why should there be any variation in how the condition is manifested?

Studies of mutated alleles tell us something not only about the causes of certain diseases and (by inference) about the functioning of the normal allele but also about how genes actually function. Genes do not exist in isolation; they function in a chemical environment produced by the activities of other genes and by the external environment, and they are often influenced by these environmental effects. The percentage of individuals carrying a particular genotype who actually show the expected phenotype is called the **penetrance** of the condition. For example, a condition that develops in about 90 percent of the individuals who carry the appropriate allele would have a penetrance of 90 percent.

Penetrance is a quantitative measure: the condition either occurs or it does not. **Expressivity,** on the other hand, is qualitative: it is the *degree* of phenotypic expression. A polydactylous individual, for example,

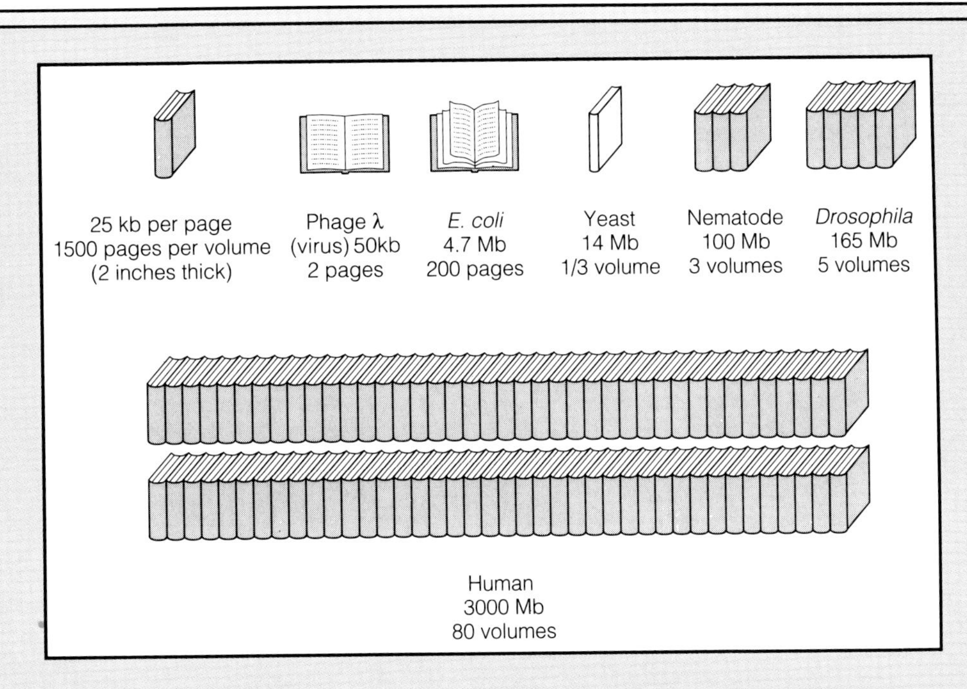

Relative Size of Genomes

By way of visualizing the magnitude of the human genome project, imagine that genomes could be printed at 15,000 characters per page and that the pages were bound in 1500 page volumes. The genome of the bacterium *E. coli* would be 200 pages in length. *Drosophila*, by comparison, would be five volumes, and the human genome, would be 80 volumes. (Mb = megabase; one million nitrogenous bases)

take place effectively until the gene responsible is located and identified—thus some fraction of the human genome will be mapped in any event, although the cost of doing so one disease at a time is very high. Moreover, safeguards are already being discussed that will protect the privacy of individuals with deleterious genetic conditions from the prying eyes of employers and insurance companies.

Nonetheless, just as is true of any major scientific project, the opportunities for misuse of information abound, and science will again pin its hopes on the will of society to ensure that such misuse does not happen.

may show only the beginning of an extra finger or toe on just one hand or foot—or he may have extra fingers and toes on both hands and feet. This range of phenotypic expression demonstrates expressivity—one measure of the outcome of the interaction of a particular mutant allele with its environment.

Autosomal Recessive Disorders

More than 500 human traits are known to be—and at least as many strongly suspected to be—the result of autosomal homozygous recessive alleles. It is estimated that each of us has from three to eight very deleterious recessive alleles, but because those alleles are masked by a dominant allele, they are not expressed. The danger comes, of course, when two individuals who are both carrying the same deleterious allele marry and have children (see Box 19.2). There are many examples of traits caused by homozygous recessive alleles.

Albinism. There are actually three different genes that, in mutated form, can lead to albinism. Albinism (L. *alba*, "white") has a frequency of one in 10,000 in the United States, although in closely inbred populations, the frequency is much greater. Among the Hopi Indians, for example, the frequency of this condition is 1 in 200. Albinism results from a failure to produce the skin pigment *melanin*, due to a defective form of the enzyme *tyrosinase*. As a consequence, no skin, hair, or eye pigment is produced (see Figure 19.3). In addition, because melanin normally protects the outer tissues from damage by ultraviolet radiation, albinos have greatly increased likelihood of skin cancer and visual problems.

Alkaptonuria. Alkaptonuria results from a missing enzyme, and the resulting accumulation of *homogentisic acid*, a breakdown product of the amino acid *tyrosine*. This acid turns black upon exposure to light and air. Because homogentisic acid is excreted in the urine, people suffering from alkaptonuria produce a urine that promptly turns black. However, other than causing quizzical expressions in public restrooms, alkaptonuria has no serious long-term consequences (see Figure 19.4). The condition has particular interest

FIGURE 19.3

Albinism

A homozygous recessive condition characterized by the absence of pigment in the skin or hair.

to geneticists because it was the first human trait identified as following a Mendelian pattern of inheritance—in 1902 (although the missing enzyme was not identified until 1958).

Cystic Fibrosis. Cystic fibrosis is the most common of all serious single gene defects, and it is also the most common cause of death in Caucasian infants. More than 40 different mutations of the cystic fibrosis gene have been identified, although not all of the mutations have equally serious consequences. Although mutated alleles are rare in blacks and Asians, more than 10 million Caucasian Americans carry a defective allele, and one in 1600 is homozygous. The locus of the gene is now known to be on chromosome 7, and the gene itself has been identified, as has the protein coded for by the gene. However, at this point, cystic fibrosis is still a fatal condition, with few individuals surviving past their twentieth birthday (but genetic engineering, discussed at the end of this chapter, may provide a future cure).

Galactosemia. Approximately one in 40,000 children born in this country suffer from galactosemia—the accumulation of toxic amounts of the 6-carbon sugar *galactose,* which, along with glucose, constitutes the disaccharide *lactose* (milk sugar). Galactose accumulates because the enzyme that normally converts it to glucose is missing (see Figure 19.5). The consequences are profound: cirrhosis of the liver, cataracts, and mental retardation, all of which can be avoided if the individual is placed on a galactose-free diet early in infancy.

Phenylketonuria. Phenylketonuria, or PKU, results from elevated levels of the amino acid *phenylalanine,* again because of a missing enzyme. In the United States, roughly one child in 14,000 is born with this disease (it is three times more common in Scotland and Ireland), and a check for phenylalanine in the urine of newborns is now mandatory in most

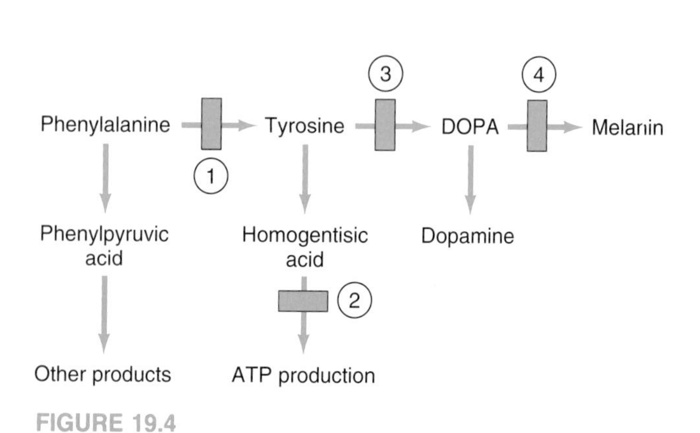

FIGURE 19.4

Phenylalanine Metabolism

Missing or defective enzymes, due to defective genes, create a variety of conditions, including (1) PKU, (2) alkaptonuria, and (3) and (4) two forms of albinism.

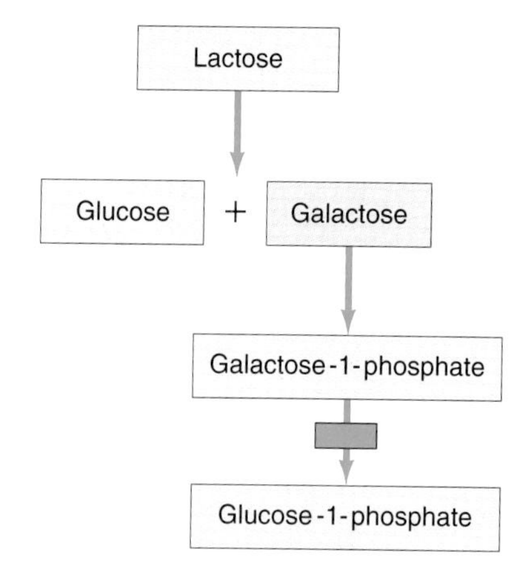

FIGURE 19.5

Galactosemia

A homozygous recessive condition characterized by the presence of a defective form of the enzyme that converts galactose to glucose.

BOX 19.2

Consanguinity

Consanguinity (L. *consanguineus,* "of the same blood") refers to matings between individuals who share a recent common ancestor. In the United States, marriages between individuals as closely related as first cousins are illegal in most states, although such marriages are permitted in some states and in many foreign countries. First cousin marriages were once common in Japan, exceeding 4 percent of all marriages, and uncle-niece marriages exceed 10 percent of all marriages in some areas of India. Brother-sister marriages were practiced by the Ptolemies of ancient Egypt.

What are the genetic implications of consanguinity? The principal concern is that homozygosity will increase. Since humans are generally thought to carry from three to seven alleles that would be lethal in the homozygous state (and many more that are deleterious when homozygous), the perils of consanguinity quickly become obvious. From 3 to 4 percent of children born to unrelated parents have serious genetic defects. This rate doubles when the parents are first cousins, and rises by a factor of ten in parent-child or sibling-sibling matings. Perhaps it is just as well that the incest taboo is so strongly upheld in most societies!

Correlation Between Recessive Disorders and First Cousin Marriages

GENETIC DISORDERS	FIRST-COUSIN MARRIAGES AMONG THE PARENTS
Phenylketonuria	5
Color blindness	11
Tay-Sachs disease	15
Albinism	17
Alkaptonuria	33

states. If untreated, it leads to mental retardation and convulsions. However, even today, approximately 1 percent of institutionalized Caucasians in this country are individuals in whom PKU was not detected early enough.

Treatment simply involves a diet that is low in phenylalanine. (Phenylalanine is one of the essential amino acids, and cannot be synthesized by the body.) Although the primary danger from excess phenylalanine ends with the maturation of the nervous system (at about the age of six), women with PKU who become pregnant must again severely restrict their phenylalanine intake, or the high levels of phenylalanine in their blood will cause severe retardation in the baby, even though the baby itself may not have PKU.

Sickle-Cell Anemia. Almost 300 variant forms of hemoglobin have been discovered, but only about 20 percent create any significant problems. The two major conditions are sickle-cell anemia (see Figure 19.6) and **thalassemia.** Individuals heterozygous for the sickle-cell or thalassemia allele have considerable protection from malaria, but homozygotes often die. Thalassemia is relatively common in Americans of Greek or Italian descent, and world-wide about 4 percent of the human population are carriers for the disease. Approximately one American black in 500 has sickle-cell anemia, and two million Americans carry the allele. Sickle-cell anemia is discussed at greater length in Chapter 22.

Tay-Sachs Disease. Tay-Sachs is rare in the general population (one in 360,000), but is 100 times more common in Ashkenazic Jews from Eastern Europe, among whom one in 30 is a carrier. Because of a defective lysosomal enzyme, *gangliosides,* a type of lipid found in neuronal membranes, begin to accumulate in the interior of neurons within a few months of birth. As the nucleus and other organelles become crowded by the accumulating gangliosides, the neurons slowly cease functioning, leading to blindness, deafness, paralysis, mental retardation, and death (usually from pneumonia, because of inhibition of the cough reflex) by age five.

Sex-Linked Disorders

Because males possess only one X chromosome, and because there is very little genetic information on the Y chromosome, (but see Box 19.3), all of the alleles on the X chromosome—whether dominant or recessive—are expressed in males. Consequently, males suffer from a variety of conditions that are rare or

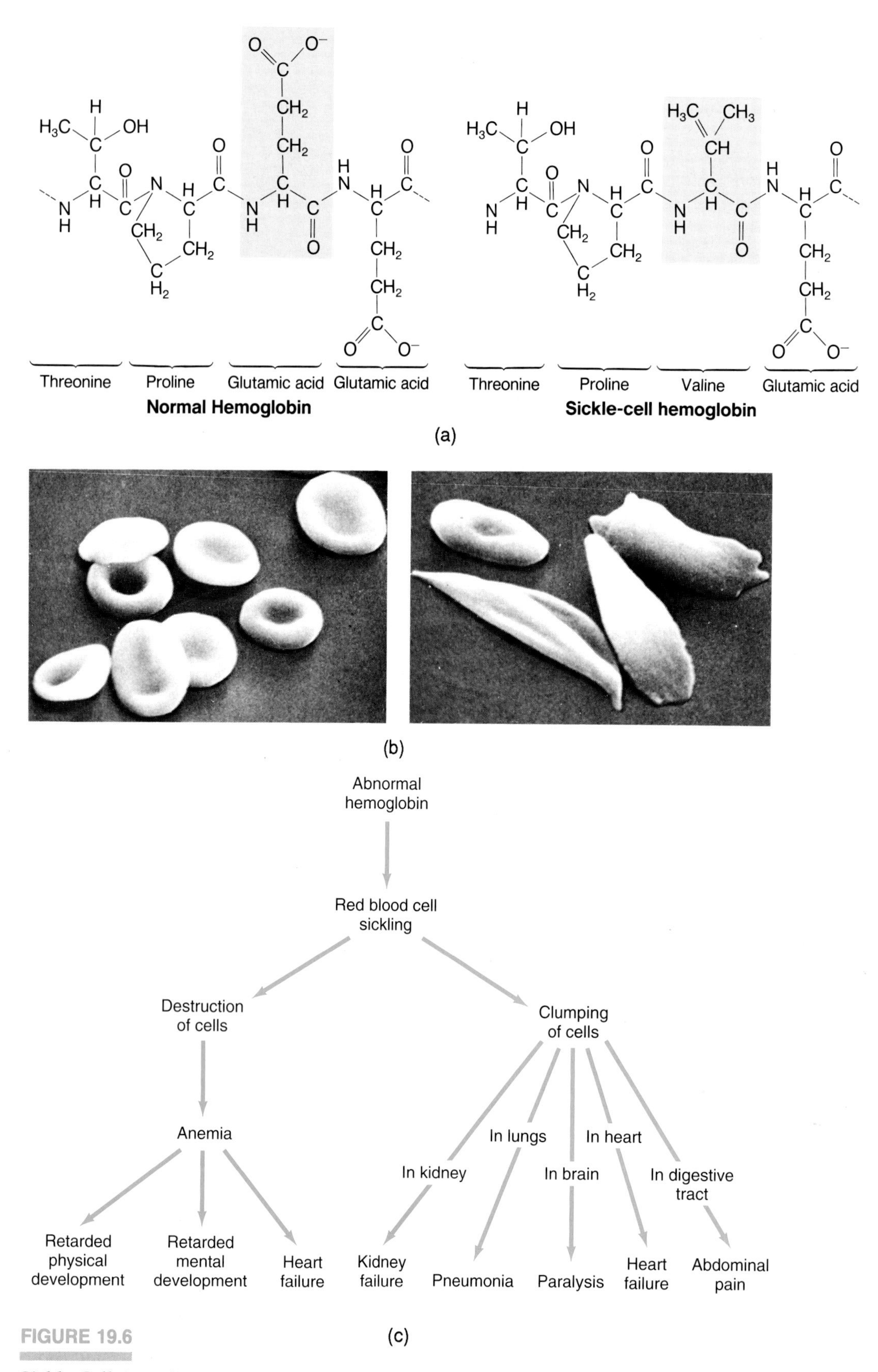

FIGURE 19.6

Sickle-Cell Anemia

A homozygous recessive condition characterized by the substitution of the amino acid valine for glutamic acid in the hemoglobin molecule (a). The consequence is that red blood cells assume a sickle shape (b), and that, in turn, has body-wide ramifications (c).

BOX 19.3

The Genetics of Gender

We are now in a position to link the fact that the reproductive organs of males and females develop from a common beginning (Chapter 20) and the fact that males and females differ chromosomally, cells from males being XY and cells from females being XX (Chapter 18).

The sexual phenotype characteristic of males develops in response to the production of a protein produced by a gene on chromosome Y called **SRY** (*Sex-determining region of the Y chromosome*). In the presence of the SRY protein, the gonads develop into testes during the seventh week of development, and the testes secrete male hormones known collectively as **androgens.** The androgens stimulate the development of the penis, scrotum, and the internal reproductive tract characteristic of males. In the absence of the SRY protein, the gonads develop into ovaries and the reproductive tract develops along characteristically female lines. Interestingly, SRY is a highly conservative gene. It has essentially the same nucleotide sequence in species as diverse as humans, rabbits, tigers, horses, cattle, and pigs—but it is found only in males.

It is intriguing to think that all of the differences characterizing males and females are ultimately attributable to a single gene. Further evidence of this fact is provided by the condition known as **testicular feminization syndrome (Tfm)**. Individuals with Tfm have cells with a functional Y chromosome, and their gonads are testes (which usually do not descend)—but they also have a vagina and well-developed breasts.

How do we explain such anomalies? Tfm individuals possess a dominant allele on the X chromosome that prevents the production of the plasma membrane receptor for the androgen responsible for initiating the development of the male genitalia. Thus, even though Tfm individuals produce normal amounts of SRY protein and testosterone, the male genitalia never develop, and such individuals are therefore assumed to be female. Moreover, with their well-developed breasts and the long legs characteristic of males, Tfm individuals are often employed as fashion models!

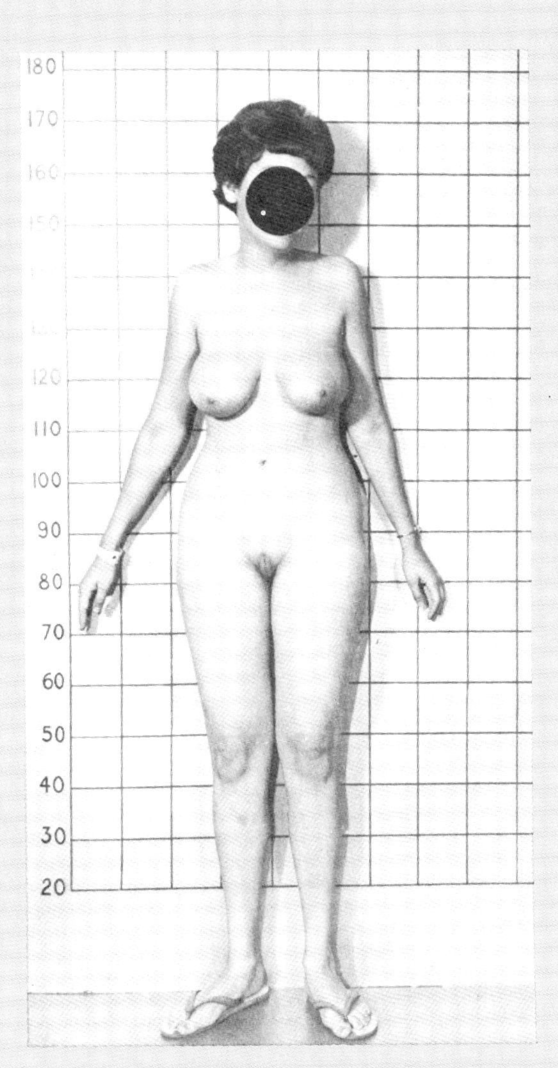

Testicular Feminization Syndrome

This individual is genotypically male, with XY sex chromosomes and a total of 46 chromosomes. This syndrome results from a dominant allele on the X chromosome.

unknown in females, including three forms of deafness, two forms of color blindness, and other, more serious, diseases.

Color Blindness. Perhaps the most common sex-linked condition is red-green color blindness (blue color blindness—see Chapter 15—is carried on an autosome). There are two closely linked genes on the X chromosome, one controlling the formation of green-sensitive cones in the retina, and the other controlling red-sensitive cones. A recessive allele that prevents normal cone formation exists for both genes. Males with either allele cannot distinguish red from green effectively. In the United States, about 8 percent of the male population, but less than 0.5 percent of the female population, are red-green color blind.

Duchenne Muscular Dystrophy. About three males in 10,000 are born with this condition, which results in a gradual wasting of the muscles. Symptoms usually appear by the age of six, and because death typically occurs before the age of 20 (that is, generally before the individual can reproduce), this disease is unknown in women.

Fragile X Syndrome. The most common form of mental retardation is caused by a breakage of the X chromosome at a specific point. In May 1991 the gene affected by this breakage was identified, a finding which promises to be of great significance in the diagnosis and treatment of this condition.

Lesch-Nyhan Disease. Again, a missing enzyme results in profound consequences—mental retardation, cerebral palsy, and a compulsive urge to self-mutilation. Children must be restrained, or they will chew off their lips, fingers, and so forth. Mercifully, death occurs at a young age.

Hemophilia. Probably the most famous of the sex-linked diseases is hemophilia, a disease in which the blood clots very poorly due to the improper formation of an essential component of the clotting process, a substance known as *factor VIII* (see Chapter 9). This disease can now be treated with some effectiveness, and more hemophiliacs are surviving to adulthood than ever before. However, although the disease is fortunately very rare, it has a decided interest to historians because of its impact on the royal houses of Europe in the decades just before and after the turn of the century (see Figure 19.7).

Victoria, Queen of England from 1837 to 1901, had nine children, one of whom, Leopold, Duke of Albany, died of hemophilia at the age of 31 (but not before siring two children). Neither Victoria, nor any of her immediate ancestors, nor her other children showed any sign of the disease, but it later became apparent that at least two of her daughters were *carriers* (meaning that they carried the recessive allele on one of their X chromosomes). The inference is that a mutation must have occurred in either the sperm or the egg that together became Victoria (although the mutation could have arisen earlier in her lineage and simply been unexpressed).

One of Victoria's daughters, Beatrice, had two hemophiliac sons, both of whom died in their early twenties, and a daughter, Victoria, who married Alfonso XIII of Spain. As were her mother and grandmother, Victoria of Spain was a carrier of the disease, and two of her sons died of it as young men—including the crown prince. Alfonso XIII was ultimately forced to abdicate, and the crown prince, though still alive at that time, was in no condition to take over the throne. Spain became a republic, and only in recent years has a limited monarchy, from a nonhemophiliac branch of the family, been restored.

Alice, another of Queen Victoria's daughters, married Grand Duke Louis IV of Germany. Alice's daughter Alexandra, a carrier of the disease as were her mother and grandmother, married Nicholas II, Tsar of Russia. After four daughters, the crown prince Alexis was born in 1904. He was a hemophiliac, and suffered excruciating pain from the swelling of what would normally be minor bumps and bruises. The doctors of the day could do little for him, and, in desperation, Alexandra turned to a mystical monk, named Rasputin, who soon became the power behind the throne, and who hastened the end (which may well have been inevitable at that point) of the Romanovs. The rest, as they say, is history.

MULTIPLE GENE EFFECTS

Things become substantially more complicated when we consider traits or conditions that are controlled by more than one gene. First, the very fact that more than one gene is involved means that the potential exists for various kinds of interactions among the particular genes and between any of those genes and other genes that affect the cellular environment. This fact further blurs whatever line we might try to draw between genetically controlled and environmentally controlled traits. Second, the nature of interaction among genes can take several forms: the effects may be additive, as appears to occur with skin color; alternatively, they can interfere with each other, as sometimes occurs with eye color (hazel or green eyes are variants of brown). Third, the fact that there are several genes involved permits more potential types

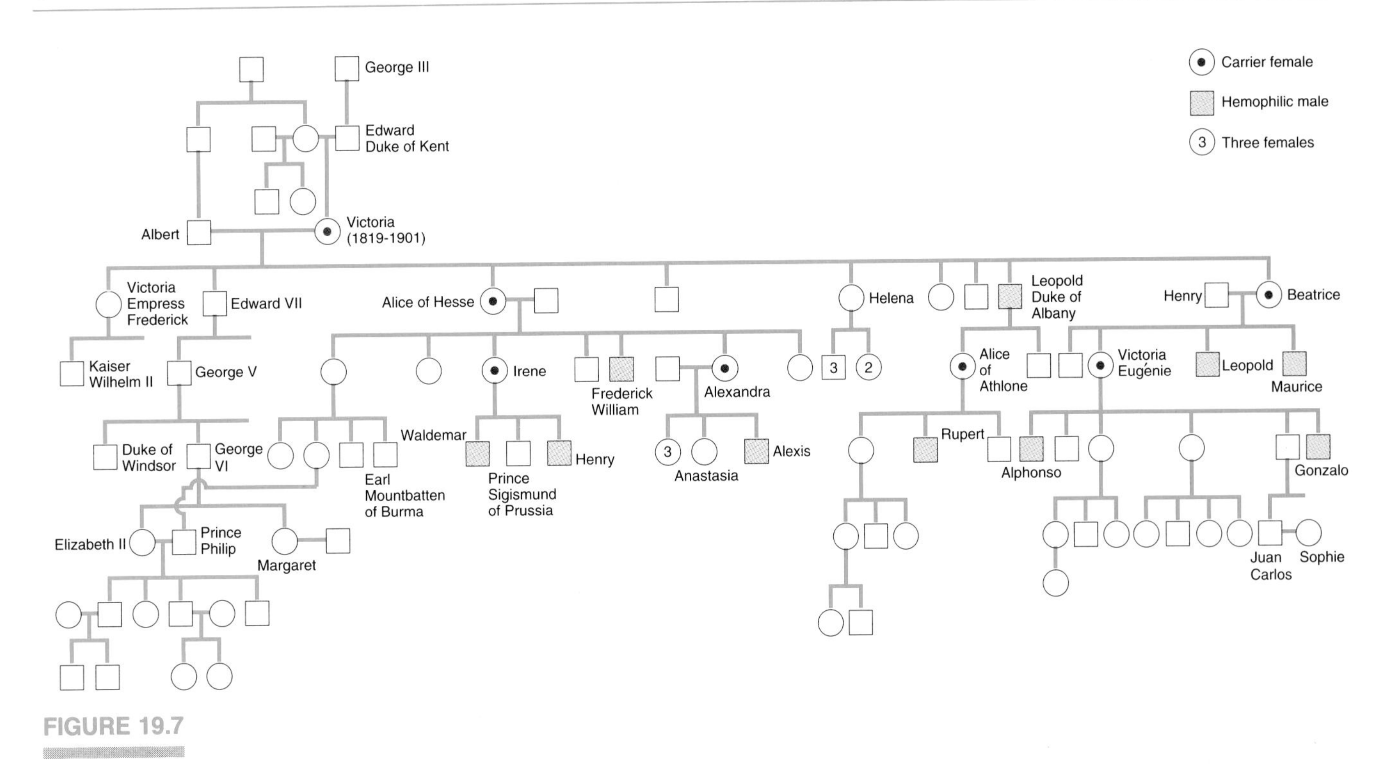

FIGURE 19.7

Hemophilia

Hemophilia, a disease in which the blood fails to clot properly, has had profound effects on the royal houses of Europe. Several of Queen Victoria's children either had the disease or were carriers. Victoria's grandaughter, the Tsarina Alexandra, was a carrier, and Crown Prince Alexis suffered from the disease.

of interactions with outside environmental factors. Thus such conditions as *spina bifida, anencephaly, cleft palate,* and *club foot* all appear to be the result of a series of genes operating against a variable environmental background.

How do we sort our way through these complexities? The primary answer is by use of the concept of **heritability.** Heritability expresses the percentage of the total phenotypic variance of a trait that is due to genetic differences, and it is often estimated by comparing variance among identical versus fraternal twins. For example, verbal aptitude has a heritability of 68 percent, but arithmetic aptitude has a heritability of only 12 percent (meaning that it is largely environmental). Among diseases, *asthma* has a heritability of 80 percent, *hypertension* is 62 percent, and *Type II diabetes* is 35 percent.

One problem with the concept of heritability is that it seems to promise more than it delivers. That is, what exactly does it mean when we say that the heritability of Type II diabetes is 35 percent? That precise-sounding figure tells us nothing about the nature of the genes involved, how they operate, or in what ways they interact with the environment. On the other hand, from the standpoint of predicting the onset of Type II diabetes, knowing its heritability value permits us to offer counsel to the children of diabetics, who are at greater risk of developing the disease than are children of nondiabetics.

These types of associations work in other ways as well. For example, certain types of diseases are highly correlated with other easily measured traits. It may be that the gene or genes responsible for the disease are closely linked to the gene for the marker trait; it may even be that the same gene is responsible (in presumably very different ways) for both the disease and the marker trait. In any event, such correlations give some clues as to the origin—and possible treatment—of particular diseases, in addition to providing the ability to predict the development of these diseases.

For example, *ankylosing spondylitis* (see Chapters 9 and 12) is rare in the general population, but is 175 times more common in individuals with a particular HLA antigen (B27). Similarly, the onset of *multiple sclerosis* is correlated with the presence of HLA antigen B7; *Hodgkin's disease* with antigen B18; *rheumatoid arthritis* with antigen DR4 and B27; *Type I diabetes* with antigens DR2, DR3, DR4 and BfF; *stomach cancer* with blood group A; and *gastric ulcers* with blood group O. In all, more than 80 diseases have been linked with marker traits.

CHROMOSOMAL ABERRATIONS

We have seen many deleterious conditions resulting from the mutation of a single gene. What happens when the error involves a portion of a chromosome, or even a whole chromosome?

Errors in Chromosome Number

Oddly enough, given their central role in inheritance, a correct chromosome count in humans was not accomplished until 1956. Prior to that time, it was believed that humans had 48 chromosomes, as do our closest relatives, the chimpanzee and gorilla.

Chromosome errors collectively are found in about one in 200 live births and in a much higher percentage of spontaneous abortions and stillbirths (see Table 19.1). Such errors may be of several types. One of the most common errors is an entire extra set of chromosomes, a condition known as **triploidy.** Triploidy generally results from the fusion of two sperm with a single egg, but can also result because of a meiotic failure in the development of the egg. The triploid fetus is unable to develop normally, and spontaneously aborts early in pregnancy. Triploidy accounts for almost 20 percent of all lethal chromosomal errors.

Tetraploidy, or two extra sets of chromosomes, generally results from a mitotic failure following fertilization. The significance of this error is a function of when during development the error occurs. If it occurs late in pregnancy, only a small amount of tissue may be involved, with the bulk of the tissue having the normal diploid number of chromosomes. Such individuals are said to be **genetic mosaics,** because, with respect to chromosome count, their cells are of two or more types. About 5 percent of all lethal chromosome errors result from tetraploidy.

Aneuploidy is the term given to cells which have one or more chromosomes missing or added. The cause of abnormal numbers of chromosomes is **nondisjunction,** the failure either of pairs of homologous chromosomes to separate during meiosis I or of chromatid pairs to separate during meiosis II. Since we normally possess two complete sets of chromosomes, we might expect that adding or losing one chromosome from one set would not be too significant. In reality, however, such additions or losses are generally very serious indeed. The loss of any of the autosomes is invariably lethal in humans, and for the most part, the possession of one or more extra autosomes is also lethal. The only exceptions appear to be one additional chromosome 13, 18, or 21, all of which nevertheless cause profound abnormalities.

Trisomy-13. An extra chromosome 13 (**Patau syndrome**) occurs in about one birth in 5000. The infant suffers from severe retardation, an underdeveloped head, cleft palate and lips, congenital heart problems, and eye defects, and usually dies shortly after birth.

Trisomy-18. An extra chromosome 18 (**Edwards syndrome**) occurs in one birth in 4500. It also causes mental retardation and congenital heart disease. More than 90 percent of affected infants die in the first year, and only one in 100 lives past the age of ten.

Trisomy-21. An extra chromosome 21 (**Down syndrome**) occurs in one birth in 600, with the rate increasing rapidly as a function of maternal age (see Figures 19.8 and 19.9). Named for its discoverer, J. Langdon Down, who first described the condition in 1866, Down syndrome includes heart defects, mental retardation, and the characteristic facial features that account for its earlier (and now discredited) name of *mongolism.* It is sufficiently common as to account for 30 percent of all mentally retarded children in the United States.

Interestingly, aneuploidy involving the sex chromosomes is, by comparison with the autosomes, much less serious. Given the small amount of genetic material on chromosome Y, such an outcome may not be surprising—but the X chromosome is larger than half the autosomes, including 13, 18, and 21. Although loss of the X chromosome is lethal, individuals can survive (albeit with increasing developmental abnormalities) with as many as four extra X chromosomes. This may be due to the fact that, in the normal

TABLE 19.1 Chromosome Abnormalities Among 100,000 Pregnancies

CHROMOSOME CONSTITUTION	SPONTANEOUSLY ABORTED FETUSES	LIVE BIRTHS
Normal	7500	84,450
Trisomy		
13	128	17
18	223	13
21	350	113
Other autosomes	3176	0
Sex chromosomes		
XYY	4	46
XXY	4	44
X	1350	8
XXX	21	44
Translocations	239	216
Polyploid		
Triploid	1275	0
Tetraploid	450	0
Others	280	49
Total	15,000	85,000

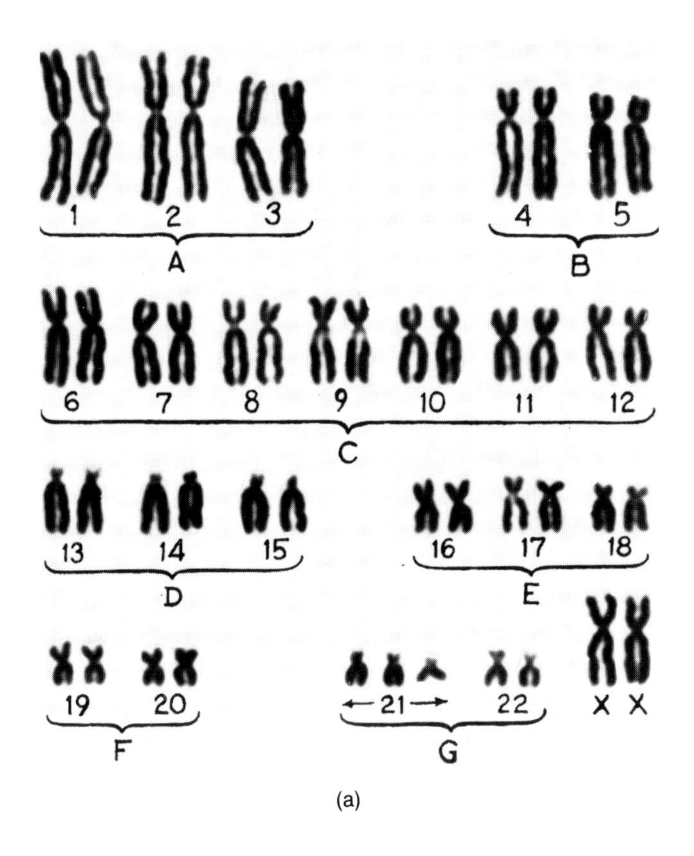

(a)

(b)

FIGURE 19.8

Down Syndrome

The most common condition resulting from an error in chromosome number is Down syndrome, which generally results from an extra chromosome 21. Sometimes this chromosome has translocated to chromosome 15; sometimes it is simply represented in triplicate (a). A child with Down syndrome (b).

XX female, only one X chromosome is functional, the other X chromosome remaining in condensed form and visible in the nucleus as the **Barr body** (named for Murray Barr, its discoverer; see Figure 19.10). Indeed, the presence of the Barr body is the diagnostic test for females, and it is used in the cell testing employed in various sporting events such as the Olympic Games. Individuals with more than two X chromosomes have a proportional increase in the number of Barr bodies. However, in the earliest stages of development all of the X chromosomes are active, and even in the Barr body state, some genetic activity remains—thus extra X chromosomes do have some impact on the individual's development.

Specific instances of sex chromosome aneuploidy include the following.

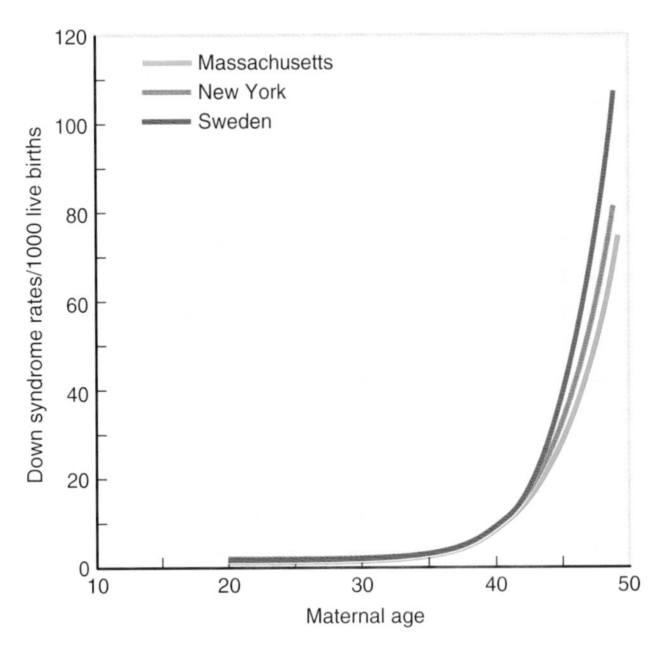

FIGURE 19.9

Incidence of Down Syndrome

Down syndrome becomes increasingly prevalent in older mothers. This graph presents the results of three different studies; the results are very similar in each instance.

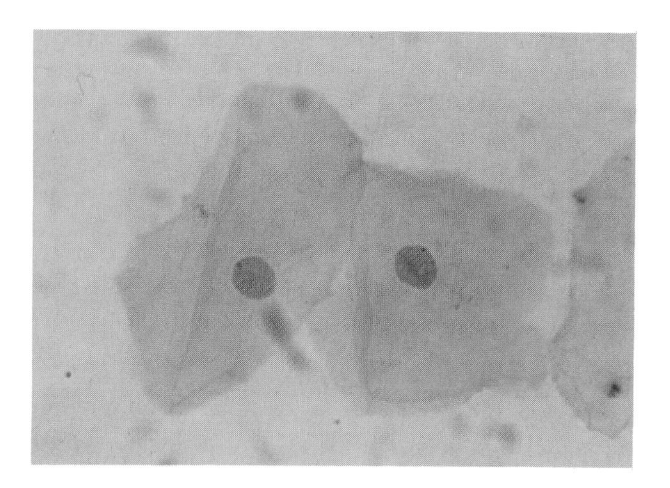

FIGURE 19.10

Barr Body

The Barr body (the dark spot at the edge of the nucleus) is a condensed form of the X chromosome found only in cells in which there are at least two X chromosomes. As such, it is the diagnostic test for gender.

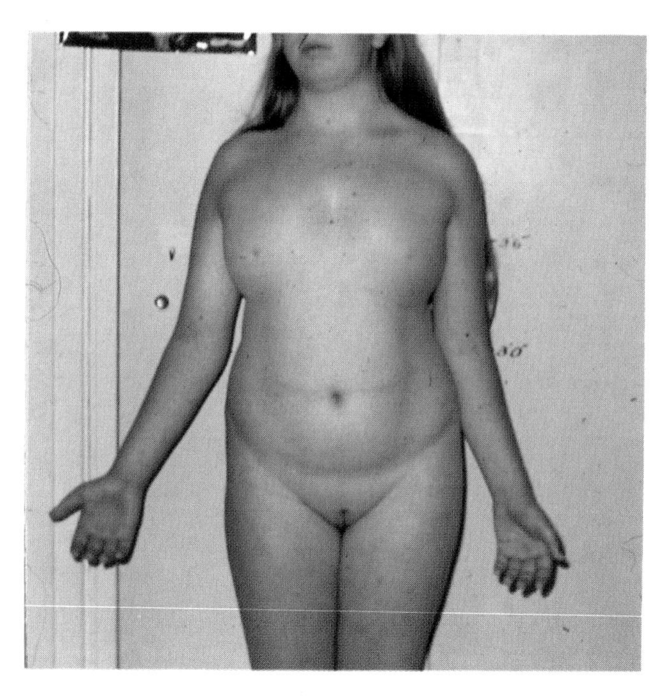

FIGURE 19.11

Extra or Missing Sex Chromosomes

A young woman with a missing X chromosome (Turner Syndrome). Note the widely spaced breasts and webbing at the neck.

Turner Syndrome. Turner syndrome was first described in 1938, although the fact that it results from a missing sex chromosome was not established until 1959. Turner syndrome individuals are XO, and are therefore female, although their ovaries are undeveloped and they are sterile. In addition, the phenotype includes short stature, webbing of the neck, wide displacement of the nipples, and a narrowing of the aorta (see Figure 19.11). Although the syndrome is relatively benign in adults, the spontaneous abortion rate of XO embryos is very high (in excess of 95 percent), for reasons which are presently unexplained.

Klinefelter Syndrome. Klinefelter syndrome was first described in 1942, but its cause (an extra sex chromosome) was not determined until 1959. Individuals with Klinefelter syndrome have an XXY genotype, but are phenotypically male, albeit with some breast development. The testes are poorly developed and the individual is sterile. Klinefelter syndrome is relatively common—about one in 800 male births—and the frequency increases with the age of the mother.

XYY Syndrome. About one in 1000 live male births have an XYY genotype, although the phenotype is only subtly different from normal. XYY males appear to be somewhat more prone to antisocial behavior, notably violence against property, than is characteristic for XY males, but this contention is still under active debate.

Errors Within the Chromosome

Another class of chromosome error results from faulty crossing over (Chapter 17). Fragments of chromosomes may *invert*, or *translocate* to another chromosome, or be inactivated, or be *duplicated*, or even be lost (*deleted*) altogether. Each of these types of chromosomal errors may cause clinical manifestations.

Inversions create enormous problems for the pairing of homologous chromosomes during meiosis, but are nonetheless found in 1 to 2 percent of the human population. The phenotypic effects vary enormously, ranging from no observable impact to severe malformation.

Translocations often create no problems, since the total amount of genetic material remains the same. However, if the break occurs within a gene, that gene will be affected—and imbalance in genetic material may in any case occur following meiosis. *Chronic granulocytic leukemia* results from a translocation from chromosome 22 to chromosome 9, presumably because a proto-oncogene (see Chapter 17) is activated in the process.

Duplications often have no effect, but duplication in the small arm of chromosome 9 is known to cause severe mental retardation and distinctive facial abnormalities.

Deletions, by comparison, are usually much more serious. Kidney cancer in children (*Wilms' tumor*) results from a loss of a section of chromosome 11; *cri-du-chat syndrome,* characterized by severe retardation and characteristic facial features, results from a delection of a section of chromosome 5.

MUTAGENS, TERATOGENS, AND DEVELOPMENT

Although the process of DNA replication is amazingly accurate, errors do still occur. These errors, which generally involve either a substitution of one base for another, or the addition or elimination of a base, are collectively called **mutations** (L. *mutare,* "to change"). *Spontaneous* mutations occur under natural conditions; *induced* mutations occur in response to specific environmental agents called **mutagens.**

Mutagens

The two principal types of mutagens are radiation and certain kinds of chemicals.

Radiation. Radiation exists both as *electromagnetic radiation* (as exemplified by ultraviolet radiation and X-rays) and *particle radiation* (as exemplified by the subatomic particles released by radioactive isotopes of elements such as uranium, radium, and radon). Both types of radiation may induce changes in the DNA of our cells by knocking electrons out of orbit and thereby initiating new, and potentially destructive chemical reactions.

Biologically significant radiation is today usually measured in **rems,** one rem being the amount of radiation necessary to generate the same effect as one *roentgen* of X-rays. (A roentgen, in turn, is defined in electrical terms.) The point is that, because they possess differing amounts of energy, various radiation sources have unequal effects on biological tissue. By converting this energy to a common unit of measure (rems), we can more easily compare the potential hazard to living tissue.

We are constantly bathed in low levels of radiation. Ultraviolet radiation from the sun, and the natural breakdown of radioactive isotopes in our environment ensure this fact, whether we like it or not. Indeed, it has been speculated that these normally moderate amounts of radiation actually operate to the biological good, since they induce the mutations that are the raw materials for natural selection (see Chapter 22). (Of course, that point tends to ignore the fact that what might be beneficial at the population level is not necessarily all that desirable at the level of the individual—mutations are often deleterious, and the affected individual may well be selected against.)

In any event, it would be wrong to conclude that natural radiation is "good" and only human-created radiation is "bad." Radiation is radiation, and its effects are cumulative. Radon gas, for example, a gas which is reasonably abundant in the soil in many parts of the world, is thought to be the cause of many cases of cancer each year.

Chemicals. The fact that certain chemicals can cause mutations has been known since 1942. The several hundred chemicals that have now been identified as mutagenic operate in only a handful of ways. Some chemicals mimic the effects of radiation, in that they cause breaks in the DNA molecule. Others mimic the nitrogenous bases that comprise the DNA message, and substitute for such bases during DNA assembly. Still others modify the bases themselves, effectively changing the DNA code. Finally, some mutagens actually bind with DNA, causing bases to be deleted or added.

The great concern with mutagens in general is that they will either cause mutations in the gametes, leading to potentially devastating results in the offspring, or that they will affect proto-oncogenes, leading to the development of cancer (see Chapter 17).

Teratogens

Teratogens (Gr. *teras,* "monster" and *gen,* "born") are environmental agents that interfere with the normal functioning of genes during embryonic development but are not themselves necessarily mutagenic. Radiation can function as a teratogen, which is one reason why X-rays should be used as little as possible during pregnancy. Certain viruses (e.g., *chickenpox, mumps, hepatitis,* and *measles*) are teratogens, but the most serious is *rubella* (German measles). As a consequence of one outbreak of this disease in the United States in 1964, more than 20,000 children were born with birth defects, often involving the heart. The effects of the disease are particularly insidious because they are the most profound during the first weeks of pregnancy, before a woman may know she is pregnant. The fact that many viruses can cause birth defects is a primary reason for the continued insistence on universal vaccination for the common viral diseases of childhood.

Finally, many chemicals can be teratogenic. *Thalidomide,* a drug widely prescribed in England and

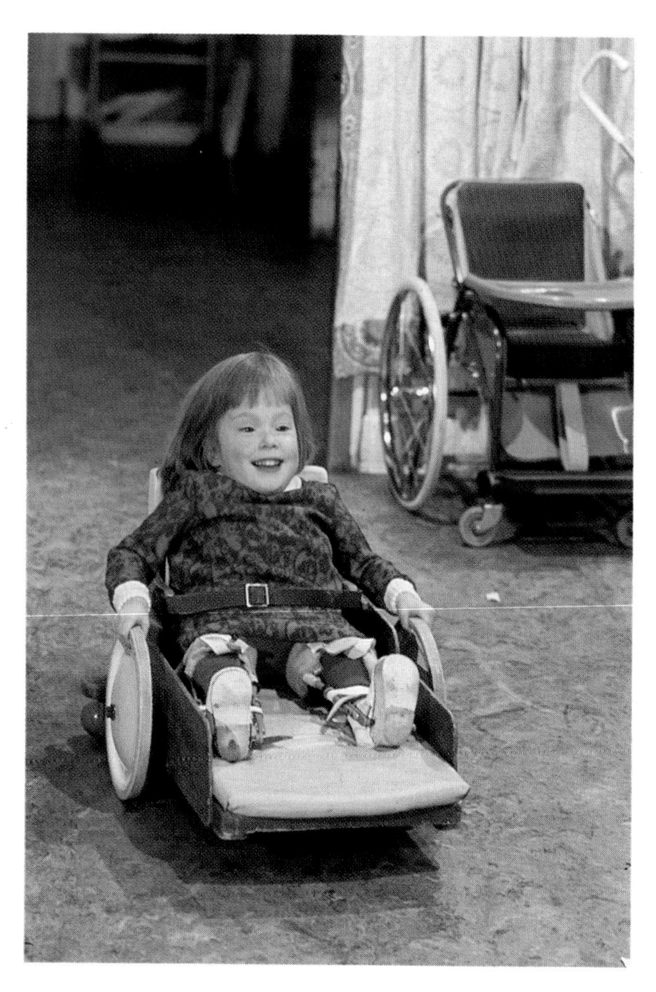

FIGURE 19.12

Thalidomide Child
In the early 1960s pregnant women who took the drug thalidomide for morning sickness bore children with incompletely developed arms or legs. These conditions mimic a rare genetic condition, but they are not themselves inherited, further proof of the germ plasm theory of inheritance. (Pangenesis would predict a heritable condition.)

FIGURE 19.13

Fetal Alcohol Syndrome
Consumption of alcohol by pregnant women can result in fetal alcohol syndrome in their children. Such children have a variety of physical and mental defects.

Germany during the late 1950s and early 1960s for morning sickness, caused several thousand birth defects, most of which involved incomplete formation of the arms or legs (see Figure 19.12). Since that time, physicians have been reluctant to prescribe virtually any medication to women during the early stages of pregnancy, and even advise against more than occasional use of such common drugs as aspirin and caffeine (although there is no conclusive evidence that these substances are teratogenic). The use of tobacco during pregnancy, however, has been linked to babies of lower than average birth weight (which accounts for the current campaign against smoking during pregnancy). Alcohol is a particularly potent teratogen (see Figure 19.13), causing a variety of physical and mental defects now collectively characterized as **fetal alcohol syndrome.** In this country, the incidence of fetal alcohol syndrome is between one and two per 1000 births, making it one of the most common of all birth defects.

Teratogens often mimic the effects of genetic mutations. Thalidomide, for example, may create a phenotype identical with the genetic condition **phocomelia** (Gr. *phoke,* "seal" and *melos,* "limb"). Such environmentally induced phenotypes are called **phenocopies.** It is of great importance to distinguish birth defects caused by teratogens from birth defects caused by mutagens. In the former instance, the defect is environmentally induced and is not transmitted to the offspring; in the latter, the defect is genetically induced and may be transmitted to the offspring. Thus thalidomide children can expect their children to have normal arms and legs, whereas children with phocomelia may have children who share their condition.

GENETIC COUNSELING, SCREENING, AND DIAGNOSIS

As we have learned more about the nature of genetic diseases and the pattern of their inheritance, and with our improved understanding of the chemistry of the gene, it has become possible to advise prospective parents on the likelihood of their child being born with a particular genetic disease or condition. It is also possible, in many instances, to determine early in embryonic development if the embryo possesses a particular genetic abnormality, thereby permitting prospective parents to decide whether or not to terminate a pregnancy.

Genetic Counseling

Genetic counseling is much more than a mere recitation of Mendelian ratios. Many traits vary in their expressivity and penetrance. Parental age, ethnicity, a history of miscarriages, the presence or absence of a particular genetic condition in other children, among many other factors, all require a unique assessment for each individual case.

Genetic Screening

Genetic counseling is highly desirable for couples who have reason to suspect that a prospective child might have a birth defect. However, for certain conditions, mass screening may be desirable even when couples have no familial history of the condition. Similarly, screening of newborns for defects that have not yet materialized is highly desirable.

For example, voluntary screening of people of Ashkenazic Jewish descent began in 1971, and led to a reduction in the number of children born with Tay-Sachs disease of up to 85 percent. The high percentage of carriers of sickle-cell anemia among black Americans makes them good candidates for similar voluntary screening.

Screening of newborns for PKU began in 1960 and is now mandatory in most states. New York state now tests routinely for eight different genetic conditions in newborns, using a single blood sample, in an effort to provide intervention in these conditions before their consequences become manifested.

On the other hand, screening may have several negative consequences. Identified carriers have sometimes been denied health insurance; many genetic conditions do not permit effective screening at this point, and some provide only ambiguous results; in the minds of many people, genetic screening raises the spectre of **eugenics,** and kindles memories both of the atrocities committed by the Nazis during World War II and the policy of mandatory sterilization for certain conditions in this country between 1917 and 1935. These social and ethical dilemmas will only increase as we learn more about the role of genes in disease.

Diagnosis

Genetic counseling and screening provide a risk assessment prior to conception. What about those couples who decide to take the risk, or who simply want to know if their prospective child possesses some particularly nasty genetic defect?

Several techniques are now in use which permit a diagnosis for many genetic conditions early in pregnancy (see Figure 19.14).

Amniocentesis. Amniocentesis involves the removal of a small amount of *amniotic fluid* (the fluid that bathes the fetus; see Chapters 20 and 21) using a hypodermic needle, and the subsequent examination of this fluid (and the cells contained in the fluid) both chemically and microscopically.

Amniocentesis is normally performed early in the second trimester of pregnancy. It cannot be performed earlier, since cells are rare in the amniotic fluid until the fourteenth week of gestation. It is of little use later in pregnancy, since some of the tests take up to three weeks, and therapeutic abortions generally cannot be performed after the second trimester. Amniocentesis permits diagnosis of all of the chromosome disorders, and more than 100 biochemical disorders, with a reliability of diagnosis in excess of 99 percent. Nevertheless, the procedure is not without some risk—from 2 to 3 percent of women spontaneously abort following amniocentesis. On the other hand, that figure is not very different from spontaneous abortions and stillbirths that occur in nondiagnosed women.

Chorionic Villus Biopsy. The *chorion* is the outermost of the extra-embryonic membranes (see Chapters 20 and 21) and contributes to the formation of the placenta. Small snippets of chorionic cells can be sampled as early as the fifth week of pregnancy, and diagnosed in the same manner as cells obtained by amniocentesis. Moreover, chorionic villus biopsy is now believed to pose less risk of spontaneous abortion than does amniocentesis and is rapidly becoming the diagnostic method of choice.

Ultrasonography. X-rays are generally considered too risky to use on pregnant women, but ultrasonography uses high-frequency sound waves to create visible "echos" of the fetus, and major structural malformations can readily be detected, even of such internal organs as the heart and liver. To date, ultrasonography has not been demonstrated to pose any risk to the mother or fetus.

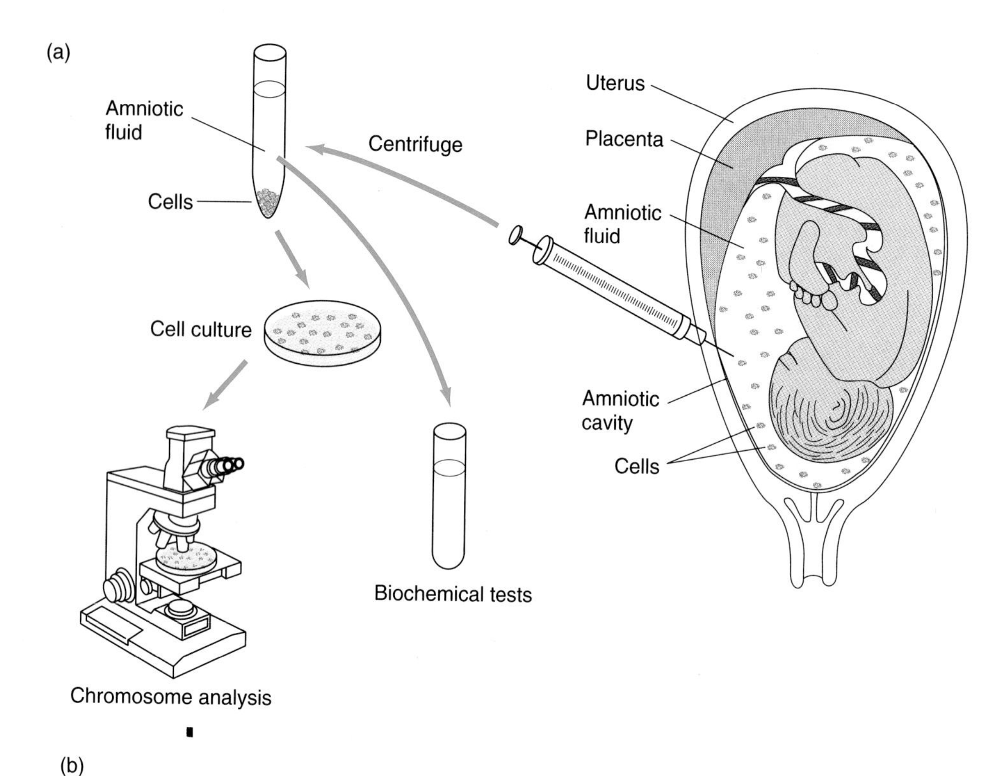

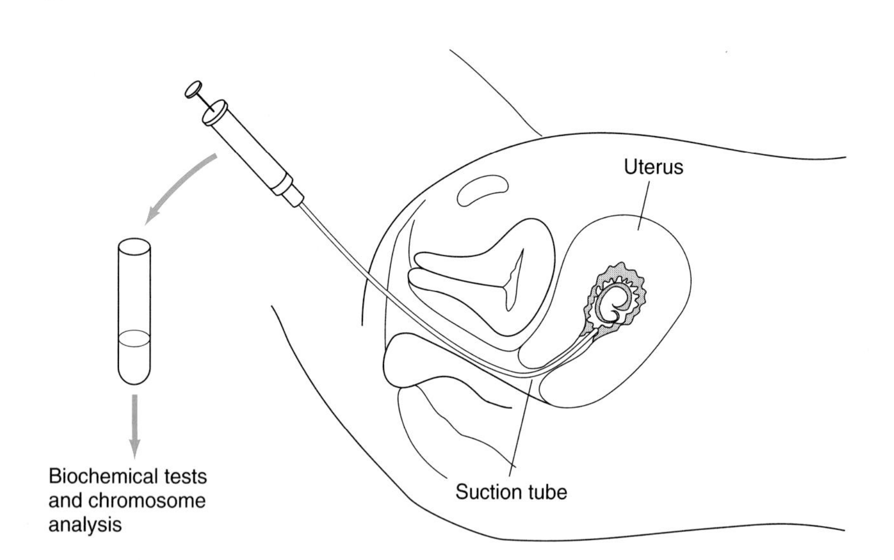

FIGURE 19.14

Diagnostic Procedures

(a) Amniocentesis involves the removal and analysis of a small amount of amniotic fluid. (b) Chorionic villus biopsy involves the removal and analysis of cells of the chorion. Ultrasonography involves the use of ultrasound to project an image of the fetus on a viewing screen (see right).

GENETIC ENGINEERING

As we saw earlier in this chapter, the loci of almost 2000 genes had been identified by 1990. However, until 1968, no genetic loci were known on autosomes. What has happened in the last 20 years to permit such an explosion of genetic knowledge—and what are the implications of this knowledge for us today?

During the past two decades, many bacterial enzymes, known as **restriction enzymes,** have been isolated and identified. These enzyme protect bacteria from being attacked by foreign DNA molecules by fragmenting these DNA molecules at specific sites. By using a combination of restriction enzymes, scientists can break human DNA into a series of fragments. Through a variety of techniques, these fragments can then be placed in the correct order, and the nucleotide sequence of each determined.

Specific genes can be determined by using a variant of this approach. Suppose that we wish to determine the locus of a gene known from other analyses (generally, the use of markers) to be on a section of chromosome 14. If we compare the restriction fragments from an individual with the mutated gene to

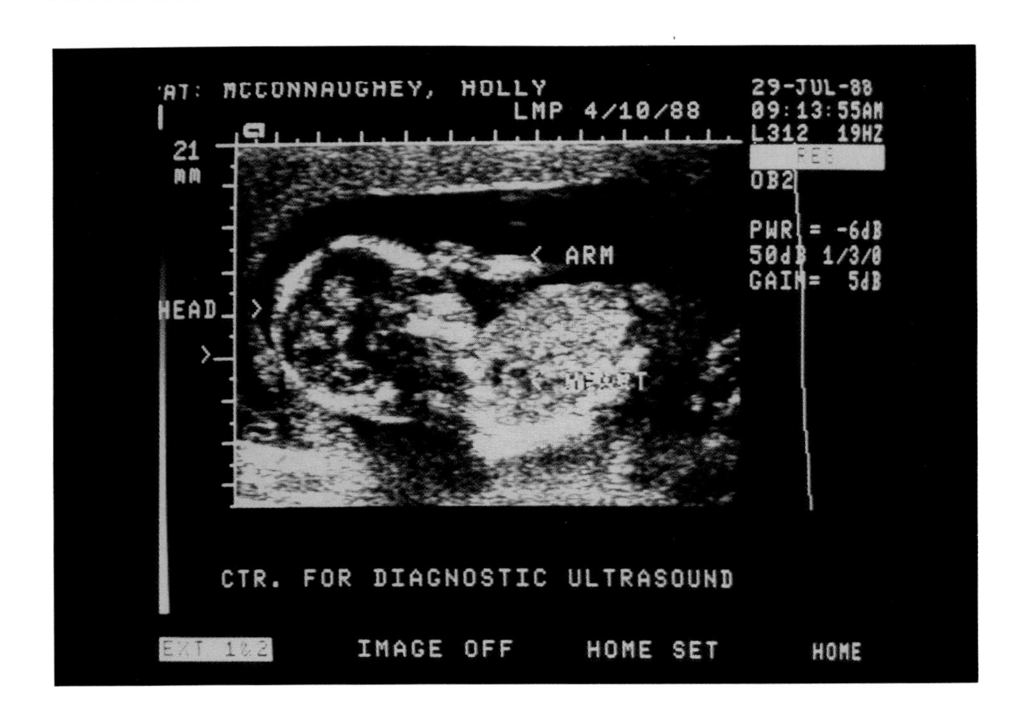

the fragments from an individual with the normal allele, we can expect to find a few fragments of different lengths (because the restriction enzymes will split the DNA at different sites owing to the different nucleotide sequence in the mutated allele). This process is known as **restriction fragment length polymorphisms (RFLPs),** and it has permitted the beginning of chromosome maps in humans (see also Box 19.1). At present, however, no chromosome is more than 2 percent mapped, and half the chromosomes are less than 0.5 percent mapped.

Recombinant DNA

The identification of gene loci by RFLP analysis led promptly to a highly practical application of this information. By using special enzymes, scientists can splice specific human genes into bacteria, thereby effectively *recombining* the DNA of humans with bacteria. The bacteria replicate the human gene along with their own as they divide, and soon a whole colony of bacteria contains the human gene—and if provided the proper substrate, the bacteria will produce the protein coded for by the gene in large amounts. Precisely such a mechanism has been used for the human genes coding for *insulin, growth hormone, interferon, tissue plasminogen activator,* and blood clotting *factor VIII,* among many others. These substances, once available only in tiny amounts following very expensive extraction techniques from human tissue, are now widely available for medical use.

Of course, as useful as these products are, they in no sense repair the defective DNA of the person suffering from a genetic disease. Moreover, many genetic diseases do not lend themselves to treatment by the injection of a genetically engineered protein.

At present, experiments are underway that take the next obvious step: Normal alleles are being installed in the cells of individuals with defective alleles. For example, the disease known as **SCID** (*severe combined immunity disorder;* see Chapter 9) is caused by a mutated recessive allele at one of three different gene loci. One of these mutations prevents the normal formation of the protein **ADA** (*adenosine deaminase*), a critical enzyme in ATP metabolism. Without this enzyme, the B and T cells of the immune system cannot survive, and thus the individual has essentially no immunity to pathogens. A proposal has recently been made to inject ten children with this form of SCID with their own T cells, altered to contain the normal allele for ADA. Similar experiments may soon be possible for *familial hypercholesterolemia* (a rabbit model has already been perfected) and *cystic fibrosis* (the normal gene has successfully been installed in defective cells in the test tube).

Another related experiment is already underway. Cells known as **TIL cells** (*tumor-infiltrating lymphocytes*) were withdrawn from 50 people suffering from advanced *melanoma* (see Chapter 17). The gene for *tumor necrosis factor* (see Chapter 9) was then spliced into the TIL cells, and they were injected back into the patients. These now very potent cells attacked tumors throughout the body, and the initial results look very promising.

Other RFLP Uses

RFLP analysis is now being used in criminal investigations, because very small amounts of tissue can now be analyzed and linked to a specific individual. More than 200 convictions (and acquittals) have been based on such evidence, although the accuracy of this technique has recently been called into question. Ideally, the use of enough restriction enzymes should yield a unique "DNA fingerprint." However, when the two commercial laboratories that perform these analyses were tested, one laboratory correctly identified 37 of 51 sample pairs, but called the other 14 pairs inconclusive. The second laboratory correctly matched 44 of 49 pairs, but erroneously identified another pair. At present, there is no set of uniform and accepted standards by which experts can agree as to how similar two samples must be to ensure a correct match.

Summary

In the last 20 years, the loci of almost 2000 human genes have been determined, and almost 5000 human traits are now known to be under genetic control. Much of our knowledge about genetics has come from identifying inherited diseases or abnormalities, and then, through the use of markers and restriction enzymes, determining the locus of the gene and the chemical composition of both the gene and its protein product.

Inherited diseases or abnormalities include single gene conditions (with either dominant or recessive defective alleles) on both autosomes and the sex chromosomes; multiple gene conditions, many of which are strongly influenced by environmental conditions; and disorders within the chromosome or the chromosome set. There is no correlation between the size of the error and the magnitude of the effect.

Radiation and various chemicals are known to increase the rate of mutations. A different type of chemical, called a teratogen, causes developmental disorders that may mimic genetic conditions, but which are not inherited.

With increased knowledge about the chemistry of particular genes, prospective parents can receive genetic counseling about their prospects for having a child with a particular genetic malady. New techniques that can be performed with the fetus *in utero* now provide extensive additional information about certain abnormalities that the fetus may have.

Finally, it is now possible to transplant individual human genes into bacteria, creating bacterial factories for the manufacture of particular proteins. The next step—transplanting normal alleles into cells with mutated alleles and then injecting those cells back into the body—is about to be taken.

Key Terms

autosome
achondroplastic dwarfism
brachydactyly
polydactyly
familial hypercholesterolemia
Huntington disease
Marfan syndrome
neurofibromatosis (NF)
penetrance
expressivity
albinism
alkaptonuria
cystic fibrosis
galactosemia
phenylketonuria (PKU)
sickle-cell anemia
thalassemia
Tay-Sachs disease
color blindness
Duchenne muscular dystrophy
Lesch-Nyhan disease
hemophilia
heritability
triploidy
tetraploidy
genetic mosaic
aneuploidy
nondisjunction
trisomy-13 (Patau syndrome)
trisomy-18 (Edwards syndrome)
trisomy-21 (Down syndrome)
Barr body
Turner syndrome
Kleinfelter syndrome
XYY syndrome
inversion
translocation
duplication
deletion
mutation
mutagen
radiation

rem
teratogen
fetal alcohol syndrome
phenocopy
eugenics
amniocentesis
ultrasonography
chorionic villus biopsy
restriction enzyme
restriction fragment length polymorphism (RFLP)
recombinant DNA
severe combined immunity disorder (SCID)
ADA
TIL cell

Box Terms

genome
consanguinity
SRY gene
androgen
testicular feminization syndrome (Tfm)

Questions

1. Describe one dominant, one recessive, and one sex-linked trait in humans.
2. What does it mean when we say that hypertension has a heritability of 62 percent?
3. What is aneuploidy? Give an example.
4. Give an example of a teratogen. Are the phenotypic consequences of teratogens inherited by the offspring? Defend your answer.
5. What techniques are currently in use for diagnosing possible genetic defects in the fetus?
6. Give three examples of genetic engineering in current use.
7. What is the human genome project? Why is it significant?
8. Many states restrict marriage between blood relatives. Geneticists could be expected to endorse such restrictions. Why do you suppose they might feel that way?

CHAPTER

20

The Reproductive System

Creating the Next Generation

Why does the growth rate of girls diminish when they enter puberty, whereas boys grow more rapidly?

Which techniques of birth control are the most effective? Why?

Are venereal diseases really serious, or just embarrassing? With all our antibiotics, why is VD on the increase?

These questions all relate to the reproductive system, the subject of this chapter.

THE NATURE OF THE PROBLEM

Unlike the other organ systems, which contribute in a variety of ways to the maintenance of homeostasis, the reproductive system is an end in itself. From the biological point of view, all the other systems are subservient to the reproductive system insofar as they collectively maintain the organism until it reaches sexual maturity and is able to reproduce. To be sure, we tend to take a less dispassionate view of things when looking at ourselves, because we generally feel that our lives have intrinsic meaning, rather than simply being a waiting period before (and after) reproduction.

Philosophical explorations aside, the effective functioning of the reproductive system requires the surmounting of significant obstacles. Consider the development of the male and female systems. While some of the other organ systems show modest sexual differences (such as skeletal system differences; see Chapter 12), no other system approaches the level of dissimilarity achieved by the reproductive system. How are such profound differences in anatomy achieved? Are they the result of differential growth of the same embryonic structures, or of the growth of entirely different structures that are present from the earliest stages of embryonic development?

Consider an entirely different set of problems. The reproductive system must begin to function only as the individual nears physical maturity. If a very young child were capable of becoming pregnant, for instance, her life would be imperiled by the growth of a fetus larger than she would physically be capable of carrying. Thus, unlike the other systems that are functional at (or even before) birth, the reproductive system must be "turned on" rather late in development. How is this "turning on" process mediated?

In humans, reproduction means pregnancy, and pregnancy is the consequence of two successive events: **fertilization,** which is the combining of the nuclei of an ovum from the female and a spermatozoan from the male, and **implantation,** which is the process whereby the fertilized egg adheres to the lining of the uterus. The male, by definition, contributes only to the first process. Consequently, the problem to be overcome in turning on the male reproductive system is relatively trivial: there need only be a mechanism whereby the organs responsible for producing sperm begin their activity. Within broad limits, the more sperm a male produces, the more likely he is to be successful in fertilizing a female (we are speaking in statistical terms at the cellular level—that is, what is the likelihood that a sperm cell will fertilize an ovum—rather than suggesting that, at the level of the individual, females choose males based on how many sperm they produce!). In any event, the male is capable of producing millions of sperm cells every day for virtually his entire adult life.

The situation in the female is far more complex. She cannot produce millions of ova—the fact that development of the embryo and fetus is internal requires that very few (in humans, typically just one) ovum be fertilized at one time. Thus the female system must be designed so as to permit the release of only one ovum during any given cycle. If fertilization is not accomplished when a particular ovum is released, there must be a process whereby, after a suitable lag period, a second ovum can be released. Conversely, if the first ovum is fertilized, release of additional ova during the pregnancy must be prevented, so as to avoid overcrowding in the uterus. In addition, the uterine lining must be properly prepared to receive the fertilized ovum, which means that the developmental stages of uterine lining development must be correlated with the timing of the release (and fertilization) of the ovum. Finally, at least in humans, the ability of a woman to become pregnant must be terminated at some point in mid to late life, in order to ensure (or at least increase the likelihood) that the woman will not only survive the pregnancy but live long enough after birth to care for her child until it can survive on its own—a period of many years. How are all these requirements achieved?

Let us consider these problems sequentially.

THE DEVELOPMENT OF THE HUMAN REPRODUCTIVE SYSTEM

Not until the seventh week of embryonic life are there discernable structural differences in the reproductive systems of males and females (discounting chromosomal differences). Prior to that time, the male and female gonads (the **testes** in the male and the **ovaries** in the female) are identical under the microscope, and the external sex organs (the **genitalia**), such as they are, are also identical (see Figure 20.1).

Male/female differences increase during fetal development, and, at birth, many of the anatomical differences by which we distinguish males from females have already occurred. However, these differences are due to the differential growth of the same initial set of structures.

The female structures resemble those of the undifferentiated embryo, albeit on a larger scale (see Figure 20.2). Just above the opening of the urethra is the **clitoris** (Gr. *kleiein,* "to hide," a reference to its hidden position). On either side of the clitoris, and surrounding the urethra and vaginal openings, are a pair of folds called the **labia minora** (L., "lesser lips").

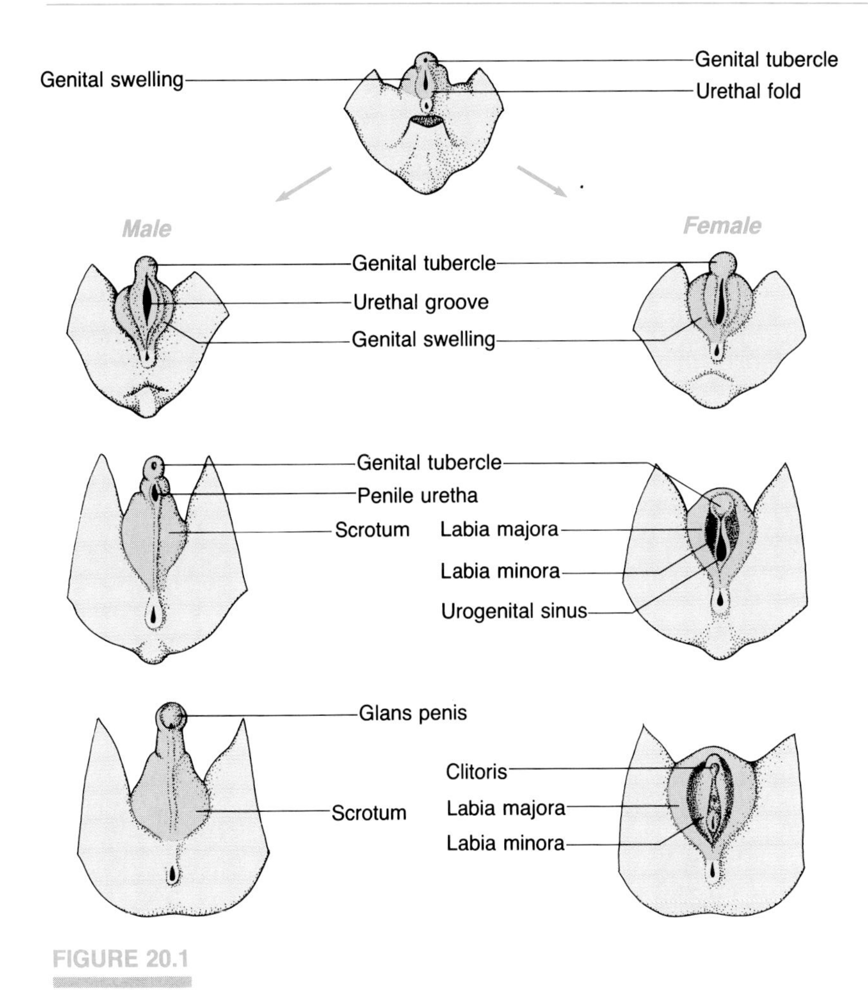

FIGURE 20.1

Embryonic Development of the Genitalia
Until the seventh week of embryonic development, the genitals are identical in the two sexes. Subsequent development is along one of two pathways, but male and female structures are derived from a common starting point.

A second, and much larger pair of folds, the **labia majora** (L., "greater lips"), lies outside the labia minora.

In males, these same structures have developed very differently (see Figure 20.3). The structure that becomes the clitoris in the female develops into the head, or **glans** (L., "acorn," a reference to its appearance) of the penis. Because the urethra must run to the end of the penis, it necessarily grows along with the glans. The portion of the urethra within the penis corresponds to the labia minora of the female. (Note that the urethra passes *through* the glans in the male, but *below* the clitoris in the female. Thus the male urethra must transport both urine and sperm, although obviously at different times and in different circumstances.) Finally, the structures that correspond to the labia majora of the female grow and fuse in the male to form the **scrotum,** the bag that holds the testes.

Internal differences between males and females are somewhat greater. The basic duct of the female—**vagina** (L., "sheath"), **uterus,** and **fallopian tubes**—is not derived from the same embryonic structure as is the male duct—the **vas deferens** (L., "duct carrying fluids away"). There is likewise no female equivalent of such male glands as the prostate and the seminal vesicles.

One of the more striking differences between the male and female systems is the relative position of the gonads. The ovaries are located in essentially the original embryonic position, whereas the testes migrate during embryonic and fetal life and come to lie outside the main body cavity by birth.

What is accomplished by this migration? In humans (and many other mammals) the slightly lower temperature within the scrotum (about 1°C lower) is required for proper development of the sperm. In individuals whose testes do not descend, the testes manufacture hormones normally, but generally fail to produce sperm.

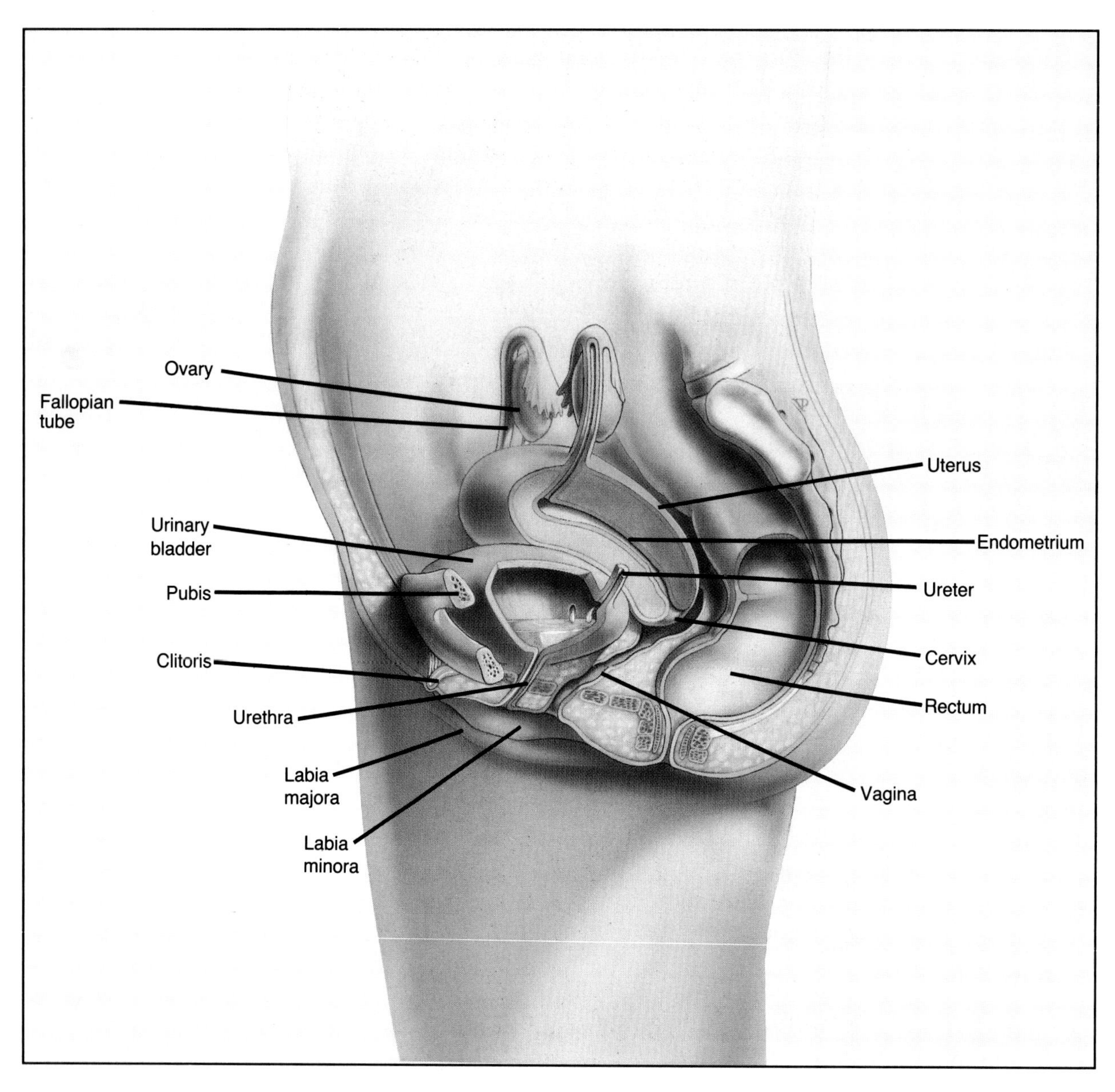

FIGURE 20.2

Female Reproductive System
The ovaries, fallopian tubes, labia majora, and labia minora are paired structures; the uterus, vagina, and clitoris are single structures.

THE ONSET OF PUBERTY IN THE FEMALE

Despite obvious differences in external genitalia, young boys and girls differ very little in their levels of hormones. However, very significant hormone differences occur at **puberty** (L. *puber*, "adult"). Puberty is the period during which sexual maturity occurs and reproduction becomes possible.

The onset of puberty in the female is triggered by factors that are largely unknown (see Box 20.1). There is some correlation with overall body growth, but it is a weak correlation. During the past century, the age at which puberty begins has declined about three years in most Western countries (from age 16 to age 13), presumably because of improved nutrition and its effects on growth (see Figure 20.4). (Those rare situations where puberty begins in children as young

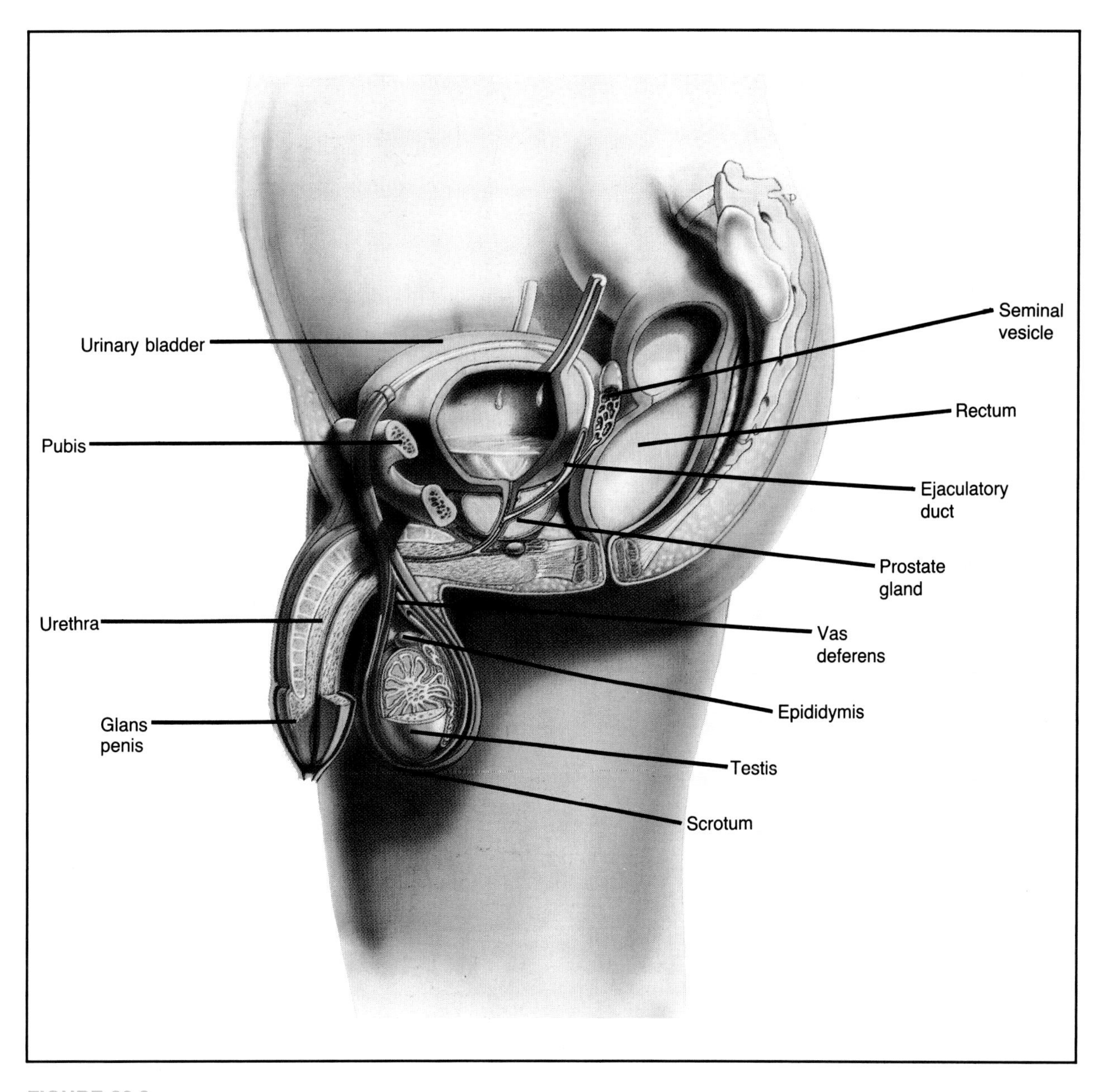

FIGURE 20.3

Male Reproductive System

The testes, vasa deferentia, and seminal vesicles are paired structures; the prostate gland, penis, and scrotum are single structures.

as age five are generally due to overproduction of hormones caused by a brain tumor.) The effects of an increasingly earlier sexual awakening in a society that "forbids" sexual experiences until marriage are, unfortunately, all too obvious.

Many of the events that are associated with puberty result from the effects of the hormone **estrogen** (see Figure 20.5), which is produced by the ovaries. (Actually, "estrogens" are a group of related hormones that, for the sake of convenience, we shall consider as a single hormone.) Among the more important of these events are the maturation of the reproductive tract and the growth of the breasts and hips. Estrogen also accelerates the activity of osteocytes, but simultaneously causes the growth areas of the long bones to fuse with the shafts. Hence, women entering puberty at a young age tend to be shorter than women who enter puberty at a later age.

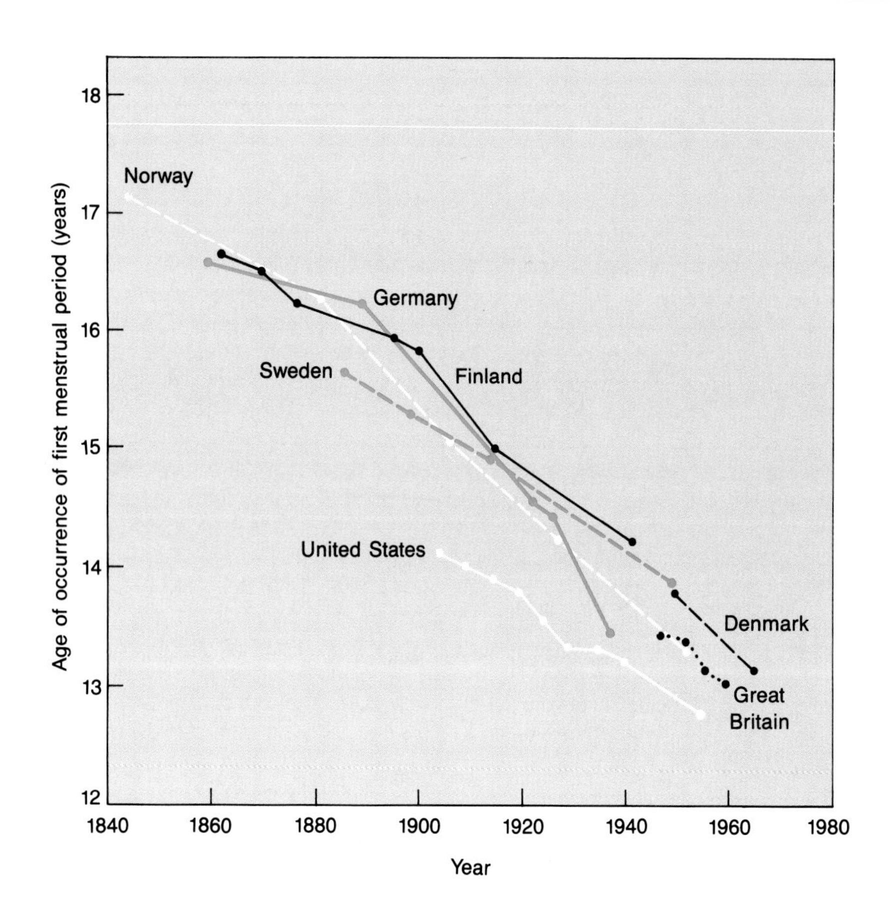

FIGURE 20.4

Age of First Menstruation

Among industrialized nations where data are available, the age of the first menstrual period has declined as much as four years in the last 120 years.

Another event associated with puberty, and one in which estrogen is intimately involved, is the onset of the **menstrual cycle** (L. *menstruus*, "monthly"), which has a periodicity of approximately 28 days. How is the menstrual cycle initiated? What is the role of estrogen?

The Menstrual Cycle

By convention, the beginning of a menstrual cycle is the first day of menstruation. A few days prior to the beginning of a new cycle, **GnRH** (**gonadotropin releasing hormone**[1]) from the hypothalamus stimulates the anterior pituitary to increase the rate of secretion of the hormones **FSH** (**follicle stimulating hormone**)

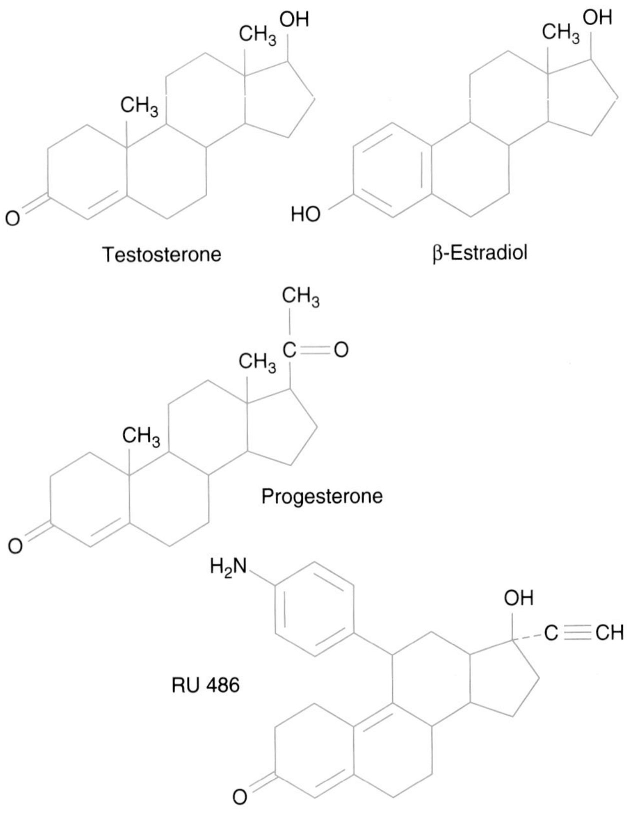

Testosterone

β-Estradiol

Progesterone

RU 486

FIGURE 20.5

Steroid Hormones

All steroids share a common structure of four rings, but differ in the nature of their side chains. Testosterone is the primary male hormone; progesterone and β-estradiol are female hormones; and RU 486 is a synthetic progesterone mimic capable of inducing menstruation.

[1]A *gonadotropin* (Gr. *trope*, "to turn or influence") is therefore a substance that influences, or stimulates, the gonads.

BOX 20.1

Menstruation and Body Fat

One of the factors associated with **menarche** (Gr. *arche*, "beginning")—the period of first menstruation that marks the beginning of puberty—is percentage of body fat. In the United States about 22 percent of the total body weight of both boys and girls prior to puberty consists of fat, but by the age of 18, body fat in women has risen to an average of 28 percent of total body weight, and dropped in men to an average of only 13 percent.

At least 17 percent of the total body weight must be in the form of fat for menarche to occur; very thin girls therefore do not begin to menstruate, or, if they have already begun puberty prior to losing weight, they cease menstruating (a condition known as **amenorrhea**). Female athletes and body builders, as well as anorexics (see Chapter 7) often cease menstruating, or do so irregularly—and do not begin menstruating regularly again until their body fat averages 22 percent of total weight (a somewhat higher level than is needed for menarche).

What does body fat have to do with puberty and menstruation—and how does the body "know" how much fat it has?

Fat is our primary means of long-term energy storage. Producing a viable infant costs between 50,000 and 80,000 kcal, and lactation requires another 500 to 1000 kcal/day. A woman without a sufficient fat reserve cannot deliver a viable infant, and is probably at risk attempting to do so. Far better that she not be able to become pregnant in the first place, and that would mean no ovulation—hence, no menstruation. The feedback loop within the body is not fully understood, but it clearly involves the brain since the absence of menstruation occurs because GnRH production is erratic and low in volume in very thin women.

This requirement for body fat explains a number of observations. The age of menarche has dropped in developed countries because of better nutrition and less disease (meaning higher levels of body fat). Women in developing countries tend to be older at menarche because malnutrition delays menarche. The fact that lactating women in developing countries do not become pregnant (whereas women in industrialized nations often do) is in part explained by the better nutritional state of the latter group of women, who will continue to menstruate (and ovulate) while nursing, unlike their more poorly nourished sisters from developing countries.

What about men? Because they do not have to carry or nurse a child, we might expect no relationship between body fat and sperm production. However, men who are more than 25 percent below the standard weight for their height—including many top marathoners and other very lean athletes—often show reduced sperm production. They, too, show reduced levels of GnRH, which suggests that GnRH production is a function of overall fat levels, although how the brain detects differences in fat levels in the body is still unknown.

---

and **LH (luteinizing hormone)**. Under the influence of these two hormones, approximately 20 **egg follicles** (ova with surrounding tissues) begin to grow and mature on the surface of the ovaries. The follicular cells (*not* the ova) produce estrogen (see Figure 20.6).

Ultimately, one follicle becomes predominant, and the other follicles regress (but see Box 20.2). The rapidly growing primary follicle produces increasing levels of estrogen. Estrogen has other effects in addition to those discussed above.

1. It inhibits GnRH, which leads to a gradual drop in the production of FSH and LH (a classic example of negative feedback).

2. It stimulates the growth of the lining, or **endometrium,** of the uterus.

For reasons not yet fully understood but apparently involving a temporary positive feedback with estrogen, at about day 12 of the menstrual cycle there is a sudden surge in the production of FSH and LH. The level of FSH doubles, and that of LH increases by as much as ten times. This surge of pituitary hormones causes the now-ripened follicle to rupture approximately two days later (day 14), releasing the ovum, which travels slowly down one of the fallopian tubes (see Figure 20.7).

The remaining follicular cells, under the influence of LH, form a scab-like structure called the **corpus luteum** (L., "yellow body") (see Figure 20.8). From day 14 to day 26, with a peak at roughly day 21, the corpus luteum produces large amounts of estrogen and the related hormone **progesterone.**

Progesterone has two primary effects:

1. It causes the endometrial cells of the uterus to swell and become secretory.

2. In conjunction with estrogen, it inhibits GnRH, leading to a sharp decline in FSH and LH production.

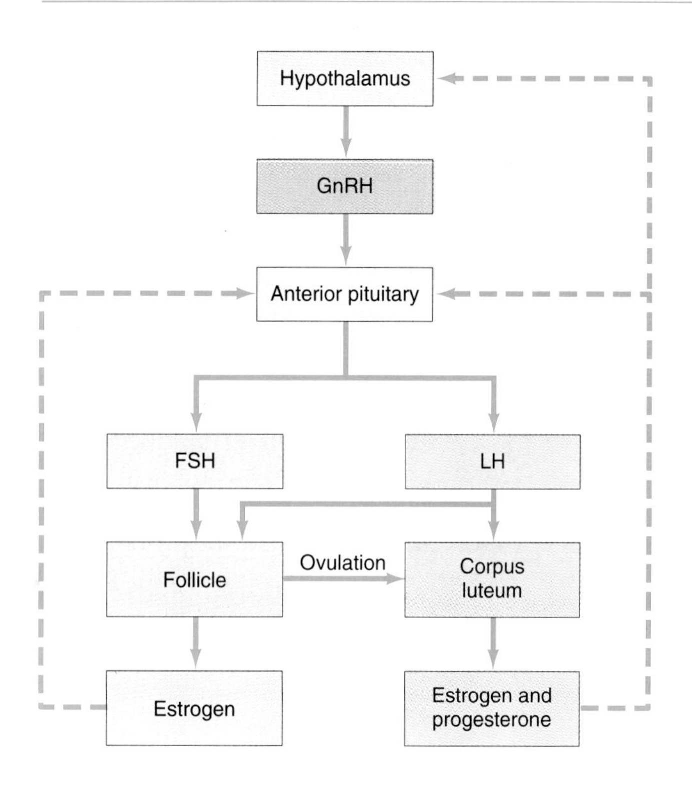

FIGURE 20.6

Negative Feedback in the Female Hormones
GnRH from the hypothalamus stimulates the anterior pituitary to secrete FSH and LH, which together stimulate the development of an egg follicle. LH stimulates ovulation and the development of the corpus luteum. Estrogen is produced by the egg follicle and the corpus luteum; progesterone is produced by the corpus luteum. Estrogen and progesterone are both in negative feedback with the anterior pituitary; progesterone is also in negative feedback with the hypothalamus.

LH must be present if the corpus luteum is to continue to function. However, when progesterone and estrogen inhibit GnRH, inevitably the rate of LH production falls, and the corpus luteum breaks down. Breakdown of the corpus luteum usually occurs by day 26 of the cycle. The scar tissue that replaces the corpus luteum produces no hormones, and consequently the production of estrogen and progesterone ceases. In the absence of estrogen and progesterone, two events occur:

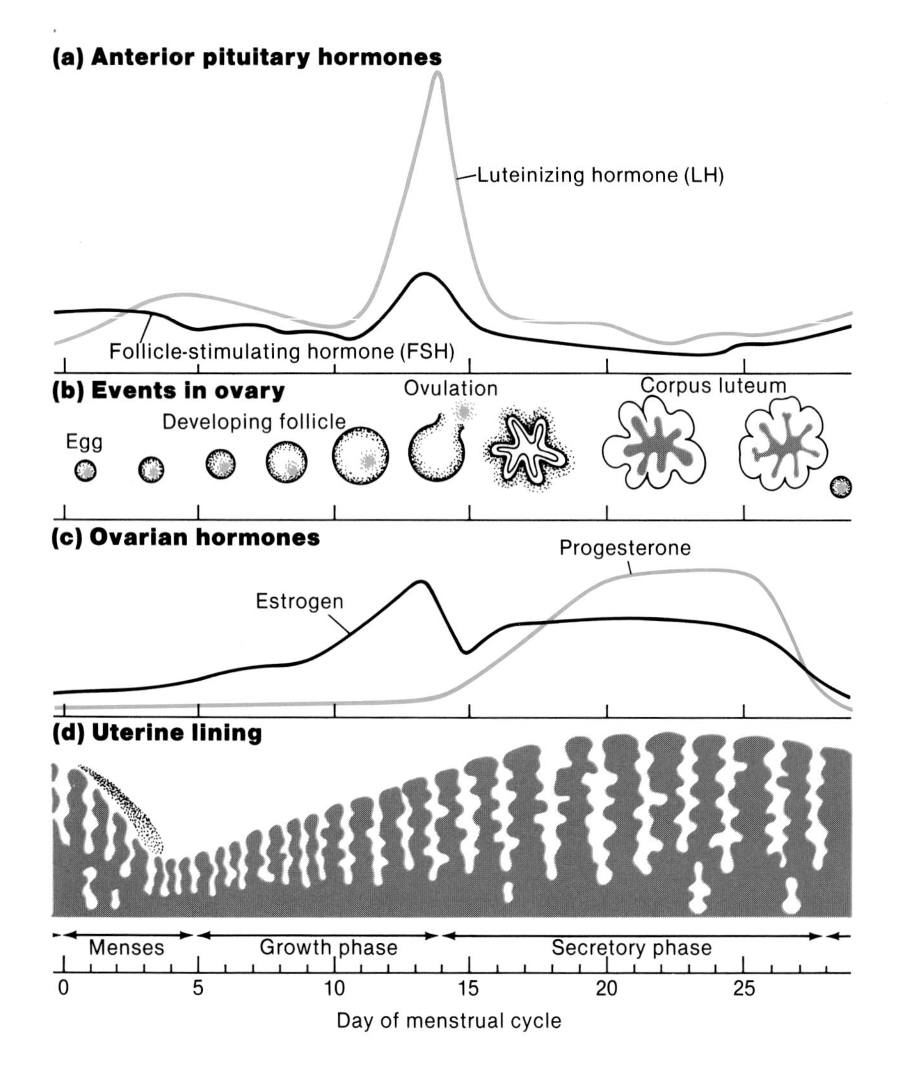

FIGURE 20.7

Menstrual Cycle
(a) The levels of FSH and especially LH peak at about day 13. (b) The egg follicle develops in response to the presence of FSH and LH, and ruptures in response to their peak rate of secretion. The corpus luteum remains as long as LH is being secreted. (c) Estrogen levels increase until ovulation, but remain relatively high until late in the cycle because the corpus luteum continues to secrete estrogen. Progesterone is abundant only in the second half of the cycle because it is produced in significant amounts only by the corpus luteum. (d) Estrogen stimulates the growth phase of the uterine lining, and progesterone stimulates the secretory phase. Menses occurs when progesterone levels drop, and the blood vessels in the uterine lining collapse. Progesterone levels drop when LH is no longer in sufficient abundance to support the corpus luteum.

BOX 20.2

Problems with the Ova

Normally, only one follicle develops to maturity in any given month. However, from 1 to 2 percent of the time, additional follicles may also develop. Fertilization and implantation of more than one ovum accounts for the birth of nonidentical twins, triplets, and so forth. "Fertility shots," now in common use by women who have difficulty conceiving, are pituitary extracts containing large amounts of FSH. (Clomid, an orally administered drug that promotes ovulation, also increases the incidence of multiple births.)

The fact that women produce no new ova during their lives may in part explain why it is that Down syndrome (see Chapter 19) becomes increasingly more common in mothers of advancing years. It seems likely that meiotic errors would be more frequent in cells that are almost half a century old than in cells half that age.

1. The cells of the endometrium, which require progesterone in order to remain functional, break down by day 28 and are passed out of the body over the next few days as menstrual flow.

2. GnRH production begins once again (because there is no estrogen and progesterone to inhibit its production). In response to GnRH, the anterior pituitary renews its production of FSH and LH—and a new menstrual cycle is initiated.

Progesterone has also been implicated in a condition that is troublesome to many women during the week or so before menstruation—**premenstrual syndrome (PMS).** A recent study found that blood zinc levels were significantly lower in a group of women suffering from PMS than in women without the syndrome. Zinc deficiency may cause lower secretion rates for progesterone and endorphins (the body's natural painkillers; see Chapter 13), both of which have been found to be lower in women suffering from PMS than in women without the syndrome.

Pregnancy

The sequence of events within the monthly menstrual cycle ensures that the release of the ovum from the ovary coincides with the onset of the secretory phase

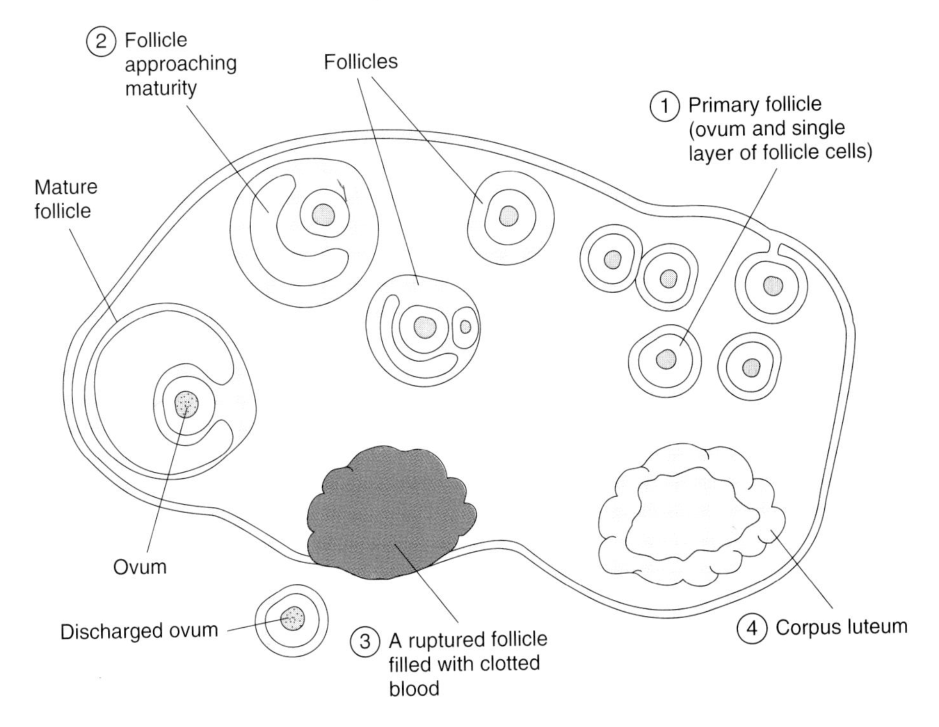

FIGURE 20.8

Life of an Egg Follicle

(1) Several primary follicles develop in response to increased levels of FSH and LH. (2) One follicle predominates and continues to develop. (3) At about the midpoint of the cycle, the follicle ruptures, and (4) gradually converts to a corpus luteum.

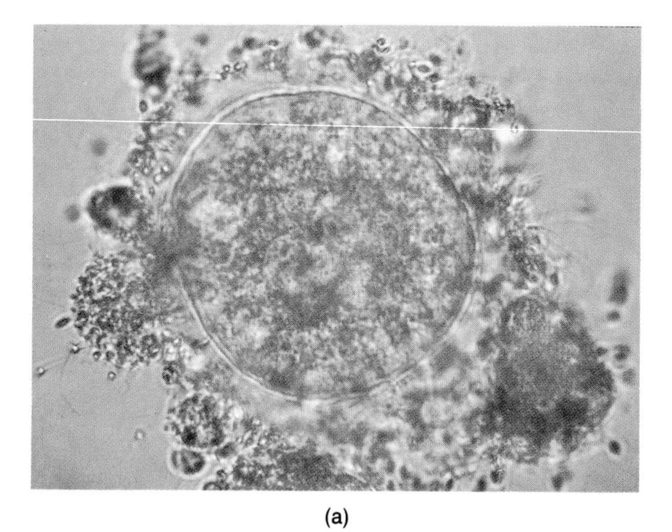

(a)

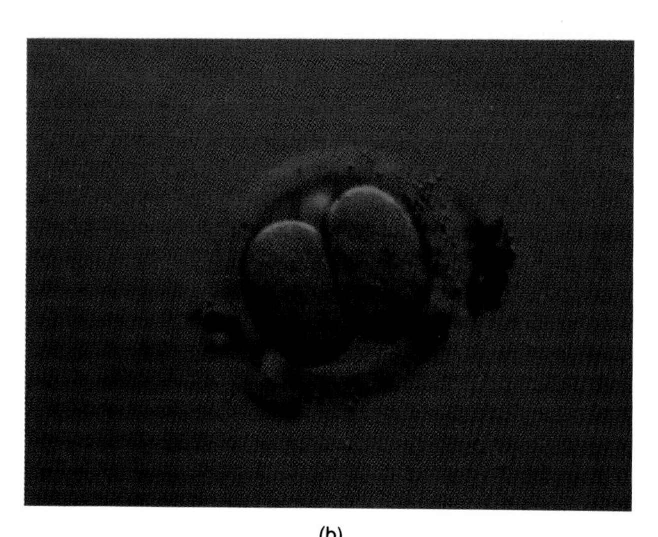

(b)

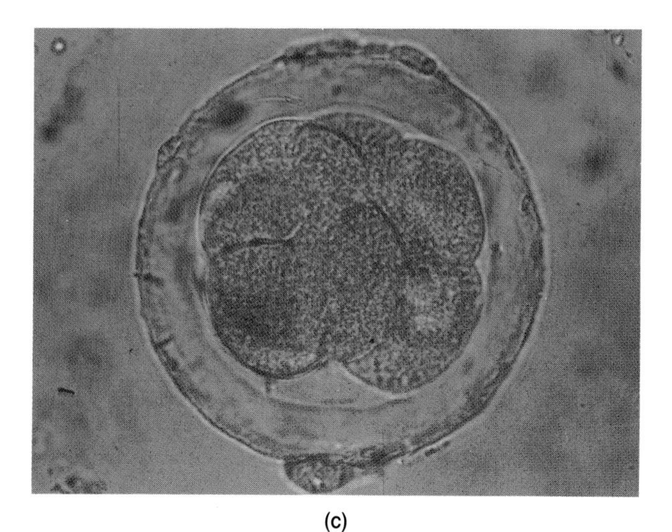

(c)

FIGURE 20.9

Events at Fertilization

Micrographs illustrating (a) the unfertilized egg; (b) the initial division after fertilization; and (c) the eight-cell stage.

of the endometrium, permitting a successful implantation if the ovum is fertilized. However, as we have seen, the menstrual cycle is self-initiating. What changes in the cycle must take place if the ovum is fertilized? (See Figure 20.9).

Fertilization occurs (certain pathological exceptions aside) in the fallopian tubes. The fertilized ovum, or **zygote** (Gr. *zygotos,* "yoked," a reference to the joining of two sex cells) takes about four days to reach the uterus, largely because the fallopian tubes are temporarily constricted at their junction with the uterus. During this interval, the zygote is nourished by secretions from the walls of the Fallopian tubes, and it divides several times into a small, solid ball of cells called a **morula** (L. *morum,* "mulberry").

For several days after entering the uterine cavity, the morula continues to be nourished by uterine secretions, and is gradually transformed into a hollow ball of cells called a **blastocyst** (Gr. *blastos,* "a sprout" and *kystis,* "sac") (See Figure 20.10). Finally, some seven or eight days after fertilization, a specialized layer of cells on the surface of the blastocyst literally digests a hole in the endometrium, and the blastocyst implants in the uterine lining (but see Box 20.3). This same layer of cells then divides rapidly, and forms the embryonic contribution to the **placenta** (Gr. *plakos,* "a flat object"), an organ formed jointly of embryonic and maternal tissues and that will function as an exchange organ between the developing embryo and the mother.

As we have seen, menstruation normally occurs some 14 days after ovulation. If menstruation were to occur in a pregnant woman, the pregnancy would be terminated, because the blastocyst would be swept away with the lining of the uterus. What happens to prevent menstruation?

Recall that the endometrium requires progesterone for its continued maintenance. However, toward the end of the typical menstrual cycle, progesterone production ceases with the loss of the corpus luteum, which itself is maintained by LH. Somehow, this successive loss of hormones must be stopped in the pregnant female—and it is. The blastocyst cells responsible for digesting a hole in the endometrium also produce a hormone, **HCG (human chorionic gonadotropin),** that is structurally and functionally very similar to LH. (This hormone is the basis of modern pregnancy tests). It not only maintains the corpus luteum but causes it to grow and to produce very large amounts of estrogen and progesterone (see Figure 20.11). These hormones, in turn, promote maintenance and growth of the endometrium. Moreover, the presence of estrogen and progesterone inhibits the production of GnRH (and, hence, FSH and LH production), with the result that normally no additional follicles mature during pregnancy.

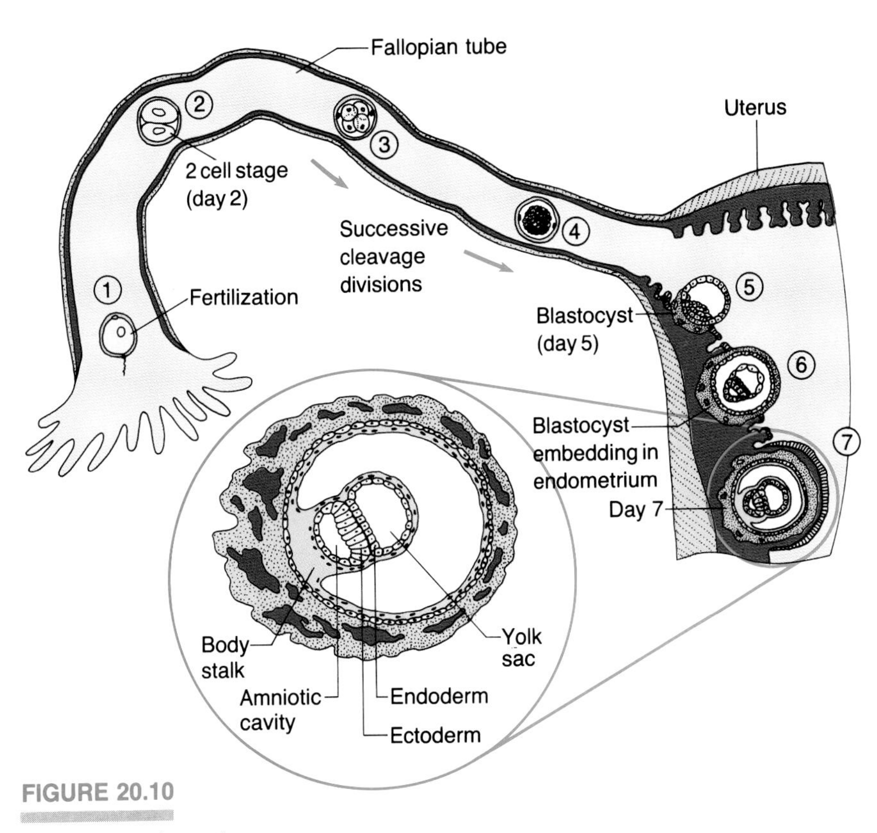

FIGURE 20.10

Steps to Implantation

(1) Fertilization normally occurs high in the fallopian tubes. (2) Cell division begins the day after fertilization. (3) Repeated cell divisions occur, leading to the development of the morula stage (4). About day 5, the morula enters the uterus, and begins to transform into a blastocyst (5 and 6). About day 7 the blastocyst implants in the uterine wall.

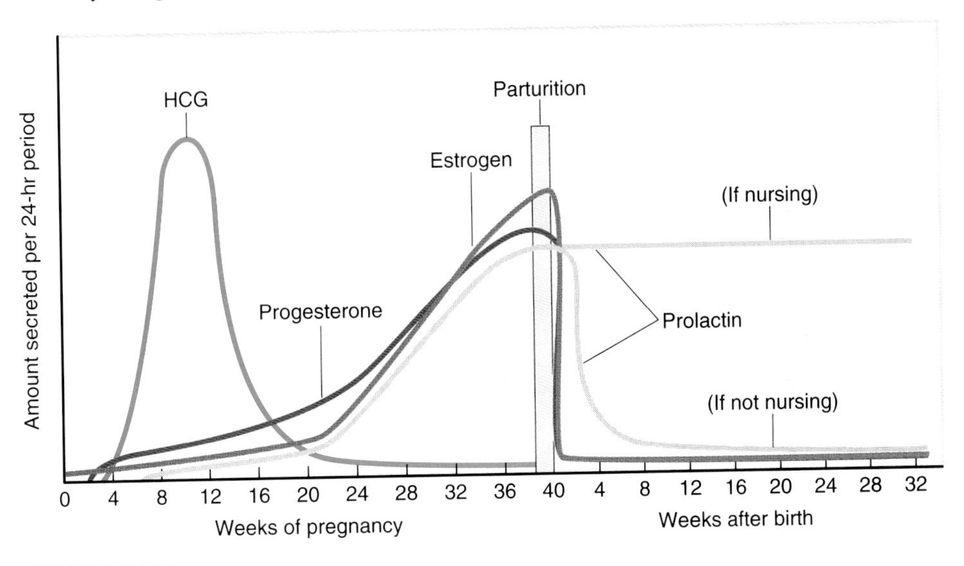

FIGURE 20.11

Hormonal Changes During and After Pregnancy

HCG peaks early in pregnancy, but is later supplanted by estrogen and progesterone, produced by the placenta. When the placenta is lost at birth, estrogen and progesterone levels drop sharply. Prolactin levels rise during pregnancy and stay high if the mother is nursing, but drop sharply if she is not.

BOX 20.3

Problems with Implantation Sites

For a variety of reasons, the fertilized egg sometimes implants in the fallopian tube rather than the uterus. Moreover, because the ovaries are not completely surrounded by the tops of the fallopian tubes, a fertilized egg can even fall into the abdominal cavity and become implanted on one of the organs of the pelvic region, such as the bladder. Implantation in sites other than the uterus is known as an **ectopic pregnancy** (Gr. *ektos*, "outside"). Since these other organs are rarely capable of sustaining a growing embryo, surgery is sometimes necessary to protect the would-be mother.

Ectopic pregnancies increase the maternal death rate by a factor of ten over normal pregnancies—and they are on the increase in the United States. In 1970, there were five ectopic pregnancies for every 1000 live births; by 1980 that number had risen to 15 per 1000 live births. The reasons for the increase are not entirely clear, but it is suspected that scarring of the fallopian tubes by such venereal diseases as *Chlamydia* and gonorrhea are at least in part responsible.

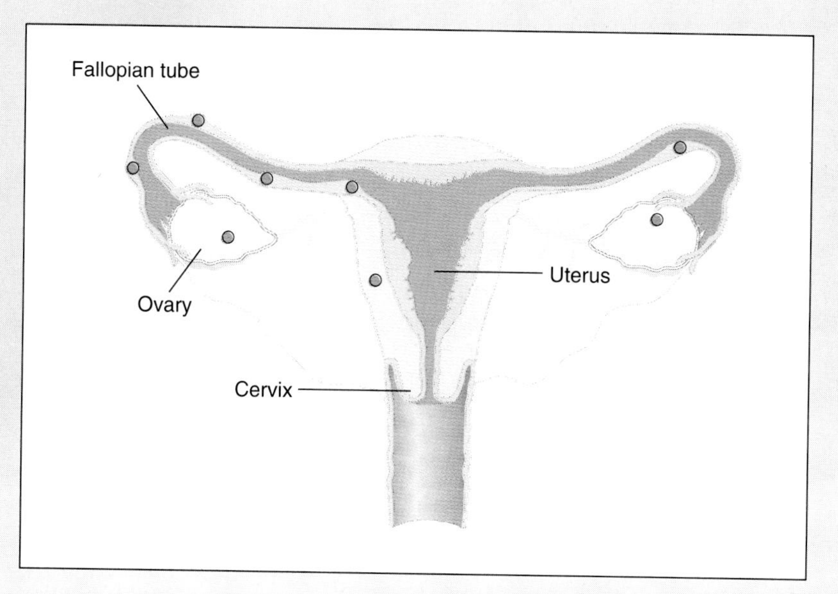

Ectopic Pregnancy Sites
Fertilized eggs may implant at any of a variety of sites in the abdominal cavity (red dots), but normally only an egg implanted in the uterus can lead to a successful pregnancy.

Of course, there are other events associated with pregnancy beyond simply the cessation of menstruation, and these, too, are primarily the result of hormonal action. The level of HCG peaks during the third month of pregnancy, but, by that time, the developing placenta is producing estrogen and progesterone in volumes never seen in the nonpregnant female. These hormones not only maintain the endometrium but cause enlargement of the breasts and uterus (see Figure 20.12). They also relax the pelvic ligaments, which will permit them to stretch extensively during birth as the baby passes through the birth canal.

Parturition

The precise physiological events that initiate **parturition** (delivery) are not completely understood. During the late stages of pregnancy, the uterus becomes increasingly more excitable until finally it begins the series of very powerful, rhythmic contractions that characterizes labor. This excitability is due both to hormonal changes and to mechanical changes in the uterus itself.

The hormonal changes are of two types. First, estrogen secretion continues to increase during the last two months of pregnancy, but progesterone secretion levels off. Since estrogen stimulates uterine contractions, whereas progesterone inhibits contractions, stimulation of contractions gradually dominates inhibition.

Second, the hormone **oxytocin,** released by the posterior pituitary, is a powerful stimulant of the uterine muscles. The volume of oxytocin produced increases sharply at the beginning of labor, possibly due

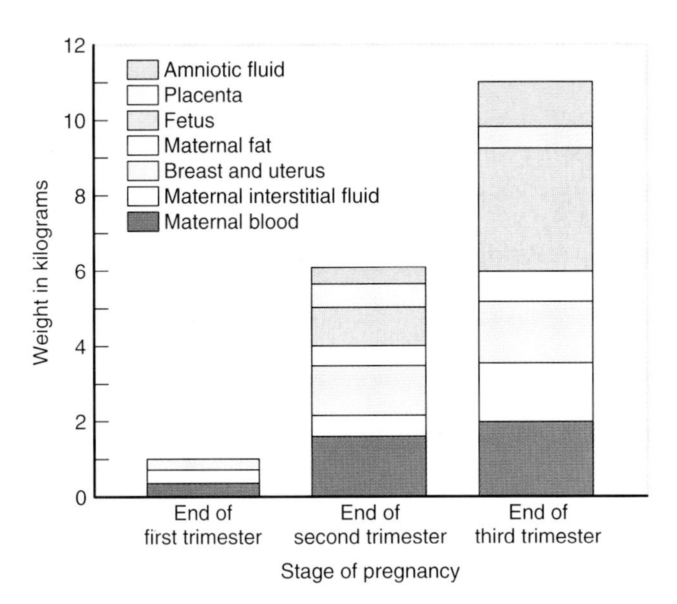

FIGURE 20.12

Weight Gain During Pregnancy
The initial weight gain during the first trimester is essentially exclusively maternal. During the second trimester, fetal weight becomes tangible, but almost half of the total weight gain of about 11 kg occurs during the last trimester, and more than half of the total weight gain occurs in the mother's tissues. Thus the birth of the baby does not mean that the mother will instantly return to her weight prior to pregnancy.

to the stretching of the opening (**cervix;** L., "neck") of the uterus by the baby (see Figure 20.13). Moreover, the uterine musculature is particularly sensitive to oxytocin at this time.

Finally, the actual stretching of the uterus increases its contractility. As the fetus grows, the volume of the uterus increases from 4 ml to 6 liters. The fact that twins are born an average of 19 days before single infants provides additional evidence that simple stretching of the uterus influences the time of delivery.

Lactation

During pregnancy, the breasts have been prepared for their role as milk producers by the combined effects of estrogen and progesterone. **Prolactin,** another hormone produced by the anterior pituitary, is also produced in increasingly larger volumes throughout the pregnancy. Prolactin stimulates not only the growth of the mammary gland but also the production of milk. However, estrogen and progesterone block the stimulatory effects of prolactin and thereby effectively prevent milk production during pregnancy.

Within an hour of the delivery of the baby, the placenta, which is the source of the enormous volumes of estrogen and progesterone produced during pregnancy, is expelled as the afterbirth. As a consequence, the level of estrogen and progesterone in the blood falls rapidly, the inhibition on prolactin is removed, and milk production begins.

The level of prolactin also falls shortly after birth, but rises dramatically in a nursing mother due to the mechanical stimulation of the infant at the mother's breast. However, if the mother stops nursing for even a few days, prolactin secretion ceases and the breasts lose their ability to produce milk.

Although prolactin is responsible for stimulating milk production, the milk produced is not automatically ejected from the mammary tissue. Oxytocin is responsible for this ejection, a process known as *milk let-down*. The release of oxytocin from the posterior pituitary is caused by stimuli from the nipples of the breast as the infant nurses. Because oxytocin stimulates the contraction of smooth muscle generally, the uterus is also affected, and, in nursing mothers, quickly returns to its pre-pregnant size.

Menopause

Menopause is the period during which menstrual cycles gradually cease, and the production of ovarian hormones decreases greatly. It usually occurs in the late 40s or early 50s, and apparently results from ovarian "burn-out."

All of the primordial follicles are present in the ovary before birth. After more than 30 years of menstrual cycles, almost 400 follicles have grown and ovulated, and thousands more have begun to grow and then regressed at the time one follicle becomes dominant. Thus, by age 50 or so, very few primordial follicles are left. In the absence of maturing follicles, no estrogen or progesterone is produced, and there is neither an FSH/LH surge at midcycle nor an inhibition on FSH and LH by estrogen and progesterone. Thus estrogen and progesterone levels fall sharply, and FSH and LH levels increase to levels much higher than those of premenopausal women (see Figure 20.14). Finally, in the absence of estrogen and progesterone, there is no cyclical development of the endometrium, and therefore no time of endometrial destruction and menstrual flow.

The hot flashes that tend to characterize menopause probably result from the effect of low estrogen levels on sensitive neurons in the temperature center of the hypothalamus. These neurons produce norepinephrine, a substance that experimentally has been shown to produce hot flashes.

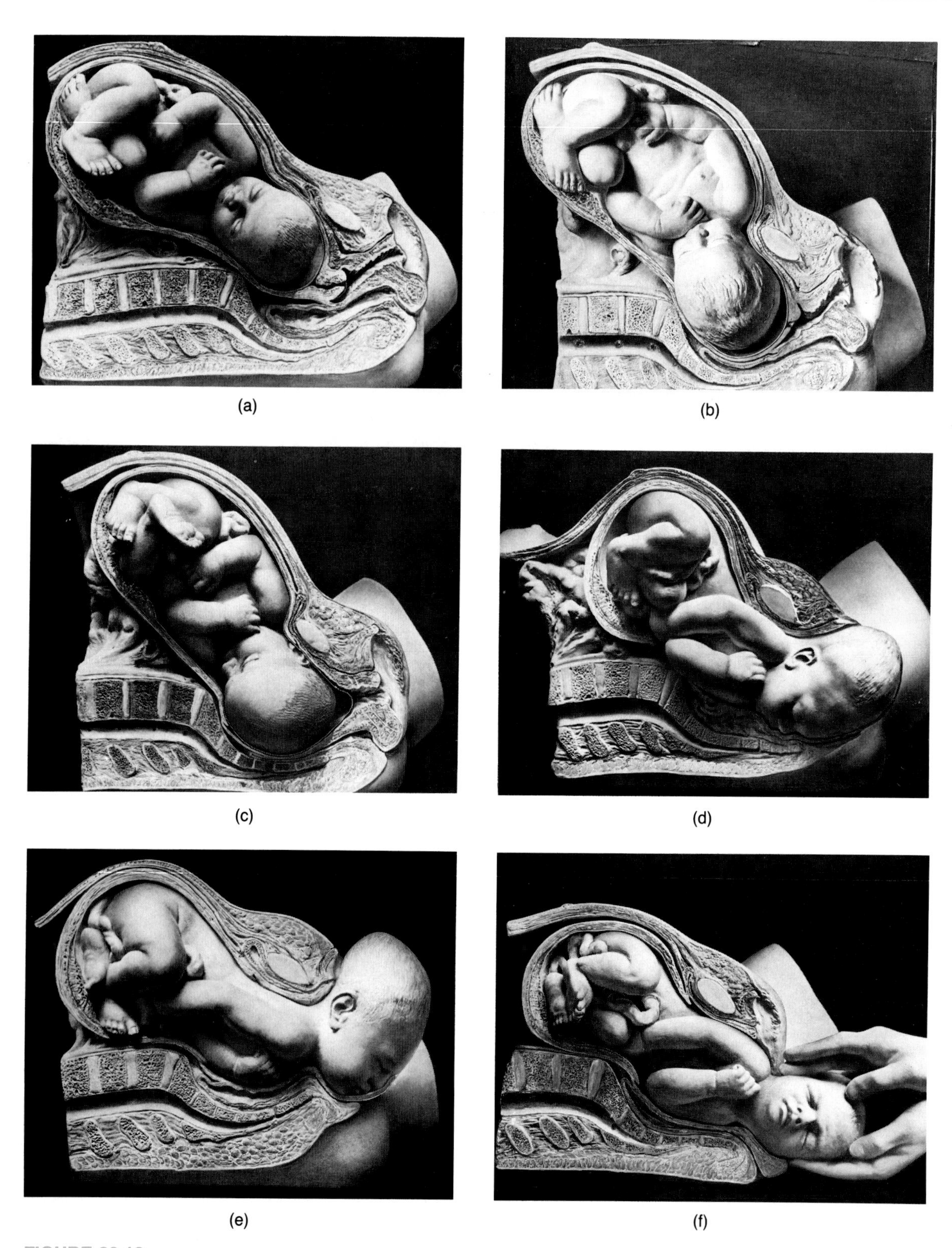

FIGURE 20.13

Birth

Babies are generally born head first (feet first is called a breech birth). As the head of the baby stretches the cervix of the uterus, the uterine muscles are stimulated to contract, pushing the baby and further dilating the cervix. The amniotic sac ("bag of water") breaks at this point, if it has not already done so. The placenta follows delivery as the afterbirth. Note the temporary distortion of the baby's soft skull as it passes through the birth canal.

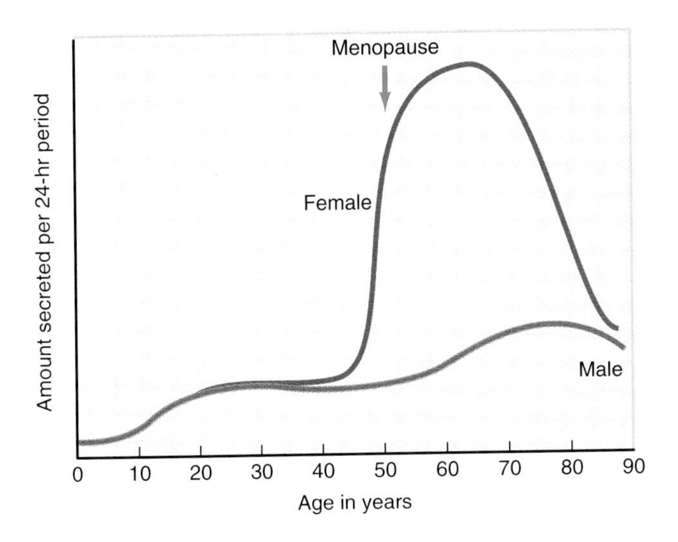

FIGURE 20.14

Secretion of FSH and LH as a Function of Age
At puberty, the secretion rates of FSH and LH in males rise gradually throughout life, with a downturn only in the elderly. Secretion rates in females rise sharply at menopause, an indication that menopause results from ovarian, not pituitary, decline.

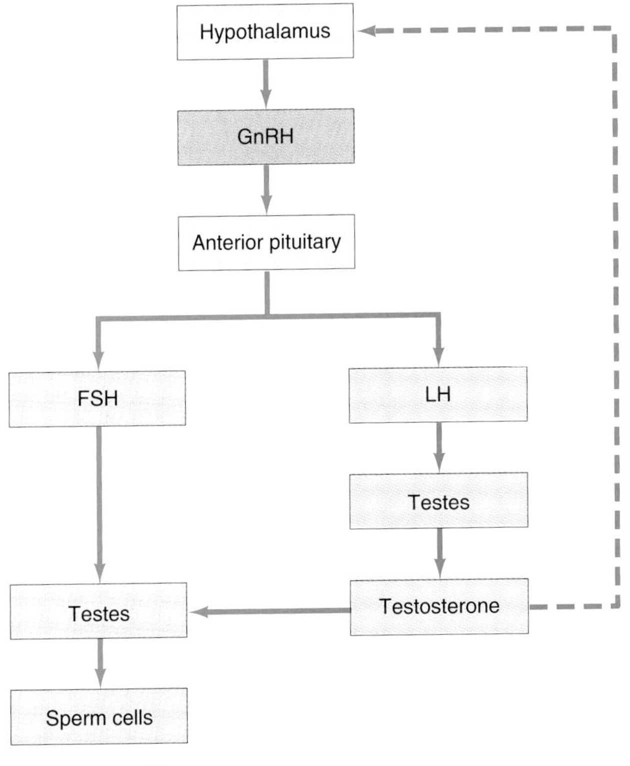

FIGURE 20.15

Negative Feedback in the Male Hormones
GnRH from the hypothalamus promotes FSH and LH secretion from the anterior pituitary. LH stimulates the testes to produce testosterone. Testosterone plus FSH result in the production of sperm cells. Testosterone is in negative feedback with the hypothalamus.

THE ONSET OF PUBERTY IN THE MALE

Compared to the female, the situation in the male is simplicity itself, although the actual cause of the onset of puberty is no better understood in males than in females. **Testosterone,** the male equivalent of estrogen, is responsible for a variety of physical changes during puberty, including the broadening of the shoulders, the growth of body and facial hair, deepening of the voice, and the increased size of the genitals. Testosterone also plays a role in baldness, as demonstrated by the fact that castrated men do not become bald. (It is unlikely, however, that such a drastic solution to the threat of baldness will ever become popular.)

Testosterone is produced by a group of cells in the testes that is not involved in the formation of sperm cells. When stimulated by GnRH, the pituitary hormone LH in turn stimulates testosterone production, and, through a negative feedback loop, the level of testosterone influences the rate of GnRH production (see Figure 20.15).

FSH plus testosterone stimulates the production of sperm cells, and about 100,000,000 of these cells are produced daily by the sexually active male. There is no monthly cycle, and this rather simple set of hormonal relationships continues, albeit at a gradually declining level, from puberty to death.

COITUS AND FERTILIZATION

Most land vertebrates, including all mammals, utilize internal fertilization as a reproductive strategem. The reproductive tracts of the human male and female provide an admirable example of the anatomy and physiology of internal fertilization.

In the sexually aroused male, the penis becomes engorged with blood as a result of increased arterial flow. This engorgement results from the filling of intricate blood sinuses in the penis. As the sinuses fill, they cause the veins draining the penis to collapse, trapping the blood within. The net result is that the penis becomes very firm and erect, often in seconds, and increases substantially in size (but see Box 20.4).

During sexual intercourse **(coitus),** the penis is inserted into the vagina of the female (see Figure 20.16). Aided by lubricating secretions produced primarily by the female, thrusting movements by the male or female (or both) stimulate the penis, causing an **emission** (the movement of the sperm to the urethra).

The path of the sperm is as follows: From their site of origin in the **seminiferous tubules** of the testes, the sperm enter the **epididymis** (Gr. *epi,* "upon" and

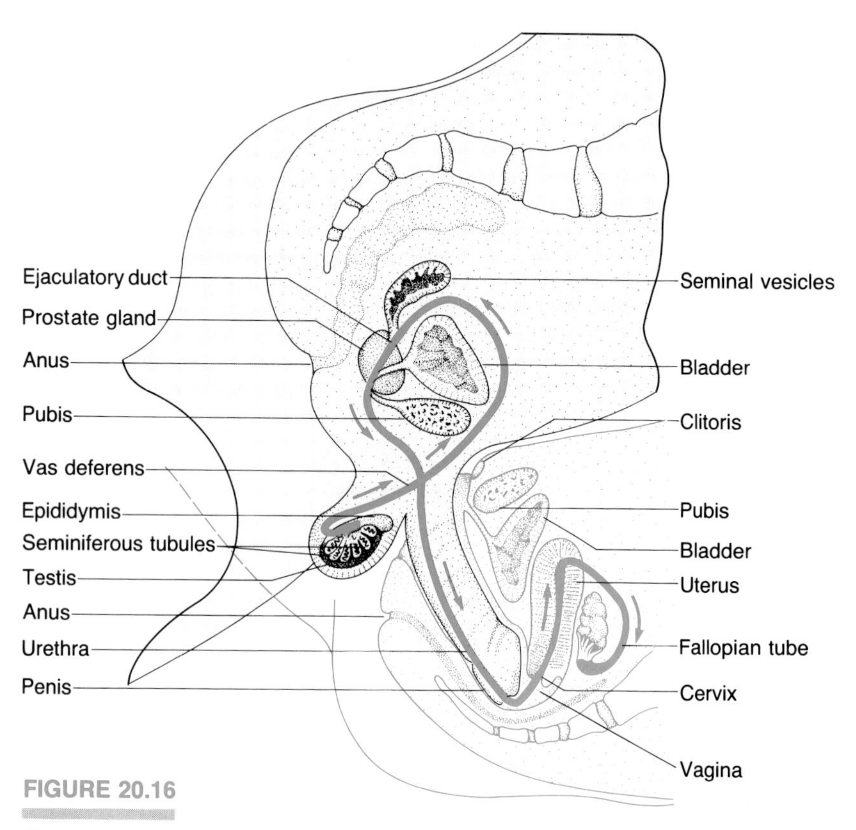

FIGURE 20.16

Coitus

Note the pathway of sperm to the site of fertilization. Early in coitus, sperm move from the epididymis (1), where they have matured after being produced by the seminiferous tubules of the testes, to the ejaculatory duct (2) by way of the vas deferens (3). During ejaculation, sperm and secretions from the seminal vesicles (4) and the prostate gland (5) are propelled through the urethra (6) and into the vagina (7). The sperm then swim up the uterus (8) to the fallopian tubes (9), where fertilization will occur if an egg is present.

didymos, "testicle"), a coiled tube atop each testis only 0.5 mm wide but more than six meters long. The sperm mature during the seven to ten days that they remain in the epididymis. They then move into the vas deferens, where they remain for a few hours, or many months, depending on the level of sexual activity.

During emission, the sperm move through the vas deferens, past the **seminal vesicles,** and into the **ejaculatory duct,** which is located in the **prostate gland.** The prostate and seminal vesicles produce a variety of secretions that both nourish the sperm and ultimately neutralize the acidic environment of the urethra and vagina. (Sperm require an alkaline environment in order to become motile, or self-propelled.)

Continued stimulation of the penis eventually results in a wave of powerful muscular contractions that propels the **semen** (the mixture of sperm and glandular secretions) out of the penis and into the vagina. This expulsion is called **ejaculation,** and is accompanied by a feeling of intense pleasure. Together, these events constitute the male **orgasm** (see Figure 20.17). During ejaculation, the sphincter to the bladder is closed, thus preventing sperm from entering the bladder and urine from entering the urethra.

The volume of an average ejaculation is about 3 ml, and contains as many as 300,000,000 sperm. Yet the sperm cells are so small that collectively they constitute only the volume of a pinhead. The bulk of the semen thus comes from the secretions of the prostate gland and seminal vesicles.

Depending on a variety of factors—including age, occurrence of last ejaculation, and psychological factors—the time between initial insertion of the penis into the vagina and ejaculation may range from less than one minute to 20 minutes or more. Following an ejaculation, however, there is a latent period of from several minutes to several hours before another erection and ejaculation can occur.

Sexual excitement in the female is typically somewhat slower to become manifest, but ultimately includes a swelling and firming of the breasts, engorgement of the clitoris and the labial folds, and

BOX 20.4

The Penis and the Testes

There is considerable variation in the size of the penis, whether erect or flaccid, partly because the amount of blood in the erectile tissues varies from moment to moment. On the average, however, an erect penis is about 15 cm long and 4 cm in diameter. These dimensions are roughly double the flaccid dimensions. Unlike many male mammals, in which a bone gives rigidity to the penis, the human male relies exclusively on blood-filled erectile tissues.

The length of the vagina is only about 8 cm, and the penis cannot enter the uterus because of the latter's angle and shape. How does a 8 cm vagina accommodate a 15 cm penis? The answer is that the vagina of the sexually aroused female balloons so as to accommodate all but the largest penis.

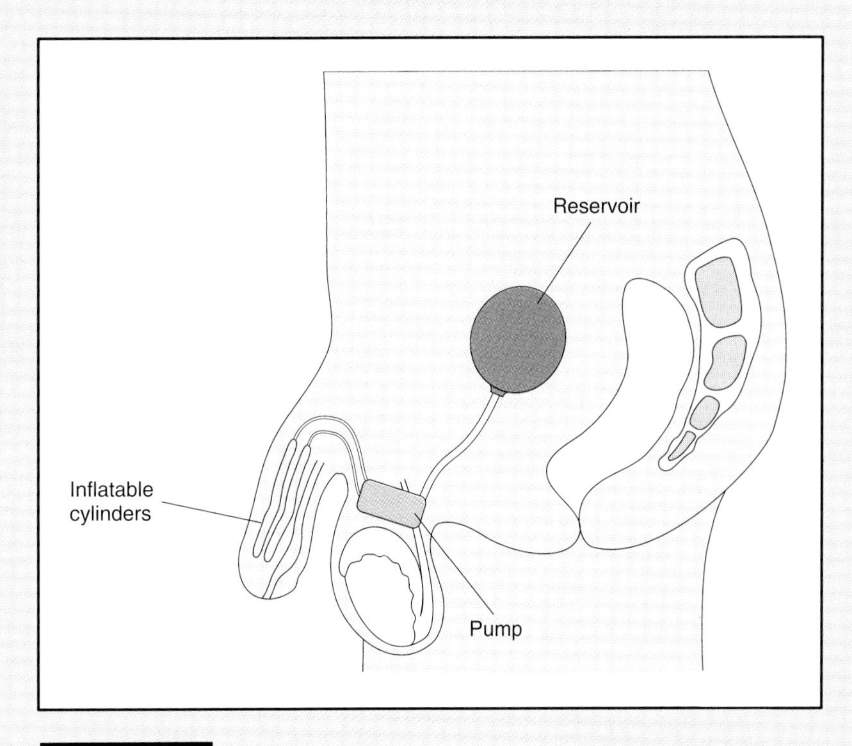

Penile Implants
For men with impotence that cannot be treated successfully, a penile implant permits the resumption of sexual relations. One form of implant functions when a pump in the scrotum is squeezed, permitting a saline solution from a reservoir situated in the abdomen to inflate cylinders implanted in the penis.

Failure to achieve an erection is called **impotence.** The causes of this condition are manifold, although these causes are more commonly psychological than physiological. Temporary impotence may be caused by fatigue or by excess alcohol, the latter apparently acting through higher brain centers. Recall the gatekeeper's speech in *MacBeth*—alcohol "provokes the desire, but it takes away the performance."

In the normal course of development, the testes must descend into the scrotum. If they fail to descend, a condition known as **cryptorchidism** (Gr., "hidden testes"—orchids derive their name from the fact that their tubers supposedly resemble testes), surgery may be required, because an undescended testis usually produces no viable sperm and can become cancerous.

The descent of the testes is not without its attendant hazards. The opening through which the testes move from the body cavity to the scrotum is called the **inguinal canal** (L. *inguinis,* "groin"). It essentially seals off following the descent, leaving only a small central area through which the vas deferens must pass. However, this secondary sealing off is frequently a point of chronic weakness, and a strain on the muscles in the area may cause a tear, resulting in an **inguinal hernia.** (The adjective is important, because hernias can occur in other parts of the body, most notably the diaphragm and the umbilicus, or navel.) Inguinal hernias may be serious, because a loop of the intestine can pass through the tear and become trapped in the scrotum. In severe cases, circulation may be cut off to this portion of the intestine, and gangrene may result. Fortunately, the scar tissue that forms once the hernia has been surgically sewn shut is usually stronger than the original muscle, making a second hernia (on the same side) uncommon. It is the descent of the testes and the resultant weakened area that accounts for the fact that inguinal hernias are far more common in men than in women.

---

the production of lubricating fluid by glands located around the orifice of the vagina, and by the walls of the vagina itself.

The female may respond to the thrusting movements of the penis with an orgasm of her own which, unlike that of the male, does not involve an ejaculation of fluid. (There is still controversy on this point, some researchers claiming that a small amount of fluid is released.) Moreover, because the female frequently does not have a significant latent period before

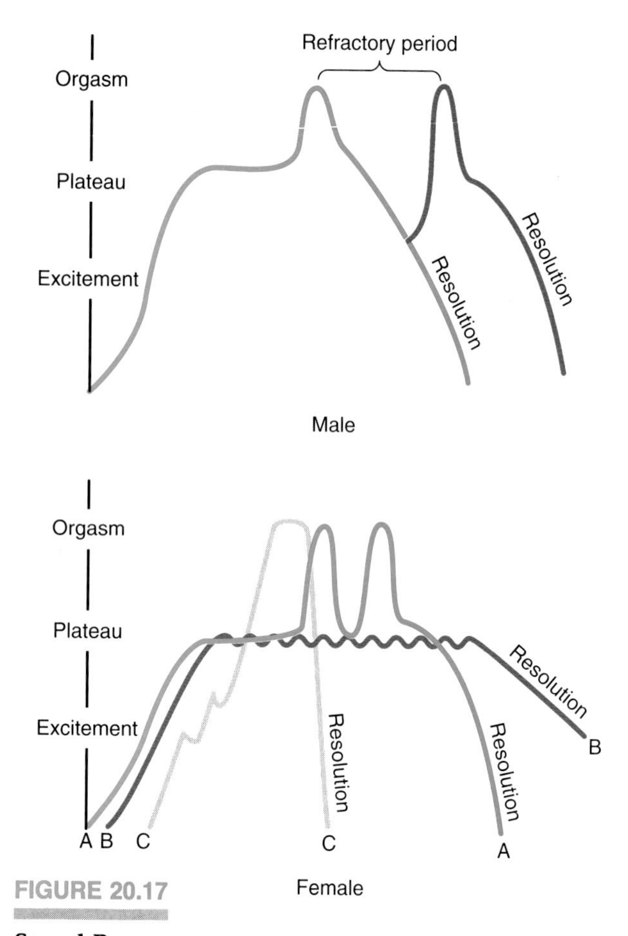

FIGURE 20.17

Sexual Response

(a) Sexual response in the male. The refractory period is variable. (b) Sexual response in the female. Note the possibilities of multiple orgasms in pattern A and of remaining at a plateau of excitement without orgasm in pattern B.

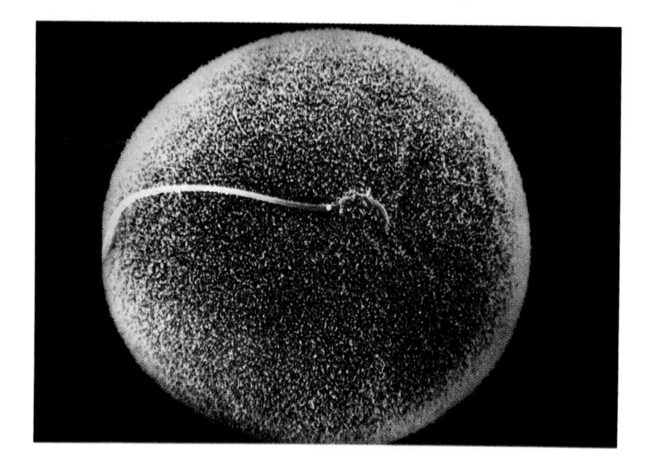

FIGURE 20.18

Fertilization

Scanning electron micrograph of a sperm cell penetrating an egg.

another orgasm can occur, she may experience multiple orgasms during a single act of coitus.

The long debate over whether there is a difference between clitoral and vaginal orgasms now seems about over. Present evidence indicates that orgasm may occur as a result of stimulation of either the clitoris or the vagina, but that the result is the same—there is only one type of orgasm. (The possible existence of a sensitive zone on the wall of the vagina—the *G spot*—is still under debate.)

The sperm are discharged near the cervix of the uterus and, despite their small size, can reach the upper regions of the fallopian tubes and fertilize an egg (if one is present) within 30 minutes of ejaculation (see Figure 20.18). To what extent the wave of muscular contractions that occurs during female orgasm assists the sperm in their migration is uncertain, although it is clear that fertilization can certainly take place in the absence of female orgasm.

The sperm may remain viable in the fallopian tubes for up to three days, but probably remain highly fertile for only a day or so. Similarly, the ovum can be fertilized only during the first eight to 12 hours of its transit through the fallopian tube.

Usually only 1000 to 3000 sperm succeed in reaching the upper fallopian tubes. The first sperm to penetrate the ovum causes the release of a fertilization membrane around the ovum, which effectively prevents any other sperm from penetrating.

BIRTH CONTROL

Depending on the circumstances, virtually everyone has, at one time or another, wished either for sterility or fertility. Partly because of our intelligence, and partly because of the unique sexual receptivity of the human female (see Chapter 16), we have long been interested in separating intercourse from conception.

There are a large number of birth control devices and techniques presently in use. Accurate data on their relative efficacy are difficult to obtain, however, because "failures" tend to be blamed on the device or method rather than on errors or lack of motivation by the users.

Male-Initiated Techniques

Probably the earliest form of birth control was **coitus interruptus** (male withdrawal). It was certainly in use more than 2000 years ago—as we know from the admonition against this technique in the Bible. Failure of this method occurs in two ways. First, the male lubricating fluid that issues forth from the penis prior to ejaculation may contain sperm, especially if there was a previous act of intercourse a short time earlier.

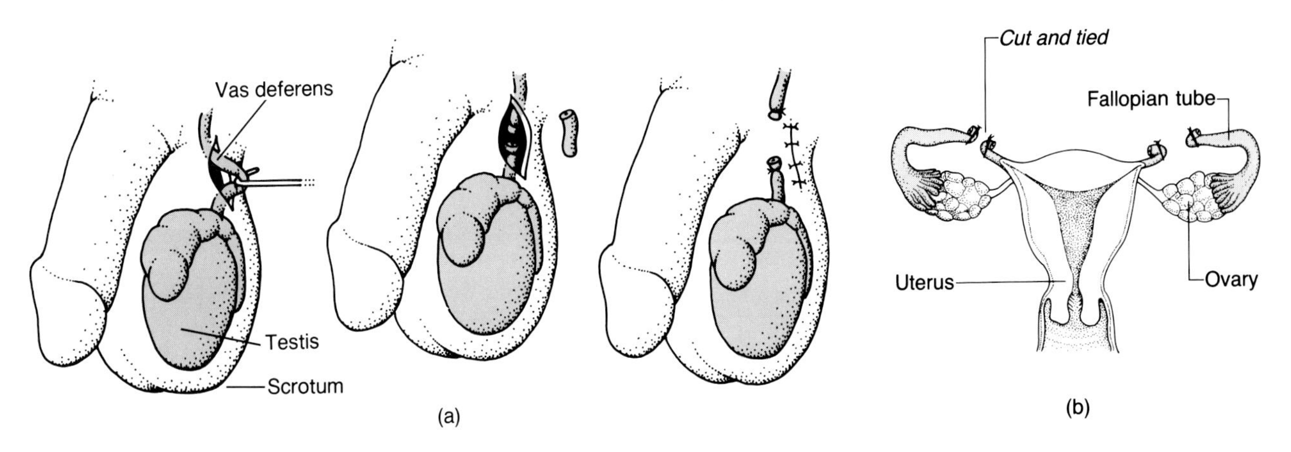

FIGURE 20.19

Surgical Sterilization

Both vasectomy (a) and tubal ligation (b) involve removal of a portion of the tube that transports the gametes. These procedures are intended to be permanent, but they can sometimes be surgically reversed.

Thus, even if the penis is withdrawn before ejaculation, some sperm may be deposited in the female tract. Second, there is the notorious "oops!" factor, which results from an error in timing. The frustration involved in withdrawing at the critical point makes this second factor very significant, and undoubtedly accounts for the high failure rate of this technique.

During the Renaissance, various types of sheaths were developed that trapped the semen and prevented its deposition in the female tract. These have been refined and are now called **condoms,** or "rubbers." Failure in this instance is less commonly due to a failure of the device (although certainly that happens) than to a failure to use them regularly and from the onset of intercourse.

Another contraceptive method used by the male is the **vasectomy,** in which the vasa deferentia are severed through an incision made in the scrotum (see Figure 20.19). Vasectomized males still ejaculate, but the semen no longer contains sperm. The production of testosterone is also unaffected by this surgery. The principal drawback is that the procedure is not reliably reversible.

Female-Initiated Techniques

There are a number of devices or techniques used by females (see Figure 20.20). Least successful of these is the **douche,** which is simply an attempt either to kill or to remove sperm after coitus. Because the sperm enter the uterus very quickly, douching must be done immediately if it is to have any benefit, and typically that is not the case.

Spermicidal agents, used prior to coitus and often in conjunction with either a **diaphragm** (a rubber disk

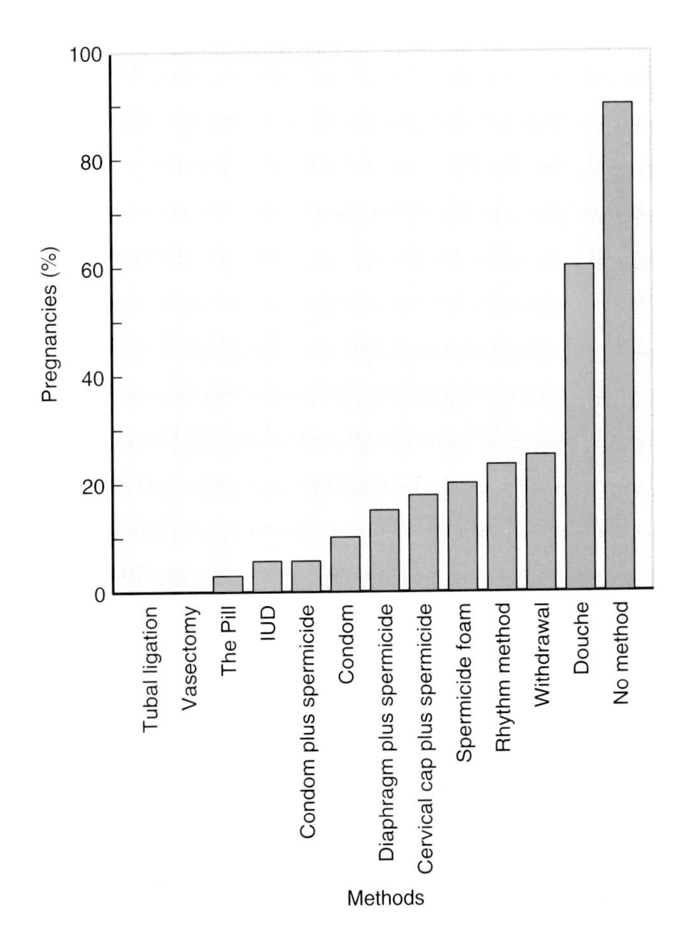

FIGURE 20.20

Effectiveness of Birth Control Methods

The likelihood of pregnancy with regular intercourse over a period of 12 months and no birth control is 90 percent. Surgical sterilization all but ensures no chance of pregnancy. Other birth control methods are of varying levels of effectiveness.

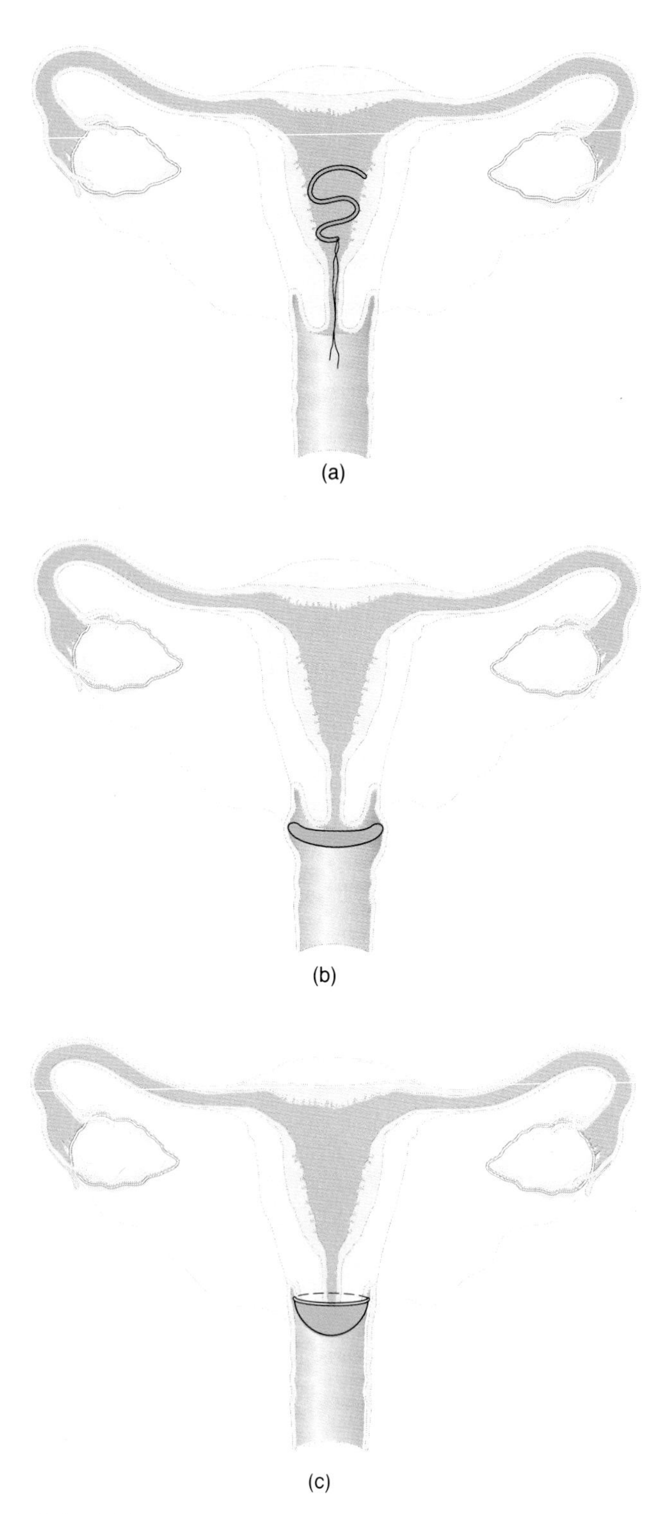

FIGURE 20.21

Three Devices for Birth Control

(a) The IUD apparently speeds up the rate of movement of the egg through the reproductive tract. The diaphragm (b) and the cervical cap (c) in theory prevent the sperm from entering the fallopian tubes. They are more effective when used with a spermicide. The cap has found favor recently because it is smaller, easier to insert, and can be left in place for longer periods of time.

that blocks the entrance to the uterus), or a **cervical cap** (a much smaller thimble-shaped device that covers only the cervix), are relatively effective, in large part because the female tends to be more motivated in matters of conception than is the male (see Figure 20.21).

Intrauterine devices (IUDs) apparently function by irritating the uterus slightly, thereby causing the ovum to move rather quickly down the Fallopian tubes, preventing either fertilization or implantation. Drawbacks include spontaneous expulsion of the device (often without the knowledge of the user, a fact that can later prove embarrassing, to say the least), occasional hemorrhaging (the device seems to work better in women who have already had a child), and increased rates of infection (possibly because the "string" which permits removal of the IUD also provides a route by which bacteria can enter the upper reproductive tract).

The favored technique today, used by more than 50,000,000 women around the world, is "the pill." There are several kinds of pills, but the major ones have both estrogen and progesterone mimics (real hormones would quickly be degraded by liver enzymes), which prevent substantial amounts of FSH and LH from being produced. As a consequence, there is no surge of these two hormones at day 12, and no ovulation. Without an ovum, conception is obviously impossible.

Despite grave concerns about the impact of artificial hormones on the female endocrine system, the pill has proved a remarkably safe drug. There are a variety of side effects, and it is clearly not the ideal method for all women, but extensive studies repeatedly confirm that benefits outweigh the risks. That is, the dangers of pregnancy considerably exceed the dangers posed by the pill.

A **tubal ligation** is the female equivalent of a vasectomy. In this procedure, the fallopian tubes are severed. The production of ova by the ovaries continues, but they cannot pass into the uterus, nor can sperm move up the tubes to fertilize the ova.

Until recently, a tubal ligation was performed under general anesthesia with all the attendant hazards of major surgery. Presently, it is possible to sever the fallopian tubes either via the vagina or the navel under local anaesthetic, and the hazards of the procedure have thus been diminished substantially.

Interestingly, the relative utilization of vasectomies and tubal ligations differs from country to country, an apparent reflection of cultural differences (see Figure 20.22). In India, for example, tubal ligations and vasectomies are performed in roughly equal numbers. In the United States, there are almost twice as many tubal ligations as vasectomies (although the ratio is diminishing). In Latin American countries, tubal

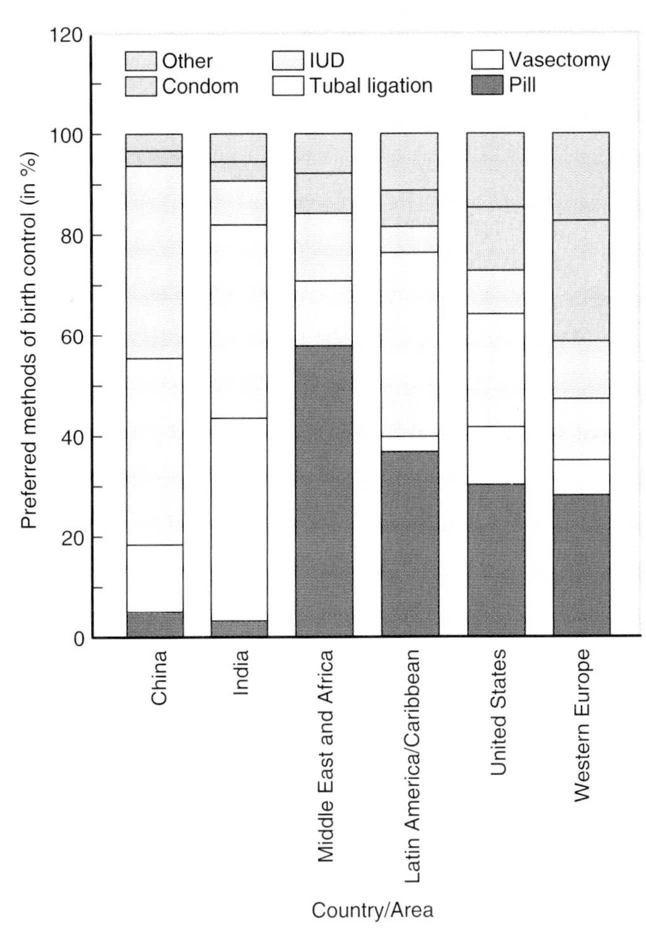

FIGURE 20.22

Preferred Methods of Birth Control

Surgical sterilization (especially of woman) and the pill are the most popular methods of birth control in those countries using modern birth control methods. (This figure does not include the percentage of individuals using no birth control.)

ligations are 15 times as common as vasectomies, and in the Middle East and Africa, vasectomies are virtually never performed.

Still in the testing stages are various types of "morning after" pills. Strictly speaking, these are not contraceptives, because fertilization and even implantation may occur. They function by inducing menstruation, which either prevents implantation or aborts the implanted blastocyst. There is still considerable debate regarding the overall safety of such pills, partly because one type contains the hormone **diethylstilbestrol (DES),** which has been found to be carcinogenic (cancer-inducing) in certain circumstances (see Box 20.5).

However, a new drug, **RU-486,** has been recently approved for use in France (see Figure 20.23). This drug competes with progesterone for receptor sites, in the same way that curare competes with acetylcholine (see Chapter 14). Thus it is as if progesterone were not present, and the endometrial lining is sloughed off. Theoretically, the pill could be used to induce menstruation once a month, or just when a woman misses her menstrual period to induce an abortion. The moral and ethical issues of such use are presently being extensively debated.

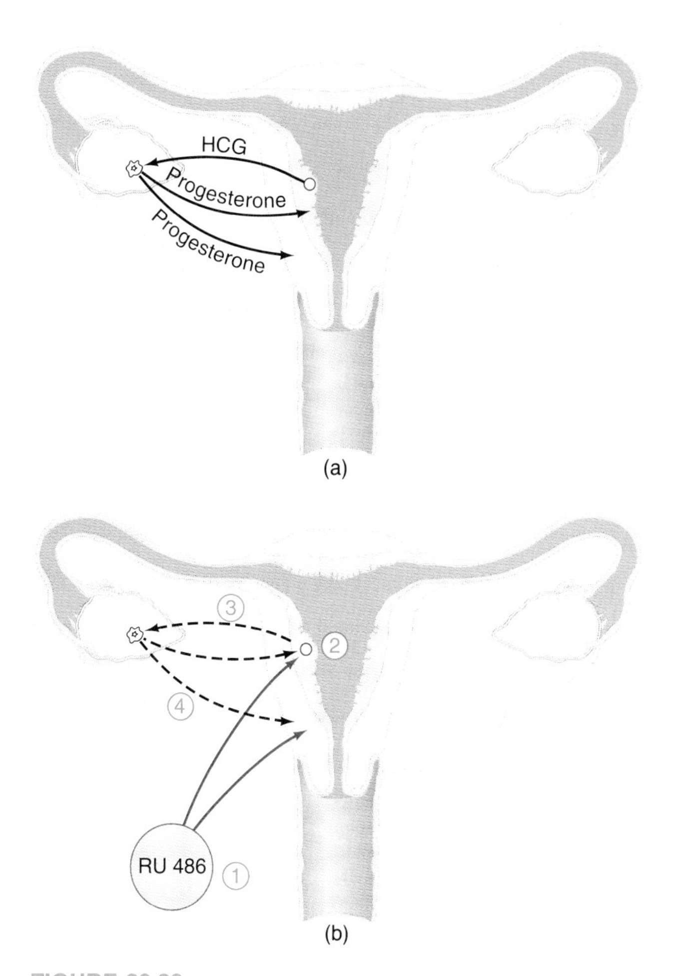

FIGURE 20.23

The Functioning of RU 486

RU 486, a contraceptive not presently approved for use in the United States but used in France and a few other countries, is controversial because it initiates menstruation and therefore can be used even after a pregnancy has occurred. Continuation of a pregnancy depends, in its early stages, on the production of progesterone by the corpus luteum (a). The corpus luteum, in turn, is maintained by HCG. RU 486 (b) competes for progesterone receptor sites and thereby thwarts its effects (1). As the uterine lining begins to break down (because progesterone cannot occupy its receptor sites), the blastocyst begins to detach (2), and its production of HCG begins to diminish (3). As HCG production falls, the corpus luteum begins to fail, and progesterone production drops (4). As this cycle continues, the uterine lining dies, and increased smooth muscle contractions expell the lining and the blastocyst.

BOX 20.5

DES and Cancer

In the late 1940s and 1950s, a hormone-like substance called **diethylstilbestrol (DES)** was widely prescribed to women who had a history of miscarriages. The treatment had limited success and was ultimately discontinued.

During the 1960s, it was discovered that DES, by then widely used in the cattle industry to promote rapid weight gain, was capable of causing tumors in mice. In the early 1970s, the Food and Drug Administration limited its use in farming operations (and some countries, such as Canada, banned its use entirely).

During the mid-1970s, there were reports of an increased incidence of vaginal cancer in teenage girls. Vaginal cancer is generally rare, especially in young girls. In analyzing the cases, researchers found that, in every instance, the girl's mother had been given DES during her pregnancy. Regrettably, the most common treatment for vaginal cancer is a hysterectomy, and permanent sterility. In addition, approximately 50 percent of the daughters born to DES mothers have structural abnormalities in their reproductive tract.

The legal repercussions of DES prescriptions continue at present.

Yet another method is simply to be a nursing mother. Especially in the first few months after birth, FSH and LH production is sharply reduced in women who nurse. However, about half the time there is a spontaneous redevelopment of ovulation despite continued nursing. Resulting conceptions have caused this "method" to fall from favor. Nevertheless, nursing is frequently an effective birth control method in many "primitive" societies, where infants may be nursed several times an hour. The almost-constant stimulation of the breast ensures high levels of prolactin production, and low FSH and LH levels. Such frequent nursing is unlikely to find favor in most Western societies. (The nutrient drain on the mother is also important in reducing FHS and LH secretion; see Box 20.1.)

Another method of birth control calls for the mutual involvement of the male and female. This is the so-called **rhythm method,** in which sexual activity is limited to those days of the menstrual cycle during which conception cannot occur. One problem associated with this method has been that many couples do not realize that "day 1" is the first day of menstruation, not the first day *following* menstruation. This error has understandably been responsible for many unexpected pregnancies.

The underlying theory behind the rhythm method is that fertilization is possible only during an interval of less than 24 hours, given the viability of the sperm and limited temporal ability of the ovum to be fertilized. Hence, if there has been no sexual activity for at least three days prior to ovulation, and there is none for at least one day following ovulation, conception cannot occur.

There are problems with this method. First, there is the difficulty of determining precisely when ovulation occurs. Even though ovulation generally occurs on day 14 of a 28-day cycle, few women are absolutely regular. More commonly, menstrual cycles vary from month to month within a range of from 26 to 31 days. Therefore, the "danger" zone on either side of ovulation must be expanded to account for such variations. (Indeed, a substantial number of women are so irregular that for them the rhythm method is essentially useless.)

This mandatory period of abstinence creates a second difficulty, which is simply perseverance with the method. Most failures of the rhythm method occur because of failed motivation rather than because of any inherent weakness in the method itself. It is actually rather effective in women who are both regular and highly motivated. Moreover, there is a small, but sudden, increase in body temperature just at the time of ovulation, and careful charting of body temperature thus can provide an accurate method of determining when ovulation occurs.

The efficiency of the rhythm method may increase as future improvements in biochemical analysis permit the regular measurement of estrogen and LH levels in the urine. Knowing these values can pinpoint the time of ovulation—information that would also be useful to women who are attempting to become pregnant.

It is ironic that, despite a broad array of contraceptive devices and techniques, unwanted pregnancies are still very much a problem in this country, even among the well-educated. (Indeed, in the United States, three million women "inadvertently" become pregnant each year.) Many of these pregnancies occur because it was the stated intention of the couple not to engage in intercourse and therefore neither partner made any plans to protect against conception. Because contraception requires advance planning, it has been criticized as detracting from the spontaneity of the event. Couples engaging in sexual activity often rationalize their failure to use contraception by suggesting that their lack of advance planning proves that they are not sexually promiscuous, that they were merely momentarily carried away. This rationalization has been responsible for many of the unplanned pregnancies that were a primary pressure behind the legalization of abortion. (Fully half of those three million "inadvertent" pregnancies end in abortion.)

The Future

Despite recent advances, most notably the pill, birth control remains in its (ahem!) infancy. A certain amount of sophistication is required to use the pill, and health workers in developing countries relate horror stories about pill misuse. One notable story involves a health worker who visited the village of a woman to whom birth control pills had been given some months previously. The worker did not find the woman, who was visiting relatives in a nearby village, but the woman's husband was quick to offer reassurance: "Don't worry about the pills. I'm taking them while my wife is away." The point of the story is simply that, although the pill may be adequate for many Western women, it is not a viable method in many medically unsophisticated cultures.

A number of other birth control methods are presently under development:

1. Long-range pills, perhaps in the form of skin implants, to thwart the common problem of "forgetting." **Norplant,** a slow-release synthetic progesterone capsule that is implanted under the skin, has recently been approved for use in the United States. Similar implants have had extensive use in several European and Asian countries. They have an effective duration of up to five years. **Depo-Provera,** a synthetic progesterone, provides three months of contraception from a single injection. Because massive doses cause breast and uterine cancer in experimental animals, it has not yet been approved by the Food and Drug Administration for use in the United States, although it has been in use in 80 countries for almost 20 years.

2. A vaccine against HCG. Preliminary studies on women bolster an earlier finding in baboons that vaccination with a portion of the HCG molecule causes the body to develop antibodies against the hormone. The vaccine causes no change in the menstrual cycle, and no side effects have yet been noted. In the baboon studies, antibody levels gradually declined after a year or so, and the females became fertile once again. In theory, this should be a very effective birth control method because it requires only one or two vaccinations, it is long-lasting (but ultimately reversible), and it interferes with a hormone that appears only during pregnancy.

3. Inhibition of FSH production in men. FSH is not involved in testosterone production, but only in sperm maturation, and would therefore be an ideal control mechanism. **Gossypol,** a chemical extracted from cottonseed oil, blocks sperm production, and is presently being tested by several developing countries. Unfortunately, it apparently causes permanent sterility in 20 percent of users.

4. The use of ultrasound to destroy sperm. Still in the experimental stage, this method has promise, because after a single painless treatment, no sperm are produced for several months.

5. The development in the woman of an immune reaction to sperm. This is only theoretical at the moment, but the possibilities of forming antibodies to sperm are very intriguing.

VENEREAL DISEASE

Improved contraceptive techniques in the last 30 years have fostered more sexual intimacy, because the fear of an unwanted pregnancy has diminished. One consequence of such intimacy has been a sharp increase in the incidence of **venereal diseases** (L. *Veneris,* "Venus, love").

Venereal diseases, by definition, are those that can be transmitted by sexual intercourse. However, in the minds of many people, the term applies only to syphilis and gonorrhea, historically the two most serious diseases, and until recently the only ones that physicians were required by law to report to state health departments. (Many states now require the reporting of herpes and AIDS as well.) For this reason, many scientists prefer the term **sexually transmitted diseases (STD)** as being less restrictive in meaning. At least 14 diseases are included under this heading, although not all are transmitted exclusively by sexual contact (see Figure 20.24).

The list includes such serious diseases as *hepatitis,* which can be transmitted sexually, although other forms of transmission are more common, as well as such minor diseases as "yeast" infections (*Candida albicans*), which are primarily in the nuisance category. It does not include diseases such as *typhus,* which is transmitted by body lice, even though sexual contact can be one means whereby the lice themselves are transmitted.

The following are among the more important sexually transmitted diseases:

1. ***Herpes simplex type II.*** Type I causes fever blisters and cold sores in the mouth, whereas type II

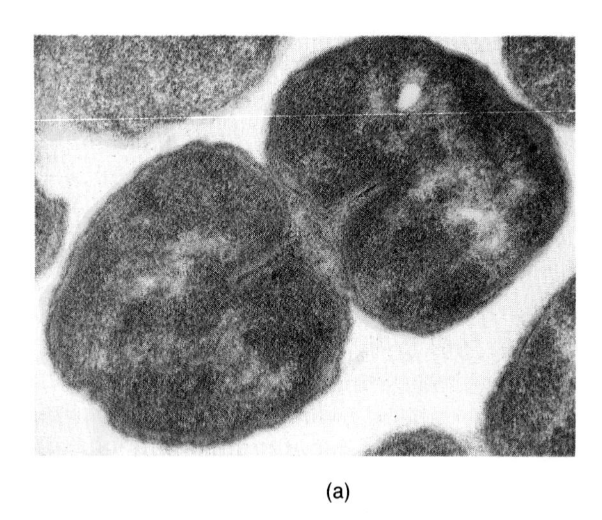

(a)

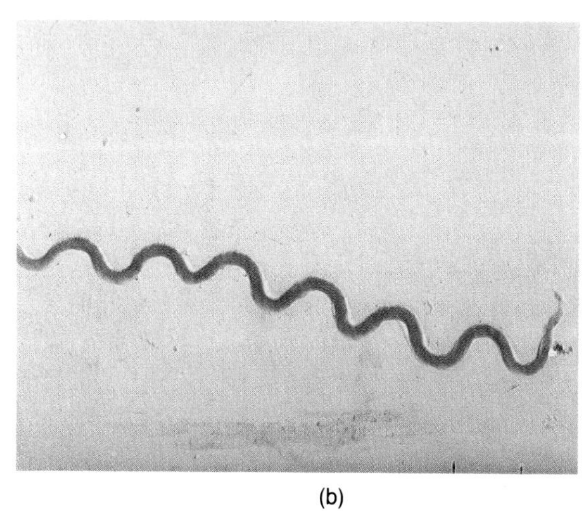

(b)

FIGURE 20.24

Agents of Venereal Disease

(a) Gonorrhea bacterium. (b) Spirochete causing syphilis.

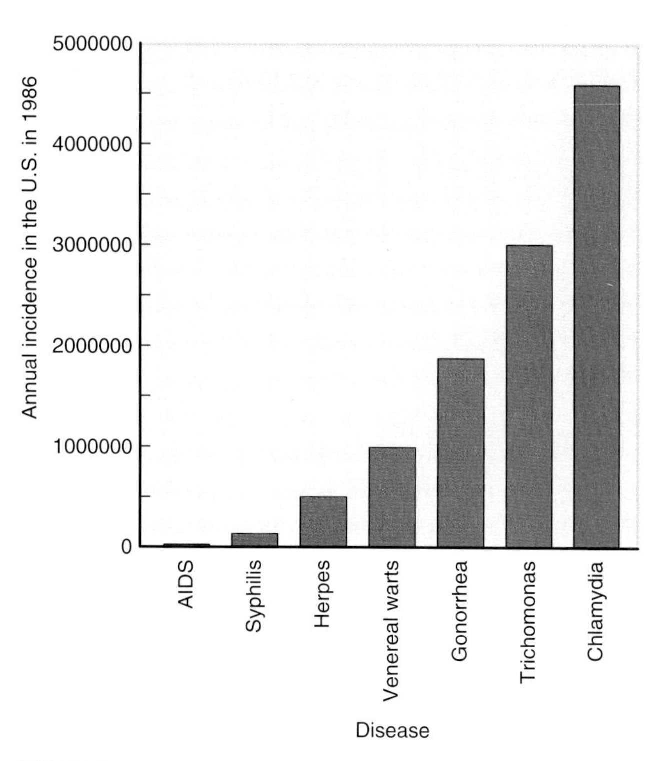

FIGURE 20.25

Incidence of Venereal Diseases

These data are from 1986. Since that time there has been a significant increase in the number of AIDS cases. AIDS is still the least common of these venereal diseases but is by far the most serious.

causes similar sores on the genitals. Oral sex permits interchange between the two. Studies on the incidence of the disease are contradictory, some suggesting an epidemic affecting tens of millions of Americans, others saying simply that the reporting practices have improved. The incidence of cervical cancer is four times greater in women with type II herpes than in women who are free of the disease. Babies may pick up the disease from their mothers during delivery, suffering possible brain damage as a consequence. There are almost 500,000 new cases of genital herpes diagnosed annually in the United States (see Figure 20.25.)

2. **Venereal warts.** Venereal warts are caused by the papilloma virus, which is transmitted during intercourse. The papilloma virus has been implicated in cancer of the cervix and penis and can also infect infants during delivery. Infants then develop warts in their respiratory system. There are one million new cases of venereal warts in the United States each year.

3. ***Trichomonas vaginalis.*** This is a protozoan infection that is symptomless in men but causes a burning sensation and vaginal discharge in women. It may be twice as common as gonorrhea, with about three million cases a year in the United States.

4. ***Chlamydia.*** *Chlamydia trachomatis* is a bacterium that often causes extreme pain during urination in men (and occasionally sterility), but that is frequently symptomless in women—until the damage is done. Arthritis, heart disease, pelvic disorders and sterility can result from this disease. Children born to infected mothers may develop eye infections and pneumonia. There are almost five million new cases diagnosed annually in the United States, making it the most common of all venereal diseases. Fortunately, if diagnosed early, it is easily and inexpensively treated.

5. **Acquired immunodeficiency syndrome.** AIDS is the newest, and the most lethal, of the sexually transmitted diseases. It is discussed in detail in Chapter 9.

Despite the recognition of other sexually transmitted diseases, **syphilis** and **gonorrhea** have historically been considered to be the two principal venereal diseases. Syphilis is caused by *Treponema pallidum,* a spirochete bacterium. Because it was widely reported shortly after Columbus' sailors returned to Europe, it is commonly assumed that the disease had a New World origin (the sailors' having presumably contracted the disease while "fraternizing" with the natives). In any case, syphilis outbreaks were commonplace throughout Europe in the years immediately after Columbus' voyages. Even today, there are more than 100,000 new cases diagnosed annually in the United States.

The disease itself passes through three stages. The *primary stage* occurs anywhere from 10 to 90 days after exposure. This prolonged latency stage compounds the difficulties in tracing the route of infection. The principal symptom in the primary stage is an open sore, called a **chancre** (pronounced "shanker"—not to be confused with the white, ulcerous spots in the mouth called *cankers,* which have no connection with venereal disease). The location of the chancre is variable, but it is always on some mucous membrane (genitals, mouth, or rectum). Actual intercourse is not essential in the transmission of syphilis, because any contact with the chancre and a mucus membrane may pass the infection.

The chancre gradually disappears, and some weeks later the *secondary stage* occurs, in the form of an extensive rash over much of the body. Again, these symptoms disappear, although they may reappear from time to time.

Most individuals never show further sign of the disease. However, approximately 30 percent enter the *tertiary stage,* often after an injury or illness, and invasion of many of the organ systems, especially the central nervous system, is common. Frequently, the severity of the tertiary stage proves lethal.

All of the above presupposes no treatment. However, syphilis is among the most easily treated of infectious diseases and is generally very susceptible to penicillin. Therefore, few cases progress to the secondary stage, let alone the tertiary stage, in this country today.

Syphilis may also be transmitted by a pregnant woman to her unborn child. The effects of this disease on the developing fetus are devastating. The bones, especially those of the face, are malformed, and normal brain development is severely impaired. As recently as 60 years ago, over 30,000 advanced syphilitics were institutionalized in the United States, many of them the children of syphilitic mothers. It was this threat to the unborn child that prompted most states to mandate a blood test (the *Wassermann test,* named for its discoverer) before issuing a marriage license. The Wassermann test is also routinely performed on pregnant women.

By comparison, gonorrhea seems at first glance an innocuous disease—but first appearances are deceiving. Gonorrhea is caused by the bacterium *Neisseria gonorrhoeae,* and unlike syphilis, manifests itself within ten days of exposure. In a minority of males, but in an overwhelming majority of females, the disease is essentially symptomless, one reason why it has proved so difficult to control. There are almost two million new cases of gonorrhea diagnosed in the United States annually.

Whether or not symptoms appear, the disease may spread throughout the reproductive tract. In the female, an inflammation of the long but narrow fallopian tubes may result in scar tissue that blocks the tubes, causing sterility. Invading bacteria may spread through the open tops of the tubes and into the pelvic cavity, causing **pelvic inflammatory disease (PID).** In the United States, approximately two million cases of PID are reported each year, and a substantial fraction end with a **hysterectomy** (surgical removal of the uterus). *Chlamydia* can also cause PID, and some researchers estimate that it is responsible for up to 50 percent of the cases of this disease.

Gonorrhea also spreads to certain other body tissues, such as the tissues surrounding the heart and the joint capsules. Although the disease is not transmitted by an infected mother to her unborn child, it was once a leading cause of infantile blindness, because, as the infant's head passed through the birth canal, the eyes became infected. The rather routine use of silver nitrate drops in the eyes of the newborn infant has largely ended this threat.

Neither syphilis nor gonorrhea promotes a permanent immunity, and, as yet, there is no effective vaccine for either disease. However, the possibilities of such a vaccine give rise to interesting policy questions. Because parents might feel that the threat of venereal disease effectively discourages premarital sex, would they permit their children to be vaccinated? Would the government have the authority, through public health laws, to require vaccinations? You may well be facing these questions by the time you are yourselves parents.

Summary

Perhaps because it represents our link with the future, or because of its obviously different organization in men and women, or because of the aura of mystery and forbidding in which society shrouds it, the reproductive system has a special fascination.

A study of the cyclical functioning of the female reproductive system provides perhaps the most elegant example of hormonal control in the human body. The pituitary hormones FSH and LH interact with the ovarian hormones estrogen and progesterone to ensure that the release of the ovum coincides precisely with the preparation of the uterus for receipt of the ovum, if it is fertilized. However, in the pregnant female, not only must the normal monthly menstrual cycle be interrupted, but the body must be prepared to nurture the baby once it is born, and the period of gestation must be ended in a timely fashion. Once again, these activities are controlled by hormones. By contrast, the male system is almost mundane, in large measure because the production of sperm cells is not cyclical.

Comparisons between the male and female systems also reveal certain societal values such as the fact that the majority of the methods of birth control now in use require involvement of the female, but not the male. This may change in the future, as new birth control techniques are perfected.

Because of the special status we afford the reproductive system, it is not surprising that we grant equal status to diseases of the system. Certainly the primary reason why these diseases are such a problem is that they are commonly transmitted by sexual activity, and sexual activity is subject to special restrictions within our society. Thus, in the minds of many people, venereal diseases are an appropriate penalty to be paid by those who choose not to follow socially proscribed norms. Unfortunately, this attitude contributes to the suffering of many innocent individuals, such as unborn children who are infected while in the uterus.

Key Terms

fertilization
implantation
testis
ovary
genitalia
clitoris
labia minora
labia majora
glans
scrotum
vagina
uterus
fallopian tube
vas deferens
puberty
estrogen
menstrual cycle
gonadotropin releasing hormone (GnRH)
follicle stimulating hormone (FSH)
luteinizing hormone (LH)
egg follicle
endometrium
corpus luteum
progesterone
premenstrual syndrome (PMS)
zygote
morula
blastocyst
placenta
HCG
parturition
oxytocin
cervix
prolactin
testosterone
coitus
seminiferous tubules
emission
epididymis
seminal vesicle
ejaculatory duct
prostate gland
semen
ejaculation
orgasm
coitus interruptus
condom
vasectomy
douche
diaphragm
cervical cap
intrauterine device (IUD)
tubal ligation
diethylstilbestrol (DES)
RU-486
rhythm method
Norplant
Depo-Provera
gossypol
venereal disease
sexually transmitted disease (STD)
Herpes simplex
venereal warts
Trichomonas vaginalis
Chlamydia
AIDS
syphilis
gonorrhea
chancre
pelvic inflammatory disease (PID)
hysterectomy

Box Terms

menarche
amenorrhea
ectopic pregnancy
impotence
cryptorchidism
inguinal canal
inguinal hernia

Questions

1. Distinguish between fertilization and implantation. Is either synonymous with pregnancy?
2. Which genital structures in the female have a homologue in the male?
3. Describe the hormonal control of the menstrual cycle. What causes the cycle to end? What causes it to reinitiate?
4. What prevents menstruation in a pregnant woman? Why is the absence of menstruation during pregnancy important?
5. Distinguish between milk production and milk letdown. What hormone or hormones are responsible for these functions?
6. Describe the hormonal control of sperm cell maturation in males.
7. Which methods are most effective in birth control? Some methods are relatively ineffective. Why, then, do you suppose that anyone wishing to avoid a pregnancy would use them?
8. List three major venereal diseases that occur in the United States. Do you find the incidence of these diseases surprising? Alarming? Why?
9. What is the relationship between menstrual cycles and body fat?

CHAPTER

21

Development

Fertilization, Birth, and Aging

What are the processes by which we are transformed from a fertilized egg to a fully developed adult? Why can't we regenerate lost limbs? Salamanders can, and many lizards can grow a new tail. How does the body know to produce new blood cells after we donate blood at a blood bank?

Dr. Sigmund Bratwurst, a brilliant but eccentric German scientist, recently announced from his aerie high in the mountains of Bavaria that he had managed to culture cells from his cheeks and, with the proper nurturing, had induced them to develop into embryos. He hopes to raise them to full adult development, thus enabling him to hold a Bratwurst festival. Is such a thing possible? Why are fertilized eggs normally the only cells that can develop into new organisms?

Finally, during the course of our lifetime, we undertake constant repair and replacement of damaged cells and tissues—but we nonetheless age. Even if we avoid death by accident or disease, eventually we do indeed die. If we are able to repair and replace damaged cells and tissues, why can't we live forever?

These questions relate to development, the subject of this chapter.

THE NATURE OF THE PROBLEM

Perhaps the most fascinating feature of life is the way in which it replicates itself. In virtually all species, a new individual begins as a single cell that divides and grows in a unique fashion, ultimately producing the adult form that is characteristic of the species. The manner by which these remarkable transformations occur remains largely a mystery even today, although, as you will recall from our discussion of epigenesis and preformationism in Chapter 1, there has been no shortage of speculation over the centuries.

Barring misfortune, the adult form survives for a period of time characteristic of the species (a period that ranges from less than a day for certain mayflies to more than 1000 years for certain trees). During this time, the organism reproduces and ultimately dies. Throughout the organism's life, damaged cells and tissues must be replaced or repaired, processes that require a type of limited development rather different from the development associated with the transformation of the zygote into the adult.

How are all these events coordinated? How does a given embryonic cell "know" that it is to form muscle, and another "know" that it is to form nerve? How do complex, multitissued organs, such as the eye, come into being during development? Surely they cannot result from a single primordial "eye" cell—or can they? In the adult, how does the body "know" when to produce a normal complement of erythrocytes, to replace those that are broken down daily, versus a greatly increased complement of erythrocytes, to replace those lost because of some serious injury? Finally, if we avoid serious injuries and illnesses during our lifetimes, why is it that we eventually die anyway?

DEVELOPMENTAL PROCESSES

The term **development** is used to describe the totality of the processes by which a zygote becomes a complete, functioning organism. Development involves two distinct processes:

1. **Differentiation,** which is the process whereby individual cells and tissues become structurally or functionally specialized; and

2. **Morphogenesis,** which is the process whereby the major organs and regions of the body develop form and structure.

Differentiation

Differentiation pertains to cells or groups of cells. At some point, as the fertilized egg undergoes successive cell divisions (a process known as **cleavage**), the resulting cells become specialized. Since by definition daughter cells possess the same chromosomal complement as the parent cell, differentiation must result not from the loss of genes but from their differential activation or suppression. That is, cells differ because their functioning genes are different. (See Box 21.1)

Morphogenesis

Morphogenesis pertains not to individual cells but rather to the individual as a whole. That is, a mature organism consists not merely of different kinds of cells but of cells that have become organized into tissues, organs, and organ systems. It is not difficult to hypothesize about differentiation, as we have just seen—but how does morphogenesis arise?

The answers are far from complete. However, it is clear that adjacent groups of cells often interact, apparently by the production and reception of chemicals called **morphogens,** and in doing so, can influence the ultimate structural outcome. This process is known as **induction.**

Perhaps the classic example of induction involves the formation of the lens of the eye (see Figure 21.1). During early embryonic development, two *optic vesicles* grow out from the brain, and ultimately come in contact with the tissue on the surface of the body. Each optic vesicle then forms a cup. The tissue at the back of the cup is destined to be the retina of the eye. However, where the vesicles touch the surface tissue, they *induce* a piece of this surface tissue to pinch off, to become globular, and ultimately to form the lens of the eye. If a thin piece of foil is placed between the vesicles and the surface tissues, there is no lens formation, suggesting that chemicals are transmitted between the two groups of tissues. However, although these classic experiments were performed more than 60 years ago, the identity of the inducing chemicals has long been a mystery.

The mystery may be nearing an end. Several morphogens have recently been identified, one of which is *retinoic acid.* (Retinoic acid is one of a group of molecules called *retinoids;* vitamin A is also a retinoid, and it seems likely that other retinoids are also inducers.) Another group of molecules critical in morphogenesis and differentiation are the **cell-adhesion molecules (CAMs),** which apparently help mediate various interactions between embryonic cells. CAMs occur in the plasma membrane. They attach cells together and are closely related to the immunoglobulins (see Chapter 9). Five different CAMs have now been identified. Because they have very different distributions in the embryo, it has been speculated that they may be critical in establishing boundaries between different tissues. In addition, cells with one type of CAM may be programmed to die, as happens in the development of fingers and toes from solid pads of tissue (see

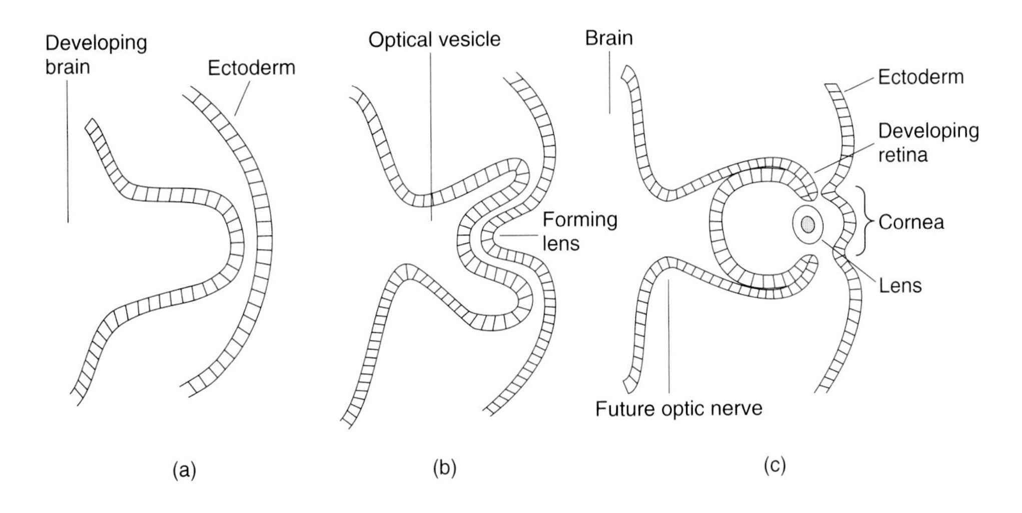

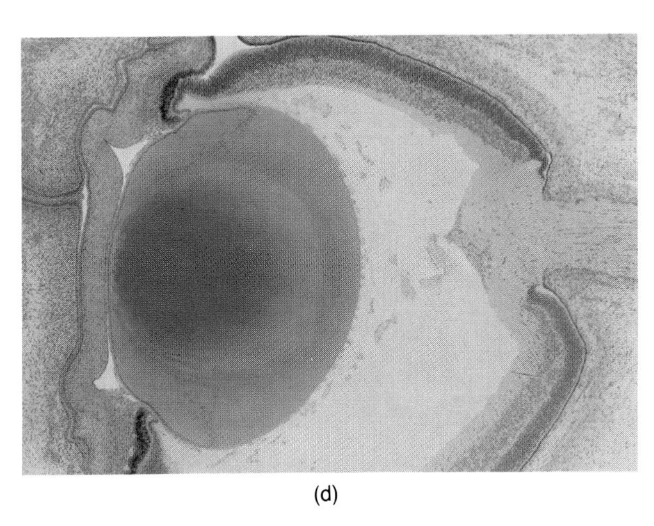

FIGURE 21.1

Induction

As the optic vesicle grows out from the brain (a), it eventually encounters the surface tissue (ectoderm) and induces the formation of the lens (b). The ectoderm grows back over the lens as the cornea, and the retina begins to develop (c). Photomicrograph of an optic vesicle and forming lens in a human embryo (d).

Figure 21.2). (It may be that the inappropriate death of nerve cells, which characterizes such diseases as Alzheimer's, Parkinson's and Huntington, results from a timing error in the activity of a gene coding for a CAM.) In other instances, cells apparently migrate along CAM trails, to assume new locations in the developing embryo. Migration of cells—presumably in response to chemical gradients—is common in morphogenesis. For example, embryonic cells move into the brain along various routes and, depending on the route they follow, mature either into various types of neurons or the supporting glial cells (see Chapter 14).

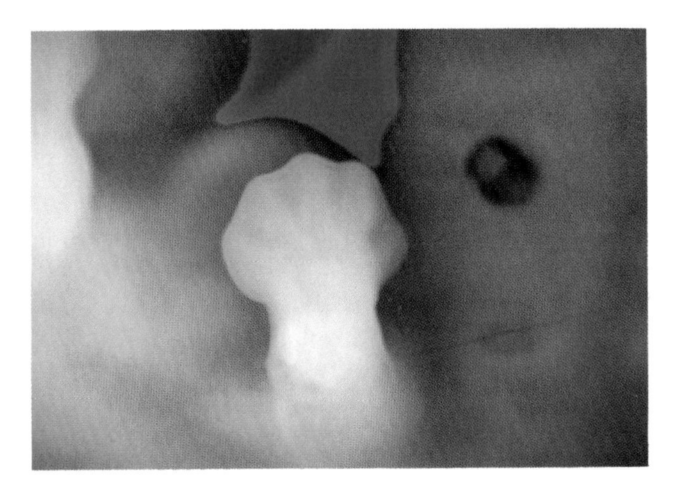

FIGURE 21.2

Programmed Cell Death

The fingers of the hand form from a flat pad of tissue by the death of cells in a programmed manner.

A recent study examined morphogenesis and differentiation in the developing peripheral nervous system. The differentiation of sympathetic neurons from embryonic nerve cells first requires exposure to fibroblast growth factor **(FGF).** Once primed by FGF, these developing nerve cells then respond to **nerve growth factor (NGF)** and become sympathetic neurons. However, the cells that have migrated to the adrenal gland are not exposed to FGF and therefore they become secretory cells of the adrenal medulla (see also Chapter 14).

Growth factors are important in the developing brain as well. One group of investigators mutated a gene thought to code for a growth factor and found that, in mice, the cerebellum failed to develop properly or was missing altogether.

BOX 21.1

Differentiation, Totipotency, and Clones

Differentiation is the specialization of cells in a structural or functional sense. Mature cells from different tissues look and perform differently—but just how different are they? All are ultimately products of the fertilized egg, and therefore all should, at least in theory, share identical sets of genes. But do they? Or do differentiated cells differ in the genes they possess?

These questions have in large measure been answered in an elegant series of experiments involving nuclear transplantation. The nucleus of a fertilized egg is removed or destroyed, and the nucleus from a differentiated cell is transplanted into the egg. With impressive regularity, the egg proceeds through normal development and becomes a functional adult, fully representative of the species. The capacity for the nuclei of differentiated cells to direct a normal development is called **totipotency,** and it indicates that the genetic makeup of differentiated cells has not been irreversibly changed by the processes of differentiation.

The adults that are produced from these nuclear transplantation experiments are genetically identical to the animals from which the transplanted nucleus was originally taken. Organisms that are genetically identical are called **clones** (Gr. *klon,* "a twig"); identical twins exemplify naturally occurring clones.

To date, the bulk of the transplantation experiments in animals have involved amphibian eggs, which are generally large and relatively easily manipulated. The eggs of mammals are much smaller (about the size of a period in this line of print), and they are much more fragile. Nonetheless, mice have been experimentally cloned, and the potential exists for cloning humans. We may eventually be facing the moral and ethical questions of such experiments. How should individuals be selected for cloning? Who will make the decisions and what criteria should be used? You have a while to think about these questions—but don't take too long. After all, it was only a generation ago that scientists were debating whether the molecules of heredity were proteins or nucleic acids.

Finally, this differentiation may occur at remarkable levels of detail. For example, secretion of the hormone **gonadotropin releasing hormone (GnRH)** is essential for the coordination of the menstrual cycle (see Chapter 20)—but this hormone is produced by only about 1500 cells (of the several hundred billion cells which comprise the brain).

How Do Morphogens Function?

Even if we allow that the developing embryo is a sea of chemicals, each influencing the differentiation of specific cells, how do these chemicals actually function? The answer to this question is only now beginning to emerge, but it is clear that they function in a manner similar to the functioning of hormones: They either bind to receptors on the surface of particular cells, thereby initiating a sequence of reactions within those cells, or they enter the cells directly and turn on and off specific genes (see Chapter 13 for a review of hormone function).

But, you may point out, there are 50,000 to 100,000 genes in each of our cells. Surely each one cannot have its own morphogen? One of the more exciting findings in biology in recent years has been the identification of genes that code for proteins that bind to specific regions of DNA, and thereby regulate the activity of other genes. In so doing, they exert influence much greater than would be supposed possible by a single gene. In other words, if a given morphogen can turn a particular regulatory gene on or off, a cascade of genetic activity can be released, all in response to one signal, in much the same way that your television set or computer becomes operational once you flip the switch and allow electricity to flow through it.

Is there evidence that morphogens and regulatory genes actually do interact during development? Various regulatory genes contain sections of DNA that in their sequence of nucleotides are remarkably alike, not only to other regulatory genes in the same species but also to regulatory genes in very different species. These common sequences of DNA are called **homeoboxes,** and the degree of similarity is stunning. For example, 59 of the 60 amino acids that comprise the portion of the protein coded for by the homeobox are identical in species as evolutionarily distant as *Drosophila* and frogs—two species that have not shared a common ancestor in the last 500 million years!

Why are these genes not more evolutionarily divergent? The answer appears to be that the precise sequence of amino acids in the protein coded for by the homeobox is needed to bind effectively with DNA. Any mutation in the DNA of the homeobox would quickly be eliminated by natural selection (see Chapter 22), because the mutated protein would not bind to DNA and would therefore be incapable of regulating gene expression.

Some 30 homeobox genes have been identified in mice, and, for many of them, their role in embryonic development has been determined. They are generally active very early in embryonic development, and most are found in tissue destined to become part of the central nervous system; some, however, are also found in limb buds, muscle tissue, and various of the prospective internal organs. Different tissues have different combinations of these genes active at any one time. Moreover, at least some of these genes are themselves turned on by retinoic acid and FGF.

We therefore have a model of morphogenesis. Morphogens create chemical gradients that differentially turn on various regulatory genes. Depending on the timing of regulatory gene activity during the development of the embryo and on the particular combination of regulatory genes that are active at any one time within a given tissue, embryonic cells differentiate along specific pathways and ultimately assume the configuration characteristic of mature tissues and organs (see Figure 21.3). Obviously, work has only begun on a complete understanding of morphogenesis, but the identification of some of the actual regulatory genes and morphogens give credence to our model.

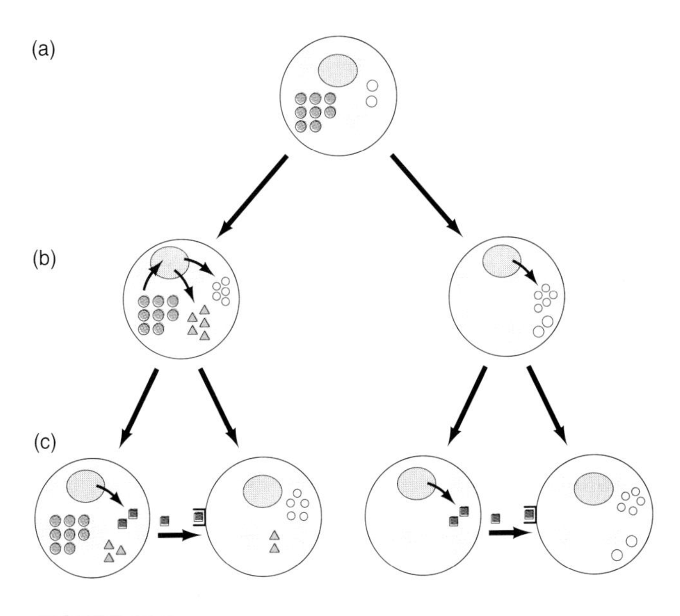

FIGURE 21.3

Model of Regulatory Mechanisms During Development

(a) The fertilized egg. As it divides, the daughter cells have unequal amounts of a particular protein (b). In the cell that receives the most protein, the protein stimulates expression of a particular gene. In the other daughter cell, the absence of this protein prevents expression of this gene. A second gene is expressed in both cells. In the second cleavage division (c), expression of a third gene creates a protein that attaches to a membrane receptor of an adjacent cell and inhibits expression of the same gene in that cell. Thus, after two divisions, the four cells are already distinct in terms of which genes are being expressed.

EARLY EMBRYONIC DEVELOPMENT

Now that we have considered some developmental principles and processes, let us turn our attention to the actual developmental history of the human embryo.

Shortly after the blastocyst implants in the uterus, the cells destined to become the embryo form first a two-layered and then a three-layered structure. These are the **germ layers,** the embryonic tissues that ultimately mature into the tissues and organs of the adult organism (see Figure 21.4). Virtually all animal species possess the same three germ layers, and marking experiments in simpler organisms have demonstrated that these germ layers invariably follow a carefully modulated developmental sequence that combines both differentiation and morphogenesis.

The outermost layer, or **ectoderm,** becomes the epidermis of the skin and the nervous system of the adult. The innermost layer, or **endoderm,** becomes the lining of the digestive system, associated glands, and the lining of the lungs and respiratory tract. The middle layer, or **mesoderm,** becomes the skeletal and muscular systems, the dermis of the skin, and the bulk of the excretory, reproductive, and circulatory systems.

How do these germ layers become distinct from one another in the first place? Recently, a molecule called *activin* has been implicated. Initial experiments suggest that the concentration of activin is the key in causing the development of the germ layers. Cells closest to the cells producing activin become ectoderm; cells somewhat farther away become mesoderm; and those most removed become endoderm.

However, not all of the cells of the blastocyst are destined to form structural components of the embryo. Instead, some of them are used to provide protection and nutrition to the growing embryo. These cells develop into the **extraembryonic membranes,** which are tissues that, as their name implies, play only an adjunct role in the development of the embryo.

THE EXTRAEMBRYONIC MEMBRANES

Initially, the developing embryo receives nutrients from the endometrial cells that are destroyed during implantation, but this nutritional source is obviously inadequate for the nine months of life in the uterus.

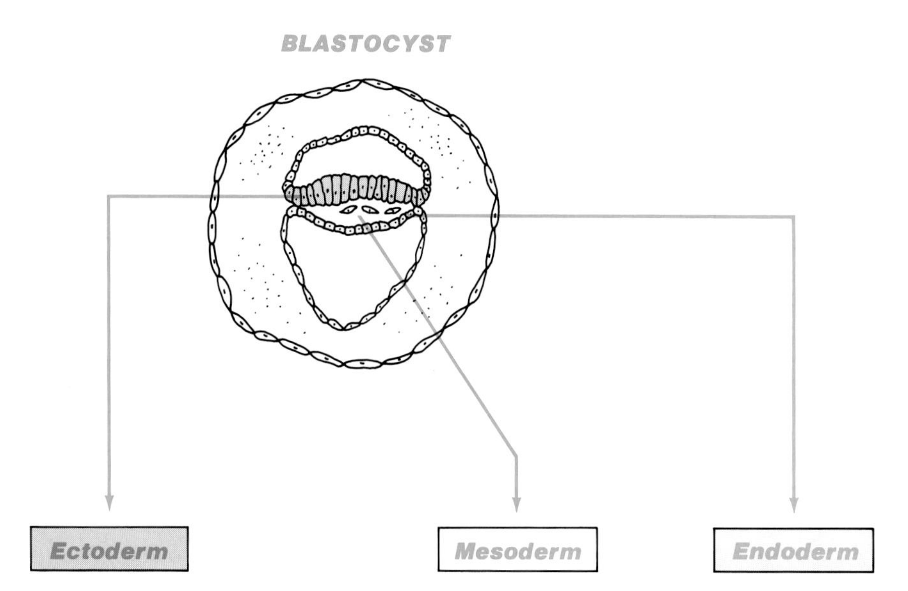

Ectoderm: Nervous Tissue	Ectoderm: External Coverings	Mesoderm	Endoderm: Linings for:
Spinal nerves	Sweat glands	Bones	Urinary bladder
Adrenal medulla	Oil glands	Dermis	Pancreas
Spinal cord	Hair	Blood	Liver
Autonomic nervous system	Milk-secreting glands	Reproductive organs	Intestines
Retina	Lens	Heart	Lungs
Brain	Mouth lining	Muscles	Pharynx
Inner ear		Kidneys	

FIGURE 21.4

Embryonic Germ Layers
Ectoderm gives rise largely to skin and nervous tissues, endoderm forms the lining of many internal organs, and mesoderm transforms into virtually all other tissues.

Amniotic fluid
Amnion
Egg shell
Chorion
Air space
Yolk sac
Allantois
Albumin (egg white)
(a)

Yolk sac
Chorion
Amniotic fluid
Amnion
Allantois
Maternal tissue
Chorionic villi
(b)

FIGURE 21.5

Extraembryonic Membranes in Reptiles and Mammals
The same four extraembryonic membranes that are found in the developing egg of reptiles (a) and birds are also found in mammals (b). In reptiles and birds, the allatois fuses with the chorion and functions as a gas exchange mechanism (the "egg lung"). In mammals the allantois fuses with the chorion to form the embryonic portion of the placenta. The yolk sac is much smaller in mammals and the stalk of the allantois becomes the primary component of the umbilical cord.

How, then, does the embryo receive nutrition from its mother?

The answer to that question requires a modest understanding of our evolutionary heritage, because the membranes that protect and nurture the human embryo are a legacy of the reptiles from which mammals evolved.

Within the reptilian egg are four extraembryonic membranes that are left behind after hatching (see Figure 21.5). An inner **amnion** (Gr. *amnus,* "lamb") encloses the developing embryo in a "bag of waters"; a **yolk sac** encloses the nutrient-laden yolk and connects it to the embryo's digestive tract; an **allantois** (Gr. *allantoeides,* "sausage-shaped") functions as the embryo's bladder by storing nitrogenous wastes; and a thick **chorion** (Gr., "leather") surrounds the other membranes. A shell is laid down on top of the chorion as the egg proceeds down the reproductive tract of the mother reptile.

These same membranes appear, in somewhat modified form, in the developing human embryo (see Figure 21.6). Within ten days of implantation, the chorion begins to develop finger-like projections, called **chorionic villi,** that sink deeply into the uterine lining. These chorionic villi soon become *vascularized* (i.e., blood-containing). Blood sinuses form in the maternal tissues opposite the villi. Nutrients from the mother diffuse from these sinuses into the blood within the chorionic villi, and metabolic wastes from the embryo diffuse in the opposite direction.

Even as the chorionic villi are forming, the amnion develops, and it grows along with the embryo, protecting the embryo just as it does in the reptiles. The yolk sac also forms about this same time, although the human ovum contains virtually no yolk. However, the yolk sac is the source of the first blood cells, a function taken over by the liver during the second month of embryonic life.

Finally, the allantois also forms at this time and grows outward to fuse with the chorion. (The same type of fusion occurs in reptiles and birds, forming a gas exchange organ known as the "egg lung.") In humans and other mammals, the allantoic stalk becomes the **umbilical cord** (L. *umbilicus,* "navel"), the connection between the embryo and the chorion.

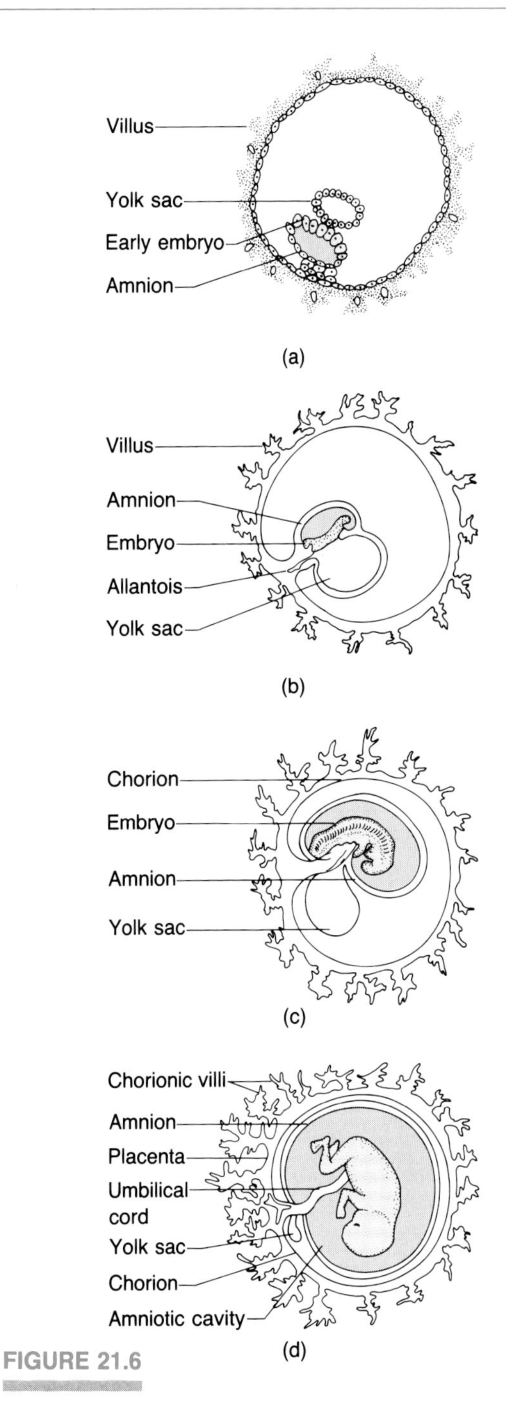

FIGURE 21.6

Extraembryonic Membranes

The relative development of the extraembryonic membranes in humans at (a) 12 days; (b) 16 days; (c) 4 weeks; and (d) 12 weeks.

The Placenta

As the embryo grows, it gradually pushes into the cavity of the uterus, remaining connected by the umbilical cord to that portion of the chorion that stays embedded in the uterine wall. The chorionic villi together with the surrounding blood sinuses and other maternal tissues form the **placenta,** a highly effective exchange organ between the embryo and its mother (see Figure 21.7). Note that the blood of the embryo and the mother normally never mix, nor do bacteria within the mother's blood gain access to the embryo. Small molecules, however, can move freely across the placenta. Unfortunately, various toxins and viruses (including both drugs of abuse and the AIDS virus) also pass across this membranous shield.

Although the total area of exchange is relatively small (the chorionic villi are not nearly as extensive

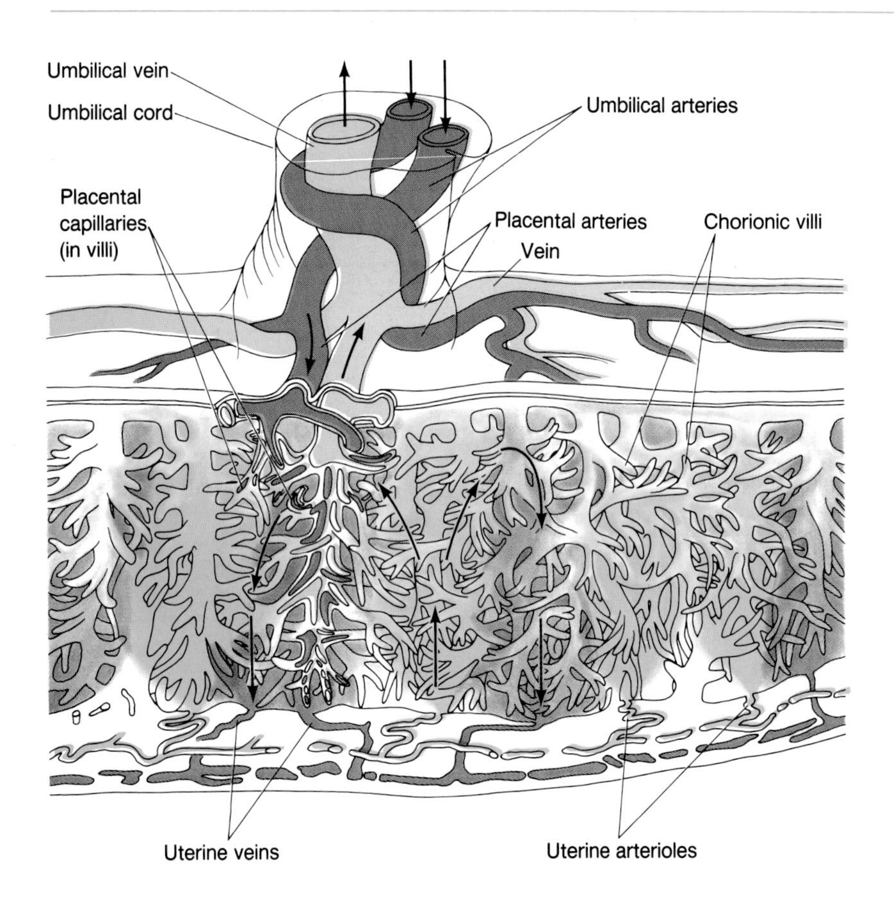

FIGURE 21.7

Placental Exchange

The placenta is responsible for exchanging materials between the mother and fetus. Blood leaves the fetus through two umbilical arteries and enters capillary beds in the chorionic villi. The villi are bathed in maternal blood from the uterine arteries. Nutrient-laden blood returns to the fetus through the umbilical vein; waste-laden blood returns to the mother's general circulation through the uterine veins. In theory, fetal and maternal bloodstreams never mix, although as a practical matter the inevitable small hemorrhages that occur generally result in some interchange of blood. That exchange can be important if the mother and fetus differ in their Rh factor (see Chapter 9).

as the villi of the small intestine; see Chapter 6), gas and nutrient exchange is entirely adequate. Transmission of oxygen from maternal blood to the blood of the embryo is aided because embryonic hemoglobin is of a different structure and has a higher affinity for oxygen than does maternal hemoglobin. Carbon dioxide pressures are slightly higher in the embryo than in the maternal blood supply, ensuring a one-way diffusion of carbon dioxide from the embryo to the mother. Reciprocally, blood glucose levels are lower in the embryo than in the mother, meaning that diffusion of this essential nutrient is always from mother to embryo. Finally, the placenta also has the capacity for active transport of calcium, phosphorus, and various amino acids.

SUBSEQUENT EMBRYONIC DEVELOPMENT

The transformation of an undifferentiated mass of cells into a recognizable embryo begins rather quickly. **Neurulation,** the process by which the **neural tube** (the primitive spinal cord) is formed occurs about three weeks after fertilization, and a definite head fold can be seen shortly thereafter. Neurulation provides another example of induction (see Figure 21.8). Folds of ectodermal tissue form along the back of the embryo, rise up, and enclose a tube of ectodermal tissue that is transformed into the neural tube. The floor of the neural tube is induced by the tissue lying immediately beneath it—the **notochord** (Gr. *noton,* "back" and *chorde,* "string"), an ancient structure in vertebrates that in ourselves ultimately becomes incorporated into the vertebral column.

Before the beginning of the fourth week, gill slits in the pharyngeal region (a characteristic trait of all vertebrates) begin to form, and, at about the same time, the heart starts to develop from the fusion of veins in the chest area (see Figure 21.9).

The Second Month

During the fifth week of development, several additional changes occur. The head becomes more clearly defined, the mouth develops, the eyes begin to form, and the heart starts to beat. The arms and legs appear as limb buds, which at this stage are just lumps of undifferentiated tissue. However, the embryo is still only about 1 cm in length.

During the sixth week, the arm and leg buds begin to differentiate, the heart divides into chambers, and the liver starts manufacturing blood cells. (The bone marrow takes over the manufacture of blood cells only later in development.) The digestive tract

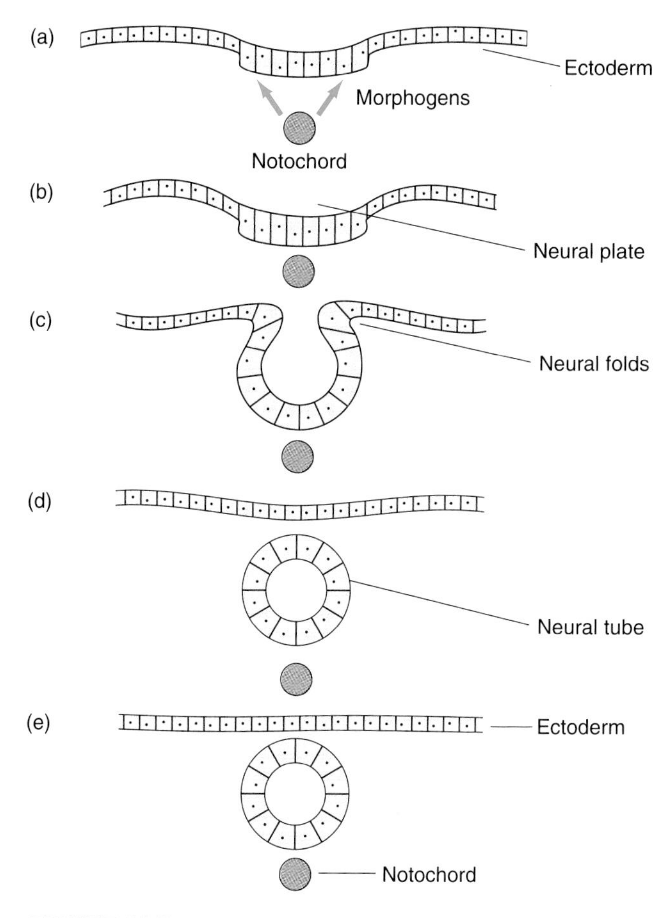

FIGURE 21.8

Neural Tube Formation

The formation of the neural tube—the structure destined to become the spinal cord and brain—provides another example of induction. The notochord induces ectodermal cells immediately above it to thicken, form ridges, and ultimately create a tube that detaches from the rest of the ectodermal tissue.

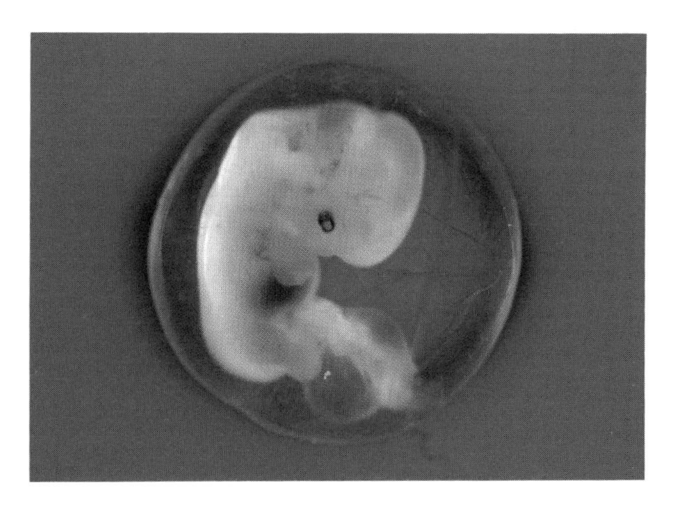

FIGURE 21.9

The Human Embryo

Embryo at six weeks, and 2 cm of body length. The arm and leg buds are well developed, but the head and face are not yet distinct.

begins to differentiate into distinct regions, although the embryo is still less than 2 cm in length.

During the seventh week, the embryo begins to assume a somewhat more human appearance, partly because the eyes shift from a lateral to a frontal position, creating a face. Most of the growth is in the head region, where the developing brain is responsible for almost half the length of the embryo. The external ears begin to develop, as do the fingers and toes. Overall length is now about 3 cm, but the embryo still weighs less than 10 g.

FETAL DEVELOPMENT

Largely because of its more human appearance, after seven weeks the embryo is referred to as a **fetus** (L., "offspring"). During the eighth week, and on through the third month of development, the fetus becomes increasingly more human in appearance. The kidneys begin to function, the nose and mouth develop, and deposition of bone begins in the hitherto exclusively cartilaginous skeleton. The fetus begins to straighten from its former C shape. The external sex organs develop, and blood begins to be formed by the bones, as well as by the liver. The nervous system matures to the point that simple reflexes occur. These movements are termed **quickening,** and were formerly thought to indicate the entry of the soul into the fetus. The mother generally cannot detect fetal movement until the fourth or fifth month, but it was arbitrarily decided by the ancients that quickening occurred on the fortieth day in males but not until the eightieth day in females. (Male chauvinism has had a long and pervasive history!) By the end of the twelfth week (the end of the *first trimester*), the fetus is about 8 cm in length and weighs approximately 30 g.

The Second and Third Trimesters

Development during the *second trimester* (the fourth through sixth months) is less dramatic. The most obvious change is the huge increase in size, to 30 cm and 1 kg by the end of that period. Reflexes such as gripping and thumbsucking occur, and the hair and eyebrows develop. However, the fetus cannot generally survive outside its mother because the lungs, digestive system, and the capacity to regulate temperature are still insufficiently developed.

In the *third trimester*, growth continues at an accelerated pace. At birth, the fetus is about 50 cm long and weighs more than 3 kg. The chances for independent survival increase steadily during the last three months, should the child be delivered prematurely. (The principal problem for such children is usually respiration, for reasons discussed below.) The amount of fat in the skin increases dramatically, and the fetus becomes less wrinkled.

EVENTS AT BIRTH

The birth process itself was discussed in Chapter 20. However, profound and immediate developmental changes occur in the infant at its birth, as it moves from an aqueous to a gaseous environment. These changes are concentrated in the respiratory and circulatory systems.

The Respiratory System

The lungs mature relatively slowly during fetal development, and the respiratory system is not fully developed until close to nine months of gestational age. **Respiratory distress syndrome** often occurs both in premature infants and in the children of diabetic mothers. Although the precise cause is unknown, it is clear that such infants do not produce a sufficient quantity of a fluid that allows the alveoli of the lungs to open with the first breaths after birth. Lung collapse is therefore common. President John F. Kennedy's third child, Patrick, died of this disease.

Even with an ample supply of this fluid, inflation of the lungs at birth requires several extremely powerful breaths, involving as much as 60 mm Hg negative pressure. (This is four times as great as the negative pressure used in normal breathing an hour or so later.) Powerful exhalations are also needed initially to clear the bronchioles of the fluid expelled from the alveoli. The initiation of breathing is usually spontaneous—the traditional slap on the buttocks is generally not required. However, if the mother has been heavily sedated during delivery, the infant also will feel the effects of the sedation and may require some prompting to begin breathing on his or her own.

The Circulatory System

During fetal life, gas exchange is accomplished through the placenta rather than through the lungs. However, with the loss of the placental connection, and the onset of normal respiration, changes in the blood circuit are imperative.

Consider the fetal circulation pattern (see Figure 21.10). Oxygenated blood flowing back to the fetus through the **umbilical vein** passes through the **ductus venosus** in the liver and connects to the inferior vena cava. Oxygen-rich and oxygen-poor blood are therefore mixed before the blood enters the right atrium. Most of this blood then passes directly into the left atrium through an opening, the **foramen ovale** (L., "egg-shaped opening"). From the left atrium, the blood flows into the left ventricle, and the left ventricle pumps the blood to the head and body. Blood returning from the head and body moves through the right atrium and into the right ventricle. However, when the right ventricle contracts, most of the blood does not travel to the lungs but instead moves through the **ductus arteriosus,** a shunt vessel that connects the pulmonary artery to the descending aorta. A pair of **umbilical arteries** branches off the descending aorta and transports the partially deoxygenated blood back to the placenta.

At birth, all of these fetal shunting vessels and openings must shut down. The umbilical arteries and veins collapse when the umbilical cord is tied, or when the placenta tears loose from the uterus. Greater pulmonary blood flow tends to force blood from the left atrium to the right atrium (the opposite pattern to that seen in fetal circulation), and when that occurs the one-way valve that guards the foramen ovale is forced closed, permanently separating the left and right atria. With the loss of the umbilical vein and with greater pulmonary blood flow, respectively, the ductus venosus and ductus arteriosus collapse, and the adult pattern of circulation begins. The entire set of events just listed is completed within a few hours of birth.

The single greatest danger in this changeover is that the foramen ovale may remain open. Infants with this problem have a bluish cast, caused by large amounts of the darker, deoxygenated blood that flows through the body. Logically enough, they are called "blue babies." Only two generations ago, these children had little hope of surviving for more than a few years, and in any event they were forced to take a vicarious view of life, as their condition prohibited any physical exertion. With the refinement of open-heart surgery, these children now can generally be restored to a totally normal life.

FROM BIRTH TO ADULTHOOD

The most visible change as we progress from birth to adulthood is growth. In a span of 20 years or so, we increase our height by a factor of three and our weight by a factor of 15 to 25 (the difference being primarily a function of gender). However, even a casual perusal of this growth demonstrates that it is not proportional growth—the legs and arms grow more than the body,

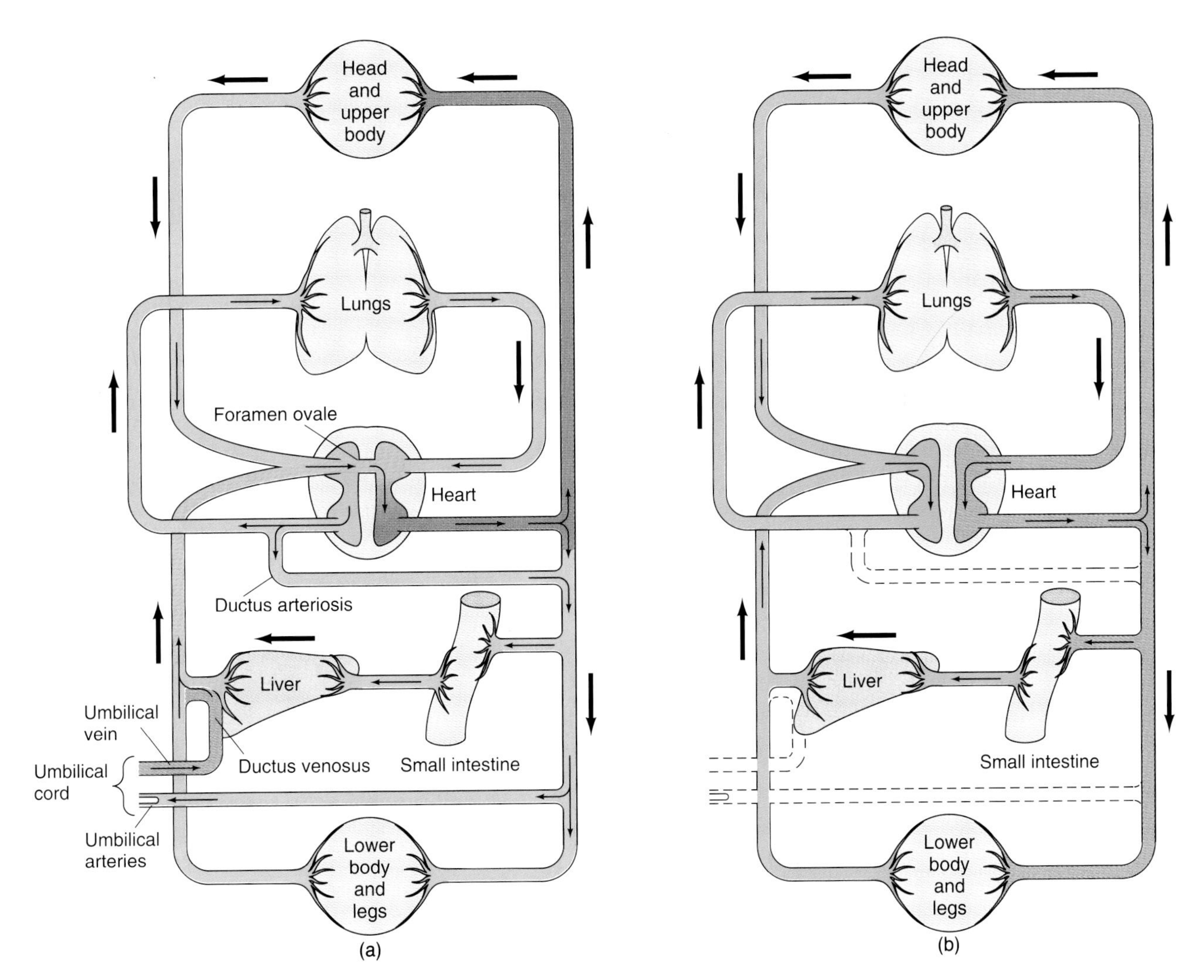

FIGURE 21.10

Fetal and Infant Circulation

In the fetus (a), the ductus venosus shunts oxygen-rich blood from the umbilical vein through the liver where it is joined by a modest flow of oxygen-poor blood from the inferior vena cava. The blood continues into the right atrium, but mixes very little with blood returning from the anterior vena cava. Instead, it passes into the right atrium by way of the foramen ovale, and then out the aorta. Because the lungs are nonfunctional in the fetus, the volume of blood returning from the lungs is modest, and therefore the blood in the aorta is still relatively oxygen-rich. Most of this blood passes to the head and brain; some continues down the aorta to the lower body. Blood returning from the brain passes through the right atrium to the right ventricle, but most of this blood does not go to the lungs but instead passes through the ductus arteriosis and then to the descending aorta. Although the lower body and legs receive some blood, most of it returns to the placenta via the umbilical arteries. At birth (b) the ductus venosus, the ductus arteriosus, the umbilical arteries and vein, and foramen ovale close, and the infant circulatory pattern takes over.

and the size of the head increases least of all. (The head is one-third to one-fourth of overall length in a baby, but only one-seventh of adult height.) This form of disproportionate growth is called **allometry.** Another example of allometry is the difference in growth of the head of apes (heavy jaw and eyebrow ridges; low-domed, crested skull; somewhat muzzle-like face) and of humans (retention of the infant pattern—flat, broad face; high-domed skull without a crest; reduced eyebrow ridges) from a juvenile form that is rather similar among hominids in general (see Chapter 23). Overall body growth is directed primarily by hormones, of which GH and TSH are the most important (see Chapter 13).

Other changes—changes that are more subtle visibly but in their own way profound—occur at the cellular level. At puberty, for instance, the growth patterns of males and females change rapidly, not only in the reproductive organs themselves but also in the muscular and skeletal systems. *Testosterone* promotes skeletal and muscular growth; *estrogen* promotes skeletal growth but also promotes the replacement of the cartilage growth zones with bone. Consequently, boys in puberty quickly grow taller than girls (in general), and substantially more muscular. Again, these growth changes are mediated by hormones—and, as we have seen (Chapter 13), hormones (especially steroid hormones) function by promoting the activity of particular genes.

During fetal and embryonic life, the process of cellular differentiation gives rise to various mature cell lines that are incapable of dividing. Mature muscle and nerve cells, for instance, cannot undergo mitosis—but then neither can many mature cells, including most of the blood cells. How, then, are these cells replaced if they are lost or damaged?

In many cell lines, some immature (embryonic) cells remain throughout adult life. The **stem cell** of blood, for instance, has the capacity to differentiate into any of the mature blood cells (see Chapter 9). Thus the loss or destruction of blood cells can be offset by the production of new cells through successive divisions of several generations of cells originating from stem cells. Epithelial tissue functions in a similar manner—thus we are able to replace lost skin tissue, or abraided intestinal lining. Conversely, there is no equivalent of the stem cell in nerve or muscle tissue, with the result that, should a nerve or muscle cell die, it cannot be replaced except by scar tissue.

Let's examine the activities of stem cells in greater detail. How does a blood stem cell "know" to differentiate into erythrocytes versus platelets—let alone any of the various types of leukocytes? Is there simply some preset proportional differentiation (i.e., 80 percent will become erythrocytes, 15 percent platelets, 5 percent leukocytes)? Proportional differentiation sounds reasonable, at first glance, especially if all we are doing is replacing half a liter of blood which has been donated to a blood center—but the number of leukocytes is known to vary considerably as a consequence of particular infections. How is that accomplished without a corresponding increase in erythrocytes or platelets?

Differentiation of the various blood cells takes place because of the interaction of hormones (see Chapter 13), interleukins (see Chapter 9), and growth factors, which in the case of blood are sometimes called **colony stimulating factors (CSFs)** or **hemopoietins** (Gr. *haima*, "blood" and *poiein*, "to make"). The distinctions among these terms may, however, be more apparent than real, as we saw in the introduction to Part 4. These substances all function like hormones, in the sense that they stimulate change within specific cells and tissues.

Stem cells may divide to produce more stem cells, or they may follow various developmental pathways leading to the maturation of any of the classes of mature blood cells. As stem cells differentiate, they pass through increasingly more specialized forms. Initially, they enter either the pathway leading toward lymphocytes (becoming **lymphoid stem cells**), or the pathway leading to all of the other types of blood cells (becoming **myeloid stem cells**). The myeloid stem cells further differentiate along five possible pathways that lead to the following:

1. Erythrocytes
2. Megakaryocytes (which produce platelets)
3. Basophils
4. Eosinophils
5. Neutrophils and monocytes

A number of the hemopoietins have now been identified, although the regulation of all of the developmental pathways has not yet been elucidated. The first hemopoietin to be isolated and identified (in 1977) was **erythropoietin.** Erythropoietin is produced by the kidney in response to a drop in the oxygen-carrying capacity of the blood (see Figure 21.11). This hormone does not actually initiate new erythrocyte production. Rather, it acts at a later point in the process by preventing the death of immature erythrocytes. Thus in the presence of erythropoietin mature erythrocytes enter the blood only a few days after major blood loss.

Other hemopoietins include **interleukin-3,** which stimulates the growth of all of the myeloid types of blood cells; **GM-CSF (granulocyte-monocyte colony stimulating factor),** which stimulates both granulocyte and monocyte colonies; **G-CSF,** which stimulates only granulocyte colonies; and **M-CSF,** which stimulates only monocyte colonies (see Figure 21.12). The genes for IL-3, GM-CSF, and M-CSF are all on chromosome 5; the G-CSF gene has been located on chromosome 17.

What cells produce these hemopoietins? GM-CSF is produced by a variety of cells, including lymphocytes, macrophages, fibroblasts, and endothelial cells. G-CSF is produced by monocytes, macrophages, and fibroblasts. M-CSF is produced by macrophages, and IL-3 is produced by activated lymphocytes. These hemopoietins are not produced in isolation from one another. To the contrary, the production of one hemopoietin often triggers the production of others by the receiving cell, in a complex network of relationships (see Figure 21.13).

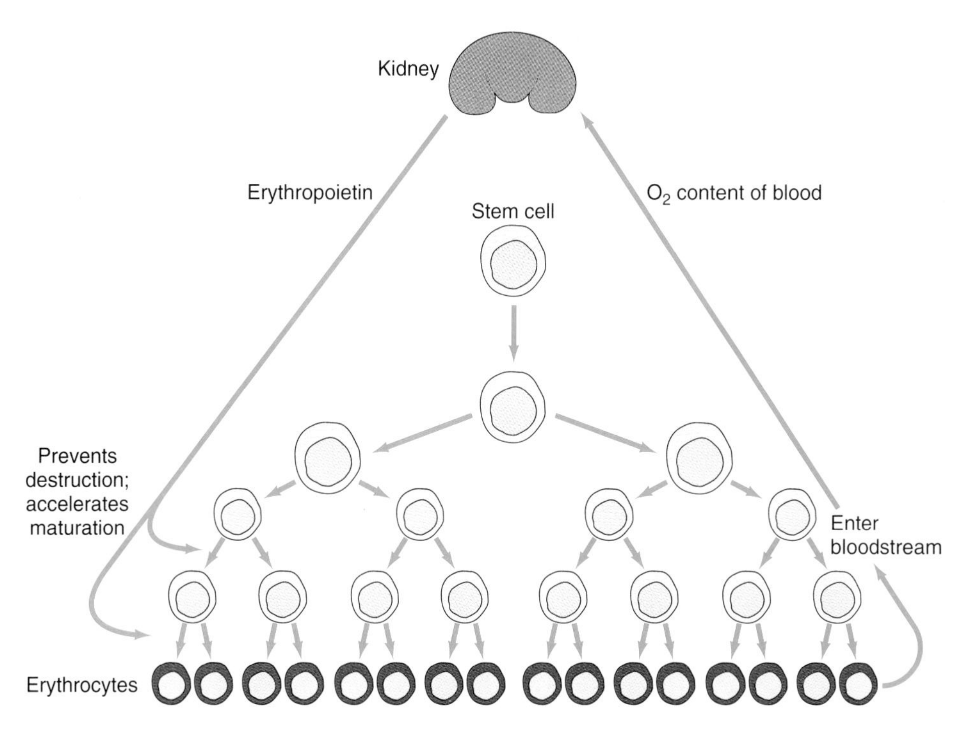

FIGURE 21.11

Erythropoietin and Red Blood Cell Formation

Erythropoietin is formed by the kidney in response to low blood oxygen levels. This growth factor prevents the destruction of erythrocytes in the bone marrow, and increases their rate of maturation.

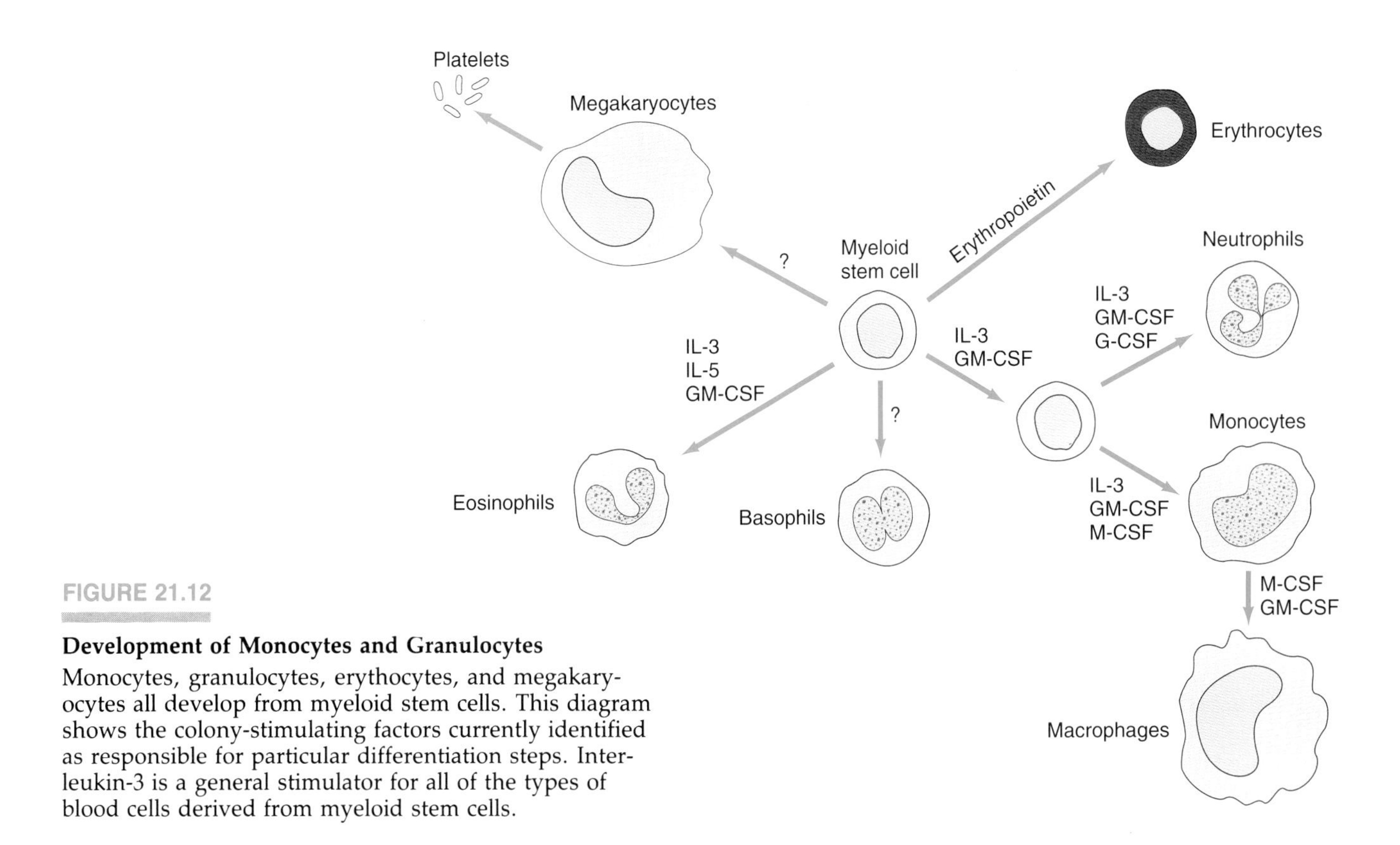

FIGURE 21.12

Development of Monocytes and Granulocytes

Monocytes, granulocytes, erythocytes, and megakaryocytes all develop from myeloid stem cells. This diagram shows the colony-stimulating factors currently identified as responsible for particular differentiation steps. Interleukin-3 is a general stimulator for all of the types of blood cells derived from myeloid stem cells.

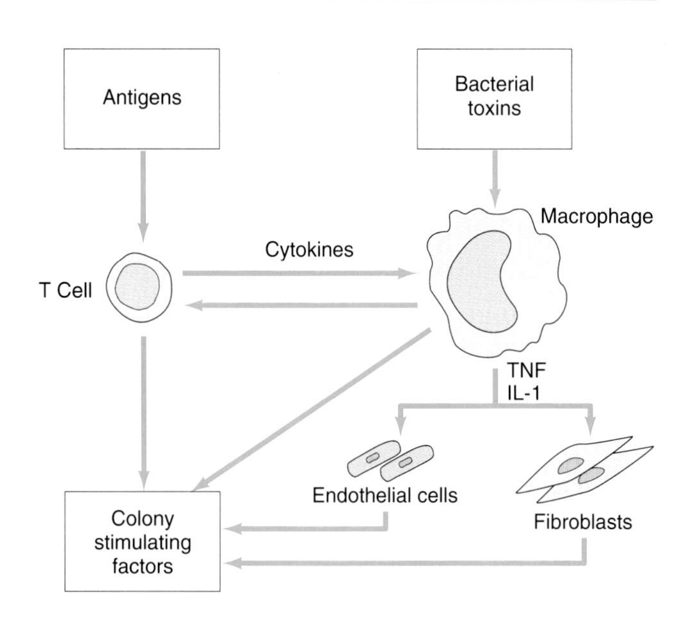

FIGURE 21.13

Production of Colony-Stimulating Factors

Antigens and bacterial toxins cause T cells and macrophages, respectively, to produce cytokines that are mutually stimulatory. In addition, both produce colony-stimulating factors, and macrophages also stimulate endothelial cells and connective tissue cells called fibroblasts to do the same.

Even though not all of the hemopoietins have been characterized, some are already being used therapeutically. Erythropoietin has the potential of increasing erythrocyte production by a factor of 10; it is being used on kidney dialysis patients, where anemia is often a side effect of dialysis. GM-CSF is being used on AIDS patients who suffer from a decrease in leukocyte production, and on certain forms of cancer and anemia. G-CSF has been shown to increase neutrophil production by a factor of 50 in monkeys, and is being used on cancer patients who have received chemotherapy (which often suppresses bone marrow function). M-CSF and IL-3 are presently awaiting clinical trials.

A QUICK REPRISE

In trying to make sense out of the details of differentiation and morphogenesis, it may be useful to remember certain points. First, the various fields of endocrinology, neuroscience, developmental biology, immunology and so forth have progressed somewhat independently of each other, and therefore have developed their own terminology. It is only now that we are beginning to appreciate that communication between cells is by chemicals—and whether we call these chemicals hormones, morphogens, growth factors, or neurotransmitters, they all function by causing a response in other cells. To be sure, morphogens, growth factors, and hormones function by activating either particular genes or genetic pathways, whereas neurotransmitters simply alter the electrical potential of plasma membranes, thereby causing only momentary changes in the receiving cell, but the principal of chemical communication between cells remains the same.

Second, although humans and other advanced vertebrates have lost the capacity for morphogenesis except during the embryonic stage of development, there is fundamentally no difference between differentiation in the embryo and fetus and differentiation in the adult. The point is that, as adults, we retain embryonic (i.e., stem) cells for various cell and tissue lines, and their periodic or continual differentiation into mature cells follows the same pathways in both adult and embryo. However, the capacity for differentiation is ultimately finite, as we shall see in the next section.

THE PROCESS OF AGING

The inevitability of the aging process has fascinated and confounded humans from the beginnings of civilization (when, presumably, people first began to live long enough to die of "old age"). During the Middle Ages, alchemists experimented with life-extending potions. Ponce de Leon, and his search for the fountain of youth, captured the imagination of the Renaissance. It was all to no avail—our life span remains little changed from the biblical "three score years and 10." (See Figures 21.14 and 21.15.) But surely now, in the closing years of the twentieth century, with our greatly improved understanding of genetics and body chemistry, science must be prepared to explain this great mystery, and even to postpone its relentless onslaught.

Don't hold your breath. We aren't quite there yet. But we're getting closer. Before we consider why we

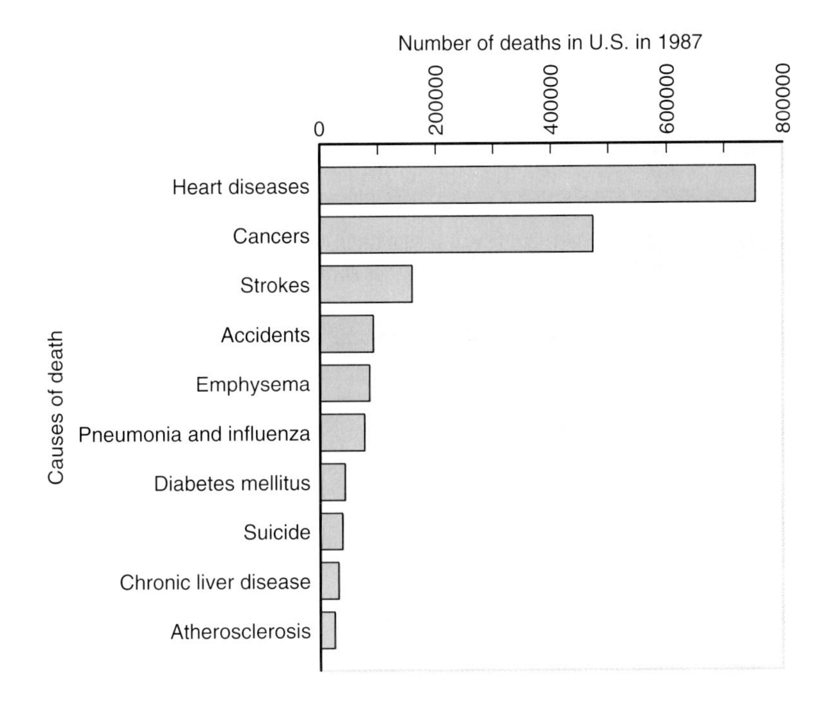

FIGURE 21.14

Causes of Death

The top 10 causes of death in the United States in 1987 are shown in this figure.

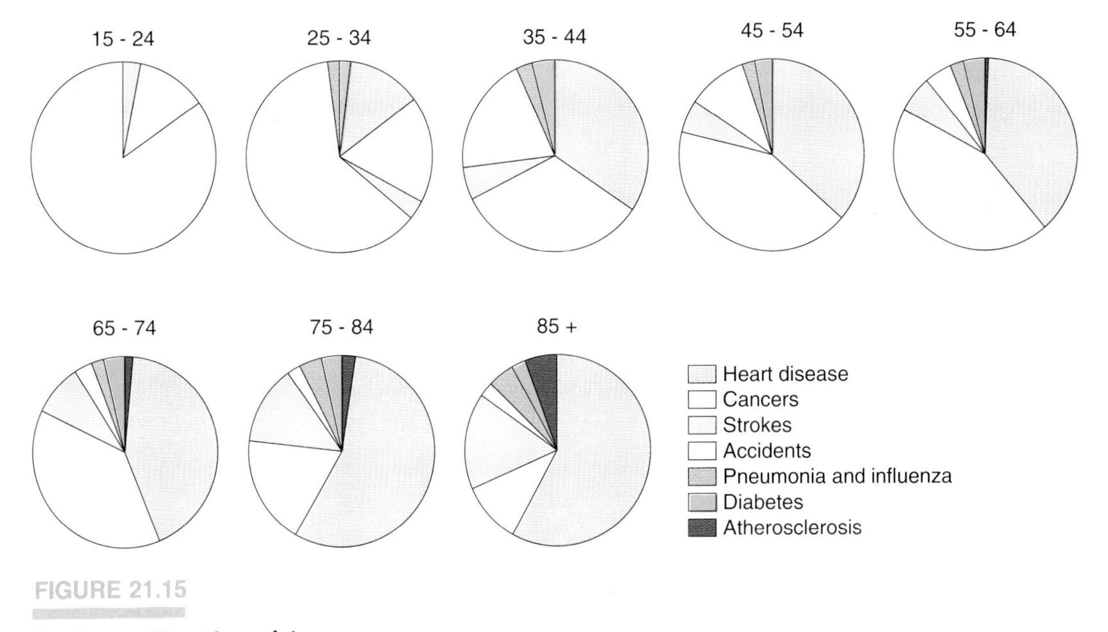

FIGURE 21.15

Death as a Function of Age

As would be expected, death rates rise sharply as a function of age. Indeed, people over the age of 85 have almost a one in five chance of dying in any given year. However, the cause of death varies greatly with age. Below the age of 35, accidents are the primary cause. Heart disease becomes increasingly more important with age, whereas cancer is concentrated primarily in the middle-aged. (Another way of looking at it is that if you don't die of cancer, then you can live long enough to die of cardiovascular disease!) Together, cancer and heart disease dominate most of the age groups.

age, however, let's briefly examine what happens in the aging process.

The Effects of Aging

We see the effects of aging both anatomically and physiologically. Structural changes are evident in the skin, for example, with the development of wrinkles. The bones decrease in density (often leading to *osteoporosis*), and the development of *osteoarthritis* is strongly correlated with age. The lens of the eye thickens, and by our mid-forties most of us need reading glasses. The blood vessels tend to harden and narrow, and *cardiovascular disease* becomes increasingly prevalent as we age. *Cancer* is primarily a disease of the middle-aged and elderly. We lose brain cells at a steady rate, and by the age of 90, our brains are fully 10 percent lighter than they were at the age of 20. The number of calcium ion channels in sympathetic neurons in the heart diminish with age, impeding signals that normally would tell the heart to work faster and harder. The walls of the alveoli of the lungs begin to collapse, and 50 percent of the area of our respiratory exchange surface is lost by the age of 70.

Physiologically, we see many changes that result from these anatomical changes (see Figures 21.16 and 21.17). In addition, hearing loss begins by the age of 50, and the sense of smell diminishes gradually as well (which explains why elderly women often use too much perfume).

With those cheery assessments of what we have to look forward to, let's examine why we age in the first place.

Why Do We Age?

The various theories of aging may be categorized under the general headings of *mechanical breakdown* theories, in which vital components simply wear out, and *planned obsolescence* theories, in which aging is, in some way, genetically programmed. Mechanical breakdown has been suggested in a number of areas: the accumulation of genetic errors, an increase in the autoimmune response (aging cells of the immune system are known to produce less interleukin-2, an essential growth factor), or the gradual collection of cellular debris and waste products, any or all of which lead to cellular death. Planned obsolescence, on the other hand, assumes the existence of specific genes that preordain the life span in ways not yet clear.

However, it now seems clear that aging, along with so many other aspects of our biology, is a product of the interaction of genes and environment. We know, for example, that each species has a characteristic life span, but that considerable individual variation in life span occurs in each species. We know that identical twins tend to have about the same life span, and that strains of *Drosophila* can be selected for longevity, facts which highlight the role of genetics in aging. We also know that mice fed diets containing about 40 percent less food than they would consume if allowed to feed to satiation live twice as long as their satiated brethren—a fact which emphasizes the environmental role. Thus current theories combine the earlier ideas of mechanical breakdown and planned obsolescence.

One line of research has focused on the products of the aging process. It has been known for many years that, with age, there is increased formation of cross-linkages between proteins, rendering the proteins useless as enzymes. High glucose levels are also correlated with the formation of such cross-linkages,

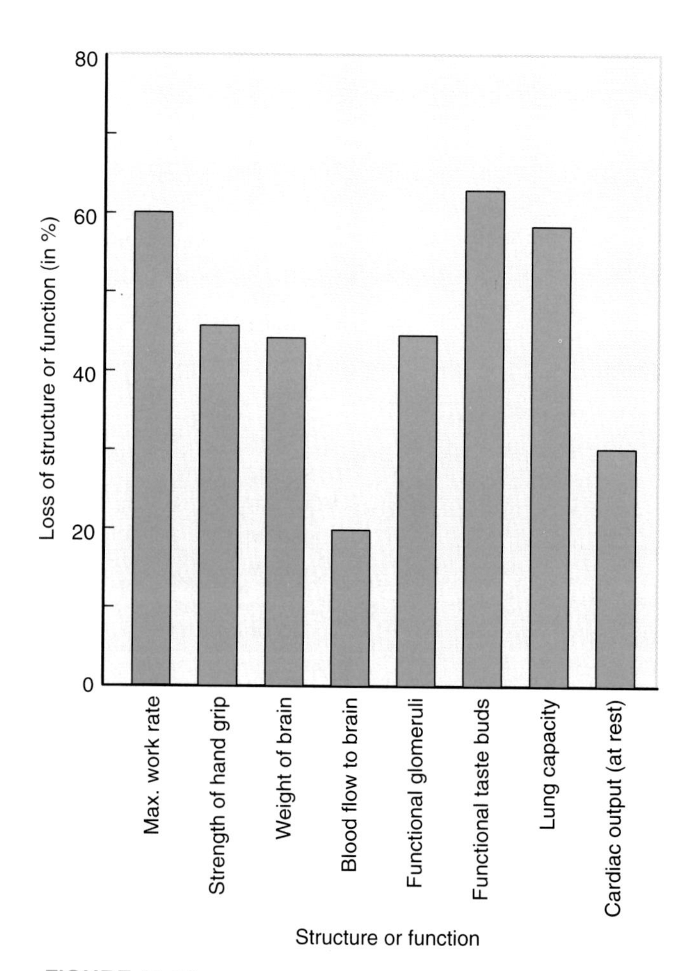

FIGURE 21.16

Functional Losses with Age I

The bars show how much tissue or function is lost in a 75-year-old man as compared to a 30-year old man.

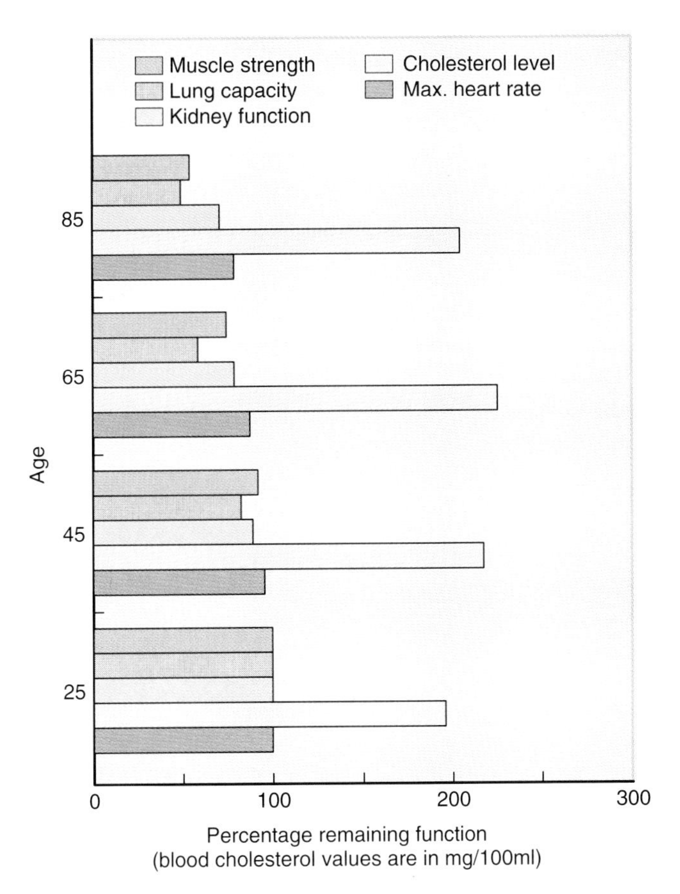

FIGURE 21.17

Functional Losses with Age II

The bars show percentage of remaining function (actual blood values in the case of cholesterol) in women as they age.

and diabetics, who have high blood glucose levels as a consequence of their illness, have an average life span that is only two-thirds of the normal life span. A medication that has been used with diabetics is now being tested on nondiabetics as well, because it appears to reduce substantially the tendency to cross-linkage formation. It remains to be seen whether this medication will slow the aging process and extend life spans.

Similarly, another correlate of aging is the increase in highly reactive oxygen-containing compounds that interfere with normal cell chemistry. Because vitamins C and E neutralize such compounds, a project is underway to determine if supplements of these two vitamins administered to the elderly will be effective in reducing their rate of aging.

A recent experimental treatment for aging is particularly intriguing. For some time it has been known that GH production diminishes after the age of 40. Preliminary studies on a group of individuals over the age of 60 suggest that administration of GH may increase the skin thickness and the muscle mass, and decrease fatty tissues. However, since prolonged use of GH may lead to hypertension, diabetes, and heart disease, it is not likely to become a standard treatment for the elderly.

Fortunately, there are simpler strategies for dealing with aging. Aging is correlated with metabolic rate—the higher the rate, the more rapid the aging process. Small birds and mammals tend to have short life spans—but hibernators live longer than nonhibernators, and flies raised at cool temperatures (18°C) live twice as long as flies raised at 30°C. Should we start sleeping in refrigerators—or all move to Fargo? An easier answer is exercise. Although metabolic rate is raised during exercise, the basal metabolic rate falls in individuals who are fit—and that may be one reason why longevity and fitness are correlated.

Some of the data on the genetic side of the issue are very intriguing. It has been known for more than 30 years that embryonic human *fibroblasts,* an important cell group in connective tissue, cannot divide successively more than about 50 times before they die. However, fibroblasts from adults can divide only about 20 times before they die. The implication is that the number of divisions a cell line can undertake is limited, and once that limit is reached, those cells (and the organism itself) die.

Studies on *endothelial cells* in culture show similar results. They, too, stop dividing after 20 to 60 divisions, during which time they accumulate interleukin-1. When the mRNA molecule that codes for IL-1 was blocked, the endothelial cells divided as many as 140 times. However, since they ultimately did stop dividing, factors other than IL-1 accumulation must be at work.

Cancer cells, on the other hand, truly are immortal. Various cancer cell lines have been maintained continuously in culture since the 1950s, and show no sign of slowing down. Cancer cells are not fully differentiated (see Chapter 17); is it the differentiation process that starts the life span clock ticking?

Recently, scientists created hybrid cells by fusing cancer cells with human fibroblasts. The hybrids followed the fibroblast pattern, and began to die after a few generations. By careful marking experiments, the scientists were able to localize the gene or genes—they are on chromosome 1—responsible for converting immortal cells to demonstrably mortal ones.

What are these "gerontogenes"? There are a variety of rare genetic conditions in humans in which individuals prematurely age. In **Werner's syndrome,** the aging process is noticeable before age 30, and death usually occurs before age 40. In another, extremely rare, condition known as **progeria,** the aging process is even more accelerated. Children become wrinkled and bald by the age of two, and stop growing when they reach a weight of about 12 kg and a height of about one meter. They generally die of cardiovascular failure before they reach their teens.

Investigators have concluded that a relatively small number of genes—probably no more than 50—are unusually active in these individuals, and assume that these highly active genes are the elusive "gerontogenes." However, none has yet been specifically identified and characterized.

As we move into the twenty-first century, the elderly will constitute an increasingly larger percentage of our population. As a consequence, it seems certain that research into the aging process will accelerate. Perhaps by the time you are elderly, we will be defining that term with an age range much different from the one used at present.

The aging process is still very incompletely understood and remains the "final frontier" in the minds of many biologists. (Actually, the entire developmental cycle is a "final frontier" at this point.) Documenting the effects of aging is easy; explaining why aging happens in the first place is far more difficult. It now seems apparent that both environmental and genetic factors are at work. Genetics appears to set the upper boundary; environmental factors determine how close to that upper boundary individuals can expect to go. However, the identification of the first "gerontogene" has yet to be accomplished.

Summary

How it is that a single cell can develop into a functioning adult organism has long been a mystery and there is still a great deal that we do not know about development.

Differentiation and morphogenesis are distinct but related phenomena that form a part of development. We now know that physical proximity to particular tissues is an important part of the developmental process, and we are beginning to characterize both the chemical messengers and the regulatory genes that are critical to development. We also know that differentiation begins very early, with the formation of three embryonic germ layers, each of which is destined to develop into different adult structures.

In the human embryo (and the embryos of many other species as well), not all of the blastocyst tissue contributes to the ultimate adult structures. Some becomes the extraembryonic membranes, tissues that help sustain the embryo during its development. The amnion encloses the embryo, the chorion provides a part of the placenta, the yolk sac produces blood cells, and the allantois becomes the umbilical cord.

Most of the basic development processes are complete within the first two months of development, at a time when the embryo is only 3 cm long. During the remaining seven months—the fetal stage—organ systems are refined and become mature, and the fetus grows dramatically in size.

Profound and immediate changes must occur at birth in the circulatory and respiratory systems of the fetus to permit the transition from an aqueous to a gaseous environment. These changes do not always occur, and are one source (along with many others) of developmental defects.

Key Terms

development
differentiation
morphogenesis
cleavage
morphogen
induction
CAMs
FGF
NGF
GnRH
homeobox
germ layer
ectoderm
endoderm
mesoderm
extraembryonic membrane
amnion
yolk sac
allantois
chorion
chorionic villi
umbilical cord
placenta
neurulation
neural tube
notochord
fetus
quickening
respiratory distress syndrome
umbilical vein
ductus venosus
foramen ovale
ductus arteriosus
umbilical artery
allometry
stem cell
CSF
hemopoietin
lymphoid stem cell
myeloid stem cell
erythropoietin
interleukin-3
GM-CSF
G-CSF
M-CSF
Werner's syndrome
progeria

Box Key Terms

totipotency
clone

Questions

1. What is induction? Give an example. What is a morphogen? Give an example.
2. What is a homeobox? What is its significance in development?
3. What are the three germ layers, and what is their ultimate destiny?
4. What is the presumed evolutionary origin of the fetal contribution to the placenta?
5. Describe the changes that must occur in the circulatory system of the fetus at birth.
6. Describe the differentiation of blood cells from a single type of stem cell.
7. Document some of the changes that occur as we age.
8. Briefly discuss current theories of aging. What are "gerontogenes"?

PART

VI

Populations

We have considered the biology of cells and of organisms, and have discussed the ways in which cells and organisms develop and reproduce. Implicit in the notion of sexual reproduction is the necessity of interrelationships among the members of one species—which means that sexual reproduction thus forms a convenient bridge from organismal biology to population biology.

We shall consider two aspects of population biology: First, how do populations change in response to environmental change? Given that all organisms ultimately share a common ancestry, how do biologists account for the presence of millions of species of organisms, each of which is reproductively isolated from all of the others? And what of humans—what is our biological heritage as a species? In short, we shall be considering evolution in some detail.

Second, how do species interact with their environment and with each other? What constraints does the environment place on species, especially in terms of energy requirements and population sizes? Finally, what lessons does biology teach us regarding our own survival as a species? In short, we shall be considering some ecological perspectives.

CHAPTER

22

Evolution

Changes in the Gene Pool

Charles Darwin first published his theory of evolution by natural selection more than a century ago, yet it remains irrevocably controversial in the minds of many nonscientists. What is it about Darwin's theory that promotes such strong feelings? What is the evidence that supports his theory?

The Irish elk was a huge animal with a rack of antlers that would have turned a moose green with envy. It became extinct around 1000 A.D.—the reason, according to some authorities, being that its antlers simply grew too large. Is such an explanation compatible with Darwin's theory of evolution by natural selection?

What is the difference between evolution and speciation? How do paleontologists *know* that two fossils represent the same or different species? If lions and tigers can be interbred in a zoo, does that mean that they are really the same species?

These and other questions will be considered in the following chapter.

THE NATURE OF THE PROBLEM

In Chapter 2, we discussed the concept of evolution and how it was that Charles Darwin's identification of natural selection as the mechanism of evolution changed biology in a fundamental way. With evolution as its main underlying principle, biology changed from a descriptive to a theoretical science. There is more than a little irony in the fact that Darwin himself was primarily a descriptive biologist—the last of the great natural historians.

However, Darwin hardly provided the last word on evolution. By identifying the role of natural selection in the evolutionary process, he fundamentally ended the debate on whether or not evolution occurs. (What remaining controversy there is on this issue comes not from science, but from religion.) Nevertheless, Darwin did not provide all the answers. He was wrong on the source of variation in the population because he was unaware of Mendel's work on genetics—and, as we have seen, it was not until 1940 that evolution and genetics were united. Darwin did not provide an adequate explanation of how new species arise (in part because the notion of what constituted a "species" has changed over the years). He did not account for other evolutionary forces besides natural selection, and his emphasis on gradualism versus catastrophism was, as we shall see, overstated.

How *does* variation in a population arise and spread? What evidence supports the notion that evolution does indeed occur? How *do* new species arise? A more complete understanding of evolution requires answers to these questions.

NEO-DARWINIAN EVOLUTION

Darwin's theory of evolution by natural selection caught the fancy—and the flak—of the nineteenth century, but, as we have seen, it really came into being only in the twentieth century, with the development of genetics as a science. Indeed, evolution is now even defined in terms of genetics:

Evolution is a change in gene or allele frequency in a population over time.

Explicit in that definition is a point often overlooked in a casual perusal of evolutionary theory. Evolution is something that happens to populations, not to individuals. Individuals obviously cannot change their allele frequency; however, populations may do so with the passage of generations. Conversely, as we shall discuss later in this chapter, individuals, not populations, are the units of natural selection.

The Source and Maintenance of Genetic Variability

As we noted in Chapter 2, only one of the three mechanisms Darwin identified as sources of variation in a population is now accepted as valid. That mechanism is mutation. However, we now appreciate that there are a number of ways in which genetic variation can be maintained in a population, and these interact with mutations in a variety of ways. A more complete picture therefore includes the following considerations.

Recurrent Mutation. Mutations occur spontaneously at a very low, but predictable, rate. Thus, even if an individual with a mutation has been selected against, that same mutation will arise again at some point in the future, and perhaps will be favored if there has been a change in the environment in the meantime. The important point is that spontaneous mutations arise because of inherent properties of DNA replication, and not because they are "needed."

Dominant-Recessive Relationships. Most mutations are recessive and are therefore not immediately subjected to the effects of natural selection, because it is the phenotype and not the genotype upon which natural selection operates (see Figure 22.1). A heterozygote with a recessive allele that would be lethal in the homozygous condition is not selected against, because the expression of the recessive allele is blocked by the normal dominant allele. Thus the recessive is passed on to half the offspring, and may become widely spread in the population before finally occurring in a homozygous (and lethal) condition. The case of Queen Victoria and hemophilia, given in Chapter 19, exemplifies this point.

Independent Assortment. Even though mutations are the sole source of genetic variation, the regrouping and recombining of chromosomes that takes place as a result of sexual reproduction and crossing over (see Chapter 17) is of much greater importance in the formation of unique genotypes. The particular genetic setting in which a mutated gene finds itself is of critical importance in determining whether it is expressed in the phenotype and thereby is subject to natural selection.

Changed Environmental Circumstances. Populations never achieve homozygous perfection in part because "perfection" is defined in terms of an ideal match with the environment, and, however subtly, the environment is constantly changing. Thus natural selection is a continuous, never-ending process.

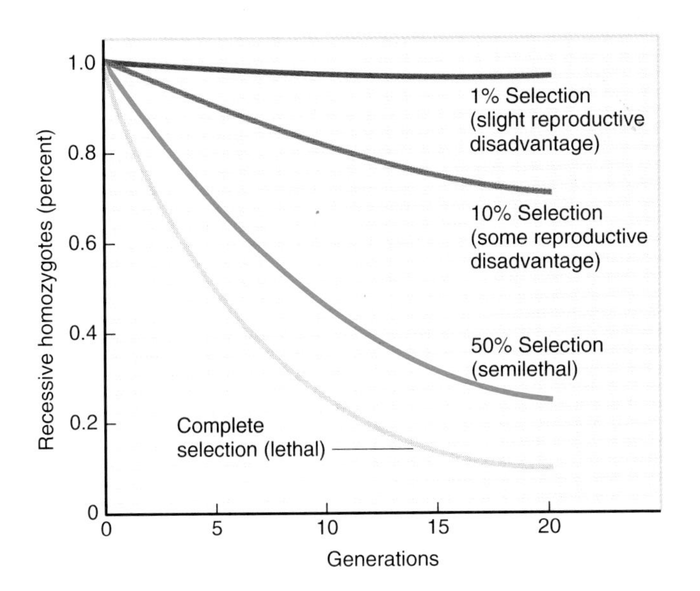

FIGURE 22.1

Dominant-Recessive Relationships

Dominant-recessive relationships preserve genetic variation in the population. Note that recessive homozygotes are never completely eliminated from the population, even when they are lethal, because in heterozygotes the recessive allele is shielded by the dominant allele.

Balanced Polymorphism. A very important method of maintaining variation in the population is **balanced polymorphism,** a phenomenon whereby multiple alleles are maintained in a population in a balanced fashion. Perhaps the classic example is the ABO series of blood types (discussed in some detail in Chapter 9). The frequency of each allele is remarkably stable within most populations, although these frequencies may differ greatly among populations.

Unfortunately, in many cases the mechanism whereby natural selection maintains these frequencies is unknown. It is known, however, for sickle-cell anemia (see Figure 22.2). This blood condition is found widely in central Africa, as well as in American blacks, most of whose ancestors came from west-central Africa. The sickle-cell allele is co-dominant with the allele for normal hemoglobin. Thus heterozygotes show some sickling of the blood cells, but not to a degree sufficient to cause all of the symptoms that characterize the disease.

How could such a condition ever become established, given the relative seriousness of the disease in the homozygous state? The answer is that the tropical regions of Africa are heavily infested with malaria, a parasite of red blood cells. Some forms of malaria are very serious, and are frequently fatal in the absence of good medical care.

Blood cells of heterozygotes change into a sickle shape when invaded by malarial parasites, and these aberrant cells are then destroyed by leukocytes, along with the parasites they contained. The result is that heterozygotes develop neither malaria nor sickle-cell anemia, whereas homozygotes develop one or the other.

Heterozygote advantage is at the heart of many polymorphisms, and the mating of heterozygotes guarantees the continued production of both heterozygotes and homozygotes, perpetuating the polymorphism. Balanced polymorphism is an exceedingly common phenomenon, and one that is gaining increased respect by geneticists as a method of maintaining genetic variation in the population.

Factors Other Than Natural Selection

A major contribution to neo-Darwinian thinking has been the recognition that natural selection, which we shall discuss in some detail shortly, is not the only factor that leads to changes in allele and gene frequencies. Heresy? Hardly—the other factor is *chance.* Chance enters the picture as **genetic drift,** the situation that occurs when, purely by chance, individuals are added to or lost from a small population and the total amount of genetic diversity in the population is significantly reduced. Let's consider some examples of genetic drift—and think about how evolution may be the result of caprice.

Population Bottlenecks. At various times, a local population of a given species may be all but exterminated—and far too often, unfortunately, human intervention has been responsible. The near-extinction of the sea otter and the elephant seal, the bison, the whooping crane, and so forth has led, in a very random way, to a marked reduction in genetic diversity in these species. Although these species have recovered in absolute numbers, there has not been any corresponding increase in genetic diversity. In the case of the cheetah, for example, which survives in a series of widely dispersed populations in Africa, there is so much genetic uniformity that there is some question about the ability of the species to survive

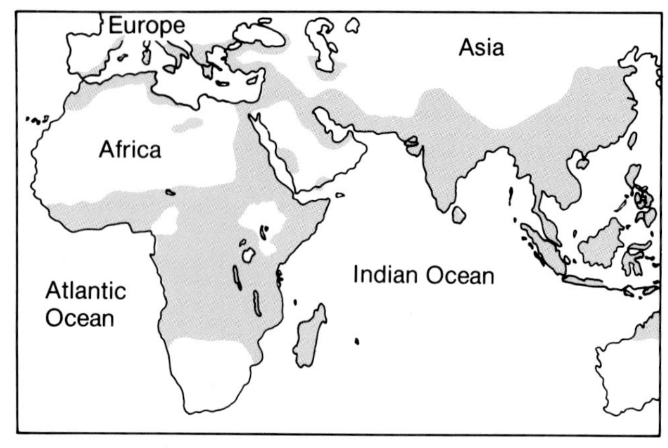

(a) Falciparum malaria

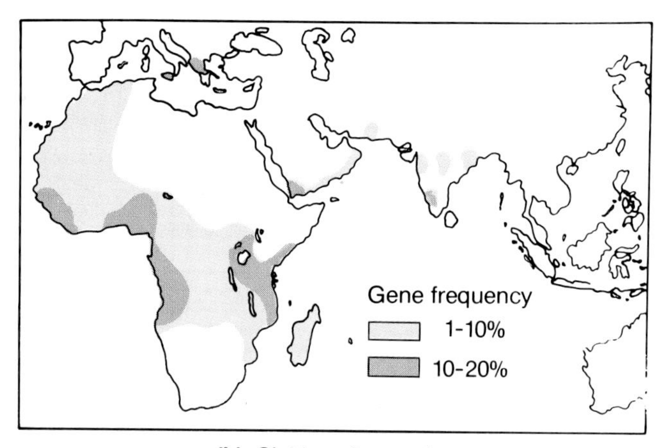

(b) Sickle cell anemia

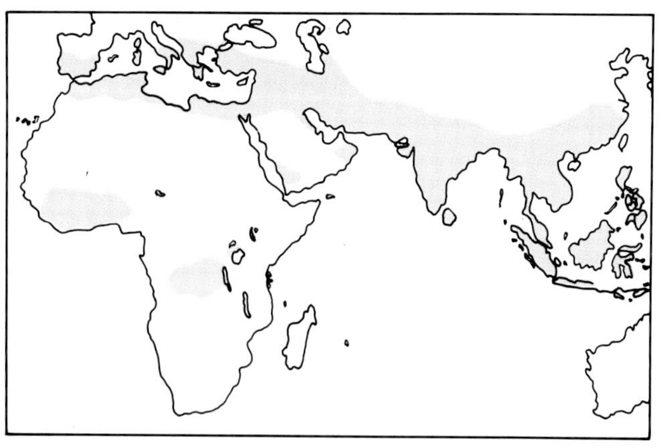

(c) Thalassemia

FIGURE 22.2

Balanced Polymorphism

The alleles for sickle-cell anemia and thalassemia are maintained in the population in those areas of the world where a particularly virulent form of malaria (falciparum malaria) is common. Individuals who are homozygous for normal hemoglobin are likely to contract malaria, whereas individuals who are homozygous for sickle-cell or thalassemia are likely to suffer the effects of those two anemia diseases. Individuals who are heterozygous are protected from both conditions, and therefore heterozygotes tend to survive while homozygotes tend to die. Heterozygote selection maintains the polymorphism.

any major disease epidemic—any pathogen that would threaten one cheetah would threaten them all.

Gene Flow. Another example of genetic drift occurs with the immigration or emigration of individuals from small populations. Many species that live in small groups drive away juvenile males as they mature. Sometimes, these males will successfully invade other groups. This mechanism, called **gene flow,** tends to increase genetic diversity for the species as a whole, but the local populations may be genetically diverse or restricted, depending on the amount of emigration versus immigration.

Founder Effect. Finally, the colonization of new habitats is often achieved by only a handful of individuals—sometimes, just one pregnant female—representing what is known as the **founder effect.** Obviously, the full range of genetic diversity in the species will not be represented by these few founders, and, as their founding population increases in number, their genetic and allelic ratios may well be very different from the original population. Island populations of many species, for instance, are often markedly different from the mainland populations.

Natural Selection and Adaptation

Natural selection, however, remains the dominant force in shaping the evolution of populations. The net effect of the phenomena that initiate and maintain genetic diversity is the production of a population of genetically very diverse, and even unique, individuals. Some of these individuals will also differ phenotypically and will therefore be acted upon by natural selection. However, our notion of natural selection has changed somewhat since Darwin's time. We now say that

Natural selection is differential reproductive success.

This means that individuals are favored by natural selection if they produce many fertile offspring and disfavored if they produce few or none. Individuals producing many offspring make larger contributions to the **gene pool** (the collective genetic information of a population) of the next generation than do individuals producing few offspring. Therefore the evolutionary direction of the population will be established by individuals favored by natural selection. Notice that we have now defined both "natural selection" and "evolution" in terms of genetics.

Darwin had problems in defining natural selection, once again because of his ignorance of genetics. He referred to the "struggle for existence" and in later editions of his work used British philosopher Herbert Spencer's phrase, "survival of the fittest." The problem with these phrases is that they focus on the survival of individuals, not of offspring. To take an extreme example, a mutation that increased the size and strength of a bull elk might allow him to win control of a harem of females, but if this same mutation caused him to be sterile, he would leave behind no offspring. By winning jousts with other bull elk, our hypothetical macho elk might well be deemed fit by Darwin, but by failing to leave behind fertile offspring he would be held unfit by today's evolutionary biologists.

In summary, natural selection is generally a matter of statistical probabilities, not a series of bloody battles. The individual who makes the greatest numerical contribution to the next generation of the population (by producing the largest number of fertile offspring), is, by definition, selected for, regardless of how scruffy and bedraggled he or she might appear to an outside observer.

The term **adaptation** refers to anything that increases the likelihood an organism will survive and multiply. Thus once a particular mutation is expressed in the phenotype, it will be deemed adaptive or nonadaptive, depending on whether it increases, or lessens, reproductive success. Natural selection generally operates to increase the overall adaptedness of a population to its environment. (See Box 23.3 in Chapter 23 for an exception in humans.)

Adaptation is also used in reference to species and populations. It is expected, for example, that a species or a population will be well adapted to its environment, assuming that the environment is relatively stable. Thus the likelihood of a new mutation's being adaptive (that is, being favored by natural selection and spreading through the population) is generally small.

The converse of that statement—only those mutations that are adaptive (or, at worst, are neutral) will spread through a population—is also generally true, although it has often been overlooked. For many years, it was widely accepted that the saber-toothed tiger became extinct because its teeth grew so large that the animal had trouble eating (see Figure 22.3). This erroneous belief should have been laid to rest long ago, but it continues to pop up occasionally in introductory texts and in the popular press. Such a situation would clearly be impossible, for any increase in tooth length that operated deleteriously would certainly be selected against, because starving animals would hardly be expected to reproduce in large numbers.

FIGURE 22.3

Adaptation
Contrary to popular belief, the saber-toothed tiger (a) and the Irish elk (b) did *not* become extinct because the teeth and antlers, respectively, grew too large.

In reality, the saber-toothed tiger was capable of opening its jaws to an angle of almost 120°, and had no trouble feeding. It evidently preyed on several species of slow-moving, thick-skinned, rhinoceros-like animals, which it killed by slashing through the thick skin of the prey with its fangs, riding about on the back of its prey until the animal bled to death. Following changes in the climate, most of these thick-skinned species became extinct, and the saber-toothed tiger followed suit, because it was less fleet than both the new herbivore species and the new carnivores. If the saber-toothed tigers starved to death, it was because of an absence of prey, not because their teeth were too large.

A similar, and equally preposterous, argument has been made about the Irish elk, which supposedly became extinct because its rack of antlers grew too large. It is much more likely that the elk became extinct because of the climatic changes that resulted from a succession of glacial advances and retreats (or from hunting pressures by early humans). The notion that the elk died out because its antlers grew too large conjures up visions of an animal suddenly becoming unable to lift its head off the ground, or periodically wedging itself between two trees, and this was certainly not the case.

Natural Selection in Action

In most circumstances, natural selection tends to preserve the status quo, since the status quo is normally well matched to the existing environment. Selection that eliminates extremes, and preserves the mean, is called **stabilizing selection** (see Figure 22.4), and it is particularly well evidenced in polygenic characters. For example, babies generally weigh between 3 kg and 3.5 kg at birth. Low weight babies tend to be at risk, because a low birth weight often means premature birth and a failure of the organ systems to mature properly. High weight babies can pose problems in delivery. Modern medicine somewhat offsets the effects of natural selection in this instance, but that is only a very recent development.

On the other hand, the preservation of one end of the continuum and the consistent loss of the other end will result in **directional selection.** The fossil record of horses, for example, indicates a rather consistent pattern of increasing size. In the human line, brain size has also consistently risen (see Chapter 23).

Finally, elimination of the norm, and preservation of the two extremes, occurs in **disruptive selection.** There are many examples—selection for small body size in the deer that inhabit islands (where food may be scarce), whereas large body size is selected for on the mainland; the separation of chimpanzees into large and pygmy species; selection for both large and small beaks in many bird groups; and so forth.

Sexual selection is in some ways the most intriguing form of selection. In sexual selection what is commonly selected for is some characteristic that increases the individual's sexual attractiveness—even though, in other situations, the trait may well be deleterious. The extreme tail length of the males of many tropical birds (e.g., the peacock) represent classic examples of traits that are advantageous in courtship but hazardous when avoiding predators (see Figure 22.5). They persist because the benefits outweigh the risks—especially in species where the male may impregnate several females.

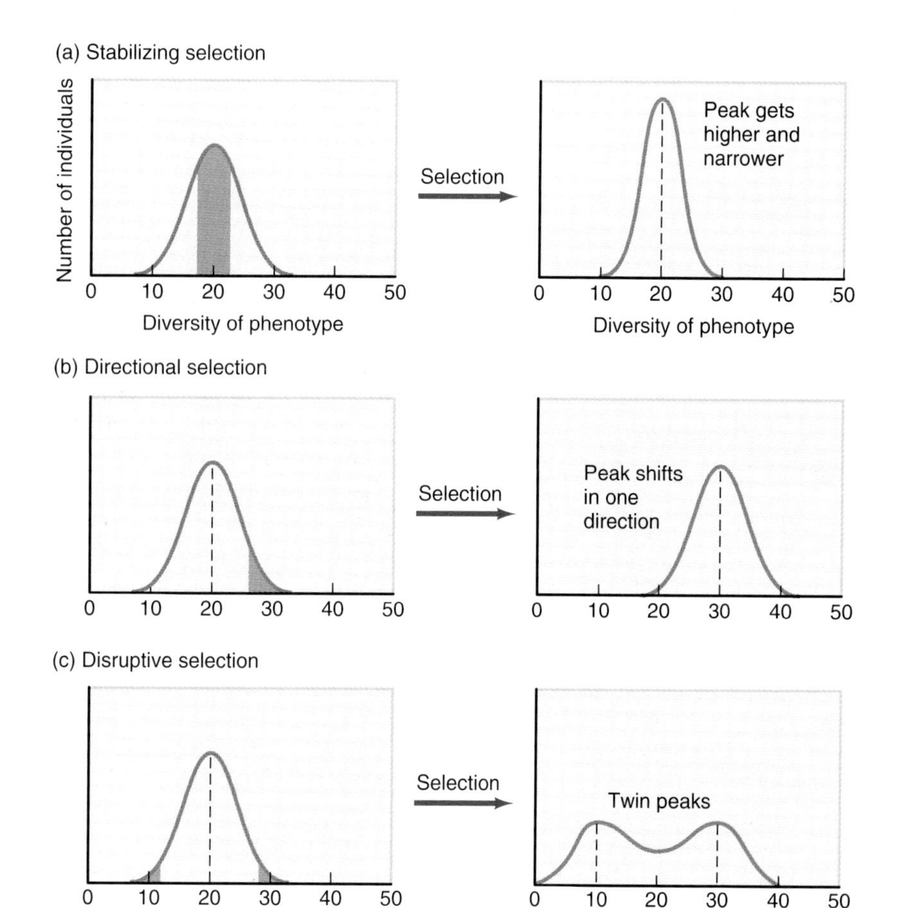

FIGURE 22.4

Type of Natural Selection

In stabilizing selection (a), individuals near the center of the range for a particular character tend to survive, and therefore the percentage of individuals near the middle of the range increases over time. In directional selection (b), individuals near one end of the range for a particular character tend to survive, and therefore the range itself tends to move over time. In disruptive selection (c), individuals near both ends of the range for a particular character tend to survive, and therefore two ranges for the character begin to emerge over time.

FIGURE 22.5

Sexual Selection
What is adaptive about the long tail of the peacock? Since it surely must be a detriment in escaping predators, why did it evolve? The answer is that it serves to attract females, and the benefits (in terms of numbers of offspring) must outweigh the risks (being eaten before the bird has a chance to reproduce).

EVIDENCE FOR EVOLUTION

The bulk of Darwin's *Origin* is devoted to a massive compilation of data that indicates the occurrence of evolution in the past and, by extrapolation, its continuing occurrence today. Since Darwin's time, even more evidence has been compiled. This evidence can be placed in six reasonably distinct categories:

1. Fossil evidence
2. Phylogenetic evidence
3. Evidence from geographic distribution
4. Evidence from comparative anatomy and embryology
5. Comparative biochemical evidence
6. Experimentally obtained evidence

Fossil Evidence

Fossils (L. *fossilis,* "dug out") may be formed in a variety of ways, but they most typically occur when an organism dies in a body of water in which the sedimentation of silt and other particulate matter in the water is occurring at a relatively rapid rate. The soft parts of the organism generally decay rather quickly and are seldom fossilized, but the skeleton may be preserved either in the form of petrified bone or as a mold, the latter condition resulting when the skeleton is initially surrounded by sediment that then hardens. Subsequently, the bony tissue is leached out by acidic water, leaving a mold of the organism in the hardened sediment (see Figure 22.6).

Although most fossils are of marine organisms that possess hard shells or skeletons, the fossil record, in total, is amazingly (but not universally) complete. For example, the fossil history of the vertebrates at the time of their presumed origin from invertebrates is not very good, in large measure because the ancestral forms seem to have been small, soft-bodied organisms, with little potential for fossilization. Most of the gaps between the major vertebrate groups, however, are well bridged by fossils. The earliest amphibians, for instance, are rather well represented in the fossil record, and they appear very much like the fossil lungfish of the same period. Similarly, the link between the amphibians and the reptiles is replete with fossil evidence. In contrast, the early history of the birds is not well represented in the fossil record,

FIGURE 22.6

Fossils as Evidence for Evolution
This fossil is of an organism called a trilobite that lived about 560 million years ago. A mold of its body has been left in ocean sediment that hardened into rock.

in large measure because birds have fragile skeletons, they seldom die in sedimenting areas, and their evolution as a group appears to have been very rapid.

The gap between the reptiles and the mammals is particularly well bridged. Mammals are distinguished from reptiles in their possession of hair, mammary glands, endothermy, and a variety of features of the circulatory and nervous systems—but none of these characteristics fossilize. Therefore the determination of whether a fossil is a reptile or a mammal must generally be based on an extrapolation from skeletal features. Modern mammals possess only one bone on each side of the lower jaw, whereas reptiles have as many as ten bones. The fossil record indicates a continuum between animals with ten bones and animals with only one. At what point along that continuum did the development of mammary glands, or any of the other mammalian characteristics, occur? We have no way of knowing. The point is simply that the fossil lineage between reptiles and mammals is so complete that many of the intermediate fossils must be classified arbitrarily, as they possess both reptilian and mammalian characteristics.

Not only are there numerous instances of rather complete lineages of fossils that link two groups, but the sequence of these fossils in sedimentary rock is very precise, rather than scattered randomly throughout the rock strata. In addition, the age of fossils generally matches well with estimated rates of evolutionary change.

The age of fossils is determined in two different ways. First, the rate of sediment deposition can be determined with reasonable accuracy, and the thickness of the overlying strata or rocks thus provides a good indication of the fossil's antiquity. Second, as we saw in Chapter 3, many elements occur in more than one atomic structure, some of which are unstable. These radioactive isotopes, as they are called, decay (give off subatomic particles in the form of radiation) at an extraordinarily regular rate. (In fact, so regular is the rate of decay that world time standards are maintained by "atomic" clocks.) Thus the amount of these radioactive isotopes still left in the fossil or in the surrounding rock can furnish a highly accurate time measure of the fossil's age.

A third aspect of the fossil record that suggests evolution is the fact that fossils are often remnants of species, or even whole lineages, that are presently extinct. However, in many instances, they are similar to living species, and that similarity is particularly evident in younger fossils.

Both facts speak eloquently and independently about evolution as a fact. Extinction without the evolution of new species would be ominous indeed. It would suggest a world that is doomed, a world once replete with color and diversity that is inevitably to become monochromatic and sterile. Yet that scenario is belied by the fossil record. With the extinction of the dinosaurs came the explosive proliferation of mammal and bird species. The incredible diversity of the flowering plants is, in an evolutionary sense, a relatively recent phenomenon—they were simply not in the fossil record until about 100 million years ago. Again, the appearance of particular species and lineages in the fossil record where they had previously been unknown strongly supports the concept of evolution.

Is it a mere coincidence that fossils of a few thousands of years ago are so often similar to, but not identical with, modern species? The Irish elk is similar to, but not identical with, the modern elk and moose. The saber-toothed tiger is similar to, but not identical with, the modern lions and cougars. As we shall see in the next chapter, our own ancestors were similar to us—but they certainly were not identical. Again, increasing similarity to modern species as one moves up in the fossil record strongly suggests evolution.

In summary, fossil evidence, although by no means complete, supports the concept of evolution in three ways:

1. Many major groups of plants and animals are tightly linked by fossils.

2. The ages of fossils, as determined by alternative methods, are usually highly consistent, and these ages are sufficiently great to have permitted the occurrence of the very slow process of evolution.

3. Many fossils are of species not present on earth today but are similar to modern species.

Phylogenetic Evidence

Since at least the time of Aristotle, there has been an interest in grouping species that are similar. The formulation of **phylogenies** (Gr. *phyle*, "tribe"), or "family trees," of related species provides evidence for evolution simply because they can be constructed at all. That is, there should be a totally random pattern of similarities and dissimilarities among different species if evolution were a myth, and it would be impossible to construct phylogenies because relationships between species would not exist. (A family tree for humans and other primates is shown in Figure 23.8.)

The fact of the matter, however, is that relationships between species do exist, and in most plant and animal groups phylogenies have been constructed that are not only in accord with the degree of similarity and dissimilarity among the various species, but that also strongly suggest the actual evolutionary patterns of relationships that interconnect these species.

Evidence from Geographic Distribution

The pattern of geographic distribution of species is also far from random, a fact strongly emphasized by Darwin. Indeed, in his initial examination of the fauna of South America, long before he had formulated his theory of evolution by natural selection, Darwin noted three facts that were later to provide the basis of much of his evidence for the existence of evolution:

1. South American fossils were structurally more similar to living South American animals than to animals of other parts of the globe.

2. There was greater similarity between South American animals in different climatic regions than between South American and African animals of the same climatic region.

3. Island fauna most closely resembled the fauna of the nearest portion of the mainland.

Geologists now realize that the earth's crust is composed of great tectonic plates, which are constantly moving relative to one another (see Figure 22.7). Hundreds of millions of years ago, all of the continents were merged in one giant land mass, but since that time they have gradually drifted apart. Africa and South America, for example, were once adjacent. They still share certain climatic conditions but have been separated so long that their plants and animals are only very distantly related. Australia, long separated from the other major land masses, also has a highly characteristic flora and fauna.

The only explanation for all these distributional patterns is the occurrence of evolution.

Comparative Anatomical and Embryological Evidence

If there were no such thing as evolution, and species remained essentially unchanged, the embryological development of a given species would be expected to show only a random similarity to the development of any other species. However, not only do species show definite similarities in their embryonic development but, as a general rule, the degree of this similarity is directly correlated with their overall similarity as adult organisms (see Figure 22.8). That is, just as adult apes are far more similar to adult humans than to adult giraffes, so, too, are embryonic apes far more similar to embryonic humans than to embryonic giraffes.

Perhaps even more dramatic is the existence of developmental similarities among species that are very different as adults. Early in the development of all vertebrates, including humans, a series of pouches forms in the throat region. In fish, these break through to the surface of the body and become gill slits. In humans, and other land vertebrates, virtually all regress, although the first remains as the eustachian tube, the canal that connects the middle ear with the throat. Why are they formed at all, only to regress before birth? Similarly, why does the human embryo have a tail? (It normally regresses before birth although occasionally babies are born with small tails that must be surgically removed.) Why does the human fetus develop a coat of hair all over the body, and then shed it all before birth?

All of these developmental features, as well as many others, make sense in light of evolution and phylogenetic relationships, but they are totally inexplicable to one who believes that species had independent beginnings and are unchanging.

An even stronger case can be made for evolution based on adult anatomy. The whole discipline of comparative anatomy is predicated on the recognition and description of anatomical similarities and differences in every organ system. Indeed, as we saw in Chapter 2, the idea of evolution, although coupled to an incorrectly designated cause, developed among anatomists of the eighteenth century who were struck by the magnitude of the similarities in the organ systems of a wide range of animals.

Consider just a single example. Primitive fish have six pairs of arteries anterior to the heart, each pair passing through one of the six pairs of gills. The number of anterior arteries is as few as four pairs in some advanced fish. Most amphibians have just three pairs. Of these, the first carries blood to the head, the second (the aorta) carries blood to the body proper, and the third carries blood to the lungs. Reptiles have a similar arrangement, except that the two aortas are of different sizes, the left being smaller than the right. In birds, which are presumed to be closely related to

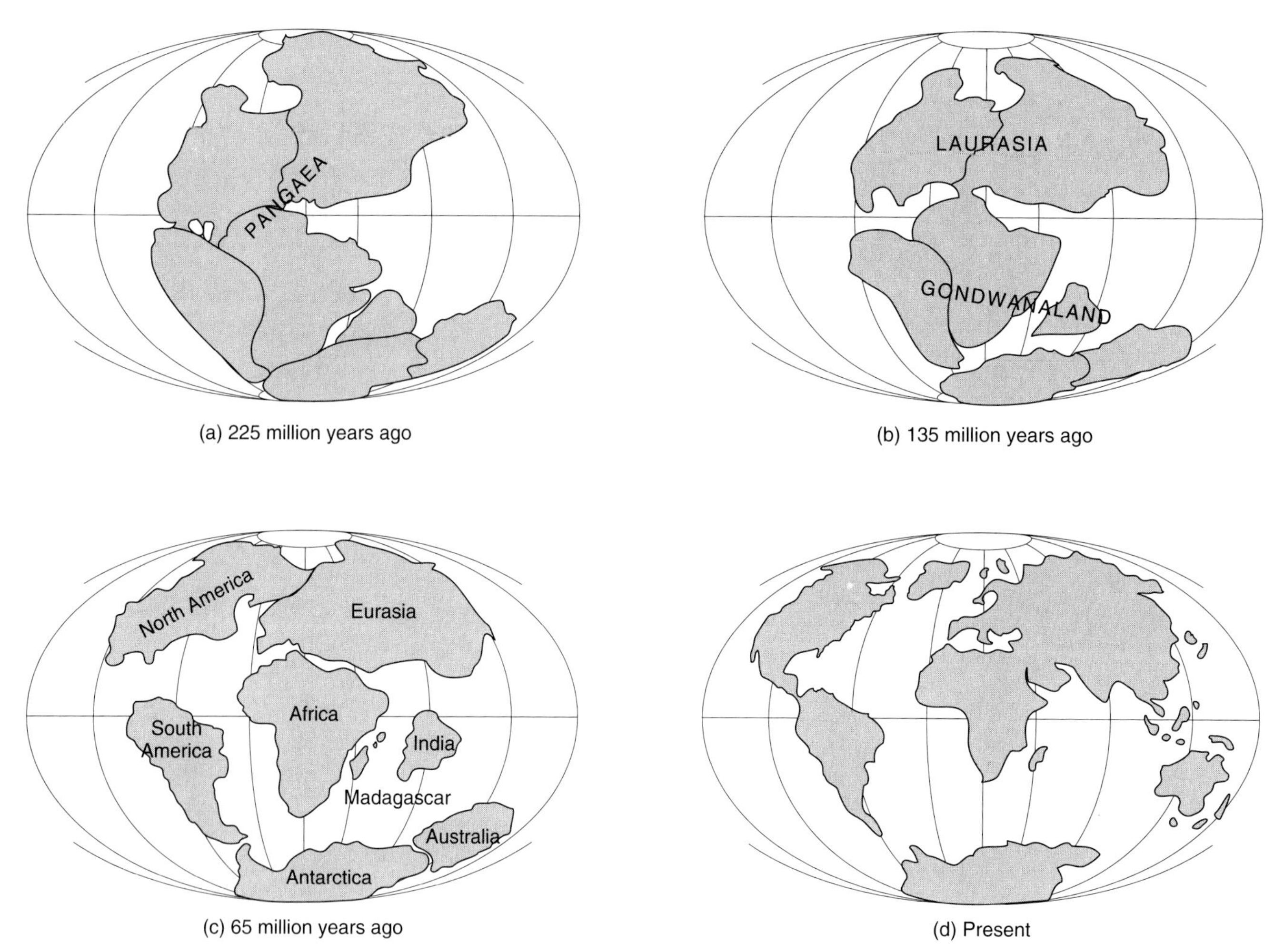

(a) 225 million years ago

(b) 135 million years ago

(c) 65 million years ago

(d) Present

FIGURE 22.7

Continental Drift

The earth is composed of great tectonic plates that continue to move to this day (volcanoes and earthquakes being obvious manifestations of these movements). More than 200 million years ago, the present continents were a part of one giant landmass. As this landmass split apart, the continents slowly drifted to their present location.

reptiles, the left aorta is lost, leaving just a single vessel. Mammals are similar, although presumed to be more distantly related to the reptiles, and they have lost the right aorta, leaving only the left vessel. In all vertebrates, all six pairs are formed during embryonic development. It is very difficult to explain away this remarkable progression of events other than by evolution.

The final piece of anatomical evidence that supports evolutionary theory is **vestigial characters** (see Figure 22.9). These are anatomical structures that apparently serve no function and are difficult to explain away except by accepting them as remnants of structures that once had a distinct purpose in our more distant ancestors. The muscles of the mid-abdominal wall are segmented, as can be seen from the washboard effect so assiduously cultivated by body-builders—but why segmented? Segmentation serves no purpose. Why do we still possess muscles to move the outer ear? We may use these muscles to help break the ice at parties but certainly never to direct the ear toward the source of a sound. Why do we possess an appendix, a structure of which we are aware only when it becomes infected and must be removed? In each case (and in many others), it is easy to see the role these structures played in ancestral species, but it is difficult to explain away their presence in humans simply by invoking chance.

Comparative Biochemical Evidence

The chemistry of life supports the concept of evolution. Why is DNA universally used as the molecule of heredity? Why is ATP universally the energy stor-

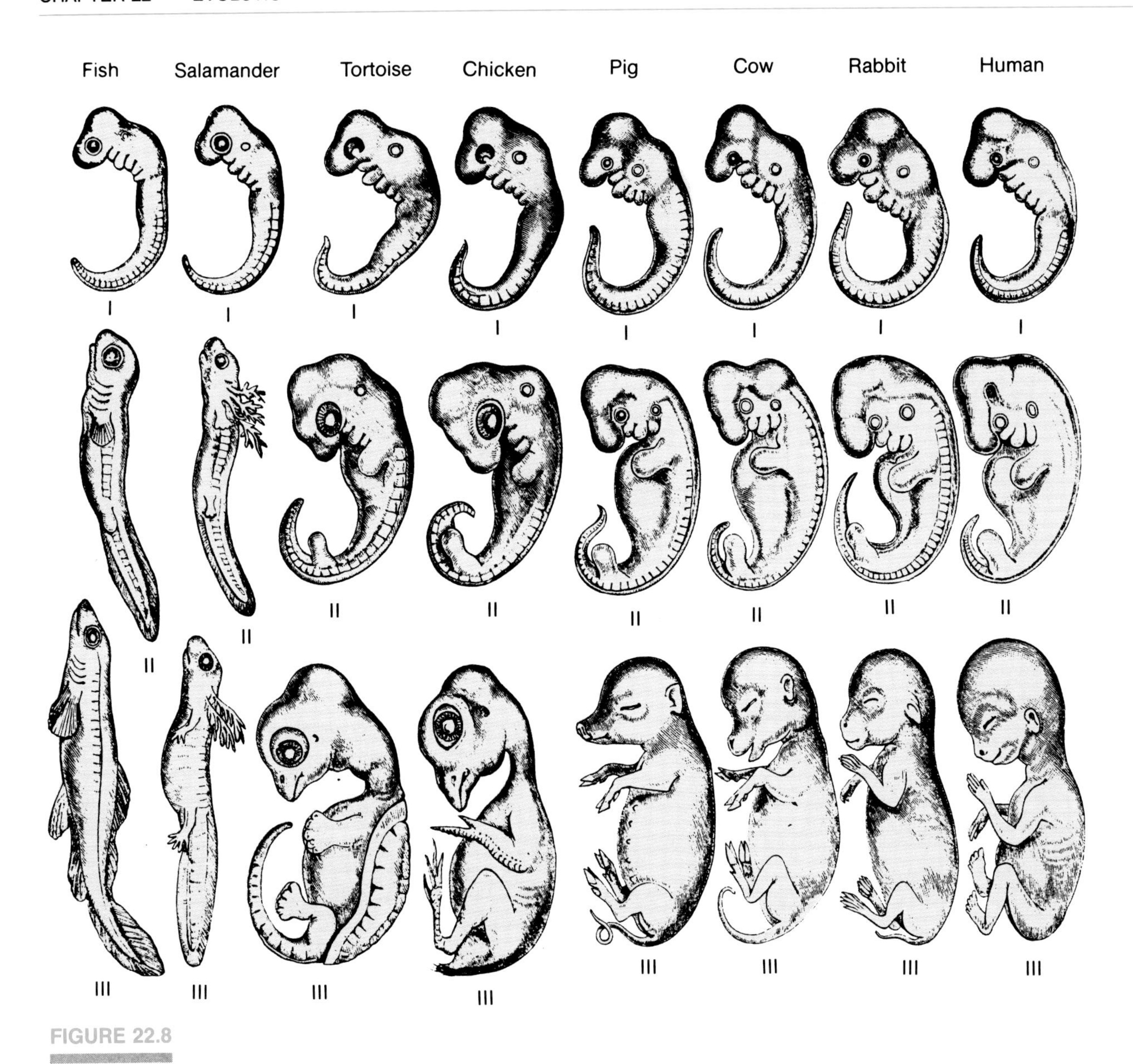

FIGURE 22.8

Comparative Embryology
Early in development, the embryos of vertebrates appear very similar. Note the presence of gill arches and a tail in embryonic humans. Species distinctiveness arises only later in embryonic development.

age molecule? Why do all vertebrates use hemoglobin as their oxygen-transport molecule, and not ***hemocyanin*** (a copper-based molecule that transports oxygen in the blood of many marine invertebrates)? Are these simply coincidences—or do they suggest a commonality of descent?

The biochemical evidence is not only strong, but extensive. Modern systematists employ very sophisticated techniques to determine just how similar or different species are. For example, the degree of protein similarity between two species can be determined by **gel electrophoresis,** which involves permitting proteins in solution to migrate along a gel strip in response to an electrical current. Each protein has a characteristic spot where it stops migrating, and a series of proteins therefore show as bands along the gel strip. The bands for two species can be compared, and the degree of similarity or dissimilarity determined at a glance.

Immunological techniques are also used. The volume of antibody that an organism forms in response to the injection of protein from another organism is a measure of the similarity, or dissimilarity, of the two organisms. Another approach to the same issue is **DNA hybridization.** DNA samples from two organisms are first fragmented into relatively short,

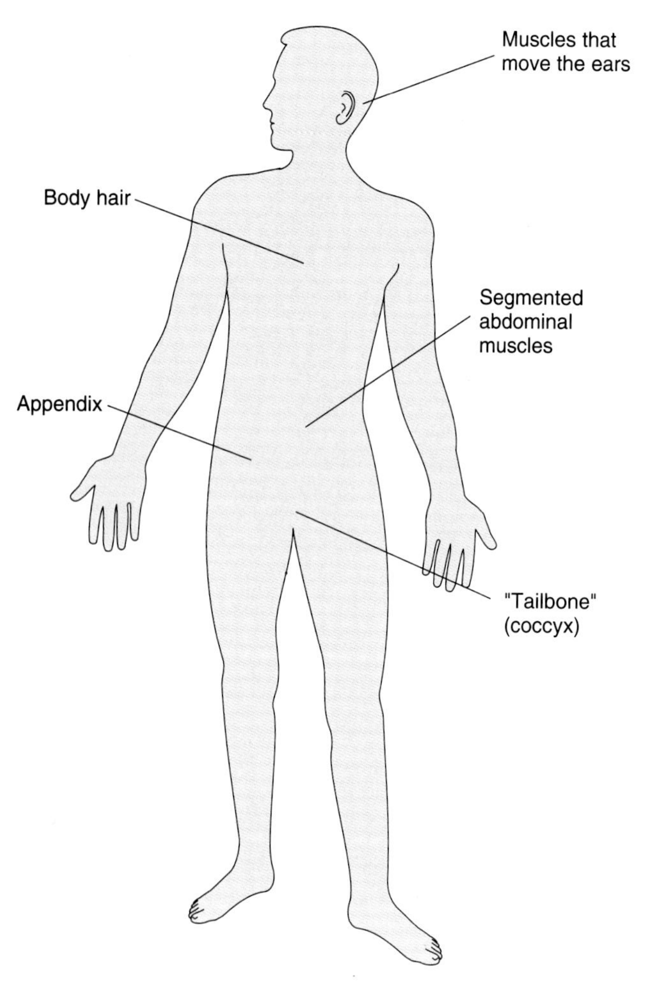

FIGURE 22.9

Vestigial Characters
Humans have many vestigial characters, some of which are indicated here. Why do we have characters that we do not use, unless they are the last remnants of an evolutionary legacy?

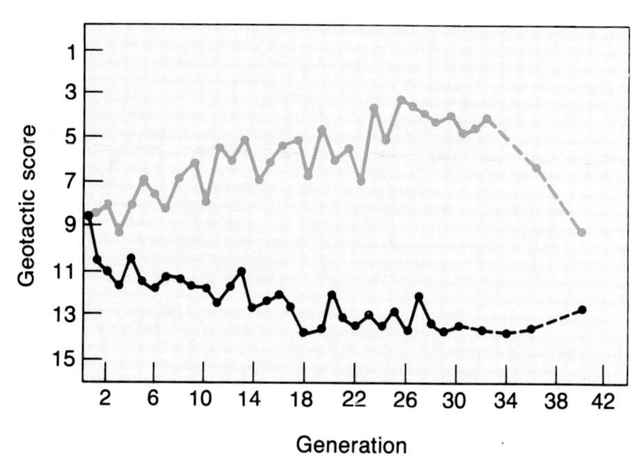

FIGURE 22.10

Disruptive Selection in *Drosophila*
For 32 generations, one group of flies was selected for negative geotaxis (a tendency to move up, away from gravity) and another for positive geotaxis (downward movement). When selection was relaxed, stabilizing selection quickly moved the two populations towards each other. Experimentally, it is possible to separate one population into two rather quickly.

single-stranded sections and then are mixed together to determine how much complementarity exists when they recombine as double strands. In closely related species, the base sequences are very similar, and the two single strands combine together evenly. In more distantly related species, there is greater dissimilarity in the bases, and the recombined double strand is very irregular. Again, the fact that this work can be done in the first place—that is, that similarity in the base sequence of two species exists at all—argues strongly in favor of evolution as fact. (See also the story of Eve in Chapter 23.)

Experimental Evidence

There are literally hundreds of experimental studies in which a species with a short generation time, such as the fruit fly, has been subjected to heavy selection pressures by the investigator (such as occurs when only those flies with the highest and lowest number of bristles on the body are permitted to breed). After 20 or 30 generations, two populations emerge that are distinctly different in bristle number (see Figure 22.10). Such experiments prove that evolution can occur in controlled experiments, but not necessarily that it occurs under natural conditions.

The problem is that, by definition, only populations evolve, and the measure of rate of evolution is the rate at which gene frequencies in the population change with time. Since natural environments normally change very slowly, natural selection pressures are normally very weak, and dramatic changes in gene frequencies would require thousands of generations. The time frame is simply beyond the capacity of direct human observation. Hence, most of the evidence for evolution is inferential.

However, there are a number of instances involving radically altered natural environments that provide evidence for evolution. One example involves the polymorphic peppered moth of England (see Figure 22.11). In collections from the last century, this moth was known almost exclusively in a single color mode (mottled gray). This color presumably served to camouflage the moth from bird predators, because the moths spend the daylight hours resting on the bark of lichen-covered tree trunks.

By 1950, virtually all the moths in the English midlands were of a different color mode (black), because industrial soot had blackened the tree trunks,

FIGURE 22.11

Industrial Melanism

Industrial melanism is a form of polymorphism in which, with the outpourings of industrial soot, and the consequent death of lichens on tree trunks, dark-colored moths are favored over light-colored forms because they are better camouflaged against bird predators.

killing the lichens and presumably rendering gray an inappropriate camouflage color. In recent years, diminished industrial pollution has allowed a return of the lichens, and the gray moths are again on the increase. This factual situation beautifully illustrates the operation of natural selection.

A similar picture can be painted for resistance. Many disease-causing bacteria have developed resistance to such drugs as penicillin, and 447 species of insects, mites, and ticks are now resistant to some or all insecticides. In each case, a powerful selective force (the lethal chemical) has been met by genetic mutations which, in various ways, allow the organisms that possess the mutations to survive and reproduce, despite the presence of the once-lethal chemical. Evolution is a change in gene or allele frequencies over time. These cases therefore exemplify evolution.

EVOLUTION AND SPECIATION

A **species** (L., "kind, variety") is defined as a population of actually or potentially interbreeding organisms that is reproductively isolated from other such populations. Speciation is the process by which new species arise.

A good deal of the confusion that has arisen in discussions of evolution and natural selection is based on a failure to distinguish between "evolution" and "speciation." As we have seen, the modern definition of evolution speaks only in terms of changed gene frequencies. Speciation always results from evolution, but evolution does not necessarily always end in speciation. DDT-resistant houseflies are genetically different from nonresistant flies, but they do not comprise a separate species.

The Species Concept

Most biologists view the species as the only truly natural category among the various categories by which plants and animals are classified. (The other categories—genus, family, order, and so forth—are held to be merely convenient pigeonholes in which scientists can arbitrarily assign clusters of species that exhibit increasingly greater differences; see Box 1.2, Chapter 1.)

However, the definition of "species" is troublesome. The geographic area of an interbreeding population is effectively determined by the size and mobility of the species. Yet there are many plants and animals that have worldwide distribution and are still referred to as a single species. An Iraqi housefly would never mate with an American housefly, for geographic, if not political, reasons. How, then, can they be held to be the same species?

Some animals that are geographically separated will interbreed if brought together, yet they are called separate species. For example, lions and tigers will interbreed, under the appropriate circumstances (usually in zoos). Does that mean that they are really a single species?

How does the species concept handle differences in time? Obviously, a population of organisms of 100 years ago will not be interbreeding with a population of organisms living today. Does that mean that each generation is a different species? If that notion strikes you as preposterous, then how different do two groups of fossils have to be before they are assigned separate species status? Human or human-like fossils date back several millions of years. How many species are represented in this chain?

These are all difficult questions, and none has a

totally satisfactory answer. Darwin himself finally gave up on the question and decided that "species" was every bit as arbitrary a term as "subspecies," "variety," "race," or any of the other categories. However, his is no longer the prevailing viewpoint on this question.

Most of the difficult questions have been arbitrarily answered. Geographically separated populations that do not appear to be structurally or behaviorally distinct are assumed to be a single species. Thus, Iraqi and American houseflies—despite their political differences—are all classified as *Musca domestica.* Animals that do not interbreed in nature, however, or that produce sterile offspring (lions and tigers; horses and donkeys) are considered separate species. Fossil species are erected when the structural differences with modern species, or with other fossils, are at least as great as the differences between modern species. Those answers are workable, if not entirely satisfactory.

Speciation by Natural Selection

There are two principal methods by which speciation is assumed to occur. First, speciation over geologic time involves the gradual transformation of a single species into a succession of newer species, although at any given moment, only one species is present. Under the influence of natural selection, the constitution of the gene pool changes to reflect adaptation to the prevailing environment, and the gradual accumulation of changes, over thousands of generations, is reflected as a new species. It is important to note that many biologists now believe that the accumulation of such changes is not steady and progressive, but may occur in fits and starts. That is, a population may remain essentially unchanged for very long periods of time, but then, in response to a dramatically changed environment, may evolve very rapidly over rather short time intervals (see Box 22.1).

The second common method of speciation occurs when populations are separated, as by the formation of a canyon, or river, or other geographic barrier. Over a sufficiently long period of time, subtle differences will accumulate to the point where the two populations cannot, or will not, interbreed, even if the geographic barrier is removed (see Box 22.2).

In some instances, this multiplication of species may occur rather rapidly. The development of flight by the ancestors of modern birds permitted a whole new way of life, and the exploitation of new food and habitats, a process known as **adaptive radiation.** The rapid multiplication of bird species exemplifies this process. A similar result may occur in the colonization of islands by animals or plants from a distant mainland.

Isolating Mechanisms

When two populations are separated for a time, what prevents interbreeding between the populations if the barrier that had separated them is removed? For that matter, what prevents species from interbreeding in the first place? The answer to both questions is the presence of **isolating mechanisms.**

Populations that were once separated but are now free to intermingle may or may not interbreed, depending on how much they diverged genetically during the period of separation. If they do not interbreed, it is presumably because of one or more of the following isolating mechanisms (see Table 22.1).

Prezygotic Isolating Mechanisms. Four isolating mechanisms can occur prior to the formation of a zygote:

1. *Behavioral isolation:* Many species of birds have elaborate courtship rituals functioning as species recognition factors. Failure to perform the ritual according to the accepted species pattern means a termination of the courtship and no interbreeding. Many insects have similar, although usually less complex, patterns. Changes in the genes which ultimately control the performance of these displays could easily ensure no interbreeding between two once-separated populations.

2. *Seasonal or temporal isolation:* Most species of animals are sexually active only at particular times of the year, or of the day. If separated populations have developed differences in the timing of their sexual behavior, then there will be no opportunity for interbreeding.

3. *Ecological isolation:* Two populations may diverge with respect to how they partition the environment. That is, one population may specialize in eating a different food, or colonizing a different site, or in dwelling in trees than on the ground, any of which would minimize the opportunities for interbreeding with the second population.

TABLE 22.1 Isolating Mechanisms

CATEGORY OF ISOLATING MECHANISMS	SPECIFIC EXAMPLES
Prezygotic Isolating Mechanisms	Behavioral isolation Seasonal/temporal isolation Ecological isolation Mechanical isolation Gamete incompatibility
Postzygotic Isolating Mechanisms	Embryonic inviability Hybrid inviability Hybrid sterility

BOX 22.1

Punctuated Equilibria and Creation Science

In the 1970s, Stephen Jay Gould and Niles Eldredge, two noted paleontologists, developed the theory of **punctuated equilibria** to explain the paucity of intermediate forms in the fossil record. Gould and Eldredge argued that evolution and speciation often—and perhaps generally—occur in fits and starts, with many long periods of stabilizing selection (equilibria), interspersed, or *punctuated,* with short periods of rapid evolution. The sudden appearance of a diversity of multicellular species at the beginning of the Precambrian, following several billion years of exclusively single-celled forms, or the explosive adaptive radiation of the birds or the flowering plants, supports the idea that evolution is not always slow and steady, but can, in fact, be uneven. Although Darwin had emphasized gradualism in his *Origin,* he also acknowledged the possibilities of periods of quiescence followed by rapid change. Nevertheless, Gould and Eldredge were perceived by some as heretics, attacking the very basis of evolution, when in fact they were fully accepting of evolution as fact but were suggesting alternative possibilities to the principal one advanced by Darwin.

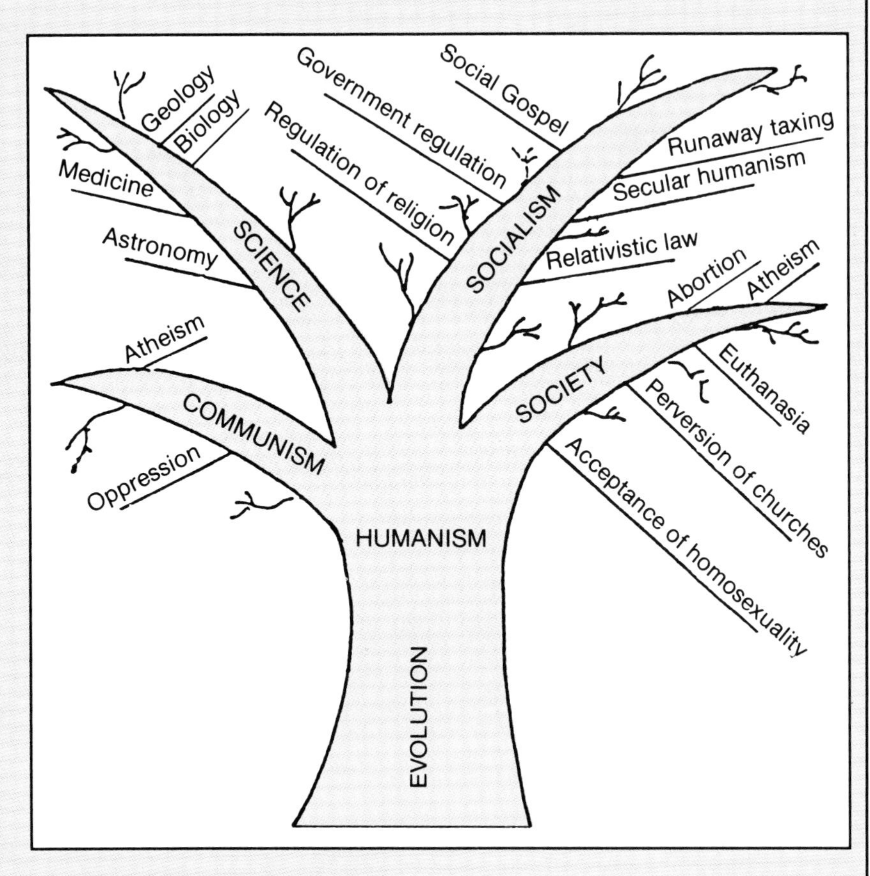

A Creationist's View of Evolution

"Evolution is the taproot which is feeding the oppressive, murderous and infidel directions we see gaining acceptance around us." This statement made in 1984 by P.A. Bartz, in the *Bible Science Newsletter,* illustrates that it is not always easy to establish a climate for rational discussion.

This subtlety was lost on (or ignored by) a small group who called themselves **creation scientists.** These individuals attempted to use Gould's and Eldredge's work as proof that evolution itself was under attack by prominent biologists. Creation science is an oxymoron, since creationism (the belief that a divine being created all of the species in the world today) is not based on scientific analysis but on faith. As we noted in Chapter 2, science cannot—and should not—investigate concepts that are outside its realm. Unfortunately, creation scientists have endeavored, under the concept of "fairness," to have creation science included in high school textbooks as an alternative to evolution—which, as former President Reagan once noted, "is only a theory." The problem is that evolution is not a theory; it is a fact, and just as much a fact as gravity, or electricity, or the shape of the earth. (In this instance, the use of the word "theory" should be linked not to evolution but to the force underlying evolution—that is, the "theory of evolution by natural selection.") Merely because some people choose not to believe the facts of science does not give them the right to have their views taught in science classes. Fortunately, the notion that creation science should be given equal time with evolution in high school texts has been rejected by most states.

4. *Mechanical isolation:* Sometimes individuals do not mate because they are anatomically incapable of doing so. The reproductive organs of many insects fit together like a lock and key. Divergence in structure of the reproductive anatomy in two populations may result in reproductive isolation.

BOX 22.2

Divergence, Convergence, Homology, and Analogy

If a population of organisms is separated into two populations, they may begin to diverge genetically (assuming the absence of any interbreeding and with different selection pressures on each population). Ultimately, they may become separate species, and, over a sufficiently long period of time, two separate and distinct lineages may evolve. Presumably mammals diverged from the reptiles in just such a fashion. **Divergent evolution**—the development of increasing amounts of distinctiveness—is a common type of evolution.

As a result of selection pressures on diverging populations, certain organs or structures may become modified for different purposes despite sharing a common ancestry. The fore and hind limbs of land vertebrates, for example, both consist of one bone in the upper portion of the limb, and two bones in the lower limb, regardless of how different the limbs themselves may appear to be. Thus the forelegs of salamanders, the wings of birds, and the human arm all contain the same arrangement of major bones. Structures that share a common ancestry, even though they may appear different, are called **homologous structures.**

Once diverged, particular lineages may converge again, to exploit a particular habitat. Thus sharks, ichthyosaurs, and porpoises share a common body form, but in fact are only very distantly related. The same is true for pterodactyls, birds, and bats—or, on a more restricted scale, of ostriches, rheas, and cassowaries, all flightless birds with superficial similarity but with very different ancestries. This type of evolution is called (not surprisingly) **convergent evolution.**

Sometimes the convergence of structure is striking indeed. Compare the vertebrate eye with the eye of an octopus or squid. Separated

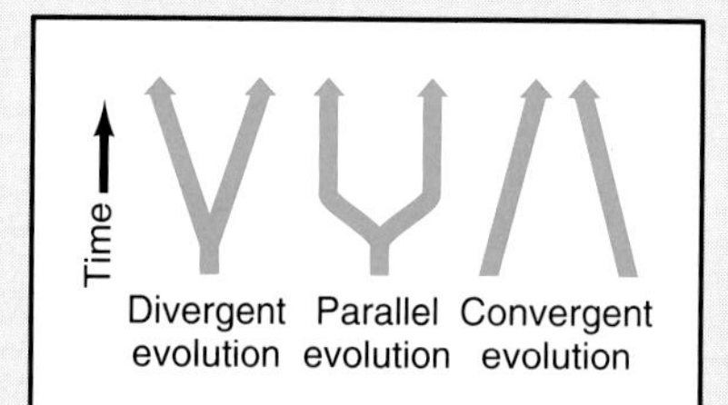

Evolutionary Patterns

In divergent evolution, two populations derived from a common ancestral population become increasingly diverse over time. In parallel evolution, the two populations diverge, but remain relatively similar over time. In convergent evolution, two very different populations begin to resemble each other (albeit superficially). In each case, the nature of the selection pressures accounts for the outcome.

Postzygotic Isolating Mechanisms. In some instances, populations remain reproductively separated even in the absence of the mechanisms just discussed. These populations may interbreed but the zygote formed may be incapable of developing (perhaps because the chromosome numbers differ, as they do between chimpanzees and humans), or because the offspring is sterile (as is the case with mules, the offspring of a horse and donkey). Sterility in the offspring prevents gene flow between two populations and is therefore ultimately an effective isolating mechanism.

Summary

After a number of false starts by others, Charles Darwin, in 1858 and 1859, correctly identified natural selection as the governing force in evolution. Although he was stymied by the formidable problem of inheritance, his identification of natural selection was amply supported by data he had amassed over a period of 25 years.

With the rediscovery of Mendel's work, evolution and genetics were united, such that natural selection is now defined as differential reproductive success and evolution as a change in gene or allele frequencies over time. Evolution, a population event, is the outgrowth of natural selection, which is an individual (and phenotypic) process. With mutation serving as the raw material of natural selection, various mechanisms are at play in maintaining substantial genetic variation in the population. Natural selection maintains those mutations that are adaptive.

Evidence for evolution comes from many sources, including fossils, phylogenies, geographic patterns, anatomical and embryological studies, and from experimental studies. Collectively, they provide broad support for a process that under normal circumstances is too slow to be within the capacity of a single scientist to observe directly.

Speciation is one end product of evolution, although the two terms are not synonymous. New species arise by replacement, over great periods of time, and by splitting, wherein two or more species

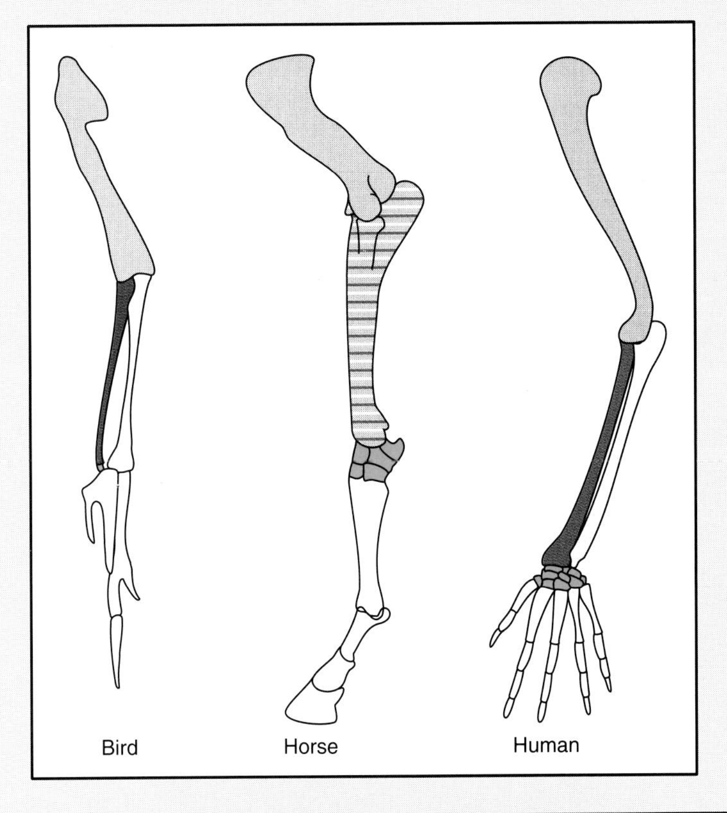

Homologous Structures
The three bones of the arm and the leg of vertebrates have been modified into many varied structures, including the wing of a bird, the foreleg of a horse, and the arm of a human, but they are homologous because they share a common ancestry.

from the bodies in which they are normally packaged, the eyes alone are so similar as to suggest a close evolutionary relationship—but no such relationship exists. Structures that do not share a common ancestry, regardless of how similar they may appear, are called **analogous structures.**

Distinguishing homologous from analogous structures is the bane of systematic biologists, because the failure to make the correct distinction leads to the misinterpretation of evolutionary relationships and the construction of erroneous phylogenies.

arise following a subdivision of a population by some event or phenomenon that prevents free interbreeding.

Key Terms

balanced polymorphism
genetic drift
gene flow
founder effect
gene pool
adaptation
stabilizing selection
directional selection
disruptive selection
sexual selection
fossil
phylogeny
vestigial character
gel electrophoresis
DNA hybridization
species
adaptive radiation
isolating mechanisms

Box Key Terms

punctuated equilibrium
creation science
divergent evolution
homologous structures
convergent evolution
analogous structures

Questions

1. How is evolution currently defined?
2. List four ways in which genetic variation is maintained in a population.
3. Is natural selection the only factor leading to changes in gene and allele frequencies in a population? Can you think of others?
4. How is natural selection currently defined?
5. What is meant by the term "adaptation"? Does it refer to individuals or to populations?
6. By category, what is the evidence for evolution?
7. What is meant by the term *species?* How does one species differ from another?
8. Is speciation the same thing as evolution? If not, how do these terms differ?
9. What factors prevent the interbreeding of species?

CHAPTER

23

Human Evolution

Our Primate Heritage

Humans are primates. Why is it that the primates are the group giving rise to the species that has come to dominate the earth, as humans have? Why not dolphins, or bears, or bats, or any of the other major groups of mammals?

What was the path of human evolution? What are the fossil links to the other primates? How far back can we trace our lineage? Were the Neanderthals our ancestors or degenerate distant relatives?

These are the kinds of questions that we shall consider in this chapter.

THE NATURE OF THE PROBLEM

Evolution has been described as the "golden thread" that unites all components of biology. It is also a subject that has direct relevance to us as humans, because we are not immune from the actions of evolution—or are we? The fuss kicked up by Darwin's publication of *On the Origin of Species by Means of Natural Selection* in 1859 was nothing compared with the furor that surrounded the 1871 publication of *The Descent of Man,* in which he unequivocally stated that modern humans were the product of evolution. That humans were related to a bunch of smelly, dirty, old monkeys! The very idea! As one proper English lady of the time said, "My dear, let us hope that it is not true, but if it is, let us pray that it will not become generally known" (see Figure 23.1).

It seems odd to us, retrospectively, that these ideas created quite the outcry they did. After all, a primary reason why the monkey house has always been among the most popular exhibits in any zoo is that the anatomy and behavior of these animals remind us so much of ourselves. In a tacit way, we are acknowledging a relationship. However, this has been a relatively recent development. In the middle of the last century, few people were willing to acknowledge anything more than vague similarity. After all, humans, were . . . well . . . human, and animals were . . . animals.

For many people, things are no different today. The concept of evolution remains threatening for visceral, not intellectual, reasons—we see ourselves as superior (and therefore apart from) the animals. Thus the question remains, "Do humans and apes share a common ancestor?" Evolution says that we must—but where is the hard evidence?

Unfortunately, the fossil record is very spotty. Our earliest ancestors were apparently never very numerous, and they displayed an annoying tendency to stay away from soft mud or shallow marine environments where they might have been fossilized. Thus we know them only from bits and pieces.

Is there enough evidence to make a convincing case about the evolution of humans? Judge for yourself.

A Darwinian Cartoon

Scientific Monkey. "Cut it off short, Tim; I can't afford to await developments before I can take my proper position in Society."

FIGURE 23.1

Cartoonist's Reaction to Darwin's Proposals

This cartoon was typical of the reaction in Victorian England to Darwin's book, *The Descent of Man.*

EARLY DISCOVERIES

A troublesome feature in postulating human evolution in Darwin's time was that virtually no fossils had been discovered that would substantiate such a theory. The earliest discovery, the **Neanderthal** fossils, occurred in 1856, just three years prior to the publication of Darwin's *Origin*. The first specimen was found during quarrying activity in Germany's Neander valley (Ger. *tal*, "valley"). The discovery created a minor sensation, not because the bones were immediately recognized as being a near relative of modern humans, for they were not, but rather because there was no agreement at all on what they were. The quarry owner thought them to be the bones of a bear; more knowledgeable individuals were sure they belonged to some poor soul who drowned in Noah's flood; another scientist proclaimed them to be the bones of a Mongolian, a member of the Russian cavalry who had deserted during the push against Napoleon in 1814; yet another anatomist thought the bones to be those of "an old Dutchman."

A few years later, in 1868, another ancient find, this time near the village of Cro-Magnon in southern France, revealed the presence of much more modern-appearing skulls. The discovery of **Cro-Magnon** fossils was greeted with a wave of relief, for it demonstrated that humans had "always" looked the way they do now.

However, when more Neanderthal skulls were discovered, this time in Belgium a few years later, concern arose once again. The first find could not now be dismissed as an aberrancy—the skeleton of a diseased individual, for example—because here were several more, several hundred kilometers from the first find.

Then, in 1891, a much more primitive but still humanlike skull was found in Java. This discovery created renewed interest in the status of the Neanderthals. In 1908, some very complete Neanderthal skeletons were found, and a series of measurements and reconstructions ensued that were to have a profound effect on the status of Neanderthals in the public eye.

Retrospectively, the magnitude of the errors was surprising. The anatomist in charge simply ignored the bones in front of him and, focusing on the heavy brow and flattened skull (see Figure 23.2), pronounced Neanderthals to be shuffling, brutish, clumsy, stupid creatures with the foot of an ape (a divergent big toe). This image has apparently been indelibly etched into the public mind. Neanderthals are usually portrayed as having spent most of their time shuffling and grunting, except when engaged in bonking nubile maidens over the head with their clubs (see Figure 23.3).

In contrast, Cro-Magnons are usually depicted as clear of mind and steady of hand; with ramrod straight posture, blond hair, and blue eyes; primitive Vikings, as it were, whose task it was to rid the earth of the smelly Neanderthals and make the world safe for humanity. Needless to say, with that type of introduction, the Neanderthals were to have a protracted wait before their role in human evolution became clear.

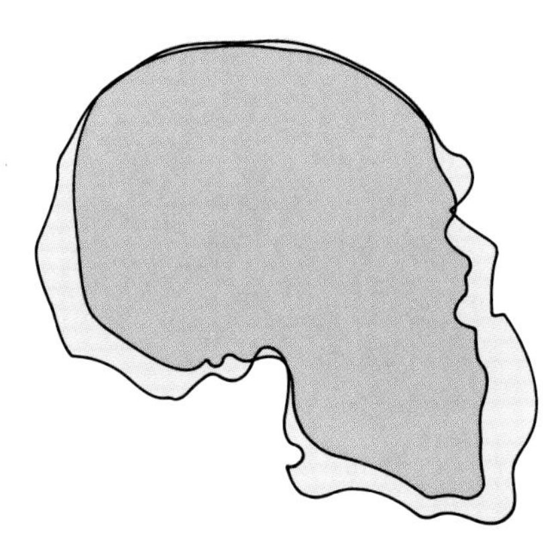

FIGURE 23.2

Neanderthal and Modern Human Skulls

The Neanderthal skull is seen to have a more elongate and heavier facial structure than does the skull of a modern human. Note, however, that the skull capacity of Neanderthals was larger than that of modern humans.

PRIMATE ORIGINS AND CHARACTERISTICS

Humans are placed in the order **Primates,** along with the apes, monkeys, and such primitive species as lemurs and tarsiers (see Box 23.1). The earliest fossil primates date back more than 53 million years, yet bear striking similarities to the tree shrews of today. Primates as a group evidently arose from insectivores, the mammalian order that includes shrews and moles.

The primates are divided into two suborders: the **Prosimii,** including the tree shrews, tarsiers, lorises, and lemurs (see Figure 23.4), and the **Anthropoidea,** including the monkeys and apes. The anthropoids, which have a fossil record dating back almost 40 million years, are further divided into three superfamilies:

1. **Ceboidea:** The new world monkeys are characterized by having a prehensile tail and flared nostrils (see Figure 23.5).

2. **Cercopithecoidea:** The old world monkeys have a nonprehensile tail and nostrils close together.

(a)

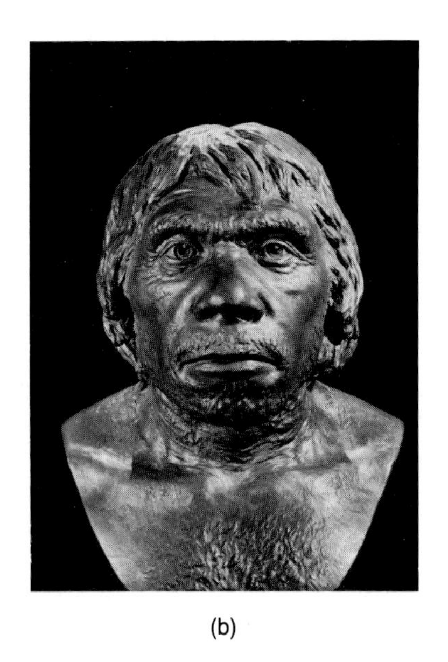

(b)

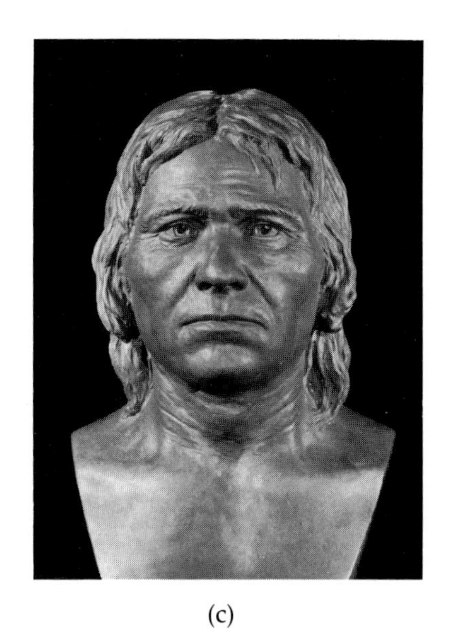

(c)

FIGURE 23.3

Reconstructions of Neanderthals and Cro-Magnons

(a) An early reconstruction of a Neanderthal compared to a more current view (b) and to a reconstruction of a Cro-Magnon. The brutish louts that Neanderthals were originally presumed to be obviously colored the way in which the first reconstruction was made.

(a)

(b)

(c)

FIGURE 23.4

Prosimians

The suborder Prosimii includes the primates most distant from humans. (a) Tree shrew, a link with insectivores; (b) a tarsier; (c) a black and white ruffed lemur.

3. **Hominoidea:** The great apes and humans are distinguished from the monkeys by their size, shape of chest, and tooth structure, among other anatomical characteristics. The lesser apes, the greater apes, and humans are placed in separate families (**Hylobatidae, Pongidae,** and **Hominidae,** respectively), that are distinguished by jaw and tooth shape, anatomy of the pelvic girdle, and the shape of the hands and feet (see Figure 23.6).

Should humans and great apes be placed in different families? That is, by standard taxonomic criteria, are humans and apes sufficiently distinct as to

BOX 23.1

Why Humans?

What is so special about the primates? Why did this group, not any other, give rise to the species that now dominates the earth (or at least likes to think that it does)?

Primates are, for the most part, very generalized in their body form. They have experienced no reduction in the numbers of fingers and toes, unlike virtually all other mammalian groups. Their largely *arboreal* (tree-dwelling) habits have favored flexible locomotory styles coupled with the ability to manipulate objects. Animals such as lions, horses, or whales have all adapted for speed and have lost versatility in limb structure.

Primates also benefited from their habitat in other ways. The need to move in a three-dimensional environment and to find fruit and other food favored stereoscopic color vision and large brains. Omnivorous diets favored generalized tooth patterns and digestive tracts. In short, humans are the product not of specialization, which tends to be evolutionarily dead-ended, but of generalization. Being jacks of all trades has given primates great flexibility in an evolutionary sense.

Dry periods in the earth's history forced our ancestors out of the shrinking forests and onto the grasslands. Our ancestors were not well equipped to survive on the grasslands—they were slow, and not terribly strong. However, there was strength in numbers, and they adopted the habit of living in groups. (A similar pattern has occurred much more recently with the baboons.) Group behavior fostered better communication skills, and changing diets required the use of tools. For both of these, larger brains were an asset.

warrant placement in separate families (as opposed to subfamilies or genera)? Or is this level of distinction chosen simply to emphasize the degree to which we see ourselves separated from (and above) all other animal species?

Interestingly, there are now objective data available that permit answers to these questions. *DNA hybridization* (see Chapter 22) shows that we are actually genetically more similar to chimpanzees than chimpanzees are to gorillas! In other words, if humans

(a)

(b)

FIGURE 23.5

New World and Old World Monkeys

(a) The New World, or ceboid, monkeys (such as this spider monkey) have broadly spaced nostrils, and most have prehensile tails. (b) The Old World, or cercopithecoid, monkeys (such as this baboon) have close-set nostrils and short, nonprehensile tails.

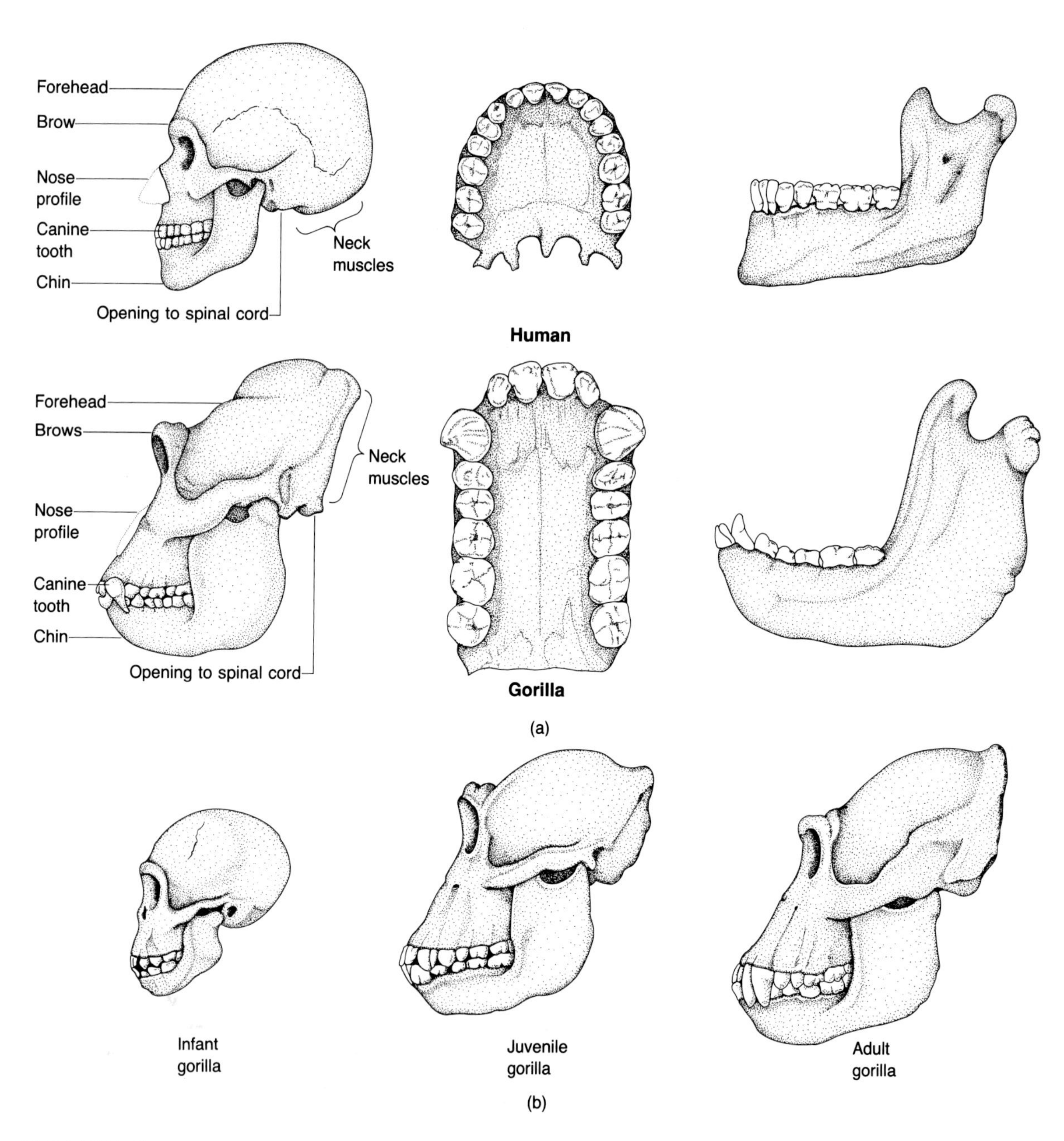

FIGURE 23.6

Skulls of Humans and Gorillas

(a) The skulls and jaws of humans and gorillas are markedly different. However, the human skull looks very similar to the skull of an infant gorilla (b). Humans retain many other juvenile traits, as compared to our primate cousins.

rate their own family, then so do both chimpanzees and gorillas (see Figure 23.7).

To be specific, the magnitude of the difference is 1.6 percent between humans and chimpanzees, and 2.1 percent between chimpanzees and gorillas. Studies of *mitochondrial DNA* (see Box 23.2) yield parallel results: The difference between humans and chimpanzees is 9.6 percent; the difference between chimpanzees and gorillas is 13.1 percent. (These figures are larger than the DNA hybridization figures because mitochondrial DNA either mutates more frequently, or selection of the mutants is less intense.)

In summary, the magnitude of the differences between humans and the great apes is not great enough to warrant separate families—but because that conclusion is undoubtedly too radical for most people, it is unlikely to be broadly accepted, although it has been proposed (see Figure 23.8).

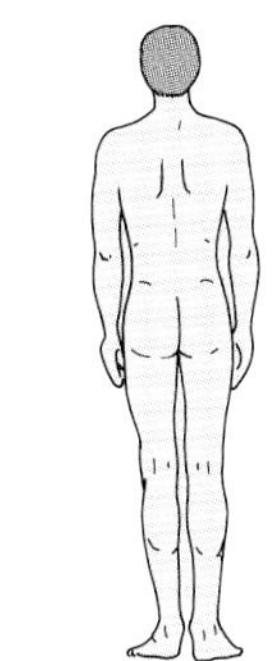

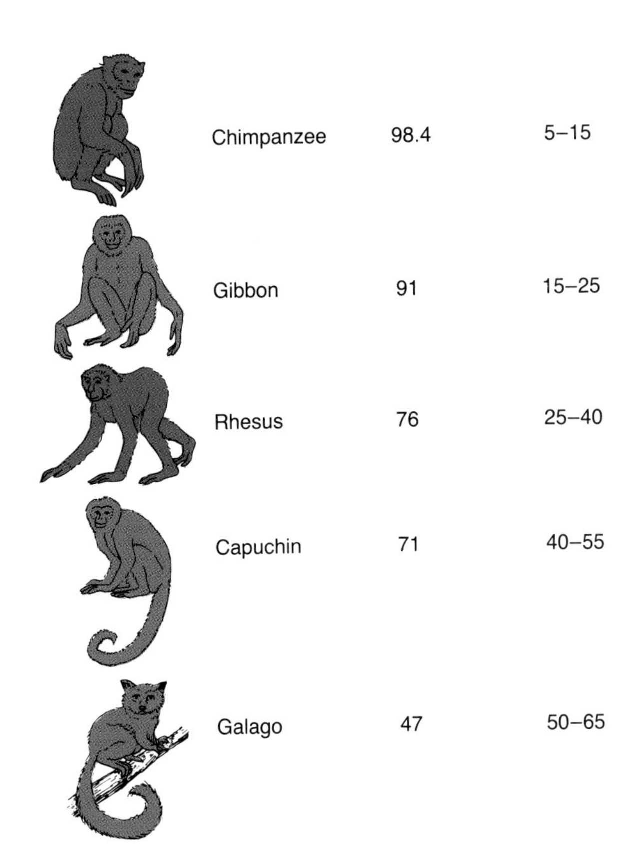

FIGURE 23.7

Genetic and Evolutionary Relationships in the Primates

Using DNA hybridization techniques, it is possible to determine the degree of similarity or dissimilarity between humans and other primates. Based on the average rate of mutational change, it is also possible to calculate the approximate time since we shared a common ancestor with various of our primate relatives.

In any event, with the exception of the problem just noted, the classification of living species of primates is a relatively easy task. What about the fossils? Where do they fit in?

THE HUMAN HERITAGE

Establishing an unequivocal fossil lineage for the primates has not yet proved possible. Although reports of new finds appear regularly in the popular press, the fact is that the total number of primate fossils is really very small. In many cases, there are only a handful of bones or bone fragments from which to draw conclusions. Trained scientists can extrapolate from a small number of bones to a complete skeleton, but how are we to know that the skeleton is representative of the population from which it came? Suppose you were on a collecting trip for the intergalactic zoo and were told to bring back a representative earthling. What would you choose—a pygmy, a Russian weightlifter, or a Chinese basketball star? It might be difficult to convince the zoo that all modern humans belong to a single species, but such is the case.

In the age of fossils, populations were few and small, and a good deal of local variation must have occurred. Undoubtedly, some of the differences among fossils that now occasion sharp debate are nothing more than local variations, but at present we have no effective way of knowing just how much variation did exist within and among populations.

Moving Backward in Time

The first fossils found–Neanderthal, Cro-Magnon, Java—were all more or less humanlike. It was possible to view them as slightly narrowing the gap between humans and apes, but they in no sense bridged the gap. Where and what was this "missing link"?

The first answer came from South Africa in 1924, although its significance was not widely recognized until after World War II. That year, Raymond Dart, a South African anthropologist, found a fossil skull that was no larger than a baboon's, but with a flat face and an even tooth row instead of the baboon's characteristic muzzle with long canine teeth. Moreover, the opening of the skull to the spinal cord was directed downward, rather than backward, suggesting that the creature must have stood erect. (Subsequent discoveries—including fossil footprints—have strongly substantiated the belief that the early hominids were bipedal.)

Was this the skull of some primitive human? No, the brain was too small—only 400 cm^3, rather than the 1200 to 1400 cm^3 of modern humans. Moreover, the creature was estimated to be only 125 cm tall and to weigh less than 30 kg. Nonetheless, it was a hominid, and certainly the most primitive then known. It was named ***Australopithecus africanus*** ("southern African ape") (see Figure 23.9).

Later excavations in South Africa by Dart's friend, Robert Broom, yielded the remains of a second, heavier-jawed type, called ***Australopithecus robustus.*** (see

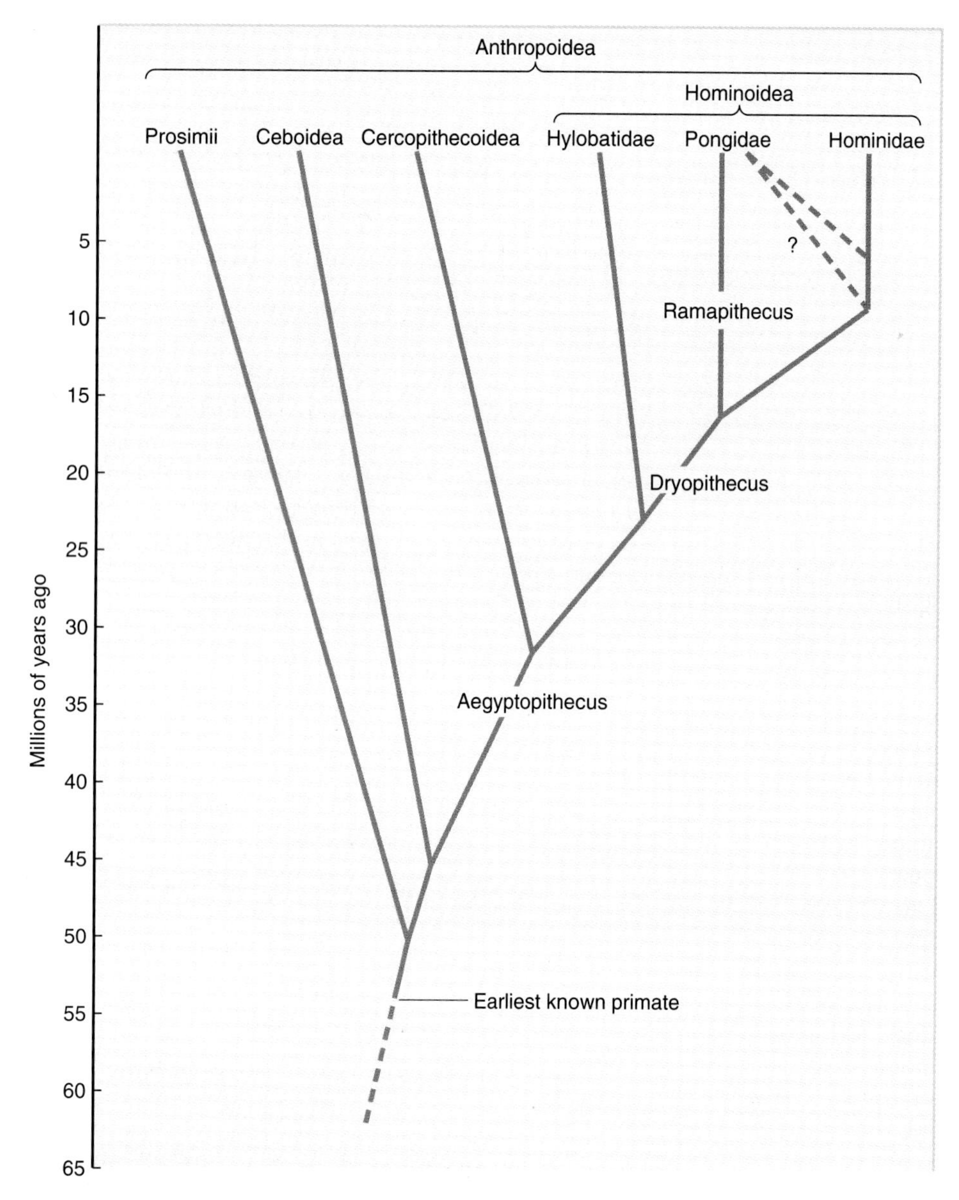

FIGURE 23.8

A Primate Phylogeny

This "family tree" of the primates shows the estimated dates of splits in the lineage and the approximate positions of several of the more important fossils. The orangutan lineage split from the rest of the hominoids about 18 million years ago. The dotted lines to the Pongidae represent both the range of estimated times since the gorilla and chimpanzee lineage split from the human lineage, and the controversy of whether to include these species in the Pongidae or the Hominidae.

Figure 23.10). This creature stood about 150 cm tall, and weighed more than 45 kg. In neither case, however, were the rocks *stratified* (laid down in layers), and, in the 1920s, stratification was the only method of dating fossils. Broom estimated their age at two million years, a figure so staggering as to shred whatever credibility he had left. The world was simply not ready to hear of human-like fossils dating back that far.

There the matter stood until 1959 when Louis and Mary Leakey discovered similar fossils in Olduvai Gorge, Tanzania. These fossils were of a creature still more robust than either of the South African types, and it was named ***Australopithecus boisei.*** It was estimated to be 175 cm tall and to weigh perhaps 75 kg. Even more important, it was discovered in well-dated strata, and its age was assigned at 1.75 million years. Suddenly, there was renewed interest in the South

FIGURE 23.9

Australopithecus africans
A photograph showing a paleontologist's reconstruction of one of our possible ancestors.

African fossils, because the figure of two million years no longer looked quite so preposterous.

Still, problems remained. Further discoveries in South Africa suggested that some of the *A. robustus* fossils were as much as one million years younger than were the more human-appearing *A. africanus*. Moreover, *A. robustus* appeared to be exclusively a plant-eater, rather like the modern gorilla—and its teeth and skull seemed to confirm this view. How, then, could this lineage lead to modern humans?

The discovery of *A. boisei* added to the problems, because it began to appear as if modern humans had evolved rather suddenly from relatively primitive creatures. However, in 1960 the Leakeys found another skull, also about 1.75 million years old, which was even more refined than that of *A. africanus*.

The correct designation of this fossil has been in some dispute. Some authorities think that it is simply a later type of *A. africanus*, but the Leakeys called it ***Homo habilis*** ("skillful man"). Certainly its brain was larger—at 700 cm^3, some 250 cm^3 larger than the typical *A. africanus*. At the very least, *Homo habilis* seems to be poised on the edge of full human status. The argument is simply to which side of the line it belongs.

More recent discoveries have muddied our understanding, not clarified it. In 1974, in the Afar region of Ethiopia, Donald Johanson discovered an amazingly complete skeleton of a female hominid. Nicknamed "Lucy," her age was determined to be approximately three million years. Although the brain size was only 500 cm^3, and the dentition was primitive, skeletal evidence indicates that "Lucy" walked erect.

As was true for all the australopithecines, "Lucy" had a body shape more like an ape than a modern

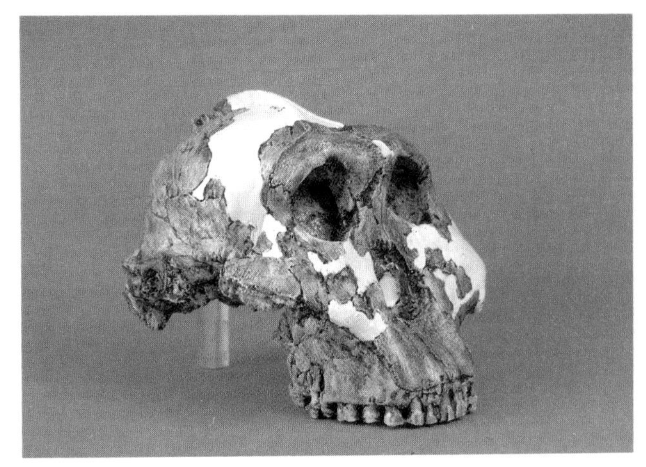

A. robustus

A. boisei

H. erectus

FIGURE 23.10

A Rogue's Gallery of Skulls
These skulls permit some general comparisons to be made between and among various fossil hominids. Note the prominant ridge on the top of the skulls of *A. robustus* and *A. boisei;* this is an attachment point for the heavy jaw muscles of these presumed plant eaters.

BOX 23.2

The Search for Adam and Eve

What can the present diversity of human genotypes tell us about our past? Because of the crossing over that occurs in meiosis, the genes of the autosomes are shuffled every generation, a fact that makes tracing a genetic lineage difficult. However, mitochondrial DNA is not subject to such shuffling. On the contrary, mitochondrial DNA is a constant from generation to generation because humans inherit mitochondria only from the mother. That is, since only the nucleus of the sperm enters the ovum at the time of fertilization, all of the mitochondria in our bodies derive from the mitochondria present in the ovum.

Mitochondrial DNA is unusual in several respects. First, it is in the form of a circle. Second, it contains 16,569 base pairs, the sequence of which has been completely mapped. Third, since there is no crossing over, the only changes that occur in mitochondrial DNA must come from mutations. If we assume that the mutation rate is constant over time (a reasonable, but not provable, assumption) then the degree of difference between the mitochondrial DNA of any two individuals is a measure of the time since those individuals shared a common ancestor.

Several studies have now been performed in which mitochondrial DNA from a diversity of individuals around the globe have been compared. Although the results of the studies are not always consistent, most suggest that we are all descended from a woman who lived about 225,000 years ago in south-central Africa. These genetic studies also suggest that her descendants fanned out of Africa about 120,000 years ago into Asia and Europe.

Is it possible that we are all descended from a single individual? Statistically, it is highly probable—lines in any species die out with the passage of generations, and if one is prepared to go back far enough, any species could, in theory, be traced back to one individual.

Does this mean that this woman was the Biblical Eve? No, in the sense that she was not the only woman of her time; yes, in the sense that there must be some universal mother at some time.

The notion of an "Eve" is not particularly controversial among anthropologists, but the methods used in the studies alluded to above are very controversial, as is the conclusion that "Eve" existed so recently (many anthropologists assume a much more ancient beginning to modern humans). Conversely, many anthropologists believe that more recent groups, such as the Neanderthals, have contributed to the collective gene pool of modern humans.

Currently, similar biochemical techniques are being employed to trace genetic changes in the Y chromosome, which is passed from father to son without engaging in crossing over. Mapping such changes is very difficult because the Y chromosome contains much more DNA than does a mitochondrion. Thus, identifying "Adam" is still several years in the future. At the very least, such a study will either lend credence to or increase doubt about the "Eve" study, since "Adam" had to be reasonably contemporaneous with "Eve"; if the study demonstrates a different place and time for the existence of "Adam," then the validity of "Eve's" identification will be questioned.

human. At slightly over one meter in height, "Lucy" was no taller than a modern five-year-old girl—but with a body weight estimated at about 30 kg, she was almost twice as heavy as a modern five-year-old.

"Lucy" was formally designated ***Australopithecus afarensis,*** and Johanson claimed that the species was ancestral both to the other australopithecine species and to genus *Homo.* More recently, Richard Leakey (the son of Mary and Louis) discovered a skull fragment dating back 2.6 million years, and with a brain size estimated at 800 cm^3—large enough to qualify as human (i.e., genus *Homo*). On the basis of this find, Leakey argued that *Homo* arose too early for *A. afarensis* to be a human ancestor. Still other researchers suggest that *A. afarensis* was ancestral to *A. africanus,* and that *Homo* arose from *A. africanus.* An even more recent discovery of a skull 2.5 million years old and variously labeled as "WT 17000," "the Black Skull," and ***"Australopithecus aethiopicus"*** is of an individual possessing both *boisei* and *afarensis* traits. Its discovery has led to even more speculation about the relationships between and among these species. It seems certain that the issues will not be completely settled until additional fossils have been discovered (see Figure 23.11).

However, clearly at least two distinct types of prehumans existed contemporaneously. Fossils found at Lake Randolph, in northern Kenya (see Figure 23.12), and at Orno, just across the border in Ethiopia (both sites are about 800 km north of Olduvai Gorge), indicate the presence of *A. boisei* from 3.7 to one million years ago. Hence, at least one, and perhaps two, robust, vegetarian species of hominids lived throughout much of eastern and southern Africa for several million years. About one million years ago, *A. robustus/boisei* apparently became extinct. Were they killed off

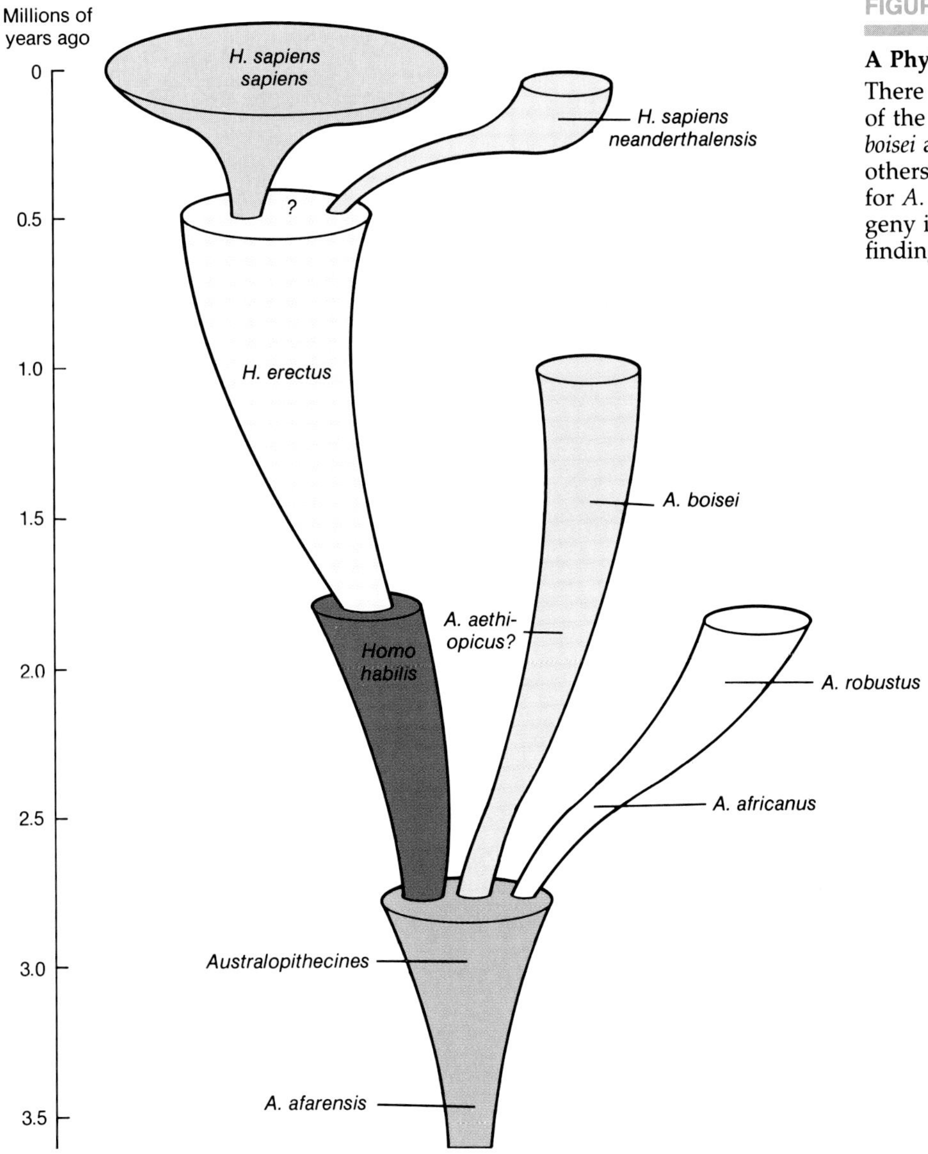

FIGURE 23.11

A Phylogeny of Recent Hominids

There are many alternative placements of the fossil hominids. Some combine *A. boisei* and *A. robustus* in a single lineage; others suggest a more central position for *A. africanus*. However, this phylogeny is consistent with the most recent findings.

FIGURE 23.12

Fossil Hominid Locale

Many of our most important fossil discoveries have come along the Rift Valley of east-central Africa. This photograph is of a site in Tanzania.

by *A. africanus/H. habilis?* We may never know for certain, but since then there has apparently been only one species of hominid on the earth at any given time.

Missing Links

During the last part of the nineteenth century and for the first half of the twentieth century, the term "missing link" was a metaphor for the creature that bridged the gap between apes and humans. As the earth grudgingly yielded a handful of fossils, it became clear that there was no "missing link." The gap between modern apes and modern humans was not to be bridged with a single link. A whole chain had been missing.

Our picture of hominid evolution during the period between one million and three and one-half million years ago has now been somewhat clarified. Nevertheless, the earliest of these fossils is still distinctly hominid. When did the hominid-pongid split occur? What does the fossil record provide?

The fossil record is unfortunately very scanty. The earliest known anthropoid was recently described from a find in Egypt. Named ***Aegyptopithecus,*** it is more than 30 million years old, and predates the split between the ceboids and cercopithecoids, on the one hand, and the hominoids on the other. In the 1930s, Leakey found a group of fossils on an island in Lake Victoria that are, at the very least, intriguing. Although the creature possessed some prosimian traits, the skull and tooth number argue persuasively that it was an anthropoid, and quite possibly a hominoid. ***Dryopithecus*** ("tree ape"), as it was called, dates back some 20 million years. Similar fossils have since been discovered in India and in Europe.

A somewhat more recent fossil, first found in India in 1932, was named ***Ramapithecus*** (for Rama, a mythical Indian prince). Its age was estimated at 12 million years. Fossils of *Ramapithecus* have more recently also been discovered in Africa, and their age is estimated at 14 million years. For many years *Ramapithecus* was generally thought of as the earliest known hominid, but the discovery of more complete fossils indicate that it is a pongid, and probably ancestral to the orangutan. Moreover, there is increasing speculation today that the hominid-pongid split occurred much more recently—perhaps only five million years ago. This speculation is based on a variety of biochemical techniques, including DNA hybridization (see Chapter 22). To date, no hominid fossils more than four million years old have been discovered. Thus the biochemical evidence has not yet been contradicted by fossil evidence, although many anthropologists still believe that the hominid-pongid split occurred much earlier than five million years ago. A jawbone found in June of 1991 may help clarify the issue. It may be ancestral to both the modern great apes and to humans, and is estimated to be ten to 15 million years old. A determination of its precise age and taxonomic position would be of great assistance in helping to pin down the timing of the hominid-pongid split.

The Genus *Homo*

As we have seen, the human lineage is very fuzzy until about three million years ago, at which point it begins to be somewhat more clear. But what about the post-australopithecine era? How well do we understand the evolution of the genus *Homo?* (See Box 23.3.)

The quick answer is, not very well. With the discovery of Java man in 1891, enormous interest was kindled in fossil humans, and for the first half of the twentieth century, a bewildering array of generic and specific names, each representing a different fossil discovery, cluttered the scientific literature. By 1960, however, there was general recognition that these various fossils were sufficiently similar that they could safely be classified as a single species. The name given to them was ***Homo erectus*** ("upright man"). They existed during a period from roughly 1.7 million years ago to about 500,000 years ago (250,000 years ago in Asia). During that time, there was a gradual increase in their brain size from about 750 cm^3 to roughly 1200 cm^3, and they were a great deal larger and heavier than *A. africanus.*

Homo erectus gave rise to ***Homo sapiens*** ("wise man") between 250,000 and 500,000 years ago. Unfortunately, the fossil record of this transition is very poor. Some fragmentary fossils exist from about 500,000 years ago, but more complete fossils are unknown until about 200,000 years ago. Even these show transitional features between *H. erectus* and *H. sapiens.*

All of this leads us back to a reconsideration of the Neanderthals, who arrived in Western Europe about 80,000 years ago, and disappeared about 32,000 years ago. What was their role in human evolution? They had a brain capacity at least the equal of modern humans, even though the skull was longer and flatter. Moreover, although they were somewhat shorter than the later Cro-Magnons, they were much more muscular. The oldest Cro-Magnon fossil dates back 34,000 years. What happened when the Neanderthals encountered the Cro-Magnons?

Before we address that question, we must further characterize the Neanderthals. During much of the time between 200,000 and 10,000 years ago, glaciers covered as much as 30 percent of the earth's surface. Throughout this period, hominid populations must

BOX 23.3

The Future of Human Evolution

The evolutionary future of humans has long been a subject of intense speculation. Even today, it is easy to find predictions that our body form will gradually change, as we rely less and less on our own locomotory capabilities, and more and more on mechanical devices. Gradually, we shall become nothing more than giant brains, with only the bare minimum of body.

EVOLUTIONARY CHANGE	SELECTIVE PRESSURE
Baldness	Hair not needed for protection from cold
Longer, more sensitive fingers	Better tool use, such as computers
Bigger feet	Needed to support a larger body
Bigger eyes	Needed to assimilate extra information

(These changes are excerpted from a recent article in a popular health magazine. Do you think the author understands evolutionary theory? Do you think his hero might be named Lamarck?)

Nonsense! This is Lamarckian evolution at its worst. We are not about to lose our legs because we ride in automobiles too much. That isn't how natural selection operates. Mutations for reduced limb size would be preserved only if they were adaptive and were therefore favored by natural selection.

Does that mean that we are no longer evolving? Certainly not as we once were. There are still changes in gene frequencies over time, and in that sense we are evolving, but we have increasingly endeavored to resist the impact of natural selection. Thus whatever evolutionary changes are now occurring are not the result of increased adaptation but are much more random—and in some instances adaptive levels are falling.

For instance, many genetic diseases that once reduced survival rates or were even lethal no longer prevent the afflicted individual from living a relatively normal life and having children. The incidence of these diseases is, in many instances, rising rapidly as a consequence, and will continue to do so until emerging techniques in gene splicing allow on-site repair of malfunctioning genes.

Of course, once that day arrives, anything is possible in an evolutionary sense, because natural selection simply will be bypassed. Genetic repair will wipe out a variety of genetic legacies—from baldness to diabetes to color blindness—and greatly increase the level of adaptiveness.

have been constantly migrating, as the glaciers alternately expanded and contracted. It is very likely that small populations were isolated by surrounding ice for many generations. We would expect that such isolated populations diverged anatomically from each other, and there is some evidence that the Neanderthals did just that.

The range of the Neanderthals extended broadly across central Europe and Asia, and even into Africa. Indeed, the earliest Neanderthal-like fossils are from Africa, more than 125,000 years ago, and true Neanderthals occurred as early as 120,000 years ago in the Middle East.

As we have seen, fossils from Western Europe were the first human-like fossils discovered—and they were significantly unlike the skulls and skeletons of modern humans. With improved dating techniques, it was also determined that these early discoveries were of individuals who lived not earlier than 80,000 years ago. The Middle East fossils are generally older, and the characteristic Neanderthal traits are less pronounced. With their discovery can we conclude that Neanderthals, far from being excluded from the modern human lineage, may have evolved into the Cro-Magnons, perhaps because of improved language skills or tool-making? In this scenario, the Western European Neanderthals were either assimilated or died out as the Cro-Magnons moved in from the east.

Many anthropologists still accept this scenario. Indeed, they see evidence of the Neanderthal legacy in the beetle brows of some Europeans. However, recent discoveries of near-modern fossils in Africa, dating back 100,000 years, coupled with biochemical evidence, suggest another scenario—namely, that yet another migration out of Africa swept the less-modern types before it, and completely replaced these earlier humans. Moreover, a re-dating of fossils in the Middle East demonstrates that from about 100,000 to about 60,000 years ago Neanderthals and modern humans coexisted, apparently without interbreeding (see Figures 23.13). Under this scenario, all modern humans derive from this central African stock, and none of the other early human stock—including the Neanderthals—made any contribution to our gene pool.

The controversy continues unabated as of this writing.

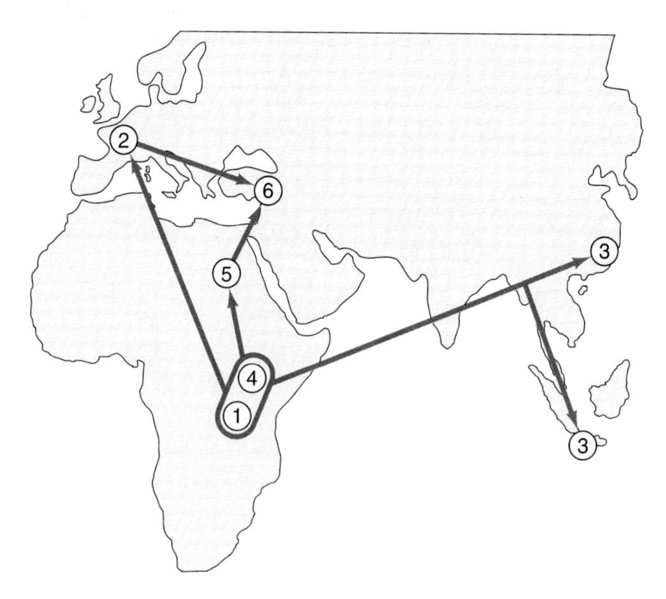

FIGURE 23.13

Out of Africa

One line of current thinking is that the genus *Homo* arose in Africa (in the form of *Homo habilis*) more than two million years ago (1) and, as *Homo erectus*, migrated to Europe (2) and Asia (3). In Europe, *Homo erectus* gave rise to the Neanderthals. Meanwhile, the original African population *Homo erectus* evolved into modern humans (4), and these humans began to move north about 150,000 years ago (5). By 100,000 years ago they had colonized the Near East, and about the same time Neanderthals migrated east ahead of the glaciers (6), the two groups eventually living in roughly the same areas for several tens of thousands of years.

HUMAN RACES

The term "race" is not widely used in biology, primarily because of the variety of interpretations that this term has suffered in other disciplines. Both biologists and social scientists are interested in cohesive groups that are largely, but not completely, reproductively isolated from other such groups. This interest exists for reasons ranging from the esoteric (measuring evolutionary rates) to the highly pragmatic (genetic counseling). However, such groupings should generally not be termed "races" for several reasons.

Poor Terminology. There are other terms readily available that are much less ambiguous. The term "population" refers to a geographically distinct group of potentially interbreeding organisms, be they sparrows or humans. The term "subspecies" designates populations that are geographically separated and anatomically distinct.

No Consensus. Anthropologists have variously distinguished anywhere from three to 30 "races" of humans. However, even the latter number is too small to recognize all of the reproductively isolated populations. "South American Indians" is a fairly restrictive category, but it is too broad to be of much use, because many of the individual Indian tribes have historically been reproductively isolated from other tribes.

Insufficient Geographic Separation. It is impossible to set geographic limits for human "races." Although there is geographic separation between some of the traditionally accepted racial groups in certain regions, there are large areas of intermixture elsewhere. For example, the Sahara Desert separates "whites" and "blacks" relatively sharply. However, to the east, there is a broad zone of mixture in the Sudan and Ethiopia. An even larger zone of mixture exists in India, and things get totally out of hand in Indonesia, the Philippines, and the other Pacific islands. Whatever utility the traditionally accepted racial classifications may have, it is lost in these areas.

Increased Genetic Mixing. Zones of mixture are becoming increasingly broader. It is likely that the major racial groups were once geographically separated. However, during historic times, and increasingly in more recent years, these groups have been mixing. To take the most obvious example—American blacks are now genetically distinct from the west African populations from which they are primarily descended. Several genes are known for which one allele is very common in West African populations, and very rare in western European populations, the homeland of most American "whites." In every case, American blacks show heterozygosity in the two alleles, with the African alleles being two or three times more common than the European alleles. This heterozygosity is a measure of interbreeding between blacks and whites in America, and indicates that American blacks are, on the average, about 30 percent "white."

In recognition of this fact, some anthropologists have proposed that American blacks are a separate "race," or an "emergent race," but such a usage has little practical application. The only valid purpose for racial designation is to identify populations that do not interbreed with other populations, and it is obvious that very considerable interbreeding has occurred and continues to occur between "blacks" and "whites" in this country.

Insufficient Biological Criteria. Perhaps most importantly, there is in no case a single characteristic that distinguishes all members of one human "race" from all members of another. The characteristics most commonly used are obvious external features such as hair and skin color, both of which are affected by many genes, and both of which exhibit a virtually continuous gradation.

Race, as it is now used, is a statistical concept, a type of averaging whereby *most* members of one population may generally be distinguished from *most* members of another population, the obvious implication being that *some* members cannot be distinguished. Unfortunately, it is in precisely this area where the term has been most abused, because of the blind insistence that everyone belongs to one race or another, and in the continued search for ways of making the identification.

The obvious fact is that even though the reproductive isolation of human populations may have once been common, and even though some populations are still reproductively isolated from other populations, there are an increasing number of people who are in no way reproductively isolated from members of the traditional racial groups.

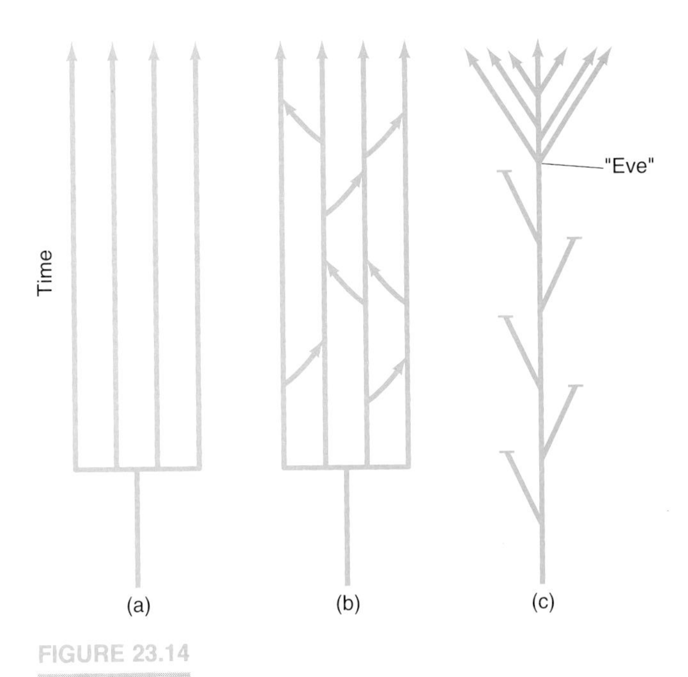

FIGURE 23.14

Path to Modern Humans

Initially, some anthropologists believed that the modern human "races" were independently derived from an ancestral stock perhaps as old as *Homo erectus* (a). Subsequently, it was suggested that, while there may have been independent origins, there was extensive mixing subsequently (b). The most popular current scenario is that all humans derive from a relatively modern African stock (c).

Thus, although there may still be justification in screening only "blacks" for sickle-cell anemia for purposes of genetic counseling, the day is not far off when, because of "mixed racial" marriages, everyone may have to be tested. By the same token, intermarriage between previously isolated populations decreases the probability of potentially dangerous homozygous recessive conditions. Thus we can expect to see a reduction in the frequency of many genetic diseases, as these "racial" barriers are broken down.

Why, then, has there been so much emphasis placed on racial categories, even among professional anthropologists? Initially, it was widely believed that racial features were very ancient and were derived from ancestral human populations—perhaps even from *H. erectus*—living in different parts of the world (see Figure 23.14). A somewhat newer hybridization model suggested ancient characters partially obscured by more recent intermixing. The newest, and currently favored, model claims that all humans derive from a relatively modern stock from Africa, which successively displaced earlier human populations (see Box 23.2). Mitochondrial DNA analysis shows that the greatest divergence among humans is never more than 2 percent, a finding suggesting that racial distinctiveness is both recent and genetically (and evolutionarily) insignificant.

Summary

The history of human evolution has been full of fits and starts, partly because we were not ready, until rather recently, to contemplate our prehuman ancestry, and partly because hominid fossils are very rare.

Discoveries during the late nineteenth century changed that, and interest in our fossil history is today very great indeed. It now appears that the earliest hominoids date back at least 20 million years, but the pongid-hominid split probably occurred much later.

There is intense debate regarding the antiquity of genus *Homo,* some authorities maintaining that *Homo* and *Australopithecus* share a common ancestry, others that *Homo* arose from *Australopithecus.* The best evidence today suggests that *Homo,* as a genus, is at least two million years old.

The evolution of the genus since that time is less controversial. The earliest fossils are called *Homo habilis,* the later fossils *Homo erectus. Homo sapiens* dates back only about 500,000 years.

The Neanderthals, the first of the hominid fossil finds, remain controversial. Some scientists believe that Neanderthals of the Near East evolved into Cro-Magnons; others believe that the Cro-Magnons were the European descendants of a group of modern humans that first arose in Africa, and that the Neanderthals made no contribution to the gene pool of present-day humans.

Key Terms

Neanderthal
Cro-Magnon
primate
Prosimii
Anthropoidea
Ceboidea
Cercopithecoidea
Hominoidea
Pongidae
Hominidae
Australopithecus africanus
Australopithecus robustus
Australopithecus boisei
Homo habilis
Australopithecus afarensis
Australopithecus aethiopicus
Aegyptopithecus
Dryopithecus
Ramapithecus
Homo erectus
Homo sapiens

Questions

1. Briefly discuss primate phylogeny. What are the major groups of living primates, and what are their presumed relationships?
2. What is the difference between hominids and hominoids? Where are the chimpanzee and gorilla placed?
3. Who, or what, are the *australopithecines?* What is their relationship to modern humans?
4. Briefly discuss *Homo habilis, H. erectus,* and the Neanderthals in terms of their presumed relationship to modern humans.
5. What is the basis of the argument that all modern humans are descended from a single female ("Eve") who lived about 225,000 years ago?
6. What is meant by the term "race"? Does the story of Eve strengthen or weaken the argument for distinguishing races in modern humans? Why?

CHAPTER

24

Principles of Ecology

Environmental Relationships

We live in comfortable, climate-controlled homes, not caves. We wear synthetic materials and eat packaged foods. We are surrounded by the marvels of our technology. Why can't this technology control our environment better? Why can't we prevent pollution?

What do DDT and radioactive fallout have in common? Why should we care if the whales become extinct? If the environment needs nitrogen and phosphorus, why does it matter if we discharge these as wastes into our rivers and lakes? What does burning more coal have to do with the melting of the polar icecaps?

These are the kinds of questions we shall consider in our discussion of ecological principles.

THE NATURE OF THE PROBLEM

Most of this text has been devoted to an understanding of how an organism—in particular, the human organism—is composed and how it functions. However, it is a truism that we do not live our lives in isolation from other members of our species, or from other species in the living world—or even from the nonliving world—although our behavior sometimes belies that fact. It is important that we spend some time examining the nature of these interactions in order to ascertain what constraints (if any) we need to consider on our collective behavior. In essence, we are asking the question, "Is the earth as a whole homeostatically controlled? If it is, are there activities of the human species that threaten to disturb that homeostasis?"

First, some terminology:

Populations are members of the same species occupying a common geographic area.

Communities are populations of different species within a common geographic area.

Ecosystems are communities plus the inorganic environment with which the organisms of the community interact.

Biosphere is the totality of the ecosystems of the world.

Ecology (Gr. *oikos,* "house") is that branch of biology devoted to the study of the interrelationships of communities and ecosystems.

Ecology is an overview science, requiring extensive knowledge in a multiplicity of disciplines, and it is still largely theoretical. The scope of ecological questions is so broad that experiments are often difficult to perform, both because of their size and because of the length of time required before definitive results can be obtained.

Three decades ago, ecological principles were all but unknown by the public. However, as public awareness of environmental problems has increased, ecology has become an area of intense interest. In fact, "ecology" has frequently been made synonymous with "the environment," and consequently it is in danger of losing its preciseness as a term. Ecology is the study of the environment, not the environment itself. Phrases such as "ecologically compatible" on the sides of detergent boxes are therefore nonsensical. "Ecology" is not an object, so how can anything be compatible with it? On the other hand, ecological principles tell us a great deal about environmental problems, and the relationship between these two topics will form the basis of our discussion in this chapter.

HOMEOSTASIS IN THE BIOSPHERE

A pervasive theme of this text has been *homeostasis*—the mechanisms and adaptations by which many organisms are able to maintain a generally stable internal environment while nonetheless functioning according to various physical and chemical laws. This same theme is equally applicable to the study of communities and ecosystems. Indeed, it is fundamental even to a cursory study of ecological principles to recall two important principles of physics:

1. According to the Second Law of Thermodynamics (see Chapter 5), no conversion from one energy state to another is ever 100 percent efficient, because some energy is always lost as heat. This law applies both to individual organisms and to the biosphere. As we have seen, it accounts for the fact that all organisms require some external source of energy in order to undertake metabolic reactions. But what of the biosphere? Dynamic processes within the biosphere involve the conversion of one form of energy to another. How does the biosphere continue to function without an external energy source? The answer is that it does not. The biosphere is not a completely closed system. Ultimately, it is powered by external energy sources, of which by far the most important is sunlight.

2. Sunlight is pure energy, unlike the organic matter that is the energy source for animals and that can also be used for the construction of new molecules. Thus in order to construct the complex organic molecules that characterize life, photosynthesizers using sunlight also require matter in the form of a particular group of elements and simple compounds. The source of this matter is not external to the biosphere; to the contrary, it is very much a part of the biosphere, and therefore the amount of this matter is limited to whatever currently exists on earth.

The point is that, unlike the energy of sunlight, which is external, continuous, and (for all practical purposes) infinite, the elements of life are internal and finite in quantity—and therefore they must be recycled through the biosphere. If required elements were to move in a dead-end fashion, the death of the biosphere would be only a matter of time, for precisely the same reasons that the failure of homeostatic controls in an organism preordains its death. Of particular interest to us will be the ways in which human intervention perturbs the homeostasis of the biosphere.

THE BIOGEOCHEMICAL CYCLES

All key elements and small compounds are recycled through the biosphere, but by looking at just four—carbon, nitrogen, phosphorus, and water—we can gain an appreciation of the complexities involved.

The Carbon Cycle

The short-term recycling of carbon between organisms and the environment primarily involves carbon dioxide (see Figure 24.1). Photosynthesizers take up carbon dioxide from the air and water, and respiring organisms release it back to the air and water. Some carbon, however, is on a much longer cycle. For example, in producing their shells, many marine organisms combine carbon with calcium and oxygen to form *calcium carbonate* ($CaCO_3$). These shells, over time, can be converted into limestone, which, if exposed to wind, freezing temperatures, or rain will gradually break down again. Organic carbon compounds can be sequestered for long periods of time and be converted to coal or oil. At present, these compounds are being reintroduced into the biosphere in great quantity.

The Nitrogen Cycle

The nitrogen cycle is somewhat more problematic (see Figure 24.2). Atmospheric nitrogen (N_2) is very abundant, but it can enter the biosphere only by first being converted to NO_3 or NH_3 by certain monerans, because it is in these forms alone that nitrogen can be taken up by plants. (Some nitrates are also formed by electrical storms and by human activities; see Box 24.1). Nitrogen wastes from animals, such as NH_3, are either taken up by plants or converted to other nitrogen compounds by bacteria; some of this nitrogen ultimately reenters the atmosphere as N_2.

The Phosphorus Cycle

Phosphorus rarely occurs in gaseous form, and therefore its cycle is largely limited to exchange with the

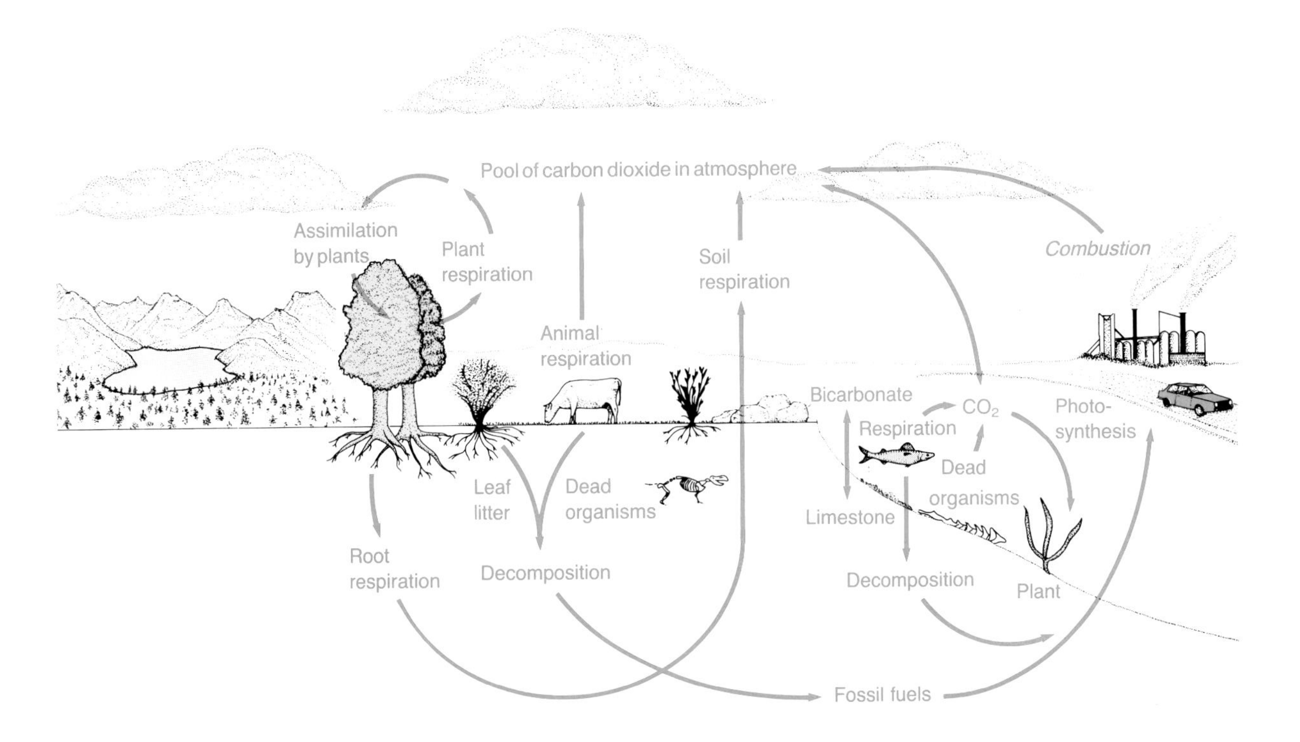

FIGURE 24.1

The Carbon Cycle

Carbon is cycled through the ecosystem when carbon dioxide is used by photosynthetic plants. Some of it then returns to the atmospheric pool when plants are eaten by animals, and carbon dioxide is expelled as a waste gas. Some carbon may be bound up for long periods of time as fossil fuels or limestone deposits.

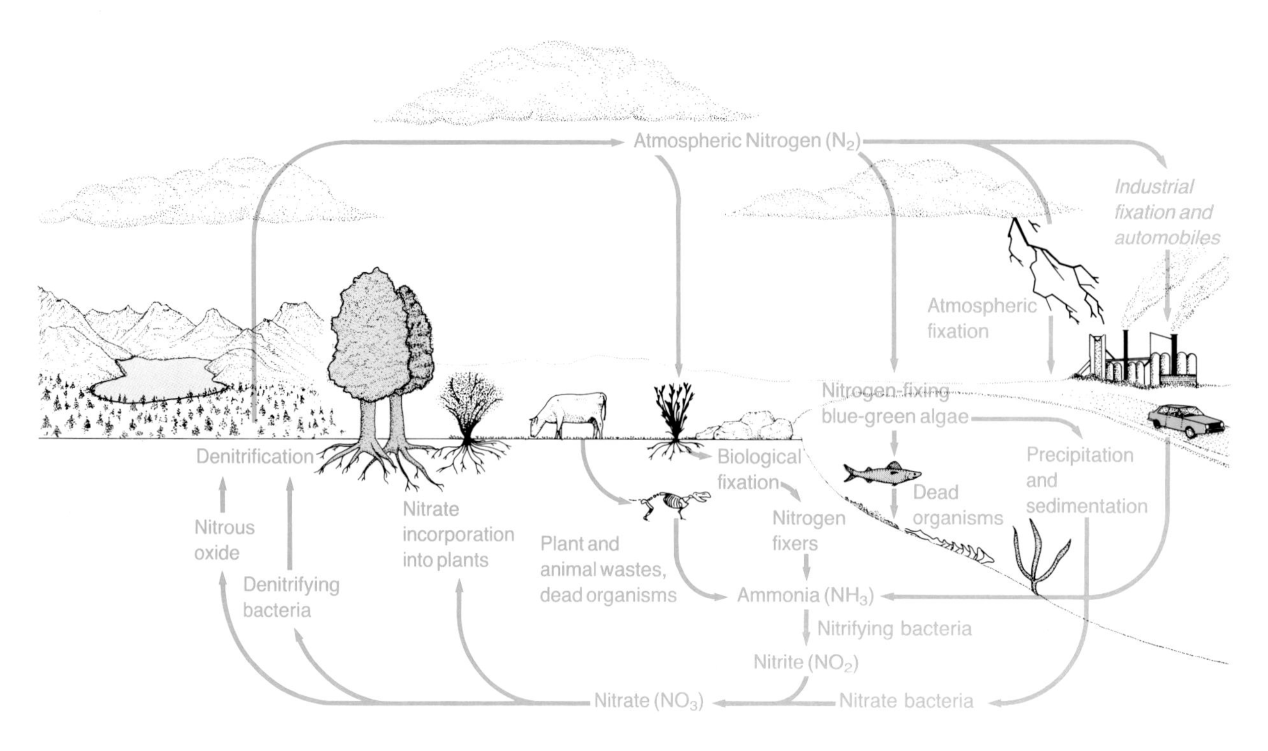

FIGURE 24.2

The Nitrogen Cycle

Nitrogen is abundant in the atmosphere, but only a few organisms can convert it for their use. In other respects nitrogen operates more like a mineral and is passed readily from plants to animals and back again.

soil and water (see Figure 24.3). Rain weathers rocks, and transports phosphorus—generally in the form of phosphate (PO_4)—into the ground water or lakes and streams where it is taken up by plants. Animals accumulate relatively large amounts of phosphates (primarily in the bones) and these phosphates are available for use by plants following the death and decomposition of the animal (see Box 24.1).

The Water Cycle

Water recycles in ways very familiar to all of us (see Figure 24.4). Evaporation from the land and oceans, transpiration from plants, and the water vapor of respiration all contribute to water in the atmosphere. Precipitation as rain or snow brings water back to the earth's surface, where it is readily taken up by living organisms.

BOX 24.1

Organic Food

Not too many years ago, organic gardeners and other "food faddists" were the objects of ridicule—extremists characterized by devotion to the cults of wheat germ and carrot juice. Such stereotypes have not totally disappeared, but organic gardening is rapidly becoming accepted as an eminently logical method of cultivation, since it involves, by definition, a recycling of materials. Fundamentally, organic gardening relies on compost, mulch, and companion planting (use of natural plant repellents) in place of synthetic fertilizers and herbicides. Organic gardening is not just for the backyard hobbyist. There are many instances of large farms successfully employing organic methods.

A present controversy centers on whether or not organic food is "healthier" than food grown with synthetic fertilizers. Whatever the answer, important is the fact that organic gardening mirrors nature's own approach to the recycling of materials within the system, rather than the interjecting of additional materials that were previously outside the system, as occurs when synthetic fertilizers are used.

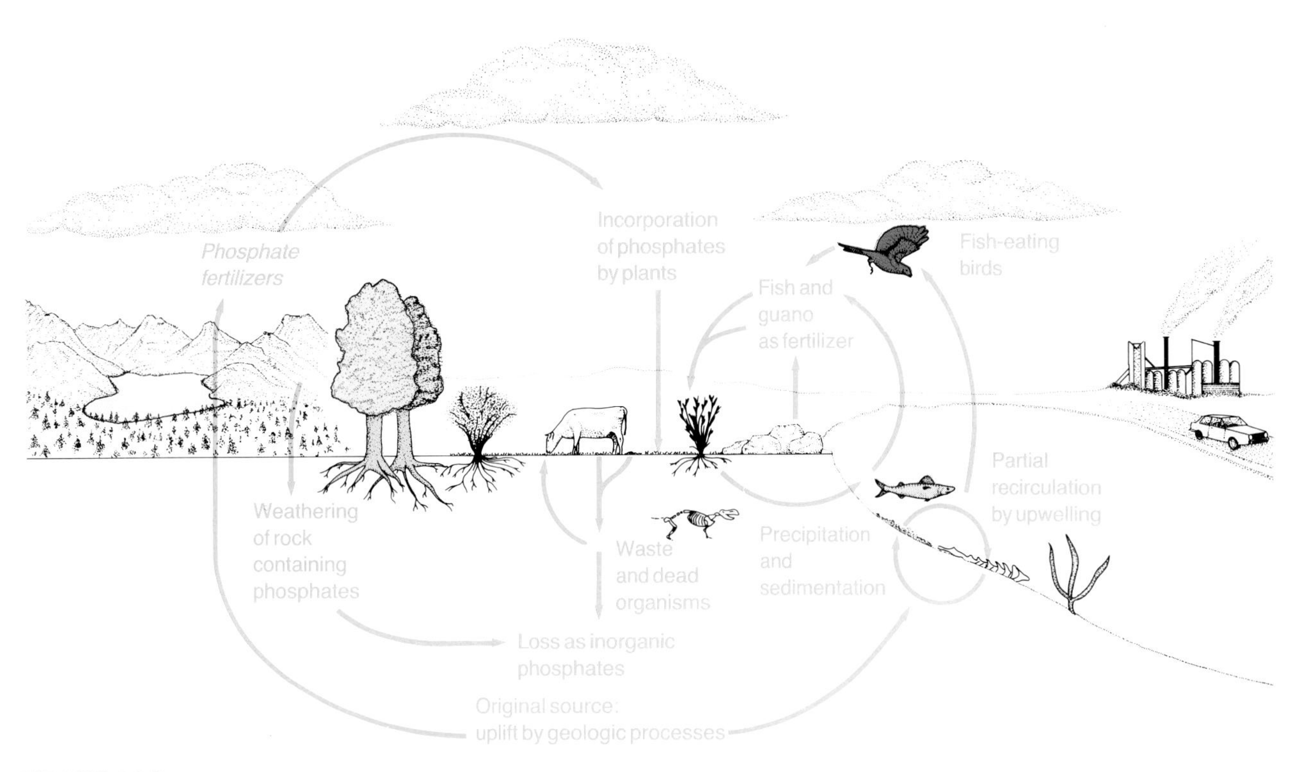

FIGURE 24.3

The Phosphate Cycle
Phosphorus is dissolved in water and used extensively by aquatic organisms and by land plants, which is why it is used in commercial fertilizers. Phosphorus does not cycle through the atmosphere.

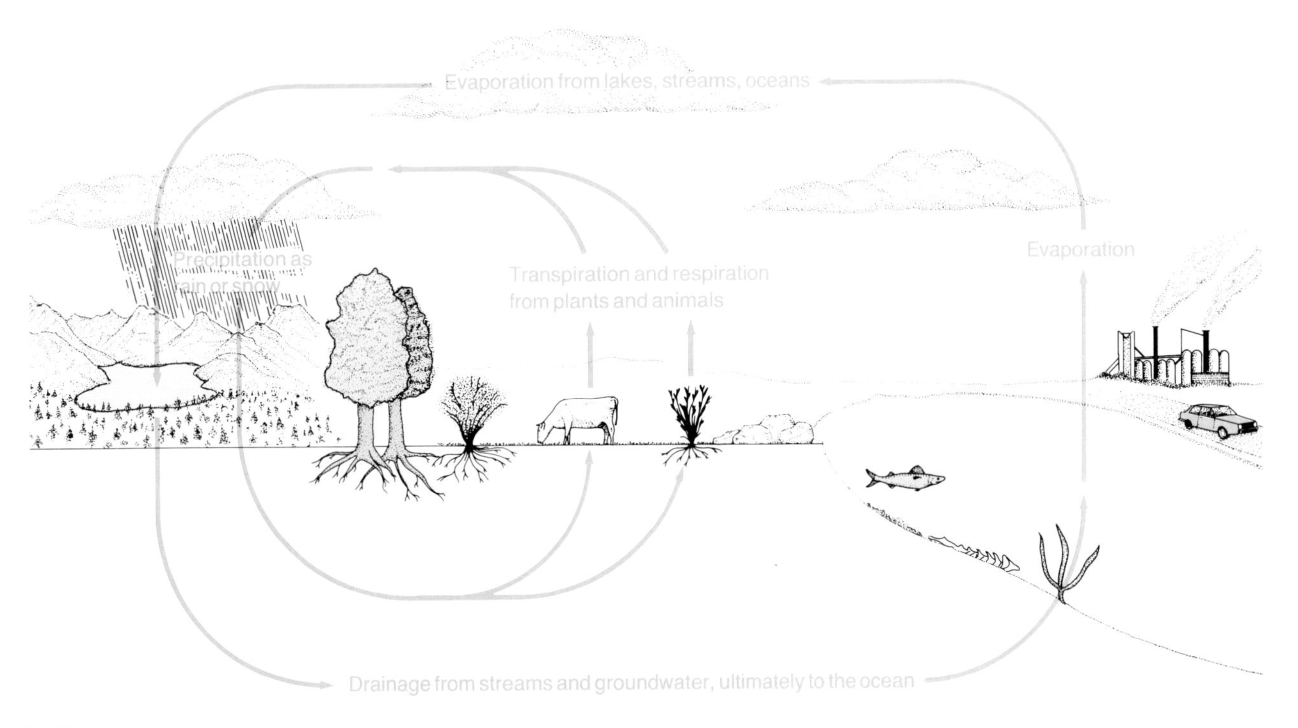

FIGURE 24.4

The Water Cycle
Water moves freely between the living and nonliving world as both a liquid and a gas.

BIOTIC RELATIONSHIPS

Ecosystems include both **biotic** (living) and **abiotic** (nonliving) components. This is a rather basic distinction, and a useful one to choose in beginning an analysis of ecological principles.

Energy Flow

By applying the Second Law of Thermodynamics to ecological relationships, we might reasonably conclude that not all of the energy stored in the molecules making up the grass that is eaten by a cow is going to be turned into additional cow (i.e., growth). A sizable proportion of the available energy is lost just in keeping the cow alive. That is, the animal requires energy to breathe, to move, to digest, to replace dead and injured cells, and so on, and all this energy must come from its food.

Moreover, a large proportion of the energy that is theoretically available is never utilized by the cow. Instead, it passes out of the animal in the form of cow pats, the volume of which is impressive, as those of you who have cleaned out cow barns well know. In fact, as a general rule only about 10 percent of the energy available in grass is actually converted into more cow (see Figure 24.5). The rest is either never used (feces), or is used for maintenance and repair.

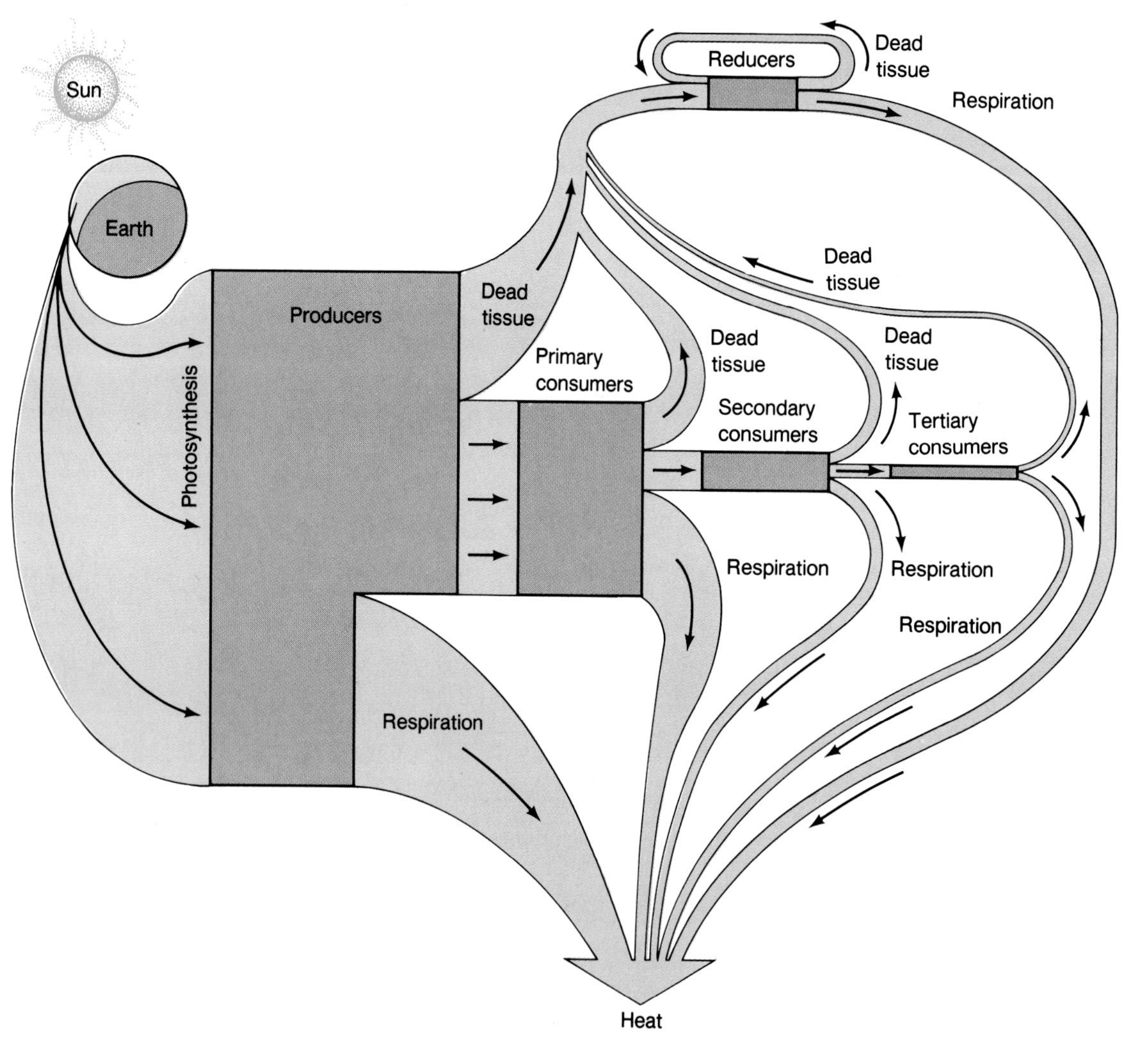

FIGURE 24.5

Energy Flow in an Ecosystem

Only about 0.5% percent of the solar energy striking the earth is used in photosynthesis, yet this tiny fraction is the source of all of the energy that moves through the biosphere. Note that only a fraction of the energy entering a given trophic level is captured and utilized for growth by the next trophic level because organisms at each level must use substantial amounts of energy for maintenance (respiration) and also because some organisms die at each level.

A similar picture could be painted for the wolf that kills and eats the cow. Not all of the cow will be eaten. Some of the blood may sink into the ground, the brain may be inaccessible within the skull, and most of the bones will be left. In addition, the wolf must use some energy for maintenance just as the cow did.

Again, only about 10 percent of the theoretically available energy will be converted into new wolf meat. However, because this represents only 10 percent of the 10 percent that the cow was able to extract from the grass for more cow meat, the wolf is converting only 1 percent of the energy originally available in the grass for its own growth. What does this fact tell us about the numbers of wolves and cows (or, more properly, the total weight of all of the wolves and all of the cows—and all of the grass, for that matter) that we might expect to find within a particular area? Read on.

Energy Pyramids

From the ecological perspective, organisms can be placed in three categories:

1. **Producers:** Organisms (such as green plants) that synthesize complex organic compounds from simple inorganic compounds;

2. **Consumers:** Organisms (such as most animals) that obtain their nutritional needs by devouring other organisms; and

3. **Reducers:** Organisms (such as most fungi and many bacteria) that break down dead organic material into its component inorganic compounds and elements.

The reducers play a critical role in the recycling of putative waste material, such as feces, bone, dead leaves, and so on, and in making the inorganic components contained in these materials available once again to the producers. We saw evidence of the activity of reducers in our earlier discussion of biogeochemical cycles. Now let us focus on the producers and consumers.

Consumers eat either producers or other consumers. Thus we can distinguish among *primary consumers* (those eating producers), *secondary consumers* (those eating primary consumers), and *tertiary consumers* (those eating secondary consumers). Many consumers, such as humans, may at any given moment be a primary, secondary, or tertiary consumer, depending on the nature of the food they are eating. However, this fact does not significantly complicate the simplified scheme of things that has just been outlined.

Based on what we have noted about the relative efficiency of energy flow from grass to cows to wolves, it stands to reason that there will be fewer primary consumers than producers, and fewer secondary consumers than primary consumers. However, raw numbers are not intuitively satisfying—of course, there are fewer cows than there are blades of grass—but we can express the relationship in terms of energy flow.

The relationships of producers and the various classes of consumers is called an **energy pyramid,** a "layer cake" in which the size of each layer represents the amount of stored energy (measured in kilocalories) in that layer (see Figure 24.6). These layers are called **trophic levels** (Gr. *trophe,* "food"). Note that the Second Law of Thermodynamics predicted the pyramidal shape of these producer-consumer energy relationships.

Predictions from the Energy Pyramid. The fact that the energy pyramid is, in fact, a pyramid, and not a rectangle, explains why large secondary consumers, such as cougars, lions, and leopards, are relatively rare. The energy pyramid also explains the virtual absence of large terrestrial tertiary consumers. For example, there are no animals that prey on lions. Lions are simply too rare to be a reliable food source, and a lion predator would have to be much less numerous. The low population density of our hypothetical lion predator would make finding a mate very

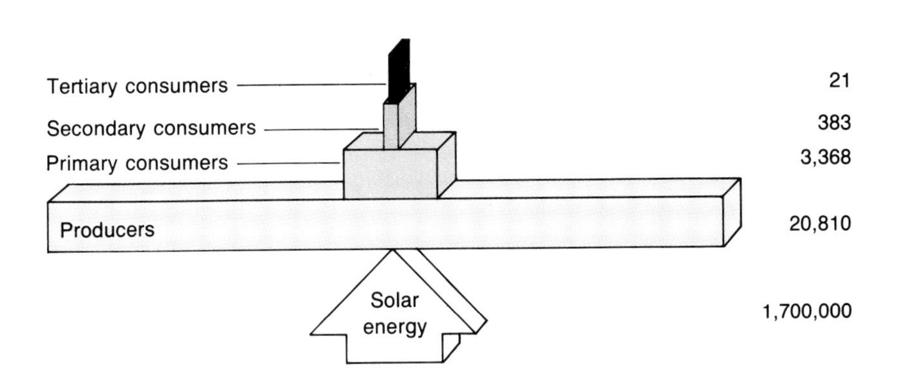

FIGURE 24.6

An Energy Pyramid

The numbers, in kilocalories per year, represent the energy stored by each trophic level. The data were calculated for a square meter of an aquatic ecosystem in Florida. It is obvious that only a small amount of the energy theoretically available to the organisms of a given trophic level is actually used by them.

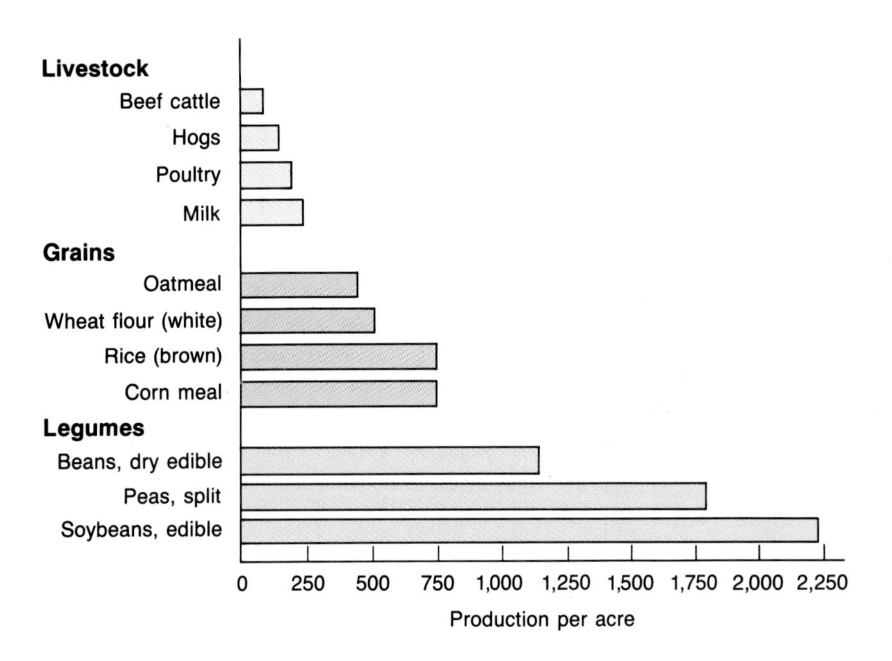

FIGURE 24.7

Protein Production in Human Foods

The scale represents the amount of protein produced per acre, expressed as the number of days it would support one person's needs. For example, in a year one acre of land will produce enough beef to support a person for a few months, but one acre will produce enough soybean protein to support a person for more than six years.

difficult, to say nothing of the energy demands involved in traversing huge areas in search of lions to eat. (Large tertiary consumers do exist in the ocean, however, where there is greater layering of trophic levels.)

Moral Dilemmas and the Energy Pyramid. Consider the human situation. As omnivores, we slide readily between various trophic levels. We can eat bread as easily as we can beef—or tuna, for that matter. (Tuna are generally tertiary consumers, meaning that we become quaternary consumers when we eat tuna.)

However, our potential flexibility in food creates a moral dilemma. More people can be supported on a diet rich in vegetables than on a diet rich in meat (see Figure 24.7). In times of food scarcity (which, on a world-wide basis, seems to be a continuous event), what are our moral and ethical obligations to become vegetarians, so more can live?

To be fair, the problem is not quite that cut and dried. Cows are actually a bit more than 10 percent efficient, in terms of conversion of grass to flesh, and pigs do even better. Moreover, many cattle are grazed on land that is too dry to support a plant crop that could serve as food for humans. However, the fact remains that huge amounts of grains, notably corn, are fed to cattle in feed lots, and this is food that could just as easily be eaten by people. Even if we are conservative and allow just half the saving—that is, projecting that only five people could live on grain for every person living on beef—do we have an obligation to eliminate beef from our diets to allow starving people in other countries to share in our grain?

Of course, this is a moral, not a biological, question. However, it is ecological theory that underlies our knowledge of the choice we face.

Biomagnification. A very different problem may also be explained by the energy pyramid. There are many elements and compounds that are not normally a part of the biosphere, and that are frequently poisonous to organisms in the biosphere (see Box 3.1, Chapter 3). These include such materials as the heavy metals, pesticides, and so on, discussed in greater detail later in this chapter. However, the relevant point here is that they are passed from trophic level to trophic level with virtually no loss. Because of the volume of food eaten by animals in the upper trophic levels, they gradually accumulate massive amounts of these poisons, a process known as **biomagnification.**

A few years ago, the brown pelican, the state bird of Louisiana, had virtually disappeared from the coastal areas of that state. The residues of DDT sprayed on interior croplands were washed out to sea via the Mississippi River, where they were picked up by microorganisms living in the Gulf of Mexico. These microorganisms are the food of small fish, and the small fish are eaten by larger fish. Pelicans, which are generally tertiary consumers, ate these large fish that were, by this point, heavily laden with DDT. The DDT levels in the pelicans reached the point where they interfered with the formation of the outer layers of the eggshell, causing the eggs to be very brittle, and leading to poor hatching.

With the banning of DDT use in the United States, and the importation of pelicans from Florida, the brown pelican has made a comeback in Louisiana.

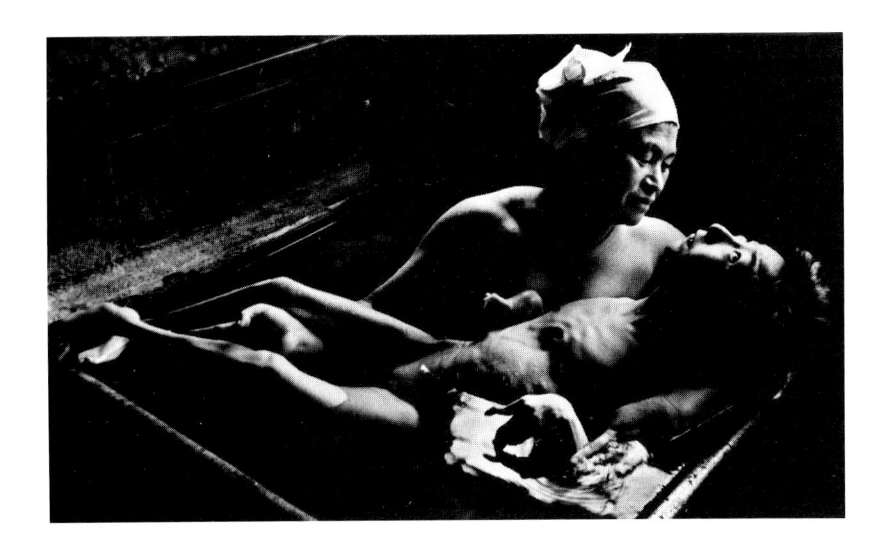

FIGURE 24.8

Mercury Poisoning

This dramatic photograph of a resident of Minamata Bay, Japan, in 1953 graphically illustrates the consequence of mercury poisoning.

However, the brown pelican story (which was also true for other tertiary bird species, including eagles and fish hawks) provides an elegant example of breakdown in the homeostasis of an ecosystem, a breakdown that was exacerbated by the nature of the energy pyramid. That is, once chemically stable, but biologically destructive, molecules are introduced into the ecosystem, they will become progressively more concentrated in the higher trophic levels, often with disastrous results.

Biomagnification can have direct impact on humans as well. In 1953, fishermen in the Minamata Bay region of Japan developed a variety of strange symptoms, beginning with loss of coordination, then paralysis, impaired intelligence, and, in more than 40 cases, death (see Figure 24.8). The cause was ultimately revealed to be mercury poisoning.

Inorganic mercury was being dumped into the bay as industrial waste. Bacteria converted this mercury into organic mercury compounds that were accumulated by microorganisms, and then passed up the food chain in increasingly higher doses. At the high levels ingested by fishermen, the organic mercury compounds began to destroy the plasma membrane of various cells of the body, with devastating consequences.

Consumption of the fish was ultimately banned, but the damage to the families of the fishermen was already done. Moreover, their means of making a living was taken away. This type of problem is not quickly resolved, even if dumping is stopped immediately, because the compounds tend to be recycled back through the ecosystem.

Food Webs

Energy pyramids group species into trophic levels of diminishing size, thereby illustrating the relative inefficiency of energy transfer from one level to another. A more fine-grained analysis can be obtained by determining predator-prey relationships for each species. The totality of these interactions is known as a **food web** (see Figure 24.9).

In complex communities, where there are large numbers of species, these relationships may be very intricate. Complexity, however, ensures stability, and stability is at the very center of homeostasis. To illustrate this point, consider the difference between communities from polar and from temperate regions.

It is not altogether clear why the number of terrestrial plant and animal species declines as one moves away from the equator toward the poles, but the fact that this decline in species number does occur is indisputable. Throughout much of the Arctic, a small mouse-like animal, the lemming, is a common herbivore that is generally an important prey species for the relative handful of secondary consumers in that ecosystem. However, the lemming population is never very stable, but passes through a series of booms and crashes, which makes life for lemming predators a succession of feasts and famines. Indeed, one reason for the lemming's population explosions may be a paucity of predators, because they have died off during the most recent lemming crash.

In contrast, the predator-prey relationships are much more complex in temperate regions. Thus the decline in the numbers of one species does not usually place a serious strain on the survival of the predator species, for there are generally several additional prey species that the predator can utilize. As a consequence, the population booms and crashes so common in polar species are uncommon in most temperate communities.

Things are even more complex in the tropics. In Malaysia a plot of land only half a square kilometer in area was found to contain 835 tree species—more

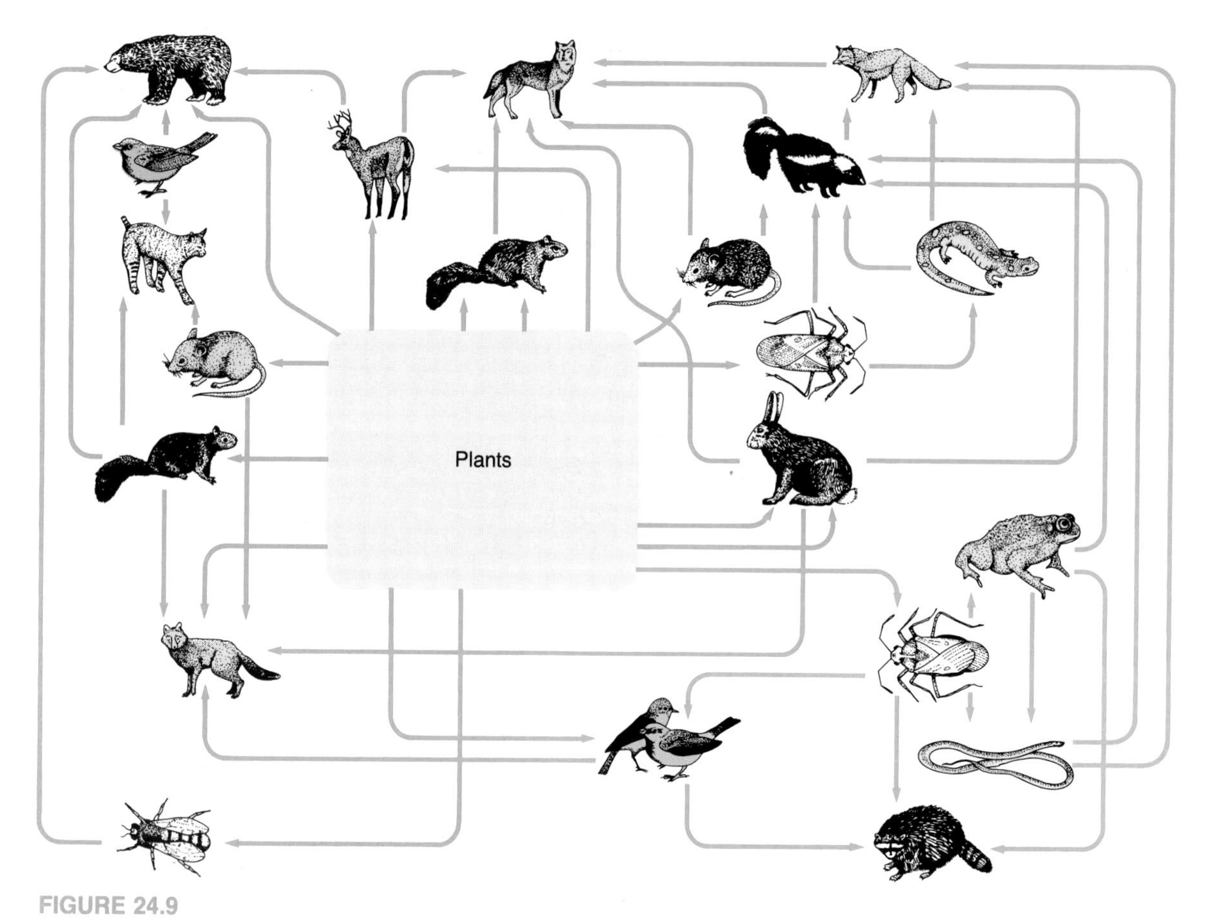

FIGURE 24.9

A Food Web

This diagram illustrates some of the major interrelationships of a North American forest food web.

species than occurs in all of the United States and Canada combined. Tropical food webs are often extraordinarily complex, but they are also fragile, because many of the species are, by temperate standards, very rare, and destruction of even modest amounts of tropical forest can result in extinction.

Monoculture. What does all this mean to humans? Consider our artificially created communities. You might well be able to walk from Ohio to Iowa and always be within spitting distance of a cornfield. Wheatfields are all but continuous from Texas to Manitoba. This uniformity represents an agricultural practice known as **monoculture.** From the standpoint of complexity, these corn and wheat fields resemble a polar community more than a temperate community, a fact well illustrated if we consider the insect pests of wheat.

The most primitive group of wasps are the *sawflies.* They have a long fossil record but are now evolutionarily on their last legs, so to speak. Some species of sawflies spend their preadult life in the stems of grasses, but most of these are rather rare. A very prominent exception is the *wheat stem sawfly,* a species that was given an evolutionary reprieve once humans decided to grow wheat in a monoculture.

In simplifying the community, we have successfully eliminated most of the species that prey on the sawfly, because the predators have few other species to eat when the sawfly is not readily available as food (that is, during its egg and pupal stages). As a consequence, we must share our wheat crop with the sawfly which, of course, ensures less wheat for us.

In some areas of the country, there has been active reduction of the practice of monoculture, through the increase of fringe areas around the crop fields, or by the planting of different crops in adjacent fields (see Figure 24.10). Such techniques reduce efficiency somewhat, as slightly less acreage is planted, and the harvesting of different crops often requires different machinery, but these techniques are ecologically more sound because they ensure greater diversity of species and, as a consequence, greater community stability.

The same "diversity equals stability" argument can also be used to justify attempts at saving endangered species from extinction. Many people are trou-

FIGURE 24.10

Strip Cropping

The intervening strips are relatively diverse communities, unlike the crops, and thus contribute to the stability of the ecosystem.

bled, perhaps out of a sense of collective guilt, over our incredible success at eliminating or endangering other species. Many other people are unconcerned, feeling that the earth is ours to do with as we please, plants and animals being here only to serve us. (One of these groups consists of enlightened idealists, whereas the other is comprised of insensitive clods, but I don't want to influence your thinking by saying which is which.)

The fact remains that it is difficult to convince doubters that *your* feeling of ill ease should change *their* thinking about *their* right to harpoon whales. What is needed is a pragmatic, nuts-and-bolts-style argument—and there is such an argument in the form of diversity equals stability. The elimination of any species cuts a series of lines in the food web of the community to which the species belonged. Thus, just as a piece of cloth is weakened when several threads are cut, so, too, is the stability of the community weakened by the extinction of a species. Because we are ourselves part of the food web, in a very real sense we are endangering our own existence by causing the extinction of other species.

This is not to say that all the inhabitants of New Jersey will suddenly turn belly up when the last whale is killed. The effects of extinctions are far more subtle. However, because we are subject to the same ecological principles as are other species, we are certainly jeopardizing ourselves by encouraging the extinction of other species—and this is hardly the best point in our history to jeopardize our future existence, since that future is already seriously threatened by other problems.

Consider a simple example. Several years ago, a Missouri farmer, tired of the eau de skunk with which his dogs kept returning home, vowed to rid his farm of skunks. Using a combination of poison, traps, and bullets, he was largely successful. However, some time later, he noticed that his ducks began disappearing in increasingly larger numbers. Mystified, he called in his agricultural extension agent who found that the source of his problem was a huge increase in the numbers of snapping turtles in the farmer's ponds. (Snapping turtles kill ducks by pulling them below the surface of the water until they drown.) What accounted for the snapping turtle increase? Skunks feed on turtle eggs, and, with virtually no skunks on the farm, the snapping turtle population was growing almost unchecked. The farmer imported some skunks, and the balance of the ecosystem was restored.

ABIOTIC RELATIONSHIPS

As was mentioned at the beginning of this chapter, it is imperative that those elements that are essential for life be recycled within the ecosystem, rather than being relegated to some ecological cul-de-sac where they are inaccessible to organisms. Problems arise when, because of human activities, recycling is impaired. Impairment may take one of two forms:

1. **Pollution:** Pollution is either the introduction of a poison into the ecosystem or the creation of an excess of a required substance to the degree that the homeostasis of the ecosystem is threatened.

2. **Depletion:** Depletion is the insufficiency of an essential substance.

Let's consider pollution and depletion as they relate to air, water, and soil.

Threats to Air Quality

We can distinguish five categories of air pollution: (1) the increase in greenhouse gases, (2) the release of toxic substances in general, (3) the production of acid rain, (4) the generation of smog, (5) and radiation.

The Greenhouse Effect. As we saw in our earlier discussion of physical laws, no energy conversion is accomplished without waste, and this waste typically takes the form of heat. However, the amounts of heat produced by the energy transformations within organisms, or even by geological events in the earth, are minuscule compared to the magnitude of the heat of the sunlight striking the earth. Put another way, the temperature of terrestrial and aquatic environments are very rarely affected significantly by anything other than climatic factors, the basis of which is heat from the sun.

However, the validity of this statement is belied by the consequences of the Industrial Revolution, which had its beginning about 250 years ago. Since that time, we have, in increasing amounts, burned fossil fuels for heat and power, and we have long since reached the point where our activities are negatively affecting the homeostatic balance of ecosystems.

Consider the **greenhouse effect** (see Figure 24.11). Certain molecules in the atmosphere, most notably carbon dioxide and water vapor, have the capacity to permit the passage of solar energy to the earth's surface, but then to absorb the radiant energy reflected back from the earth, in much the same way that the glass of a greenhouse permits the passage of solar energy but traps much of the heat within the greenhouse (see Table 24.1). The consequence of the presence of carbon dioxide and water vapor in the atmosphere has been that the earth's average surface temperature has been raised by several degrees. (Matters are complicated somewhat by the varying amounts of water vapor in the atmosphere at different times and places. Large amounts of water vapor create clouds that can prevent solar energy from reaching the earth's surface in the first place.)

The overall contribution to carbon dioxide levels by humans seems minuscule by comparison to the forces of nature. Primarily as a consequence of the burning of fossil fuels (and, more recently, because of the burning of tropical forests), human activity adds more than 6×10^{12} kg of carbon dioxide to the atmosphere annually (see Figure 24.12). On the other hand nature (through respiration, volcanic activities, and other events) adds almost 182×10^{12} kg annually. Homeostatic control mechanisms remove some 185×10^{12} kg annually (primarily into the oceans), leaving a net gain of almost 3×10^{12} kg per year. Cumulatively, over the past century, this addition has resulted in a 25 percent increase in atmospheric carbon dioxide levels (see Figure 24.13).

On a worldwide basis, it is not simply the industrialization of developing countries that is contributing to the problem; it is also the destruction of the rain forests, primarily by burning, which both introduces large amounts of carbon dioxide into the atmosphere and destroys the trees that are an important factor in the removal of carbon dioxide from the atmosphere. In 1987 alone, an area the size of Kansas was burned in the Amazon. The original area of the rain forests worldwide was about 16 million km^2. By 1990, this area had been reduced to less than 10 million km^2, and by the year 2000, less than 6 million km^2 of rain forest are expected to remain.

The increase in carbon dioxide, coupled with a small but detectable increase in average global temperatures, has promoted fears of a forthcoming global warming of unprecedented magnitude—a rise in global temperatures of perhaps 3° C by 2030, and (because of icecap melting) a corresponding increase in ocean levels of about one meter (see Figure 24.14). Moreover, the changes in temperature would not be uniform throughout the world; consequently, ocean current patterns would change, and there would be wholesale changes in the earth's climate.

On the other hand, the greenhouse effect has been actively considered by scientists for the past 30 years, and many scientists view the magnitude of current concerns as unwarranted. They point out that the increase in carbon dioxide concentrations over the past century should have already resulted in a global temperature increase of 3° C, when in fact the actual increase has been about 0.5° C. Their interpretation is that the earth's homeostatic control mechanisms are minimizing the impact of excessive carbon dioxide release.

However, a recent study on a series of lakes in northwestern Ontario showed both air and lake temperatures up 2° C over a 20-year period, with a three-week increase in the ice-free season. Moreover, in the last four years, carbon dioxide levels in the atmosphere have increased at the rate of 1.71 ppm annually—substantially above the 1.4 ppm annual average of the 15 years previously. Finally, yet another study of cores from the Antarctic ice sheets has concluded that carbon dioxide levels rose just before a previous melting of the ice sheets, some 130,000 years ago.

The problem is that no one knows for certain what the magnitude of the risk is—and the danger is that, by the time we are certain, the damage will already have been done. Even if the earth has thus far been able to adjust for the current increase in carbon dioxide levels, who is to say that will continue in the

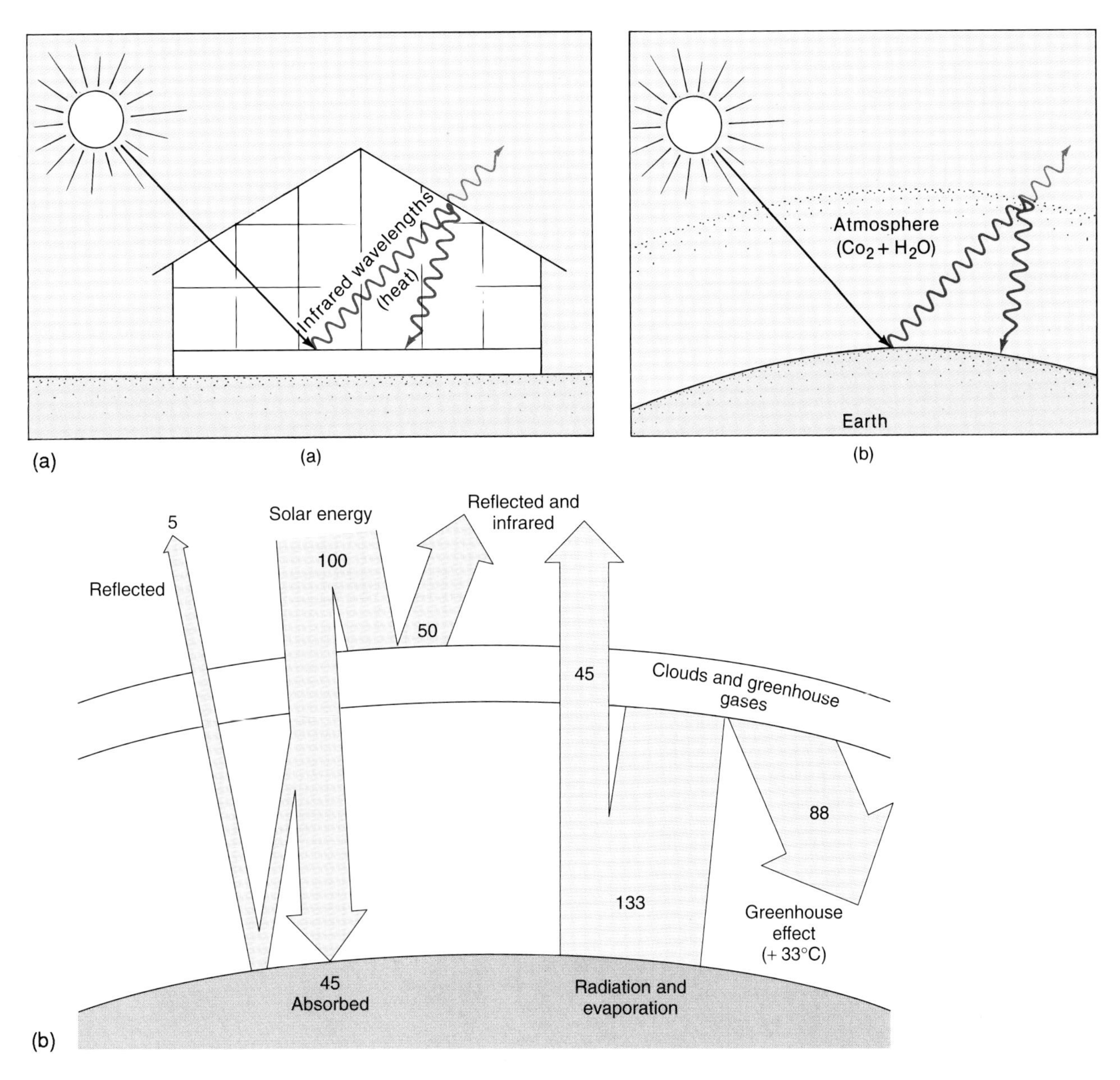

FIGURE 24.11

The Greenhouse Effect

(a) Light readily penetrates the glass of a greenhouse, but it is reradiated from the plants and soil as infrared. It is then deflected by the glass, thus raising the temperature of the greenhouse. (b) Similarly, some solar energy is reflected by clouds and gases in the atmosphere, but about half is absorbed by the earth's surface. Radiation and evaporation are responsible for energy leaving the earth's surface, but most of that energy is reflected back by clouds and gases, effectively raising the earth's temperature by 33°C over what would occur in the absence of an atmosphere. (Numbers are in percent, with solar energy arbitrarily assigned 100 percent.)

future, as carbon dioxide levels continue to rise? Would it not be prudent to attempt to reduce our present emission levels?

Unfortunately, that is easier said than done. In the United States, 34 percent of our carbon dioxide emissions are from automobiles, and, unless we develop solar automobiles, or turn to hydrogen as a fuel, there is no easy way to reduce these emissions. Another 34 percent is from industry; only a massive conversion to nuclear power plants (which do not produce carbon dioxide, but which have other obvious liabilities) would have much effect on industrial emissions.

In November 1989, an international conference on greenhouse gases was convened in Amsterdam. Of

TABLE 24.1 Atmospheric Impact of Trace Gases

GAS	HUMAN SOURCES	EMISSIONS PER YEAR (IN MILLIONS OF TONS)	DURATION IN ATMOSPHERE	CURRENT CONCENTRATION (PARTS PER BILLION)	ESTIMATED CONCENTRATION IN 2030 (PARTS PER BILLION)	GREENHOUSE EFFECT	ACID RAIN	SMOG AND DECREASED VISIBILITY
CFCs	Refrigerants, foams, aerosols	1	60–100 years	3	3–6	+	—	—
Carbon dioxide	Fossil fuels, deforestation	5500	100 years	350,000	475,000	+	—	—
Sulfur dioxide	Fossil fuels, smelting	150	Weeks	Up to 50	Up to 50	—	+	+
Methane	Landfills, cattle, rice paddies	400	10 years	1700	2300	+	—	—
Nitric oxide and nitrogen dioxide	Fossil fuels, deforestation	30	Days	Up to 50	Up to 50	—	+	+
Nitrous oxide	Fertilizers, deforestation	15	150 years	310	340	+	—	—

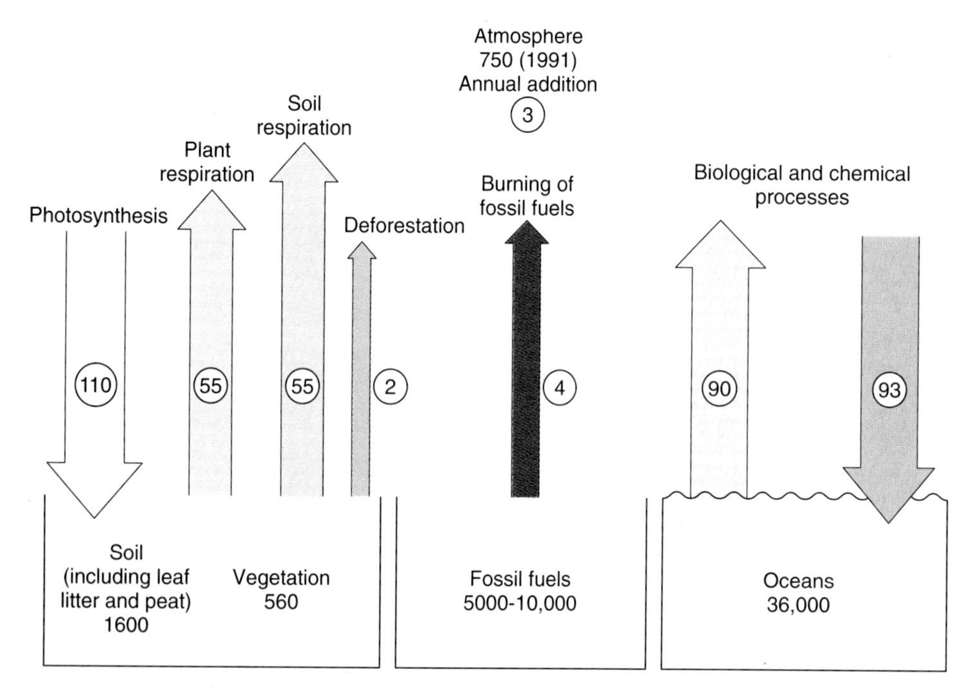

FIGURE 24.12

Carbon Dioxide and the Greenhouse Effect

Human activities, including deforestation and the burning of fossil fuels, have created a homeostatic imbalance in the recycling of carbon dioxide. As a consequence, the atmosphere is accumulating about three billion metric tons of additional carbon dioxide each year. (All figures are in billions of metric tons.)

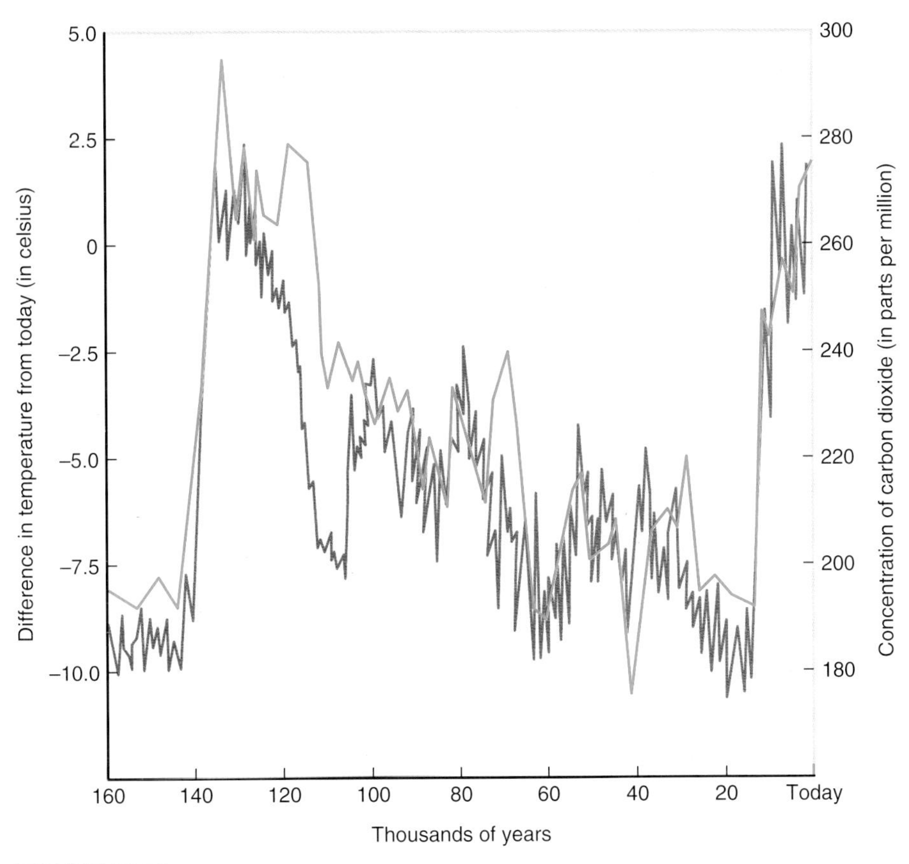

FIGURE 24.13

Historic Record of Carbon Dioxide

Analysis of the carbon dioxide content of air bubbles trapped in polar icecaps reveals that for the last 160,000 years there has been a very strong correlation between the amount of carbon dioxide in the atmosphere and the relative temperature of the earth.

the 68 nations attending, 30 agreed to a proposal to stabilize the annual emissions of greenhouse gases by the year 2000, and to reduce emissions by 20 percent from that level by 2005. The United States, the Soviet Union, China, and Japan—which collectively account for 58 percent of the emissions of such gases—were among the nations voting against the proposal. Japan agreed to standards at a subsequent meeting in Geneva in October 1990, but the United States once again balked.

Toxic Emissions and the Ozone Layer. Carbon dioxide is obviously not the only gas released from automobiles, power plants, and industrial facilities—and many of the other emissions are of substances that are inherently toxic. Indeed, recent data indicates that 1.2 billion kg of toxic materials are released into the air in the United States each year.

The news is not all bad. Auto emission controls enacted in 1970 mandated a 90 percent reduction in auto emissions by 1990—and this figure has been achieved. In fact, the typical 1990 automobile emits just 4 percent of the pollutants produced by the typical 1970 automobile—although the impact of this change is mitigated somewhat by the increase in the number of automobiles in the United States over the past two decades.

On the other hand, some of the toxic emissions are increasing, both in amount and in environmental impact. **Chlorofluorocarbons (CFCs)** are used extensively as refrigerants, industrial solvents, and in the manufacture of various foam plastics. They were also once used as the propellants in aerosol cans (because they are stable, biologically inactive, and inexpensive to manufacture). However, in 1974, scientists implicated CFCs in the destruction of the ozone layer (see Figure 24.15).

At this point, a brief digression is warranted. The **ozone layer** is a region of the upper atmosphere where oxygen gas (O_2) is converted to ozone (O_3) by the action of sunlight. The third atom of oxygen is loosely bound to the other two; consequently, ozone is a very reactive gas and is highly irritating to the human respiratory system, among many other biological tissues.

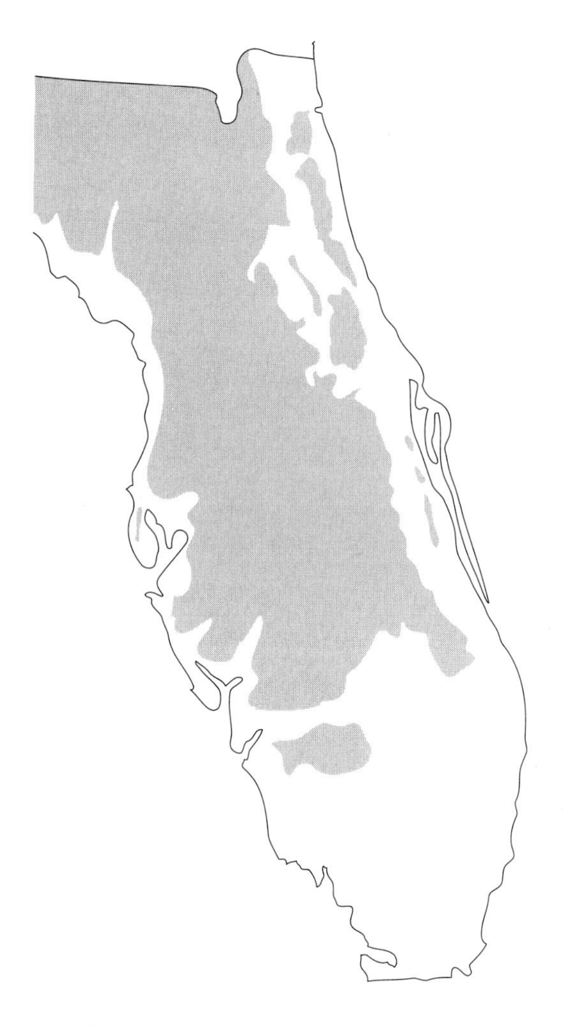

FIGURE 24.14

Effects of Global Warming

A fear of many scientists is that an increased greenhouse effect will lead to melting of the polar icecaps. Should the sea level rise by as much as 7.6 meters, the coastline of Florida (among many other places) would be dramatically changed.

Fortunately, sunlight-manufactured ozone normally remains high in the atmosphere, although on bright, windy spring days, the "bracing" feel to the air is due to downdrafts of ozone. The total amount of ozone in the atmosphere is not large—if it were all concentrated at sea level pressures, it would make a layer only one meter thick. Nevertheless, it is present in the upper atmosphere in sufficient volume as to screen out 99 percent of the ultraviolet radiation present in sunlight.

As we saw in Chapter 17, UV radiation is the primary source of skin cancer. Medical experts estimate that, for each 1 percent decrease in the ozone layer, there will be a 2 percent increase in melanoma (the most dangerous form of skin cancer, causing 5000 deaths per year in the United States) and a 5 percent increase in squamous cell cancer. (The lifetime incidence of melanoma in the United States has increased from 1 in 1500 in 1930 to 1 in 150 in 1986, primarily because of greater interest in tanning.)

What do CFCs have to do with the ozone layer? CFCs undergo complex chemical interactions with ozone, and ultimately a single molecule of CFC can destroy more than 100,000 molecules of ozone. Although CFCs were banned from aerosol sprays in this country in 1978, we still produce more than 600 million kg of CFCs every year, and there is now strong evidence that the ozone layer is indeed thinning. CFCs also contribute to global warming, since they are 100,000 times more efficient than carbon dioxide at retaining radiant energy reflected from the earth's surface. Manufacturers are looking for suitable replacements for CFCs, but even if all production were stopped tomorrow, our problems are not over—CFCs are stable, remember, and today's CFCs will remain in the atmosphere for another 75 years!

The 1990 amendment to the 1987 Montreal Protocol calls for ending CFC production by the year 2000, and establishes a $240 million fund from developed countries to assist poorer countries in shifting away from CFCs. Other solvents containing chlorine and destroying ozone are also affected by this agreement: Carbon tetrachloride production will cease by the year 2000, and methyl chloroform by 2005.

Acid Rain. Many of the small lakes that dot eastern Canada and the northeastern United States have become increasingly acidic in recent years, to the point where some are all but devoid of life (see Figure 24.16). Some tree species in these areas have been severely damaged as well—and similar effects have been observed in central Europe (see Figure 24.17). Many scientists have believed for some time that the source of these problems is acidic rain and snow, falling into the sparse, granite soils of the northeast, which have little buffering capacity. They also believed that the source of the acidity of the rain was the *sulfur dioxide* (SO_2) being emitted from coal-fed power plants, primarily in the midwest. In the atmosphere, sulfur dioxide combines with water and oxygen to form *sulfuric acid* (H_2SO_4).

These conclusions have been challenged for several years, although more on political than scientific grounds. However, new techniques have recently permitted the "fingerprinting" of effluents from particular areas, and it is now apparent that fully half of the SO_2 released into the atmosphere in the United States is from just 50 power plants. There is, as yet, no plan or agreement on when and how these power plants will reduce their emissions.

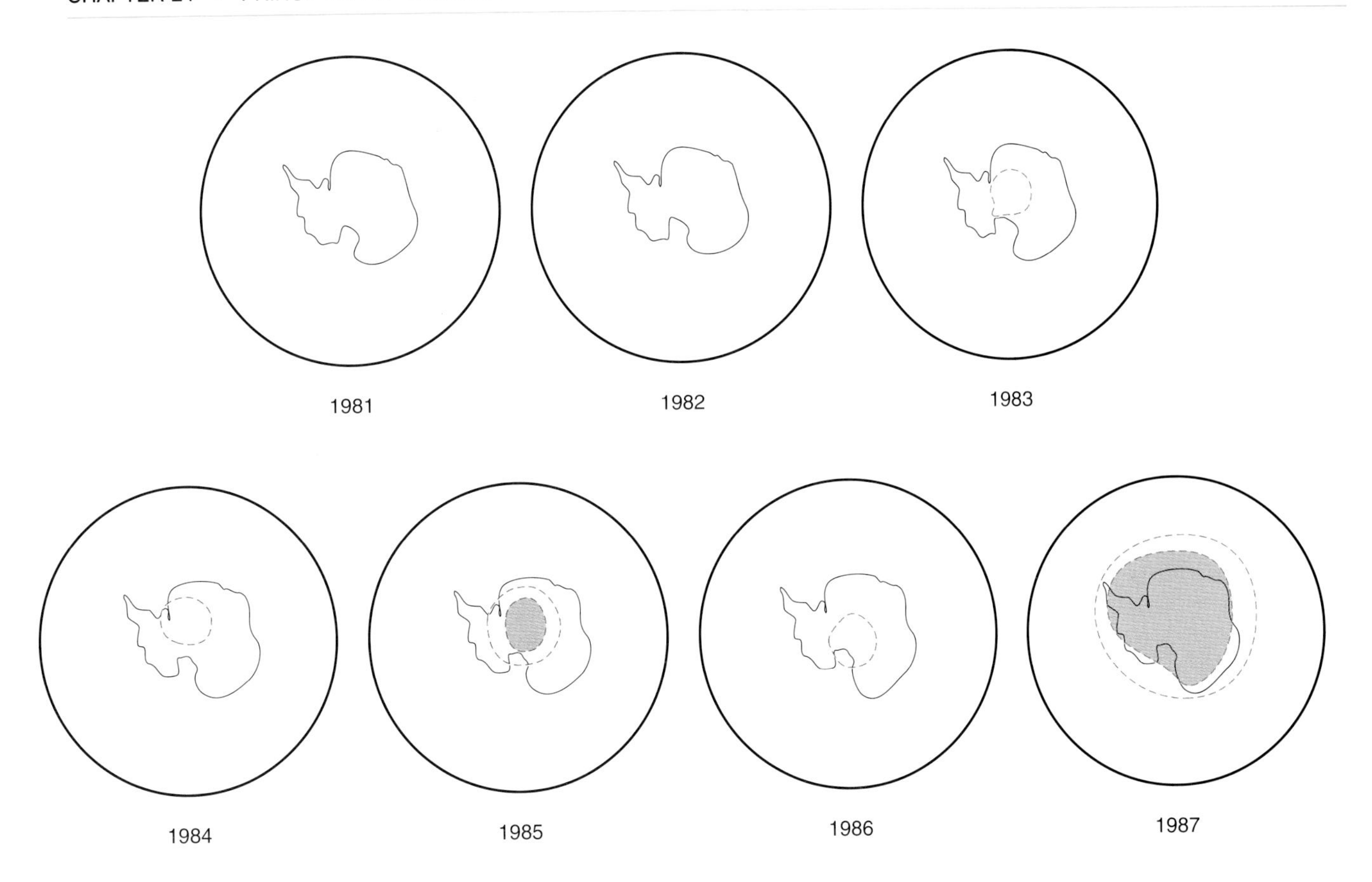

FIGURE 24.15

Holes in the Ozone Layer

In the last decade, satellite photos of the Antarctic have revealed increasingly larger losses of ozone, almost certainly as a consequence of the release of CFCs into the atmosphere.

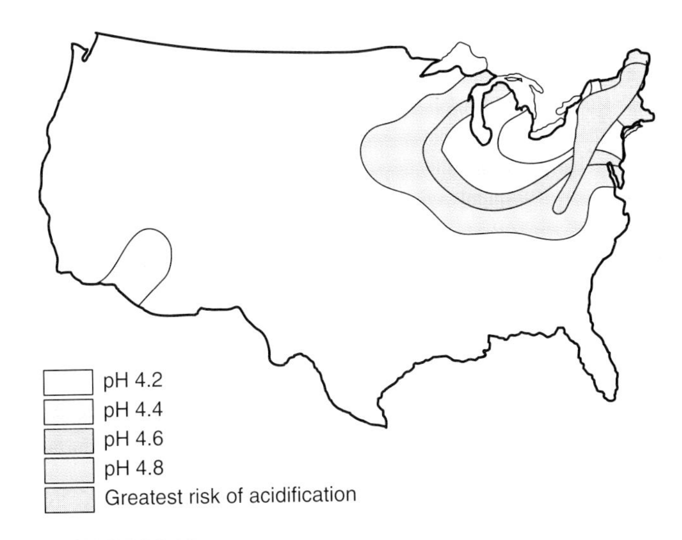

FIGURE 24.16

Dangers of Acid Rain

Sulphate-producing factories and power plants are concentrated in the Midwest and Northeast, the areas of the country with the most acid rain. Unfortunately, lakes and rivers in the Adirondacks and in New England are also at great risk from acidification.

FIGURE 24.17

Effects of Acid Rain

This forest has been severely damaged by acid rain. Trees weakened by acidic conditions are more susceptible to disease and insect attack.

(a)

(b)

FIGURE 24.18

Industrial Smog

These photographs of the Boston skyline dramatically illustrate the effects of industrial smog.

Smog. Smog is of two types: *Industrial smog,* gray in color, is characteristic of industrial cities where high sulfur coal is burned during cold, wet winters (see Figure 24.18). This type of smog consists largely of particulate matter (soot and smoke), combined with sulfur dioxide. The sulfur dioxide ultimately becomes sulfuric acid, in a smog equivalent of acid rain. *Photochemical smog,* brown in color, is more common in warm, dry climates where automobiles are abundant. It results largely from the conversion of NO (released by cars) to NO_2 in the air. In the presence of sunlight, NO_2 reacts with *hydrocarbons* (any molecule comprised exclusively of hydrogen and carbon) to form ozone and *peroxyacetyl nitrates (PANs).* Because they are highly reactive, ozone and PANs exacerbate respiratory problems and irritate moist membranes. In cities such as Los Angeles that are surrounded by mountains, warm air aloft tends to trap the smog below, a condition known as a **temperature inversion** (see Figure 24.19). In 1988, Los Angeles experienced 148 days when the ozone level exceeded the allowable federal standard.

Some 40 percent of photochemical smog is attributable to automobile exhaust, and another 40 percent to bakeries, dry cleaners, and consumer products; only 15 percent is industrial. The current plan in Los Angeles is to begin to mandate the burning of alternate fuels in vehicles (methanol is the likely choice), starting with fleet cars (rental agencies, government cars, etc.). By 2009, no new gasoline-burning cars will be permitted, and every service station will have to install at least one methanol pump. The burning of methanol produces no soot, nitrogen oxides, or hydrocarbons—but it does increase the production of *formaldehyde,* another toxic chemical, and it does nothing to reduce carbon dioxide emissions.

The one "bright spot" in the smog story is that smog filters out excess UV radiation. Thus the problem of not enough ozone high in the atmosphere is compensated for by too much ozone at ground level. Before you move to a smoggy city to avoid melanoma, however, remember that smog is highly irritating to the lungs. In 1952, 4000 people died in London from respiratory problems caused by smog.

Radiation. Radiation remains an important part of air pollution. *Radon* is a radioactive element and occurs naturally as a gas. Some eight million homes in the United States are estimated to have potentially dangerous levels of radon, due to seepage from the soil, and it is blamed for 5000 to 20,000 deaths annually (see also Chapter 17).

Radiation from radon gas is a natural pollutant (albeit one made more dangerous by our nearly airtight houses of today), and has presumably been a threat to humans for thousands of years—but we were ignorant of radiation until the discovery of X-rays in 1895 and of radium in 1898. It may seem odd to us today, but for most of this century, there was essentially no concern about the possible hazards of X-rays or radioactive elements. For example, radium-impregnated paint was at one time used to coat the numerals on watch faces so they would glow in the dark. The amount of radium was so low that the wearer was not injured—but the workers who did the painting often had the habit of moistening the paintbrushes

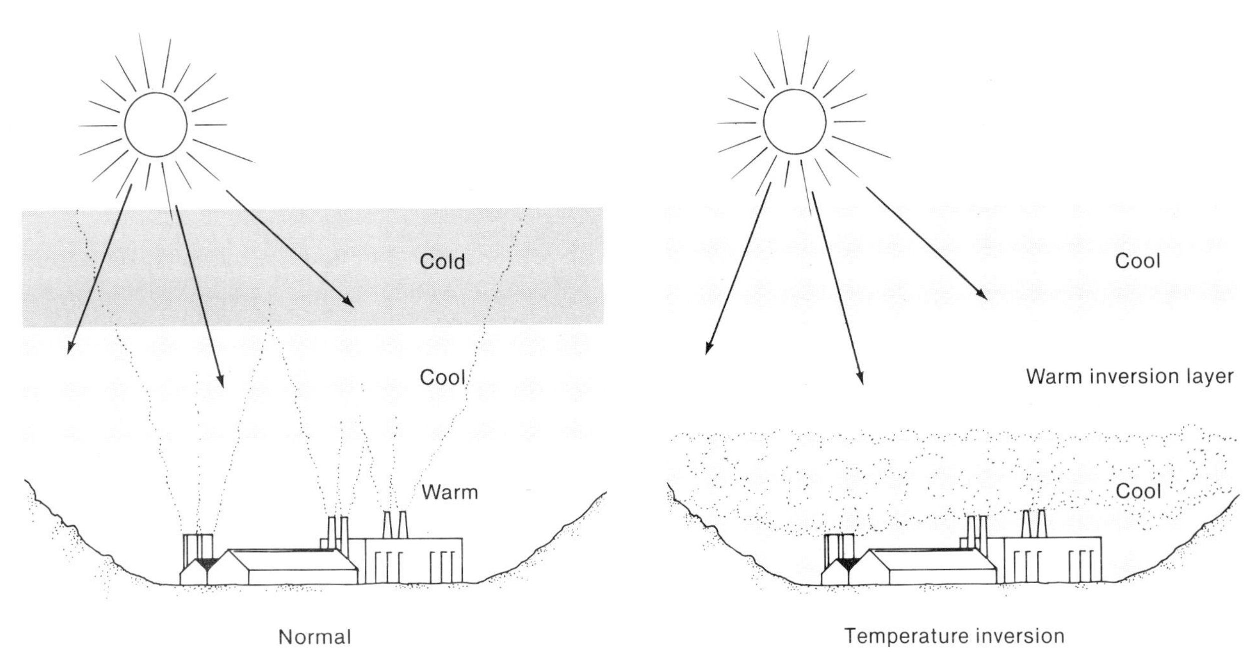

FIGURE 24.19

Temperature Inversions

When a layer of warm air develops over a layer of cool air, it prevents the air from circulating and traps pollutants and smog in place. Temperature inversions are particularly common in cities surrounded by mountains, since the mountains tend to restrict movement of the air.

in their mouths. Repeated instances of loosened teeth, falling hair, and other symptoms of radiation poisoning ultimately put a stop to that practice.

As recently as 40 years ago, shoe stores routinely used an X-ray device to examine children's feet as they were trying on new shoes. The child was told to put his newly shod feet into two openings at the bottom of a machine the size and shape of a drinking fountain. Viewholes at the top allowed the child, his mother, and the clerk all to watch as the child wiggled his toes inside his new shoes to be sure they fit properly. Although no child presumably ever had his feet fall off as he attempted to walk out of the store, the cumulative effects of X-rays are now so widely known that we cringe at this flagrant misuse of X-ray exposure.

All of this pales beside the threat posed by the release of radioactive materials from nuclear explosions. Many radioactive isotopes, most of which emit radiation for many years, were released into the atmosphere during the 1950s, when the United States and the Soviet Union competed for the largest bomb. Faced with a world community enraged by the hazards of radioactive fallout, both countries gave up above-ground testing. (France, China, and, more recently, India have attempted to fill the void, however, and several other nations are thought to be nearing the testing stages for their own bombs.)

The primary concern with fallout is the fate of *strontium-90,* a radioactive isotope of the element strontium. The atomic structure of strontium is similar to that of calcium (although the strontium atom is much larger), and it tends to be incorporated, in place of calcium, in the growing bones of young children. Although the amount of strontium-90 taken up is relatively small, it emits radiation for many years. Because radiation causes mutation and other cell damage, and because the bone marrow is a site for blood cell formation, it was feared that the continued production of strontium-90 by bomb tests would lead to an increase in leukemia. The meltdown at Chernobyl in 1986 was simply the most recent instance of a major release of radioactive isotopes into the atmosphere (see Figures 24.20 and 24.21).

Threats to Water Quality

We can consider two classes of water pollution: heat pollution and pollution by toxics.

Heat. Because of its abundance and its capacity to absorb heat, water is very frequently used as a coolant in electrical generating facilities, manufacturing plants, and so on. Typically, water is pumped in

FIGURE 24.20

The Chernobyl Reactor
This photograph shows the damaged power plant shortly after the accident in April 1986.

from a lake or river, passed through a set of cooling coils in the plant and is then pumped back into the river. Even if the river water is not exposed to plant impurities, and the quality of the water remains the same in all respects save temperature, it is still considered to be polluted. The effects may include:

1. The inability of cold-water organisms, such as trout, to survive, either because of death of the adults, or because the eggs fail to hatch;

2. The death of various organisms because of insufficient oxygen (cold water can hold considerably more dissolved oxygen than can hot water); and

3. The death of organisms due to their inability to withstand the sudden temperature fluctuations that often accompany periodic discharge of water used for cooling.

Thus, by altering abiotic factors (temperature in this instance), we are eliminating organisms and are thereby affecting the ecosystem just as surely as if we were to cause the extinction of species by any other method.

Toxic Substances. Many of the substances we discussed earlier as air pollutants also affect the water, simply because pollutants in the air often settle in the water, or are carried there by rain. The consequence is that waterways very distant from pollution sites can become contaminated. For example, Isle Royale National Park, in northern Lake Superior, has no permanent settlements except for the park's headquarters. Recently, however, *toxaphene,* a pesticide used on cotton crops, was found in the mud at the bottom of a lake on Isle Royale. Since there is no cotton grown within 1200 km of Isle Royale, toxaphene was presumably carried to Isle Royale in the air.

Heavy metals are a particular problem. We discussed mercury earlier in the chapter, but a similar picture could be painted for lead. Lead interferes with a variety of metabolic processes. For instance, the presence of lead in the blood in concentrations of less than one part per million inhibits the activity of the enzyme responsible for the synthesis of hemoglobin. In addition, lead damages the liver, kidneys, and the brain, and the damage is not always reversible.

Where does the lead come from? At one time, ead was extensively used in paint, and many children become seriously ill from eating bits of old, peeling paint. Until recent years, lead was released into the atmosphere in large amounts from the burning of gasoline, although all new cars sold in the United States are permitted to burn only lead-free gasoline.

So there's no longer a problem, right? Well, not quite. In 1986, the United States imported 873 million pieces of ceramics, at least some of which were improperly glazed and, if used for food or drink, could release lead into the body. The total number of ceramic pieces that were tested in the following year was 794 (not a very large sample!)—and 14 percent of them leached lead in amounts exceeding federal limits. Moreover, lead is still used in solder, and solder is used in the installation of water lines and the manufacture of refrigerated drinking fountains. (Indeed, the Latin word for lead—*plumbum*—is the root word for "plumbing.") Present estimates are that 20 percent of lead pollution in children comes from drinking fountains!

Interestingly, it is not simply exotic pollutants that have posed a problem to our waterways; we have also managed to pollute lakes and rivers with essential elements. For example, under normal conditions, population explosions of bacteria and algae do not occur because of the limited availability of nitrogen

FIGURE 24.21

The Chernobyl Aftermath
Measurements of radioactive iodine in the year following the Cherynobyl reactor failure show a very uneven fallout, with heavy fallout in areas as far removed as southern France and Sweden.

and phosphorus in the water. However, if they are introduced into pond or lake water as run-off from synthetic fertilizers (nitrogen) or detergents (phosphorus), both elements are capable of causing very rapid growth of bacteria and algal populations.

The algae multiply very quickly, choking waterways and preventing the penetration of sunlight to organisms living at or near the bottom of the lake. Without sunlight, these organisms die, and in decaying they promote bacterial growth, leading to an exhaustion of the dissolved oxygen. The resulting drop in oxygen promotes the extinction of many species, including most game and commercial fish. Even very large lakes, such as Lake Erie, have been affected (although thanks to pollution abatement measures, Erie has experienced something of a resurrection).

In many respects, water pollution is no longer the problem it once was. The National Pollution Discharge Elimination System requires a permit for any "point source" that empties into a river or lake, and the permit can be easily revoked. Of course, exceptions remain. Modern sewage treatment is still not in place in such cities as Boston and New York (which empty their wastes into the ocean); oil spills remain very much with us, unplanned, to be sure, but seemingly ever present; and perhaps most important, 65 percent of water pollution is from "non-point" sources, such as farm runoff. Finally, although treatment of water before discharge is now common, there remains the problem of what to do with the toxics that accumulate in the sludge at treatment plants.

Threats to Soil Quality

The major pollutants of the soil are domestic wastes (garbage) and industrial chemicals (pesticides and hazardous wastes).

Garbage. The United States produces 82 billion kg of household garbage each year (see Figure 24.22), or almost one kg per person per day, with only 11 percent of this amount being recycled. The total volume of garbage has increased by 80 percent in the past 30 years—and it is expected to double by the end of the century. More than 75 percent of our garbage is sent to landfills, but many states have virtually no landfill sites left. A few states are already sending their garbage to other states, but this is presumably a short-term solution since 80 percent of all U.S. landfills will be filled by the year 2000. Moreover, once buried, even "biodegradable" substances are slow to decompose. Hot dogs are recognizable (although somewhat less appetizing) after 25 years; newspapers may be readable after 30. Finally, run-off from dump sites have polluted underground water sources, and the sheer diversity of the contents of garbage ensures that dump sites will be potential hazards for decades, if not centuries.

What should be done? Many European countries (and some individual communities in the United States) are active in recycling, separation, and incineration of nonrecyclables. Incineration, of course, can simply trade soil pollution for air pollution if not properly managed. Ocean dumping is not a viable solution either (see Figure 24.23). For instance, plastics, which are notoriously resistant to biological breakdown, are threatening the giant leatherback turtles of the Atlantic, which swallow plastic bags in the belief that they are jellyfish. Eventually, the animals succumb to blocked intestines.

Long after our monuments have crumbled, the only remaining relics of our civilization—the *objets d'art* of our culture—may be our plastic bottles and bags.

Pesticides. Pesticides include compounds used to kill weeds, fungi, and insects. One of the major types of weedkillers (*herbicides*) is *2,4-D,* a plant hormone mimic that kills plants by disrupting their metabolism. A related compound, *2,4,5-T,* is the primary substance in *Agent Orange*—a product widely used in Vietnam as a **defoliant** (a substance that causes leaves to fall off). More than 10 percent of the entire country was defoliated, and large sections of the forests destroyed. To say that destruction of the forests deleteriously affected the ecosystem is to state the obvious. Moreover, there is continuing debate about the effects of Agent Orange on humans. Many Vietnam veterans claim a variety of side effects, ranging from cancer to birth defects in their children, from having been inadvertently sprayed with the chemical. In 1989, some 11 years after the first lawsuits were filed, and without ever acknowledging that Agent Orange was responsible for any of the claimed side effects, manufacturers amassed a fund that is being distributed to about 30,000 veterans, and about 10,000 widows and families, in amounts ranging up to $12,000 per award.

FIGURE 24.22

Garbage

This photograph, which could have been taken in almost any city in the country, is of a home appliance landfill.

The *insecticides* are of particular interest because of their extensive use. Most modern insecticides fall into one of three major groups of chemicals:

1. *Chlorinated hydrocarbons,* including DDT and chlordane, among many others;

2. *Organophosphates,* including dieldrin, malathion, and parathion; and

3. *Carbamates,* including Sevin, and certain other, relatively recent, insecticides.

Most of the organophosphates break down much more rapidly than do the chlorinated hydrocarbons, and in that sense pose less of a persisting threat to the ecosystem. However, the organophosphates are also generally a good deal more toxic to mammals. In addition, at least one of them, *endrin,* was found to induce cancer in animals, and consequently was banned. Carbamates are generally both short-lived and relatively nontoxic to humans.

Insecticides are a major industry in this country. We use about 0.5 billion kg yearly, at an annual cost of $3 billion—but only 1 percent of what is sprayed actually hits the insect. The rest becomes soil and water contaminants. More than 17 different insecticides have been identified in the ground water in some 23 states, and over 1400 wells in central California alone (where insecticide use has been particularly intensive) are contaminated.

A historical problem with insecticides has been that once a population of insects develops resistance to one insecticide within a given group, the population is usually also resistant to most of the other insecticides from that group. Thus it is difficult to read with a straight face the pronouncements of scientists at the end of World War II who declared that DDT would cause the extinction of the housefly and other pest species. As you know from looking about you, the housefly managed to *adapt* (see Chapter 22) rather well to this threat.

The housefly is by no means the only species of insect to develop resistance. Some 447 species of insects, ticks, and mites are now resistant to some or all insecticides. Crop losses to insects, which were 7 percent in the 1940s, were 13 percent in the 1980s, in part because the new hybrid crops are less resistant to insect attack than were the older types. Moreover, insect-borne diseases are killing more people than ever. Malaria alone kills more than one million children in Africa under the age of five each year. There are an estimated 40 million cases of river blindness (another insect-borne disease), mostly in West Africa, and 400 million cases of elephantiasis, which may be the world's fastest spreading disease.

The extensive use of pesticides also causes significant ecological damage. In addition to the problems of biomagnification, discussed earlier, levels of DDT as low as one or two parts per million reduce photosynthetic activity by green plants by as much as 20 percent. These are not high levels—human breast milk has been found to contain as much as five parts per million! Moreover, DDT has a half-life of 30 years, which means that it is very persistent in the environment. Finally, more than nine million kilograms are still exported annually from the United States for malaria control in tropical countries (although its use in the United States has been banned since 1972).

During the 1980s, one of the principal manufacturers of DDT paid an indemnity of $24 million to the residents of a small Alabama town where the insecticide was manufactured for 26 years, as a settlement of all personal injury and environmental claims against the company. Such recognition of the potential harm of this insecticide is a far cry from the days when manufacturers used to eat and drink DDT as a testimonial to its safety.

FIGURE 24.23

Garbage and Wildlife

This photograph speaks for itself. What is garbage to us may represent danger to wildlife.

Fortunately, as we have learned more about how insects live, other, more specific, methods of control have been discovered (see Figure 24.24). These include the use of selected parasites and predators, the application of insect viruses, and the utilization of insect hormones, all of which are far less threatening to the biosphere because each is a "natural" product, meaning one to which the biosphere has long been exposed. The use of chemical pesticides continues at a very high rate, however, primarily because of savings in time and money.

Hazardous Wastes. Few pollutants have been more obvious to the public than the dumping of hazardous wastes. During the early 1980s, illegal dump sites

FIGURE 24.24

Integrated Pest Management
Ladybugs are very effective at controlling aphids. They are also nonpolluting and aphids are never able to develop a resistance to them.

seemed to appear on television news with ominous regularity. Love Canal in New York and Times Beach in Missouri became the symbols of our mistreatment of the earth. As a consequence, hazardous wastes were placed high on the hit lists of federal and state governments. In May 1990, it became illegal to dump nearly any chemical that was untreated (i.e., not neutralized or incinerated). As a consequence some wastes are now buried by "deep injection" (wells dug far under potable water sources).

BIOLOGY AND THE LAW

One of the ironies of modern life is that, at a time when more and more individuals are desirous of increased personal freedom, laws are becoming more pervasive and restrictive. Yet there is a certain logic to that fact. The growth of laws is a function of an increasingly mobile and crowded society, for as the number of contacts we have with other individuals increases, the chances of inimical encounters also increase, making it necessary for society to provide standards to govern such encounters.

For example, we are now surrounded by laws regulating the environment, whereas 25 years ago such laws were almost unknown. The National Environmental Policy Act, which established the Environmental Protection Agency (EPA), dates only to 1969. Effective regulation of air pollution dates only to the Clean Air Amendments of 1970. Water quality laws have a longer history, but it was not until the 1972 Amendments to the Water Pollution Control Act that the EPA was given a broad charge with respect to water quality (specifically, "to restore and maintain the chemical, physical, and biological integrity of the Nation's water"). The Resource Conservation and Recovery Act of 1976 provided regulations for the disposal of solid wastes, and the Superfund (for toxic waste dumps) was established in 1980.

Laws are not usually formed in a vacuum; rather they are typically created by reasonable people attempting to find reasonable answers to widely recognized problems. The last phrase should be emphasized—particularly when esoteric concepts such as "environmental quality" are involved, an informed (and irate) citizenry is the primary force behind the establishment of laws and regulations designed to effect remedies. However, to state the obvious, there is a world of difference between having laws, or having the power to set regulations, and actually enforcing or setting them. Some observers have claimed, with more than a little validity, that regulatory agencies very often become the tools of those they were erected to regulate.

The EPA has had mixed reviews. It was the courts, not the EPA, that interpreted and set standards for the phrase "no significant deterioration" in the Clean Air Act. In 1970, Congress directed the EPA to restrict 320 different toxic air pollutants—but by 1989, the EPA had completed regulations on just seven. For example, the discharge of mercury is regulated in plants that work directly with mercury and mercury compounds, but less than half the mercury that enters the environment does so through these manufacturing plants. The huge amount of mercury that enters the atmosphere through the stacks of power plants that burn coal (which contains trace amounts of mercury) is totally unregulated. The EPA was ordered by the Federal Insecticide, Fungicide, and Rodenticide Act of 1972 to reanalyze the impacts on health and the environment of 50,000 pesticides approved prior to 1972. By 1986, the EPA had not completed a single analysis.

The EPA showed equal reluctance to a suggested ban on the use of chlorofluorocarbons, a ban widely recommended because of preliminary reports that indicated their deleterious effect on the ozone layer. The EPA argued that it was unwilling to act until the deleterious effects were proved. Would it be unreasonable to suggest that perhaps those who propose to release novel substances into the atmosphere should have the burden placed on them to prove, at least to a reasonable degree, that the substances are *not* going to cause environmental problems?

EPA's track record was not improved with its handling of toxic waste dumps. Of an estimated 14,000 such dumps nationwide, EPA designated 1224 as top priority for cleanup, but 80 percent of the Superfund spending to date has been for consultants—and by 1989, only 77 of the sites had been cleaned up. Similarly, the EPA action with respect to DDT can best be described as cautious. Most of the 4.5 million kg once used annually in the United States has been eliminated, but twice that amount is still exported. Admittedly, there is a tradeoff here. DDT is a very effective and inexpensive insecticide, and malaria, which is transmitted by mosquitoes, remains a serious and widespread disease. Nevertheless, it is interesting to note that when the developed countries took the opportunity to aid the underdeveloped countries, they chose the cheapest insecticide—and one that had been banned in the United States. Does that signify our intentions with respect to the larger problems facing the underdeveloped countries in the future?

In the final analysis, laws generally will be structured to reflect the will of the majority (as long as the rights of the minority are not infringed). Therefore it is necessary to have citizens who are informed about the tradeoffs of a given policy, and who, in that way, can make their wishes known based on knowledge, not ignorance. The importance of biological information in making these decisions in the future can hardly be overemphasized.

Summary

A principal tenet of biology is that the various levels of biological organization are subject to the same physical laws. Thus the complexly organized biosphere is just as influenced by the laws of physics as is any individual organism, for both need an external source of energy. Matter, however, is recycled within the biosphere.

The efficiency of energy capture is generally very low, averaging about 10 percent between the various feeding, or trophic, levels. Thus the trophic levels decrease sharply in size, forming a pyramid, and the number of layers is equally sharply limited.

Another parallel between individual organisms and the world's ecosystems is that each has developed homeostasis. Oscillations in population size tend to be greatest in polar ecosystems, which are relatively unstable because they have few species. Because complex ecosystems are the most stable, we have reason to be concerned when we bring about extinctions, or create artificially simple ecosystems by engaging in monoculture.

We also affect the homeostasis of ecosystems by misuse of our technology. Pollutants, poisons, pesticides, and radiation all result in disturbances of the homeostasis of ecosystems, and many of our efforts to compensate for their effects simply compound the problems.

Key Terms

population
community
ecosystem
biosphere
ecology
biotic
abiotic
producer
consumer
reducer
energy pyramid
trophic level
biomagnification
food web
monoculture
pollution
depletion
greenhouse effect
chlorofluorocarbon (CFC)
ozone layer
acid rain
smog
temperature inversion
defoliant

Questions

1. Explain how the concept of homeostasis applies to an understanding of ecology.
2. Give an example of a biogeochemical cycle. Is sunlight an appropriate example? Why or why not?
3. Why are there always relatively few levels in an energy pyramid?
4. Explain the phenomenon of biomagnification. Give an example of a substance that illustrates this phenomenon.
5. How do the concepts of food webs and monoculture interrelate? Are monocultures stable or unstable? Why?
6. Explain the greenhouse effect. Name two gases that contribute to this effect.
7. What risks may occur from a partial destruction of the ozone layer?
8. List two threats to each of: air quality, water quality, and soil quality.
9. Are current laws, in your opinion, adequate to deal with the threats to environmental quality? Defend your answer.

CHAPTER

25

Population Ecology

Bending the Rules

Do all species have population explosions? If not, why are we special? How many people *can* live on the earth at one time? How many people *should* there be? The earth's population, at its present rate of increase, will double in the next 40 years to something like ten billion. Are we destined to have a population crash, as the lemmings do periodically?

In 1982, 43,000 children under the age of four died every day. However, there were approximately 122 million babies born that same year, a number that exceeds the entire population of all but a handful of countries. Should we try to save the lives of those children? Do we have the resources to feed them if they survive? What should our policy be toward countries that have a rapidly growing population but need our aid and assistance?

Are things getting better? In the 1940s, the skies of Pittsburgh were dark at noon, the sign of a healthy industrial economy. Lethal smogs in London killed thousands in the 1950s. The INCO smelter at Sudbury, Ontario (which closed in the late 1970s), spewed out 1 percent of all of the sulfur emitted in the world each year, and no trees grew within a 15 km radius. Those stories are all relics of our past—but in Eastern Europe things have not changed much. In Russia, Lake Baikal, which contains 20 percent of the world's fresh water, is now polluted by chemical and paper plants. The Sea of Azov is polluted by pesticides, and, as more water is diverted for irrigation, is becoming increasingly more salty. The Aral Sea is biologically dead. More than 25 percent of the farmland of Byelorussia was rendered unusable by the Chernobyl disaster. Pollution in Poland may cost $25 billion to clean up, and in what was formerly East Germany the cost may be six times that amount. Birth defects occur in 10 percent of infants in northern Bohemia, and Romania is even more polluted. Are things simply hopeless?

These and other issues are considered in this chapter.

THE NATURE OF THE PROBLEM

It should be obvious to all of us that humans have had a profound effect on the biosphere—an effect that has not usually been for the better. The cycling of essential elements, for example, takes place on such a large scale that it is tempting to think that we can function without regard to the impact our activities may have on these cycles. We have learned, from bitter experience, that such thinking is folly. We have been tempted to alter whole ecosystems to serve our needs, only to find ourselves engaged, like Sisyphus with his rock, in a never-ending struggle to prop up an ecosystem we have thrown completely out of balance.

As painful and expensive as these lessons have been to learn, however, we can take comfort in the fact that we now take heed of ecological laws in planning the activities of our society, right? Well, no, perhaps we are not ready to concede just yet. Consider our present efforts to "go forth and multiply."

REPRODUCTIVE STRATEGIES

Living organisms employ a variety of reproductive strategies, although they can be lumped in one or two major categories (see Figure 25.1). We might characterize these categories as being "Don't put all your eggs in one basket" versus, "Put all your eggs in one basket, but then watch that basket!"

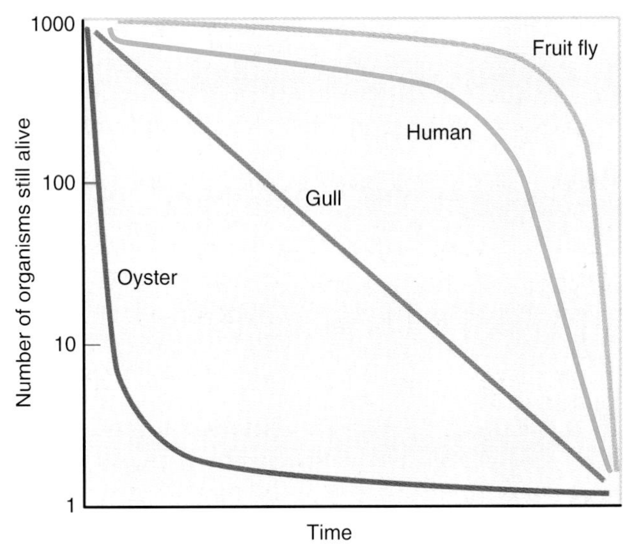

FIGURE 25.1

Survivorship Curves

Some organisms, such as oysters, have very high mortality rates early in life, and only a few reach adulthood. Other organisms, including humans and fruit flies, have high mortality rates only late in their lifespan.

Many species adopt the first strategy. Animals as evolutionarily diverse as oysters and tapeworms produce literally millions of eggs, only a very few of which ever reach adulthood. These species beat the long odds against surviving by producing so many eggs that some survive just by chance.

Most birds and mammals, and especially the primates, follow the second strategy. The number of offspring is small, often only one or two at a time, but these offspring are carefully nurtured, and a much larger percentage of them reach adulthood.

As different as these two strategies seem, in one important respect they are identical. Each permits population explosions.

Reproductive Potential

The **reproductive potential** of a species is its theoretical reproductive capacity, free from any constraints. Animals that produce many offspring and have short generation times are usually used to illustrate reproductive potential. For example, if all the offspring from each generation survived, one pair of houseflies would have more than five trillion descendants at the end of one year. That's about 1000 for each person on the face of the earth! One bacterium, dividing every 15 minutes, would produce, at the end of a year, a ball of bacteria the size of the known universe that was expanding outward at the speed of light!

Those are dramatic examples, to be sure, but a similar result would occur with any species, given a long enough time, be it aardvarks, elephants—or humans.

The Human Reproductive Potential

Although the evidence is scanty, it would appear that the total human population was roughly stable, or at most grew very slowly, until the introduction of agriculture about 10,000 years ago (see Figure 25.2). From that point, there was a gradual increase until the beginning of the Industrial Revolution in the eighteenth century. Growth since then has been increasingly rapid.

Estimates of population size over this period place the numbers of humans at about four million at the beginning of agriculture, about 500 million at the beginning of the Industrial Revolution, and about five billion today.

How was it that our species (or our immediate ancestors) took half a million years to reach a population of four million, and only 10,000 years to increase to 500 million? Why did the human population not reach one billion until about 1800 but is expected to top six billion by the year 2000?

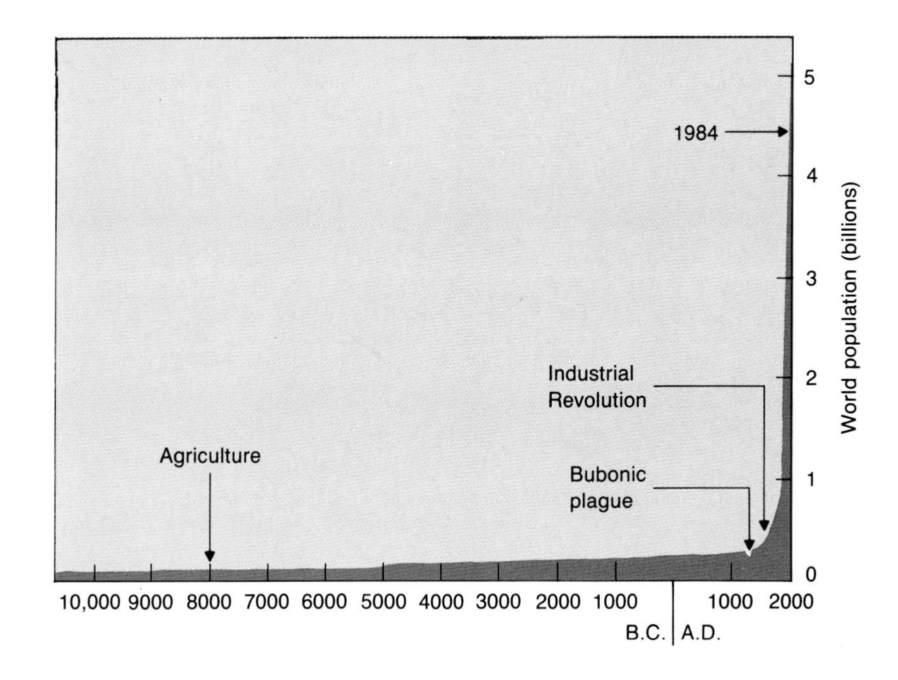

FIGURE 25.2

Growth Curve for the Global Human Population

The world population was relatively small until the Industrial Revolution. Since then, it has increased by a factor of ten. Wars and plagues have had almost no effect on total population size.

At least five factors have been at work:

1. *Agriculture:* It is believed that, for most of our history as a species, we humans were hunter-gatherers, a lifestyle employed by a number of technologically primitive societies, such as the !Kung bushmen of the Kalahari region of Africa, the aborigines of Australia, and many of the North American Indian tribes before the coming of the Europeans. Because hunter-gatherers must move with the herds of game animals they hunt, they have no permanent settlements, and the groups are generally small, consisting of fewer than 50 individuals.

Agriculture entirely changed the social organization. In order to grow crops, or raise livestock, permanent (or at least semipermanent) dwellings were desirable, even necessary. The food supply became a good deal more stable and predictable, and small settlements developed because more people could help raise more crops.

2. *Urbanization and Trade:* With the stabilization of food supplies and the growth of settlements, some individuals began to act as traders and merchants, rather than as farmers. Small market towns developed, where farmers could trade or sell their produce for other goods and services. The human population density began to rise markedly.

3. *Colonization:* In the sixteenth and seventeenth centuries, many of the then rather crowded European countries established colonies in the New World. Great mineral wealth was returned to Europe, and many colonists left Europe to find their fortune in North and South America. These rapidly growing colonies became natural trading partners of the mother countries, and the prosperity—and populations—of both continued to rise.

4. *Industrialization:* The Industrial Revolution substituted machines for human labor, and fostered the growth of very large cities. Farming and distribution of food became more efficient, and increasingly fewer individuals were needed to grow the crops and raise the herds of farm animals, permitting (or forcing) many people to become factory workers.

5. *Medicine:* The impact of medicine and public health on the size of the world's population, especially in the last 150 years, has been profound. The increasing availability of food over the centuries may have prompted some increase in birth rate, but medicine was responsible for a precipitous decline in death rate. Growth rate is simply birth rate minus the death rate. Thus, with sharp drops in the rate of infant mortality and death from childhood diseases, many more children survived to become reproductive-age adults, and that fact, as much as any other, has accounted for the recent explosion in population.

Is the human population destined to continue to grow at our present growth rates? Things are not entirely bleak. There is some evidence that, with increased prosperity, there is a drop in the birth rate—at least, this is what has happened in many of the developed countries (see Figure 25.3). In the United States, the birth rate was 26 per 1000 population in 1947, but it dropped to less than 15 per 1000 by the mid-1970s and has been roughly stable since then. Some of this decline has been due to better methods

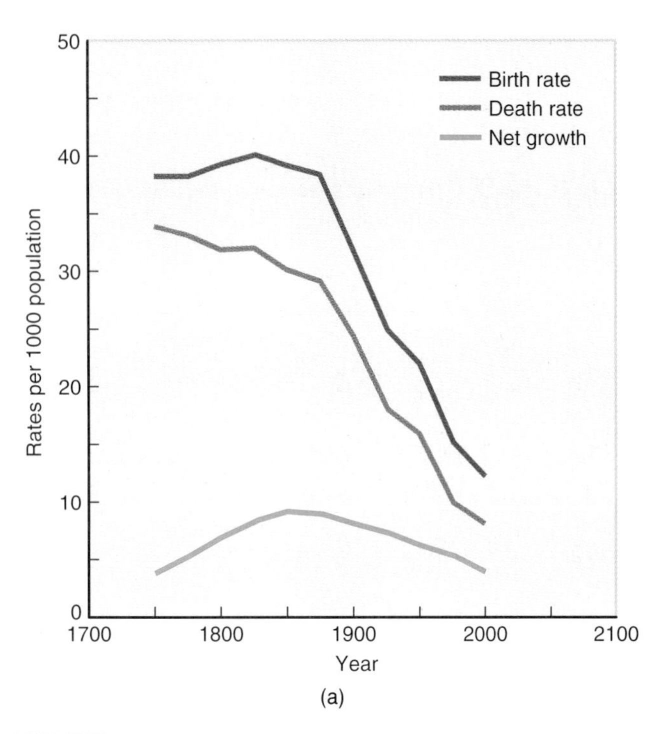

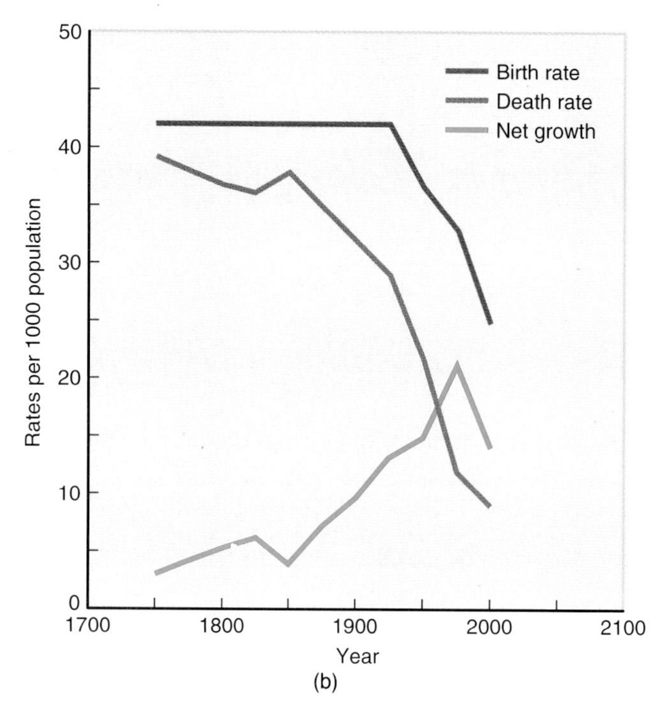

FIGURE 25.3

Birth Rates, Death Rates, and Growth Rates

These two graphs, which are based on estimates from the United Nations, show that in industrialized nations (a), both birth rates and death rates have been falling for the past century, and net growth has slowed measurably in that time. On the other hand, the birth rate in developing countries (b) did not begin to drop until well into the twentieth century, and the rate of decline has not kept pace with the precipitous drop in death rates. As a consequence, the growth rates of these countries are not expected to show a decline until the beginning of the next century and even then they will remain extremely high.

of contraception, and to the legalization of abortion, but some can also be attributed to greater social acceptance of couples who choose to have few (or no) children, and to increased awareness of the population explosion itself.

However, in the developing countries, the population explosion continues unabated. The current growth rate in Africa, for example, is 2.8 percent, which means a doubling of the population in 26 years (see Figure 25.4). (By comparison, the growth rate in Great Britain is so low that its doubling time is 460 years.) In some countries, such as Kenya, the growth rate is even higher, and the doubling rate is only 17 years. The impact of such rapid growth rates on the environment, on social services, and on economic development efforts in general is enormous (see Figure 25.5).

What should the developed countries do? Do we have the obligation to change from a heavily animal protein diet to a largely vegetarian diet in order to send grain that would otherwise be used in feedlots to countries facing starvation? Should we make relief efforts contingent upon adequate population control in these countries? Do we have the right to legislate their social programs? These are the tough kinds of policy questions that your generation is going to have to help decide—and they all arise from the fact that, as a species, our population growth is unchecked.

Why don't we see similar population explosions in every species?

CARRYING CAPACITY

No species ever reaches its reproductive potential for more than brief periods, because of various **limiting factors** in the environment. Our hypothetical housefly population faces risks from larval habitats that dry out, food shortages, spider and bird predators, and even people armed with rolled-up newspapers. The balance between reproductive potential and limiting factors represents the **carrying capacity** of the environment (see Figure 25.6).

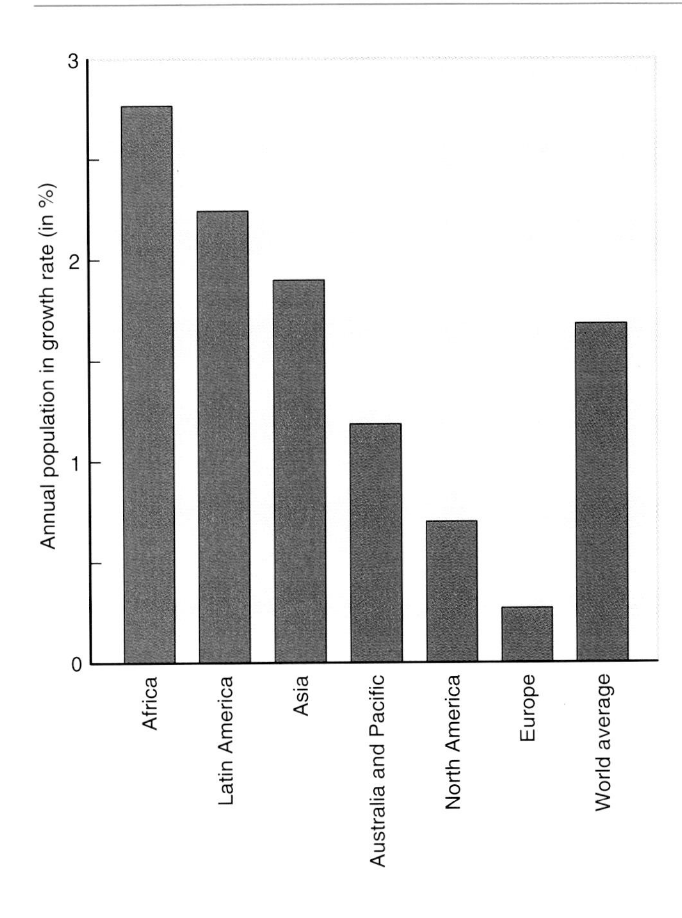

FIGURE 25.4

Current Population Growth Rates by Continent
Annual growth rates range from almost 3 percent annually in Africa to less than 0.5 percent in Europe. With a growth rate of 3 percent, the population of Africa will double in less than 24 years!

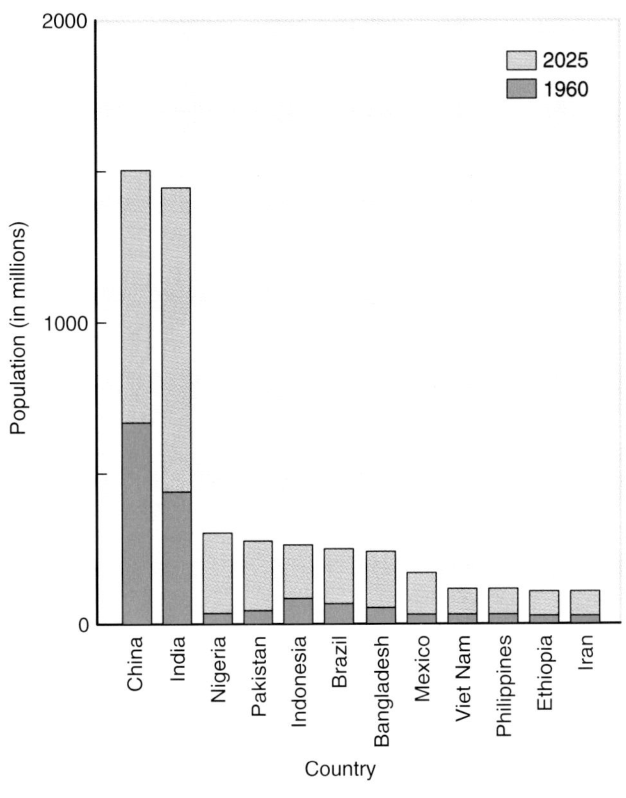

FIGURE 25.5

The Hundred Million Club
In 1960 the only countries in the world with populations in excess of 100 million were China, India, the Soviet Union, and the United States—two developing countries, and two industrialized countries. By the year 2025, ten additional countries will exceed the 100 million mark—and all of them are developing countries. Would you expect development of a country to be easier or more difficult when it is growing rapidly? Do you think these countries will succeed in becoming developed?

Carrying capacity and limiting factors are not constants, but rather are subject to considerable variation over time. For instance, the housefly population is larger in the summer than in the winter, because the carrying capacity of the environment is greater in the summer (there are fewer limiting factors). Food is available in quantity, and shelter from the elements is less of a necessity.

Similarly, the algal "blooms" that appear every spring are a reflection of the ending of cold temperatures as a limiting factor. Artificially created blooms occur when nitrogen or phosphorous are dumped into a pond or lake, demonstrating that levels of nitrogen and phosphorus are limiting factors in "unfertilized" ponds.

Other species exhibit variation in their population size over longer periods of time. However, over a sufficient period of time, the average population size of most species is relatively constant, and booms and crashes are rare. Instead, there are rather gentle oscillations at or near the carrying capacity.

In species that regularly experience population explosions, there is typically an equally dramatic crash, and the crash carries the population not just back to the carrying capacity but considerably below it. For example, after a three year build-up in population size in southern New England between 1979 and 1982, the gypsy moth population crashed sharply in the summer of 1983. The population went from a size capable of stripping whole forests, and literally blackening the sides of houses as the caterpillars crawled upwards in search for food, to a point where they were virtually absent from the local fauna.

The gypsy moth is an imported species, with few natural enemies. It is therefore able to reach and maintain its reproductive potential for extended periods of

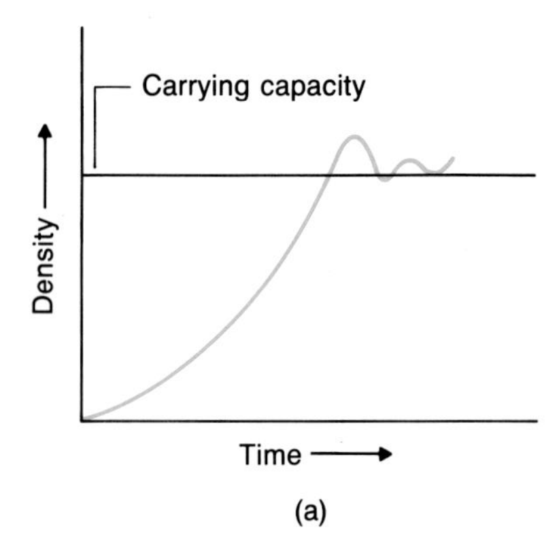

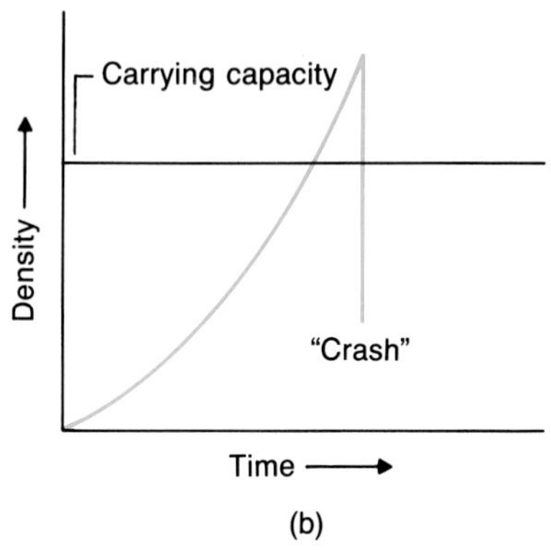

FIGURE 25.6

Carrying Capacity

(a) Many species have populations that oscillate about the carrying capacity. (b) Other species overshoot the carrying capacity and then "crash" well below it. Compare these curves with Fig. 25.2. Which of the two most looks like the human population growth curve?

time. However, the crashes in gypsy moth populations do not occur primarily because of an absence of food, or for some other resource-limited reason. Rather, the primary factor is plague. Gypsy moths are susceptible to a virus that kills the caterpillars before they molt to form pupae. In very high population densities, the virus spreads rapidly, and virtually all the caterpillars die. Thus it is clear that limiting factors are not always constraints by the environment. Sometimes they are created by the very success of the species itself in attaining huge population sizes.

What do these findings suggest about our own population explosion?

THE HUMAN CARRYING CAPACITY

One of the most important debates of modern times has an ecological basis. In everyday terms, it involves the question of whether or not the world (or the United States) is overpopulated. In ecological parlance, the question is, "What is the carrying capacity for humans?" (See Figure 25.7.) An admission that there is, in fact, a carrying capacity short of the point where we are literally piled on top of each other is an important first step, but even this concession is not made willingly by some adherents of unlimited growth. Let's examine the evidence.

It is useful to use ecological models for our discussion. Let's assume, for the moment, that current projections for the world's population are accurate, and that, by the year 2030, we will number some 10 billion. Is that number above or below the carrying

FIGURE 25.7

Human Carrying Capacity

The idea of the human carrying capacity means different things for different people. To many people, life under the conditions illustrated here demonstrates that our population is already beyond carrying capacity, at least in some countries.

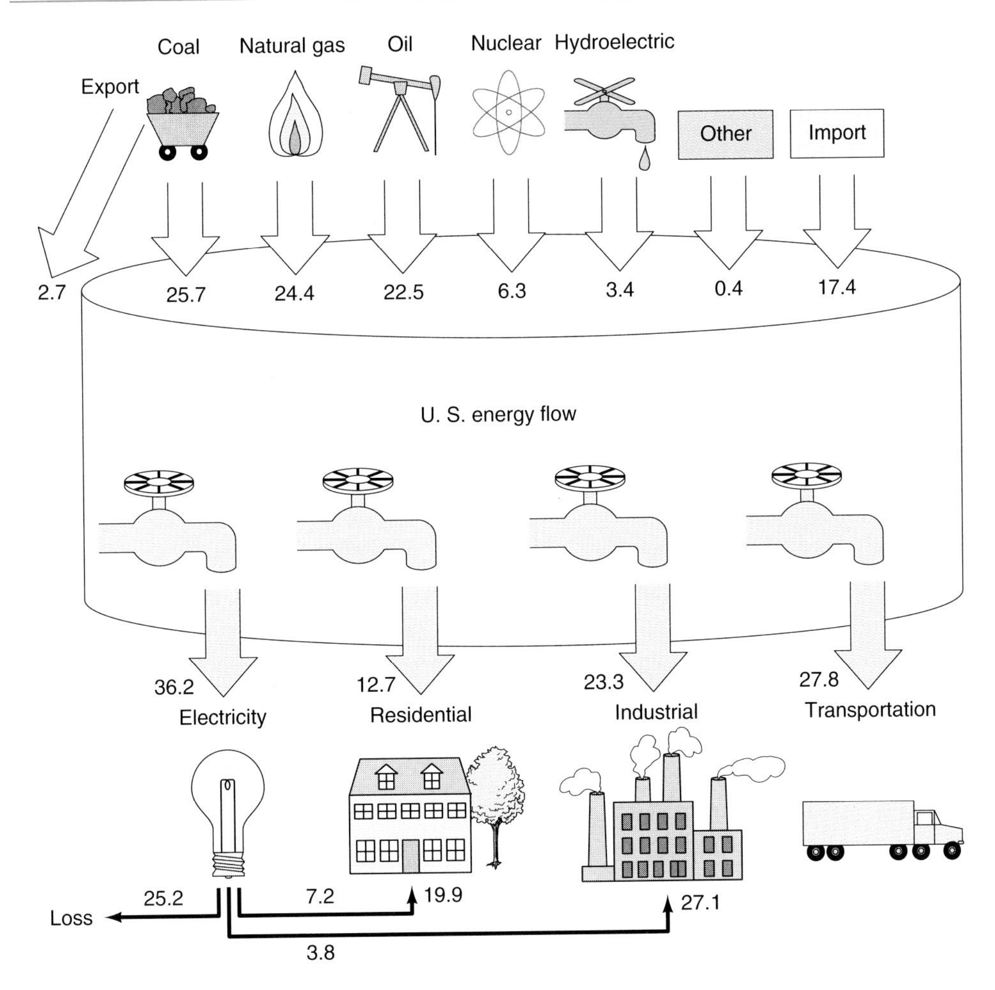

FIGURE 25.8

Energy Flow in the United States

The percentage contribution of each of the major energy sources utilized by this country is shown at the top of the diagram. Note that we export some coal. Note also that, at present, we import more than 17 percent of our energy needs, primarily in the form of oil and gas. Percentages are also given at the bottom for our uses of this energy. Note that more than 36 percent of our energy sources are used to generate electricity—and more than two-thirds of this electricity is lost in transmission. The balance is split between residential and industrial use.

capacity of the earth? What are the impediments to achieving a population of ten billion and what accommodations would have to be made to sustain a population of that size—that is, what are the limiting factors, and can they be stretched or circumvented?

As we address these questions, we must pay attention to ecological principles—or justify why we can ignore them without peril. Ecosystems import energy but recycle matter. Can we import enough energy to sustain a population of ten billion? Can we recycle matter with enough efficiency so as to avoid exhausting some vital resource, thereby imperiling our ability to sustain a population of ten billion? Let's examine the evidence.

Energy

Developed nations, with about 20 percent of the world's population, consume 70 percent of the energy, although their per capita use has slowly fallen in recent years. In developing countries, by comparison, per capita use is rising—and at a rate about twice as fast as the population is increasing.

Will it be possible, in 2030, for our population of ten billion to consume energy at the current U.S. per capita rate? (See Figure 25.8.) Barring some breakthrough in energy production, the answer is almost certainly no. Consider the following statistics: In 1900, annual global energy consumption was the equivalent of 3.6 billion barrels of oil. In 1988, this figure had

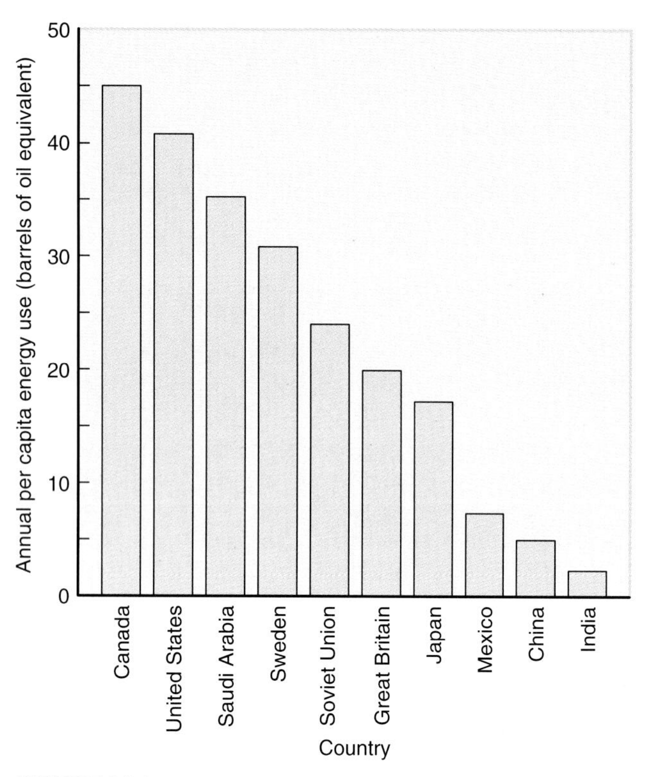

FIGURE 25.9

Differences in per Capita Use of Energy

By converting all energy uses to equivalent barrels of oil, it is a simple matter to demonstrate that northern industrial countries use a great deal more energy than tropical developing countries. However, there is considerable variation even among industrialized countries (compare, for example, Canada and Japan). Will the rate of use of energy by industrialized nations drop more quickly than the use by developing nations will rise?

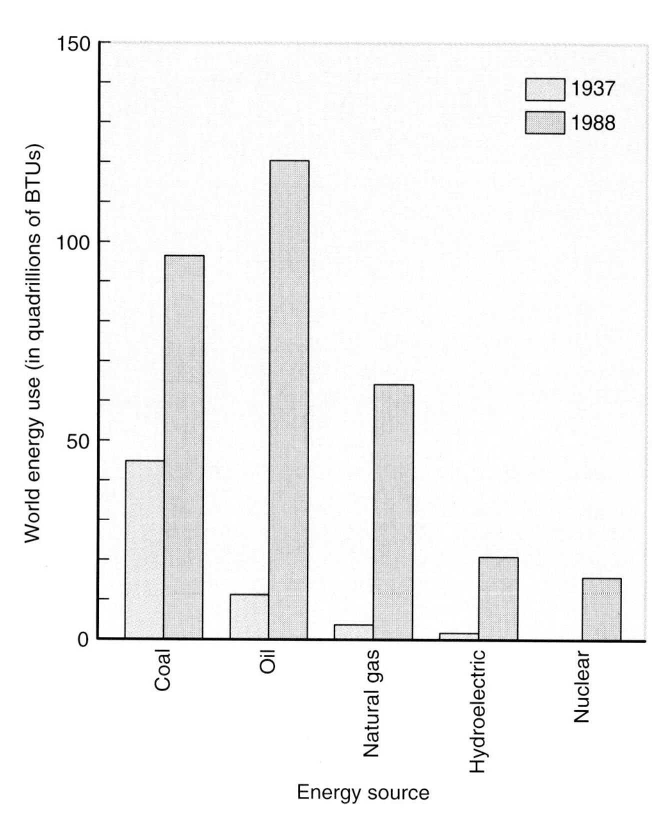

FIGURE 25.10

Changes in Energy Sources

In 1937 the world used 60 quadrillion BTUs of energy, more than 70 percent of which came from the burning of coal. In 1988 the world's energy use increased by a factor of five, to more than 320 quadrillion BTUs, and oil and natural gas together accounted for almost 60 percent of this energy.

risen to 54 billion barrels of oil. Extrapolating to 2030 at current U.S. rates would give us an annual figure of about 500 billion barrels of oil—an amount that is simply not available as an energy source (see Figure 25.9).

Currently, 38 percent of the world's energy comes from the burning of oil (and its derivatives), and another 20 percent comes from natural gas. Coal provides about 20 percent of our energy, and hydroelectric and nuclear make up virtually all the balance. (Hydroelectric and nuclear are used for electrical generation only.) Expanding our reliance on any one of these will be problematic, for different reasons (see Figure 25.10).

Although our ecological model permits us to import energy in order to sustain our population of ten billion, the *source* of this energy must either be renewable or, for all intents and purposes, infinite (such as sunlight). Oil and coal do not meet these criteria. Current estimates are that at present rates of use we will have between three and seven years' supply of oil left in the year 2030. Coal is much more abundant—we will have from 30 to 100 years' supply in 2030 (perhaps even more, depending on how aggressively we wish to rip up the landscape to extract it). Thus, to the extent that we must rely on these fuels—and recall that they presently are responsible for almost 60 percent of our energy production—we cannot sustain a population of ten billion. Moreover, both contribute significantly to air and water pollution, coal being a particularly dirty fuel. Even with pretreatment and the use of "scrubbers" or precipitators, burning of these fuels will continue to add huge amounts of carbon dioxide to the atmosphere, with the attendant risk of global warming.

What about alternative sources of energy? Known reserves of natural gas will be exhausted by 2030—but since natural gas is largely methane and ethane, and since methane is readily produced by bacterial decomposition of organic matter, it can be considered a renewable resource. Natural gas is, in many respects, an ideal fuel, since it produces much less pollution than coal or even oil (although burning it does produce carbon dioxide). The only problem is that we cannot possibly produce more than a tiny fraction of our energy needs in this manner—the earth simply does not produce enough biomass per unit time to permit the volume of recycling that would be required.

Hydroelectric power is renewable, since it simply takes advantage of the water cycle. However, there are several drawbacks to hydroelectric power. First, it generates only electricity, and much of our current energy use is not geared to the use of electricity (e.g., automobiles). Second, most of the obvious dam sites have already been exploited, at least in the developed countries. Third, hydroelectric dams tend to silt up over the span of a few decades, sharply reducing their output. Fourth, dams are often very destructive of the environment, inasmuch as they flood the land behind the dam and completely change the character of the river. Hydroelectric power is, on the other hand, relative nonpolluting and produces no greenhouse gases.

What about nuclear power? High grade uranium reserves are finite, in the same way that oil and coal reserves are finite, but breeder reactors, which convert nonfissionable uranium and thorium to plutonium and then use the plutonium for fuel, expand the fissionable fuel reserve by a factor of several hundred times—long enough to suggest long-term stability for our population of ten billion. However, some of the same problems associated with hydroelectric plants are also a part of nuclear power. First, it is electrical only, meaning that it could not serve some of our present energy needs. Second, public sentiment (after Three Mile Island and Chernobyl) is currently very much against expanded nuclear facilities—indeed, no new plants have been ordered in this country since 1978. Third, there is the very real problem of finding safe and effective storage sites for radioactive wastes. However, they produce no greenhouse gases, and, with the very important exception of radioactive wastes, are not significant pollution sources (although heat pollution can be a problem).

What about other sources—wind, geothermal, solar? (see Figure 25.11). Those are all renewable and nonpolluting—but at present supply less than 0.5 percent of this country's energy needs. Solar power in particular is becoming more competitive with other energy sources—a plant that opened in California in 1989 is producing electricity for less than eight cents per kilowatt-hour, as compared with three cents for gas or coal-fueled plants—but how much of the earth's surface do we want to devote to solar collectors? What do we do for power at night, or on rainy days, or during winter months in northern latitudes? Geothermal power is of use only in certain geological environments; wind power has similar limitations. On the other hand, these sources are renewable and relatively nonpolluting. (There is some venting of toxics in the water and air near geothermal plants, and wind or solar "farms" preempt most other uses for the land on which they sit.)

There is always the possibility of technological breakthroughs occurring prior to our exhaustion of existing energy sources. Solar power is certainly going to be used much more, and there is still the hope that

FIGURE 25.11

Solar Power

Light from acres of mirrors is used to vaporize water. The resulting steam causes a turbine to turn, thereby producing electricity.

fusion reactors will one day be practical—but it would be folly to assume that technology will always advance quickly enough to sustain our growing population.

Let's attack the problem from the other end. Can we reduce the per capita use of energy to the point where our energy production will be substantially less than our projections for 2030?

There are several opportunities. Prototypes of most of our major electrical appliances are three or four times as efficient as current models. Prototype houses are almost 20 times as energy efficient as the average house of today, and prototype automobiles are about five times more efficient than the average new car. Even larger savings are possible if we want to change certain of our present habits. One-third of the world's oil production is used to power automobiles. If we decide that no one will have cars in 2030, or that they can drive solar vehicles only, we could cut our projected energy needs by 12 percent. Almost 70 percent of electricity produced today is lost in transmission lines. If we could eliminate this loss (perhaps by the use of superconductors now being developed), we could save another 14 percent of our projected total energy needs.

Through a combination of efforts, we might be able to reduce our projected energy needs in 2030 by 50 percent or even more. However, even a 50 percent savings would still require the consumption of the equivalent of 250 billion barrels of oil, an amount more than four times greater than our present usage. Barring the technological breakthroughs we alluded to earlier, energy generation at this magnitude does not seem achievable, let alone sustainable. In short, from the standpoint of energy generation, a human population of ten billion is beyond the carrying capacity of the planet.

Minerals and Manufacturing

There are three fundamental questions that must be addressed in a consideration of the problems posed by a mineral and manufacturing environment capable of supporting the needs of ten billion people in the year 2030:

1. Is there a sufficient resource and manufacturing base to provide for ten billion people on an ongoing basis?

2. Can the processes of extraction of minerals and the manufacture of goods be accomplished without wholesale destruction of the environment?

3. What do we do with the waste products of the manufacturing process, and with the manufactured goods once they are worn out?

Let's consider these questions successively.

Resource and Manufacturing Base. At current rates of use, and with a population of ten billion in 2030, many key minerals will be very scarce (see Figure 25.12). We would, for example, expect to have essentially no more copper, nickel, molybdenum, or cobalt by the year 2040, and platinum would be exhausted by 2050. Even such abundant minerals as aluminum would be exhausted by the middle of the following century. Thus it seems clear that our supply of certain key resources is such that we cannot maintain a population of ten billion using present practices.

The manufacturing base is more difficult conceptually. In principle, there is no reason not to expect the manufacturing base to continue to keep pace with consumer demands, since (in most countries) there is a market-driven economy. Assuming continued development of the developing counties (a questionable assumption, to be sure), and a concomitant increase in wealth, the manufacturing base will presumably expand to meet demands. The limitations on the manufacturing base are not inherent to the process itself but are imposed by other considerations: availability of raw materials, trained workers, and energy, pol-

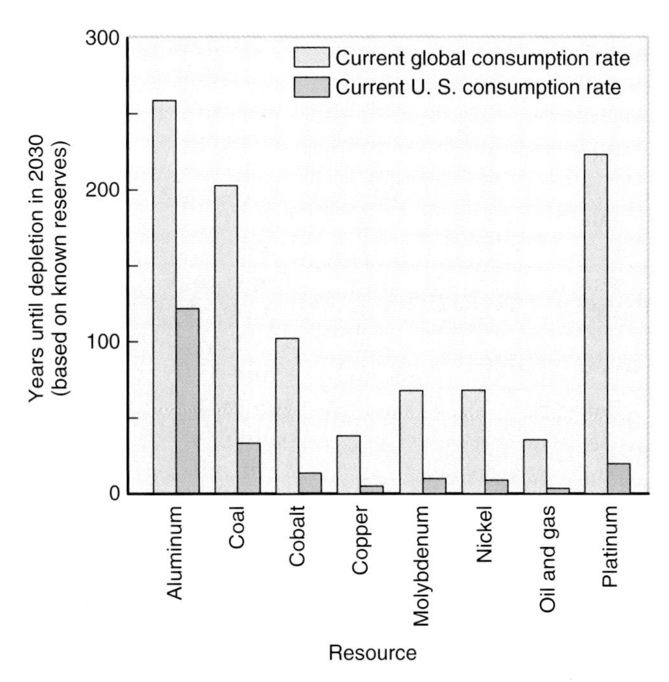

FIGURE 25.12

Lifetimes of Global Resources

Based on known reserves, in 2030 the lifetimes of many key resources will be measured in a handful of years if the world continues to use these resources at present rates. If the world is using these resources in 2030 at the rates currently used in the United States, these already brief lifetimes become dramatically shorter.

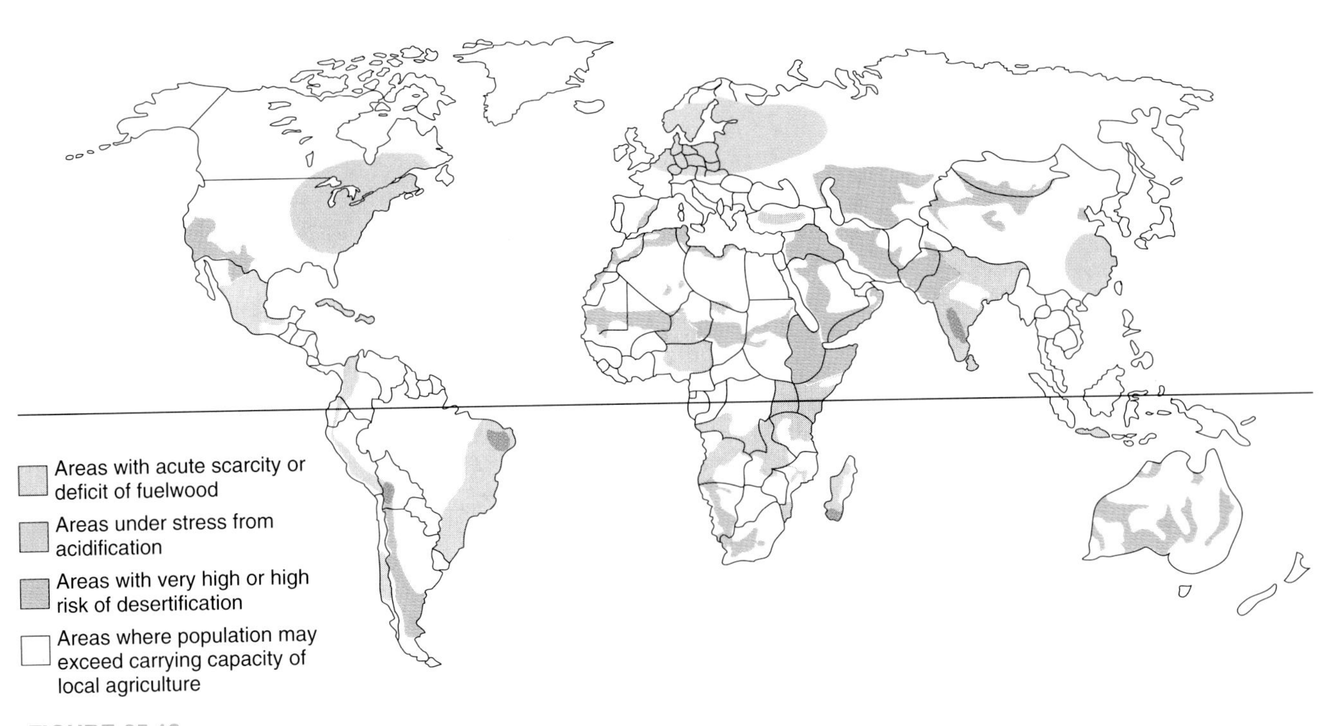

FIGURE 25.13

Global Land Degradation
This world view, based on data from the Food and Agriculture Organization of the United Nations and from the Scientific Committee on Problems of the Environment, illustrates four types of degradation of land, all of which are attributable to human activities—and all of which may be considered evidence of a population that has exceeded its carrying capacity.

lution abatement costs, and so on. Barring the development of new social constraints, these factors will limit the extent of manufacturing, rather than any self-imposed limitation by manufacturers themselves. Thus manufacturing in the strict sense is not a limiting factor in the growth of the human population.

Extraction, Manufacture, and the Environment. Historically, the extraction of minerals has resulted in extensive environmental damage (see Figure 25.13). Some experts feel that open pit and strip mining, which are the prevalent methods of extraction for many minerals, have caused more environmental damage than any other human activity. Certainly the magnitude of our activities are impressive. A single copper mine in Utah covers over 7 km^2, and is almost 800 meters deep. The total amount of material removed from this mine exceeds the volume of material moved to create the Panama Canal by a factor of seven! Almost half of the coal mined in the United States has historically been removed by strip mining, and much of our most retrievable reserves are in surface deposits in the western states where strip mining is a more logical method of extraction than is underground mining. Yet to strip mine these fields will require moving surface soil from an area equivalent to Pennsylvania and West Virginia combined! Finally, the estimated current annual movement of soil and rock worldwide is 3×10^{18} kg—a staggering number that is a hundred times more than the volume of sediment carried by all of the earth's rivers each year. That figure would have to be increased by an order of magnitude in 2030, to support the earth's predicted population in that year—and, over time, the environmental impact of such activity would be devastating.

Stripmined land can be reclaimed, of course—but the cost of doing so is often several times more than the value of the coal or minerals being extracted. Moreover, there are serious problems with the availability of water for mining, with the pollution of ground water from the effects of mining, and with the destruction of underground aquifers (see Figure 25.14).

The process of manufacture is generally less damaging to the environment than is the process of extraction. However, manufacturing uses more than 25 percent of U.S. energy production, and as we have seen earlier, there are enormous problems associated with the long-term expansion of world energy production, both in terms of generating the energy in the

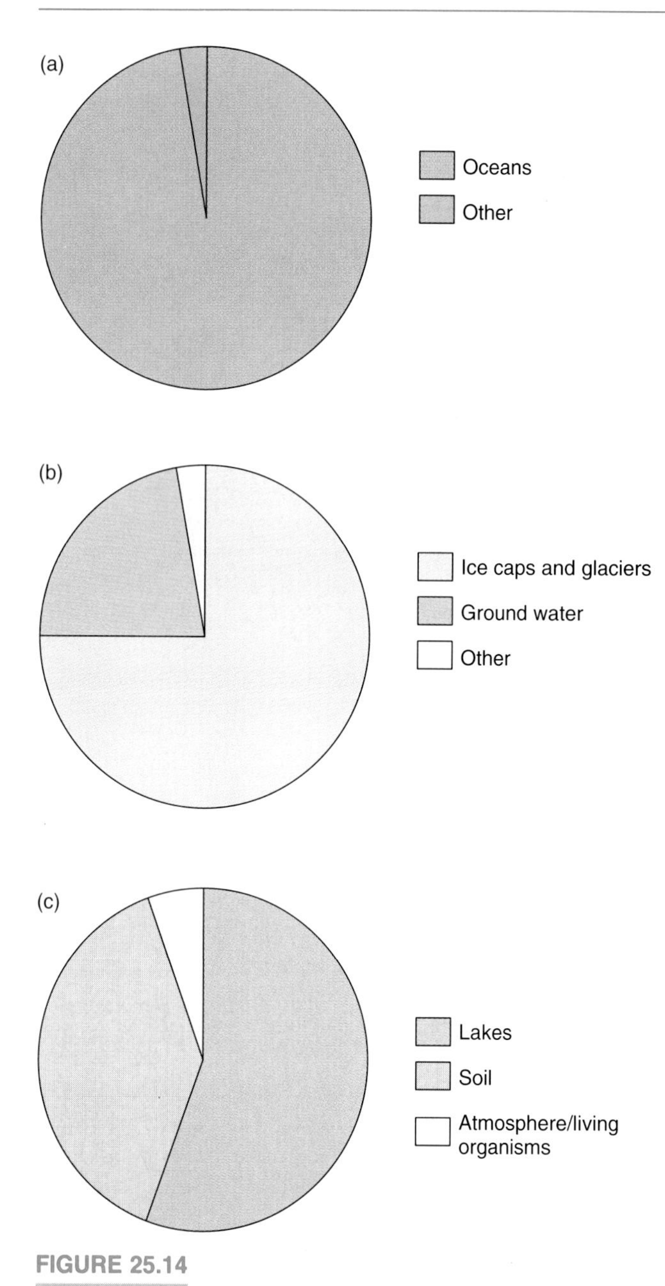

FIGURE 25.14

Global Water Supply

More than 97 percent of the earth's water is in the oceans (a). Of the remainder (b), most is inaccessible to living organisms because it is frozen. Ground water is accessible to humans through the construction of wells, but other species must rely on the 0.015 percent of the water that is present in the soil, lakes, and the atmosphere (c).

first place and in terms of the pollution that results from the generation of energy. Thus the energy costs alone make a great expansion of world manufacturing highly problematic.

Wastes and Disposal. At current U.S. rates of production, in the year 2030 the world population will generate enough solid waste every year to cover greater Los Angeles to a depth of 100 meters. Whether or not you think that is a good idea, you must acknowledge that it is an impressive volume of waste—and it is obviously unthinkable to continue to generate such a large volume year after year. Much of this waste comes from the throw-away society in which we currently live—a throw-away society that is the antithesis of the recycling that we will have to do.

Solutions. Although expanding the world's manufacturing base might well prove possible to meet the needs and demands of an expanding population, the resources needed to manufacture things that are ultimately thrown away, plus land needed for storage sites for the things thrown away are very definitely finite and will prevent our sustaining a population of ten billion. Manufacturers and economists are, interestingly enough, now looking at an ecological model for manufacturing wherein virtually everything is recycled—including the waste products of manufacture itself. It will presumably never be possible to recycle literally everything, but it will certainly be essential to recycle those resources that are in short supply or are energy-intensive to manufacture.

Some recycling, of course, is taking place even now. Iron and steel are recycled rather well, as is aluminum. Certain plastics are potentially very recyclable, but at the moment, only about 1 percent of **polyvinyl chloride (PVC)**—the type of plastic used in bleach or shampoo bottles—is recycled. Because the United States presently manufactures almost two billion kilograms of this plastic each year, an amount that is about one-sixth of the total plastic manufactured in this country, and because it is enormously persistent in the environment, it simply must be recycled.

By comparison, another type of plastic, **polyethylene terephthalate (PET)** is being recycled at a much higher rate. This is the type of plastic used in large containers of carbonated beverages and is subject to mandatory deposit laws in nine states. Moreover, recyclers pay more than ten cents per kilogram, an amount exceeded only by aluminum among solid wastes. As a consequence, about 20 percent of the 350 million kg of PET manufactured each year in the United States is now being recycled.

Platinum is used in the catalytic converters of new automobiles, and because it is so scarce in nature, and the extraction process is so expensive, there is strong economic incentive for recycling this metal. Indeed, in most industrial operations, 85 percent or more of the platinum used is recycled. However, less than 15 percent of the platinum used in catalytic converters is presently being recycled, in part because there has yet to evolve an efficient means of locating and collecting catalytic converters from the thousands of scrapyards and junkyards that receive wrecked or abandoned autos in this country. This situation will presumably change once more cars with catalytic converters reach the junk stage.

Many municipalities are now mandating separation and recycling of garbage components. Aluminum cans, plastics, glass, and paper, at a minimum, are separated from other garbage and are being recycled. Since many of these materials are the products of manufacturers—but are no longer controlled by manufacturers—it is essential that consumers also participate in the recycling process. To the extent that the products (and byproducts) of manufacture are recycled, our society emulates an ecosystem, and that points the way to a population of a sustainable size.

Finally, another possibility is injecting water and bacteria into the garbage at landfills in order to reduce the active biological life of the landfill from 50 years to fewer than ten years. The danger is the increased risk of ground water contamination.

Disease

Historically, disease was an important factor in the control of human populations. As long as we existed as small, dispersed, and largely isolated groups, there was no potential for the rapid spread of disease. However, with the beginnings of agriculture, followed by commerce and urbanization, the potential for disease increased dramatically. *Leprosy*, one of the least contagious of diseases, was nevertheless the scourge of the biblical lands. *Malaria* forced the Romans to leave Rome during the summers, and it remains a serious disease in much of the tropical and semitropical world even today. *Smallpox* was a repeated visitor throughout Europe and Asia all through the Middle Ages and beyond. *Bubonic plague* killed almost 25 percent of the people of Western Europe between 1346 and 1352, and was a common affliction until the middle of the seventeenth century. *Syphilis* (great pox) was even more feared than smallpox (though possibly because it afflicted adults, whereas smallpox was more commonly a disease of children), and it was omnipresent in Europe from the time of Columbus. *Tuberculosis* was the great leveler in the nineteenth century.

Our knowledge of infectious disease agents and basic sanitation has changed all this, at first in the western world, and more recently, as we exported our medical technology, to the rest of the world as well. Sewage is no longer dumped into the streets; surgical instruments are sterilized between operations; food and health workers wash their hands at regular intervals. As commonplace as these practices now seem, they were not routinely done as recently as a century ago and are not faithfully done in many parts of the world even today.

On one level, small scale plagues are still very much with us. Every year, 14 million children under the age of five die of infectious diseases, most of which are readily preventable. A global immunization of children is now underway for six major diseases: *measles, diphtheria, pertussis, tetanus, polio*, and *tuberculosis*. In 1974, only 5 percent of the children in the developing world were immunized against these diseases; by 1988, 50 percent were immunized.

By comparison, death in the western world is much less commonly caused by infectious diseases than it was even 50 years ago. Instead, death comes more commonly from some internal malfunction (cancer, cardiovascular ailments). However, many authorities feel we are overdue for a plague. Crowded conditions and rapid travel already cause regular epidemics of such minor diseases as influenza, new strains of which circle the globe virtually every year. Could the same thing happen with a more serious disease?

AIDS (acquired immunodeficiency syndrome) might be a candidate (see Chapter 9). Since it was first identified in 1981, it has caused hundreds of thousands of deaths worldwide, and the virus has infected millions more, many of whom are expected to develop the disease in the future. Because it renders the body virtually helpless to resist a variety of secondary infections, constant (and expensive) medical monitoring—of a type often not available in developing countries—is required to keep AIDS victims alive.

However, the likelihood is that AIDS will not, in fact, become the plague of the twenty-first century, if only because it is rather difficult to transmit. Moreover, an effective vaccination for AIDS is likely before the disease becomes much more widespread. Perhaps we now are sufficiently knowledgeable about diseases and medicine in general that plagues are a thing of the past. Nevertheless, our dense social structure renders us highly vulnerable, and there is a much greater chance of such diseases spreading rapidly through less developed countries.

In short, current diseases will not prevent our reaching and maintaining a population of a given size. However, new diseases continue to arise, and the larger and denser the population, the more the opportunity exists for a devastating population crash caused by disease.

Food

Can we provide a diet of at least sustenance level to a population of ten billion? Over the past 20 years, world production of grains has been increasing at an annual rate of 2.9 percent, whereas the world's population has been increasing at an annual rate of less than 2 percent. Since the population growth rate is continuing to slow, it would appear as if we could supply the food needs of a population of ten billion, especially because food is a renewable resource.

However, there are several other factors that must also be considered (see Figure 25.15). More than half of the increase in grain production has resulted from irrigation of marginal land—and past irrigation projects have lowered ground water levels and have led to **salinization** (an increase in soil salt levels due to evaporation of water), which sharply reduces yields. Erosion and **desertification** (the loss of habitat to desert, usually because of overgrazing) are other problems that diminish, rather than increase, productivity. At present, for example, one estimate is that 20 million hectares (one hectare equals 2.4 acres) are lost to desertification each year. Moreover, in many parts of the world, water itself is scarce and cannot continue to be used in current irrigation practices. Desertification, salinization, and a drop in water tables suggest that our current agricultural practices are not self-renewing, meaning that they are not sustainable over extended periods of time.

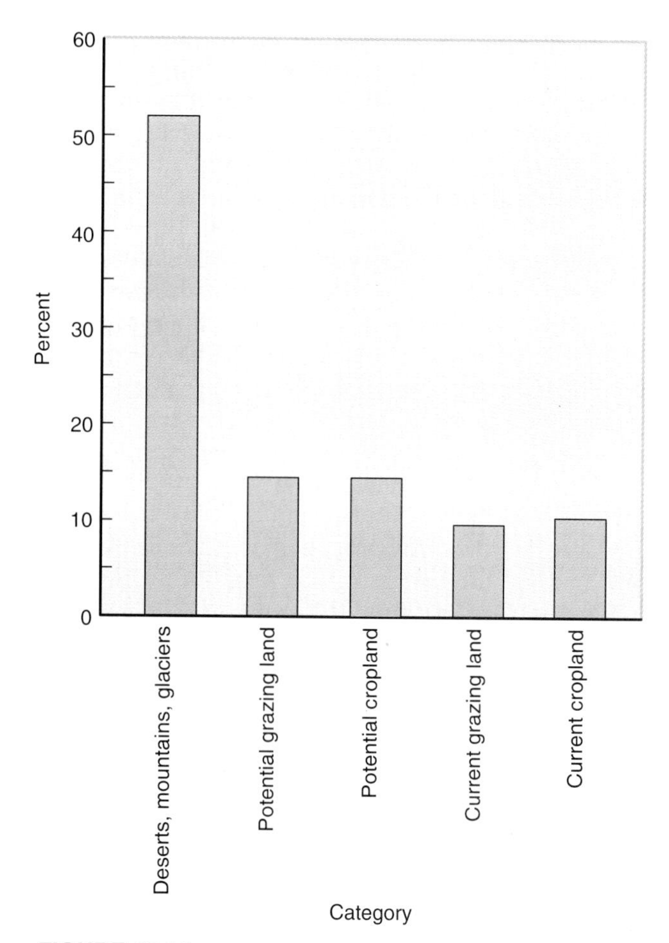

FIGURE 25.15

Global Land Supply

Slightly more than half the surface of the earth is unavailable for human crops and livestock. Of the remainder, slightly over more than 20 percent is used for grazing and crops, an amount that could be doubled if all the remaining tropical and temperate forests were cleared, and if all marginal land were fully utilized. Should we undertake such activities in order to permit a doubling of the world's population? Do we have any choice?

The other major factor that has led to the dramatic increase in grain production in the last 20 years has been the use of fertilizers and pesticides. Fertilizer use in North America has doubled during the past two decades but has increased by a factor of ten in Asia. As we saw in the previous chapter, fertilizers and pesticides can contaminate drinking water, and there is considerable pressure to reduce the dependence that agriculture has on these products. The use of these products again illustrates that present agricultural practices cannot be sustained for long periods.

Can these problems be resolved even as we continue to increase food production? The use of drip irrigation methods would reduce salinization and the demand for irrigation water. Integrated pest control methods, with greater reliance on biological control, would reduce the need for pesticides. Genetically engineered plants, which could fix nitrogen directly from the atmosphere, would reduce the need for fertilizers. The growing of crops better suited for the particular environment would reduce desertification.

The problem is that there is currently no global consensus that these kinds of things must happen—and that is a major flaw in the arguments we have been making. It is one thing to determine that the earth has a theoretical carrying capacity of ten billion humans (or whatever number you choose); it is quite another to achieve that carrying capacity in the real world.

The reality today, for example, is that outmoded and destructive agricultural practices continue, and at great cost. At the current rate of destruction, the tropical forests will be gone by 2030, primarily the victim of slash and burn agriculture (see Figures 25.16 and 25.17). Since the bulk of the nutrients in the tropics is in living organisms, burning the trees leaves an impoverished soil that cannot sustain crops for more than a few years. Moreover, the burning of the tropical forests creates other problems. It increases ozone levels (which are now near the level of plant toxicity in many tropical countries); it contributes to acid rain; and it contributes between one-fourth and one-half of all of the excess carbon dioxide added to the atmosphere each year.

Finally, to what extent do we maintain a stable population at the expense of the bulk of the other

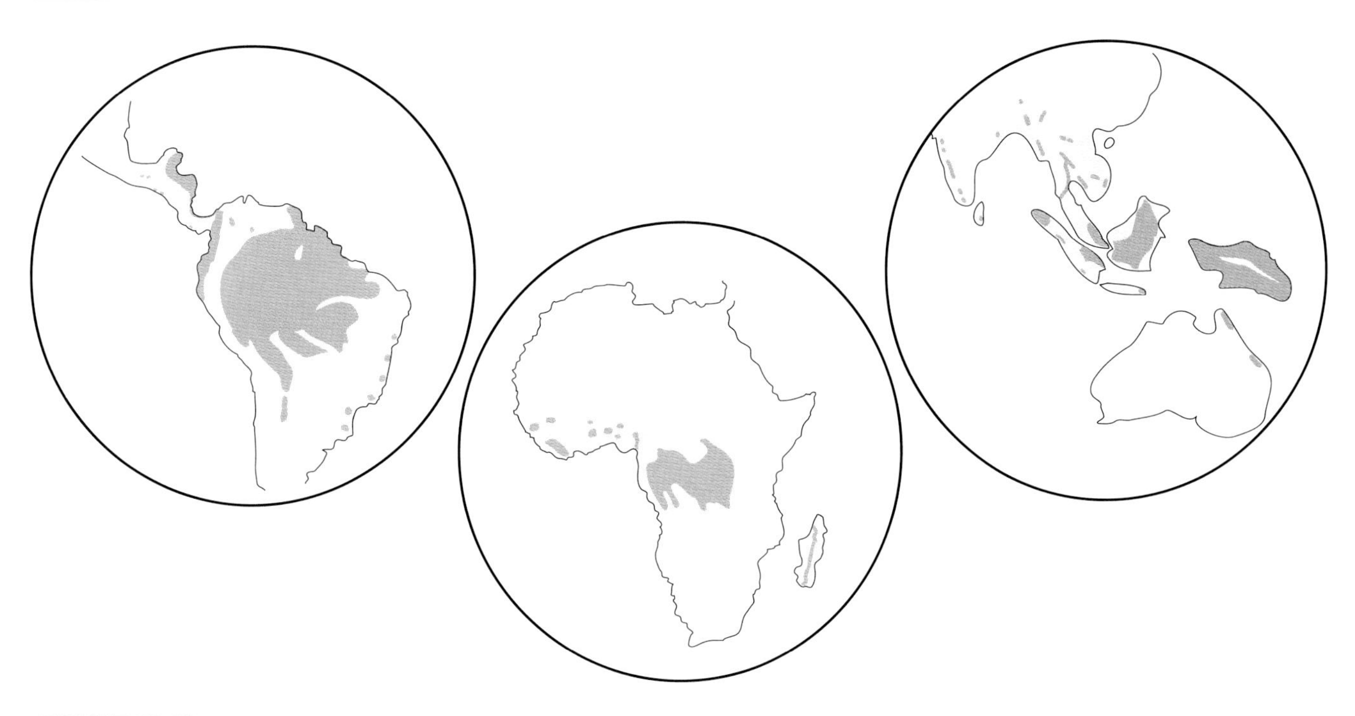

FIGURE 25.16

The Loss of Tropical Rain Forests
These maps show the original extent of tropical rain forests, and their current extent.

species of organisms on the planet? Unlike many other aspects of our population growth and ecological impact, biological diversity, in terms of genotypes, is totally nonreversible—once a species is extinct, we cannot bring it back. More than half of all species live in the 6 percent of the earth's land surface comprising the tropical rain forests. However, these species are becoming extinct as the rain forests are destroyed to create open land for farming, or jobs for loggers. People often have the impression that as a particular habitat is diminished, the species simply reduce their population sizes accordingly, but that extinction of a species is a rare phenomenon. That is not how things work. Instead, the size of the habitat is directly correlated with the number of species, not just the number of individuals (see Figure 25.18). Large islands, for instance, have more species than do smaller, but adjacent, islands—and very often the population sizes

FIGURE 25.17

The Undisturbed Tropical Rain Forest
This photograph of a tropical rain forest in Costa Rica gives only a hint of the enormous biological diversity present in these regions of the world.

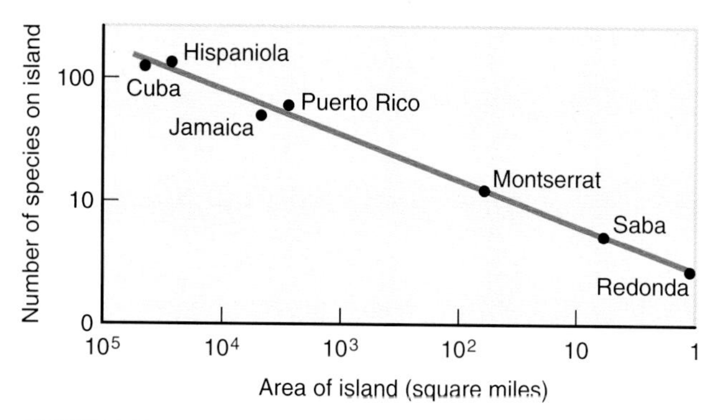

FIGURE 25.18

Species Diversity and Land Area

The number of species of island reptiles and amphibians is directly correlated with the size of the island, as these data demonstrate.

in rain forest species are not large in the first place, which means that any significant reduction threatens the continued viability of the species.

The tropical rain forests now occupy only 55 percent of their original area, and each year another 2 percent—or about 200,000 square kilometers—is lost. For every 1 percent loss of rain forest, there is a loss of 0.2 to 0.3 percent of the indigenous species. Thus, at the present rates of forest destruction, it is conservatively estimated that some 4000 to 6000 species are now being obliterated every year (see Figure 25.19). This rate of extinction is unprecedented in the history of the earth. Moreover, since 45 percent of the medicines used by humans are derived in whole or in part from other species—indeed the income from medicines derived from just one species of periwinkle, a plant from Madagascar, exceeds $100 million per year—the wholesale destruction of any species is ultimately detrimental to our own self-interest. In any case, if we cannot maintain our human population without the continued destruction of other species, we clearly are not in an equilibrium with our environment.

The Human Carrying Capacity Revisited

We began by asking if a human population of ten billion in the year 2030 represented the carrying capacity of the planet for our species. We have examined some of the essential factors underlying this question, and it seems clear that 10 billion is well beyond the carrying capacity of the planet, at least in terms of the types of demands the developed countries presently place on the environment. We must either settle for a population well below ten billion, or change many aspects of how we live—or both. The question is whether we can do so voluntarily, or whether we are destined to expand like lemmings, until we outstrip

FIGURE 25.19

Deforestation

Deforestation is not just a problem in tropical rain forests. It is also occurring in the United States, as this photograph taken in Washington state demonstrates.

our resources and suffer a monumental population crash. The difference, of course, is that the lemmings' habitat regenerates while their population is small. In our case, we are likely to render our habitat incapable of regeneration before we suffer a population crash, in which case the ultimate size of the human population is likely to be very much smaller than it is at present.

Summary

Although humans have a very limited reproductive potential, even species with limited potential can experience population explosions, as we are only too painfully aware. In other species, limiting factors come into play that place a ceiling on population size, but with our usual display of arrogance toward the laws of ecology, we have chosen to ignore these factors wherever possible. Now the question has become, "At what point are we likely to see a crash—and from what source?"

Our present awareness of the dangers we are causing to ecosystems and to ourselves is reflected by stringent constraints on how business and industry can exploit the environment, and, at least to some degree, by declining birth rates in much of the western world. However, in the less developed regions of the world, the population explosion continues, not only unabated but in a crescendo. A burgeoning population possessed with rising expectations exacerbates all environmental problems. Fewer people with fewer demands may be the only way to save the ecosystems of our earth as we now know them—if we do not run out of time.

Key Terms

reproductive potential
limiting factor
carrying capacity
polyvinyl chloride (PVC)
polyethylene terephthalate (PET)
AIDS
salinization
desertification

Questions

1. What is meant by "reproductive potential," and how does it relate to the growth in human populations?
2. What factors account for the sharp increase in the human population over the past two hundred years?
3. What is meant by "carrying capacity," and how does it relate to the growth in human populations?
4. Can our earth sustain a population of 10 billion humans? Defend your answer.
5. What problems exist in trying to increase energy production?
6. Give three examples of minerals for which the reserves are sufficiently small, thus posing a threat to our manufacturing abilities.
7. Will it be possible to continue to expand food production? What factors influence your answer?
8. What are the moral and ethical dilemmas created when we consider whether or not to cut down the rain forests to provide cropland to feed a growing population? How do you think these dilemmas should be resolved?

Glossary

A band In striated muscle, the region of the sarcomere composed of thick filaments of myosin molecules.

abiotic Nonliving.

ABO system A genetically controlled series of antigens on the surface of red blood cells.

absorption The uptake of digested materials by cells of the digestive tract.

accommodation Adjustment, as occurs when the lens of the eye changes shape for focusing on objects at various distances, or when touch or olfactory receptors cease responding after repeated stimulation.

ACE inhibitor A chemical that reduces blood pressure by inhibiting the production of angiotensin.

acetyl CoA The combination of a two-carbon derivative of pyruvate with coenzyme A; the entry point to the Krebs cycle in the metabolic breakdown of glucose.

acetylcholine The most common of the neurotransmitters, found at both synapses and neuromuscular junctions.

Achilles tendon The tendon connecting the gastrocnemius muscle to the heel.

achondroplastic dwarfism The most common type of dwarfism, caused by a dominant allele that negatively affects the production of collagen.

acid A substance that can release a hydrogen ion, thereby lowering the pH of a solution.

acid rain Acid pollution caused when sulfur dioxide from coal-burning power plants combines with moisture in the air, forming sulfuric acid that then falls to earth as natural precipitation (rain, snow, fog).

acromegaly Enlargement of the bones of the face, hands, and feet in adults, caused by the overproduction of growth hormone by the pituitary gland.

acquired immune deficiency syndrome *See* **AIDS.**

ACTH A hormone from the anterior pituitary that stimulates hormone production by the adrenal cortex.

actin One of the two major contractile proteins of muscle; found in lesser amounts in most types of cells.

action potential A wave of depolarization in a neuron, resulting in the transmission of a nerve impulse.

activation energy The energy required to initiate a chemical reaction.

active immunity A form of immunity caused by exposure to disease agents or vaccines and the consequent production of antibodies; contrasted with *passive immunity*.

active site The portion of an enzyme molecule where amino acids are arranged in a configuration specific to a particular substrate.

active transport Movement of materials across the plasma membrane against a concentration gradient and requiring the expenditure of energy.

acute leukemia A form of leukemia in which lymphatic tissue is invaded by abnormal leukocytes.

ADA 1. American Dietetic Association. 2. Adenosine deaminase, a critical enzyme in ATP metabolism.

adaptation 1. An adjustment to local conditions. 2. An inherited change in the anatomy, physiology, or behavior of an organism that results from natural selection and that increases the organism's ability to survive and reproduce.

adaptive radiation The evolution from a single species of many species specialized for a variety of ways of life.

Addison's disease A condition caused by the failure of the adrenal cortex to produce an adequate supply of hormones.

adenoma Benign tumors having a gland-like appearance.

adenosine diphosphate *See* **ADP.**

adenosine triphosphate *See* **ATP.**

ADH A hormone produced by the hypothalamus and released by the posterior pituitary that causes the collecting ducts of the kidney to reabsorb water.

ADP A molecule of adenine to which the sugar ribose and two phosphate groups are linked, and from which ATP is formed by the addition of a third phosphate.

adrenal cortex The outer layer of the adrenal gland.

adrenal gland A pair of small glands that, in humans, are located above the kidneys and that produce a variety of hormones in both the cortical and medullary portions of the gland.

adrenal medulla The inner layer of the adrenal gland.

adrenal sex hormone Any of several steroid hormones produced by the adrenal medulla that influences the development of secondary sexual traits.

adrenaline *See* **epinephrine.**

adrenocorticotropic hormone *See* **ACTH.**

adrenogenital syndrome A masculizing effect occurring in women and children when excessive amounts of adrenal sex hormones are produced.

aerobic Metabolic reactions requiring the use of oxygen gas.

AIDS A fatal viral disease resulting from the failure of the body's immune mechanisms.

albinism A genetic condition caused by a recessive allele in which the skin and other superficial tissues contain no melanin.

alcohol Any of a group of organic chemicals having a characteristic molecular structure; usually refers to ethyl alcohol, a two carbon molecule produced by the fermentation of glucose.

aldosterone A steroid hormone produced by the adrenal cortex that affects the rate at which potassium and sodium are reabsorbed by the tubules of the kidney.

alkaptonuria A homozygous recessive genetic condition resulting in the accumulation of homogentisic acid.

allantois An extraembryonic membrane that is responsible for storing nitrogenous wastes during embryonic development in reptiles and birds and that contributes to the development of the placenta in mammals.

allele Any of the alternate forms of a particular gene.

allergen A substance that triggers an allergic reaction.

allergy Overproduction of immunoglobulin E in response to an allergen, and the consequent release of excessive amounts of histamine by basophils and mast cells either locally or body-wide.

allometry Differential growth rates of particular body parts.

all-or-none rule The rule that states that a stimulus to a neuron is either above threshold (in which case an action potential will be generated) or is below threshold (in which case no action potential will be generated).

altruism Behavior in which the welfare of others has priority over the welfare of the individual.

alveoli In the mammalian lung, the tiny air sacs in which gas exchange occurs with the blood.

Alzheimer's disease A chronic and progressive disease of the brain, characterized by accumulation of amyloid protein in the brain, destruction of brain cells, and memory loss.

amacrine cell A class of cells in the retina that synapses with bipolar cells and ganglion cells.

amebic dysentery A disease caused by pathogenic amebae in the large intestine and characterized by extreme diarrhea and excessive water and salt loss.

amenorrhea Cessation of monthly menstrual cycles.

amino acid Any of twenty nitrogen-containing subunits of proteins.

ammonia A colorless gas composed of nitrogen and hydrogen (NH_3); the primary waste product of protein metabolism.

amniocentesis A medical procedure whereby fluid is withdrawn from the amniotic cavity of a pregnant woman and analyzed to determine possible developmental defects in the fetus.

amnion The innermost of the extraembryonic membranes and the one containing the developing embryo or fetus.

amphetamine Any of a class of drugs that stimulates the central nervous system.

amygdala A structure in the brain that forms a part of the limbic system.

amylase A starch-digesting enzyme produced both by the salivary glands and by the pancreas.

amyloid protein A brain protein found in excessive amounts in patients with Alzheimer's disease.

amyotrophic lateral sclerosis Lou Gehrig's disease; a progressive autoimmune disease characterized by loss of myelin from motor neurons.

anabolic steroids Steroids that promote muscle growth.

anaerobic respiration Respiration that does not require atmospheric oxygen; used by yeasts, bacteria and by very active muscle tissue.

analogous structures Structures that are superficially similar in function or appearance but that have different embryological origins and evolutionary history.

anaphase The stage in mitosis at which the chromatids separate and in meiosis at which the paired chromosomes separate.

anaphylactic shock A body-wide allergic reaction, potentially fatal because of a sudden drop in blood pressure.

androgen Any of several male sex hormones, including testosterone.

anemia Low levels of erythrocytes or hemoglobin.

aneuploidy A condition in which one or more chromosomes is missing or added, relative to the number characteristic for the species.

aneurysm A weakened area in an artery.

ANF *See* **atrial natriuretic factor.**

angina pectoris Pain in the chest caused by insufficient blood flow to the muscles of the ventricular walls, generally as a result of atherosclerosis of the coronary arteries.

angioplasty The clearing of a blockage in a coronary artery (or other vessel) by a balloon that is inflated at the blockage site.

angiotensin I The active form of angiotensinogen; changed into angiotensin II by converting enzyme.

angiotensin II A transformed plasma protein that stimulates the adrenal cortex to produce aldosterone.

angiotensinogen A plasma protein that is produced by the liver and is transformed into angiotensin I in the presence of renin.

ankylosing spondylitis A form of arthritis affecting the vertebral column.

anorexia nervosa An eating disorder characterized by the willful suppression of appetite due to the person's perception that he/she is overweight.

Anthropoidea The suborder of primates that includes monkeys, apes, and humans.

anterior In front of; opposite of *posterior*.

antibiotic Organic molecules capable of inhibiting or destroying pathogenic bacteria.

antibody An immunoglobulin produced by a plasma cell that is specific for, and can bind to, an antigen.

anticoagulant Any substance that interferes with the formation of blood clots.

anticodon A portion of a transfer RNA molecule composed of three nitrogenous bases that can form a complementary pairing with three bases in a messenger RNA molecule.

antidiuretic hormone *See* **ADH.**

antigen Any substance that stimulates the production of antibodies specific to it.

antigen presenting cell (APC) Leukocytes, generally macrophages, that display antigens from destroyed pathogens on their plasma membrane surface and thereby activate helper T cells.

antihistamine A chemical that blocks the effects of histamine, thus preventing the swelling and reddening of tissues.

antisense mRNA A form of messenger RNA that binds with sense mRNA and stops the production of protein synthesis.

anus The posterior opening of the digestive tract through which feces are expelled.

aorta The largest of the arteries in vertebrates.

aplastic anemia Anemia resulting from the destruction of the bone marrow, commonly caused by radiation poisoning.

appendicitis Inflammation of the appendix.

appendicular skeleton Bones that comprise the arms, legs, and limb girdles.

appendix A small, fingerlike, blind sac of the large intestine near its connection with the small intestine, having no known function in humans.

aqueous humor A watery solution found in the anterior chamber of the eye, helping to maintain the shape of the eye.

Archaebacteria The kingdom comprised of the most ancient bacterial lineages.

arrhythmia *See* **heart block.**

arteriole A small artery.

artery A blood vessel that carries blood away from the heart.

arthritis Inflammation of a joint.

articulation A joint between two or more bones.

artificial pacemaker A mechanical contrivance that provides regular electrical stimulation to a heart with heart block.

ascorbic acid Vitamin C—a water-soluble vitamin that functions in the formation and maintenance of connective tissues and in iron and calcium absorption.

aster Either of a pair of structures formed during meiosis and mitosis, consisting of a spiral of microtubules and microfilaments arrayed about a centriole.

asthma An allergic reaction in the bronchioles of the lungs, characterized by great difficulty in breathing.

astigmatism Lack of one point of sharp focus in the eye, caused by a misshapen cornea or lens.

atherosclerosis A condition in which irregular deposits of fat accumulate along the lining of an artery.

atom The smallest unit of an element still possessing the characteristics of that element.

atomic number The number of protons in the nucleus of the atoms of a particular element.

ATP A molecule involved in powering many cellular reactions, consisting of the nitrogenous base adenine, the sugar ribose, and three phosphate groups.

ATPase An enzyme that forms a part of the sodium/potassium pump.

atrial natriuretic factor A hormone produced by the atria that increases the rate of sodium excretion by the kidneys, thereby reducing blood volume and, consequently, blood pressure.

atrioventricular node *See* **AV node.**

atrium Either of two chambers of the heart that receives venous blood and passes it to the ventricles.

atropine A poisonous drug used in small amounts in certain medical procedures, such as dilating the pupil of the eye.

auditory canal The ear passage that leads from the outer ear to the eardrum.

Australopithecus aethiopicus A fossil hominid from about 2.5 million years ago possessing traits of both *A. boisei* and *A. afarensis*.

Australopithecus afarensis A fossil hominid from about 3 million years ago that may have been a direct ancestor of modern humans.

Australopithecus africanus The first of the ancient fossil hominids to be discovered, dating to about 2 million years ago.

Australopithecus boisei A fossil hominid from South Africa that is more heavy set than *A. africanus* or *A. robustus*.

Australopithecus robustus A heavy-jawed South African hominid fossil.

autoimmunity The development of an immune reaction against the tissues of one's own body.

autonomic nervous system The portion of the nervous system that controls the actions of the visceral organs and the diameter of blood vessels, and that comprises both the sympathetic and parasympathetic systems.

autosome Any chromosome other than the sex chromosomes.

autotroph An organism capable of synthesizing all of its required organic compounds from inorganic sources and an external energy source such as sunlight (e.g., most plants).

AV node A small mass of highly specialized muscle fibers located near the junction of the atria and the ventricles that receives impulses from the SA node and conducts them throughout the ventricles.

axial skeleton Bones aligned along the primary axis of the body, including the skull, vertebral column, and ribs.

axon The long process of a neuron.

AZT The first of the drugs approved for use in the treatment of AIDS.

B cell A type of lymphocyte that matures in the bone marrow and that can transform into an antibody-producing plasma cell.

barbiturate Any of a class of drugs that functions as a sedative.

balanced polymorphism *See* **polymorphism.**

Barr body A darkly-staining object characteristic of the nucleus of cells from females that is actually a condensed X chromosome.

baroreceptor A blood pressure monitor located in the carotid sinus and aorta.

basal body A cytoplasmic organelle found at the base of cilia and flagella and composed of microtubules.

base A substance that releases hydroxyl (OH^-) ions when dissolved in water, causing the pH to rise.

base substitution A form of mutation in which one base is erroneously substituted for another during DNA replication.

basilar fibers A series of thin filaments in the cochlea of the ear that vibrates in response to sound waves.

basilar membrane A membrane that divides the cochlea of the ear longitudinally.

basophil A category of granular leukocytes responsible for the release of histamine and for attack on parasitic worms.

benign Non-malignant.

beriberi A muscle-weakening disease caused by thiamin deficiency.

beta blocker A type of drug used by some heart patients to reduce heart activity and to dilate the coronary arteries.

bicep A two-headed muscle of the arm, responsible for closing the elbow joint.

bicuspid valve The heart valve between the left atrium and the left ventricle that prevents blood from returning to the atrium when the ventricle contracts.

bile A fat-emulsifying liquid produced by the liver and stored in the gall bladder.

binomial The two word system of naming species.

biomagnification A process whereby organisms in the upper trophic levels accumulate increasingly greater amounts of various poisons such as DDT and other insecticides.

biopsy The examination of excised tumor tissue for the presence of cancerous cells.

biosphere The air, land, and water of the earth in which organisms live.

biotic Pertaining to any aspect of life.

biotin A water-soluble vitamin required in the formation of antibodies and in carbon dioxide transfers in cellular metabolism.

bipedality The capacity to walk on two feet.

bipolar cell Cells in the retina that transmit action potentials to the ganglion cells.

bladder *See* **urinary bladder.**

blastocyst An early stage in the embryonic development of a mammal.

blood A liquid tissue contained within the vessels of the circulatory system.

blood-brain barrier The tight junctions found between the cells of the choroid plexus as well as those comprising the walls of capillaries in the brain that greatly limit the passage of materials between the blood and the interstitial fluid of the brain.

blood clot A dense network of fibrin molecules and blood cells that forms in response to injury to blood vessels.

blood doping The use, by athletes, of transfusion of their own blood shortly before an athletic event, to increase the number of erythrocytes and theoretically to provide more oxygen-carrying capacity.

bond Any of a number of chemical interactions between atoms, including covalent, ionic, and hydrogen bonds.

bone A rigid connective tissue comprising the bulk of the skeletal system of vertebrates.

bone matrix The salts and proteins that make up the non-cellular portion of bone.

botulism A disease in which a toxin produced by the bacterium *Clostridium botulinum* blocks the release of acetylcholine at neuromuscular junctions.

Bowman's capsule The portion of the nephron in which the glomerulus is imbedded.

brachial Pertaining to the arm, as in brachial artery, the principal artery of the arm.

brachydactyly The genetic condition of having short fingers and toes.

bradykinin A pain-producing molecule formed from the splitting of an interstitial fluid protein by the enzyme kallikrein.

brain The anterior enlargement of the spinal cord, surrounded by the skull.

brainstem The region in the floor of the brain that leads directly from the spinal cord.

Broca's area The area of the brain responsible for grammatical refinement.

bronchiole A subdivision of a bronchus.

bronchitis Inflammation of the bronchi or bronchioles.

bronchus Either of the two main branches of the trachea.

buffer A substance capable of neutralizing either an acid or a base, and thus also capable of maintaining a constant pH.

bulimia An eating disorder characterized by bingeing followed by purging (through the use of laxatives or induced vomiting).

bulk flow The overall movement of a fluid induced by hydrostatic pressure.

bundle of His A cluster of specialized muscle fibers that conducts impulses from the AV node throughout the ventricles.

bursa A fluid-filled sac that allows free movement of joints and tendons.

bursitis Inflammation of a bursa.

caffeine A stimulant found in coffee, tea, and cola that enhances synaptic transmission in the CNS by blocking the breakdown of cyclic AMP.

calcitonin A hormone produced by the thyroid gland that lowers blood calcium levels.

calcium channel blocker A drug used by angina patients to prevent spasms of the coronary arteries.

Calorie A measure of energy in food; one Calorie is the amount of heat energy needed to raise one kilogram of water one degree Celsius.

Calvin cycle The principal set of chemical reactions of the light-independent reactions of photosynthesis during which carbon dioxide is incorporated in the synthesis of carbohydrates.

CAMs Cell-adhesion molecules, a group of plasma membrane molecules that are critical in morphogenesis.

cancelleous bone The spongy inner layer of bone.

cancer A malignant tumor or the disease it causes.

caniculi Tiny canals that link the lacunae of bone.

capillary A blood vessel that links an arteriole with a venule and in which exchange of materials occurs with the surrounding interstitial fluid.

capillary pores The tiny gaps that exist between the cells that comprise the capillary walls and through which materials may flow between the blood and the interstitial fluid.

carbohydrate Any of a class of organic compounds characterized by being composed of carbon, hydrogen, and oxygen in roughly a 1:2:1 ratio.

cardiac muscle A type of muscle tissue found only in the heart.

cardiac output The amount of blood pumped by the heart in a given time; stroke volume times heart rate.

caries Tooth decay.

carnivore An animal that feeds on other animals.

carotenoid Any of a class of lipids characterized by its yellow, orange, or red color.

carotid artery Either of a pair of major arteries in the neck that carries blood to the brain.

carotid sinus A swelling at the point where the common carotid artery divides into internal and external vessels and in which blood pressure is monitored by baroreceptors.

carpals The eight bones of the wrist.

carrying capacity The largest population of a particular organism that can be supported indefinitely by a given habitat.

cartilage An elastic but strong type of connective tissue found in the nose, ears, and intervertebral discs, and on the ends of bones with movable articulations.

catalyst A substance that accelerates the rate of a chemical reaction without itself being permanently altered.

catalyze Chemical reactions that are speeded up by the use of enzymes or other agents that reduce activation energy.

cataract A clouding or opacity of the lens of the eye.

cause and effect Utilization of the scientific method requires distinguishing events related by cause and effect from events linked only by correlation or coincidence.

Ceboidea The primate superfamily that includes the New World monkeys.

cell The basic structural unit of life.

cell cycle The life cycle of a cell, from its formation until it undergoes mitosis.

cell division Mitosis plus cytokinesis.

cell theory The concept that the cell is the basic organizational unit of life; that living organisms are composed of cells and cell products; and that cells arise only from preexisting cells.

cell wall A thick coat, variously comprised, that surrounds the plasma membrane in moneran, fungal, and plant cells (and some protists).

cellular respiration The series of chemical reactions by which cells break down organic molecules in the presence of atmospheric oxygen and produce energy in the form of ATP.

central nervous system The spinal cord and brain.

centriole One of a pair of organelles in animal cells that evidently organizes the aster and spindle fibers during mitosis and meiosis.

centromere The portion of the chromosome at which sister chromatids are linked and the point at which spindle fibers attach during mitosis and meiosis.

Cercopithecoidea The primate superfamily that includes the Old World monkeys.

cerebellum The portion of the brain, located behind the cerebrum, that coordinates motor activity.

cerebral hemorrhage Rupture of an artery in the brain, causing a stroke.

cerebral palsy A condition resulting from destruction of certain motor centers of the cerebrum and leading to impaired motor function.

cerebral thrombosis A blood clot in an artery of the brain, causing a stroke.

cerebrospinal fluid A fluid produced by the choroid plexus and in which the brain and spinal cord float.

cerebrum The largest portion of the human brain, consisting of two lobes that together control most conscious activities.

cervical cap A small, thimble-shaped birth control device that covers the opening to the cervix.

cervical vertebrae The seven vertebrae of the neck.

cervix The neck of the uterus, just above the vagina.

chancre An ulcer-like sore that forms on mucous membranes in the primary stage of syphilis.

chemautotroph Bacteria that assemble organic molecules from inorganic molecules, using energy from inorganic reactions.

chemical evolution Natural selection in non-living systems.

chemoreceptor A sense organ that responds to the presence of various types of chemicals in air, food, or internal body fluids.

chemotaxis Orientation of a cell (such as a leukocyte) to a chemical gradient (as from an infection).

Chlamydia A bacterium causing an STD characterized by painful urination in men and possible pelvic disorders, arthritis, and heart disease in women.

chiasma Crossover points of nonsister chromatids during prophase I of meiosis.

chlorofluorocarbon (CFC) An ozone-destroying chemical used as a coolant in refrigerators and air conditioners.

chlorophyll A green photosynthetic pigment found in the chloroplasts of the leaf cells of most plants.

chloroplast A specialized type of plant cell organelle, containing chlorophyll.

cholecystokinin A hormone produced by the small intestine that stimulates contraction of the gall bladder.

cholera A bacterial disease of the intestines, characterized by extreme diarrhea, vomiting, and loss of water and salts.

cholesterol A steroid common to animal tissues, excess amounts of which may build up on arterial walls, contributing to the development of atherosclerosis.

cholinesterase An enzyme, found in the synaptic cleft between two neurons and in neuromuscular junctions, that catalyzes the breakdown of acetylcholine.

chorion The outermost of the extraembryonic membranes and the source of most of the fetal contribution to the placenta.

chorionic villi Tiny fingerlike outgrowths of the chorion that imbed deeply in the uterine lining, thereby increasing the surface exchange area between the circulatory system of the fetus and that of the mother.

chorionic villus biopsy A process by which samples of chorionic cells are taken from the placenta in order to permit examination for possible genetic defects.

choroid A membrane of the eye, located between the sclera and the retina.

choroid plexus A specialized tissue located at the top of the brainstem that secretes cerebrospinal fluid.

chromatid One of the two identical strands of the replicated chromosome, visible during prophase and metaphase and separated during anaphase of mitosis.

chromosome Nuclear bodies that carry genetic information in a specific order; comprised of two intertwined DNA molecules and proteins.

chronic leukemia A form of leukemia in which abnormally high numbers of lymphocytes persist for many years.

chyme The thick, semifluid substance that results from the actions of the stomach on food.

chylomicron Tiny packages of protein and triglycerides that are transformed into VLDLs in the liver.

ciliary muscle The muscle that controls the shape of the lens of the eye.

cilium A short, hairlike organelle, capable of movement, found on the surface of some cells, and generally present in large numbers.

cirrhosis Liver disease characterized by the death of liver cells and the formation of scar tissue.

classical conditioning A type of learning that substitutes one stimulus for another, as exemplified by Pavlov's dogs.

clavicle The collarbone, linking the sternum with the scapula and humerus.

cleavage The division of a fertilized egg into a series of smaller cells.

cleavage furrow The cytoplasmic depression that begins cytokinesis.

cleft palate A birth defect in which the bones of the hard palate do not fuse in the midline, leaving an opening in the roof of the mouth.

clitoris A small, sensitive organ composed of erectile tissue, lying just above the urethral opening in female mammals, and homologous to the head of the penis.

clone Any of a group of genetically identical individuals arising by asexual means from a single ancestor.

clot retraction The compaction of a blood clot that occurs when microfilaments from platelets pull the fibrin molecules in a blood clot closer together.

CNS *See* **central nervous system.**

coacervate droplet A lipid-coated droplet in a water solution that may have represented an intermediate stage in the evolution of life from non-life.

cobalamin Vitamin B_{12}; a water-soluble vitamin required in the formation of red blood cells and nerve cells.

cocaine A stimulant that blocks sodium channels in the brain and inhibits the uptake of dopamine, allowing continued stimulation of the pleasure centers of the hypothalamus.

coccyx The tailbone, comprised of the four most posterior vertebrae.

cochlea A spiral-shaped portion of the inner ear of mammals.

cochlear duct The area between the vestibular canal and the tympanic canal in the cochlea.

codominance A situation in which both alleles of a gene are expressed, neither being recessive to the other.

codon A series of three nitrogenous bases in a molecule of messenger RNA that matches a corresponding anticodon of transfer RNA.

coenzyme A small organic molecule that is essential for the functioning of a particular enzyme.

cofactor An ion that is necessary for the functioning of a particular enzyme.

coitus Sexual intercourse.

coitus interruptus Male withdrawal during intercourse prior to ejaculation.

collagen A gelatinous protein found in the dermis and in most types of connective tissues.

collecting duct Any of a large number of tiny tubes in the kidney into which the products of the nephrons pass, prior to entering the ureters as urine.

colon The large intestine.

color blindness A genetic defect in the cones of the retina that prevents a distinction being made between red and green (rarely, blue).

coma A state of deep and prolonged unconsciousness, typically caused by an injury to the head.

community The totality of the species living within a common geographic area.

compact bone The solid outer layer of bone.

complement system A group of about 20 plasma proteins that, in various ways, assist in the destruction of bacteria.

compound A substance composed of two or more elements, the atoms of which are linked by chemical bonds.

condensation A chemical reaction in which the components of water molecules are removed thereby linking monomers to form polymers.

condom A thin sheath of rubber or animal membrane worn over the penis during intercourse to prevent conception or transmission of venereal disease.

conduction deafness Deafness caused by impairment of the transmission of sounds from the eardrum to the cochlea.

cone One of a class of light-sensitive cells found in the retina of the eye.

congestive heart failure Inadequate functioning of the bicuspid valve, allowing blood to be forced back up the pulmonary veins into the lungs when the left ventricle contracts.

consanguinity Matings between individuals who share a recent common ancestor.

consumer A member of one of the three classes of organisms used in ecological parlance, in contrast to producer and reducer; includes animals and some other organisms.

contact dermatitis An allergic reaction characterized by a reddening and itching of the skin.

convergent evolution Increased similarity between two evolutionarily diverse lineages.

converting enzyme An enzyme located on the capillary walls of the lungs that transforms angiotensin I into angiotensin II.

cornea The transparent outer coat of the eye, covering the iris and lens.

coronary Relating to the heart, as in coronary artery or vein.

coronary artery Either of a pair of vessels that supplies blood to the ventricular walls.

coronary bypass The surgical creation of alternate pathways around blockage sites in coronary arteries.

coronary thrombosis A blood clot in a coronary artery, generally causing a myocardial infarction.

coronary transplant Removal of the heart and replacement with the heart of a donor.

corpus callosum A broad band of neural tissue that links the left and right cerebral lobes.

corpus luteum The yellowish remains of an egg follicle following ovulation, consisting of secretory cells that produce the hormones estrogen and progesterone.

cortex The outer layer of such organs as the brain, kidney, and adrenal glands.

corticotropic releasing hormone *See* **CRH.**

cortisol The primary glucocorticoid hormone secreted by the adrenal gland.

cortisone A synthetic glucocorticoid used in the treatment of inflammation.

countercurrent mechanism In urine formation, the mechanism operating in the peritubular capillaries that illustrates why exchange between two fluids is greatest when they flow in opposite directions.

covalent bond A type of chemical bond in which electrons are shared between two atoms.

cramp A muscle spasm wherein the muscle remains in a continuously contracted state.

cranial nerve Any of the nerves emanating directly from the brain.

cranium The braincase, composed of eight bones.

creation science A non-scientific, anti-evolutionary belief that a divine being created all of the species in the world today just as they are.

cretinism A type of mental retardation caused by deficient amounts of thyroxin, either due to a genetic condition or to the absence of iodine in the food.

Creutzfeldt-Jacob disease (CJD) One of two diseases in humans known to be caused by a slow virus.

CRH A releasing factor from the hypothalamus that stimulates the anterior pituitary to produce ACTH.

critical period The brief developmental period during which imprinting occurs.

Crohn's disease An autoimmune disease of the small intestine; a form of inflammatory bowel disease.

Cro-Magnon The most modern of the fossil humans, dating from about 40,000 years ago.

crossing over The exchange of chromatid segments between homologous chromosomes during prophase I of meiosis.

cryptorchidism The condition in which the testes fail to descend into the scrotum during development.

CSF *See* **cerebrospinal fluid.**

cultural learning The transmission of specific information from generation to generation.

curare A paralytic poison derived from certain South American plants.

Cushing's disease Overproduction of the hormones of the adrenal cortex.

cuteness response A benevolent reaction to the offspring of a species.

Cyanobacteria Blue-green photosynthetic bacteria.

cyclic AMP The second messenger of many hormones, the presence of which sets off a cascade of enzymatic activity.

cyclic GMP A compound found in the rods and cones of the eye that maintains sodium channels open in the plasma membrane.

cyclin The molecular trigger that initiates mitosis.

cyclosporine An extract from a fungus that largely prevents the rejection of tissue and organ transplants by the body.

cystic fibrosis A crippling and ultimately fatal genetic disease affecting the lungs and digestive system.

cystitis Inflammation of the urinary bladder.

cytokine Small protein molecules that function as chemical signals between cells.

cytokinesis Division of the cytoplasm of a cell following mitosis.

cytoplasm The contents of a cell, exclusive of the nucleus and organelles.

cytoskeleton A constantly changing arrangement of linear proteins that gives rigidity to a cell.

decibel A unit for measuring the volume of a sound, equal to one-tenth of a bel.

decibel scale A logarithmic scale measuring the intensity of sound.

deductive logic Reasoning from the general to the specific.

defoliant A chemical that causes plants to lose their leaves.

deletion A type of chromosomal mutation involving the loss of a segment of a chromosome.

denaturation A generally irreversible consequence of the loss of the secondary and tertiary structures of a protein because of excessive heat or a change in pH.

dendrite One of the short, numerous branches of a neuron.

deoxyribonucleic acid *See* **DNA.**

dependent variable Any change in a test population not found in the control population that can logically be linked to the independent variable.

depletion The insufficiency of an essential substance for the growth of a population.

depolarization The process occurring during the generation of an action potential in which the interior of a neuron becomes electrically positive relative to the surrounding fluids, rather than electrically negative as it is during the resting potential.

Depo-Provera A synthetic progesterone that provides three months of contraception from a single injection.

depressant Any of a series of drugs that decreases neurological or physiological activity.

dermis The inner layer of the skin, beneath the epidermis.

DES A steroid implicated in the development of vaginal cancer in young women whose mothers took the drug during their pregnancies.

desertification The loss of habitat to desert, usually because of overgrazing.

development The processes by which a zygote becomes a complete and functioning organism.

diabetes insipidus A condition caused by the failure of the hypothalamus to produce ADH, resulting in excessive urine production.

diabetes mellitus A condition caused by insufficient insulin, resulting in elevated blood glucose levels and loss of glucose in the urine.

dialysis A process employed in the artificial kidney, whereby waste materials are allowed to pass from the blood through a porous membrane into a fluid bath that is later discarded.

diapedesis Movement of leukocytes through capillary pores and into the surrounding tissues.

diaphragm 1. The horizontal muscular wall that separates the chest cavity from the abdominal cavity. 2. A rubber cup that is placed over the cervix as a birth control device.

diastolic pressure Blood pressure during the interbeat period; contrasted with *systolic pressure*.

diethylstilbestrol *See* **DES.**

differentiation A developmental process in which a cell line becomes structurally or functionally specialized.

diffusion Movement of materials in a fluid from regions of high concentration to regions of low concentration.

digestion The fragmentation of food by mechanical and chemical means into sizes small enough to permit uptake by the cells of the digestive tract.

digestive tract A series of interconnected organs stretching from mouth to anus.

dihybrid cross Mating between individuals from two inbred lines that are homozygous dominant and recessive, respectively, for two different traits.

diploid A cell with two sets of chromosomes, one set from each parent.

directional selection The preservation of one end of the continuum of variation, and the consistent loss of the other end of the continuum.

disaccharide A sugar, such as sucrose, that is composed of two monosaccharides.

displacement activity The expression of a fixed action pattern in an incongruous situation.

disruptive selection Elimination of the norm, and preservation of the two extremes of variation in a population over time.

distal convoluted tubule The portion of the nephron located between the ascending arm of the loop of Henle and the collecting duct.

diuretic A substance that increases the rate and volume of urine formation.

divergent evolution The development of an increasing amount of distinctiveness between two evolutionary lines.

diverticulitis Inflammation of outpocketings of the colon.

diverticulosis The presence of outpocketings in the colon.

DNA The molecule responsible for carrying coded genetic information in cells.

DNA hybridization The amount of complementarity that exists in the DNA of two organisms.

DNA polymerase The enzyme that selects bases during the replication of DNA.

dominant An allele that is phenotypically expressed.

dorsal root The upper branch of a spinal nerve, through which sensory neurons enter the spinal cord.

douche Washing of some internal body part, especially the vagina.

drive Motivation; what underlies the performance of a particular fixed action pattern.

Dryopithecus A fossil primate having both prosimian traits and anthropoid (and possibly hominoid) traits.

Duchenne muscular dystrophy A genetic disease characterized by a wasting away of skeletal muscles.

ductus arteriosus A shunt vessel that connects the pulmonary artery to the descending aorta in the fetus.

ductus venosus A shunt vessel in the liver that permits blood to flow from the umbilical vein to the inferior vena cava in the fetus.

duodenum The first portion of the small intestine, just posterior to the stomach.

duplication A chromosomal mutation in which a segment of a chromosome is repeated and therefore is present twice.

dwarfism Any of several developmental conditions in which the bones of the arms and legs fail to grow properly.

dynein A cytoplasmic protein believed to be responsible for helping to move organelles within the cell; see *kinesin*.

ear One of a pair of sense organs located on the sides of the head that function in hearing and balance.

eardrum The tympanum, a membrane that separates the outer ear from the middle ear.

ecology The branch of biology that deals with the relationships between organisms and their environment.

ecosystem A community of organisms plus the nonliving environment in which the community resides.

ectoderm One of the embryonic germ layers from which the epidermis and the nervous system develop.

ectopic pregnancy A pregnancy in which implantation has occurred outside the reproductive tract (e.g., on the wall of the bladder).

edema The accumulation of abnormally high levels of interstitial fluid.

egg follicle On the surface of the ovary, the developing ova with surrounding tissue.

ejaculation The ejection of semen from the penis during the male orgasm.

ejaculatory duct A structure in the prostate gland to which the sperm move during emission.

elastin A protein found in the dermis and in connective tissue in the form of long, elastic fibers.

electromagnetic spectrum The continuum of radiation ranging from gamma rays to radio waves.

electron A negatively charged subatomic particle found at some distance from the nucleus of the atom.

electron acceptor An atom that takes electrons from another atom, thereby filling its own outer ring.

electron donor An atom with only a few electrons in its outermost ring that becomes more stable when the outer electrons are lost to an electron acceptor.

electron transport system A series of enzymes found within mitochondria that receive electrons from the Krebs cycle and use their energy in the manufacture of ATP.

electrophoresis Separation of proteins or nucleic acids in a solution or on a gel through the application of an electric current.

element A substance that cannot be separated into simpler materials by chemical means alone.

emergent property A characteristic or ability that is not intrinsic in the component parts of an entity, but is true only for the entity as a whole; contrasted with *inherent property*.

emission Movement of sperm to the ejaculatory duct just prior to ejaculation.

empiricism Reliance on direct observation rather than on speculation.

emphysema A chronic lung disease, characterized by the gradual breakdown of the alveoli.

emulsify To form an emulsion, a stable suspension of tiny fat particles in a water medium.

endocrine gland A ductless gland, producing one or more types of hormones.

endocytosis The process whereby a cell engulfs substances; includes both phagocytosis and pinocytosis.

endoderm The innermost of the three embryonic germ layers, from which the linings of many of the internal organs are derived.

endometrium The lining of the uterus.

endoplasmic reticulum A complex series of cytoplasmic membranes, which may be covered with ribosomes (rough ER, the site of protein synthesis) or be ribosome-free (smooth ER, the site of lipid synthesis).

endorphins Small proteins produced by the brain and released during stress that reduce pain and enhance the feeling of well-being.

endothelium A tissue that lines many organs, including blood vessels.

energy pyramid The ecological relationships of producers and the various classes of consumers.

enkephalins Small chains of amino acids produced by the brain that enhance the perception of pain.

enzyme A globular protein capable of catalyzing one or more chemical reactions.

eosinophil A class of granular leukocytes that bind readily with the acidic stain eosin.

epidermis The outer layer of the skin, lying over the dermis.

epididymis A coiled tube in which the sperm mature, located atop the testis.

epigenesis A now broadly accepted theory that the fertilized egg is essentially homogeneous and that development consists of gradual specialization; contrasted with *preformationism*.

epiglottis A thin flap of tissue that covers the opening to the trachea during swallowing.

epilepsy A disease of the brain, characterized by periodic abnormal levels of electrical activity and seizures.

epinephrine A hormone produced by the adrenal medulla that causes rapid and powerful ventricular contractions and various changes in the blood flow.

epistasis The masking of the expression of a gene by a nonallelic gene located at a different locus.

epithelium A type of tissue that covers surfaces or lines organs.

equational division A type of cell division in which the number of chromosomes is not reduced from parent to daughter cells.

erythrocyte A red blood cell.

erythropoietin A hemopoietin produced by the kidneys in response to a drop in the oxygen-carrying capacity of the blood that stimulates the maturation of erythrocytes.

esophagus The tube linking the pharynx with the stomach.

essential amino acids Amino acids that cannot be synthesized by the body and are therefore required in the diet.

essential fatty acids Fatty acids that cannot be synthesized by the body and are therefore required in the diet.

essential hypertension A form of high blood pressure where the cause is unknown; contrasted with *organic hypertension*.

estrogen Any of a group of feminizing steroid hormones involved in the development of secondary sexual characteristics and in governing the menstrual cycle.

ethanol A two-carbon alcohol (grain alcohol).

ethology The branch of biology that concerns the biological bases of behavior.

Eubacteria The kingdom containing the more evolutionary recent lineages of bacteria.

eugenics The notion that the human species can be improved through the elimination or control of certain deleterious genes by selective breeding.

eukaryote Any cell with a distinct nucleus and a variety of cytoplasmic organelles.

eustachian tube In mammals, a canal linking the pharynx with the middle ear; derived from one of the gill clefts of fishes.

evolution A change in the allele or gene frequencies of a population of organisms over time as a consequence of natural selection and adaptation.

excretion The elimination of metabolic by-products from a cell or from the body as a whole.

exhalation The act of breathing out.

exocrine gland A ducted gland, such as sweat or sebaceous glands.

exocytosis Elimination of a waste vacuole from a cell; the opposite of endocytosis.

exon A region of DNA that is expressed in the formation of a messenger RNA molecule.

expressivity The degree of phenotypic expression of a particular gene.

extraembryonic membrane Any of four membranes that are produced by the embryo but are not destined to become incorporated as adult structures.

facilitated diffusion Diffusion of certain molecules across the plasma membrane as loosely bound complexes attached to a carrier molecule in the membrane.

fallopian tube The oviduct, a narrow canal between an ovary and the uterus.

familial hypercholesterolema A genetic disease characterized by extremely high blood cholesterol levels.

FAP *See* **fixed action pattern.**

fat Any of a class of lipids consisting of a molecule of glycerol to which three fatty acid molecules are bound.

feces Intestinal waste material expelled through the anus.

femur The thighbone.

fermentation The conversion of pyruvate to ethanol and water, or to lactic acid, in the absence of oxygen.

fertilization The union of a sperm nucleus with an egg nucleus.

fetal alcohol syndrome A series of developmental problems in a child caused by excessive alcohol consumption by the mother during pregnancy.

fetus The stage of development from two months after fertilization until birth.

FGF Fibroblast growth factor, a morphogen that, in conjunction with NGF, causes differentiation of sympathetic neurons from embryonic nerve cells.

fibrin A linear plasma protein that is an important component of blood clots.

fibrinogen An inactive plasma protein that is converted to fibrin in the presence of thrombin.

fibula A long, slender bone of the lower leg.

filtration The first step in the formation of urine, wherein some of the blood serum passes through the walls of the glomerulus and Bowman's capsule and into the cavity of the nephron.

First Law of Thermodynamics Energy cannot be created or destroyed, but it can be converted from one form to another.

fission reaction A type of nuclear reaction wherein atomic nuclei (generally, uranium) are split, releasing enormous amounts of energy.

fixed action potential Highly stereotyped genetically determined behavior of a type characteristic for a given species that is triggered by a releaser and that varies little from performance to performance.

flagellum A long, whiplike organelle used for propulsion; longer and fewer in number than cilia.

fluid homeostasis The maintenance of the blood and interstitial fiuid in essentially constant volume and composition.

foam cell A form of macrophage that can accumulate along arterial walls and that has been implicated in the development of atherosclerosis.

folic acid One of the water-soluble vitamins; important in DNA and RNA synthesis.

follicle stimulating hormone *See* **FSH.**

food vacuole The small membranous sac that is formed around ingested material during phagocytosis.

food web The complex of predator-prey relationships within a given ecological community.

foramen ovale The opening between the fetal atria, which normally closes at birth.

formula Information about the kind and number of atoms that make up a molecule.

fossil The remains or impression of an organism from prehistoric times.

founder effect A limited gene pool in a population created when the original colonization of an area is by a very small number of individuals.

fovea The most sensitive portion of the retina.

fracture Injury to a bone ranging from a minor crack to major damage to the bone and surrounding tissue.

FSH A hormone produced by the anterior pituitary that stimulates the growth of an egg follicle in the ovary (and the production of sperm in males).

fulcrum The pivot point in a lever system.

Fungi The kingdom of heterotrophic, vaguely plant-like organisms that includes mushrooms, yeast, molds, and their relatives.

fusion reaction A type of nuclear reaction wherein atomic nuclei (generally, isotopes of hydrogen) are combined to form helium, with the release of enormous amounts of energy.

G spot A region of tissue on the upper wall of the vagina that is reportedly highly erogenous in some women.

G-CSF A hemopoietin that stimulates the development of granulocyte colonies.

galactosemia A genetic condition characterized by the accumulation of toxic amounts of galactose in children, causing cirrhosis of the liver, cataracts, and mental retardation.

gall bladder A small accessory organ located near the liver that stores and concentrates bile.

gallstones Precipitates of cholesterol and bile salts that may accumulate in the gall bladder and cause abdominal pain.

gamete A cell capable of fusing with another cell to form a zygote; sperm and egg cells.

ganglion A swelling of the dorsal root of spinal nerves, containing the cell bodies of sensory neurons.

ganglion cell A class of cells in the retina of the eye that receives stimuli from bipolar and amacrine cells.

gastrin A hormone produced by the stomach that increases the rate of stomach secretion.

gastrocnemius The calf muscle.

gastrocolic reflex The wave of peristaltic contractions that moves food residue from the colon to the rectum.

gel electrophoresis *See* **electrophoresis.**

gene The basic unit of heredity in a chromosome, consisting of a section of the DNA molecule that codes for a single protein.

gene flow A type of genetic drift that occurs with the immigration or emigration of individuals from small populations.

gene pool The collective genetic information in a population.

genetic drift Changes in the genetic diversity of a population (especially a small population) due to chance, rather than to natural selection.

genetic mosaic Individuals with cells that mutated very early in embryonic development, and therefore having genetically distinct zones of the body.

genetics The science of heredity at the organismal or population level (Mendelian genetics) or at the cellular level (molecular genetics).

genitalia The organs of the reproductive system.

genome The set or sets of chromosomes possessed by each cell of an organism.

genotype The genetic constitution of an organism.

germ layer Any of the three embryonic tissues that give rise to all of the adult tissues and organs.

germ plasm The cells of reproduction; gametes; sperm and ova.

GH A hormone produced by the anterior pituitary that promotes increased metabolic activity and growth.

GHRH A releasing factor produced by the hypothalamus that stimulates the anterior pituitary to produce growth hormone.

glans The head of the penis.

glaucoma An eye disease in which high pressure in the fluids of the eye causes the collapse of the artery that supplies the retina and the retinal cells die.

glial cell A class of cells of the brain that supports neurons.

globular protein Proteins with primary, secondary and tertiary structures.

glomerular filtrate The portion of the serum that is forced through the glomerular walls and into the cavity of Bowman's capsule.

glomerulonephritis An autoimmune disease characterized by the inadvertent destruction of the cells of the glomeruli by phagocytes.

glomerulus A cluster of arterial capillaries imbedded in Bowman's capsule.

glottis The opening into the larynx and trachea from the pharynx.

glucagon A pancreatic hormone that raises blood glucose levels by promoting the conversion of glycogen into glucose.

glucocorticoid Any of a group of adrenocortical steroid hormones that increases blood glucose levels and reduces inflammation.

gluconeogenesis The production of "new" glucose through the conversion of amino acids and glycerol to glucose.

gluteus maximus The largest of the three pairs of muscles that comprise the buttocks.

glycogen A polysaccharide storage molecule in animals, composed of long, branched chains of glucose.

glycolipid A lipid to which a carbohydrate is attached.

glycolysis Literally, the breakdown of glucose; more specifically, that portion of glucose metabolism accomplished without oxygen.

glycoprotein A protein to which a carbohydrate is attached.

GM-CSF A hemopoietin that simulates the development of both granulocyte and monocyte colonies.

GnRH A hormone produced by the hypothalamus that stimulates the anterior pituitary to produce FSH and LH.

goiter A greatly expanded thyroid gland, usually attributable to a diet deficient in iodine.

Golgi apparatus An organelle responsible for the packaging of secretory materials.

Golgi tendon organ Multibranched nerve fibers located in the tendons just beyond the end of muscle fibers that measure the amount of muscle tension.

gonad The ovary or testis.

gonadotropin A hormone that stimulates the gonads, such as FSH or LH.

gonadatropin releasing hormone *See* **GnRH.**

gonorrhea A venereal disease caused by gonococcal bacteria.

gossypol A chemical used as a male contraceptive that functions by blocking sperm production.

gout An ailment caused by the precipitation of uric acid in the joints.

granulocyte The class of leukocytes that includes basophils, neutrophils, and eosinophils.

Graves' disease An autoimmune disease in which a goiter forms because of the destruction of TSH receptors in the cells of the thyroid gland.

greenhouse effect The retention of the heat of the sun by the glass of a greenhouse or by the presence of water vapor, carbon dioxide, and certain other gases in the atmosphere.

growth factor Any of a class of cytokines produced by leukocytes that stimulates the differentiation of stem cells along particular lines.

growth hormone *See* **GH.**

growth hormone releasing hormone *See* **GHRH.**

H zone In striated muscle, the space between the ends of the thin filaments of sarcomeres.

habituation The capacity of the nervous system to screen out irrelevant stimuli.

hair end-organ A nerve fiber wrapped around the base of every hair.

hallucinogen Any of a class of drugs that affects the central nervous system and promotes hallucinations.

haploid A cell or organism with a single set of chromosomes.

harelip A congenital deformity in which the upper lip is cleft.

Haversian canal A small canal found in bone through which a nerve and blood vessel run.

hay fever A type of allergic reaction caused by the pollen of certain plants.

HCG A hormone produced by the placenta that maintains the endometrium intact.

HDL *See* **high density lipoprotein.**

heart A muscular organ at the base of the sternum that propels blood through the vessels of the circulatory system.

heart block Failure of a portion of the pacemaker system of the heart, usually due to a preceding myocardial infarction.

heartburn Irritation of the esophageal lining by acidic stomach secretions regurgitated into the esophagus.

heart murmur Any irregular sound produced by the heart.

heat capacity The amount of energy necessary to raise or lower the temperature of a substance.

heat of vaporization The amount of heat energy necessary to convert a liquid to a gas.

hemoglobin An iron-containing respiratory pigment found in great abundance in erythrocytes.

hemophilia A genetic condition in which one of the agents needed for the clotting cascade is missing.

hemopoietin Any of several growth factors that stimulate the development of particular lines of blood cells.

hemorrhage Loss of blood through a ruptured vessel.

hemorrhoid A varicosity of one of the hemorrhoidal veins that surround the anus.

hepatic portal vein A vein that transports most of the materials absorbed by the intestine from capillary beds in the small intestine to capillary beds in the liver.

herbivore An animal that feeds on plants.

heritability The percentage of the total phenotypic variance of a trait that is due to genetic differences.

hernia The protusion of all or part of an organ through a torn muscle or other abnormal opening.

heroin A narcotic drug derived from morphine.

Herpes simplex An STD that causes sores on the genitals (type II).

heterotroph An organism dependent on other organisms for at least some of its organic needs (e.g., all animals).

heterozygous A diploid cell or organism with two different alleles for the same gene.

hiccups Noises made by the bursts of air that enter the larynx when the diaphragm spasms.

high density lipoprotein Lipoprotein with no more than 15% cholesterol by weight that transports cholesterol from the cells of the body to the liver for conversion into bile.

hippocampus A part of the limbic system that routes recent memories to storage centers in the cerebrum.

histamine A protein found within various cells that, when released, causes tissue swelling.

HLA system *See* **human leukocyte antigen system.**

homeostasis The capacity of an organism to maintain an essentially stable internal environment.

Hominidae A family of primates, including *Australopithecus* and *Homo.*

Hominoidea A superfamily of primates, including humans and apes.

Homo erectus "Upright man"—the direct ancestor of *H. sapiens,* living from about 1.7 million years ago to as recently as 250,000 years ago.

Homo habilis "Skillful man"—the species intermediate between *Australopithecus* and *H. erectus,* living about 1.75 million years ago.

homologous structures Structures that share a common evolutionary ancestry, despite what may be profound differences in appearance.

homologue In diploid cells, either member of a chromosome pair.

Homo sapiens "Wise man"—modern humans, dating from about 250,000 years ago.

homozygous A diploid cell or organism with the same two alleles for a particular gene.

hormone A small organic molecule secreted in one part of an organism that travels through the blood to regulate the function of a tissue or organ in another part of the organism.

human chorionic gonadotropin *See* **HCG.**

humanism The study of the interests and ideals of people.

human leukocyte antigen system The MHC complex in humans, responsible for the proteins used in identification of self.

humerus The bone of the upper arm.

Huntington disease A genetic disease that causes deterioration of nerve cells with a corresponding loss of motor control and memory.

hybridization The crossing of two species or two genetically distinct lines within a single species.

hydrogen bond A weak chemical bond between a hydrogen atom and an oxygen or nitrogen atom of another part of the molecule or an adjacent molecule.

hydrolysis A chemical reaction whereby polymers are split into monomers with the addition of the components of water.

hydrophilic Water-loving; said of substances that dissolve readily in water.

hydrophobic Water-hating; said of substances that do not dissolve readily in water.

hyperglycemia Abnormally high blood glucose levels, most commonly the result of diabetes mellitus.

hyperopia Farsightedness.

hyperpolarize A neuron with a resting potential greater (i.e., more negative) than normal.

hypertension High blood pressure.

hyperventilation Consciously engaging in deep breathing, an activity that increases blood oxygen levels and lowers blood carbon dioxide levels.

hypoglycemia Abnormally low blood glucose levels.

hypothalamus A portion of the floor of the brain that is responsible for governing most of the activities of the pituitary gland.

hypothesis A proposition set forth to be tested that tentatively offers an explanation for a set of observations and suggests some general principle or cause-and-effect relationship.

hypothyroid dwarf An individual of small stature and with mental retardation owing to a deficiency of thyroxin.

hypoventilation Reducing the oxygen level of the blood and raising the carbon dioxide level by holding one's breath.

hypoxia The failure of sufficient oxygen to reach the cells of the body.

hysterectomy Surgical removal of the uterus.

I band In striated muscle, the space between the ends of the thick filaments in sarcomeres.

ileum The most posterior portion of the small intestine, leading into the large intestine.

immune system The various processes and mechanisms used by the body to repel or destroy foreign matter, including disease agents.

immunity Resistance to a particular disease.

immunoglobulin Any of a group of circulating blood proteins that functions as an antibody.

implantation The attachment of the blastocyst to the wall of the uterus.

impotence The inability of a man to achieve or maintain an erection.

imprinting A form of learning that occurs only during certain critical periods in the life of some species and, once learned, remains highly stereotyped.

inclusive fitness The concept that fitness is not limited just to the reproductive success of an individual but also includes the reproductive success of the individual's close relatives.

incomplete dominance Alleles not in a dominant-recessive relationship, meaning that both are expressed in the phenotype.

incus The central of the three bones in the middle ear, shaped somewhat like an anvil.

independent variable In a well designed experiment, the one factor that varies between the experimental group and the control group.

induction The embryological process wherein the developmental fate of a tissue is determined by chemical interactions with surrounding tissues.

inductive logic Reasoning from particular facts to a general conclusion.

inferior vena cava The main vein draining blood from the abdomen to the right atrium.

inflammatory bowel disease An autoimmune disease of the intestines, including both Crohn's disease and ulcerative colitis.

inguinal canal The canal through which the testes descend into the scrotum.

inguinal hernia A rupture of the abdominal muscles in the region of the inguinal canal.

inhalation The act of breathing in.

inherent property A characteristic or ability that is intrinsic to one or more of the component parts of an entity; contrasted with *emergent property*.

inheritance of acquired characteristics The now-discredited belief that traits or capacities developed during an individual's lifetime can be genetically transmitted to the offspring.

innate Genetically determined as opposed to learned; said of certain behavioral patterns such as fixed action patterns.

innate releasing mechanism A neural circuit in the brain that issues the motor signals to the muscles responsible for performing a fixed action pattern.

inner ear The cochlea and the organs of the ear associated with the maintenance of balance and posture.

insertion A mutation in which an extra base is added, creating one or more new codons.

insulin A hormone produced by the pancreas that lowers blood glucose levels.

insulin shock Unconsciousness from an overdose of insulin, causing blood glucose levels to drop below the level required by the brain.

intention movement A fixed action pattern that signals another, subsequent behavior, often used for communicative purposes and frequently ritualized.

intercalated disc The highly complex double cell membrane between two adjacent cells in cardiac muscle.

interferon Any of several cytokines that stimulates macrophage and natural killer cell activity.

interleukin Any of several cytokines produced primarily by T cells that regulates the growth and differentiation of B and T cells.

interneuron A neuron linking motor and sensory neurons, located in the brain or spinal cord.

interstitial fluid Fluid that fills the tiny spaces between the cells of the body.

intervertebral disc Shock-absorbing cartilaginous rings located between the vertebrae.

intrauterine device *See* **IUD.**

intron A section of DNA that is not expressed following the editing of a messenger RNA molecule.

inversion A mutation in which a chromosomal segment is reversed.

ion An electrically charged atom or group of atoms.

ionic bond A chemical bond formed between ions of opposite charges.

iris The round, pigmented membrane surrounding the pupil of the eye.

iron-deficiency anemia Insufficient hemoglobin due to an inadequate supply of iron in the diet.

IRM *See* **innate releasing mechanism.**

irritable bowel syndrome Chronic sensitivity of the colon, believed to be caused primarily by stress and nervous tension.

isolating mechanism Circumstances that prevent the interbreeding of individuals from two populations or species.

isotope Any of two or more forms of an element, each of which has atoms with the same atomic number (i.e., same number of protons) but different mass numbers (i.e., different numbers of neutrons).

IUD A coil or loop, generally of plastic, which is placed in the uterus as a birth control device.

jaundice A yellowish cast to the skin caused by excessive amounts of bile pigments being retained by the body.

jejunum The middle portion of the small intestine, between the duodenum and the ileum.

kallikrein An interstitial fluid enzyme involved in pain perception and activitated by acidity, as occurs when cells are damaged.

keratin A linear protein forming the major component of the epidermis, hair, and nails.

kidney Either of a pair of organs responsible for the excretion of nitrogenous wastes and the control of water, salt, and pH balance in the blood.

kidney stone Precipitates of uric acid that may lodge in the ureters, causing intense pain.

killer T (T_C) cell A class of T lymphocytes that are activated by T helper cells and in turn that attack virally infected cells by producing perforin.

kinesin A molecule of the cytoplasm believed to be responsible for helping to move organelles within the cell; see also *dynein*.

Kleinfelter syndrome Individuals with an XXY genotype; phenotypically male, but with some breast development.

Krebs cycle A series of biochemical reactions in the mitochondria that form an integral part of cellular respiration and through which pyruvate is reduced to carbon dioxide.

kuru One of two diseases of humans known to be caused by a slow virus.

kwashiorkor A malnutrition disease caused by insufficient protein in the diet and characterized by an accumulation of interstitial fluid in the abdomen.

labia majora The outer folds of skin surrounding the vagina, clitoris, and urethral opening.

labia minora The inner folds of skin surrounding the vagina.

labor The period of intense muscular contractions responsible for parturition.

lacteal A lymphatic vessel within the villi of the small intestine.

lacuna A small cavity in bone occupied by an osteocyte.

Langerhans cell An antigen-presenting cell found in the epidermis that plays a key role in contact dermatitis.

larynx The bony structure at the top of the trachea containing the vocal cords; Adam's apple.

Law of Independent Assortment The second of Mendel's laws, stating that the genes for different traits are distributed independently from each other during meiosis.

Law of Segregation The first of Mendel's laws, stating that the pair of alleles governing a particular trait are separated during meiosis.

Law of Use and Disuse A now-discredited belief that excessive use of a structure during one's lifetime enhances the inheritance of the structure in one's offspring, whereas non-use of the structure causes it to be reduced or absent in one's offspring.

LDL *See* **low density lipoprotein.**

learning The gradual acquisition of particular types of behavior through extensive interaction with the environment.

lens The clear spheroid structure in the front of the eye, responsible for focusing light on the retina.

Lesch-Nyhan disease A genetic disease caused by a missing enzyme, characterized by mental retardation, cerebral palsy, and self-mutilation.

leukemia A form of cancer characterized by the uncontrolled production of leukocytes.

leukocyte Any of the white blood cells.

LH A hormone produced by the anterior pituitary that is largely responsible for inducing ovulation in females and stimulating testosterone production in males.

ligament A band of tough, inflexible tissue holding two or more bones in place.

light dependent reactions Those photosynthetic reactions involving chlorophyll and requiring light.

light independent reactions The chemical reactions of photosynthesis that use the products of the light dependent reactions but that do not themselves require light.

light receptor A sense organ that responds to electromagnetic radiation of particular wavelengths in the visible spectrum.

limbic system A ring-like system of brain centers controlling various emotional responses.

limited penetrance Incomplete phenotypic expression of a trait because of the effects of other genes or the environment.

limiting factor Any factor in an environment that, by its presence or its scarcity, limits the growth of a population.

linear protein A protein possessing primary and secondary structures only; contrasted with *globular protein.*

lipid Any member of a group of organic molecules characterized by being composed largely of carbon and hydrogen with some oxygen and by being more soluble in organic solvents than in inorganic solvents.

lipoprotein A protein to which a lipid is attached.

liver The largest organ of the body and, but for the brain, the most complex, with a host of regulatory functions.

loop of Henle That portion of the nephron that often passes deep into the medulla; located between the proximal and distal convoluted tubules.

low density lipoprotein A lipoprotein with about 45% cholesterol by weight that transports cholesterol from the liver through the blood to the cells of the body.

LSD A synthetic but very powerful hallucinogen.

lumbar vertebrae The five vertebrae of the abdominal region.

lung cancer A malignancy of the lungs, caused primarily by cigarette smoking.

lungs Internal gas exchange organs found in all terrestrial vertebrates.

lupus Systemic lupus erythematosus, an autoimmune disease involving the attack by agents of the immune system on many different tissues of the body at one time.

luteinizing hormone *See* **LH.**

Lyme disease A form of arthritis caused by a tick-borne bacterium.

lymphatic system The network of vessels, nodes, and organs responsible for accumulating and filtering excess interstitial fluid.

lymph node A round mass of lymphatic tissue filled with lymphocytes and phagocytes through which lymph vessels pass.

lymphocyte Any of several types of leukocytes involved in the immune response; see also *B cell, T cell.*

lymphoid stem cell Stem cells destined to differentiate into lymphocytes.

lymphokine A cytokine produced by lymphocytes.

lymphokine-activated killer cell (LAK) A lymphocyte cultured with interleukin-2, then transferred back into the body to attack tumors.

lysergic acid diethylamide *See* **LSD.**

lysosome An enzyme-containing organelle responsible for digestion or destruction of phagocytized materials.

lysozyme An enzyme produced by various cells, including those in the mouth, that splits the bonds of one of the chemical components of bacterial cell walls.

macroelements Minerals found in relatively large amounts in the body.

macromolecule Any of the large biochemical polymers, including polysaccharides, proteins, nucleic acids, and certain of the lipids.

macrophage A large phagocytic leukocyte, derived from a monocyte.

macular degeneration Destruction of the most sensitive region of the retina due to the proliferation of blood vessels.

major histocompatibility (MHC) complex The genes responsible for the protein markers on cells that permit an identification of self.

malignant In reference to a tumor, one which is very dangerous or life threatening because of its capacity to metastasize; contrasted with *benign.*

malleus The outermost of the three bones of the middle ear, somewhat fancifully viewed as hammer shaped.

mandible The lower jaw bone.

manic depression A disorder characterized by alternating extreme emotional peaks and valleys.

Marfan syndrome A genetic disorder characterized by weakened connective tissue, often resulting in aneurysms.

margination The tendency for certain leukocytes to become attached to the walls of the capillaries, especially in areas of infection.

mass A measure of quantity determined by dividing the weight of an object by the acceleration due to gravity.

masseter Either of a pair of large muscles in the jaw angles responsible for closing the jaws.

mass number The number calculated for each element that represents the sum of all protons and neutrons in the atomic nucleus.

mast cell A basophil-like cell that does not circulate in the blood but that releases histamine when immunoglobulins bind to its plasma membrane.

mastoid process A projection of the temporal bone near the opening of the ear.

M-CSF A hemopoietin that stimulates development of monocyte colonies.

mechanist One who believes that the processes of life can be explained by adequate knowledge of the laws of chemistry and physics.

mechanoreceptor A sense organ that responds to deformation of the receptor itself or of surrounding cells.

medulla The inner region of various organs such as the adrenal gland, kidney, or brain.

medulla oblongata The most posterior part of the brain stem, joining with the spinal cord, and responsible for controlling many vital functions such as breathing.

megakaryocyte A type of blood cell that is the source of platelets.

megavitamin Massive doses of a given vitamin.

meiosis The process by which gametes are formed, involving a chromosomal reduction from the diploid to the haploid number.

Meissner's corpuscle An expanded series of terminal nerve filaments surrounded by a capsule, common in non-hairy areas of the skin.

melanin A black pigment of the skin and hair.

melanocytes Pigment-containing cells found in the stratum germinativum of the epidermis.

melanoma Cancer of the skin.

memory cell A cell of helper T, killer T, or B cell lines, produced in response to a particular antigen, and capable of becoming mobilized quickly in response to a reinvasion by the same antigen.

menarche The first menstruation, marking the onset of puberty.

Mendelian genetics The branch of genetics that deals with the inheritance of traits as units.

meninges Any of three membranes that envelops the brain and spinal cord.

meningitis Inflammation of the meninges, caused by either viruses or bacteria.

menopause The permanent cessation of menstrual cycles.

menstrual cycle The cycle of hormonal, ovarian, and uterine events involved with the production and release of an egg and the preparation of the uterus for the egg's implantation, and having a periodicity of roughly once a month.

Merkel's disc A type of encapsulated expanded nerve ending common in both hairy and non-hairy areas of the skin.

mesoderm One of the three embryonic germ layers, responsible for producing the bulk of the internal organs in the adult organism.

messenger RNA A form of RNA that transfer protein synthesis information from the DNA of the nucleus to ribosomes in the cytoplasm.

metabolic water Water produced as an end product of cellular respiration.

metacarpal The bones that form the back of the hand.

metaphase The stage of mitosis or meiosis in which the chromosomes line up along the midline of the cell.

metastasis The invasion of a tissue or organ by the cells of a cancer tumor located in a different part of the body.

metatarsal The bones that form the arch of the foot.

micelle In water, a clustering of large molecules with both polar and non-polar ends, created because of the positive interaction of the polar ends with water molecules.

microelements Minerals found in small amounts in the body.

microfilament A linear organelle, composed largely of actin, and responsible for changes in cell shape.

microsphere Small cell-like structures formed by heating dry amino acids and placing the resulting large molecules in water; a possible intermediate step in the evolution of life.

microtubule A long, thin cylinder composed of tubulin that is found within cilia, flagella, and spindle fibers.

microvilli The brushlike fringe on the cells forming the intestinal villi.

middle ear Three small, articulated bones that link the eardrum to the cochlea of the inner ear.

midget A pituitary dwarf.

milk letdown The release of milk from the glandular portions of the breast to the nipple, caused by the hormone oxytocin.

mineral An inorganic ion, element, or compound needed for normal growth.

mineralocorticoid Any of a class of steroid hormones produced by the adrenal cortex responsible for regulating the level of various salts in the blood.

mitochondria Large organelles of eukaryotic cells that are the site for cellular respiration and most ATP production.

mitosis Nuclear division of eukaryotic cells characterized by duplication of chromosomes prior to division such that each daughter cell retains a full chromosome complement.

mitral valve *See* **bicuspid valve.**

molecular genetics The study of the biochemical activity of the molecules involved in heredity.

molecule The smallest unit of a compound that possesses the properties characteristic of that compound.

Monera The kingdom of prokaryotic organisms, including bacteria and their relatives.

monoclonal antibody A type of antibody produced by a hybrid cell composed of an activated B lymphocyte and a cancer cell.

monoculture The growing of a single crop over large areas with a resulting decrease in species diversity.

monocyte A type of leukocyte that can transform into a highly phagocytic macrophage.

monohybrid cross Mating between individuals from two inbred lines, differing only in that one line is homozygous dominant for a particular trait whereas the other is homozygous recessive.

monomer The single unit of a polymer (e.g., the amino acids that constitute a protein).

monosaccharide Any sugar not composed of smaller subunits of sugar (e.g., glucose).

morphine A powerful narcotic derived from opium.

morphogen A chemical that induces morphogenesis.

morphogenesis The development of structure and form of the major organs in the embryo.

morula A zygote that has divided several times into a small, solid ball of cells.

motor neuron A neuron carrying information from the brain or spinal cord to an effector organ.

multiple sclerosis A complex of autoimmune diseases that creates lesions on myelinated neurons.

muscle fiber A bundle of long, multinucleated, cylindrical cells found in skeletal muscle.

muscle spindle Tiny, specialized muscle fibers that lack contractile proteins but act as sense organs in responding to changes in body position.

mutagen A substance that induces a mutation.

mutation An inheritable change in the structure or sequence of DNA.

myasthenia gravis An autoimmune disease in which acetylcholine receptor sites on skeletal muscles are destroyed.

myelin A white, fatty substance that surrounds the axons of many neurons in the peripheral nervous system and augments the speed of impulse transmission in those neurons.

myeloid stem cell Stem cells destined to differentiate into any type of blood cell other than lymphocytes.

myocardial infarction An area of tissue death in the ventricles, generally resulting from a coronary thrombosis.

myofibril The contractile unit of a muscle fiber.

myopia Nearsightedness.

myosin Along with actin the principal protein of muscles.

myxedema A disease caused by insufficient thyroxin and characterized by a swelling of the face and a reduction in metabolic activity.

narcotic Any drug that induces profound sleep and relief of pain.

nasal cavity Either of a pair of hollows above the nostrils in which inhaled air is warmed, moistened, cleaned, and monitored for odors.

nasal septum A membrane in the nose that separates the two nostrils.

natural killer (NK) cell Any of several types of leukocytes that produces perforin and destroys tumor cells or virally-infected cells.

natural selection The differential survival and reproduction of different phenotypes in a population; the central tenet in Darwin's theory of evolution.

Neanderthal A highly diverse form of human existing from about 200,000 to about 80,000 years ago whose relationship to modern humans is highly controversial.

negative feedback An important mechanism of homeostasis wherein an increase in the output of a process causes the rate of the process to be diminished, thus insuring a roughly stable level of output.

negative pressure breathing The type of breathing used by mammals wherein a negative pressure is created in the lungs by raising the rib cage and flattening the diaphragm.

nephritis Inflammation of the kidney.

nephron The basic functional unit of the kidney.

nephrosclerosis A narrowing of the arteries supplying blood to the kidneys (usually because of hypertension or diabetes) that leads to the degeneration of the kidneys.

nerve A bundle of axons connecting the central nervous system with a peripheral organ or tissue.

nerve deafness Deafness caused by impairment of the cochlea or auditory nerve.

neural tube The embryonic spinal cord.

neurofibromatosis A genetic disease characterized by growth of small tumors of the central and peripheral nervous systems.

neuromuscular junction The motor end-plate, where there is contact between a motor neuron and a muscle fiber.

neuron A nerve cell.

neurotransmitter Any of a group of chemicals responsible for transmitting a nerve impulse between neurons or between neurons and muscles (e.g., acetylcholine).

neurulation The developmental process by which the embryonic spinal cord is formed.

neutron A particle in the atomic nucleus having mass but no electrical charge.

neutrophil A granular phagocytic leukocyte that can be stained with a neutral stain.

NGF Nerve growth factor, a morphogen that, in conjunction with FGF, causes the differentiation of sympathetic neurons from embryonic nerve cells.

niacin Vitamin B_3, a water-soluble vitamin that assists in hydrogen transport in cellular respiration.

nicotine A stimulant found in large amounts in tobacco, and formerly used as an insecticide.

nodes Gaps in the myelin sheath of neurons that occur between adjoining Schwann cells.

nondisjunction The failure of either of a pair of homologous chromosomes to separate during anaphase I of meiosis, or of chromatid pairs to separate during anaphase II.

noradrenaline *See* **norepinephrine.**

norepinephrine One of two closely related hormones produced by the adrenal medulla that have largely stimulatory effects on the circulatory system and heart.

Norplant A slow-release synthetic progesterone capsule implanted under the skin to prevent pregnancy.

nostrils The openings to the respiratory system, located above the mouth.

nuclear pore An opening in the double membrane surrounding the cell nucleus.

nucleic acid Any of a class of biological macromolecules that have particular importance in heredity and protein formation (e.g., DNA, RNA).

nucleolus A dark-staining body composed of RNA and protein found within the nucleus.

nucleoplasm The undifferentiated contents of the cell nucleus.

nucleotide A single unit of a DNA or RNA molecule, consisting of a nitrogenous base, a sugar, and a phosphate group.

nucleus 1. A prominent organelle of eukaryotic cells containing the chromosomes. 2. The center of an atom containing protons and neutrons.

obsessive-compulsive disorder A disease characterized by prolonged bouts of bizarre behavior, such as extensive and repeated washing of the hands.

olfactory membrane The epithelial tissue at the top of the nostrils that is responsible for detecting odors.

omnivore An animal that eats both plants and animals.

oncogene A gene that is responsible for a cell's becoming cancerous.

operant conditioning Learning by trial and error.

operator A regulatory gene that triggers messenger RNA synthesis by structural genes.

opsin A multi-looped protein of 348 amino acids imbedded in the discs of the rod cells of the eye and containing a molecule of retinal.

optic nerve The nerve that runs from the eye to the brain.

optic vesicle An outgrowth from the side wall of the brain during embryonic development that becomes the back of the eye and that induces the epithelium to form a lens.

organ A bodily structure composed of several types of tissues and adapted for the performance of one or more specific functions.

organ of Corti The portion of the cochlea involved in the perception of sound.

organ system A group of functionally interrelated organs.

organelle A structurally and functionally organized part of a cell.

organic A chemical containing both carbon and hydrogen.

organic hypertension High blood pressure with a known cause; contrasted with *essential hypertension*.

orgasm Sexual climax.

osmosis The diffusion of water across the plasma membrane from regions of high concentration to regions of low concentration.

osmotic pressure Pressure induced by the presence of dissolved materials in solution that retain or attract water by osmosis.

osteoarthritis A condition found primarily in the elderly in which a wearing down of the cartilage surfaces of a joint causes painful movement of the joint.

osteoblast An immature bone cell that develops into an osteocyte.

osteoclast A large multinuclear cell that absorbs bone.

osteocyte A cell that is responsible for the deposition of bone.

osteoporosis A weakening of the bones with age, due primarily to calcium loss.

outer ear Two flaps of tissue on the side of the head that help direct sound to the ear drum.

oval window An area of the cochlea against which the stapes rests and that is deflected when the stapes moves, creating waves of pressure in the cochlear fluid.

ovary The female organ responsible for producing eggs.

oxygenated blood Blood that has passed through the lungs and is rich in oxygen; arterial blood.

oxytocin A hormone released from the posterior pituitary that is responsible for smooth muscle contractions in the uterus and breast.

ozone layer A thin layer of O_3 molecules located high in the atmosphere that acts as a filter for ultraviolet radiation.

pacemaker The source of heartbeat regulation; may be natural (SA node) or a medically implanted mechanical device (artificial pacemaker).

Pacinian corpuscle A mechanoreceptor of the skin, consisting of concentric layers of tissue surrounding a nerve ending.

pain receptor A sense organ that responds to physical or chemical damage to the tissues of the body.

palate A horizontal plate of bone and tissue that separates the nasal cavities from the mouth.

pancreas A large, elongate gland located just below the stomach, responsible for both exocrine and endocrine functions.

pangenesis The now-discredited notion that all of the tissues and organs of the body contribute material to the formation of the gametes.

Plate 1

THE SKELETAL SYSTEM (front view)

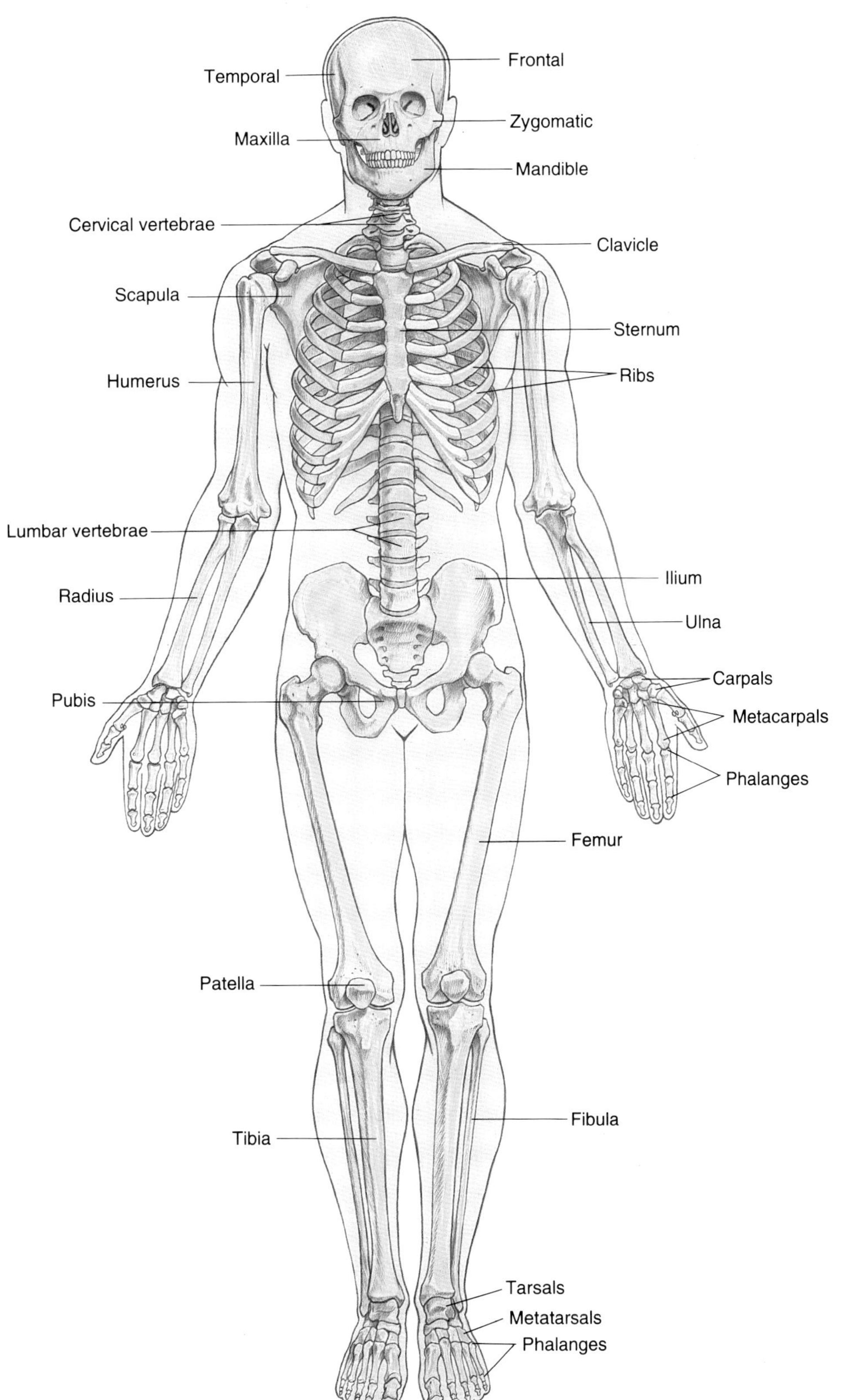

PLATE 2

THE SKELETAL SYSTEM (side and back views)

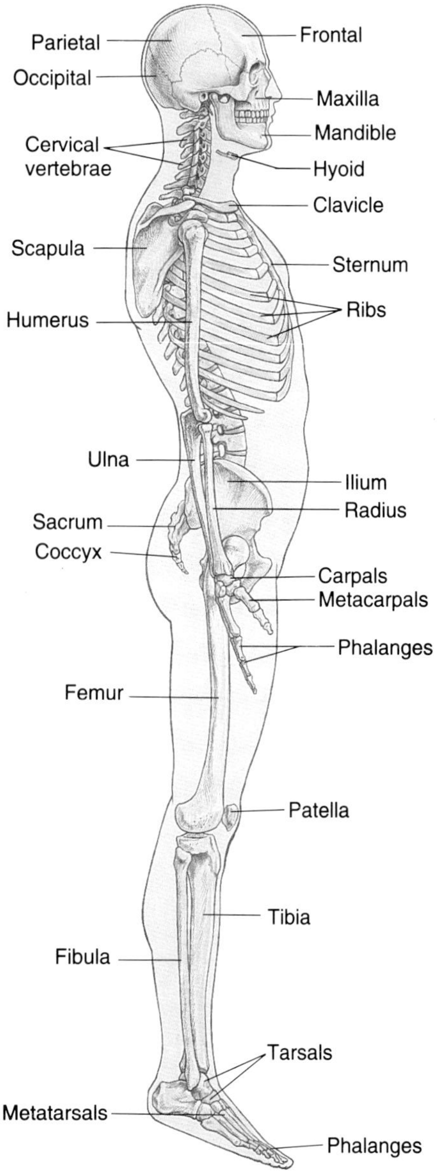

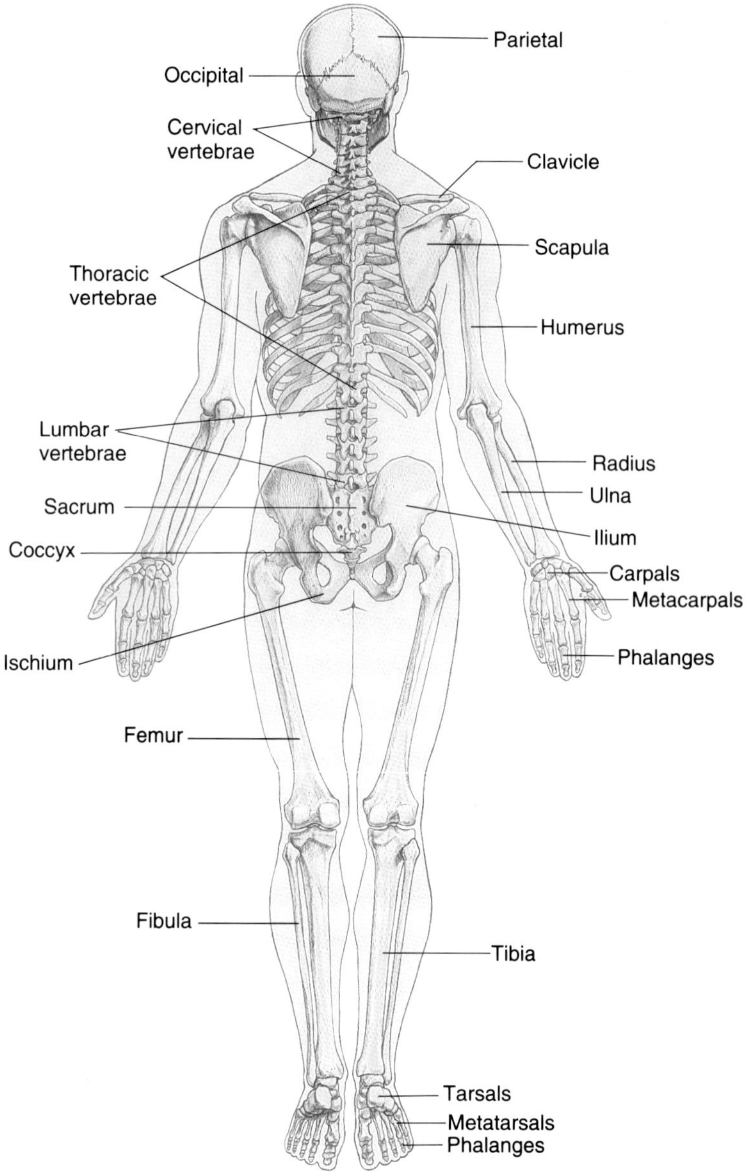

Parietal
Frontal
Occipital
Maxilla
Mandible
Cervical vertebrae
Hyoid
Clavicle
Scapula
Sternum
Humerus
Ribs
Ulna
Ilium
Radius
Sacrum
Coccyx
Carpals
Metacarpals
Phalanges
Femur
Patella
Tibia
Fibula
Tarsals
Metatarsals
Phalanges

Parietal
Occipital
Cervical vertebrae
Clavicle
Scapula
Thoracic vertebrae
Humerus
Lumbar vertebrae
Radius
Ulna
Sacrum
Ilium
Coccyx
Carpals
Metacarpals
Ischium
Phalanges
Femur
Fibula
Tibia
Tarsals
Metatarsals
Phalanges

PLATE 3

THE MUSCULAR SYSTEM (front view)

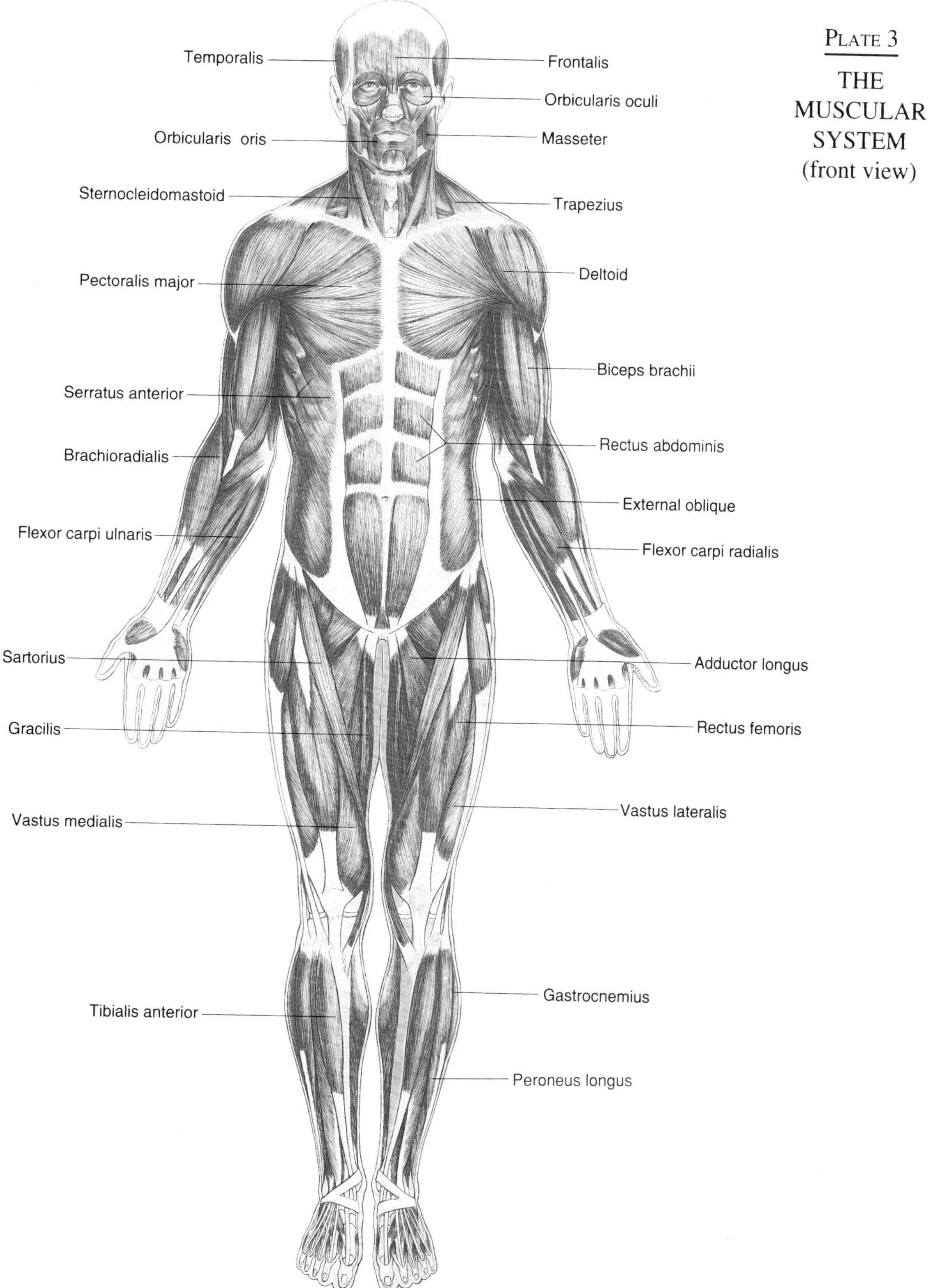

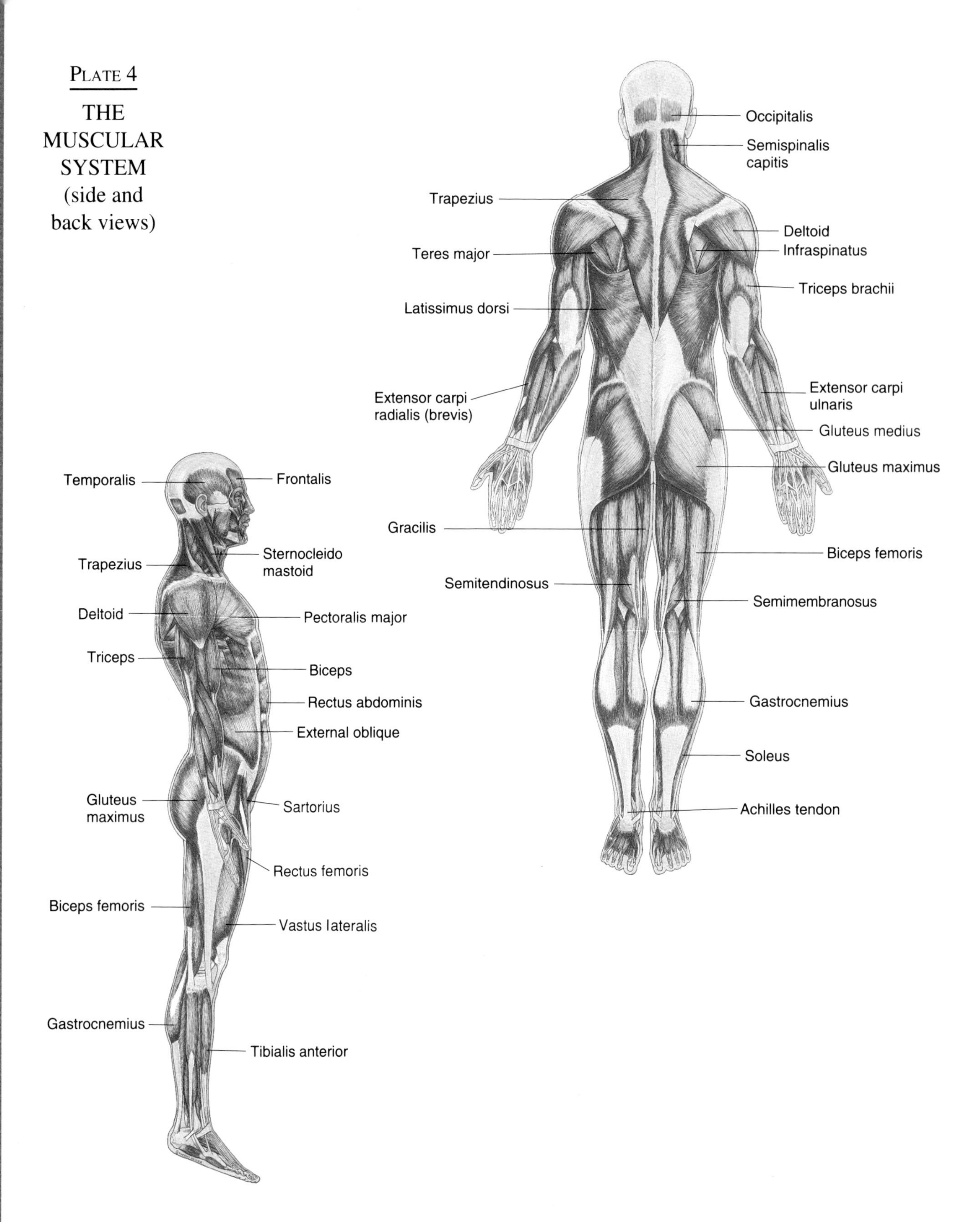
PLATE 4
THE MUSCULAR SYSTEM (side and back views)
Occipitalis
Semispinalis capitis
Trapezius
Deltoid
Teres major
Infraspinatus
Triceps brachii
Latissimus dorsi
Extensor carpi radialis (brevis)
Extensor carpi ulnaris
Gluteus medius
Gluteus maximus
Gracilis
Biceps femoris
Semitendinosus
Semimembranosus
Gastrocnemius
Soleus
Achilles tendon
Temporalis
Frontalis
Trapezius
Sternocleido mastoid
Deltoid
Pectoralis major
Triceps
Biceps
Rectus abdominis
External oblique
Gluteus maximus
Sartorius
Rectus femoris
Biceps femoris
Vastus lateralis
Gastrocnemius
Tibialis anterior

PLATE 5

THE SKULL and MUSCLES OF THE HEAD

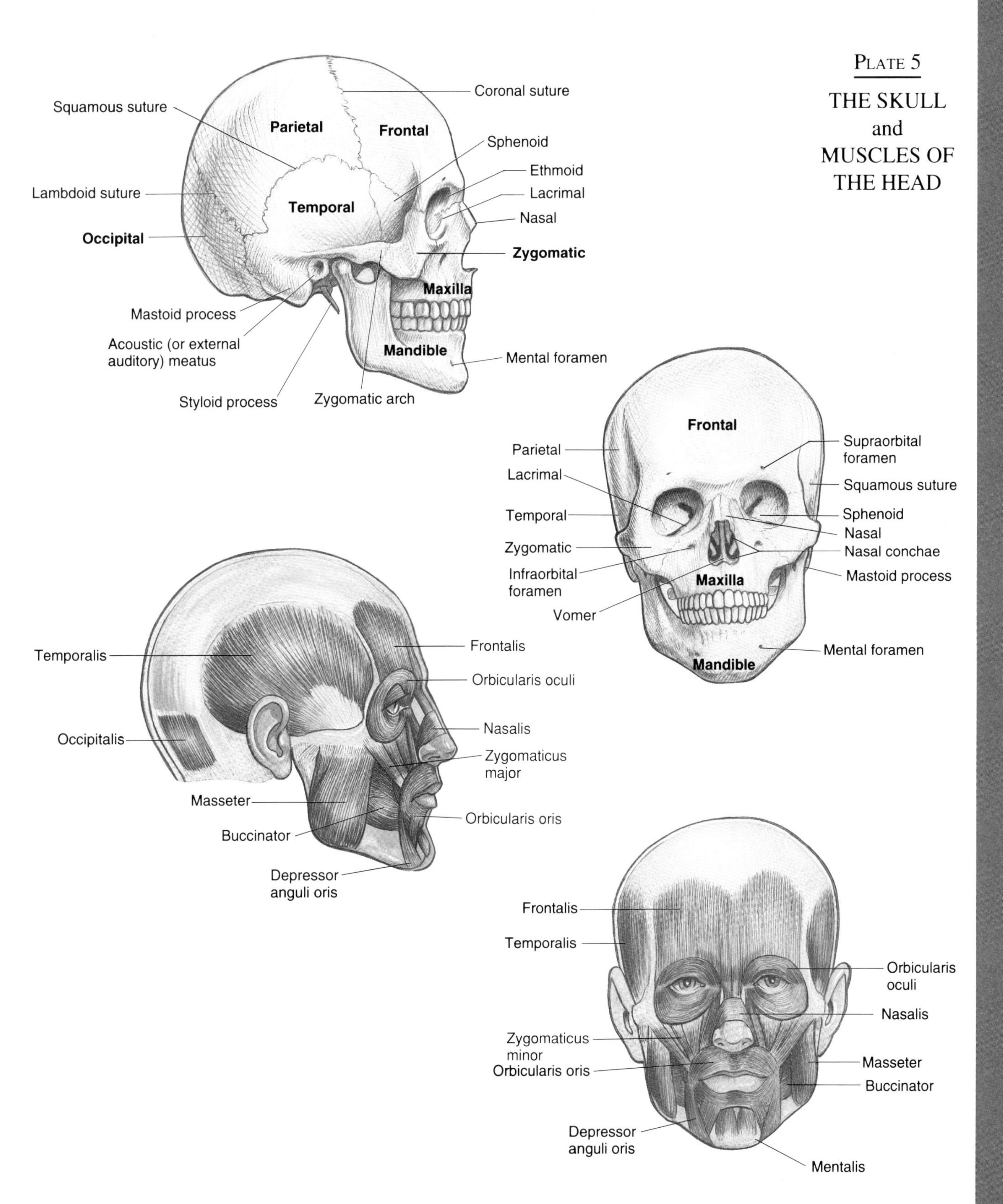

PLATE 6

THE ARTERIAL SYSTEM and THE VENOUS SYSTEM

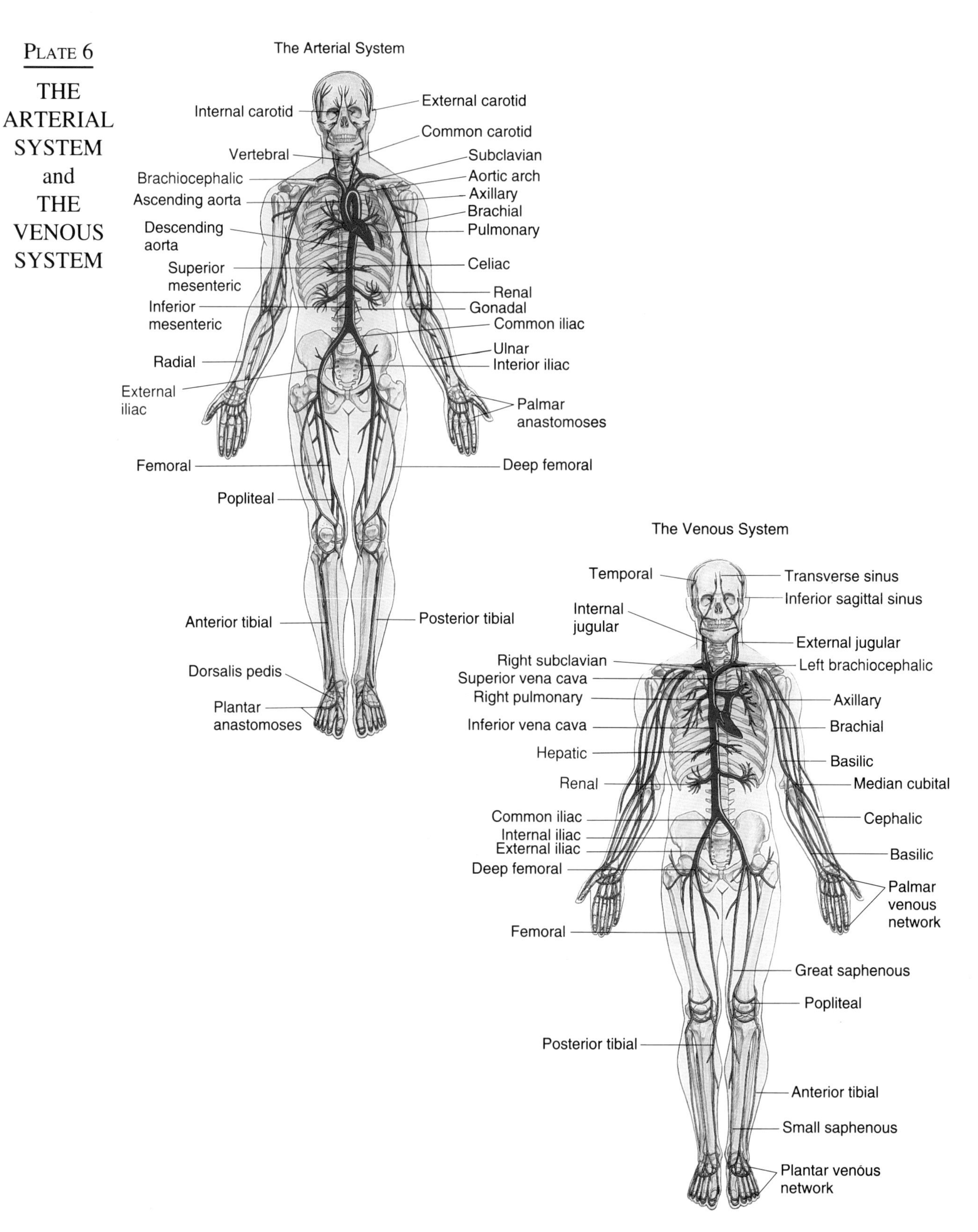

PLATE 7

THE ARTERIAL-VENOUS SYSTEM

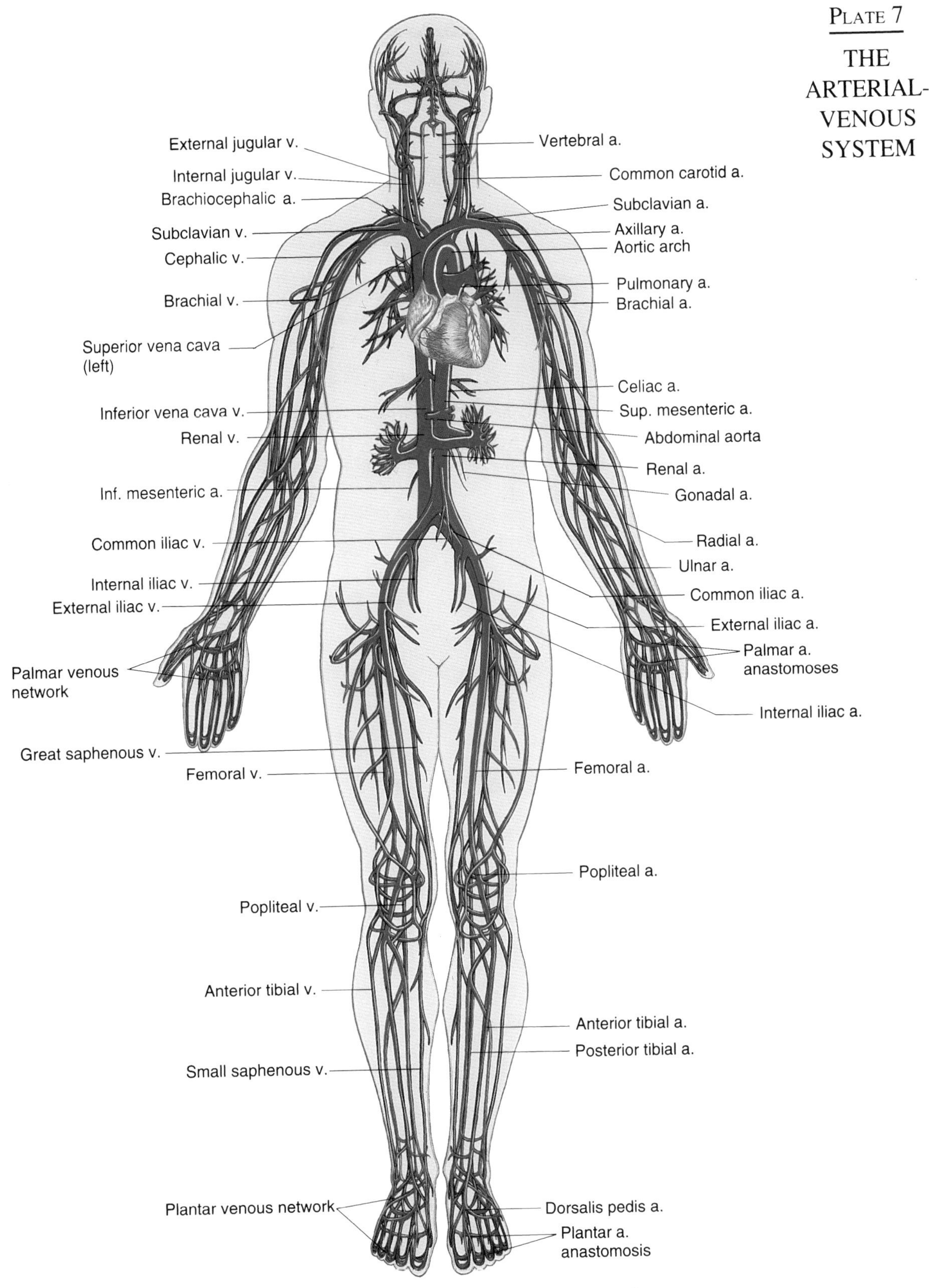

PLATE 8

THE HEART and BLOOD CIRCUITS

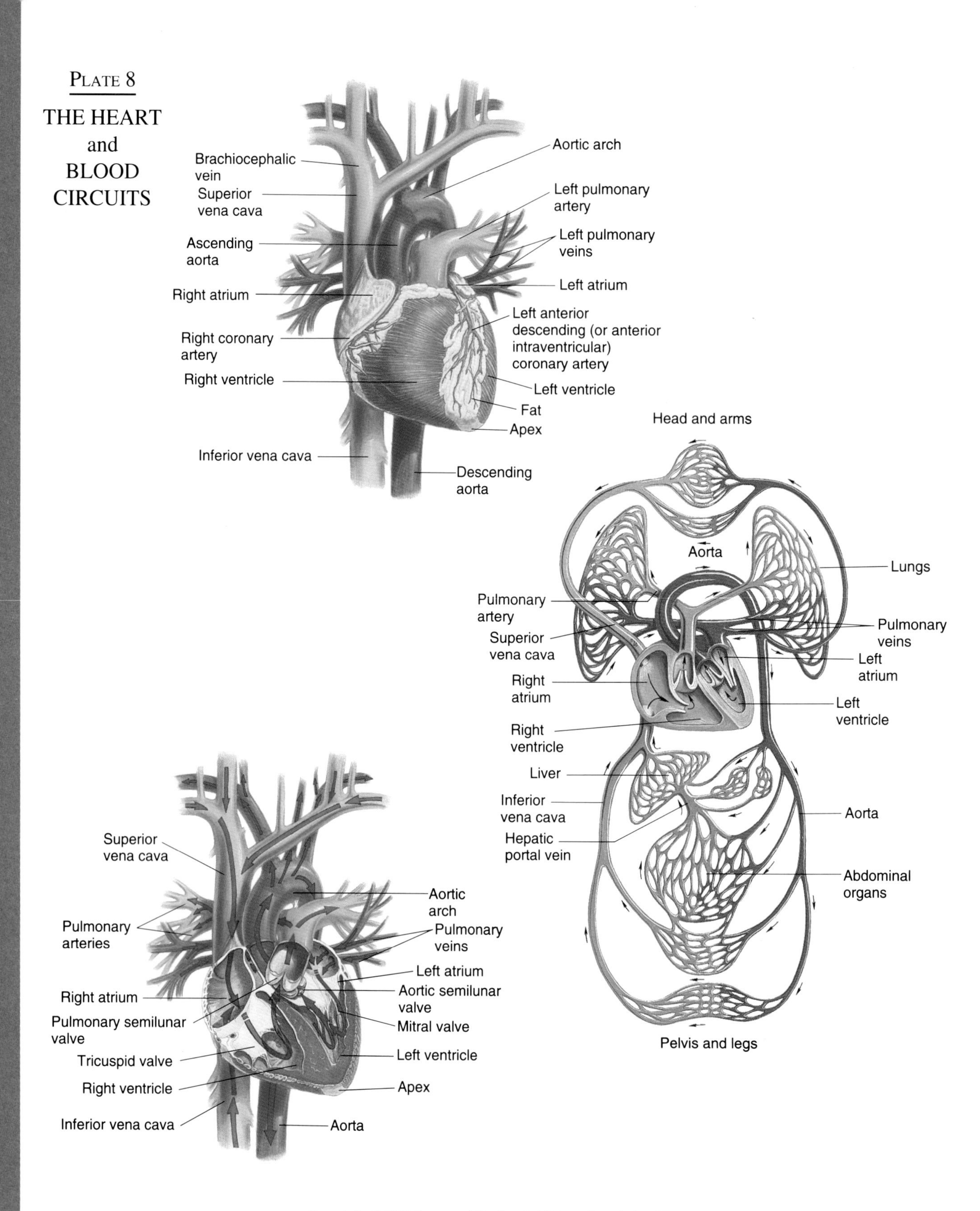

PLATE 9

THE ENDOCRINE SYSTEM

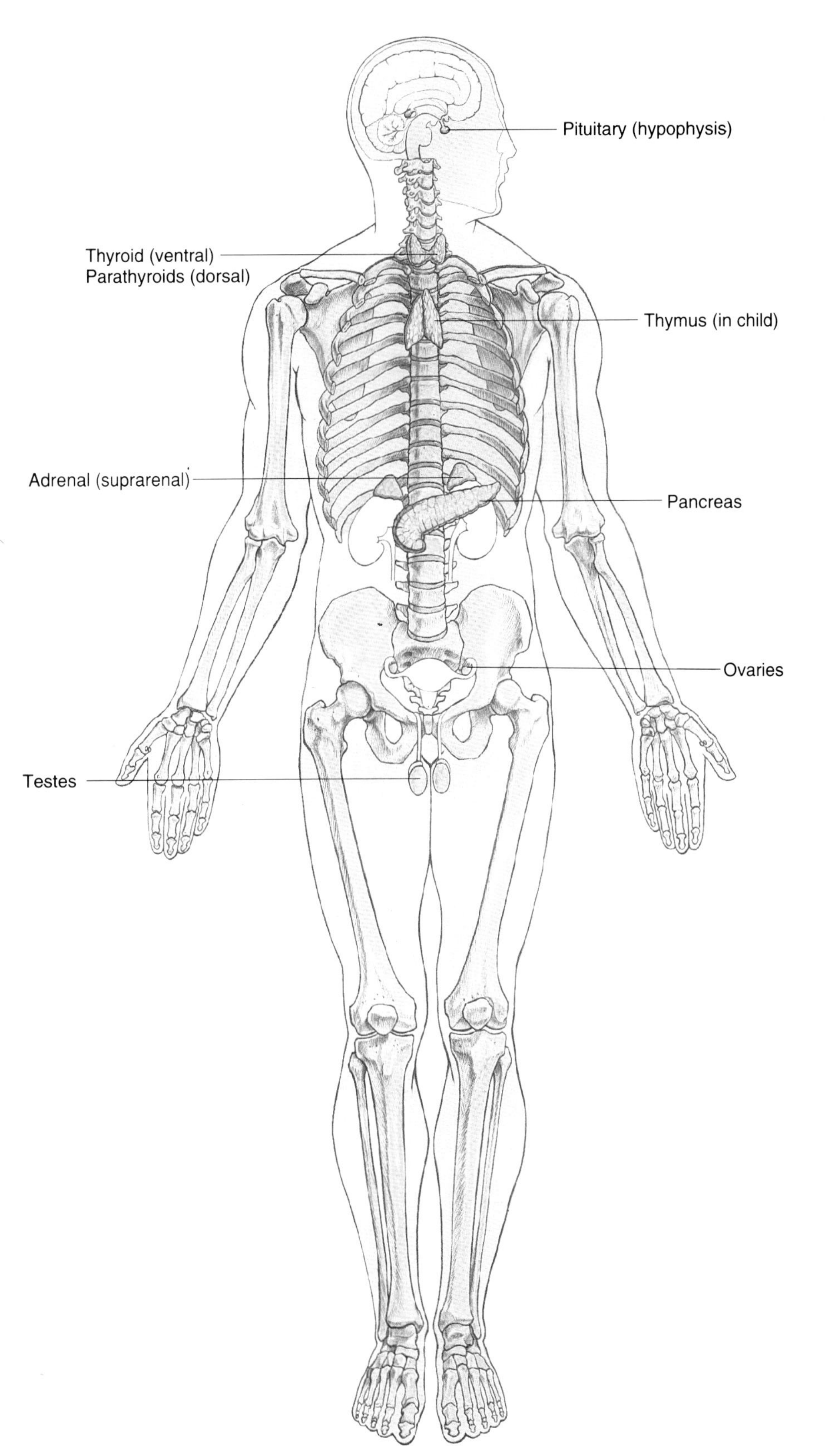

PLATE 10

THE LYMPHATIC SYSTEM

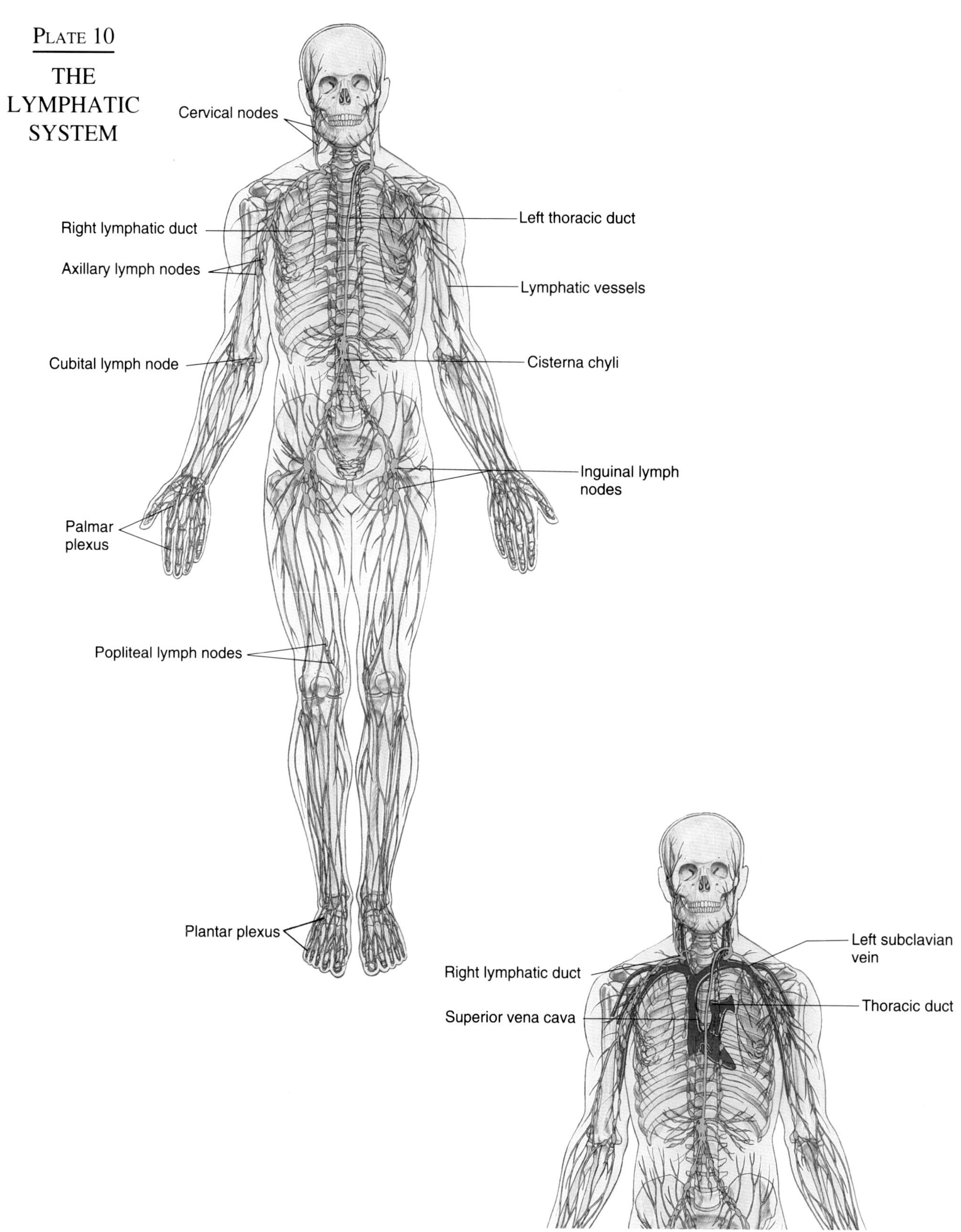

Plate 11

THE NERVOUS SYSTEM and THE BRAIN (surface and midsagittal views)

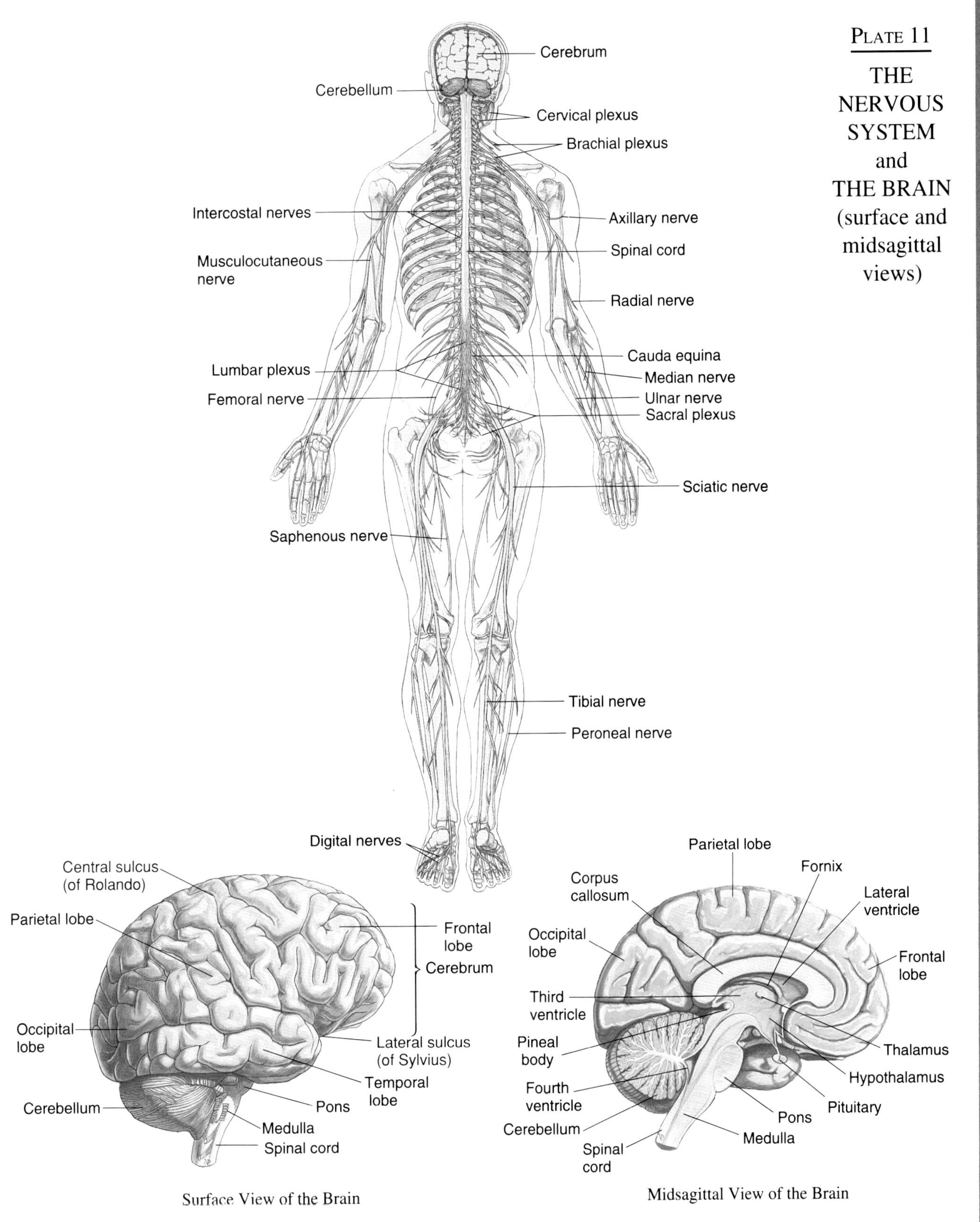

Surface View of the Brain

Midsagittal View of the Brain

Plate 12

THE VISCERA

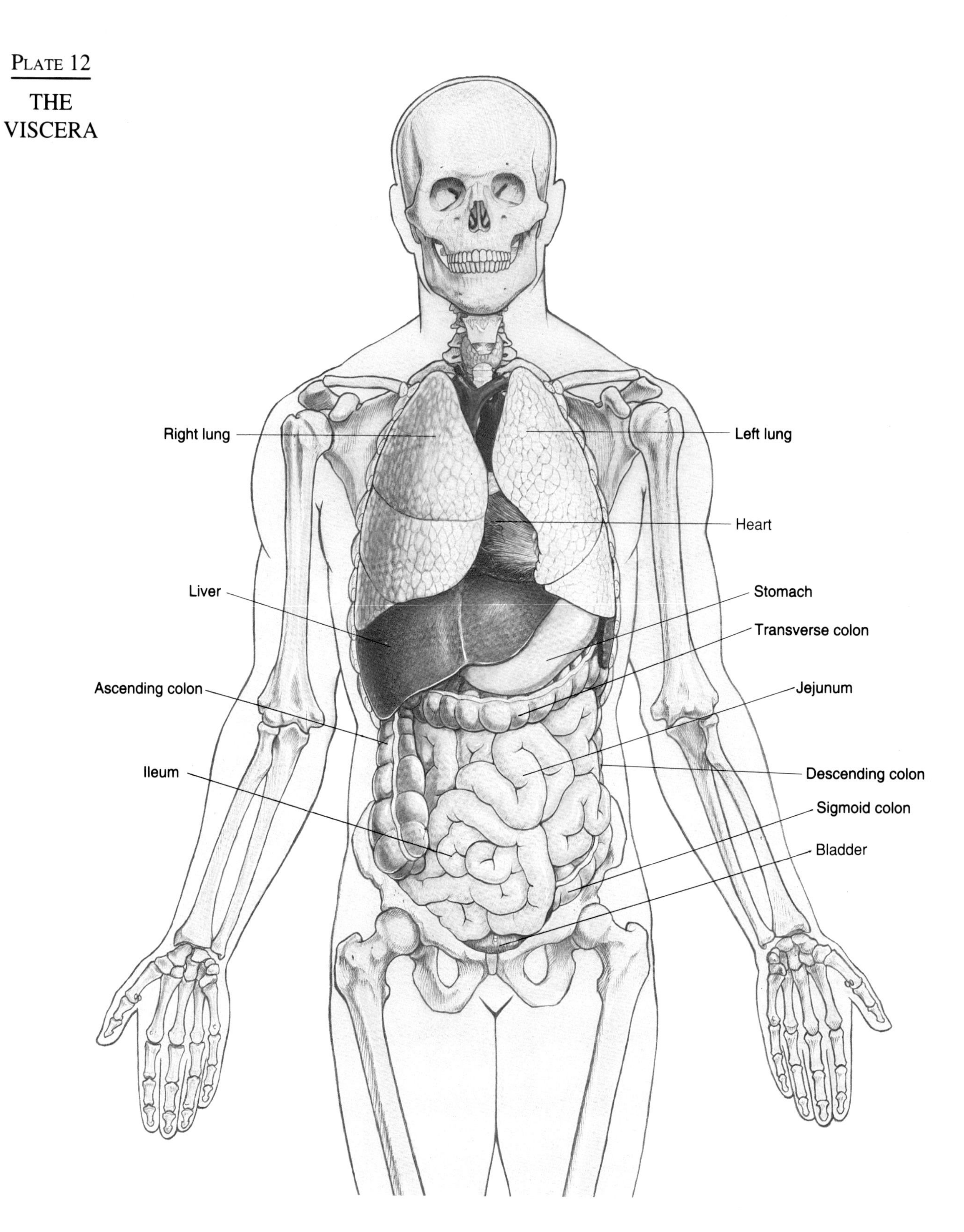

THE RESPIRATORY SYSTEM

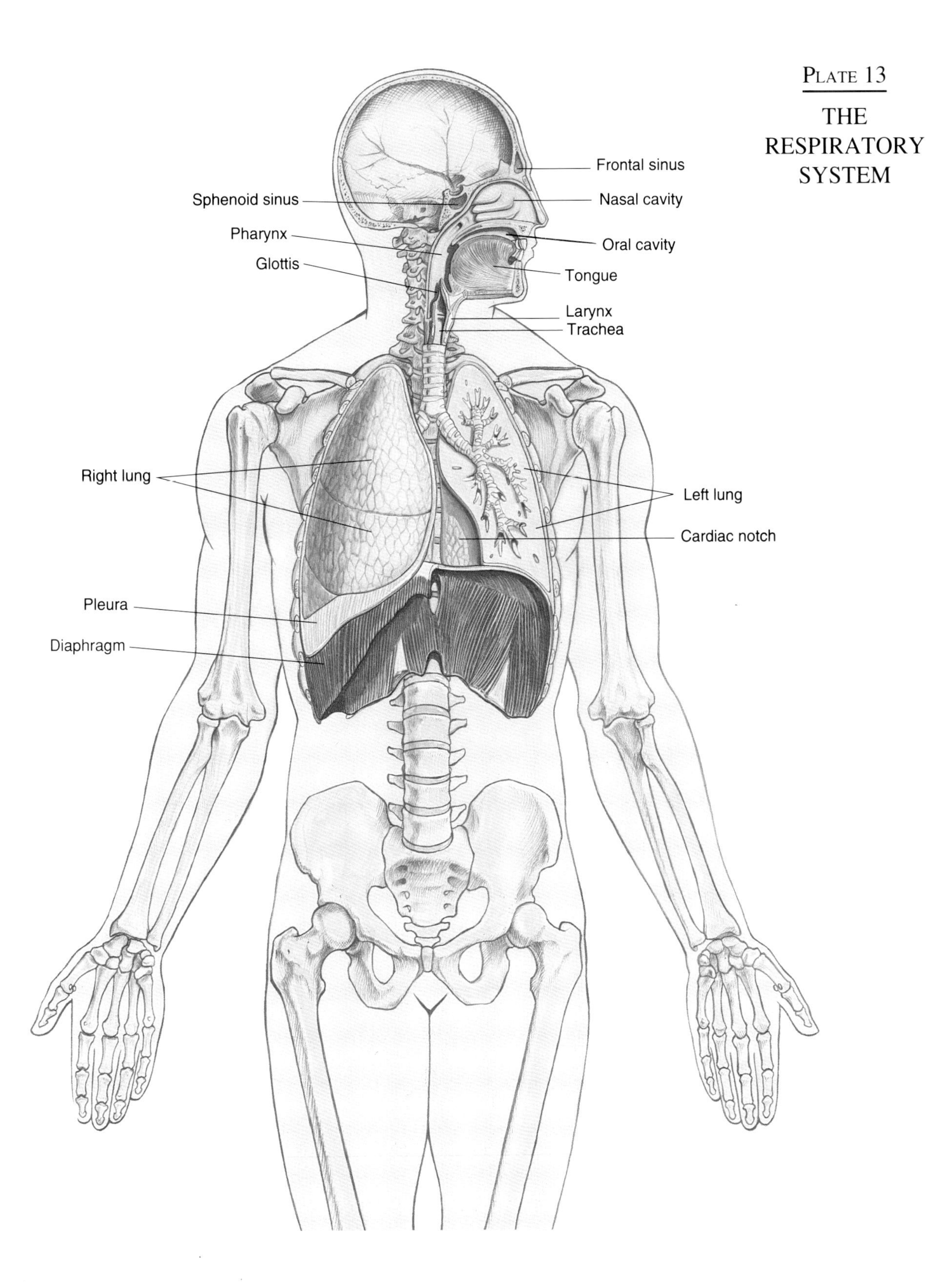

PLATE 14

THE DIGESTIVE SYSTEM

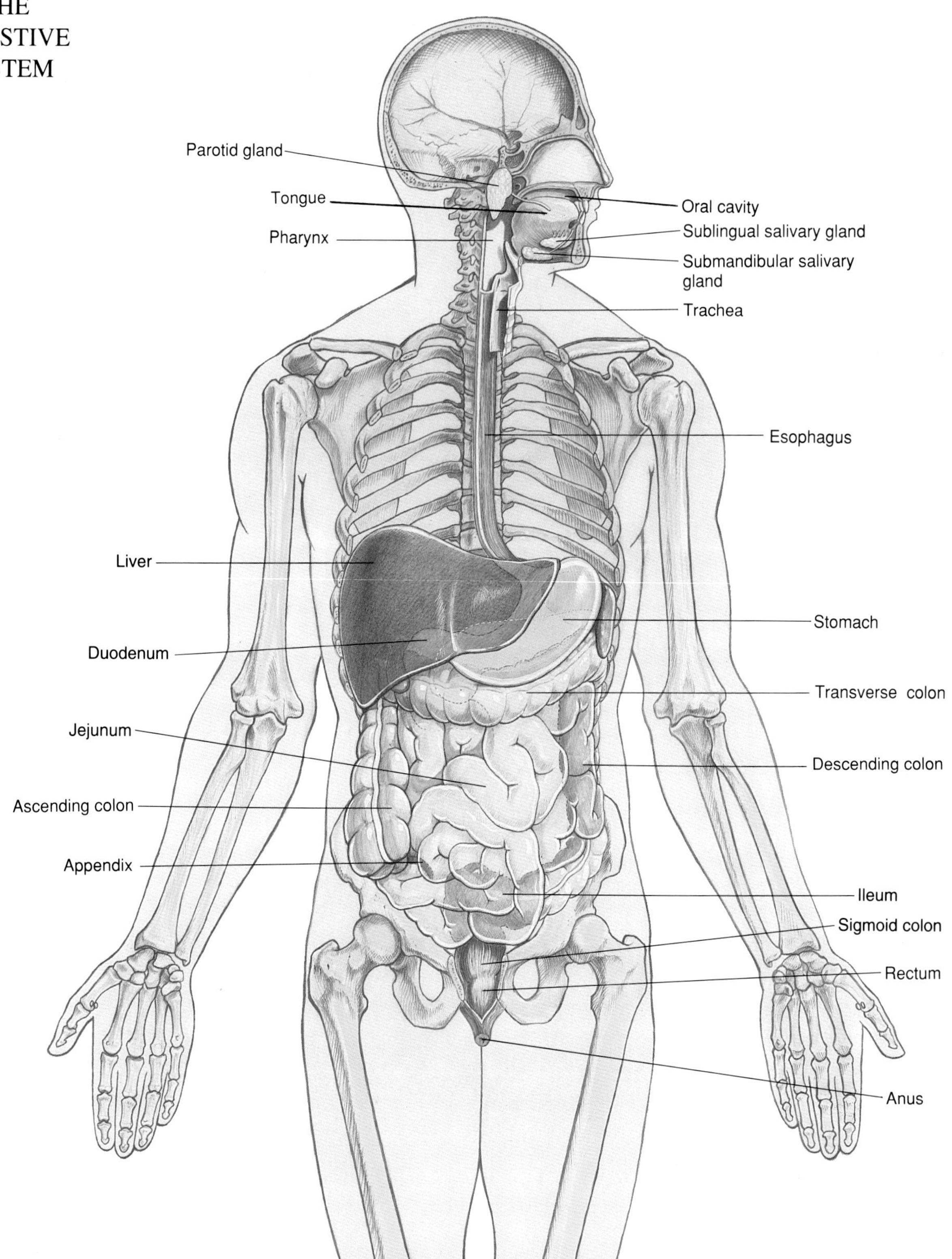

Plate 15

THE URINARY SYSTEM and THE FEMALE AND MALE REPRODUCTIVE SYSTEMS

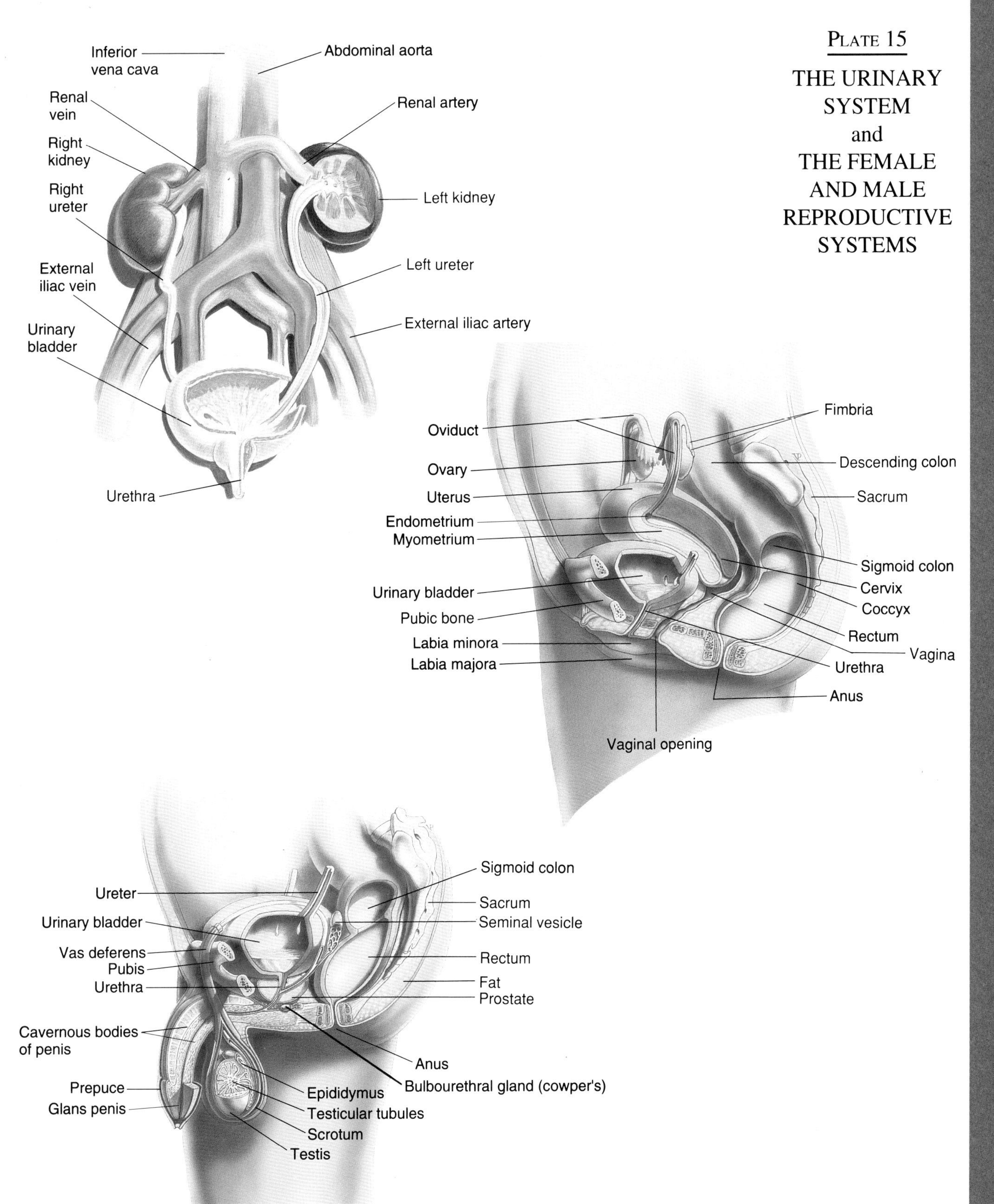

PLATE 16

THE SENSES

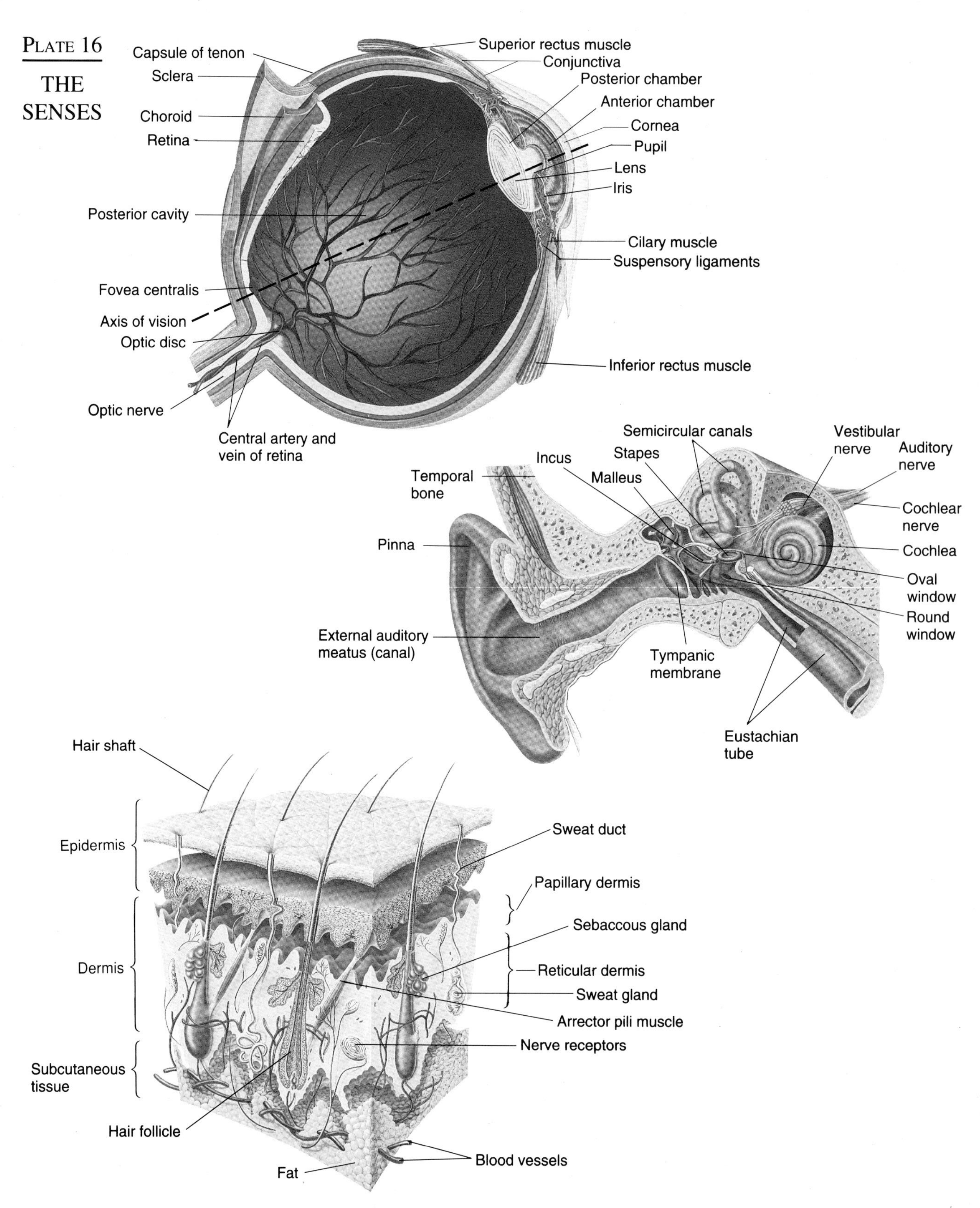

pantothenic acid Vitamin B_5—a water-soluble vitamin essential in glucose metabolism.

papillae Raised areas of the tongue that house the taste buds.

parasympathetic nervous system A set of cranial and spinal nerves responsible for increasing the activity of digestive organs and reducing the activity of the heart.

parathion An organophosphate insecticide originally developed as a nerve gas.

Parkinson's disease A disease characterized by rigidity and tremor, and resulting from the destruction of certain motor centers in the brain stem.

parathormone A hormone produced by the parathyroid glands that elevates blood calcium levels.

parathyroid glands A set of four glands imbedded in the thyroid gland that produces the hormone parathormone.

parturition The act of giving birth.

passive immunity Temporary resistance to disease that results from the injection of antibodies; contrasted with *active immunity*.

pathogen Any microorganism or virus that causes disease.

PDGF Platelet-derived growth factor, a substance produced by foam cells that stimulates the growth of smooth muscle cells in the walls of arteries, a step in the development of atherosclerosis.

pectoral girdle The bones of the shoulder, connecting the arm to the vertebral column and ribs.

pellagra A malnutrition disease caused by a deficiency of niacin, and affecting the skin, digestive system, and central nervous system.

pelvic girdle The bones of the hip, connecting the leg to the vertebral column.

pelvic inflammatory disease *See* **PID.**

penis The male organ of sexual intercourse.

pepsin A protein-splitting enzyme secreted as the inactive molecule pepsinogen by cells in the stomach and activated by hydrochloric acid.

pepsinogen The inactive form of the protein-digesting enzyme pepsin.

peptic ulcer Irritation of the lining of the stomach or small intestine caused by pepsin and hydrochloric acid.

peptide bond A chemical bond formed between two amino acids through a dehydration reaction.

perforin A molecule produced by several types of leukocytes that forms holes in the plasma membrane of attacked cells, causing their destruction.

periosteum A tough layer of connective tissue that covers bone except at the joints.

peripheral nervous system The nerves and ganglia of the nervous system excluding the brain and spinal cord.

peristalsis The rhythmic contraction of visceral muscle in the walls of the digestive tract and certain other hollow organs.

peritonitis Inflammation of the abdominal cavity.

pernicious anemia An inadequate supply of erythrocytes, due to a deficiency of vitamin B_{12}.

peroxisome An organelle consisting of a membrane-bound sac of detoxifying enzymes.

pH scale A logarithmic scale of hydrogen ion concentration running from O to 14; values below 7 are increasingly acidic, values above 7 are increasingly basic.

phagocyte Any of several types of leukocytes that engulfs foreign matter or damaged cells.

phagocytosis The process by which a cell engulfs material and consequently forms digestive vacuoles.

pharynx The region behind and below the oral and nasal cavities and above the esophagus and larynx.

phenocopy An environmentally-induced phenotype.

phenotype The observable properties of an organism, resulting from the interaction of the environment with the genotype.

phenylketonuria (PKU) A genetic disorder in which a missing enzyme causes elevated levels of the amino acid phenylalanine, that, if untreated, can lead to mental retardation.

phlebitis Inflammation of the veins.

phospholipid A fat in which a phosphate group substitutes for one of the three fatty acids.

phosphorylation The attachment of phosphate groups to organic molecules.

photophosphorylation The attachment of a free phosphate group to ADP, thereby creating ATP as occurs in the light dependent reactions of photosynthesis.

photosynthesis The conversion of light energy to chemical energy, most commonly accomplished by the plant pigment chlorophyll.

Photosystem I A portion of the photosynthetic pathway employing chlorophyll with maximum light absorbance at 700 nm and believed to be of relatively ancient origin.

Photosystem II A portion of the photosynthetic pathway employing chlorophyll with maximum light absorbance at 680 nm and believed to be of relatively recent origin.

phototroph Autotrophs that use sunlight as their energy source.

phylogeny The evolutionary history of a species or group of species.

PID An extensive pathological involvement of organs in the pelvic region, generally in women and generally as a consequence of an untreated venereal disease such as gonorrhea.

PIH A hormone produced by the hypothalamus that inhibits the production and release of prolactin by the anterior pituitary.

pinocytosis The formation and uptake by the cell of tiny pieces of the plasma membrane that surround substances in solution.

pituitary gland The major gland of the endocrine system located at the base of the brain just above the mouth and consisting of anterior and posterior portions, each with a different embryological origin.

pituitary dwarf An individual showing very slow growth as a child, due to underproduction of growth hormone.

placenta The organ of exchange between fetus and mother, formed jointly by tissues of each and located on the wall of the uterus.

plasma Whole blood minus the blood cells.

plasma cell An antibody-producing cell resulting from the division and differentiation of an activated B lymphocyte.

plasma membrane The selectively permeable membrane that encloses a cell.

platelet A fragment of a type of blood cell called a megakaryocyte that plays a key role in the formation of blood clots.

platelet factor-4 (PF 4) A substance produced by platelets that inhibits the growth of new blood vessels.

platelet plug A cluster of platelets that forms immediately at the site of blood vessel injury.

pleura A thin, moist membrane lining the chest cavity and covering the lungs.

pleurisy Inflammation of the pleura.

PMS A period of irritability in some women in the week prior to menstruation.

pneumonia A lung inflammation caused by various agents.

pneumothorax A collapsed lung.

point mutation Errors in one or two bases during DNA replication.

polar body A daughter cell formed during the meiotic formation of an ovum that is little more than a nucleus and that plays no role in reproduction.

polarity The possession of unequally distributed electrical charges by a molecule.

polio A disease caused by a virus that destroys the cell bodies of motor neurons in the spinal cord.

pollution The introduction of a poison into an ecosystem or the creation of an excess of a required substance to the degree that the homeostasis of the ecosystem is threatened.

polydactyl A genetic condition in which a person has extra fingers and/or toes.

polyethylene terephthalate (PET) A type of plastic used for plastic soft drink bottles, among other uses.

polygenic inheritance Inheritance involving many independent genes, each of which has a small but cumulative effect on the phenotype (e.g., weight, height).

polymer A large organic molecule (such as a protein) made up of many individual units (such as amino acids) linked together.

polymorphism The existence within a population of two or more phenotypically discrete and genetically determined forms of an organism, at a frequency such that the rarest cannot be maintained solely by spontaneous mutation.

polysaccharide A long chain carbohydrate composed of a series of individual subunits (monosaccharides).

polyvinyl chloride A type of plastic with many industrial applications.

Pongidae A primate family that includes the great apes.

pons A broad band of nerve fibers in the floor of the vertebrate brain connecting the medulla oblongata to the rest of the brain stem.

population All of the members of a given species within a common geographic area.

positive feedback A type of relationship where the presence of an interaction increases the product, and vice versa (e.g., nursing behavior causes continued prolactin and oxytocin production and milk production and release—and the presence of milk causes nursing behavior)

positive pressure breathing The form of breathing used by frogs wherein air is forced into the lungs by the pressure produced by contractions of the floor of the mouth.

posterior Behind; toward the rear; opposite of *anterior*.

postsynaptic neuron The second neuron is the transmission of an action potential between two neurons.

power arm That portion of a lever between the fulcrum and the force or power.

preformationism A now-discredited belief that one or the other of the gametes (usually the sperm) contains a complete organism and that development consists merely of growth of the parts.

premenstrual syndrome *See* **PMS.**

presbyopia Farsightedness in middle-aged people caused when the lens of the eye loses the ability to accommodate and cannot focus on objects near at hand.

presynaptic neuron The first neuron in the transmission of an action potential between two neurons.

primary structure In proteins, the sequence of amino acids.

primate Any member of the mammalian order that includes prosimians, monkeys, apes, and humans.

prion Particular types of proteins correlated with certain diseases, although whether they are the cause or the product of the disease is a matter of intense debate.

producer In ecological parlance, green plants or other photosynthetic organisms.

progeria A genetic condition in which the aging process is greatly accelerated.

progesterone An ovarian hormone produced by the corpus luteum and responsible for maintenance of the endometrium of the uterus.

prokaryote A cell or organism without an organized nucleus or membrane-bound organelles.

prolactin An anterior pituitary hormone that promotes growth of the mammary glands and the production of milk.

prolactin relase inhibiting hormone *See* **PIH.**

promotor A regulatory gene that stimulates the activity of structural genes.

prophase The first stage of mitosis and meiosis during which the chromosomes condense and become visible.

prosimian A member of the most primitive group of primates, including lemurs and tarsiers.

prostaglandin Any of a group of hormone-like lipids produced by most, and perhaps all, cells.

prostate A large gland at the base of the urethra in males, secretions from which form the bulk of the semen.

protease Any protein-digesting enzyme.

protein Any of a class of biological macromolecules composed of serially linked amino acids.

proteinoid Proteins produced synthetically, by heating amino acids.

prothrombin A plasma protein that, when catalyzed by thromboplastin, is converted to thrombin, a key enzyme in the formation of blood clots.

Protista The kingdom of single-celled eukaryotes.

proton A positively charged particle found in the nucleus of the atom.

proto-oncogene Normal genes that, if mutated, become oncogenes.

proximal convoluted tubule The region of the nephron located between Bowman's capsule and the descending arm of the loop of Henle.

proximate cause Something that occurs immediately before, and produces, an effect; contrasted with *ultimate cause*.

puberty The developmental stage during which the reproductive organs mature.

pulmonary Pertaining to the lungs, as in pulmonary circulation or pulmonary thrombosis.

pulmonary artery Either of a pair of vessels that carries blood from the right ventricle to the lungs.

pulmonary embolus A blood clot in a pulmonary artery.

pulmonary vein Any of the four vessels that carries blood from the lungs to the left atrium.

punctuated equilibrium The theory that evolution often consists of many long periods of stabilizing selection, interspersed with short periods of rapid change.

pupil The hole in the iris of the eye through which light passes before being focused on the retina by the lens.

pupillary reflex Changes in the size of the pupil based on the relative presence or absence of light.

Purkinje fibers Cardiac muscle fibers specialized for conducting impulses from the bundle of His to the ventricular walls.

PVC *See* **polyvinyl chloride.**

pygmy Individuals of short stature owing to a deficiency in plasma membrane receptors for growth hormone.

pyloric sphincter muscle A ring of visceral muscle located between the stomach and the duodenum.

pyridoxine Vitamin B_6—a water-soluble vitamin essential in protein metabolism and in the healing of wounds.

quickening An archaic word used to denote the time of entry of the soul into the developing fetus.

radiation Energy emitted in the form of rays or particles from unstable atoms.

radioactive An element capable of giving off radiant energy in the form of particles or rays.

radius One of two bones in the forearm.

radon A radioactive element that occurs in nature as a gas and that releases radiation as it breaks down.

Ramapithecus A fossil primate from about 12–14 million years ago, long thought to be a hominid but now believed to be ancestral to the orangutan.

reabsorption In the formation of urine, the process whereby certain substances are extracted from the glomerular filtrate by the cells of the nephron.

recessive An allele that is not expressed in the phenotype of an organism whenever a dominant allele is present.

recombinant DNA What is created when human genes are spliced into bacterial DNA.

rectum The most posterior portion of the large intestine.

redirection activity A fixed action pattern that is expressed not to the releaser but to some other object in the environment.

red marrow A specialized blood-producing tissue found in cancellous bone.

reducer Any of the fungi or microorganisms responsible for the decomposition of dead plant or animal matter.

reduction division The first of the meiotic divisions which reduces the chromosome complement in the daughter cells from the diploid to the haploid number.

reflex A type of simple behavior characterized by its speed, because of the involvement of only two or three neurons.

reflex arc The series of neurons involved in the receipt of a stimulus and the performance of a reflex act.

refractory period The brief span of time following the generation of an action potential during which the neuron is unresponsive to additional stimuli.

releaser Specific signals that prompt the performance of a fixed action pattern.

rem The amount of radiation necessary to generate the same effect as one roentgen of X-ray radiation.

renin An enzyme produced by the kidney as a consequence of declining sodium levels in the blood that catalyzes the conversion of the plasma protein angiotensinogen into angiotensin I.

replication The process whereby a DNA molecule duplicates itself.

repolarization The resumption of resting potential values by a neuron following an action potential.

repressor A protein that binds to an operator and thereby prevents RNA polymerase from attaching and initiating messenger RNA synthesis by adjacent structural genes.

reproductive potential The theoretical maximum rate at which a population of a particular species can reproduce, disregarding any environmental constraints.

residual volume The volume of air left in the lungs after maximum exhalation.

respiration The cellular utilization of oxygen in the manufacture of ATP through the degradation of glucose to carbon dioxide and water.

respiratory distress syndrome A condition often found in prematurely-born infants whose respiratory system is not fully developed.

resting potential The electrical difference existing across the neuronal membrane in a neuron that is at rest (i.e., not transmitting an impulse).

restriction enzyme A bacterial enzyme that protects against invasion by foreign DNA by fragmenting the DNA at specific sites.

restriction fragment length polymorphism (RFLP) A process by which gene loci and chromosome maps can be prepared, involving the use of restriction enzymes on chromosomes with normal genes and chromosomes with mutated genes and comparing the resulting pieces.

reticular activating system A network of neurons and structures in the brain stem that is responsible for consciousness and alertness.

retina The photosensitive layer at the back of the eye composed largely of rods and cones plus supporting cells.

retinal A vitamin A derivative found within an opsin molecule in the discs of rod and cone cells in the eye.

retinitis pigmentosa Loss of vision due to the gradual degeneration of the retina and the replacement of rods and cones with cells containing abundant melanin.

retinoblastoma Cancer of the retina.

retrovirus A virus in which the hereditary molecule is RNA, not DNA.

Rh system A polymorphic series of blood antigens first discovered in rhesus monkeys.

rheumatic fever An autoimmune disease triggered by streptococcal bacteria, and often resulting in damage to heart valves.

rheumatoid arthritis An autoimmune disease characterized by a destruction of the cartilaginous ends of bones at the joints.

rhodopsin The combination of opsin and retinal, two molecules critical in the detection of light by the light-sensitive cells of the eye.

rhythm method A method of birth control involving abstinence near the time of ovulation.

riboflavin Vitamin B_2—a water-soluble vitamin required for the formation of erythrocytes and the synthesis of DNA, glycogen, and the hormones of the adrenal cortex.

ribonucleic acid *See* **RNA.**

ribosomal RNA The primary constituent of ribosomes.

ribosome A cytoplasmic organelle composed largely of ribosomal RNA and protein; the site for protein synthesis (translation).

rickets A bone-softening malnutrition disease caused by a deficiency of vitamin D.

rigor mortis Muscle rigidity that occurs within a few hours of death.

ritualization The transformation of intention movements and other activities into communicative displays.

RNA A nucleic acid formed of the sugar ribose, a phosphate group, and four different nitrogenous bases, occurring as messenger RNA, transfer RNA, and ribosomal RNA.

RNA polymerase The enzyme responsible for guiding the transcription of RNA from a DNA template.

rod One of two types of photosensitive cells in the retina of the eye.

rough ER *See* **endoplasmic reticulum.**

round window A membranous aperture in the cochlea that bulges out when the oval window is pressed inward, allowing movement of the cochlear fluid.

RU-486 A synthetic progesterone mimic that occupies progesterone receptor sites and induces menstruation.

Ruffini's end-organ Multi-branched nerve endings located deep in the skin and some other tissues.

SA node A small mass of tissue in the wall of the right atrium that functions as the heart's pacemaker.

saccule The smaller of the two chambers in the labyrinth of the inner ear involved in balance and equilibrium.

sacrum The five fused vertebrae that articulate with the pelvic girdle.

salinization An increase in the salt levels of soil due to water evaporation.

salivary amylase *See* **amylase.**

salivary glands Three pairs of glands that secrete saliva into the mouth.

saltatory conduction Nervous transmission in myelinated fibers, with depolarization "jumping" from node to node.

sarcolemma The plasma membrane of a muscle fiber.

sarcomere The basic contractile unit of striated muscle, bounded at each end by the Z line and containing the thick and thin filaments.

sarcoplasmic reticulum The endoplasmic reticulum of a muscle fiber.

scapula The shoulder blade.

schizophrenia A mental disorder characterized by withdrawal, indifference, hallucinations, and paranoia.

scholasticism A system of philosophy and logic of medieval university scholars, based upon Aristotelian logic and early Christian writings.

Schwann cell The source of the myelin sheath in myelinated neurons.

scientific method The general method of scientific investigation involving the statement and testing of a hypothesis, followed by a conclusion.

sclera The outer white coat of the eye.

scrotum The pouch of skin containing the testes.

scurvy A degenerative disease of the connective tissues caused by a deficiency of vitamin C.

sebaceous gland Oil-producing glands of the skin, associated with hair follicles.

Second Law of Thermodynamics When energy is changed from one form to another, the amount of usable energy decreases.

second messenger Cyclic AMP, the chemical produced in the cytoplasm of cells that are targets for most proteinaceous hormones in response to binding of the hormone to the plasma membrane.

secondary structure The alpha helix or pleated sheet configuration of linear proteins.

secretin A hormone produced by the small intestine that promotes bicarbonate secretion by the pancreas.

secretion In urine formation, the process whereby certain cells of the nephron actively transport certain substances from the blood and interstitial fluid into the glomerular filtrate.

segmentation The alternating contraction and relaxation of bands of smooth muscle of the small intestine during digestion.

semen Ejaculatory fluid made up of a variety of glandular secretions plus sperm.

semicircular canal Any of three loops of the inner ear involved in balance and orientation.

semilunar valve Either of the half-moon valves at the base of the pulmonary artery and aorta that prevents blood from flowing backward into the ventricle.

seminal vesicles A pair of outgrowths of the vas deferens producing glandular secretions that are a part of the semen.

seminiferous tubules Coiled tubules in the testes in which sperm are produced.

sensitization An abnormally vigorous reaction to a stimulus previously associated with pain or reward.

sensory neuron A neuron carrying information from a sense receptor to the brain or spinal cord.

serum Whole blood minus both the cells and the clotting proteins.

severe combined immunity disorder (SCID) A genetic condition in which very few lymphocytes are produced, resulting in virtually no immunity to pathogens.

sex chromosome An X or Y chromosome; contrasted with *autosome*.

sex linkage A trait the gene for which is carried on a sex chromosome.

sexual selection Selection based on traits that increase the individual's sexual attractiveness to the opposite sex.

sexually transmitted disease Any venereal disease.

shell The orbit (and, hence, the energy level) of a particular electron or group of electrons around the nucleus of an atom.

sickle-cell anemia A form of anemia caused by a mutation of the hemoglobin gene that provides resistance to malaria but that creates malformed erythrocytes that are prone to rupture.

sinoatrial node *See* **SA node.**

sinus A cavity in any of several bones of the face, linked to the nasal cavities by narrow canals.

sinus venosus A chamber in the fish heart that is the evolutionary ancestor of the SA node in mammals.

skeletal muscle Muscle associated with the skeletal system; striated muscle.

skeleton An organized framework of bones in the body.

slime mold Any member of a group of organisms having attributes both of Fungi and of Protista.

slow virus Viruses that take many years from the time of exposure to the onset of disease, kuru and Creutzfeldt-Jacob disease being the only two known examples in humans.

small intestine The narrow but lengthy section of the digestive tract located between the stomach and the colon.

smog Air pollution principally caused by the burning of high-sulfur coal (industrial smog) or by automobile emissions (photochemical smog).

smooth ER *See* **endoplasmic reticulum.**

sociobiology The study of the biological bases of social behavior.

sodium-potassium pump An active transport system in the plasma membrane that accumulates potassium and removes sodium.

solute The dissolved substance in a solution.

solvent A substance that dissolves a solute, forming a solution.

somatic nervous system The system of nerves serving the skeletal muscles and under voluntary control; contrasted with the *autonomic nervous system.*

somatomedin A collection of related proteins produced by the liver in response to growth hormone stimulation that, in turn, stimulates growth of cartilage and bone.

somatostatin A hormone produced by the hypothalamus that inhibits the formation and release of growth hormone by the anterior pituitary gland.

speciation The process or processes by which new species are formed.

species A group of actually or potentially interbreeding organisms that are reproductively isolated from other such groups.

sphincter A ring-shaped muscle that surrounds a natural opening in the body and can open or close it by relaxing or contracting.

spinal cord A thick, elongate bundle of axons and neurons extending posteriorly from the brain through the vertebral column.

spinal nerve Any of the nerves emanating directly from the spinal cord.

spinal tap The drawing off of a small amount of cerebrospinal fluid by a needle inserted at the base of the vertebral column.

spindle apparatus The functional unit that at anaphase draws the chromosomes to either side of the cell.

spleen An organ located near the liver that serves as a reservoir for, and modifier of, blood.

spontaneous generation The now-refuted belief that the spontaneous creation of life from nonlife is possible under present environmental conditions.

sprain An injury to a joint involving either the joint capsule, associated tendons, or both.

SRY gene Sex-determining region of the Y chromosome; the gene responsible for determining the male phenotype.

stabilizing selection Selection that eliminates extreme variants and favors those closest to the mean.

stapes A small, stirrup-shaped bone that is the innermost of the three bones of the middle ear.

STD *See* **sexually transmitted disease**.

stem cell The type of cell from which all of the classes of blood cells are derived.

sternum The breastbone.

steroid Any of a group of lipids characterized by having four linked carbon rings with various side units.

stomach A food storage organ located between the esophagus and the duodenum.

stratum germinativum The living layer of epidermal cells that produces the outer layers of dead cells.

streptokinase An enzyme that dissolves blood clots.

stroke Injury or death of brain tissue due to a failure in the blood supply (either a vessel rupture or a blood clot).

stroke volume The amount of blood the heart pumps with each contraction of the ventricles.

subatomic particle Any proton, neutron, or electron.

subcutaneous tissue The tissue that connects the skin to the underlying muscles.

substance P A substance released by nerve endings in damaged tissue that binds to mast cells, causing the release of histamine, and in general increases the sensation of pain.

substrate A molecule that is chemically altered by enzymatic action.

superior vena cava The major vein draining blood from the head and arms into the right atrium.

suppressor T (T_s) cell Any of a class of T lymphocytes that modulates the response of helper T cells and B cells.

surface tension The amount of attraction—typically in the form of hydrogen bonds—exerted downward and sideways on the surface molecules of a liquid by molecules forming the body of a liquid.

sweat gland A gland that release water and salt onto the surface of the skin.

syllogism A form of deductive reasoning in which two statements of fact, or premises, are made and from these a logical conclusion is drawn.

sympathetic nervous system A set of spinal nerves that forms a part of the autonomic nervous system and that is responsible for increasing heartbeat and decreasing digestive activity.

synapse The junction between two neurons, where chemical transmission of an impulse replaces electrical transmission.

synapsis The pairing of homologous chromosomes during prophase I of meiosis.

synaptic cleft The gap that separates two neurons at a synapse.

syncytium A multinucleate mass of cytoplasm formed by the fusion of cells.

synovial membrane The membrane surrounding a joint that produces the lubricant known as synovial fluid.

syphilis A type of venereal disease caused by a spirochete.

systolic The blood pressure during contraction of the ventricles; contrasted with *diastolic*.

T cell A type of lymphocyte that matures in the thymus gland and that can attack virally-infected cells displaying particular antigens.

target organ An organ that is responsive to a particular hormone.

tarsals The bones that collectively form the ankle and heel.

taste bud A sensory organ responsive to sweet, sour, salt, or bitter substances and found primarily on the surface of the tongue.

taxa The particular name assigned to a given phylogenetic category.

Tay-Sachs disease A fatal genetic disease characterized by the accumulation of particular lipids within neurons, leading to a gradual collapse of the nervous system.

tectorial membrane A part of the organ of Corti that moves in response to sound waves, causing deflection of hair cells and the initiation of action potentials.

telophase The final stage of mitosis and meiosis, during which the chromosomes become diffuse and the nuclear membrane begins to reform.

temperature inversion A situation in which a layer of warm air sits above a region of cold air and prevents pollutants in the cold air from dissipating.

temporalis A muscle in the temporal area of the skull that is responsible for closing the jaws.

tendon A strap-like band of collagen that connects bone to bone.

teratogen Any environmental agent that interferes with the normal functioning of genes during embryonic and fetal development.

tertiary structure The folded and twisted appearance of globular proteins.

testicular feminization syndrome (TFM) Genotypic males possessing a gene that creates a female phenotype.

testis The male gonad, the site of sperm production.

testosterone The primary male hormone, produced by cells of the testis in response to the presence of LH.

tetany Extreme muscular spasms, especially of the muscles of the arms and legs.

tetraploidy Four sets of chromosomes within a cell as a consequence of a mitotic failure.

thalamus A region of the brain wall located just above the hypothalamus, and serving as an important link between the cerebrum and the rest of the brain.

thalassemia A disease resulting from a mutated form of hemoglobin, providing some resistance to malaria in heterozygotes, but at the cost of death from anemia for many of the homozygotes.

T helper (T_H) cell Any of a class of T lymphocytes that is activated by antigen-presenting cells and that in turn activates killer T cells and B cells.

thalidomide A medication used widely in many countries during the early 1960s and later found to cause severe birth defects.

theory A hypothesis that has withstood repeated experimental testing.

thermodynamics The branch of physics concerned with interconversions of energy.

thermoreceptor A sense organ that responds either to heat or to cold.

thiamin Vitamin B_1—a water-soluble vitamin required for carbohydrate metabolism, deficiencies of which cause beriberi.

thick filament Strands of myosin that are found in the A band region of the sarcomere of striated muscles.

thin filament Strands of actin that are anchored at one end to the Z line of the sarcomere of striated muscles.

thoracic vertebrae The 12 vertebrae of the chest.

thrombin A blood enzyme that catalyzes the conversion of fibrinogen to fibrin in the formation of blood clots.

thromboplastin An enzyme released by damaged cells and ruptured platelets that catalyzes the conversion of the plasma protein prothrombin to thrombin, and begins the series of reactions leading to the formation of a blood clot.

through-flow channels Vessels in capillary beds through which blood can flow directly from arterioles to venules when the bed is metabolically inactive.

thymus gland A gland located beneath the sternum in which T cells mature.

thyroid An endocrine gland located just below the larynx.

thyroid stimulating hormone *See* **TSH.**

thyroid stimulating hormone releasing hormone *See* **TRH.**

thyroxin An iodine-containing hormone produced by the thyroid gland and responsible for elevating metabolic rate.

tibia The shinbone.

tidal volume The volume of air exchanged during normal breathing.

tight junction A structure joining adjacent cells closely together, eliminating any intercellular spaces and preventing the movement of materials except by diffusion and active transport through the plasma membrane.

TIL cell Tumor-infiltrating lymphocyte, a type of killer T cell that attacks cancerous cells.

tissue A group of similar cells organized into a structural and functional unit.

tissue plasminogen activator (TPA) A substance that dissolves blood clots.

tonsil Either of a pair of outgrowths of the wall of the upper pharynx that contains lymphoid tissue.

tonus Muscles in a state of partial contraction; normal muscle tone.

totipotency The capacity of the nucleus of a body cell to direct differentiation and development when substituted for the nucleus of a fertilized egg.

toxin Any of the poisonous compounds produced by microorganisms.

trachea The windpipe, lying between the pharynx and the bronchi.

tracheotomy A procedure in which the trachea is opened surgically; generally done because of an obstruction in the upper respiratory passages.

tranquilizer Any of a variety of compounds used to control anxiety and tension.

transcription The synthesis of RNA from a DNA template.

transducer Sense organs that can convert particular kinds of environmental changes into a neural response (i.e., an action potential).

transfer RNA A molecule consisting of about 90 bases that conveys amino acids to the messenger RNA-ribosome complex where proteins are being synthesized.

transforming growth factor (TGF) A substance produced by heart muscle cells that neutralizes tumor necrosis factor and thereby minimizes the damage to heart muscle following a heart attack.

translation The reading of a messenger RNA molecule at a ribosome and the consequent formation of a protein.

translocation A chromosomal mutation in which a section of one chromosome is attached to another chromosome.

transplant A tissue or organ from one individual inserted into the body of another individual.

transposon A particular sequence of DNA that can move from chromosome to chromosome; a "jumping gene."

TRH A hormone produced by the hypothalamus that stimulates the anterior pituitary to produce TSH.

triceps A large muscle on the back of the upper arm responsible for opening the elbow joint.

Trichomonas vaginalis An STD protozoan infection, generally symptomless in men but producing vaginal discharge in women.

tricuspid valve A valve between the right atrium and the right ventricle that prevents blood from flowing back into the atrium during ventricular contraction.

triiodothyronine An iodine-containing hormone produced by the thyroid gland; closely related in structure and function to thyroxin.

trisomy Possession of an extra chromosome by a diploid cell or organism.

trisomy-13 (Patau syndrome) A fatal genetic disease caused by an extra chromosome 13 and characterized by an underdeveloped head, severe retardation, and congenital heart problems.

trisomy-18 (Edwards syndrome) A generally fatal genetic disease caused by an extra chromosome 18 and characterized by mental retardation and heart defects.

trisomy-21 (Down syndrome) A genetic disease caused by an extra chromosome 21 and characterized by mental retardation and heart defects.

trophic level An ecological term referring to any level of a food pyramid.

tropomyosin A protein molecule that masks the myosin-binding sites on the actin molecules of sarcomeres in resting muscle.

troponin A protein molecule that holds tropomyosin in place in resting muscle.

TSH A hormone produced by the anterior pituitary that stimulates the thyroid gland to produce thyroxin and triiodothyronine.

tubal ligation The severing of the fallopian tubes as a permanent method of birth control.

tuberculosis A bacterial disease primarily of the lungs.

tubulin The globular protein of which microtubules are composed.

tumor An uncontrolled growth of tissue; it may be either benign or malignant.

tumor necrosis factor (TNF) Either of two cytokines produced by macrophages and lymphocytes that attacks tumor cells and virally-infected cells and that stimulates granulocytes and natural killer cells.

Turner syndrome An individual with an XO genotype; phenotypically female but with underdeveloped ovaries and breasts.

tympanic canal A canal in the cochlea of the inner ear.

type specimen The reference specimen for a particular species.

typology The now-outmoded belief in the fixity of species; more specifically, a belief that essentially no variation occurs within the members of a species.

ulcerative colitis An autoimmune disease of the colon; a form of inflammatory bowel disease.

ulna One of two bones in the forearm.

umbilical artery Either of two vessels carrying deoxygenated blood from the fetus to the placenta.

umbilical cord The connection between the fetus and the placenta through which pass two umbilical arteries and one umbilical vein.

umbilical vein The vein by which oxygenated blood flows back to the fetus from the mother.

urea A nitrogenous waste product consisting of two molecules of ammonia linked by a molecule of carbon dioxide.

uremia Excessive amounts of urea in the blood.

ureter Either of a pair of tubes that conveys urine from the kidneys to the bladder.

urethra A tube that conveys urine from the bladder to the outside environment.

uric acid A nitrogenous waste molecule of very low toxicity.

urinary bladder A muscular sac located in the lower abdomen that receives urine from the kidneys and ureters and stores it until the urine is voided through the urethra.

urine The liquid waste produced from the blood by the kidneys.

uterus A chamber in the female reproductive tract in which the blastocyst implants.

utricle The larger of the two chambers in the labyrinth of the inner ear into which the semicircular canals open.

vaccination The injection of dead or weakened pathogens or toxins into the body in order to induce an immune response.

vacuole A membrane-bound, fluid-filled sac in the cytoplasm.

vacuum activity The performance of a fixed action pattern in the absence of a releaser.

vagina The canal leading from the labia minora to the uterus.

vagus nerve The main nerve of the parasympathetic nervous system, running from the medulla of the brain to most of the organs of the chest and upper abdomen.

valvular heart disease Damage to one of the heart valves, most commonly the bicuspid valve, often because of rheumatic fever.

varicose vein A vein that has become enlarged and prominent because of failure of the venous valves.

vas deferens The sperm duct leading from the testis to the urethra.

vasectomy A surgical procedure involving the severing of the vas deferens to induce sterility.

vasoconstriction Contraction of the smooth muscles in blood vessels, resulting in narrowed diameters and reduced blood flow.

vein Any blood vessel carrying blood to the heart.

venereal disease A sexually transmitted disease.

venereal warts Genital warts caused by the papilloma virus that is transmitted during sexual intercourse.

ventral root The lower branch of a spinal nerve, through which motor neurons exit the spinal cord.

ventricle Either of two chambers of the heart that receives blood from the atria and pumps it to the lungs (right ventricle) and body (left ventricle).

ventricular fibrillation A life-threatening situation wherein the ventricles twitch rapidly instead of contracting slowly; often the result of a myocardial infarction.

venule A small vein.

vertebral column The array of bones comprising the backbone and through which the spinal cord passes.

very low density lipoprotein (VLDL) A lipoprotein synthesized by the liver with about 19% cholesterol by weight that is transformed into an LDL as it moves through the blood.

vesicle A small sac of materials transported within the cell to and from the plasma membrane.

vestibular apparatus The primary organ of equilibrium; part of the inner ear.

vestibular canal The upper canal of the cochlea of the inner ear.

vestibular membrane The upper membrane of the cochlea of the inner ear.

vestigial A degenerate or rudimentary organ, no longer having the function it had at an earlier point in evolution.

villus Any of the small, fingerlike projections of the small intestine that collectively vastly increase the surface area for absorption.

viral encephalitis A disease of the brain transmitted by viruses carried by mosquitoes.

viroid Plant pathogens consisting of small circles of DNA but without the protein coat characteristic of viruses.

virus A small, pathogenic entity comprised entirely of a protein coat and a strand or coil of nucleic acid.

visceral muscle The smooth muscle characteristic of the internal organs.

vital capacity The maximum volume of air that can be exchanged with one breath.

vitamin Any of a chemically diverse variety of organic molecules that cannot be synthesized by the organism but that is essential in small amounts for normal growth and development.

vitamin A A fat-soluble vitamin necessary for bone development, cell differentiation, and night vision.

vitamin D A fat-soluble vitamin that stimulates the uptake of calcium and phosphate absorption by the small intestine.

vitamin E A fat-soluble vitamin that assists in cellular respiration and in the synthesis of hemoglobin.

vitamin K A fat-soluble vitamin essential in the formation of blood clotting proteins.

vitreus humor The transparent, jellylike substance that fills the eye between the lens and the retina.

vocal cords Membranous structures within the larynx that vibrate to produce speech.

vomiting Regurgitation of food, caused by a relaxation of the esophageal sphincter and violent contractions of the diaphragm and abdominal muscles.

weight arm That portion of a lever between the weight and the fulcrum.

Werner's syndrome A genetic condition causing premature aging.

Wernicke's area The area of the brain responsible for the storage of written and spoken language.

whole blood Blood that contains all its components; contrasted with *plasma* or *serum*.

XYY syndrome A hereditary condition involving possession of an extra Y chromosome.

yolk sac An extraembryonic membrane in which the yolk of the egg is stored during early embryonic development.

Z line In striated muscle, the boundaries of a sarcomere and the point at which the thin filaments attach.

zygote A fertilized ovum.

Index

f indicates figure or illustration; t indicates table

continued from page iv

S. Friend, MD, University of California, San Francisco. Fig. 4.18 p. 86: photo © Omikron, Photo Researchers. Fig. 4.19 p. 87: © Eugene L. Vigil, Seed Research Laboratory. Fig. 4.20 p. 87: © Dr. Jeremy Burgess, Science Photo Library, Photo Researchers. Fig. 4.21 p. 88: © Omikron, Photo Researchers. Fig. 4.22 p. 89: © Dr. Daniel Branton, Harvard University. Fig. 4.23 p. 89: © Biophoto Associates/Science Source, Photo Researchers.

CHAPTER 5: Fig. 5.4 p. 97 left, top right photos: © Dr. Marion I. Barnhart, Wayne State University Medical School; bottom photo: © Dr. Peck-Sun Lin, New England Medical Center. Fig. 5.7 p. 99: Courtesy Drs. John J. Bozzola and Lonnie D. Russell. From *Electron Microscopy: Principles and Techniques for Biologists* © 1992, Jones and Bartlett Publishers, Boston. Fig. 5.21 p. 113: Photo © Dr. Jeremy Burgess, Science Photo Library/Photo Researchers.

CHAPTER 6: Fig. 6.4b p. 128: © CNRI/Science Photo Library, Photo Researchers. Fig. 6.5b p. 130: © Cameron Thatcher/The National Audubon Society Collection, Photo Researchers. Photo, Box 6.1 p. 131: From Crowley, *Introduction to Human Disease, Third Edition* © 1992, Jones and Bartlett Publishers, Boston. Fig. 6.10 p. 137: Photos © Jeanne M. Riddle, Ph.D., Henry Ford Hospital.

CHAPTER 7: Fig. 7.1 p. 150: From J.R. Paulsrud et al., *Amer. J. Clin. Nutr.*, 25 [1972]: 897. Fig. 7.2 p. 152: Lester V. Bergman, NY. Fig. 7.3 p. 152: © Nancy Durrell McKenna/Science Source, Photo Researchers. Fig. 7.4 p. 155: From W. Varavithya et al., Clin. Pediatr. *14 [1975]: 1083. Fig. 7.7 p. 159: © John Radcliffe Hospital/Science Photo Library, Photo Researchers.*

CHAPTER 8: Photos, Box 8.2 p. 179: © American Heart Association. Fig. 8.8 p. 180: Supplied by Carolina Biological Supply Company. Photo, Box 8.2 p. 181: © Salem Hospital. Photos, Box 8.3 p. 184: From Crowley, Introduction to Human Disease, Third Edition © 1992, Jones and Bartlett Publishers.

CHAPTER 9: Fig. 9.3 p. 197: Photo by Emil Bernstein and Eila Kairinen, Gillette Company Research Institute, from *Science* 1973:Cover, 27 August 1971. Copyright © 1971 by the American Association for the Advancement of Science. Photos, Box 9.1 p. 199: (a) © Mayo Foundation; (b) © The Government of Kenya, issued by FAO. Fig. 9.7 p. 203: © Ralph M. Albrecht, University of Wisconsin. Fig. 9.17 p. 213: © David Greenwood, *Science* 163:1076–1077; Copyright © 1969 by the American Association for the Advancement of Science. Fig. 9.19 p. 216: © Bruce Iverson, BSc. Fig. 9.22 p. 219: © NIBSC/Science Photo Library, Photo Researchers.

CHAPTER 11: Fig. 11.3(a) p. 243: © Biophoto Associates/Science Source, Photo Researchers. Fig. 11.5 p. 247: Photos © Biophoto Associates/Science Source, Photo Researchers.

CHAPTER 12: Fig. 12.6 p. 265: Photo © Bruce Iverson, BSc. Fig. 12.9 p. 267: Photos supplied by Carolina Biological Supply Company. Fig. 12.11 p. 270: Photo © Hugh E. Huxley, Clinical Research Center, Cambridge. Box 12.3: p. 273 photo © CNRI/Science Photo Library, Photo Researchers; p. 274 photo © Larry Mulvehill/Science Source, Photo Researchers.

CHAPTER 13: Fig. 13.5 p. 297: Photo by Lester V. Bergman, NY.

CHAPTER 14: Fig. 14.1 p. 311: (b) Photo © Bruce Iverson, BSc. (c) © R. Coggeshall, Photo Researchers. Fig. 14.8 p. 317: © E.R. Lewis/Omikron, Photo Researchers. Fig. 14.9 p. 318: © D.W. Fawcett/Desaki & Venara, Photo Researchers.

CHAPTER 15: Box 15.1 p. 348: Photos © Joseph E. Hawkins, Kresge Hearing Research Institute, University of Michigan Medical School. Fig. 15.5 p. 349: Photo © Dr. G. Bredberg, Science Photo Library/Photo Researchers. Fig. 15.7 p. 352: Photo © Omikron/Photo Researchers. Fig. 15.12 p. 357: Photo © Omikron/Photo Researchers.

CHAPTER 16: Fig. 16.1 p. 369: © Ken Robert Buck/Stock, Boston. Fig. 16.2 p. 370: Thomas McAvoy, *Life* Magazine © Time Warner Inc. Fig. 16.3: © Cary S. Wolinsky/Stock, Boston. Fig. 16.4 p. 371: I. DeVore/Anthro-Photo. Fig. 16.5 p. 373: Photo © I. Eibl-Eibesfeldt. Fig. 16.6 p. 374: (a) Zoological Society of San Diego; (b) AP Wirephoto from Wide World Photo. Fig. 16.7 p. 374: (a), (b) Zoological Society of San Diego; (c) © Sergio Larrain, Magnum Photos. Fig. 16.8 p. 375: Photos by Paul Foley, © Barry M. Lester, Ph.D. Child Development Unit, Children's Hospital Medical Center. Fig. 16.9 p. 376: © I. Eibl-Eibesfeldt. Box 16.2 p. 378: Photo © Mark D. Phillips/Photo Researchers.

CHAPTER 17: Fig. 17.2 p. 387: © K.R. Porter/Photo Researchers. Fig. 17.4 p.s 388–389: Photos © Bruce Iverson, BSc. Fig. 17.6 p. 392 © Leonard Lessin/Photo Researchers.

CHAPTER 18: Fig. 18.1 p. 410: The Bettmann Archive. Fig. 18.9 p. 421: The Bettmann Archive. Fig. 18.16 p. 427: © Sung-Hou Kim.

CHAPTER 19: Fig. 19.1 p. 437: © Biophoto Associates/Science Source, Photo Researchers. Fig. 19.4 p. 437. Photo © Lester V. Bergman, NY. Fig. 19.3 p. 440: © Richard Dranitzke/Science Source, Photo Researchers. Fig. 19.6 p. 442: Photos © Marion I. Barnhart, Wayne State University School of Medicine. Photo, Box 19.3 p. 443: from Money, *Man and Woman, Boy and Girl,* © 1973 by Johns Hopkins University Press. Fig. 19.8 p. 447: (a) © Biophoto Associates/Photo Researchers; (b): © Elaine Rebman/Photo Researchers. Fig. 19.10 p. 448: © Martin M. Rotker/Photo Researchers. Fig. 19.11 p. 448: Photo © Lester V. Bergman, NY. Fig. 19.12 p. 450: © John Moss/Photo Researchers. Fig. 19.13 p. 450: © Andy Levin/Photo Researchers. Fig. 19.14 p. 453: Photo © Fred McConnaughey/Photo Researchers.

CHAPTER 20: Fig. 20.9 p. 466 (a): © Alexander Tsiaras/Science Source, Photo Researchers; (b): © Dan McCoy/Rainbow; (c): © Petit Format/Nestle/Science Source, Photo Researchers. Fig. 20.13 p. 470: Photos © Maternity Center Association, from *The Birth Atlas*. Fig. 20.18 p. 474: © D.W. Fawcett/D. Phillips, Photo Researchers. Fig. 20.24 p. 480: (a): © CNRI/SPL, Photo Researchers; (b): © PR/CDC.

CHAPTER 21: Fig. 21.1 p. 487: Photo supplied by Carolina Biological Supply Company. Fig. 21.2 p. 487: Supplied by Carolina Biological Supply Company. Fig. 21.9 p. 493: Supplied by Carolina Biological Supply Company.

CHAPTER 22: Fig. 22.2 p. 510: From Strickberger, *Evolution,* © 1990 by Jones and Bartlett Publishers. Fig. 22.5 p.

513: Supplied by Carolina Biological Supply Company. Fig. 22.6 p. 514: © Dan McCoy/Rainbow. Fig. 22.8, p. 517: From Strickberger, *Evolution*, © 1990 by Jones and Bartlett Publishers. Fig. 20-9 p. 519: Photos reprinted from Colin Patterson: *Evolution*. © Trustees of the British Museum (Natural History) 1978. Publisher: Cornell University Press (permission pending). Fig., Box 22.1 p. 521: From Strickberger, *Evolution*, © 1990 by Jones and Bartlett Publishers.

CHAPTER 23: Fig. 23.1 p. 526. Culver Pictures. Fig. 23.3 p. 528: (a) Photo Neg. No. 260298, © Library Services Department, American Museum of Natural History, photo: R.N. Wegner; (b) Photo Neg. No. 313686, and (c) Photo Neg. No. 313684, © Library Services Department, American Museum of Natural History, photos: H.S. Rice. Fig. 23.4 p. 528: Photos © Zoological Society of San Diego. Fig. 23.5 p. 529: (a) Photo © D.J. Chivers/Anthro-Photo; (b) Photo © I. DeVore/Anthro-Photo. Fig. 23.9 p. 533: Supplied by Carolina Biological Supply Company. Fig. 23.10 p. 533: Photos supplied by Carolina Biological Supply Company. Fig. 23.11 p. 535: From Strickberger, *Evolution*, © 1990 by Jones and Bartlett Publishers. Fig. 23.12 p. 535: Supplied by Carolina Biological Supply Company.

CHAPTER 24: Fig. 24.8 p. 551: W. Eugene Smith © Magnum Photos, Inc. Fig. 24.10 p. 553: Supplied by Carolina Biological Supply Company. Fig. 24.18 p. 559: Supplied by Carolina Biological Supply Company. Fig. 24.18 p. 560: Photos by Allen H. Morgan © Massachusetts Audubon Society. Fig. 24.20 p. 562: The Bettmann Archive. Fig. 24.22 p. 564: Supplied by Carolina Biological Supply Company. Fig. 24.23 p. 565: © Randall Hyman/Stock, Boston. Fig. 24.24 p. 566: © Tom Branch, The National Audubon Society Collection/ Photo Researchers.

CHAPTER 25: Fig. 25.7 p. 574: © Peter Menzel/Stock, Boston. Fig. 25.11 p. 577: © Sandia Laboratories. Fig. 25.17 p. 583: © Gregory G. Dimijian, MD/The National Audubon Society, Photo Researchers. Fig. 25.19 p. 584: © Earl Roberge, Photo Researchers.

METRIC-ENGLISH CONVERSIONS

LENGTH

English (USA)	*= Metric*
inch	= 2.54 cm
foot	= 0.30 m
yard	= 0.91 m
mile (5,820 ft)	= 1.61 km
Metric	*= English (USA)*
millimeter	= 0.039 in
centimeter	= 0.39 in
meter	= 3.28 ft.
kilometer	= 0.62 mi.

WEIGHT

English (USA)	*= Metric*
ounce	= 28.35 g
pound	= 0.45 kg
ton (short—2000 lb)	= 0.91 metric tons
Metric	*= English (USA)*
gram	= 0.03502 oz
kilogram	= 2.20 lb
metric ton (1000 kg)	= 1.10 tons

VOLUME

English (USA)	*= Metric*
ounce	= 0.03 l
pint	= 0.47 l
quart	= 0.95 l
gallon	= 3.79 l
Metric	*= English (USA)*
milliliter*	= 0.03 oz
liter	= 2.12 pt

*Note: 1 ml = 1 cc